Third Edition

Microprocessors and Microcontrollers

(Includes 8085 Microprocessor, 8086 Microprocessor,
Advance Microprocessors and Microcontrollers)

Third Edition

Microprocessors and Microcontrollers

(Includes 8085 Microprocessor, 8086 Microprocessor,
Advance Microprocessors and Microcontrollers)

A Nagoor Kani

Founder, RBA Educational Group
Chennai

CBS

CBS Publishers & Distributors Pvt Ltd

New Delhi • Bengaluru • Chennai • Kochi • Kolkata • Lucknow • Mumbai
Hyderabad • Jharkhand • Nagpur • Patna • Pune • Uttarakhand

Third Edition

Microprocessors and Microcontrollers

(Includes 8085 Microprocessor, 8086 Microprocessor, Advance Microprocessors and Microcontrollers)

ISBN: 978-93-5466-092-4

Third Edition: 2022
Reprint 2023
First Edition: 2005
Second Edition: 2012

Published by Satish Kumar Jain and produced by Varun Jain for
CBS Publishers & Distributors Pvt Ltd
4819/XI Prahlad Street, 24 Ansari Road, Daryaganj, New Delhi 110 002, India
Ph: 011–23289259, 23266861   Website: www.cbspd.com
                                 e-mail: delhi@cbspd.com

Corporate Office: 204 FIE, Industrial Area, Patparganj, Delhi 110 092, India
Ph: 011–49344934   Fax: 011–49344935   e-mail: publishing@cbspd.com; publicity@cbspd.com

Branches

- **Bengaluru:** Seema House 2975, 17th Cross, K.R. Road, Banasankari 2nd Stage, Bengaluru 560 070, Karnataka, India
  Ph: +91-80-26771678/79   Fax: +91-80-26771680   e-mail: bangalore@cbspd.com
- **Chennai:** 7, Subbaraya Street, Shenoy Nagar, Chennai 600 030, Tamil Nadu, India
  Ph: +91-44-26680620, 26681266   Fax: +91-44-42032115   e-mail: chennai@cbspd.com
- **Kochi:** 42/1325, 1326, Power House Road, Opposite KSEB, Power House, Ernakulam 682018, Kochi, Kerala, India
  Ph: +91-484-4059061-67   Fax: +91-484-4059065   e-mail: kochi@cbspd.com
- **Kolkata:** 147, Hind Ceramics Compound, 1st Floor, Nilgunj Road, Belghoria, Kolkata 700056, West Bengal, India
  Ph: +91-33-25330055/56   e-mail: kolkata@cbspd.com
- **Lucknow:** Basement, Khushuma Complex, 7 Meerabai Marg (behind Jawahar Bhawan), Lucknow 226001, UP, India
  Ph: +91-522-4000032   e-mail: tiwari.lucknow@cbspd.com
- **Mumbai:** PWD Shed, Gala No. 25/26, Ramchandra Bhatt Marg, Next JJ Hospital Gate No. 2, Opp. Union Bank of India, Noorbaug, Mumbai 400009, Maharashtra, India
  Ph: +91-22-66661880/89   e-mail: mumbai@cbspd.com

Representatives

• **Hyderabad**	0-9885175004	• **Jharkhand**	0-9811541605	• **Nagpur**	0-9421945513
• **Patna**	0-9334159340	• **Pune**	0-9923910676	• **Uttarakhand**	0-9716462459

Printed at: Mudrak, Noida, UP, India.

to

My Father-in-law Capt R Chandrasekaran and
Mother-in-law Mrs Amala Chandra

PREFACE

The main objective of this book is to explore the basic concepts of popular microprocessors and microcontrollers programming and interfacing techniques in a simple and easy-to-understand manner. This text on microprocessors and microcontrollers has been crafted and designed to meet student's requirements of various universities in India.

Considering the complex technical nature of this subject, equal emphasis has been given to programming and design aspects. Considerable effort has been made to explain the assembly language programs with step-by-step algorithm and flowchart. The peripheral interfacing techniques have been explained with simple sketches clearly showing the necessary signals and supported by programming examples. Short-answer questions with varied difficulty levels are given in the text to help students get an intuitive grasp on the subject. This book with its lucid writing style and germane pedagogical features will prove to be a master text for engineering students and practitioners in industry.

Chapter 1 briefs about evolution of microprocessors and basics of microprocessor-based system. The architecture of popular microprocessors 8085, 6800, Z80, 8086 and 68000 are also presented in Chapter 1. The architecture of popular microcontrollers 8051 and 8096 are discussed in Chapter 2. An introduction to PIC and ARM controllers is also presented in Chapter 2.

The instruction set of microprocessors 8085 and 8086 are explained in detail in Chapters 3 and 4 respectively. The instructions of an 8051 microcontroller are explained in Chapter 5. A brief discussion about semiconductor memory and their interfacing with microprocessors and microcontrollers are presented in Chapter 6. Memory interface design examples are also included for better understanding of the concepts of the memory and IO interfacing with 8085, 8086, and 8051.

The importance of interrupts and the various interrupts of 8085 and 8086 processors are discussed in Chapter 7. The implementation of interrupt scheme and the expansion of interrupts in 8085 and 8086 systems are also presented in Chapter 7. The interrupts of 8051 microcontroller are also explained with programming examples in Chapter 7.

The concepts of assembly language programming are discussed in Chapter 8. A number of assembly language programs using 8085, 8086 and 8051 instructions are included in Chapter 8. The concepts of DOS and BIOS services, and writing assembly language programs using these services are also presented in Chapter 8. The example programs presented in this book are assembled using Microsoft assemblers and verified in RBA trainer kits and personal computer.

A brief discussion about peripheral devices and their interfacing with 8085, 8086 and 8051 are presented in Chapter 9. The popular peripheral devices 8255, 8279, 8251, 8237, 8257, 8254, ADC0809 and DAC0800 and their interfacing with 8085/8086/8051 along with programming examples are discussed in Chapter 9.

The architectures of 80×86 and Pentium family of microprocessors are presented in Chapter 10. The microprocessor and microcontroller-based system design concepts with some application examples of 8085/8086/8051 based systems are presented in Chapter 11. The instruction set of 8051, 8085 and 8086 are listed in the Appendices-II to VI, and the BIOS and DOS interrupts are listed in Appendix VII, for the use of assembly language programmers.

I have taken care to present the concepts of microprocessors and microcontrollers in a simple manner and hope that the teaching and student community will welcome the book. The readers can feel free to convey their criticism and suggestions to nagoorkani65@yahoo.com for further improvement of the book.

A Nagoor Kani

ACKNOWLEDGEMENTS

I express my heartfelt thanks to my wife Ms C Gnanaparanjothi Nagoor Kani and my sons N Bharath Raj Alias Chandrakani Allaudeen and N Vikram Raj for the support, encouragement and cooperation they have extended to me throughout my career. I thank Ms T A Benazir, Manager, RBA Group, and all my office-staff for their cooperation in carrying out my day-to-day activities.

It is my pleasure to acknowledge the contributions of our technical editors, Ms S Saranya and Ms P Kanimozhi and Ms K G Sathyapriya for editing, proof-reading and type-setting of the manuscript and preparing the layout of the book.

My sincere thanks to all reviewers for their valuable suggestions and comments which helped me to explore the subject to a greater depth.

I am also grateful to Mr Satish K Jain, CMD, CBS Publishers & Distributors, for his keen interest in publishing this work in CBS banner. My sincere thanks to all team members of CBS Publishers & Distributors, for their concern and care in publishing this work.

Finally, a special note of appreciation is due to my sisters, brothers, relatives, friends, students and the entire teaching community for their overwhelming support and encouragement to my writing.

A Nagoor Kani

CONTENT

CHAPTER 3: Instruction Set of 8085   3.1–3.68

CHAPTER 4: Instruction Set of 8086   4.1–4.94

CHAPTER 7: Interrupts 7.1–7.48

CHAPTER 8: Assembly Language Programming 8.1–8.214

CHAPTER 9: Peripheral Devices and Interfacing   9.1–9.180

CHAPTER 11: Microprocessor and Microcontroller-Based System Design   11.1–11.68

APPENDICES   A.1–A.24

GENERAL INDEX   I.1–I.10

CHIP INDEX   I.1–I.3

LIST OF ABBREVIATIONS

Abbreviations

ADC	-	Analog to Digital Converter
AF	-	Auxiliary carry Flag
ALE	-	Address Latch Enable
ALU	-	Arithmetic Logic Unit
ASCII	-	American Standard Code for Information Interchange
BCD	-	Binary Coded decimal
CF	-	Carry Flag
CMOS	-	Complementary Metal Oxide Semiconductor
CPU	-	Central Processing Unit
CRT	-	Cathode Ray Tube
DAC	-	Digital to Analog Converter
DIP	-	Dual In-line Package
DMA	-	Direct Memory Access
DRAM	-	Dynamic Random Access Memory
EEPROM	-	Electrically Erasable Programmable Read Only Memory
EOI	-	End Of Interrupt
EPROM	-	Erasable-Programmable Read Only Memory
FIFO	-	First In First Out
HMOS	-	High density Metal Oxide Semiconductor
IC	-	Integrated circuit
IO	-	Input-Output
IP	-	Instruction Pointer
IR	-	Instruction Register
IRR	-	Interrupt Request Register
ISP	-	In-System Programmable
ISR	-	Interrupt Service Routine/ In-Service Register
ISS	-	Interrupt Service Subroutine
LCD	-	Liquid Crystal Display
LDT	-	Local Descriptor Table
LED	-	Light Emitting Diode
LIFO	-	Last In First Out
LRU	-	Least Recently Used
LSB	-	Least Significant Byte
LW	-	Lower Word
MSB	-	Most Significant Byte
MSW	-	Machine Status Word
NDRO	-	Non-Destructive Read Out Memory
NMI	-	Non Maskable Interrupt
NMOS	-	N-type Metal Oxide Semiconductor
NVRAM	-	Non Volatile Random Access Memory
OCW	-	Operational Common Word
OF	-	Overflow Flag
PC	-	Program Counter/ Personal Computer
PCB	-	Printed Circuit Board
PF	-	Parity Flag
PGA	-	Pin Grid Array
PMOS	-	P-type Metal Oxide Semiconductor
PROM	-	Programmable Read Only Memory
PS	-	Priority Resolver
PSW	-	Program Status Word
PWM	-	Pulse Width Modulation
RAM	-	Random Access Memory
RI	-	Receive Interrupt
SEC	-	Single Edge Connector
SF	-	Sign Flag
SFR	-	Special Function Registers
SI	-	Source Index
SIMD	-	Single Instruction Multiple Data
SP	-	Stack Pointer
SPP	-	Speed Power Product
TI	-	Transmit Interrupt
TLB	-	Translation Look aside Buffer
TTL	-	Transistor Transistor Logic
UART	-	Universal Asynchronous Receiver Transmitter
USART	-	Universal Synchronous Asynchronous Receiver Transmitter
UV	-	Ultra Violet
VLSI	-	Very Large Scale Integration
WDT	-	Watchdog Timer
ZF	-	Zero Flag

Symbols

&	-	Logical AND
|	-	Logical OR
^	-	Logical Exclusive OR
~	-	Logical NOT
$	-	Denote the end of a String

INTRODUCTION TO MICROPROCESSORS

1.1 TERMS USED IN MICROPROCESSOR LITERATURE

Bit	: A digit of the binary number or code is called a bit.
Nibble	: The 4-bit (4-digit) binary number or code is called a nibble.
Byte	: The 8-bit (8-digit) binary number or code is called a byte.
Word	: The 16-bit (16-digit) binary number or code is called a word.
Double Word	: The 32-bit (32-digit) binary number or code is called a double word.
Multiple Word	: The 64, 128, ... bit/digit binary numbers or codes are called multiple words.
Data	: The quantity (binary number/code) operated by an instruction of a program is called data. The size of data is specified as bit, byte, word, etc.
Address	: Address is an identification number (in binary) for memory locations. The 8086 processor uses a 20-bit address for memory.
Memory Word Size (or Addressability)	: The memory word size or addressability is the size of binary information that can be stored in a memory location. The memory word size for an 8086 processor-based system is 8-bit.

[Address and program codes in a microprocessor system are given in binary (i.e., as a combination of "0" and "1"). With n-bit binary we can generate 2^n different binary codes or addresses.]

Microprocessor	: The microprocessor is a program-controlled semiconductor device (IC), which fetches instruction and data (from memory), decodes and executes instructions. It is used as CPU (**C**entral **P**rocessing **U**nit) in computers.
	The basic functional blocks of a microprocessor are ALU (**A**rithmetic **L**ogic **U**nit), an array of registers and a control unit. The microprocessor is identified with the size of data the ALU of the processor can work with at a time. The 8085 processor has a 8-bit ALU; hence, it is called a 8-bit processor. The 8086 processor has a 16-bit ALU; hence, it is called a 16-bit processor.
Bus	: A bus is a group of conducting lines that carries data, address and control signals. Buses can be classified into Data bus, Address bus and Control bus.
	The group of conducting lines that carries data is called a data bus.
	The group of conducting lines that carries address is called an address bus.
	The group of conducting lines that carries control signals is called a control bus.

CPU Bus   :   The group of conducting lines that are directly connected to the microprocessor is called a CPU bus. In a CPU bus, the signals are multiplexed, i.e., more than one signal is passed through the same line but at different timings.

System Bus   :   The group of conducting lines that carries data, address and control signals in a microcomputer system is called System bus. Multiplexing is not allowed in a system bus.

[In microprocessor-based systems, each bit of information (data/address/control signal) is sent through a separate conducting line. Due to practical limitations, the manufacturers of microprocessors may provide multiplexed pins, i.e., one pin is used for more than one purpose. This leads to a multiplexed CPU bus. For example, in an 8086 processor, the address and data are sent through the same pins but at different timings. But when the system is formed, the multiplexed bus lines should be demultiplexed by using latches, ports, transceivers, etc. The demultiplexed bus lines are called system bus. In a system bus, separate conducting lines will be provided for each bit of data, address and control signals.]

Clock   :   A clock is a square wave used to synchronize various devices in the microprocessor and in the system. Every microprocessor system requires a clock for its functioning. The time taken for the microprocessor and the system to execute an instruction or program are measured only in terms of the time period of its clock.

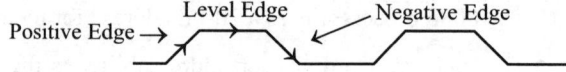

A clock has three edges: rising edge (positive edge), level edge and falling edge (negative edge). The device is made sensitive to any one of the edges for better functioning (it means that the device will recognize the clock only when the edge is asserted or arrived).

Tristate Logic   :   Almost all the devices used in a microprocessor-based system use tristate logic. In devices with tristate logic, three logic levels will be available: **High** state, **Low** state and **High impedance** state.

The **high** and **low** level states are normal logic levels for data, address or control signals. The **high impedance** state is an electrical open-circuit condition. The **high impedance** state is provided to keep the device electrically isolated from the system. The tristate devices will normally remain in the **high impedance** state and their pins are physically connected in the system bus but electrically isolated. In the **high impedance** state, they cannot receive or send any signal or information. These devices are provided with chip enable/chip select pins. When the signal at this pin is asserted to the right level, they come out from the **high impedance** state to normal levels.

1.2 EVOLUTION OF MICROPROCESSORS

History tells us that it was the ancient Babylonians who first began using the abacus (a primitive calculator made of beads) in about 500 BC. This simple calculating machine eventually sparked the human mind into the development of calculating machines that use gears and wheels (Blaise Pascal in 1642). The giant computing machines of the 1940s and 1950s were constructed with relays and vacuum tubes. Next, the transistor and solid-state electronics were used to build the mighty computers of the 1960s. Finally, the advent of the Integrated Circuit (IC) led to the development of the microprocessor and microprocessor-based computer systems.

In 1971, INTEL Corporation released the world's first microprocessor the INTEL 4004, a 4-bit microprocessor. It addresses 4096 memory locations of 4-bit word size. The instruction set consists of 45 different instructions. It is a monolithic IC employing large-scale integration in PMOS technology. The INTEL 4004 was soon followed by a variety of microprocessors, with most of the major semiconductor manufacturers producing one or more types.

First-Generation Microprocessors

The microprocessors introduced between 1971 and 1973 were the first-generation processors. They were designed using PMOS technology. This technology provided low cost, slow speed and low output currents and was not compatible with TTL (**T**ransistor **T**ransistor **L**ogic) levels.

The first-generation processors required a lot of additional support ICs to form a system, sometimes as high as 30 ICs. The 4-bit processors are provided with only 16 pins, but 8-bit and 16-bit processors are provided with 40 pins. Due to limitations of pins, the signals are multiplexed. A list of first-generation microprocessors are as follows:

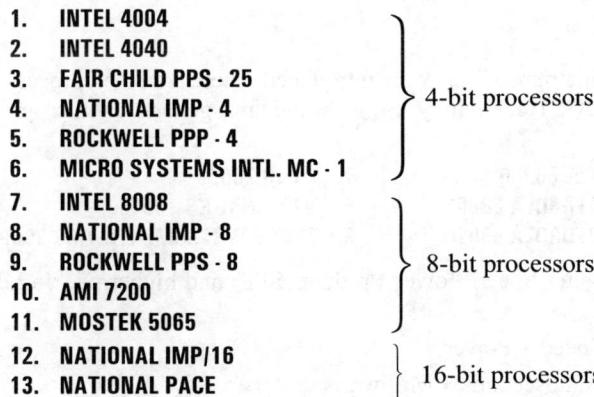

1.	INTEL 4004	
2.	INTEL 4040	
3.	FAIR CHILD PPS - 25	
4.	NATIONAL IMP - 4	4-bit processors
5.	ROCKWELL PPP - 4	
6.	MICRO SYSTEMS INTL. MC - 1	
7.	INTEL 8008	
8.	NATIONAL IMP - 8	
9.	ROCKWELL PPS - 8	8-bit processors
10.	AMI 7200	
11.	MOSTEK 5065	
12.	NATIONAL IMP/16	16-bit processors
13.	NATIONAL PACE	

Second-Generation Microprocessors

The second-generation microprocessors appeared in 1973 and were manufactured using the NMOS technology. The NMOS technology offers faster speed and higher density than PMOS and it is TTL compatible. Some of the second-generation processors are as follows:

1. INTEL 8080
2. INTEL 8085
3. FAIRCHILD F - 8
4. MOTOROLA M6800
5. MOTOROLA M6809
6. NATIONAL CMP -8   } 8-bit processors
7. RCA COSMAC
8. MOS TECH. 6500
9. SIGNETICS 2650
10. ZILOG Z80

11. INTERSIL 6100   } 12-bit processors
12. TOSHIBA TLCS - 12

13. TI TMS 9900
14. DEC - W.D. MCP - 1600
15. GENERAL INSTRUMENT CP 1600   } 16-bit processors
16. DATA GENERAL μN601

Characteristics of Second-Generation Microprocessors

1. Larger chip size (170 × 200 mil).   [1mil = 10^{-3} inch]
2. 40 pins.
3. More numbers of on-chip decoded timing signals.
4. The ability to address large memory spaces.
5. The ability to address more IO ports.
6. Faster operation.
7. More powerful instruction set.
8. A greater number of levels of subroutine nesting.
9. Better interrupt-handling capabilities.

Third-Generation Microprocessors

After 1978, the third-generation microprocessors were introduced. These are 16-bit processors and designed using HMOS (**H**igh density **MOS**) technology. Some of the third generation microprocessor are given below:

1. INTEL 8086   4. INTEL 80286   7. ZILOG Z8000
2. INTEL 8088   5. MOTOROLA 68000   8. NATIONAL NS 16016
3. INTEL 80186   6. MOTOROLA 68010   9. TEXAS INSTRUMENTS TMS 99000

The HMOS technology offers better **S**peed **P**ower **P**roduct (SPP) and higher packing density than NMOS.

Speed Power Product (SPP) = Speed × Power

Unit of SPP   = Nanoseconds × Milliwatts

                        = Picojoules

1. **Speed Power Product of HMOS is four times better than NMOS.**
    SPP of NMOS = 4 picojoules (pJ)
    SPP of HMOS = 1 picojoules (pJ)
2. **Circuit densities provided by HMOS are approximately twice those of NMOS.**
    Packing density of NMOS = 1852.5 gates/mm²
    Packing density of HMOS = 4128 gates/mm² (1 mm = 10^{-6} metre)

Characteristics of Third-Generation Microprocessors

- **Provided with 40/48/64 pins.**
- **High speed and very strong processing capability.**
- **Easier to program.**
- **Allow for dynamically relocatable programs.**
- **Size of internal registers are 8/16/32 bits.**
- **The processor has multiply/divide arithmetic hardware.**
- **Physical memory space is from 1 to 16 megabytes.**
- **The processor has segmented addresses and virtual memory features.**
- **More powerful interrupt-handling capabilities.**
- **Flexible IO port addressing.**
- **Different modes of operations (e.g., user and supervisor modes of M68000).**

Fourth-Generation Microprocessors

The fourth-generation microprocessors were introduced in the year 1980. These generation processors are 32-bit processors and are fabricated using the low-power version of the HMOS technology called HCMOS. These 32-bit microprocessors have increased sophistications that compete strongly with mainframes. Some of the fourth-generation microprocessors are given below:

1. **INTEL 80386**	4. **MOTOROLA M68020**	7. **MOTOROLA MC88100**
2. **INTEL 80486**	5. **BELLMAC - 32**	
3. **NATIONAL NS16032**	6. **MOTOROLA M68030**	

Characteristics of Fourth-Generation Microprocessors

1. **Physical memory space of 2^{24} bytes = 16 MB (megabytes).**
2. **Virtual memory space of 2^{40} bytes = 1 TB (terabytes).**
3. **Floating-point hardware is incorporated.**
4. **Supports increased number of addressing modes.**

Fifth-Generation Microprocessors

In microprocessor technology, INTEL has taken a leading edge and is developing more and more new processors. The latest processor by INTEL is the **pentium** which is considered a fifth-generation processor. The pentium is a 32-bit processor with 64-bit data bus and is available in a wide range of clock speeds from 60 MHz to 3.2 GHz. With improvement in semiconductor technology, the processing speed of microprocessors has increased tremendously. The 8085 released in the year 1976 executes 0.5 Million Instructions Per Second (0.5 MIPS). The 80486 executes 54 Million Instructions Per Second. The pentium is optimized to execute two instructions in one clock period. Therefore, a pentium processor working at 1 GHz clock can execute 2000 **Million Instructions Per Second (2000 MIPS)**. The various processors released by INTEL are listed in Appendix I.

1.3 FUNCTIONAL BUILDING BLOCKS OF A MICROPROCESSOR

A microprocessor is a programmable IC which is capable of performing arithmetic and logical operations. The basic functional block diagram of a microprocessor is shown in Fig. 1.1.

The basic functional blocks of a microprocessor are ALU, Flag register, Register array, Program Counter (PC)/Instruction Pointer (IP), Instruction decoding unit, and the Timing and Control unit.

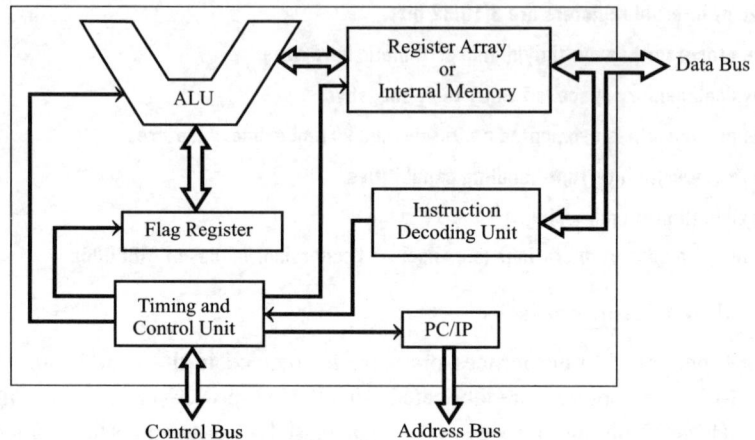

Fig. 1.1: Block diagram showing functional blocks of a microprocessor.

ALU is the computational unit of the microprocessor which performs arithmetic and logical operations on binary data. The various conditions of the result are stored as status bits called flags in the flag register. For example, consider sign flag. One of the bit positions of the flag register is called sign flag and it is used to store the status of sign of the result of the ALU operation (output data of ALU). If the result is negative then "1" is stored in the sign flag and if the result is positive then "0" is stored in the sign flag.

The register array is the internal storage device and so it is also called internal memory. The input data for ALU, the output data of ALU (result of computations) and any other binary information needed for processing are stored in the register array.

For any microprocessor, there will be a set of instructions given by its manufacturer. For doing any useful work with the microprocessor, we have to first write a program using these instructions, and store them in a memory device external to the microprocessor.

The instruction pointer generates the address of the instructions to be fetched from the memory and sends it through the address bus to the memory. The memory will send the instruction codes and data through the data bus. The instruction codes are decoded by the decoding unit and it sends information to the timing and control unit. The data is stored in the register array for processing by the ALU.

The control unit will generate the necessary control signals for internal and external operations of the microprocessor.

1.4 MICROPROCESSOR-BASED SYSTEM (ORGANIZATION OF A MICROCOMPUTER)

A microprocessor is a semiconductor device (or integrated circuit) manufactured by using the VLSI (Very Large Scale Integration) technique. It includes the ALU, the register arrays and the control circuit on a single chip. To perform a function or useful task we have to form a system by using the microprocessor as a CPU (Central Processing Unit) and interfacing the memory, input and output devices to it. A system designed by using a microprocessor as its CPU is called a microcomputer or a single board microcomputer. A microprocessor-based system consists of a microprocessor as the CPU, semiconductor memories like EPROM and RAM, an input device, an output device and interfacing devices. The memories, input devices, output devices and interfacing devices are called peripherals.

The commonly used EPROM and static RAM in microcomputers are given below:

EPROM	Static RAM
INTEL 2708 (1 kB)	MOTOROLA 6208 (1kB)
INTEL 2716 (2 kB)	MOTOROLA 6216 (2kB)
INTEL 2732 (4kB)	MOTOROLA 6232 (4kB)
INTEL 2764 (8kB)	MOTOROLA 6264 (8kB)

Note: kB refers to kilo bytes.

Popular input devices are keyboard, floppy disk, etc., and output devices are printer, LED and LCD displays, CRT monitor, etc.

The block diagram of a microprocessor-based system (or organization of microcomputer) is shown in Fig. 1.2. In this system the microprocessor is the master and all other peripherals are slaves. The master controls all the peripherals and initiates all operations.

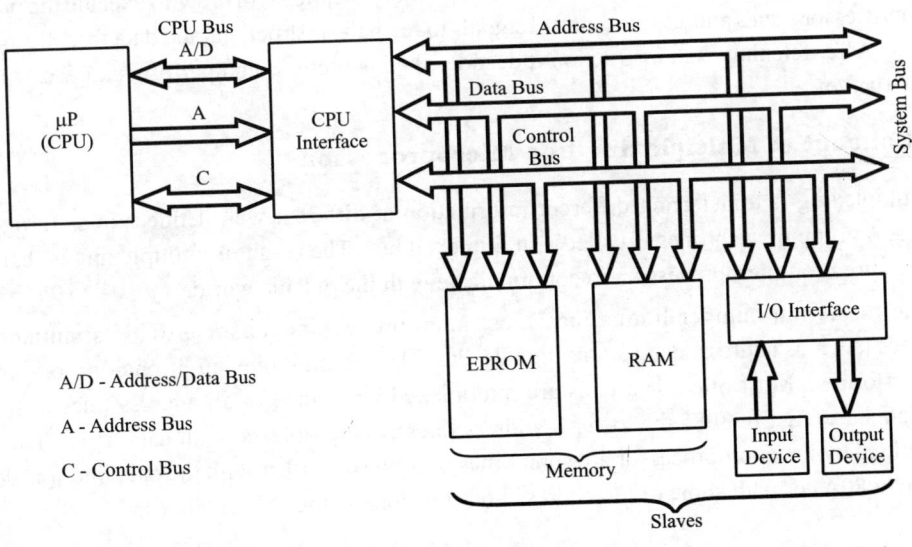

Fig. 1.2: Microprocessor-based system (organization of microcomputer).

Buses are groups of lines that carry data, addresses or control signals. The CPU bus has multiplexed lines, i.e., the same line is used to carry different signals. The CPU interface is provided to demultiplex the multiplexed lines to generate chip select signals and additional control signals. The system bus has separate lines for each signal.

All the slaves in the system are connected to the same system bus. But communication takes place between the master and one of the slaves at any one time. All the slaves have tristate logic and hence normally remain in **high impedance** state. The processor selects a slave by sending an address. When a slave is selected, it comes to the normal logic and communicates with the processor.

The EPROM memory is used to store permanent programs and data. The RAM memory is used to store temporary programs and data. The input device is used to enter the program, data and to operate the system. The output device is used for examining the results. Since the speed of IO devices does not match with the speed of the microprocessor, an interface device is provided between the system bus and the IO devices. Generally IO devices are slow devices.

The work done by the processor can be classified into the following three groups:

    1. Work done internal to the processor.

    2. Work done external to the processor.

    3. Operations initiated by the slaves or peripherals.

The work done internal to the processor are additions, subtractions, logical operations, data transfer within registers, etc. The work done external to the processor are reading/writing the memory and reading/writing the IO devices or the peripherals. If the peripheral requires the attention of the master, then it can interrupt the master and initiate an operation.

The microprocessor is the master which controls all the activities of the system. To perform a specific job or task, the microprocessor has to execute a program stored in the memory. The program consists of a set of instructions stored in consecutive memory locations. In order to execute the program, the microprocessor issues address and control signals to fetch the instructions and data from the memory one by one. After fetching each instruction it decodes the instructions and performs the task specified by the instruction.

1.4.1 Concept of Multiplexing in a Microprocessor

Multiplexing is transferring different information at different well-defined times through the same lines. A group of such lines is called a multiplexed bus. The result of multiplexing is that fewer pins are required for microprocessors to communicate with the outside world.

Due to the pin number limitations, most microprocessors cannot provide simultaneously similar lines (such as address, data, status signals, etc.). Hence multiplexing of one or more of these buses is performed. Most often data lines are multiplexed with some or all address lines to form an address/data bus. (e.g., In 8085 the lower 8-address lines are multiplexed with data lines.) The status signals emitted by the microprocessor are sometimes multiplexed either with the data lines (as done in the INTEL 8080A) or with some of the address lines (as done in the INTEL 8086).

Whenever multiplexing is used, the CPU interface of the system must include the necessary hardware to demultiplex those lines to produce separate address, data and control buses required for the system. Demultiplexing of a multiplexed bus can be handled either at the CPU interface or locally at appropriate points in the system. Besides a slower system operation, a multiplexed bus also results in additional interface hardware requirements.

1.4.2 Demultiplexing of Address/Data Lines in an 8085 Microprocessor

In order to demultiplex the address/data lines (of the processor), the processor provides a signal called the ALE (Address Latch Enable). The ALE is asserted **high** and then **low** by the processor at the beginning of every machine cycle. At the same time the low byte address is given out through the AD_0 - AD_7 lines. The demultiplexing of address/data lines using an 8-bit D-latch 74LS373 is shown in Fig. 1.3.

The ALE is connected to the enable pin (EN) of an external 8-bit latch. When the ALE is asserted **high** and then **low**, the addresses are latched into the output lines of the latch. It holds the low byte of the address until the next machine cycle. After latching the address, the AD_0 - AD_7 lines are free for data transfer. The first T-state of every machine cycle is used for address latching in 8085 and the remaining T-states are used for reading or writing operation.

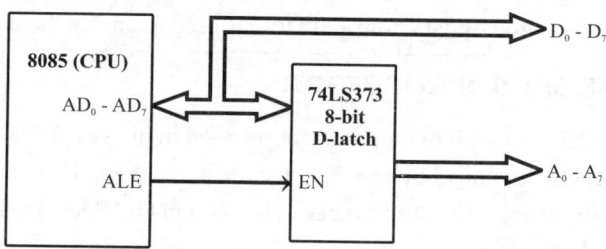

Fig. 1.3: Demultiplexing of address and data lines in an 8085 processor.

1.4.3 Demultiplexing of Address/Data Lines in an 8086 Microprocessor

In order to demultiplex the address/data lines (of the processor), the processor provides a signal called ALE (Address Latch Enable). The ALE is asserted **high** and then **low** by the processor at the beginning of every bus cycle. At the same time, the address is given out through AD_0-AD_{15} lines and A_{16}-A_{19}/status lines. Demultiplexing of address/data lines and address/status lines using 8-bit D-latch 74LS373 is shown in Fig. 1.4.

The ALE is connected to the **E**nable Pin (EN) of the external 8-bit latches. When ALE is asserted **high** and then **low**, the addresses are latched into the output lines of the latch. It holds the address until the next bus cycle. After latching the address, the AD_0- AD_{15} lines are free for data transfer and A_{16} -A_{19}/status lines are free for carrying status information. The first T-state of every bus cycle is used for address latching in 8086 and the remaining T-states are used for reading or writing operation.

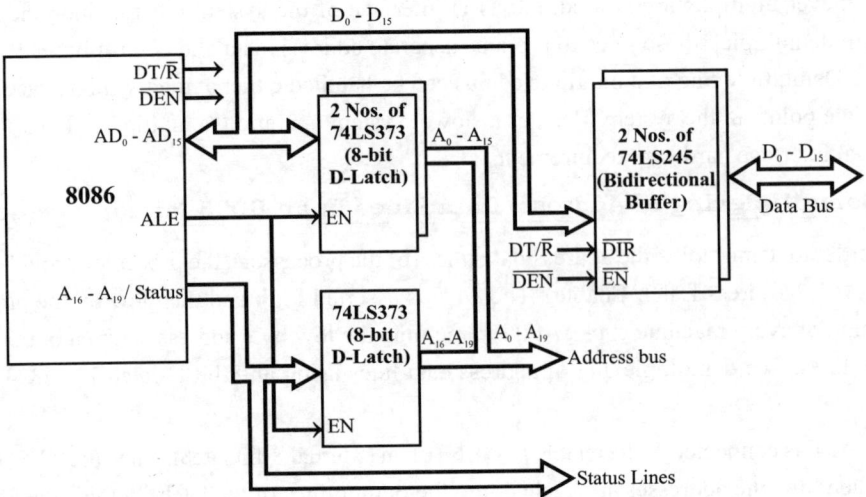

Fig. 1.4: Demultiplexing of address and data lines in an 8086 processor.

The data bus is provided with a bidirectional buffer in order to drive the data to a longer distance in the bus. The 8086 provides two control signals, $DT/\overline{R}$ and $\overline{DEN}$, for controlling the data buffers. $DT/\overline{R}$ is used to decide the direction of data flow and $\overline{DEN}$ is used to enable the data buffer.

1.5 INTEL 8085 MICROPROCESSOR

The INTEL 8085 is an 8-bit microprocessor released in the year 1976. The 8085 was originally designed using NMOS technology but now it is manufactured using HMOS technology and contains approximately 6500 transistors. The 8085 is packed in a 40-pin DIP (**D**ual **I**n-line **P**ackage) and requires a single 5V supply.

The 8085 has an internal clock oscillator. It generates a clock signal internally and divides by two for use as internal clock. This internal clock is also given out through the CLK pin for the clock requirement of peripheral devices.

The NMOS 8085 is available in two versions: 8085A and 8085A-2, with a maximum internal clock frequency of 3.03 MHz and 5 MHz respectively. The enhanced version of the 8085 is designed with HMOS transistors. It is available in three versions: 8085AH, 8085AH-2 and 8085AH-1 with maximum internal clock of 3 MHz, 5 MHz and 6 MHz respectively.

The basic data size of an 8085 is 8-bit. Therefore the memory word size of the memories interfaced with a 8085 processor is also 8-bit or byte. The 8085 uses a 16-bit address to access memory and hence it can address up to $2^{16} = 65,536_{10} = 64$ k memory locations. Since, one byte of information can be stored in one memory location, the maximum memory capacity of an 8085-based system is 64 kilobytes. For accessing IO-mapped devices, the 8085 uses a separate 8-bit address and so it can generate $2^8 = 256_{10}$ IO addresses.

1.5.1 Pin Configuration of an 8085 Microprocessor

The pin configuration of an 8085 microprocessor is shown in Fig. 1.5. The signals of the 8085 are listed in Table 1.1. The 8085 has 8 pins AD_0 to AD_7 for data transfer, which are multiplexed with low byte of address. The 8085 provides a signal ALE (**Address Latch Enable**) to demultiplex the low byte address and data using an external latch. The demultiplexing of address and data lines in an 8085 is shown in Fig. 1.3 in Section 1.4.2.

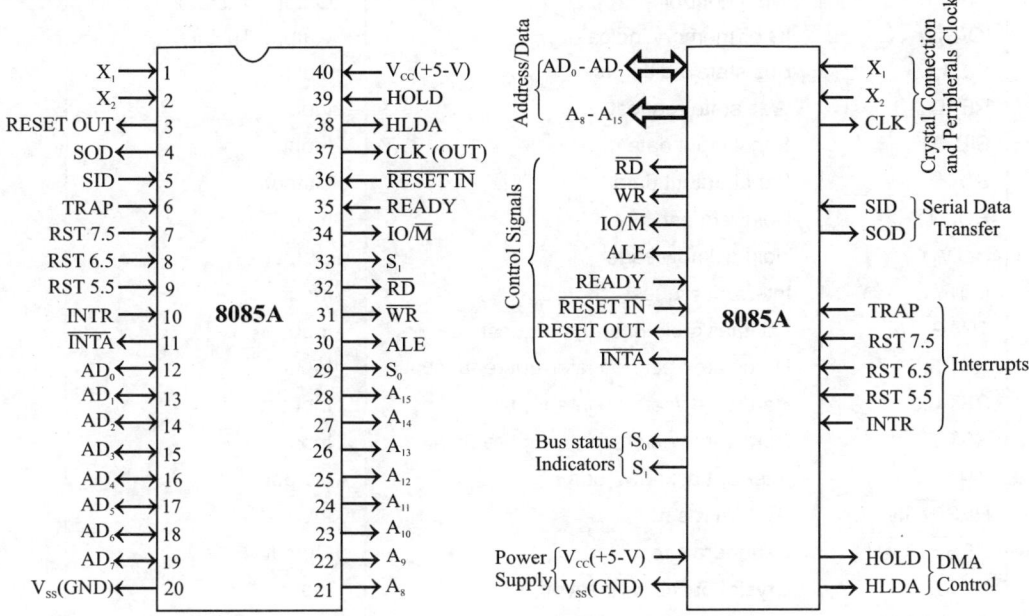

Fig. 1.5: 8085 microprocessor signals and pin assignment.

During memory access, the 16-bit memory address are output on AD_0 to AD_7 and A_8 to A_{15} lines. During IO access of IO-mapped devices the 8-bit IO address are output on both AD_0 to AD_7 and A_8 to A_{15} lines.

The 8085 processor differentiates the memory and IO address using the signal $IO/\overline{M}$. When the processor outputs a memory address, the $IO/\overline{M}$ is asserted **low** and when the processor outputs an IO address, the $IO/\overline{M}$ is asserted **high**.

The $\overline{RD}$ signal is asserted **low** by the processor during a memory or IO read operation. The $\overline{WR}$ signal is asserted **low** by the processor during a memory or IO write operation. The S_0 and S_1 are bus status indicators. The output signals on these lines during various bus activity (or machine cycles) are listed in Table 1.2.

Table 1.1: 8085 Signal Description Summary

Pin Name	Description	Type
AD_0 - AD_7	Address/Data	Bidirectional, Tristate
A_8 - A_{15}	Address	Output, Tristate
ALE	Address latch enable	Output, Tristate
$\overline{RD}$	Read control	Output, Tristate
$\overline{WR}$	Write control	Output, Tristate
IO/$\overline{M}$	IO or memory indicator	Output, Tristate
S_0, S_1	Bus state indicators	Output
READY	Wait state request	Input
SID	Serial input data	Input
SOD	Serial output data	Output
HOLD	Hold request	Input
HLDA	Hold acknowledge	Output
INTR	Interrupt request	Input
TRAP	Nonmaskable interrupt request	Input
RST 5.5	Hardware vectored interrupt request	Input
RST 6.5	Hardware vectored interrupt request	Input
RST 7.5	Hardware vectored interrupt request	Input
$\overline{INTA}$	Interrupt acknowledge	Output
$\overline{RESET\ IN}$	System reset	Input
$\overline{RESET\ OUT}$	Peripherals reset	Output
X_1, X_2	Crystal or RC connection	Input
CLK (OUT)	Clock signal	Output
V_{cc}	+5 V	Power supply
V_{ss}	Ground	Power supply

> **Note:** *A overbar on the signal, indicates that it is active low. (i.e., the signal is normally high and when the signal is activated it is low).*

Table 1.2: Bus Status Signals

IO/$\overline{M}$	S_1	S_0	Operation performed by the 8085
0	0	1	Memory write
0	1	0	Memory read
1	0	1	IO write
1	1	0	IO read
0	1	1	Opcode fetch
1	1	1	Interrupt acknowledge

READY is an input signal that can be used by slow peripherals to get extra time in order to communicate with the 8085. The 8085 will work only when READY is tied to logic **high**. Whenever READY is tied to logic **low**, the 8085 will enter a wait state. When the system has slow peripheral devices, additional hardware is provided in the system to make the READY input **low** during the required extra time while executing a machine cycle, so that the processor will remain in wait state during this extra time.

The HOLD and HLDA signals are used for **D**irect **M**emory **A**ccess (DMA) type of data transfer. This type of data transfers are achieved by employing a DMA controller in the system. When DMA is required, the DMA controller will place a **high** signal on the HOLD pin of the 8085. When the HOLD input is asserted **high**, the processor will enter a wait state and drive all its tristate pins to a **high impedance** state and send an acknowledgement signal to the DMA controller through the HLDA pin. Upon receiving the acknowledgement signal, the DMA controller will take control of the bus and perform DMA transfer and at the end it asserts HOLD signal **low**. When HOLD is asserted **low** the processor will resume its execution.

The 8085 has five interrupt pins. The order of priority of the interrupts is TRAP, RST 7.5, RST 6.5, RST 5.5 and INTR. The interrupts TRAP, RST 7.5, RST 6.5 and RST 5.5 are hardware vectored interrupts and are enabled by appropriate signals at the appropriate pins of the 8085. When a vectored interrupt is enabled and if it is accepted then the program execution branches to the vector addresses specified by INTEL. The interrupts RST 7.5, RST 6.5 and RST 5.5 are maskable interrupts by software.

The INTR is enabled by appropriate signals at its pin. In order to service the INTR, one of the eight opcodes (RST 0 to RST 7) has to be provided on the AD_0 - AD_7 bus by external logic. The 8085 then executes this instruction and vectors to the appropriate address to service the interrupt. The vector address for an interrupt RST n is given by $(08 \times n)_H$. The vector addresses of the interrrupts of 8085 are listed in Table 1.3. (The interrupt TRAP is RST 4.5.)

Table 1.3: Vector Addresses of Interrupts

Interrupt	Vector address	Interrupt	Vector address
RST 0	0000_H	RST 5	0028_H
RST 1	0008_H	RST 5.5	$002C_H$
RST 2	0010_H	RST 6	0030_H
RST 3	0018_H	RST 6.5	0034_H
RST 4	0020_H	RST 7	0038_H
TRAP	0024_H	RST 7.5	$003C_H$

The 8085 has the clock generation circuit on the chip but an external quartz crystal or LC circuit or RC circuit should be connected at the pins X_1 and X_2. The frequency at X_1 and X_2 is divided by two internally and used as an internal clock. The frequency of the output clock signal at the CLK(OUT) pin is same as that of the internal clock.

$\overline{\text{RESET IN}}$ is the system reset input signal and it is used to bring the processor to a known state. For proper reset, the $\overline{\text{RESET IN}}$ pin should be held **low** for at least three clock periods. When pin is asserted **low**, the program counter, instruction register, interrupt mask bits and all internal registers are cleared/reset. Also the RESET OUT signal is asserted **high** to clear/reset all the peripheral devices in the system. After a reset, the content of the program counter will be 0000_H and so the processor will start executing the program stored at 0000_H.

The pins SID and SOD can be used for serial data communication between the 8085 and any serial device under software control.

1.5.2 Driving X_1 and X_2 Inputs in an 8085 Microprocessor

The X_1 and X_2 pins of an 8085 processor are provided to connect an external quartz crystal or LC circuit. It can also be driven by an RC circuit or an external clock source. This connection is necessary for the internal oscillator to generate the clock signal for the processor. An oscillator consists of an amplifier and a feedback circuit. The feedback circuit of an oscillator can be of RC type, LC type or quartz crystal (a quartz crystal is electrically equivalent to an RLC circuit.) Also the feedback circuit decides the frequency of the signal generated by the oscillator.

In an 8085 processor, the oscillator circuit is provided internally except the feedback circuit. This feature facilitates the system designer to choose his own frequency for clock signals. But this frequency should not exceed the maximum clock frequency specified by the manufacturer.

Another reason for keeping feedback circuit external to the processor is that the high Q circuits (quartz crystal or large values of L) cannot be fabricated by IC technology.

In an 8085, the frequency generated by the oscillator circuit will be double that of the internal clock frequency. (The maximum clock frequencies specified by the manufacturer are internal clock frequencies.) In other words, the frequency at X_1 - X_2 pins of an 8085 is divided by two internally. This means that in order to obtain an internal clock of 3.03 MHz, a clock source of 6.06 MHz must be connected to X_1 - X_2. (Crystal/LC/RC should be designed for double the internal frequency.)

Quartz crystals are the best choice for connecting at X_1 - X_2, because they are less expensive, highly stable, have a large Q, occupy a very small space and frequencies do not drift with ageing. For crystals with less than 4 MHz, a capacitor of 20 pF should be connected between X_2 and ground to ensure the starting up of the crystal at the right frequency.

When an LC circuit is used, the value of L_{ext} and C_{ext} can be chosen using the formula,

$$f = \frac{1}{2\pi L_{ext}(C_{ext} + C_{int})}$$

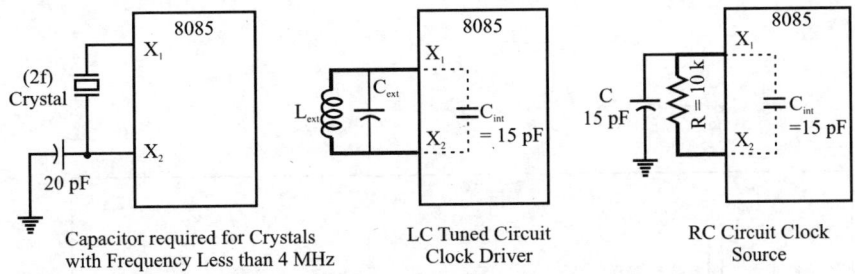

Fig. 1.6: Clock driver circuits for an 8085 microprocessor.

To minimize the variations in frequency, it is recommended that the value for C_{ext} should be chosen which is twice that of C_{int} or 30 pF. The use of LC circuit is not recommended for external frequencies higher than 5 MHz.

An RC circuit may also be used as the clock source for the 8085A if an accurate clock frequency is of no concern. Its advantage is the low component cost. The values shown in Fig. 1.6 are for generating an approximate external frequency of 3 MHz. Note that frequencies higher or lower than 3 MHz should not be attempted on this circuit.

1.5.3 Hardware Architecture of an 8085 Microprocessor

The architecture of an 8085 is shown in Fig. 1.8. The 8085 includes an ALU, a timing and control unit, a instruction register and a decoder, a register array, an interrupt control and a serial IO control.

The ALU performs the arithmetic and logical operations. The operations performed by the ALU of an 8085 are **addition, subtraction, increment, decrement, logical AND, OR, EXCLUSIVE-OR, compare, complement and left/right shift.** The accumulator and temporary register are used to hold the data during an arithmetic/logical operation. After an operation the result is stored in the accumulator and the flags are set or reset according to the result of the operation. The accumulator and flag register together is called the **P**rogram **S**tatus **W**ord (PSW).

There are five flags in an 8085: **Sign Flag (SF), Zero Flag (ZF), Auxiliary Carry Flag (AF), Parity Flag (PF) and Carry Flag (CF).** The bit positions reserved for these flags in the flag register are shown in Fig. 1.7.

D_7	D_6	D_5	D_4	D_3	D_2	D_1	D_0
SF	ZF		AF		PF		CF

Fig. 1.7: Bit positions of various flags in the flag register of 8085.

After an ALU operation if the most significant bit of the result is 1, the sign flag is set. The zero flag is set if the ALU operation results in zero and it is reset if the result is nonzero. In an arithmetic operation, when a carry is generated by the lower nibble, the auxiliary carry flag is set. After an arithmetic or logical operation if the result has an even number of 1's, the parity flag is set, otherwise it is reset.

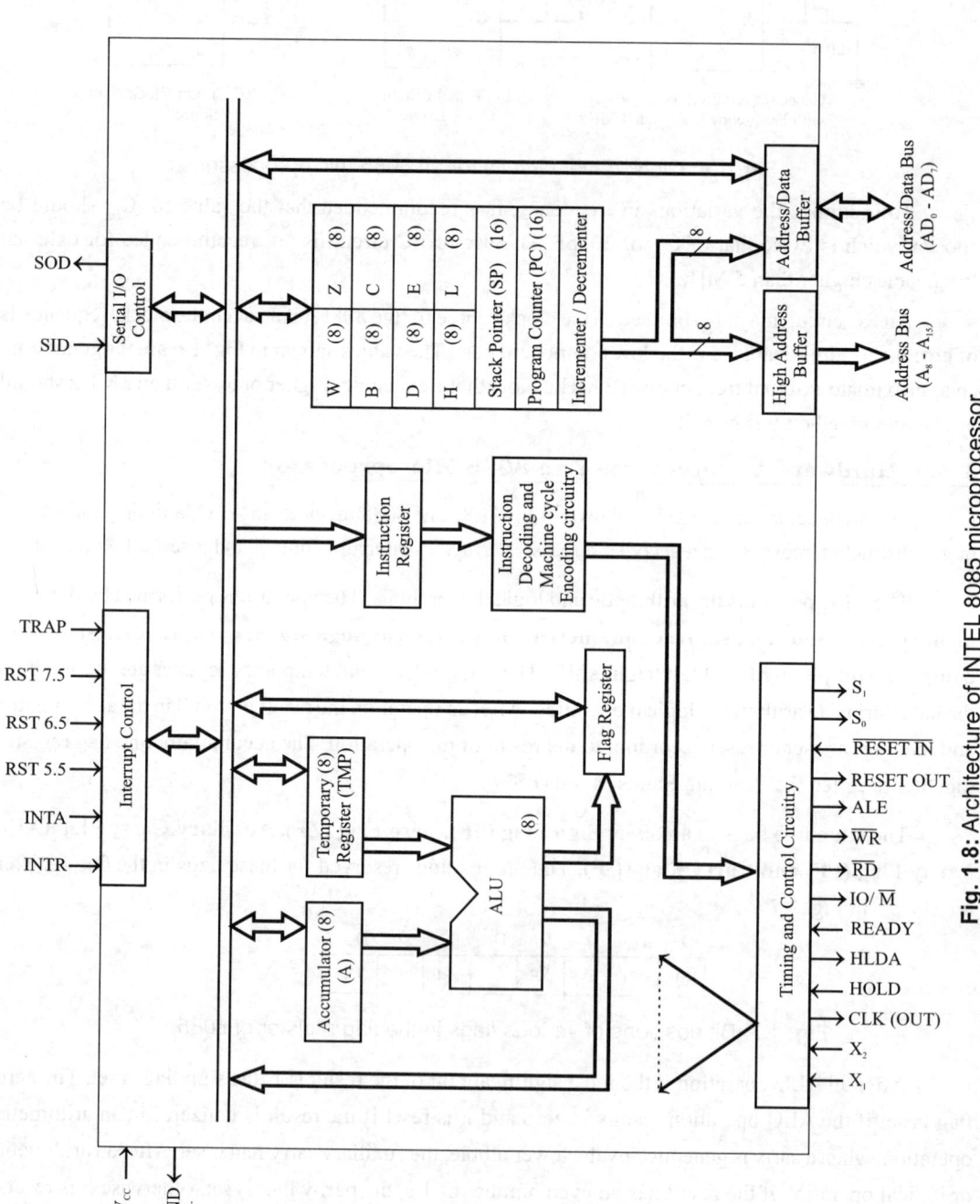

Fig. 1.8: Architecture of INTEL 8085 microprocessor.

If an arithmetic operation results in a carry, the carry flag is set, otherwise it is reset. Among the five flags, the AF flag is used internally for BCD arithmetic and other four flags can be used by the programmer to check the conditions of the result of an operation.

The **timing and control unit** synchronizes all the microprocessor operations with the clock, and generates the control signals necessary for communication between the microprocessor and the peripherals.

When an instruction is fetched from the memory it is placed in the instruction register. Then it is decoded and encoded into various machine cycles. Apart from the **Accumulator** (A-register) there are six general purpose programmable registers B, C, D, E, H and L.

They can be used as 8-bit registers or paired to store 16-bit data. The allowed pairs are BC, DE and HL. The temporary registers TMP, W and Z cannot be used by the programmer.

The **S**tack **P**ointer SP holds the address of the stack top. The stack is a sequence of RAM memory locations defined by the programmer. The stack is used to save the content of the registers during the execution of a program.

The **P**rogram **C**ounter (PC) keeps track of program execution. To execute a program the starting address of the program is loaded in the program counter. The PC sends out an address to fetch a byte of instruction from memory and increment its content automatically. Hence when a byte of instruction is fetched, the PC holds the address of the next byte of the instruction or the next instruction.

1.5.4 Instruction Execution and Data Flow in an 8085 Microprocessor

The program instructions are stored in the memory, which is an external device. In order to execute a program in an 8085, the starting address of the program should be loaded in the program counter. The 8085 outputs the contents of the program counter to the address bus and asserts the read control signal **low**. Also, the program counter is incremented.

The address and the read control signal enables the memory to output the content of the memory location on the data bus. Now the content of the data bus is the opcode of an instruction.

The read control signal is made **high** by the timing and control unit after a specified time. At the rising edge of read control signals, the opcode is latched into the microprocessor internal bus and placed in the instruction register.

The instruction decoding unit decodes the instructions and provides information to the timing and control unit to take further action.

1.6 ZILOG Z80

The ZILOG Z80 is an 8-bit microprocessor manufactured using NMOS technology. The Z80 is available in a 40-pin DIP (**D**ual **I**n-line **P**ackage). It requires a single external clock and a single 5-V power supply. The maximum internal clock of standard Z80 is 2.5 MHz and for Z80-A it is 4 MHz. The Z80 provides more registers, extra addressing modes and a much larger instruction set than an 8085. It also has a built-in logic to refresh its dynamic RAM memories.

The signals of Z80 microprocessor and its functional block diagram (architecture) are shown in Figs.1.9 and 1.10 respectively. The Z80 communicates with other system modules via three functionally separate buses: data, address and control buses.

The Z80 has separate pins for data and address. It operates on an 8-bit data and uses a 16-bit memory address. The physical memory size of the Z80 system is 64 kB. The IO devices can be mapped by memory-mapping or IO-mapping similar to that of 8085. For IO-mapped devices an 8-bit address is allotted. During memory refresh time, the seven lower-order bits of the address bus (A_0 - A_6) contain a valid refresh address.

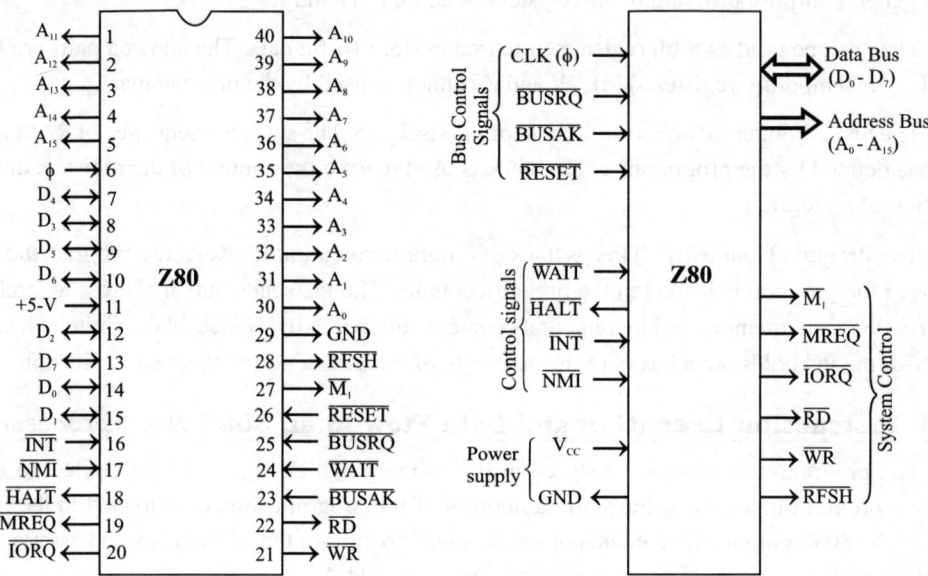

Fig. 1.9: Signals of Z80.

The control bus has three types of control signals. They are listed below:

1. System control signals

$\overline{M}_1$	-	First machine cycle of an instruction
$\overline{MREQ}$	-	Memory request
$\overline{IORQ}$	-	IO request
$\overline{RD}$	-	Read control
$\overline{WR}$	-	Write control
$\overline{RFSH}$	-	Refresh cycle

2. CPU control signals

$\overline{WAIT}$	-	Wait request
$\overline{HALT}$	-	Halt request
$\overline{INT}$	-	Interrupt request
NMI	-	Non-maskable interrupt

3. Bus control signals

$\overline{\text{BUSRQ}}$	-	**BUS request**
$\overline{\text{BUSAK}}$	-	**BUS acknowledge**
$\overline{\text{RESET}}$	-	**System reset**
$\overline{\text{CLK}}$ (ϕ)	-	**Clock input**

The ALU is eight bits wide and performs similar functions to those of the 8085 ALU. The Z80 has two independent 8-bit accumulators A and A′ and two independent flag registers F and F′. The ALU operation involving accumulator A affects the flag register F. The ALU operation involving accumulator A′ affects the flag register F′.

The flag registers have six flags: sign (S and S′), zero (Z and Z′), carry (C and C′), parity/over flow (P/V and P′/V′), half carry (H and H′) and subtract (N and N′).

The Z80 has two sets of 8-bit general purpose register. Each set has 6 registers. They are B, C, D, E, H & L and B′, C′, D′, E′, H′, & L′. They can be used individually as 8-bit registers or as 16-bit register pairs. The allowed pairs are BC, DE & HL and B′C′, D′E′ & H′L′.

At any time instant the programmer can select and work with either the main register set or the alternate register set. To work in an alternate register set, the programmer has to use a single **Ex**change **I**nstruction (EXI) for the entire set of instructions. This alternate set allows the background mode of operation or handling fast interrupt response requirements while servicing an interrupt or executing a subroutine. While executing a program if one set of registers is not sufficient then we need not push them to stack, alternatively we can deactivate them without destroying its contents and switch to an alternate set of registers through exchange instructions.

The 16-bit **P**rogram **C**ounter (PC) and **S**tack **P**ointer (SP) registers are same as that of a 8085 microprocessor and operate in exactly the same way. The registers IX and IY allow two independent indexed addressing modes.

The Z80 includes an 8-bit interrupt vector (I). It is used in one of the interrupt response modes of the processor. It holds the upper eight bits of a memory pointer (or vector address). The lower eight bits of this pointer are supplied (as a vector number) by the interrupting device that requests service. The CPU then uses this 16-bit vector address to make an indirect call to the memory location that holds the first instruction of the interrupt service routine. This feature allows the vector table to be located anywhere in the memory.

The Z80 also contains an 8-bit memory refresh register (R) that contains the current memory refresh address, thus providing for automatic, totally transparent refresh of external dynamic RAM memories. Although the programmer can load this register for testing purposes, the R register is not normally used by the programmer.

The Z80 can execute 158 instruction types. The microprocessor includes all the instructions of an 8080A microprocessor with total software compatibility at the machine code level.

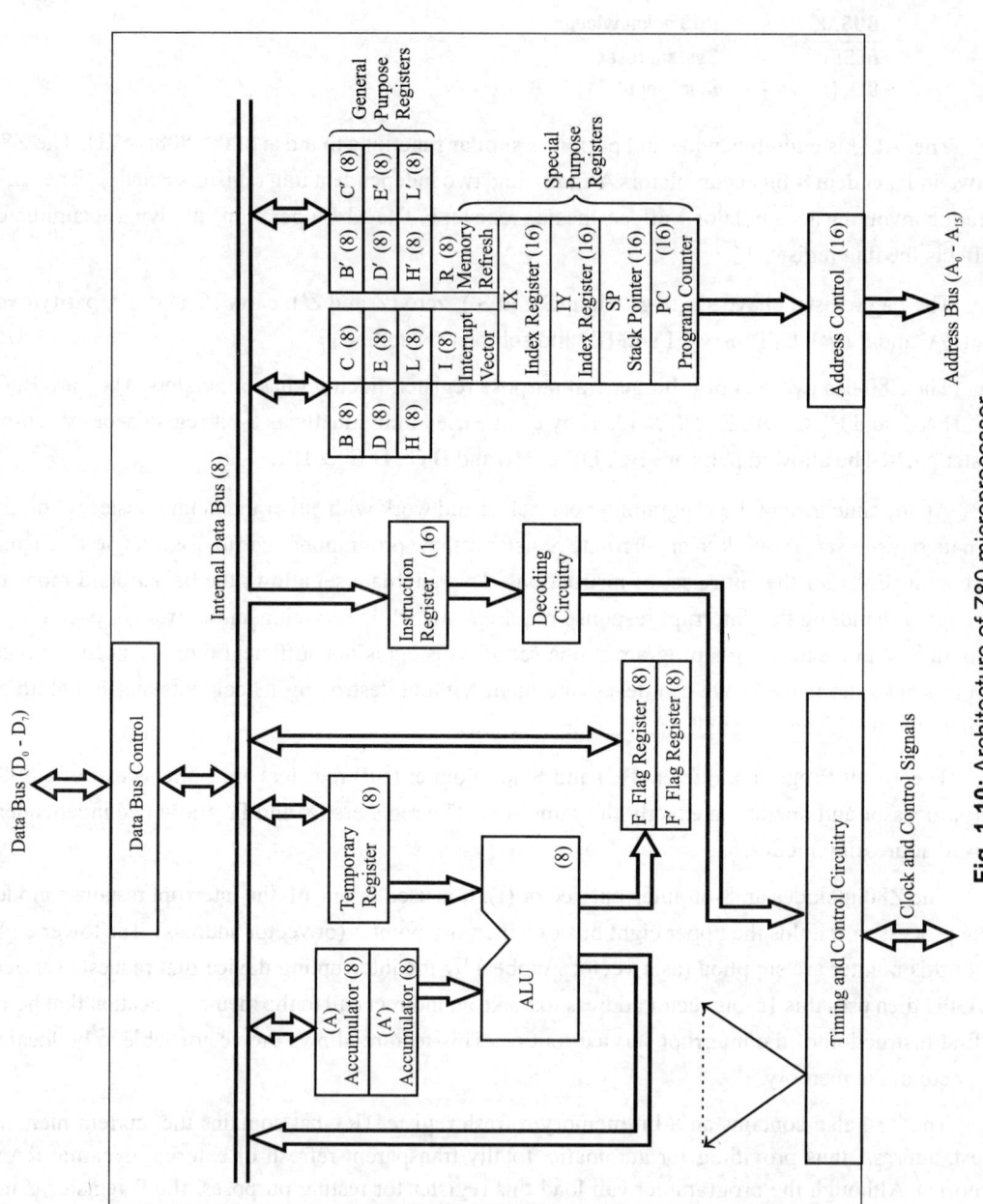

Fig. 1.10: Architecture of Z80 microprocessor.

> **Note:** The 8085 has same instructions of 8080 except two new instructions SIM and RIM. Hence 8085 is also software compatible with Z80.

The new instructions in Z80 include 1/4/8/16-bit operations, exchange instructions, block-transfer and block-search instructions and a full set of rotate and shift instructions applicable to any register, rather than just to the accumulator.

The size of a Z80 instruction is one to four bytes. A 1-byte instruction has just one-byte opcode. A 2-byte instruction has one or two byte opcode plus data-byte/device-number/displacement.

In multibyte instructions the opcode is one or two bytes. The remaining bytes are data/device-number/displacement/address.

The device-number is an 8-bit IO port address. The data-byte is the immediate operand. The displacement is a signed 2's complement number which is added to a 16-bit number residing in an index register, during indexed addressing.

Every Z80 instruction consists of one to six machine cycles. All types of machine cycles consist of either three or four states. Some Z80 instructions always insert wait states (T_w) between the states T_2 and T_3.

The basic operation of the Z80 is analogous to that of the INTEL 8085. The main difference is that instead of IO/$\overline{M}$ of 8085, the Z80 has $\overline{MREQ}$ and $\overline{IORQ}$. They are activated along with $\overline{RD}$ and $\overline{WR}$ for the memory or IO access.

1.7 MOTOROLA 6800

The Motorola 6800 product family was originally introduced in 1974. The 6800 microprocessor CPU is manufactured in NMOS technology on a 40-pin chip, has TTL compatible pins and it is the first 8-bit single chip microprocessor to exploit a single 5-V power supply.

The 6800 CPU can drive from seven to ten 6800 family devices without buffering. A two-phase external clock (1 MHz, maximum) must be externally supplied.

The signals of a motorola 6800 and its simplified functional block diagram are shown in Figs. 1.11 and 1.12 respectively.

The 6800 CPU has three buses to communicate with the other system modules, they are data, address and control buses. The data bus is bidirectional and has 8-lines, D_0-D_7.

The address bus has 16-lines, A_0-A_{15}. The processor operates on 8-bit data and uses a 16-bit address for memory and IO devices.

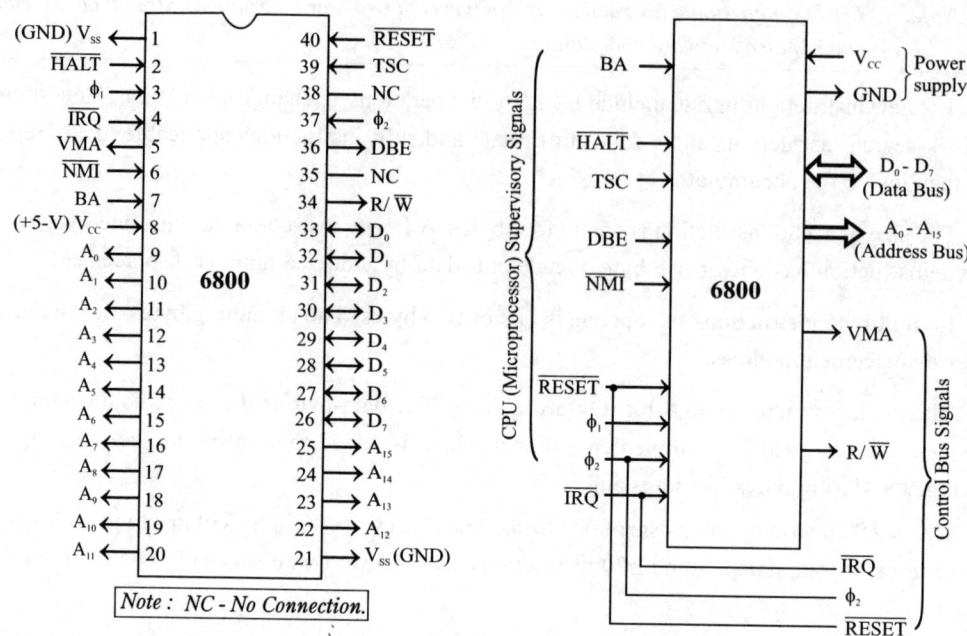

Fig.1.11: Signals of Motorola 6800.

This microprocessor does not distinguish between memory and peripheral addresses. Therefore some of the 64 k addresses must be reserved for peripheral addresses. The control bus carries two types of signals called **control bus signals** and **CPU (microprocessor) supervisory signals.**

Control bus signals :	VMA	-	**Valid memory address**
	R/$\overline{W}$	-	**Read/Write control**
	$\overline{IRQ}$	-	**Interrupt request**
	ϕ_2	-	**Phase-2 of clock**
	$\overline{RESET}$	-	**System reset**
CPU supervisory signals :	BA	-	**Bus acknowledge**
	$\overline{HALT}$	-	**Halt request**
	TSC	-	**Tristate control**
	DBE	-	**Data bus enable**
	$\overline{NMI}$	-	**Non-maskable interrupt**
	ϕ_1	-	**Phase-1 of clock**
	$\overline{IRQ}$	-	**Interrupt request**
	$\overline{RESET}$	-	**System reset**

The $\overline{HALT}$ pin is used for DMA data transfer in block-transfer mode or a cycle-stealing mode. When $\overline{HALT}$ is asserted **low** the microprocessor halts all its activity at the completion of the current instruction.

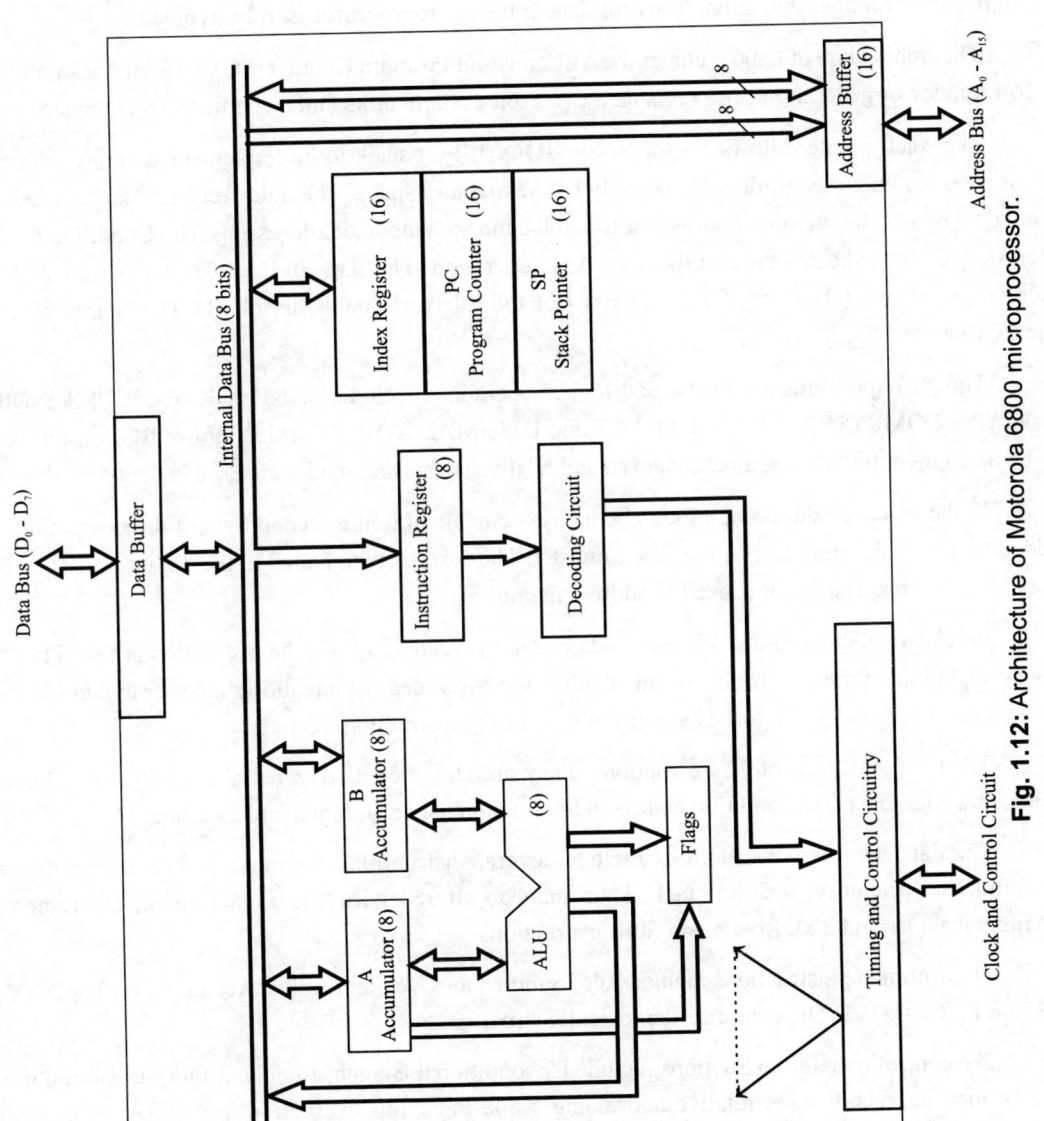

Fig. 1.12: Architecture of Motorola 6800 microprocessor.

The **T**ristate **C**ontrol (TSC) may be used to implement DMA on a cycle-stealing basis. If TSC is placed in a **high** state, the address bus and the R/W line go to a **high impedance** state 500 ns later. The data bus is not affected by TSC and has its own enable (DBE). This approach assures rapid response to the DMA request. Since the internal memory of the 6800 is dynamic, the TSC terminal cannot be held in **high** state for longer than 5 μs, if loss of data in the microprocessor is to be avoided.

The architecture of 6800 includes the ALU, a 16-bit **P**rogram **C**ounter (PC), a 16-bit stack pointer, a 16-bit index or general purpose register, a two 8-bit accumulators and a condition code register.

The stack pointer allows a **L**ast-**I**n-**F**irst-**O**ut (LIFO) stack to be implemented at any address in the memory and to be limited in size only by the memory space. The index register may be used to store data or a 16-bit memory address for use in the indexed mode of addressing. The **C**ondition **C**ode **R**egister (CCR) indicates the results of an ALU operation. The flags in the CCR are **N**egative (N), **Z**ero(Z), **O**verflow (O), **C**arry(C), **H**alf carry (H) and **I**nterrupt enable/disable (I). The unused bits of the CCR are the 1's.

The ALU performs arithmetic and logical operations including AND, OR, EXCLUSIVE-OR, NEGATE, COMPARE, ADD, SUBTRACT and DECIMAL ADJUST which allows BCD arithmetic to be performed. Immediate, direct, indexed and relative addressing modes are used in 6800.

In the indexed addressing mode, the address contained in the second byte of the instruction is added to the lowest eight bits of the index register. The carry is then added to the higher order bits of the index register. The result is used to address memory.

In relative addressing the address contained in the second byte of the instruction is added to the lowest eight bits of the PC. To this result, a value of **+2** is added, which allows the user to address the data within a range of – 125 to + 129 bytes of the present instruction.

The 6800 has a set of 72 instructions. They are classified as data handling, arithmetic, logic, control transfer, data test, condition codes, address maintenance and interrupt handling.

The data handling instructions include several instructions for moving data between two accumulators, memory and the stack. Data may be altered with Clear, Increment, Decrement, Complement (1's and 2's), Rotate and Shift instructions.

The arithmetic instructions include Add, Subtract and Decimal Adjust Accumulator. The AND, OR and EXCLUSIVE-OR comprise the logical instructions.

The control transfer instructions include Unconditional Branch, Jump and Jump-to-subroutine. The Branch instruction uses relative addressing while the Jump instruction uses direct or indirect addressing. A number of conditional branches are available which test the condition of one or more bits of the condition code register.

The data test instructions set the condition codes (alter the flags) without altering the data. They include Bit Test (for comparing individual bits of accumulator A or B with a memory word), Compare and Test (for determining the sign of a number).

Condition code instructions are provided which enable the programmer to set or reset directly the Carry, Interrupt or Over flow flags. The entire contents of the condition code register may be moved to or from the accumulator A with a single instruction. Eleven instructions are provided for address maintenance. These instructions allow operations on the index register, e.g., Compare, Increment, Decrement and Transfer to or from the memory or the stack pointer. Similar instructions are available for operation on addresses stored in the stack pointer.

The interrupt handling instructions include a software interrupt (SWI) which stores the status of the processor in the stack before processing the interrupt and a **Return** from **Interrupt** (RTI) instruction which restores the status of the microprocessor after an interrupt is processed. A **Wa**it for **Interrupt** (WAI) instruction causes the status to be stored in the stack and places the processor in a halt condition until a hardware interrupt occurs.

A 6800 instruction may be one, two or three bytes long, its length being closely related to the addressing mode used.

Usually every 6800 instruction cycle consists of two to eight machine cycles, all of which are identical in length (except interrupt instructions which require longer instruction execution cycles.) In the 6800, a machine cycle is one and the same thing as a clock cycle (or state.)

The operation of the 6800 is very simple, since it consists of only three types of machine cycles: a read machine cycle (during which a byte of data is input into the CPU), a write machine cycle (during which a byte of data is output by the CPU) and an interrupt operation machine cycle (during which the CPU is busy and no activity occurs on the system buses.) The timing of any 6800 instruction is simply a concatenation of these three basic machine cycle types.

The control signals required to access a memory location are $R/\overline{W}$, VMA and DBE. Under normal circumstances, DBE is identical to ϕ_2. The signal $R/\overline{W}$ controls the reading or writing operation. For read operation $R/\overline{W}$ is asserted **high** and for write operation $R/\overline{W}$ is asserted **low**.

1.8   INTEL 8086 MICROPROCESSOR

The INTEL 8086 is the first 16-bit processor released by INTEL in the year 1978. The 8086 is designed using the HMOS technology and now it is manufactured using HMOS III technology and contains approximately 29,000 transistors. The 8086 is packed in a 40-pin DIP and requires a single 5-V supply.

The 8086 does not have an internal clock circuit. The 8086 requires an external asymmetric clock source with 33% duty cycle. An 8284 clock generator is used to generate the required clock for 8086. The maximum internal clock of 8086 is 5 MHz. The other versions of 8086 with different clock rates are 8086-1, 8086-2 and 8086-4 with maximum internal clock frequency of 10 MHz, 8 MHz and 4 MHz respectively.

The 8086 uses a 20-bit address to access memory and hence it can directly address up to one megabytes (2^{20} = 1 Mega) of memory space. The one megabyte (1MB) of addressable memory space of 8086 are organized as two memory banks of 512 kilobytes each (512 kB + 512 kB = 1MB).

The memory banks are called even (or lower) bank and odd (or upper) bank. The address line A_0 is used to select even bank and the control signal $\overline{BHE}$ is used to select odd bank.

For accessing IO-mapped devices, the 8086 uses a separate 16-bit address, and so the 8086 can generate 64k (2^{16}) IO addresses. The signal $M/\overline{IO}$ is used to differentiate the memory and IO addresses. For memory address the signal $M/\overline{IO}$ is asserted **high** and for IO address the signal M/IO is asserted **low** by the processor.

The 8086 can operate in two modes: minimum mode and maximum mode. The mode is decided by a signal at $MN/\overline{MX}$ pin. When the $MN/\overline{MX}$ is tied **high** it works in minimum mode and the system is called a uniprocessor system. When $MN/\overline{MX}$ is tied **low** it works in maximum mode and the system is called a multiprocessor system. Usually the pin $MN/\overline{MX}$ is permanently tied to **low** or **high** so that the 8086 system can work in any one of the two modes. The 8086 can work with an 8087 coprocessor in maximum mode. In this mode an external bus controller 8288 is required to generate bus control signals.

The 8086 has two families of processors. They are 8086 and 8088. The 8088 uses 8-bit data bus externally but 8086 uses 16-bit data bus externally. The 8086 access memory is in words but 8088 access memory is in bytes. IBM designed its first **P**ersonal **C**omputer (PC) using an INTEL 8088 microprocessor as the CPU.

1.8.1 Pins and Signals of INTEL 8086

The 8086 pins and signals are shown in Fig. 1.13. The 8086 is a 40-pin IC and all the 8086 pins are TTL compatible. The signal assigned to pins 24 to 31 is different for minimum and maximum mode of operation. The signal assigned to all the other pins are common for minimum and maximum mode of operation.

Table 1.4: Common Signals

Name	Description/Function	Type
$AD_{15} - AD_0$	Address/Data	Bidirectional, Tristate
$A_{19}/S_6 - A_{16}/S_3$	Address/Status	Output, Tristate
$\overline{BHE}/S_7$	Bus high enable/Status	Output, Tristate
$MN/\overline{MX}$	Minimum/Maximum mode control	Input
$\overline{RD}$	Read control	Output, Tristate
$\overline{TEST}$	Wait on test control	Input
READY	Wait state control	Input
RESET	System reset	Input
NMI	Non-maskable interrupt request	Input
INTR	Interrupt request	Input
CLK	System clock	Input
V_{cc}	+ 5-V	Power supply input
GND	Ground	Power supply ground

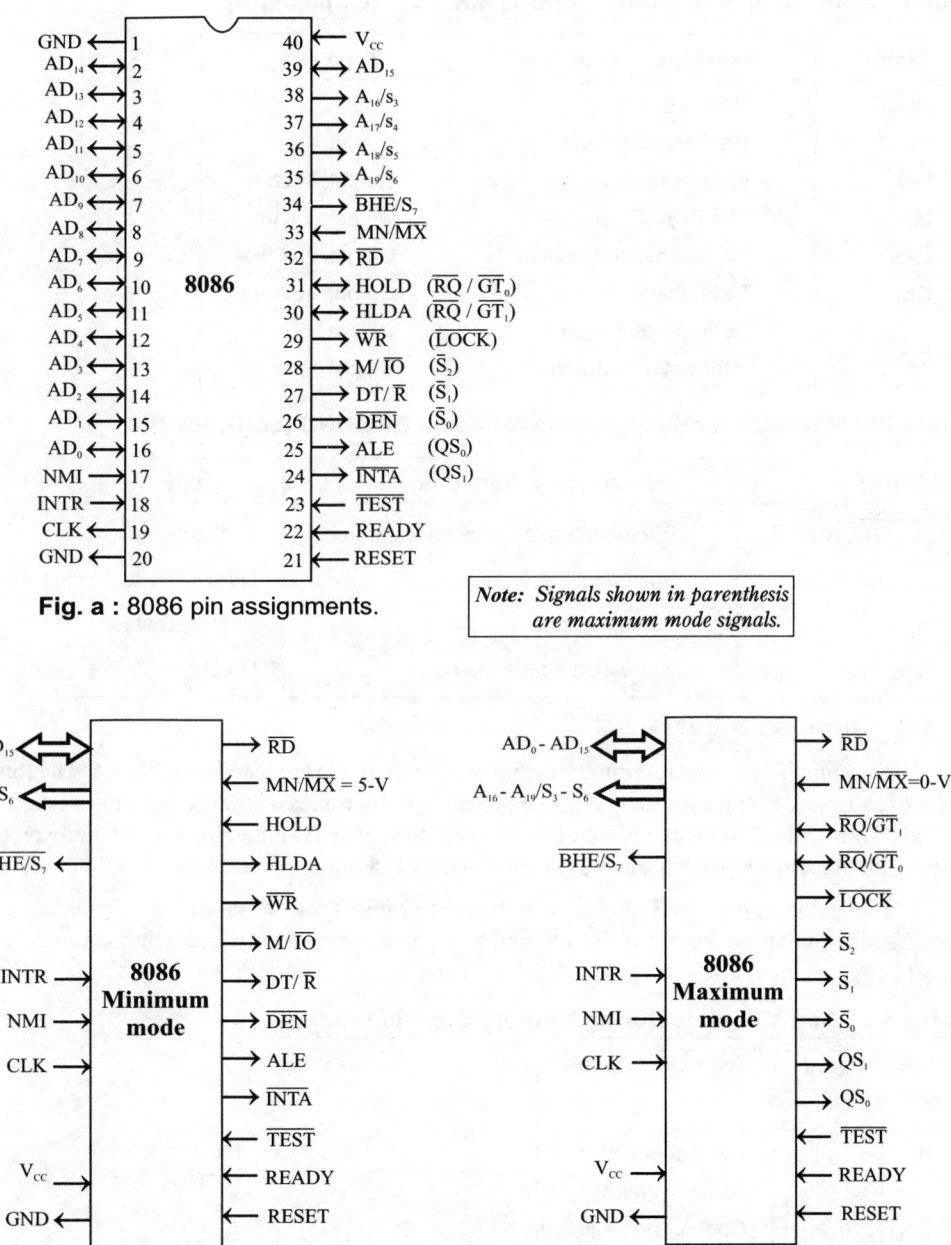

Fig. a : 8086 pin assignments.

Note: *Signals shown in parenthesis are maximum mode signals.*

Fig. b: 8086-Minimum mode.

Fig. c: 8086-Maximum mode.

Fig. 1.13: 8086 signals and pin assignment.

Table 1.5: Minimum Mode Signals [MN / $\overline{MX}$ = V$_{cc}$ (Logic high)]

Name	Description / Function	Type
HOLD	Hold request	Input
HLDA	Hold acknowledge	Output
$\overline{WR}$	Write control	Output, Tristate
M/$\overline{IO}$	Memory / IO control	Output, Tristate
DT/$\overline{R}$	Data transmit / Receive	Output, Tristate
$\overline{DEN}$	Data enable	Output, Tristate
ALE	Address latch enable	Output
$\overline{INTA}$	Interrupt acknowledge	Output

Table 1.6: Maximum Mode Signals [MN / $\overline{MX}$ = GROUND (Logic low)]

Name	Description / Function	Type
$\overline{RQ/GT}_1$, $\overline{RQ/GT}_0$	Request / Grant bus access control	Bidirectional
$\overline{LOCK}$	Bus priority lock control	Output, Tristate
$\overline{S}_2$, $\overline{S}_1$, $\overline{S}_0$	Bus cycle status	Output, Tristate
QS$_1$, QS$_0$	Instruction queue status	Output

1.8.2 Common Signals

The signals common for minimum and maximum mode are listed in Table 1.4. The lower sixteen lines of address are multiplexed with data and the upper four lines of address are multiplexed with status signals. During the first clock period of a bus cycle the entire 20-bit address is available on these lines. During all other clock periods of a bus cycle, the data and status signals will be available on these lines.

The status signals on S$_3$ and S$_4$ specifies the segment register used for calculating the physical address. The output on the status lines S$_3$ and S$_4$ when the processor is accessing various segments are listed in Table 1.7.

Table 1.7: Status Signals During Memory Segment Access

Status signal		Segment register
S$_4$	S$_3$	
0	0	Extra segment
0	1	Stack segment
1	0	Code or no segment
1	1	Data segment

The status lines S$_3$ and S$_4$ can be used to expand the memory up to 4 Mb. The status line S$_5$ indicates the status of an 8086 interrupt enable flag. A **low** on the line S$_6$ indicates that 8086 is on the bus (i.e., it indicates that 8086 is the bus master) and during hold acknowledge this pin is driven to **high impedance** state. The output signal $\overline{BHE}$ on the first T-state of a bus cycle is maintained as status signal S$_7$ on the same pin.

The 8086 outputs a **low** on $\overline{BHE}$ pin during read, write and interrupt acknowledge cycles when the data is to be transferred to the high-order data bus. The $\overline{BHE}$ can be used in conjunction with AD_0 to select memory banks.

When the processor reads from memory or an IO location it asserts $\overline{RD}$ **low**. The $\overline{TEST}$ input is tested by the WAIT instruction. The 8086 will enter a wait state after execution of the WAIT instruction, and it will resume execution only when $\overline{TEST}$ is made **low** by an external hardware. This is used to synchronize an external activity to the processor internal operation. $\overline{TEST}$ input is synchronized internally during each clock cycle on the leading edge of the clock signal.

INTR is the maskable interrupt and INTR must be held **high** until it is recognized to generate an interrupt signal. NMI is the non-maskable interrupt input activated by a leading edge signal.

RESET is the system reset input signal. For power-ON reset it is held **high** for 50 micro-second. For reset while working, it is held **high** for at least four clock cycles. When the processor is resetted, the DS, SS, ES, IP and flag register are cleared, **C**ode **S**egment (CS) register is initialized to $FFFF_H$ and queue is emptied. After reset the processor will start fetching instruction from 20-bit physical address $FFFF0_H$.

READY is an input signal to the processor, used by the memory or IO devices to get extra time for data transfer or to introduce **wait states** in the bus cycles. Normally READY is tied **high**. If the READY is tied **low**, the 8086 introduces wait states after second T-state of a bus cycle and it will complete the bus cycle only when READY is made **high** again.

CLK input is the clock signal that provides basic timing for the 8086 and bus controller. The 8086 does not have an on-chip clock generation circuit. Hence the 8284 clock generator chip is used to generate the required clock. A quartz crystal whose frequency is thrice that of the internal clock of 8086 must be connected to the 8284. The 8284 generates the clock at crystal frequency. It divides the generated clock by three and modifies the duty cycle to 33% and output on the CLK pin of the 8284. This CLK output of the 8284 must be connected to the 8086 CLK pin. The 8284 also provides the RESET and READY signal to an 8086.

1.8.3 Minimum Mode Signals

The minimum mode signals of an 8086 are listed in Table 1.5. For minimum mode of operation the $MN/\overline{MX}$ pin is tied to V_{cc} (logic **high**). In minimum mode, the 8086 itself generates all bus control signals. The minimum mode signals are explained below:

$DT/\overline{R}$ - [*Data Transmit / Receive*] It is an output signal from the processor to control the direction of data flow through the data transceivers.

$\overline{DEN}$ - (*Data Enable*) - It is an output signal from the processor used as output enable for the data transceivers.

ALE - (*Address Latch Enable*) - It is used to demultiplex the address and data lines using external latches.

$M/\overline{IO}$ - It is used to differentiate memory access and IO access. For IN and OUT instructions it is asserted **low**. For memory reference instructions it is asserted **high.**

$\overline{\text{WR}}$ -   It is a write control signal and it is asserted **low** whenever the processor writes data to memory or IO port.

$\overline{\text{INTA}}$ -   (*Interrupt Acknowledge*) - The 8086 output is asserted **low** on this line to acknowledge when the interrupt request is accepted by the processor.

HOLD   -   It is an input signal to the processor from other bus masters as a request to grant control of the bus. It is usually used by the DMA controller to get control of the bus.

HLDA   -   (*Hold Acknowledge*) - It is an acknowledge signal by the processor to the master requesting the control of the bus through HOLD. The acknowledge is asserted **high** when the processor accepts the HOLD. [*On accepting the hold the processor drives all the tristate pins to* **high** **impedance** *state and sends an acknowledgement to the device which requested HOLD. On receiving the acknowledgement the other master will take control of the bus.*]

1.8.4 Maximum Mode Signals

The maximum mode signals of an 8086 are listed in Table 1.6. An 8086-based system can be made to work in maximum mode by grounding the MN/$\overline{\text{MX}}$ pin (i.e., MN/$\overline{\text{MX}}$ is tied to logic **low**). In maximum mode, the pins 24 to 31 are redefined as follows:

$\overline{S_0}, \overline{S_1}, \overline{S_2}$ -   These are status signals and they are used by the 8288 bus controller to generate bus timing and control signals. The status signals are decoded as shown in Table 1.8.

Table 1.8: Status Signals During Various Machine Cycles

Status Signal			Machine Cycle
S_2	S_1	S_0	
0	0	0	Interrupt acknowledge
0	0	1	Read IO port
0	1	0	Write IO port
0	1	1	Halt
1	0	0	Code access
1	0	1	Read memory
1	1	0	Write memory
1	1	1	Passive/Inactive

$\overline{\text{RQ/GT}_0}$, -   (Bus Request/Bus Grant) These requests are used by the other local bus masters to force
$\overline{\text{RQ/GT}_1}$ the processor to release the local bus at the end of the processor's current bus cycle These pins are bidirectional. The request on GT_0 will have higher priority than GT_1.

The bus request to an 8086 works as follows:

1. When a local bus master requires an system bus control, it sends a low pulse to the 8086.
2. At the end of the current bus cycle, the processor (8086) drives its pins to high impedance state and sends an acknowledgement as a low pulse on the same pin to the device which had requested the bus control.

3. On receiving the acknowledgement the local master will take control of the system bus. After completing its work, at the end, the local bus master sends a low signal on the same pin to 8086 to inform the end of control. Now 8086 regains the control of the bus.

$\overline{LOCK}$ - It an output signal activated by the LOCK prefix instruction and remains active until the completion of the instruction prefixed by LOCK. The 8086 asserts the $\overline{LOCK}$ pin low while executing an instruction prefixed by LOCK to prevent other bus masters from gaining control of the system bus.

QS_1, QS_0 - (Queue Status) - The processor provides the status of queue on these lines. The queue status can be used by the external device to track the internal status of the queue in an 8086. The QS_0 and QS_1 are valid during the clock period following any queue operation. The output on QS_0 and QS_1 can be interpretted as shown in Table 1.9.

Table 1.9: Queue Status

Queue status		Queue operation
QS_1	QS_0	
0	0	No operation
0	1	First byte of an opcode from the queue
1	0	Empty the queue
1	1	Subsequent byte from the queue

1.8.5 Architecture of INTEL 8086

The 8086 has a pipelined architecture. In pipelined architecture the processor will have a number of functional units and the execution time of the functional units are overlapped. Each functional unit works independently most of the time. The simplified block diagram of the internal architecture of an 8086 is shown in Fig. 1.14. The architecture of the 8086 can be internally divided into two separate functional units: **B**us Interface Unit (BIU) and **E**xecution Unit (EU).

The BIU fetches instructions, reads data from memory and IO ports and writes data to memory and IO ports. The BIU contains segment registers, an instruction pointer, an instruction queue, an address generation unit and a bus control unit. The EU executes instructions that have already been fetched by the BIU. The BIU and EU function independently.

The instruction queue is a FIFO (**F**irst-**I**n-**F**irst-**O**ut) group of registers. The size of the queue is 6 bytes. The BIU fetches the instruction code from memory and stores it in queue. The EU fetches the instruction codes from the queue.

The BIU has four 16-bit segment registers: **C**ode **S**egment (CS) register, **D**ata **S**egment (DS) register, **S**tack **S**egment (SS) register and **E**xtra **S**egment (ES) register. The 8086 memory space can be divided into segments of 64 kB. The 4-segment registers are used to hold four segment base addresses. Hence 8086 can directly address 4 segments of 64 kB at any time instant (4 × 64 = 256 kB within 1MB memory space). This feature of the 8086 allows the system designer to allocate separate areas for storing program codes and data.

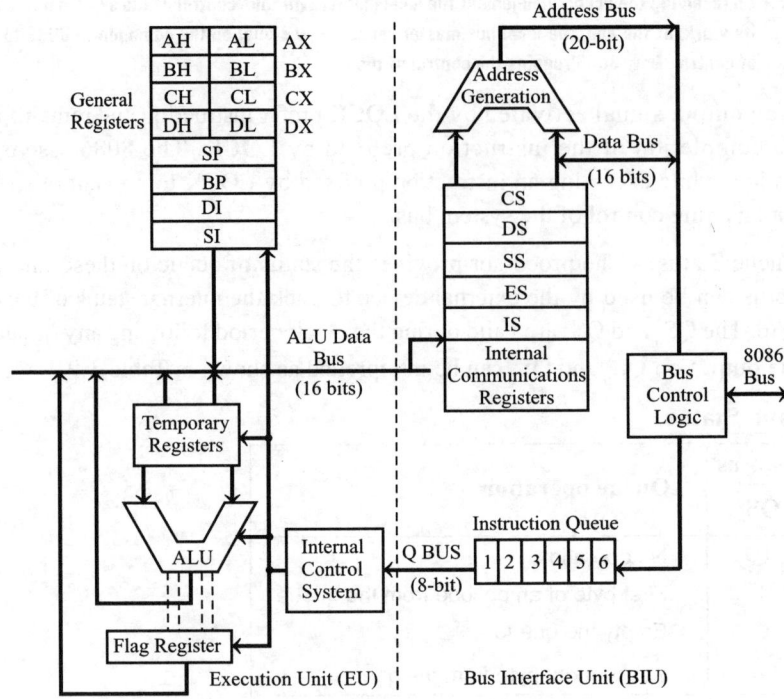

Fig. 1.14: Internal architecture of 8086.

The contents of the segment registers are programmable. Hence the processor can access the code and data in any part of the memory by changing the contents of the segment registers. The memory segment can be continuous, partially overlapped, fully overlapped or disjointed.

> **Note:** *Since segment registers are programmable it is possible to design multitasking and multiuser systems using 8086. The program code and data for each task/user can be stored in separate segments. The program execution can be switched from one task/user to another by changing the contents of the segment registers.*

The dedicated address generation unit generates a 20-bit physical address from the segment base and an offset or effective address. The segment base address is logically shifted left four times and added to the offset. *[logically shifting left four times is equal to multiplying it by 16_{10}.]*

The address for fetching instruction codes is generated by logically shifting the content of the CS to the left four times and then adding it to the content of the IP (**I**nstruction **P**ointer). The IP holds the offset address of the program codes. The content of the IP gets incremented by two after every bus cycle. *[In one bus cycle the processor fetches two bytes of the instruction code.]*

The data address is computed by using the content of the DS or ES as the base address and an offset or effective address specified by the instruction.

The stack address is computed by using the content of the SS as the base address and the content of the SP (Stack **P**ointer) as the offset address or effective address.

The bus control logic of the BIU generates all the bus control signals such as read and write signals for memory and IO. The EU consists of the ALU, the flag register and the general purpose registers. The EU decodes and executes the instructions. A decoder in the EU control system translates the instructions.

The EU has a 16-bit ALU to perform arithmetic and logical operations. The EU has eight numbers of 16-bit general purpose registers. They are AX, BX, CX, DX, SP, BP, SI and DI.

Some of the 16-bit registers can also be used as two numbers of 8-bit registers as given below:

AX - can be used as AH and AL

BX - can be used as BH and BL

CX - can be used as CH and CL

DX - can be used as DH and DL

The general purpose registers can be used for data storage when they are not involved in any special functions assigned to them. These registers are named after special functions carried out by each one of them as given in Table 1.10.

Table 1.10: Special Functions of 8086 Registers

Register	Name of the register	Special function
AX	16-bit Accumulator	Stores the 16-bit result of certain arithmetic and logical operations.
AL	8-bit Accumulator	Stores the 8-bit result of certain arithmetic and logical operations.
BX	Base Register	Used to hold the base value in base addressing mode to access memory data
CX	Count Register	Used to hold the count value in SHIFT, ROTATE and LOOP instructions.
DX	Data Register	Used to hold data for multiplication and division operations.
SP	Stack Pointer	Used to hold the offset address of top of stack memory.
BP	Base Pointer	Used to hold the base value in base addressing using stack segment register to access data from stack memory.
SI	Source Index	Used to hold the index value of source operand (data) for string instructions.
DI	Destination Index	Used to hold the index value of destination operand (data) for string instruction.

1.8.6 8086 Flag Register

The size of an 8086 flag register is 16 bits and in this nine bits are defined as flags. The six flags are used to indicate the status of the result of the arithmetic or logical operations. Three flags are used to control the processor operation and so they are also called control bits. The various flags of an 8086 processor and their bit position in flag register are shown in Fig. 1.15.

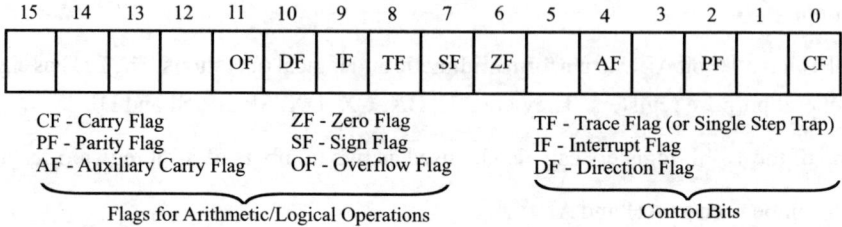

Fig. 1.15: Bit positions of various flags in the flag register of 8086.

The **C**arry **F**lag (CF) is set if there is a carry from the addition or borrow from the subtraction. Auxiliary carry **F**lag (AF) is set if there is a carry from low nibble to high nibble of the low order 8-bit of a 16-bit number.

The **O**verflow **F**lag (OF) is set to **one** if there is an arithmetic overflow, that is, if the size of the result exceeds the capacity of the destination location. **S**ign **F**lag (SF) is set to **one** if the most significant bit of the result is **one** and SF is cleared to **zero** for non-negative result. The **P**arity **F**lag (PF) is set to **one** if the result has even parity and PF is cleared to **zero** for odd parity of the result. The **Z**ero **F**lag (ZF) is set to **one** if the result is zero and ZF is cleared to **zero** for a non zero result.

The three control bits in the flag register can be set or reset by the programmer. The **D**irection **F**lag (DF) is set to **one** for autodecrement and reset to **zero** for autoincrement of the SI and DI registers during string data accessing. Setting **I**nterrupt **F**lag (IF) to **one** causes the 8086 to recognize the external maskable interrupts, and clearing IF to **zero** disables the interrupts.

Setting **T**race **F**lag (TF) to **one**, places the 8086 in the single step mode. In this mode the 8086 generates an internal interrupt after execution of each instruction. The single stepping is used for debugging a program.

1.8.7 Instruction and Data Flow in 8086

The 8086 microprocessor allows the user to define different memory areas for storing the program and data. The program memory can be accessed using CS-register and the data memory can be accessed using DS, ES and SS registers.

The program instructions are stored in the program memory which is an external device. To execute a program in 8086, the base address and offset address of the first instruction of the program should be loaded in CS-register and IP respectively.

The 8086 computes the 20-bit physical address of the program instruction by multiplying the content of the CS-register by 16_{10} and adding to the content of the IP. The 20-bit physical address is given out on the address bus.

Then $\overline{RD}$ is asserted **low**. Also other control signals necessary for program memory read operation are asserted. The IP is incremented by two to point the next instruction or the next word of the same instruction.

The address and control signals enable the memory to output one word (two bytes) of program memory on the data bus. After a predefined time the $\overline{RD}$ is asserted **high** and at this instant the content of the data bus is latched into two empty locations of the instruction queue. Then the BIU starts fetching the next word, of the program code as explained above. The BIU keeps on fetching the program codes, word by word, from consecutive memory locations whenever two locations of queue is empty. When a branch instruction is encountered, the queue is emptied and then filled with program codes from the new address loaded in the CS and IP by the branch instruction.

The EU reads the program instructions from the queue, decodes and executes them one by one. If the execution of an instruction requires data from memory (or to store data in memory) then BIU is interrupted to read (or write) data in memory. When BIU is interrupted it completes the fetching of the current instruction word and then starts reading/writing the data by generating a 20-bit data memory address. The 20-bit data memory address is obtained by multiplying the content of the segment base register specified by the instruction by 16_{10} and adding to an effective or offset address specified by the instruction.

1.9 MOTOROLA MC68000

The MC68000 is Motorola's first 16-bit microprocessor. It has a 16-bit data bus and 24-bit address bus to address up to 16 MB of physical memory space. The 16 MB physical memory space is organized as two banks of 8 MB each. The MC68000 is designed using HMOS transistors and requires a single +5-V supply.

In the 68000 family Motorola has released a number of processors which share a common base architecture but differ in data bus size, address bus size, instruction set, operating system support and performance. The 68000 family of processors are listed in Table 1.11.

The 68000 does not have an on-chip clock circuitry and hence it requires an external clock generator to generate the required clock and supply to the processor. Its maximum internal clock is 25 MHz. The 68000 is available with a maximum clock rating of 6, 8, 10, 12.5, 16.67 and 25 MHz.

Table 1.11: 68000 Family of Processors

Processor	Address bus size	Data bus size
MC68000	24	16
MC6808	20	8
MC68010	24	16
MC68012	31	16
MC68020	32	8/16/32
MC68030	32	8/16/32

The 68000 has a general register based architecture and in this architecture any register can be used as the accumulator or scratch pad register. The internal address and data registers of 68000 are 32-bit wide and its ALU is 16-bit wide. It operates on five different data types: 4-bit BCD, 8-bit, 16-bit and 32-bit binary data. The 68000 has 56 basic instructions and has more than 1000 opcodes. It has 14 addressing modes and supports only memory-mapped IO.

The 68000 has two operating modes: the supervisor mode and the user mode. The supervisor mode is also called the operating system mode, and in this mode the processor can execute all the instructions. Upon hardware reset the 68000 enters the supervisor mode. The processor can switch from the supervisor mode to the user mode by clearing the S-bit in the status register. The processor can switch from the user mode to the supervisor mode by recognition of a trap/reset/interrupt.

1.9.1 Pins and Signals of 68000

The MC68000 microprocessor is a 64-pin IC available in **D**ual **I**n-line **P**ackage (DIP). It is also available as a 68-pin IC in quad or **P**in **G**rid **A**rray (PGA) package or available as a 68-terminal chip carrier. In the 68-pin version the extra four pins are either ground or NC (**N**o **C**onnection).

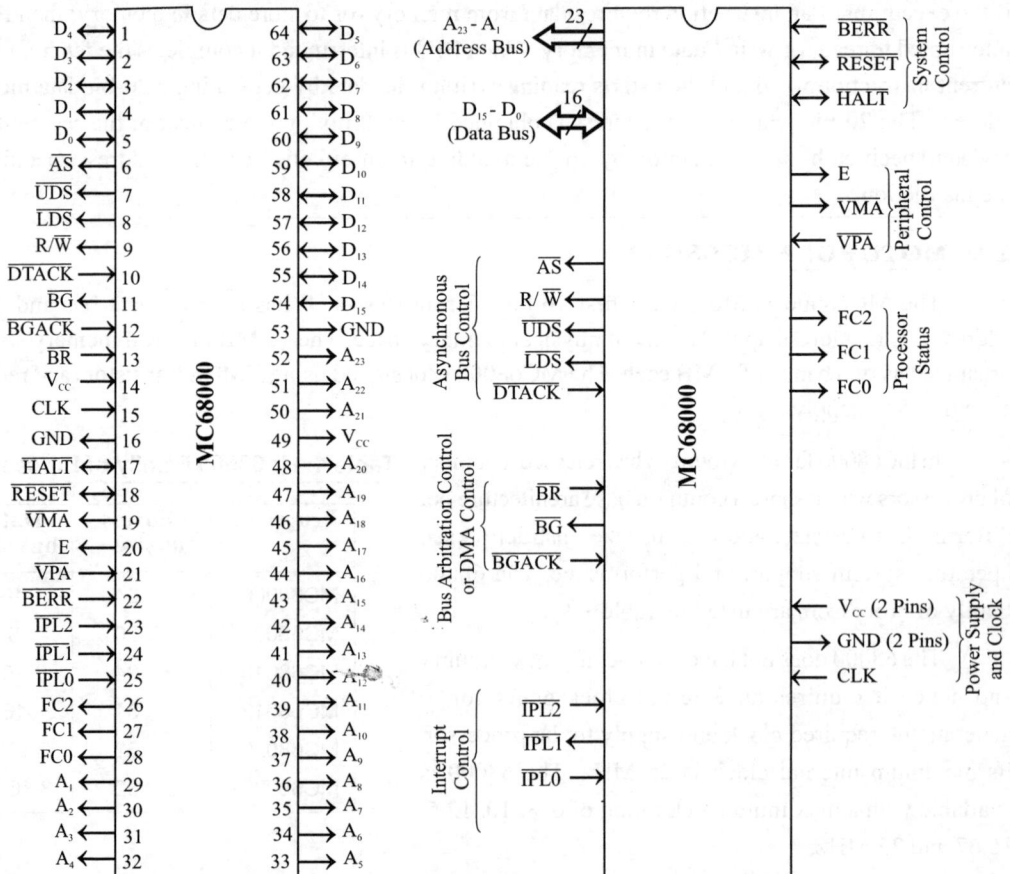

Fig. 1.16: Pin description of MC68000 microprocessor.

The pin configuration of the 64-pin version of MC68000 in DIP is shown in Fig. 1.16. The signals of a 68000 processor are listed in Table 1.12.

The 68000 has 16 data pins D_{15} - D_0 to form a 16-bit data bus and 23 address pins A_{23} - A_1 to address up to 8 MB of physical memory space. The address bit A_0 is internally decoded and supplied as two memory bank select signals $\overline{UDS}$ and $\overline{LDS}$. Hence, 68000-based systems can have two memory banks with a capacity of 8 MB in each bank and so the total addressable memory space is 16 MB. The two memory banks are called odd bank and even bank. The odd bank is enabled by $\overline{LDS}$ and the even bank is enabled by $\overline{UDS}$. The data lines D_{15} to D_8 are connected to the even bank and the data lines D_7-D_0 are connected to the odd bank.

The processor can access bytes from the even bank via D_{15} - D_8 lines by asserting $\overline{UDS}$ as **low** or can access bytes from the odd bank via D_7 - D_0 lines by asserting $\overline{LDS}$ as **low**. For word access both the banks are simultaneously enabled so that one byte from each bank forms a word.

The MC68000 can perform synchronous or asynchronous data transfer with peripherals. The synchronous data transfer can be employed for peripherals having compatible timing with the processor and asynchronous data transfer can be employed for slow peripherals.

The synchronous data transfer between the processor and the peripheral involves the following operations:

1. The 68000 initiates read/write operation by sending an address and asserting $\overline{AS}$ as low to inform the peripheral that a valid address is on the bus.
2. The peripheral will assert $\overline{VPA}$ as low to inform the processor that it requires synchronization with clock.
3. The 68000 synchronizes read/write operation with E-clock when it is low.
4. Then 68000 asserts $\overline{VMA}$ as low to inform the peripheral that it is synchronized.
5. When E-clock is high the peripheral transfer the data.

The asynchronous data transfer between the processor and the peripheral involves the following operations:

1. The 68000 initiates read/write operation by sending as address and asserting $\overline{AS}$ as low to inform the peripheral that a valid address is on the bus.
2. Then 68000 remains in wait state until it gets a $\overline{DTACK}$ signal from the peripheral.
3. When the data is ready, the peripheral will assert $\overline{DTACK}$ as low.
4. On receiving the $\overline{DTACK}$ signal the processor will accomplish the read/write operation using R/W, UDS and $\overline{LDS}$.

The $\overline{BERR}$ is an error signal sent by the external timer/peripheral device to inform the processor that some error has occurred in the machine cycle. (Usually in a 68000-based system an external timer is used to monitor the timing of the machine cycle.) When the 68000 receives an error signal it either returns the instruction cycle which caused the error or executes an error service routine.

Table 1.12: Signals of a 68000 Processor

Signal	Type	Description
A_{23}-A_1	Output	23 pins for address. Used to form 23-bit address bus to address up to 8 MB memory space.
$\overline{UDS}$	Output	Upper Data Strobe. Used to select even memory bank.
$\overline{LDS}$	Output	Lower Data Strobe. Used to select odd memory bank.
D_{15}- D_0	Bidirectional	16 pins for data. Used to form 16-bit data bus.
$\overline{AS}$	Output	Address strobe. It is asserted **low** by the processor to indicate a valid address whenever an address is output on the address bus.
R/$\overline{W}$	Output	Read/Write control signal. For a read operation it is asserted **high** and for write operation it is asserted **low**.
$\overline{DTACK}$	Input	Data acknowledge. Used for asynchronous data transfer between processor and peripheral. It is an acknowledge signal supplied by the peripheral to complete the bus cycle.
E	Output	Enable or E-clock. It is a clock output at one-tenth(1/10) of processor clock.
$\overline{VMA}$	Output	Valid Memory Address. An acknowledge from processor in response to $\overline{VPA}$.
$\overline{VPA}$	Input	Valid Peripheral Address. A request from peripheral to synchronize with clock.
$\overline{BERR}$	Input	Bus error. Input signal from peripheral to indicate bus error.
$\overline{RESET}$	Bidirectional	Reset input/output. As input it is used to bring the processor to known state. As output it is used to reset the peripherals.
$\overline{HALT}$	Bidirectional	Halt control. Used to drive the processor to **high impedance** state during single step or error condition or DMA.
$\overline{BR}$	Input	Bus Request. Used by other bus master to make a request for the bus.
$\overline{BG}$	Output	Bus Grant. It is an acknowledge signal from the processor in response to $\overline{BR}$ signal.
$\overline{BGACK}$	Input	Bus Grant Acknowledge. It is an acknowledgement from the other master controlling the bus.
$\overline{IPL2}$ to $\overline{IPL0}$	Input	Interrupt Privilege Level (or priority level).
FC2-FC0	Output	Function code.
CLK	Input	Clock input.
V_{cc}	Input	Power supply (+5-V).
GND	Output	Power supply ground (0-V).

The $\overline{HALT}$ input signal is used to drive the processor to **high impedance** state. The $\overline{HALT}$ is asserted **low** in order to perform single stepping or to indicate double error. It can also be used by other bus masters (such as the DMA controller) to take control of the system bus. The $\overline{HALT}$ signal can also be used as the output signal. During a major failure the processor asserts the $\overline{HALT}$ signal as **low** to inform the peripheral and goes to **high impedance** state.

In order to reset the system both $\overline{RESET}$ and $\overline{HALT}$ pins must be asserted **low** at the same time. The 68000 $\overline{RESET}$ pin can also be used as an output signal. In the supervisor mode, when the RESET instruction is executed, the processor outputs **low** signal on the $\overline{RESET}$ line in order to reset the peripherals connected to this pin. During a hardware reset through $\overline{RESET}$ pin the **P**rogram **C**ounter (PC), **S**tack **P**ointer (SP) and **S**tatus **R**egister (SR) are initialized, and other registers are not altered.

The input signals $\overline{IPL2}$, $\overline{IPL1}$ and $\overline{IPL0}$ are used to initiate seven hardware interrupts. The input signal 000_2 to 110_2 will initiate interrupt level-7 to level-1. (The complement of the input gives the interrupt level.) The interrupt level-7 has the highest priority and level-1 has the lowest priority. Level-7 is a nonmaskable interrupt while the other interrupts are maskable.

The signals $\overline{BR}$, $\overline{BG}$ and $\overline{BGACK}$ are used as bus arbitration signals in multi-master systems. In a 68000 processor-based multimaster system, the 68000 processor will be the main/primary master and the other masters such as DMA controller will be the secondary master. The $\overline{BR}$ signal is used by the secondary master to make a request for the control of the system bus.

On receiving the $\overline{BR}$ signal the processor will send an acknowledge via the $\overline{BG}$ line to the secondary master which made a request for the bus. The secondary master has to check for completion of the current bus cycle and then assert $\overline{BGACK}$ signal as **low** to drive the processor to **high impedance** state. The processor will come to a normal state only when $\overline{BGACK}$ is asserted **high** by the secondary master.

The 68000 processor outputs the bus status signal on pins FC2 to FC0 to indicate whether the processor is fetching data/program from user/supervisor memory or the processor is servicing an interrupt.

The status signals of a 68000 processor are listed in Table 1.13. In a 68000 processor the status signals are also known as function codes.

Table 1.13: Status Signals of the MC68000

FC2	FC1	FC0	Operation performed by bus
0	0	1	User data access
0	1	0	User program access
1	0	1	Supervisor data access
1	1	0	Supervisor program access
1	1	1	Interrupt acknowledge

Note: Other codes are not defined.

1.9.2 Architecture of a MC68000 Microprocessor

The architecture of a MC68000 microprocessor is shown in Fig. 1.17. It has a general register-based architecture in which any data register can be used as the accumulator or scratch-pad register. The 68000 processor has seventeen numbers of 32-bit registers in which eight registers are called data registers and the remaining nine registers are called address registers. The data registers are denoted as D0 - D7 and the address registers are denoted as A0 - A7, A7′.

The data registers are used to store 8/16/32 bit data. The address registers are used to store 24-bit address. (The upper 8-bits of address registers are ignored in 68000 processor.) The address register A7 is the **Stack Pointer (SP)** or **User Stack Printer (USP)** and the address register A′ is the **Supervisor Stack Pointer (SSP)**.

The 68000 processor has a 16-bit ALU **(Arithmetic Logic Unit)** which can operate on 8, 16 or 32-bit data. The architecture includes a 32-bit temporary register which is used to hold the operand or intermediate result during the execution of an instruction.

The 68000 processor has a 16-bit **Instruction Register (IR)** which holds the first word of the currently executing instruction. The control unit decodes the instructions and controls the other blocks to fetch and execute the instructions.

The architecture includes a 32-bit **Program Counter (PC)** in which the lower 24-bit portion is used to hold the 24-bit address of instruction being executed. When an instruction word is fetched the PC will hold the address of next word of same instruction or the address of next instruction.

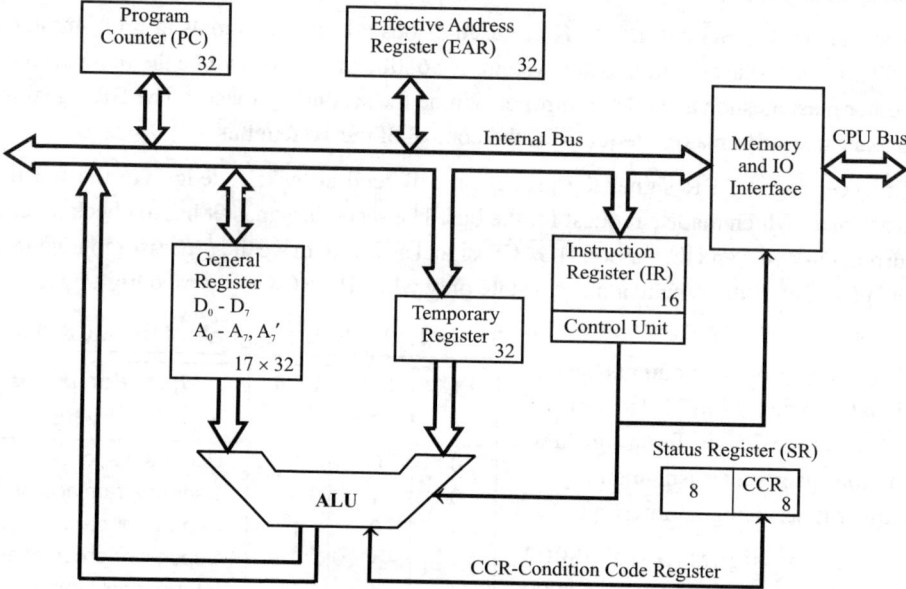

Fig. 1.17: Architecture of MC68000 microprocessor.

The memory and IO interface takes care of the read/write operation with memory or IO device as commanded by the control unit. The 68000 processor has a 16-bit status register in which the lower 8-bit portion is also known as **Condition Code Register (CCR)**.

The format of the status register of a 68000 microprocessor is shown in Fig. 1.18. The content of the status register is also known as status word in which the lower byte is called user byte and upper byte is called system byte.

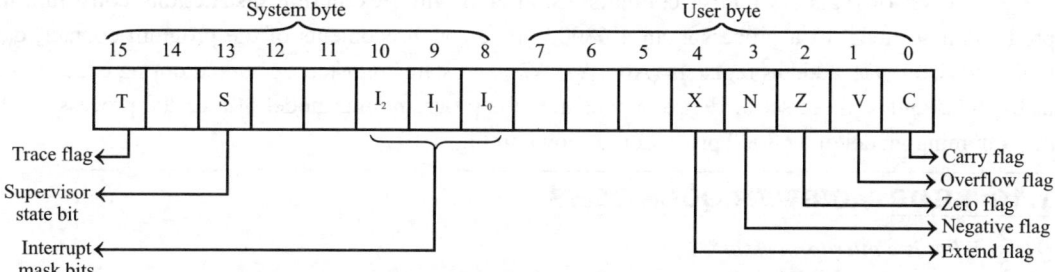

Fig. 1.18: Status register of a MC68000 microprocessor.

The user byte consists of arithmetic flags like carry, overflow, negative, zero and extend flags. During arithmetic operations the carry and extend flag are affected in an identical manner. The instruction set of a 68000 processor includes instructions which uses extend flag for addition with carry or subtraction with borrow.

The system byte consists of trace flag, supervisor state bit and interrupt mask bits. When the trace flag is set to one, the 68000 processor generates an internal interrupt called trap after execution of each instruction and so the trace flag can be used for single step execution of programs for debugging.

The supervisor state bit is used to switch the operating mode from the supervisor mode to the user mode or vice versa. When S-bit is set to one, the processor operates in the supervisor mode and when S-bit is cleared to zero, the processor operates in the user mode. The interrupt mask bits provide the status of 68000 interrupt input pins IPL2, IPL1 and IPL0. The signals on these pins are inverted and then stored as I_2, I_1 and I_0 respectively.

1.9.3 Programming Model of a 68000 Microprocessor

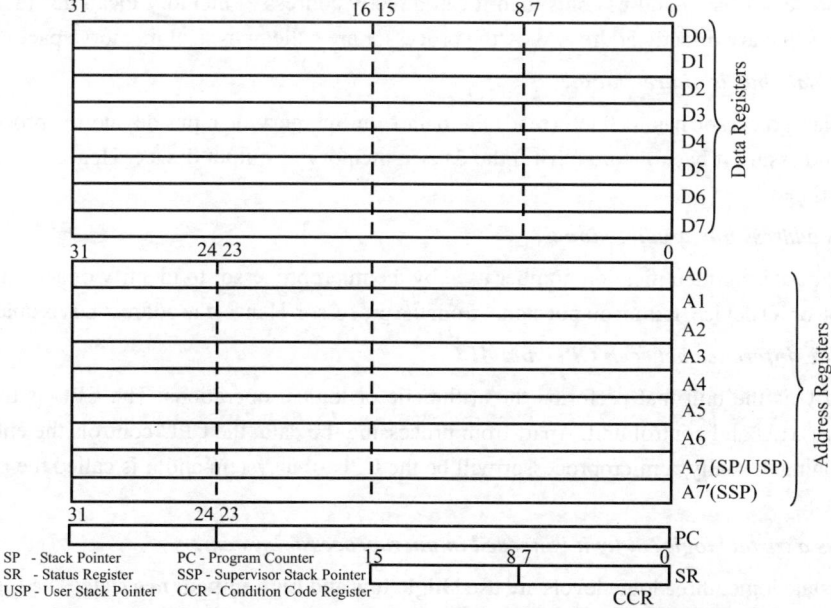

Fig. 1.19: Programming model of a MC68000 microprocessor.

The set of registers whose contents are altered while executing instructions constitute the programming model of a processor. In a 68000 processor the contents of the program counter, data registers, (D0 - D7), address registers (A0 - A7, A7′) and status register are altered during execution of instructions and so these set of registers constitute the programming model of a 68000 processor. The programming model of a 68000 processor is shown in Fig. 1.19.

1.10 SHORT-ANSWER QUESTIONS

Q1.1 *What is a microprocessor ?*

A microprocessor is a program controlled semiconductor device (IC), which fetches, decodes and executes instructions.

Q1.2 *What are the basic functional blocks of a microprocessor ?*

The basic functional blocks of a microprocessor are ALU, an array of registers and control unit.

Q1.3 *What is a bus ?*

Bus is a group of conducting lines that carries data, addresses and control signals.

Q1.4 *Define bit, byte and word.*

A digit of the binary number or code is called bit. The bit is also the fundamental storage unit of computer memory.

The 8-bit (8-digit) binary number or code is called byte and 16-bit binary number or code is called word. (Some microprocessor manufacturers refer to the basic data size operated by the processor as word.)

Q1.5 *State the relation between the number of address pins and physical memory space?*

The size of the binary number used to address the memory decides the physical memory space. If a microprocessor has n-address pins then it can directly address 2^n memory locations. (The memory locations that are directly addressed by the processor are called physical memory space.)

Q1.6 *Why is data bus is bidirectional?*

The microprocessor has to fetch (read) the data from memory or input device for processing and after processing it has to store (write) the data in memory or output device. Hence, the data bus is bidirectional.

Q1.7 *Why is address bus unidirectional?*

The address is an identification number used by the microprocessor to identify or access a memory location or IO device. It is an output signal from the processor. Hence, the address bus is unidirectional.

Q1.8 *State the difference between CPU and ALU.*

The ALU is the unit that performs the arithmetic or logical operations. The CPU is the unit that includes ALU and control unit. Apart from processing the data, the CPU controls the entire system functioning. Usually, a microprocessor will be the CPU of a system and it is called the brain of the computer.

Q1.9 *What is a tristate logic? Why it is needed in microprocessor system?*

In a tristate logic, three logic levels are used **high, low** and **high impedance** state. The **high** and **low** are normal logic levels and **high impedance** state is electrical open circuit condition.

In a microprocessor system, all the peripheral/slave devices are connected to a common bus. But communication (data transfer) takes place between the master (microprocessor) and one slave (peripheral) at any time instant. During this time instant, all other devices should be isolated from the bus. Therefore, normally all the slaves (peripherals) will remain in **high impedance** state (i.e., in electrical isolation). The master will select a slave by sending address and chip select signal. When the slave is selected, it comes to normal logic and it can communicate with the master.

Q1.10 What is HMOS and HCMOS.

The HMOS is High density n-type Metal Oxide Silicon field effect transistors. The third generation microprocessors are fabricated using HMOS transistors.

The HCMOS is High density n-type Complementary Metal Oxide Silicon field effect transistors. It is the low power version of HMOS and the fourth generation microprocessors are fabricated using HCMOS transistors.

Q1.11 What are the drawbacks of first generation microprocessors.

The first generation processors are fabricated using PMOS technology and it has the drawbacks like slow speed, provides low output currents and was not compatible with TTL logic levels.

Q1.12 What is a microcomputer? Explain the difference between a microprocessor and a microcomputer.

A system designed using a microprocessor as its CPU is called microcomputer. The term microcomputer refers to the whole system, whereas the microprocessor is the CPU of the system.

Q1.13 What is the function of microprocessor in a system?

The microprocessor is the master in the system, which controls all the activity of the system. It issues address and control signals and fetches the instruction and data from memory. Then it executes the instruction to take appropriate action.

Q1.14 List the components of microprocessor-based (single board microcomputer) system.

The microprocessor-based system consist of microprocessor as CPU, semiconductor memories like EPROM and RAM, input device, output device and interfacing devices.

Q1.15 Why interfacing is needed for IO devices?

Generally IO devices are slow devices. Therefore, the speed of IO devices does not match with the speed of microprocessor. And so an interface is provided between system bus and IO devices.

Q1.16 What is the difference between CPU bus and system bus?

The CPU bus has multiplexed lines but the system bus has separate lines for each signal. (The multiplexed CPU lines are demultiplexed by the CPU interface circuit to form system bus.)

Q1.17 What is multiplexing and what is its advantage?

Multiplexing is transferring different information at different well-defined times through same lines. A group of such lines is called multiplexed bus. The advantage of multiplexing is that fewer pins are required for microprocessors to communicate with the outside world.

Q1.18 How the address and data lines are demultiplexed in 8085?

The low order address and data lines of 8085 are demultiplexed using an external 8-bit D-Latch (74LS373) and the ALE signal of 8085, as shown in Fig. Q1.18.

At the beginning of every machine cycle, ALE is asserted **high** and then **low**. Also, the low byte of address is given out through AD_0 - AD_7 lines. Since, the ALE is connected to enable of latch, whenever ALE is asserted **high** and then **low,** the addresses are latched into the output lines of the latch then the lines AD_0 - AD_7 are free for data transfer.

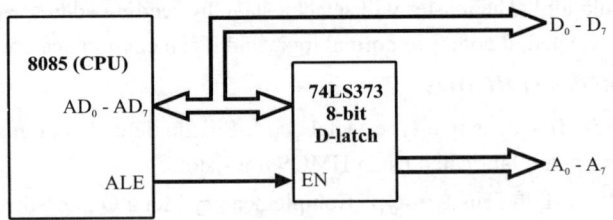

Fig. Q1.18: Demultiplexing of address and data lines in 8085.

Q1.19 *What do you mean by 16 and 8-bit processors? Mention a few 8-bit and 16-bit processors.*

The processors are classified into 8-bit or 16-bit depending on the basic data size handled by the ALU of the processor.

8-bit microprocessors : 8085, Z80, Motorola 6800.

16-bit microprocessors : 8086, Z8000, MC68000.

Q1.20 *What is the fabrication technology used for 8085?*

The 8085A is fabricated used NMOS technology and 8085AH is fabricated using HMOS technology.

Q1.21 *What is the physical memory space in 8085?*

The 8085 uses 16-bit address to access memory locations. Hence, it can directly address 64 k memory locations (2^{16} = 65,536 = 64 k). Since 8085 has 8 data lines, it can read or write 8-data bits from a memory address. Therefore, the physical memory space is 64 k × 1byte = 64 kilobytes (64 kB).

Q1.22 *What is ALE?*

The ALE (**A**ddress **L**atch **E**nable) is a signal used to demultiplex the address and data lines using an external latch. It is used as enable signal for the external latch.

Q1.23 *Explain the function of IO/$\overline{M}$ in 8085.*

The IO/$\overline{M}$ is used to differentiate memory access and IO access. For IN and OUT instruction it is asserted **high**. For memory reference instructions it is asserted **low**.

Q1.24 *How the READY signal is used in microprocessor system?*

The READY is an input signal that can be used by slow peripherals to get extra time in order to communicate with 8085. The 8085 will work only when READY is tied to logic **high**. Whenever READY is tied to logic **low**, the 8085 will enter a wait state. When the system has slow peripheral devices, additional hardware is provided in the system to make the READY input **low** during the required extra time while executing a machine cycle, so that the processor will remain in wait state during this extra time.

Q1.25 What is HOLD and HLDA? How is it used?

The HOLD and HLDA signals are used for the **D**irect **M**emory **A**ccess (DMA) type of data transfer. These type of data transfers are achieved by employing a DMA controller in the system. When DMA is required, the DMA controller will place a **high** signal on the HOLD pin of 8085. When HOLD input is asserted **high,** the processor will enter a wait state and drive all its tristate pins to **high impedance** state and send an acknowledge signal to DMA controller through HLDA pin. Upon receiving the acknowledge signal, the DMA controller will take control of the bus and perform DMA transfer and at the end it asserts HOLD signal **low**. When HOLD is asserted **low,** the processor will resume its execution.

Q1.26 How clock signals are generated in 8085 and what is the frequency of the internal clock?

The 8085 has the clock generation circuit on the chip but an external quartz crystal or LC circuit or RC circuit should be connected at the pins X_1 and X_2 in order to generate a clock signal. The 8085 clock generation circuit, generate a clock whose frequency is double as compared to that of internal clock. The generated clock is divided by two and then used as internal clock. The maximum internal clock frequency of 8085A is 3.03 MHz.

Q1.27 What happens to the 8085 processor when it is resetted?

When $\overline{\text{RESET IN}}$ pin is asserted **low**, the program counter, instruction register, interrupt mask bits and all internal registers are cleared or resetted. Also the RESET OUT signal is asserted **high** to clear or reset all the peripheral devices in the system. After a reset, the content of program counter will be 0000_H and so the processor will start executing the program stored at 0000_H.

Q1.28 What are the operations performed by ALU of 8085?

The operations performed by ALU of 8085 are addition, subtraction, logical AND, OR, Exclusive-OR, compare, complement, increment, decrement and left/right shift.

Q1.29 Mention the names of various registers in 8085 along with its size.

Register		Size (bits)	Register		Size (bits)
Accumulator (A)	-	8	Stack pointer	-	16
Temporary register	-	8	Program counter	-	16
Instruction register	-	8			
SGeneral purpose register	-	8			
(B, C, D, E, H and L)					

Q1.30 What is a flag?

Flag is a flip-flop used to store the information about the status of the processor and the status of the instruction executed most recently.

Q1.31 List the flags of 8085.

There are five flags in 8085. They are sign flag, zero flag, auxiliary carry flag, parity flag and carry flag.

Q1.32 What are the hardware interrupts of 8085?

The hardware interrupts in 8085 are TRAP, RST 7.5, RST 6.5 and RST 5.5.

Q1.33 Which interrupt has highest priority in 8085? What is the priority of other interrupts?

The TRAP has the highest priority, followed by RST 7.5, RST 6.5, RST 5.5 and INTR.

Q1.34 Show the bit positions of various flags in 8085 flag register.

The bit positions of various flags in the flag register of 8085 is shown in Fig. Q1.34.

D_7	D_6	D_5	D_4	D_3	D_2	D_1	D_0
SF	ZF		AF		PF		CF

SF - Sign Flag
PF - Parity Flag
ZF - Zero Flag
AF - Auxiliary Carry Flag
CF - Carry Flag

Fig. Q1.34: Bit positions of various flags in the flag register of 8085.

Q1.35 Define stack.

Stack is a sequence of RAM memory locations defined by the programmer.

Q1.36 What is program counter? How is it useful in program execution?

The program counter keeps a track of program execution. To execute a program, the starting address of the program is loaded in program counter. The PC sends out an address to fetch a byte of instruction from memory and increment its content automatically.

Q1.37 How is the microprocessor synchronized with peripherals ?

The timing and control unit synchronizes all the microprocessor operations with clock and generates control signals necessary for communication between the microprocessor and peripherals.

Q1.38 What are the additional features in Z80, as compared to an 8085?

The Z80 has separate pins for data and address. The Z80 provides more register, extra addressing modes, a larger instruction set than 8085 and it has built-in logic to refresh dynamic RAM memories. The Z80 has an indexed addressing mode.

Q1.39 What are shadow registers of Z80?

Each register of Z80 has an alternate register. The set of alternate registers are called shadow registers.

Q1.40 How are the control signals classified in Z80?

The control signals of Z80 are classified into bus control, CPU control and system control signals.

Q1.41 List the register pairs of Z80.

The registers pairs of Z80 are BC, DE, HL, B'C', D'E' and H'L'.

Q1.42 List the flags of Z80.

The Z80 has six flags:

1. **Sign flag (S and S')**   4. **Parity/Overflow flag (P/V and P'/V')**
2. **Zero flag (Z and Z')**   5. **Half carry flag (H and H')**
3. **Carry flag (C and C')**   6. **Subtract flag (N and N')**

Q1.43 What are the common features of 8085 and Z80?

The common features of 8085 and Z80 are the following:

1. Both are fabricated using NMOS technology and have 40 pins.
2. Memory is accessed by a 16-bit address and the IO device by an 8-bit address.
3. The 8085 is software compatible with Z80.

Q1.44 List the difference between 8085 and Z80.

8085	Z80
1. Low order address and data lines are multiplexed.	1. Separate lines are provided for address and data.
2. A single signal is IO/$\overline{M}$ used to differentiate IO access and memory access.	2. Separate signals are used to differentiate the memory address and the IO address.
3. The instruction size is one to three bytes.	3. The instruction size is one to four bytes.
4. The flag register has five flags	4. The flag register has six flags.
5. It has five hardware interrupts.	5. The Z80 has two hardware interrupts.
6. It has 74 types of instructions.	6. It has 156 types of instructions.

Q1.45 What is the data and address size in Motorola 6800?

In Motorola 6800 the data size is 8-bit and address size is 16-bit.

Q1.46 How are IO devices addressed in M6800?

The Motorola 6800 does not have a separate address for its memory and IO devices. Hence some of the memory addresses are used to address IO devices.

Q1.47 How are the control signals of M6800 classified?

The control signals of M6800 are classified into control bus signals and CPU (microprocessor) supervisory signals.

Q1.48 What is the clock requirement of M6800?

The Motorola 6800 requires an external 2-phase clock whose maximum frequency can be 1 MHz.

Q1.49 What is CCR or what is the name of the flag register in M6800?

The flag register in the Motorola processor is called Condition Code Register (CCR).

Q1.50 What are the flags of M6800?

The M6800 has six flags:

1. Negative (N)   4. Carry (C)
2. Zero (Z)   5. Half Carry (H)
3. Overflow (V)   6. Interrupt enable/disable (I).

Q1.51 What are the addressing modes available in Motorola 6800?

The addressing modes of Motorola 6800 are immediate, direct, indexed and relative addressing.

Q1.52 List the differences between 8085 and M6800.

8085	M6800
1. Low order address and data lines are multiplexed.	1. Separate lines are provided for address and data.
2. It has a 16-bit address for memory and an 8-bit address for IO-mapped devices.	2. It does not have a separate address for memory and IO-mapped devices.
3. The flag register has five flags.	3. The flag register has six flags.
4. It has five hardware interrupts.	4. It has two hardware interrupts.

Q1.53 What are the different types of instructions available in Motorola 6800?

Data handling, arithmetic, logic, control transfer, data test, condition codes, address maintenance and interrupt handling are the different types of instructions available in Motorola 6800.

Q1.54 What are the modes in which 8086 can operate?

The 8086 can operate in two modes and they are minimum (or uniprocessor) mode and maximum (or multiprocessor) mode.

Q1.55 What is the data and address size in 8086?

The 8086 can operate on either 8-bit or 16-bit data. The 8086 uses 20-bit address to access memory and 16-bit address to access IO devices.

Q1.56 What is the difference between 8086 and 8088?

The external data bus in 8086 is 16-bit and that of 8088 is 8-bit, i.e., the 8086 access memory in words but 8088 access memory in bytes.

Q1.57 Explain the function of M/$\overline{IO}$ in 8086.

The signal M/$\overline{IO}$ is used to differentiate memory address and IO address. When the processor is accessing memory locations M/$\overline{IO}$ is asserted **high** and when it is accessing IO-mapped devices it is asserted **low**.

Q1.58 What are the hardware interrupts of 8086?

The hardware interrupts of 8086 are INTR and NMI. The INTR is general maskable interrupt and NMI is non-maskable interrupt.

Q1.59 How is clock signal generated in 8086? What is the maximum internal clock frequency of 8086?

The 8086 does not have on-chip clock generation circuit. Hence the clock generator chip, 8284 is used to generate the required clock. The frequency of clock generated by 8284 is thrice that of internal clock frequency of 8086. The 8284 divides the generated clock by three and modifies the duty cycle to 33% and then supply as clock signal to 8086. The maximum internal clock frequency of 8086 is 5 MHz.

Q1.60 What is pipelined architecture?

In pipelined architecture, the processor will have the number of functional units and the execution time of functional units overlapped. Each functional unit works independently most of the time.

Q1.61 What are the functional units available in 8086 architecture?

The **B**us **I**nterface **U**nit (BIU) and **E**xecution **U**nit (EU) are the two functional units available in 8086 architecture.

Q1.62 List the segment registers of 8086.

The segment registers of 8086 are **C**ode **S**egment (CS), **D**ata **S**egment (DS), **S**tack **S**egment (SS) and **E**xtra **S**egment (ES) registers.

Q1.63 What is the difference between segment register and general purpose register?

The segment registers are used to store 16-bit segment base address of the four memory segments. The general purpose registers are used as the source or destination register during data transfer and computation, as pointers to memory and as counters.

Q1.64 What is queue? How is queue implemented in 8086?

A data structure which can be accessed on the basis of first in first out is called queue. The 8086 has six numbers of 8-bit FIFO registers, which are used as instruction queue.

Q1.65 Write the flags of 8086.

The 8086 has nine flags. They are:

1.	Carry Flag (CF)	6.	Overflow Flag (OF)
2.	Parity Flag (PF)	7.	Trace Flag (TF) (or Single step trap)
3.	Auxiliary carry Flag (AF)	8.	Interrupt Flag (IF)
4.	Zero Flag (ZF)	9.	Direction Flag (DF)
5.	Sign Flag (SF)		

Q1.66 Write the special functions carried out by the general purpose registers of 8086.

The special functions carried out by the registers of 8086 are the following:

Register	Name of the register	Special function
AX	16-bit Accumulator	Stores the 16-bit result of certain arithmetic and logical operations.
AL	8-bit Accumulator	Stores the 8-bit result of certain arithmetic and logical operations.
BX	Base Register	Used to hold the base value in base addressing mode to access memory data
CX	Count Register	Used to hold the count value in SHIFT, ROTATE and LOOP instructions.
DX	Data Register	Used to hold data for multiplication and division operations.
SP	Stack Pointer	Used to hold the offset address of top of stack memory.
BP	Base Pointer	Used to hold the base value in base addressing using stack segment register to access data from stack memory.
SI	Source Index	Used to hold the index value of source operand (data) for string instructions.
DI	Destination Index	Used to hold the index value of destination operand (data) for string instruction.

Q1.67 What are control bits?

The flags TF, IF and DF of 8086 are used to control the processor operation and hence are called control bits.

Q1.68 List the internal registers of a 68000 processor.

The 68000 processor has eight numbers of 32-bit data registers D_0-D_7 and nine numbers of 32-bit address registers A_0-A_7, A'$_7$. Also it has a 32-bit **P**rogram **C**ounter (PC) and a 16-bit **S**tatus **R**egister (SR). For addressing, the 68000 processor uses only lower 24-bit position of address registers and PC.

Q1.69 What is status word in a 68000 processor?

In a 68000 processor, the content of the status register is called the status word, in which the lower byte is called the user byte and the upper byte is called the system byte.

The user byte consists of arithmetic flags like carry, overflow, negative, zero and extend flags. The system byte consists of the trace flag, the supervisor state bit and the interrupt mask bits.

Q1.70 List few differences between 8086 and 68000 processor?

8086	68000
1. The address pins are multiplexed with data and status signals.	1. Separate (or Non-multiplexed) pins are provided for address and data.
2. Uses a 20-bit address for memory and so physical address space is 1MB.	2. Uses a 24-bit address for memory and so the physical address space is 16 MB.
3. Separate IO address space is available. Therefore both memory mapping and IO mapping of IO devices are possible.	3. The Motorola processor does not have a separate IO address space and so only memory-mapped IO is possible
4. All the internal registers are 16 bits wide.	4. All the internal registers (except SR) are 32 bits wide.

1.11 EXERCISES

I. Fill in the blanks with appropriate words

1. A digit of the binary number or code is callled _____ .

2. The group of conducting lines that carry control signals is called _____ bus.

3. The _____ state is used to keep the device electrically isolated from the system.

4. INTEL 8088 is an _____ bit processor.

5. The third generation microprocessors were designed using _____ technology.

6. Transfering different information at different well defined times through the same lines is called _____.

7. 1 mil is equivalent to _____ inch.

8. The _____ and _____ signals of 8085 are used for serial data communication between 8085 and any serial device.

9. The _____ register of 8085 points to the next instruction to be executed.

10. The two independent 8-bit accumulators of Z80 processor are _____ and _____.

11. The two registers of Z80 processor used for indexed addressing mode are _____ and _____.

12. _____ register is used to indicate the results of an ALU operation in motorola 6800.

13. _____ signal is used to differentiate the minimum mode and maximum mode operation in 8086 processor.

14. _____ flag is set to 1 if the most significant bit of the result is one.

15. _____ register is used to hold the upper 16 bits of the starting address of the code segment in 8086 processor.

16. _____ pins are used to track the internal status of the instruction queue in 8086.

17. _____ register is used as the counter register in 8086.

18. _____ flag is used for single step execution in 8086.

19. The address line _____ is used to select the even bank of 8086.

20. _____ pin of 8086 is made low by external hardware when the instruction WAIT is executed.

21. The status signals _____ and _____ are use to indicate the selection of segments in 8086.

22. The maximum internal clock frequency of motorola MC68000 is _____.

23. The _____ bit in the status register of MC68000 is reset to zero to switch the processor from supervisor mode to user mode.

24. The _____ pin is asserted low by MC68000 to indicate a valid address whenever an address is output on the address bus.

Answers

1. bit	7. 10^{-3}	13. MN/$\overline{\text{MX}}$	19. A_0
2. control	8. SID, SOD	14. Sign	20. TEST
3. high impedance	9. Program Counter(PC)	15. CS	21. S_3, S_4
4. eight	10. A,A′	16. QS_1 and QS_0	22. 25 MHz
5. High Density MOS(HMOS)	11. IX,IY	17. CX	23. S
6. multiplexing	12. Condition Code	18. TF	24. $\overline{\text{AS}}$

II. State whether the following statements are True/False.

1. The address bus is bidirectional.
2. The CPU bus is directly connected to the microprocessor.
3. The high impedance state is an electrical open-circuit condition .
4. The NMOS technology offers faster speed and higher density than HMOS technology.
5. Registers can be read/written faster than memory chips.
6. The 8085 has an internal clock oscillator.
7. The 8085 interrupts RST 7.5, RST 6.5 and RST 5.5 are non-maskable interrupts.
8. To reset the 8085 microprocessor the $\overline{\text{RESET IN}}$ pin should be held low for atleast three clock pulses.
9. The stack pointer (SP) holds the address of the stack top.
10. Z80 processor does not require a separate circuit to refresh dynamic RAM.
11. Z80 is software compatible with 8085.
12. The physical memory size of Z80 is larger than that of 8085.
13. The motorola 6800 uses I/0 mapping technique to interface the peripherals.
14. The 8086 processor does not have an internal clock circuit
15. The 8086 uses 16-bit address to access memory.
16. The 8088 uses 8-bit data bus externally.
17. The 8086 processor executes single bus cycle to read/write a word from/to the memory address 50013_H.
18. The 8086 and 8088 have a common instruction set.
19. The minimum mode signals of 8086 are used in multiprocessor environment.
20. The memory segments in 8086 cannot be overlapped.
21. SP register of 8086 always points to the top of the stack.
22. Data segment is the default segment in 8086.
23. The motorola MC68000 can address up to 16 Mb of physical memory space.
24. The Motorola MC68000 enters the supervisor mode when it is reset.
25. MC68000 consists of only one accumulator.

Answers

1. False	6. True	11. True	16. True	21. True
2. True	7. False	12. False	17. False	22. True
3. True	8. True	13. False	18. True	23. True
4. False	9. True	14. True	19. False	24. True
5. True	10. True	15. False	20. False	25. False

III. Choose the right answer for the following questions.

1. *Microprocessors are intended to be a _____ computer.*

   a) general-purpose   *b)* special purpose   *c)* hybrid   *d)* analog

2. *Group of 4-bits is called*

   a) byte   *b)* nibble   *c)* word   *d)* double word

3. *What would be the total memory capacity of a microprocessor with 10 address lines*

   a) 1 MB   *b)* 1 GB   *c)* 1 KB   *d)* 512 MB

4. *The first 8-bit processor introduced by INTEL is*

   a) 8080   *b)* 8008   *c)* 8085   *d)* 8086

5. *Which of the following is used to store temporary programs and data?*

   a) EPROM   *b)* ROM   *c)* RAM   *d)* all the three

6. *The 1-bit register provided to store the results of certain program instructions is*

   a) status register   *b)* instruction register   *c)* program counter   *d)* flag

7. *Which of the following is not a 16-bit processor?*

   a) 8086   *b)* 80186   *c)* 8088   *d)* 8096

8. *Which of the following signals is used to demultiplex the address/data lines in 8085 and 8086 processors?*

   a) $DT/\bar{R}$ *b)* ALE   *c)* SOD   *d)* READY

9. *The maximum internal clock frequency of 8085A microprocessor is*

   a) 5.03 MHz   *b)* 6 MHz   *c)* 10.6 MHz   *d)* 3.03 MHz

10. *The total memory capacity of 8085 processor is*

    a) 64 KB   *b)* 1 MB   *c)* 64 MB   *d)* 10 MB

11. *Which of the following 8085 signals are used for DMA operation?*

    a) $\overline{RD}, \overline{WR}$ *b)* $\overline{RESET\ IN}, \overline{RESET\ OUT}$ *c)* HOLD, HLDA   *d)* SOD, SID

12. *When initiated, certain interrupts can be delayed or rejected but when allowed, the program execution starts from a fixed location. Such an interrupt is known as,*

    a) non maskable and non vectored   *b)* maskable and non vectored
    c) non maskable and vectored   *d)* maskable and vectored

13. *The vector address for the 8085 interrupt TRAP is,*

    a) 0028_H *b)* 0020_H *c)* 0024_H *d)* 0000_H

14. *Which of the following flag bit is available in Z80 processor but not in 8085 processor*

    a) sign flag   *b)* zero flag   *c)* subtract flag   *d)* carry flag

15. *Which of the following is the first 8-bit single chip microprocessor to exploit a single 5-V power supply?*

    a) 8080   *b)* 8085   *c)* motorola 6800   *d)* Z80

16. What is the total memory capacity of 8086 microprocessor?

   a) 1 GB   *b)* 1 MB   *c)* 2 MB   *d)* 64 KB

17. Which of the following is the maximum mode signal of 8086?

   a) HOLD   *b)* ALE   *c)* DT/$\overline{\text{R}}$ *d)* $\overline{\text{LOCK}}$

18. Which of the following signal is used to select the odd memory bank of 8086?

   a) READY   *b)* A_0 *c)* $\overline{\text{BHE}}$ *d)* ALE

19. What is the size of the instruction queue in 8086?

   a) 4 bytes   *b)* 6 bytes   *c)* 8 bytes   *d)* 5 bytes

20. What will be the content of CS register if the 8086 is reset?

   a) FFF0_H *b)* FF00_H *c)* FFFF_H *d)* 0000_H

21. Which of the following 8086 registers is used to hold the index value of destination operand for string instructions?

   a) SP   *b)* BP   *c)* SI   *d)* DI

22. Which of the following signals is used by slow peripherals to get extra time in order to communicate with 8086?

   a) TEST   *b)* READY   *c)* DEN   *d)* none of the above

23. What is the significance of instruction queue in 8086?

   a) overlapping   *b)* multitasking   *c)* pipeling   *d)* multiprogramming

24. Which of the following signals of MC68000 are used to enable the odd bank and even bank respectively?

   a) $\overline{\text{UDS}}, \overline{\text{LDS}}$ *b)* $\overline{\text{LDS}}, \overline{\text{UDS}}$ *c)* $A_0, \overline{\text{BHE}}$ *d)* $\overline{\text{BHE}}, A_0$

25. Which of the following signals of MC68000 is used to indicate the bus error from peripheral to the processor?

   a) $\overline{\text{BERR}}$ *b)* $\overline{\text{BR}}$ *c)* BG   *d)* $\overline{\text{DTACK}}$

Answers

1. a	5. c	9. d	13. c	17. d	21. d	25. a
2. b	6. d	10. a	14. c	18. c	22. b	
3. c	7. d	11. c	15. c	19. b	23. c	
4. b	8. b	12. d	16. b	20. c	24. b	

IV. Answer the following questions.

E1.1 What is meant by addressability of a microprocessor?

E1.2 State the significance of clock pulse in microprocessor based system.

E1.3 List some of the 4-bit microprocessors with specifications.

E1.4 Define the term speed power product(SPP).

E1.5 Mention the packing density of NMOS and HMOS technology

E1.6 Define the term MIPS.

E1.7   Draw the basic functional blocks of a microprocessor.

E1.8   What is meant by active low signal? Explain with an example.

E1.9   Write down the significance of bus status signals $IO/\overline{M}$, S_0 and S_1.

E1.10   What is the use of signals SOD and SID in 8085?

E1.11   Define maskable and non-maskable interrupts.

E1.12   What is meant by hardware interrupt?

E1.13   What is meant by software interrupt?

E1.14   What is meant by vectored and non-vectored interrupts?

E1.15   Write down the vector address of all the interrupts of 8085.

E1.16   When power on, how does the CPU know the starting address of the first instruction it has to execute? What is that first instruction? why?

E1.17   What is the use of stack pointer in 8085 microprocessor?

E1.18   Why the program counter and stack pointer of 8085 are 16-bit registers?

E1.19   How is the dynamic RAM refreshed in Z80?

E1.20   What is the use of shadow registers in Z80?

E1.21   How is the interrupt service routine serviced in Z80?

E1.22   Mention the clock frequency of different versions of 8086 processor.

E1.23   How are the even and odd bank of memory accessed in 8086?

E1.24   What is meant by memory segmentation? What are its advantages?

E1.25   Name the different segments of 8086 and mention the segment size.

E1.26   How is the physical address generated in 8086?

E1.27   Write down the status of all the six status flag after execution of an 8086 instruction that adds $AL = 28_H$ and $BL = 1D_H$.

E1.28   How many bus cycles are required for accessing a word which is stored at memory location 20013_H? Which (higher or lower) data lines are used for placing lower byte and higher byte?

E1.29   Is it possible for a segment to begin at a memory address that is not divisible by 16 in 8086? Why?

E1.30   An 8-bit data 50_H is stored at the address 75380_H in a data segment with base address 7500_H. What will be the physical address if the same data is stored in another data segment with same offset address but the base address is changed to 6000_H?

E1.31   Which pin of 8086 is used to synchronize the slow peripherals with 8086? How?

E1.32   Show the bit positions of various flags in an 8086 flag registers.

E1.33   Differentiate the 8086 signals READY and $\overline{TEST}$.

E1.34   What is the significance of WAIT instruction in 8086?

E1.35   How does the 8086 processor read the status of instruction queue?

E1.36   What is meant by supervisor mode? Also state its importance in MC68000.

E1.37   What is meant by scratch pad register?

E1.38   How is the motorola MC68000 processor reset?

E1.39   Draw the programming model of MC68000 processor.

E1.40   Differentiate conditional and control flags.

CHAPTER 2

INTRODUCTION TO MICROCONTROLLERS

2.1 INTRODUCTION TO MICROCONTROLLERS

Since the invention of microprocessors, different companies have started manufacturing more and more sophisticated processors with improved features such as large data bus, large address bus, sophisticated memory management techniques and instruction set, capability of handling a wide range of integer and floating-point data, parallel processing of instructions, etc. But these sophisticated processors are not necessary for small applications such as controlling a motor, monitoring/controlling temperature, switching ON/OFF traffic lights, etc. In the 1980s, the manufacturers of microprocessors realized that there is a need for low cost, compact, single-chip programmable systems for small dedicated applications and so started manufacturing another class of programmable ICs called microcontrollers.

2.1.1 Comparison of Microprocessors and Microcontrollers

Microcontrollers are similar to microprocessors, but they are designed to work as a true single-chip system by integrating all the devices needed for a system on a single chip. The basic functional units of a microprocessor will be ALU, a set of registers, timing and control unit. The microcontroller will have these functional blocks and in addition may have IO ports, a programmable timer, RAM memory and EPROM/EEPROM memory. Some microcontrollers may even have internal ADC and/or DAC.

Table 2.1: Comparison of Microprocessor and Microcontroller

Microprocessor	Microcontroller
1. The functional blocks of a microprocessor are the ALU, registers, and timing and control unit.	1. The microcontroller includes the functional blocks of a microprocessor and in addition has a timer, a parallel IO port, a serial IO port, internal RAM and EPROM/EEPROM memory. Some controllers even have ADC and/or DAC.
2. A microprocessor is concerned with rapid movement of code and data between the external memory and the microprocessor. Hence it has a large number of instructions for moving data between the external memory and the microprocessor.	2. A microcontroller is concerned with rapid movement of code and data within the microcontroller. Hence it has few instructions for data transfer between the external memory and the microcontroller.
3. Microprocessors mostly operate on byte/word data and so have very few bit manipulating instructions.	3. Microcontrollers often manipulate with bits and so have a large number of bit manipulating instructions.

Table 2.1 continued...

Microprocessor	Microcontroller
4. Microprocessors usually require interfacing of a large number of additional ICs to form a microcomputer-based system. Hence the PCB of a microprocessor-based system will be large and so the system will be costly.	4. Microcontrollers can be used to form a single chip microcomputer-based system without any additional ICs. Hence the PCB of a microcontroller-based system will be small and so the system will be cheap.
5. The microprocessors are used for designing general purpose digital computing system (or computers).	5. Microcontrollers are used for designing application specific dedicated systems.

2.1.2 Functional Building Blocks of a Microcontroller

The microcontroller is a programmable IC manufactured by VLSI (**V**ery **L**arge **S**cale **I**ntegration) technique, and capable of performing arithmetic and logical operations. The various functional blocks of a typical microcontroller are shown in Fig. 2.1.

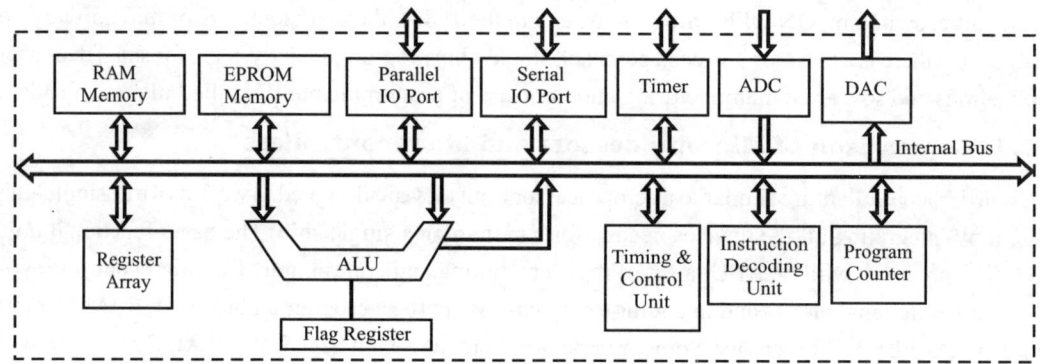

Fig. 2.1: Functional block diagram of a microcontroller.

The basic functional blocks of a microcontroller are the ALU, Flag register, Register array, **P**rogram **C**ounter (PC), Instruction Decoding Unit, Timing and control unit, RAM memory, EPROM/EEPROM memory, Parallel IO port, Serial IO port, Programmable timer, ADC and DAC. All microcontrollers may not have all the blocks shown in Fig. 2.1. Some of the functional blocks shown in Fig. 2.1, may not be available in certain microcontrollers.

The ALU is the computational unit of the microcontroller which performs arithmetic and logical operations. The various conditions of the result are stored as status bits called flags in the flag register. The register array and internal RAM memory are used as a temporary storage device for storing temporary data during execution of a program.

The program codes and permanent data are stored in EPROM/EEPROM. In microcontroller-based systems, an external memory is provided only when the internal memory is not sufficient and so in most of the microcontroller-based systems, the program and data are stored in the internal memory of the microcontroller itself.

The program counter generates the address of the instructions to be fetched from the memory and sends it through the internal bus to the memory. (If the instruction to be fetched is stored in the external memory then the address is sent through IO ports to the external memory. Because the microcontrollers communicate with the external world only through IO ports.) The memory will send the instruction codes, which are decoded by the instruction decoding unit and send the information to the timing and control unit. The timing and control unit will generate the necessary control signals for internal and external operation of the microcontroller.

The parallel and serial IO ports are used for interfacing IO devices like switches, keyboard, LCD/ LED, ADC, DAC, etc., and also for any other input/output operations.

Microcontrollers do not have a dedicated external address and data bus. Therefore, for interfacing any additional peripheral devices, the external address and data buses are formed only by using port lines.

Microcontrollers with internal ADC can directly accept analog signals for processing. Likewise, microcontrollers with internal DAC can directly generate analog signals for controlling analog devices. A programmable timer can be used for time-based operations and it can also be used as a counter.

2.2 INTEL 8051 MICROCONTROLLER

The 8051 family of microcontrollers was originally released by INTEL and later on licensed to many semiconductor companies like PHILIPS, ATMEL, SIEMENS, HARRIS, etc. These companies have developed a family of 8X5X microcontrollers with the same base architecture but with different internal memory capacity and internal devices. Some of the popular members of 8X5X family of microcontrollers are listed in Table 2.2.

Table 2.2: Some Members of the 8X5X Family of Microcontrollers

Microcontroller	Internal memory capacity				Number of timers
	RAM	ROM	EPROM	FLASH EPROM	
8031	128 bytes	-	-	-	2
8032	256 bytes	-	-	-	3
8051	128 bytes	4 kB	-	-	2
8052	256 bytes	8 kB	-	-	3
8751	128 bytes	-	4 kB	-	2
8752	256 bytes	-	8kB	-	3
8951	128 bytes	-	-	4kB	2
8952	256 bytes	-	-	8 kB	3

2.2.1 Pins and Signals of 8051

The INTEL 8051 is an 8-bit microcontroller with 128 byte internal RAM and 4 kB internal ROM. The 8051 is a 40-pin IC available in **D**ual **I**n-line **P**ackage (DIP) and it requires a single power supply of +5 V. Its maximum internal clock frequency rating is 12 MHz.

The 8X5X family members listed in Table 2.2 are pin-to-pin compatible with 8051. The pin configuration of 8051 microcontroller is shown in Fig. 2.2 and the signals of the controller are listed in Table 2.3. Some of the port pins of a 8051 microcontroller have alternate functions and they are listed in Table 2.4.

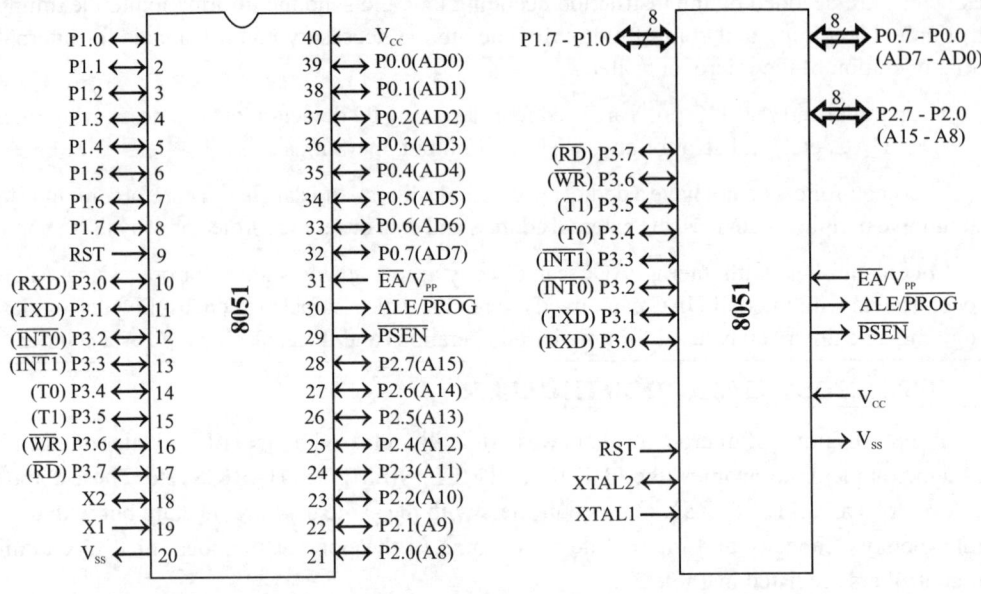

Note: The signals shown within brackets are alternate functions of port pins.

Fig. 2.2: Pin configuration of an INTEL 8051 microcontroller.

Table 2.3: Signals of an 8051 Microcontroller

Pins/Signal	Description
P0.7 - P0.0	Port-0 input/output pins.
P1.7 - P1.0	Port-1 input/output pins.
P2.7 - P2.0	Port-2 input/output pins.
P3.7 - P3.0	Port-3 input/output pins.
RST	Reset input
X1, X2	Pins for crystal connection. The signal at X2 can be used as clock signal for peripherals.
PSEN	Program store enable. Used as read control or enable for external program memory.
ALE/PROG	Address Latch Enable or program pulse input during EPROM/ROM programming.
EA/V$_{PP}$	External Access or Programming voltage.
V$_{CC}$	Power supply (+5 V)
V$_{SS}$	Power supply ground (0 V)

Table 2.4: Alternate Functions of Port Pins

Port pins	Alternate signal	Description
P0.7-P0.0	AD7-AD0	Multiplexed low byte address/data.
P2.7-P2.0	A15-A8	High byte address
P3.7	$\overline{RD}$	External memory read control signal
P3.6	$\overline{WR}$	External memory write control signal
P3.5	T1	External input to timer 1
P3.4	T0	External input to timer 0
P3.3	$\overline{INT1}$	External interrupt 1
P3.2	$\overline{INT0}$	External interrupt 0
P3.1	TxD	Serial data output
P3.0	RxD	Serial data input

The 8051 microcontroller has 32 IO pins and they are organized as four numbers of an 8-bit parallel port. The ports are denoted as port-0, port-1, port-2 and port-3. Each port can be used either as an 8-bit parallel port or 8 numbers of 1-bit port (i.e., individual pins of each port can be used as 1-bit IO line independently). When used as 1-bit port, the port pins are denoted as PX.Y, where X can take values 0 to 3 and Y can take values 0 to 7. For example, the bit-0 of port-1 is denoted as P1.0. The ports behave as latches during output operation and as buffers during input operation. Except port-0 all other ports are provided with internal pull up. Hence, while using port-0 for IO operation, external pull up should be provided.

Except port-1 all other ports have alternate functions. (Port-1 can be used only for IO operation.) When external memory is employed, port-0 functions as multiplexed low byte address or data lines, and port-2 functions as high byte address lines. Therefore, for accessing external memory the microcontroller uses a 16-bit address and accesses the memory in bytes.

Hence, the addressable memory space is 64 kB ($2^{16}= 64$ k). The 8051 allows the external memory to be organized as two banks of 64 kB, one for the program/code and the other for the data. The signal $\overline{PSEN}$ is used as read control/enable for program memory. The port pin P3.7 functions as read control ($\overline{RD}$) and the port pin P3.6 functions as write control ($\overline{WR}$) for data memory. When two external memory banks are not desirable, the $\overline{PSEN}$ and $\overline{RD}$ should be externally ANDed to provided a single read control signal. In such cases, the controller will access a common memory space (of maximum capacity 64 kB) for program and data.

In systems with external memory, the signal ALE is used to demultiplex the low byte address or data using an external latch. The output signal on the ALE pin is the clock signal with a frequency one-sixth of a crystal or internal clock frequency. The controller will output the ALE signal at a constant rate (i.e., at one-sixth of internal clock) even when there is no external memory.

Therefore, the ALE can also be used for external timing and clock source for peripherals or IO devices. In EPROM/ROM version of 8051 family controllers, the programming pulse can be input through ALE during programming of EPROM/ROM.

Signal $\overline{EA}$ is used as an external program memory access control. The microcontroller will access the program from external memory if $\overline{EA}$ pin is grounded. During programming mode of internal EPROM/ROM, this pin is used to supply the programming voltage (+12 V).

> *Note:* *For programming the internal EPROM/ROM of 8051 family of microcontroller, a separate programmer should be employed. The controllers listed in Table 2.2, do not have ISP (in-system programmable) facility.*

The X1 and X2 pins are provided for external quartz crystal connection, in order to generate the required clock for the microcontroller. The maximum frequency of a quartz crystal that can be connected to an 8051 microcontroller is 12 MHz. (There are higher clock versions of the 8051 family of microcontrollers. For details please refer to manufacturers data sheet.) Alternatively, the external clock can be supplied through an X1 pin.

The internal clock frequency of an 8051 microcontroller is same as a crystal frequency or externally supplied clock frequency. When a crystal is connected between X1 and X2, the controller will output a clock signal through X2 whose frequency is same as crystal frequency and this clock signal can be used for peripheral or IO devices.

The RST signal is used to reset the microcontroller in order to bring the controller to a known state. For proper reset the RST pin should be held **low** for at least two machine cycles. When the 8051 controller is reset, all the internal registers are cleared except the port latches, stack pointer and SBUF register. The internal RAM is not affected by reset. The content of the various registers of 8051 after a reset are listed in Table 2.5.

Table 2.5: Contents of Registers After a Reset

Register	Content after reset	Register	Content after reset
PC	00_H	SP	07_H
ACC	00_H	TCON	00_H
B	00_H	TH0	00_H
PSW	00_H	TL0	00_H
DPTR	0000_H	TH1	00_H
P0-P3	FF_H	TL1	00_H
IP	$xxx00000_B$	SCON	00_H
IE	$0xx00000_B$	SBUF	Indeterminate
TMOD	00_H	PCON	$0xxx00000_B$

The 8051 has five interrupts. In this two interrupts are external interrupts and the remaining three are internal interrupts. The two external interrupts are interrupts initiated by applying appropriate signals through the pins $\overline{INT0}$ and $\overline{INT1}$, and they are called external interrupt-0 and external interrupt-1 respectively. The internal interrupts are initiated by timer-0, timer-1 and the serial port. All the interrupts of 8051 are maskable and vectored interrupts. The vector address and the priorities of the interrupts of 8051 are listed in Table 2.6. (The priorities of the interrupts can also be altered by programming the IP register.)

Table 2.6: Vector Address and Priority of an 8051

Interrupt	Vector address	Normal priority
External interrupt-0	0003_H	highest
Timer-0 interrupt	$000B_H$	
External interrupt-1	0013_H	
Timer-1 interrupt	$001B_H$	
Serial port interrupt	0023_H	lowest

2.2.2 Architecture of 8051

The architecture of 8051 is shown in Fig. 2.3. The various functional blocks of 8051 are the ALU, **S**pecial **F**unction **R**egisters (SFRs) listed in Table 2.7, **I**nstruction **R**egister (IR), **P**rogram **C**ounter (PC), 128 bytes RAM, 4 kB ROM, Port latches and drivers, Oscillator, Timing and Control Unit.

The 8051 has **Harvard architecture** in which the same address in different memory devices or banks is used for program (or code) and data. Therefore, the architecture has two dedicated 16-bit address pointers, namely, **P**rogram **C**ounter (PC) and **D**ata **P**ointer (DPTR). The PC is used as the address pointer to access program instructions and it is automatically incremented after every byte of instruction fetch. The DPTR is used as address pointers to read/write data in data memory and it is programmable using instructions.

Since, the size of the address pointers are 16-bit they can address up to $2^{16} = 64k$ memory locations. Hence, 8051 supports two memory banks of 64 kB each, one for program and the other for data. In 8051, when the $\overline{EA}$ pin is tied to the V_{CC} (logic-1), the first 4 kB of program memory address space refers to 4 kB internal ROM and the remaining 60 kB refer to external (EPROM/RAM) memory. In 8051, when the $\overline{EA}$ pin is tied to the ground (logic-0), the entire 64 kB of program address space refers to external (EPROM/RAM) memory.

The 8051 has separate 256 bytes internal RAM accessed by using an 8-bit address. In this 256 bytes address space, the first 128 addresses are allotted to internal RAM and the next 128 bytes are allotted to SFR. The internal RAM/SFR can be accessed by using MOV instructions and external data memory (RAM) can be accessed by using MOVX instruction.

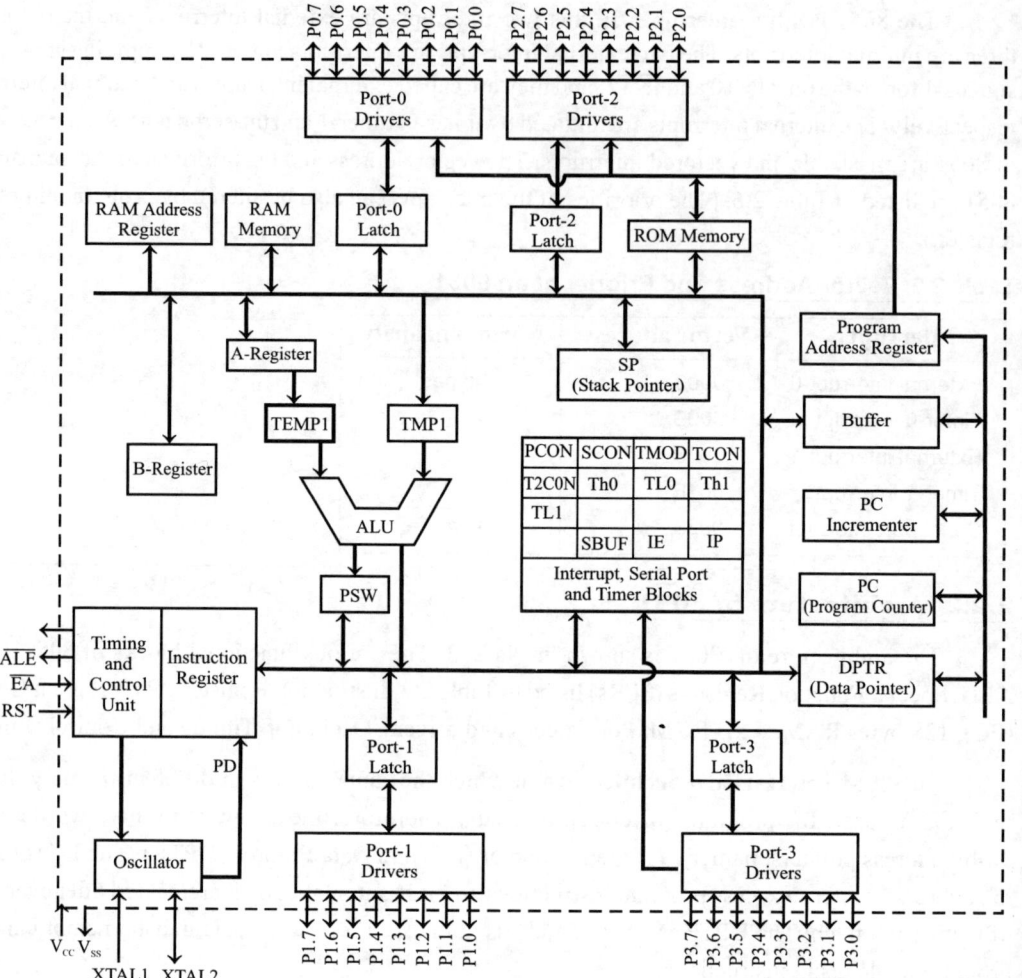

Fig. 2.3: Architecture of an 8051 microcontroller.

The 8051 has four 8-bit ports, namely, port-0, port-, port-2 and port-3. Each port has a latch and driver (or buffer). When external memory is employed the port-0 lines will function as multiplexed low byte address/data lines and port-2 lines will function as high byte address lines. Also, the port pins P3.7 and P3.6 are used to output read and write control signals respectively. In fact, each pin of port-3 has an alternate function which are listed in Table 2.4.

Port-1 is a dedicated IO port and does not have any alternate function. The ports are also mapped as internal memory in the controller and so they can be addressed as memory locations for 8-bit operation.

The SFRs include 21 internal registers listed in Table 2.7. (A detailed explanation of SFRs is presented in the Section 2.2.4.) The SFRs are mapped as internal data memory.

The data memory address space 80_H to FF_H are reserved for SFRs. Each register of SFR has one-byte address. Some of the registers are both byte and bit-addressable. (The registers whose address ends with 0_H or 8_H are bit-addressable.)

The 8051 has an 8-bit ALU which performs arithmetic and logical operations on binary data. The A and B registers are used to hold the input data and the result of the ALU operation. Starting from the address stored in the PC, the controller will fetch the instructions one by one, and store in IR, which decodes the instructions and give information to the timing and control unit. Using the information supplied by the IR unit, the control signals necessary for internal and external operations are generated by the timing and control unit. The 8051 has an internal oscillator and so it is sufficient if an external quartz crystal is connected for clock generation.

The 8051 has two 16-bit programmable timers/counters, namely, timer-1 and timer-0. In the counter mode of operation they can count the number of high to low transitions of the signal applied to the timer pins. In the timer mode of operation, they can be independently programmed to work in any one of the four operating modes. The timer operating modes are called mode-0, mode-1, mode-2 and mode-3.

Table 2.7: Special Function Registers (SFR)

Special function register		Byte address in hexa
Symbol	Name	
A or ACC	A-register or accumulator	E0
B	B-register	F0
DPH	Data pointer higher order register	83
DPL	Data pointer lower order register	82
IE	Interrupt enable register	A8
IP	Interrupt priority register	B8
P0	Port-0	80
P1	Port-1	90
P2	Port-2	A0
P3	Port-3	B0
PCON	Power control register	87
PSW	Program status word	D0
SCON	Serial port control register	98
SBUF	Serial port data buffer	99
SP	Stack pointer	81
TMOD	Timer/Counter mode control register	89
TCON	Timer/Counter control register	88
TL0	Timer-0 low order register	8A
TH0	Timer-0 high order register	8C
TL1	Timer-1 low order register	8B
TH1	Timer-1 high order register	8D

The 8051 family of microcontrollers has a full duplex serial port which can be programmed to work in any one of the four operating modes, namely, mode-0, mode-1, mode-2 and mode-3. In mode-0 the serial port can either receive or transmit at a fixed baud rate. In mode-2, it can simultaneously transmit and receive at any one of the two selectable baud rates. In mode-1 and mode-3, it can work as a full duplex serial port with variable baud rates which is programmed using timer-1.

2.2.3 Programming Model of 8051

The programming model of the 8051 microcontroller is shown in Fig. 2.4. The model will show the programmable internal devices of a processor or controller. During execution of a program the contents of the registers and RAM memory locations shown in the programming model are altered.

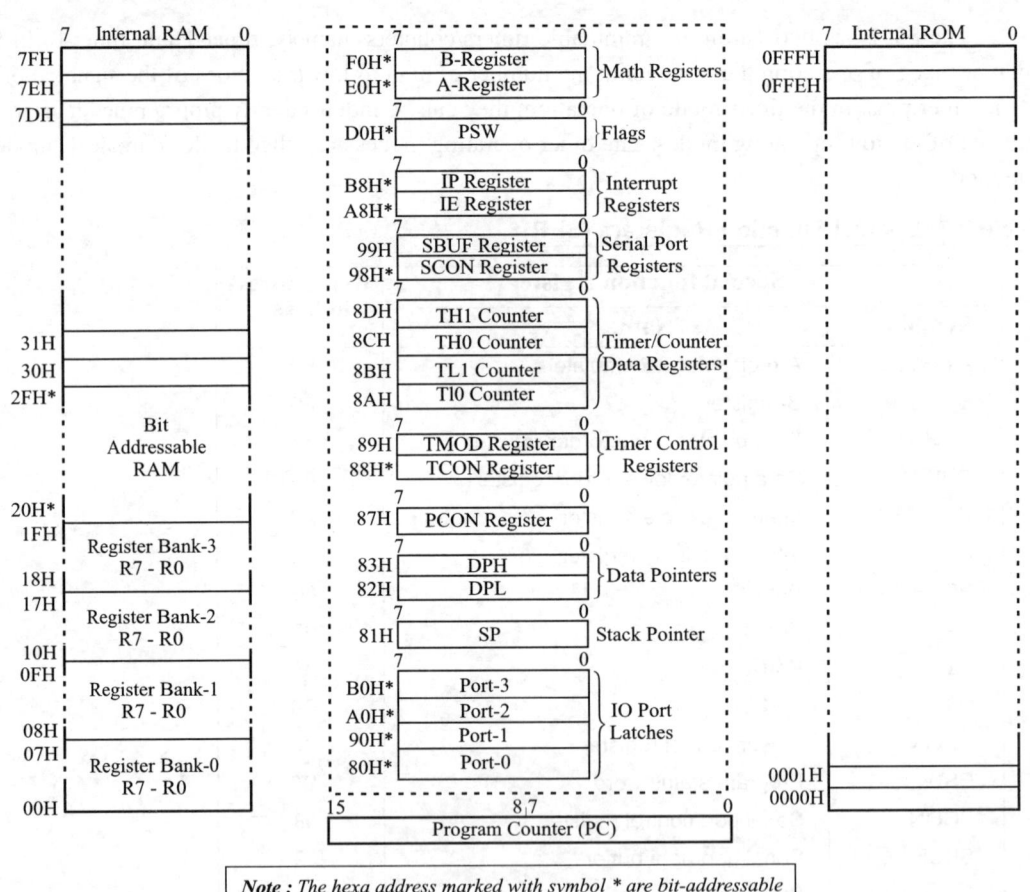

Fig. 2.4: Programming model of an 8051 microcontroller.

The programming model of an 8051 microcontroller includes a 128 byte internal RAM and all the Special Function Registers (SFRs). In addition, the programming model of an 8051 includes a 4 kB internal ROM. The internal RAM locations are separate data memory address space and accessed either by direct or indirect addressing using an 8-bit address.

The SFRs are also separate data memory address spaces and are accessed by direct addressing using an 8-bit address. The 4 kB internal ROM in an 8051 controller is mapped as program memory and accessed using a 16-bit address.

The first 32 bytes of internal RAM are organized as four groups of eight registers. Each group is called a **register bank** and denoted as bank-0, bank-1, bank-2 and bank-3. The eight registers of a bank are denoted as R_n where n takes values from 0 to 7. At any one time, the controller can use any one of the register bank as general purpose registers (or scratch pad registers). The selection of the register bank depends on the value of bits RS0 and RS1 in the PSW register. After a reset the PSW register is cleared and so the controller works with register bank-0.

The internal RAM locations in the address range 20_H to $2F_H$ are bit addressable. The internal RAM locations in the address range 30_H to $7F_H$ can be used as general purpose RAM. The hexa address of the bit-addressable RAM are listed in Table 2.8. The 16 RAM locations in the address range 20_H to $2F_H$ has a capacity of 128 bits ($16 \times 8 = 128$). Each bit in this RAM area can be addressed by an 8-bit address in the range 00_H to $7F_H$, as shown in Table 2.8.

Table 2.8: Byte Address of Bit-Addressable RAM

Byte address	Hexa address of bit position							
	B_7	B_6	B_5	B_4	B_3	B_2	B_1	B_0
20	07	06	05	04	03	02	01	00
21	0F	0E	0D	0C	0B	0A	09	08
22	17	16	15	14	13	12	11	10
23	1F	1E	1D	1C	1B	1A	19	18
24	27	26	25	24	23	22	21	20
25	2F	2E	2D	2C	2B	2A	29	28
26	37	36	35	34	33	32	31	30
27	3F	3E	3D	3C	3B	3A	39	38
28	47	46	45	44	43	42	41	40
29	4F	4E	4D	4C	4B	4A	49	48
2A	57	56	55	54	53	52	51	50
2B	5F	5E	5D	5C	5B	5A	59	58
2C	67	66	65	64	63	62	61	60
2D	6F	6E	6D	6C	6B	6A	69	68
2E	77	76	75	74	73	72	71	70
2F	7F	7E	7D	7C	7B	7A	79	78

2.2.4 Special Function Registers (SFR) OF 8051

The 8051 is provided with 21 special function registers and they are used for selecting various programmable features of the microcontroller. The special functions of most of the SFR are distinguishable. Each SFR has an internal one-byte address assigned to it.

Some of the registers are both byte and bit-addressable. The SFR along with their byte address are listed in Table 2.7. The bit-addressable register along with address for each bit are listed in Table 2.9.

Table 2.9: Bit Address of SFR

SFR	Hexa address of bit position							
	B_7	B_6	B_5	B_4	B_3	B_2	B_1	B_0
B	F7	F6	F5	F4	F3	F2	F1	F0
A or ACC	E7	E6	E5	E4	E3	E2	E1	E0
PSW	D7	D6	D5	D4	D3	D2	D1	D0
IP	BF	BE	BD	BC	BB	BA	B9	B8
P3	B7	B6	B5	B4	B3	B2	B1	B0
IE	AF	AE	AD	AC	AB	AA	A9	A8
P2	A7	A6	A5	A4	A3	A2	A1	A0
SCON	9F	9E	9D	9C	9B	9A	99	98
P1	97	96	95	94	93	92	91	90
TCON	8F	8E	8D	8C	8B	8A	89	88
P0	87	86	85	84	83	82	81	80

A and B Registers

The A and B registers are called CPU registers. They are used to hold the data for most of the CPU (ALU) operations. The sizes of A and B registers are 8-bit and they are mapped as on-chip data memory with byte address $E0_H$ and $F0_H$ respectively. These registers are also bit-addressable.

In most of the ALU operations, the result is stored in the A-register and so, it is also known as the **accumulator.**

Data Pointer (DPTR)

The data pointer is a 16-bit register used to hold the 16-bit address of data memory. The 16-bit data pointer can also be used as two numbers of 8-bit data pointer, namely, DPH and DPL. The 8-bit data pointers are used for accessing internal RAM and SFR. The 16-bit data pointer is used for accessing external data memory.

The 8-bit data pointer DPH and DPL are mapped as internal memory with byte address 83_H and 82_H respectively. The contents of the data pointers are programmable using instructions.

Program Status Word (PSW)

The program status word stores the status of the result of the ALU operations and some of the status of the processor by means of a 1-bit status called flag. The PSW is also known as a **flag register.** The flags are useful for the programmer to test the condition of the result and make decisions.

The format of the PSW of an 8051 microcontroller is shown in Fig. 2.5. The PSW consists of four math flags, two register bank select bits and two user flag bits. The math flags are carry, auxiliary carry, overflow and parity flags. These flags are altered after arithmetic and logical operations depending on the result. The carry flag is set when the result has a carry. When there is a carry from the lower nibble to the upper nibble, the auxiliary carry is set. When the result has even parity, the parity flag is set. In signed mathematical operations if the size of the result exceeds the maximum range then the overflow flag is set.

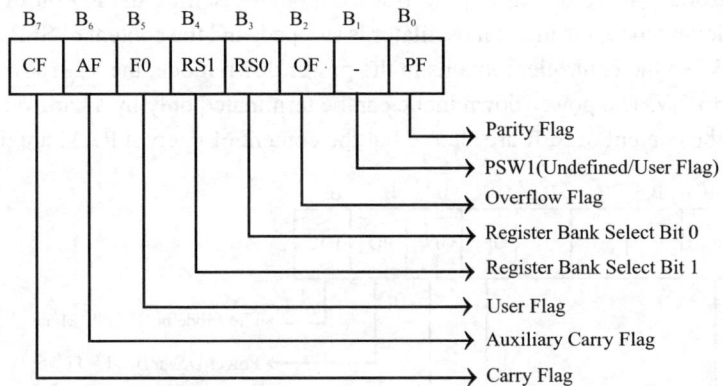

Fig. 2.5: Format of PSW of the 8051 family of microcontrollers.

The register bank select bits RS1 and RS0 are used to select any one of the four register banks of the internal RAM. At any one time, the microcontroller can work with (or access) only one register bank selected by these bits. The bank select bits are programmable and after reset the controller defaults to bank-0. The selection of a register bank using the RS1 and RS0 bits are listed in Table 2.10. The user flag bits can be used by the programmer to indicate the status of certain events during program execution.

Table 2.10: Selection of a Register Bank

Bank select bits		Selected register bank	Range of hexa address of the selected bank
RS1	RS0		
0	0	Bank-0	00_H - 07_H
0	1	Bank-1	08_H - $0F_H$
1	0	Bank-2	10_H - 17_H
1	1	Bank-3	08_H - $1F_H$

Stack Pointer (SP)

The stack pointer always holds the 8-bit address at the top of stack. The 8051 microcontroller supports the LIFO (**L**ast-**I**n-**F**irst-**O**ut) stack, and the stack may reside anywhere in on-chip RAM (i.e., the programmer can reserve any portion of on-chip RAM as stack.) After a reset, the stack pointer is initialized to 07_H. The stack can be accessed using PUSH and POP instructions. During a PUSH operation, the stack pointer is automatically incremented by one and during POP operation the stack pointer is automatically decremented by one.

Power Control Register (PCON)

The PCON register is used for power control and baud rate selection. It also consists of general purpose user flags. The format of a PCON register is shown in Fig. 2.6.

The controller can be driven to the idle mode by setting the IDL bit of the PCON register. When the idle mode is activated, the clock signal is stopped to CPU(ALU), but the clock signal is supplied to interrupt, timer and serial port blocks. The idle mode can be terminated either by an interrupt or by hardware reset.

The controller can be driven to power-down mode by setting the PD bit of a PCON register. During power-down mode, the internal oscillator is stopped, and the content of SFR and internal RAM are preserved. When the controller remains in the power down mode, the V_{CC} (power supply voltage) can be reduced to 2 V. The power-down mode can be terminated only by a hardware reset and during hardware reset the content of SFR are altered but the content of internal RAM are preserved.

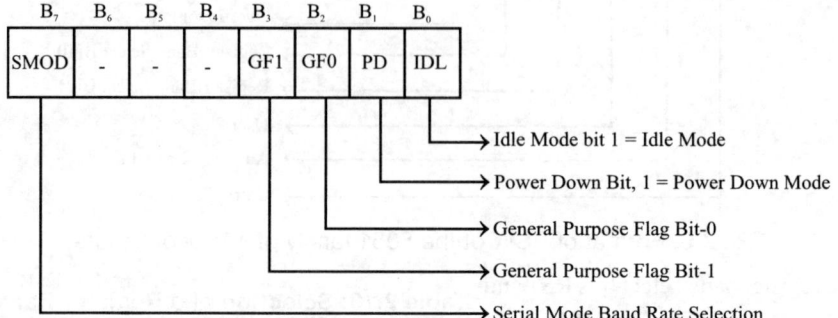

Fig. 2.6: Format of a PCON register of the 8051 family of microcontrollers.

Note: The idle and power-down mode are not available in NMOS version of 8051.

The SMOD bit is used to decide the baud rate in serial port operating modes 1, 2 or 3. In mode 2, if SMOD = 0, then the baud rate is 1/64 of oscillator frequency and if SMOD = 1 then the baud rate is 1/32 of oscillator frequency.

The general purpose flag bits GF1 and GF0 can be used by the programmer to indicate the status of certain events during program execution.

Serial Data Buffer (SBUF) Register

The SBUF register is used to hold the parallel data during transmission and reception. During serial reception, the serial data is received via RxD pin and converted to parallel data and stored in the receive buffer. During serial transmission, the parallel data is stored in the transmit buffer and then converted to serial data to transmit via TxD pin.

The transmit and receive buffers are assigned the same internal address 99_H but the transmit buffer can be accessed only for write operation and the receive buffer can be accessed only for read operation. When data is written to SBUF, it goes to transmit buffer and when data is read from SBUF it comes from the receive buffer.

Serial Port Control Register (SCON)

The format of a SCON register is shown in Fig. 2.7. The SCON register consists of mode selection bits, the 9th data bit (bit-B_8) for transmit and receive, and the serial port interrupt bits TI and RI. The bits SM0 and SM1 are used to select any one of the four operating modes for serial transmission and reception. The four modes of a serial port are mode-0, mode-1, mode-2 and mode-3.

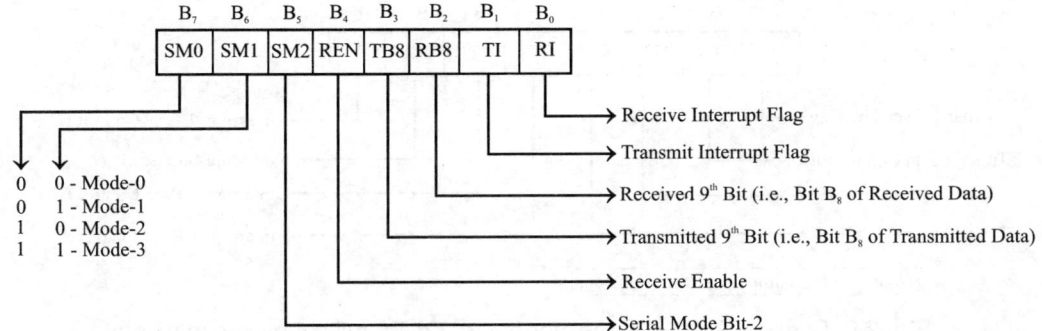

Fig. 2.7: Format of a SCON register of the 8051 family of microcontrollers.

Timer Mode Control (TMOD) Register

The TMOD register is used to select the operating mode and the timer/counter operation of the timers. The format of a TMOD register is shown in Fig. 2.8. The lower four bits of the TMOD register is used to control timer-0 and the upper four bits are used to control timer-1. The two timers can be independently programmed to operate in various modes. The register has two separate two bit field M0 and M1 to program the operating mode of timers. The operating modes of timers are mode-0, mode-1, mode-2 and mode-3. In all these operating modes, the oscillator clock is divided by 12 and applied as the input clock to the timer.

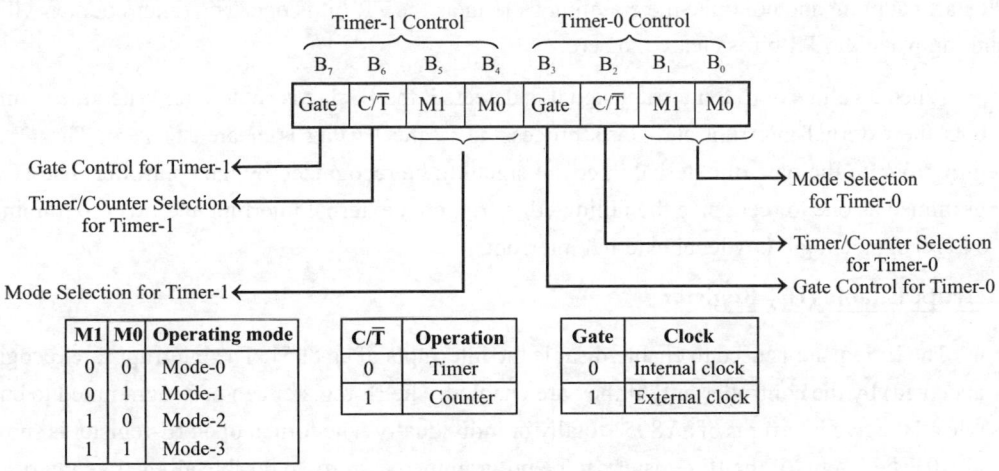

Fig. 2.8: Format of the TMOD register of the 8051 family of microcontrollers.

Timer Control (TCON) Register

The TCON register consists of timer overflow flags, timer run control bits, external interrupt flags and external interrupt-type control bits. The format of a TCON register is shown in Fig. 2.9.

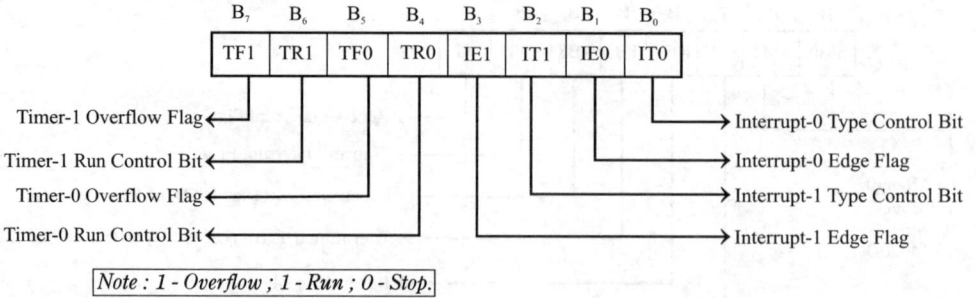

Fig. 2.9: Format of a TCON register of the 8051 family of microcontrollers.

The timers in an 8051 microcontroller are upcounters and keep on incrementing as long as the clock is applied. Therefore, when the clock is applied after reaching the maximum value (i.e., the content of the counter is all 1s), the content of the counter will become zero (i.e., all 0 s). This condition is called timer overflow and it is also the end of timing which a program wants to maintain by using the timer. The TCON register has a 1-bit flag, TF for each timer to indicate the timer overflow or end of timing. Whenever the timer/counter overflows, the TF flag is set to one. The TF flag is also used as an interrupt signal to initiate the execution of a subroutine. When the controller vectors to subroutine, the TF flag is cleared.

The TR bit is used to start/stop the timer/counter. When the TR bit is set to one, the timer/counter will start counting and continue the counting as long as the TR bit is one. The timer/counter will stop counting when the TR bit is cleared to zero.

When a valid external interrupt signal is detected, the IE flag is set to one. When the controller accepts the external interrupt and starts processing it, the IE flag is cleared to zero. The IT bit is used to program the type of external interrupt signal to be recognized by the controller. The IT bit is programmed as one to recognize the falling edge-triggered external interrupt and it is programmed as zero to recognize logic **low** level external interrupt.

Interrupt Enable (IE) Register

The IE register is used to enable/disable the interrupts of an 8051. The interrupts are recognized (or accepted) by the controller only if they are enabled. The IE register can be programmed to enable/disable all the five interrupts of an 8051 totally or individually. The format of an IE-register is shown in Fig. 2.10. The EA bit of the IE register can be programmed as zero, to disable all the five interrupts of 8051. When EA bit is programmed as one, the interrupts are enabled provided their individual enable bits are programmed as one. (The EA bit is also called global enable.)

Each interrupt has one-bit field to enable or disable it individually. When EA = 1, if the enable bit of a particular interrupt is programmed as one then it is enabled and if the enable bit is programmed as zero, then it is disabled.

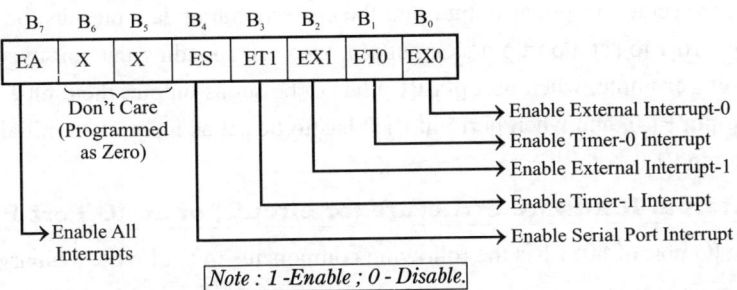

Fig. 2.10: Format of the IE register of the 8051 family of microcontrollers.

Interrupt Priority (IP) Register

The 8051 has five interrupts and the normal priority of these interrupts from highest to lowest are external interrupt-0, Timer-0 interrupt, External interrupt-1, Timer-1 interrupt and serial port interrupt.

The IP register can be programmed to make the priority of any of the interrupt as highest. The format of an IP register is shown in Fig. 2.11. The IP register has one-bit field for the priority of each interrupt. When the priority bit of a particular bit is programmed as one then its priority will be highest. In 8051, while servicing a lower priority interrupt a higher priority interrupt will be recognized but another lower priority interrupt will not be recognized.

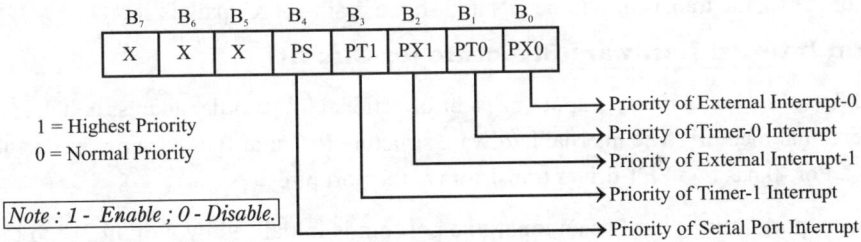

Fig. 2.11: Format of the IP register of the 8051 family of microcontrollers.

2.2.5 IO Pins, Ports and Circuits

The 8051 microcontroller has 32 IO pins and they are organized as four numbers of 8-bit parallel port. The ports are denoted as Port-0, Port-1, Port-2 and Port-3. Each port can be used either as an 8-bit parallel port or 8 numbers of 1-bit port (i.e., individual pins of each port can be used as 1-bit IO line independently).

When used as 1-bit port, the port pins are denoted as PX.Y, where X can take values 0 to 3 and Y can take values 0 to 7. For example, the bit-0 of port-1 is denoted as P1.0 (Least Significant Bit of Port-1) and the bit-7 of port-1 is denoted as P1.7 (Most Significant Bit of Port-1).

Most of these pins are used to connect to IO devices or external data and code memory. The ports behave as latches during output operation and as buffers during input operation. Except port-0 all other ports are provided with internal pull up. Hence, while using port-0 for IO operation, external pull up should be provided.

In order to set a port pin as output pin, the corresponding data bit must be cleared in the port register and in order to set a port pin as input pin, the corresponding data bit must be set high in the port register. For example, when port pin P1.0 has to be set as output, then initialize the port pin by writing "0" to pin P1.0, and when port pin P1.0 has to be set as input, then initialize the port pin by writing "1" to pin P1.0.

General Internal Hardware Structure (or circuit) of an IO Port Pin of 8051

Each bit of an IO port of 8051 has the following components to read/write a binary bit.

1. An internal data bus line for data transfer between the port pin and internal bus.

2. A D-latch to store the value of the data bit during write operation and is controlled by "Write to latch" control signal. When "Write to latch" = 1, the data on the internal bus line is latched into the output of D latch.

3. Two tri-state buffers, are employed for read operation. One buffer is used to read the data on the output of D-latch and it is controlled by "Read latch" control signal. Another buffer is used to read the data on the port pin and it is controlled by "Read pin" control signal. When "Read pin" = 1, the data present at the port pin is read and when "Read latch" = 1, the data in the output of D-latch is read.

4. FET transistor to drive signal at the port pin. When the signal at the gate of transistor is "0", the transistor will be OFF and so the drain-source path is open. When the signal at the gate of the transistor is "1", the transistor will be ON and so the drain-source path is short.

Port-0 Pin Internal Hardware Structure (or Circuit)

The Port-0 serves as input, output or as a bi-directional lower order address and data bus (AD0-AD7) for external memory. The internal hardware structure (or circuit) of each pin in Port-0 is shown in Fig. 2.12. Port-0 has two FET driver transistors at the port pin.

When a port-0 pin is used as an input, the port pin is initialized by writing "1" to the D-latch, which will make the signal at transistor gate as "0". Therefore, both the transistors will turn OFF, which in turn causes the pin to "float" in a high impedance state, and so the port pin is connected only to the input buffer.

When a port-0 pin is used as an output, the port pin is initialized by writing "0" to D-latch, which will make the signal at transistor gate as "1". Therefore, both the transistors will turn ON, which in turn causes the port pin to be connected to ground. Now, during a write operation when a "0" is written the same condition exists, but when a "1" is written then transistors are OFF and the port pin floats, and so an external pull-up resistor should be connected to maintain logic **high** at the port pin.

When port-0 is used as an address bus to external memory, internal control signal switches the address lines to the transistor gates. Logic "1" on an address bit will turn the upper transistor ON and the lower transistor OFF to provide logic **high** at the port pin. When the address bit is a "0", the lower transistor is ON and the upper transistor is OFF to provide logic **low** at the port pin. After the address output, the port-0 pins are internally initialized for data transfer. Hence, for normal address/data interfacing (or for external memory access), no pull-up resistors are required.

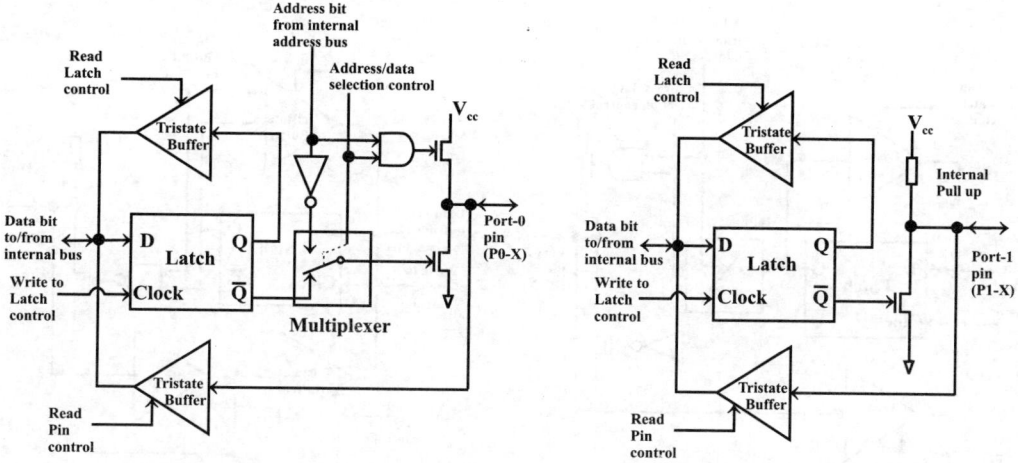

Fig. 2.12: Circuit of port-0 bit   **Fig. 2.13:** Circuit of port-1 bit

Port-1 Pin Internal Hardware Structure (or Circuit)

The Port-1 can serve as either input or output and does not have any alternate function. The internal hardware structure (or circuit) of each pin in Port-1 is shown in Fig. 2.13. Since the Port-1 does not have alternate function, the output of D-latch is connected directly to the gate of the transistor which has an internal pull-up resistor as shown in Fig. 2.13. This internal pull-up resistor help to maintain a stable logic level when used as output port.

When a Port-1 pin is used as an input, the port pin is initialized by writing "1" to the D-latch, which will make the signal at transistor gate as "0".

Therefore, the transistor will turn OFF, which in turn causes the port pin and input of buffer to be pulled high by the internal pull up. An external circuit can overcome the high impedance pull-up and drive the pin **low** or **high** to input '0' or '1', respectively.

When a Port-1 pin is used as an output, the port pin is initialized by writing "0" to D-latch, which will make the signal at transistor gate as "1". Therefore, the transistor will turn ON, which in turn causes the port pin to be connected to ground. Now, during a write operation, when a "0" is written the same condition exists, but when a "1" is written then the transistor will be OFF, and so the port pin is pulled **high** by the internal pull up.

Port-2 Pin Internal Hardware Structure (or Circuit)

Port-2 serves as input, output or as a higher order address bus (A8-A15) for external memory. The internal hardware structure (or circuit) of each pin in port-2 is shown in Fig. 2.14. The Port-2 pins are also provided with internal pull up and so the input/output operation of Port-2 pins are similar to that of Port-1.

When Port-2 is used as an address bus to external memory, the internal control signal switches the address lines to the transistor gates.

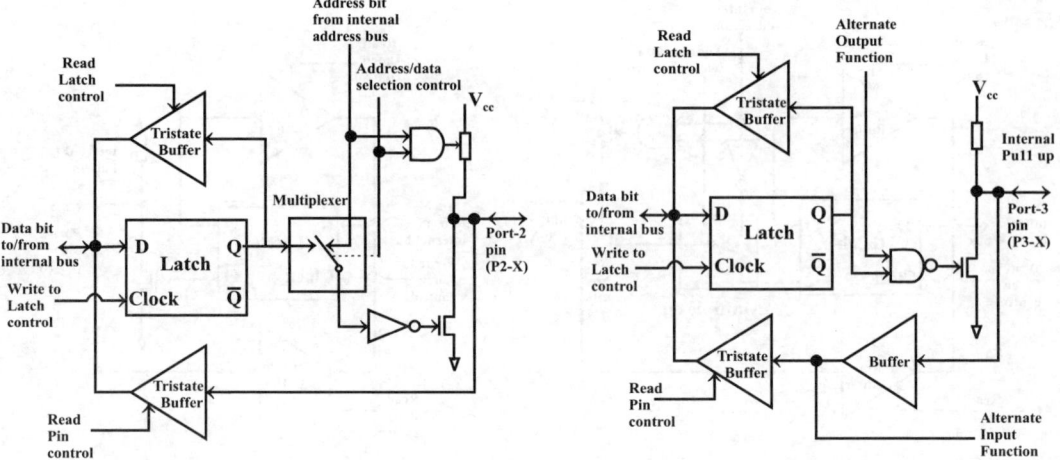

Fig. 2.14: Circuit of port-2 bit   **Fig. 2.15:** Circuit of Port-3 bit

Port-3 Pin Internal Hardware Structure (or Circuit)

Port-3 serves as input or output and also each pin of Port-3 has an alternate function. The internal hardware structure (or circuit) of each pin in Port-3 is shown in Fig. 2.15. Port-3 pins are also provided with internal pull up and so the input/output operation of Port-3 pins are similar to that of Port-1.

In order to select the alternate functions of port-3 pins, the port pin has to be initialized by writing "1" to it and the alternate function of port-3 pins are controlled by various other special function registers.

2.3   INTEL 8096 MICROCONTROLLER

The INTEL 8096 microcontroller is a 16-bit microcontroller with dedicated IO subsystems and a set of 16-bit arithmetic instructions including multiply and divide operations. INTEL has designed the 8096 microcontroller for high speed/high performance control applications.

The salient features of a 8096 microcontroller are,

- **16-bit ALU**
- **16-bit arithmetic instructions including multiply and divide**
- **256 bytes internal RAM called register file**
- **PWM output equivalent to DAC**
- **ADC with 8 multiplexed inputs and 10-bit resolution**
- **Serial port with own baud rate generation**
- **A 16-bit timer (Timer-1)**
- **A 16-bit counter (Timer-2)**
- **Four numbers of input programmable edge detector**
- **Six numbers of programmable event generators**
- **Four numbers of software timers**
- **100 instructions operating on bits and bytes (some operating on words and long words.)**

2.3.1 Pins and Signals of INTEL 8096

The INTEL 8096 is 68-pin IC available in PLCC package. The signals of an 8096 microcontroller are shown in Fig. 2.16. The description of these signals are listed in Table 2.11.

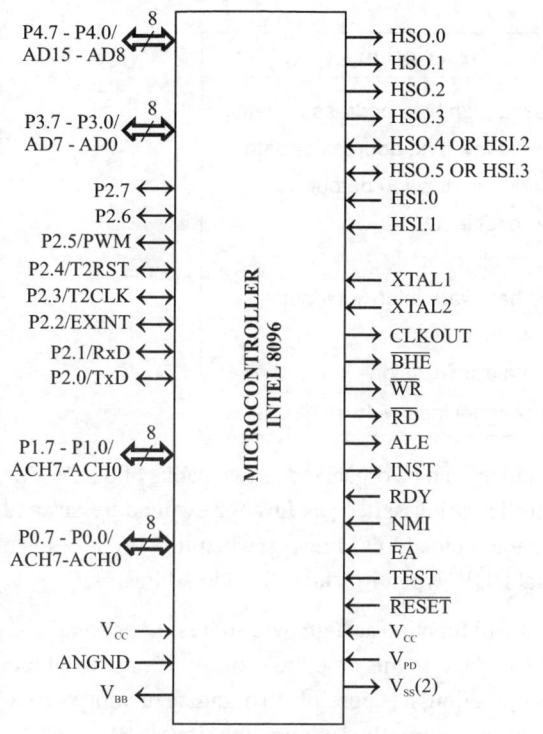

Fig. 2.16: Signals of an INTEL 8096 microcontroller.

Table 2.11: Signals of an 8096 Microcontroller

Signal	Description
P0.7-P0.0	Port-0 IO lines
P1.7-P1.0	Port-1 IO lines
P2.7-P2.0	Port-2 IO lines
P3.7-P3.0	Port-3 IO lines
P4.7-P4.0	Port-4 IO lines
V_{ref}	Reference voltage for internal ADC
ANGND	Analog ground
V_{BB}	Back bias voltage output
HSO.0-HSO.5	High speed output lines
HSI.0-HSI.3	High speed input lines
XTAL1-XTAL2	Pins for quartz crystal connection
CLKOUT	Output clock signal
$\overline{BHE}$	Bus High Enable
$\overline{WR}$	Write Control Signal
$\overline{RD}$	Read Control Signal
ALE	Address Latch Enable
INST	Instruction fetch indication
$\overline{EA}$	External Access
RDY	Ready synchronization
NMI	Non-maskable Interrupt
TEST	Test control input
$\overline{RESET}$	Controller reset input
V_{PD}	Power Down voltage
V_{CC}	Power supply, +5-V
V_{SS}	Power supply ground, 0-V

The 8096 has five numbers of 8-bit parallel port. The port pins are bit-addressable and so they can also be used as individual IO lines. Except port-1 all other port pins have alternate functions. The alternate functions of port pins are listed in Table 2.12.

When external memory is employed, the port-4 pins function as multiplexed high byte address/ data bus and the port-3 pins function as multiplexed **low** byte address/data bus.

The size of the external address bus in an 8096 is 16-bit and so the total addressable memory space is 64 kB (2^{16} = 64 k.) The external data bus in an 8096 can be 16-bit or 8-bit and this feature is programmable.

When an 8096 is programmed for a 16-bit data bus, the memory is accessed in words. In this case, memory is organized as two banks of 32 kB, so that the controller can access one-byte from each bank to form a word.

When an 8096 is programmed for an 8-bit data bus, the memory is accessed in bytes and so the memory is organized as a single bank of 64 kB (2^{16} = 64 k.) The pin $\overline{EA}$ should be tied **low** for accessing external memory. (If $\overline{EA}$ is tied **high,** the controller will work only with internal memory.)

Table 2.12: Alternate Functions of Port Pins

Port pins	Alternate function	Description
P4.7-P4.0	AD15-AD8	Multiplexed high byte address or data
P3.7-P3.0	AD7-AD0	Multiplexed low byte address or data
P2.5	PWM	Pulse width modulation output
P.2.4	T2RST	Timer-2 reset input
P2.3	T2CLK	Timer-2 clock input
P2.2	EXINT	External hardware interrupt input
P2.1	RxD	Serial data reception
P2.0	TxD	Serial data transmission
P0.7-P0.0	ACH7-ACH0	Analog channel input-7 to input-0

When a 16-bit external bus is employed the memory can be organized as two banks of 32 kB. The banks are called even bank and odd bank. The controller will assert A_0 as **low** for even addresses and $\overline{BHE}$ as **low** for odd addresses. The even bank is enabled by address line A_0 (when it is **low**) for even addresses. The odd bank is enabled by control signal $\overline{BHE}$ (when it is **low**) for odd addresses.

When an 8-bit external bus is employed, port-4 will function as high byte address lines and port-3 will function as a multiplexed low byte address or 8-bit data. In this case the memory is organized like that of an 8051 microcontroller. Also, in an 8-bit bus operation, it is possible to organize memory as two banks and access the data from the even and odd banks by externally decoding the signals $\overline{BHE}$ and A_0.

The signal ALE is used to demultiplex the multiplexed address/data lines using external latches. The $\overline{WR}$ and $\overline{RD}$ are asserted **low** for write and read operations respectively. The signal INST is asserted **low** during instruction read (or opcode fetch) cycles in order to indicate the code access. The signal RDY can be tied **low** for introducing wait states in machine cycle in order to extend the read/write time while accessing slow peripherals. The RDY should be tied **high** for normal timings.

The 8096 microcontroller has an internal oscillator. Pins XTAL1 and XTAL2 are provided for crystal connection to the internal oscillator for clock generation. Alternatively, the external clock signal can be applied through the XTAL1 pin. The crystal/external clock frequency can be in the range of 6 to 12 MHz. The crystal/external oscillator frequency is divided by 3 to generate the three internal timing phases called phase-A, phase-B and phase-C, as shown in Fig. 2.17.

Each phase has a duty cycle of 33% and one period of each phase is equal to three periods of crystal/external clock. The one period of a phase is called one state time. Most of the internal operations are synchronized to any one of these three phases. The phase-A clock is output through the CLKOUT pin for use by the peripherals. Therefore the signal at the CLKOUT pin of the microcontroller will have a frequency one-third of the crystal frequency and it is an asymmetric clock with a duty cycle of 33%.

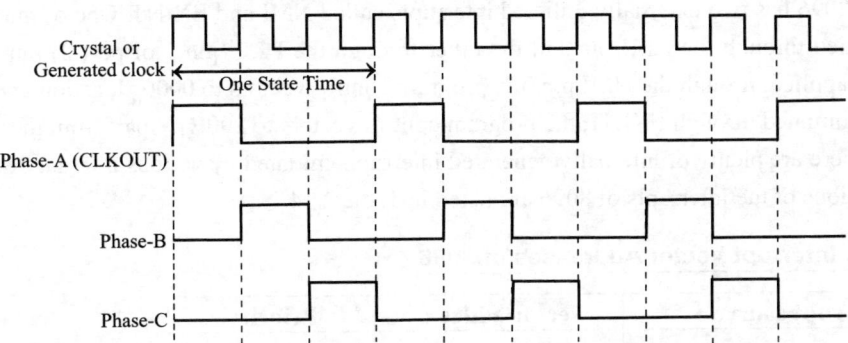

Fig. 2.17: Clock signals of an 8096 microcontroller.

The $\overline{\text{RESET}}$ pin is used to reset the microcontroller and bring the controller to a known state. Whenever the controller is resetted the content of the memory address 2018_H is loaded in the CCR (**C**hip **C**onfiguration **R**egister). The content of all other internal registers after a reset are listed in Table 2.13. For proper reset, the $\overline{\text{RESET}}$ pin should be held **low** for at least 10 state times. (i.e., for 30 clock periods).

Table 2.13: Reset Value of Internal Registers

Register	Reset value	Register	Reset value
Port-1	xxxx xxxx	Timer-1	0000 0000 0000 0000
Port-2	xx0x xxx1	Timer-2	0000 0000 0000 0000
Port-3	1111 1111	Watchdog timer	0000 0000 0000 0000
Port-4	1111 1111	HSI mode	xxxx xxxx
PWM control	0000 0000	HSI status	Undefined
Serial port (Transmit)	Undefined	IOS0	0000 0000
Serial port (Receive)	Undefined	IOS1	0000 0000
Baud rate register	Undefined	IOC0	x0x0 x0x0
Serial port control	xxxx 0xxx	IOC1	x0x0 xxx1
Serial port status	x00x xxxx	HSI FIF0	empty
A/D command	Undefined	HSO CAM	empty
A/D result	Undefined	HSO SFR	0000 0000
Interrupt pending	Undefined	PSW	0000 0000 0000 0000
Interrupt mask	0000 0000	Stack pointer	Undefined
		Program counter	0010 0000 1000 0000

The 8096 requires a single power supply of +5-V connected between V_{CC} and V_{SS}. The V_{PD} pin should also be tied to +5-V in order to activate the power down internal RAM. During power failure condition, the supply in the V_{PD} pin alone be maintained through a backup battery, in order to retain the contents of the power down RAM.

The 8096 has an internal 8-channel, 10-bit ADC. The analog inputs for eight channels are applied through port-0 pins. The reference voltage for ADC is applied between V_{ref} and ANGND. The maximum value of V_{ref} is +5-V.

The 8096 has two externally initiated interrupts called NMI and EXINT. One of the interrupts can be applied through the NMI pin and the other through the P2.2 (pin 2 of port-2) pin. When an interrupt is applied through the NMI pin, the program control vectors to 0000_H location and when an interrupt is initiated through EXINT, the program control vectors to $200E_H$. Apart from these external interrupts there are plenty of internally generated interrupts initiated by various internal devices. The vector locations of the interrupts of 8096 are listed in Table 2.14.

Table 2.14: Interrupt Vector Addresses of 8096

Interrupt source	Vector address	Priority
Software	2010_H-2011_H	Not applicable
External interrupt	$201E_H$-$200F_H$	7 (Highest)
Serial port	$200C_H$ -$200D_H$	6
Software Timers	$200A_H$-$200B_H$	5
HSI.0	2008_H-2009_H	4
High speed outputs	2006_H-2007_H	3
HSI data available	2004_H-2005_H	2
A/D conversion complete	2002_H-2003_H	1
Timer overflow	2000_H-2001_H	0 (Lowest)

The 8096 microcontroller has one serial port which can operate in four modes referred to as mode-0, mode-1, mode-2 and mode-3. For serial transmission and reception, the port-2 pins P2.0 and P2.2 are used as TxD and RxD respectively.

The 8096 has two timers, namely, timer-1 and timer-2. The timer-1 is a free running timer and it is incremented once in every eight state time.

The timer-1 is cleared only by reset. The timer-2 is controlled by an external clock signal applied through T2CLK or HSI.1. The timer-2 will increment by one for every transition (i.e., for every **low** to **high** and for every **high** to **low**) of either T2CLK or HSI.1.

The 8096 microcontroller has four **H**igh **S**peed **I**nput (HSI) pins denoted as HSI.0, HSI.1, HSI.2 and HSI.3. These input pins can be used to record events and the time at which the event occurs with respect to timer-1. There are four possible modes of operation for each of the HSI pins.

The 8096 microcontroller has four numbers of independent **H**igh **S**peed **O**utput (HSO)pins denoted as HSO.0, HSO.1, HSO.2 and HSO.3. The HSI.2 and HSI.3 can be used either as input or output lines. When used as output lines they are referred to as HSO.4 and HSO.5.

The six high speed output lines can be used to trigger six time-based events. For timings, the HSO lines can refer to either timer-1 or timer-2.

2.3.2 Architecture of an 8096 Microcontroller

The 8096 is a 16-bit microcontroller and has a register-based architecture in which any of the internal register can be the source or destination of the instructions. The architecture (or functional block diagram) of an 8096 microcontroller is shown in Fig. 2.18. The architecture consists of two major sections: CPU section and IO section.

The CPU section consists of a 16-bit ALU called the Register ALU (RALU), which uses a 256 byte register file as sources and destinations of the operand, instead of the accumulator. The IO section consists of a watchdog timer, a PWM generator equivalent to DAC, an 8-channel ADC, a serial port, a two 16-bit timers, high-speed inputs, high-speed outputs and buffers for IO ports.

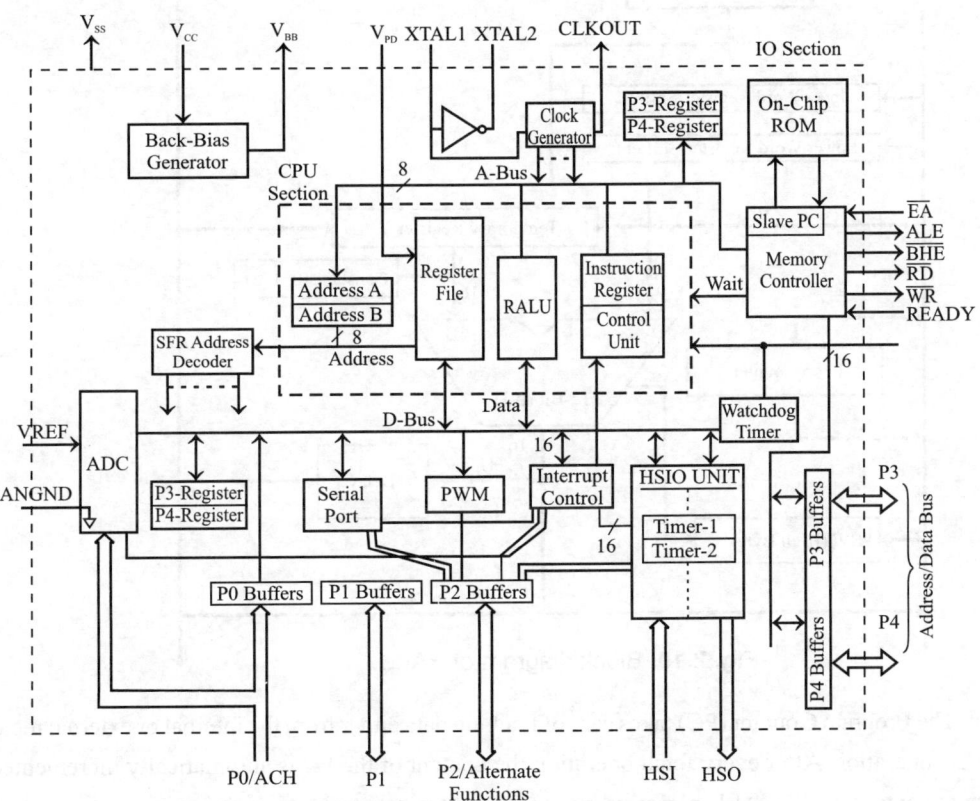

Fig. 2.18: Block diagram of an 8096 microcontroller.

The architecture of the 8096 includes two internal buses called A-bus and D-bus. The D-bus is 16-bit wide and used to transfer data between RALU and the register file or SFR (**S**pecial **F**unction **R**egister). The A-bus is 8-bit wide and used as address bus for the data transfer between RALU and register file. Also the A-bus is used as a multiplexed address/data bus for data transfer between the memory controller and the register file.

Register ALU (RALU)

The RALU consists of a 17-bit ALU (16-bit + 1 sign extension bit), a high byte of **Program Status Word (PSW)**, a **Program Counter (PC)**, an incrementer, a loop counter, three temporary registers, a delay unit and stored constants. The block diagram of RALU is shown in Fig. 2.19.

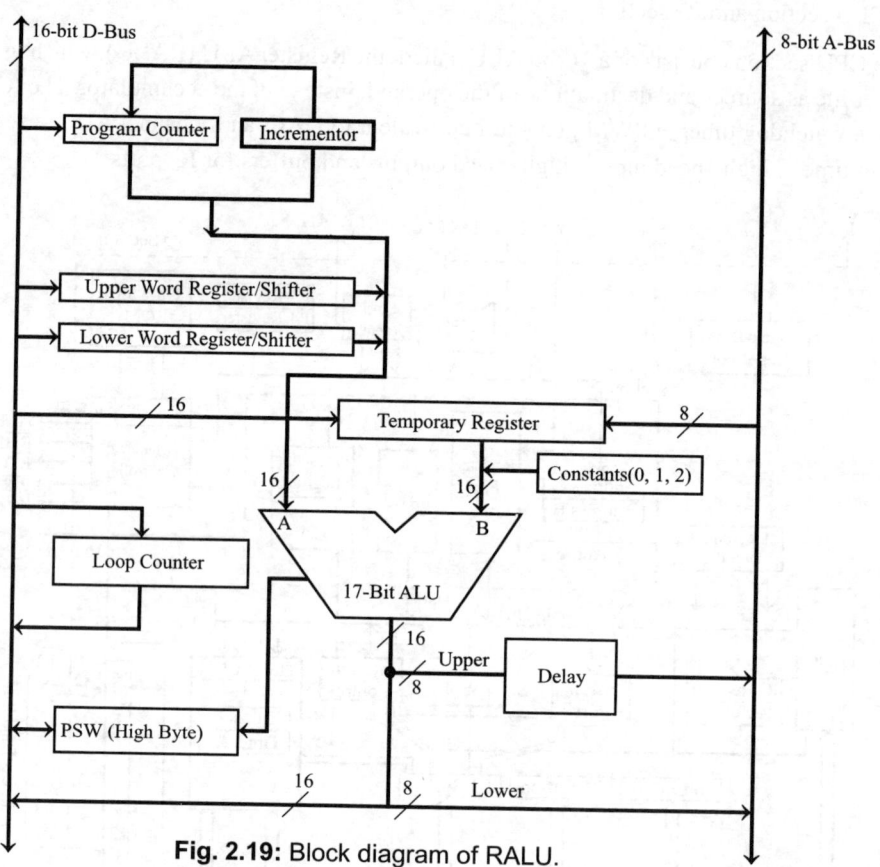

Fig. 2.19: Block diagram of RALU.

The **Program Counter (PC)** takes care of fetching data either from the internal or external memory for ALU operation. After every fetch operation the content of the PC is automatically incremented by the incrementor unit. The ALU performs most of the arithmetic and logical operations. Two temporary registers are capable of performing the shift operations required for Normalize, Multiply and Divide.

For shift of 16-bit data (word data) only the upper word register is used and for shift of 32-bit data (double word data) both upper and lower word registers are used. The loop counter will function as a 5-bit counter for repetitive shifts.

The upper or high byte of the PSW consists of the arithmetic flags which are altered by the ALU operation to indicate various condition of ALU result. The zero flag is set, when the result of the ALU operation is zero. The negative flag is set, when the result of the ALU operation is negative.

The overflow flag is set, when the size of the result exceeds the size of the destination. The overflow trap flag is also set whenever overflow is set, but the overflow trap flag can be cleared only by specific instructions.

The carry flag is set to indicate a carry over from the most significant bit. The sticky flag is set to indicate that during a right shift, a 1 has been shifted first into the carry flag and then has been shifted out. The sticky flag can be used along with the carry flag to control rounding after a right shift. The format of the upper byte of the PSW is shown in Fig. 2.20. (The lower byte of the PSW register is called interrupt mask register and it is mapped as internal memory.)

The delay unit in RALU is used to transfer the data and address between the 16-bit and 8-bit bus. The RALU has several stored constants such as 0, 1 and 2 which are used for certain calculations.

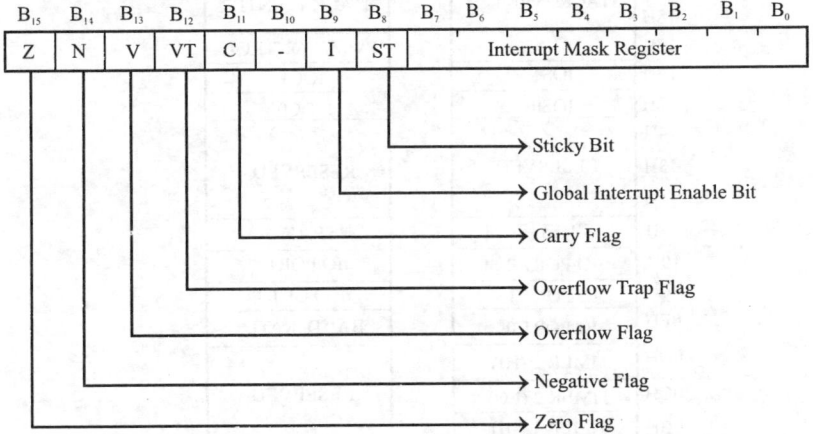

Fig. 2.20: Format of the upper word of a PSW register.

Register File

The register file is an array of 256 registers of size 8-bit each and they are generally referred to as internal RAM. The 256 bytes of register file are mapped in the memory space in the address range 0000_H to $00FF_H$ and they can be addressed with 8-bit addresses in the range 00_H to FF_H. This address space can be addressed only as data memory and cannot be used for code/program storage. The 8096 has a provision for external program memory in the address range 0000_H to $00FF_H$ and if an attempt is made to execute instruction in this address space, the controller will fetch instructions only from the external memory. Actually this address is used by INTEL development tools for various procedures, including interrupt service procedures for NMI and TRAP.

The lower 26 bytes of a register file (in the address range 00_H to 19_H) are Special Function Registers (SFRs) and they are used to control or program various IO features of 8096. The upper 16 bytes of register file (in the address range $F0_H$ to FF_H) are called power down RAM, because the content of these memory location receives power through a separate pin called V_{PD} and so during power down condition the content of these RAM locations can be preserved by providing backup supply from the V_{PD} pin.

The remaining 213 bytes register file in the address range $1A_H$ to EF_H are called general purpose RAM and can be used for general purpose data storage. The layout of a register file is shown in Fig. 2.21. Most of the SFR serve two functions, one if they are read from and another if they are written to.

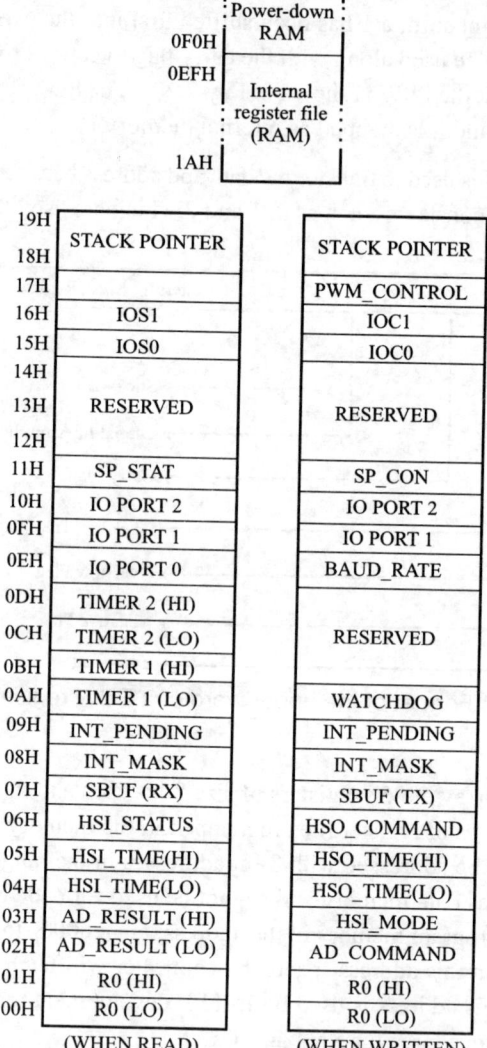

Address	(WHEN READ)	(WHEN WRITTEN)
19H / 18H	STACK POINTER	STACK POINTER
17H		PWM_CONTROL
16H	IOS1	IOC1
15H	IOS0	IOC0
14H / 13H / 12H	RESERVED	RESERVED
11H	SP_STAT	SP_CON
10H	IO PORT 2	IO PORT 2
0FH	IO PORT 1	IO PORT 1
0EH	IO PORT 0	BAUD_RATE
0DH	TIMER 2 (HI)	
0CH	TIMER 2 (LO)	RESERVED
0BH	TIMER 1 (HI)	
0AH	TIMER 1 (LO)	WATCHDOG
09H	INT_PENDING	INT_PENDING
08H	INT_MASK	INT_MASK
07H	SBUF (RX)	SBUF (TX)
06H	HSI_STATUS	HSO_COMMAND
05H	HSI_TIME(HI)	HSO_TIME(HI)
04H	HSI_TIME(LO)	HSO_TIME(LO)
03H	AD_RESULT (HI)	HSI_MODE
02H	AD_RESULT (LO)	AD_COMMAND
01H	R0 (HI)	R0 (HI)
00H	R0 (LO)	R0 (LO)

Memory map (top): 0FFH–0F0H Power-down RAM; 0EFH–1AH Internal register file (RAM).

Fig. 2.21: Layout of a register file.

Internal ROM and EPROM

The 8x9x family of microcontrollers has an internal ROM version as well as an internal EPROM version of microcontrollers. Most of the family members have an 8 kB internal ROM/EPROM mapped in the address range 2000_H to $3FFF_H$ and some of the family members have a 16 kB internal ROM/EPROM mapped in the address range from 2000_H to $5FFF_H$.

In both these family members the address space 2000_H to $207F_H$ are reserved for various specific functions and the remaining address space 2080_H to $3FFF_H$ (or 2080_H to $5FFF_H$ in case 16 kB internal ROM/EPROM) can be used as internal program memory by the user. After a reset, the **Program Counter** (PC) is initialized with the address 2080_H and so after a reset the PC will start fetching instructions from the address 2080_H. The specific usage of the memory locations in the address range 2000_H to $207F_H$ are listed in Table 2.15.

Table 2.15: Function of Reserved Memory Space

Memory address	Function/Usage
2000_H-2011_H	Interrupt vectors
2012_H-2017_H	Reserved
2018_H	Chip configuration register
2019_H	Reserved
$201A_H$-$201B_H$	Jump to self opcode
$201C_H$-$201F_H$	Reserved
2020_H-$202F_H$	Security key
2030_H-$207F_H$	Reserved

The reserved locations should not be used by the programmer and all reserved locations except 2019_H should be filled with FF_H to ensure compatibility with future devices. The location 2019_H must be filled with 20_H.

Memory Space of an 8096 Microcontroller

The 8096 microcontroller has 64 kB address space. In this, the first 256 bytes are allotted to the internal register file and their hexa address range is 0000_H to $00FF_H$.

The address space from 0100_H to $1FFF_H$ can be used as an external memory address space or IO address space and in this address range the addresses $1FFE_H$ and $1FFF_H$ are reserved for port-3 and port-4 respectively.

The addresses 2000_H to $3FFF_H$ are allotted to internal ROM/EPROM. The remaining address space from 4000_H to $FFFF_H$ refers to external memory or IO. The 8096 also has 256 byte external program memory address space in the range 0000_H to $00FF_H$ which can be used only by INTEL development systems.

The memory map of an 8096 microcontroller is shown in Fig. 2.22.

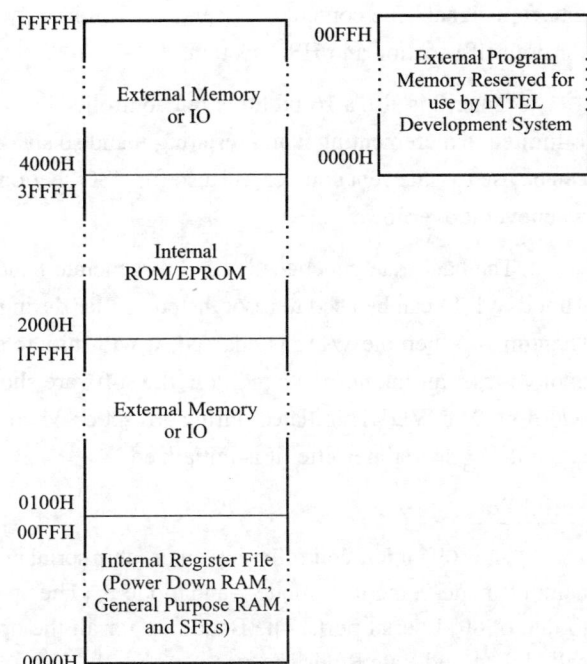

Fig. 2.22: Memory map of an 8096 microcontroller.

Memory Controller

The memory controller takes care of accessing the program codes from the internal ROM/EPROM and accessing program codes and data from the external memory or IO. The memory controller communicates with the RALU through the A-bus and several control lines. The memory controller has a slave PC which holds the address of the program memory (i.e., the content of the PC is used to fetch program codes and not to fetch data.) The content of the slave PC will always be same as that of the PC in the RALU.

The slave PC is loaded with the same value as that of the PC in the RALU during reset and jump or call. During all other times the slave PC is incremented automatically after every instruction fetch similar to that of the PC in a RALU. Thus, the slave PC avoids the overhead time for getting the address of instruction code through the 8-bit A-bus for every instruction fetch.

Timers

The 8096 microcontroller has two general purpose timers: Timer-1 and Timer-2, and two dedicated timers: Baud rate generator and Watchdog timer.

Timer-1 is a 16-bit free running timer and it is incremented once in every 8-state time. The content of the timer-1 cannot be modified or stopped. Whenever the timer-1 overflows, it generates an internal interrupt signal. The content of timer-1 is cleared only during a reset. Timer-1 is used as a reference for both HSI section and HSO section.

Timer-2 is also a 16-bit timer but controlled by an external clock source. The control of Timer-2 is limited to incrementing it and resetting it and so specific values cannot be written to it. The Timer-2 can be used as an event counter and also for HSO section. Timer-2 generates an internal interrupt signal whenever it overflows.

The baud rate generator is used to generate baud rate clock for the serial port. The Watchdog Timer (WDT) can be used to reset the controller during software malfunction, which is referred to as "hanging". When the WDT is enabled, it will initiate a hardware reset in every 64k state times. To avoid a reset during normal condition, the software should take care of clearing WDT well before its overflow. The WDT is initiated during a reset or when it is cleared and it overflows between 65280_{10} and 65535_{10} state times after it is initialized.

Serial Port

The 8096 microcontroller has an on-chip serial port which can function in four operating modes, namely, mode-0, mode-1, mode-2 and mode-3. (The operating modes of serial port of 8096 are similar to that of 8051 serial port.) The Baud rate for all the operating modes are controlled through an SFR called the Baud Rate Register.

The mode-0 is synchronous mode and in this mode, the serial port functions as the shift register. The shift clocks output through a TxD pin and the data is either transmitted or received through an RxD pin. In modes 1, 2 and 3 the serial port functions as a full duplex asynchronous serial port. In mode-1, one data character can have 8 data bits, one start bit and one stop bit.

In mode-2 and mode-3, one data character can have 8 data bits, one control/parity bit called 9th data bit, one start bit and one stop bit. The difference in modes 2 and 3 is the initiation of the interrupt. In mode-2, the serial port interrupt is activated only when the 9th bit is a one and in mode-3 the serial port interrupt is activated whenever a byte is received.

Analog to Digital Converter (ADC)

The 8096 controller has an internal 8-channel, 10-bit ADC with an internal multiplexer. The conversion of analog to digital signal is performed by successive approximation techniques and one conversion requires 168 state times. The SFR, which is used for ADC are the A/D command register to control the conversion process and the A/D result register to store the output of the ADC.

Pulse Width Modulation (PWM) Generator

The 8096 controller has an internal PWM generator which can generate a fixed frequency, variable duty cycle waveforms. The frequency is fixed at 15.625 kHz (the time period being 64 microseconds) for a 12 MHz clock speed. The duty cycle can be varied from 0% to 99.6% by writing an 8-bit value to one of the SFR called the PWM register.

This PWM output can be easily converted to analog signal using an external hardware and used for various applications like temperature control, speed control of motor, etc.

The block diagram of a PWM generator and its output waveform for various duty cycles are shown in Fig. 2.23 and Table 2.16 respectively. The PWM counter is an 8-bit counter and it is incremented by one in every state time. When the counter overflows (i.e., changes from FF_H to 00_H) the PWM output is set **high**.

When the content of the counter is equal to the content of the PWM register, the PWM output is made **low**. Whenever the counter overflows the content of the temporary latch is reloaded into the PWM register and so the PWM generator does not recognize a new value before a overflow.

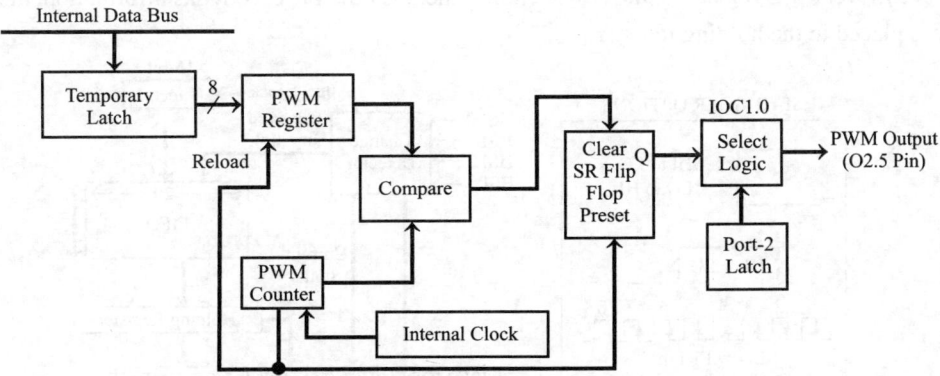

Fig. 2.23: Block diagram of a PWM generator.

Table 2.16: Output Waveform of PWM Generator for Various Duty Cycles

Duty cycle	8-bit value loaded in PWM register	PWM output waveform
0%	00_H	HI LO
10%	19_H (25_{10})	HI LO
50%	$7F_H$ (127_{10})	HI LO
90%	$E6_H$ (230_{10})	HI LO
99.6%	FF_H (255_{10})	HI LO

High Speed Input/Output (HSIO) Unit

The HSIO unit consists of four **High Speed Input** (HSI) lines, four **High Speed Output** (HSO) lines and two high speed input or output lines. The HSI lines are used to record events and the time at which the event occurred. The HSI lines can be independently programmed to recognize any one of the following four events at any one time:

1. **Eight positive transitions**
2. **Each positive transition**
3. **Each negative transition**
4. **Every transition (Positive and negative)**

The block diagram of an HSI section is shown in Fig. 2.24. Whenever a valid transition is detected the content of timer-1 along with the inputs which recognizes the transition are stored as a 20-bit value in a seven-level FIFO register in the HSI section. Whenever the FIFO contains information, the earliest entry is placed in the holding register.

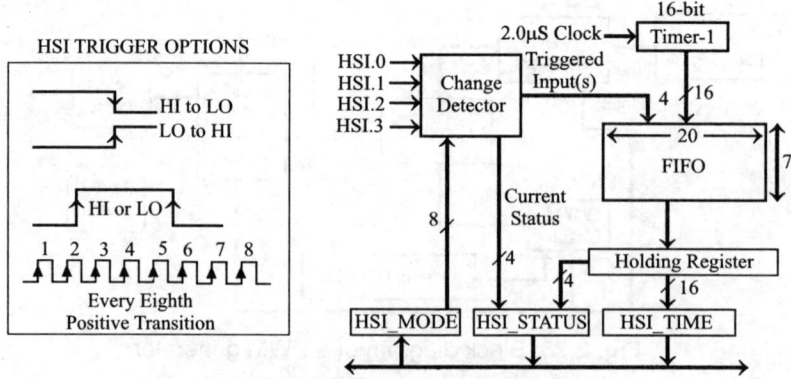

Fig. 2.24: High speed input (HSI) section.

After each read operation of the holding register, the next entry in FIFO is placed in the holding register on the basis of **First-In-First-Out** (FIFO). An internal interrupt is also generated by the HSI unit whenever the holding register is loaded or when the FIFO has six or more entries.

The HSO lines can be used to generate time-based signals to initiate various programmable time-based activities like turning the ON/OFF device, start or reset a process, etc. The HSO lines can use either timer-1 or timer-2 as reference for their time-based signals.

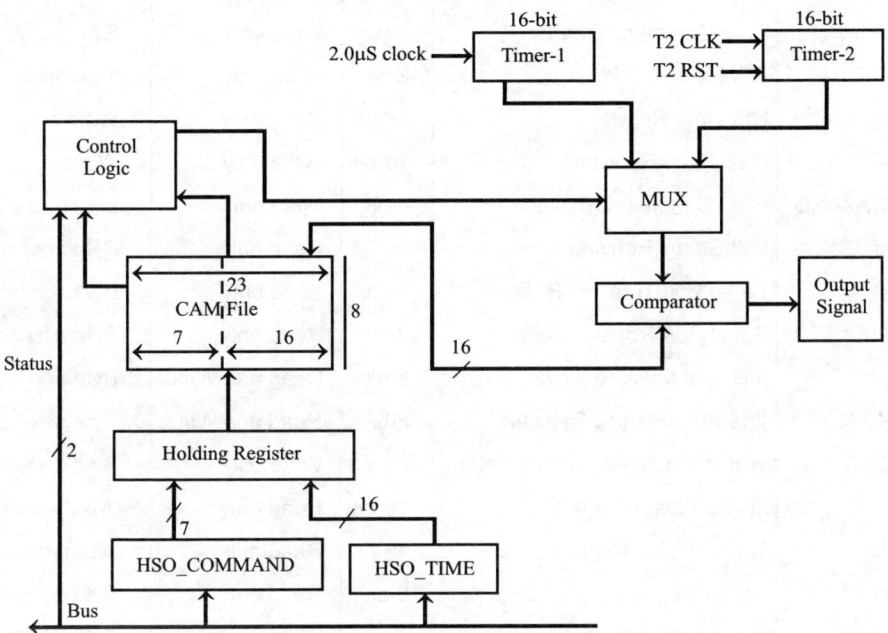

Fig. 2.25: High speed output (HSO) section.

The block diagram of an HSO section is shown in Fig. 2.25. The main component of the HSO is the **C**ontent **A**ddressable **M**emory (CAM) file. The CAM file stores up to eight events which are pending to occur. In every state time one location of the CAM is compared with the two timers.

In a time of 8 state times, the entire CAM file has been searched for time matches. If a match occurs, the specified event will be triggered and that location of the CAM will be made available for another pending event.

2.3.3 Special Function Registers (SFR) of 8096

The **S**pecial **F**unction **R**egisters (SFRs) of 8096 are listed in Table 2.17. Most of the SFRs serve two functions, one when they are read and another when they are written to. The SFRs are mapped as internal memory in the address range 0000_H to 0019_H. The address of a specific SFR can be obtained from the SFR layout shown in Fig. 2.21.

Table 2.17: Special Function Registers (SFR) of an 8096 Microcontroller

SFR	Description	Size	Possibility of read/write operation	Byte/word access
R0	Zero Register	16-bit	Read and Write	Byte/word access
AD_RESULT	ADC Result Register	16-bit	Read only	Byte access
AD_COMMAND	ADC Command Register	8-bit	Write only	Byte access
HSI_MODE	HSI Mode Register	8-bit	Write only	Byte access
HSI_TIME	HSI Time Register	16-bit	Read only	Word access
HSO_TIME	HSO Time Register	16-bit	Write only	Word access
HSO_COMMAND	HSO Command Register	8-bit	Write only	Byte access
HSI_STATUS	HSI Status Register	8-bit	Read only	Byte access
SBUF(Tx)	Serial Port Transmit Buffer	8-bit	Write only	Byte access
SBUF (Rx)	Serial Port Receive Buffer	8-bit	Read only	Byte access
INT_MASK	Interrupt Mask Register	8-bit	Read and Write	Byte access
INT_PENDING	Interrupt Pending Register	8-bit	Read and Write	Byte access
WATCHDOG	Watchdog Timer Data Register	8-bit	Write only	Byte access
TIMER1	Timer-1 Data Register	16-bit	Read only	Word access
TIMER2	Timer-2 Data Register	16-bit	Read only	Word access
IOPORT0	Port-0 Register	8-bit	Read only	Byte access
BAUD_RATE	Baud Rate Register	8-bit	Write only	Byte access
IO PORT1	Port-1 Register	8-bit	Read and Write	Byte access
IO PORT 2	Port-2 Register	8-bit	Read and Write	Byte access
SP_STAT	Serial Port status Register	8-bit	Read only	Byte access
SP_CON	Serial Port Control Register	8-bit	Write only	Byte access
IOS0	IO Status Register-0	8-bit	Read only	Byte access
IOS1	IO Status Register-1	8-bit	Read only	Byte access
IOC0	IO Control Register-0	8-bit	Write only	Byte access
IOC1	IO Control Register-1	8-bit	Write only	Byte access
PWM_CONTROL	PWM Control Register	8-bit	Write only	Byte access
SP	Stack Pointer	16-bit	Read and Write	Word access

Zero Register (R0)

When the R0-register is read, it will always give a zero value and this zero can be used as a base for indexing and as a constant for calculations and comparisons.

AD_Result and AD_Command Registers

The ADC command register is used to select a channel and initiate conversion. The format of an ADC command register is shown in Fig. 2.26. The controller will store the converted digital value in the ADC result register along with the channel number. Hence, the digital value of the analog signal can be read from the ADC result register. The result register also has a one-bit status to indicate whether the ADC is idle or a conversion is in progress. This status bit can be tested by the programmer before giving any command for start of conversion.

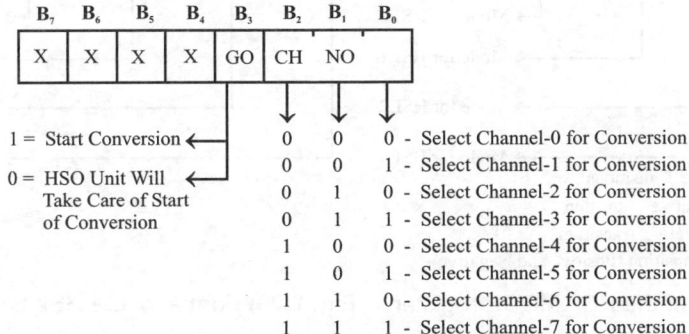

Fig. 2.26: Format of the AD_COMMAND register.

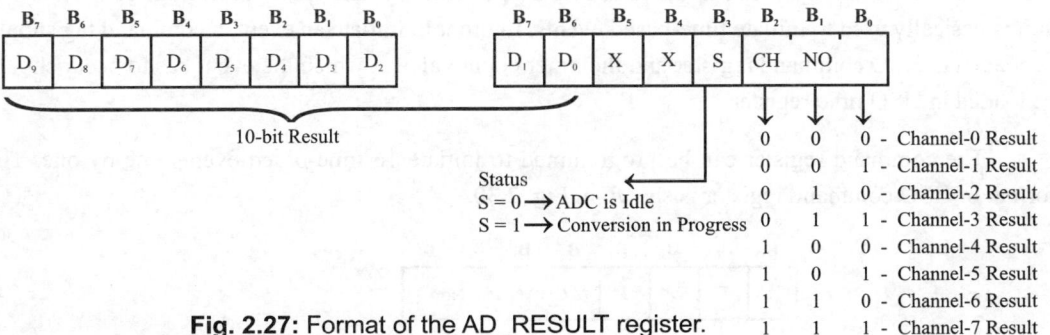

Fig. 2.27: Format of the AD_RESULT register.

SFRs in High Speed Input (HSI) Section

The HSI section has three SFRs, one for mode, one for status and another for time. The HSI mode register is used to program the operating mode of the HSI input lines. The format of the HSI_MODE register is shown in Fig. 2.28. The HSI mode register has a two-bit field for each HSI line, to program any one of the four possible mode of operation.

The status of the HSI lines can be obtained from the HSI status register. The format of the HSI_STATUS register is shown in Fig. 2.29.

The HSI status register has a two-bit field for each HSI line in which the lower bit indicates whether an event has occurred or not and the upper bit indicates the current status of the HSI line.

The HSI_time register holds the count value of Timer-1 at which the high speed input is triggered. (Actually it will be loaded from FIFO of the HSI section into the HSI_time register.) The programmer can read the time at which the HSI was triggered from the HSI_time register.

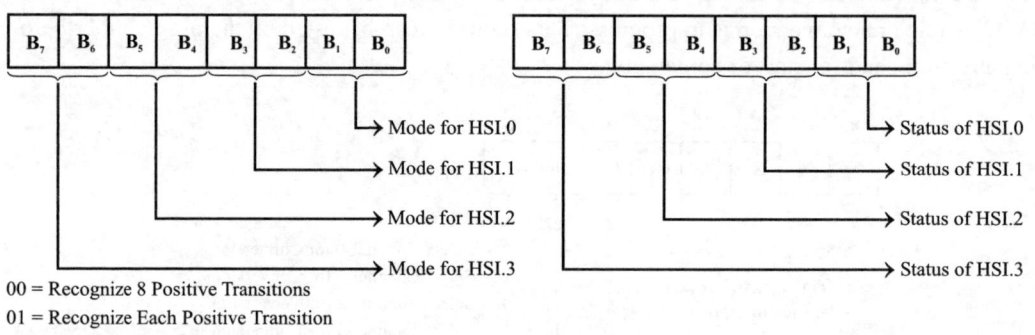

Fig. 2.28: Format of the HSI_MODE register.   **Fig. 2.29:** Format of the HSI_STATUS register.

SFRs in High Speed Output (HSO) Section

The HSO section has two SFRs called HSO_COMMAND and HSO_TIME registers. The HSO unit is basically used to initiate time-based events. In order to initiate an event, a command tag should be loaded in HSO command register and the timer count value at which the event has to occur should be loaded in HSO time register.

The command register can be programmed to initiate 16 time-based events one by one. The format of HSO command register is shown in Fig. 2.30.

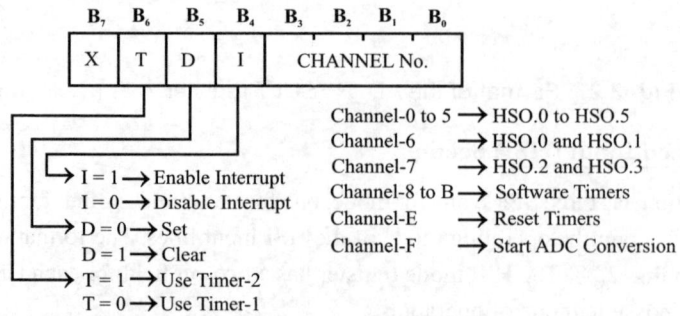

Fig. 2.30: Format of the HSO_COMMAND register.

SFRs of a Serial Port

The SFRs used by the serial port are the Transmit buffer (SBUF(Tx)), the Receive buffer (SBUF(Rx)), the Baud rate register (BAUD_RATE), the Serial port status register (SP_STAT) and the Serial port control register (SP_CONT). The received serial data is converted to parallel data and loaded in SBUF(Rx) to be read by the controller. The parallel data to be transmitted as serial data is loaded in SBUF (Tx), from which the serial port reads the parallel data and converts it to serial for transmission.

The serial port has its own baud rate generator for all the modes of operation. In order to generate the required baud rate, a 16-bit value should be loaded in the baud rate register. (The 16-bit value is loaded such that low byte first and then high byte.) The most significant digit of this 16-bit value selects one of the two clock sources for the baud rate generator.

If this bit is 1, then the frequency at the XTAL1 pin is selected or if it is 0, then the signal applied to the T2CLK pin is used for baud rate generation. The lower 15 bits of the baud rate register is an unsigned integer in the range 0_{10} to 32767_{10} and it is denoted as **B**. The baud rate for serial transmission/reception for all the operating modes will be decided by the value of **B** and the clock frequency at XTAL1 or T2CLK pin as given below:

USING XTAL1	USING T2CLK
Mode-0:	Mode-0:
$\text{Baud rate} = \dfrac{\text{XTAL1 frequency}}{4 \times (\mathbf{B}+1)} \quad ; \quad \mathbf{B} \neq 0$	$\text{Baud rate} = \dfrac{\text{T2CLK frequency}}{\mathbf{B}} \quad ; \quad \mathbf{B} \neq 0$
Other modes:	Other modes:
$\text{Baud rate} = \dfrac{\text{XTAL1 frequency}}{64 \times (\mathbf{B}+1)} \quad ; \quad \mathbf{B} \neq 0$	$\text{Baud rate} = \dfrac{\text{T2CLK frequency}}{16 \times \mathbf{B}} \quad ; \quad \mathbf{B} \neq 0$

The serial port control register is used to select the mode of operation for the serial port. It is also used to enable the parity, program the 9th bit to be transmitted in modes 2 and 3, and to enable/disable the reception. The format of a serial port control register is shown in Fig. 2.31.

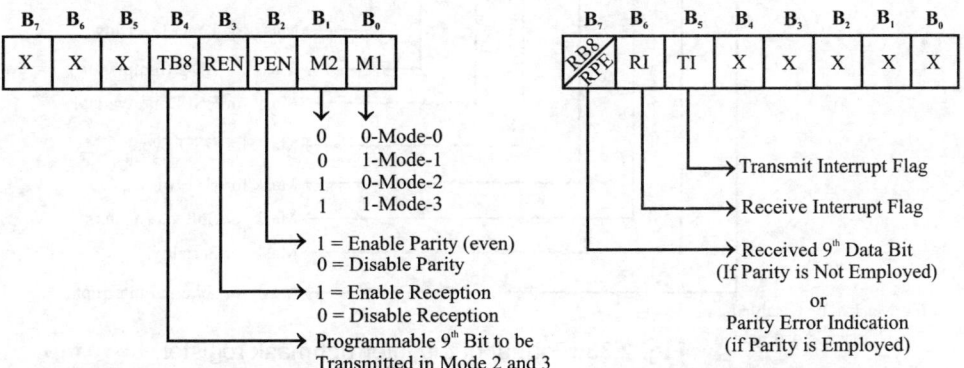

Fig. 2.31: Format of the SP_CON register.

Fig. 2.32: Format of the SP_STAT register.

The format of a serial port status register is shown in Fig. 2.32. It has two interrupt flags, one for transmit and the other for receive. The **Transmit Interrupt** (TI) flag is set during transmission of the last bit of a data character. The **Receive Interrupt** (RI) flag is set during reception of the last bit of a data character.

The TI and RI flags are logically ORed internally and used to generate the serial port interrupts. These flags are cleared whenever the status register is read. Also, the status register stores the 9th data bit received which can also be a parity error flag, if parity is enabled.

Timer Data Register

The SFRs of an 8096 includes two 16-bit registers, one for timer-1 and the other for timer-2, and one 8-bit register for Watchdog Timer (WDT). The timer-1 data register is cleared, whenever the controller is resetted or during overflow. The timer-2 data register is cleared by applying a reset signal to P2.4 pin of port-2. Both these timer registers can only be read at any time, but cannot be written.

The WDT has an 8-bit register which is cleared during a reset and incremented once in 256 state times. The overflow of this timer register will initiate an internal reset, which is useful to overcome any software upset. But during normal program execution, the WDT reset should be avoided by clearing the WDT register well before its overflow.

Interrupt Mask and Pending Register

The SFRs include two 8-bit registers for interrupts, one for individually masking the interrupts and the other for monitoring the pending interrupts. The interrupt mask register can also be accessed as low byte of PSW (**Program Status Word**). (*Note : The high byte of PSW is flag register.*) The interrupt mask register has a 1-bit field for each of the eight interrupt sources.

A zero in the particular interrupt will mask (or disable) that interrupt and a 1 will unmask (or enable) that interrupt. The format of an interrupt mask register is shown in Fig. 2.33.

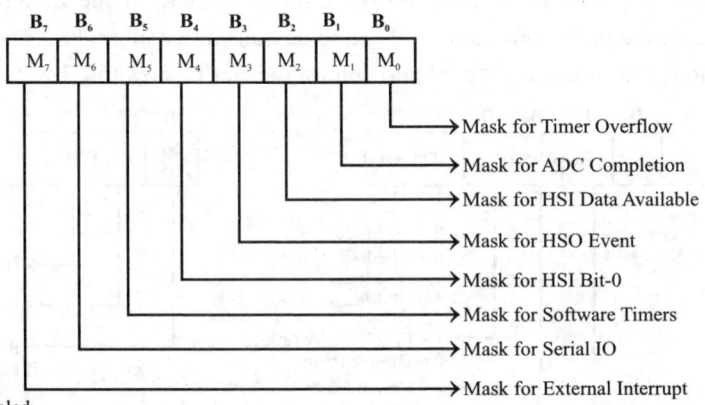

0 = Masked or Disabled
1 = Unmasked or Enabled

Fig. 2.33: Format of the interrupt mask register.

Similar to that of a mask register, the interrupt pending register also has a 1-bit field for each of the eight interrupt sources to indicate the pending status of the interrupts. Whenever an interrupt signal is generated, the 1-bit field corresponds to that interrupt in the pending register is set to 1 and when that interrupt is serviced, the pending status bit is cleared to zero. The format of an interrupt pending register is shown in Fig. 2.34.

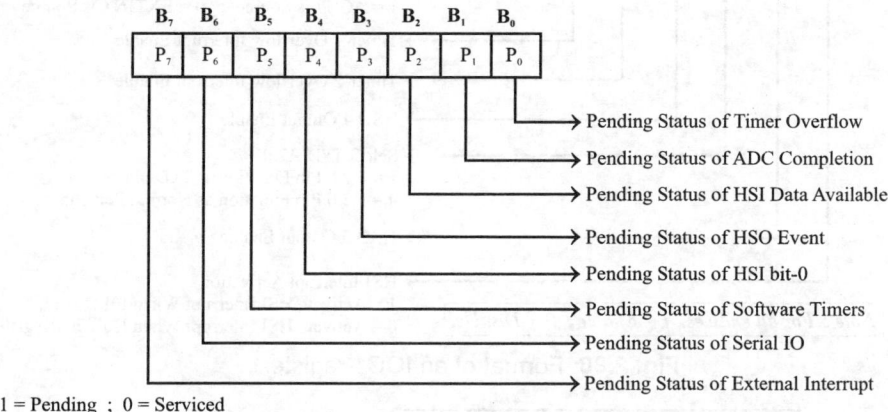

Fig. 2.34: Format of the interrupt pending register.

IO Control and Status Registers

The SFRs includes two IO control registers, namely, IOC0 and IOC1, and two status registers, namely, IOS0 and IOS1. The IOC0 can be programmed to control the HSI lines and timer-2. The format of an IOC0 is shown in Fig. 2.35. The IOC1 can be programmed to select the functions of some of the port pins and to control some of the interrupts and HSO lines. The format of an IOC1 is shown in Fig. 2.36.

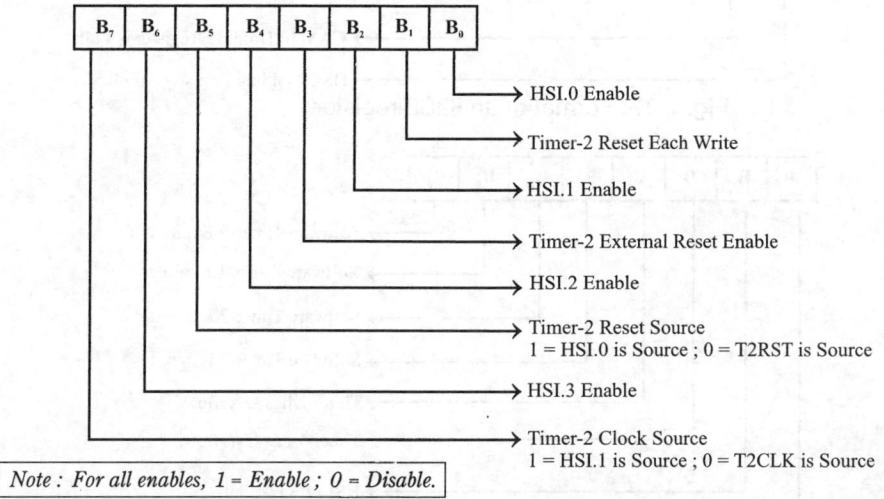

Fig. 2.35: Format of an IOC0 register.

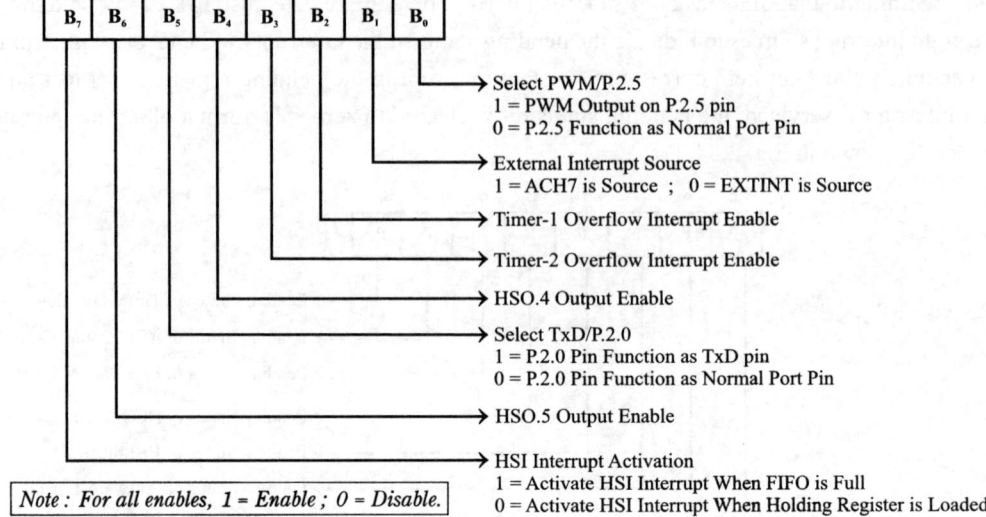

Fig. 2.36: Format of an IOC1 register.

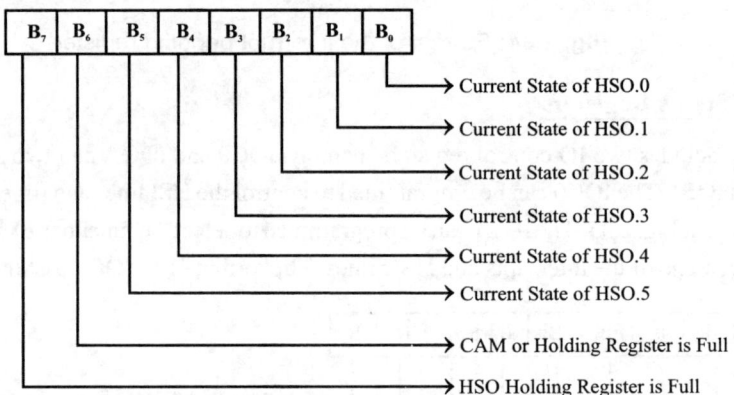

Fig. 2.37: Format of an IOS0 register.

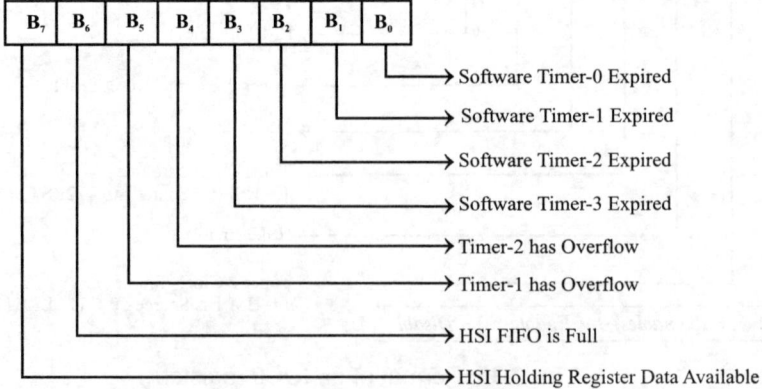

Fig. 2.38: Format of an IOS1 register.

The IOS0 register holds the current status of HSO lines and has status bits to indicate whether the CAM or holding register is full or not. The format of IOS0 is shown in Fig. 2.37. The IOS1 contains the status bits for the timers and HSI unit FIFO full indication. The format of an IOS1 register is shown in Fig. 2.38. Whenever IOS1 is read, all the status of timers (time-related flags) are cleared.

2.4 INTRODUCTION TO PIC MICROCONTROLLER

History: **PIC** is a family of microcontrollers made by Microchip Technology, a company formed by acquiring General Instruments Microelectronics Division in 1987. The name PIC initially referred to **Peripheral Interface Controller,** later it was renamed as **Programmable Intelligent Computer.** The first parts of the family were available in 1976 made by General Instruments.

Features of PIC microcontroller

- Harvard architecture with separate code and data memory spaces.
- RISC based instruction set with a small number of fixed-length instructions.
- Most instructions are single-cycle execution.
- One accumulator which is the implied source of one operand.
- All RAM locations function as registers as both source and/or destination.
- A hardware stack for storing return addresses.
- A small amount of addressable data space (32, 128, or 256 bytes, depending on the family), extended through banking.
- Data-space mapped CPU, port and peripheral registers.
- ALU status flags are mapped into the data space.
- The program counter is also mapped into the data space and writeable (this is used to implement indirect jumps).

The entire family of PIC microcontrollers has Harvard architecture, in which instructions and data come from separate sources which simplifies timing and microcircuit design and in turn benefits clock speed, price and power consumption.

In PIC microcontrollers, there is no distinction between memory space and register space because the RAM serves the job of both memory and registers. External data memory is not directly addressable.

In PIC microcontrollers there is no provision for storing code in external memory due to the lack of an external memory interface. The code space is generally implemented in internal memory or on-chip ROM, EPROM or flash ROM.

PIC handles data and address of data memory space in 8-bits. But, the unit of addressability of the code memory space is not same as the data space, the instruction width is 12 or 14 or 16 bits.

Advantages of PIC Microcontroller are as follows:

- **Small instruction set to learn**
- **RISC architecture**
- **Built-in oscillator with selectable speeds**
- **Easy entry level, in-circuit programming plus in-circuit debugging**
- **Inexpensive microcontrollers**
- **Wide range of interfaces including I²C, SPI, USB, USART, A/D, programmable comparators, PWM, LIN, CAN, PSP, and Ethernet**
- **Availability of processors in DIL (Dual-In-Line) package makes them easy for hobby use**

2.4.1 PIC16F877A Microcontroller

The PIC16F877A is a popular and most widely used member of PIC microcontroller family. The PIC16F877A is an 8-bit microcontroller released by Microchip Technology. The PIC is packed with 40 pins and requires single 5 volts supply. The PIC instruction set consists of 35 instructions. The PIC has an internal clock oscillator. For clock generation, it uses either low cost RC circuit or quartz crystal. This internal clock is also given out for the clock requirement of peripheral devices.

Features of PIC16F877A

- **Instruction set with 35 single-word instructions.**
- **Instructions execute in single-cycle except program branches which takes two-cycles.**
- **20 MHz clock input with instruction cycle time of 200 ns.**
- **8K x 14 words of Flash Program Memory, 368 x 8 bytes of RAM Data Memory, 256 x 8 bytes of EEPROM Data Memory.**
- **Three Timers: Timer0 and Timer2 (8-bit timer/counter), Timer1 (16-bit timer/counter).**
- **Two Capture, Compare, PWM modules.**
- **Synchronous Serial Port (SSP).**
- **Universal Synchronous Asynchronous Receiver Transmitter (USART).**
- **10-bit, 8-channel Analog-to-Digital Converter.**
- **Five bidirectional IO ports: Port A, B, C, D and E with bit addressable feature.**

2.4.2 Architecture of PIC16F877A

The architecture of PIC16F877A microcontroller is shown in Fig. 2.39. PIC16F877A has Havard architecture so that it has separate program and data memory address space. PIC architecture includes 8-bit ALU, RAM file register, instruction register, program counter, instruction decoder and control unit, stack, timing circuit and internal peripheral devices timers, analog to digital convertor, synchronous serial port, IO ports and USART.

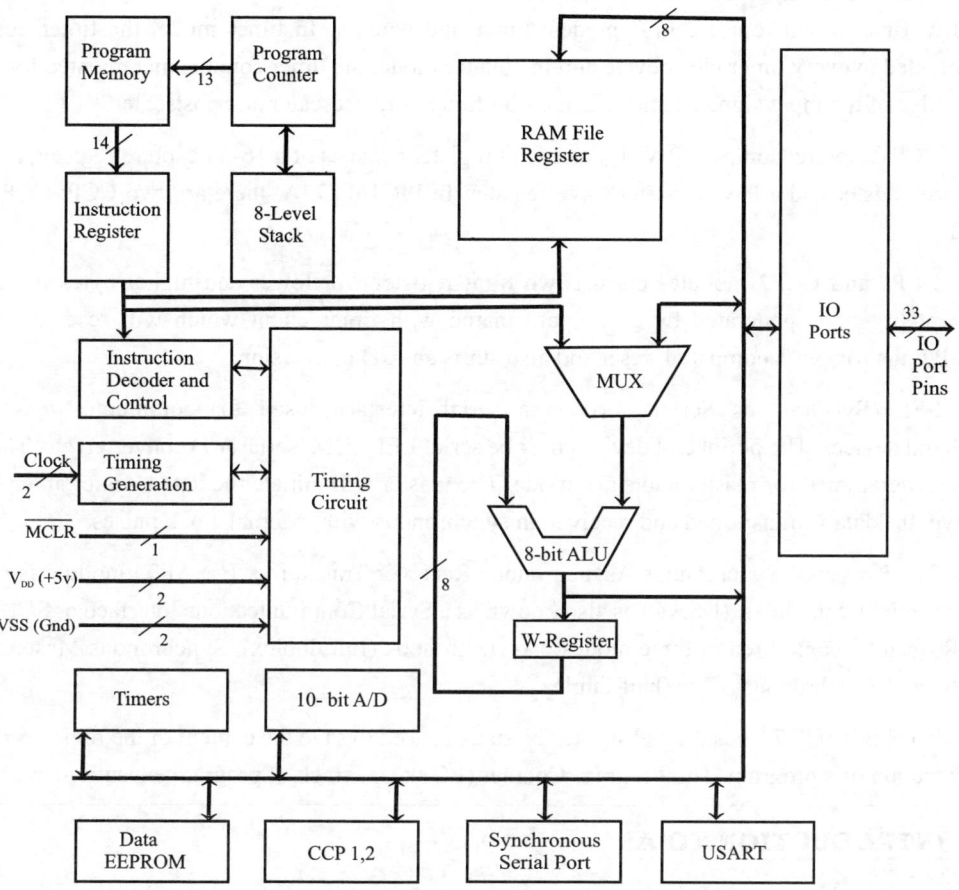

Fig. 2.39: Architecture of PIC16F877A microcontroller.

The ALU performs the arithmetic and logical operations. The operations performed by ALU are addition, subtraction, increment, decrement, logical AND, OR, XOR, compare and complement. The W-register is used to store the data during an ALU operation.

PIC16F877A has five bidirectional IO ports: Port A, B, C, D and E. Many port pins are multi-function pins. When special functions of port pins are not used they can be used as IO pins. The port pins can be used either in group or individually for IO operations.

The PIC16F877A consists of three timers: Timer 0, 1 and 2. Timer 0 consists 8-bit timer/counter along with 8 bit software programmable prescaler. The clock is selected among internal or external clock.

Timer 1 is a 16-bit timer with two 8-bit register which are readable and writeable. Usually, Timer 1 will operate in 2 modes: timer and counter. In timer mode, the timer count is incremented in every timer clock cycle and in counter mode, the timer count is incremented for every rising edge of the input signal. Timer 2 is an 8-bit timer with prescaler and postscaler.

CCP (Capture/Compare/PWM) is a 16-bit register consists of a 16-bit capture register, a 16-bit compare register and a PWM master/slave register. In PIC16F877A, there are two CCP: CCP1 and CCP2.

CCP1 and CCP2 modules consist two 8-bit registers for lower and higher bytes. In CCP1, an event trigger is generated by a compare match with timer count which will reset the timer. In CCP2, the trigger is compared, reset and also starts an A/D conversion.

SSP (**S**ynchronous **S**erial **P**ort) is a serial interface, used for communicating among peripheral devices. The peripheral devices may be serial EEPROM, serial A/D converter, etc. The SSP can be programmed for master and slave mode. The master will initiate the data transfer at any time. In slave, the data is transmitted and received in synchronous with external clock pulses.

The Universal Synchronous Asynchronous Receiver Transmitter (USART) module is one of the two serial IO modules. (USART is also known as a Serial Communications Interface or SCI.). The USART can be configured in three modes: Asynchronous (full-duplex), Synchronous Master (half-duplex) and Synchronous Slave (half-duplex).

The PIC16F877A has 8-level stack. The stack is used to save the content of the registers during the execution of a program. The **P**rogram **C**ounter (PC) keeps track of program execution.

2.5 INTRODUCTION TO ARM

History: In 1983 a British company named Acorn Computers started developing RISC (Reduced Instruction Set Computer) processors and later in 1990 joined with Apple computers and VLSI technology to start a company named ARM (**A**dvanced **R**ISC **M**achine) Ltd and in 1998 the company was renamed as ARM Holdings.

ARM Holdings develops the ARM architecture and licenses it to other companies, who design their own microprocessor/microcontroller that implement one of those architectures as their CPU core. The ARM architectures are IP (**I**ntellectual **P**roperty) architecture cores of ARM Holdings. ARM Holdings provides to all licensees hardware description of the ARM core as well as complete software development toolset like compiler, debugger, software development kit, etc. ARM is the most licensed processor cores in the world used in portable devices due to its high performance with low power consumption.

ARM is a family of RISC architectures that can be employed as CPU core for microcontrollers and microprocessors configured for various applications.

The major design features of ARM core are as follows:

- **Reduced set of instructions that execute in single cycle**
- **Pipelined architecture that decode instructions in one stage**
- **A large set of general-purpose registers**
- **Load/store architecture which restricts data processing instructions to registers only and load/ store instructions to transfer data between registers and memory.**

2.5.1 ARM7 Processor Core

The ARM7 series is the first popular and highly successful core of the ARM Holdings. ARM7 is a general purpose 32-bit microprocessor core with a 32-bit ALU, 32-bit data bus and 32-bit address bus. ARM7 has Von Neumann architecture, with a single 32-bit data bus carrying both instructions and data. Only load, store and swap instructions can access data from memory and so this architecture is also called load/store architecture. ARM7 supports three types of data: 8-bit bytes, 16-bit halfwords, or 32-bit words. Words must be aligned to 4-byte boundaries. Halfwords must be aligned to 2-byte boundaries.

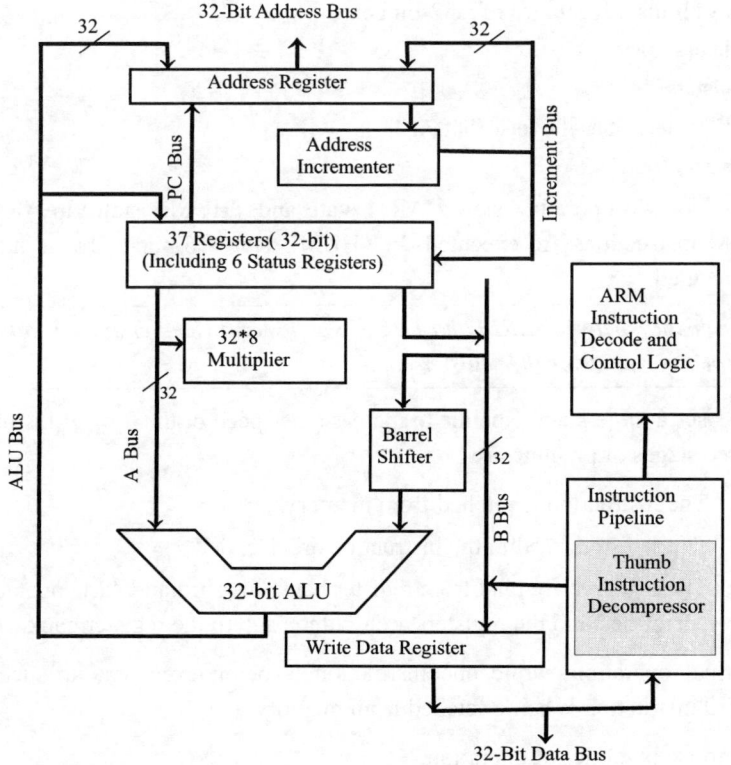

Fig. 2.40: Architecture of ARM7 processor core.

The architecture of ARM7 core is shown in Fig 2.40. ARM7 core has a 32-bit ALU (**A**rithmetic and **L**ogic Unit), 32x8 multiplier and 32-bit barrel shifter. It has five independent internal buses: ALU bus, PC bus, Increment bus, A and B-buses. These independent buses help in high degree of operational parallelism in instruction execution.

ARM7 employs a unique architectural strategy known as THUMB, which supports a variable-length instruction set that provides both 32 and 16-bit instructions for improved code density.

ARM7 processor has two instruction sets:

1. **The standard 32-bit ARM instruction set.**

2. **A 16-bit THUMB instruction set.**

The THUMB instruction set is a subset of the most commonly used 32-bit ARM instructions. THUMB instructions are each 16 bits long and each THUMB instruction has a corresponding 32-bit ARM instruction. THUMB instruction set operates on 16-bit data in 32-bit core. THUMB code use the same 32-bit register set as ARM code and by this, THUMB code is able to provide up to 65% of the code size of 32-bit ARM and 160% of the performance of an equivalent 16-bit memory system.

THUMB has all the advantages of a 32-bit core:

*   **32-bit address space**
*   **32-bit registers**
*   **32-bit shifter and Arithmetic Logic Unit (ALU)**
*   **32-bit memory transfer.**

The ARM7 has two operating states: ARM state and THUMB state. In ARM state 32-bit, word-aligned ARM instructions are executed. In THUMB state 16-bit, halfword-aligned THUMB instructions are executed.

> **Note:** *Transition between ARM and THUMB states and vice-versa does not affect the processor mode or the register contents.*

The ARM7 uses a three-stage pipeline to increase the speed of the flow of instructions to the processor. The three stages of pipeline are

Fetch : The instruction is fetched from memory.

Decode : The registers used in the instruction are decoded.

Execute : The registers are read from register bank, the shift and ALU operations are performed and the registers are written back to the register bank.

During normal operation, while one instruction is being executed, its successor is being decoded, and a third instruction is being fetched from memory.

ARM7 supports six modes of operation.

User mode (usr) : It is the normal program execution mode.

FIQ mode (fiq) : It is designed to support a data transfer or channel process.

IRQ mode (irq)   :   It is used for general purpose interrupt handling.

Supervisor mode (svc) :   It is a protected mode for the operating system.

Abort mode (abt)   :   This mode is entered after a data or instruction prefetch abort.

Undefined mode (und) :   This mode is entered when an undefined instruction is executed.

Most application programs will execute in User mode. Modes other than user mode are collectively known as privileged modes. Privileged modes are used to service interrupts, exceptions, or access protected resources. Mode changes may be made under software control or may be brought about by external interrupts or exception processing.

The ARM7 has a total of 37 registers and can be classified into 31 general purpose 32-bit registers and 6 status registers. These registers are not all accessible at the same time. The processor state and operating mode determine which registers are available to the programmer. At any one time 16 general registers (R0 to R15) and one or two status registers are accessible to the programmer.

Registers R0 to R13 are general-purpose registers used to hold either data or address. Register R14 is called link register that holds the return address during execution of subroutine. Register R15 is the Program Counter that keeps track of the address of the instruction being executed. In ARM state, bits [1:0] of R15 are zero and bits [31:2] contain the address of instruction. In THUMB state, bit [0] of R15 is zero and bits [31:1] contain the address of instruction.

The six status registers of ARM7 are a CPSR (**C**urrent **P**rogram **S**tatus **R**egister) and five SPSR (**S**aved **P**rogram **S**tatus **R**egister) for the use of exception handlers. The CPSR is accessible in user mode and one SPSR is accessible in each privilege mode. SPSR contains the condition code flags, and the mode bits saved as a result of the exception that caused entry to the current mode.

CPSR contains condition code flags and the control bits to enable or disable interrupts and select or determine the operating mode. The arrangement of bits of CPSR / SPSR is shown in Fig. 2.41.

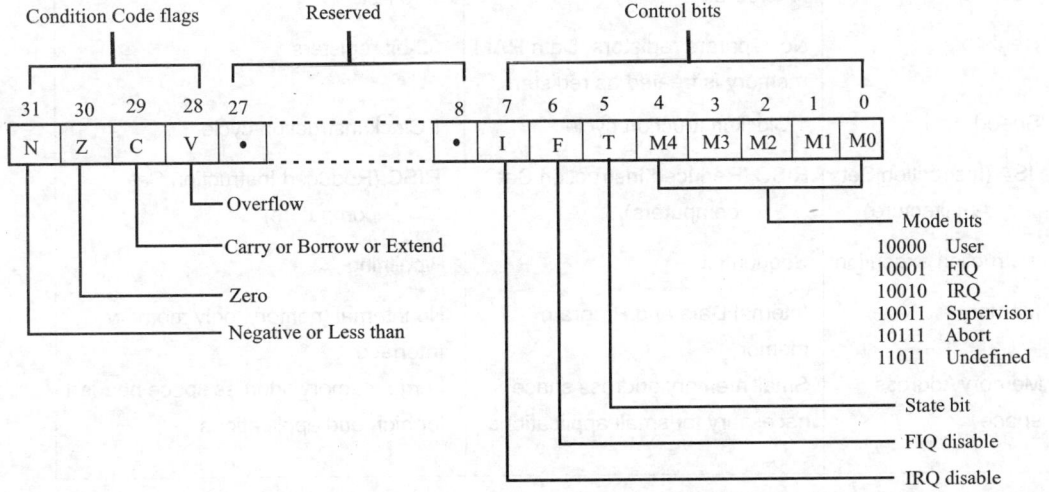

Fig. 2.41: Format of CPSR / SPSR.

The N, Z, C and V bits are the condition code flags modified by arithmetic and logical operations. ARM7 also has instructions to set or clear condition code flags.The lower eight bits of a CPSR are called control bits.

The I and F bits are the interrupt disable bits, when the I bit is set, IRQ interrupts are disabled and when the F bit is set, FIQ interrupts are disabled. The T bit reflects the operating state, when the T bit is set, the processor is executing in Thumb state and when the T bit is clear, the processor executing in ARM state. The M4, M3, M2, M1, and M0 bits (M[4:0]) are the mode bits. These bits determine the processor operating mode.

2.5.2 Comparison of PIC and ARM Controller

PIC microcontrollers include processor core and peripherals required for single chip embedded system. Whereas ARM controllers are processor cores that can be used to design a microprocessor or microcontroller by adding memory and required peripherals. PIC microcontrollers can be used as such to design a system. Whereas a microcontroller itself has to be manufactured by getting the license of ARM core and then system has to be designed using the microcontroller with ARM core. There are a number of commercial microcontrollers are available with ARM core. One such example is LPC2148 commercial microcontroller manufactured with ARM7 core by Philips semiconductors.

Table 2.18: Comparison of PIC and ARM Controller

Feature	PIC	ARM
Device type	Complete microcontroller	Processor core
Architecture	Harvard architecture	Von Neumann (ARM7) and Modified Harvard architecture (Higher versions)
Bus width	8/16/32-bit	32/64-bit
Registers	No separate registers. Data RAM memory is treated as registers.	32-bit registers
Speed	4 Clock/instruction cycle	1 Clock/instruction cycle
ISA (Instruction Set Architecture)	RISC (Reduced Instruction Set computers)	RISC (Reduced Instruction Set computers)
Instruction execution	Sequential	Pipelining
Memory	Internal Data and Program memory.	No internal memory, only memory interface.
Memory Address space	Small memory address space necessary for small applications	Large memory address space needed for high end applications.

2.6 SHORT-ANSWER QUESTIONS

Q2.1   *What is a microcontroller?*

A microcontroller is a programmable semiconductor device available as IC and capable of performing arithmetic and logical operations.

Q2.2   *What are the basic units of a microcontroller?*

The basic units of a microcontroller are the ALU, a set of registers, IO ports, memory, timing and control unit. In addition some of the controllers may have timers, ADC and DAC.

Q2.3   *Mention few differences between a microprocessor and a microcontroller?*

1.   A microprocessor is concerned with the rapid movement of code and data between the external memory and the processor, whereas a microcontroller is concerned with the rapid movement of code and data within the controller.

2.   A microprocessor will have few bit manipulating instructions, whereas a microcontroller will have a large number of bit manipulating instructions.

3.   Microprocessors are generally used for designing general purpose systems, whereas microcontrollers are used for designing dedicated application specific systems.

Q2.4   *List the features of an 8051 microcontroller?*

The features of an 8051 microcontroller are:

- 8-bit controller operating on bit and byte operand
- Provides separate code and data memory address space
- 256 bytes internal RAM and 4 kB internal ROM
- 64/60 kB external program memory address space
- 64 kB external data memory address space
- Four numbers of 8-bit parallel ports
- One number of programmable serial port
- Two numbers of programmable timers
- Five members of vectored interrupts

Q2.5   *List the alternate functions of the ports of an 8051 microcontroller.*

Port pins	Alternate signal	Description
P0.7 - P0.0	AD7 - AD0	Multiplexed low byte address/data.
P2.7 - P2.0	A15 - A8	High byte address
P3.7	$\overline{RD}$	External memory read control signal
P3.6	$\overline{WR}$	External memory write control signal
P3.5	T1	External input to timer 1
P3.4	T0	External input to timer 0
P3.3	$\overline{INT1}$	External interrupt 1
P3.2	$\overline{INT0}$	External interrupt 0
P3.1	TxD	Serial data output
P3.0	RxD	Serial data input

Q2.6   *List the interrupts of an 8051 microcontroller.*

The 8051 microcontroller has five interrupts and they are (in the order of higher to lower priority) External interrupt-0, Timer-0 interrupt, External interrupt-1, Timer-1 interrupt and Serial port interrupt.

Q2.7   What are dedicated address pointers in an 8051?

The 8051 has two dedicated address pointers: Program Counter (PC) and Data Pointer (DPTR). The PC is used as an address pointer for programs and DPTR is used as an address pointer for data.

Q2.8   What are SFRs?

SFRs (Special Function Registers) are internal registers of a microcontroller dedicated for specific functions. These registers can be used only for their specified/defined functions and cannot be used for any other function. In microcontrollers, the SFRs are mapped as internal data memory and can be accessed by direct addressing.

Q2.9   What are register banks in an 8051?

The register banks are internal RAM locations of 8051 which can be used as general purpose registers or scratch pad registers. The first 32 bytes of internal RAM of 8051 are organized as four register banks with each bank consisting of eight locations. At any one time, the processor can work with only one register bank depending on the value of bits RS0 and RS1 in the PSW register.

Q2.10   What is PSW in an 8051?

The flag register of an 8051 is called PSW (Program Status Word). The PSW consists of four math flags and two register bank select bits. The math flags are carry, auxiliary carry, overflow and parity flag. The register bank select bits are RS0 and RS1.

Q2.11   How is stack implemented in an 8051?

The 8051 supports a LIFO (Last-In-First-Out) stack and the stack can reside anywhere in the internal RAM. The 8051 has an 8-bit Stack Pointer (SP) to indicate the top of stack. The stack can be accessed using PUSH and POP instructions. During PUSH, the SP is automatically incremented by one and during POP, the SP is automatically decremented by one.

Q2.12   What are the operating modes of the serial port of an 8051?

The operating modes of the serial port of an 8051 are mode-0, mode-1, mode-2 and mode-3. In mode-0, the serial port functions as a half duplex serial port at fixed baud rate and one data character is framed as 8 bits. In modes 1, 2 and 3, the serial port can function as full duplex serial port. In modes 1 and 3, the baud rate is variable and in mode-1, the baud rate is either 1/32 or 1/64 of oscillator frequency. In mode-1, one data character should be framed as 10 bits, and in modes 2 and 3, one data character is framed as 11 bits.

Q2.13   How is the baud rate decided in modes 1 and 3 of serial transmission in an 8051?

In serial transmission modes 1 and 3 of 8051, the baud rate depends on the SMOD bit of PCON register and the timer-1 overflow rate as shown below:

$$\text{The baud rate in mode 1 or 3} = \frac{2^{\text{SMOD}}}{32} \times (\text{Timer-1 overflow rate})$$
(in modes 1 and 3)

Q2.14   What are the operating modes of the timer of an 8051?

The operating modes of the timers of an 8051 are mode-0, mode-1, mode-2 and mode-3. In mode-0 the timers will function as 13-bit timers and in mode-1 the timers will function 16-bit timers. In mode-2 the timers will function as 8-bit timers with auto reload feature. Timer-0 alone can work in mode-3 and in this mode the TL0 will function as an 8-bit timer controller by standard timer-0 control bits and TH0 will function as 8-bit timer controlled by timer-1 control bits.

Q2.15 List the features of an 8096 microcontroller.

The features of 8096 microcontroller are:

- 16-bit controller operating on bit, byte and word operands
- 16-bit arithmetic operations including multiply and divide
- 256 bytes internal RAM and 8 kB internal RAM/EPROM
- Total address space of 64 kB organised either as one bank of 64 kb or two banks of 32 kB each
- 8-channel, 10-bit ADC
- PWM output
- One number of 16-bit timer, one number of 16-bit counter and four numbers of software timers
- Serial port with own baud rate generation
- High speed input and output lines

Q2.16 What is power down RAM?

In an 8096 microcontroller, the 16 bytes of internal RAM in the address range $F0_H$ to FF_H will receive power through a separate pin called V_{PD}. These memory locations are called power down RAM because supply to these RAM locations can be maintained through a separate small battery backup supply so that the content of these RAM can be preserved during power failure conditions.

Q2.17 What are the functional blocks of RALU in an 8096?

The functional blocks of RALU (Register ALU) of an 8096 microcontroller are 17-bit ALU, high byte of PSW, Program Counter (PC), incrementer, loop counter, three temporary registers, delay unit and stored constants.

Q2.18 What is Watchdog Timer (WDT)?

The Watchdog Timer is a dedicated timer to take care of system malfunction. It can be used to reset the controller during software malfunction, which is referred to as "hanging". A microcontroller will have facility for enabling or disabling the WDT. When the WDT is enabled, it will initiate a hardware reset, whenever it overflows. In order to avoid a reset during normal conditions, the software should take care of clearing the WDT well before it overflows.

Q2.19 Write a short note on the PWM generator of an 8096 controller?

The 8096 controller has an internal PWM generator which can generate fixed frequency and variable duty cycle waveforms. The frequency is fixed at 15.625 kHz for a 12 MHz clock speed. The duty cycle can be varied from 0% to 99.6% by writing an 8-bit value in the PWM register.

Q2.20 What are HSI and HSO lines?

The **H**igh **S**peed **I**nput (HSI) lines are input lines to an 8096 microcontroller that can be used to record external events and the time at which the event has occurred. The **H**igh **S**peed **O**utput (HSO) lines are output from microcontrollers through which the controller can generate time-based signals to initiate various programmable events like turning the ON/OFF a device, start/stop a process, etc.

2.7 EXERCISES

I. Fill in the blanks with appropriate words

1. The _____ architecture uses separate address for code and data.

2. The size of the two external memory banks of 8051 is _____.

3. The interrupt vector address of second highest priority interrupt of 8051 is _____.

4. The _____ and _____ registers of 8051 are used to access the program memory and external data memory respectively.

5. The special function registers(SFRs) of 8051 include _____ number of internal registers.

6. The _____ and _____ bits of PSW registers are used to select the register bank of 8051.

7. The address range of 16 RAM locations of internal 128 byte RAM of 8051 are from _____ to _____.

8. After the reset, the stack pointer of 8051 is initialized to the address _____.

9. The _____ bit of SCON register is used to enable multiprocessor communication using 8051 controllers.

10. The _____ bit of TCON register is used to start/stop the timer/counter in 8051.

11. The _____ bit of Interrupt Enable (IE) register of 8051 is called as global enable.

12. The _____ signal of 8096 is tied low to introduce wait states in machine cycle when slow peripherals are interfaced.

13. The _____ pin of 8096 is used to provide backup power during power failure.

14. The vector address of the 8096 interrupt EXINT is _____.

15. The _____ pins of 8096 are used to record events.

16. The _____ unit of 8096 produces DAC output.

17. The 5-bit counter present in RALU of 8096 to perform repetitive shifts is called as _____.

18. The _____ flag is used along with carry flag to control rounding after a right shift in 8096 microcontroller.

19. The upper 16 bytes of 8096 register file are called _____.

20. When 8096 microcontroller is reset, the program counter(PC) is intialized with the address _____.

21. The _____ timer can be used to reset the 8096 microcontroller based system during software malfunction.

22. The _____ special function register of 8096 microcontroller can be used as a base for indexing and as a constant for calculations and comparisons.

23. The _____ register (special function register) is used to read the status of ADC conversion in 8096 controller.

24. In 8096, the low byte of program status word is also called as _____.

Answers

1. Harvard	7. 20_H, $2F_H$	13. V_{PP}	19. power down RAM
2. 64 KB	8. 07_H	14. $200E_H$	20. 2080_H
3. $000B_H$	9. SM2	15. high speed input	21. Watch dog
4. Program Counter(PC) and Data Pointer(DPTR)	10. TR	16. PWM generator	22. zero register(R0)
5. 21	11. EA	17. loop counter	23. AD_RESULT
6. RS_0, RS_1	12. RDY	18. sticky	24. interrupt mask register.

II. State whether the following statements are True/False.

1. Microcontrollers are general purpose systems.
2. Microcontrollers can have on-chip timers.
3. The 8031 has 128 byte internal RAM and 4 kB internal ROM.
4. The port-1 of 8051 is a dedicated I/O port.
5. The 8051 family of microcontrollers are built with Harvard architecture.
6. The 8051 family microcontrollers come with in system programmers.
7. The internal clock frequency of 8051 is same as crystal frequency.
8. All the interrupts of 8051 are maskable and vectored interrupts.
9. The 8051 special function registers(SFRs) whose address ends with 0_H (or) 8_H are bit addressable.
10. The Data Pointer (DPTR) register of 8051 is automatically incremented.
11. All SFRs of 8051 are bit addressable.
12. After the reset, the 8051 always works with register bank-0.
13. The stack memory of 8031/8051 is always placed in external data memory.
14. The 8051 timer is an up-counter.
15. In 8051 the TF flag of TCON register is reset to zero whenver the timer/counter overflows.
16. The priorities of 8051 interrupts are fixed.
17. INTEL 8096 is a 16-bit microcontroller.
18. Timer and counter are separtely available in 8096 microcontroller.
19. The RDY signal of 8096 should always be tied high.
20. One state time of 8096 is equal to three clock pulses.
21. The timer-1 of 8096 is cleared only by reset.
22. There is no accumulator register in 8096.
23. The watch dog timer of 8096 initiates a hardware reset for every 64k state times.
24. The High Speed Output (HS0) section of 8096 microcontroller can trigger maximum of eight events at a time.
25. The timer-1 and timer-2 registers of 8096 are read only.
26. In 8096, the watch dog timer should not overflow during normal operation.

Answers

1. False	6. False	11. False	16. False	21. True	26. True
2. True	7. True	12. True	17. True	22. True	
3. False	8. True	13. False	18. True	23. True	
4. True	9. True	14. True	19. False	24. False	
5. True	10. False	15. False	20. True	25. True	

III. Choose the right answer for the following questions.

1. *Which of the following is called as ROM less version of 8051?*

   a) 8031 *b)* 8032 *c)* 8751 *d)* 8752

2. Which of the following pin of 8051 is externally ANDed with $\overline{RD}$ pin, to produce read control signal when there is only one external memory?

   a) ALE   *b)* $\overline{EA}$ *c)* $\overline{PSEN}$ *d)* RxD

3. The following 8051 pin is used as programming pulse for programming 8051

   a) $\overline{EA}$ *b)* ALE   *c)* $\overline{WR}$ *d)* TxD

4. Which of the following interrupts of 8051 has been assigned the lowest priority?

   a) Timer-0 interrupt   *b)* Timer-1 interrupt   c) External interrupt-0   *d)* Serial port interrrupt

5. Which of the following condition is required to store a 8051 program of size 15 kB in external code memory without ignoring internal 4 kB ROM?

   a) $\overline{EA}$ = 1   *b)* $\overline{EA}$ = 0   *c)* $\overline{PSEN}$ = 1   *d)* $\overline{PSEN}$ = 0

6. Which of the following data memory address space are alloted for 8051 SFRs?

   a) 00_H to $7F_H$ *b)* 80_H to FF_H *c)* 00_H to FF_H *d)* 20_H to $2F_H$

7. The following register of 8051 is used to select the operating modes of timer/counter.

   a) TCON   *b)* TMOD   *c)* PCON   *d)* SCON

8. Which of the following mode of serial port function with fixed baud rate in 8051 controller?

   a) Mode-0   *b)* Mode-1   *c)* Mode-2   *d)* Mode-3

9. The mode-2 of timer/counter in 8051 is also called

   a) 13-bit timer/counter   *b)* 16-bit timer/counter

   c) autoreload 8-bit timer/counter   *d)* autoreload 16-bit timer/counter

10. An external pulse is applied to $\overline{INT}$ pin of 8051. What should be the content of TMOD register to count the pulse using timer/counter-1 in mode-0?

    a) 00_H *b)* $C0_H$ *c)* 80_H *d)* $0C_H$

11. What should be the content of Interrupt Enabler (IE) register to enable the serial port interrupt in 8051 controller?

    a) 80_H *b)* 81_H *c)* 01_H *d)* 82_H

12. Which of the following is internal interrupt of 8051?

    a) Timer-0 interrupt   *b)* Timer-1 interrupt   *c)* Serial port interrupt   *d)* all the three

13. The following register of 8051 is a bit-addressable register.

    a) TCON   *b)* PCON   *c)* SP   *d)* TMOD

14. Which of the following flag is not supported by 8051?

    a) carry   *b)* auxilary carry   *c)* sign   *d)* parity

15. The program counter in 8051 is a _____ bit register.

    a) 8-bit   *b)* 16-bit   *c)* 32-bit   *d)* 64-bit

16. Which of the following pins are common to 8051 and 8096 microcontroller?

    a) HSO.0   *b)* HSI.0   *c)* $\overline{EA}$ *d)* INST

17. **Which of the following clock pulse of 8096 can be used to clock the peripherals?**

    a) phase-A   *b)* Phase-B   *c)* Phase-C   *d)* all the three

18. **The crystal frequency of 8096 is 10 MHz. How long the $\overline{RESET}$ pin should be held low for proper reset?**

    a) 0.1 µs   *b)* 1 µs   *c)* 0.3 µs   *d)* 3 µs

19. **In 8096, the two events analog to digital conversion and timer overflow occurs simultaneously .**
    The microcontroller services the _____ interrupt first.

    a) external interrupt *b)* software timers *c)* timer overflow   *d)* A/D conversion complete

20. **Which of the following signals are used to record and trigger time-based events n 8096 controller?**

    a) HSI.1   *b)* HSI.2   *c)* HSO.1   *d)* HSO.2

21. **The following unit is not part of I/O section of 8096 controller**

    a) ADC   *b)* PWM   *c)* serial port   *d)* RALU

22. **The size of the 8096 internal RAM is _____ bytes.**

    a) 128   *b)* 256   *c)* 512   *d)* 1024

23. **The special function registers of 8096 are mapped in the address range**

    a) 00_H to $2F_H$ *b)* 00_H to 19_H *c)* 00_H to 29_H *d)* $F0_H$ to FF_H

24. **What value should be loaded in the 8096 PWM register to generate a PWM output waveform with 50% duty cycle?**

    a) 00_H *b)* 19_H *c)* $7F_H$ *d)* FF_H

25. **Which of the following events cannot be detected by High Speed Input (HSI) unit of 8096?**

    a) Eight positive transition   *b)* Eight negative transition

    c) Every positive transition   *d)* Every negative transition

26. **The serial port control word to be written in the 8096 SP_CON register to receive an 8-bit serial data is_____.**

    a) 08_H *b)* 09_H *c)* $0C_H$ *d)* $0D_H$

Answers						
1. a	5. a	9. c	13. a	17. a	21. d	25. b
2. c	6. b	10. b	14. c	18. c	22. b	26. a
3. b	7. b	11. b	15. b	19. d	23. b	
4. d	8. a	12. d	16. c	20. b	24. c	

IV. Answer the following questions.

E2.1   Compare 8051 and 8052 microcontrollers

E2.2   How is the external program memory accessed in 8051? Explain with diagram.

E2.3   How are the interrupts of 8051 classified?

E2.4   List the vector address and priority of 8051 interrupts.

E2.5   Why is 8031 called as ROM less version of 8051?

E2.6   Find the contents of the condition flags in the 8051 after adding the binary numbers $A2_H$ and $C8_H$.

E2.7   Write the binary word to be written into the PSW of the 8051, to select the register bank-3 in the internal RAM.

E2.8   Specify the size of various memory supported by 8051.

E2.9   How many ports are bit-addressable in 8051?

E2.10   What happens in power down mode of 8051?

E2.11   What happens in idle down mode of 8051?

E2.12   Compare power down and idle mode of 8051.

E2.13   What is the significance of $\overline{EA}$ line of 8051?

E2.14   Draw the format of PSW register of 8051. Explain each flag

E2.15   What is the significance of SMOD bit of PCON register in 8051?

E2.16   Draw the format of SCON register of 8051. Explain each bit.

E2.17   How is multiprocessor communication implemented using 8051?

E2.18   What is meant by autoreload counter? How is it implemented in 8051?

E2.19   How is the timer/counter used to count the internal and external pulses in 8051?

E2.20   When does the timer/counter start and stop counting in 8051?

E2.21   How are the interrupts of 8051 enabled?

E2.22   How are the priority of interrupts of 8051 changed?

E2.23   What is state time in 8096 controller? Draw the waveform

E2.24   The crystal/external frequency of 8096 microcontroller is 10 MHz. Draw the waveform of three internal timing waveforms

E2.25   What is meant by Register ALU(RALU)?

E2.26   Draw and discuss the program status word of 8096.

E2.27   Write short notes on register file of 8096.

E2.28   How is the memory of 8096 mapped?

E2.29   Write the uses of memory controller of 8096.

E2.30   Differenciate timer-1,timer-2 and watch dog timer of 8096 controller.

E2.31   What is zero register(R0) of 8096 controller?

E2.32   What are the 8096 special function registers used in analog to digital conversion? Draw the format and explain.

E2.33   The crystal frequency of 8096 is 8 MHz. Calculate the baud rate required for transmitting an 8-bit serial data. The baud rate count, $B = 999_{10}$.

E2.34   Write short notes on interrupt mask register of 8096.

E2.35   Discuss about IO control registers of 8096 controller.

E2.36   What are the significance of IO status registers of 8096 controller?

CHAPTER 3

INSTRUCTION SET OF 8085

3.1 TIMING DIAGRAM OF 8085

The timing diagram provides information about the various condition (**high** state or **low** state or **high impedance** state) of the signals while a machine cycle of an instruction is executed. The timing diagrams are supplied by the manufacturer of the microprocessor. The timing diagrams are essential for a system designer. Only from the knowledge of timing diagrams, the matched peripheral devices like memories, ports, etc., can be selected to form a system with microprocessor as CPU.

3.1.1 Processor Cycles

The sequence of operations that a processor has to carry out while executing the instruction is called instruction cycle. Each instruction cycle of a processor in turn consists of a number of machine cycles. The machine cycles are the basic operations performed by the processor. To execute an instruction, the processor executes one or more machine cycles in a particular sequence. The machine cycles of a processor are also called processor cycles. The manufacturers of microprocessors define the timings and status of various signals during the processor cycles.

In general, the instruction cycle of an instruction can be divided into two sub cycles: Fetch cycle and Execute cycle. The fetch cycle is executed to fetch the opcode from memory and the execute cycle is executed to decode the instruction and to perform the work specified by the instruction.

3.1.2 Machine Cycles of 8085

The 8085 microprocessor has seven basic machine cycles. They are as follows:

1. **Opcode fetch cycle (4T or 6T)**
2. **Memory read cycle (3T)**
3. **Memory write cycle (3T)**
4. **IO read cycle (3T)**
5. **IO write cycle (3T)**
6. **Interrupt acknowledge cycle (6T or 12T)**
7. **Bus idle cycle (2T or 3T)**

Each instruction of the 8085 processor consists of one to five machine cycles, i.e., when the 8085 processor executes an instruction, it will execute some of the machine cycles in a specific order. The processor takes a definite time to execute the machine cycles. The time taken by the processor to execute a machine cycle is expressed in T-states. One T-state is equal to the time period of the internal clock signal of the processor. The T-state starts at the falling edge of a clock.

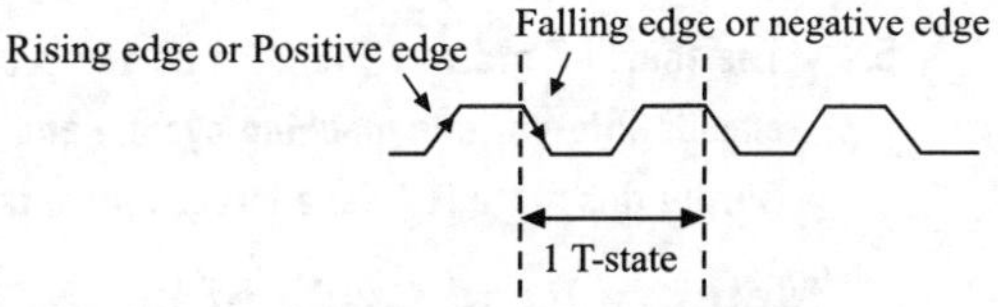

Rising edge or Positive edge   Falling edge or negative edge

1 T-state

Fig. 3.1: Clock signal.

> **Note:** *Time period, $T = 1/f$; where f = Internal clock frequency.*

The T-states required by the 8085 processor to execute each machine cycle are mentioned within brackets in the list of machine cycles given above.

3.1.3 Opcode Fetch Cycle of 8085

Each instruction of the processor has one-byte opcode. The opcodes are stored in memory. The opcode fetch machine cycle is executed by the processor to fetch the opcode from memory. Hence, every instruction starts with opcode fetch machine cycle.

The time taken by the processor to execute the opcode fetch cycle is either 4T or 6T. In this time, the first 3T states are used for fetching the opcode from memory and the remaining T-states are used for internal operations by the processor. The timings of various signals during opcode fetch cycle are shown in Fig. 3.2.

1. At the falling edge of first T-state (T_1), the microprocessor outputs the low byte address on AD_0-AD_7 lines and high byte address on A_8 to A_{15} lines. ALE is asserted high to enable the external address latch. The other control signals are asserted as follows:

   $IO/\overline{M}=0$, $S_0 = 1$, $S_1 = 1$. ($IO/\overline{M}$ is asserted low to indicate memory access.)

2. At the middle of T_1, the ALE is asserted low and this enables the external address latch to take low byte of the address and keep on its output lines.

3. In the second T-state (T_2), the memory is requested for read by asserting read line low. When read is asserted low, the memory is enabled for placing the opcode on the data bus. The time allowed for memory to output the opcode is the time during which read remains low.

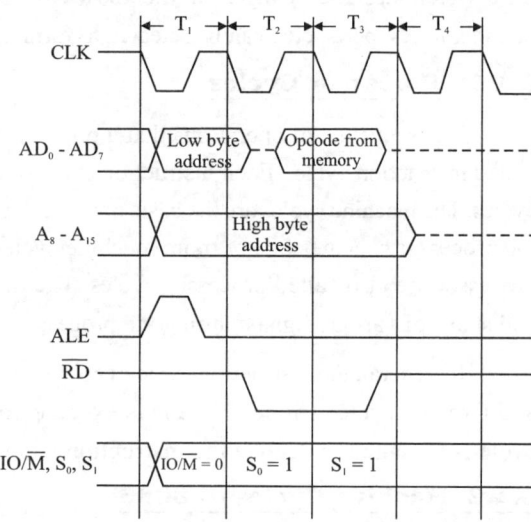

(WR will be **high**; READY is tied **high** either permanently or temporarily in the system.)

Fig. 3.2: Opcode fetch cycle of 8085.

4. In the third T-state (T_3), the read signal is asserted high. On the rising edge of read signal, the opcode is latched into microprocessor. Other control signals remain in the same state until the next machine cycle.

5. The fourth T-state (T_4) is used by the processor for internal operations to decode the instruction and encode into various machine cycles, and also for completing the task specified by 1-byte instruction. During this state (T_4) the address and data bus will be in high impedance state.

3.1.4 Memory Read Cycle of 8085

The memory read machine cycle is executed by the processor to read a data byte from memory. The processor takes 3T states to execute this cycle. The timings of various signals during memory read cycle are shown in Fig. 3.3.

1. At the falling edge of T_1, the microprocessor outputs the low byte address on AD_0 - AD_7 lines and high byte address on A_8 to A_{15} lines. ALE is asserted high to enable the external address latch. The other control signals are asserted as follows:

    $IO/\overline{M}=0$, $S_0 = 0$, $S_1 = 1$. ($IO/\overline{M}$ is asserted low to indicate memory access.)

2. At the middle of T_1, the ALE is asserted low and this enables the external address latch to take low byte of address and keep on its output lines.

3. In the second T-state (T_2), the memory is requested for read by asserting read line low. When read is asserted low, the memory is enabled for placing the data on the data bus. The time allowed for memory to output the data is the time during which read remains low.

4. At the end of T_3, the read signal is asserted high. On the rising edge of read signal, the data is latched into microprocessor. Other control signals remain in the same state until the next machine cycle.

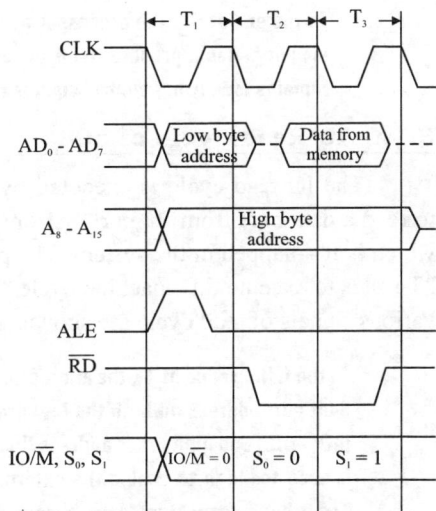

($\overline{WR}$ will be **high** ; READY is tied **high** either permanently or temporarily in the system.)

Fig. 3.3: Memory read cycle of 8085.

3.1.5 Memory Write Cycle of 8085

The memory write machine cycle is executed by the processor to write a data byte in a memory location. The processor takes 3T states to execute this machine cycle. The timings of various signals during memory write cycle are shown in Fig. 3.4.

1. At the falling edge of T_1, the microprocessor outputs the low byte address on AD_0 - AD_7 lines and high byte address on A_8 to A_{15} lines. ALE is asserted high to enable the external address latch. The other control signals are asserted as follows:

    $IO/\overline{M}=0$, $S_0 = 1$, $S_1 = 0$. ($IO/\overline{M}$ is asserted low to indicate memory access.)

2. At the middle of T_1, the ALE is asserted low and this enables the external address latch for latching the low byte address into its output lines.

3. In the falling edge of T_2, the processor output data on AD_0 to AD_7 lines and then request memory for write operation by asserting the write control signal $\overline{WR}$ to low.

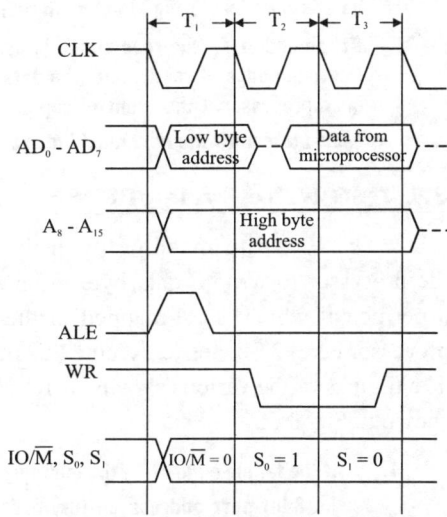

($\overline{RD}$ will be **high** ; READY is tied **high** either permanently or temporarily in the system.)

Fig. 3.4: Memory write cycle of 8085.

4. At the end of T_3, the processor asserts $\overline{WR}$ high. This enables the memory to latch the data into it. The memory should prepare itself to accept the data within the time duration in which write control signal remains low. Other control signals remain in the same state until the next machine cycle.

3.1.6 IO Read Cycle of 8085

The IO read cycle is executed by the processor to read a data byte from IO port or from the peripheral which is IO-mapped in the system. The processor takes 3T states to execute this machine cycle. The timings of various signals of read cycle are shown in Fig. 3.5.

1. At the falling edge of T_1, the microprocessor output the 8-bit port address on both the low order address lines (AD_0 - AD_7) and high order address lines (A_8 to A_{15}). ALE is asserted high to enable the external address latch. The other control signals are asserted as follows: $IO/\overline{M}$ = 1, S_0 = 0 and S_1 = 1. ($IO/\overline{M}$ is asserted high to indicate IO access.)

2. At the middle of T_1, the ALE is asserted low and this enables the external address latch to take the port address and keep on its output lines.

3. In the second T-state (T_2) the IO device is requested for read by asserting read line low. When read is asserted low, the IO port is enabled for placing the data on the data bus. The time allowed for IO port to output the data is the time during which read remains low.

4. At the end of T_3, the read signal is asserted high. On the rising edge of read signal the data is latched into microprocessor. Other control signals remains in the same state until the next machine cycle.

3.1.7 IO Write Cycle of 8085

The IO write machine cycle is executed by the processor to write a data byte in an IO port or to a peripheral which is IO-mapped in the system. The processor takes 3T states to execute this machine cycle. The timings of the various signals of IO write cycle are shown in Fig. 3.6.

1. At the falling edge of T_1, the microprocessor outputs the 8-bit port address on low order address line (AD_0 - AD_7) and high order address lines (A_8 to A_{15}).

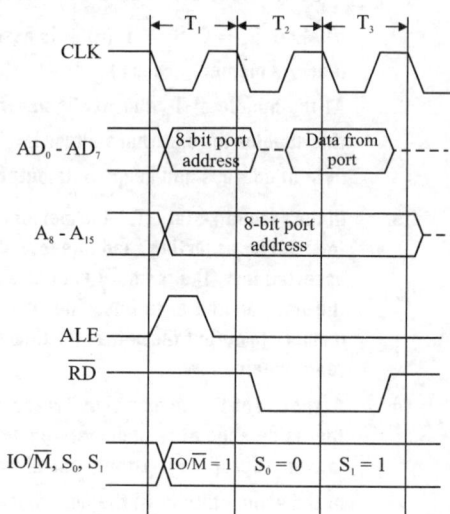

(WR will be **high** ; READY is tied **high** either permanently or temporarily in the system.)

Fig. 3.5: I/O read cycle of 8085.

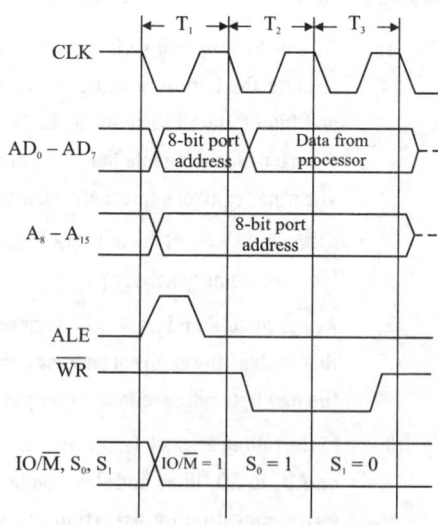

(RD will be **high** ; READY is tied **high** either permanently or temporarily in the system.)

Fig. 3.6: I/O write cycle of 8085.

ALE is asserted high to enable the external address latch. The other control signals are asserted as follows: $IO/\overline{M}=1$, $S_0 = 1$ and $S_1 = 0$. ($IO/\overline{M}$ is asserted high to indicate IO access.)

2. At the middle of T_1, the ALE is asserted low and this enables the external address latch for latching the port address into its output lines.

3. In the falling edge of T_2, the processor output data on AD_0 - AD_7 lines and then request IO port for write operation by asserting the write control signal $\overline{WR}$ to low.

4. At the end of T_3, the processor asserts $\overline{WR}$ high. This enables the IO port to latch the data into it. The IO port should prepare itself to accept the data within the time duration in which write control signal remains low. Other control signals remains in the same state until the next machine cycle.

3.1.8 Interrupt Acknowledge Cycle of 8085

The interrupt acknowledge machine cycle is executed by the processor to service an interrupt when an interrupt request is made through INTR pin of the processor.

The 8085 processor checks for an interrupt at the second T-state of the last machine cycle of every instruction. If there is a valid interrupt request and if INTR is enabled then the processor completes the current instruction execution and then executes an interrupt acknowledge machine cycle. The interrupt acknowledge machine cycle is executed to get either a **RST n** instruction from the interrupting device or to get a CALL instruction with CALL address from the interrupting device. It also stores the content of program counter (return address) in stack.

3.1.9 Interrupt Acknowledge Cycle of 8085 with RST n Instruction

The timings of various signals during interrupt acknowledge cycle of 8085 when **RST n** instruction is supplied by the interrupting device are shown in Fig. 3.7.

1. In the first T-state of interrupt acknowledge cycle, the address is placed on the AD_0 - AD_7 and A_8-A_{15} lines and ALE is asserted high. But the address is not used to read from memory. The other control signals are asserted as follows:
$IO/\overline{M}=1$, $S_0 = 1$ and $S_1 = 1$.

In the middle of T_1, ALE is asserted low. The INTR signal can remain high or it can go low once the interrupt is accepted.

2. In the second T-state (T_2), $\overline{INTA}$ is asserted low, and this enables the interrupting device to place the opcode of RST n instruction on the data bus.

3. At the end of T_3, the $\overline{INTA}$ is asserted high and the RST n opcode is latched into the processor. The time allowed for the external hardware to place the RST n opcode is the time during which $\overline{INTA}$ remains low.

4. The next three T states T_4, T_5 and T_6 are used for internal operations. The internal operations performed are decoding the instruction and encoding into various machine cycles and generation of vector address for the RST n interrupt.

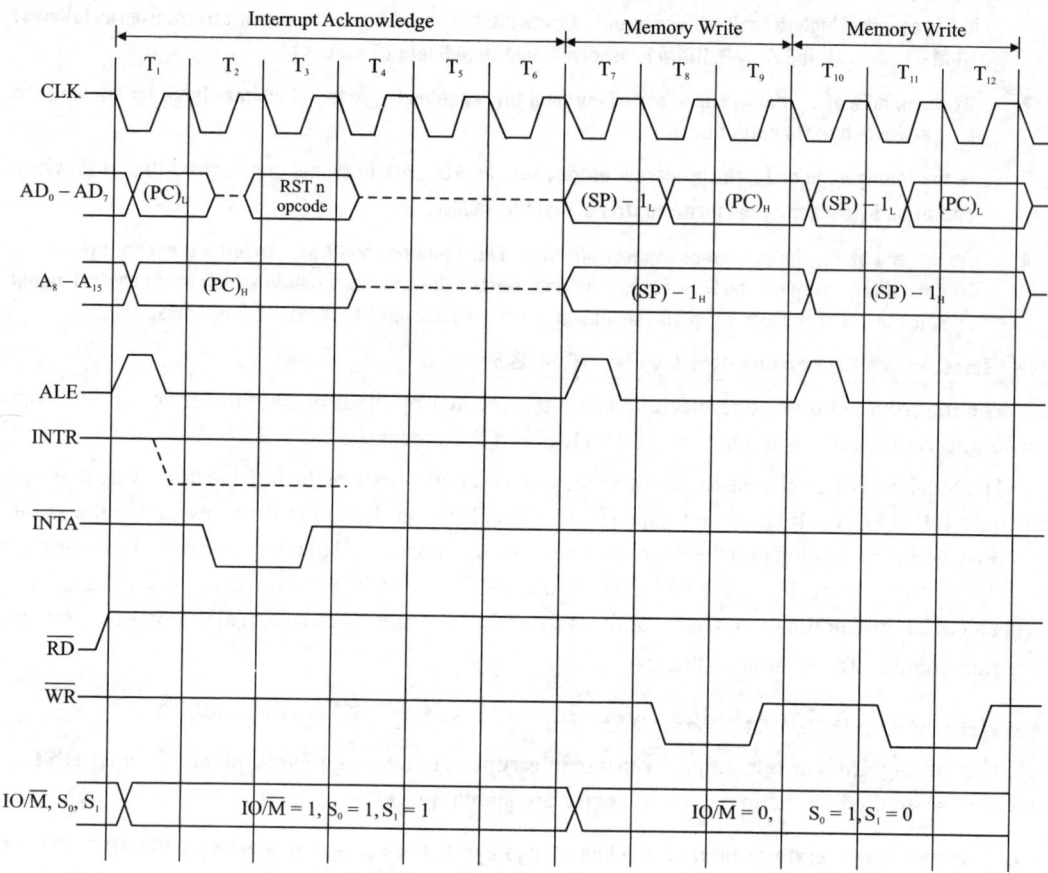

Fig. 3.7: Interrupt acknowledge cycle with **RST n** opcode.

5. The T states T_7, T_8 and T_9 are used to store the high byte of the Program Counter (PC) in stack (using the content of Stack Pointer (SP) as address).

    In T_7, the content of SP is decremented by one and placed on AD_0-AD_7 and A_8-A_{15} lines. ALE is asserted high and then low, to latch the low byte of address into external latch. The status signals are asserted as $IO/\overline{M}=0$, $S_0 = 1$ and $S_1 = 0$.

    In T_8, the high byte of PC is placed on AD_0 - AD_7 lines and $\overline{WR}$ is asserted low to enable the stack memory for write operation. At the end of T_9, $\overline{WR}$ is asserted high.

6. The T states T_{10}, T_{11} and T_{12} are used to store the low byte of the program counter into stack.

    In T_{10}, the content of SP is again decremented by one and placed on AD_0-AD_7 and A_8 - A_{15} lines. ALE is asserted high and then low, to latch the low byte of address into external latch. The status signals are asserted as $IO/\overline{M}=0$, $S_0 = 1$ and $S_1 = 0$.

    In T_{11}, the low byte of PC is placed on AD_0 - AD_7 lines and $\overline{WR}$ is asserted low to enable the stack memory for write operation. At the end of T_{12} $\overline{WR}$ is asserted high.

After the interrupt acknowledge machine cycle, the PC will have the vector address of **RST n** instruction and so the processor starts servicing the interrupt by executing the interrupt service subroutine stored at this address.

3.1.10 Interrupt Acknowledge Cycle of 8085 with CALL Instruction

This cycle is executed by the machine to service an interrupt, when an interrupt request is made through 8259 (Interrupt Controller) to the INTR pin of 8085. The INTEL 8259 can accept 8 interrupt request and allow one by one to the INTR pin of the 8085 processor. It also supplies CALL opcode and CALL address, when it receives INTA signal from the processor.

The processor checks for an interrupt at the second T-state of the last machine cycle of every instruction. If there is a valid interrupt request and if INTR is enabled then the processor completes the current instruction execution and then executes an interrupt acknowledge machine cycle.

The timings of various signals during interrupt acknowledge cycle when CALL instruction is supplied by the interrupting device are shown in Fig. 3.8.

1. At the falling edge of T_1 the address is placed on AD_0 - AD_7 and A_8 - A_{15} lines and ALE is asserted high. But the address is not used to read from memory. The other control signals are asserted as $IO/\overline{M} = 1$, $S_0 = 1$ and $S_1 = 1$.

   In the middle of T_1, ALE is asserted low. The INTR signal can remain high or it can go low once the interrupt is accepted by executing acknowledge cycle.

2. In T_2, $\overline{INTA}$ is asserted low and this enables the interrupt controller 8259 to place a CALL opcode on the data bus.

3. At the end of T_3, the $\overline{INTA}$ is asserted high and the CALL opcode is latched into the processor.

4. The T states T_4, T_5 and T_6 are used for internal operations. The internal operations performed are decoding the opcode and encoding into various machine cycles.

5. The T states T_7, T_8 and T_9 are used to fetch the low byte of call address from 8259. In T_7, the content of Program Counter (PC) is placed on address bus but not used for memory operation. In T_8 the $\overline{INTA}$ is asserted low and this enables the interrupt controller 8259 to place the low byte of call address on data bus. At the end of T_9 the $\overline{INTA}$ is asserted high and the low byte call address on the data bus is latched into the processor.

6. The T states T_{10}, T_{11} and T_{12} are used to fetch the high byte of call address from 8259. In T_{10} the content of PC is placed on address bus, but not used for memory operation. In T_{11} the INTA is asserted low and 8259 is enabled for placing the high byte of call address on data bus. At the end of T_{12}, the $\overline{INTA}$ is asserted high and the high byte call address on the data bus is latched into the processor.

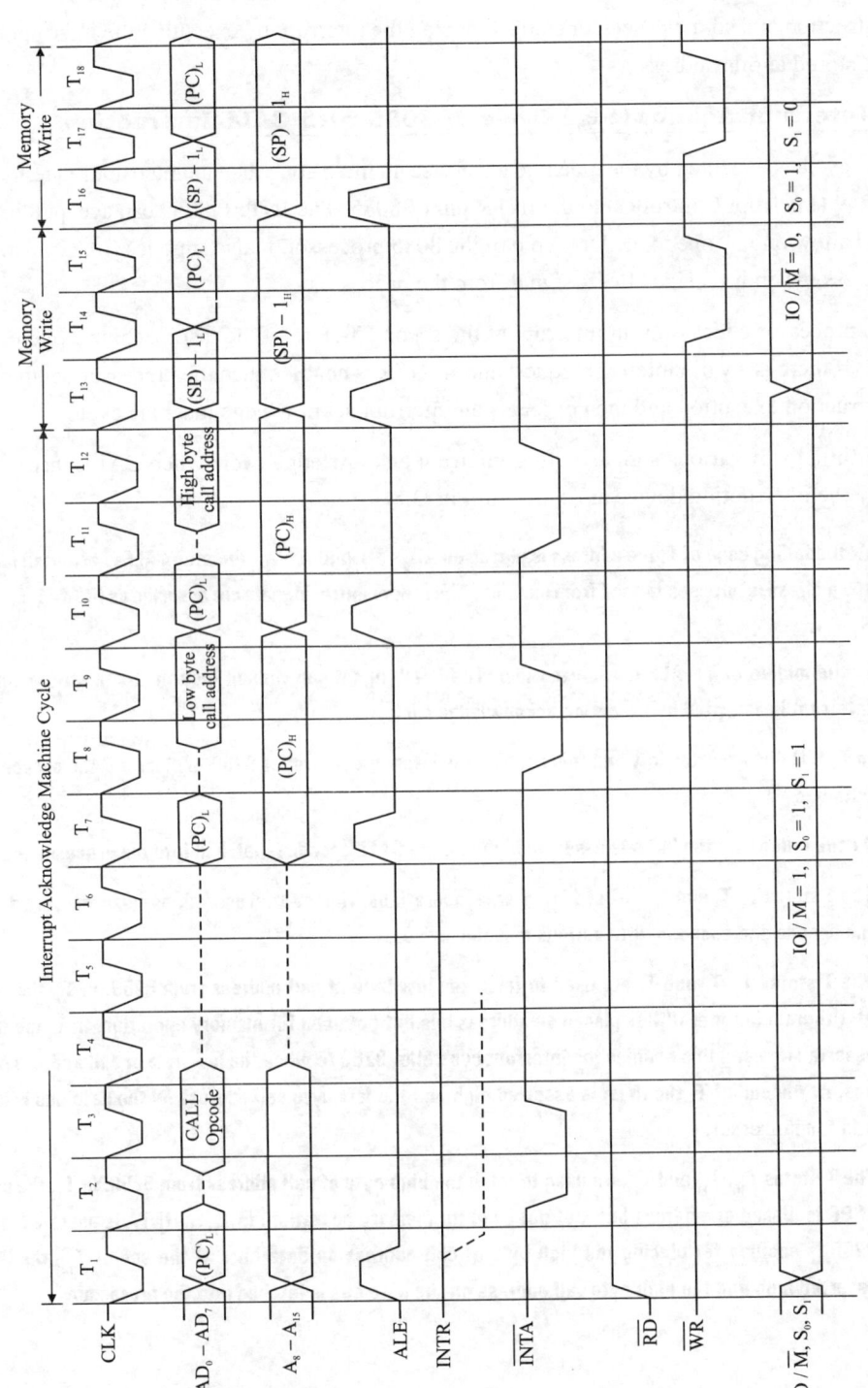

Fig. 3.8: Interrupt acknowledge cycle with CALL opcode.

7. The T states T_{13}, T_{14} and T_{15} are used to store the high byte of the program counter in stack memory. In T_{13}, the content of Stack Pointer (SP) is decremented by one and placed on address bus. ALE is asserted high and then low, to latch the low byte of address into external latch. The other control signals are asserted as $IO/\overline{M}=0$, $S_0 = 1$ and $S_1 = 0$. In T_{14}, the high byte of PC is placed on AD_0 - AD_7 lines and $\overline{WR}$ is asserted low to enable the stack memory for write operation. At the end of T_{15}, $\overline{WR}$ is asserted high.

8. The T states T_{16}, T_{17} and T_{18} are used to store the low byte of the program counter in stack memory.

In T_{16}, the content of SP is again decremented by one and placed on address bus. ALE is asserted high and then low, to latch the low byte of address into external latch. The other control signals are asserted as $IO/\overline{M}=0$, $S_0 = 1$ and $S_1 = 0$.

In T_{17}, the low byte of PC is placed on AD_0 - AD_7 lines and $\overline{WR}$ is asserted low to enable the stack memory for write operation. At the end of T_{15} $\overline{WR}$ is asserted high.

After the interrupt acknowledge machine cycle, the PC will have the call address and so the processor starts servicing the interrupt by executing the interrupt service subroutine stored at this address.

3.1.11 Bus Idle Cycle

The bus idle machine cycle is executed, when extra time or more time is needed for an internal operation of the processor. During this cycle, the status signals S_0 and S_1 are asserted **low**. The data, address and control pins are driven to **high impedance** state. The READY signal will not be sampled by the processor during this cycle.

3.1.12 Machine Cycle with Wait States

Wait states can be introduced in any machine cycle except bus idle cycle between T_2 and T_3. The wait states are introduced in the machine cycle if READY pin is tied **low** at the second T-state of a machine cycle. The processor samples (or check) the READY signal at the second T-state of every machine cycle. If READY is tied **low** at this time, then the processor keeps on introducing wait state until the READY is again tied **high**. This facility is used by the slow memories, IO devices and peripherals to get extra time for read or write operations.

In the system when the peripheral timings are matched with processor timings, then the READY pin is permanently tied **high**. If the system peripherals require more time for read or write cycles, then using additional hardware the READY pin should be tied **low** for the required number of T-states.

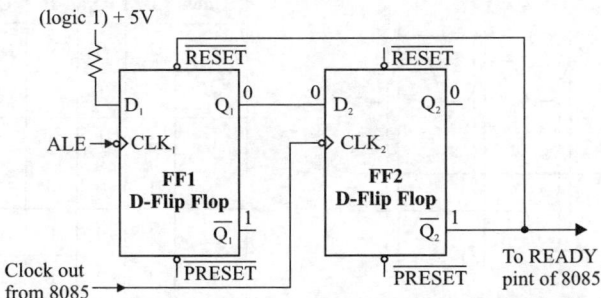

(The values shown at the input and output of the flip-flops are initial conditions)

Fig. 3.9: Circuit to introduce one wait state in 8085 machine cycle.

The circuit shown in Fig. 3.9 can be used to introduce one wait state in the machine cycles. The working of the circuit shown in Fig. 3.9 can be explained as follows :

1. Initially Q_2 = 0 and $\overline{Q}_2$ = 1. The input D_1 is permanently tied high. The flip-flops are negative edge sensitive and so they are clocked (recognizes the clock) at the falling edges.

2. In the beginning of every machine cycle (except bus idle), ALE is asserted high and then low. At the falling edge of ALE, FF1 is clocked and its output Q_1 changes to 1. Also the input to FF2, D_2 changes to 1.

3. Now D_1 = 1, Q_1 = 1, D_2 = 1, Q_2 = 0, $\overline{Q}_2$ = 1 and $\overline{RESET}$ = 1.

4. At the falling edge (beginning) of T_2, FF2 is clocked and so its output Q_2 changes to 1 and changes to 0.

5. Now, D_1 = 1, Q_1 = 1, D_2 = 1, Q_2 = 1, $\overline{Q}_2$ = 0 and $\overline{RESET}$ = 0.

6. Since $\overline{Q}_2$ is connected to READY pin of 8085, the READY will be tied low. The $\overline{Q}_2$ is also used to reset FF1 and so when $\overline{Q}_2$ goes to 0 the FF1 is resetted or cleared. Now Q_1 = 0 and since Q_1 = D_2, the D_2 is also equal to 0.

7. Now, D_1 = 1, Q_1 = 0, D_2 = 0, Q_2 = 1, $\overline{Q}_2$ = 0 and $\overline{RESET}$ = 0.

8. At the falling edge of next T-state (i.e., in wait state) again FF2 is clocked and so the output of FF2 will change.

9. Now, D_1 = 1, Q_1 = 0, D_2 = 0, Q_2 = 0, $\overline{Q}_2$ = 1 and $\overline{RESET}$ = 1.

10. Since $\overline{Q}_2$ = 1, again READY is tied high. When the processor checks the READY at the falling edge of next cycle (T_3), it will be high and it will continue the machine cycle.

Thus, the hardware shown in Fig. 3.9 introduces one wait state in the machine cycles. A machine cycle with one wait state is shown in Fig. 3.10.

Truth Table of D-flip-flop

Clock	Input	Output
	D	$\overline{Q}$
↓	1	1 0
↓	0	0 1

Preset and reset/clear facility in D-flip-flop

PRESET	$\overline{RESET}$	Q	$\overline{Q}$
0	1	1	0
1	0	0	1
1	1	Clock and D input decide the output	
0	0	Should not occur	

Time	D_1	Q_1	D_2	Q_2	$\overline{Q}_2$	$\overline{RESET}$
t_1	1	0	0	0	1	1
t_2	1	1	1	0	1	1
t_3	1	1	1	1	0	0
	1	0	0	1	0	0
t_4	1	0	0	0	1	1
t_5	1	0	0	0	1	1
t_6	1	0	0	0	1	1

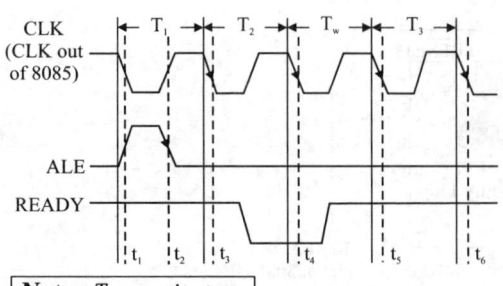

Note : T_w = wait state.

Fig. 3.10: Machine cycle with one wait state.

3.2 INSTRUCTION FORMAT OF 8085

The 8085 has 74 basic instructions and 246 total instructions. The instruction set of 8085 is defined by the manufacturer INTEL Corporation. Each instruction of 8085 has one-byte opcode. With 8-bit binary code, we can generate 256 different binary codes. In this, 246 codes have been used for opcodes of 8085 instructions. The instructions of 8085 in hexadecimal order are listed in Appendix IV and in alphabetical order are listed in Appendix V.

The size of 8085 instruction can be one-byte, two-byte or three-byte. The one-byte instruction has an opcode alone and the two-byte instruction has an opcode followed by an eight bit address or data. The 3-byte instruction has an opcode followed by 16-bit address or data. While storing the 3-byte instruction in memory, the sequence of storage is, opcode first followed by low byte of address or data and then high byte of address or data. The data or address specified in the instruction is also known as operand. The format of 8085 instructions are shown in Fig. 3.11.

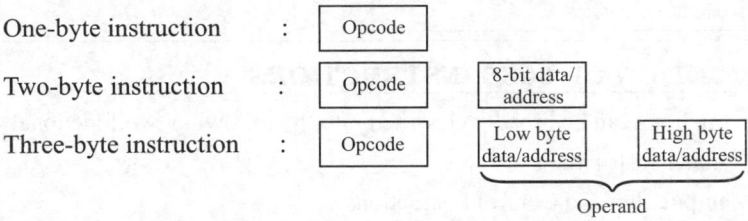

Fig. 3.11: Format of 8085 instructions.

3.3 ADDRESSING MODES

Every instruction of a program has to operate on a data. The method of specifying the data to be operated by the instruction is called Addressing. The 8085 supports the following five addressing modes:

1. **Immediate Addressing**
2. **Direct Addressing**
3. **Register Addressing**
4. **Register Indirect Addressing**
5. **Implied Addressing**

Immediate Addressing

In immediate addressing mode, the data is specified in the instruction itself. The data will be a part of the program instruction.

> Example :   MVI B, $3E_H$
>
> Move the data $3E_H$ given in the instruction to B-register.

Direct Addressing

In direct addressing mode, the address of the data is specified in the instruction. The data will be in memory. In this addressing mode, the program instructions and data can be stored in different memory blocks.

> Example :   LDA 1050_H
>
> Load the data available in memory location 1050_H in accumulator.

Register Addressing

In register addressing mode, the instruction specifies the name of the register in which the data is available.

Example :   MOV A, B
Move the content of B-register to A-register.

Register Indirect Addressing

In register indirect addressing mode, the instruction specifies the name of the register in which the address of the data is available. Here the data will be in memory and the address will be in a register pair.

Example :   MOV A, M
The memory data addressed by HL pair is moved to A-register.

Implied Addressing

In implied addressing mode, the instruction itself specifies the data to be operated.

Example :   CMA
Complement the content of accumulator.

3.4   CLASSIFICATION OF 8085 INSTRUCTIONS

The 8085 instructions can be broadly classified into the following two functional groups.

1. Data Transfer Instructions

2. Data Manipulation and Control Instructions.

The data manipulation and control instructions can be classified as follows:

(a) Arithmetic Instructions.

(b) Logical Instructions.

(c) Branching Instructions.

(d) Machine Control Instructions.

Data Transfer Instructions : Includes the instructions that moves (copies) data between registers or between memory location and register. In all data transfer operations, the content of source register or memory is not altered. Hence the data transfer is copying operation.

Arithmetic Instructions : Includes the instructions which performs addition, subtraction, increment or decrement operations. The flag conditions are altered after execution of an instruction in this group.

Logical Instructions : The instructions which performs the logical operations like AND, OR, EXCLUSIVE-OR, complement, compare and rotate instructions are grouped under this heading. The flag conditions are altered after execution of an instruction in this group.

Branching Instructions : The instructions that are used to transfer the program control from one memory location to another memory location are grouped under this heading.

Machine Control Instructions : Includes the instructions related to interrupts and the instruction used to halt program execution.

The 74 basic instructions of 8085 are listed in Table 3.1. The opcode of each instruction, size, machine cycles, number of T-state and the total number of instructions in each type are also shown in Table 3.1. The instructions affecting the status flag are listed in Table 3.2.

Table 3.1: Summary of 8085 Instruction Set

S.No.	Mnemonic	Opcode	Number of bytes	Machine cycles	Number of T-states	Total number of instructions
Data transfer instructions						
1.	MOV Rd, Rs	`0 1 D D D S S S`	1	F	4T	49
2.	MOV Rd, M	`0 1 D D D 1 1 0`	1	F, R	7T	7
3.	MOV M, Rs	`0 1 1 1 0 S S S`	1	F, W	7T	7
4.	MVI Rd, d8	`0 0 D D D 1 1 0`	2	F, R	7T	7
5.	MVI M, d8	`0 0 1 1 0 1 1 0`	2	F, R, W	10T	1
6.	LDA addr16	`0 0 1 1 1 0 1 0`	3	F, R, R, R	13T	1
7.	LDAX rp	`0 0 R P 1 0 1 0`	1	F, R	7T	2
8.	LXI rp, d16	`0 0 R P 0 0 0 1`	3	F, R, R	10T	4
9.	LHLD addr16	`0 0 1 0 1 0 1 0`	3	F,R,R,R,R	16T	1
10.	STA addr16	`0 0 1 1 0 0 1 0`	3	F, R, R,W	13T	1
11.	STAX rp	`0 0 R P 0 0 1 0`	1	F, W	7T	2
12.	SHLD addr16	`0 0 1 0 0 0 1 0`	3	F,R,R,W,W	16T	1
13.	SPHL	`1 1 1 1 1 0 0 1`	1	S	6T	1
14.	XCHG	`1 1 1 0 1 0 1 1`	1	F	4T	1
15.	XTHL	`1 1 1 0 0 0 1 1`	1	F,R,R,W,W	16 T	1
16.	PUSH rp	`1 1 R P 0 1 0 1`	1	S, W, W	12T	3
17.	PUSH PSW	`1 1 1 1 0 1 0 1`	1	S, W, W	12T	1
18.	POP rp	`1 1 R P 0 0 0 1`	1	F, R, R	10T	3
19.	POP PSW	`1 1 1 1 0 0 0 1`	1	F, R, R	10T	1
20.	IN addr8	`1 1 0 1 1 0 1 1`	2	F, R, I	10T	1
21.	OUT addr8	`1 1 0 1 0 0 1 1`	2	F, R, O	10T	1
Arithmetic instructions						
22.	ADD reg	`1 0 0 0 0 S S S`	1	F	4T	7
23.	ADD M	`1 0 0 0 0 1 1 0`	1	F, R	7T	1

Table 3.1 continued...

S.No.	Mnemonic	Opcode	Number of bytes	Machine cycles	Number of T-states	Total number of instructions
24.	ADI d8	1 1 0 0 0 1 1 0	2	F, R	7T	1
25.	ADC reg	1 0 0 0 1 S S S	1	F	4T	7
26.	ADC M	1 0 0 0 1 1 1 0	1	F, R	7T	1
27.	ACI d8	1 1 0 0 1 1 1 0	2	F, R	7T	1
28.	DAA	0 0 1 0 0 1 1 1	1	F	4T	1
29.	DAD rp	0 0 R P 1 0 0 1	1	F, B, B	10T	4
30.	SUB reg	1 0 0 1 0 S S S	1	F	4T	7
31.	SUB M	1 0 0 1 0 1 1 0	1	F, R	7T	1
32.	SUI d8	1 1 0 1 0 1 1 0	2	F, R	7T	1
33.	SBB reg	1 0 0 1 1 S S S	1	F	4T	7
34.	SBB M	1 0 0 1 1 1 1 0	1	F, R	7T	1
35.	SBI d8	1 1 0 1 1 1 1 0	2	F, R	7T	1
36.	INR reg	0 0 S S S 1 0 0	1	F	4T	7
37.	INR M	0 0 1 1 0 1 0 0	1	F, R, W	10T	1
38.	INX rp	0 0 R P 0 0 1 1	1	S	6T	4
39.	DCR reg	0 0 S S S 1 0 1	1	F	4 T	7
40.	DCR M	0 0 1 1 0 1 0 1	1	F, R, W	10T	1
41.	DCX rp	0 0 R P 1 0 1 1	1	S	6T	4
Logical instructions						
42.	ANA reg	1 0 1 0 0 S S S	1	F	4T	7
43.	ANA M	1 0 1 0 0 1 1 0	1	F, R	7T	1
44.	ANI d8	1 1 1 0 0 1 1 0	2	F, R	7T	1
45.	ORA reg	1 0 1 1 0 S S S	1	F	4T	7
46.	ORA M	1 0 1 1 0 1 1 0	1	F, R	7T	1
47.	ORI d8	1 1 1 1 0 1 1 0	2	F, R	7T	1

Table 3.1 continued...

S.No.	Mnemonic	Opcode	Number of bytes	Machine cycles	Number of T-states	Total number of instructions
48.	XRA reg	1 0 1 0 1 S S S	1	F	4T	7
49.	XRA M	1 0 1 0 1 1 1 0	1	F,R	7T	1
50.	XRI d8	1 1 1 0 1 1 1 0	2	F, R	7T	1
51.	CMP reg	1 0 1 1 1 S S S	1	F	4T	7
52.	CMP M	1 0 1 1 1 1 1 0	1	F, R	7T	1
53.	CPI d8	1 1 1 1 1 1 1 0	2	F, R	7T	1
54.	CMA	0 0 1 0 1 1 1 1	1	F	4T	1
55.	CMC	0 0 1 1 1 1 1 1	1	F	4T	1
56.	STC	0 0 1 1 0 1 1 1	1	F	4T	1
57.	RLC	0 0 0 0 0 1 1 1	1	F	4T	1
58.	RAL	0 0 0 1 0 1 1 1	1	F	4T	1
59.	RRC	0 0 0 0 1 1 1 1	1	F	4T	1
60.	RAR	0 0 0 1 1 1 1 1	1	F	4T	1

Branching instructions

S.No.	Mnemonic	Opcode	Number of bytes	Machine cycles	Number of T-states	Total number of instructions
61.	JMP addr16	1 1 0 0 0 0 1 1	3	F,R,R	10T	1
62.	J<condition> addr16	1 1 C C C 0 1 0	3	F,R/F,R,R	7T/10T	8
63.	CALL addr16	1 1 0 0 1 1 0 1	3	S,R,R,W,W	18T	1
64.	C<condition> addr16	1 1 C C C 1 0 0	3	S, R or S,R,R,W,W	9T/18T	8
65.	RET	1 1 0 0 1 0 0 1	1	F,R,R	10T	1
66.	R<condition>	1 1 C C C 0 0 0	1	S/S,R,R	6T/12T	8
67.	RST n	1 1 N N N 1 1 1	1	S,W,W	12T	8
68.	PCHL	1 1 1 0 1 0 0 1	1	S	6T	1

Table 3.1 continued...

S.No.	Mnemonic	Opcode	Number of bytes	Machine cycles	Number of T-states	Total number of instruction
Machine control instructions						
69.	SIM	0 0 1 1 0 0 0 0	1	F	4T	1
70.	RIM	0 0 1 0 0 0 0 0	1	F	4T	1
71.	DI	1 1 1 1 0 0 1 1	1	F	4T	1
72.	EI	1 1 1 1 1 0 1 1	1	F	4T	1
73.	HLT	0 1 1 1 0 1 1 0	1	F,B	5T	1
74.	NOP	0 0 0 0 0 0 0 0	1	F	4T	1
						246

Meanings of various symbols used in Table 3.1

Symbol	Meaning
rp, RP	Register pair
Rs, SSS	Source register
Rd, DDD	Destination register
M	Memory
d8	8-bit data
d16	16-bit data
addr8	8-bit address
addr16	16-bit address
reg	Register
PSW	Program status word
n, NNN	Type number of restart instruction
<condition>, CCC	Flag condition
F	4T-Opcode fetch cycle
S	6T-Opcode fetch cycle
R	Memory read cycle
W	Memory write cycle
I	IO read cycle
O	IO write cycle
B	Bus idle cycle

Flag condition can be any one of the conditions given below

Z	→	Zero flag = 1	M	→	Sign flag = 1
NZ	→	Zero flag = 0	P	→	Sign flag = 0
C	→	Carry flag = 1	PE	→	Parity flag = 1
NC	→	Carry flag = 0	PO	→	Parity flag = 0

The binary codes for the symbols used in opcode of 8085 instructions are given below:

Register	DDD or SSS
B	0 0 0
C	0 0 1
D	0 1 0
E	0 1 1
H	1 0 0
L	1 0 1
A	1 1 1

Register	RP
BC	0 0
DE	0 1
HL	1 0
SP	1 1

Flag condition	C C C
NZ	0 0 0
Z	0 0 1
NC	0 1 0
C	0 1 1
PO	1 0 0
PE	1 0 1
P	1 1 0
M	1 1 1

n	N N N
0	0 0 0
1	0 0 1
2	0 1 0
3	0 1 1
4	1 0 0
5	1 0 1
6	1 1 0
7	1 1 1

Table 3.2: 8085 Instructions Affecting the Status Flags

Instructions	Status flags				
	CF	AF	ZF	SF	PF
ACI d8	+	+	+	+	+
ADC reg	+	+	+	+	+
ADC M	+	+	+	+	+
ADD reg	+	+	+	+	+
ADD M	+	+	+	+	+
ADI d8	+	+	+	+	+
ANA reg	0	1	+	+	+
ANA M	0	1	+	+	+
ANI d8	0	1	+	+	+
CMC	+				
CMP reg	+	+	+	+	+
CMP M	+	+	+	+	+
CPI d8	+	+	+	+	+
DAA	+	+	+	+	+
DAD rp	+				
DCR reg		+	+	+	+
DCR M		+	+	+	+
INR reg		+	+	+	+
INR M		+	+	+	+
ORA reg	0	0	+	+	+
ORA M	0	0	+	+	+

Table 3.2 continued...

Instructions	Status flags				
	CF	AF	ZF	SF	PF
ORI d8	0	0	+	+	+
RAL	+				
RAR	+				
RLC	+				
RRC	+				
SBB reg	+	+	+	+	+
SBB M	+	+	+	+	+
SBI d8	+	+	+	+	+
STC	+				
SUB reg	+	+	+	+	+
SUB M	+	+	+	+	+
SUI d8	+	+	+	+	+
XRA reg	0	0	+	+	+
XRA M	0	0	+	+	+
XRI d8	0	0	+	+	+

Note:

+ → *Indicates that the particular flag is affected.*

0 → *Indicates that the particular flag is always zero.*

1 → *Indicates that the particular flag is always one.*

Table 3.3: Meaning/Expansion of Mnemonics used in an 8085 Instruction Set

S.No.	Mnemonic	Meaning
1.	ACI	Add the immediate data and the carry to the accumulator.
2.	ADC	Add the register/memory and the carry to the accumulator.
3.	ADD	Add the register/memory to the accumulator.
4.	ADI	Add the immediate data to the accumulator.
5.	ANA	AND register/memory with the accumulator.
6.	ANI	AND immediate data with the accumulator.
7.	CALL	Call a subroutine/procedure.
8.	CC	Call on carry.
9.	CM	Call on minus.
10.	CMA	Complement accumulator.
11.	CMC	Complement carry.
12.	CMP	Compare register/memory with accumulator.
13.	CNC	Call on no carry.
14.	CNZ	Call on not zero.

Table 3.3 continued...

S.No.	Mnemonic	Meaning
15.	CP	Call on positive.
16.	CPE	Call on parity even.
17.	CPI	Compare immediate data with the accumulator.
18.	CPO	Call on parity odd.
19.	CZ	Call on zero.
20.	DAA	Decimal adjust accumulator after addition.
21.	DAD	Double addition.
22.	DCR	Decrement the register/memory.
23.	DCX	Decrement the register pair.
24.	DI	Disable interrupt.
25.	EI	Enable interrupt.
26.	HLT	Halt program execution.
27.	IN	Input data from specified port to accumulator.
28.	INR	Increment the register/memory.
29.	INX	Increment the register pair.
30.	JC	Jump on carry.
31.	JM	Jump on minus.
32.	JMP	Jump to specified address to get the next instruction.
33.	JNC	Jump on no carry.
34.	JNZ	Jump on not zero.
35.	JP	Jump on positive.
36.	JPE	Jump on parity even.
37.	JPO	Jump on parity odd.
38.	JZ	Jump on zero.
39.	LDA	Load the accumulator.
40.	LDAX	Load accumulator indirectly using the address in the specified register pair.
41.	LHLD	Load HL direct.
42.	LXI	Load the immediate data in the register pair.
43.	MOV	Move (copy) the content of register/memory to another register/memory.
44.	MVI	Move the immediate data to register/memory.
45.	NOP	No operation.
46.	ORA	OR register/memory with accumulator.
47.	ORI	OR immediate data with accumulator.
48.	OUT	Output the content of accumulator to specified port.
49.	PCHL	Move the content of HL to PC.

Table 3.3 continued...

S.No.	Mnemonic	Meaning
50.	POP	Move the top of stack to the specified register pair.
51.	PUSH	Push the content of the specified register pair to top of stack.
52.	RAL	Rotate the accumulator left along with carry.
53.	RAR	Rotate the accumulator right along with carry.
54.	RC	Return on carry.
55.	RET	Return from subroutine/procedure to calling program.
56.	RIM	Read interrupt mask status.
57.	RLC	Rotate accumulator left to carry.
58.	RM	Return on minus.
59.	RNC	Return on no carry.
60.	RNZ	Return on not zero.
61.	RP	Return on positive.
62.	RPE	Return on parity even.
63.	RPO	Return on parity odd.
64.	RRC	Rotate accumulator right to carry.
65.	RST	Restart the program execution from the specified vector address.
66.	RZ	Return on zero.
67.	SBB	Subtract register/memory and the carry (borrow) from accumulator.
68.	SBI	Subtract the immediate data and the carry (borrow) from accumulator.
69.	SHLD	Store HL direct.
70.	SIM	Set interrupt mask.
71.	SPHL	Move HL to SP.
72.	STA	Store accumulator.
73.	STAX	Store accumulator indirectly by using the address in specified register pair.
74.	STC	Set carry.
75.	SUB	Subtract register/memory from accumulator.
76.	SUI	Subtract the immediate data from accumulator.
77.	XCHG	Exchange DE and HL.
78.	XRA	Exclusive-OR register/memory with accumulator.
79.	XRI	Exclusive-OR the immediate data with accumulator.
80.	XTHL	Exchange the top of stack and HL.

3.5 DATA TRANSFER INSTRUCTIONS

1.   MOV Rd, Rs   (Rd) ← (Rs)

The content of source register (Rs) is copied to the destination register (Rd). The registers Rd and Rs can be any one of the general purpose registers A, B, C, D, E, H or L. No flags are affected.

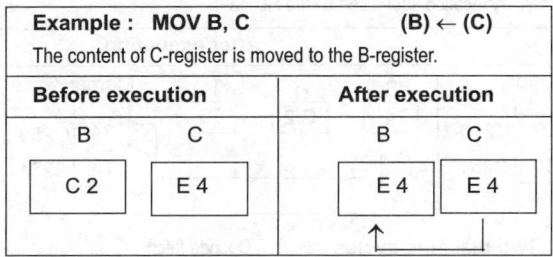

One-byte instruction   **One machine cycle :** Opcode fetch - 4T

Register addressing

Total number of instructions = 49

MOV A, A	MOV B, A	MOV D, A	MOV H, A
MOV A, B	MOV B, B	MOV D, B	MOV H, B
MOV A, C	MOV B, C	MOV D, C	MOV H, C
MOV A, D	MOV B, D	MOV D, D	MOV H, D
MOV A, E	MOV B, E	MOV D, E	MOV H, E
MOV A, H	MOV B, H	MOV D, H	MOV H, H
MOV A, L	MOV B, L	MOV D, L	MOV H, L
	MOV C, A	MOV E, A	MOV L, A
	MOV C, B	MOV E, B	MOV L, B
	MOV C, C	MOV E, C	MOV L, C
	MOV C, D	MOV E, D	MOV L, D
	MOV C, E	MOV E, E	MOV L, E
	MOV C, H	MOV E, H	MOV L, H
	MOV C, L	MOV E, L	MOV L, L

2.   MOV Rd, M   (Rd) ← (M)   or   (Rd) ← ((HL))

The content of memory (M) addressed by the HL pair is moved to the destination register (Rd). The register Rd can be any one of the general purpose registers A, B, C, D, E, H or L. No flags are affected.

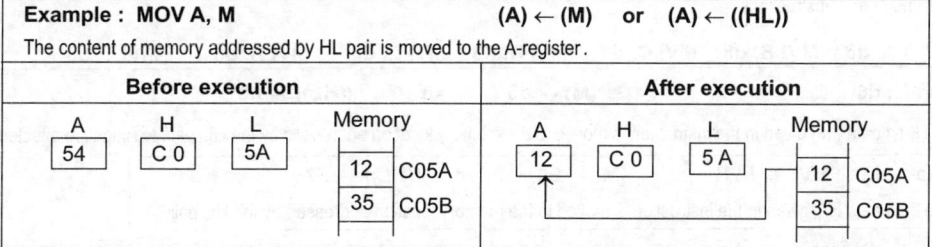

One-byte instruction   **Two machine cycles:** Opcode fetch - 4T

Register indirect addressing   Memory read - 3T

                                                                                                        7T

Total number of instructions = 7

MOV A, M	MOV B, M	MOV C, M	MOV D, M	MOV E, M	MOV H, M	MOV L, MOV

3.   **MOV M, Rs**   **(M) ← (Rs)**   **or**   **((HL)) ← (Rs)**

The content of source register (Rs) is moved to the memory location addressed by HL pair. The register Rs can be any one of the general purpose registers A, B, C, D, E, H or L. No flags are affected.

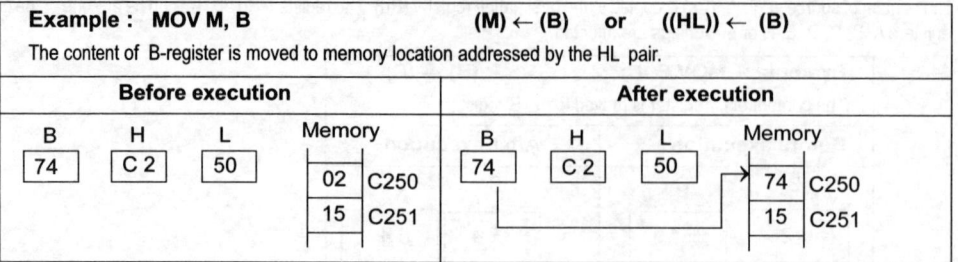

Example : MOV M, B	**(M) ← (B) or ((HL)) ← (B)**
The content of B-register is moved to memory location addressed by the HL pair.	

One-byte instruction   **Two machine cycles :**   Opcode fetch - 4T

Register indirect addressing   Memory write - 3T

                                                                           7 T

Total number of instructions = 7

MOV M,A   MOV M,B   MOV M, C   MOV M, D   MOV M, E   MOV M, H   MOV M, L

4.   **MVI Rd, d8**   **(Rd) ← d8**

The 8-bit data (d8) given in the instruction is moved to the destination register (Rd). The register Rd can be any one of the general purpose registers A, B, C, D, E, H or L. No flags are affected.

Example : MVI D,09H	**(D) ← 09$_H$**
The 8-bit data 09$_H$ given in the instruction is moved to the D-register.	

Before execution	**After execution**
D	D
C2	09

Two-byte instruction   **Two machine cycles:**   Opcode fetch - 4T

Immediate addressing   Memory read - 3T

                                                                           7 T

Total number of instructions = 7

MVI A, d8   MVI B, d8   MVI C, d8   MVI D, d8   MVI E, d8   MVI H, d8   MVI L, d8

5.   **MVI M, d8**   **(M) ← d8   or   ((HL)) ← d8**

The 8-bit data (d8) given in the instruction is moved to the memory location addressed by the HL pair. No flags are affected.

Example : MVI M, E7H **(M) ← E7$_H$ or ((HL)) ← E7$_H$**	
The 8-bit data E7$_H$ given in the instruction is moved to the memory location addressed by the HL pair.	

Before execution			**After execution**		
		Memory			Memory
H	L	28 205C	H	L	E 7 205C
20	5 C	3A 205D	20	5 C	3A 205D

Two-byte instruction

Register indirect addressing or

Immediate addressing

Three machine cycles : Opcode fetch - 4T

Memory read - 3T

Memory write - 3T

—————

10 T

Total number of instructions = 1

---

6.   **LDA addr16**   **(A) ← (M)   or   (A) ← (addr16)**

The content of the memory location whose address is given in the instruction, is moved to accumulator. No flags are affected.

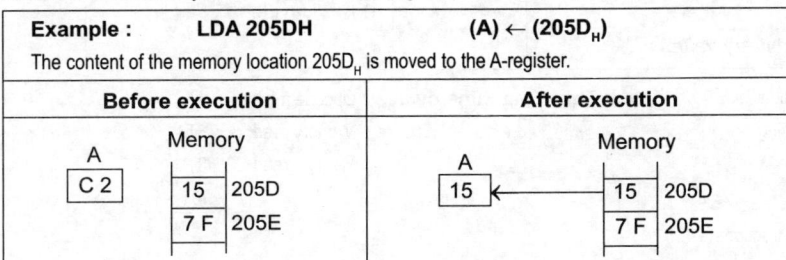

Example :   LDA 205DH   (A) ← (205D_H)

The content of the memory location 205D_H is moved to the A-register.

Three-byte instruction

Direct addressing

Four machine cycles :   Opcode fetch - 4T

Memory read - 3T

Memory read - 3T

Memory read - 3T

—————

13T

Total number of instructions = 1

---

7.   **LHLD addr16**   **(L) ← (M)   or   (L) ← (addr16)**

(H) ← (M)   (H) ← (addr16 + 01)

The content of the memory location whose address is given in the instruction, is moved to the L-register. The content of the next memory location is moved to the H-register. No flags are affected.

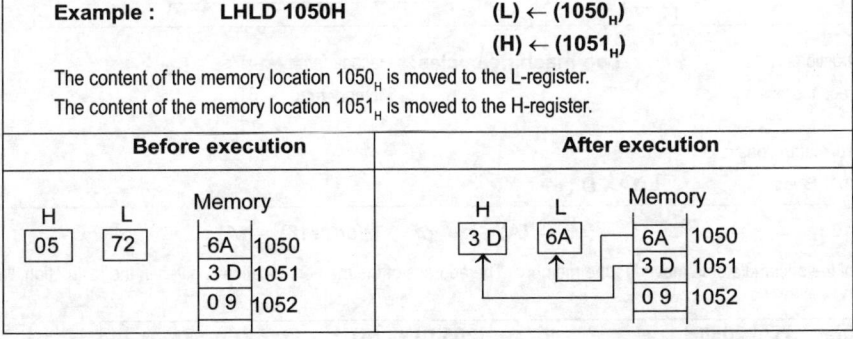

Example :   LHLD 1050H   (L) ← (1050_H)

(H) ← (1051_H)

The content of the memory location 1050_H is moved to the L-register.

The content of the memory location 1051_H is moved to the H-register.

Three-byte instruction

Direct addressing

Five machine cycles: Opcode fetch - 4T

Memory read - 3T

Memory read - 3T

Memory read - 3T

Memory read - 3T

—————

16T

Total number of instructions = 1

8.   LXI rp, d16   **(rp) ← d16**

The 16-bit data given in the instruction is moved to the register pair (rp). The register pair can be BC, DE, HL or SP.

Example : LXI H, 1050H	(L) ← 50$_H$
	(H) ← 10$_H$

The 16-bit data 1050$_H$ given in the instruction is moved to the HL register pair.

Before execution	After execution
H L	H L
xx yy	10 50
(some arbitrary value)	

Three-byte instruction   **Three machine cycles :** Opcode fetch   -   4T

Immediate addressing   Memory read   -   3T

                                                                                      Memory read   -   3T

                                                                                      _____

                                                                                      10T

Total number of instructions = 4

       LXI B, d16   **LXI D, d16**   **LXI H, d16**   **LXI SP, d16**

9.   LDAX rp   **(A) ← (M)**   **or**   **(A) ← ((rp))**

The content of the memory addressed by the register pair (rp) is moved to the accumulator. (The content of the register pair is the memory address). The register pair can be either BC or DE.

Example : LDAX B	(A) ← (M) or (A) ← ((BC))

The content of the memory location addressed by the BC pair is moved to the A-register.

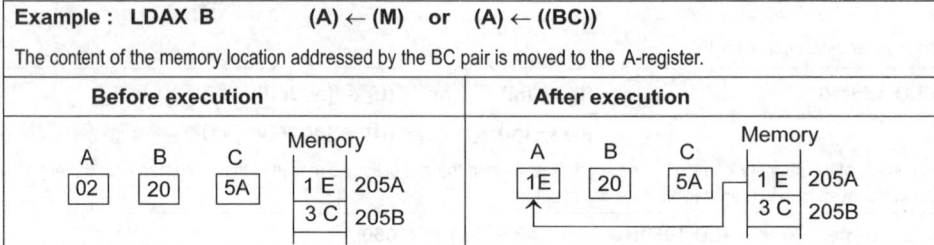

One-byte instruction   **Two machine cycles:**   Opcode fetch   - 4T

Register indirect addressing   Memory read   - 3T

                                                                            _____

                                                                            7T

Total number of instructions = 2

       LDAX B   **LDAX D**

10.   STA addr16   **(M) ← (A)**   **or**   **(addr16) ← (A)**

The content of the accumulator is moved to the memory. The address of the memory location is given in the instruction. No flags are affected.

Example : STA 2050H	(2050$_H$) ← (A)

The content of the accumulator is moved to memory location 2050$_H$.

Before execution	After execution
A Memory	A Memory
F4 0 6 2050	F4 → F4 2050
7A 2051	7A 2051

Three-byte instruction	**Four machine cycles:**	Opcode fetch	-	4T
Direct addressing		Memory read	-	3T
		Memory read	-	3T
		Memory write	-	3T
				13T

Total number of instructions = 1

---

11.   **STAX rp**   **(M) ← (A)**   **or**   **((rp)) ← (A)**

The content of the accumulator is moved to the memory addressed by the register pair (rp). (The content of the register pair is the memory address.) The register pair can be either BC or DE.

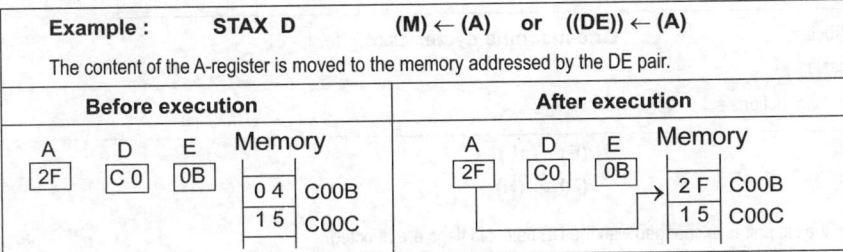

Example :	**STAX D**	**(M) ← (A)** **or** **((DE)) ← (A)**

The content of the A-register is moved to the memory addressed by the DE pair.

One-byte instruction	**Two machine cycles:**	Opcode fetch	-	4 T
Register indirect addressing		Memory write	-	3 T
				7 T

Total number of instructions = 2

         STAX B   **STAX D**

---

12.   **SHLD addr16**   **(M) ← (L)**   **or**   **(addr16) ← (L)**
                               (M) ← (H)   **(addr16+1) ← (H)**

The content of the L-register is stored in the memory location, whose address is given in the instruction. The content of the H-register is stored in the next memory location. No flags are affected.

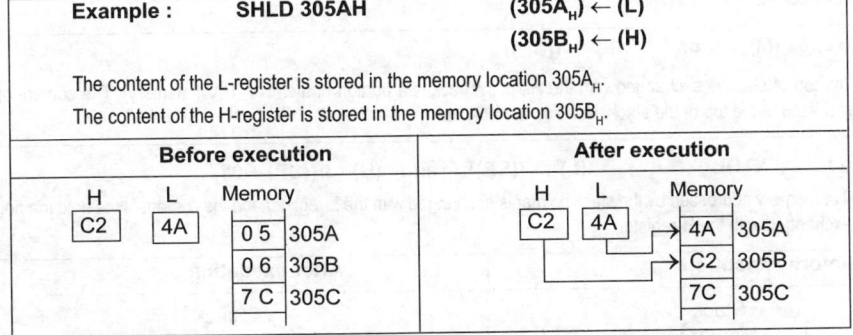

Example :	**SHLD 305AH**	**($305A_H$) ← (L)**
		($305B_H$) ← (H)

The content of the L-register is stored in the memory location $305A_H$.
The content of the H-register is stored in the memory location $305B_H$.

Three-byte instruction	**Five machine cycles:**	Opcode fetch	-	4T
Direct addressing		Memory read	-	3T
		Memory read	-	3T
		Memory write	-	3T
		Memory write	-	3T
				16T

Total number of instructions = 1

13.   SPHL   **(SP) ← (HL)**

The content of the HL pair is moved to the Stack Pointer (SP). No flags are affected.

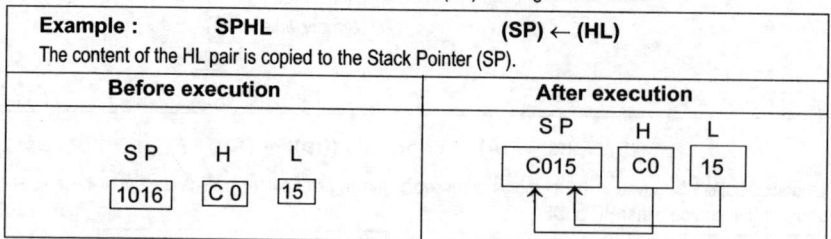

Example : **SPHL**	**(SP) ← (HL)**

The content of the HL pair is copied to the Stack Pointer (SP).

One-byte instruction   **One machine cycle:** Opcode fetch - 6T

Implied addressing

Total number of instructions = 1

---

14.   XCHG   **(E) ↔ (L)**

(D) ↔ (H)

The content of the HL pair is exchanged with the DE pair. No flags are affected.

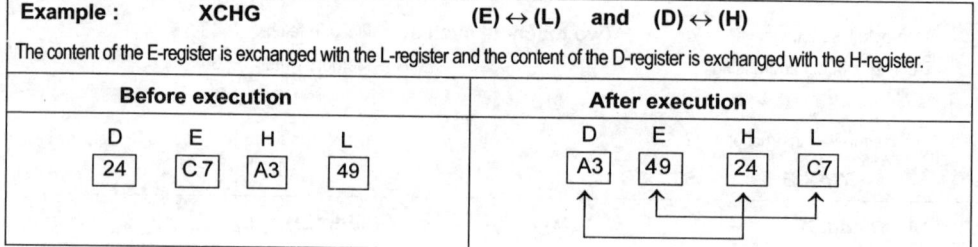

Example : **XCHG**	**(E) ↔ (L) and (D) ↔ (H)**

The content of the E-register is exchanged with the L-register and the content of the D-register is exchanged with the H-register.

One-byte instruction   **One machine cycle:** Opcode fetch - 4T

Implied addressing

Total number of instructions = 1

---

15.   XTHL   (HL) ↔ (M)   or   (HL) ↔ ((SP))

The content of the top of stack is exchanged with the HL pair. Stack is a portion of memory (RAM memory). The content of the Stack Pointer (SP) is the address of the top of the stack. No flags are affected.

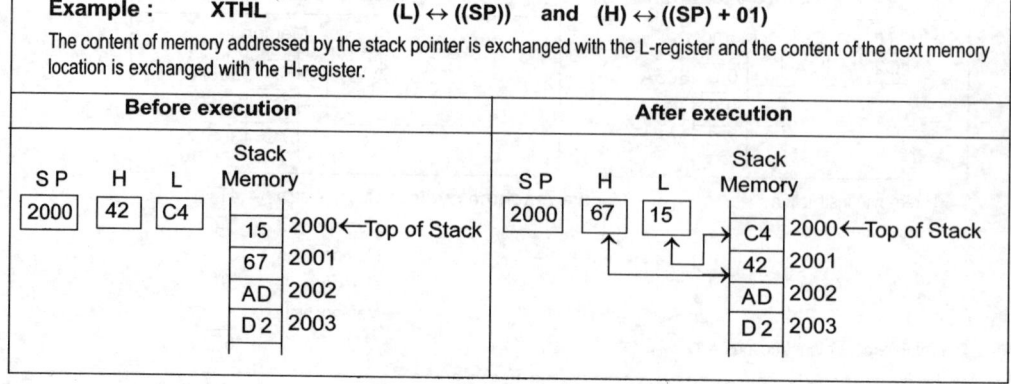

Example : **XTHL**	**(L) ↔ ((SP)) and (H) ↔ ((SP) + 01)**

The content of memory addressed by the stack pointer is exchanged with the L-register and the content of the next memory location is exchanged with the H-register.

One-byte instruction	**Five machine cycles:** Opcode fetch	-	4T
Implied addressing	Memory read	-	3T
	Memory read	-	3T
	Memory write	-	3T
	Memory write	-	3T
			16T

Total number of instructions = 1

16.   PUSH rp $(SP) \leftarrow (SP) - 1$; $((SP)) \leftarrow (rp)_H$
         $(SP) \leftarrow (SP) - 1$; $((SP)) \leftarrow (rp)_L$

The content of the register pair (rp) is pushed to the stack. After execution of this instruction, the content of the Stack Pointer (SP) will be 02 less than the earlier value. The register pairs can be BC, DE , HL and PSW. No flags are affected.

[PSW (Program Status Word) : Accumulator and Flag register together called PSW. Accumulator is high order register and Flag register is low order register.]

The instruction is executed as follows:

(i) The content of the SP is decremented by one.

(ii) The content of the high order register is moved to memory addressed by SP.

(iii) The content of the SP is decremented by one.

(iv) The content of the low order register is moved to memory addressed by SP.

One-byte instruction	**Three machine cycles:** Opcode fetch	-	6T
Register indirect addressing	Memory write	-	3T
	Memory write	-	3T
			12T

Total number of instructions = 4

PUSH PSW   PUSH B   PUSH D   PUSH H

Example : PUSH B $(SP) \leftarrow (SP) - 01$
                             $((SP)) \leftarrow (B)$
                             $(SP) \leftarrow (SP) - 01$
                             $((SP)) \leftarrow (C)$

(i)   The content of the SP is decremented by one.

(ii)   The content of the B-register is moved to the memory addressed by the Stack Pointer (SP).

(iii)   Again the content of SP is decremented by one.

(iv)   The content of the C-register is moved to the memory addressed by SP.

Before execution	**After execution**

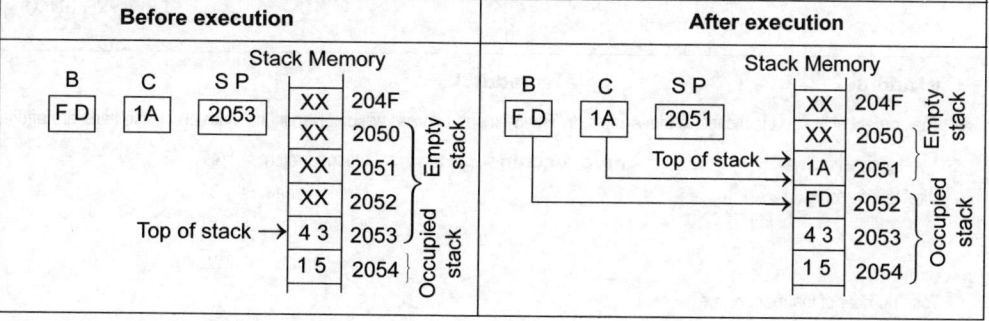

17. **POP r p** $(rp)_L \leftarrow ((SP))$; $(SP) \leftarrow (SP) + 1$

$(rp)_H \leftarrow ((SP))$; $(SP) \leftarrow (SP) + 1$

The content of top of stack memory is moved to the register pair. After execution of this instruction the content of the Stack Pointer (SP) will be 02 greater than the earlier value. The register pairs can be BC, DE , HL and PSW. No flags are affected. [PSW (Program Status Word) : Accumulator and Flag register are together called PSW. The accumulator is a high order register and the flag register is a low order register.]

The pop instruction is executed as follows:

  (i)   The content of the memory addressed by the SP is moved to the low order register.

  (ii)   The content of the SP is incremented by one.

  (iii) The content of the memory addressed by the SP is moved to the high order register.

  (iv) The content of the SP is incremented by one.

One-byte instruction	**Three machine cycles**: Opcode fetch - 4T
Register indirect addressing	Memory read - 3T
	Memory read - 3T
	10T

Total number of instructions = 4

POP PSW   **POP B**   **POP D**   **POP H**

Example : POP D $(E) \leftarrow ((SP))$

                                  $(SP) \leftarrow (SP) + 01$

                                  $(D) \leftarrow ((SP))$

                                  $(SP) \leftarrow (SP) + 01$

  (i)   The content of the memory addressed by the SP is moved to the E-register.

  (ii)   The content of the SP is incremented by one.

  (iii) The content of the memory addressed by the SP is moved to the D-register.

  (iv) The content of the SP is incremented by one.

Before execution	After execution

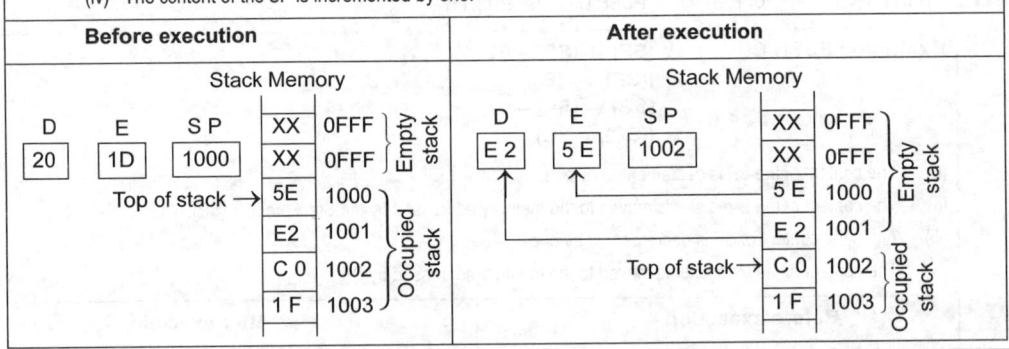

---

18. **IN addr8**   **(A)← (addr8)**

The content of the port is moved to the A-register. The 8-bit port address will be given in the instruction. No flags are affected.

Two-byte instruction	**Three machine cycles:** Opcode fetch - 4T
Direct addressing	Memory read - 3T
	IO read - 3T
	10T

Total number of instructions = 1

19.   OUT addr8   (addr8) ← (A)

The content of the A-register is moved to the port. The 8-bit port address will be given in the instruction. No flags are affected.

Two-byte instruction	**Three machine cycles:**	Opcode fetch - 4T
Direct addressing		Memory read - 3T
		IO write - 3T
		$\overline{10T}$

Total number of instructions = 1

> **Note:** *In an 8085 processor-based system when the IO devices are mapped by IO mapping then the processor can communicate with these IO devices only by using IN and OUT instructions. The processor uses an 8-bit address to select IO-mapped IO devices. With 8-bit address the processor can generate $2^8 = 256_{10}$ IO addresses.*

3.6 DATA MANIPULATION AND CONTROL INSTRUCTIONS

3.6.1 Arithmetic Instructions

1.   ADD reg   (A) ← (A) + (reg)

The content of the register is added to the content of the accumulator (A-register). After addition the result is stored in the accumulator. All flags are affected. The register can be any one of the general purpose register A, B, C, D, E, H or L.

Example : ADD E	(A) ← (A) + (E)

The content of the E-register is added to the content of the A-register.

The result will be in the A-register. All flags are affected.

Before execution	Addition	After execution
A E [C 2] [B8] C F = 0 P F = 0 A F = 0 Z F = 0 S F = 0	$C2_H$ = 1100 0010 $B8_H$ = 1011 1000 $\overline{\boxed{1}\ 0111\ 1010}$ Sum = 0111 1010 =$7A_H$ Carry = 1 (Addition is performed in ALU)	A E [7A] [B8] C F = 1 P F = 0 A F = 0 Z F = 0 S F = 0

One-byte instruction	**One machine cycle:** Opcode fetch - 4T
Register addressing	

Total number of instructions = 7

ADD A   ADD B   ADD C   ADD D   ADD E   ADD H   ADD L

2.   ADI d8   (A) ← (A) + d8

The 8-bit data given in the instruction is added to the content of the A-register (Accumulator). After addition, the result is stored in the accumulator. All flags are affected.

Two-byte instruction	Two machine cycles:	Opcode fetch - 4T
Immediate addressing		Memory read - 3T
		7T

Total number of instructions = 1

3.   ADD M   (A) ← (A) + (M)   or   (A) ← (A) + ((HL))

The content of memory addressed by HL pair is added to the content of the A-register. After addition, the result is stored in the A-register. All flags are affected.

| **Example : ADD M** | **(A) ← (A) + (M)** | **or** | **(A) ← (A) + ((HL))** |

Let the content of A be 44_H.

Let the content of memory location $C00A_H$ be 73_H.

The content of the memory location $C00A_H$ is added to the content of the A-register. The result is put back in the A-register.

Before execution	Addition	After execution

A   H L   Memory
┌──┐ ┌────┐ ┌──┐
│44│ │C00A│ │73│ C00A
└──┘ └────┘ ├──┤
C F = 0   │14│ C00B
P F = 0   ├──┤
             │27│ C00C
AF = 0   └──┘
ZF = 0
SF = 0
	44_H = 0100 0100
	73_H = 0111 0011
	―――――――――
	       1011 0111
	Sum = B7
	Carry = 0
	(Addition is performed in ALU)
		A H L Memory
		┌──┐ ┌────┐ ┌──┐
		│B7│ │C00A│ │73│ C00A
		└──┘ └────┘ ├──┤
		C F = 0 │14│ C00B
		P F = 1 ├──┤
		             │27│ C00C
		AF = 0 └──┘
		ZF = 0
		SF = 1

One-byte instruction   **Two machine cycles:**   Opcode fetch - 4T

Register indirect addressing   Memory read - 3T

                                                                7T

Total number of instructions = 1

---

4.   ACI d8   **(A) ← (A) + d8 + CF**

The 8-bit data given in the instruction and the carry flag (the value of carry flag before executing this instruction) are added to the content of the A-register (Accumulator). After addition, the result is stored in the accumulator. All flags are affected.

Two-byte instruction   **Two machine cycles :**   Opcode fetch - 4T

Immediate addressing   Memory read - 3T

                                                                7T

Total number of instructions = 1

---

5.   ADC reg   **(A) ← (A) + (reg) + CF**

The content of the register and the carry flag are added to the content of the A-register. After addition, the result is stored in the A-register. All flags are affected. The register can be any one of the general purpose register A, B, C, D, E, H or L.

| **Example : ADC H** | **(A) ← (A) + (H) + CF** |

The content of the H-register and the value of the carry flag (before executing this instruction) are added to the content of the A-register. After addition, the result will be in the A-register.

Before execution	Addition	After execution
A H ┌──┐ ┌──┐ │43│ │7A│ └──┘ └──┘ C F = 1 P F = 0 AF = 0 ZF = 0 S F = 1	43_H = 0100 0011 $7A_H$ = 0111 1010 CF = 1 ―――――――― 1011 1110 Sum = BE_H Carry = 0 (Addition is performed in the ALU)	A H ┌──┐ ┌──┐ │BE│ │7A│ └──┘ └──┘ C F = 0 P F = 1 AF = 0 ZF = 0 S F = 1

One-byte instruction   **One machine cycle :** Opcode fetch - 4T

Register addressing

Total number of instructions = 7

ADC A	ADC B	ADC C	ADC D	ADC E	ADC H	ADC L

6.   **ADC M** $(A) \leftarrow (A) + (M) + CF$ **or** $(A) \leftarrow (A) + ((HL)) + CF$

The content of the memory addressed by the HL pair and the value of the carry flag (before executing this instruction) are added to the content of A-register. After addition, the result is stored in the A-register. All flags are affected.

One-byte instruction   **Two machine cycles:**   Opcode fetch -   4T

Register indirect addressing   Memory read -   3T

                                                                     7T

Total number of instructions = 1

7.   **SUB reg** $(A) \leftarrow (A) - (reg)$

The content of the register is subtracted from the content of the accumulator (A-register). After subtraction the result is stored in the A-register. All flags are affected. The register can be any one of the general purpose register A, B, C, D, E, H or L.

Example: **SUB C** $(A) \leftarrow (A) - (C)$

The content of the C-register is subtracted from A-register. The result will be in the A-register.

Case i

Before execution	Subtraction
A C $\boxed{C4}$ $\boxed{89}$ CF = 0 PF = 0 AF = 0 ZF = 0 SF = 1	$C4_H = 1100\ 0100$ $89_H = 1000\ 1001$ 1's complement of $89_H = 0111\ 0110$ 2's complement of $89_H = 0111\ 0110 + 1$ $= 0111\ 0111 = 77_H$
A C $\boxed{3B}$ $\boxed{89}$ CF = 0 PF = 0 AF = 0 ZF = 0 SF = 0	$\begin{array}{r} C4_H = 1100\ 0100 \\ +77_H = 0111\ 0111 \\ \hline \boxed{1}\,0011\ 1011 \end{array}$ Complement Carry $\downarrow$ 3 B $\boxed{0}$ Result = $3B_H$ CF = 0

Case ii

Before execution	Subtraction
A C $\boxed{89}$ $\boxed{C4}$ CF = 0 PF = 0 AF = 0 ZF = 1 SF = 1	$89_H = 1000\ 1001$ $C4_H = 1100\ 0100$ 1's complement of $C4_H = 0011\ 1011$ 2's complement of $C4_H = 0011\ 1011 + 1$ $= 0011\ 1100 = 3C_H$

Case ii continued ...	
After execution	**Subtraction**
A C $\boxed{C\,5}$ $\boxed{C\,4}$ C F = 1 P F = 1 A F = 1 Z F = 0 S F = 1	89_H = 1000 1001 $+3C_H$ = 0011 1100 $\overline{\quad\boxed{0}\ 1100\ 0101\quad}$ $\underset{\text{Carry}}{\text{Complement}}\downarrow$ C 5 $\boxed{1}$ Result = $C5_H$ CF = 1 $\boxed{\textit{Note : 2's complement of } C5_H = 3B_H}$

Note: *The 8085 microprocessor performs 2's complement subtraction. But after subtraction, it will complement the carry alone. In 2's complement subtraction, if CF =1, then the result is positive and if CF =0, then the result is negative. Since, the 8085 processor complements the carry after subtraction, here if CF = 0, then the result is positive and if CF = 1, then the result is negative. If the result is negative, then it will be in 2's complement form.*

One-byte instruction   **One machine cycle:** Opcode fetch - 4T

Register addressing

Total number of instructions = 7

SUB A   SUB B   SUB C   SUB D   SUB E   SUB H   SUB L

8.   SUI d8 $(A) \leftarrow (A) - d8$

The 8-bit data given in the instruction is subtracted from the A-register (accumulator). After subtraction, the result is stored in the A-register. All flags are affected.

Two-byte instruction   **Two machine cycles :**   Opcode fetch - 4T

Immediate addressing   Memory read - 3T

$\overline{\qquad\qquad\qquad\qquad\ 7T}$

Total number of instructions = 1

9.   SUB M $(A) \leftarrow (A) - (M)$ or $(A) \leftarrow (A) - ((HL))$

The content of the memory addressed by the HL pair is subtracted from the A-register. After subtraction, the result is stored in the A-register. All flags are affected.

One-byte instruction   **Two machine cycles :**   Opcode fetch - 4T

Register indirect addressing   Memory read - 3T

$\overline{\qquad\qquad\qquad\qquad\ 7T}$

Total number of instructions = 1

10.   SBB reg $(A) \leftarrow (A) - (reg) - CF$

The content of the register and the value of carry (before executing this instruction) are subtracted from the accumulator (A-register). After subtraction, the result is stored in the accumulator. All flags are affected. The register can be any one of the general purpose register A, B, C, D, E, H or L.

One-byte instruction   One machine cycle : Opcode fetch - 4T

Register addressing

Total number of instructions = 7

SBB A   SBB B   SBB C   SBB D   SBB E   SBB H   SBB L

11.   SBI d8   **(A) ← (A) – d8 – CF**

The 8-bit data given in the instruction and the value of carry (before executing this instruction) are subtracted from accumulator. After subtraction, the result is stored in the accumulator. All flags are affected.

Two-byte instruction	**Two machine cycles :**	Opcode fetch	-	4T
Immediate addressing		Memory read	-	3T
				7T

Total number of instructions = 1

---

12.   SBB M   **(A) ← (A) – (M) – CF   or   (A) ← (A) – ((HL)) – CF**

The content of the memory addressed by HL and the value of carry (before executing this instruction) are subtracted from accumulator (A-register). After subtraction, the result is stored in the A-register. All flags are affected.

One-byte instruction	**Two machine cycles:**	Opcode fetch	-	4T
Register indirect addressing		Memory read	-	3T
				7T

Total number of instructions = 1

---

13.   DAA

(DAA - Decimal Adjust Accumulator)

After BCD addition, the DAA instruction is executed to get the result in BCD. When DAA instruction is executed, the content of the accumulator is altered or adjusted as explained below :

i)   If the sum of the lower nibbles exceeds 09_H or auxiliary carry is set, then a correction 06_H (0110) is added to sum of lower nibbles.

ii)   If the sum of the upper nibbles exceeds 09_H or carry is set, then a correction 06_H (0110) is added to sum of upper nibble.

After executing this instruction all flags are modified to indicate the status of the result.

One-byte instruction	**One machine cycle:** Opcode fetch - 4T
Implied addressing	

Total number of instructions = 1

---

14.   DAD rp   **(HL) ← (HL) + (rp)**

(DAD - Double Addition)

The content of the register pair is added to the content of the HL pair. After addition, the result is stored in the HL pair. Only the carry flag is affected. The register pair can be BC, DE, HL or SP.

One-byte instruction	**Three machine cycles:**	Opcode fetch	-	4T
Register addressing		Bus idle	-	3T
		Bus idle	-	3T
				10T

Total number of instructions = 4

       DAD B   **DAD D**   **DAD H**   **DAD SP**

15. **INR reg** $(reg) \leftarrow (reg) + 01$

The content of the register is incremented by one. Except carry flag, all other flags are affected. The register can be any one of the general purpose register A, B, C, D, E, H or L.

Example : **INR B** $(B) \leftarrow (B) + 01$

The content of the B-register is incremented by one. The increment opertation is performed by adding 01_H to the content of B-register.

Before execution	Increment Operation	After execution
B 4 A	CF = 0 PF = 0 AF = 0 ZF = 0 SF = 0	$4A_H$ = 0100 1010 + 01_H = 0000 0001 ─────────── 0100 1011 4 B	B 4B	CF = 0 PF = 1 AF = 0 ZF = 0 SF = 0

One-byte instruction

Register addressing

One machine cycle: Opcode fetch - 4T

Total number of instructions = 7

INR A	INR B	INR C	INR D	INR E	INR H	INR L

16. **INR M** $(M) \leftarrow (M) + 01$ or $((HL)) \leftarrow ((HL)) + 01$

The content of the memory addressed by the HL pair is incremented by one. Except carry, all other flags are affected.

Example : **INR M** $(M) \leftarrow (M) + 01$

Let the content of the HL pair be $C00A_H$. Let the content of memory location $C00A_H$ be $C5_H$. The content of the memory location $C00A_H$ is incremented by one. The increment operation is performed by adding 01_H to the content of the memory.

Before execution	Increment Operation	After execution
H L C00A CF = 0 PF = 0 AF = 0 ZF = 0 SF = 0	Memory C 5 │ C00A A2 │ C00B 0 7 │ C00C	$C5_H$ = 1100 0101 + 01_H = 0000 0001 ─────────── 1100 0110 C 6	H L C00A CF = 0 PF = 1 AF = 0 ZF = 0 SF = 1	Memory C 6 │ C00A A2 │ C00B 0 7 │ C00C

One-byte instruction

Register indirect addressing

Three machine cycles : Opcode fetch - 4T

Memory read - 3T

Memory write - 3T

──────

10T

Total number of instructions = 1

17. **DCR reg** $(reg) \leftarrow (reg) - 01$

The content of the register is decremented by one. Except carry , all other flags are affected. The register can be A, B, C, D, E, H or L.

Example : DCR D $(D) \leftarrow (D) - 01$

The content of the D-register is decremented by one. The decrement operation is performed by subtracting 01_H from the content of the D-register.

Before execution	Decrement operation
D 60 CF = 0 PF = 0 AF = 0 ZF = 0 SF = 0	01_H = 0000 0001 1's complement of 01_H = 1111 1110 2's complement of 01_H = 1111 1110 + 1 = 1111 1111 = FF_H
After execution	60_H = 0110 0000 + FF_H = 1111 1111 ⎯⎯⎯⎯⎯⎯⎯⎯ 1 0101 1111 5 F Carry is discarded
D 5F CF = 0 PF = 1 AF = 0 ZF = 0 SF = 0	

One-byte instruction   **One machine cycle :** Opcode fetch - 4T

Register addressing

Total number of instructions = 7

DCR A	DCR B	DCR C	DCR D	DCR E	DCR H	DCR L

18. **DCR M**   **(M) ← (M) − 01   or   ((HL)) ← ((HL)) − 01**

The content of memory addressed by the HL pair is decremented by one. Except carry, all other flags are affected.

Example: DCR M   **(M) ← (M) − 01**

Let the content of the HL pair be 2010_H. Let the content of memory location 2010_H be FA_H. The content of memory location 2010_H is decremented by one.

Before execution	Decrement operation
H L 2010 Memory FA 2010 02 2011 CF = 0 PF = 0 AF = 0 ZF = 0 SF = 0	01_H = 0000 0001 1's complement of 01_H = 1111 1110 2's complement of 01_H = 1111 1110 + 1 = 1111 1111 = FF_H
After execution	
H L 2010 Memory F9 2010 02 2011 CF = 0 PF = 1 AF = 1 ZF = 0 SF = 1	FA_H = 1111 1010 + FF_H = 1111 1111 ⎯⎯⎯⎯⎯⎯⎯⎯ 1 1111 1001 F 9 Carry is discarded

One-byte instruction   **Three machine cycles :** Opcode fetch   - 4T

Register indirect addressing   Memory read   - 3T

                                                                     Memory write - 3T
                                                                                ⎯⎯⎯⎯
                                                                                   10T

Total number of instructions = 1

19.   INX rp   **(rp) ← (rp) + 01**

The content of the register pair is incremented by one. The register pair can be BC, DE, HL or SP. No flags are affected.

Example : INX H (HL) ← (HL) + 01	
The content of the HL pair is incremented by one.	
Before execution	**After execution**
H L	H L
00FF	0100

One-byte instruction   **One machine cycle :** Opcode fetch - 6T

Register addressing

Total number of instructions = 4

     INX B   INX D   INX H   INX SP

---

20.   DCX rp   **(rp) ← (rp) − 01**

The content of the register pair is decremented by one. The register pair can be BC, DE, HL or SP. No flags are affected.

Example : DCX SP (SP) ← (SP) − 01	
The content of the stack pointer is decremented by one.	
Before execution	**After execution**
S P	S P
1000	0FFF

One-byte instruction   **One machine cycle :** Opcode fetch - 6T

Register addressing

Total number of instructions = 4

     DCX B   DCX D   DCX H   DCX SP

3.6.2   Logical Instructions

1.   ANA reg   **(A) ← (A) & (reg)**

(& is the symbol used for logical AND operation)

The content of the register is logically ANDed bit by bit with the content of the accumulator. In bit by bit AND operation, the bit D_0 of register is ANDed with the bit D_0 of A-register, the bit D_1 of register is ANDed with bit D_1 of A-register, and so on. The register can be any one of the general purpose register A, B, C, D, E, H or L. After execution of the instruction, carry flag is always reset and auxiliary carry flag is always set. Other flags are altered (according to the results). After AND operation, result is stored in accumulator.

Example : ANA E (A) ← (A) & (E)		
The content of E-register is logically ANDed bit by bit with the content of accumulator.		
Before execution	**AND operation**	**After execution**

A   E   C F   =   0
 15   E 2   P F   =   0
         AF   =   0
         Z F   =   0
         S F   =   0
	15_H = 0001 0101
	
	 $E2_H$ = 1110 0010
	 ——————————
	      0000 0000
	       0 0
		A E C F = 0
		 00 E 2 P F = 1
		         AF = 1
		         Z F = 1
		         S F = 0

One-byte instruction   **One machine cycle:** Opcode fetch - 4T

Register addressing

Total number of instructions = 7

ANA A	ANA B	ANA C	ANA D	ANA E	ANA H	ANA L

2.   **ANI d8**   **(A) ← (A) & d8**

The 8-bit data given in the instruction is logically ANDed bit by bit with the content of the accumulator. The result is stored in the accumulator. After execution of this instruction, CF = 0 and AF = 1. Other flags are affected.

Two-byte instruction   **Two machine cycles :**   Opcode fetch - 4T

Immediate addressing   Memory read - 3T

                                                                  7T

Total number of instructions = 1

3.   **ANA M**   **(A) ← (A) & (M)**   or   **(A) ← (A) & ((HL))**

The content of the memory addressed by the HL pair is logically ANDed bit by bit with the content of the accumulator. The result is stored in the accumulator. After execution, CF = 0 and AF = 1. Other flags are affected.

Example : ANA M   **(A) ← (A) & (M)**

Let the content of HL be $105A_H$. Let the content of the memory location $105A_H$ be $4C_H$. The content of the memory location $105A_H$ is logically ANDed bit by bit with the content of the accumulator. The result is stored in the accumulator.

Before execution	AND operation	After execution
A = 27, HL = 105A, Memory: 14|1059, 4C|105A CF = 0 PF = 0 AF = 0 ZF = 0 SF = 0	27_H = 0010 0111 $4C_H$ = 0100 1100 0000 0100 0 4	A = 04, HL = 105A, Memory: 14|1059, 4C|105A CF = 0 PF = 0 AF = 1 ZF = 0 SF = 0

One-byte instruction   **Two machine cycles:**   Opcode fetch - 4T

Register indirect addressing   Memory read - 3T

                                                                   7T

Total number of instructions = 1

4.   **ORA reg**   **(A) ← (A) | (reg)**

(| is the symbol used for logical OR operation)

The content of the register is logically ORed bit by bit with the content of the accumulator. In bit by bit OR operation, the bit D_0 of the register is ORed with bit D_0 of the A-register, the bit D_1 of the register is ORed with bit D_1 of the A-register, and so on. The register can be any one of the general purpose register A, B, C, D, E, H or L. After execution of the instruction, both the carry and auxiliary flags are always reset (AF = 0, CF = 0). Other flags are modified (according to the result). After OR operation, the result is stored in the accumulator.

One-byte instruction   **One machine cycle:** Opcode fetch - 4T

Register addressing

Example : ORA B   **(A) ← (A) | (B)**

The content of the B-register is logically ORed bit by bit with the content of the accumulator.

Before execution	OR operation	After execution

A   B   C F = 0
04 7A P F = 0
          A F = 0
          Z F = 0
          S F = 0
	04_H = 0000 0100
	$7A_H$ = 0111 1010
	———————
	0111 1110
	———————
	7 E
		A B C F = 0
		7E 7A P F = 1
		          A F = 0
		          Z F = 0
		          S F = 0

Total number of instructions = 7

ORA A   ORA B   ORA C   ORA D   ORA E   ORA H   ORA L

5.   **ORA M**   **(A) ← (A) | (M)   or   (A) ← (A) | ((HL))**

The content of the memory addressed by the HL pair is logically ORed bit by bit with the content of the accumulator. The result is stored in the accumulator. After execution, CF = AF = 0. Other flags are affected.

Example :   ORA M   **(A) ← (A) | (M)**

Let the content of the HL pair be 2050_H. Let the content of memory location 2050_H be $1B_H$. The content of the memory location 2050_H is logically ORed bit by bit with the content of the accumulator. The result is stored in the accumulator.

Before execution	OR operation	After execution

A   HL   Memory
45   2050   1B 2050
             07 2051
C F = 0
P F = 0
A F = 0
Z F = 0
S F = 0
	45_H = 0100 0101
	$1B_H$ = 0001 1011
	———————
	0101 1111
	———————
	5 F
		A HL Memory
		5F 2050 1B 2050
		             07 2051
		C F = 0
		P F = 1
		A F = 0
		Z F = 0
		S F = 0

One-byte instruction   **Two machine cycles:**   Opcode fetch -   4T

Register indirect addressing   Memory read -   3T
                                                          ————
                                                          7 T

Total number of instructions = 1

6.   **ORI d8**   **(A) ← (A) | d8**

The 8-bit data given in the instruction is logically ORed bit by bit with the content of the accumulator. The result is stored in the accumulator. After execution of this instruction, CF = AF = 0. Other flags are affected.

Two-byte instruction   **Two machine cycles :**   Opcode fetch -   4T

Immediate addressing   Memory read -   3T
                                                          ————
                                                          7 T

Total number of instructions = 1

7.   **XRA reg   (A) ← (A) ^ (reg)**

(^ is the symbol used for logical EXCLUSIVE-OR operation).

The content of the register is logically EXCLUSIVE-ORed bit by bit with the content of the accumulator. In bit by bit EXCLUSIVE-OR operation, the bit D_0 of register is EXCLUSIVE-ORed with bit D_0 of A-register, the bit D_1 of register is EXCLUSIVE-ORed with bit D_1 of A-register, and so on. The result is stored in the accumulator. The register can be any one of the general purpose register A, B, C, D, E, H or L. After execution AF = CF = 0. Other flags are modified (according to the result).

Example : XRA A	$(A) \leftarrow (A) \wedge (A)$	
The content of the A-register is EXCLUSIVE-ORed bit by bit with the content of the A-register itself.		
Before execution	**EXCLUSIVE-OR operation**	**After execution**
A CF = 1 74 PF = 0 AF = 1 ZF = 0 SF = 1	74_H = 0111 0100 74_H = 0111 0100 <u>0000 0000</u>	A CF = 0 00 PF = 1 AF = 0 ZF = 1 SF = 0

One-byte instruction   **One machine cycle**: Opcode fetch - 4T

Register addressing

Total number of instructions = 7

XRA A	XRA B	XRA C	XRA D	XRA E	XRA H	XRA L

8.   **XRI d8** $(A) \leftarrow (A) \wedge d8$ or $(A) \leftarrow (A) \wedge d8$

The 8-bit data given in the instruction is logically EXCLUSIVE-ORed bit by bit with the content of the accumulator. The result is stored in the accumulator. After execution of this instruction, CF = AF = 0. Other flags are affected.

Two-byte instruction   **Two machine cycles :**   Opcode fetch - 4T

Immediate addressing   Memory read - <u>3T</u>

                                                      7T

Total number of instructions = 1

9.   **XRA M** $(A) \leftarrow (A) \wedge (M)$ or $(A) \leftarrow (A) \wedge ((HL))$

The content of the memory addressed by the HL pair is logically EXCLUSIVE-ORed bit by bit with the content of accumulator. The result is stored in accumulator. After execution, CF = AF = 0. Other flags are affected.

Example : XRA M	$(A) \leftarrow (A) \wedge (M)$	
Let the content of the HL pair be $805A_H$. Let the content of memory location $805A_H$ be $C4_H$. The content of the memory location $805A_H$ is logically EXCLUSIVE-ORed bit by bit with the content of the accumulator. The result will be in the accumulator.		
Before execution	**Exclusive-OR operation**	**After execution**
A H L Memory B7 805A 1 C 8059 CF = 1 C 4 805A PF = 1 2 0 805B AF = 1 51 805C ZF = 0 SF = 1	$B7_H$ = 1011 0111 $C4_H$ = 1100 0100 <u>0111 0011</u> 7 3	A H L Memory 73 805A 1 C 8059 CF = 0 C 4 805A PF = 0 2 0 805B AF = 0 51 805C ZF = 0 SF = 0

One-byte instruction   **Two machine cycles :** Opcode fetch - 4T

Register indirect addressing   Memory read - <u>3T</u>

                                                      7T

Total number of instructions = 1

10.   CMP reg   (A) – (reg) ⇒ **Modify flags**

The content of the register is compared with the accumulator. The comparison is performed by subtracting the content of register from the A-register. The subtraction is performed in the ALU, and the result is used to modify flags and then the result is discarded (i.e., it is not stored in any register). After execution of this instruction, the content of accumulator and the register are not altered. All flags are affected by this instruction. The register can be any one of the general purpose register A, B, C, D, E, H or L.

The status of carry and zero flag after comparison are given below :

   i)   If (A) < (reg) then the carry flag is set (i.e., CF = 1)

   ii) If (A) > (reg) then the carry flag is reset or cleared (i.e., CF = 0)

   iii) If (A) = (reg) then the zero flag is set (i.e., ZF = 1).

Example : CMP B	**(A) – (B) ⇒ Modify flags.**	

The content of the B-register is compared with the accumulator. The comparison is performed by subtracting the content of the B-register from the content of the accumulator. The subtraction is performed in the ALU and the result is used to modify the flags and then discarded. The content of the accumulator and the B-register are not altered.

Before execution	**Comparison**	**After execution**

A   B
[15] [C2]
C F = 0
P F = 0
A F = 0
Z F = 0
S F = 0
	$C2_H = 1100\ 0010$
	1's complement of $C2_H$ = 0011 1101
	2's complement of $C2_H$ = 0011 1101+1
	         = 0011 1110 = $3E_H$
	
	15_H = 0001 0101
	+$3E_H$ = 0011 1110
	Complement [0] 0101 0011
	Carry ↓ 5 3
	       [1]
		A B
		[15] [C2]
		
		C F = 1
		P F = 1
		A F = 1
		Z F = 0
		S F = 0

One-byte instruction   **One machine cycle:** Opcode fetch - 4T

Register addressing

Total number of instructions = 7

CMP A   CMP B   CMP C   CMP D   CMP E   CMP H   CMP L

11.   CPI d8   (A) – d8 ⇒ **Modify flags.**

The 8-bit data given in the instruction is compared with the accumulator. The comparison is performed by subtracting the 8-bit data from the A-register. The subtraction is performed in ALU and the result is used to modify flags and then discarded. After execution of the instruction, the content of the accumulator is not altered. All flags are affected.

The status of carry and zero flag after comparision are given below :

   i)   If (A) < d8 then the carry flag is set (i.e., CF = 1)

   ii) If (A) > d8 then the carry flag is reset or cleared (i.e., CF = 0)

   iii) If (A) = d8 then the zero flag is set (i.e., ZF = 1).

Two-byte instruction   **Two machine cycles :**   Opcode fetch - 4T

Immediate addressing   Memory read   - 3T

                                                                              ————
                                                                               7T

Total number of instructions = 1

12.   **CMP M**   **(A) – (M) $\Rightarrow$ Modify flags or (A) – ((HL)) $\Rightarrow$ Modify flags.**

The content of the memory addressed by HL pair is compared with the accumulator. The comparison is performed by subtracting the content of memory from the A-register. The subtraction is performed in the ALU and the result is used to modify flags and then discarded. After execution of the instruction, the content of the accumulator and the memory are not altered. All flags are affected by this instruction.

The status of carry and zero flag after comparison are given below:

    i)   If (A) < (M) then the carry flag is set (i.e., CF = 1).

    ii)   If (A) > (M) then the carry flag is reset or cleared (i.e., CF = 0).

    iii) If (A) = (M) then the zero flag is set (i.e., ZF = 1).

Example : CMP M

Let the content of the HL pair be C050$_H$. Let the content of the memory location C050$_H$ be 7A$_H$. The content of the memory location C050$_H$ is compared with the content of the accumulator. Only flags are altered. The content of the accumulator and the memory remains the same.

Before execution	Comparison	After execution
A HL \[25\] \[C050\] Memory \[7A\] C050 \[10\] C051 C F = 0 P F = 0 A F = 0 Z F = 0 S F = 0	$25_H = 0010\ 0101$ $7A_H = 0111\ 1010$ 1'complement of $7A_H = 1000\ 0101$ 2'complement of $7A_H = 1000\ 0101 +1$ $= 1000\ 0110 = 86_H$ $25_H = 0010\ 0101$ $+86_H = 1000\ 0110$ ——————— \[0\]$1010\ 1011$ Complement $\quad\downarrow$ Carry $\quad$ A B \[1\]	A HL \[25\] \[C050\] Memory \[7A\] C050 \[10\] C051 C F = 1 P F = 0 A F = 0 Z F = 0 S F = 1

One-byte instruction   **Two machine cycles:** Opcode fetch   -   4 T

Register indirect addressing   Memory read   -   <u>3 T</u>

                                                        <u>7 T</u>

Total number of instructions = 1

---

13.   **CMA**   **(A) ← $\left(\overline{\text{A}}\right)$**

(CMA - Complement Accumulator)

The content of the accumulator is complemented. No flags are affected.

One-byte Instruction   **One machine cycle:** Opcode fetch - 4T

Implied addressing

---

14.   **STC**   **(CF) ←1**

(STC - Set Carry)

The carry flag is set to 1. Only carry flag is affected by this instruction.

One-byte instruction   **One machine cycle :** Opcode fetch - 4T

Implied addressing

15.   CMC $(CF) \leftarrow (\overline{CF})$

(CMC - Complement Carry)

The carry flag is complemented. Only the carry flag is affected by this instruction.

One-byte instruction   **One machine cycle:** Opcode fetch - 4T

Implied addressing

---

16.   RLC $D_{n+1} \leftarrow D_n$; $D_0 \leftarrow D_7$ and $(CF) \leftarrow D_7$

(RLC - Rotate Accumulator Left to carry)

The content of the A-register is rotated left by one bit and the left most bit of A-register is rotated to the carry. [The left most bit is most significant bit.] Only the carry flag is affected.

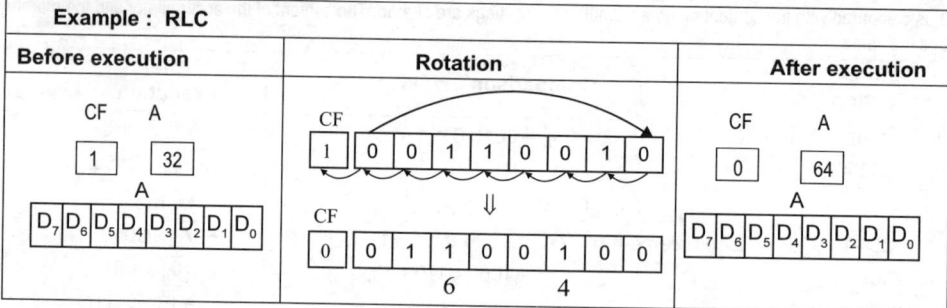

One-byte instruction   **One machine cycle:** Opcode fetch - 4T

Implied addressing

---

17.   RRC $D_n \leftarrow D_{n+1}$; $D_7 \leftarrow D_0$ and $(CF) \leftarrow D_0$

(RRC - Rotate Accumulator Right to Carry)

The content of A-register is rotated right by one bit and the right most bit of A-register is rotated to carry. [The right most bit is least significant bit.] Only carry flag is affected.

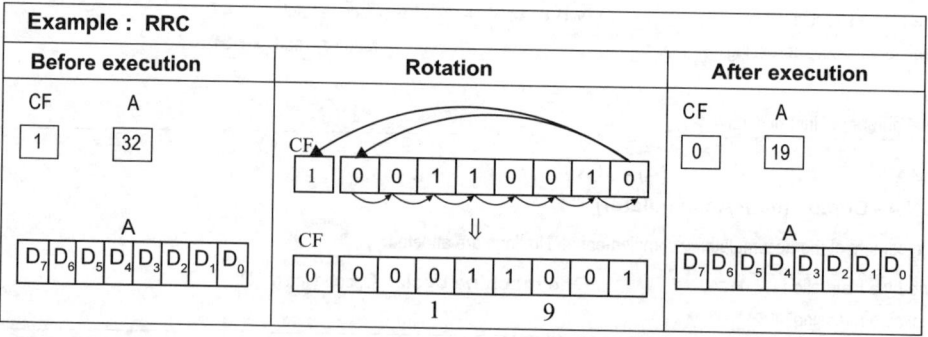

One-byte instruction   **One machine cycle:** Opcode fetch - 4T

Implied addressing

---

18.   RAR $D_n \leftarrow D_{n+1}$; $D_7 \leftarrow (CF)$ and $(CF) \leftarrow D_0$

(RAR - Rotate Accumulator Right through carry)

The content of the A-register along with the carry is rotated right by one bit. Here the carry is moved to the most significant bit position (D_7) and the least significant bit (D_0) is moved to the carry. Only the carry flag is affected.

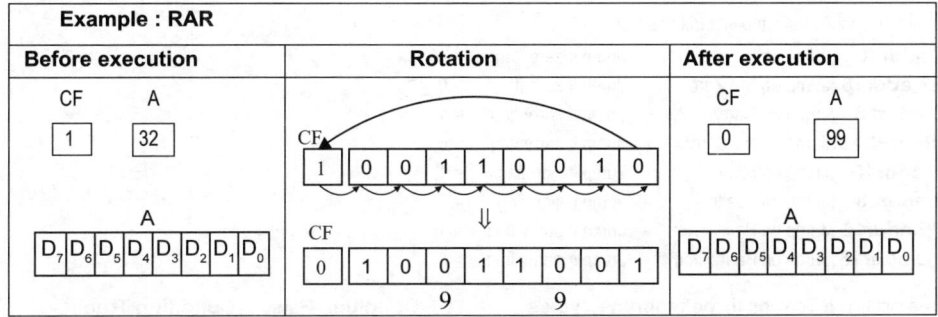

Example : RAR		
Before execution	**Rotation**	**After execution**

One-byte instruction

Implied addressing

One machine cycle: Opcode fetch - 4T

---

19. **RAL** $D_{n+1} \leftarrow D_n$; $D_0 \leftarrow (CF)$ and $(CF) \leftarrow D_7$

(RAL - Rotate Accumulator Left through carry)

The content of the A-register along with the carry is rotated left by one bit. Here the carry is moved to the least significant bit position (D_0) and the most significant bit (D_7) is moved to the carry. Only the carry flag is affected.

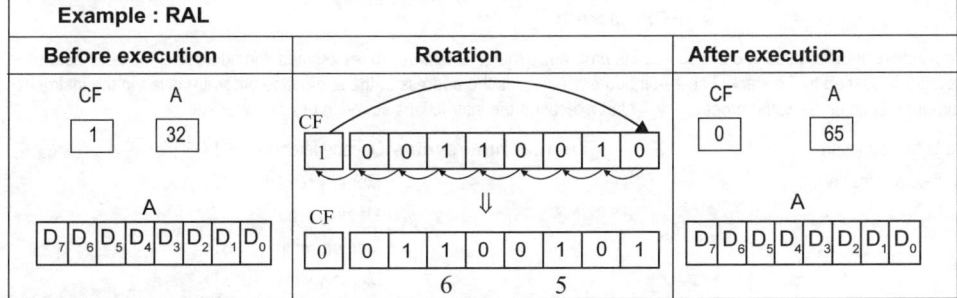

Example : RAL		
Before execution	**Rotation**	**After execution**

One-byte instruction

Implied addressing

One machine cycle: Opcode fetch - 4T

3.6.3 Branching Instructions

1. **JMP addr16** **(PC) ← addr16**

It is unconditional jump instruction. When this instruction is executed, the address given in the instruction is moved to the program counter. Now, the processor starts executing the instructions stored in this address.

Three-byte instruction

Immediate addressing

Three machine cycles:	Opcode fetch -	4 T
	Memory read -	3 T
	Memory read -	3 T
		10T

2. **J <condition> addr16**

If <condition> is TRUE then,

(PC) ← addr16

It is conditional jump instruction. The conditional jump instruction will check a flag condition. If the flag condition is true, then the address given in the instruction is moved to the program counter. Thus the program control is branched to the jump address. If the flag condition is false, then the next instruction is executed.

There are eight conditional jump instructions.

i)   **JZ addr16**   ;Jump on Zero   -   Jump if zero flag   = 1.
ii) **JNZ addr16** ;Jump on Not Zero   -   Jump if zero flag   = 0.
iii) **JC addr16**   ;Jump on Carry   -   Jump if carry flag   = 1.
iv) **JNC addr16** ;Jump on No Carry   -   Jump if carry flag   = 0.
v)   **JM addr16**   ;Jump on Minus   -   Jump if sign flag   = 1.
vi) **JP addr16**   ;Jump on Positive   -   Jump if sign flag   = 0.
vii) **JPE addr16** ;Jump on Parity Even -   Jump if parity flag = 1.
viii) **JPO addr16** ;Jump on Parity Odd -   Jump if parity flag = 0.

Three-byte instruction **Two or three machine cycles:**	**Condition False**	**Condition True**
Immediate addressing	Opcode fetch - 4T	Opcode fetch - 4T
	Memory read - 3T	Memory read - 3T
		Memory read - 3T
	7T	10T

3.   CALL addr16 $(SP) \leftarrow (SP) - 1$; $((SP)) \leftarrow (PC)_H$

$(SP) \leftarrow (SP) - 1$; $((SP)) \leftarrow (PC)_L$

$(PC) \leftarrow addr16$

It is unconditional CALL used to call a subroutine program. When this instruction is executed, the address of the next instruction in the program counter is pushed to the stack. The 16-bit address (which is the address of the subroutine program) given in the instruction is loaded in the program counter. Now, the processor will start executing the instructions stored in this call address.

Three-byte instruction	**Five machine cycles:** Opcode fetch	-	6T
Immediate addressing	Memory read	-	3T
	Memory read	-	3T
	Memory write	-	3T
	Memory write	-	3T
			18T

4.   C<condition>   addr16

If <condition> is TRUE then,

$(SP) \leftarrow (SP) - 1$; $((SP)) \leftarrow (PC)_H$

$(SP) \leftarrow (SP) - 1$; $((SP)) \leftarrow (PC)_L$

$(PC) \leftarrow addr16$

It is conditional subroutine call instruction. The conditional CALL instruction will check for a flag condition. If the flag condition is true, then the address of the next instruction is pushed to the stack and the call address (address given in the instruction) is loaded in the program counter. Now, the processor will start executing the instructions stored in this address. If the flag condition is false, then the next instruction is executed.

There are eight conditional CALL instructions. These are:

i)   **CZ addr16**   ;Call on Zero   -   Call if zero flag   = 1.
ii) **CNZ addr16** ;Call on Not Zero   -   Call if zero flag   = 0.
iii) **CC addr16**   ;Call on Carry   -   Call if carry flag   = 1.
iv) **CNC addr16** ;Call on No Carry   -   Call if carry flag   = 0.
v)   **CM addr16**   ;Call on Minus   -   Call if sign flag   = 1.
vi) **CP addr16**   ;Call on Positive   -   Call if sign flag   = 0.
vii) **CPE addr16** ;Call on Parity Even -   Call if parity flag = 1.
viii) **CPO addr16** ;Call on Parity Odd -   Call if parity flag = 0.

Three-byte instruction **Two or five machine cycles:**

Immediate addressing

	Condition False		Condition True	
	Opcode fetch	- 6T	Opcode fetch	- 6T
	Memory read	- 3T	Memory read	- 3T
		9T	Memory read	- 3T
			Memory write	- 3T
			Memory write	- 3T
				18T

5.   **RET** $(PC)_L \leftarrow ((SP))$; $(SP) \leftarrow (SP) + 1$

                       $(PC)_H \leftarrow ((SP))$; $(SP) \leftarrow (SP) + 1$

(RET - Return to the main program)

It is an unconditional return instruction. This instruction is placed at the end of the subroutine program, in order to return to the main program. When this instruction is executed, the top of the stack is poped to (loaded in) the program counter .

> **Note:** *While calling the subroutine using CALL instruction, the return address of the main program is pushed to the stack. The return instruction, (RET) pops that to the program counter. Thus the processor resumes the execution of main program.*

One-byte instruction

Register indirect addressing

Three machine cycles:	Opcode fetch	- 4 T
	Memory read	- 3 T
	Memory read	- 3 T
		10 T

6.   **R<condition>**

If <condition> is TRUE then,

$(PC)_L \leftarrow ((SP))$; $(SP) \leftarrow (SP) + 1$

$(PC)_H \leftarrow ((SP))$; $(SP) \leftarrow (SP) + 1$

It is conditional return instruction.

In a conditional return instruction a flag condition is tested. If the flag condition is true, then the program control return to main program by poping the top of the stack to the program counter . If the flag condition is false, then the next instruction is executed.

There are eight conditional return instructions:

i)	**RZ**	;Return on Zero	-	Return if zero flag	= 1.
ii)	**RNZ**	;Return on Not Zero	-	Return if zero flag	= 0.
iii)	**RC**	;Return on Carry	-	Return if carry flag	= 1.
iv)	**RNC**	;Return on No Carry	-	Return if carry flag	= 0.
v)	**RM**	;Return on Minus	-	Return if sign flag	= 1.
vi)	**RP**	;Return on Positive	-	Return if sign flag	= 0.
vii)	**RPE**	;Return on Parity Even	-	Return if parity flag	= 1.
viii)	**RPO**	;Return on Parity Odd	-	Return if parity flag	= 0.

One-byte instruction   **One or three machine cycles:**

Register indirect addressing

	Condition False		Condition True	
	Opcode fetch - 6T		Opcode fetch	- 6T
			Memory read	- 3T
			Memory read	- 3T
				12T

7.   RST n

It is a restart instruction. The restart instructions are also called software interrupts. Each restart instruction has a vector address. The vector address is fixed by the manufacturer (INTEL).

When a restart instruction is executed, the content of the program counter is pushed to the stack and the vector address is loaded in the program counter. The vector address is internally generated (computed) by the processor. The vector address for RST n is obtained by multiplying n by 8. Thus the program control is branched to a subroutine program stored in this vector address.

One-byte instruction	**Three machine cycles:** Opcode fetch - 6 T
Register indirect addressing	Memory write - 3 T
	Memory write - 3 T
	12T

There are eight restart instructions.

RST 0   RST 1   RST 2   RST 3   RST 4   RST 5   RST 6   RST 7

The vector addresses for the restart instructions are listed in the table given below :

Restart instruction	Vector address	Computation of vector address
RST 0	0000_H	$0 \times 8 = \quad 0_{10} \quad = \quad 0_H$
RST 1	0008_H	$1 \times 8 = \quad 8_{10} \quad = \quad 8_H$
RST 2	0010_H	$2 \times 8 = \quad 16_{10} \quad = \quad 10_H$
RST 3	0018_H	$3 \times 8 = \quad 24_{10} \quad = \quad 18_H$
RST 4	0020_H	$4 \times 8 = \quad 32_{10} \quad = \quad 20_H$
RST 5	0028_H	$5 \times 8 = \quad 40_{10} \quad = \quad 28_H$
RST 6	0030_H	$6 \times 8 = \quad 48_{10} \quad = \quad 30_H$
RST 7	0038_H	$7 \times 8 = \quad 56_{10} \quad = \quad 38_H$

8.   PCHL   **(PC) ← (HL)**

The content of the HL register pair is moved to the program counter. Since this instruction alters the content of the program counter, the program control is transferred to a new address. This instruction is used by the system designer to implement the system subroutine to execute a program.

One-byte instruction	**One machine cycle:** Opcode fetch - 6T
Implied addressing	

3.6.4 Machine Control Instructions

1.   DI

(DI - Disable Interrupts)

When this instruction is executed, all the interrupts except TRAP are disabled. [When the interrupts are disabled the processor will not accept or recognize the interrupt request made by the external devices through the interrrupt pins. When the processor is doing an emergency work, it can execute DI instruction to prevent the interrupts from interrupting the processor.]

One-byte instruction   **One machine cycle:** Opcode fetch - 4T

2.   **EI**

     (EI - Enable Interrupts)

         This instruction is used (or executed) to allow the interrupts after disabling. (The interrupts except TRAP are disabled after processor reset or after execution of DI instruction. When we want to allow the interrupts, we have to execute EI instructions.)

         One-byte instruction   **One machine cycle:** Opcode fetch - 4T

---

3.   **SIM**

     (SIM - Set Interrupt Mask)

         The SIM instruction is used to mask the hardware interrupts RST 7.5, RST 6.5 and RST 5.5. It is also used to send data through the SOD line. (SOD: Serial Output Data pin of the 8085 processor.) The execution of SIM instruction uses the content of the accumulator to perform the following functions:

         i)   Program the interrupt mask for the hardware interrupts RST 5.5, RST 6.5 and RST 7.5.

         ii)   Reset the edge-triggered RST 7.5 input latch.

         iii)   Load the SOD output latch.

         The bits in the accumulator before execution of the SIM instruction are defined as shown in the Fig. 3.12.

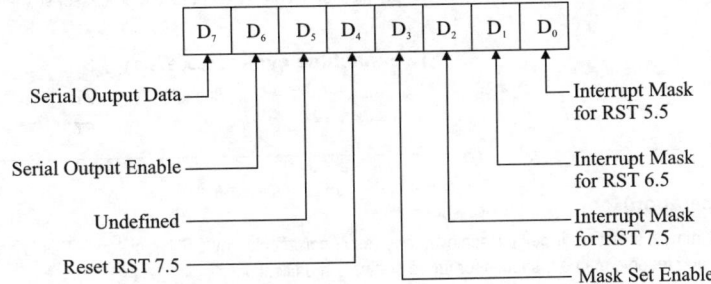

Fig. 3.12: Accumulator content before execution of SIM instruction.

         If the mask set enable bit is set to "1" then the interrupt mask bits for RST 7.5, RST 6.5 and RST 5.5 (D_0, D_1 and D_2) are recognized and if it is "0" then these bits are not recognized by the processor. The interrupt mask bits D_0, D_1 and D_2 can be independently set to "1" to mask the particular interrupt and reset to "0" to unmask the particular interrupt.

         If the bit D_4 is set to "1", then an internal flip-flop is reset to "0" in order to disable the RST 7.5 interrupt. If the serial output enable is "1", the serial output data is sent to the SOD pin.

         One-byte Instruction   **One machine cycle:** Opcode fetch - 4T

---

4.   **RIM**

     (RIM - Read Interrupt Mask)

         The RIM instruction is used to check whether an interrupt is masked or not. It is also used to read data from the SID line. (SID: Serial Input Data pin of 8085 processor).

         When a RIM instruction is executed, the accumulator is loaded with 8-bit data. The 8-bit data in the accumulator (content of accumulator) can be interpretted as shown in Fig. 3.13.

         Bits D_0, D_1 and D_2 provide the mask status of the RST 5.5, RST 6.5 and RST 7.5 interrupts respectively. If the mask bit corresponding to a particular RST is "1", then the interrupt is masked and if the mask bit is "0" then the interrupt is unmasked.

         If the interrupt enable bit (D_3) is "0", the 8085's maskable interrupts are disabled. The interrupts are enabled if this bit is "1".

         A "1" in a particular interrupt pending bit indicates that an interrupt is being requested on the identified RST line. When this bit is "0", no interrupt is waiting to be serviced. The serial input data (bit D_7) indicate the value of the signal at the SID pin.

         One-byte instruction   **One machine cycle:** Opcode fetch - 4T

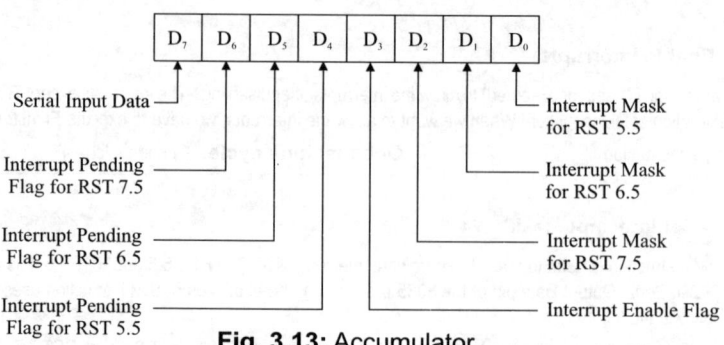

Fig. 3.13: Accumulator.

5.   **HLT**
      (HLT - Halt program Execution)

      This instruction is placed at the end of the program. When this instruction is executed, the processor suspends program execution and bus will be in idle state.

One-byte instruction   **Two machine cycle:** Opcode fetch - 3T
                                                    Bus idle   - 2T
                                                                   ——
                                                                   5T

6.   **NOP**
      (NOP - No operation)

      The NOP is a dummy instruction, it neither achieves any result nor affects any CPU registers. This is an useful instruction for producing software delay and reserve memory spaces for future software modifications.

One-byte instruction   **One machine cycle :** Opcode fetch - 4T

3.7 TIMING DIAGRAM OF 8085 INSTRUCTIONS

The 8085 instructions is one to five machine cycles. (Refer Table 3.1 for the machine cycles of instructions.) Actually, the execution of an instruction is the execution of the machine cycles of that instruction in a predefined order. Therefore, from the knowledge of the timing diagrams of machine cycles, the timing diagram of an instruction can be obtained.

The machine cycles of an 8085 instuction can be divided into two parts as shown below:

Machine cycles of an instruction

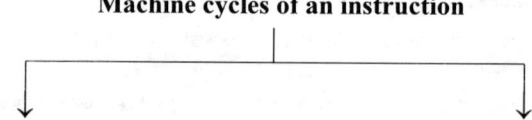

Machine cycles to fetch instruction
          bytes from memory.

One-byte instruction   : **Opcode fetch**

Two-byte instruction   : **Opcode fetch**
                              + memory read

Three-byte instruction   : **Opcode fetch**
                              + memory read
                              + memory read

Additional machine cycles for external read/write with memory/IO in order to complete instruction execution. These machine cycles depend on instruction execution logic.

Based on the execution of the machine cycles, the instructions can be classified as shown below:

Case(i)   :   *1-byte, 1-cycle*   -   Opcode fetch.

Case(ii)   :   *1-byte, 2-cycle*   -   Opcode fetch + memory read/write.

Case(iii)   :   *1-byte, 3-cycle*   -   Opcode fetch + memory read/write (or Bus idle)
                                         +   memory read/write (or Bus idle).

Case(iv)   :   *1-byte, 5-cycle*   -   Opcode fetch + memory read + memory read
                                           + memory write + memory write.

Case(v)   :   *2-byte, 2-cycle*   -   Opcode fetch + memory read (read second byte of instruction.)

Case(vi)   :   *2-byte, 3-cycle*   -   Opcode fetch + memory read (read second byte of instruction)
                                           + memory read/write/or IO read/write.

Case(vii)   :   *3-byte, 3-cycle*   -   Opcode fetch + memory read (read second byte of instruction)
                                           + memory read (read third byte of instruction.)

Case(viii)   :   *3-byte, 4-cycle*   -   Opcode fetch + memory read (read second byte of instruction)
                                           + memory read (read third byte of instruction)
                                           + memory read/write.

Case(ix)   :   *3-byte, 5-cycle*   -   Opcode fetch + memory read (read second byte of instruction)
                                           + memory read (read third byte of instruction)
                                           + memory read/write + memory read/write.

The timing diagram of an instruction is obtained by drawing the timing diagrams of the machine cycles of that instruction one by one in the order of execution. The timing diagrams of few instructions are presented from Figs. 3.14 to 3.20.

3.7.1 Timing Diagram of STA Instruction

The **"STA addr16"** instruction is used to store the content of the accumulator to a memory location. This instruction employs direct addressing. Let the content of the accumulator be $C7_H$ and it is desired to store the content of the accumulator to a memory location $526A_H$.

The STA addr16 instruction is a three byte instruction. The first byte is the opcode of the instruction 32_H. The second byte is low byte address $6A_H$ and the third byte is high byte address 52_H. Let the three bytes of the instructions be stored in memory locations $41FF_H$, 4200_H and 4201_H.

In order to execute this instruction, the 8085 microprocessor will first execute opcode fetch machine cycle to get the opcode, followed by two memory read cycles to read the address of data (i.e., to read second and third byte of instruction).

Then, the processor executes the memory write cycle to store the content of the accumulator in the memory. The status of various signals during execution of this instruction are shown in Fig. 3.14.

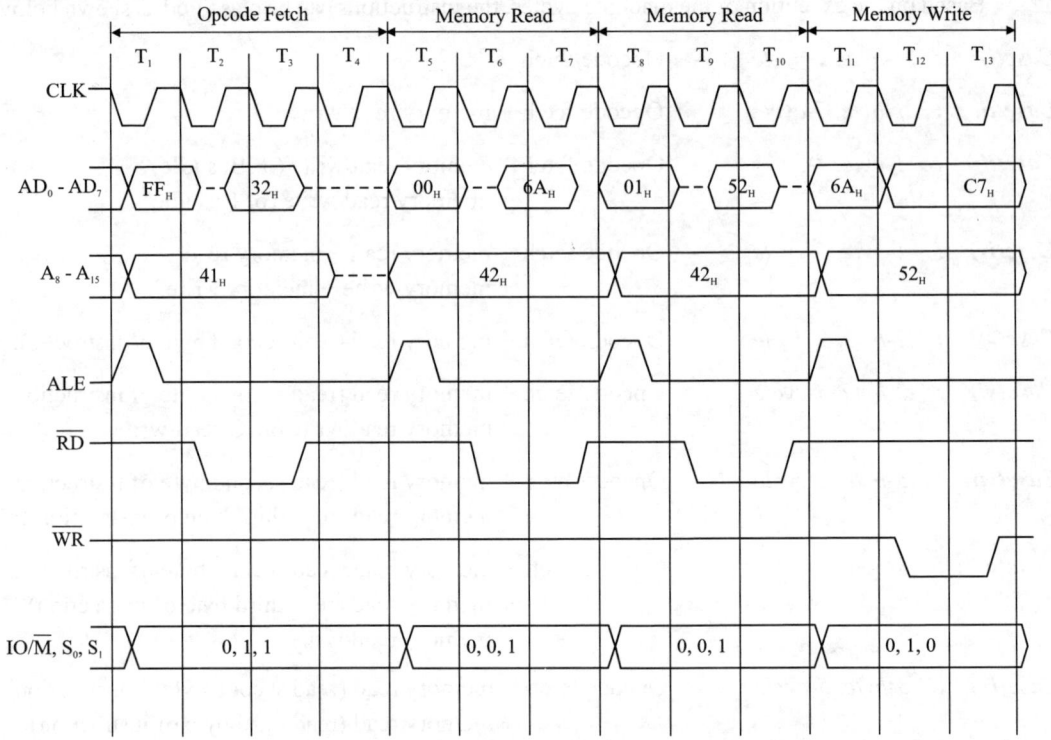

Fig. 3.14: Timing diagram of STA 526AH instruction.

3.7.2 Timing Diagram of PUSH Instruction

The **"PUSH rp"** instruction is used to store the content of a register pair in the stack memory. This instruction employs register indirect addressing using Stack Pointer (SP). Let us consider PUSH B instruction. On execution of this instruction, the content of the BC pair is pushed to the stack. Let the content of the BC pair be $E25D_H$ and the content of SP be $A100_H$.

The PUSH rp is one-byte instruction and it is the opcode of the instruction. The opcode of PUSH B instruction is $C5_H$ and let it be stored in memory location $C010_H$. In order to execute this instruction, the processor will first execute the opcode fetch cycle to get the opcode $C5_H$. Then the processor executes two memory write cycles to store the content of the BC pair in the stack memory. The status of various signals during execution of this instruction are shown in Fig. 3.15.

During the memory write cycles in PUSH rp instruction, the content of the SP is used as the memory address. In the first write cycle, the content of the SP is decremented by one ($A100_H - 1 = A0FF_H$) and output on the address lines and in this address, the content of B-register ($E2_H$) is stored. In the second write cycle, the content of the SP is again decremented by one ($A0FF_H - 1 = A0FE_H$) and output on the address lines and in this address, the content of C-register ($5D_H$) is stored.

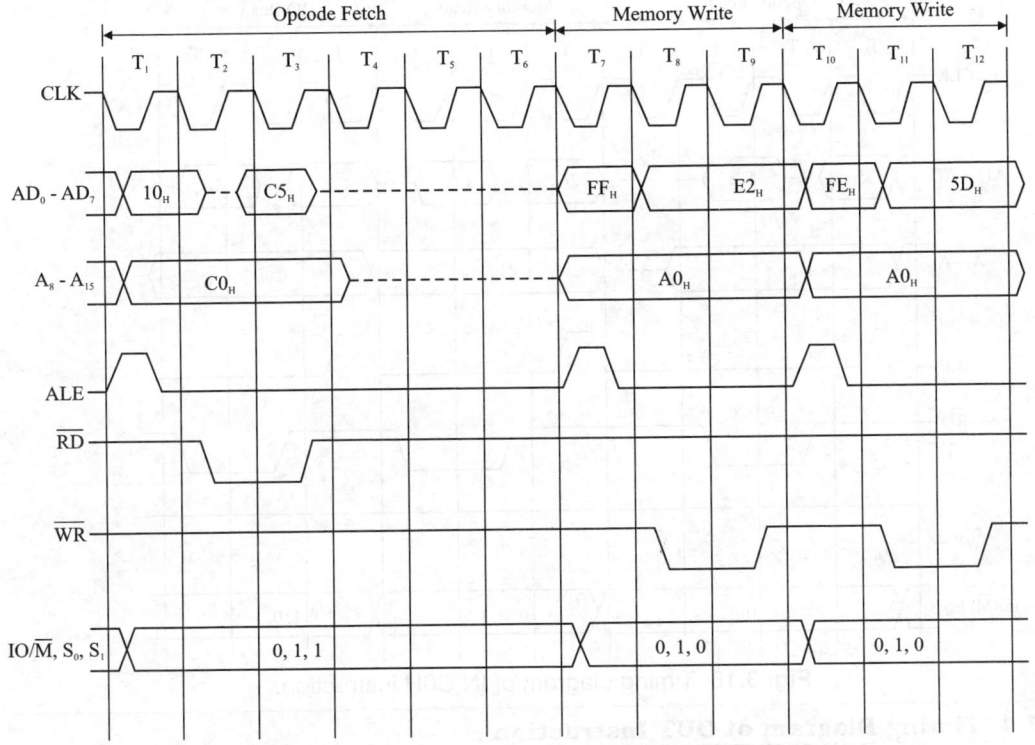

Fig. 3.15: Timing diagram of PUSH B instruction.

3.7.3 Timing Diagram of IN Instruction

The **"IN addr8"** instruction is used to read the content of an IO-mapped device/port and store in the accumulator. For addressing IO-mapped devices the 8085 microprocessor employs 8-bit address. Let the 8-bit address of the IO port be $C0_H$ and the content of IO port be $5E_H$.

The IN addr8 instruction, is a two-byte instruction. The first byte is the opcode of the instruction DB_H and the second byte is the IO port address $C0_H$. Let the two bytes of the instruction be stored in memory locations 4125_H and 4126_H.

In order to execute this instruction, the 8085 microprocessor will first execute the opcode fetch machine cycle to get the opcode, followed by the memory read cycle to read the IO port address (i.e., to read the second byte of the instruction.)

Then, the processor executes IO read cycle to read the content of the IO port. The status of various signals during execution of this instruction are shown in Fig. 3.16.

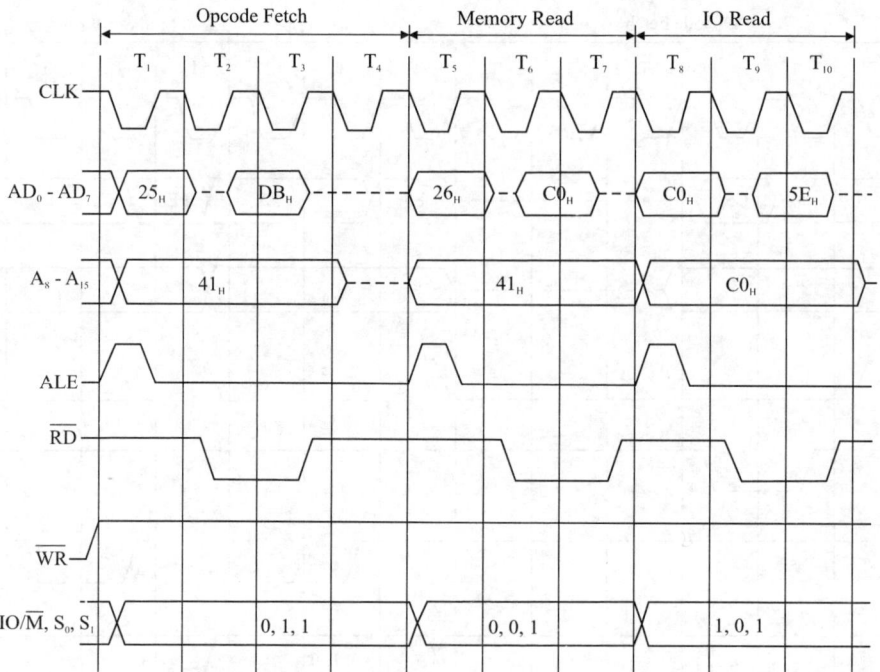

Fig. 3.16: Timing diagram of IN C0H instruction.

3.7.4 Timing Diagram of OUT Instruction

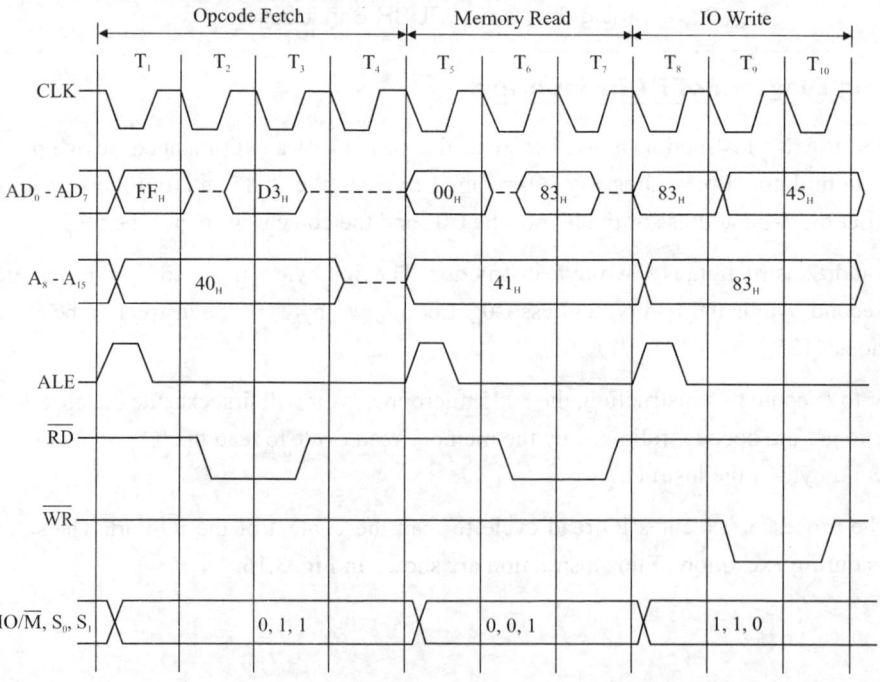

Fig. 3.17: Timing diagram of OUT 83H instruction.

The **"OUT addr8"** instruction is used to output the content of the accumulator to the IO-mapped device/port. For addressing IO-mapped devices, the 8085 microprocessor employs 8-bit address. Let the 8-bit address of the IO port be 83_H and the content of the accumulator be 45_H.

The OUT addr8 instruction is a two-byte instruction. The first byte is the opcode of the instruction $D3_H$, and the second byte is the IO port address 83_H. Let the two bytes of instruction be stored in memory locations $40FF_H$ and 4100_H.

In order to execute this instruction, the 8085 microprocessor will first execute the opcode fetch machine cycle to get the opcode, followed by the memory read cycle to read the IO port address. (i.e., to read the second byte of the instruction.) Then the processor executes the IO write cycle to write the content of the accumulator to the IO port. The status of various signals during execution of this instruction are shown in Fig. 3.17.

3.7.5 Timing Diagram of INR M Instruction

The **"INR M"** instruction is used to increment the content of a memory location. This instruction employs register indirect addressing using an HL pair. Let the content of the HL pair be 4250_H and let the content of the memory location 4250_H be 12_H.

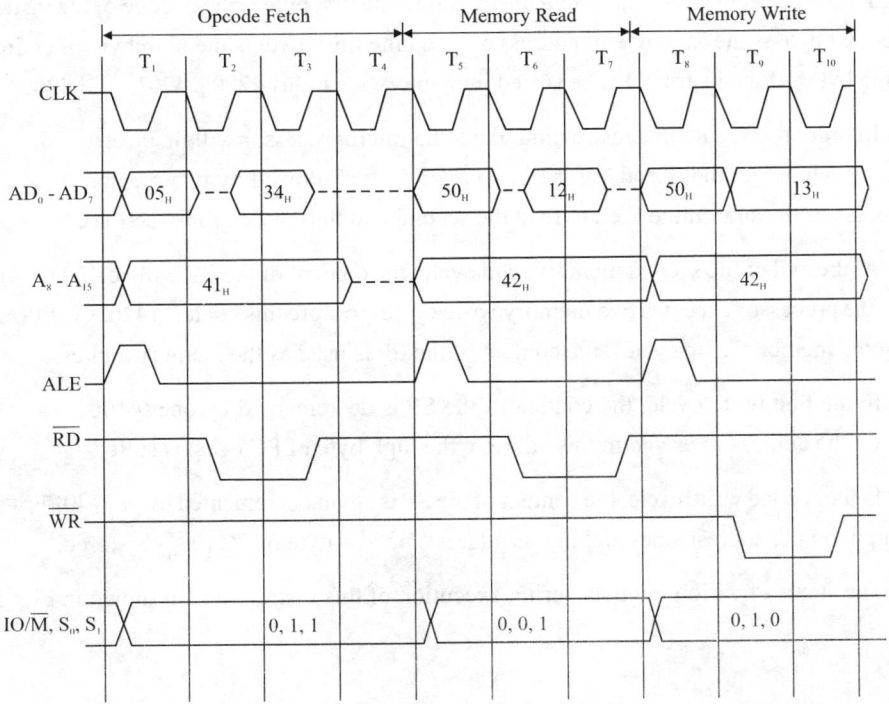

Fig. 3.18: Timing diagram of INR M instruction.

The INR M is one-byte instruction and it is the opcode of instruction 34_H. Let this instruction be stored in the memory location 4105_H.

In order to execute the instruction, the 8085 microprocessor will first execute the opcode fetch cycle to get the opcode 34_H. Then, it executes the memory read cycle to read the content of the memory location 4250_H.

The content (12_H) of the memory location is incremented by one in the ALU and then, the processor executes the memory write cycle to store the result (13_H) of the ALU operation in the same memory location 4250_H. The status of various signals during execution of this instruction are shown in Fig. 3.18.

3.7.6 Timing Diagram of CALL Instruction

The **"CALL addr16"** instruction is used to execute a subroutine/procedure stored at addr16, after saving the address of the next instruction in the stack memory.

On execution of this instruction, the addr16 is loaded in the **Program Counter (PC)** and the previous value of the PC is stored in the stack memory pointed by the **Stack Pointer (SP)**.

Let the address of the subroutine be $4F50_H$ and the content of SP be 4100_H.

The CALL addr16 is a three-byte instruction. The first byte is the opcode of the instruction CD_H. The second byte is the low byte of address 50_H and the third byte is the high byte of address $4F_H$. Let the three bytes of the instructions be stored in memory locations 4200_H, 4201_H and 4202_H.

In order to execute this instruction, the 8085 microprocessor will first execute the opcode fetch machine cycle to get the opcode of the instruction CD_H, followed by two memory read cycles to get the address of the subroutine (i.e., to read the second and third byte of instruction).

At the end of the second memory read cycle, the content of the PC will be 4203_H. After the read cycles, the processor executes two memory write cycles to store this content (4203_H) of PC in the stack. During the memory write cycles, the content of the SP is used as the memory address.

In the first write cycle, the content of the SP is decremented by one $(4100_H - 1 = 40FF_H)$ and output on the address lines and in this address, the high byte of PC (42_H) is stored.

In the second write cycle, the content of the SP is again decremented by one $(40FF_H - 1 = 40FE_H)$ and output on the address lines and in this address the low byte of PC (03_H) is stored.

The status of various signals during execution of this instruction are shown in Fig. 3.19.

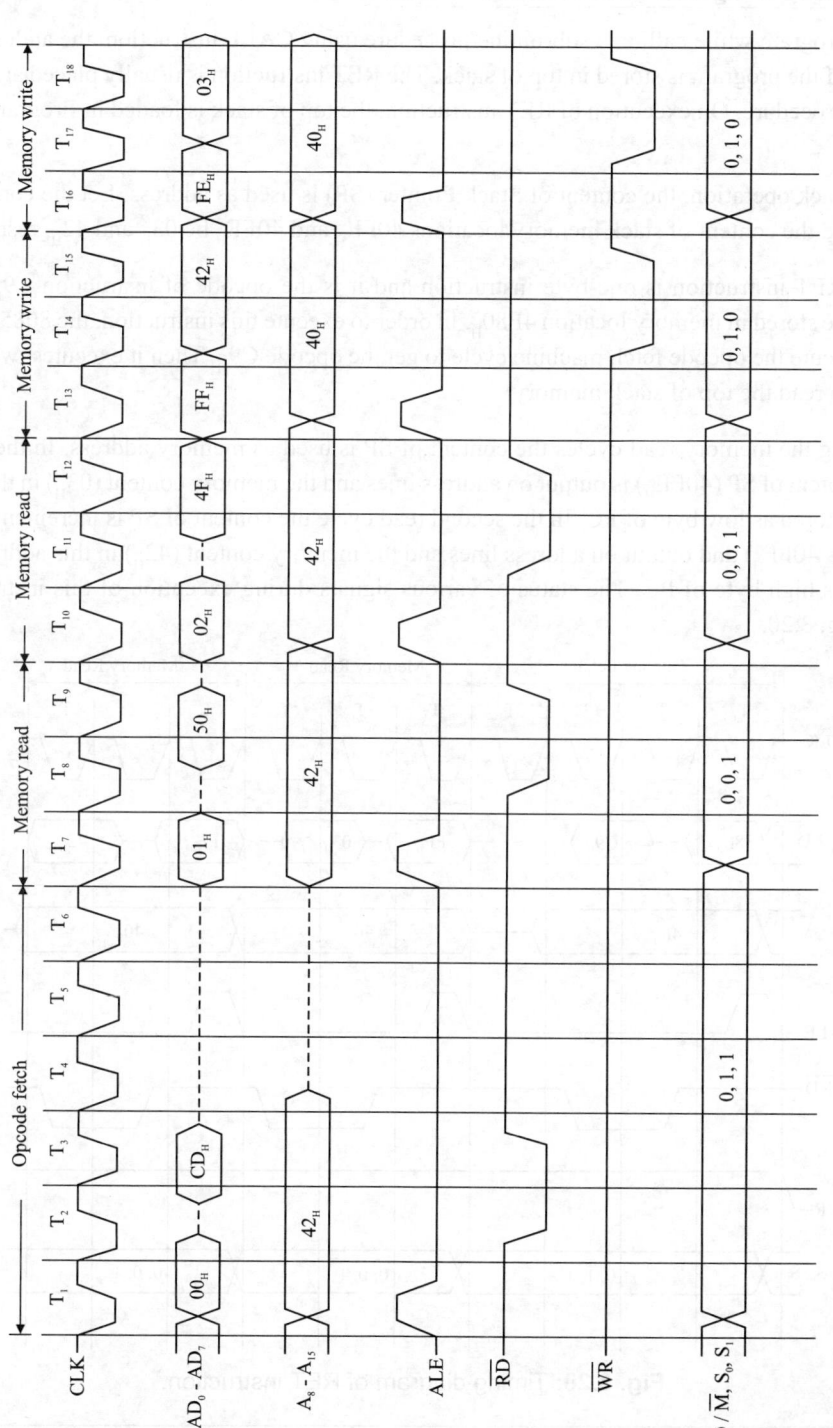

Fig. 3.19: Timing diagram of CALL 4F50H instruction.

3.7.7 Timing Diagram of RET Instruction

In a program while calling a subroutine/procedure using CALL instruction, the address of next instruction of the program is stored in top of stack. The RET instruction is usually placed at the end of subroutine/procedure. On execution of RET instruction, the top of stack is loaded in **P**rogram **C**ounter (PC).

For stack operation, the content of **S**tack **P**ointer (SP) is used as address. Let the content of SP be $40FE_H$ and the content of stack memory locations $40FE_H$ and $40FF_H$ be 03_H and 42_H, respectively.

The RET instruction is one-byte instruction and it is the opcode of instruction $C9_H$. Let this instruction be stored in memory location $4F80_H$. In order to execute this instruction, the 8085 processor will first execute the opcode fetch machine cycle to get the opcode $C9_H$. Then it executes two memory read cycle to read the top of stack memory.

During the memory read cycles the content of SP is used as memory address. In the first read cycle the content of SP $(40FE_H)$ is output on address lines and the memory content (03_H) in this address is read and stored as low byte of PC. In the second read cycle the content of SP is incremented by one $(40FE_H + 1 = 40FF_H)$ and output on address lines and the memory content (42_H) in this address is read and stored as high byte of PC. The status of various signals during execution of this instruction are shown in Fig. 3.20.

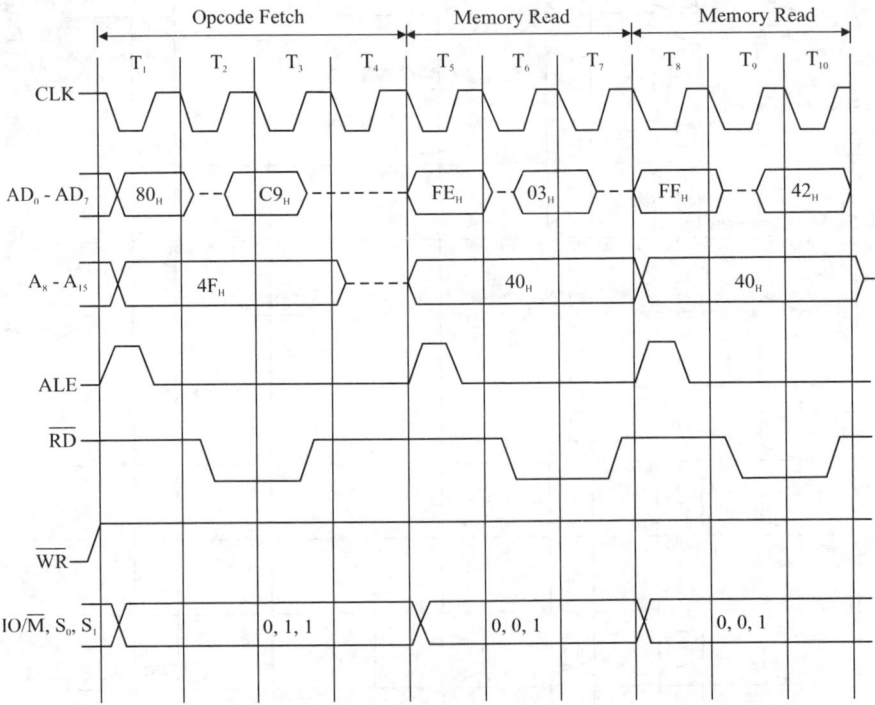

Fig. 3.20: Timing diagram of RET instruction.

3.8 SHORT-ANSWER QUESTIONS

Q3.1 What is Software and Hardware?

The software is a set of instructions or commands needed for performing a specific task by a programmable device or a computing machine.

The hardware refers to the components or devices used to form computing machine in which the software can be run and tested. Without software the hardware is an idle machine.

Q3.2 What is machine language?

The language that can be understood by a programmable machine is called machine language. The machine language program are developed using 1s and 0 s.

Q3.3 What is assembly language?

The language in which the mnemonics (short-hand form of instructions) are used to write a program is called assembly language. The mnemonics are given by the manufacturers of microprocessor.

Q3.4 What are machine language and assembly language programs?

The software developed using 1s and 0 s are called machine language programs. The software developed using mnemonics are called assembly language programs.

Q3.5 What is the drawback in machine language and assembly language programs?

The machine language and assembly language programs are machine dependent. The programs developed using these languages for a particular machine cannot be directly run on another machine (But after conversion using suitable conversion software it can be run on another machine.)

Q3.6 Define mnemonics.

The short-hand form of describing the instructions are called mnemonics. The mnemonics are given by the manufacturers of microprocessors and programmable devices.

Q3.7 What is processor cycle (machine cycle)?

The processor cycle or machine cycle is the basic external operation performed by the processor. To execute an instruction, the processor will run one or more machine cycles in a particular order.

Q3.8 What is instruction cycle?

The sequence of operations that a processor has to carry out while executing an instruction is called instruction cycle. Each instruction cycle of a processor in turn consists of a number of machine cycles.

Q3.9 What is fetch and execute cycle?

In general, the instruction cycle of an instruction can be divided into fetch and execute cycles. The fetch cycle is executed to fetch the opcode from memory. The execute cycle is executed to decode the instruction and to perform the work instructed by the instruction.

Q3.10 List the various machine cycles of 8085.

The various machine cycles of 8085 are as follows:

1. **Opcode fetch cycle**
2. **Memory read cycle**
3. **Memory write cycle**
4. **IO read cycle**
5. **IO write cycle**
6. **Interrupt acknowledge cycle**
7. **Bus idle cycle.**

Q3.11 What is the need for timing diagram?

The timing diagram provides information regarding the status of various signals, when a machine cycle is executed. The knowledge of timing diagram is essential for system designer to select matched peripheral devices like memories, latches, ports, etc., to form a microprocessor system.

Q3.12 What is T-state?

The T-state is the time period of the internal clock signal of the processor. The time taken by the processor to execute a machine cycle is expressed in T-state.

Q3.13 How many machine cycles constitute one instruction cycle in 8085?

Each instruction of the 8085 processor consist of one to five machine cycles.

Q3.14 Define opcode and operand.

Opcode (**Op**eration **Code**) is the part of an instruction/directive that identifies a specific operation. Operand is a part of an instruction/directive that represents a value on which the instruction acts.

Q3.15 What is opcode fetch cycle?

The opcode fetch cycle is a machine cycle executed to fetch the opcode of an instruction stored in memory. The first machine cycle of every instruction is opcode fetch machine cycle.

Q3.16 What operation is performed during first T-state of every machine cycle in 8085?

In 8085, during the first T-state of every machine cycle the low byte address is latched into an external latch using ALE signal.

Q3.17 Why status signals are provided in microprocessor?

The status signals can be used by the system designer to track the internal operations of the processor. Also, it can be used for memory expansion (by providing separate memory banks for program and data, and selecting the banks using status signals).

Q3.18 How the 8085 processor differentiates memory access (read/write) and IO access (read/write)?

The memory access and IO access is differentiated using IO/$\overline{\text{M}}$ signal. The 8085 processor asserts IO/$\overline{\text{M}}$ **low** for memory read/write operation and IO/$\overline{\text{M}}$ is asserted **high** for IO read/write operation.

Q3.19 *In which lines the 8085 processor gives the output of IO port address during IO read/write operation?*

When the processor executes an IO read or write cycle, 8-bit port address is sent out both on low order address bus and high order address bus. This facility offers a flexibility for system designer to use either low-order address lines or high-order address lines for addressing ports and generating chip select signals for IO devices.

Q3.20 *When the 8085 processor checks for an interrupt?*

In the second T-state of the last machine cycle of every instruction, the 8085 processor checks whether an interrupt request is made or not.

Q3.21 *What is interrupt acknowledge cycle?*

The interrupt acknowledge cycle is a machine cycle executed by 8085 processor after acceptance of the interrupt to get the address of the interrupt service routine in-order to service the interrupting device.

Q3.22 *What will be the status of the processor during bus idle cycle?*

During bus idle cycle, the status signals S_0 and S_1 are both asserted **low** and data, address and control pins are driven to **high impedance** state. Also, the processor will not sample the READY signal.

Q3.23 *How the slow peripherals are interfaced with 8085 processor?*

The slow peripherals require longer read/write time than allowed by the processor. Hence to interface slow peripherals, an extra hardware should be designed so that it introduces required number of wait states in machine cycles between T_1 and T_2. An alternate solution is to interface the slow peripherals using ports.

Q3.24 *When is the READY signal sampled by the processor?*

The 8085 processor samples or checks the READY signal at the second T-state of every machine cycle.

Q3.25 *What are wait states?*

The T-states introduced between T_2 and T_3 of a machine cycle by the slow peripherals (to get extra time for read/write operation) are called wait states.

Q3.26 *When the 8085 processor will enter wait state?*

The 8085 processor will check the READY signal at the second T-state of a machine cycle. If the READY is tied **low** at this time, then it will enter into wait state (i.e., after second T-state). The processor will come out of wait state only when READY is again made **high**.

Q3.27 *What is the difference between wait state and bus idle condition?*

During bus idle condition, the tristate pins of the processor are driven to **high impedance** state, but during wait state they are in normal states (either **low** or **high**). The READY is not sampled during bus idle condition but it is sampled during wait state.

Q3.28 *How many instructions are available in 8085 instruction set?*

The 8085 instruction set consists of 74 basic instructions and 246 total instructions.

Q3.29 *What is the instruction format of 8085?*

The size of 8085 instruction is 1 to 3 bytes. Each instruction has one-byte opcode. The remaining bytes are either data or address. The format of 8085 instructions are shown below :

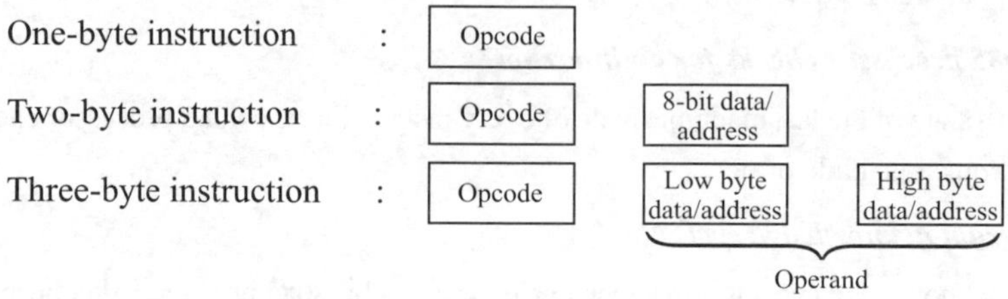

One-byte instruction	:	Opcode		

Two-byte instruction	:	Opcode	8-bit data/ address	

Three-byte instruction	:	Opcode	Low byte data/address	High byte data/address

Operand

Fig. Q3.29: Format of 8085 instructions.

Q3.30 *What is addressing?*

The method of specifying the data to be operated (operand) by the instruction is called addressing.

Q3.31 *What are the addressing modes available in 8085?*

The 8085 has the following five different modes of addressing.

1. **Immediate addressing**
2. **Direct addressing**
3. **Register addressing**
4. **Register indirect addressing**
5. **Implied addressing.**

Q3.32 *Explain the immediate addressing with an example.*

In immediate addressing mode, the data is specified in the instruction itself. The data will be a part of the program instruction.

Example: MVI B, 3EH - Move the data $3E_H$ given in the instruction to B-register.

Q3.33 *What is direct addressing? Give an example.*

If the address of the data is directly specified in the instruction then the addressing mode is called direct addressing.

Example: LDA 1050H - Load the data available in memory location 1050_H in accumulator.

Q3.34 *Explain register addressing with an example.*

In register addressing mode, the instruction specifies the name of the register in which the data is available.

Example: MOV A, B - Move the content of B-register to A-register.

Q3.35 *Explain register indirect addressing with an example.*

In register indirect addressing mode, the instruction specifies the name of the register in which the address of the data is available. Here the data will be in memory and the address will be in the register pair.

Example: MOV A, M - The content of memory whose address is available in HL pair is moved to A-register.

Q3.36 What is implied or implicit addressing mode?

If the instruction operates on a data available in the register defined by the opcode then the addressing mode is called implied or implicit addressing mode.

Example: CMA - Complement the content of accumulator.

Q3.37 What are the functions performed by data transfer instruction? Give an example and explain.

The data transfer instructions can copy the content of one register to another and copy the content of register to memory or vice versa.

Example: MOV B, C - The content of C-register is moved (copied) to B-register.

Q3.38 What are the functions performed by arithmetic instructions? Give an example and explain.

The functions performed by arithmetic instructions are Addition, Subtraction, Increment and Decrement.

Example: ADD E - The content of E-register is added to accumulator.

Q3.39 What are the operations performed by logical instructions? Give an example and explain.

The operations performed by logical instructions are AND, OR, EXCLUSIVE-OR, Complement, Compare and Shift (Rotate).

Example: ANA D - The content of D-register is logically ANDed with accumulator.

Q3.40 In which unit the arithmetic and logical operations are performed. Which unit is the destination of result.

The arithmetic and logical operations are performed in ALU. After the operation, the result will be stored in accumulator.

Q3.41 Which group of instruction affects the flags?

The flags are altered after execution of arithmetic and logical instructions.

Q3.42 What are the arithmetic instructions that do not affect the flag?

The 16-bit increment and decrement instructions (INX rp and DCX rp) will not affect any flags.

Q3.43 What are the flags affected by 8-bit increment and decrement instructions?

Except carry, all other flags are affected by 8-bit increment and decrement instructions.

Q3.44 What will be condition of flags after logical AND and OR operations?

After logical AND operation the carry flag is RESET (0), auxiliary carry flag is SET (1) and depending on the result of AND operation other flags are altered.

After logical OR operation the carry flag and auxiliary carry flag are RESET (0). Depending on the result of OR operation other flags are altered.

Q3.45 List the instructions that affect only carry flag.

The instructions that affect only carry flag are the following:

CMC	RAR	STC
DAD rp	RLC	
RAL	RRC	

Q3.46 What is DAA ?

DAA - **D**ecimal **A**djust **A**ccumulator.

After BCD addition, this instruction is executed to get the result in BCD. When DAA instruction is executed, the content of the accumulator is altered or adjusted as explained below:

1. If the sum of lower nibbles exceeds 09_H or auxiliary carry is set, a correction 06_H (0110) is added to lower nibble.

2. If the sum of upper nibbles exceeds 09_H or carry is set, a correction 06_H (0110) is added to upper nibble.

Q3.47 What is DAD and what are the flags affected by this instruction?

DAD refers to Double Addition. This instruction is used to perform addition of two 16-bit data.

> *Syntax:* *DAD* *rp*

The content of **r**egister **p**air (rp) is added to the content of HL pair. After addition, the result will be in HL pair. The register pair can be BC, DE, HL, or SP. On execution of this instruction, only carry flag is affected.

Q3.48 List the various instructions that can be used to clear accumulator ?

The accumulator can be cleared by the following instructions:

1. MVI A,00_H

2. SUB A

3. ANI 00_H

4. XRA A.

Q3.49 What is the similarity and difference between subtract and compare instruction?

Similarity : Both the subtraction and comparison are performed by subtracting two data in ALU and flags are altered depending upon the result.

Difference: After subtract instruction is executed, the result is stored in accumulator, but after the execution of compare instruction the result is discarded (i.e., the subtract instruction alters the content of destination register (accumulator), but the compare instruction will not alter the content of any register or memory).

Q3.50 List the IO instruction in 8085.

1. The IO instruction of 8085 are IN addr8 and OUT addr8.

2. The IN instruction is used to input a data byte from the IO-mapped device or port. The OUT instruction used to output data byte to IO-mapped device or port.

Q3.51 State the difference between LDA and LDAX.

The LDA instruction uses direct addressing mode to load a data byte from memory to accumulator, but LDAX instruction uses register indirect addressing for the same operation.

In LDA instruction, the content of memory location whose address is given in the instruction is moved to accumulator.

In LDAX instruction, a register pair contains the address of memory location. The content of memory location whose address is available in register pair is moved to accumulator.

Q3.52 Explain DI and EI.

DI - Disable Interrupt. When this instruction is executed all the interrupts except TRAP are disabled. When the interrupts are disabled the processor will not accept or recognize the interrupt.

EI - Enable Interrupt. This instruction is used or executed to allow the interrupts after disabling.

Q3.53 What is the function performed by SIM instruction?

SIM - Set Interrupt Mask. The SIM instruction is used to mask the hardware interrupts RST 7.5, RST 6.5 and RST 5.5. The execution of SIM instruction output and the content of the accumulator to program interrupt mask bits are also used to output serial data on the SOD line.

Q3.54 What is the function performed by RIM instruction?

RIM - Read Interrupt Mask. The RIM instruction is used to check whether an interrupt is masked or not. It is also used to read data from SID line.

Q3.55 What will be the state of the processor after executing HLT instruction?

When the HLT instruction is executed, the processor suspends program execution and the bus will be in idle state (i.e., the processor keeps on executing bus idle cycle until a reset or interrupt).

Q3.56 What is NOP? State its importance.

The NOP is a dummy instruction, it neither achieves any result nor affects any CPU register. This is used for producing software delay and reserve memory spaces for future software modifications.

Q3.57 What is PSW?

PSW - Program Status Word. The flag register and accumulator together is called PSW. Flag register is low order register. Accumulator is high order register.

Q3.58 Explain RET instruction.

RET - Return to main program. This instruction is placed at the end of subroutine program in order to return to the main program. When this instruction is executed, the top of stack is poped to program counter.

Q3.59 Explain the difference between the conditional and unconditional return instructions.

In a conditional return instruction a flag condition is tested. If the flag condition is true, then the program control returns to main program. If the flag condition is false, then the next instruction is executed. In unconditional return instruction, the program control returns to the main program irrespective of the condition of the flag.

Q3.60 State the difference between STA and STAX instructions.

The STA instruction uses direct addressing mode to store the content of accumulator to a memory location, but the STAX instruction uses indirect addressing mode for the same operation.

Q3.61 What will be the content of SP (Stack Pointer) after execution of PUSH and POP instructions?

1. After execution of PUSH instruction, the content of Stack Pointer (SP) will be 02 less than the earlier value.

2. After execution of POP instruction, the content of Stack Pointer (SP) will be 02 greater than the earlier value.

Q3.62 What is the difference between ADD and ADC instruction?

The ADD instruction will not consider the value of carry flag for addition, but the ADC instruction will consider the value of carry flag (before executing this instruction) for addition. In ADC instruction the content of register or memory and the carry flag are added to the content of accumulator.

Q3.63 How is the subtraction performed in 8085?

The 8085 processor performs 2's complement subtraction and after subtraction, it complements the carry flag.

Q3.64 How the result of subtract operation can be interpreted?

1. After subtract operation, if the carry flag is SET (1), then the result is negative will be in 2's complement form.

2. After subtract operation, if the carry flag is RESET (0), then the result is positive.

Q3.65 What is the difference in 2's complement subtraction and 8085 subtraction?

In 2's complement subtraction, the result is positive if carry is equal to one (1) and negative if carry is equal to zero (0). But in 8085, the result is negative if carry is equal to one (1) and positive if carry is equal to (0).

Q3.66 What is the difference between CALL and JUMP instruction?

In CALL instruction, the address of next instruction is pushed to stack (i.e., stored in stack memory) before transferring the program control to call address. But in JUMP instruction, the address of next instruction is not saved.

Q3.67 What is the difference between conditional and unconditional branch instructions?

In unconditional branch instructions, the program control is transferred to branch address without

3.9 EXERCISES

I. Fill in the blanks with appropriate words

1. The software which is used to convert assembly language programs into machine language program is called _____ .

2. The number of T-states required to execute the 8085 memory read cycle is _____.

3. The various ways of specifying data are called _____.

4. The addressing mode used in the 8085 instruction MOV A,M is _____ addressing.

5. The 8085 instruction _____ is used to load the accumulator with content of memory.

6. The 8085 assembly language instruction that stores the content of H and L registers into the memory locations 2050$_H$ and 2051$_H$, respectively is _____.

7. A single byte 8085 instruction which is used to push the memory content into stack memory is

   _____.

8. The _____ and _____ instructions of 8085 are used to communicate with peripheral devices.

9. The _____ instruction of 8085 is used for BCD addition.

10. The content of A register is equal to 23$_H$. The _____ instruction of 8085 is used to mask the lower nibble of A register.

11. The status of carry flag, CF = _____ and zero flag, ZF = _____ after executing the 8085 instruction CMP A,B where A = 25$_H$ and B = 52$_H$.

12. The 8085 instruction XCHG exchanges the content of _____ and _____ register pairs.

13. The single byte 8085 instruction which is used to find the one's complement of the accumulator content is _____.

14. The content of Accumulator, A = 12$_H$. The execution of 8085 instruction RLC and RRC change the A register as _____ and _____ respectively.

15. The _____ instruction of 8085 is used to transfer the program control back to the main program.

16. The vector address for the 8085 restart instruction RST 5 is _____.

17. When the 8085 instruction DI is executed, all the interrupts except _____ are disabled.

18. The _____ instruction of 8085 is used to read the status of interrupt masks.

19. The _____ instruction performs nothing but simply waste clock cycles in 8085 processor.

Answers

1. assembler	5. LDA	9. DAA	13. CMA	17. TRAP
2. three	6. SHLD 2050H	10. ANI 0FH	14. 24$_H$, 09$_H$	18. RIM
3. addressing mode	7. SPHL	11. 1, 0	15. RET	19. NOP
4. register indirect	8. IN, OUT	12. HL, DE	16. 0028$_H$	

II. State whether the following statements are True/False.

1. The assembly language programs are machine dependent.

2. One T-state is equal to twice the time period of internal clock signal of the 8085 processor.

3. The T-state of 8085 always starts at falling edge of the clock.

4. The READY signal of 8085 will not be sampled by the processor during bus idle sycle.

5. In implied addressing mode, the operand is always in accumulator.

6. The source remains unchaned after executing data transfer instructions.

7. The SP register of 8085 processor always points to top of the stack.

8. The register pair DE of 8085 processor is used to point to the memory.

9. The content of SP register of 8085 processor is incremented by 02 after the execution of PUSH instruction.

10. The DAD instruction of 8085 modifies only the carry flag.

11. The 8085 processor performs subtraction by using 1's complement method.

12. There are no direct instructions to perform multiplication and division in 8085.

13. The source and destination contents remain unchanged after executing CMP instruction of 8085.

14. The CALL instruction of 8085 is a conditional branching instruction.

15. The logical operation AND is generally used to mask the bits.

16. The 8085 instruction INX will not alter any flags.

17. The STC instruction of 8085 will always clear the carry flag.

Answers					
1. True	4. True	7. True	10. True	13. True	16. True
2. False	5. False	8. False	11. False	14. False	17. False
3. True	6. True	9. False	12. True	15. True	

III. Choose the right answer for the following questions.

1. *Which of the following is the high level language?*
   a) BASIC   b) COBOL   c) C   d) all the three

2. *During which T-state of opcode fetch cycle the higher order address/data lines are demultiplexed in 8085 processor?*
   a) T_1 b) T_2 c) T_3 d) T_4

3. *The data, address and control pins are driven to high impedance state during _____ machine cycle of 8085.*
   a) interrupt acknowledge   b) bus idle   c) wait state   d) all the three

4. *The 8085 instruction used for providing direct value to a register through the instruction itself is*
   a) MOV   b) MVI   c) ADD   d) JNZ

5. *Logical operation in an 8085 processor include _____ operation?*
   a) AND,OR,EXOR   b) rotate   c) compare & complement   d) all of the above

6. *The binary code for the registers A and B are 111 and 000 respectively. what is the 8-bit opcode of the 8085 instruction MOV A,B?*
   a) 47_H b) 78_H c) 38_H d) 37_H

7. *The _____ instruction is used to store the accumulator content into memory pointed by register pair DE in 8085 processor.*
   a) STA D   b) STA E   c) STAX D   d) STAX E

8. *The number of 8085 machine cycles required to execute the instruction LDA 2000H is*
   a) 2T   b) 3T   c) 4T   d) 6T

9. *Which of the following is a two byte 8085 instruction?*
   a) STA 1200H   b) LDA 3000H   c) MVI A,12H   d) MOV A,B

10. *In which of the following 8085 instructions, the operand is specified by the instruction itself?*
    a) CMA   b) SPHL   c) XCHG   d) all the three

11. *Which of the following is an invalid 8085 instruction?*
    a) STAX D   b) STAX H   c) STA 2000H   d) LDAX D

12. *The content of HL register and DE register are 1234_H and 2567_H respectively. A single byte 8085 instruction which is used to add the two 16 bit data is*
    a) DAA   b) DAD D   c) DAD H   d) ADD D

13. Which of the following 8085 instructions change the program control to new location?

   a) SPHL   *b)* PCHL   *c)* LDA   *d)* STA

14. The _____ instruction of 8085 is used to implement the 8-bit full adder.

   a) ADD B   *b)* ACI 35H   *c)* ADC B   *d)* all the three

15. The following set of 8085 instructions are used to multiply the content of HL register pair by three

   a) DAA   b) DAD H   c) DAD H   d) ADD H

                                      DAD H   ADD H

16. The content of accumulator, A = FF_H and carry flag CY = 0. What would be the content of A register and carry flag after executing the 8085 instructions (i) INR A and (ii) ADD 01H respectively?

   a) (i) A=00H ; CY = 1   *b)* (i) A=00H ; CY = 0

      (ii) A=00H ; CY = 1   (ii) A=00H ; CY = 1

   c) (i) A = 00H ; CY = 1   *d)* (i) A=00H ; CY = 0

      (ii) A = 00H ; CY = 0   (ii) A=00H ; CY = 0

17. Which of the following 8085 instruction is used to clear the accumulator?

   a) MOV A, 00H   *b)* XRA A   *c)* ANI A,00H   *d)* all the three

18. Which of the following 8085 instructions use the auxilary carry flag for its execution?

   a) ADC   *b)* SBB   *c)* CMP   *d)* DAA

19. The two single byte 8085 instructions which are used (i) to load SP register with HL pair and (ii) to exchange the content of HL pair and top of the stack respectively are

   a) SPHL and XTHL   *b)* XTHL and SPHL   *c)* SPHL and XCHG   *d)* XTHL and XCHG

20. Which of the following 8085 instruction sequences are used to swap the nibbles of A register?

   a) RAL   *b)* RAR   *c)* RRC   *d)* RLC

       RAL   RAR   RRC   RLC

                       RAR   RRC

                       RAR   RRC

21. Given the sequence of 8085 instructions

Address	Instructions	Address	Instructions
2000:	MVI A, 12H	3000:	INR A
2002:	CPI 05H		RET
2004:	CZ 3000H		
2007:	ADI A,05H		
2009:	HLT		

What would be the content of A register after the execution of the 8085 program given above.

   a) 17_H b) 18_H c) 12_H d) 05_H

22. The following 8085 instruction forces the bus to idle state

   a) NOP   b) HLT   c) RET   d) all the three

23. An 8085 instruction used for generating software delay is

   a) HLT   b) RET   c) NOP   d) EI

24. *The order of 8085 machine cycle executed in executing the instruction STA 1200H are*
   a) opcode fetch, memory read, memory write
   b) opcode fetch, memory write, memory read
   c) opcode fetch, memory read, memory read, memory read
   d) opcode fetch, memory read, memory read, memory write

Answers

1. d	4. b	7. c	10. d	13. b	16. b	19. a	22. b
2. a	5. d	8. c	11. b	14. c	17. d	20. c	23. c
3. b	6. b	9. c	12. b	15. b	18. d	21. a	24. d

IV.   Answer the following questions.

E3.1   How many T-states are required to execute interrupt acknowledge machine cycle of 8085? Justify the answer

E3.2   Identify the operations performed and operand in the following 8085 instructions?
   a) MVI A, 12H   b) CMA   c) RLC

E3.3   Write the number of bytes and machine code of the following 8085 instructions.
   a) LDAX B   b) JC 2000H   c) RST 4   d) NOP

E3.4   How fast the 8085 processor with clock frequency 1 MHz executes the following instructions
   a) MVI A, 00H   b) XRA A. Comment on the operation.

E3.5   Identify the machine cycles involved in the following 8085 instructions
   a) SHLD 4000H   b) OUT 80H   c) INX B   d) DAD B

E3.6   In an 8085 microprocessor based system, it is desired to increment the contents of memory location whose address is available in DE register pair and store the result in same location. Write the sequence of instruction to perform the above mentioned action.

E3.7   In an 8085 microprocessor, what would be the contents of accumulator after the following instructions are executed?

   XRA A

   MVI B, F0H

   SUB B

E3.8   Write the sequence of 8085 instructions to add two decimal numbers 12 and 39 which are stored in A and B registers respectively.

E3.9   The content of register B and accumulator A of 8085 microprocessor are 49_H and $3A_H$ respectively. What would be the content of A and the status of all the flags after execution of SUB B instruction?

E3.10   Assume that the accumulator contains the data 65_H and the 8085 instruction MOV C,A ($4F_H$) is fetched. List the operations performed in various T-states while executing the instruction.

CHAPTER 4

INSTRUCTION SET OF 8086

4.1 BUS CYCLES AND TIMING DIAGRAM

The 8086 processor has two functional units called **Bus Interface Unit (BIU)** and **Execution Unit (EU)**. Most of the time each unit works independently. The BIU takes care of fetching instruction codes from memory, and data from memory and IO devices. The EU takes care of executing the instructions prefetched by BIU. The BIU initiates all external operations which are also called **bus activity**. The external bus activities are repetitions of certain basic operations. The basic operations performed by the CPU bus are called bus cycles. Depending on the activities of 8086, the bus cycles can be classified as follows:

1. Memory read cycle (Four T-states)
2. Memory write cycle (Four T-states)
3. IO read cycle (Four T-states)
4. IO write cycle (Four T-states)
5. Interrupt acknowledge cycle (Eight T-states)

The processor takes a definite time to perform a bus cycle. The time taken to perform a bus cycle is specified in terms of T-states. In an 8086 processor, the time duration of one T-state is equal to one time period of the internal clock of the processor. The T-state starts in the middle of falling edge of the clock signal as shown in Fig. 4.1.

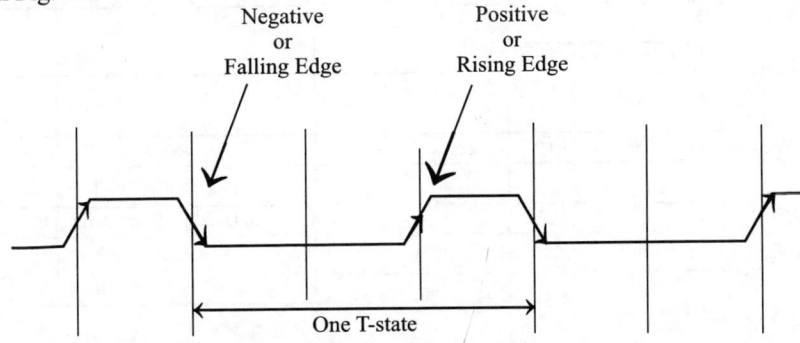

Fig. 4.1: Clock signal and one T-state of 8086.

The normal time taken by 8086 to perform read/write cycle is four T-states. The processor also has the facility to extend the timing of bus cycles by introducing extra T-states called wait states using the READY control signal. If READY is permanently tied **high** then the bus cycles are executed in normal timing.

The memory or IO access time allowed by the 8086 processor with a 5 MHz clock is 400 ns. If the memory or IO devices used in the system has access time more than 400 ns then wait states have to be introduced in the bus cycles, between the second and third T-states (T_2 and T_3) to extend the timing of the bus cycle.

In order to introduce wait states, the READY is made **low** by an external hardware in the beginning of second T-state and then made **high** after required time delay. The processor samples READY signal at the end of second T-state of every bus cycle. If READY is **low** at this time then the processor introduces one wait state.

Again, it samples READY signal at the end of wait state and if READY is still **low** then it introduces another wait state. This process is continued until READY is **high**. Once READY is made **high** the processor will resume the bus activity and complete the bus cycle.

4.1.1 Timing Diagram

The timing diagram provides information about the various conditions (**high** state or **low** state or **high impedance** state) of the signals while a bus cycle is executed. The timing diagrams are supplied by the manufacturer of the microprocessor. The timing diagrams are essential for a system designer. Only from the knowledge of timing diagrams, the matched peripheral devices like memories, ports, etc., can be selected to form a system with a microprocessor as the CPU.

4.1.2 Memory Read Cycle of 8086

The memory read cycle is initiated by BIU of 8086 to read a program code or data from memory. The normal time taken by memory read cycle is four clock periods. The timing of various signals involved in reading a word (16-bit) starting from even address of memory in minimum mode are shown in Fig. 4.2. The activities of the bus in each T-state are given in the following list.

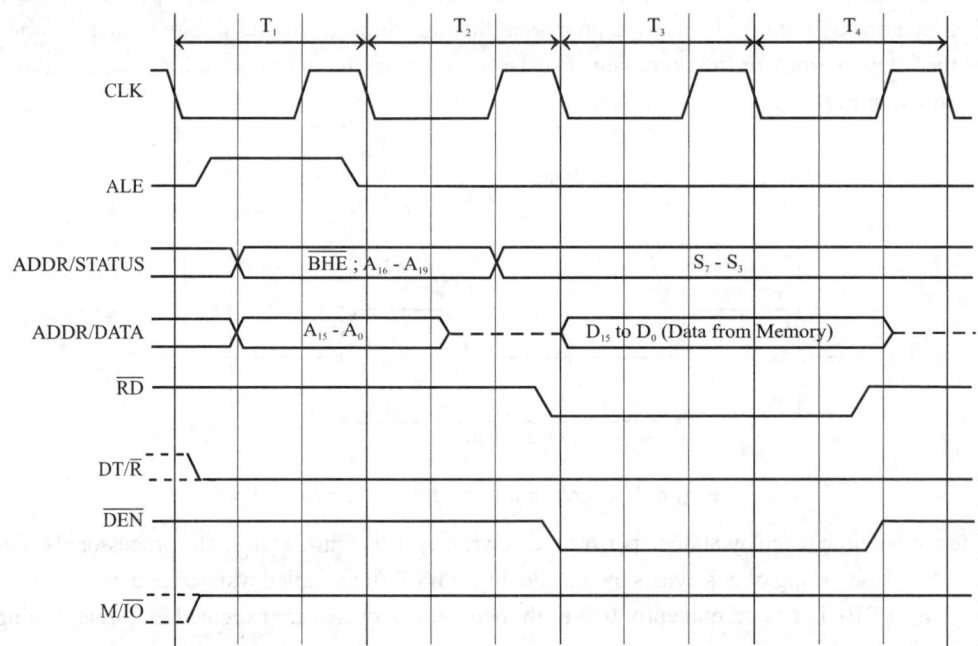

($\overline{\text{WR}}$ is **high** ; READY is tied **high** permanently or temporarily in the system.)

Fig. 4.2: Memory read cycle of 8086.

Activities during T_1

1. The 8086 outputs a 20-bit memory address on AD_0-AD_{15} lines and ADDR/STATUS lines.

2. The address latch enable signal ALE is asserted high in the beginning of T_1 and then asserted low at the end of T_1. This enables the external address latches to latch the address (at the falling edge of ALE) and keep on their output lines.

3. The direction control signal DT/$\overline{R}$ is asserted low to inform the external bidirectional data buffer that the processor has to receive data. (If DT/$\overline{R}$ is already low in the previous bus cycle then it remains low as such.)

4. The M/$\overline{IO}$ signal is asserted high to indicate memory access. (If M/$\overline{IO}$ is already high then it remains high as such.)

5. The $\overline{BHE}$ is asserted low to enable the odd/upper memory bank.

Activities during T_2

1. The AD_0 - AD_{15} lines becomes inactive.

2. The address is with drawn from the ADDR/STATUS lines and status signals S_7 - S_3 are issued on these lines. (The $\overline{BHE}$ becomes the status signal S_7.)

3. At the end of T_2 the read control signal $\overline{RD}$ is asserted low to enable the output buffer of memory. The time during which $\overline{RD}$ remains low is the time allowed for memory to load data in the data bus.

4. The $\overline{DEN}$ signal is asserted low to enable the external bidirectional data buffers.

5. The 8086 samples READY signal during T_2. (If READY is high then T_3 and T_4 are executed otherwise wait states are introduced.)

Activities during T_3

No activities are performed during T_3. The status of the signals at the end of T_2 are maintained throughout T_3.

Activities during T_4

1. The $\overline{RD}$ is asserted high and at this time (i.e., at the rising edge of $\overline{RD}$) the data is latched into 8086.

2. The $\overline{DEN}$ is made high to disable the data buffer.

4.1.3 Memory Write Cycle of 8086

The memory write cycle is initiated by BIU of 8086 to write a data in memory. The normal time taken by memory write cycle is four clock periods. The timing of various signals involved in writing a word (16-bit) starting from even address of memory in the minimum mode are shown in Fig. 4.3. The activities of the bus in each T-state are explained in the following list.

Activities during T_1

The activities during T_1 are same as that of the read cycle except DT/$\overline{R}$ signal. In a memory write cycle, the DT/$\overline{R}$ signal is asserted high to inform the external bidirectional data buffer that the processor is going to transmit data. (If DT/$\overline{R}$ is already high in the previous bus cycle then it remains high as such.)

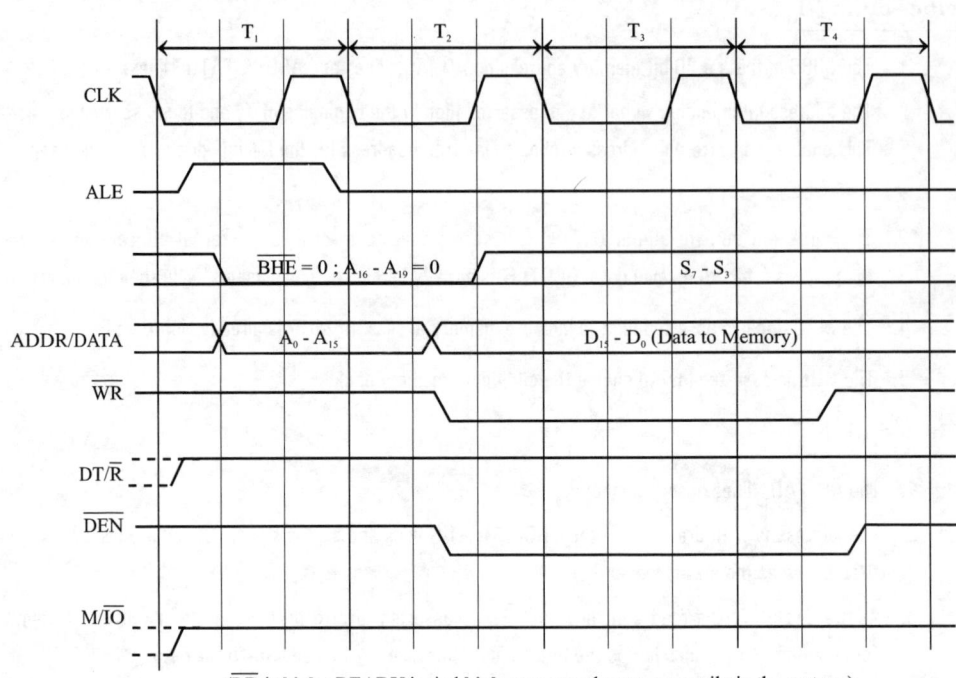

(RD is **high** ; READY is tied **high** permanently or temporarily in the system.)

Fig. 4.3: Memory write cycle of 8086.

Activities during T_2

1. The address is withdrawn from AD_0- AD_{15} lines and data is output on these lines.

2. The address is withdrawn from ADDR/STATUS lines and status signals are issued on these lines (The **BHE** becomes the status signal S_7.)

3. When data is output on a data bus, the control signals **WR** and **DEN** are also asserted low to enable the input buffer of memory and external data buffer on the data bus, respectively.

4. The 8086 samples READY signal during T_2. (If READY is high then T_3 and T_4 are executed otherwise wait states are introduced.)

Activities during T_3

No activities are performed during T_3. The status of the signals at the end of T_2 are maintained throughout T_3.

Activities during T_4

1. The **WR** is asserted high and at this time (i.e., at the rising edge of **WR**), the data is latched into memory.

2. The **DEN** is made high to disable the data buffers.

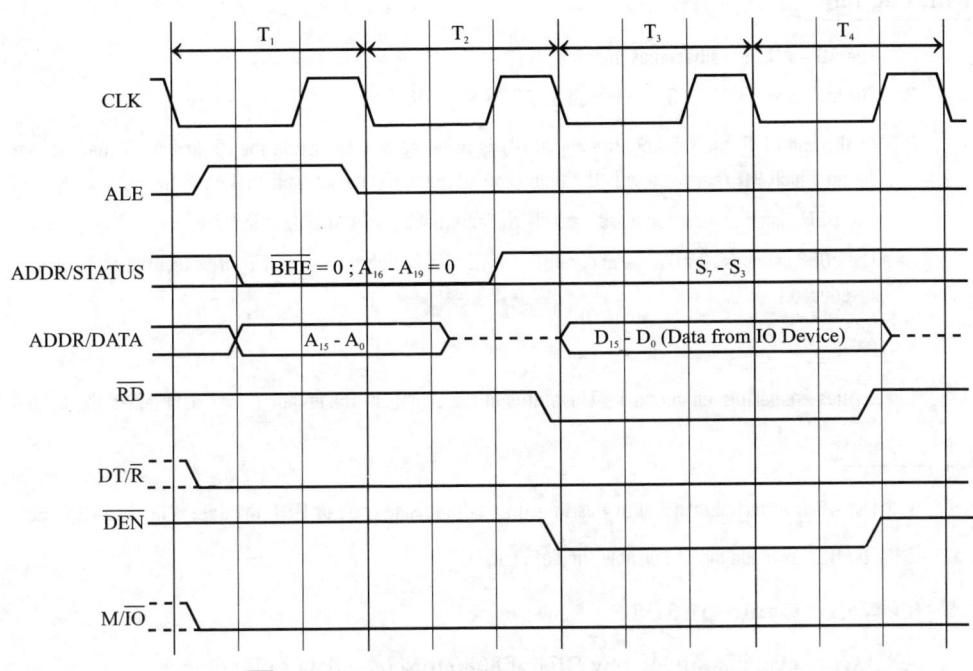

(WR is **high** ; READY is tied **high** permanently or temporarily in the system.)

Fig. 4.4: IO read cycle of 8086.

4.1.4 IO Read Cycle of 8086

The IO read cycle is initiated by BIU of 8086 to read a data from an IO-mapped device or IO port. The normal time taken by an IO read cycle is four clock periods. The timing of various signals involved in reading an IO port in the minimum mode are shown in Fig. 4.4. The activities of the bus in each T-state are given in the following list.

Activities during T_1

1. The 8086 outputs a 16-bit IO address on AD_0-AD_{15} lines. Logic low is output on the $\overline{BHE}$ and ADDR/STATUS lines.

2. The ALE is asserted high and then low. This enables the external address latches to latch the address (at the falling edge of ALE) and keep on their output lines.

3. The DT/$\overline{R}$ signal is asserted low to inform the external bidirectional data buffer that the processor has to receive data. (If DT/$\overline{R}$ is already low in the previous bus cycle then it remains low as such.)

4. The M/$\overline{IO}$ signal is asserted low to indicate IO access. (If M/$\overline{IO}$ is already low then it remains low as such.)

Activities during T_2

1. The AD_0 - AD_{15} lines becomes inactive.

2. The status signals S_7 - S_3 are issued on ADDR/STATUS lines.

3. At the end of T_2 the read control signal $\overline{RD}$ is asserted low to enable the IO device for read operation. The time during which $\overline{RD}$ remains low is the time allowed for the IO device to load data in the data bus.

4. The $\overline{DEN}$ signal is asserted low to enable the external bidirectional data buffers.

5. The 8086 samples READY signal during T_2. (If READY is high then T_3 and T_4 are executed otherwise wait states are introduced.)

Activities during T_3

No activities are performed during T_3. The status of the signals at the end of T_2 are maintained throughout T_3.

Activities during T_4

1. The $\overline{RD}$ is asserted high and at this time (i.e., at the rising edge of $\overline{RD}$), the data is latched into 8086.

2. The $\overline{DEN}$ is made high to disable the data buffer.

4.1.5 IO Write Cycle of 8086

The IO write cycle is initiated by BIU of 8086 to send a data to IO device. The normal time taken by an IO write cycle is four clock periods. The timing of various signals involved in sending a word to the IO device in the minimum mode are shown in Fig. 4.5.

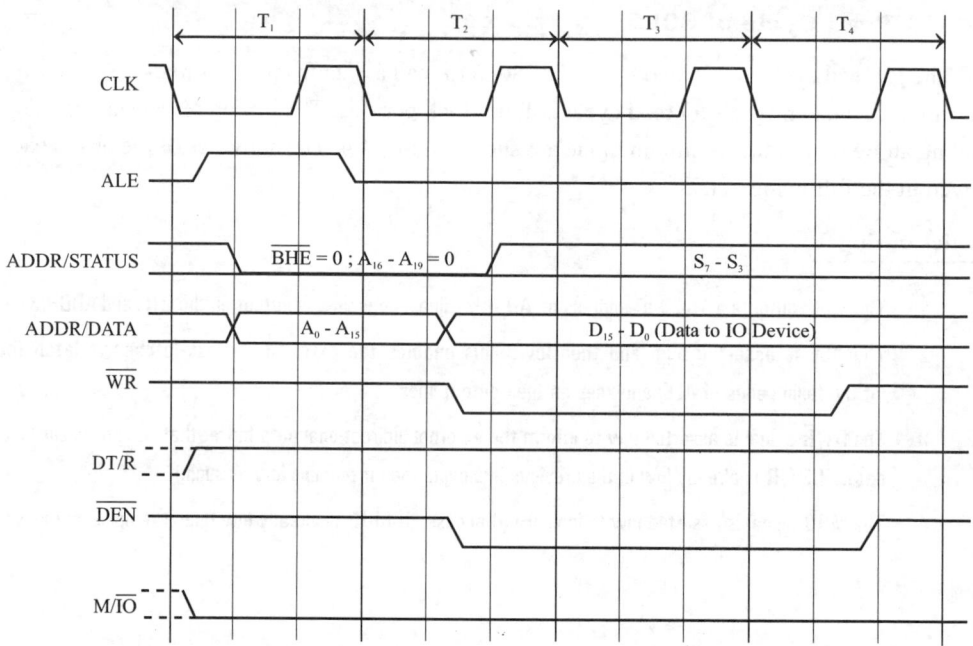

($\overline{RD}$ is **high** ; READY is tied **high** permanently or temporarily in the system.)

Fig. 4.5: IO write cycle of 8086.

The activities during an IO write cycle will be same as an IO read cycle except the following:

1. During T_1, DT/$\overline{R}$ is asserted high to inform the external bidirectional data buffer that the processor is going to transmit data.

2. During T_2, the address is withdrawn from AD_0 - AD_{15} lines and data is output on these lines. At the same time, $\overline{WR}$ is asserted low to enable the IO device for a write operation and $\overline{DEN}$ is asserted low to enable the data buffer on the bus. Here, $\overline{RD}$ remains high.

3. During T_4, $\overline{WR}$ is asserted high and at this time (i.e., at the rising edge of $\overline{WR}$), the data is latched into the IO device.

4.1.6 Interrupt Acknowledge Cycle of 8086

The interrupt acknowledge cycle is executed in response to an interrupt request through the INTR pin of 8086.

The 8086 samples the status of the INTR pin during the last T-state of an instruction (or at the end of instruction execution).

If INTR is **high** at the time of sampling and the Interrupt flag is enabled (i.e., IF = 1) then the processor saves (or pushes) the content of flag register, CS-register and IP in stack, and clears IF and TF flags, and then executes interrupt acknowledge cycle.

The time taken by 8086 to execute an interrupt acknowledge cycle is eight T-states. It is actually two cycles with each cycle extending for 4T states.

In the first cycle, the processor sends $\overline{INTA}$ to the interrupting device to inform the acceptance of interrupt. In the second cycle, the processor requests the interrupting device to supply interrupt type number or pointer and read type number from the interrupting device by using an $\overline{INTA}$ signal.

The timing of various signals during interrupt acknowlegde cycle in minimum mode are shown in Fig. 4.6. During T_1 of both the cycles, ALE is made **high** and **low** which results in loading a junk value in address latches.

During T_1 of the first cycle DT/$\overline{R}$, M/$\overline{IO}$ and S_5 are asserted **low**. The DT/$\overline{R}$ is asserted **low** to inform the data buffer that the processor has to receive data. The M/$\overline{IO}$ is asserted **low** to indicate IO operation.

The S_5 is asserted **low** to inform the peripheral devices that the interrupt system is disabled. (Actually, S_5 is the status of the interrupt flag.)

In both the cycles, the $\overline{INTA}$ is asserted **low** during T_2 and then **high** during T_4. In the second cycle, when $\overline{INTA}$ is **low,** the processor expects an 8-bit interrupt pointer on the lower eight lines (AD_0 - AD_7) of the data bus.

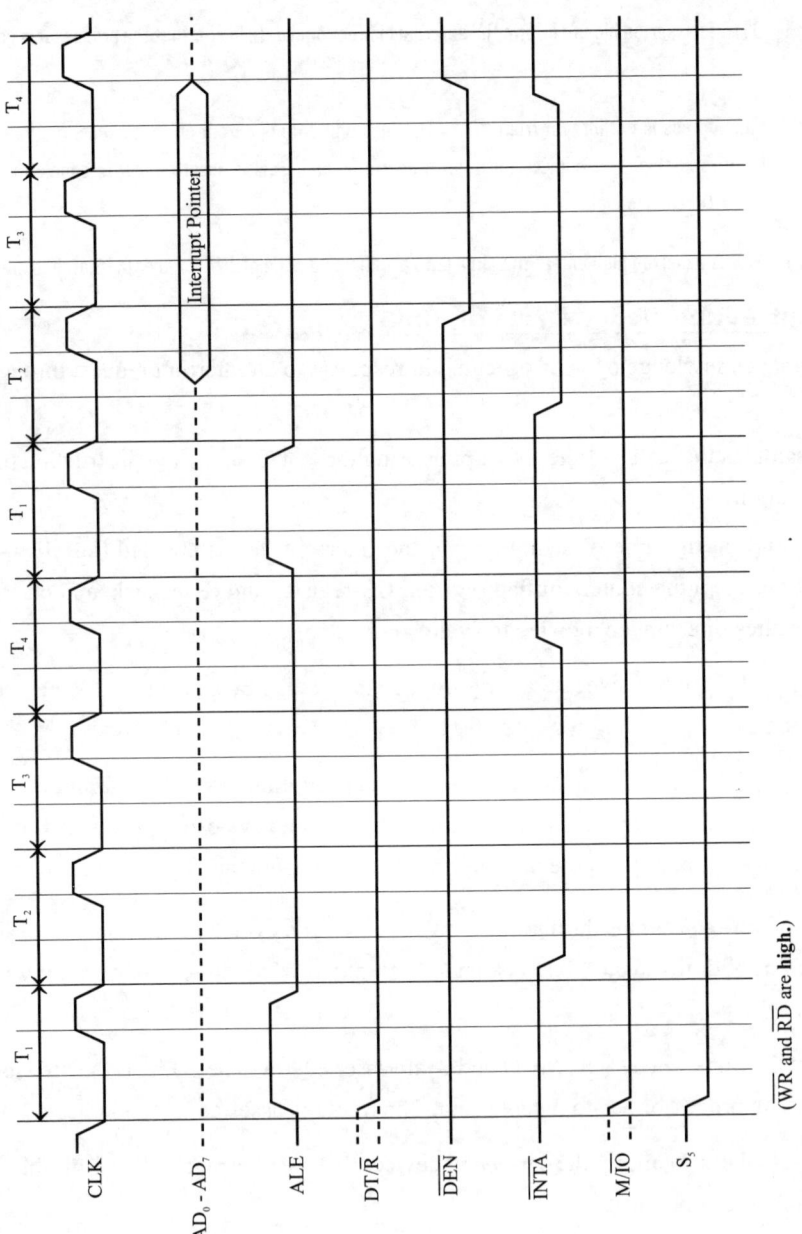

Fig. 4.6: Interrupt acknowledge cycle of 8086.

The time allowed to the interruping device to load the pointer is the time during which $\overline{\text{INTA}}$ remains **low**. The processor samples the interrupt pointer on the rising edge of INTA signal in the second cycle.

4.2 CLASSIFICATION OF 8086 INSTRUCTIONS

The 8086 instructions can be classified into the following six groups.

1. Data-transfer instructions

2. Arithmetic instructions

3. Logical instructions

4. String-manipulating instructions

5. Control-transfer instructions

6. Processor-control instructions

Data-Transfer Instructions : Includes instructions for moving data between registers, register and memory, register and stack memory, and accumulator and IO devices.

Arithmetic Instructions : Includes instructions for addition and subtraction of binary, BCD and ASCII data, and instructions for multiplication and division of signed and unsigned binary data.

Logical Instructions : Includes instructions for performing logical operations like AND, OR, Exclusive-OR, Complement, Shift, Rotate, etc.

String-Manipulation Instructions : Includes instructions for moving string data between two memory locations and comparing string data word by word or byte by byte.

Control-Transfer Instructions : Includes instructions to call a procedure/subroutine in the main program. It also includes instructions to jump from one part of a program to another part either conditionally (after checking flags) or unconditionally (without checking flags).

Processor Control Instructions : Includes instructions to set/clear the flags, to delay and halt the processor execution.

4.3 FORMAT OF 8086 INSTRUCTIONS

The size of an 8086 instruction is one to six bytes. Some examples of 8086 instruction formats are shown in Fig. 4.7 and the general format of 8086 instruction is shown in Fig. 4.8. The exact format of individual instructions are summarized in Appendix VI : *Templates for 8086 Instructions*.

In general, the first byte of the instruction will have a 6-bit opcode and two special bit indicators d-bit and w-bit or (s-bit and w-bit) or (v-bit and w-bit). Some instructions will have an 8-bit opcode and some instructions will have a 7-bit opcode followed by a special bit indicator, w-bit or z-bit.

Some of the instructions will have a 2-bit or 3-bit register field in the first byte of the instruction. The usage of special one-bit indicators are as follows:

w-bit : This bit appears in the format of instructions which can operate on both byte and word data.

If w = 0 then the data operated by the instruction is 8-bit/byte.

If w = 1 then the data operated by the instruction is 16-bit/word.

d-bit : This bit appears in the format of instructions which has a double operand. In double-operand instructions, one of the operands should be a register specified by a reg field. The d-bit is used to specify whether the register specified by the reg field is the source operand or destination operand.

If d = 0 then the register specified by the reg field is the source operand.

If d = 1 then the register specified by the reg field is the destination operand.

s-bit : This bit appears in the format of arithmetic instructions which operate on immediate data. If s = 1, w = 1 and immediate data is 8-bit then the immediate data is sign extended to 16-bit and used for arithmetic operation.

sw = 00 → 8-bit operation with an 8-bit immediate data.

sw = 01 → 16-bit operation with a 16-bit immediate data.

sw = 11 → 16-bit operation with a sign extended 8-bit immediate operand.

v-bit : This bit appears in the format of shift and rotate instructions.

If v = 0 then the shift/rotate operation is performed one time.

If v = 1 then the content of CL is count value for the number of shift/rotate operations.

z-bit : This bit appears in the format of REP prefix for string instructions and is used for comparing with zero flag.

If z = 0 then repeat execution of string instruction until the zero flag is zero.

If z = 1 then repeat execution of string instruction until the zero flag is one.

In multi-byte instructions, the second byte will specify the addressing mode of the operands. The second byte usually has three fields: mod, reg and r/m.

The mod field is 2-bit wide and it defines the method of addressing the operand specified by the r/m field. The r/m field is 3-bit wide and it is used to indicate the source or destination operand in the memory/register.

The reg field is 3-bit wide and it is used to indicate the source or destination operand in the register. If the register specified by the reg field is the source operand then the r/m field is used to indicate the destination operand or vice versa.

The mod and r/m fields are used to calculate the effective address of the memory operand as shown in Table 4.3.

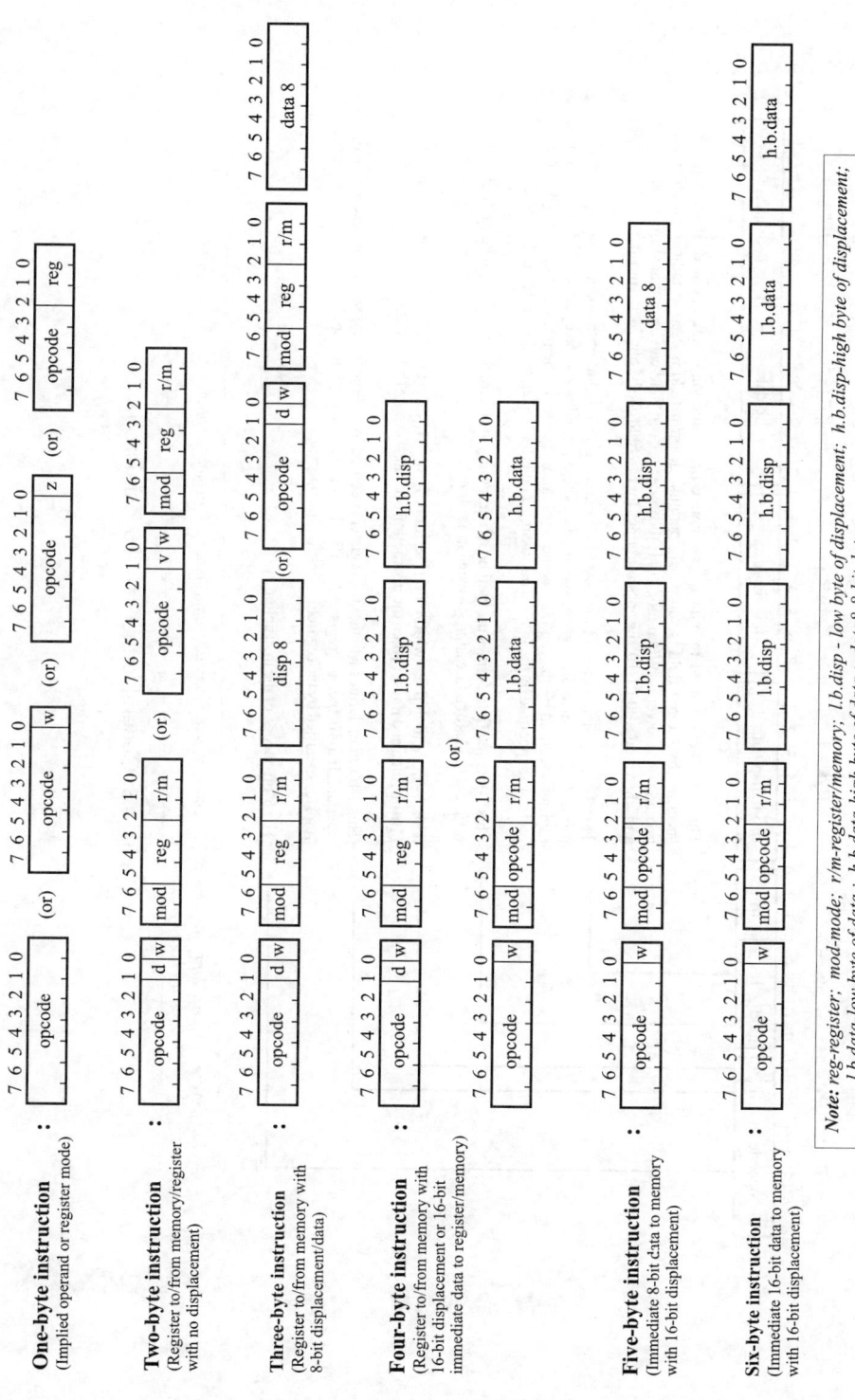

Fig. 4.7: Examples of 8086 instruction formats.

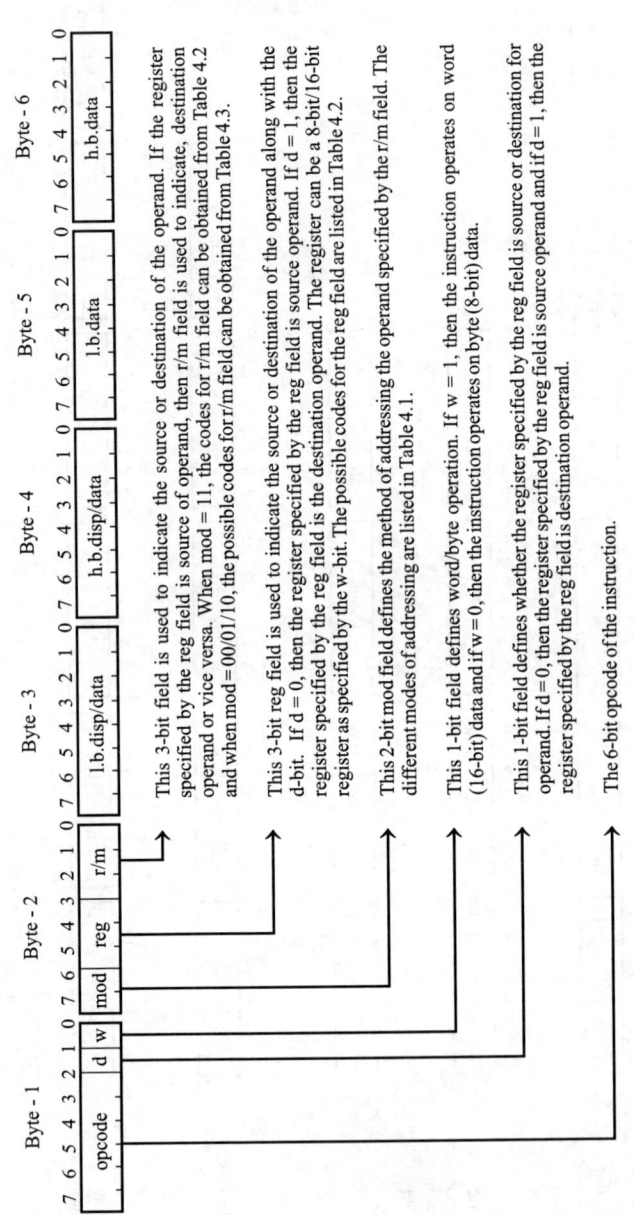

Fig. 4.8: General format of 8086 instruction.

Note: *l.b.disp/data - low byte displacement or data ; h.b. disp/data-high byte displacement or data ; l.b. data-low byte data ;*
h.b. data - high byte data.

In multi-byte instructions, the bytes following the opcode and address mode bytes (1st and 2nd bytes) may be any one of the following:

1. **No additional bytes.**
2. **Two-byte effective address.**
3. **One-byte (8-bit) signed displacement or two-byte (16-bit) unsigned displacement.**
4. **One-byte (8-bit) immediate data or two-byte (16-bit) immediate data (operand).**
5. **One/two byte displacement followed by one/two-byte immediate data (operand).**
6. **Two-byte effective address followed by two-byte segment address.**

Note: If a displacement or immediate data is two bytes long then the low-order byte always appears first.

Table 4.1: Codes for mod Field

Code for mod field	Name of the mode
00	Memory mode with no displacement
01	Memory mode with 8-bit signed displacement
10	Memory mode with 16-bit unsigned displacement
11	Register mode

Table 4.2: Codes for reg Field

Code for reg field	Name of the register represented by the code when w = 0 or 1	
	When w = 0	When w = 1
000	AL	AX
001	CL	CX
010	DL	DX
011	BL	BX
100	AH	SP
101	CH	BP
110	DH	SI
111	BH	DI

Table 4.3: Codes for r/m Field

Code for r/m field	Effective address calculation when mod = 00/01/10		
	mod = 00	mod = 01	mod = 10
000	[BX + SI]	[BX + SI + disp8]	[BX + SI + disp16]
001	[BX + DI]	[BX + DI + disp8]	[BX + DI + disp16]
010	[BP + SI]	[BP + SI + disp8]	[BP + SI + disp16]
011	[BP + DI]	[BP + DI + disp8]	[BP + DI + disp16]
100	[SI]	[SI + disp8]	[SI + disp16]
101	[DI]	[DI + disp8]	[DI + disp16]
110	[disp16]	[BP + disp8]	[BP + disp16]
111	[BX]	[BX + disp8]	[BX + disp16]

Note: disp 8 → 8-bit signed displacement ; disp16 → 16-bit unsigned displacement.

Table 4.4: Memory Address Calculation in 8086 Using Default Segment Register

S.No.	Addressing mode	Effective address EA	Physical address MA/MA_S
1.	[BX + SI]	EA = (BX) + (SI)	$MA = (DS) \times 16_{10} + EA$
2.	[BX + SI + disp8]	disp8 $\xrightarrow{\text{sign extend}}$ disp16 EA = (BX) + (SI) + disp16	$MA = (DS) \times 16_{10} + EA$
3.	[BX + SI + disp16]	EA = (BX) + (SI) + disp16	$MA = (DS) \times 16_{10} + EA$
4.	[BX + DI]	EA = (BX) + (DI)	$MA = (DS) \times 16_{10} + EA$
5.	[BX + DI + disp8]	disp8 $\xrightarrow{\text{sign extend}}$ disp16 EA = (BX) + (DI) + disp16	$MA = (DS) \times 16_{10} + EA$
6.	[BX + DI + disp16]	EA = (BX) + (DI) + disp16	$MA = (DS) \times 16_{10} + EA$
7.	[BP + SI]	EA = (BP) + (SI)	$MA_S = (SS) \times 16_{10} + EA$
8.	[BP + SI + disp8]	disp8 $\xrightarrow{\text{sign extend}}$ disp16 EA = (BP) + (SI) + disp16	$MA_S = (SS) \times 16_{10} + EA$
9.	[BP + SI + disp16]	EA = (BP) + (SI) + disp16	$MA_S = (SS) \times 16_{10} + EA$
10.	[BP + DI]	EA = (BP) + (DI)	$MA_S = (SS) \times 16_{10} + EA$
11.	[BP + DI + disp8]	disp8 $\xrightarrow{\text{sign extend}}$ disp16 EA = (BP) + (DI) + disp16	$MA_S = (SS) \times 16_{10} + EA$
12.	[BP + DI + disp16]	EA = (BP) + (DI) + disp16	$MA_S = (SS) \times 16_{10} + EA$
13.	[SI]	EA = (SI)	$MA = (DS) \times 16_{10} + EA$
14.	[SI + disp8]	disp8 $\xrightarrow{\text{sign extend}}$ disp16 EA = (SI) + disp16	$MA = (DS) \times 16_{10} + EA$
15.	[SI + disp16]	EA = (SI) + disp16	$MA = (DS) \times 16_{10} + EA$
16.	[DI]	EA = (DI)	$MA = (DS) \times 16_{10} + EA$
17.	[DI + disp8]	disp8 $\xrightarrow{\text{sign extend}}$ disp16 EA = (DI) + disp16	$MA = (DS) \times 16_{10} + EA$
18.	[DI + disp16]	EA = (DI) + disp16	$MA = (DS) \times 16_{10} + EA$
19.	[disp16]	EA = disp16	$MA = (DS) \times 16_{10} + EA$
20.	[BP + disp8]	disp8 $\xrightarrow{\text{sign extend}}$ disp16 EA = (BP) + disp16	$MA_S = (SS) \times 16_{10} + EA$
21.	[BP + disp16]	EA = (BP) + disp16	$MA_S = (SS) \times 16_{10} + EA$
22.	[BX]	EA = (BX)	$MA = (DS) \times 16_{10} + EA$
23.	[BX + disp8]	disp8 $\xrightarrow{\text{sign extend}}$ disp16 EA = (BX) + disp16	$MA = (DS) \times 16_{10} + EA$
24.	[BX + disp16]	EA = (BX) + disp16	$MA = (DS) \times 16_{10} + EA$

Note: *Segment registers used in address calculation can be modified using segment override prefix.*
$MA \rightarrow$ Memory address of data segment $MA_s \rightarrow$ Memory address of stack segment.

4.3.1 Addressing Modes of 8086

Every instruction of a program has to operate on a data. The method of specifying the data to be operated by the instruction is called addressing. The 8086 has 12 addressing modes and they can be classified into the following five groups.

1. Register addressing
2. Immediate addressing   **Group I : Addressing modes for register and immediate data**

3. Direct addressing
4. Register indirect addressing
5. Based addressing   **Group II : Addressing modes for memory data**
6. Indexed addressing
7. Based index addressing
8. String addressing

9. Direct IO port addressing
10. Indirect IO port addressing   **Group III : Addressing modes for IO ports**

11. Relative addressing   **Group IV : Relative addressing mode**

12. Implied addressing   **Group V : Implied addressing mode**

Notes: *1. The "register" or "register + constant" enclosed by square brackets in the operand field of instructions refer to the method of effective address calculation of memory. The 16-bit constant enclosed by square brackets in the operand field of instructions refers to the effective address of memory data. The 8-bit/16-bit constants which are not enclosed by square brackets in the operand field refer to immediate data.*

    2. The term MA used in the symbolic description of instructions refer to physical memory address of data segment memory, MA_S refers to physical memory address of stack segment memory; and MA_E refers to physical memory address of extra segment memory.

    3. The register/memory enclosed by brackets in symbolic description refers to the content of register/memory.

    4. For hexadecimal constant (data/address), the letter H is included at the end of 8-bit/16-bit constants(data/address), and the numeral 0 is included in the front of the hexadecimal constant starting with A through F.

Register Addressing

In register addressing, the instruction will specify the name of the register which holds the data to be operated by the instruction.

Examples:

(a) MOV CL,DH   (CL) ← (DH)

    The content of the 8-bit register DH is moved to another 8-bit register CL.

(b) MOV BX,DX   (BX) ← (DX)

    The content of the 16-bit register DX is moved to another 16-bit register BX.

Immediate Addressing

In immediate addressing mode, an 8-bit or 16-bit data is specified as part of the instruction.

Examples:

(a) MOV DL,08H   (DL) ← 08$_H$

   The 8-bit data (08$_H$) given in the instruction is moved to DL-register.

(b) MOV AX,0A9FH   (AX) ← 0A9F$_H$

   The 16-bit data (0A9F$_H$) given in the instruction is moved to AX-register.

Direct Addressing

*In direct addressing, an unsigned 16-bit displacement or signed 8-bit displacement will be specificied in the instruction. The displacement is the **E**ffective **A**ddress (EA) or offset. In case of 8-bit displacement, the effective address is obtained by sign extending the 8-bit displacement to 16-bit*

   The 20-bit physical address of memory is calculated by multiplying the content of DS-register by 16$_{10}$ (or 10$_H$)and adding to effective address. When segment override prefix is employed, the content of the segment register specified in the override prefix will be used for segment base address calculation instead of DS-register.

Examples:

(a) MOV DX,[08H]

   EA = 0008$_H$ (sign extended 8-bit displacement)
   BA = (DS) ×16$_{10}$;   MA = BA + EA
   (DX) ← (MA)   or   DL ← (MA)
                           DH ← (MA+1)

   *The **E**ffective **A**ddress (EA) is obtained by sign extending the 8-bit displacement given in the instruction to 16-bit. The segment **B**ase **A**ddress (BA) is computed by multiplying the content of DS by 16$_{10}$. The **M**emory **A**ddress (MA) is computed by adding the **E**ffective **A**ddress (EA) to segment **B**ase **A**ddress (BA).*

   The content of memory whose address is calculated as explained above is moved to DL-register and the content of the next memory location is moved to DH-register.

(b) MOV AX, [089DH]

   EA = 089D$_H$;   BA = (DS) × 16$_{10}$;   MA = BA + EA
   (AX) ← (MA)   or   (AL) ← (MA)
                           (AH) ← (MA+1)

   *Here, the 16-bit displacement given in the instruction is the effective address. The segment **B**ase **A**ddress (BA) is computed by multiplying the content of DS by 16$_{10}$. The **M**emory **A**ddress (MA) is computed by adding the **E**ffective **A**ddress (EA) to segment **B**ase **A**ddress (BA).*

   The content of memory whose address is calculated as explained above is moved to AL-register and the content of the next memory location is moved to AH-register.

Register Indirect Addressing

*In register indirect addressing, the name of the register which holds the **E**ffective **A**ddress (EA) will be specified in the instruction. The registers used to hold the effective address are BX, SI and DI. The content of DS is used for segment base address calculation. When segment override prefix is employed, the content of segment register specified in the override prefix will be used for base address calculation instead of DS-register.*

   The base address is obtained by multiplying the content of the segment register by 16$_{10}$. The 20-bit physical address of the memory is computed by adding the effective address to the base address.

Examples:

(a) MOV CX, [BX]

   EA = (BX)   ;   BA = (DS) × 16$_{10}$;   MA = BA + EA
   (CX) ← (MA)   or   (CL) ← (MA)
                           (CH) ← (MA+1)

The content of BX is the **E**ffective **A**ddress (EA). The segment **B**ase **A**ddress (BA) is computed by multiplying the content of DS by 16_{10}. The **M**emory **A**ddress (MA) is obtained by adding BA and EA.

The content of memory whose address is calculated as explained above is moved to CL-register and the content of the next memory location is moved to CH-register.

(b) MOV AX,[SI]

$$EA = (SI) \quad ; \quad BA = (DS) \times 16_{10} \quad ; \quad MA = BA + EA$$

$$(AX) \leftarrow (MA) \quad or \quad (AL) \leftarrow (MA)$$

$$(AH) \leftarrow (MA + 1)$$

The content of SI is the **E**ffective **A**ddress (EA). The segment **B**ase **A**ddress (BA) is computed by multiplying the content of DS by 16_{10}. The memory address is obtained by adding BA and EA.

The content of memory whose address is calculated as explained above is moved to AL-register and the content of the next memory location is moved to AH-register.

Based Addressing

In this addressing mode, the BX or BP-register is used to hold a base value for effective address and a signed 8-bit or unsigned 16-bit displacement will be specified in the instruction. The displacement is added to the base value in BX or BP to obtain the **E**ffective **A**ddress(EA). In case of 8-bit displacement, it is sign extended to 16-bit before adding to base value.

When BX is used to hold the base value for EA, the 20-bit physical address of memory is calculated by multiplying the content of DS by 16_{10} adding to EA.

When BP is used to hold the base value for EA, the 20-bit physical address of memory is calculated by multiplying the content of SS by 16_{10} and adding to EA.

Example:

MOV AX, [BX+08H]

$$0008_H \qquad\qquad 08_H \; ; \; EA = (BX) + 0008_H$$

$$BA = (DS) \times 16_{10} \; ; \; MA = BA + EA$$

$$(AX) \leftarrow (MA) \quad or \quad (AL) \leftarrow (MA)$$

$$(AH) \leftarrow (MA+1)$$

The effective address is calculated by sign extending the 8-bit displacement given in the instruction to 16-bit and adding to the content of BX-register. The Base Address (BA) is obtained by multiplying the content of DS by 16_{10}. The Memory Address (MA) is obtained by adding BA and EA.

The content of memory whose address is calculated as explained above is moved to AL-register and the content of the next memory is moved to AH-register.

Indexed Addressing

In this addressing mode, the SI or DI-register is used to hold an index value for memory data and a signed 8-bit displacement or unsigned 16-bit displacement will be specified in the instruction. The displacement is added to index value in SI or DI-register to obtain the **E**ffective **A**ddress(EA). In case of 8-bit displacement, it is sign extended to 16-bit before adding to the index value.

The 20-bit memory address is calculated by multiplying the content of **D**ata **S**egment(DS) by 16_{10} and adding to EA.

Note: *In general, Effective Address = Reference + Modifier.*

In this context, the based and indexed addressing looks similar, but in based addressing, the base value is the reference and displacement is the modifier, whereas in indexed addressing, displacement is the reference and index value is the modifier.

Example:

MOV CX, [SI+0A2H]

$FFA2_H \xleftarrow{\text{sign extend}} A2_H$; $EA = (SI) + FFA2_H$

$BA = (DS) \times 16_{10}$; $MA = BA + EA$

$(CX) \leftarrow (MA)$ or $(CL) \leftarrow (MA)$

$(CH) \leftarrow (MA + 1)$

The effective address is calculated by sign extending the 8-bit displacement given in the instruction to 16-bit and adding to the content of SI-register. The Base Address (BA) is obtained by multiplying the content of DS by 16_{10}. The Memory Address (MA) is obtained by adding BA and EA.

The content of memory whose address is calculated as explained above is moved to CL-register and the content of next memory is moved to CH-register.

Based Indexed Addressing

In this addressing mode, the effective address is given by sum of base value, index value and an 8-bit or 16-bit displacment specified in the instruction. The base value is stored in BX or BP-register. The index value is stored in SI or DI-register. In case of 8-bit displacement it is sign extended to 16-bit before adding to base value. This type of addressing will be useful in addressing two-dimensional arrays where we require two modifiers.

When BX is used to hold base value for EA, the 20-bit physical address of memory is calculated by multiplying the content of DS by 16_{10} and adding to EA.

When BP is used to hold the base value for EA, the 20-bit physical address of memory is obtained by multiplying the content of SS-register by 16_{10} and adding it to EA.

Example:

MOV DX, [BX+SI+0AH]

$000A_H \xleftarrow{\text{sign extend}} 0A_H$; $EA = (BX) + (SI) + 000A_H$

$BA = (DS) \times 16_{10}$; $MA = BA + EA$

$(DX) \leftarrow (MA)$ or $(DL) \leftarrow (MA)$

$(DH) \leftarrow (MA + 1)$

The Effective Address (EA) is calculated by sign extending the 8-bit displacement given in the instruction to 16-bit and adding it to the content of BX and SI-register. The Base Address (BA) is obtained by multiplying the content of DS by 16_{10}. The 20-bit Memory Address (MA) is obtained by adding BA and EA.

The content of memory whose address is calculated as explained above is moved to DL-register and the content of the next memory location is moved to DH-register.

String Addressing

This addressing mode is employed in string instructions to operate on string data. In string addressing mode, the Effective Address (EA) of source data is stored in SI-register and the EA of destination data is stored in DI-register.

The segment register used for calculating base address for source data is DS and can be overridden. The segment register used for calculating base address for destination is ES and cannot be overridden.

This addressing mode also supports auto increment/decrement of index registers SI and DI depending on Direction Flag (DF). If DF = 1 then the content of index registers are decremented to point to the previous byte/word of the string after execution of a string instruction. If DF = 0 then the content of index registers are incremented to point to the next byte/word of the string after execution of a string instruction. (For word operand, the content of index registers are incremented/decremented by two and for byte operand, the content of index registers are incremented/decremented by one.)

Example:

MOVS BYTE

     $EA = (SI)$; $BA = (DS) \times 16_{10}$; $MA = BA + EA$

     $EA_E = (DI)$; $BA_E = (ES) \times 16_{10}$; $MA_E = BA_E + EA_E$

     $(MA_E) \leftarrow (MA)$

     If DF = 1, then $(SI) \leftarrow (SI) - 1$ and $(DI) \leftarrow (DI) - 1$

     If DF = 0, then $(SI) \leftarrow (SI) + 1$ and $(DI) \leftarrow (DI) + 1$

     This instruction moves a byte of string data from one memory location to another memory location. The address of source memory location is calculated by multiplying the content of DS by 16_{10} and adding to SI. The address of destination memory location is calculated by multiplying the content of ES by 16_{10} and adding to DI.

     After the move operation, if DF = 1 then the content of index registers DI and SI are decremented by one. If DF = 0 then the content of index registers DI and SI are incremented by one.

Direct IO Port Addressing

     This addressing mode is used to access data from standard IO mapped devices or ports. In the direct port addressing mode, an 8-bit port address is directly specified in the instruction.

Example:

IN AL, [09H]

         $PORT_{addr} = 09_H$

         $(AL) \leftarrow (PORT)$

     The content of the port with the address 09_H is moved to AL-register.

Indirect IO Port Addressing

     This addressing mode is used to access data from standard IO mapped devices or ports. In the indirect port addressing mode, the instruction will specify the name of the register which holds the port address. In 8086, the 16-bit port address is stored in DX-register.

Example:

OUT [DX], AX

$PORT_{addr} = (DX)$; $(PORT) \leftarrow (AX)$

     The content of AX is moved to the port whose address is specified by DX-register.

Relative Addressing

     In this addressing mode, the effective address of a program instruction is specified relative to the Instruction Pointer (IP) by an 8-bit signed displacement.

Example:

JZ 0AH

     $000A_H \xleftarrow{\text{sign extend}} 0A_H$

     If ZF = 1, then $(IP) \leftarrow (IP) + 000A_H$; $EA_C = (IP) + 000A_H$

                    $BA_C = (CS) \times 16_{10}$; $MA_C = BA_C + EA_C$

Note: Suffix C refers to code memory

     If ZF = 1 then the program control jumps to a new code address as calculated above. If ZF = 0 then the next instruction of the program is executed.

Implied Addressing

     In implied addressing mode, the instruction itself will specify the data to be operated by the instruction.

Example:

CLC - Clear carry ; CF $\leftarrow$ 0

     Execution of this instruction will clear the Carry Flag (CF).

4.4 INSTRUCTION EXECUTION TIME

The execution time of each instruction of 8086 is specified by INTEL in terms of clock cycles or periods of the processor. The execution time can be determined by multiplying the number of clock cycles needed to execute the instruction by the time period of the processor clock. The execution time of 8086 instructions are listed in the Table 4.5: *INTEL 8086 instruction set*. The execution time specified in Table 4.5, assumes that the instruction to be executed has already been fetched and stored in the instruction queue.

When the instruction execution involves memory access then extra time is needed for memory address calculation. This extra time is denoted as EACT (**E**ffective **A**ddress **C**alculation **T**ime) in Table 4.5. The time required for address calculation (EACT) depends on addressing mode used in the instruction. The clock cycles required for address calculation for various memory addressing modes are listed in Table 4.6.

Instructions like multiply, divide, shift and rotate will have variable execution time depending on the type of data operated by the instruction.

The conditional branch instructions will have different timings for TRUE and FALSE condition. When the condition is TRUE, branch operation takes place which involves modifying the IP and CS, clearing the queue and then filling the queue with instruction codes from branch address, and so instruction execution takes a longer time. When the condition is FALSE, the next instruction is executed and so the instruction takes lesser time.

The execution time specified in Table 4.5 also assumes that word operand is located in an even address. If the word operand is located in an odd address then the processor executes two bus cycles for each memory access and so four extra clock cycle has to be added to execution time of the instruction for each memory access. The number of memory access while executing the instruction is also listed in Table 4.5.

Table 4.5: INTEL 8086 Instruction Set

S.No.	Mnemonic	Size of instruction (Number of bytes)	Clock period (or T-state) needed for instruction execution	Number of memory access (or transfer)
GROUP I : DATA TRANSFER INSTRUCTIONS				
a)	MOV reg2, reg1	2	2	-
b)	MOV mem, reg1	2 to 4	9 + EACT	1
c)	MOV reg2, mem	2 to 4	8 + EACT	1
2.	MOV reg/mem, data			
a)	MOV reg, data	3 to 4	4	-
b)	MOV mem, data	3 to 6	10 + EACT	1
3.	MOV reg, data	2 to 3	4	-
4.	MOV A, mem			
a)	MOV AL, mem	3	10	1
b)	MOV AX, mem	3	10	1

Table 4.5 continued...

S.No.	Mnemonic	Size of instruction (Number of bytes)	Clock period (or T-state) needed for instruction execution	Number of memory access (or transfer)
5.	MOV mem, A			
a)	MOV mem, AL	3	10	1
b)	MOV mem, AX	3	10	1
6.	MOV segreg, reg16/mem			
a)	MOV segreg, reg16	2	2	-
b)	MOV segreg, mem	2 to 4	8 + EACT	1
7.	MOV reg16/mem, segreg			
a)	MOV reg16, segreg	2	2	-
b)	MOV mem, segreg	2 to 4	9 + EACT	1
8.	PUSH reg16/mem			
a)	PUSH reg16	2	11	1
b)	PUSH mem	2 to 4	16 + EACT	2
9.	PUSH reg16	1	11	1
10.	PUSH segreg	1	10	1
11.	PUSHF	1	10	1
12.	POP reg16/mem			
a)	POP reg16	2	8	1
b)	POP mem	2 to 4	17 + EACT	2
13.	POP reg16	1	8	1
14.	POP segreg	1	8	1
15.	POPF	1	8	1
16.	XCHG reg2/mem, reg1			
a)	XCHG reg2, reg1	2	4	-
b)	XCHG mem, reg1	2 to 4	17 + EACT	2
17.	XCHG AX, reg16	1	3	-
18.	XLAT	1	11	1
19.	IN A, [DX]			
a)	IN AL, [DX]	1	8	1 ⎫ IO
b)	IN AX, [DX]	1	8	1 ⎭ access

Table 4.5 continued...

S.No.	Mnemonic	Size of instruction (Number of bytes)	Clock period (or T-state) needed for instruction execution	Number of memory access (or transfer)
20.	IN A, addr8			
a)	IN AL, addr8	2	10	1 ⎫
b)	IN AX, addr8	2	10	1 ⎬ IO access
21.	OUT [DX], A			
a)	OUT [DX], AL	1	8	1 ⎫
b)	OUT [DX], AX	1	8	1 ⎬ IO access
22.	OUT addr8, A			
a)	OUT addr8, AL	2	10	1 ⎫
b)	OUT addr8, AX	2	10	1 ⎬ IO access
23.	LEA reg16, mem	2 to 4	2 + EACT	-
24.	LDS reg16, mem	2 to 4	16 + EACT	2
25.	LES reg16, mem	2 to 4	16 + EACT	2
26.	LAHF	1	4	-
27.	SAHF	1	4	-
GROUP II : ARITHMETIC INSTRUCTIONS				
28.	ADD reg2/mem, reg1/mem			
a)	ADD reg2, reg1	2	3	-
b)	ADD reg2, mem	2 to 4	9 + EACT	1
c)	ADD mem, reg1	2 to 4	16 + EACT	2
29.	ADD reg/mem, data			
a)	ADD reg, data	3 to 4	4	-
b)	ADD mem, data	3 to 6	17 + EACT	2
30.	ADD A, data			
a)	ADD AL, data8	2	4	-
b)	ADD AX, data16	3	4	-
31.	ADC reg2/mem, reg1/mem			
a)	ADC reg2, reg1	2	3	-
b)	ADC reg2, mem	2 to 4	9 + EACT	1
c)	ADC mem, reg1	2 to 4	16 + EACT	2

Table 4.5 continued...

S.No.	Mnemonic	Size of instruction (Number of bytes)	Clock period (or T-state) needed for instruction execution	Number of memory access (or transfer)
32.	ADC reg/mem, data			
a)	ADC reg, data	3 to 4	4	-
b)	ADC mem, data	3 to 6	17 + EACT	2
33.	ADC A, data			
a)	ADC AL, data8	2	4	-
b)	ADC AX, data16	3	4	-
34.	AAA	1	4	-
35.	DAA	1	4	-
36.	SUB reg2/mem, reg1/mem			
a)	SUB reg2, reg1	2	3	-
b)	SUB reg2, mem	2 to 4	9 + EACT	1
c)	SUB mem, reg1	2 to 4	16 + EACT	2
37.	SUB reg/mem, data			
a)	SUB reg, data	3 to 4	4	-
b)	SUB mem, data	3 to 6	17 + EACT	2
38.	SUB A, data			
a)	SUB AL, data8	2	4	-
b)	SUB AX, data16	3	4	-
39.	SBB reg2/mem, reg1/mem			
a)	SBB reg2, reg1	2	3	-
b)	SBB reg2, mem	2 to 4	9 + EACT	1
c)	SBB mem, reg1	2 to 4	16 + EACT	2
40.	SBB reg/mem, data			
a)	SBB reg, data	3 to 4	4	-
b)	SBB mem, data	3 to 6	17 + EACT	2
41.	SBB A, data			
a)	SBB AL, data8	2	4	-
b)	SBB AX, data16	3	4	-
42.	AAS	1	4	-
43.	DAS	1	4	-

Table 4.5 continued...

S.No.	Mnemonic	Size of instruction (Number of bytes)	Clock period (or T-state) needed for instruction execution	Number of memory access (or transfer)
44.	MUL reg/mem			
a)	MUL reg			
	i) MUL reg8	2	70 to 77	-
	ii) MUL reg16	2	118 to 133	-
b)	MUL mem			
	i) MUL mem8	2 to 4	(76 to 83) + EACT	1
	ii) MUL mem16	2 to 4	(124 to 139)+EACT	1
45.	IMUL reg/mem			
a)	IMUL reg			
	i) IMUL reg8	2	80 to 98	-
	ii) IMUL reg16	2	128 to 154	-
b)	IMUL mem			
	i) IMUL mem8	2 to 4	(86 to 104) + EACT	1
	ii) IMUL mem16	2 to 4	(134 to 160) + EACT	1
46.	AAM	2	83	-
47.	DIV reg/mem			
a)	DIV reg			
	i) DIV reg8	2	80 to 90	-
	ii) DIV reg16	2	144 to162	-
b)	DIV mem			
	i) DIV mem8	2 to 4	(86 to 96) + EACT	1
	ii) DIV mem16	2 to 4	(150 to 168)+EACT	1
48.	IDIV reg/mem			
a)	IDIV reg			
	i) IDIV reg8	2	101 to 112	-
	ii) IDIV reg16	2	165 to 184	-
b)	IDIV mem			
	i) IDIV mem8	2 to 4	(107 to 118)+EACT	1
	ii) IDIV mem16	2 to 4	(171 to 190)+EACT	1
49.	AAD	2	60	-
50.	NEG mem/reg			
a)	NEG reg	2	3	1
b)	NEG mem	2 to 4	16 + EACT	2
51.	INC reg8/mem			
a)	INC reg8	2	3	-
b)	INC mem	2 to 4	15 + EACT	2

Table 4.5 continued...

S.No.	Mnemonic	Size of instruction (Number of bytes)	Clock period (or T-state) needed for instruction execution	Number of memory access (or transfer)
52.	INC reg16	1	2	-
53.	DEC reg8/mem			
a)	DEC reg8	2	3	-
b)	DEC mem	2 to 4	15 + EACT	2
54.	DEC reg16	1	2	-
55.	CBW	1	2	-
56.	CWD	1	5	-
57.	CMP reg2/mem, reg1/mem			
a)	CMP reg2, reg1	2	3	-
b)	CMP reg2, mem	2 to 4	9 + EACT	1
c)	CMP mem, reg1	2 to 4	9 + EACT	1
58.	CMP reg/mem, data			
a)	CMP reg, data	3 to 4	4	-
b)	CMP mem, data	3 to 6	10 + EACT	1
59.	CMP A, data			
a)	CMP AL, data8	2	4	-
b)	CMP AX, data16	3	4	-
GROUP III : LOGICAL INSTRUCTIONS				
60.	AND reg2/mem, reg1/mem			
a)	AND reg2, reg1	2	3	-
b)	AND reg2, mem	2 to 4	9 + EACT	1
c)	AND mem, reg1	2 to 4	16 + EACT	2
61.	AND reg/mem, data			
a)	AND reg, data	3 to 4	4	-
b)	AND mem, data	3 to 6	17 + EACT	2
62.	AND A, data			
a)	AND AL, data8	2	4	-
b)	AND AX, data16	3	4	-

Table 4.5 continued...

S.No.	Mnemonic	Size of instruction (Number of bytes)	Clock period (or T-state) needed for instruction execution	Number of memory access (or transfer)
63.	OR reg2/mem, reg1/mem			
a)	OR reg2, reg1	2	3	-
b)	OR reg2, mem	2 to 4	9 + EACT	1
c)	OR mem, reg1	2 to 4	16 + EACT	
64.	OR reg/mem, data			
a)	OR reg, data	3 to 4	4	-
b)	OR mem, data	3 to 6	17 + EACT	2
65.	OR A, data			
a)	OR AL, data8	2	4	-
b)	OR AX, data16	3	4	-
66.	XOR reg2/mem, reg1/mem			
a)	XOR reg2, reg1	2	3	-
b)	XOR reg2, mem	2 to 4	9 + EACT	1
c)	XOR mem, reg1	2 to 4	16 + EACT	2
67.	XOR reg/mem, data			
a)	XOR reg, data	3 to 4	4	-
b)	XOR mem, data	3 to 6	17 + EACT	2
68.	XOR A, data			
a)	XOR AL, data8	2	4	-
b)	XOR AX, data16	3	4	-
69.	TEST reg2/mem, reg1/mem			
a)	TEST reg2, reg1	2	3	-
b)	TEST reg2, mem	2 to 4	9 + EACT	1
c)	TEST mem, reg1	2 to 4	9 + EACT	1
70.	TEST reg/mem, data			
a)	TEST reg, data	3 to 4	5	-
b)	TEST mem, data	3 to 6	11 + EACT	1
71.	TEST A, data			
a)	TEST AL, data8	2	4	-
b)	TEST AX, data16	3	4	-
72.	NOT reg/mem			
a)	NOT reg	2	3	-
b)	NOT mem	2 to 4	16 + EACT	2

Table 4.5 continued...

S.No.	Mnemonic	Size of instruction (Number of bytes)	Clock period (or T-state) needed for instruction execution	Number of memory access (or transfer)
73.	SHL reg/mem (or SAL reg/mem)			
a)	SHL reg (or SAL reg)			
	i) SHL reg, 1 (or SAL reg, 1)	2	2	-
	ii) SHL reg, CL (or SAL reg, CL)	2	8 + 4 per bit	-
b)	SHL mem (or SAL mem)			
	i) SHL mem, 1 (or SAL mem, 1)	2 to 4	15 + EACT	2
	ii) SHL mem, CL (or SAL mem, CL)	2 to 4	20 + EACT + 4 per bit	2
74.	SHR reg/mem			
a)	SHR reg			
	i) SHR reg, 1	2	2	-
	ii) SHR reg, CL	2	8 + 4 per bit	-
b)	SHR mem			
	i) SHR mem, 1	2 to 4	15 + EACT	2
	ii) SHR mem, CL	2 to 4	20 + EACT + 4 per bit	2
75.	SAR reg/mem			
a)	SAR reg			
	i) SAR reg, 1	2	2	-
	ii) SAR reg, CL	2	8 + 4 per bit	-
b)	SAR mem			
	i) SAR mem, 1	2 to 4	15 + EACT	2
	ii) SAR mem, CL	2 to 4	20 + EACT + 4 per bit	2
76.	ROL reg/mem			
a)	ROL reg			
	i) ROL reg, 1	2	2	-
	ii) ROL reg, CL	2	8 + 4 per bit	-
b)	ROL mem			
	i) ROL mem, 1	2 to 4	15 + EACT	2
	ii) ROL mem, CL	2 to 4	20 + EACT + 4 per bit	2
77.	RCL reg/mem			
a)	RCL reg			
	i) RCL reg, 1	2	2	-
	ii) RCL reg, CL	2	8 + 4 per bit	-

Table 4.5 continued...

S.No.	Mnemonic	Size of instruction (Number of bytes)	Clock period (or T-state) needed for instruction execution	Number of memory access (or transfer)
b)	RCL mem			
	i) RCL mem, 1	2 to 4	15 + EACT	2
	ii) RCL mem, CL	2 to 4	20+EACT+4 per bit	2
78.	ROR reg/mem			
a)	ROR reg			
	i) ROR reg, 1	2	2	-
	ii) ROR reg, CL	2	8 + 4 per bit	-
b)	ROR mem			
	i) ROR mem, 1	2 to 4	15 + EACT	2
	ii) ROR mem, CL	2 to 4	20+EACT+4 per bit	2
79.	RCR reg/mem			
a)	RCR reg			
	i) RCR reg, 1	2	2	-
	ii) RCR reg, CL	2	8 + 4 per bit	-
b)	RCR mem			
	i) RCR mem, 1	2 to 4	15 + EACT	2
	ii) RCR mem, CL	2 to 4	20+EACT+4 per bit	2
GROUP IV : STRING MANIPULATION INSTRUCTIONS				
80.	REP			
a)	REPZ/REPE	1	2	-
b)	REPNZ/REPNE	1	2	-
81.	MOVS		18 or	
a)	MOVSB	1	9 + 17 per	2
b)	MOVSW		repetition	
82.	CMPS		22 or	
a)	CMPSB	1	9 + 22 per	2
b)	CMPSW		repetition	
83.	SCAS		15 or	
a)	SCASB	1	9 + 15 per	1
b)	SCASW		repetition	
84.	LODS		12 or	
a)	LODSB	1	9 + 13 per	1
b)	LODSW		repetition	
85.	STOS		11 or	
a)	STOSB	1	9 + 10 per	1
b)	STOSW		repetition	

Table 4.5 continued...

S.No.	Mnemonic	Size of instruction (Number of bytes)	Clock period (or T-state) needed for instruction execution	Number of memory access (or transfer)
GROUP V : CONTROL TRANSFER INSTRUCTIONS				
86.	CALL disp16	3	19	1
87.	CALL reg/mem			
a)	CALL reg	2	16	1
b)	CALL mem	2 to 4	21 + EACT	2
88.	CALL addr$_{offset}$, addr$_{base}$	5	28	2
89.	CALL mem	2 to 4	37 + EACT	4
90.	RET (Return from call within segment)	1	8	1
91.	RET data16 (Return from call within segment adding immediate data to SP)	3	12	1
92.	RET (Return from intersegment call)	1	18	2
93.	RET data16 (Return from intersegment call adding immediate data to SP)	3	17	2
94.	JMP disp16 (Unconditional jump near-direct within segment)	3	15	-
95.	JMP disp8 (Unconditional jump short-direct within segment)	2	15	-
96.	JMP reg/mem (Unconditional jump near-indirect within segment)			
a)	JMP reg	2	11	-
b)	JMP mem	2 to 4	18 + EACT	1
97.	JMP addr$_{offset}$, addr$_{base}$ (Unconditional jump far-direct intersegment)	5	15	-
98.	JMP mem (Unconditional jump far-indirect intersegment)	2 to 4	24 + EACT	2
99.	JE/JZ disp8	2	16 or 4	-
100.	JL/JNGE disp8	2	16 or 4	-

Table 4.5 continued...

S.No.	Mnemonic	Size of instruction (Number of bytes)	Clock period (or T-state) needed for instruction execution	Number of memory access (or transfer)
101.	JLE/JNG disp8	2	16 or 4	-
102.	JB/JNAE/JC disp8	2	16 or 4	-
103.	JBE/JNA disp8	2	16 or 4	-
104.	JP/JPE disp8	2	16 or 4	-
105.	JNB/JAE/JNC disp8	2	16 or 4	-
106.	JNBE/JA disp8	2	16 or 4	-
107.	JNP/JPO disp8	2	16 or 4	-
108.	JNO disp8	2	16 or 4	-
109.	JNS disp8	2	16 or 4	-
110.	JO disp8	2	16 or 4	-
111.	JS disp8	2	16 or 4	-
112.	JNE/JNZ disp8	2	16 or 4	-
113.	JNL/JGE disp8	2	16 or 4	-
114.	JNLE/JG disp8	2	16 or 4	-
115.	JCXZ disp8	2	18 or 6	-
116.	LOOP disp8	2	17 or 5	-
117.	LOOPZ/LOOPE disp8	2	18 or 6	-
118.	LOOPNZ/LOOPNE disp8	2	19 or 5	-
119.	INT type	2	51	5
120.	INT 3	1	52	5
121.	INTO	1	53 or 4	5
122.	IRET	1	24	3
GROUP VI : PROCESSOR CONTROL INSTRUCTIONS				
123.	CLC	1	2	-
124.	CMC	1	2	-
125.	STC	1	2	-
126.	CLD	1	2	-
127.	STD	1	2	-
128.	CLI	1	2	-
129.	STI	1	2	-
130.	HLT	1	2	-
131.	WAIT	1	3 + 5n	-
132.	ESC opcode, mem/reg			
a)	ESC opcode, mem	2 to 4	8 + EACT	1
b)	ESC opcode, reg	2	2	-
133.	LOCK	1	2	-
134.	NOP	1	3	-

Table 4.6: Effective Address Calculation Time

S.No.	Method of addressing memory	Number of clock cycles for EACT
1.	Direct addressing	6
2.	Register indirect addressing [BX] or [SI] or [DI]	5
3.	Based addressing [BX + disp] or [BP + disp]	9
4.	Indexed addressing [SI + disp] or [DI + disp]	9
5.	Based indexed addressing	
	a) Without displacement	
	i) [BP + DI] or [BX + SI]	7
	ii) [BP + SI] or [BX + DI]	8
	b) With displacement	
	i) [BP + DI + disp] or [BX + SI + disp]	11
	ii) [BP + SI + disp] or [BX + DI + disp]	12

Table 4.7: Meanings of Various Terms used in Operand Field of Instructions

Term	Meaning	Term	Meaning
reg/reg1/reg2	8-bit or 16-bit register	data8	8-bit data
reg8	8-bit register	data16	16-bit data
reg16	16-bit register	addr8	8-bit address
segreg	segment register	$addr_{offset}$	16-bit offset/effective
mem	8-bit or 16-bit memory		address
mem8	8-bit memory	$addr_{base}$	16-bit base address
mem16	16-bit memory	disp8	8-bit displacement
data	8-bit or 16-bit data	disp16	16-bit displacement

Note: *1. Possible choice for reg/reg1/reg2 are AL, AH, BL, BH, CL, CH, DL, DH, AX, BX, CX, DX, SI, DI, SP and BP.*

       2. Possible choice for reg8 are AL, AH, BL, BH, CL, CH, DL and DH.

       3. Possible choice for reg16 are AX, BX, CX, DX, SI, DI, SP and BP.

       4. Possible choice for segreg are DS, ES and SS.

       5. The term mem stands for the 24 different methods of addressing memory data as shown in Table 4.4.

Table 4.8: Meaning/Expansion of Mnemonics Used in 8086 Instruction Set

S.No.	Mnemonic	Meaning
1.	AAA	ASCII adjust after addition.
2.	AAD	ASCII adjust before division.
3.	AAM	ASCII adjust after multiply.
4.	AAS	ASCII adjust after subtraction.
5.	ADC	Add two specified data along with carry.
6.	ADD	Add two specified data.
7.	AND	AND two specified data bit by bit.
8.	CALL	Call a procedure/subroutine.
9.	CBW	Convert byte to word (Sign extend byte to word).
10.	CLC	Clear carry flag (CF = 0).
11.	CLD	Clear direction flag (DF = 0).
12.	CLI	Clear interrupt enable flag (IF = 0).
13.	CMC	Complement the state of the carry flag (CF = $\overline{\text{CF}}$).
14	CMP	Compare two specified data.
15.	CMPS/CMPSB/CMPSW	Compare two string byte or two string word.
16.	CWD	Convert word to double word. (Sign extend the word to double word.)
17.	DAA	Decimal adjust after addition.
18.	DAS	Decimal adjust after subtraction.
19.	DEC	Decrement specified data.
20.	DIV	Divide unsigned word by byte, or unsigned double word by word.
21.	ESC	Escape to external coprocessor such as 8087.
22.	HLT	Halt until interrupt.
23.	IDIV	Divide signed word by byte, or signed double word by word.
24.	IMUL	Multiply signed byte by byte or signed word by word.
25.	IN	Copy a byte/word from specified port to accumulator.
26.	INC	Increment specified data.
27.	INT	Interrupt program execution. (Call interrupt service procedure.)
28.	INTO	Interrupt program execution, if OF = 1.
29.	IRET	Interrupt return.
30.	JA/JNBE	Jump if above/Jump if not below nor equal.
31.	JAE/JNB	Jump if above or equal/Jump if not below.
32.	JB/JNAE	Jump if below/Jump if not above nor equal.
33.	JBE/JNA	Jump if below or equal/Jump if not above.
34.	JC	Jump if CF = 1.
35.	JCXZ	Jump if CX = 0.
36.	JE/JZ	Jump if equal/Jump if ZF = 1.

Table 4.8 continued...

S.No.	Mnemonic	Meaning
37.	JG/JNLE	Jump if greater/Jump if not less than nor equal.
38.	JGE/JNL	Jump if greater than or equal/Jump if not less than.
39.	JL/JNGE	Jump if less than/Jump if not greater than.
40.	JLE/JNG	Jump if less than or equal/Jump if not greater than.
41.	JMP	Jump to specified address to get next instruction.
42.	JNC	Jump if no carry (Jump if CF = 0).
43.	JNE/JNZ	Jump if not equal/Jump if not zero (ZF = 0).
44.	JNO	Jump if no overflow (Jump if OF = 0).
45.	JNP/JPO	Jump if not parity/Jump if parity odd. (PF = 0).
46.	JNS	Jump if not sign (Jump if sign flag = 0).
47.	JO	Jump if OF = 1.
48.	JP/JPE	Jump if parity/Jump if parity even (PF = 1).
49.	JS	Jump if SF = 1.
50.	LAHF	Load AH with the low byte of the flag register.
51.	LDS	Load DS-register and other specified register from memory.
52.	LEA	Load effective address of operand into specified register.
53.	LES	Load ES-register and other specified register from memory.
54.	LOCK	An instruction prefix which prevents another processor from taking bus while the adjacent instruction (i.e., instruction prefixed with lock) executes.
55.	LODS/LODSB /LODSW	Load string byte into AL or string word into AX.
56.	LOOP	Loop through a sequence of instructions until CX = 0.
57.	LOOPE/LOOPZ	Loop through a sequence of instructions while ZF = 1 and CX $\neq$ 0.
58.	LOOPNE/LOOPNZ	Loop through a sequence of instructions while ZF = 0 and CX $\neq$ 0.
59.	MOV	Move (copy) a byte/word from specified source to specified destination.
60.	MOVS/MOVSB /MOVSW	Move a byte/word from one string to another.
61.	MUL	Multiply two specified unsigned data.
62.	NEG	Negative of specified data (2's complement value of specified data).
63.	NOP	No action (operation) except fetch and decode.
64.	NOT	Invert/complement each bit of specified data.
65.	OR	OR two specified data bit by bit.
66.	OUT	Copy a byte/word from accumulator to specified port.
67.	POP	Move the top of stack to specified location.

Table 4.8 continued...

S.No.	Mnemonic	Meaning
68.	POPF	Move the top of stack to flag register.
69.	PUSH	Push (copy) the specified register to top of stack.
70.	PUSHF	Push (copy) the flag register to top of stack.
71.	RCL	Rotate left through carry.
72.	RCR	Rotate right through carry.
73.	REP	An instruction prefix. Repeat adjacent instruction (i.e., instruction prefixed with REP) until CX = 0.
74.	REPE/REPZ	An instruction prefix. Repeat adjacent instruction until CX = 0 or ZF ≠ 1.
75.	REPNE/REPNZ	An instruction prefix. Repeat adjacent instruction until CX = 0 or ZF = 1.
76.	RET	Return from procedure to calling program.
77.	ROL	Rotate left to carry.
78.	ROR	Rotate right to carry.
79.	SAHF	Store (copy) AH-register to low byte of flag register.
80.	SAR	Arithmetic Right shift.
81.	SBB	Subtract specified data and carry flag from another specified data.
82.	SCAS/SCASB/ SCASW	Scan (compare) a string byte/word with accumulator.
83.	SHL/SAL	Logical left shift/Arithmetic left shift.
84.	SHR	Logical right shift.
85.	STC	Set carry flag (CF = 1).
86.	STD	Set direction flag (DF = 1).
87.	STI	Set interrupt enable flag (IF = 1).
88.	STOS/STOSB/ STOSW	Store byte from AL or word from AX into string.
89.	SUB	Subtract a specified data from another specified data.
90.	TEST	Test by performing logical AND operation of specified operands and modify flags.
91.	WAIT	Wait until signal on the test pin is **low**.
92.	XCHG	Exchange bytes or exchange words.
93.	XLAT	Translate a byte in AL using a table in memory.
94.	XOR	Exclusive-OR two specified data bit by bit.

4.5 INSTRUCTIONS AFFECTING FLAGS OF 8086

The 8086 microprocessor has 9 flags. In this, six flags are altered by arithmetic and logical instructions, and three flags are used to control the processor operation.

The flags which are altered by arithmetic and logical instructions are carry flag, auxiliary carry flag, parity flag, zero flag, sign flag and overflow flag. The flags which are used to control the processor operation are trace flag (or single step trap), interrupt flag and direction flag.

The status of various flags after execution of arithmetic and logical instructions are listed in Table 4.9. The 8086 processor has instructions to directly set or clear the interrupt flag, direction flag and carry flag.

While servicing an interrupt the 8086 processor, save the status of flags in stack and the status of the flags are restored at the end of service procedure by executing IRET instruction.

The 8086 also has instruction to directly save the flags in stack (PUSHF) and to restore the saved flags (POPF).

Table 4.9: 8086 Instructions Affecting Flags

Instruction	Flags								
	O	D	I	T	S	Z	A	P	C
AAA	u	-	-	-	u	u	+	u	+
AAD	u	-	-	-	+	+	u	+	u
AAM	u	-	-	-	+	+	u	+	u
AAS	u	-	-	-	u	u	+	u	+
ADC	+	-	-	-	+	+	+	+	+
ADD	+	-	-	-	+	+	+	+	0
AND		0	-	-	-	+	+	u	+
CLC	-	-	-	-	-	-	-	-	0
CLD	-	0	-	-	-	-	-	-	-
CLI	-	-	0	-	-	-	-	-	-
CMC	-	-	-	-	-	-	-	-	+
CMP	+	-	-	-	+	+	+	+	+
CMPS	+	-	-	-	+	+	+	+	+
DAA	u	-	-	-	+	+	+	+	+
DAS	u	-	-	-	+	+	+	+	+
DEC	+	-	-	-	+	+	+	+	-
DIV	u	-	-	-	u	u	u	u	u
IDIV	u	-	-	-	u	u	u	u	u

Table 4.9 continued...

Instruction	Flags								
	O	D	I	T	S	Z	A	P	C
IMUL	+	-	-	-	u	u	u	u	+
INC	+	-	-	-	+	+	+	+	-
INT	-	-	0	0	-	-	-	-	-
INTO	-	-	0	0	-	-	-	-	-
IRET	r	r	r	r	r	r	r	r	r
MUL	+	-	-	-	u	u	u	u	+
NEG	+	-	-	-	+	+	+	+	+
OR	0	-	-	-	+	+	u	+	0
POPF	r	r	r	r	r	r	r	r	r
RCL	+	-	-	-	-	-	-	-	+
RCR	+	-	-	-	-	-	-	-	+
ROL	+	-	-	-	-	-	-	-	+
ROR	+	-	-	-	-	-	-	-	+
SAHF	-	-	-	-	r	r	r	r	r
SAL/SHL	+	-	-	-	+	+	u	+	+
SAR	+	-	-	-	+	+	u	+	+
SBB	+	-	-	-	+	+	+	+	+
SCAS	+	-	-	-	+	+	+	+	+
SHR	+	-	-	-	+	+	u	+	+
STC	-	-	-	-	-	-	-	-	1
STD	-	1	-	-	-	-	-	-	-
STI	-	-	1	-	-	-	-	-	-
SUB	+	-	-	-	+	+	+	+	+
TEST	0	-	-	-	+	+	u	+	0
XOR	0	-	-	-	+	+	u	+	0

Note: "+" → *Flag is altered and defined (i.e., set or cleared according to the result).*

"u" → *Flag is undefined (i.e., altered but not defined).*

"–" → *Flag is not altered/affected.*

"r" → *The flag is restored from previous saved value.*

"1" → *Set to 1.*

"0" → *Cleared to 0.*

4.6 EXPLANATION OF 8086 INSTRUCTIONS

The 8086 instructions can be classified into the following six groups:

1. Data transfer instructions
2. Arithmetic instructions
3. Logical instructions
4. String-manipulation instructions
5. Control-transfer instructions
6. Processor-control instructions

A brief explanation about each instruction is given in the following sections. The list of symbols/abbreviations used in the instruction set are listed as follows:

Symbols/Abbreviations used in Instruction Set

reg, reg1, reg2	–	8-bit or 16-bit register	Flags –	Flag register
reg8	–	8-bit register	& –	logical AND
reg16	–	16-bit register	| –	logical-OR
mem	–	8-bit or 16-bit memory	^ –	logical Exclusive-OR
mem8	–	8-bit memory	~ –	logical NOT
mem16	–	16-bit memory	disp –	8-bit or 16-bit displacement
data	–	8-bit or 16-bit immediate data	disp8 –	8-bit displacement
data8	–	8-bit immediate data	disp16 –	16-bit displacement
segreg	–	segment register (Excluding CS)	CF –	Carry Flag
data16	–	16-bit immediate data	AF –	Auxiliary Carry Flag

OF	–	Overflow Flag
TF	–	Trace Flag
IF	–	Interrupt Flag
DF	–	Direction Flag
SF	–	Sign Flag
ZF	–	Zero Flag
PF	–	Parity Flag
addr8	–	8-bit port address
$addr_{offset}$	–	16-bit offset/effective address
$addr_{base}$	–	16-bit base address

Note :

1. *Possible choices for reg/reg1/reg2 are AL, AH, BL, BH, CL, CH, DL, DH, AX, BX, CX, DX, SI, DI, SP and BP.*

2. *Possible choices for reg8 are AL, AH, BL, BH, CL, CH, DL and DH.*

3. *Possible choices for reg16 are AX, BX, CX, DX, SI, DI, SP and BP.*

4. *Possible choices for segreg are DS, ES and SS.*

5. *The term mem stands for the 24 different methods of addressing memory data as shown in Table 4.4.*

4.6.1 Data Transfer Instructions

The instruction set of the 8086 microprocessor includes a variety of instructions to transfer data/address into registers, memory locations and IO ports. The various mnemonics used for data-transfer instructions are MOV, XCHG, PUSH, POP, IN, OUT, etc., and they perform any one of the following operations:

1. Copy the content of a register to another register.

2. Copy the content of a register to memory or vice versa.

3. Load the immediate operand to memory/register.

4. Copy the content of a register/memory to segment register (excluding CS-register) or vice versa.

5. Exchange the content of two registers or register and memory.

6. Copy the content of the accumulator to port or vice versa.

7. Load the effective address in segment registers.

The data-transfer instructions generally involve two operands: source operand and destination operand. The source and destination operands should be of same size, both the operand sizes should be either byte or word. This means that only 8-bit data can be moved to an 8-bit register/memory and only 16-bit data can be moved to a 16-bit register/memory. Moving the content of an 8-bit register to a 16-bit register/memory or vice versa is illegal.

The source can be a register or a memory location or an immediate data. The destination can be a register or a memory location. In double-operand instructions, the source and destination cannot refer to memory locations in the same instruction. Therefore, copying the content of one memory location to another memory location in a single instruction is not possible (except PUSH instruction).

The data-transfer instructions (except POPF and SAHF instructions) do not affect the flags of 8086. While executing the POPF instruction, the previously stored status of the flag is restored in the flag register. The instruction SAHF is used to modify the content of the flag register.

The data-transfer instructions of 8086 are listed in Table 4.10, with a brief description about each instruction.

Note: 1. The terms MA, MA_S and MA_E used in symbolic description of instructions refer to physical memory address of data segment, stack segment and extra segment respectively.

2. The register or memory enclosed by brackets in symbolic description refer to content of register or memory.

Table 4.10: Data Transfer Instructions

S.No.	Instruction	Symbolic representation	Explanation
1.	MOV reg2/mem, reg1/mem		
a)	MOV reg2, reg1	(reg2) ← (reg1)	The content of register1 is transferred to register2.
b)	MOV mem, reg1	(mem) ← (reg1)	The content of register1 is transferred to memory.
c)	MOV reg2, mem	(reg2) ← (mem)	The content of memory is transferred to register2.
2.	MOV reg/mem, data		
a)	MOV reg, data	(reg) ← data	The data given in the instruction is transferred to register.
b)	MOV mem, data	(mem) ← data	The data given in the instruction is transferred to memory.
3.	MOV reg, data	(reg) ← data	The data given in the instruction is transferred to register.
4.	MOV A, mem		
a)	MOV AL, mem	(AL) ← (mem)	The content of (8-bit) memory is transferred to 8-bit accumulator (AL).
b)	MOV AX, mem	(AX) ← (mem)	The content of (16-bit) memory is transferred to accumulator (AX).
5.	MOV mem, A		
a)	MOV mem, AL	(mem) ← (AL)	The content of 8-bit accumulator (AL) is transferred to memory.
b)	MOV mem, AX	(mem) ← (AX)	The content of 16-bit accumulator (AX) is transferred to memory.
6.	MOV segreg, reg16/mem		
a)	MOV segreg, reg16	(segreg) ← (reg16)	The content of 16-bit register is transferred to segment register.
b)	MOV segreg, mem	(segreg) ← (mem)	The content of (16-bit) memory is transferred to segment register.

Table 4.10 continued...

S.No.	Instruction	Symbolic representation	Explanation
7.	MOV reg16/mem, segreg		
a)	MOV reg16, segreg	$(reg16) \leftarrow (segreg)$	The content of segment register is transferred to 16-bit register.
b)	MOV mem, segreg	$(mem) \leftarrow (segreg)$	The content of segment register is transferred to 16 bit memory.
8.	PUSH reg16/mem		
a)	PUSH reg16	$(SP) \leftarrow (SP) - 2$ $MA_s = (SS) \times 16_{10} + (SP)$ $(MA_s ; MA_s + 1) \leftarrow (reg16)$	The stack pointer is decremented by 2 and the content of 16-bit register is pushed to stack memory pointed by SP.
b)	PUSH mem	$(SP) \leftarrow (SP) - 2$ $MA_s = (SS) \times 16_{10} + (SP)$ $(MA_s ; MA_s + 1) \leftarrow (mem)$	The stack pointer is decremented by 2 and the content of (16-bit) memory is pushed to stack memory pointed by SP.
9.	PUSH reg16	$(SP) \leftarrow (SP) - 2$ $MA_s = (SS) \times 16_{10} + (SP)$ $(MA_s ; MA_s + 1) \leftarrow (reg16)$	The stack pointer is decremented by 2 and the content of 16-bit register is pushed to stack memory pointed by SP.
10.	PUSH segreg	$(SP) \leftarrow (SP) - 2$ $MA_s = (SS) \times 16_{10} + (SP)$ $(MA_s ; MA_s + 1) \leftarrow (segreg)$	The stack pointer is decremented by 2 and the content of segment register is pushed to stack memory pointed by SP.
11.	PUSHF	$(SP) \leftarrow (SP) - 2$ $MA_s = (SS) \times 16_{10} + (SP)$ $(MA_s ; MA_s + 1) \leftarrow (Flags)$	The stack pointer is decremented by 2 and the content of 16-bit flag register is pushed to stack memory pointed by SP.
12.	POP reg16/mem		
a)	POP reg16	$MA_s = (SS) \times 16_{10} + (SP)$ $(reg16) \leftarrow (MA_s ; MA_s + 1)$ $(SP) \leftarrow (SP) + 2$	The content of stack memory pointed by SP is moved to 16-bit register and the stack pointer is incremented by 2.
b)	POP mem	$MA_s = (SS) \times 16_{10} + (SP)$ $(mem) \leftarrow (MA_s ; MA_s + 1)$ $(SP) \leftarrow (SP) + 2$	The content of (16-bit) stack memory pointed by SP is moved to memory and the stack pointer is incremented by 2.

Table 4.10 continued...

S.No.	Instruction	Symbolic representation	Explanation
13.	POP reg16	$MA_s = (SS) \times 16_{10} + (SP)$ $(reg16) \leftarrow (MA_s ; MA_s + 1)$ $(SP) \leftarrow (SP) + 2$	The content of (16-bit) stack memory pointed by SP is moved to 16-bit register and the stack pointer is incremented by 2.
14.	POP segreg	$MA_s = (SS) \times 16_{10} + (SP)$ $(segreg) \leftarrow (MA_s ; MA_s + 1)$ $(SP) \leftarrow (SP) + 2$	The content of (16-bit) stack memory pointed by SP is moved to segment register and the stack pointer is incremented by 2.
15.	POPF	$MA_s = (SS) \times 16_{10} + (SP)$ $(Flags) \leftarrow (MA_s ; MA_s + 1)$ $(SP) \leftarrow (SP) + 2$	The content of (16-bit) stack memory pointed by SP is moved to flag register and the stack pointer is incremented by 2.
16.	XCHG reg2/mem, reg1		
a)	XCHG reg2, reg1	$(reg2) \leftrightarrow (reg1)$	The content of two registers are exchanged.
b)	XCHG mem, reg1	$(mem) \leftrightarrow (reg1)$	The content of memory and register are exchanged.
17.	XCHG AX, reg16	$(AX) \leftrightarrow (reg16)$	The content of accumulator and 16-bit register are exchanged.
18.	XLAT	$MA = (DS) \times 16_{10} + (BX) + (AL)$ $(AL) \leftarrow (MA)$	The content of (8-bit) memory is transferred to AL. The effective address of memory is given by sum of BX and AL.
19.	IN A, [DX]	$PORT_{addr} = (DX)$	
a)	IN AL, [DX]	$(AL) \leftarrow (PORT)$	The content of (8-bit) port whose address is specified by DX-register is transferred to 8-bit accumulator (AL).
b)	IN AX, [DX] $(AX) \leftarrow (PORT)$	$PORT_{addr} = (DX)$	The content of (16-bit) port whose address is specified by DX-register is transferred to accumulator (AX).
20.	IN A, addr8		
a)	IN AL, addr8	$(AL) \leftarrow (addr8)$	The content of (8-bit) port whose address is given in the instruction is transferred to 8-bit accumulator (AL).
b)	IN AX, addr8	$(AX) \leftarrow (addr\ 8)$	The content of (16-bit) port whose address is given in the instruction is transferred to accumulator (AX).

Table 4.10 continued...

S.No.	Instruction	Symbolic representation	Explanation
21.			
a)	OUT [DX], A OUT [DX], AL	PORT$_{addr}$ = (DX) (PORT) ← (AL)	The content of 8-bit accumulator (AL) is transferred to the (8-bit) port whose address is specified by DX-register.
b)	OUT [DX], AX	PORT$_{addr}$ = (DX) (PORT) ← (AX)	The content of 16-bit accumulator (AX) is transferred to the (16-bit) port, whose address is specified by DX-register.
22.	OUT addr8, A		
a)	OUT addr8, AL	(addr8) ← (AL)	The content of 8-bit accumulator (AL) is transferred to the (8-bit) port whose address is given in the instruction.
b)	OUT addr8, AX	(addr8) ← (AX)	The content of 16-bit accumulator (AX) is transferred to the (16-bit) port whose address is given in the instruction.
23.	LEA reg16, mem	(reg16) ← EA	The 16-bit register is loaded with the **Effective Address** (EA) of the memory location specified by the instruction.
24.	LDS reg16, mem	(reg16) ← (mem) (DS) ← (mem+2)	The word from first two memory locations is moved to the 16-bit register and the word from next two memory locations is moved to DS-register.
25.	LES reg16, mem	(reg16) ← (mem) (ES) ← (mem+2)	The word from first two memory locations is transferred to the 16-bit register and the word from next two memory locations is moved to ES-register.
26.	LAHF	(AH) ← (lower byte flag register)	The content of the lower byte flag register is transferred to the higher byte register of the accumulator.
27.	SAHF	(lower byte flag register) ← (AH)	The content of the higher byte register of the accumulator is moved to lower byte flag register.

4.6.2 Arithmetic Instructions

The arithmetic group includes instructions for performing the following operations:

1. Addition or subtraction of binary, BCD or ASCII data.

2. Multiplication or division of signed or unsigned binary data.

3. Increment or decrement or comparison of binary data.

The mnemonics used for arithmetic instructions are ADD, ADC, SUB, SBB, INC, DEC, MUL, DIV, CMP, etc.

The arithmetic instructions generally involve two operands: source operand and destination operand. The source can be a register or a memory location or an immediate data. The destination can be a register or memory. The result of an arithmetic operation is stored in the destination register or memory except in case of comparison. (In comparison, the result is used to modify the flags and then the result is discarded.)

In double-operand arithmetic instructions, the source and destination cannot refer to memory locations in the same instruction. Therefore, performing arithmetic operation directly on two memory data is not possible.

In double-operand arithmetic instructions, except division, the source and destination operand should be of same size, both the operand sizes should be either byte or word. In all arithmetic instructions employing immediate addressing mode, if the immediate operand/data is 8-bit and the size of register/memory is 16-bit then the 8-bit immediate operand is sign extended to 16-bit. The arithmetic operation is performed between the sign-extended data and the content of register/memory.

The arithmetic instructions alter the flags of 8086. The processor uses the result of an arithmetic operation to alter the flag. The flags reflect the status of the result (for example, whether the result is zero or not; result has carry or not, etc.).

The arithmetic instructions of 8086 are listed in Table 4.11, with a brief description about each instruction.

Table 4.11: Arithmetic Instructions

S.No.	Instruction	Symbolic representation	Explanation
28.	ADD reg2/mem, reg1/mem		
a)	ADD reg2, reg1	(reg2) ← (reg1) + (reg2)	The content of two registers are added and the result is stored in register2.
b)	ADD reg2, mem	(reg2) ← (reg2) + (mem)	The content of register2 and memory are added and the result is stored is register2.
c)	ADD mem, reg1	(mem) ← (mem) + (reg1)	The content of register1 and memory are added and the result is stored in memory.
29.	ADD reg/mem, data		
a)	ADD reg, data	(reg) ← (reg) + data	The data given in the instruction is added to the content of register and the result is stored in register.
b)	ADD mem, data	(mem) ← (mem) + data	The data given in the instruction is added to the content of memory and the result is stored in memory.
30.	ADD A, data		
a)	ADD AL, data8	(AL) ← (AL) + data8	The 8-bit data given in the instruction is added to the content of 8-bit accumulator and the result is stored in 8-bit accumulator (AL).
b)	ADD AX, data16	(AX) ← (AX) + data16	The 16-bit data given in the instruction is added to the content of 16-bit accumulator and the result is stored in 16-bit accumulator (AX).
31.	ADC reg2/mem, reg1/mem		
a)	ADC reg2, reg1	(reg2) ← (reg2) + (reg1) + CF	The content of registers and carry flag are added and the result is stored in register2.
b)	ADC reg2, mem	(reg2) ← (reg2) + (mem) + CF	The carry flag and the content of memory are added to register2 and the result is stored in register2.
c)	ADC mem, reg1	(mem) ← (mem) + (reg1) + CF	The carry flag and the content of register1 are added to memory and the result is stored in memory.

Table 4.11 continued...

S.No.	Instruction	Symbolic representation	Explanation
32. a)	ADC reg/mem, data ADC reg, data	$(reg) \leftarrow (reg) + data + CF$	The data given in the instruction and the carry flag are added to the content of register and the result is stored in register.
b)	ADC mem, data	$(mem) \leftarrow (mem) + data + CF$	The data given in instruction and the carry flag are added to the content of memory and the result is stored in memory.
33. a)	ADC A, data ADC AL, data8	$(AL) \leftarrow (AL) + data8 + CF$	The 8-bit data given in instruction and the carry flag are added to content of 8-bit accumulator(AL) and the result is stored in 8-bit accumulator(AL).
b)	ADC AX, data16	$(AX) \leftarrow (AX) + data16 + CF$	The 16-bit data given in instruction and the carry flag are added to content of accumulator(AX) and the result is stored in 16-bit accumulator (AX).
34.	AAA	Adjust AL to unpacked BCD 1. $(AL) \leftarrow (AL) \ \& \ 0F_H$ 2. If AL > 9 or AF = 1 then $(AL) \leftarrow (AL) + 6$ $(AH) \leftarrow (AH) + 1$ $CF \leftarrow 1 ; AF \leftarrow 1$ $(AL) \leftarrow (AL) \ \& \ 0F_H$	This instruction is executed after addition of two ASCII data to convert the result in AL to correct unpacked BCD.
35.	DAA	Adjust AL to packed BCD. 1. If lower nibble of AL>9 or AF=1 then $(AL) \leftarrow (AL)+06$; AF $\leftarrow 1$ 2. If higher nibble of AL>9 or CF=1 then $(AL) \leftarrow (AL) + 60$; CF $\leftarrow 1$	This instruction is executed after addition of two packed BCD data to convert the result in AL to packed BCD data.
36. a)	SUB reg2/mem, reg1/mem SUB reg2, reg1	$(reg2) \leftarrow (reg2) - (reg1)$	The content of register1 is subtracted from the register2 and result is stored in register2.
b)	SUB reg2, mem	$(reg2) \leftarrow (reg2) - (mem)$	The content of memory is subtracted from the content of register2 and result is stored in register2.
c)	SUB mem, reg1	$(mem) \leftarrow (mem) - (reg1)$	The content of register1 is subtracted from the content of memory and the result is stored in memory.

Table 4.11 continued...

S.No.	Instruction	Symbolic representation	Explanation
37.	SUB reg/mem, data		
a)	SUB reg, data	(reg) ← (reg) – data	The data given in the instruction is subtracted from the register and the result is stored in register.
b)	SUB mem, data	(mem) ← (mem) – data	The data given in the instruction is subtracted from the content of memory and the result is stored in memory.
38.	SUB A, data		
a)	SUB AL, data8	(AL) ← (AL) – data8	The 8-bit data given in the instruction is subtracted from AL and the result is stored in AL-register.
b)	SUB AX, data16	(AX) ← (AX) – data16	The 16-bit data given in the instruction is subtracted from accumulator (AX) and the result is stored in accumulator (AX).
39.	SBB reg2/mem, reg1/mem		
a)	SBB reg2, reg1	(reg2) ← (reg2) – (reg1) – CF	The carry flag and the content of register1 are subtracted from register2 and the result is stored in register2.
b)	SBB reg2, mem	(reg2) ← (reg2) – (mem) – CF	The carry flag and the content of memory are subtracted from register2 and the result is stored in register2.
c)	SBB mem, reg1	(mem) ← (mem) – (reg1) – CF	The carry flag and the content of register1 are subtracted from the content of memory and the result is stored in memory.
40.	SBB reg/mem, data		
a)	SBB reg, data	(reg) ← (reg) – data – CF	The carry flag and the data given in instruction are subtracted from register and the result is stored in register.
b)	SBB mem, data	(mem) ← (mem) – data – CF	The carry flag and the data given in the instruction are subtracted from the content of memory and the result is stored in memory.

Table 4.11 continued...

S.No.	Instruction	Symbolic representation	Explanation
41. (a)	SBB A, data SBB AL, data8	(AL) ← (AL) – data8 – CF	The carry flag and the 8-bit data given in the instruction are subtracted from AL-register and the result is stored in AL-register.
(b)	SBB AX, data16	(AX) ← (AX) – data16 – CF	The carry flag and the 16-bit data given in the instruction are subtracted from the accumulator and the 16-bit result is stored in accumulator.
42.	AAS	Adjust AL to unpacked BCD. 1. (AL) ← (AL) & 0F$_H$ 2. If (AL) > 9 or AF = 1 then, (AL) ← (AL) – 6 ; (AH) ← (AH) – 1 AF ← 1 ; CF ← 1 ; AL ← (AL) & 0F$_H$	This instruction is executed after subtraction of ASCII data to convert the result in AL to correct unpacked BCD.
43.	DAS	Adjust AL to packed BCD. 1. If lower nibble of AL>9 or AF = 1 then, (AL) ← (AL) –6 ; AF ← 1 2. If upper nibble of AL>9 or CF=1 then, (AL) ← (AL) –60 ; CF ← 1.	This instruction is executed after subtraction of packed BCD data to convert the result in AL to packed BCD data.
44. (a)	MUL reg/mem MUL reg	For byte (AX) ← (AL) × (reg8) For word (DX)\(AX) ← (AX) × (reg16)	It is unsigned multiplication. While using this instruction the content of accumulator and register should be unsigned binary and the result is also unsigned binary. For byte operand The content of 8-bit accumulator(AL) is multiplied by the content of 8-bit register and the product is stored in AX-register. For word operand The content of 16-bit accumulator(AX) is multiplied by the content of 16-bit register.The lower word of the result is stored in AX-register and the upper word in DX-register.
(b)	MUL mem	For byte (AX) ← (AL) × (mem8) For word (DX)\(AX) ← (AX) × (mem16)	This instruction is same as MUL reg, except that one of the source operands is in memory instead of register.

Table 4.11 continued...

S.No.	Instruction	Symbolic representation	Explanation
45. (a)	IMUL reg/mem IMUL reg	For byte (AX) ← (AL) × (reg8) For word (DX)(AX) ← (AX) × (reg16)	It is signed multiplication. While using this instruction, the content of accumulator and register should be sign extended binary in 2's complement form and the result is also sign extended binary. For byte operand The content of AL is multiplied by the content of 8-bit register and the sign extended result is stored in AX. For word operand The content of AX is multiplied by the content of 16-bit register. The lower word of sign-extended result is stored in AX-register and the upper word in DX-register.
(b)	IMUL mem	For byte (AX) ← (AX) × (mem8) For word (DX)(AX) ← (AX) × (mem16)	This instruction is same as IMUL reg, except that one of the source operands is in memory instead of register.
46.	AAM	Adjust AH to unpacked BCD data. (AH) = (AL) ÷ 0A$_H$ in AX (AL) = (AL) MOD 0A$_H$ *Note:* $0A_H = 10_{10}$	After multiplication of two 8-bit unpacked BCD data, the result will be in binary. This instruction can be executed after multiplication to convert the result in AX to unpacked BCD.

Table 4.11 continued...

S.No.	Instruction	Symbolic representation	Explanation
47. (a)	DIV reg/mem DIV reg	For 16-bit ÷ 8-bit (AL)← (AX) ÷ (reg8) Quotient (AH) ← (AX) MOD (reg8) Remainder For 32-bit ÷ 16-bit (AX) ← (DX)(AX) ÷ (reg16) Quotient (DX) ← (DX)(AX) MOD (reg16) Remainder	It is unsigned division. While using this instruction, the content of accumulator and register should be an unsigned binary. The result is also an unsigned binary. This instruction divides the content of accumulator by the content of register. Division by zero will generate a type-0 interrupt. For 16-bit ÷ 8-bit The quotient is stored in AL-register and the remainder is stored in AH-register. For 32-bit ÷ 16-bit The quotient is stored in AX (accumulator) while the remainder is stored in DX-register.
(b)	DIV mem	For 16-bit÷8-bit (AL)← (AX) ÷ (mem8) Quotient (AH) ← (AX) MOD (mem8) Remainder For 32-bit ÷ 16-bit (AX) ← (DX)(AX) ÷ (mem16) Quotient (DX) ← (DX)(AX) MOD (mem16) Remainder	This instruction is same as DIV reg except that the divisor is stored in memory instead of register.

Table 4.11 continued...

S.No.	Instruction	Symbolic representation	Explanation
48. (a)	IDIV reg/mem IDIV reg	For 16-bit ÷ 8-bit (AL) ← (AX) ÷ (reg8) Quotient (AH) ← (AX) MOD (reg8) Remainder For 32-bit ÷ 16-bit (AX) ← (DX)(AX) ÷ (reg16) Quotient (DX) ← (DX)(AX) MOD (reg16) Remainder	It is signed division.While using this instruction the content of accumulator and register should be sign extended binary. The sign of quotient depends on the sign of the dividend and divisor. The sign of the remainder will be same as that of the dividend. Division by zero generates a type-0 interrupt. For 16-bit ÷ 8-bit The quotient is stored in AL-register and the remainder is stored in AH-register. For 32-bit ÷ 16-bit The quotient is stored in AX(accumulator), while the remainder is stored in DX-register.
(b)	IDIV mem	For 16-bit ÷ 8-bit (AL) ← (AX) ÷ (mem8) Quotient (AH) ← (AX) MOD (mem8) Remainder For 32-bit ÷ 16-bit (AX) ← (DX)(AX) ÷ (mem16) Quotient (DX) ← (DX)(AX) MOD (mem16) Remainder	This instruction is same as IDIV reg, except that the signed divisor is stored in memory instead of register.
49.	AAD	(AL) ← (AH) × 16₁₀ + (AL) (AH) ← 00ₕ	The unpacked BCD digit in AH and AL registers are converted to equivalent binary data and stored in AL-register. This instruction should be used before the use of division instruction
50. (a)	NEG mem/reg NEG reg	(reg) ← 0-(reg)	Changes the sign of the register content. (The register content will be replaced by its 2's complement value.)
(b)	NEG mem	(mem) ← 0-(mem)	Changes the sign of the memory content. (The memory content is replaced by its 2's complement value.)

Table 4.11 continued...

S.No.	Instruction	Symbolic representation	Explanation
51.	INC reg8/mem		
a)	INC reg8	(reg8) ← (reg8) + 1	The content of the 8-bit register is incremented by 1.
b)	INC mem	(mem) ← (mem) + 1	The content of the memory is incremented by 1.
52.	INC reg16	(reg16) ← (reg16) + 1	The content of the 16-bit register is incremented by 1.
53.	DEC reg8/mem		
a)	DEC reg8	(reg8) ← (reg8) −1	The content of the 8-bit register is decremented by 1.
b)	DEC mem	(mem) ← (mem) −1	The content of memory is decremented by 1.
54.	DEC reg16	(reg16) ← (reg16)−1	The content of the 16-bit register is decremented by 1.

55.	CBW	Bit-7 of AL is moved to all the bits of AH-register.
		1. If AL = 1xxx xxxx (ie., ≥80$_H$)
		   then AH ← 1111 1111(FF$_H$)
		2. If AL = 0xxx xxxx (ie., <80$_H$)
		   then AH ← 0000 0000 (00$_H$)
			Sign extends the content of AL to AH-register by copying the sign bit of AL to all the bits in AH-register.

56.	CWD	Bit-15 of AX is moved to all the bits of DX-register.
		1. If AX = 1xxx xxxx xxxx xxxx
		   (ie., ≥8000$_H$) then
		    DX←1111 1111 1111 1111 (FFFF$_H$)
		2. If AX = 0xxx xxxx xxxx xxxx
		   (ie., <8000$_H$) then
		    DX←0000 0000 0000 0000 (0000$_H$)
			Sign extends the content of AX to DX-register by copying the sign bit of AX to all the bits in DX-register.

Table 4.11 continued...

S.No.		Instruction	Symbolic representation	Explanation
57.		CMP reg2/mem, reg1/mem	Modify flags ← (reg2)–(reg1)	

	a)	CMP reg2, reg1	If (reg2) > (reg1) then CF=0 ; ZF=0 ; SF=0
			If (reg2) < (reg1) then CF=1 ; ZF=0 ; SF=1
			If (reg2) = (reg1) then CF=0 ; ZF=1 ; SF=0
				The content of two registers are compared by subtraction and the result is used to modify the flags. The content of the two registers are not altered.

	b)	CMP reg2, mem	Modify flags ← (reg2) – (mem)
			If (reg2) > (mem) then CF=0 ; ZF=0 ; SF=0
			If (reg2) < (mem) then CF=1 ; ZF=0 ; SF=1
			If (reg2) = (mem) then CF=0 ; ZF=1 ; SF=0
				The content of memory and register2 are compared by subtraction and the result is used to modify the flags. The content of memory and register are not altered.

	c)	CMP mem, reg1	Modify flags ← (mem) – (reg1)
			If (mem) > (reg1) then CF=0 ; ZF=0 ; SF=0
			If (mem) < (reg1) then CF=1 ; ZF=0 ; SF=1
			If (mem) = (reg1) then CF=0 ; ZF=1 ; SF=0
				The content of memory and register1 are compared by subtraction and the result is used to modify flags. The content of memory and register are not altered.

58.		CMP reg/mem, data	Modify flags ← (reg)–data	

	a)	CMP reg, data	If (reg) > data then, CF=0 ; ZF=0 ; SF=0
			If (reg) < data then, CF=1 ; ZF=0 ; SF=1
			If (reg) = data then, CF=0 ; ZF=1 ; SF=0
				The content of reg/mem is compared with data given in the instruction by subtraction and the result is used to modify the flags. The content of reg/mem is not altered.

	b)	CMP mem, data	Modify flags ← (mem) –data
			If (mem) > data then, CF=0 ; ZF=0 ; SF=0
			If (mem) < data then, CF=1 ; ZF=0 ; SF=1
			If (mem) = data then, CF=0 ; ZF=1 ; SF=0	

Table 4.11 continued...

S.No.	Instruction	Symbolic representation	Explanation
59.	CMP A, data	Modify flags ← (AL) –data 8	The content of accumulator is compared with data given in the instruction by subtraction and the result is used to modify the flags. The content of accumulator is not altered.

(a)	CMP AL, data8	If (AL)>data8 then CF=0 ; ZF=0 ; SF=0
		 If (AL)<data8 then CF=1 ; ZF=0 ; SF=1
		 If (AL) = data8 then CF=0 ; ZF=1 ; SF=0	

(b)	CMP AX, data16	Modify flags ← (AX) –data16
		 If (AX)>data16 then CF=0 ; ZF=0 ; SF=0
		 If (AX)<data16 then CF=1 ; ZF=0 ; SF=1
		 If (AX) =data16 then CF=0 ; ZF=1 ; SF=0	

4.6.3 Logical Instructions

The logical group includes instructions for performing AND, OR, Exclusive-OR, complement, shift and rotate operations on binary data. The mnemonics used for logical instructions are AND, OR, XOR, TEST, SHR, SHL, RCR, RCL, etc.

The logical instructions except shift and rotate involve two operands: source operand and destination operand. The source operand can be a register or memory location or immediate data. The destination can be a register or memory. The result of a logical operation is stored in the destination register or memory except in case of TEST. (In a TEST operation, the result is used to modify the flags and then the result is discarded.)

In double-operand logical instructions, the source and destination cannot refer to memory locations in the same instruction. Therefore, performing logical operation directly on two memory data is not possible.

In double-operand logical instructions, the source and destination operand should be of same size, both the operand size should be either byte or word.

The logical instructions alter the flags of 8086. The processor uses the result of a logical operation to alter the flag. The flags reflect the status of the result (for example, whether the result is zero or not).

The logical instructions of 8086 are listed in Table 4.12, with a brief description about each instruction.

Table 4.12: Logical Instructions

S.No.	Instruction	Symbolic representation	Explanation
60.	AND reg2/mem, reg1/mem		
(a)	AND reg2, reg1	(reg2) ← (reg2) & (reg1)	The content of registers are logically ANDed bit by bit and the result is stored in register2.
(b)	AND reg2, mem	(reg2) ← (reg2) & (mem)	The content of register2 and memory are logically ANDed bit by bit and the result is stored in register2.
(c)	AND mem, reg1	(mem) ← (mem) & (reg1)	The content of the memory and register1 are logically ANDed bit by bit and result is stored in memory.
61.	AND reg/mem, data		
(a)	AND reg, data	(reg) ← (reg) & data	The content of register and the data given in the instruction are logically ANDed bit by bit and result is stored in register.
(b)	AND mem, data	(mem) ← (mem) & data	The content of memory and the data given in the instruction are logically ANDed bit by bit and result is stored in memory.
62.	AND A, data		
(a)	AND AL, data8	(AL) ← (AL) & data8	The content of 8-bit accumulator(AL) and 8-bit data given in the instruction are logically ANDed bit by bit and result is stored in AL.
(b)	AND AX, data16	(AX) ← (AX) & data16	The content of accumulator(AX) and 16-bit data given in the instruction are logically ANDed bit by bit and the result is stored in AX.

Table 4.12 continued...

S.No.	Instruction	Symbolic representation	Explanation
63.	OR reg2/mem, reg1/mem		
(a)	OR reg2, reg1	(reg2) ← (reg2) | (reg1)	The content of registers are logically ORed bit by bit and result is stored in register2.
(b)	OR reg2, mem	(reg2) ← (reg2) | (mem)	The content of register2 and memory are logically ORed bit by bit and result is stored in register2.
(c)	OR mem, reg1	(mem) ← (mem) | (reg1)	The content of memory and register1 are logically ORed bit by bit and result is stored in memory.
64.	OR reg/mem, data		
(a)	OR reg, data	(reg) ← (reg) | data	The content of register and the data given in the instruction are logically ORed bit by bit and the result is stored in register.
(b)	OR mem, data	(mem) ← (mem) | data	The content of memory and the data given in the instruction are logically ORed bit by bit and the result is stored in memory.
65.	OR A, data		
(a)	OR AL, data8	(AL) ← (AL) | data8	The content of 8-bit accumulator (AL) and 8-bit data given in the instruction are logically ORed bit by bit and the result is stored in AL.
(b)	OR AX, data16	(AX) ← (AX) | data16	The content of 16-bit accumulator (AX) and 16-bit data given in the instruction are logically ORed bit by bit and the result is stored in AX.

Table 4.12 continued...

S.No.	Instruction	Symbolic representation	Explanation
66.	XOR reg2/mem, reg1/mem		
(a)	XOR reg2, reg1	(reg2) ← (reg2) ^ (reg1)	The content of registers are Exclusive-ORed bit by bit and the result is stored in register2.
(b)	XOR reg2, mem	(reg2) ← (reg2) ^ (mem)	The content of register and memory are Exclusive-ORed bit by bit and result is stored in register2.
(c)	XOR mem, reg1	(mem) ←(mem) ^ (reg1)	The content of memory and register1 are Exclusive-ORed bit by bit and the result is stored in memory.
67.	XOR reg/mem, data		
(a)	XOR reg, data	(reg) ← (reg) ^ data	The content of register and the data given in the instruction are Exclusive-ORed bit by bit and the result is stored in register.
(b)	XOR mem, data	(mem) ← (mem) ^ data	The content of memory and the data given in the instruction are Exclusive-ORed bit by bit and results stored in memory.
68.	XOR A, data		
(a)	XOR AL, data8	(AL) ← (AL) ^ data8	The content of 8-bit accumulator (AL) and 8-bit data given in the instruction are Exclusive-ORed bit by bit and result is stored in AL.
(b)	XOR AX, data16	(AX) ← (AX) ^ data16	The content of 16-bit accumulator (AX) and 16-bit data given in the instruction are Exclusive-ORed bit by bit and result is stored in AX.

Table 4.12 continued...

S.No.	Instruction	Symbolic representation	Explanation
69.	TEST reg2/mem, reg1/mem		
(a)	TEST reg2, reg1	Modify flags ← (reg2) & (reg1)	The content of registers are ANDed and the result is used to modify flags. The content of registers are not altered.
(b)	TEST reg2, mem	Modify flags ← (reg2) & (mem)	The content of register and memory are ANDed and the result is used to modify flags. The content of register/memory are not altered.
(c)	TEST mem, reg1	Modify flags ← (mem) & (reg1)	
70.	TEST reg/mem, data		
(a)	TEST reg, data	Modify flags ← (reg) & data	The content of register and the data given in the instruction are ANDed and the result is used to modify flags. The content of register is not altered.
(b)	TEST mem, data	Modify flags ← (mem) & data	The content of memory and the data given in the instruction are ANDed and the result is used to modify flags. The content of memory is not altered.
71.	TEST A, data		
(a)	TEST AL, data8	Modify flags ← (AL) & data8	The content of accumulator and the data given in the instruction are logically ANDed and the result is used to modify flags. The content of accumulator is not altered.
(b)	TEST AX, data16	Modify flags ← (AX) & data16	
72.	NOT reg/mem		
(a)	NOT reg	(reg) ← ~ (reg)	The content of the register is complemented.
(b)	NOT mem	(mem) ← ~ (mem)	The content of memory is complemented.

Table 4.12 continued...

S.No.	Instruction	Symbolic representation	Explanation
73. (a)	SHL reg/mem or SAL reg/mem SHL reg or SAL reg (i) SHL reg, 1 or SAL reg, 1 (ii) SHL reg, CL or SAL reg, CL	$CF \leftarrow B_{MSD}$; $B_{n+1} \leftarrow B_n$; $B_{LSD} \leftarrow 0$ reg 8/mem8 reg16/mem16	The content of register/memory is shift left, the MSD is shifted to carry flag while the LSD is filled with zero. For **SHL reg/mem, 1** the content of the register/memory is shifted left once. For **SHL reg/mem, CL**, the number of times the content of register/memory has to be shifted left is specified by a count value (1 to 255_{10}) stored in CL-register.
(b)	SHL mem or SAL mem (i) SHL mem, 1 or SAL mem, 1 (ii) SHL mem, CL or SAL mem, CL		
74. (a)	SHR reg/mem SHR reg (i) SHR reg, L (ii) SHR reg, CL	$CF \leftarrow B_{LSD}$; $B_n \leftarrow B_{n+1}$; $B_{MSD} \leftarrow 0$ reg 8/mem8 reg16/mem16	The content of register/memory is shifted right, the LSD is shifted to carry flag while the MSD is filled with zero. For **SHR reg/mem, 1** the content of register/memory is shifted right once. For **SHR reg/mem, CL**, the number of times the content of register/memory has to be shifted right is specified by a count value (1 to 255_{10}) stored in CL-register.
(b)	SHR mem (i) SHR mem, 1 (ii) SHR mem, CL		

Note: MSD - Most Significant Digit ; LSD - Least Significant Digit.

Table 4.12 continued...

S.No.	Instruction	Symbolic representation	Explanation
75. (a)	SAR reg/mem SAR reg (i) SAR reg, 1 (ii) SAR reg, CL	$C \leftarrow B_{LSD}$; $B_n \leftarrow B_{n+1}$; $B_{MSD} \leftarrow B_{MSD}$ reg 8/mem8 LSD MSD B_7 B_6 B_5 B_4 B_3 B_2 B_1 B_0 CF	The content of register/memory is shifted right, the LSD is shifted to carry flag while the MSD is retained. For **SAR reg/mem, 1**, the conten of register/ memory is shifted right once. For **SAR reg/mem, CL**, the number of times the content of register/memory has to be shifted right is specified by a count value (0 to 255₁₀) stored in CL-register.
(b)	SAR mem (i) SAR mem, 1 (ii) SAR mem, CL	reg16/mem16 LSD MSD B_{15} B_{14} .... B_2 B_1 B_0 CF	
76. (a)	ROL reg/mem ROL reg (i) ROL reg, 1 (ii) ROL reg, CL	$B_{n+1} \leftarrow B_n$; $CF \leftarrow B_{MSD}$; $B_{LSD} \leftarrow B_{MSD}$ reg 8/mem8 LSD MSD CF B_7 B_6 B_5 B_4 B_3 B_2 B_1 B_0	The content of register/memory is rotated left, while the MSD is moved to both LSD and carry flag. For **ROL reg/mem, 1**, the content of register/ memory is rotated left once. For **ROL reg/mem, CL**, the number of times the content of register/memory has to be rotated left is specified by a count value (0 to 255₁₀) stored in CL-register.
(b)	ROL mem (i) ROL mem, 1 (ii) ROL mem, CL	reg16/mem16 LSD MSD CF B_{15} B_{14} B_{13} .... B_2 B_1 B_0	

Table 4.12 continued...

S.No.	Instruction	Symbolic representation	Explanation
77. (a)	RCL reg/mem RCL reg (i) RCL reg, 1 (ii) RCL reg, CL	$B_{n+1} \leftarrow B_n$; $B_{LSD} \leftarrow CF$; $CF \leftarrow B_{MSD}$	The content of register/memory is rotated left, the carry flag is moved to the LSD, while the MSD is moved to carry flag. For **RCL reg/mem, 1**, the content of register/memory is rotated left once. For **RCL reg/mem, CL**, the number of times the content of register/memory has to be rotated left is specified by a count value (0 to 255_{10}) stored in CL-register.
(b)	RCL mem (i) RCL mem, 1 (ii) RCL mem, CL		
78. (a)	ROR reg/mem ROR reg (i) ROR reg, 1 (ii) ROR reg, CL	$B_n \leftarrow B_{n+1}$; $B_{MSD} \leftarrow CF$; $CF \leftarrow B_{LSD}$	The content of register/memory is rotated right, the LSD is moved both to MSD and carry flag. For **ROR reg/mem, 1**, the content of register/memory is rotated right once. For **ROR reg/mem, CL**, the number of times the content of register/memory has to be rotated right is specified by a count value (0 to 255_{10}) stored in CL-register.
(b)	ROR mem (i) ROR mem,1 (ii) ROR mem, CL		

Table 4.12 continued...

S.No.	Instruction	Symbolic representation	Explanation
79. (a)	RCR reg/mem RCR reg (i) RCR reg, 1 (ii) RCR reg, CL	$B_n \leftarrow B_{n+1}$; $B_{MSD} \leftarrow$ CF ; CF $\leftarrow B_{LSD}$ reg 8/mem8 MSD B_7 B_6 B_5 B_4 B_3 B_2 B_1 B_0 LSD CF	The content of register/memory is rotated right, the carry flag is moved to MSD while the LSD is moved to carry flag. For **RCR reg/mem, 1**, the content of register/memory is rotated right once. For **RCR reg/mem, CL**, the number of times the content of register/memory has to be rotated right is specified by a count value (0 to 255_{10}) stored in CL-register.
(b)	RCR mem (i) RCR mem, 1 (ii) RCR mem, CL	reg16/mem16 MSD B_{15} B_{14} B_{13} .... B_2 B_1 B_0 LSD CF	

4.6.4 String-Manipulation Instructions

A string is a sequence of bytes or words. The 8086 instruction set includes instructions for string movement, comparison, scan, load and store. It also consists of the REP instruction prefix which is used to repeat the execution of string instructions.

The string instructions end with "S" or "SB" or "SW". Here, "S" represents **String**, "SB" represents **String Byte** and "SW" represents **String Word**.

All string instructions have an implied source and destination operand (i.e., the operands are not specified as a part of the instruction). The string instructions MOVS and CMPS assume that the source operand is in the data segment memory and the destination is in the extra segment memory. The string instructions STOS and SCANS assume that the source operand is in the accumulator and the destination is in the extra segment memory. The string instruction LODS assumes that the source operand is in the data segment memory and the destination is the accumulator.

For string operations, the offset, or effective address, of the source operand is stored in SI-register and that of the destination operand is stored in DI-register. On execution of a string instruction depending on **D**irection **F**lag (DF), SI and DI registers are automatically updated to point to the next byte/word of the source and destination. If DF = 0 then SI and DI are incremented by one for byte and incremented by two for word. If DF = 1 then SI and DI are decremented by one for byte and decremented by two for word.

The string instructions of 8086 are listed in Table 4.13, with a brief description about each instruction.

Table 4.13: String-Manipulation Instructions

S.No.	Instruction	Symbolic representation	Explanation
80. (a)	REP REPZ/REPE	While CX ≠ 0 and ZF = 1, repeat execution of string instruction and (CX) ← (CX) − 1	It is a prefix used to compare or scan a string instruction. When a string instruction is prefixed with **REPZ/ REPE**, the instruction execution is repeated if CX ≠ 0 and ZF = 1. After each execution of a string instruction, the content of CX is decremented by 1. The repeat operation is terminated if CX = 0 or ZF = 0.
(b)	REPNZ/REPNE	While CX ≠ 0 and ZF = 0, repeat execution of string instruction and (CX) ← (CX) − 1	It is a prefix used to compare or scan string instructions. When a string instruction is prefixed with **REPNZ/ REPNE**, the instruction execution is repeated if CX ≠ 0 and ZF = 0. After each execution of a string instruction, the content of CX is decremented by 1. The repeat operation is terminated if CX = 0 or ZF = 1.

Table 4.13 continued...

S.No.	Instruction	Symbolic representation	Explanation
81. (a)	MOVS MOVSB	$MA = (DS) \times 16_{10} + (SI)$ $MA_E = (ES) \times 16_{10} + (DI)$ $(MA_E) \leftarrow (MA)$ If DF=0 then $(DI) \leftarrow (DI)+1$; $(SI) \leftarrow (SI)+1$ If DF=1 then $(DI) \leftarrow (DI)-1$; $(SI) \leftarrow (SI)-1$	One byte of a string data stored in the data segment is copied into the extra segment, and SI and DI are automatically incremented/decremented by 1 depending on **D**irection **F**lag (DF).
(b)	MOVSW	$MA = (DS) \times 16_{10} + (SI)$ $MA_E = (ES) \times 16_{10} + (DI)$ $(MA_E ; MA_E + 1) \leftarrow (MA ; MA + 1)$ If DF = 0 then $(DI) \leftarrow (DI)+2$; $(SI) \leftarrow (SI) + 2$ If DF = 1 then $(DI) \leftarrow (DI)-2$; $(SI) \leftarrow (SI) + 2$	One word of a string data stored in the data segment is copied into the extra segment, and SI and DI are automatically incremented/decremented by 2 depending on **D**irection **F**lag (DF).
82. (a) (b)	CMPS CMPSB CMPSW	$MA = (DS) \times 16_{10} + (SI)$ $MA_E = (ES) \times 16_{10} + (DI)$ Modify flags $\leftarrow$ $(MA) - (MA_E)$ If $(MA) > (MA_E)$ then CF = 0 ; ZF = 0 ; SF = 0 If $(MA) < (MA_E)$ then CF = 1 ; ZF = 0 ; SF = 1 If $(MA) = (MA_E)$ then CF = 0 ; ZF = 1 ; SF = 0 <u>For byte operation</u> If DF=0 then $(DI) \leftarrow (DI)+1$; $(SI) \leftarrow (SI)+1$ If DF=1 then $(DI) \leftarrow (DI)-1$; $(SI) \leftarrow (SI)-1$ <u>For word operation</u> If DF=0 then $(DI) \leftarrow (DI)+2$; $(SI) \leftarrow (SI)+2$ If DF=1 then $(DI) \leftarrow (DI)-2$; $(SI) \leftarrow (SI)-2$	One byte/word of a string data in the extra segment is subtracted from one byte/word of a string data in the data segment and the result is used to modify flags. The contents of DI and SI are automatically incremented/decremented depending on **D**irection **F**lag (DF). For byte operation, the contents of DI and SI are incremented/decremented by 1. For word operation, the contents of DI and SI are incremented/decremented by 2.

Table 4.13 continued...

S.No.	Instruction	Symbolic representation	Explanation

83.
(a)
	SCAS
	SCASB
		$MA_E = (ES) \times 16_{10} + (DI)$
		Modify flags ← (AL) − (MA_E)
		If (AL) > (MA_E) then CF=0 ; ZF=0 ; SF=0
		If (AL) < (MA_E) then CF=1 ; ZF=0 ; SF=1
		If (AL) = (MA_E) then CF=0 ; ZF=1 ; SF=0
		If DF = 0 then (DI) ← (DI) + 1
		If DF = 1 then (DI) ← (DI) − 1
			One byte of string data in the extra segment is subtracted from the content of AL and the result is used to modify flags. The content of DI and SI are automatically incremented/decremented by 1 depending on **Direction Flag** (DF).

(b)	SCASW	$MA_E = (ES) \times 16_{10} + (DI)$
		Modify flags ← (AX) − $(MA_E ; MA_E + 1)$
		If (AX) > $(MA_E ; MA_E + 1)$ then CF=0 ; ZF=0 ; SF=0
		If (AX) < $(MA_E ; MA_E + 1)$ then CF=1 ; ZF=0 ; SF=1
		If (AX) = $(MA_E ; MA_E + 1)$ then CF=0 ; ZF=1 ; SF=0
		If DF=0 then (DI) ← (DI) + 2
		If DF=1 then (DI) ← (DI) − 2
			One word of string data in the extra segment is subtracted from the content of AX and the result is used to modify flags. The content of DI and SI are automatically incremented/decremented by 2 depending on **Direction Flag** (DF).

84. (a)	LODS LODSB	$MA = (DS) \times 16_{10} + (SI)$ (AL) ← (MA) If DF=0 then (SI) ←(SI) + 1 If DF=1 then (SI) ← (SI) − 1	One byte of a string data stored in the data segment is copied into the AL-register and SI is automatically incremented/decremented by 1, depending on **Direction Flag** (DF).
(b)	LODSW	$MA = (DS) \times 16_{10} + (SI)$ (AX) ← (MA ; MA + 1) If DF=0 then (SI) ← (SI) + 2 If DF=1 then (SI) ← (SI) − 2	One word of a string data stored in the data segment is copied into the accumulator. SI is automatically incremented/decremented by 2, depending on **Direction Flag** (DF).

Table 4.13 continued...

S.No.	Instruction	Symbolic representation	Explanation
85. (a)	STOS STOSB	$MA_E = (ES) \times 16_{10} + (DI)$ $(MA_E) \leftarrow (AL)$ If DF=0 then $(DI) \leftarrow (DI) + 1$ If DF=1 then $(DI) \leftarrow (DI) - 1$	The content of AL-register is stored as one byte of string data in the extra segment. DI is automatically incremented/decremented by 1 depending on Direction Flag (DF).
(b)	STOSW	$MA_E = (ES) \times 16_{10} + (DI)$ $(MA_E ; MA_E + 1) \leftarrow (AX)$ If DF=0 then $(DI) \leftarrow (DI) + 2$ If DF=1 then $(DI) \leftarrow (DI) - 2$	The content of AX-register is stored as one word of string data in the extra segment. DI is automatically incremented/decremented by 2 depending on Direction Flag (DF).

4.6.5 Control-Transfer Instructions

The control-transfer group consists of call, jump, loop and software interrupt instructions. Normally, a program is executed sequentially (i.e., the program instructions are executed one after the other). When a branch instruction is encountered, the program execution control is transferred to the specified destination or target instruction. The transfer of program execution control is done either by changing the content of IP or by changing the content of IP and CS. When the content of IP alone is modified, the program control branches to a new memory location in the same segment. When the content of both IP and CS are modified, the program control branches to a new memory location in another memory segment.

The control-transfer instructions do not affect the flags of 8086. The jump and loop instructions can be classified into conditional and unconditional instructions. In conditional instructions, the status of one or more flags are checked and control transfer takes place only if the specified condition is satisfied.

The control transfer instructions are listed in Table 4.14 - Table 4.19, with a brief description about each instruction.

CALL and RET Instructions

The CALL instructions transfer control to a subprogram or subroutine or a procedure after saving the return address in the stack memory. There are two types of call instructions: Intrasegment or near call and Intersegment or far call. A **near call** refers to calling a procedure stored in the same code segment memory in which the main program (or calling program) resides. A **far call** refers to calling a procedure stored in a different code segment memory than that of the main program.

While executing a near call, the content of IP alone is pushed to stack. While executing a far call, the content of CS and IP are pushed to stack. Every procedure or subroutine ends with a RET instruction. The execution of a RET instruction at the end of a subroutine or procedure will pop the content of the top of the stack to the IP in case of a near call or to IP and CS in case of a far call. Thus, the program control returns back to the main program.

The call and return instructions are listed in Table 4.14, with a brief description about each instruction.

Table 4.14: CALL and RET Instructions

S.No.	Instruction	Symbolic representation	Explanation
86.	CALL disp16 (Call near-direct within segment)	$(SP) \leftarrow (SP) - 2$ $MA_s = (SS) \times 16_{10} + (SP)$ $(MA_s) \leftarrow (IP)$ $(IP) \leftarrow disp16$	This instruction is a near-direct call in which the program control is transferred within the same segment. The stack pointer is decremented by 2, the Instruction Pointer (IP) is pushed into stack and the effective address (disp16) of the subroutine/procedure to be executed is loaded in IP.

Table 4.14 continued...

S.No.	Instruction	Symbolic representation	Explanation
87.	CALL reg/mem (Call near-indirect within segment)		
(a)	CALL reg	$(SP) \leftarrow (SP) - 2$ $MA_S = (SS) \times 16_{10} + (SP)$ $(MA_S) \leftarrow (IP)$ $(IP) \leftarrow (reg)$	This instruction is a near-indirect call in which the control transfer is within same segment and the effective address of the subroutine/procedure to be called is stored in the register/memory. The stack pointer is decremented by 2, the Instruction **Pointer** (IP) is pushed into the stack and the effective address of the subroutine/procedure to be executed is loaded in IP from the register memory.
(b)	CALL mem	$(SP) \leftarrow (SP) - 2$ $MA_S = (SS) \times 16_{10} + (SP)$ $(MA_S) \leftarrow (IP)$ $(IP) \leftarrow (mem)$	
88.	CALL addr$_{offset}$, addr$_{base}$ (Call far-direct intersegment)		This instruction is a far-direct call in which the program control is transferred to another segment. The offset and segment base address of the procedure to be executed are directly given in the instruction. The stack pointer is again decremented by 2 and CS is pushed into stack and the base address of the procedure to be executed is loaded in CS. The stack pointer is again decremented by 2, the Instruction **Pointer** (IP) is pushed into stack. The offset address of the procedure to be executed is loaded in IP.

Table 4.14 continued...

S.No.	Instruction	Symbolic representation	Explanation
89.	CALL mem (Call far - indirect intersegment)		This instruction is far-indirect call in which the program control is transferred to different segment, and the offset and segment base address of procedure to be executed are stored in memory. The stack pointer is decremented by 2 and CS is pushed to stack. The base address available in memory is moved to CS. The stack pointer is again decremented by 2 and IP is pushed to stack. The offset address available in memory is moved to IP.
90.	RET (Return from intersegment call)	$MA_s = (SS) \times 16_{10} + (SP)$ $(IP) \leftarrow (MA_s)$ $(SP) \leftarrow (SP) + 2$	Return the control back to calling procedure from the called procedure within the segment. The content of top of the stack is transferred to IP. The stack pointer is incremented by 2.
91.	RET data16 (Return from call within segment adding immediate value to SP)	$MA_s = (SS) \times 16_{10} + (SP)$ $(SP) \leftarrow (SP) + 2$	Return the control back to the calling procedure from the called procedure within the segment. The content of top of the stack is transferred to IP and the stack pointer is incremented by a value (data16) specified in the instruction.
92.	RET (Return from intersegment call)	$MA_s = (SS) \times 16_{10} + (SP)$ $(IP) \leftarrow (MA_s)$ $MA_s = (SS) \times 16_{10} + (SP)$ $(SP) \leftarrow (SP) + 2$ $(CS) \leftarrow (MA_s)$ $(SP) \leftarrow (SP) + 2$	Return the control back to calling procedure from the called procedure which is in different segment. The content of top of the stack is moved to IP, and the stack pointer is incremented by 2. Next the content of current top of the stack is moved to CS and SP is incremented by 2.

Table 4.14 continued...

S.No.	Instruction	Symbolic representation	Explanation
93.	RET data16 (Return from intersegment call adding immediate data to SP)	$MA_s = (SS) \times 16_{10} + (SP)$ $(IP) \leftarrow (MA_s)$ $(SP) \leftarrow (SP) + 2$ $MA_s = (SS) \times 16_{10} + (SP)$ $(CS) \leftarrow (MA_s)$ $(SP) \leftarrow (SP) + data16$	Return the control back to calling procedure from the called procedure which is in a different segment. The content of top of the stack is moved to IP and the stack pointer is incremented by 2. Next the content of the current top of the stack is moved to CS and the stack pointer is incremented by a value (data16) specified in the instruction.

Unconditional Jump Instructions

The unconditional jump instructions does not check for any flag condition. When the unconditional jump instruction is executed, the program control is transferred to a new memory location either in the same segment or in another segment. In a near jump instruction, the program control is transferred to the new memory location in the same segment by modifying the content of Instruction Pointer (IP). In a far jump instruction, the program control is transferred to a new memory location in another segment by modifying the content of Instruction Pointer (IP) and Code Segment (CS) register.

Table 4.15: Near Jump Instructions

S.No.	Instruction	Symbolic representation	Explanation
94.	JMP disp16	$(IP) \leftarrow (IP) + disp16$	The 16-bit value (disp16) given in the instruction is added to Instruction Pointer (IP).
95.	JMP disp8	$disp16 \xleftarrow[\text{extend}]{\text{Sign}} disp8$ $(IP) \leftarrow (IP) + disp16$	The 8-bit value (disp8) given in the instruction is sign extended to 16-bit and added to Instruction Pointer (IP).

Table 4.15 continued...

S.No.	Instruction	Symbolic representation	Explanation
96.	JMP reg/mem		
(a)	JMP reg	(IP) ← (reg)	The 16-bit value stored in the register/memory is moved to Instruction Pointer (IP).
(b)	JMP mem	(IP) ← (mem)	

Table 4.16: Far Jump Instructions

S.No.	Instruction	Symbolic representation	Explanation
97.	JMP addr$_{offset}$, addr$_{base}$	(IP) ← addr$_{offset}$ (IP) ← addr$_{base}$	The offset address given in the instruction is loaded in IP and the base address given in the instruction is loaded in CS-register.
98.	JMP mem	(IP) ← (mem) (CS) ← (mem + 2)	The content of (16-bit) memory is moved to IP and the next word in the memory is moved to CS-register.

Conditional Jump Instructions

In a conditional jump instruction, one or more flag conditions are checked. If the conditions are TRUE then the program control is transferred to the new memory location in the same segment by modifying the content of the IP. All conditional instructions are only near jump (or short jump); hence, the content of CS is not altered.

In all conditional jump instructions, an 8-bit value (disp8) will be directly specified in the instruction which is sign extended to 16-bit and added to IP. The new value in IP is the effective address of the instruction where the program control is transferred if the condition is TRUE.

Instruction Format

J <condition> disp8

If <condition> is TRUE then   disp16 ← $\xleftarrow{\text{Sign extend}}$ disp8 ;   (IP) ← (IP) + disp16

Note: If the condition specified by the instruction is FALSE then the content of IP is not altered.

Table 4.17: Conditional Jump Instructions

S.No.	Instruction	Explanation
99.	JE disp8 (JZ disp8)	Jump if ZF=1
100.	JL disp8 (JNGE disp8)	Jump if SF≠OF
101.	JLE disp8 (JNG disp8)	Jump if SF≠OF or ZF = 1
102.	JB disp8 (JNAE/JC disp8)	Jump if CF=1
103.	JBE disp8 (JNA disp8)	Jump if CF=1 or ZF=1
104.	JP disp8 (JPE disp8)	Jump if PF=1
105.	JNB disp8 (JAE/JNC disp8)	Jump if CF=0
106.	JNBE disp8 (JA disp8)	Jump if CF=0 and ZF=0

S.No.	Instruction	Explanation
107.	JNP disp8 (JPO disp8)	Jump if PF=0
108.	JNO disp8	Jump if OF=0
109.	JNS disp8	Jump if SF=0
110.	JO disp8	Jump if OF=1
111.	JS disp8	Jump if SF=1
112.	JNE disp8 (JNZ disp8)	Jump if ZF=0
113.	JNL disp8 (JGE disp8)	Jump if SF=OF
114.	JNLE disp8 (JG disp8)	Jump if CF=OF and ZF=0
115.	JCXZ disp8	Jump if (CX) = 0

Note: The execution of the instruction JCXZ is similar to that of any conditional jump instruction except that the content of CX-register is checked to make a decision instead of a flag.

LOOP Instructions

LOOP instructions are used to execute a group of instructions, a number of times, as specified by a count value stored in CX-register. The number of instructions to be looped will be specified directly in the instruction as a signed 8-bit number (displacement or disp8). For positive displacement, the instructions below the LOOP instructions are executed and for negative displacement the instructions above the LOOP instructions are executed. The content of CX-register is decremented by one after each execution of looped instructions. The effective address of the first instruction of the loop is obtained by sign extending the disp8 to 16-bit and adding it to IP.

Table 4.18: Loop Instructions

S.No.	Instruction	Symbolic representation	Explanation
116.	LOOP disp8	Loop if (CX) ≠ 0 (CX) ← (CX) − 1	Repeat execution of the group of instructions until the content of CX is zero. After each execution, CX is decremented by one.
117.	LOOPZ disp8	Loop if (CX)≠0 and ZF=1 (CX) ← (CX) − 1	Repeat execution of the group of instructions, if the content of CX is not zero and ZF=1. After each execution, CX is decremented by one.
118.	LOOPNZ disp8 (LOOPNE disp8)	Loop if (CX)≠0 and ZF=0 (CX) ← (CX) − 1	Repeat execution of the group of instructions, if the content of CX is not zero and ZF=0. After each execution, the CX is decremented by one.

Software Interrupts

The INT instructions are called software interrupts. The INT instruction is used to call a procedure or subroutine on interrupt basis. Hence, the procedure executed on interrupt basis is called **Interrupt Service Routine** (ISR).

The INT instruction is accompanied by a type number, which can be in the range of 0 - 255. Therefore, in 8086 processor, 256 types of software interrupts can be implemented. These software interrupts are used to implement the system call service of the operating system.

In order to execute an ISR, a 16-bit effective address for IP and a 16-bit base address for CS are needed. Therefore, for each INT instruction four memory locations are reserved in the first 1k address space of memory. In the reserved locations, the first two locations are used to store the effective address (to be loaded in IP), and the next two locations are used to store the base address (to be loaded in CS-register).

The address of the reserved memory location is called **vector address**. The vector address of an interrupt is obtained by multiplying the type number by 4.

Before executing ISR, the content of IP, CS and flag register are pushed to stack. Each ISR is terminated by IRET (**Interrupt return**) instruction. On executing IRET instruction, the top of the stack are popped to IP, CS and flag register. Thus, the program control return back to the main program after executing ISR.

Table 4.19: Software Interrupt

S.No.	Instruction	Symbolic representation	Explanation
119.	INT type	$(SP) \leftarrow (SP) - 2$; $(MA_s) \leftarrow$ Flags $IF \leftarrow 0$; $TF \leftarrow 0$ $(SP) \leftarrow (SP) - 2$; $(MA_s) \leftarrow (CS)$ $(SP) \leftarrow (SP) - 2$; $(MA_s) \leftarrow (IP)$ $(IP) \leftarrow (0000 : (type \times 4))$ $(CS) \leftarrow (0000 : (type \times 4) + 2)$ For each push operation the stack memory address is calculated as shown below.	This instruction is a software interrupt and used to call a service procedure (or subroutine) on interrupt basis. The type number is from 0 to 255. On execution of this instruction, the content of flag register, CS-register and IP are pushed to stack one by one after decrementing SP by 2 before each push operation. The flags TF and IF are also cleared. The effective vector address is calculated by multiplying the type number by 4. The memory location pointed by the vector address contains the address of interrupt service routine. The first word pointed by the calculated vector address is moved to IP and the next word is moved to CS-register.
120.	INT 3	$(SP) \leftarrow (SP) - 2$; $(MA_s) \leftarrow$ Flags $IF \leftarrow 0$; $TF \leftarrow 0$ $(SP) \leftarrow (SP) - 2$; $(MA_s) \leftarrow (CS)$ $(SP) \leftarrow (SP) - 2$; $(MA_s) \leftarrow (IP)$ $(IP) \leftarrow (0000C_H)$; $(CS) \leftarrow (0000E_H)$ ***Note:*** $3 \times 4 = 12_{10} = 0C_H$; $12 + 2 = 14_{10} = 0E_H$ For each push operation, the MA_s is given by $MA_s = (SS) \times 16_{10} + (SP)$	This instruction is a special type of software interrupt which has the single byte code of CC_H. Many systems use this as a break point instruction. The operations performed by this instruction is same as that of a type3 interrupt.

Table 4.19 continued...

S.No.	Instruction	Symbolic representation	Explanation
121.	INTO	If OF=1, then following operations are performed. $(SP) \leftarrow (SP) - 2$; $(MA_S) \leftarrow$ Flags $IF \leftarrow 0$; $TF \leftarrow 0$ $(SP) \leftarrow (SP) - 2$; $(MA_S) \leftarrow (CS)$ $(SP) \leftarrow (SP) - 2$; $(MA_S) \leftarrow (IP)$ $(IP) \leftarrow (00010_H)$; $(CS) \leftarrow (00012_H)$ *Note:* $4 \times 4 = 16_{10} = 10_H$; $16 + 2 = 18_{10} = 12_H$ For each push operation, the MA_S is given by $MA_S = (SS) \times 16_{10} + (SP)$	If **Overflow Flag** (OF) is 1 then the type-4 interrupt is performed.
122.	IRET	$(IP) \leftarrow (MA_S)$; $(SP) \leftarrow (SP) + 2$ $(CS) \leftarrow (MA_S)$; $(SP) \leftarrow (SP) + 2$ Flag $\leftarrow (MA_S)$; $(SP) \leftarrow (SP) + 2$ For each pop operation, the stack memory address is calculated as shown below: $MA_S = (SS) \times 16_{10} + (SP)$	This instruction is used to terminate an interrupt service procedure and transfer the program control back to the main program. On execution of this instruction, the contents of top of the stack (Pointed by SP) are moved (poped) to IP, CS and flag registers one by one. After every pop operation, the SP is incremented by 2.

4.6.6 Processor-Control Instructions

The processor-control group includes instructions to set or clear the carry flag, direction flag and interrupt flag. It also includes HLT, NOP, LOCK and ESC instructions which controls the processor operation.

The processor control instructions are listed in Table 4.20 with a brief description about each instruction.

Table 4.20: Processsor-Control instructions

S.No.	Instruction	Symbolic representation	Explanation
123.	CLC	CF ← 0	The carry flag is reset to zero.
124.	CMC	CF ← ~ CF	The carry flag is complemented.
125.	STC	CF ← 1	The carry flag is set to one.
126.	CLD	DF ← 0	The direction flag is reset to zero.
127.	STD	DF ← 1	The direction flag is set to one.
128.	CLI	IF ← 0	The interrupt flag is reset to zero.
129.	STI	IF ← 1	The interrupt flag is set to one.
		Explanation	
130.	HLT		Halt program execution. This instruction is used to terminate a program. On execution of this instruction, the processor enters into an idle state and performs no operation until an interrupt occurs.
131.	WAIT		Wait for test line active. This instruction causes the processor to enter into an idle state or a wait state and continue to remain in that state until a signal is asserted on TEST input pin or until a valid interrupt signal is received on the INTR or NMI interrupt input pin.
132. (a) (b)	ESC opcode, mem/reg ESC opcode, mem ESC opcode, reg		This instruction is used to pass instructions to a coprocessor which shares the address and data bus with the 8086. The lower 6 bits of the opcode is the opcode of 8087 instruction and the upper two bits are zeros. For **ESC opcode, mem** the data is accessed by 8087 from memory; and for **ESC opcode reg**, the data is accessed by 8087 from the 8086 register specified in the instruction.
133.	LOCK		The **LOCK** is used as a prefix to a critical instruction which has to be executed without any disturbance from other bus masters. When **LOCK** prefix is used in an instruction then during execution of this instruction, the **LOCK** prefix ensures that the shared system resources are not taken over by other bus masters in the middle of instruction execution.
134.	NOP		**No operation** is performed for 3 clock periods when this instruction is executed. The processor waits for 3 clock periods and then the next instruction is executed.

4.7 EXAMPLES OF 8086 ASSEMBLY LANGUAGE INSTRUCTIONS

Notes: 1. *The register or register + constant enclosed by square brackets in the operand field of instructions refers to the method of effective address calculation of memory. The 16-bit constant enclosed by square brackets in the operand field of instructions refers to the effective address of memory data. The 8-bit/16-bit constants which are not enclosed by brackets in the operand field refers to immediate data.*

2. *The term MA used in the symbolic description of instructions refer to physical Memory Address of memory. MA_E refers to physical Memory Address of data segment memory and MA_S refers to physical Memory Address of Stack segment memory and MA_E refers to physical Memory Address of Extra segment memory.*

3. *The register/memory enclosed by brackets in symbolic description refers to the content of register/memory.*

4. *For hexadecimal constant (data/address) the letter H is included at the end of 8-bit/16-bit constants(data/address). When a hexadecimal constant start with A, B, C, D, E or F, a zero should be placed before the constant, otherwise the assembler will treat the constant as a variable.*

Table 4.21: 8086 Assembly Language Instructions

Instruction	Symbolic description	Explanation
MOV AX, SI	$(AX) \leftarrow (SI)$	The content of SI-register is moved to AX-register.
MOV CH, CL	$(CH) \leftarrow (CL)$	The content of CL-register is moved to CH-register.
MOV [BX + 08H], AX	$MA = (DS) \times 16_{10} + (BX) + 0008_H$ $(MA) \leftarrow (AL) ; (MA+1) \leftarrow (AH)$	The contents of AX-register is moved to two consecutive memory locations.
MOV AX, [BP + SI + 07H]	$MA_S = (SS) \times 16_{10} + (BP) + (SI) + 0007_H$ $(AL) \leftarrow (MA_S) ; (AH) \leftarrow (MA_S+1)$	The contents of two consecutive memory locations from stack memory are moved to AX-register.
MOV CX, 150AH	$(CX) \leftarrow 150A_H$	The 16-bit data ($150A_H$) given in the instruction is moved to CX-register.
MOV CX, [150AH]	$MA = (DS) \times 16_{10} + 150A_H$ $(CL) \leftarrow (MA) ; (CH) \leftarrow (MA + 1)$	The contents of two consecutive memory locations addressed by the instruction are moved to CX-register.

Table 4.21 continued...

Instruction	Symbolic description	Explanation
MOV ES, CX	$(ES) \leftarrow (CX)$	The content of CX-register is moved to ES-register.
MOV ES, [SI + 0008H]	$MA = (DS) \times 16_{10} + (SI) + 0008_H$ $(ES) \leftarrow (MA ; MA+1)$	The contents of two consecutive memory locations are moved to ES-register.
MOV DX, SS	$(DX) \leftarrow (SS)$	The content of SS-register is moved to DX-register.
MOV [BX + 0C0H], SS	$FFC0_H \xleftarrow{\text{Sign extend}} C0_H$ $MA = (DS) \times 16_{10} + (BX) + FFC0_H$ $(MA ; MA+1) \leftarrow (SS)$	The content of SS-register is moved to two consecutive memory locations.
PUSH CX	$(SP) \leftarrow (SP) - 2$ $MA_s = (SS) \times 16_{10} + (SP)$ $(MA_s ; MA_s+1) \leftarrow (CX)$	The content of CX-register is pushed to top of the stack. (The content of CX-register is moved to two consecutive locations in stack memory, whose effective address is obtained by decrementing SP by two.)
PUSH [BX + 05H]	$0005_H \xleftarrow{\text{Sign extend}} 05_H$ $MA = (DS) \times 16_{10} + (SI) + 0005_H$ $(SP) \leftarrow (SP) - 2$ $MA_s = (SS) \times 16_{10} + (SP)$ $(MA_s) \leftarrow (MA) ; (MA_s+1) \leftarrow (MA+1)$	The content of data memory specified by the instruction is pushed to top of the stack memory. (The content of two consecutive memory locations in data memory are moved to two consecutive location in stack memory. The effective address of stack memory is obtained by decrementing SP by two.)
POP BX	$MA_s = (SS) \times 16_{10} + (SP)$ $(BX) \leftarrow (MA_s ; MA_s+1)$ $(SP) \leftarrow (SP) + 2$	The content of top of the stack is moved to BX-register. (The content of two consecutive locations in stack memory are moved to BX-register.) After this move operation, SP is incremented by two.

Table 4.21 continued...

Instruction	Symbolic description	Explanation
POP [SI + 05H]	$0005_H \xleftarrow{\text{Sign extend}} 05_H$ $MA = (DS) \times 16_{10} + (SI) + 0005_H$ $MA_s = (SS) \times 16_{10} + (SP)$ $(MA) \leftarrow (MA_s)$; $(MA+1) \leftarrow (MA_s+1)$	The content of top of the stack memory is moved to data memory specified by the instruction. (The content of two consecutive locations in stack memory are moved to two consecutive locations in data memory.) Then SP is incremented by two.
XCHG CX, SI	$(CX) \leftrightarrow (SI)$	The content of SI-register is exchanged with the content of CX-register.
XCHG DH, CL	$(DH) \leftrightarrow (CL)$	The content of CL-register is exchanged with the content of DH-register.
XCHG [DI+07H], DX	$0007_H \xleftarrow{\text{Sign extend}} 07_H$ $MA = (DS) \times 16_{10} + (DI) + 0007_H$ $(MA ; MA + 1) \leftrightarrow (DX)$	The content of DX-register is exchanged with the content of memory.
IN AX, [DX]	$PORT_{addr} = (DX)$ $(AX) \leftarrow (PORT)$	The content of port (whose address is in DX-register) is moved to AX-register.
IN AX, 0C0H	$PORT_{addr} = C0_H$ $(AX) \leftarrow (PORT)$	The content of port (whose address is specified in the instruction) is moved to AX-register.
OUT [DX], AL	$PORT_{addr} = (DX)$ $(PORT) \leftarrow (AL)$	The content of AL-register is moved to port addressed by DX-register.
OUT 0F2H, AX	$PORT_{addr} = F2_H$ $(PORT) \leftarrow (AX)$	The content of AX-register is moved to port whose address is specified in the instruction.

Table 4.21 continued...

Instruction	Symbolic description	Explanation
LEA CX, [BX + DI]	EA = (BX) + (DI) (CX) ← (EA) *Note: EA - Effective Address*	The instruction LEA determines the **Effective Address** (EA) of source operand in memory and loads the address in CX-register.
LDS BX, [420AH]	$MA = (DS) \times 16_{10} + 420A_H$ (BX) ← (MA ; MA + 1) (DS) ← (MA+2 ; MA+3)	This instruction copies a word from memory to BX-register and copies the next word in memory to DS-register.
LES BP, [0C00FH]	$MA = (DS) \times 16_{10} + C00F_H$ (BP) ← (MA ; MA + 1) (ES) ← (MA + 2 ; MA + 3)	This instructio copies a word from memory to BP-register and copies the next word in memory to ES-register
ADD CX, DX	(CX) ← (CX) + (DX)	The contents of CX and DX registers are added and the result is stored in CX-register.
ADC BH, AL	(BH) ← (BH) + (AL) + CF	The content of BH-register, AL-register and the carry flag are added. The result is stored in BH-register.
ADD CX, [BX + 05H]	$0005_H \xleftarrow{\text{Sign extend}} 05_H$ $MA = (DS) \times 16_{10} + (BX) + 0005_H$ (CX) ← (CX) + (MA ; MA + 1)	The content of CX-register and a word from memory are added. The result is stored in CX-register.
ADC [DI], C2H	$MA = (DS) \times 16_{10} + (DI)$ (MA) ← (MA) + $C2_H$ + CF	The 8-bit data ($C2_H$) given in the instruction and the carry flag are added to memory. The result is stored in memory.
ADD DX, 0FA5H	(DX) ← (DX) + $0FA5_H$	The 16-bit data ($0FA5_H$) given in the instruction is added to the content of DX-register. The result is stored in DX-register.
SUB DI, SI	(DI) ← (DI) – (SI)	The content of SI-register is subtracted from DI-register. The result is stored in DI-register.

Table 4.21 continued...

Instruction	Symbolic description	Explanation
SUB [BP+DI], AH	$MA_S = (SS) \times 16_{10} + (BP) + (DI)$ $(MA_S) \leftarrow (MA_S) - (AH)$	The content of AH-register is subtracted from the content of a location in stack memory. The result is stored in stack memory location.
SUB SP, 0500H	$(SP) \leftarrow (SP) - 0500_H$	The 16-bit data (0500_H) given in the instruction is subtracted from SP-register. The result is stored in SP-register.
SBB AX, DI	$(AX) \leftarrow (AX) - (DI) - CF$	The content of DI-register and carry flag are subtracted from the content of AX-register. The result is stored in AX-register.
SBB [BX + 08H], DL	$0008_H \xleftarrow{\text{Sign extend}} 08_H$ $MA = (DS) \times 16_{10} + (BX) + 0008_H$	The content of DL-register and carry flag are subtracted from memory. The result is story in memory.
MUL BX	$(DX)\ (AX) \leftarrow (AX) \times (BX)$	The content of AX and BX registers are multiplied. The lower word of the result is stored in AX-register and the upper word of the result is stored in DX-register.
MUL DL	$(AX) \leftarrow (AL) \times (DL)$	The content of AL and DL registers are multiplied. The 16-bit result is stored in AX-register.

Table 4.21 continued...

Instruction	Symbolic description	Explanation
MUL [BX + 08H]	$0008_H \xleftarrow{\text{Sign extend}} 08_H$ $MA = (DS) \times 16_{10} + (BX) + 0008_H$ $(AX) \leftarrow (AL) \times (MA)$ (or) $(DX)(AX) \leftarrow (AX) \times (MA ; MA + 1)$	In 8-bit multiplication, the content of AL and 8-bit memory are multiplied. The result is stored in AX-register. In 16-bit multiplication, the content of AX and 16-bit memory are multiplied. The result is stored in AX and DX registers. The 8-bit or 16-bit multiplication is defined by w-bit in the instruction template.
DIV CH	$(AL) \leftarrow (AX) \div (CH)$ Quotient $(AH) \leftarrow (AX) \text{ MOD } (CH)$ Remainder	The content of AX-register is divided by the content of CH-register. The quotient is stored in AL-register and the remainder in AH-register.
DIV BX	$(AX) \leftarrow (DX)(AX) \div (BX)$ Quotient $(DX) \leftarrow (DX)(AX) \text{ MOD } (BX)$ Remainder	The content of AX and DX registers are divided by the content of BX-register. The quotient is stored in AX-register and the remainder in DX-register.
DIV [SI + 0C002H]	$MA = (DS) \times 16_{10} + (SI) + C002_H$ 16-bit ÷ 8-bit $(AL) \leftarrow (AX) \div (MA)$ Quotient $AH \leftarrow (AX) \text{ MOD } (MA)$ Remainder 32-bit ÷ 16-bit $AX \leftarrow (DX)(AX) \div (MA ; MA + 1)$ Quotient $DX \leftarrow (DX)(AX) \text{ MOD } (MA ; MA + 1)$ Remainder	In 16-bit by 8-bit division, the content of AX-register is divided by 8-bit memory. The quotient is stored in AL-register and the remainder is stored in AH-register. In 32-bit by 16-bit division, the contents of AX and DX register are divided by 16-bit memory. The quotient is stored in AX-register and the remainder is stored in DX-register. The 8-bit or 16-bit divisor is defined by w-bit in the instruction template.

Table 4.21 continued...

Instruction	Symbolic description	Explanation
INC DL	$(DL) \leftarrow (DL) + 1$	The content of DL-register is incremented by one.
INC DX	$(DX) \leftarrow (DX) + 1$	The content of DX-register is incremented by one.
INC [BP + SI + 0F5H]	$FFF5_H \xleftarrow{\text{Sign extend}} F5_H$ $MA_s = (SS) \times 16_{10} + (BP) + (SI) + FFF5_H$ $(MA_s) \leftarrow (MA_s) + 1$ (or) $(MA_s ; MA_s + 1) \leftarrow (MA_s ; MA_s + 1) + 1$	In 8-bit operation, the content of 8-bit stack memory is incremented by one. In 16-bit operation, the content of 16-bit stack memory is incremented by one. The 8-bit or 16-bit operation is defined by w-bit in the instruction template.
DEC CH	$(CH) \leftarrow (CH) - 1$	The content of CH-register is decremented by one.
DEC BP	$(BP) \leftarrow (BP) - 1$	The content of BP-register is decremented by one.
DEC [DI + 0007H]	$MA = (DS) \times 16_{10} + (DI) + 0007_H$ $(MA) \leftarrow (MA) - 1$ (or) $(MA_s ; MA_s + 1) \leftarrow (MA_s ; MA_s + 1) + 1$	In 8-bit operation, the content of 8-bit memory is decremented by one. In 16-bit operation, the content of 16-bit memory is decremented by one. The 8-bit or 16-bit operation is defined by w-bit in the instruction template.
CMP DL, CH	Modify flags $\leftarrow (DL) - (CH)$ If $(DL) = (CH)$; then CF = 0, SF = 0, ZF = 1 If $(DL) > (CH)$; then CF = 1, SF = 1, ZF = 0 If $(DL) > (CH)$; then CF = 0, SF = 0, ZF = 0	The comparison is performed by subtracting the content of CH register from the content of DL register. The result is used to modify flags. The content of DL and CH registers are not altered.

Table 4.21 continued...

Instruction	Symbolic description	Explanation

CMP CX, SI	Modify flags ← (CX) – (SI)
	If (CX) = (SI) ; then CF = 0, SF = 0, ZF = 1
	If (CX) < (SI) ; then CF = 1, SF = 1, ZF = 0
	If (CX) > (SI) ; then CF = 0, SF = 0, ZF = 0
		The comparison is performed by subtracting the content of SI-register from the content of CX-register. The result is used to modify flags. The content of SI and CX registers are preserved.

CMP [BX], CL	MA = (DS) × 16_{10} + (BX)
	Modify flags ← (MA) – (CL)
	If (MA) = (CL) ; then CF = 0, SF = 0, ZF = 1
	If (MA) < (CL) ; then CF = 1, SF = 1, ZF = 0
	If (MA) > (CL) ; then CF = 0, SF = 0, ZF = 0
		The comparison is performed by subtracting the content of CL-register from 8-bit memory. The result is used to modify flags. The contents of CL-register and memory are preserved.

CMP [DI], 00FFH	MA = (DS) × 16_{10} + (DI)
	Modify flags ← (MA ; MA + 1) – $00FF_H$
	If (MA ; MA + 1) = $00FF_H$; then CF = 0, SF = 0, ZF = 1
	If (MA ; MA + 1) < $00FF_H$; then CF = 1, SF = 1, ZF = 0
	If (MA ; MA + 1) > $00FF_H$; then CF = 0, SF = 0, ZF = 0
		The comparison is performed by subtracting the 16-bit data ($00FF_H$) given in the instruction from16-bit memory. The result is used to modify flags. The content of memory is preserved.

AND CX, DX	(CX) ← (CX) & (DX)	The contents of CX and DX registers are ANDed and the result is stored in CX-register.
AND [BX + SI], AX	MA = (DS) × 16_{10} + (BX) + (SI) (MA ; MA + 1) ← (MA ; MA + 1) & (AX)	The contents of 16-bit memory and AX-register are ANDed and the result is stored in memory.
AND CL, 0FH	(CL) ← (CL) & $0F_H$	The contents of CL-register and the 8-bit data ($0F_H$) given in the instruction are ANDed. The result is stored in CL-register.

Table 4.21 continued...

Instruction	Symbolic description	Explanation
OR AH, DL	$(AH) \leftarrow (AH) \mid (DL)$	The contents of AH and DL registers are ORed and the result is stored in AH-register.
OR CX, [BP + DI + 05H]	$0005_H \xleftarrow{\text{Sign extend}} 05_H$ $MA_s = (SS) \times 16_{10} + (BP) + (DI) + 0005_H$ $(CX) \leftarrow (CX) \mid (MA_s ; MA_s + 1)$	The contents of 16-bit stack memory and CX-register are ORed and the result is stored in CX-register.
OR DI, 0F0F0H	$(DI) \leftarrow (DI) \mid F0F0_H$	The contents of DI-register is ORed with 16-bit data ($F0F0_H$) given in the instruction. The result is stored in DI-register.
XOR BX, DX	$(BX) \leftarrow (BX) \wedge (DX)$	The contents of BX and DX registers are Exclusive-ORed. The result is stored in BX-register.
XOR SP, [SI + 0A00AH]	$MA = (DS) \times 16_{10} + (SI) + A00A_H$ $(SP) \leftarrow (SP) \wedge (MA ; MA + 1)$	The contents of SP-register and 16-bit memory are Exclusive-ORed. The result is stored in SP-register.
XOR [DI + 05H], 0F0FH	$0005_H \xleftarrow{\text{Sign extend}} 05_H$ $MA = (DS) \times 16_{10} + (DI) + 0005_H$ $(MA ; MA + 1) \leftarrow (MA ; MA + 1) \wedge 0F0F_H$	The contents of 16-bit memory is Exclusive-ORed with 16-bit data ($0F0F_H$) given in the instruction. The result is stored in memory.
TEST CX, DI	Modify flags $\leftarrow (CX) \& (DI)$	The TEST operation is performed by logically ANDing he contents of CX and DI registers. The result is used to modify flags. The contents of CX and DI registers are not altered.
TEST [SI + 0F0H], 0C000H	$FFF0_H \xleftarrow{\text{Sign extend}} F0_H$ $MA = (DS) \times 16_{10} + (SI) + FFF0_H$ Modify flags $\leftarrow (MA ; MA + 1) \& C000_H$	The TEST operation is performed by logically ANDing the content of 16-bit memory and 16-bit data given in the instruction. The result is used to modify flags. The content of memory is not altered.

4.8 SHORT-ANSWER QUESTIONS

Q4.1 **What is the size of 8086 instructions?**

The size of 8086 instruction is one to six bytes. The first byte consists of opcode and special bit indicators. The second byte will specify the addressing mode of the operands. The subsequent bytes will specify immediate data or address.

Q4.2 **Write the general format of 8086 instructions?**

The general format of 8086 instruction is shown in Fig. Q4.2.

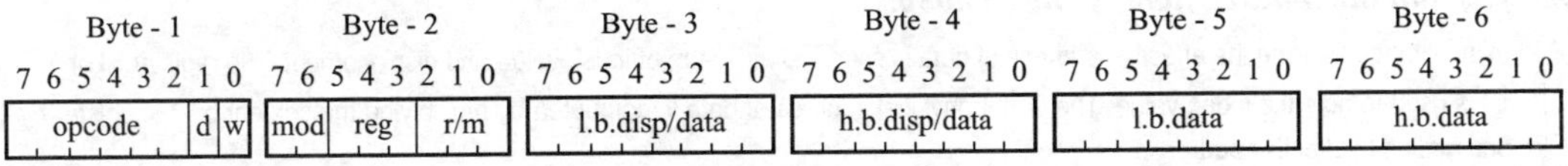

Byte - 1	Byte - 2	Byte - 3	Byte - 4	Byte - 5	Byte - 6
7 6 5 4 3 2 1 0	7 6 5 4 3 2 1 0	7 6 5 4 3 2 1 0	7 6 5 4 3 2 1 0	7 6 5 4 3 2 1 0	7 6 5 4 3 2 1 0
opcode d w	mod reg r/m	l.b.disp/data	h.b.disp/data	l.b.data	h.b.data

Fig. Q4.2: General format of 8086 instruction.

Q4.3 **What is addressing?**

The method of specifying the data to be operated (operand) by the instruction is called addressing.

Q4.4 **What are the addressing modes available in 8086?**

The 8086 has the following 12 addressing modes.

1. Register addressing	5. Based addressing	9. Direct IO port addressing
2. Immediate addressing	6. Indexed addressing	10. Indirect IO port addressing
3. Direct addressing	7. Based index addressing	11. Relative addressing
4. Register indirect addressing	8. String addressing	12. Implied addressing

Q4.5 **What is register addressing? Give example.**

In register addressing the instruction will specify the name of the register which holds the data to be operated by the instruction.

Example : MOV CX, DX - The content of DX-register is moved to CX-register.

Q4.6 **What is immediate addressing? Give example.**

In immediate addressing mode an 8-bit or 16-bit data is specified as part of the instruction.

Example : MOV BX, 0CA5H - The 16-bit data ($0CA5_H$) given in the instruction is moved to BX-register.

Q4.7 **Explain the direct addressing in 8086.**

In direct addressing an unsigned 16-bit displacement or signed 8-bit displacement will be specified in the instruction. The displacement is the effective address of the data. The 20-bit physical address of the data is computed by multiplying the content of DS-register by 16_{10} and adding to effective address.

Example : MOV CL, [0F2AH] - The memory address is computed by multiplying the content of DS-register by 16_{10} and adding the 16-bit displacement ($0F2A_H$) given in the instruction. Then the content of memory is moved to CL-register.

Q4.8 **Explain the register indirect addressing in 8086.**

In register indirect addressing the name of the register which holds the effective address of data will be specified in the instruction. The register used to hold the effective address are BX, SI or DI. The 20-bit physical address of data is obtained by multiplying the content of DS-register by 16_{10} and adding to effective address.

Example : MOV DX, [DI] - The memory address of the data is obtained by multiplying the content of DS-register by 16_{10} and adding the content of DI-register. The content of memory is moved to DX-register.

Q4.9 Explain the based addressing in 8086.

In based addressing the effective address of data is specified as a sum of base value and displacement. The register BX or BP is used to hold the base value. When BX holds the base value, the 20-bit physical address of data is calculated by multiplying the content of DS-register by 16_{10} and adding to effective address. When BP holds the base value, the 20-bit physical address of data is calculated by multiplying the content of SS register by 16_{10} and adding to effective address.

Example : MOV CX, [BP + 00A2H] - The effective address is computed by adding the 16-bit displacement ($00A2_H$) given in the instruction to the content of BP-register. The 20-bit physical address is obtained by multiplying the content of SS-register by 16_{10} and adding to effective address. The content memory is moved to CX-register.

Q4.10 Explain the indexed addressing in 8086.

In indexed addressing the effective address of data is specified as a sum of index value and displacement. The register SI or DI is used to hold the index value. The 20-bit physical address of data is computed by multiplying the content of DS-register by 16_{10} and adding to effective address.

Example : MOV AX, [DI + 04H] - The effective address is computed by sign extending the 8-bit displacement given in the instruction to 16-bit and adding to the content of DI. The 20-bit physical address of memory is computed by multiplying the content of DS-register by 16_{10} and adding to effective address. The content of memory is moved to AX-register.

Q4.11 What is based indexed addressing? Give an example.

In based indexed addressing, the effective address is specified as a sum of base value, index value and displacement. The base value is stored in BX or BP and the index value is stored in SI or DI-register. When BX holds the base value of effective address, the content of DS-register is considered as segment base address and when BP holds the base value, the content of SS-register is considered as segment base address.

The 20-bit physical address is computed by multiplying the segment base address by 16_{10} and adding to effective address .

Example : MOV CX, [BP + DI + 01A0H] - The effective address is computed by adding the contents of BP-register, DI-register and the 16-bit displacement ($01A0_H$) given in the instruction. The 20-bit physical address of memory is computed by multiplying the content of SS-register by 16_{10} and adding to effective address. The content of memory is moved to CX-register.

Q4.12 Explain string addressing in 8086.

In 8086, string addressing is used by string instructions to address the source and destination operand data. In this mode the SI-register is used to hold the effective address of source data and DI-register is used to hold the effective address of destination. The memory address of source is obtained by multiplying the content of DS-register by 16_{10} and adding to effective address (content of SI-register). The memory address of destination is obtained by multiplying the content of ES-register by 16_{10} and adding to effective address (content of DI-register). After execution of string instruction the content of SI and DI are incremented or decremented depending on direction flag.

Q4.13 How are IO ports addressed in 8086?

The IO ports in 8086-based system can be addressed either by direct addressing or by indirect addressing. In direct addressing an 8-bit port address is directly specified in the instruction. In indirect addressing a 16-bit port address is stored in DX-register and the name of the register (DX) is specified in the instruction.

Q4.14 What is relative addressing?

In relative addressing, the effective address of a program instruction is specified relative to instruction pointer by an 8-bit signed displacement.

Example : JC 0F2H - If carry flag is one, then a new effective address is calculated and loaded in instruction pointer. The new effective address is obtained by sign extending the 8-bit displacement ($F2_H$) given in the instruction and adding to the content of instruction pointer.

Q4.15 What is implied addressing?

In implied addressing mode, the instruction itself will specify the data to be operated by the instruction.

Example : CLD - Clear Direction Flag.

Q4.16 List the data transfer instructions that affects the flags in 8086.

In 8086, the data transfer instructions affecting the flags are POPF and SAHF. The POPF instruction is used to restore the previously stored status of the flag. The instruction SAHF is used to modify the content of flag register.

Q4.17 List the instructions of 8086 that affect only carry flag.

The instructions that affect only carry flag are CLC, CMC and STC.

Q4.18 List the instructions of 8086 that affect direction flag.

The instructions that affect direction flag are CLD, POPF and STD.

Q4.19 List the instructions of 8086 that affects interrupt flag.

The instructions that affects interrupt flag are CLI, INT, INTO, IRET, POPF and STI.

Q4.20 What are the operations performed by data transfer instructions?

The operations performed by the data transfer instructions are :

1. Copy the content of a register to another register.
2. Copy the content of a register/segment register to memory or vice versa.
3. Copy the content of accumlator to port or vice versa.
4. Exchange the content of two registers or register and memory.
5. Load an immediate operand to register/memory.
6. Load effective address in segment registers.

Q4.21 What are the operations performed by arithmetic instructions?

The operations performed by arithmetic instructions are:

1. Addition or subtraction of binary, BCD or ASCII data.
2. Multiplication or division of signed or unsigned binary data.
3. Increment or decrement or comparison of binary data.

Q4.22 What are the operations performed by logical instructions?

The operations performed by logical instructions are AND, OR, Exclusive-OR, complement, arithmetic shift and logical shift.

Q4.23 What are the operations performed by string instructions?

The operations performed by string instructions are :

1. Copy a byte/word of a string data from data segment to extra segment.
2. Compare the content of two memory locations or accumulator and a memory location.
3. Load a byte/word of a string data from memory to accumulator or vice versa.

Q4.24 List the string instructions of 8086.

The string instructions of 8086 are REPZ/REPE, REPNZ/REPNE, MOVSB, MOVSW, CMPSB, CMPSW, SCASB, SCASW, LODSB, LODSW, STOSB and STOSW.

Q4.25 What will be the content of stack pointer (SP) after a PUSH operation and after a POP operation?

The PUSH operation will decrement the content of SP by two and so after a PUSH operation the content of SP will be less by two than earlier value.

The POP operation will increment the content of SP by two and so after a POP operation the content of SP will be greater by two than earlier.

Q4.26 List the IO instructions of 8086.

The IO instructions of 8086 are:

1. IN A, addr8
2. IN A, [DX]
3. OUT addr8, A
4. OUT [DX], A

The IN instruction is used to load a byte/word from IO port to accumlator and the OUT instruction is used to send a byte/word from accumlator to IO port.

Q4.27 Explain the instruction LEA reg16, mem.

The instruction LEA is used to load the effective address of the memory operand to the register specified in the instruction (i.e., this instruction will not load the content of memory in register but the calculated effective address of memory data is loaded in the register).

Q4.28 How can the low byte flag register be modified in 8086?

The low byte of flag register can be modified by moving an 8-bit data to AH-register and then moving the content of AH to low byte flag register using SAHF instruction.

Q4.29 How can the 16-bit flag register be modified in 8086?

The steps involved in modifying the 16-bit flag register are given below:

1. First move a 16-bit data to a 16-bit register.
2. Second save the content of register in stack using PUSH instruction.
3. Finally move the top of stack to flag register using POPF instruction.

Q4.30 How is subtraction performed in 8086 and how can the result be interpreted?

The 8086 processor performs 2's complement subtraction and after subtraction the carry flag is complemented. Therefore the result of subtraction can be interpreted as follows:

1. After subtraction if carry flag is set (i.e., CF = 1) then the result is negative and the result will be in 2's complement form.
2. After subtraction if carry flag is cleared/reset (i.e., CF = 0) then the result is positive.

Q4.31 What is the similarity and difference between subtract and compare instructions?

Similarity: Both the subtraction and comparison are performed by subtracting two data in ALU and flags are altered depending upon the result.

Difference: After subtract operation, the result is stored in destination register/memory, but after compare operation the result is discarded.

Q4.32 What will be the status of flags after division and multiplication operations?

The division and multiplication operation will modify all the six arithmetic flags (CF, AF, PF. ZF, SF and OF flags) but all these flags will be in undefined state after the division and multiplication operations.

Q4.33 What is the difference between Compare and Test operations in 8086?

In Compare operation the content of register or memory is subtracted from the content of another register and the result is used to modify the flags.

In Test operation the content of register or memory is bit by bit ANDed with the content another register and the result is used to modify the flags.

In both Compare and Test operations the content of source and destination are not altered.

Q4.34 What is the difference between arithmetic shift and logical shift?

In logical shift operation zero is inserted in the shifted location (i.e., zero is inserted in LSD position for left shift and for right shift zero is inserted in MSD position).

The arithmetic left shift operation is same as logical left shift, whereas in arithemetic right shift operation, the sign bit is copied into shifted location i.e., after every right shift the old value of MSD (**M**ost **S**ignificant **D**igit) is copied into the current MSD location.

Q4.35 What is the difference between shift and rotate operation?

In shift operation either zero or one is inserted in the shifted location, whereas in rotate operation ony the content of register/memory with or without carry are rotated (i.e., in rotate operation there is no insertion of extra bit in the shifted position).

Q4.36 What is near call and far call?

Near call refers to calling a procedure stored in the same code segment memory in which main program (or calling program)resides. Far call refers to calling a procedure stored in different code segment memory than that of main program.

While executing near call instructions the content of IP alone is pushed to stack. While executing far call the content of CS and IP are pushed to stack.

Q4.37 What is the difference between CALL and JUMP instruction?

In CALL instruction, the address of next instruction is pushed to stack (i.e., stored in stack memory) before transferring the program control to call address. But in JUMP instruction the address of next instruction is not saved.

Q4.38 What is the difference between conditional and unconditional branch instructions?

In unconditional branch instruction, the program control is transferred to branch address without checking any flag condition. But in conditional branch instructions, a flag condition is checked and only if the flag condition is true, program control is transferred to branch address, otherwise next instruction is executed.

Q4.39 What is near jump and far jump?

In near jump, the program control is transferred to new memory location in the same segment by modifying the content of **I**nstruction **P**ointer (IP). In far jump the program control is transferred to new memory location in another segment by modifying the content of **I**nstruction **P**ointer (IP) and **C**ode **S**egment (CS) register.

Q4.40 What is difference between CALL and INT instruction?

While executing CALL instruction only IP and CS are saved in stack. But, While executing INT instruction IP,CS and flag register are saved in stack.

4.9 EXERCISES

I. Fill in the blanks with appropriate words

1. The number of T states(clock periods) required to execute an 8086 memory read cycles is _____.

2. The _____ signal of 8086 is sampled at the end of second T-state of every bus cycle.

3. The _____ machine cycle is executed whenever the 8086 receives the interrupt through INTR pin

4. The _____ segment is used with register indirect addressing mode that uses BP or SP register to address the memory in 8086.

5. _____ addressing mode is used in the instruction CLC of 8086.

6. The _____ instruction of 8086 is used to store the content of program status word register into stack memory.

7. The _____ and _____ instructions of 8086 are used to read from and write into I/O ports respectively.

8. _____ instruction of 8086 is used in code conversion to read the look up table.

9. The _____ and _____ instructions of 8086 can be used to load both segment registers and offset value into specified registers.

10. The content of AL register is 80_H. The value of AX register after the execution of CBW instruction of 8086 is _____

11. An 8086 instruction that is used to convert hexadecimal number into packed BCD number after addition is _____

12. The 8086 instruction used for BCD division is _____

13. The 8086 instruction used to perform signed multiplication is _____

14. The status of zero flag is _____ after comparing AL = 45H and BL = 45H using the 8086 instruction CMP AL,BL.

15. The _____ instruction of 8086 is used to find 1's complement of the special register/memory content.

16. The content of AL register after execution of the 8086 instructuion XRA AL, AL is _____

17. In 8086, the counter register used in all string operation is _____

18. Short jump in 8086 refers to jump within _____

19. The _____ and _____ instructions of 8086 are used to execute the subroutine and return to the main program respectively.

20. The _____ instruction of 8086 is used to invert the carry flag.

21. The 8086 instruction _____ disables all the interrupt.

22. The 8086 instruction with _____ prefix is used to differentiate the 8086 and 8087 instructions.

Answers

1. 4	7. IN, OUT	13. IMUL	19. CALL, RET
2. READY	8. XLAT	14. 1	20. CMC
3. interrupt acknowledge	9. LDS, LES	15. NOT	21. CLI
4. stack	10. $FF80_H$	16. 00_H	22. ESC
5. Implied	11. DAA	17. CX	
6. PUSHF	12. AAD	18. same segment	

II. State whether the following statements are True/False.

1. The T state of 8086 always starts in the middle of falling edge of the clock signal.

2. The READY signal of 8086 is permanently tied high.

3. The M/$\overline{\text{IO}}$ signal of 8086 is asserted high during execution of an IO read machine cycle.

4. The 8086 instruction MOV AX,[2000H] uses immediate addressing mode to get the operand.

5. The 8086 instruction JMP 5000:0032 specifies far jump.

6. The segment registers of 8086 cannot be initialized directly.

7. The PUSH instruction of 8086 is used to retrieve data from stack.

8. Stack grows downwards in 8086.

9. The stack memory is first in first out.

10. The instruction AAA of 8086 is used to convert the result in AL into packed BCD data after BCD addition.

11. In string operations, the source and destination are always data segment and extra segment respectively.

12. The 8086 instruction AAD is always used before division.

13. The AND and TEST instructions are same in 8086.

14. The CALL instruction of 8086 is conditional control transfer instruction.

15. Conditional jump instructions of 8086 are always near jump instructions.

16. The LOOP instruction of 8086 terminates the execution when CX = 0.

17. INTO instruction is type-4 interrupt in 8086.

18. STI instruction of 8086 enables the interrupt.

Answers

1. True	4. False	7. False	10. False	13. False	16. True
2. False	5. True	8. True	11. True	14. False	17. True
3. False	6. True	9. False	12. True	15. True	18. True

III. Choose the right answer for the following questions.

1. *If the 8086 clock frequency is 10MHz, what is the time for a memory write cycle if there are 2 wait states and no wait states respectively?*

    a) 600 ns and 400 ns *b)* 400 ns and 600 ns *c)* 500 ns and 300 ns *d)* 300 ns and 500 ns

2. *What are the 8086 machine cycles executed for the instruction SUB [BX], AX*

    a) opcode fetch *b)* memory read *c)* memory write *d)* all the three

3. *The addressing mode used in the 8086 instruction MOV AL,12H is*

    a) immediate addressing mode *b)* register addressing mode

    c) direct addressing mode *d)* relative addressing mode

4. *If the data is available in a memory location pointed by a register, then it is*

    a) direct addressing mode *b)* register addressing mode

    c) register indirect addressing mode *d)* immediate addressing mode

5. *The addressing mode that is used in conditional jump instructions of 8086 is*

    a) intrasegment direct addressing mode

    b) intersegment direct addressing mode

    c) intrasegment direct and indirect addressing mode

    d) intersegment direct addressing and indirect mode

6. *Which of the following 8086 instruction uses relative addressing mode?*

    a) ADD *b)* AAA *c)* LES *d)* POP

7. Which one of the following 8086 instructions use intrasegment direct addressing mode?

   a) JMP [BX]   b) JMP 50H   c) JMP 5000:0002   d) JMP [2000H]

8. In a machine instruction format of 8086, s-bit is the

   a) status bit   b) sign bit   c) sign extension bit d) none of the above

9. If w-bit value in machine instruction format of 8086 is '1' then the operand is of

   a) 8-bits   b) 4-bits   c) 16-bits   d) 2-bit

10. Which of the following is not an 8086 data transfer instruction?

    a) MOV   b) PUSH   c) DAS   d) POP

11. The following 8086 instruction affects carry flag

    a) PUSH   b) POP   c) PUSHF   d) SAHF

12. Which of the following 8086 instruction is not valid?

    a) MOV AL,DL   b) MOV 2000H,3000H c) PUSH BX   d) ADD CX, 01H

13. The 8086 instruction that uses table look up procedure for data access is

    a) XCHG   b) XLAT   c) XOR   d) JCXZ

14. In general, the destination operand of an instruction cannot be

    a) memory location b) register   c) immediate data   d) memory location and register

15. The 8086 instruction used for multibyte addition is

    a) ADD   b) ADC   c) AAA   d) DAA

16. Given that AX = 2FFF$_H$, BX = 0001$_H$. What will be the status of flags after execution of an 8086 instruction ADD AX, BX

    a) PF = 1 ; OF = 0 ; SF = 0 ; ZF = 0 ; AC = 1 ; CF = 0

    b) PF = 0 ; OF = 1 ; SF = 0 ; ZF = 1 ; AC = 0 ; CF = 0

    c) PF = 1 ; OF = 1 ; SF = 0 ; ZF = 0 ; AC = 1 ; CF = 1

    d) PF = 0 ; OF = 0 ; SF = 0 ; ZF = 0 ; AC = 0 ; CF = 0

17. Which of the following is 8086 mnemonic of one byte instruction?

    a) ADD   b) ADC   c) AAA   d) all the above

18. The 8086 instructions used to produce the result AL = 15 when AL = 09$_H$ and BL = 06$_H$ are

    a) ADD AL, BL   b) ADD AL, BL   c) AAA   d) DAA
       AAA   DAA   ADD AL, BL   ADD AL, BL

19. The _____ instruction of 8086 is used before BCD division.

    a) AAA   b) AAS   c) AAD   d) DAA

20. The following 8086 instruction is used to mask the lower nibble of AL register

    a) OR AL, 00H   b) AND AL, 00H   c) XOR AL, 0F0H   d) AND AL, 0F0H

21. The 8086 instruction that performs logical AND operation and modifies flag register without changing the source and destination operands is

    a) AAA   b) AND   c) TEST   d) XOR

22. Which one of the following logical operations performed with FF_H will invert the register

    a) AND reg, 0FFH   *b)* OR reg, 0FFH   *c)* XOR reg, 0FFH   *d)* TEST reg, 0FFH

23. Which of the following is not a bit manupulating instruction of 8086

    a) CMA   *b)* CLC   *c)* CMC   *d)* STC

24. An 8086 instuction used to return to the main program after executing subroutine

    a) CALL   *b)* HLT   *c)* RET   *d)* END

25. The following 8086 instuction is used to load the string byte into AL register

    a) LODSB   *b)* LODSW   *c)* STOSB   *d)* STOSW

26. Which of the following is not a 8086 machine control instruction?

    a) HLT   *b)* CLC   *c)* LOCK   *d)* ESC

27. Which of the following is not a 8086 machine control instruction?

    a) STI   *b)* SIT   *c)* CLI   *d)* RTI

Answers

1. a	5. c	9. c	13. b	17. c	21. c	25. a
2. d	6. b	10. c	14. c	18. b	22. c	26. b
3. a	7. c	11. d	15. b	19. c	23. a	27. c
4. c	8. c	12. b	16. a	20. d	24. c	

IV. Answer the following questions.

E4.1 Write the 8086 machine cycles executed for the following instructions. Also calculate the time required to execute each instruction. Assume that procesor works at 8 MHz frequency.

    a. MOV [BX], 1234H

    b. XOR AL, BL

    c. NOP

E4.2 Identify the addressing modes used in each instruction of 8086.

    a. ADD AX, BX   d. SUB BX, [CX]   g. ADD AL, [BX][SI]   j. JMP [BX]

    b. OR AL, 0FH   e. MOV AX, 02 [BX]   h. JMP 2000H   k. IN AL, [89H]

    c. MOV AX, [2000H] f. CLC   i. CALL 5000:0002

E4.3 Identify and correct the syntax errors if any in the following instructions of 8086

    a. MOV DS, 2000H   b. MOV SI, 0005H   c. MOV [0015],[0102]

    d. XCHG AL, 20H   e. IN BL, 30H   f. MOV BL, AX

E4.4 Differntiate the 8086 instructions (a) MOV SI, [1000H] and (b) LEA SI, [1000H]. Assume the memory location DS: 1000 contains the data 34_H and DS: 1001 contains the data 12_H.

E4.5 Write an equivalent 8086 instruction for the following sequence of instructions. Justify the answer

    MOV SI, [2000H]

    MOV AX, [2002H]

    MOV DS, AX

E4.6 What will be the status of AX, BX registers and carry, zero and sign flags after the execution of the two 8086 instructions (a) SUB AX, BX and (b) CMP AX, BX if AX = 2131_H and BX = 5131_H and CF = ZF = SF = 0.

E4.7 Assume that AL = 20_H and BL = FE_H. What will be the output when AL and BL are multiplied by MUL and IMUL instructions of 8086. Specify the location of output.

E4.8 Assume AX = 1234_H and BL = $E0_H$. What will be the quotient and remainder when AX is divided by BL using DIV and IDIV instruction of 8086. Also specify the output location.

E4.9 Write the 8086 instructions used to change the magnitude of data stored at AL = 02_H.

E4.10 Assume that AL = - 07H. What will be the output of AX after execution of the following set of 8086 instructions?

MOV AL, - 07H

CBW

E4.11 Write the 8086 instructions to perform the following operations

a. Load the data 1234_H into stack memory

b. Load DS register with data 3945_H

c. To invert the upper nibble of CL register

d. To mask the lower nibble of BL register

e. To swap the nibbles of AL register

f. To increment the content of DL register by one without affecting carry flag

g. To copy the status flags to BL register

E4.12 What is the symbol used to differentiate immediate data and direct address in 8086 instruction?

E4.13 What is the significance of MOD field in machine language of 8086?

E4.14 Mention the exception conditions which occur during division in 8086 processor.

E4.15 Differentiate the operations inversion and negation. Also write the 8086 instructions used for performing those operations.

CHAPTER 5

INSTRUCTION SET OF 8051

5.1 MACHINE CYCLES AND TIMING DIAGRAM

The timing diagram provides information about the various conditions of the signal while a machine cycle is executed.

The external basic operations performed by a microcontroller are called machine cycles. The executive of an instruction involves execution of one or more machine cycles in a specified order. The 8051 microcontroller takes one to four machine cycles to execute an instruction. The basic timing of the 8051 machine cycle is shown in Fig. 5.1.

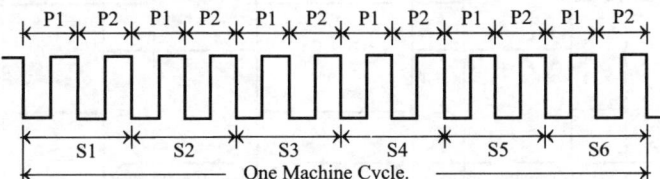

Fig. 5.1: Basic timing of a machine cycle.

The entire timing of a machine cycle of 8051 is divided into 6 states and they are denoted as S1, S2, S3, S4, S5 and S6. The timing of each state is two clock periods and they are denoted as P1 and P2.

A state in a machine cycle is a basic time interval for discrete operation of the microcontroller such as fetching an opcode byte, decoding an opcode, executing an opcode, writing a data, etc. The time taken to execute a machine cycle is 12 clock periods and so the time taken to execute an instruction is obtained by multiplying the number of machine cycles of that instruction by 12 clock periods.

Instruction execution time $= C \times 12 \times T$

$$= C \times 12 \times \frac{1}{f}$$

where, C = Number of machine cycles of an instruction

T = Time period of crystal frequency in seconds

f = Crystal frequency in Hz

The 8051 microcontroller has four machine cycles. These are:

1. External program memory fetch cycle

2. External data memory read cycle

3. External data memory write cycle

4. Port operation cycle

5.1.1 External Program Memory Fetch Cycle

The External program memory fetch machine cycle is executed by the 8051 to fetch the opcode and subsequent instruction bytes from the memory. The timing diagram of an external memory fetch cycle is shown in the Fig. 5.2. During one machine cycle (6 states), two consecutive bytes of program memory are read. In one-byte instruction, the second byte is discarded. The timing of various signals involved in the fetch operation are shown in the timing diagram in Fig. 5.2.

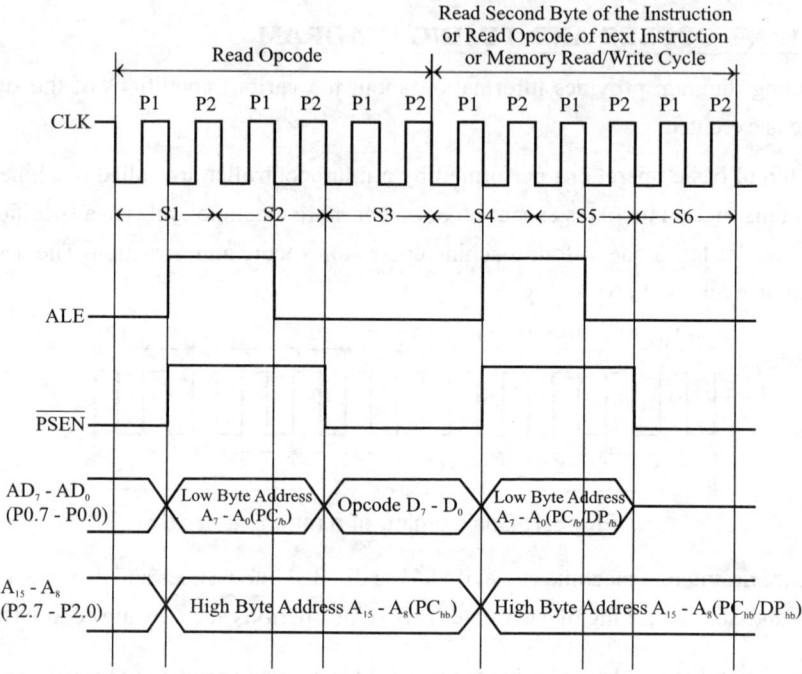

Fig. 5.2: Timing diagram of an external program memory fetch cycle.

1. At the falling edge of phase P2 of first state S1, the microcontroller outputs the low byte address on AD_7-AD_0 lines and high byte address on A_{15}-A_8 lines. For program memory fetch, the content of the Program Counter (PC) is the address of the program code. The ALE is asserted high to enable the address Latch.

2. At the middle of state S2, the ALE is asserted low and this enables the Latch to take low byte of the address and keep on its output lines.

3. The program store enable PSEN is asserted low to fetch the opcode from the memory and load on the AD_7-AD_0 lines at the state S3.

4. The microcontroller utilizes the first three states (S1, S2 and S3) to fetch the opcode from the memory.

5. During the remaining three states of one machine cycle (S4, S5 and S6), the microcontroller fetches the second byte of the same instruction or opcode of the next instruction or executes the external memory read/write cycle.

6. When executing the one-byte instruction, the states S4, S5 and S6 are used by the processor for internal operations to decode the instructions and for completing the task specified by the one-byte instruction.

5.1.2 External Data Memory Read cycle

The memory read cycle is executed by the 8051 to read a data from the external data memory. The data memory read cycle is executed immediately after an opcode fetch if the instruction execution requires an external data memory access. The timing diagram of the external memory read cycle is shown in Fig. 5.3. The timing of various signals involved in read operation are shown in the timing diagram.

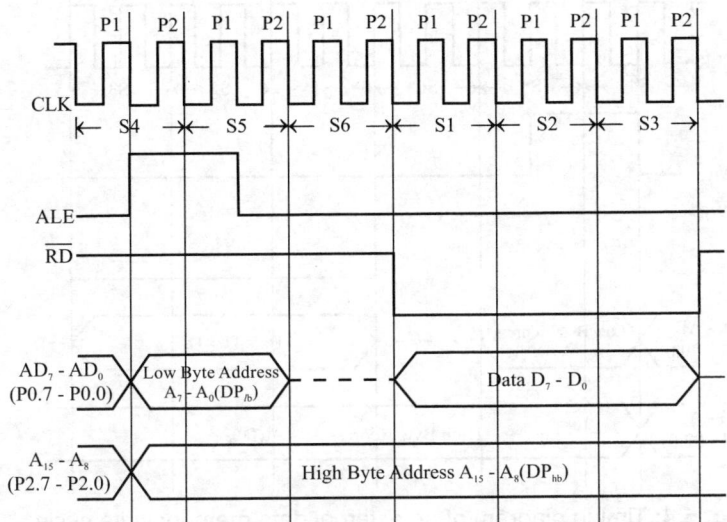

Fig. 5.3: Timing diagram of an external data memory read cycle.

1. External memory read operation needs six states. In one-byte instruction, when the external memory access is required, these six states are S4, S5 and S6 of the first machine cycle and S1, S2 and S3 of the second machine cycle. (In two-byte instruction, when external memory access is required these six states are S1 to S6 of the second machine cycle.)

2. At the first half of the state S4, the ALE is asserted high to enable the address latch. In this state, the microcontroller outputs the content of the Data Pointer low (DP_{lb}) on AD_7- AD_0 lines and the content of the Data Pointer high (DP_{hb}) on A_{15}-A_8 lines.

3. At the S1 state of the second machine cycle, the memory read signal is asserted low and the data in the specified address can be read and placed on the AD_7-AD_0 lines. The read signal is asserted low for the three states S1, S2 and S3 of the second machine cycle.

4. At the end of the S3 state of the second cycle, the RD signal is asserted low and at this time the data is latched into the microcontroller.

5. The high byte of address A_{15}-A_8 is valid for six states, i.e., from the S4 of the first machine cycle to S3 of second machine cycle.

6. At the last three states of the second machine cycle, the microcontroller reads the next byte of program memory and discards it.

5.1.3 External Data Memory Write Cycle

The memory write cycle is executed by the 8051 to store the data to the external data memory. The timing diagram of the memory write cycle is shown in Fig. 5.4. The timings of various signals involved in write operation are shown in the timing diagram.

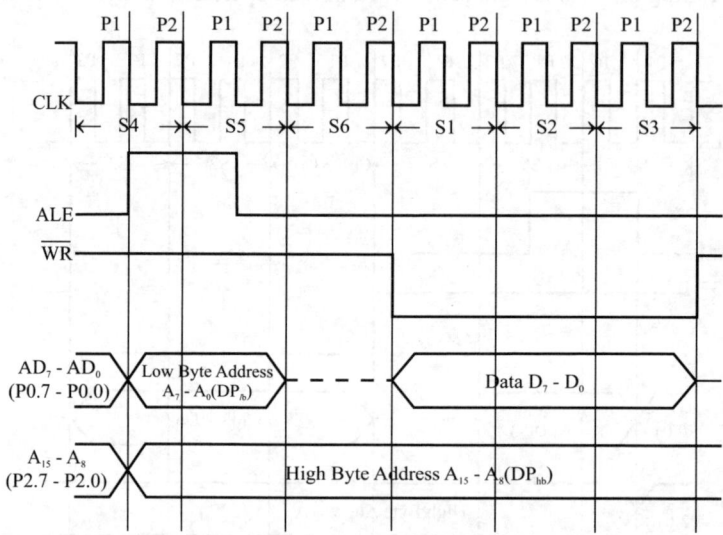

Fig. 5.4: Timing diagram of an external data memory write cycle.

1.   This write operation needs six states. In one-byte instruction, when the external memory access is required, these six states are S4, S5 and S6 of the first machine cycle and S1, S2 and S3 of the second machine cycle. (In two-byte instruction, when external memory access is required these six states are S1 to S6 of the second machine cycle.)

2.   At the first half of the state S4, the ALE is asserted high to enable the address latch. In this state, the microcontroller outputs the content of the Data Pointer low (DP$_{lb}$) on the AD$_7$- AD$_0$ lines and the content of Data Pointer high (DP$_{hb}$) on the A$_{15}$-A$_8$ lines.

3.   At the S1 state of the second machine cycle, the memory write signal is asserted low and the data is output on the AD$_7$-AD$_0$ lines. The write signal is asserted low by the microcontroller for three states S1, S2 and S3 of the second machine cycle.

4.   At the end of the S3 state of the second cycle, the WR signal is asserted low and at this time the data is latched into the external memory.

5.   The high byte of address A$_8$-A$_{15}$ valid for six states, i.e., from S4 of the first machine cycle to S3 of the second machine cycle.

6.   At the last three states of the second machine cycle, the microcontroller reads the next byte of the program memory and discards it.

5.1.4 Port Operation Cycle

The port value can be changed during the port operation machine cycle. The timing diagram of the port operation cycle is shown in the Fig. 5.5. The various signals during the port operation cycle are shown in the timing diagram. By using some instructions we can change the port value immediately. For this, the 8051 executes the port operation cycle.

1. **The port operation cycle needs six states. When port operation is required in 1-byte instruction, these six states are S4, S5 and S6 of the first machine cycle and S1, S2 and S3 of the second machine cycle.**

2. **In the fifth state S5 of the every machine cycle, all the port values are sampled.**

3. **The ports P0 and P1 are sampled at phase P1 of state S5. The ports P2, P3, RST are sampled at phase P2 of the state S5 in the first machine cycle.**

4. **The port values are changed by placing the new value on the specified port.**

5. **The changes in the new value takes place at the S1 state of the second machine cycle.**

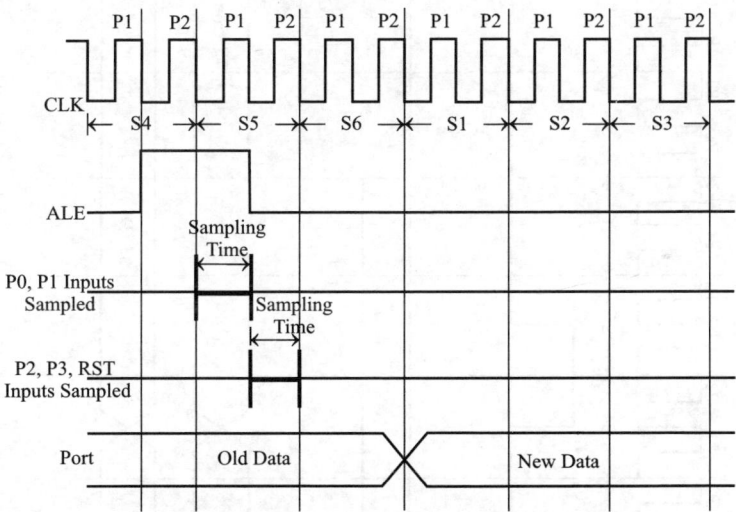

Fig. 5.5: Timing diagram of a port operation cycle.

5.2 TIMING DIAGRAM OF 8051 INSTRUCTIONS

The size of an 8051 instruction is one to three bytes. The first byte is opcode and the subsequent bytes are address or data. The 8051 microcontroller executes the instructions in one to four machine cycles.

Based on the method of execution of the machine cycles, the instructions can be classified as shown below. The various operations performed during execution is also shown below.

Case (i)   : *1-byte, 1-cycle* - Opcode fetch (3 states) + Dummy program memory fetch (3 states)

Case (ii) : *2-byte, 1-cycle* - Opcode fetch (3 states) + Fetch second byte of instruction (3 states)

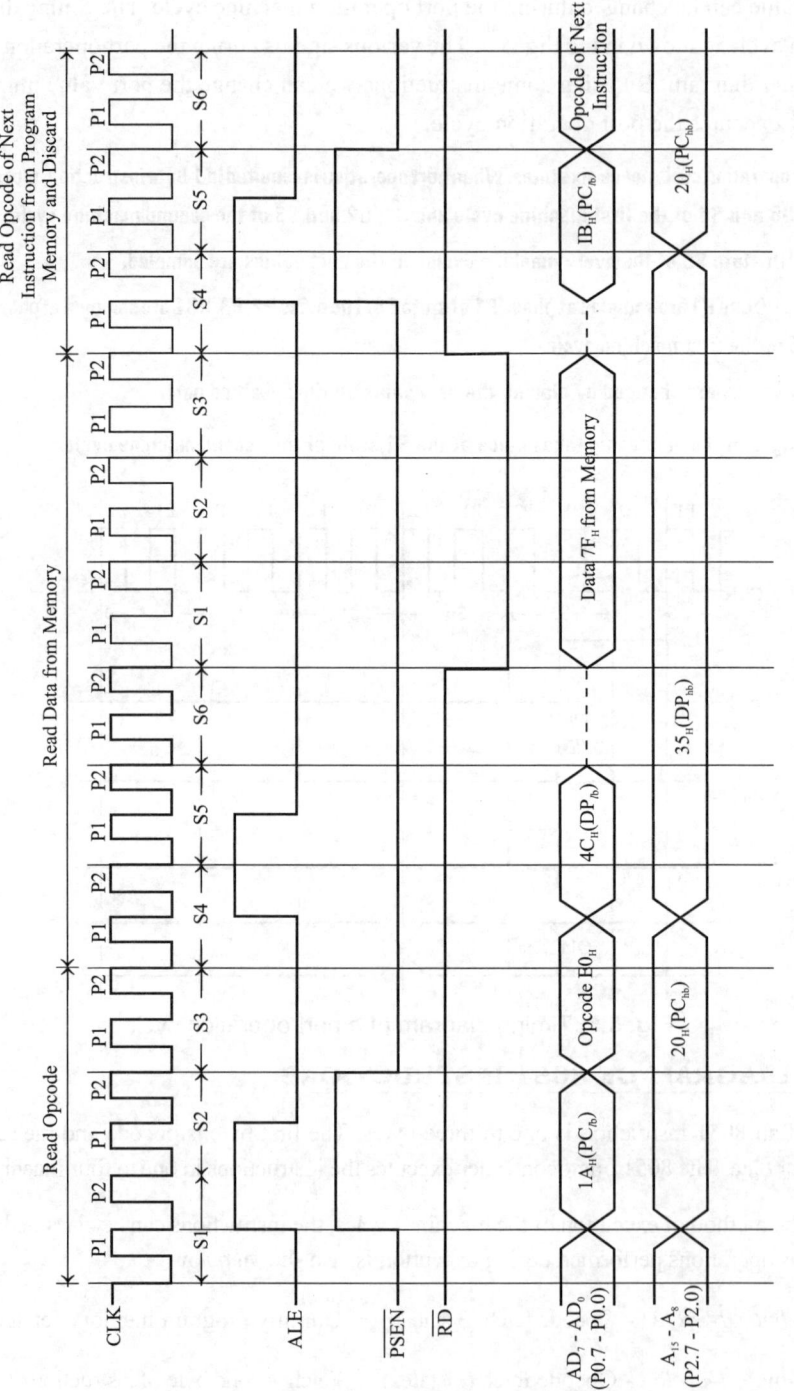

Fig. 5.6: Timing diagram of MOVX A, @DPTR.

Case (iii) : 1-byte, 2-cycle - Opcode fetch (3 states) + Data memory read/write (6 states) + dummy program memory fetch (3 states)

Case (iv) : 2-byte, 2-cycle - Opcode fetch (3 states) + Fetch second byte of instruction (3 states) + Data memory read/write (6 states)

Case (v) : 3-byte, 2-cycle - Opcode fetch (3 states) + Fetch second byte (3 states) + fetch third byte (3 states) + dummy program memory fetch (3 states)

Case (vi) : 1-byte, 4-cycle - Opcode fetch (3 states) +Dummy fetches (7 × 3 states)

The timing diagram of MOVX and ADD instructions are shown in Fig. 5.6 and Fig. 5.7 respectively:

5.2.1 Timing Diagram of MOVX A,@DPTR

The MOVX A,@DPTR instruction is used to move the content of the data memory addressed by the DPTR to the A-register (or accumulator). The timing diagram of this instruction is shown in Fig. 5.6.

The MOVX A,@DPTR is a one-byte instruction and executed in two machine cycles. In the first three states (S1, S2 and S3) of the first macshine cycle, the opcode is fetched and the next six states (S4, S5 and S6 of the first machine cycle and S1, S2 and S3 of the second machine cycle) are used for data memory read operation.

During the last three states (S4, S5 and S6) of the second machine cycle, a dummy program memory fetch is performed and it is discarded. The controller will not increment the program counter for this dummy fetch.

In the timing diagram shown in Fig. 5.6 it is assumed that $201A_H$ is the address of the program memory where the instruction is stored. Also, it is assumed that the content of the DPTR (**Data Pointer**) is $354C_H$ and the content of the data memory location with address $354C_H$ be $7F_H$. The opcode of the MOVX A,@ DPTR instruction is $E0_H$.

5.2.2 Timing Diagram of ADD A, #DATA

The instruction ADD A,#data is used to add an 8-bit immediate data to the A-register (accumulator). The timing diagram of this instruction is shown in Fig. 5.7. The ADD A,#data is a two-byte instruction and executed in one machine cycle.

In the first three states (S1, S2 and S3) of the machine cycle, the opcode is fetched from the program memory. In the next three states (S4, S5 and S6), the immediate data (which is the second byte of instruction) is fetched from the program memory.

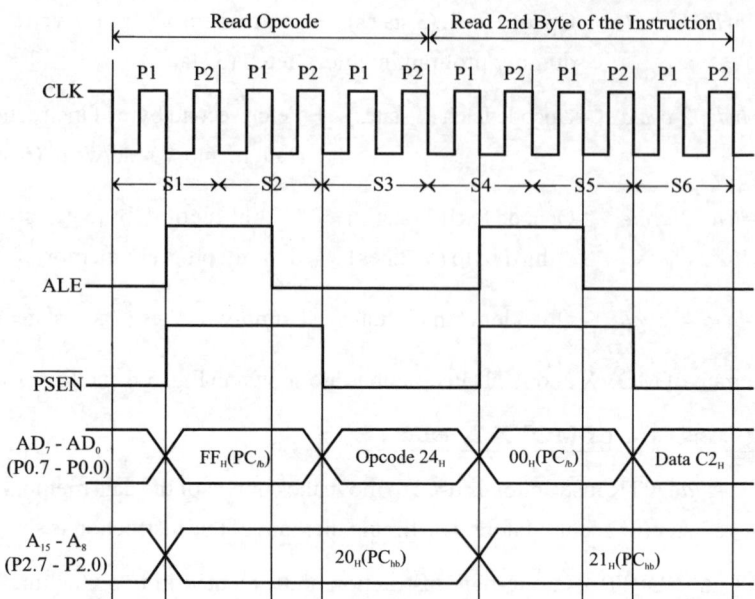

Fig. 5.7: Timing diagram of ADD A, #10$_H$.

In the timing diagram shown in Fig. 3.23, it is assumed that 20FF$_H$ and 2100$_H$ are the address of the program memory where the two bytes of the instruction are stored. Also, it is assumed that the immediate data in the instruction is C2$_H$. The opcode of ADD A,#data instruction is 24$_H$.

5.3 ADDRESSING MODES

Every instruction of a program has to operate on a data. The method of specifying the data to be operated by the instruction is called addressing. The 8051 has the following types of addressing:

       1. **Immediate addressing**

       2. **Direct addressing**

       3. **Register addressing**

       4. **Register indirect addressing**

       5. **Implied addressing**

       6. **Relative addressing**

Immediate Addressing

In immediate addressing mode, an 8/16-bit immediate data/constant is specified in the instruction itself.

Example:

MOV A, #6CH

Move the immediate data 6C$_H$ given in the instruction to the A-register. (Accumulator).

Example:

MOV DPTR, #0100H

Load the immediate 16-bit constant given in the instruction in the DPTR (Data pointer). This constant will be an address of the data memory location.

Direct Addressing

In direct addressing mode, the address of the data is directly specified in the instruction. The direct address can be the address of an internal data RAM location (00_H to $7F_H$) or the address of a special function register (80_H to FF_H).

Example:

MOV A,07 H

The address of the R7-register of bank-0 is 07. This instruction will move the content of the R7-register to the A-register (Accumulator).

Register Addressing

In register addressing mode, the instruction will specify the name of the register in which the data is available.

Example:

MOV R2,A

The content of the A-register (accumulator) is moved to register R2 of the currently selected memory bank.

Register Indirect Addressing

In this mode, the instruction specifies the name of the register in which the address of the data is available. The internal data RAM locations (00_H to $7F_H$) can be addressed indirectly through registers R1 and R0. The registers R3 to R7 cannot be used for register indirect addressing. The external RAM can be addressed indirectly through DPTR.

Example:

MOV A,@R0

The internal RAM location R0 holds the address of the data. The content of the RAM location addressed by R0 is moved to the A-register (Accumulator).

Implied Addressing

In implied addressing mode, the instruction itself specifies the data to be operated by the instruction.

Example:

CPL C

Complement carry flag.

Relative Addressing

In relative addressing mode, the instruction specifies the address relative to the program counter. The instruction will carry an offset whose range is -128_{10} to $+127_{10}$. The offset is added to the PC to generate the 16-bit physical address.

Example:

JC Offset

     If carry is one, the program control jumps to an address obtained by adding the content of the program counter and offset value in the instruction.

5.4 INSTRUCTIONS AFFECTING FLAGS OF 8051

     The 8051 microcontroller has 4 flags. These flags are carry flag, auxiliary carry flag, overflow flag and parity flag. The flags are set or cleared after arithmetic and logical operations. The status of various flags after execution of arithmetic and logical instructions are listed in Table 5.1.

Table 5.1: Instructions Affecting Flags of 8051

INSTRUCTION	CF	AF	OF	PF
ADD	+	+	+	+
ADDC	+	+	+	+
ANL C,bit	+	-	-	-
ANL C,/bit	+	-	-	-
CJNE	+	-	-	-
CLR C	0	-	-	-
CPL C	+	-	-	-
DIV	0	-	0	-
DA	+	+	-	+
MUL	0	-	+	-
MOV C, bit	+	-	-	-
ORL C, bit	+	-	-	-
ORL C,/bit	+	-	-	-
RRC A	+	-	-	-
RLC A	+	-	-	-
SETB C	1	-	-	-
SUBB	+	+	+	+

> *Note:*   *"+" Indicate that the flag is modified*
>
>   *"0" Indicates that the flag is always 0*
>
>   *"1" Indicates that the flag is always 1*

5.5 CLASSIFICATION OF 8051 INSTRUCTIONS

The 8051 instructions can be classified into the following five groups:

1. **Data transfer instructions**
2. **Arithmetic instructions**
3. **Logical instructions**
4. **Branching instructions**
5. **Boolean instructions**

The summary of 8051 instructions are provided in Table 5.2. The meaning of mnemonics used in the instruction set are listed in Table 5.3.

Table 5.2: Summary of 8051 Instruction Set

S.No.	Mnemonic	Opcode	Size of instruction in bytes	Time taken to execute the instruction in clock periods
I. DATA TRANSFER INSTRUCTIONS				
1.	MOV A,Rn	1 1 1 0 1 r r r	1	12
2.	MOV A,direct	1 1 1 0 0 1 0 1	2	12
3.	MOV A,@Ri	1 1 1 0 0 1 1 i	1	12
4.	MOV A, #data	0 1 1 1 0 1 0 0	2	12
5.	MOV Rn,A	1 1 1 1 1 r r r	1	12
6.	MOV Rn,direct	1 0 1 0 1 r r r	2	24
7.	MOV Rn,#data	0 1 1 1 1 r r r	2	12
8.	MOV direct,A	1 1 1 1 0 1 0 1	2	12
9.	MOV direct,Rn	1 0 0 0 1 r r r	2	24
10.	MOV direct,direct	1 0 0 0 0 1 0 1	3	24
11.	MOV direct,@ Ri	1 0 0 0 0 1 1 i	2	24
12.	MOV direct,#data	0 1 1 1 0 1 0 1	3	24
13.	MOV @Ri,A	1 1 1 1 0 1 1 i	1	12
14.	MOV @Ri,direct	1 0 1 0 0 1 1 i	2	24
15.	MOV @Ri,#data	0 1 1 1 0 1 1 i	2	12
16.	MOV DPTR,#data16	1 0 0 1 0 0 0 0	3	24
17.	MOVC A,@A+ DPTR	1 0 0 1 0 0 1 1	1	24
18.	MOVC A,@A+PC	1 0 0 0 0 0 1 1	1	24
19.	MOVX A,@ Ri	1 1 1 0 0 0 1 i	1	24
20.	MOVX A,@DPTR	1 1 1 0 0 0 0 0	1	24

Table 5.2 continued ...

S.No.	Mnemonic	Opcode	Size of instruction in bytes	Time taken to execute the instruction in clock periods
21.	MOVX @Ri,A	1 1 1 1 0 0 1 i	1	24
22.	MOVX @DPTR,A	1 1 1 1 0 0 0 0	1	24
23.	PUSH direct	1 1 0 0 0 0 0 0	2	24
24.	POP direct	1 1 0 1 0 0 0 0	2	24
25.	XCH A,Rn	1 1 0 0 1 r r r	1	12
26.	XCH A,direct	1 1 0 0 0 1 0 1	2	12
27.	XCH A,@Ri	1 1 0 0 0 1 1 i	1	12
28.	XCHD A,@Ri	1 1 0 1 0 1 1 i	1	12
II. ARITHMETIC INSTRUCTIONS				
29.	ADD A,Rn	0 0 1 0 1 r r r	1	12
30.	ADD A,direct	0 0 1 0 0 1 0 1	2	12
31.	ADD A, @Ri	0 0 1 0 0 1 1 i	1	12
32.	ADD A,#data	0 0 1 0 0 1 0 0	2	12
33.	ADDC A,Rn	0 0 1 1 1 r r r	1	12
34.	ADDC A,direct	0 0 1 1 0 1 0 1	2	12
35.	ADDC A,@Ri	0 0 1 1 0 1 1 i	1	12
36.	ADDC A,#data	0 0 1 1 0 1 0 0	2	12
37.	SUBB A,Rn	1 0 0 1 1 r r r	1	12
38.	SUBB A,direct	1 0 0 1 0 1 0 1	2	12
39.	SUBB A,@Ri	1 0 0 1 0 1 1 i	1	12
40.	SUBB A,#data	1 0 0 1 0 1 0 0	2	12
41.	INC A	0 0 0 0 0 1 0 0	1	12
42.	INC Rn	0 0 0 0 1 r r r	1	12
43.	INC direct	0 0 0 0 0 1 0 1	2	12
44.	INC @Ri	0 0 0 0 0 1 1 i	1	12
45.	DEC A	0 0 0 1 0 1 0 0	1	12
46.	DEC Rn	0 0 0 1 1 r r r	1	12
47.	DEC direct	0 0 0 1 0 1 0 1	2	12
48.	DEC @Ri	0 0 0 1 0 1 1 i	1	12

Table 5.2 continued ...

S.No.	Mnemonic	Opcode	Size of instruction in bytes	Time taken to execute the instruction in clock periods
49.	INC DPTR	1 0 1 0 0 0 1 1	1	24
50.	MUL AB	1 0 1 0 0 1 0 0	1	48
51.	DIV AB	1 0 0 0 0 1 0 0	1	48
52.	DA A	1 1 0 1 0 1 0 0	1	12

III. LOGICAL INSTRUCTIONS

S.No.	Mnemonic	Opcode	Size of instruction in bytes	Time taken to execute the instruction in clock periods
53.	ANL A,Rn	0 1 0 1 1 r r r	1	12
54.	ANL A,direct	0 1 0 1 0 1 0 1	2	12
55.	ANL A,@ Ri	0 1 0 1 0 1 1 i	1	12
56.	ANL A,#data	0 1 0 1 0 1 0 0	2	12
57.	ANL direct,A	0 1 0 1 0 0 1 0	2	12
58.	ANL direct,#data	0 1 0 1 0 0 1 1	3	24
59.	ORL A,Rn	0 1 0 0 1 r r r	1	12
60.	ORL A,direct	0 1 0 0 0 1 0 1	2	12
61.	ORL A,@Ri	0 1 0 0 0 1 1 i	1	12
62.	ORL A,#data	0 1 0 0 0 1 0 0	2	12
63.	ORL direct,A	0 1 0 0 0 0 1 0	2	12
64.	ORL direct,#data	0 1 0 0 0 0 1 1	3	24
65.	XRL A,Rn	0 1 1 0 1 r r r	1	12
66.	XRL A,direct	0 1 1 0 0 1 0 1	2	12
67.	XRL A,@ Ri	0 1 1 0 0 1 1 i	1	12
68.	XRL A,#data	0 1 1 0 0 1 0 0	2	12
69.	XRL direct,A	0 1 1 0 0 0 1 0	2	12
70.	XRL direct,#data	0 1 1 0 0 0 1 1	3	24
71.	CLR A	1 1 1 0 0 1 0 0	1	12
72.	CPL A	1 1 1 1 0 1 0 0	1	12
73.	RL A	0 0 1 0 0 0 1 1	1	12
74.	RLC A	0 0 1 1 0 0 1 1	1	12
75.	RR A	0 0 0 0 0 0 1 1	1	12

Table 5.2 continued ...

S.No.	Mnemonic	Opcode	Size of instruction in bytes	Time taken to execute the instruction in clock periods
76.	RRC A	0 0 0 1 0 0 1 1	1	12
77.	SWAP A	1 1 0 0 0 1 0 0	1	12

IV. BRANCHING INSTRUCTIONS

S.No.	Mnemonic	Opcode	Size of instruction in bytes	Time taken to execute the instruction in clock periods
78.	ACALL addr11	a_{10} a_9 a_8 1 0 0 0 1	2	24
79.	LCALL addr16	0 0 0 1 0 0 1 0	3	24
80.	RET	0 0 1 0 0 0 1 0	1	24
81.	RETI	0 0 1 1 0 0 1 0	1	24
82.	AJMP addr 11	a_{10} a_9 a_8 0 0 0 0 1	2	24
83.	LJMP addr 16	0 0 0 0 0 0 1 0	3	24
84.	SJMP offset	1 0 0 0 0 0 0 0	2	24
85.	JMP @A+ DPTR	0 1 1 1 0 0 1 1	1	24
86.	JZ offset	0 1 1 0 0 0 0 0	2	24
87.	JNZ offset	0 1 1 1 0 0 0 0	2	24
88.	CJNE A,direct,offset	1 0 1 1 0 1 0 1	3	24
89.	CJNE A,#data,offset	1 0 1 1 0 1 0 0	3	24
90.	CJNE @Rn,#data,offset	1 0 1 1 1 r r r	3	24
91.	CJNE @Ri,#data,offset	1 0 1 1 0 1 1 i	3	24
92.	DJNZ Rn,offset	1 0 1 1 1 r r r	2	24
93.	DJNZ direct,offset	1 1 0 1 0 1 0 1	3	24
94.	NOP	0 0 0 0 0 0 0 0	1	12

V. BOOLEAN INSTRUCTIONS

S.No.	Mnemonic	Opcode	Size of instruction in bytes	Time taken to execute the instruction in clock periods
95.	CLR C	1 1 0 0 0 0 1 1	1	12
96.	CLR bit	1 1 0 0 0 0 1 0	2	12
97.	SETB C	1 1 0 1 0 0 1 1	1	12
98.	SETB bit	1 1 0 1 0 0 1 0	2	12
99.	CPL C	1 0 1 1 0 0 1 1	1	12
100.	CPL bit	1 0 1 1 0 0 1 0	2	12
101.	ANL C,bit	1 0 0 0 0 0 1 0	2	24
102.	ANL C,/bit	1 0 1 1 0 0 0 0	2	24

Table 5.2 continued ...

S.No.	Mnemonic	Opcode	Size of instruction in bytes	Time taken to execute the instruction in clock periods
103.	ORL C,bit	0 1 1 1 0 0 1 0	2	24
104.	ORL C,/bit	1 0 1 0 0 0 0 0	2	24
105.	MOV C,bit	1 0 1 0 0 0 1 0	2	12
106.	MOV bit,C	1 0 0 1 0 0 1 0	2	24
107.	JC offset	0 1 0 0 0 0 0	2	24
108.	JNC offset	0 1 0 1 0 0 0 0	2	24
109.	JB bit,offset	0 0 1 0 0 0 0 0	3	24
110.	JNB bit,offset	0 0 1 1 0 0 0 0	3	24
111.	JBC bit,offset	0 0 0 1 0 0 0 0	3	24

Symbols/Abbreviations Used in the Instruction Set

Rn	Register R_7 to R_0 of currently selected register bank	bit	Address of bit-addressable RAM/SFR
direct	8-bit address of internal RAM/SFR	&	Logical AND
A	Accumulator	|	Logical OR
@A+	Memory addressed indirectly through the accumulator	~	Complement/Logical NOT
		^	Logical Exclusive-OR
@Ri	Internal RAM addressed indirectly through R_0 or R1	CF	Carry flag
		B_n	n^{th} bit of register/memory
@DPTR	External data memory addressed indirectly through DPTR	B_{n+1}	$(n + 1)^{th}$ bit of register/memory
		$(A)_{3-0}$	Lower nibble of accumulator
#data	8-bit immediate data/constant	$(A)_{7-4}$	Upper nibble of accumulator
#data16	16-bit immediate data/constant	$(PC)_{7-0}$	Lower byte of program counter
addr11	11-bit address	$(PC)_{15-8}$	Upper byte of program counter
addr16	16-bit address	$(PC)_{10-0}$	Lower 11 bits of program counter
offset	8-bit signed offset value in the range -128_{10} to $+ 127_{10}$	$(RAM)_{3-0}$	Lower nibble of RAM.

Symbols Used in Opcode

rrr	3-bit register field representing the 8 registers of a bank. The code 000 to 111 represents the registers R_0 to R_7 respectively.
i	1-bit register field representing R_0 or R_1. For R_0, i = 0 and for R_1, i = 1.
a_{10} a_9 a_8	Upper 3 bits of 11-bit address and can take eight possible values 000 to 111

Table 5.3: Meaning of Mnemonics Used in the Instruction Set

Mnemonic	Meaning/Expansion
ACALL	Absolute Subroutine call
ADD	Add
ADDC	Add including carry
AJMP	Absolute jump
ANL	AND Logic Operation
CJNE	Compare and jump if not equal
CLR	Clear
CPL	Complement
DA	Decimal adjust
DEC	Decrement
DIV	Divide
DJNZ	Decrement and jump if not zero
INC	Increment
JB	Jump if bit is set
JBC	Jump if bit is set and clear
JC	Jump on carry
JMP	Jump
MOV C	Move code byte
MOV X	Move to/from external RAM
MOV	Move to/from internal RAM
MUL	Multiply
NOP	No operation
ORL	OR logic operation
POP	Retrieve from stack (POP from stack)
PUSH	Store in stack (PUSH to stack)
RET	Return from subroutine
RETI	Return from interrupt

Table 5.3 continued...

Mnemonic	Meaning/Expansion
RL	Rotate left
RLC	Rotate left carry
RR	Rotate right
RRC	Rotate right carry
SETB	Set bit
SJUMP	Short jump
SUBB	Subract with borrow
SWAP	Swap/Exchange lower and upper nibble
XCH	Exchange
XCHD	Exchange digit
XRL	Exclusive-OR logic

5.6 DATA TRANSFER INSTRUCTIONS

The instruction set of the 8051 microcontroller includes a variety of instructions for data transfer between the registers and the memory locations.

The various mnemonics used for data transfer instructions are MOV, MOVC, MOVX, PUSH, POP, XCH and XCHD, and they perform any one of the following operations:

1. Copy the content of an SFR to the internal memory or vice versa.

2. Load an immediate operand to the SFR/internal memory.

3. Exchange the content of the SFR/internal memory with the accumulator.

4. Copy the content of the program memory to the accumulator.

5. Copy the content of the data memory to the accumulator or vice versa.

The data transfer instructions of 8051 are listed in Table 5.4 with a brief explanation about each instruction.

Table 5.4: Data Transfer Instructions

S.No.	Instruction	Symbolic representation	Explanation
1.	MOV A,Rn	(A) ← (Rn)	The content of register Rn is moved to the accumulator (A-register). The Rn can be any one of the 8 registers of the currently selected bank.
2.	MOV A,direct	direct = 8-bit address of internal RAM/SFR (A) ← (RAM/SFR)	The content of internal RAM/SFR (whose address is specified directly in the instruction) is moved to the accumulator (A-register).
3.	MOV A,@Ri	(Ri) = Internal RAM address (A) ← (RAM)	The content of internal RAM memory (whose address is specified by the Ri-registeris moved to the accumulator (A-register)). The register Ri can be either R0 or R1 or the currently selected register bank.
4.	MOV A,#data	(A) ← data	The data given in the instruction is moved to the accumulator (A-register).
5.	MOV Rn,A	(Rn) ← (A)	The content of the accumulator is moved to register Rn, where Rn is any one of the 8 registers of the currently selected register bank.
6.	MOV Rn,direct	direct = 8-bit address of internal RAM/SFR (Rn)← (RAM/SFR)	The content of internal RAM/SFR (whose address is directly specified in the instruction) is moved to register Rn, where Rn is any one of the 8 registers of the currently selected register bank.
7.	MOV Rn,#data	(Rn) ← data	The immediate data given in the instruction is moved to register Rn, where Rn is any one of the 8 registers of the currently selected register bank.
8.	MOV direct,A	direct = 8-bit address of internal RAM/SFR (RAM/SFR) ← (A)	The content of the accumulator is moved to internal RAM/SFR whose address is directly specified in the instruction.
9.	MOV direct,Rn	direct = 8-bit address of internal RAM/SFR (RAM/SFR) ← (Rn)	The content of register Rn is moved to internal RAM/SFR whose address is directly specified in the instruction.
10.	MOV direct,direct	direct = 8-bit address of internal RAM/SFR (RAM/SFR) ←(RAM/SFR)	The content of one internal RAM/SFR is moved to another internal RAM/SFR. The address of the source and destination are directly specified in the instruction.
11.	MOV direct,@Ri	(Ri) = Internal RAM address of source operand direct = Internal RAM/SFR address of destination operand (RAM/SFR) ← (RAM)	The content of internal RAM whose address is specified by Ri is moved to another internal RAM/SFR whose address is directly specified in the instruction. The register Ri can be either R0 or R1.

Table 5.4 continued...

S.No.	Instruction	Symbolic representation	Explanation
12.	MOV direct,#data	direct = Address of internal RAM/SF (RAM/SFR) ← data	The immediate data given in the instruction is moved to the internal RAM/SFR, whose address is directly specified in the instruction.
13.	MOV @Ri,A	(Ri) = Internal RAM address (RAM) ← (A)	The content of the accumulator is moved to an internal RAM location whose address is specified by the Ri-register. The register Ri can be either R0 or R1.
14.	MOV @Ri,direct	direct = Internal RAM/SFR address of source operand (Ri) = Internal RAM address of destination operand. (RAM) ← (RAM/SFR)	The content of the internal RAM/SFR whose address is directly specified in the instruction is moved to another internal RAM location whose address is specified by the Ri-register. The register Ri can be either R0 or R1.
15.	MOV @Ri,#data	(Ri) = Internal RAM address (RAM) ← data	The immediate data given in the instruction is moved to an internal RAM location, whose address is specified by the Ri-register. The register Ri can be R0 or R1.
16.	MOV DPTR,#data16	(DPTR) ← data16	The 16-bit constant (data16) given in the instruction is moved to the DPTR. (The content of the DPTR is used as address of external data memory in the subsequent instruction.)
17.	MOVC A,@A+DPTR	(A) + (DPTR) = Address of program memory (A) ← (program memory)	This instruction will copy a byte from the code/program memory to the accumulator. The address of the program memory is given by the sum of the content of the DPTR and accumulator before the move operation.
18.	MOVC A,@A+PC	(PC) ← (PC)+1 (A) + (PC) = Address of program memory (A) ← (program memory)	This instruction will copy a byte from the code/program memory to the accumulator. The address of the program memory is given by the sum of the PC and the accumulator. Here, the content of the PC is incremented before adding to A to get the address of the code memory.
19.	MOVX A,@Ri	(Ri) = 8-bit address external data RAM (A) ← (RAM)	The content of external data RAM is moved to the accumulator. The content of register Ri is the 8-bit address of the external memory. The register Ri can be either R0 or R1 of the currently selected register bank.
20.	MOVX A,@DPTR	(DPTR) = 16-bit address of external data RAM (A) ← (RAM)	The content of external data RAM is moved to the accumulator. The content of DPTR is the 16-bit address of the external RAM.

Table 5.4 continued...

S.No.	Instruction	Symbolic representation	Explanation
21.	MOVX @Ri,A	(Ri) = 8-bit address of external data RAM (RAM) ← (A)	The content of the accumulator is moved to the external data RAM. The content of the Ri is the 8-bit address of external RAM. The register Ri can be either R0 or R1 of the currently selected register bank.
22.	MOVX @DPTR,A	(DPTR) = 16-bit address of external data RAM (RAM) ← (A)	The content of the accumulator is moved to the external data RAM. The content of the DPTR is the 16-bit address of the external RAM.
23.	PUSH direct	(SP) ← (SP) + 1 direct = 8-bit address of internal RAM/SFR ((SP)) ← (RAM/SFR)	The stack pointer is incremented by one. The content of the internal RAM/SFR (whose address is directly specified in the instruction) is moved to the internal RAM memory pointed by the SP.
24.	POP direct	direct = 8-bit address of internal RAM/SFR (RAM/SFR) ← ((SP)) (SP) ← (SP) − 1	The content of the internal RAM memory pointed by the SP is moved to the internal RAM/SFR (whose address is directly specified in the instruction). Then the stack pointer is decremented by one.
25.	XCH A,Rn	$(A) \underset{\leftarrow}{\overset{\rightarrow}{}} (Rn)$	The content of the register Rn is exchanged with the accumulator. The register Rn can be any one of the eight registers of the currently selected register bank.
26.	XCH A,direct	direct = 8-bit address of internal RAM/SFR $(A) \underset{\leftarrow}{\overset{\rightarrow}{}} (RAM/SFR)$	The content of the internal RAM /SFR whose address is directly specified in the instruction is exchanged with the accumulator.
27.	XCH A,@Ri	(Ri) = 8-bit address of internal RAM $(A) \underset{\leftarrow}{\overset{\rightarrow}{}} (RAM)$	The content of the internal RAM whose address is specified by the Ri-register is exchanged with the accumulator. The register Ri can be either R0 or R1 of the currently selected register bank.
28.	XCHD A,@Ri	(Ri) = 8-bit address of internal RAM $(A)_{3-0} \underset{\leftarrow}{\overset{\rightarrow}{}} (RAM)_{3-0}$	The lower nibble of the internal RAM addressed by Ri-register is exchanged with the lower nibble of the accumulator. The content of the upper nibble of RAM and accumulator are not altered. The Register Ri can be either R0 or R1 of the currently selected register bank.

5.7 DATA MANIPULATION INSTRUCTIONS

The instructions that manipulate data either by arithmetic or logical instructions can be called data manipulating instructions. These instructions can be further classified into arithmetic instructions and logical instructions.

5.7.1 Arithmetic Instructions

The arithmetic group includes instructions for performing addition, subtraction, multiplication, division, increment and decrement operation on the binary data. The mnemonics used in arithmetic instructions are ADD, ADDC, SUBB, INC, DEC, MUL, DIV and DA. The results of most of the arithmetic operations are stored in the accumulator except a few decrement and increment operations. The arithmetic instructions except increment and decrement instructions modify the flags of 8051. The arithmetic instructions of 8051 are listed in Table 5.5.

Table 5.5: Arithmetic Instructions

S.No.	Instruction	Symbolic representation	Explanation
29.	ADD A,Rn	$(A) \leftarrow (A) + (Rn)$	The content of the register Rn and the accumulator are added. The result is stored in the accumulator. The register Rn can be any one of the eight registers of the currently selected register bank.
30.	ADD A,direct	direct = 8-bit address of internal RAM/SFR $(A) \leftarrow (A) + (RAM/SFR)$	The content of the internal RAM/SFR and the accumulator are added. The result is stored in the accumulator. The address of internal RAM/SFR is directly specified in the instruction.
31.	ADD A,@Ri	(Ri) = Address of internal RAM $(A) \leftarrow (A) + (RAM)$	The content of the internal RAM and the accumulator are added. The result is stored in the accumulator. The register Ri holds the address of the internal RAM and Ri can be either R0 or R1 of the currently selected register bank.
32.	ADD A,#data	$(A) \leftarrow (A) + data$	The immediate data given in the instruction is added to accumulator.
33.	ADDC A,Rn	$(A) \leftarrow (A) + CF + (Rn)$	This instruction is same as **ADD A, Rn** except that the current value of carry flag (i.e., previous carry) is also added to the sum.
34.	ADDC A,direct	direct = address of internal RAM/SFR $(A) \leftarrow (A) + CF + (RAM/SFR)$	This instruction is same as **ADD A, direct** except that the current value of carry flag (i.e., previous carry) is also added to the sum.
35.	ADDC A,@Ri	(Ri) = address of internal RAM $(A) \leftarrow (A) + CF + (RAM)$	This instruction is same as **ADD A, @Ri** except that the current value of the carry flag (i.e., previous carry) is also added to sum.

Table 5.5 continued...

S.No.	Instruction	Symbolic representation	Explanation
36.	ADDC A,#data	$(A) \leftarrow (A) + CF + data$	The immediate data given in the instruction, the carry flag and the content of the accumulator are added. The result is stored in the accumulator.
37.	SUBB A,Rn	$(A) \leftarrow (A) - CF - (Rn)$	The carry flag and the content of the Rn-register are subtracted from the content of the accumulator. The result is stored in the accumulator. The register Rn can be any one of the 8 registers of the currently selected register bank. The 8051 perform 2's complement subtraction and then complement carry.
38.	SUBB A,direct	direct = Address of internal RAM/SFR $(A) \leftarrow (A) - CF- (RAM/SFR)$	The carry flag and the content of RAM/SFR (specified by direct address) are subtracted from the content of the accumulator. The result is stored in the accumulator.
39.	SUBB A,@Ri	(Ri) = Address of internal RAM $(A) \leftarrow (A) - CF - (RAM)$	The carry flag and the content of RAM (specified by the Ri-register) are subtracted from the content of the accumulator. The result is stored in the accumulator. The register Ri can be either R0 or R1 of the currently selected register bank.
40.	SUBB A,#data	$(A) \leftarrow (A) - CF - data$	The carry flag and the data given in the instruction are subtracted from the accumulator. The result is stored in the accumulator.
41.	INC A	$(A) \leftarrow (A) + 1$	The content of the accumulator is incremented by one.
42.	INC Rn	$(Rn) \leftarrow (Rn) +1$	The content of the register Rn is incremented by one. The Rn can be any one of the eight registers of the currently selected register bank.
43.	INC direct	direct = Address of internal RAM/SFR $(RAM/SFR) \leftarrow (RAM/SFR) +1$	The content of RAM/SFR (whose address is directly given in the instruction) is incremented by one.
44.	INC @Ri	(Ri) = Address of internal RAM/SFR $(RAM) \leftarrow (RAM) +1$	The content of RAM (whose address is specified by Ri) is incremented by one. The Ri can be either R0 or R1 of the currently selected register bank.
45.	DEC A	$(A) \leftarrow (A) - 1$	The content of the accumulator is decremented by one.
46.	DEC Rn	$(Rn) \leftarrow (Rn) - 1$	The content of register Rn is decremented by one. The Rn can be any one of the eight registers of the currently selected register bank.

Table 5.5 continued...

S.No.	Instruction	Symbolic representation	Explanation
47.	DEC direct	direct = Address of internal RAM/SFR $(RAM/SFR) \leftarrow (RAM/SFR) - 1$	The content of RAM/SFR (whose address is directly given in the instruction) is decremented by one.
48.	DEC @Ri	(Ri) = Address of internal RAM $(RAM) \leftarrow (RAM) - 1$	The content of RAM (whose address is specified by Ri) is decremented by one. The Ri can be either R0 or R1 of the currently selected register bank.
49.	INC DPTR	$(DPTR) \leftarrow (DPTR) + 1$	The 16-bit content of the DPTR (**Data Pointer**) is incremented by one.
50.	MUL AB	(B) $(A) \leftarrow (A) \times (B)$ high low byte byte	The contents of A and B registers are multiplied. The low byte of the product is stored in the A-register and the high byte of the product is stored in the B-register.
51.	DIV AB	$(A) \leftarrow (A) \div (B)$ Quotient $(B) \leftarrow (A)$ MOD (B) Remainder	The content of the A-register is divided by the content of the B-register. The quotient is stored in the A-register and the remainder is stored in B-register.
52.	DAA	i) If $(A)_{3-0} > 9$ or AF = 1 then $(A)_{3-0} \leftarrow (A)_{3-0} + 06$ ii)If $(A)_{7-4} > 9$ or CF = 1 then $(A)_{7-4} \leftarrow (A)_{7-4} + 06$	This instruction is executed after the addition of two packed BCD data, to convert the result in the accumulator to the packed BCD data. If the lower nibble of the accumulator is greater than 09 or the AF is set, then it is corrected by adding 06. If the upper nibble of the accumulator is greater than 09 or the CF is set, then it is corrected by adding 06.

5.7.2 Logical Instructions

The logical group includes instructions for performing logical AND, OR, Exclusive-OR, and Complement operations, and instructions for right and left rotation. The mnemonics used in logical operations are ANL, ORL, XRL, CLR, CPL, RL, RLC, RR, RRC and SWAP. The logical operations except rotate through carry do not modify the flags of 8051. In rotate through carry, the carry flag alone is modified. In most of the logical instructions, the result is stored in the accumulator and in some instructions the result is stored in the internal RAM/SFR. The logical instructions of 8051 are listed in Table 5.6 with a brief explanation about each instruction.

Table 5.6: Logical Instructions

S.No.	Instruction	Symbolic representation	Explanation
53.	ANL A,Rn	$(A) \leftarrow (A)$ & (Rn)	The content of the register Rn and the accumulator are bit by bit logically ANDed, and the result is stored in the accumulator. The register Rn can be any one of the 8 registers of the currently selected register bank.
54.	ANL A,direct	direct = Address of internal RAM/SFR $(A) \leftarrow (RAM/SFR)$ & (A)	The content of the RAM/SFR (whose address is directly given in the instruction) and the accumulator are bit by bit logically ANDed, and the result is stored in the accumulator.
55.	ANL A,@Ri	(Ri) = Address of internal RAM $(A) \leftarrow (RAM)$ & (A)	The content of the RAM (whose address is specified by Ri) and the accumulator are bit by bit logically ANDed, and the result is stored in the accumulator. The register Ri can be either R0 or R1 of the currently selected register bank.
56.	ANL A,#data	$(A) \leftarrow (A)$ & data	The data given in the instruction and the content of the accumulator are bit by bit logically ANDed, and the result is stored in the accumulator.
57.	ANL direct, A	direct = Address of internal RAM/SFR $(RAM/SFR) \leftarrow (RAM/SFR)$ & (A)	The content of the accumulator and the RAM/SFR are bit by bit logically ANDed, and the result is stored in the RAM/SFR. The address of the RAM/SFR is directly specified in the instruction.
58.	ANL direct,#data	direct = Address of internal RAM/SFR $(RAM/SFR) \leftarrow (RAM/SFR)$ & data	The data given in the instruction and the content of RAM/SFR are bit by bit logically ANDed, and the result is stored in RAM/SFR. The address of the RAM/SFR is directly specified in the instruction.
59.	ORL A,Rn	$(A) \leftarrow (A) \mid (Rn)$	The content of the register Rn and the accumulator are bit by bit logically ORed, and the result is stored in the accumulator. The register Rn can be any one of the 8 registers of the currently selected register bank.
60.	ORL A,direct	direct = Address of internal RAM/SFR $(A) \leftarrow (RAM/SFR) \mid (A)$	The content of the RAM/SFR (whose address is directly given in the instruction) and the accumulator are bit by bit logically ORed, and the result is stored in the accumulator.
61.	ORL A,@Ri	(Ri) = Address of internal RAM $(A) \leftarrow (RAM) \mid (A)$	The contents of the RAM (whose address is specified by Ri) and the accumulator are bit by bit logically ORed, and the result is stored in the accumulator. The register Ri can be either R0 or R1 of the currently selected register bank.

Table 5.6 continued...

S.No.	Instruction	Symbolic representation	Explanation
62.	ORL A,#data	(A) ← (A) | data	The data given in the instruction and the content of the accumulator are bit by bit logically ORed, and the result is stored in the accumulator.
63.	ORL direct,A	direct = Address of internal RAM/SFR (RAM/SFR) ← (RAM/SFR) | (A)	The content of the accumulator and the RAM/SFR are bit by bit logically ORed, and the result is stored in the RAM/SFR. The address of the RAM/SFR is directly specified in the instruction.
64.	ORL direct,#data	direct = Address of internal RAM/SFR (RAM/SFR) ← (RAM/SFR) | data	The data given in the instruction and the content of the RAM/SFR are bit by bit logically ORed, and the result is stored in the RAM/SFR. The address of the RAM/SFR is directly specified in the instruction.
65.	XRL A,Rn	(A) ← (A) ^ (Rn)	The contents of the register Rn and the accumulator are bit by bit logically exclusive-ORed, and the result is stored in the accumulator. The register Rn can be anyone of the 8 registers of the currently selected register bank.
66.	XRL A,direct	direct = Address of internal RAM/SFR (A) ← (RAM/SFR) ^ (A)	The content of the RAM/SFR (whose address is directly given in and instruction) and the accumulator are bit by bit logically exclusive-ORed, and the result is stored in the accumulator.
67.	XRL A,@Ri	(Ri) = Address of internal RAM (A) ← (RAM) ^ (A)	The content of the RAM (whose address is specified by Ri) and the accumulator are bit by bit logically exclusive-ORed, and the result is stored in the accumulator. The register Ri can be either R0 or R1 of the currently selected register bank.
68.	XRL A,#data	(A) ← (A) ^ data	The data given in the instruction and the content of the accumulator are bit by bit logically exclusive-ORed, and the result is stored in the accumulator.
69.	XRL direct,A	direct = Address of internal RAM/SFR (RAM/SFR) ← (RAM/SFR) ^ (A)	The content of the accumulator and RAM/SFR are bit by bit logically exclusive-ORed, and the result is stored in the RAM/SFR. The address of the RAM/SFR is directly specified in the instruction.
70.	XRL direct,#data	direct = Address of internal RAM/SFR (RAM/SFR) ← (RAM/SFR) ^ data	The data given in the instruction and the content of RAM/SFR are bit by bit logically exclusive-ORed, and the result is stored in the RAM/SFR. The address of the RAM/SFR is directly specified in the instruction.

Table 5.6 continued...

S.No.	Instruction	Symbolic representation	Explanation
71.	CLR A	$(A) \leftarrow 0$	The content of the accumulator is cleared.
72.	CPL A	$(A) \leftarrow \sim (A)$	The content of the accumulator is complemented.
73.	RL A	$B_{n+1} \leftarrow B_n \; ; \; B_0 \leftarrow B_7$	The content of the accumulator is rotated left by one bit. The most significant digit (B_7) is moved to the least significant digit (B_0) position.
74.	RLC A	$B_{n+1} \leftarrow B_n \; ; \; B_0 \leftarrow CF \; ; \; CF \leftarrow B_7$	The content of the accumulator along with the carry is rotated left by one bit. The carry is moved to the least significant digit position and the most significant digit is moved to the carry.
75.	RR A	$B_n \leftarrow B_{n+1} \; ; \; B_7 \leftarrow B_0$	The content of the accumulator is rotated right by one bit. The least significant digit (B_0) is moved to the most significant digit (B_7) position.
76.	RRC A	$B_n \leftarrow B_{n+1} \; ; \; B_7 \leftarrow CF \; ; \; CF \leftarrow B_0$	The content of the accumulator along with the carry is rotated right by one bit. The carry is moved to the most significant digit (position) and the least significant digit (B_0) is moved to carry.
77.	SWAP A	$(A)_{3-0} \longleftrightarrow (A)_{7-4}$	The higher nibble of the accumulator is exchanged with the lower nibble of the accumulator.

5.8 CONTROL INSTRUCTIONS

The instructions that changes the proram flow are called control instructions. Normally, program instructions are executed sequencially so that instructions are executed one by one. But when a control instruction is encountered the sequential execution is modified depending on the condition specified by the control instruction. The control instructions can be classified into program branching and boolean instructions.

5.8.1 Program Branching Instructions

Normally, a program is executed sequentially and the PC (Program Counter) keeps track of the address of the instructions and it is incremented appropriately after each fetch operation. The program branching instructions will modify the content of the PC so that, the program control branches to a new address. The program branching instructions of 8051 includes conditional and unconditional branching instruction. In conditional branching instructions, the content of the PC is modified, only if the condition specified in the instruction is true, whereas in unconditional branching instruction, the PC is always modified. The instructions like ACALL and LCALL will save the previous value of the PC in the stack before modifying the PC. The program branching instructions of 8051 are listed in Table 5.7 with a brief explanation about each instruction.

Table 5.7: Program Branching Instructions

S.No.	Instruction	Symbolic representation	Explanation
78.	ACALL addr11	$(PC) \leftarrow (PC) + 2$ $(SP) \leftarrow (SP) + 1$ $((SP)) \leftarrow (PC)_{7-0}$ $(SP) \leftarrow (SP) + 1$ $((SP)) \leftarrow (PC)15-8$ $(PC)_{10-0} \leftarrow$ addr11	This instruction is used to unconditionally call a subroutine which resides within the same 2k block of the program memory in which the instruction following ACALL is stored. This instruction first increments the PC by two, to point to the address of the instruction next to ACALL. Next, the content of the SP is incremented by one and the low byte of PC is saved in the stack memory pointed by SP. Again, content of the SP is incremented by one and then, high byte of PC is saved in the stack memory pointed by the SP. Then, the 11-bit address given in the instruction is moved to the lower 11-bit position of the PC. (The 11- bit address is the second byte of the instruction and upper 3 bits of the opcode.) Now, the controller starts fetching the instructions from this new address.
79.	LCALL addr16	$(PC) \leftarrow (PC) + 3$ $(SP) \leftarrow (SP) + 1$ $((SP)) \leftarrow (PC)_{7-0}$ $(SP) \leftarrow (SP) + 1$ $((SP)) \leftarrow (PC)_{15-8}$ $(PC) \leftarrow$ addr16	This instruction is used to unconditionally call a subroutine anywhere in the 64 k space. First, the PC is incremented by three to point to the next instruction. Next, the SP is incremented and the content of the PC is saved in the stack memory pointed by the SP. Then, the 16-bit address given in the instruction is moved to the PC and so the controller fetching the instruction from this new address. starts
80.	RET	$(PC)_{15-8} \leftarrow ((SP))$ $(SP) \leftarrow (SP) - 1$ $(PC)_{7-0} \leftarrow ((SP))$ $(SP) \leftarrow (SP) - 1$	This instruction is used to terminate a subroutine. On execution of this instruction, the content of the stack memory pointed by the SP is moved to the high byte of PC and SP is decremented by one. Then, the content of the stack memory pointed by SP is moved to the low byte of PC and again the SP is decremented by one.
81.	RETI	$(PC)_{15-8} \leftarrow ((SP))$ $(SP) \leftarrow (SP) - 1$ $(PC)_{7-0} \leftarrow ((SP))$ $(SP) \leftarrow (SP) - 1$	This RETI instruction is used to terminate an interrupt service subroutine. This instruction moves the top of the stack to the PC similar to that of RET instruction and in addition restores the interrupt logic to accept additional interrupts of the same priority level as the one just processed.

Table 5.7 continued…

S.No	Instruction	Symbolic representation	Explanation
82.	AJMP addr11	(PC) ← (PC) + 2 (PC)$_{10-0}$ ← addr11	This instruction is used to unconditionally jump to a memory location within the same 2k block of program memory in which the instruction following AJMP is stored. First, the PC is incremented by two to point to the address of next instruction and then the 11-bit address given in the instruction is moved to the lower 11-bit position of the PC. The 11-bit address is the second byte of the instruction and upper 3 bits of opcode
83.	LJMP addr16	(PC) ← addr16	This instruction is used to unconditionally jump to any location in the 64 k memory space. Upon execution of this instruction, the 16-bit address given in the instruction is moved to the PC, and so the controller starts fetching the instruction from this new address.
84.	SJMP offset	(PC) ← (PC) + 2 (PC) ← (PC) + offset	This instruction is used to unconditionally transfer the program control to a new address obtained by adding the 8-bit signed offset to the content of the PC. The offset will be in the range of −128$_{10}$ to +127$_{10}$.
85.	JMP @A+DPTR	(A) + (DPTR) = Address (PC) ← Address	This instruction computes the address to which the program control has to be transferred and loads this address in the PC. The address is given by the sum of the signed 8-bit in the accumulator and the 16-bit content of the DPTR.
86.	JZ offset	(PC) ← (PC) + 2 If (A) = 0 then (PC) ← (PC) + offset	First, the content of the PC is incremented by two. Next, the content of accumulator is checked. If the content of the accumulator is zero, then the 8-bit signed offset given in the instruction is added to the PC, so that the program control branches to new address. If the accumulator is not zero then PC is not modified, so that the next instruction of the program is fetched and executed.
87.	JNZ offset	(PC) ← (PC) + 2 If (A) ≠ 0 then (PC) ← (PC) + offset	First, the content of the PC is incremented by two. Next, the content of the accumulator is checked. If the content of the accumulator is not zero, then the 8-bit signed offset given in the instruction is added to the PC, so that the program control branches to a new address. If the accumulator is zero, then PC is not modified so that the next instruction of the program is fetched and executed.

Table 5.7 continued...

S.No.	Instruction	Symbolic representation	Explanation

88.	CJNE A,direct,offset	(PC) ← (PC) + 3
		direct = Address of internal RAM/SFR
		If (A) ≠ (RAM/SFR) then
		(PC) ← (PC) + offset
		If (A) < (RAM/SFR) then, CF ← 1
		If (A) > (RAM/SFR) then, CF ← 0
			First, the PC is incremented by three to point to the next instruction. The content of the accumulator and the internal RAM/SFR (whose address is directly specified in the instruction) are compared. If the contents are not equal then the program control is transferred to a new address. The new address is the sum of the PC and offset given in the instruction. Also, if the content of the accumulator is less than RAM/SFR, then the carry flag is set, otherwise it is cleared.

89.	CJNE A,#data,offset	(PC) ← (PC) + 3
		If (A) ≠ data then
		(PC) ← (PC) + offset
		If (A) < data then, CF ← 1
		If (A) > data then, CF ← 0
			This instruction is same as **CJNE A,direct,offset** except that the comparison is performed with the immediate data given in the instruction and the accumulator.

90.	CJNE Rn,#data,offset	(PC) ← (PC) + 3
		If (Rn) ≠ data then
		(PC) ← (PC) + offset
		If (Rn) < data then, CF ← 1
		If (Rn) > data then, CF ← 0
			This instruction is same as **CJNE A,direct,offset** except that the comparison is performed with the content of the Rn and the immediate data. The Rn can be any one of the eight registers of the currently selected register bank.

91.	CJNE @Ri,#data,offset	(PC) ← (PC) + 3
		(Ri) = Address of internal RAM
		If (RAM) ≠ data then
		(PC) ← (PC) + offset
		If (Ri) < data then, CF ← 1
		If (Ri) > data then, CF ← 0
			This instruction is same as **CJNE A,direct,offset** except that the comparison is performed between the RAM (whose address is specified by Ri) and the immediate data given in the instruction. The register Ri can be either R0 or R1 of the currently selected register bank

92.	DJNZ Rn,offset *[AU May'15, 2 marks]*	(PC) ← (PC) + 2 (Rn) ← (Rn) − 1 If (Rn) ≠ 0 then (PC) ← (PC) + offset	First, the PC is incremented by two to point to the address of the next instruction. Then, the content of register Rn is decremented by one. If the content of Rn (after decrement) is not equal to zero then, the offset given in the instruction is added to the PC so that, the program control branches to a new address. The register Rn can be any one of the eight registers of the currently selected register bank

Table 5.7 continued...

S.No.	Instruction	Symbolic representation	Explanation
93.	DJNZ direct,offset	(PC) ← (PC) + 2 direct = Address of RAM/SFR (RAM/SFR) ← (RAM/SFR) −1 If (RAM/SFR) ≠ 0 then (PC) ← (PC) + offset	This instruction is same as **DJNZ Rn, offset** except that the content of the RAM/SFR is decremented and compared. The address of the RAM/SFR is directly specified in the instruction.
94.	NOP	(PC) ← (PC) + 1	This instruction will not perform any operation, except that the PC is incremented by one to point to the next instruction. Execution of NOP will produce a delay of one machine cycle time and so, this instruction can be used to create small delays in multiplies of machine cycle time.

5.8.2 Boolean Instructions

The boolean instructions operate on a particular bit of a data. This group includes instructions which clear, complement or move a particular bit of bit-addressable RAM/SFR or carry flag. It also include jump instructions, which transfers the program control to a new address, if a particular bit is set or cleared. The boolean instructions of the 8051 are listed in Table 5.8 with a brief explanation about each instruction.

Table 5.8: Boolean Instructions

S.No.	Instruction	Symbolic representation	Explanation
95.	CLR C	CF ← 0	Clear carry flag.
96.	CLR bit	bit = Address of particular bit of RAM/SFR (bit) ← 0	The particular bit of RAM/SFR whose address is specified in the instruction is cleared to zero.
97.	SETB C	CF ← 1	The carry flag is set to one.

Table 5.8 continued ...

S.No.	Instruction	Symbolic representation	Explanation
98.	SETB bit	bit = Address of particular bit of RAM/SFR (bit) ← 1	The particular bit of RAM/SFR whose address is specified in the instruction is set to one.
99.	CPL C	CF ← ~ CF	The carry flag is complemented.
100.	CPL bit	bit = Address of particular bit of RAM/SFR (bit) ← ~ (bit)	The particular bit of RAM/SFR whose address is specified in the instruction is complemented.
101.	ANL C,bit	bit = Address of particular bit of RAM/SFR CF ← CF & (bit)	The particular bit of RAM/SFR is logically ANDed with the carry flag and the result is stored in the carry flag.
102.	ANL C,/bit	bit = Address of particular bit of RAM/SFR (bit) ← ~ (bit) CF ← CF & (bit)	The complement of the particular bit of RAM/SFR is logically ANDed with the carry flag and the result is stored in the carry flag.
103.	ORL C,bit	bit = Address of particular bit of RAM/SFR CF ← CF | (bit)	The particular bit of RAM/SFR is logically ORed with the carry flag and the result is stored in the carry flag.
104.	ORL C,/bit	bit = Address of particular bit of RAM/SFR (bit) ← ~ (bit) CF ← CF | (bit)	The complement of the particular bit of RAM/SFR is logically ORed with the carry flag and the result is stored in the carry flag.
105.	MOV C,bit	bit = Address of particular bit of RAM/SFR CF ← (bit)	The particular bit of RAM/SFR is moved to the carry flag. The address of the bit is directly given in the instruction.
106.	MOV bit,C	bit = Address of particular bit of RAM/SFR (bit) ← CF	The carry flag is moved to a particular bit of RAM/SFR whose address is directly specified in the instruction.

Table 5.8 continued...

S.No.	Instruction	Symbolic representation	Explanation
107.	JC offset	(PC) ← (PC) + 2 If CF = 1 then, (PC) ← (PC) + offset	The content of the PC is incremented by two to point to the next instruction. Then the carry flag is checked. If the carry flag is one then the offset given in the instruction is added to the PC so that the program control is transferred to a new address.
108.	JNC offset	(PC) ← (PC) + 2 If CF = 0 then, (PC) ← (PC) + offset	Same as JC offset, except that the branching will take place if the carry flag is zero.
109.	JB bit, offset	bit = Address of particular bit of RAM/SFR (PC) ← (PC) + 3 If If (bit) = 1, then (PC) ← (PC) + offset	The content of the PC is incremented by three to point to the next instruction. The particular bit of RAM/SFR is tested. If the bit is one, then the offset given in the instruction is added to PC so that the program control is transferred to a new address.
110.	JNB bit, offset	bit = Address of particular bit of RAM/SFR (PC) ← (PC) + 3 If (bit) = 0, then (PC) ← (PC) + offset	The content of the PC is incremented by three to point to the next instruction. The particular bit of RAM/SFR is tested. If the bit is zero then the offset given in the instruction is added to the PC so that the program control is transferred to a new address.
111.	JBC bit, offset	bit = Address of particular bit of RAM/SFR (PC) ← (PC) + 3 If If (bit) = 1, then bit ← 0 ; (PC) ← (PC) +offset	The content of the PC is incremented by three to point to the next instruction. The particular bit of RAM/SFR is tested. If the bit is one, then clear the bit and the offset given in the instruction is added to the PC, so that the program control is transferred to a new address.

5.9 IO INSTRUCTIONS

The IO ports of 8051 are mapped as RAM memory and so that content of port can be accessed by direct addressing. Therefore the instructions that employs direct addressing can be called IO instructions.

Example: Mov A, diorect; Mov direct, A; Add A, direct; DJNZ direcr,offset; CJNE A, direct,offset; ORL direct,#data.

5.10 COMPARISON OF 8085 AND 8051 ASSEMBLY LANGUAGE PROGRAMMING

The 8085 has **von Neumann architecture** in which there will be a single memory bank with a common address space for program and data. The system designer has to divide or allot the space for program and data depending on the need.

The 8051 has **Harvard architecture** which support two different memory banks, one for program and another for data, with independent or separate address space. The 8051 microcontrollers have more advanced features than 8085 microprocessor and some of the features are listed here.

1. Program codes are stored in program address space and accessed using program address pointer called program counter (PC).

2. Data can be stored in either in data memory space or program memory space. When data is stored in external data memory space it is accessed using data memory pointers R0, R1 and DPTR. When data is stored in internal memory space it can be accessed by direct addressing or using data memory pointer DPTR. When data is stored in program memory space it can be accessed using program memory pointer PC.

3. Jump address can be 3 types: 16-bit address, 11-bit address, 8-bit signed offset address

4. Subtraction is always with previous borrow and so carry should be set or cleared depending upon program logic before executing the subtract instruction.

5. In 8051 combined decrement, compare and jump instructions using zero flag are available which makes programming task easier.

6. 8051 has number of bit manipulating instructions. So that individual bits of data can modified more efficiently and also can be used for decision making.

7. In 8051, 8-bit multiplication can be performed using single instruction MUL. Similarly, 8-bit division can be performed using single instruction DIV.

8. Microcontrollers are designed to run continuously and so program execution cannot be halted by any instruction. Alternatively, program execution can be made to remain in an idle loop or simple loop without doing any work, until reset or interrupt.

5.11 SHORT-ANSWER QUESTIONS

Q5.1 *What is state in an 8051 microcontroller?*

The state is the basic time unit for discrete operation of the controller such as fetching an opcode, executing an opcode, writing a data, etc. A machine cycle consists of six states and the timing of each state is 2 oscillator clock periods.

Q5.2 *How many machine cycles are needed to execute an instruction in an 8051 controller ?*

The 8051 microcontroller executes an instruction in one to four machine cycles.

Q5.3 *How can the time taken to execute an instruction be estimated in an 8051 controller?*

The time taken to execute an instruction by an 8051 controller is obtained by multiplying the time to execute a machine cycle by the number of machine cycles of the instruction. The time to execute a machine cycle is 12 clock periods.

$\therefore$ Time to execute an instruction $= C \times 12 \times T = C \times 12 \times \dfrac{1}{f}$

where, C = Number of machine cycles of an instruction.

T = Time period of crystal frequency in seconds.

f = Crystal frequency in Hz.

Q5.4 *What is the size of 8051 instructions?*

The size of 8051 instructions is one to three bytes. The first byte is an opcode and the subsequent bytes are the address or data.

Q5.5 *List the various machine cycles of an 8051 controller.*

The various machine cycles of 8051 microcontroller are:

1. **External program memory fetch cycle.**

2. **External data memory read cycle.**

3. **External data memory write cycle.**

4. **Port operation cycle.**

Q5.6 *How does an 8051 microcontroller differentiate between external program memory access and data memory access?*

With external program memory, the controller can perform only read operations but with external data memory, the controller can perform both read and write operations. For reading program memory, the controller asserts $\overline{PSEN}$ as **low**, for reading data memory the controller asserts $\overline{RD}$ as **low**, and for writing data memory the controller asserts $\overline{WR}$ as **low**.

Q5.7 ***What are the addressing modes available in an 8051 controller?***

The addressing modes available in the 8051 microcontroller are

1.	Immediate addressing	4. Register indirect addressing
2.	Direct addressing	5. Implied addressing
3.	Register addressing	6. Relative addressing

Q5.8 ***Explain register indirect addressing in an 8051.***

In register indirect addressing, the instruction specifies the name of the register in which the address of the data is available. The internal data RAM locations can be addressed indirectly through registers R1 and R0. The external RAM can be addressed indirectly through the DPTR (Data pointer).

Example : **MOV A,@R0** - The content of the RAM location addressed by the R0 is moved to the A-register.

Q5.9 ***Explain relative addressing in an 8051.***

In relative addressing mode, the instruction specifies the address relative to the Program Counter (PC). The instruction will carry an offset whose range is -128_{10} to $+127_{10}$. The offset is added to the PC to generate the 16-bit physical address.

Example : **JC offset** - If carry is one, then the program control jumps to an address obtained by adding the content of the PC and the offset value in the instruction.

Q5.10 ***How can the 8051 instructions be classified ?***

The 8051 instructions can be classified into the following five groups:

1. Data transfer instructions
2. Arithmetic instructions
3. Logical instructions
4. Branching instructions
5. Boolean instructions

Q5.11 ***List the instructions of 8051 that affect all the flags of 8051.***

The 8051 instructions that affect all the flags are ADD, ADDC, and SUBB.

Q5.12 ***List the instructions of 8051 that affect the overflow flag in 8051.***

The 8051 instructions that affect the overflow flag are ADD, ADDC, DIV, MUL and SUBB.

Q5.13 ***List the instructions of 8051 that affect only the carry flag.***

The 8051 instructions that affect only the carry flag are,

ANL C,bit	CPL C	RRC A
ANL C,/bit	MOV C,bit	RLC A
CJNE	ORL C,bit	SETB C
CLR C	ORL C,/bit	

Q5.14 List the instructions of 8051 that always clear the carry flag.

The instructions that always clear the carry flag are CLR C, DIV and MUL.

Q5.15 What are the operations performed by the Boolean variable instructions of an 8051?

The Boolean variable instructions can clear or complement or move a particular bit of bit-addressable RAM/SFR or carry flag. They can also transfer the program control to a new address if a particular bit is set or cleared.

5.12 EXERCISES

I. Fill in the blanks with appropriate words

1. One machine cycle of 8051 consists of _____ number of clock pulses.

2. _____ machine cycle of 8051 is executed to load the data from external data memory into the internal register.

3. The 8051 instruction MOVX A, @ DPTR is executed in _____ number of machine cycles.

4. The _____ addressing mode is used in the 8051 instruction ADD R1, 20H.

5. The 8051 instruction that copies the data of external RAM pointed by R_1 register into A register is _____ .

6. The _____ instruction of 8051 is used to exchange the lower nibbles of two operands.

7. The borrow which results from subtraction is always reflected on _____ flag.

8. The 8051 instruction SUB A, R1 is _____ instruction.

9. The quotient and remainder are stored in _____ and _____ registers respectively after executing the 8051 instruction DIV AB.

10. The _____ logical instruction of 8051 is used to mask the bits.

11. The instructions that change the sequence of execution are called _____ instructions.

12. The _____ instruction of 8051 is executed to jump to a location within the same page of 2K memory space unconditionally.

13. The _____ instruction of 8051 is used to clear the 0th bit of port 1.

14. The JNB C, offset instructioin of 8051 is equivalent to _____ .

15. The result of 8051 instruction RL A, if A = 01_H is _____ .

16. The 8051 instruction CJNE stands for _____ .

Answers

1. 12	5. MOV A, @R1	9. A, B	13. CLRP1.0
2. data memory read	6. XCHD	10. ANL	14. JNC offset
3. two	7. carry	11. control transfer	15. 02_H
4. direct	8. invalid	12. AJMP	16. compare and jump if not equal

II. State whether the following statements are True/False.

1. One state of 8051 consists of two clock pulses.

2. The 8051 microcontroller utilizes all the six states to fetch the opcode from code memory.

3. The 8051 special function registers can be accessed using direct addressing mode.

4. MOV A, 07H instruction of 8051 uses immediate addressing mode.

5. Data transfer between two internal RAM/SFR is not possible in 8051.

6. External data memory of 8051 controller can be read/written only through A register.

7. Internal RAM memory of 8051 can be pointed only by R_0 and R_1 register.

8. The 8051 instructions INC and DEC do not affect carry flag.

9. The destination register is always the A register in 8051 addition and subtraction.

10. The 8051 instruction CPL A changes the magnitude of A register.

11. The execution of 8051 instruction ANL A does not affect carry flag.

12. The 8051 always consider previous carry (borrow) while performing subtraction.

13. The rotate instructions of 8051 are used to transmit data serially.

14. The program counter(PC) of 8051 is always modified by unconditional jump instruction.

15. The 8051 instruction ACALL is a three byte instruction.

16. All the 8051 conditional jumps are short jumps.

17. The 8051 instruction "JNB bit, offset " jumps to the specified relative address if specified bit is 1.

Answers

1. True	4. False	7. True	10. False	13. True	16. True
2. False	5. False	8. True	11. True	14. True	17. False
3. True	6. True	9. True	12. True	15. False	

III. Choose the right answer for the following questions.

1. *The time required to execute one machine cycle of 8051 is*

   a) 6 states   b) 12 clock pulses   c) a or b   d) none

2. *Which of the following machine cycles are executed by 8051 to execute the instruction CLR A?*

   a) External program memory fetch cycle

   b) External data memory read cycle

   c) External data memory write cycle

   d) Port operation cycle

3. *Identify the 8051 instruction which uses immediate addressing mode*

   a) ADD A, 20H   *b)* MOV R1, #20H   *c)* both a and b   *d)* none

4. *Identify the invalid 8051 instrution*

   a) MOV A, #75H   *b)* ADD A, 80H   *c)* MOV A, R1   *d)* MOV A, @R4

5. *This 8051 instruction copies the data from code memory into accumulator*

   a) MOVX A, @PC *b)* MOVX A, @A+DPTR *c)* MOVC A, @PC *d)* MOVC A, @A+PC

6. *The 8051 instruction used to perform multibyte addition is*

   a) ADD   *b)* ADDC   *c)* both a and b   *d)* neither a nor b

7. *Which of the following 8051 instruction is used to multiply the contents of internal RAM E0$_H$ and F0$_H$?*

   a) MUL F0H   *b)* MUL E0H   *c)* MUL AB   *d)* MUL B

8. *What is the status of CY and overflow flag after executing the 8051 instruction SUBB A, 09H? Assume A = 12$_H$ and CY = 0 before execution of instruction.*

   a) CY = 0 ; OF = 0   *b)* CY = 0 ; OF = 1   *c)* CY = 1 ; OF = 0   *d)* CY = 1 ; OF = 1

9. *Identify the invalid 8051 instruction.*

   a) ADD A, 12H   *b)* ADD A, #12H   *c)* ADD A, R1   *d)* ADD R1, A

10. *The _____ instruction of 8051 is used to set the least significant bit of the register 'A'*

   a) ANL A, #01H   *b)* ANL A, 01H   *c)* ORL A, 01H   *d)* ORL A, #01H

11. *Which of the following 8051 instruction is equivalent to MOV A, #00H?*

   a) CPL A   *b)* CLR A   *c)* both a and b   *d)* neither a nor b

12. *The content of A register is 95$_H$. What is the status of A after executing the 8051 instruction SWAP A?*

   a) 90H   *b)* 05H   *c)* 00H   *d)* 59H

13. Which of the following is not a valid jump instruction of 8051?

   a) JMP 2000H   b) JNZ 12H   c) JC 1000H   d) JNC F2H

14. The _____ instruction of 8051 is used to terminate an interrupt service subroutine

   a) RET   b) RETI   c) either a or b   d) neither a nor b

15. Which of the following instruction is used to set the carry flag when the port-1 pin P1.4 is 0? Assume carry flag is intially reset.

   a) ANL C, P1.4   b) ANL C, /P1.4   c) ORL C, P1.4   d) ORL C, /P1.4

16. What is the status of CY flag after execution of the following 8051 codes

   MOV A, #82H

   CJNE A, #20H, NEXT

   a) CY = 0   b) CY = 1   c) carry does not change   d) CY = $\overline{CY}$

17. Assume CY = 1 and P1.2 = 0. What is the status of CY flag after the execution of the 8051 instruction ANL C, /P1.2?

   a) CY = 0   b) CY = 1   c) CY = $\overline{CY}$ d) CY is not altered

18. In 8051 short jump instruction, if MSB of relative address = 1 then it is

   a) forward jump   b) backward jump   c) a or b   d) neither a nor b

Answers								
1. c	3. a	5. d	7. c	9. d	11. b	13. c	15. d	17. b
2. a	4. d	6. b	8. a	10. d	12. d	14. b	16. a	18. b

IV. Answer the following questions.

E5.1 The 8051 microcontroller works at 10 MHz. How much time is taken by 8051 to execute the instruction ADD A, #12H?

E5.2 Where the program control is transferred when the instruction JMP 06H is executed by 8051? Assume PC = 2100_H.

E5.3 Identify the addressing mode for the following 8051 instructions

   a) MOV A,R2   b) MOVX A,@DPTR   c) MUL AB

   d) XRL A,20H   e) CJNE A,#04H,07H

E5.4 Write down the steps involved in executing the 8051 instruction PUSH 07H. Assume SP = 3070_H and the $R_7 = 25_H$.

E5.5 Differentiate XCH and XCHD instructions of 8051.

E5.6 Write down the 8051 instsruction sequences to transfer the data stored at internal RAM location 45_H into external data memory 36_H.

E5.7 Name the 8051 instruction used to increment the external data memory pointer by one

E5.8 Identify the syntax errors if any in the following 8051 instructions

    a. MOV A,@R4   b. INC PC   c. MUL A,#25H   d. SUB R2,R4

E5.9 The content of the register A = 78_H. What do the 8051 instructions RL A and RLC A produce? Assume CF = 1.

E5.10 Calculate the physical jumping address after the execution of the instruction SJMP F2H if it is stored in the address 2536_H.

E5.11 Write down the single equivalent 8051 instruction of the following

    DEC R1

    JNZ 06_H

E5.12 Find the status of CY flag after execution of the following 8051 instructions sequentially

    CLR   C

    SETB C

    SETB P1.4

    CPL   C

    ANL   C,P1.4

E5.13 Write down the 8051 instructions used to read the status of LED which is connected to P1.6. If the LED is ON switch it OFF and jump three steps forward in the program.

CHAPTER 6

MEMORY AND IO INTERFACING

6.1 INTRODUCTION TO MEMORY

A memory unit is an integral part of any microcomputer system and its primary purpose is to store programs and data. In a broad sense, a microcomputer memory system can be logically divided into three groups. They are as follows:

1. **Processor memory**
2. **Primary or main memory**
3. **Secondary memory**

Processor memory refers to registers inside the microprocessor. These registers are used to hold data and results temporarily when computation is in progress. Since the registers of the processor are fabricated using the same technology as that of a microprocessor, there is no speed disparity between these registers and a microprocessor. However, the cost involved in this approach forces a manufacturer to include only a few registers in the microprocessor.

Primary or main memory refers to the storage area which can be directly accessed by the microprocessor. Therefore, all programs and data must be stored only in primary memory prior to execution. In primary memory the access time should be compatible with the read/write time of the processor. Therefore, only semiconductor memories are used as primary memories and they (the latest versions) are fabricated using CMOS technology. Primary memory normally includes ROM, EPROM, static RAM, DRAM and NVRAM.

Secondary memory refers to the storage medium which comprises of slow devices such as magnetic tapes and disks (hard disk, floppy disc and Compact Disc (CD)). They are called as auxiliary or backup storage devices. These devices are used to hold large data files and huge programs such as operating systems, compilers, data bases, permanent programs, etc. The microcomputer system copies the required programs and data from secondary memory to main memory and work directly with main memory only.

6.2 SEMICONDUCTOR MEMORY

The main or primary memory elements are semiconductor devices, because the semiconductor devices alone can work at high speeds and consume less power. Moreover, they can be fabricated as ICs and so they occupy less space.

A typical semiconductor memory IC will have n address pins (lines) and m data pins (lines). The capacity of the memory will be $2^n \times m$ bits. Figure 6.1 shows a simplified functional block diagram of semiconductor memory. The functional blocks of semiconductor memory are Row address decoder, Column address decoder, Memory array, Input buffer and Output buffer.

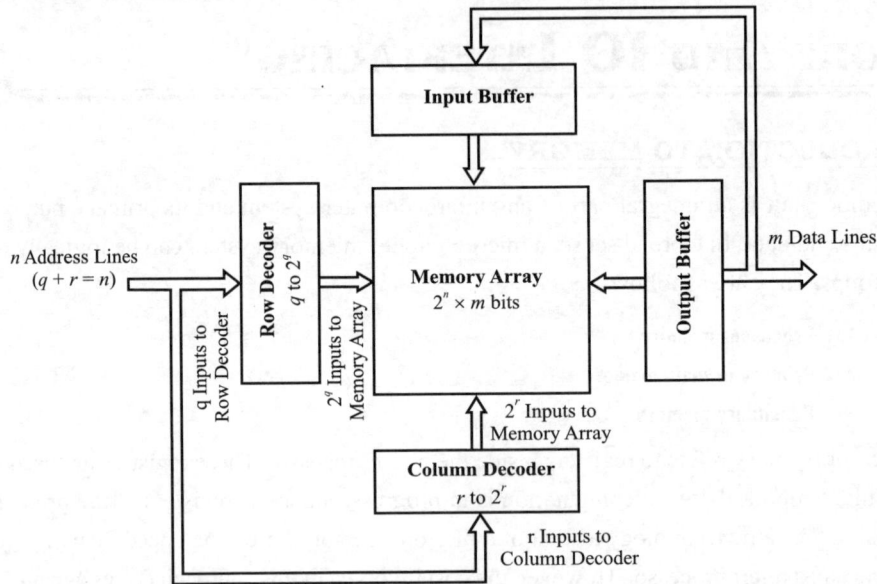

Fig. 6.1: A simplified functional block diagram of a typical semiconductor memory.

The input and output buffers are used to hold the data until the valid time and also takes care of the signal current level matching (Impedance matching). The n address lines are split into q lines and r lines, such that $q + r = n$. The q address lines are applied as input to the row decoder and r address lines are applied as input to the column decoder.

The output lines of the row and column decoders are used to form a matrix array of size, $2^q \times 2^r$ consisting of 2^n crossing points as shown in Fig. 6.2. Each crossing point is called **memory cell** and can store one bit of binary information. A typical memory array will consist of m layers of matrix array as that of Fig. 6.2 and all of them are wired in parallel.

When an address is send to memory IC, the row and column decoder will select one line each, which in turn select one memory cell in each layer. Thus, m memory cells are selected by an address. Then using read or write control signals, the data can be read or stored in the selected memory cells.

In the first version of semiconductor memory, the memory cells were made of passive elements like resistors and capacitors. Later, diodes were used instead of passive elements. With advancement in semiconductor technology, bipolar and MOS transistors were used to form memory cells. The latest technology used for fabricating memory cells are CMOS and HMOS which offer very low power and high-speed operation.

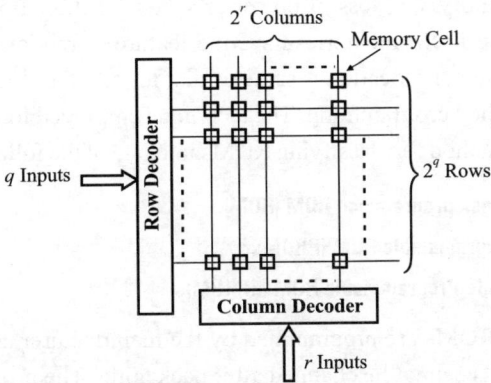

Fig. 6.2: One layer of memory array.

The different types of semiconductor memory are ROM, PROM, EPROM, static RAM, DRAM and NVRAM. These semiconductor memories can be classified into volatile and nonvolatile memory. If the information stored in a semiconductor memory is lost when the power supply to that IC is switched OFF then the memory is called volatile.

On the other hand, if the stored information is retained even if the power supply is switched OFF then the memory is called nonvolatile. The ROM, PROM, EPROM and NVRAM are nonvolatile memories. The static RAM and DRAM are volatile memories.

The semiconductor memories can also be classified into read-only memory and read/write memory. In read-only memories, information is stored permanently either during manufacturing or after manufacturing and then interfaced to the microcomputer system. The processor can only read the stored information from these memories and cannot write into it. But in read/write memory, the processor can store (write) the information as well as read from it. The ROM, PROM and EPROM are read-only memories. The NVRAM, static RAM and DRAM are read/write memories.

The other features of semiconductor memories are random access and nondestructive readout. In random-access memory, the memory access time is independent of the memory location being accessed (i.e., the access time will be same for the first or last location). All semiconductor memories are random-access memories. In semiconductor memories, a read operation by the processor will not destroy the stored information and for this reason, the semiconductor memory is also called the NDRO memory (**N**ondestructive **R**ead-**O**ut memory).

6.2.1 ROM and PROM

ROM is a semiconductor memory which permits only a read access. ROM functions as a memory array whose contents once programmed, are permanently fixed and cannot be altered by the microprocessor to which the memory is interfaced. Other names for this type of memory are dead memory, fixed memory, permanent memory and **R**ead-**O**nly **S**tore (ROS). In ROM memory, the memory cell (storage unit) will have a MOS transistor either with open gate or closed gate. The transistors with closed gate represent **1**'s and with open gate represent **0**'s. Since the configuration is fixed, they permanently store **1**'s and **0**'s.

ROM is a nonvolatile memory, i.e., loss of power or system malfunction does not change the contents of the memory. Also, the ROM memories have the feature of random access, which means that the access time for a given memory location is same as that for all other locations. The process of storing information in ROM is called programming. The technique employed for storing information in the ROM provides a convenient method for classifying ROMs into one of the following three categories:

1. **Custom programmed or Mask programmed ROM (ROM).**
2. **Programmable or Field programmable ROM (PROM).**
3. **Reprogrammable or Erasable-Programmable ROM (EPROM).**

The custom programmed ROMs are programmed by the manufacturer as specified by the user during fabrication and the contents cannot be changed after packaging. The programmable ROMs are one-time programmable by the user.

The reprogrammable ROMs have facilities for programming as well as for erasing its content and reprogramming the memory. The reprogrammable ROMs are erased either by passing electrical current or Ultraviolet light.

The programming of ROMs can be carried out using ROM (EPROM) programmer. Usually the ROM programmer is a digital system interfaced to a **P**ersonal **C**omputer (PC). The information to be programmed are first stored as a file in the PC and converted to the required binary format using a conversion software. Then the information is transferred from the PC to ROM programmer.

6.2.2 EPROM

The **R**ead **O**nly **M**emory (ROM) which has reprogrammable features is called EPROM (**E**rasable-**P**rogrammable **R**ead **O**nly **M**emory). The EPROM memory is nonvolatile and also has the feature of random access. In an EPROM, the binary informations are entered using electrical impulses and the stored information is erased using ultraviolet rays. Typical erase time varies between 10 to 30 minutes.

In EPROM, the memory cell (storage location of a bit) consists of a MOS transistor with isolated gate. The isolated gate is located between the normal control gate and the source/drain region of a MOS transistor.

This gate may be charged with electrons during the programming operation and when charged with electrons, the transistor is permanently turned OFF. The state of the floating gate, charged or uncharged, is permanent because the gate is isolated in an extremely pure oxide.

The charge on the isolated gate may be removed if the device is irradiated with ultraviolet light. The ultraviolet light allows the electrons to recombine and discharge through the control gate. The process of charging and discharging is repeatable.

The EPROM is programmed by inserting the EPROM chip into the socket of a PROM programmer and providing addresses and voltage pulses at the appropriate pins of the chip. Usually, the PROM programmer is interfaced to a **P**ersonal **C**omputer (PC) and the information to be programmed is downloaded from PC.

EPROMs are manufactured by many semiconductor industries like INTEL, Hitachi, Toshiba, Cypress, etc. The manufacturers have a common industry standard, so that a product from different industry will be pin-to-pin compatible and slightly differ in electrical and switching characteristics. The various features of 2764 (8 kB EPROM) manufactured by Cypress semiconductor corporation are discussed in this section.

CY27C64 (Cypress Make CMOS 2764)

The CY27C64 is a high performance 8192 byte (8 kB) CMOS EPROM. It has power down mode, in which the device will enter a low-power standby mode when it is not enabled (or deselected).

The logic block diagram of CY27C64 is shown in Fig. 6.3 and the pin configuration during read mode is shown in Fig. 6.4. [The pin configuration of CY27C64 will be different to that of Fig. 6.4 during programming or write mode.] The chip has thirteen address inputs denoted as A_0-A_{12}. The address is used to access any one of the 8 kilo (8192) locations within the chip.

The eight output lines, O_0 to O_7 are used to output data from the chip. The chip will be in standby mode when $\overline{CE}$ is inactive. The $\overline{CE}$ is activated for selecting the chip and $\overline{OE}$ is activated for enabling the output buffer during the read operation.

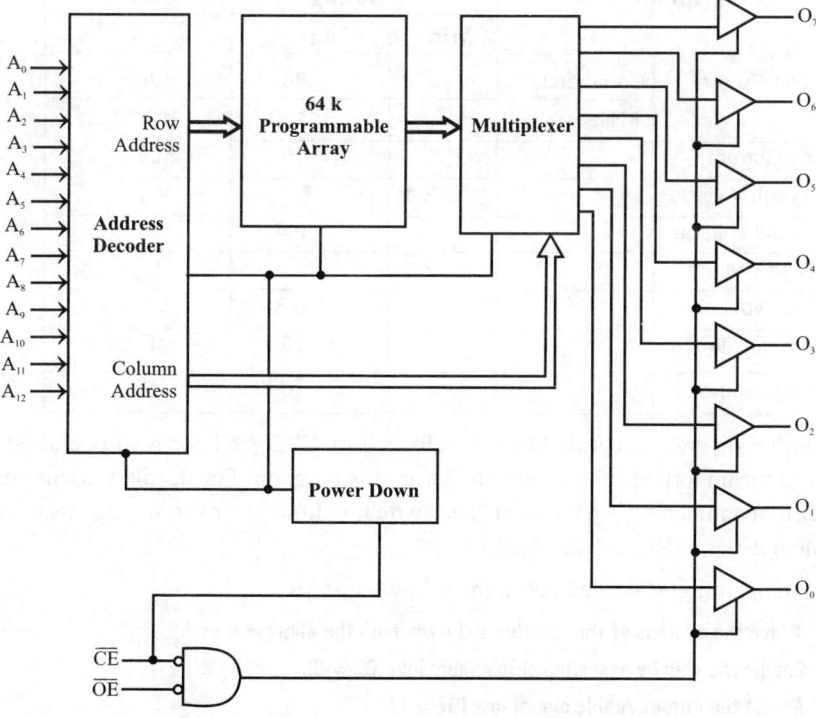

Fig. 6.3: Logic block diagram of a CY27C64.

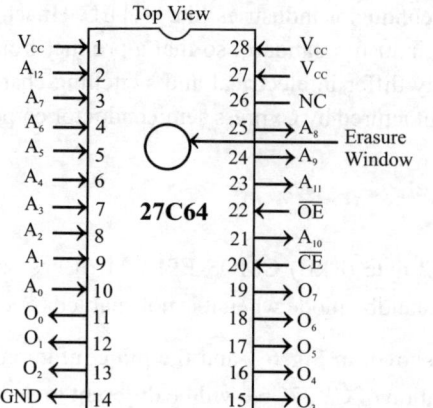

Fig. 6.4: Pin configuration of a CY27C64 in read mode.

Pin	Description
A_0 - A_{12}	Address
O_0 - O_7	Output/Data
$\overline{CE}$	Chip Enable
$\overline{OE}$	Output Enable
V_{CC}	Power supply, +5-V
GND	Ground (0-Volt)
NC	No Connection

The CY27C64 EPROM is available with maximum access time of 70, 90, 120, 150 or 200 ns (nanosecond). The electrical characteristics or ratings of the EPROM are listed in Table 6.1.

Table 6.1: Electrical Characteristics of CY27C64

Description		Rating		Unit
		Min	**Max**	
Operating Current	Commercial		80	mA
	Military		100	mA
Standby Current			15	mA
Output High Voltage		2.4		V
Output Low Voltage			0.4	V
Input High Voltage		2.0		V
Input Low Voltage			0.8	V
Output Capacitance			10	pF
Input Capacitance			10	pF

The timing diagram (or switching waveforms) of CY27C64 for read operation is shown in Fig. 6.5. Only four important timings are shown in this diagram. For detailed discussions on timing diagram refer to manufacturer's data sheet. The switching timings of various signals of CY27C64 are listed in Table 6.2.

The read operation is carried out in the following steps:

1. **Place the address of the location to be read, on the address pins A_0 - A_{12}.**
2. **Enable the chip by asserting chip enable low ($\overline{CE}$ = 0).**
3. **Assert the output enable signal low ($\overline{OE}$ = 0).**
4. **The data can be read from the output lines (O_0 to O_7) after a delay time of t_{OE} (40 or 50 ns) after asserting $\overline{OE}$ signal low.**

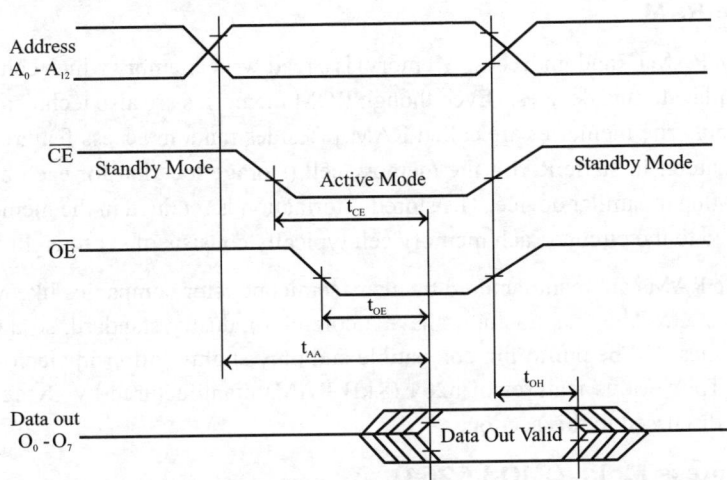

Fig. 6.5: Timing diagram of a CY27C64 for read operation.

Table 6.2: Switching Characteristics of CY27C64

Parameter	Description	Time		Unit
		Min	**Max**	
t_{AA}	Address to output valid		70 to 200	ns
t_{OE}	Output enable active to output valid		40 or 50	ns
t_{CE}	Chip enable active to output valid		70 to 200	ns
t_{OH}	Data hold from address change	3		ns

When the address is placed on the address lines, the memory will take a time of t_{AA} to place the data on the output lines, provided the $\overline{CE}$ and $\overline{OE}$ are both asserted **low**, at the appropriate time.

The CY27C64 EPROM is equipped with an erasure window. When the window is exposed to UV light, the contents of EPROM are erased and then it can be reprogrammed. Wavelengths of light less than 4000A° (Angstrom unit) begin to erase the EPROM. Hence, an opaque label should be placed over the window if the EPROM is exposed to sunlight or fluorescent lighting for a very long time.

The recommended dose of UV light for erasure is a wavelength of 2537 A° for a minimum dose (UV intensity multiplied by exposure time) of 25 W-sec/cm². For an UV lamp with a 12 mW/cm² power rating, the exposure time would be approximately 35 minutes.

The EPROM has to be placed within a distance of 1 inch from the lamp during erasure. Permanent damage may result if EPROM is exposed to high intensity UV light for a very long time. (Maximum dosage is 7258 W-sec/cm²).

6.2.3 Static RAM

The static RAM (**R**andom **A**ccess **M**emory) is a read/write memory which consists of an array of flip-flops or similar storage devices. [Even though ROM memories are also technically random access memory, the read/write memories are called RAM.] Besides random access feature, the static RAMs are volatile in nature. In static RAM, the memory cell (storage location for each bit of information) consist of a flip-flop or similar device. The stored information is retained in the memory cell as long as power is supplied to the circuit. Each memory cell typically consists of six to eight MOS transistors.

The static RAMs are manufactured by many semiconductor companies like Motorola, Hitachi, Toshiba, Cypress, etc. The manufacturers have a common industry standard, so that a product from different companies will be pin-to-pin compatible and olny slightly differ in electrical and switching characteristics. The various features of 6264 (8 kB RAM) manufactured by Cypress semiconductor corporation are discussed in this section.

CY6264 (Cypress Make CMOS 6264)

The CY6264 is a high performance CMOS static RAM organized as 8192 bytes (8 kB). The device has a power down mode. When CY6264 is not enabled (deselected), it will enter the power down mode and in this mode the power consumed is reduced to 30% of active mode power.

The logic block diagram and the pin configuration of CY6264 are shown in Figs. 6.6 and 6.7. The chip has 13 address inputs denoted as A_0-A_{12}. The address is used to access any one of the 8 kilo (8192) locations within the chip. It has eight IO pins for reading/writing the data and they are denoted as IO_0-IO_4.

The chip has four control signals $\overline{CE}_1$, CE_2, $\overline{WE}$ and $\overline{OE}$. When $\overline{CE}_1$ and $\overline{WE}$ inputs are both **low** and is **high**, data on the eight data pins (IO_0 through IO_7) is written into the memory location addressed by the address pins (A_0 through A_{12}).

When $\overline{CE}_1$ and $\overline{OE}$ are both **low** and CE_2 is **high**, the content of the memory location addressed by the address pins will be loaded on the eight data pins, IO_0 to IO_4.

The CY6264 RAM is available with maximum access time of 55 or 70 ns (nanosecond). The electrical characteristics or ratings of the RAM are listed in Table 6.3.

Table 6.3: Electrical Characteristics of CY6264

Description	Value		Unit
	Min	**Max**	
Operating Current		100	mA
Standby Current		15 or 20	mA
Output High Voltage	2.4		V
Output Low voltage		0.4	V
Input High Voltage	2.2	V_{cc}	V
Input Low Voltage	−0.5	0.8	V
Output Capacitance		7	pF
Input Capacitance		7	pF

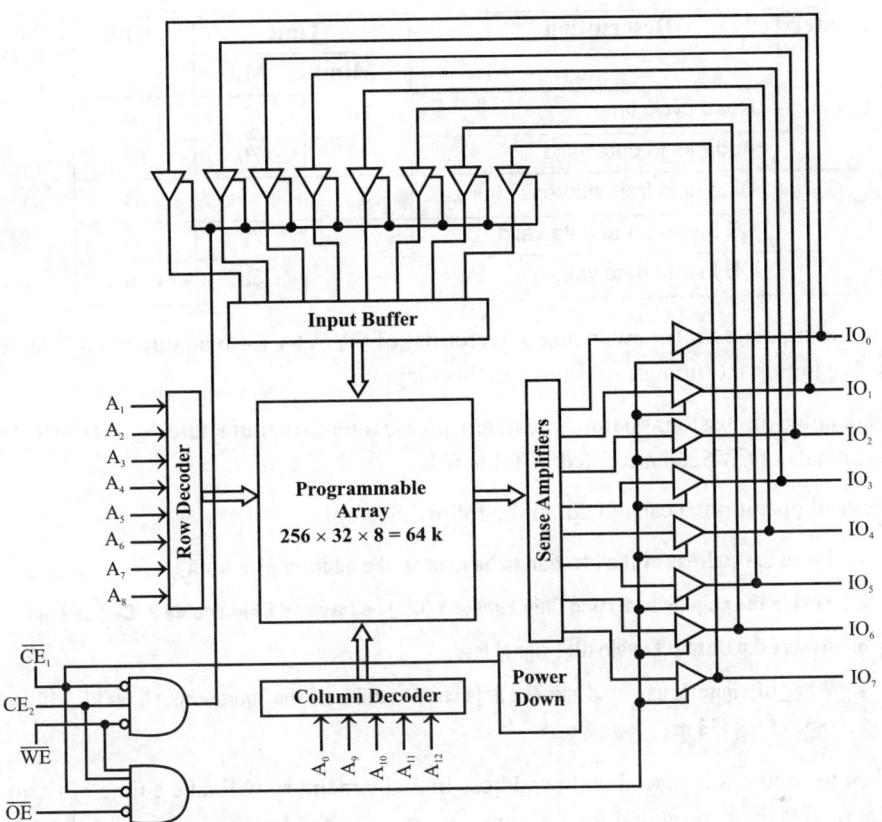

Fig. 6.6: Logic block diagram of a CY6264.

Top View

Pin	Description
A_0 - A_{12}	Address
IO_0 - IO_7	Input/Output Data
$\overline{CE_1}$	Active **low** Chip Enable
CE_2	Active **high** Chip Enable
$\overline{WE}$	Active **low** Write Enable
$\overline{OE}$	Active **low** Output Enable
V_{cc}	Power supply, +5-V
GND	Ground (0-Volt)
NC	No Connection

Fig.6.7: Pin configuration of a CY6264.

Table 6.4: Read Cycle Timings of CY6264

Parameter	Description	Time		Unit
		Min	Max	
t_{RC}	Read cycle time	70		ns
t_{AA}	Address to data valid		70	ns
t_{OHA}	Data hold from address change	5		ns
t_{ACE}	CE **low/high** to data valid		70	ns
t_{DOE}	OE **low** to data valid		35	ns

The timing diagram (or switching waveforms) of CY6264 for read operation is shown in Fig. 6.8. Only five important timings are shown in this diagram.

For detailed discussions on timing diagram, please refer to manufacturer's data sheet. The timings of various signals of CY6264 are listed in Table 6.4.

The read operation is carried out in the following steps:

1. Place the address of the location to be read on the address pins A_0 - A_{12}.
2. Enable the chip by asserting Chip Enable-1 ($\overline{CE}_1$) as low and Chip Enable-2 (CE_2) as high.
3. Assert the Output Enable ($\overline{OE}$) signal low.
4. When $\overline{OE}$ signal is asserted low, the data can be read from the input/output lines (IO_0 - IO_7) after a delay time of t_{DOE} (35 ns).

When the address is placed on the address line, the memory will take a time of t_{AA} to place the data on the output lines, provided the $\overline{CE}_1$ and $\overline{OE}$ are asserted **low** and CE_2 is asserted **high** at the appropriate time.

The timing diagram (or switching waveform) of CY6264 for write operation is shown in Fig. 6.9. The diagram shows some important timings of write cycle.

For detailed discussions on timing diagram, refer to the manufacturer's data sheet. The timings of various signals are listed in Table 6.5.

The write operation is carried in the following steps:

1. Place the address of the location to be written on the address pins A_0 - A_{12}.
2. Enable the chip by asserting $\overline{CE}_1$ signal as low and after a small delay assert CE_2 signal as high.
3. Assert the write enable $\overline{WE}$ signal as low.
4. Place the data to be written on the IO_0-IO_7 lines immediately after $\overline{WE}$ is asserted low.

After the address is placed on the address lines and $\overline{CE}_1$, CE_2 and $\overline{WE}$ are asserted appropriately, the data has to be placed on the data lines within the time t_{SD} (data set-up to write end).

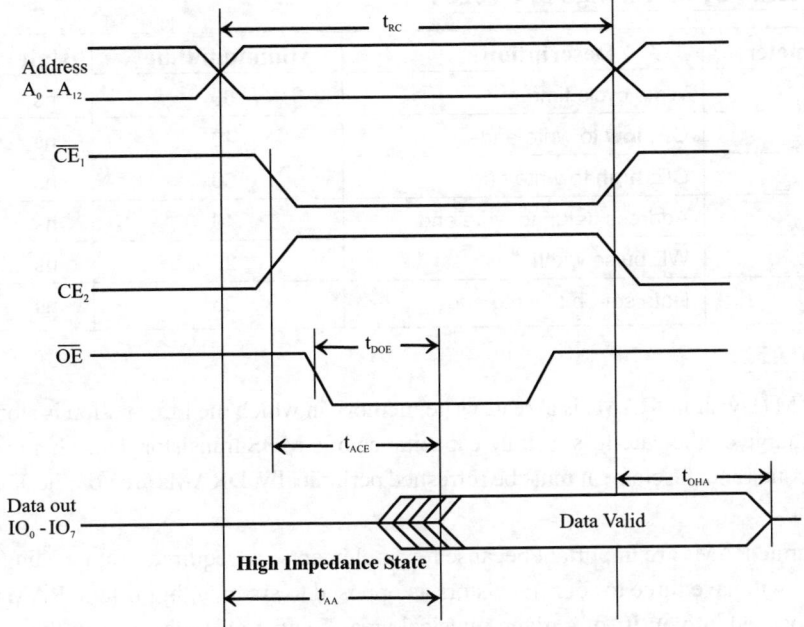

Fig. 6.8: Read cycle timings of a CY6264.

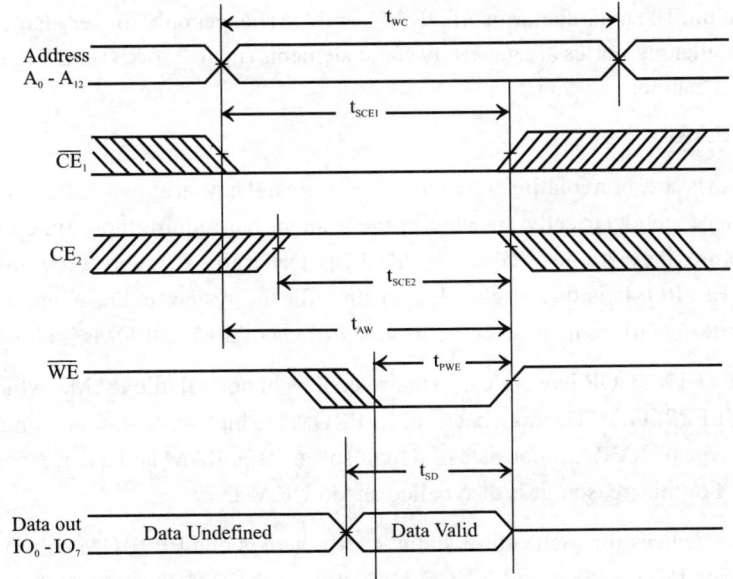

Fig. 6.9: Write cycle timings of CY6264.

Table 6.5: Write Cycle Timings of CY6264

Parameter	Description	Minimum time	Unit
t_{WC}	Write cycle time	50	ns
t_{SCE1}	$\overline{CE}_1$ **low** to write end	40	ns
t_{SCE2}	CE_2 **high** to write end	30	ns
t_{AW}	Address setup to write end	40	ns
t_{PWE}	$\overline{WE}$ pulse width	25	ns
t_{SD}	Data set-up to write end	25	ns

6.2.4 DRAM

DRAM (**D**ynamic **RAM**) is a read/write memory in which the information is stored in the form of electric charge on the gate-to-substrate capacitance of a MOS transistor. This charge dissipates in a few milliseconds and the element must be refreshed periodically. DRAMs are volatile and have random access feature.

Dynamic RAMs are important because fewer elements are required to store a bit (typically each memory cell will have three to four transistors as opposed to six to eight in static RAM), so that more bits can be packed into an IC of a given physical area. They are also faster than the static RAM and consume less power in the quiescent state.

Refreshing of DRAMs needs extra circuitry and so the interfacing of DRAMs to the microprocessor are more complex than the interfacing of static RAMs. The recent versions of DRAMs have internal refreshing circuit. The manufacturing of DRAMs will be cheaper only for very large capacity memories. Therefore, smaller memories are generally static elements (up to 256 kB) and large memories (> 1MB) are typically dynamic.

6.2.5 NVRAM

NVRAMs are nonvolatile read/write memories. They are also called flash memory. These memory devices are electrically erasable in the system, but require more time to erase than a static RAM. Therefore, they are also called EEPROM (**E**lectrically **E**rasable **P**rogrammable **ROM**). The drawback in EEPROMs is that it takes longer time for the system to erase and write. The maximum number of write operations that can be performed in most of the EEPROMs is about 10,000 operations.

INTEL and XYCOR have released their versions of nonvolatile RAMs, which does not have the drawbacks of EEPROMs. (The drawbacks of EEPROM are high write time and limited number of write cycles.) This type of NVRAM consists of a high speed static RAM and a corresponding EEPROM on a single chip. For this reason, it is also called shadow RAM.

In these devices for each cell of static RAM, there is one EEPROM cell. A typical example of shadow RAM is INTEL 2004 and XYCOR's X2004. The 2004 has a special pin called nonvolatile enable (NE). Normally, this pin is **high** and read or write operation is performed with static RAM. When NE is asserted **low**, the data in the static RAM cells are written into the corresponding EEPROM cells.

6.3 INTERFACING STATIC RAM AND EPROM

The primary function of memory interfacing is that the microprocessor should be able to read from and write into a set of semiconductor memory IC chips. Generally, EPROM is interfaced for read operations and RAM is interfaced for read and write operations. The procedure for interfacing SRAM for read/write operation and EPROM for read operation are similar. So, they are dealt commonly in this section.

In order to perform the read/write operation the memory access time should be less than the read/write time of processor, chip select signals should be generated for selecting a particular memory IC, suitable control signals have to be generated for read/write operation and a specific address should be allotted to each memory location. Hence, memory interfacing deals with choosing memories with suitable access time, designing address decoding circuit to generate chip select signals, generating control signals for read/write operation and allocation of addresses to various memory ICs and their locations.

6.3.1 Typical EPROM and Static RAM

A typical semiconductor memory IC will have **n** address pins, **m** data pins (or output pins) and a minimum of two power supply pins (one for connecting required supply voltage (V_{CC}) and the other for connecting ground). The control signals needed for static RAM are chip select (chip enable), read control (output enable) and write control (write enable). The control signals needed for read operation in EPROM are chip select (chip enable) and read control (output enable). A typical static RAM and EPROM are shown in Figs. 6.10 and 6.11 respectively.

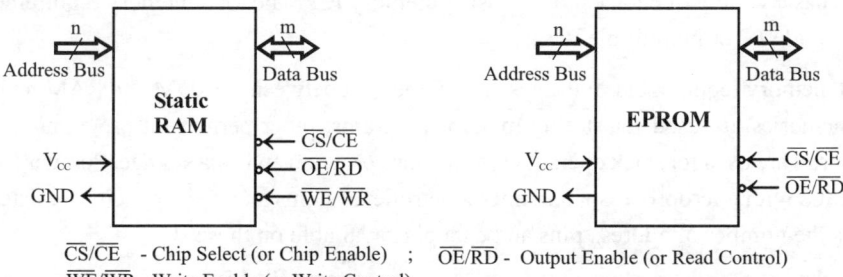

$\overline{CS}/\overline{CE}$ - Chip Select (or Chip Enable)   ; $\overline{OE}/\overline{RD}$ - Output Enable (or Read Control)
$\overline{WE}/\overline{WR}$ - Write Enable (or Write Control)

Fig. 6.10: A typical static RAM IC.   **Fig. 6.11:** A typical EPROM IC in read mode.

> *Note:*   *The pins of EPROM are redefined for write operation. An EPROM requires a different hardware setup and high supply voltage for write operation.*

6.3.2 Memory Capacity

A semiconductor memory IC will have **n** address pins and **m** data pins. Such a memory has 2^n locations and each location can store **m**-bit data. The size of data stored in each memory location is called memory word size. In INTEL 8085-based systems normally memories with word size of 1-byte are used. (But we can even interface memories with word size 1-bit, 2-bit and 4-bit.) The memory capacity is specified in kilobytes. If the memory IC has **m** data pins and **n** address pins, then the memory IC will have a capacity of $2^n \times m$ bits. When m = 8, the memory capacity is 2^n bytes. One kilobyte is 1024_{10} (= 400_H) bytes. The relation between address pins and capacity of memory ICs are listed in Table 6.6.

Table 6.6: Relation between Number of Address Pins and Memory Capacity

Number of address pins	Memory capacity			Range of address in hexa
	in decimal	in kilo	in hexa	
10	$2^{10} = 1024$	1k	400	000 to 3FF
11	$2^{11} = 2 \times 2^{10} = 2048$	2k	800	000 to 7FF
12	$2^{12} = 2^2 \times 2^{10} = 4 \times 2^{10} = 4096$	4k	1000	000 to FFF
13	$2^{13} = 2^3 \times 2^{10} = 8 \times 2^{10} = 8192$	8k	2000	0000 to 1FFF
14	$2^{14} = 2^4 \times 2^{10} = 16 \times 2^{10} = 16384$	16k	4000	0000 to 3FFF
15	$2^{15} = 2^5 \times 2^{10} = 32 \times 2^{10} = 32768$	32k	8000	0000 to 7FFF
16	$2^{16} = 2^6 \times 2^{10} = 64 \times 2^{10} = 65536$	64k	10000	0000 to FFFF
17	$2^{17} = 2^7 \times 2^{10} = 128 \times 2^{10} = 131072$	128k	20000	00000 to 1FFFF
18	$2^{18} = 2^8 \times 2^{10} = 256 \times 2^{10} = 262144$	256k	40000	00000 to 3FFFF
19	$2^{19} = 2^9 \times 2^{10} = 512 \times 2^{10} = 524288$	512k	80000	00000 to 7FFFF
20	$2^{20} = 2^{10} \times 2^{10} = 1024 \times 2^{10} = 1048576$	1024k = 1M	100000	00000 to FFFFF

6.3.3 Choice of Memory ICs and Address Allocation

The memory requirement of a system depends on the application for which it is designed. A system designer has a variety of choices for choosing memory ICs. The total memory requirement can be realized in a single IC or in multiple ICs.

The total memory requirement of the system will be split between EPROM and RAM memories. The EPROM memories are used for storing monitor programs, other permanent programs and data. The RAM memories are used for stack operations, temporary program and data storage. Popular EPROM and static RAM ICs with microprocessor and microcontroller systems and their capacity are listed here. Table 6.7 shows the number of address pins and data pins available on these ICs.

EPROM	Static RAM
2708 (1k × 8 = 8 kilobits/1kB)	6208 (1k × 8 = 8 kilobits/1kB)
2716 (2k × 8 = 16 kilobits/2 kB)	6216 (2k × 8 = 16 kilobits/2 kB)
2732 (4k × 8 = 32 kilobits/4 kB)	6232 (4k × 8 = 32 kilobits/4 kB)
2764 (8k × 8 = 64 kilobits/8 kB)	6264 (8k × 8 = 64 kilobits/8 kB)
27256 (32k × 8 = 256 kilobits/32 kB)	62256 (32k × 8 = 256 kilobits/32 kB)
27512 (64k × 8 = 512 kilobits/64 kB)	62512 (64k × 8 = 512 kilobits/64 kB)
27010 (128k × 8 = 1 megabits/128 kB)	62128 (128k × 8 = 1 megabit/128 kB)
27020 (256k × 8 = 2 megabits/256 kB)	62138 (256k × 8 = 2 megabits/256 kB)
27040 (512k × 8 = 4 megabits/512 kB)	62148 (512k × 8 = 4 megabits/512 kB)

Note: In this book kB refers to kilobytes.

Table 6.7: Number of Address and Data Pins in Memory ICs

Memory IC EPROM/RAM	Capacity	Number of address pins	Number of data pins
2708/6208	1kB	10	8
2716/6216	2 kB	11	8
2732/6232	4 kB	12	8
2764/6264	8 kB	13	8
27256/62256	32 kB	15	8
27512/62512	64 kB	16	8
27010/62128	128 kB	17	8
27020/62138	256 kB	18	8
27040/62148	512 kB	19	8

Note: 16 kB memory is not available as a standard product.

6.3.4 Generation of Chip Select Signals

Decoders are used to generate chip select signals. The 2-to-4 decoder will give four chip select signals. The 3-to-8 decoder will give eight chip select signals. The 4-to-16 decoder will give sixteen chip select signals.

Decoder is a logic circuit that identifies each combination of the signals present at its input. Decoders have **n** input lines and 2^n output lines. In logic **low** decoder, at any one time one of the 2^n outputs will remain **low** and all other outputs will remain **high**.

The output which remains **low** depends on the input signal. Hence if the decoder outputs are connected to chip select pins of ICs in the microprocessor system at any one time, only one chip will be selected. The input to the decoders are unused address lines or high order address lines. While interfacing memories, low order address lines are connected to memory ICs. The remaining unused address lines (or high order address lines) are connected to the input of the decoder. The outputs of the decoder are connected to CS or CE pins of memory ICs.

In a microprocessor-based system, all the memory ICs and peripheral ICs are connected to a common system bus. Therefore, the data, address and control lines are connected to all the slaves (memory/peripheral ICs). But all the slaves remain in **high impedance** state. So, they cannot communicate with the master (processor) through bus (i.e., they are physically connected but electrically isolated).

When the address is given out by the processor for read/write operation, only one of the memory ICs is selected and the selected memory IC will come to normal logic. The selection logic depends on address decoding logic. All other memory ICs will remain in **high impedance** state. So, they are electrically isolated from the system. The read/write operation is performed by the processor with the selected memory IC.

6.3.5 Decoder

Popular decoders used in the microprocessor-based system are 74LS138 and 74LS139. The 74LS138 is a 3-to-8 decoder and 74LS139 is dual 2-to-4 decoder.

The 74LS138 decoder consists of 3-input lines, 8-output lines (logic **low**) and three enables or ground. In the three enables, two are logic **low** and one is a logic **high** enable. The pin configuration of 3-to-8 decoder (74LS138) is shown in Fig. 6.12. The truth table of the decoder is given in Table 6.8.

The 74LS139 decoder consists of two numbers of 2-to-4 decoder packed in a single IC package. Each decoder has two input pins, four output lines and a logic **low** enable. The pin configuration of 74LS139 is shown in Fig. 6.13. The truth table of 2-to-4 decoder is given in Table 6.9. In the 74LS139 each decoder can work independently.

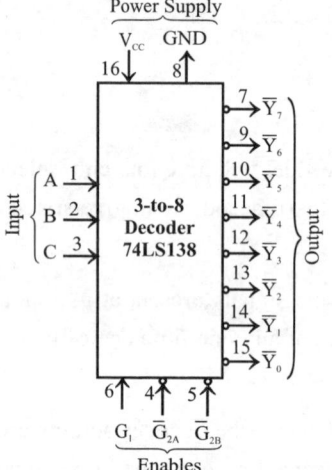

Fig. 6.12: Signals of 74LS138.

Table 6.8: Truth Table of 3-to-8 Decoder

Enables			Input			Output							
G_1	$\overline{G}_{2A}$	$\overline{G}_{2B}$	C	B	A	$\overline{Y}_7$	$\overline{Y}_6$	$\overline{Y}_5$	$\overline{Y}_4$	$\overline{Y}_3$	$\overline{Y}_2$	$\overline{Y}_1$	$\overline{Y}_0$
1	0	0	0	0	0	1	1	1	1	1	1	1	0
1	0	0	0	0	1	1	1	1	1	1	1	0	1
1	0	0	0	1	0	1	1	1	1	1	0	1	1
1	0	0	0	1	1	1	1	1	1	0	1	1	1
1	0	0	1	0	0	1	1	1	0	1	1	1	1
1	0	0	1	0	1	1	1	0	1	1	1	1	1
1	0	0	1	1	0	1	0	1	1	1	1	1	1
1	0	0	1	1	1	0	1	1	1	1	1	1	1
0	1	1	X	X	X	H	H	H	H	H	H	H	H

Table 6.9: Truth Table of The 2-to-4 Decoder

Enable	Input		Output			
$\overline{E}$	B	A	$\overline{Y}_3$	$\overline{Y}_2$	$\overline{Y}_1$	$\overline{Y}_0$
0	0	0	1	1	1	0
0	0	1	1	1	0	1
0	1	0	1	0	1	1
0	1	1	0	1	1	1
1	X	X	H	H	H	H

Fig. 6.13: Signals of 74LS139.

6.4 EVEN AND ODD MEMORY BANKS IN AN 8086-BASED SYSTEM

The 8086 microprocessor uses a 20-bit address to access memory. With a 20-bit address the processor can generate $2^{20} = 1$ mega address. The basic memory word size of the memories used in an 8086 system is 8-bit or 1-byte (i.e., in one memory location an 8-bit binary information can be stored). Hence, the physical memory space of 8086 is 1 MB (1 mega-byte).

For the programmer, the 8086 memory address space is a sequence of one mega-byte in which one location stores an 8-bit binary code/data and two consecutive locations stores 16-bit binary code/data. But physically (i.e., in the hardware), the 1 MB memory space is divided into two banks of 512 kB (512 kB + 512 kB = 1 MB). The two memory banks are called even (or lower) bank and odd (or upper) bank. The organization of even and odd memory banks in an 8086-based system is shown in Fig. 6.14.

The 8086-based system will have two sets of memory IC's: One set for the even bank and another set for the odd bank. The data lines D_0 - D_7 are connected to the even bank and the data lines D_8 - D_{15} are connected to the odd bank.

The even memory bank is selected by the address line A_0 and the odd memory bank is selected by the control signal $\overline{BHE}$. The memory banks are selected when these signals are **low** (active low). Any memory location in the memory bank is selected by the address lines A_1 to A_{19}.

The organization of memory into two banks and providing bank select signals allows the programmer to read/write the byte (8-bit) operand in any memory address through a 16-bit data bus. Also, it allows the programmer to read/write the word (16-bit) operand starting from even address or odd address.

The memory access for byte and word operand from the even and odd bank by the 8086 processor will be as follows:

Case i : Byte access from the even bank

For read/write operation of a byte in the even memory address, A_0 is asserted **low** and $\overline{BHE}$ is asserted **high** (i.e., $A_0 = 0$ and $\overline{BHE} = 1$). Now, the even bank alone is enabled and the data transfer take place through D_0-D_7 data lines.

Case ii : Byte access from odd bank

For read/write operation of a byte in the odd memory address, A_0 is asserted **high** and $\overline{BHE}$ is asserted **low** (i.e., $A_0 = 1$ and $\overline{BHE} = 0$). Now, the odd bank alone is enabled and the data transfer take place through D_8 - D_{15} data lines.

Case iii : Word access from even boundary

For read/write operation of a word (16-bit) in even boundary (i.e., low byte in even address and high byte in next address (odd address)), both A_0 and $\overline{BHE}$ are asserted **low** (i.e., $A_0 = 0$ and $\overline{BHE} = 0$). Now, both the memory banks are enabled simultaneously and the processor read/write the 16-bit operand in one bus cycle through D_0 - D_{15} data lines.

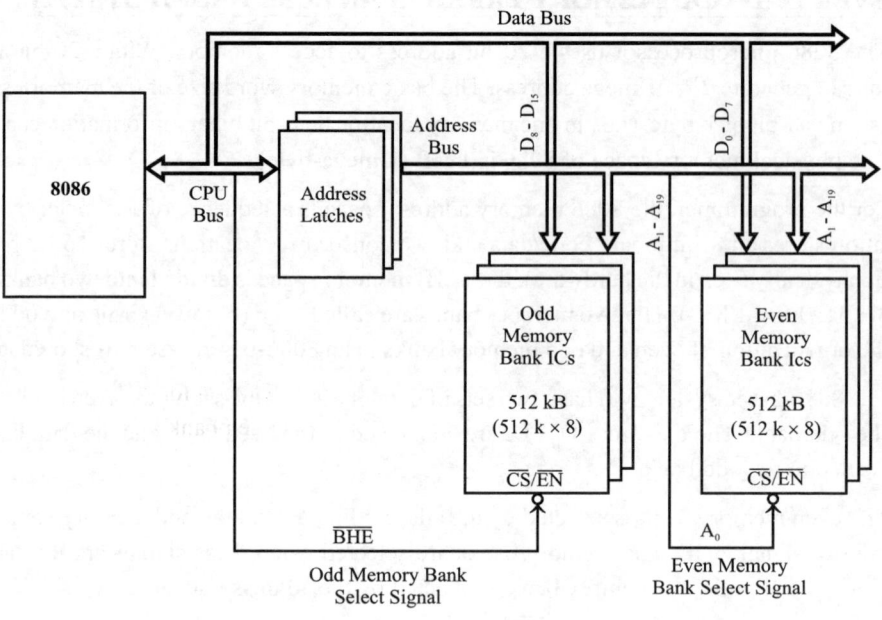

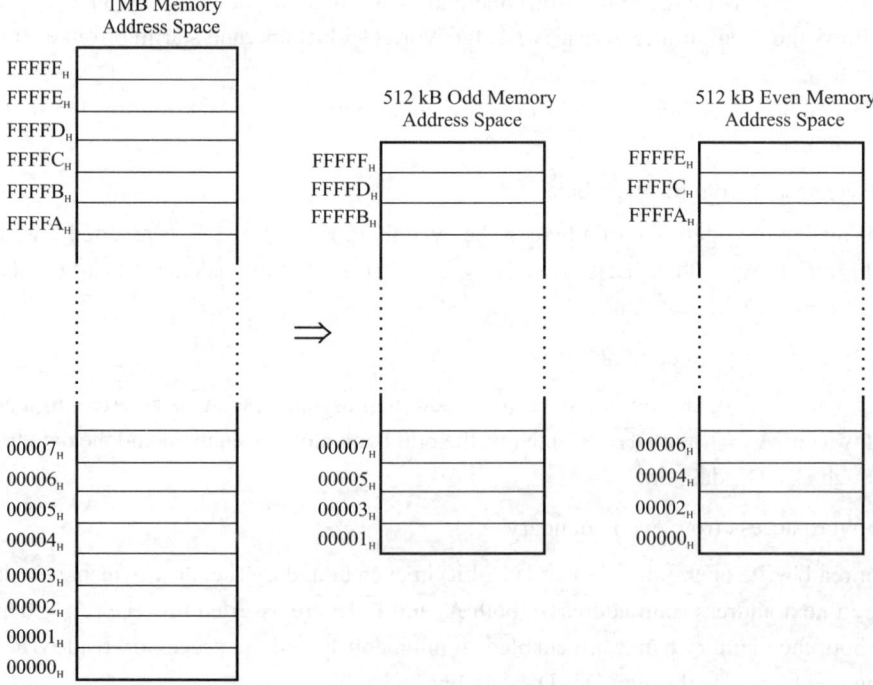

Fig. 6.14: Organization of even and odd memory banks in an 8086-based system.

Case iv : Word access from odd boundary

For read/write operation of a word (16-bit) in odd boundary (i.e., low byte in odd address and high byte in next address (even address)), the processor executes two bus cycles to read/write the word (16-bit) operand. In the first bus cycle, A_0 is asserted **high** and $\overline{BHE}$ is asserted **low** (i.e., $A_0 = 1$ and $\overline{BHE} = 0$). Now, the odd bank alone is enabled and the low byte of the 16-bit operand is read/write through D_8- D_{15} data lines.

In the second bus cycle, A_0 is asserted **low** and $\overline{BHE}$ is asserted **high** (i.e., $A_0 = 0$ and $\overline{BHE} = 1$). Now, the even bank alone is enabled and the high byte of the 16-bit operand is read/write through D_0-D_7 data lines.

The status of A_0 and $\overline{BHE}$ for byte and word memory access are listed in Table 6.10.

Table 6.10: Status of A_0 and $\overline{BHE}$ During Memory Access

Boundary	Operand type	Status of A_0	Status of $\overline{BHE}$	Data lines used for memory access	No. of bus cycle
Even	Byte	0	1	D_0 - D_7	One
Odd	Byte	1	0	D_8 - D_{15}	One
Even	Word	0	0	D_0 - D_{15}	One
Odd	Word	1	0	D_8 - D_{15}	First cycle
		0	1	D_0 - D_7	Second cycle

Note: 1. *The processor may access the low byte operand via the upper data lines and high byte operand via the lower data lines, but while placing the operand in the registers the processor places it in the appropriate locations.*

2. *When word operand's are accessed from the odd boundary, the instruction execution will take extra time due to two bus cycle memory access.*

6.5 MEMORY ORGANIZATION IN A MICROCOMPUTER SYSTEM

A microprocessor-based system requires both EPROM and RAM. Hence the available memory space has to be divided between EPROM and RAM. This choice depends on the system designer as well as on the application for which the system is designed. In 16-bit system when memory is organized as two banks, for proper system functioning, the system designer should allot equal address space in the odd and even banks for both EPROM and RAM.

Some systems may require a large memory space and so full memory space is utilized. But in some systems, the memory requirement may be less and in this case the full memory space is not utilized.

When full memory space is not utilized for memory, then the unused memory addresses can be used for addressing IO devices. Such IO devices are called memory-mapped IO devices and they can be accessed similar to that of a memory device.

The required EPROM memory capacity of the system can be implemented in one IC in an 8-bit system and two ICs in a 16-bit system (one for even and the other for odd bank) or in multiple ICs. Similarly, the RAM capacity of the system can be implemented in one IC in an 8-bit system and two ICs in a 16-bit system or in mulitple ICs. This choice depends on the availability of memory IC and the system designer. Some examples of memory organizations for 8085, 8086 and 8051 processor-based system are discussed in the following section.

6.5.1 Memory Organization in an 8085-Based System

A microprocessor-based system requires both EPROM and RAM. Hence the available memory space has to be divided between EPROM and RAM. The 8085 has 64 kB of addressable memory space and allotting this address space for EPROM and RAM depends on the system designer as well as the application for which the system is designed.

In 8085 system, the EPROM is mapped at the beginning of memory space (i.e., 0000_H address is allotted to EPROM memory location). Whenever the power supply is switched ON, the microprocessor chip will be reset. This power-on reset will be implemented by the system designer. When the processor is reset all the internal registers, flag register and program counter will be cleared. Hence, after a reset, the program counter will have an address 0000_H and so the processor starts fetching and executing the instruction stored at 0000_H.

The system designer will store the monitor program starting from the address 0000_H. The monitor program should be executed to initialize system peripherals whenever the system is switched ON. To enable automatic execution of monitor program, whenever the system is switched ON, the EPROM should be mapped from 0000_H location in 8085-based system. Monitor program is a permanent program written by the system designer to take care of system initializations. System initializations includes the following:

1.   **Programming 8279 for keyboard scanning and display refreshing.**

2.   **Programming peripheral ICs 8259, 8257, 8255, 8251, 8254, etc.**

3.   **Initializing stack.**

4.   **Display a message on display (output) device.**

5.   **Initializing interrupt vector table.**

Note: *8279 - Programmable keyboard/display controller.*	*8257 - DMA controller.*
8259 - Programmable interrupt controller.	*8251 USAR*
8255 - Programmable peripheral interface.	*8254 - Programmable timer.*

The total address space (64 kB) in an 8085 microprocessor and its allocation is shown in Fig. 6.15.

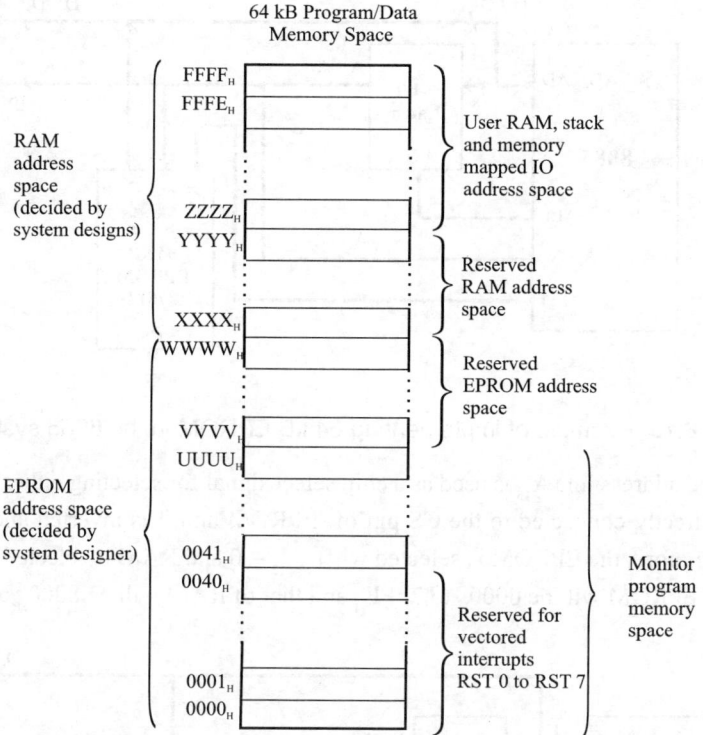

Fig. 6.15: Memory address space in an 8085 microprocessor.

The required EPROM memory capacity of the system can be implemented in one IC or in multiple ICs. Similarly the RAM capacity of the system can be implemented in one IC or in multiple ICs. This choice depends on the availability of memory IC and the system designer. Some examples of memory organizations for 8085 microprocessor-based system are discussed in this section.

Consider a system in which the full memory space 64 kB is utilized for EPROM memory. In this system the entire 16 address lines of the processor are connected to address input pins of memory IC in order to address the internal locations of memory and Chip Select ($\overline{CS}$) pin of EPROM is permanently tied to logic **low** (i.e., tied to ground) as shown in Fig. 6.16. Now the range of address for EPROM is 0000_H to $FFFF_H$.

Consider a system in which the available 64 kB memory space is equally divided between EPROM and RAM. Let us implement 32 kB memory capacity of EPROM using single IC 27256. Similarly, 32 kB RAM capacity is implemented using single IC 62256.

The 32 kB memory requires 15 address lines and so the address lines A_0 - A_{14} of the processor are connected to 15 address pins of both EPROM and RAM as shown in Fig. 6.16.

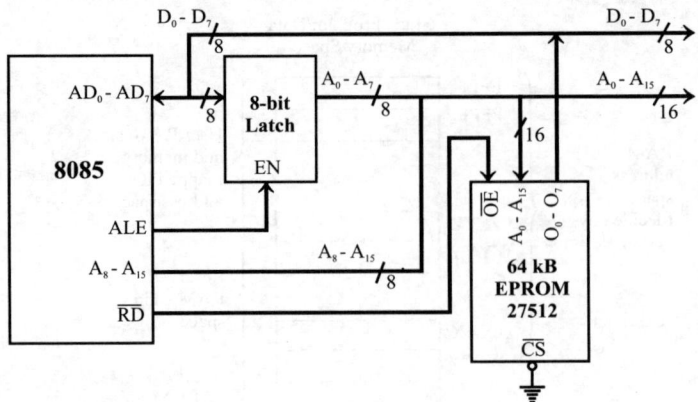

Fig. 6.16: Example of implementing 64 kB EPROM in the 8085 system.

The unused address line A_{15} is used as a chip select signal for selecting either EPROM or RAM. The A_{15} line is directly connected to the $\overline{CS}$ pin of EPROM and it is inverted and connected to $\overline{CS}$ pin of RAM. Therefore, the EPROM is selected when $A_{15} = 0$ and RAM is selected when $A_{15} = 1$. The address range of EPROM will be 0000_H to $7FFF_H$ and that of RAM will be 8000_H to $FFFF_H$.

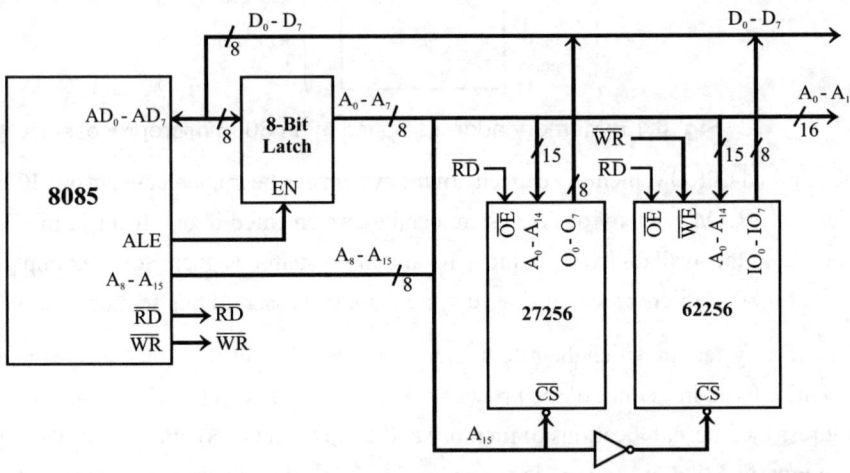

Fig. 6.17: Example of implementing 32 kB EPROM and 32 kB RAM in an 8085 system.

Consider a system in which 32 kB memory space is implemented using four 8 kB memory. Let two 8 kB memory be EPROM and the remaining two be RAM. Each 8 kB memory requires 13 address lines.

So, the address lines A_0 - A_{12} of the processor are connected to 13 address pins of all the memory ICs. The address lines A_{13} and A_{14} can be decoded using a 2-to-4 decoder to generate four chip select signals.

Table 6.11: Address Allocation for Memory ICs Shown in Fig. 6.18

Device	Binary address					Hexa address
	Decoder enable/input	Input to address pins of memory IC				
	A_{15} A_{14} A_{13} A_{12}	A_{11} A_{10} A_9 A_8	A_7 A_6 A_5 A_4	A_3 A_2 A_1 A_0		
8 kB EPROM - I	0 0 0 0	0 0 0 0	0 0 0 0	0 0 0 0		0000
	0 0 0 0	0 0 0 0	0 0 0 0	0 0 0 1		0001
	0 0 0 0	0 0 0 0	0 0 0 0	0 0 1 0		0002
	. . . .	. . . .	. . . .	. . . .		.
	0 0 0 1	1 1 1 1	1 1 1 1	1 1 1 1		1FFF
8 kB EPROM - II	0 0 1 0	0 0 0 0	0 0 0 0	0 0 0 0		2000
	0 0 1 0	0 0 0 0	0 0 0 0	0 0 0 1		2001
	0 0 1 0	0 0 0 0	0 0 0 0	0 0 1 0		2002
	. . . .	. . . .	. . . .	. . . .		.
	0 0 1 1	1 1 1 1	1 1 1 1	1 1 1 1		3FFF
8 kB RAM - I	0 1 0 0	0 0 0 0	0 0 0 0	0 0 0 0		4000
	0 1 0 0	0 0 0 0	0 0 0 0	0 0 0 1		4001
	0 1 0 0	0 0 0 0	0 0 0 0	0 0 1 0		4002
	. . . .	. . . .	. . . .	. . . .		.
	0 1 0 1	1 1 1 1	1 1 1 1	1 1 1 1		5FFF
8 kB RAM -II	0 1 1 0	0 0 0 0	0 0 0 0	0 0 0 0		6000
	0 1 1 0	0 0 0 0	0 0 0 0	0 0 0 1		6001
	0 1 1 0	0 0 0 0	0 0 0 0	0 0 1 0		6002
	. . . .	. . . .	. . . .	. . . .		.
	0 1 1 1	1 1 1 1	1 1 1 1	1 1 1 1		7FFF

These four chip select signals can be used to select one of the four memory IC at any one time. The address line A_{15} is used as an enable for the decoder. The simplified schematic of this memory organization is shown in Fig. 6.18 and address allotted to each memory IC is shown in Table 6.11.

Consider a system in which the 64 kB memory space is implemented using eight numbers of 8 kB memory. Each 8 kB memory requires 13 address lines and so the address line A_0-A_{12} of the processor are connected to 13 address pins of all the memory ICs.

The address lines A_{13}, A_{14} and A_{15} are decoded using a 3-to-8 decoder to generate eight chip select signals. These eight chip select signals can be used to select one of the eight memory IC at any one time. Design example-2 given at the end of this chapter is an example of implementing 64 kB address space using 8 numbers of 8 kB memory.

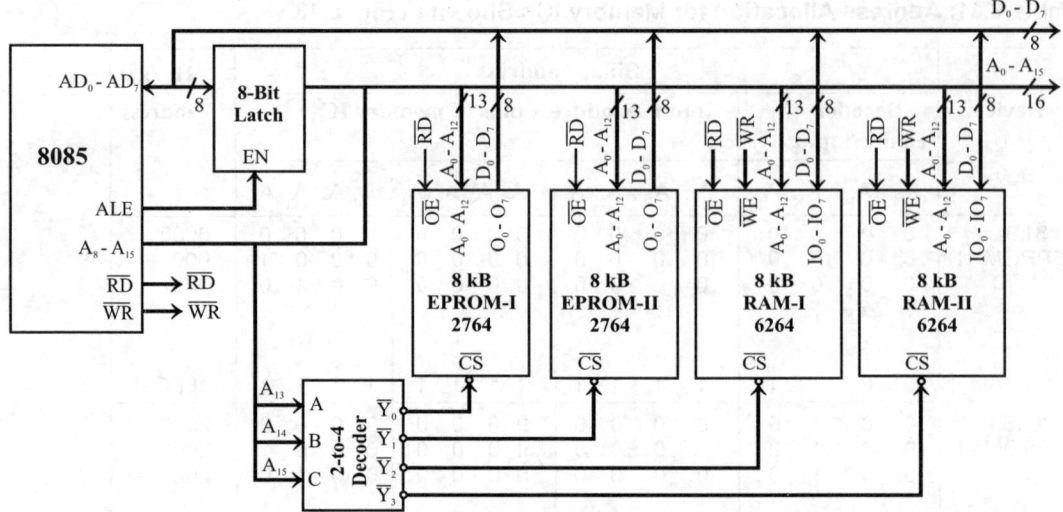

Fig. 6.18: Example of implementing 16 kB EPROM and 16 kB RAM in an 8085 system.

6.5.2 Memory Organization in an 8086-Based System

In 8086, the one megabyte (1MB) of addressable memory space is divided into two banks:Even memory bank and Odd memory bank. Each bank will have an addressable space of 512 kilobytes (512 kB). In an 8086 based system, the lower eight lines of data bus, D_0- D_7 are connected to even bank memory ICs and the upper eight lines of data bus, D_8- D_{15} are connected to the odd bank memory ICs. The even bank is selected by the address line A_0 and the odd bank is selected by the control signal $\overline{BHE}$

A microprocessor-based system requires both EPROM and RAM. Hence, the available memory space has to be divided between EPROM and RAM. This choice depends on the system designer as well as on the application for which the system is designed. For proper system functioning, the system designer should allot equal address space in odd and even banks for both EPROM and RAM.

In the 8086 system, the EPROMs are mapped at the end of memory space and RAMs are mapped at the beginning of memory space (i.e., 00000_H address is alloted to RAM and $FFFFF_H$ is alloted to EPROM). This organization will facilitate automatic execution of monitor program and creation of an interrupt vector table in RAM upon reset.

Whenever the power supply is switched ON, the microprocessor chip will be reset. This power-on reset will be implemented by the system designer. When the processor is reset, except CS-register all other internal registers, flag register and instruction pointer will be cleared. The CS-register is initialized with $FFFF_H$. Hence, after a reset, the processor starts fetching and executing the instruction stored at $FFFF0_H$. $[(CS) \times 16_{10} + (IP) = FFFF_H \times 16_{10} + 0000_H = FFFF0_H$.

The system designer will store the monitor program starting from the address $FFFF0_H$. The monitor program should be executed to initialize system peripherals whenever the system is switched ON. To enable automatic execution of monitor program, whenever the system is switched ON, the EPROM should be mapped at the end of memory space in an 8086-based system. The monitor program is a permanent program written by the system designer to take care of system initializations. The system initializations is similar to that of 8085-based system discussed in Section 6.5.1. The total address space (1 MB) in an 8086 microprocessor and its allocation is shown in Fig. 6.19.

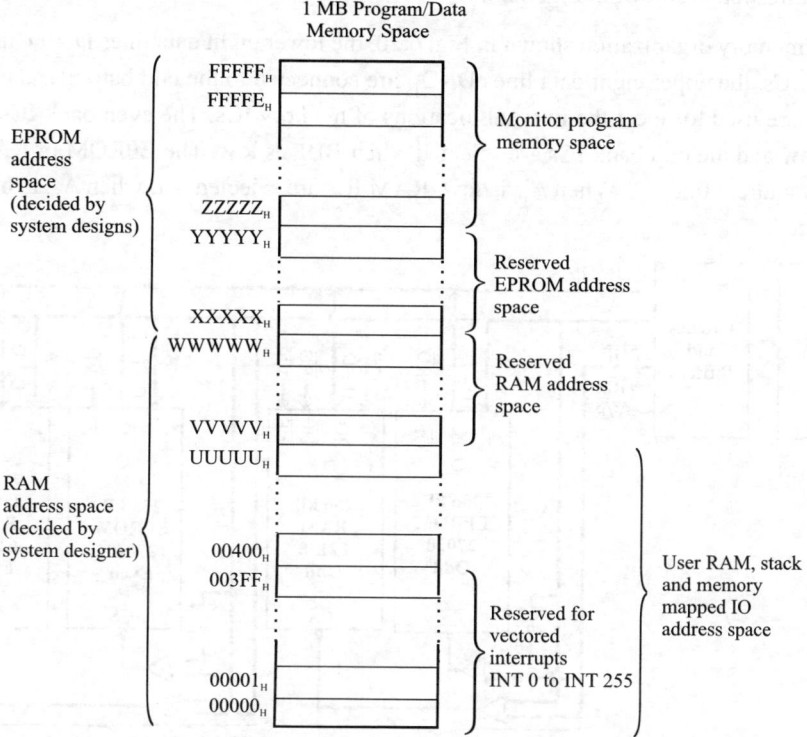

Fig. 6.19: Memory address space in an 8086 microprocessor.

Some systems may require large memory space and so full memory space is utilized. But in some systems, the memory requirement may be less and in this case, the full memory space is not utilized.

When full memory space is not utilized for memory then the unused memory addresses can be used for addressing IO devices. Such IO devices are called memory-mapped IO devices and they can be accessed similar to that of memory device.

The required EPROM memory capacity of the system can be implemented in two ICs (one for even and the other for odd bank) or in multiple ICs. Similarly, the RAM capacity of the system can be implemented in two ICs or in multiple ICs.

This choice depends on the availability of memory IC and the system designer. Some example memory organizations for the 8086-processor-based system are discussed in this section.

Consider a system in which the full memory space is utilized and the memory space is equally divided between EPROM and RAM. For this system, in each memory bank, 256 kB of EPROM memory space and 256 kB of RAM memory space is available. If the 256 kB memory space is implemented in single IC then we require two numbers of 256 kB RAM and two numbers of 256 kB EPROM. This memory organization is shown in Fig. 6.20.

In the memory organization shown in Fig. 6.20, the lower eight data lines D_0 - D_7 are connected to even bank ICs, the upper eight data lines D_8-D_{15} are connected to the odd bank ICs and the address lines A_1 - A_{18} are used to select the internal locations of memory ICs. The even bank ICs are selected when A_0 is **low** and the odd bank ICs are selected when $\overline{BHE}$ is **low**. The EPROM or RAM selection is decided by address line A_{19}. When A_{19} is **low**, RAM ICs are selected and when A_{19} is **high** EPROM ICs are selected.

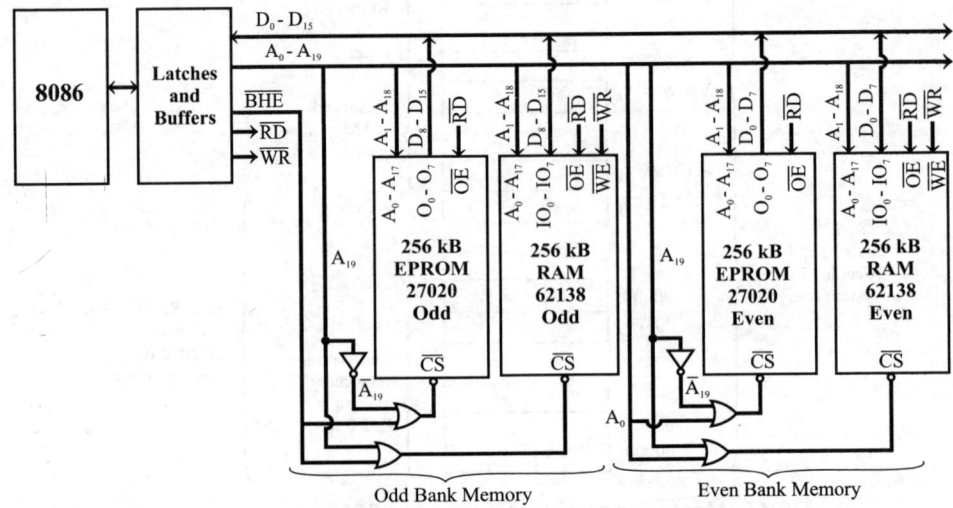

Fig. 6.20: Example of implementing full memory space in 8086-based system.

In this organization, the first half address space (i.e., first 512 kB) is implemented in RAM. The address range of RAM memory is 00000_H to $7FFFF_H$. The second half address space (i.e., second 512 kB) is implemented in EPROM. The address range of EPROM memory is 80000_H to $FFFFF_H$. The address allocation for RAM and EPROM locations are shown in Table 6.12.

> **Note:** *While alloting binary address to odd bank memory, A_0 is considered as 1.*

Let us discuss another example of memory organization with unequal space for EPROM and RAM and utilized only half the available memory space. Consider a system with 128 kB (2×64 kB = 128 kB) EPROM and 384 kB (6×64 kB = 384 kB) RAM.

For this system in each bank, 64 kB of EPROM and 192 kB of RAM is available. If the memory system is designed using 64 kB memory IC then we may require two 64 kB EPROM and six 64 kB RAM. This memory organization showin in Fig. 6.21.

Table 6.12: Address Allocation Table for Memory Organization Shown in Fig. 6.19

Memory device	Binary address							Hexa address	
	Input to memory address pins								
	A_{19}	A_{18} A_{17} A_{16}	A_{15} A_{14} A_{13} A_{12}	A_{11} A_{10} A_9 A_8	A_7 A_6 A_5 A_4	A_3 A_2 A_1	A_0		
RAM 256 kB Even	0	0 0 0	0 0 0 0	0 0 0 0	0 0 0 0	0 0 0	0	0 0 0 0 0	
	0	0 0 0	0 0 0 0	0 0 0 0	0 0 0 0	0 0 1	0	0 0 0 0 2	
	0	0 0 0	0 0 0 0	0 0 0 0	0 0 0 0	0 1 0	0	0 0 0 0 4	
	0	0 0 0	0 0 0 0	0 0 0 0	0 0 0 0	0 1 1	0	0 0 0 0 6	
	·	·	·	·	·	·	·	·	RAM Address Range 00000_H to $7FFFF_H$
	0	1 1 1	1 1 1 1	1 1 1 1	1 1 1 1	1 1 1	0	7 F F F E	
RAM 256 kB Odd	0	0 0 0	0 0 0 0	0 0 0 0	0 0 0 0	0 0 0	1	0 0 0 0 1	
	0	0 0 0	0 0 0 0	0 0 0 0	0 0 0 0	0 0 1	1	0 0 0 0 3	
	0	0 0 0	0 0 0 0	0 0 0 0	0 0 0 0	0 1 0	1	0 0 0 0 5	
	0	0 0 0	0 0 0 0	0 0 0 0	0 0 0 0	0 1 1	1	0 0 0 0 7	
	·	·	·	·	·	·	·	·	
	0	1 1 1	1 1 1 1	1 1 1 1	1 1 1 1	1 1 1	1	7 F F F F	
EPROM 256 kB Even	1	0 0 0	0 0 0 0	0 0 0 0	0 0 0 0	0 0 0	0	8 0 0 0 0	
	1	0 0 0	0 0 0 0	0 0 0 0	0 0 0 0	0 0 1	0	8 0 0 0 2	
	1	0 0 0	0 0 0 0	0 0 0 0	0 0 0 0	0 1 0	0	8 0 0 0 4	
	1	0 0 0	0 0 0 0	0 0 0 0	0 0 0 0	0 1 1	0	8 0 0 0 6	
	·	·	·	·	·	·	·	·	EPROM Address Range 80000_H to $FFFFF_H$
	1	1 1 1	1 1 1 1	1 1 1 1	1 1 1 1	1 1 1	0	F F F F E	
EPROM 256 kB Odd	1	0 0 0	0 0 0 0	0 0 0 0	0 0 0 0	0 0 0	1	8 0 0 0 1	
	1	0 0 0	0 0 0 0	0 0 0 0	0 0 0 0	0 0 1	1	8 0 0 0 3	
	1	0 0 0	0 0 0 0	0 0 0 0	0 0 0 0	0 1 0	1	8 0 0 0 5	
	1	0 0 0	0 0 0 0	0 0 0 0	0 0 0 0	0 1 1	1	8 0 0 0 7	
	·	·	·	·	·	·	·	·	
	1	1 1 1	1 1 1 1	1 1 1 1	1 1 1 1	1 1 1	1	F F F F F	

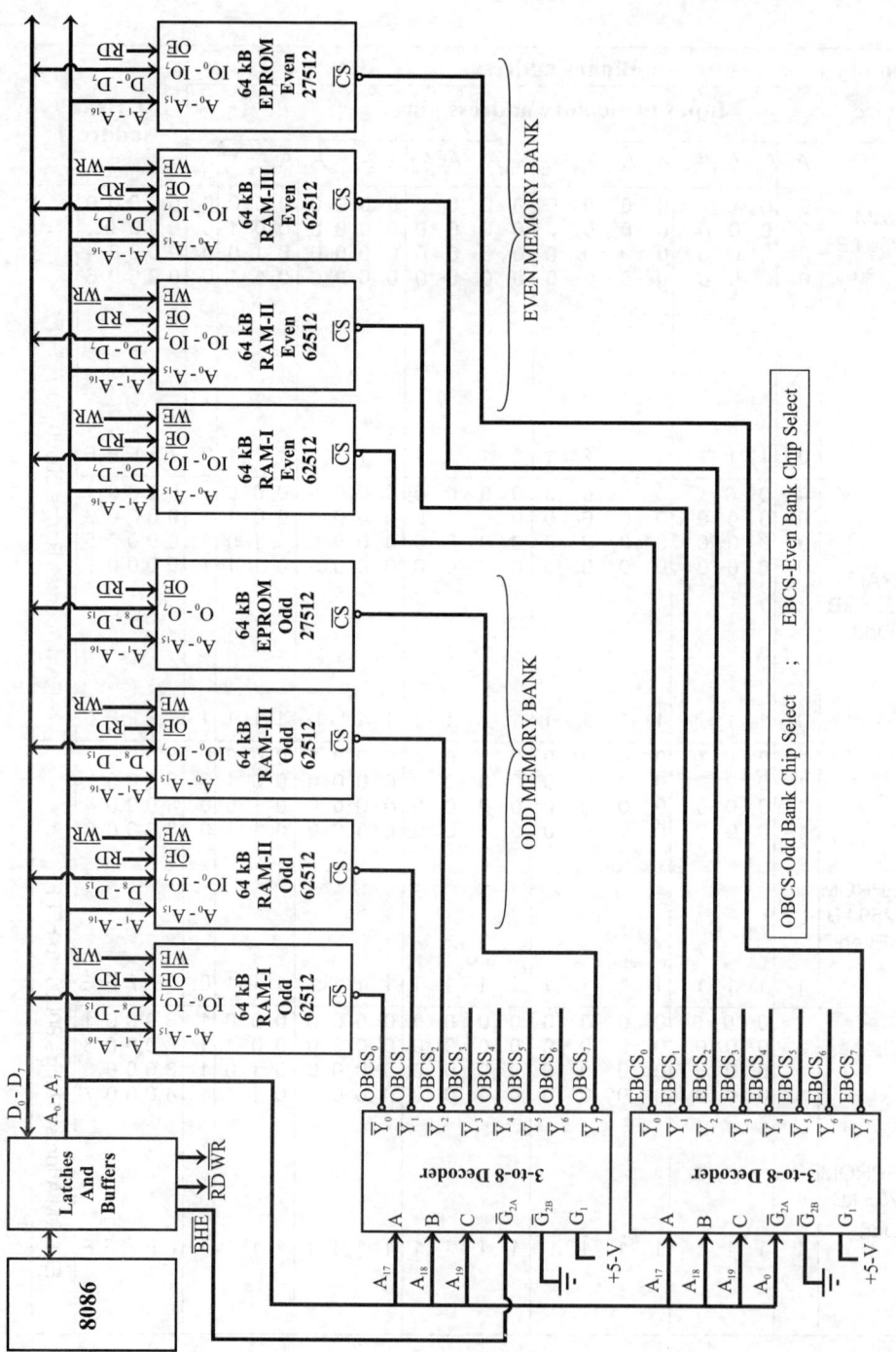

Fig. 6.21: Example of implementing 512 kB memory space in an 8086-based system.

Table 6.13: Address Allocation Table for Memory Organisation Shown in Fig. 6.20

Memory device	Decoder input A_{19} A_{18} A_{17}	Input to memory address pins A_{16}	A_{15} A_{14} A_{13} A_{12}	A_{11} A_{10} A_9 A_8	A_7 A_6 A_5 A_4	A_3 A_2 A_1	A_0	Hexa Address	
64 kB RAM-I Even	0 0 0 0 0 0 0 0 0 ⋮ 0 0 0	0 0 0 ⋮ 1	0 0 0 0 0 0 0 0 0 0 0 0 ⋮ 1 1 1 1	0 0 0 0 0 0 0 0 0 0 0 0 ⋮ 1 1 1 1	0 0 0 0 0 0 0 0 0 0 0 0 ⋮ 1 1 1 1	0 0 0 0 0 1 0 1 0 ⋮ 1 1 1	0 0 0 ⋮ 0	0 0 0 0 0 0 0 0 0 2 0 0 0 0 4 ⋮ 1 F F F E	RAM-I Address Range : 00000$_H$ to 1FFFF$_H$
64 kB RAM-I Odd	0 0 0 0 0 0 0 0 0 ⋮ 0 0 0	0 0 0 ⋮ 1	0 0 0 0 0 0 0 0 0 0 0 0 ⋮ 1 1 1 1	0 0 0 0 0 0 0 0 0 0 0 0 ⋮ 1 1 1 1	0 0 0 0 0 0 0 0 0 0 0 0 ⋮ 1 1 1 1	0 0 0 0 0 1 0 1 0 ⋮ 1 1 1	1 1 1 ⋮ 1	0 0 0 0 1 0 0 0 0 3 0 0 0 0 5 ⋮ 1 F F F F	
64 kB RAM-II Even	0 0 1 0 0 1 0 0 1 ⋮ 0 0 1	0 0 0 ⋮ 1	0 0 0 0 0 0 0 0 0 0 0 0 ⋮ 1 1 1 1	0 0 0 0 0 0 0 0 0 0 0 0 ⋮ 1 1 1 1	0 0 0 0 0 0 0 0 0 0 0 0 ⋮ 1 1 1 1	0 0 0 0 0 1 0 1 0 ⋮ 1 1 1	0 0 0 ⋮ 0	2 0 0 0 0 2 0 0 0 2 2 0 0 0 4 ⋮ 3 F F F E	RAM-II Address Range : 20000$_H$ to 3FFFF$_H$
64 kB RAM-II Odd	0 0 1 0 0 1 0 0 1 ⋮ 0 0 1	0 0 0 ⋮ 1	0 0 0 0 0 0 0 0 0 0 0 0 ⋮ 1 1 1 1	0 0 0 0 0 0 0 0 0 0 0 0 ⋮ 1 1 1 1	0 0 0 0 0 0 0 0 0 0 0 0 ⋮ 1 1 1 1	0 0 0 0 0 1 0 1 0 ⋮ 1 1 1	1 1 1 ⋮ 1	2 0 0 0 1 2 0 0 0 3 2 0 0 0 5 ⋮ 3 F F F F	
64 kB RAM-III Even	0 1 0 0 1 0 0 1 0 ⋮ 0 1 0	0 0 0 ⋮ 1	0 0 0 0 0 0 0 0 0 0 0 0 ⋮ 1 1 1 1	0 0 0 0 0 0 0 0 0 0 0 0 ⋮ 1 1 1 1	0 0 0 0 0 0 0 0 0 0 0 0 ⋮ 1 1 1 1	0 0 0 0 0 1 0 1 0 ⋮ 1 1 1	0 0 0 ⋮ 0	4 0 0 0 0 4 0 0 0 2 4 0 0 0 4 ⋮ 5 F F F E	RAM-III Address Range : 40000$_H$ to 5FFFF$_H$
64 kB RAM-III Odd	0 1 0 0 1 0 0 1 0 ⋮ 0 1 0	0 0 0 ⋮ 1	0 0 0 0 0 0 0 0 0 0 0 0 ⋮ 1 1 1 1	0 0 0 0 0 0 0 0 0 0 0 0 ⋮ 1 1 1 1	0 0 0 0 0 0 0 0 0 0 0 0 ⋮ 1 1 1 1	0 0 0 0 0 1 0 1 0 ⋮ 1 1 1	1 1 1 ⋮ 1	4 0 0 0 1 4 0 0 0 3 4 0 0 0 5 ⋮ 5 F F F F	
64 kB EPROM Even	1 1 1 1 1 1 1 1 1 ⋮ 1 1 1	0 0 0 ⋮ 1	0 0 0 0 0 0 0 0 0 0 0 0 ⋮ 1 1 1 1	0 0 0 0 0 0 0 0 0 0 0 0 ⋮ 1 1 1 1	0 0 0 0 0 0 0 0 0 0 0 0 ⋮ 1 1 1 1	0 0 0 0 0 1 0 1 0 ⋮ 1 1 1	0 0 0 ⋮ 0	E 0 0 0 0 E 0 0 0 2 E 0 0 0 4 ⋮ F F F F E	EPROM Address Range : E0000$_H$ to FFFFF$_H$
64 kB EPROM Odd	1 1 1 1 1 1 1 1 1 ⋮ 1 1 1	0 0 0 ⋮ 1	0 0 0 0 0 0 0 0 0 0 0 0 ⋮ 1 1 1 1	0 0 0 0 0 0 0 0 0 0 0 0 ⋮ 1 1 1 1	0 0 0 0 0 0 0 0 0 0 0 0 ⋮ 1 1 1 1	0 0 0 0 0 1 0 1 0 ⋮ 1 1 1	1 1 1 ⋮ 1	E 0 0 0 1 E 0 0 0 3 E 0 0 0 5 ⋮ F F F F F	

Total RAM Address Range : 00000$_H$ to 5FFFF$_H$

In the memory organization shown in Fig. 6.21, the lower eight data lines, D_0-D_7 are connected to even bank ICs, the upper eight data lines, D_8-D_{15} are connected to the odd bank ICs and the address lines A_1-A_{16} are used to select the internal locations of memory ICs.

The address lines A_{17}, A_{18} and A_{19} are decoded to generate chip select signals. Here, two 3-to-8 decoders are employed for generating separate chip select signals for even and odd bank ICs. Each 3-to-8 decoder will generate eight chip select signals and in this, four signals are used as chip select signals for four memory ICs of a bank. The remaining four signals are reserved for future expansion (or can be used for IO devices).

In the memory organization shown in Fig. 6.21, the first 384 kB of memory space is implemented in RAM and the last 128 kB of memory space is implemented in EPROM. The address range of RAM is 00000_H to $5FFFF_H$ and the address range of EPROM is $E0000_H$ to $FFFFF_H$. The address allocation for RAM and EPROM locations are shown in Table 6.13.

6.5.3 Memory Organization in an 8051-Based System

A microcontroller-based system requires both EPROM and RAM. The EPROM is required for permanent program and permanent data storage. The RAM is required for temporary data storage and stack. The 8051 has 64 kB program memory address space and 64 kB data memory address space.

The microcontroller can only read from the program memory and the signal PSEN is used as read control for reading the program memory. Therefore read only memories like ROM/EPROM/EEPROM can be employed as the program memory.

The microcontroller can read and write with data memory. It has a separate read control signal, $\overline{RD}$ and write control signal, $\overline{WR}$ for reading and writing with data memory respectively. Hence read-write memories like static RAM can be employed as data memory. The interfacing of external memory to the 8051 microcontroller-based system bus and memory organization of an 8051 microcontroller are shown in Figs. 6.21 and 6.22 respectively.

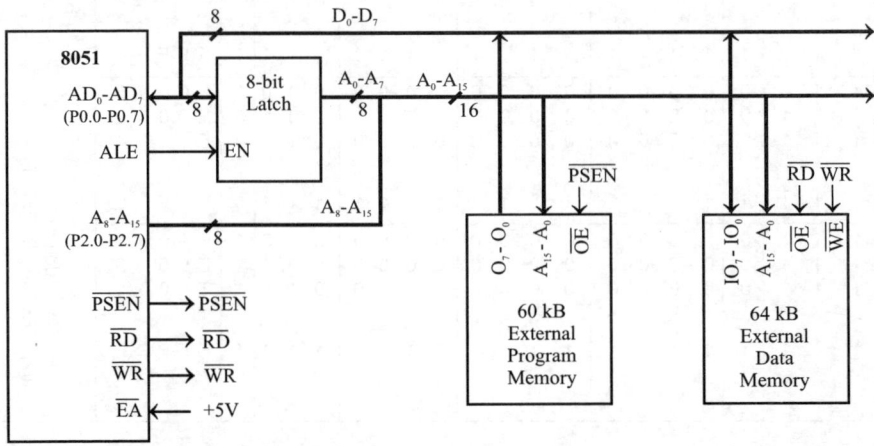

Fig. 6.22: Interfacing external memory to an 8051 microcontroller.

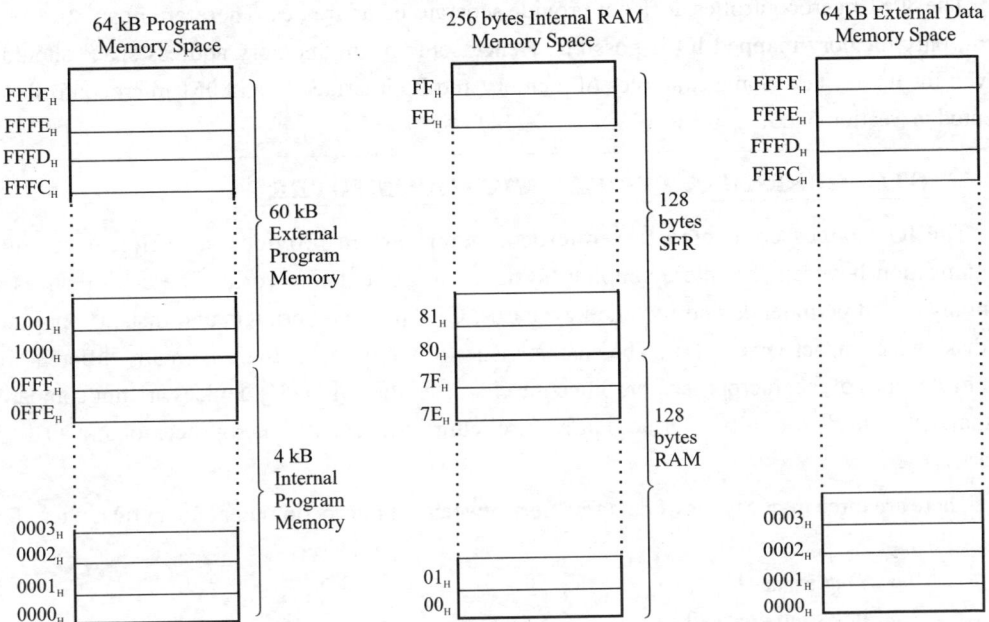

Fig. 6.23: Memory organization in an 8051 microcontroller.

In an 8051 microcontroller the entire 64 kB data memory space is external. The address range of the external data memory is 0000_H to $FFFF_H$. Apart from the external data memory the 8051 has 256 bytes of internal data memory in which the first 128 bytes are called RAM and the next 128 byte are called SFR. The address range of SFRs and internal RAM are 00_H to FF_H. The organization of 256 bytes of internal data memory (SFRs and internal RAM) are discussed in Chapter-2, Section 2.2.3.

The 8051, has a 4 kB internal ROM which can be mapped to the first 4 kB address space of the program memory if the $\overline{EA}$ pin is tied **high** or tied to V_{CC} (+5V). This means that the usage of internal ROM in the 8051 is optional. When $\overline{EA}$ is tied **high** or tied to V_{CC} (+5V) the internal 4 kB ROM is mapped as the first 4 kB of the program memory address space and when $\overline{EA}$ is tied **low** or grounded (0V), the internal ROM is ignored or cannot be accessed.

When $\overline{EA}$ is tied **high**, the internal 4 kB ROM will be mapped as the program memory in the address range 0000_H to $0FFF_H$ and the external program memory 60 kB will have the address range 1000_H to $FFFF_H$. When $\overline{EA}$ is tied to the ground, the entire 64 kB program memory address space is external with an address range 0000_H to $FFFF_H$. In an 8051-based system, it is possible to have a single memory bank for both program and data, to minimize the cost of the system where memory requirement is less. When a single memory bank is provided, the combined read control signal is generated by logically ANDing the $\overline{PSEN}$ and $\overline{RD}$ signals. When a single bank is provided the total memory capacity is 64 kB and this address space is common to the program memory and the data memory.

The system designer has to partition the address space for program and data. Apart from this, the 256 bytes internal memory can be accessed as data memory using an 8-bit address.

The 8051 microcontroller does not provide separate IO addresses. Therefore in an 8051 based system, only memory-mapped IO is possible. Hence some of the memory address space should be reserved for IO devices. Some examples of memory and IO interface to an 8051 microcontroller are presented in Section 6.7.3.

6.6   IO STRUCURE OF A TYPICAL MICROCOMPUTER

The IO devices connected to a microcomputer system provides an efficient means of communication between the microcomputer system and the outside world. These IO devices are commonly called peripherals and include keyboards, CRT displays, printers and disks (floppy disk, hard disk and Compact Disc (CD)). The characteristics of the IO devices are normally different from the characteristics of the microprocessor. Since the characteristics of the IO devices are not compatible with that of the microprocessor, interface hardware circuitry between the microprocessor and IO device are necessary.

There are three major types of data transfer between the microcomputer and an IO device. They are as follows:

- **Programmed IO**
- **Interrrupt driven IO**
- **Direct memory access (DMA)**

In programmed IO the data transfer is accomplished through an IO port and controlled by software. In interrupt driven IO, the IO device will interrupt the processor and initiate data transfer. In DMA, the data transfer between memory and IO can be performed by bypassing the microprocessor. Each type of data transfer scheme mentioned above, includes different methods of data transfer schemes. Fig. 6.24 shows all the types of data transfer schemes in a microcomputer and it can also be called **IO structure of a microcomputer**.

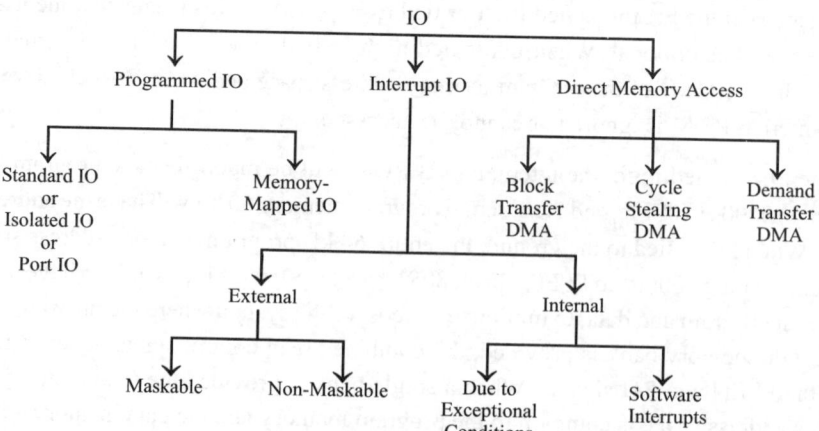

Fig. 6.24: IO structure of a typical microcomputer.

6.6.1 Interfacing IO and Peripheral Devices

The IO devices are generally slow devices. So, they are connected to the system bus through ports. The ports are buffer IC which is used to temporarily hold the data transmitted from the microprocessor to IO device or to hold the data transmitted from IO device to the microprocessor.

To data transfer from the input device to the processor the following operations are performed:

1. **The input device will load the data to the port.**
2. **When the port receives the data, it sends message to the processor to read the data.**
3. **The processor will read the data from the port.**
4. **After the data has been read by the processor the input device will load the next data into the port.**

To data transfer from the processor to the output device the following operations are performed:

1. **The processor will load the data to the port.**
2. **The port will send a message to the output device to read the data.**
3. **The output device will read the data from the port.**
4. **After the data has been read by the output device the processor can load the next data to the port.**

6.6.2 INTEL IO Port Devices

The various INTEL IO port devices are 8212, 8155/8156, 8255, 8355 and 8755.

INTEL 8212

The 8212 is a 24-pin IC. It consists of eight number of D-type latches, each followed by a tristate buffer. It has 8-input lines DI_1 to DI_8 and 8-output lines DO_1 to DO_8. The 8212 can be used as an input or output device and the function is determined by the mode pin. However, it cannot be used simultaneously for input and output in the same circuit, since its mode pin is hardwired. It has 2-device select signals $\overline{DS}_1$ and DS_2. The port is selected by the processor by sending appropriate address to device select pins.

Output Port : When $MD = 1$, $\overline{DS}_1 = 0$ and $DS_2 = 1$

Input Port : When $MD = 0$, $\overline{DS}_1 = 0$ and $DS_2 = 1$

INTEL 8155

INTEL 8155 has 256×8 static RAM, two numbers of 8-bit parallel IO port (ports A and B), one number of 6-bit parallel IO port (port-C) and 14-bit timer. The ports A and B can be programmed to work as simple or handshake input or output port. If port-A and port-B are simple ports then port-C can be used as input or output port. The timer can be programmed to operate in four different modes. INTEL 8155 requires six internal addresses and has one logic **low** Chip Select pin ($\overline{CS}$). The addresses of internal devices of 8155 are listed in Table 6.14.

Table 6.14: Internal Address of 8155/8156

Internal device	A_2	A_1	A_0
Control Register/ Status Register	0	0	0
Port-A	0	0	1
Port-B	0	1	0
Port-C	0	1	1
LSB of Timer	1	0	0
MSB of Timer	1	0	1

INTEL 8156

INTEL 8156 is same as 8155, but it has logic **high** Chip Select ($\overline{CS}$), i.e., the chip is selected when $\overline{CS} = 1$.

INTEL 8255

It has 3 numbers of 8-bit parallel IO ports (ports A, B and C). Port-A can be programmed in mode 0, mode1 or mode-2 as input or output port. Port-B can be programmed in mode-1 and mode-2 as IO port. When ports A and B are in mode-0, port-C can be used as IO port. The individual pins of port-C can be set or reset. INTEL 8255 requires four internal addresses and has one logic **low** Chip Select ($\overline{CS}$) pin. The address of internal devices of 8255 are listed in Table 6.15.

Table 6.15: Internal Address of 8255

Internal device	A_1	A_0
Port-A	0	0
Port-B	0	1
Port-C	1	0
Control Register	1	1

INTEL 8355

It has 2 k × 8 ROM and two numbers of 8-bit port (Ports A and B). The individual pins of ports A and B can be programmed as input or output lines by sending a control word to DDR (**D**ata **D**irection **R**egister). The address of internal devices of 8355 are listed in Table 6.16. The 8355 requires four internal addresses and has one logic **low** Chip Select ($\overline{CS}$) pin.

Table 6.16: Internal Address of 8355/8755

Internal device	A_1	A_0
Port-A	0	0
Port-B	0	1
DDR A	1	0
DDR B	1	1

INTEL 8755

Same as 8355 but has 2 k × 8 EPROM.

6.6.3 INTEL Peripheral Devices

Apart from port ICs, dedicated programmable controller/peripheral ICs are used in the system for various activities. Some of the controller/peripheral devices used in the 8085 system and their functions and internal addresses are listed in Table 6.17.

Table 6.17: Functions and Internal Addresses of Peripheral Devices

Device	Function	Internal addresses
INTEL 8279	Keyboard/display controller. Used for keyboard scanning and display refreshing.	**Two-internal addresses** $A_0 = 0 \rightarrow$ Data register $A_0 = 1 \rightarrow$ Control register
INTEL 8257 or INTEL 8237	DMA controller. Used for supporting DMA access to the IO device. It acts as a master during the DMA mode. It is a slave device during programming mode.	**Sixteen-internal addresses** A_3 A_2 A_1 A_0 $\quad$ 0 $\quad$ 0 $\quad$ 0 $\quad$ 0 $\quad$ 0 $\quad$ 0 $\quad$ 0 $\quad$ 1 $\quad$. $\quad$. $\quad$. $\quad$. $\quad$ 1 $\quad$ 1 $\quad$ 1 $\quad$ 1
INTEL 8259	Interrupt controller. Used to expand the hardware interrupt INTR to eight interrupts in an 8085-based system and 256 interrupts in an 8086-based system.	**Two-internal addresses** $A_0 = 0$ $A_0 = 1$
INTEL 8253/ 8254	Programmable timer. Used in the system to produce various timing signals. It has three independent counters and can be programmed in six operating modes.	**Four-internal addresses** $\quad\quad\quad A_1$ A_0 Counter-0 $\quad$ 0 $\quad$ 0 Counter-1 $\quad$ 0 $\quad$ 1 Counter-2 $\quad$ 1 $\quad$ 0 Control Register $\quad$ 1 $\quad$ 1
INTEL 8251 (USART)	Universal **S**ynchronous/**A**synchronous **R**eceiver **T**ransmitter. Used for serial data communication.	**Two-internal addresses** C/D = 0 $\rightarrow$ Data register C/D = 1 $\rightarrow$ Control register

6.6.4 IO Mapping

The port and peripheral devices will have one logic **low/high** chip select pin. The processor can access the port/peripheral device by supplying internal address and chip select signals. Therefore, the port and peripheral device interfacing (IO interfacing) deals with allocation of various internal addresses and generation of chip select signals.

There are two ways of interfacing IO devices in 8085-based system.

1. **Memory-mapped IO device.**

2. **Standard IO-mapped IO device or Isolated IO mapping.**

> *Note:*   *The interfacing of IO ports and controller/peripheral ICs are commonly referred as IO device mapping.*

The 8085 and 8086 microprocessors supports both memory-mapped IO and IO-mapped IO. The 8051 microcontroller supports only memory-mapped IO. Hence in a 8051-based system, some of the memory addresses should be reserved for IO devices, and in these systems the IO devices are interfaced similar to that of memory devices.

6.6.5 IO Mapping in an 8085 Microprocessor-Based System

The two methods of interfacing IO devices in an 8085-based system are: Memory mapping of IO device and Standard IO mapping of IO device or Isolated IO mapping.

In memory mapping of IO devices the ports are allotted a 16-bit address like that of the memory location. Some of the chip select signals generated to select memory ICs are used for selecting the IO port devices.

Hence, the processor treats the IO ports as memory locations for reading and writing (i.e., the devices which are mapped by memory mapping are accessed by executing memory read cycle or memory write cycle).

In standard IO mapping or isolated IO mapping, a separate 8-bit address is allotted for the IO ports and the peripheral ICs. The processor differentiates the IO-mapped devices, from the memory-mapped devices in the following ways:

1. **For accessing the IO-mapped devices the processor executes IO read or write cycle.**

2. **During IO read or write cycle, the 8-bit address is placed on both low order address lines and the high order address lines.**

3. **IO/$\overline{\text{M}}$ is asserted high to indicate the IO operation (for read as well as write).**

A 8085 processor does not provide separate read ($\overline{\text{RD}}$) and write ($\overline{\text{WR}}$) signals for memory and IO devices. But it differentiates the memory and IO device accessed by IO/$\overline{\text{M}}$ signal.

The three signals $\overline{\text{RD}}$, $\overline{\text{WR}}$ and IO/$\overline{\text{M}}$ can be decoded as shown in Fig. 6.25 to provide separate read and write control signals for IO devices and memory devices.

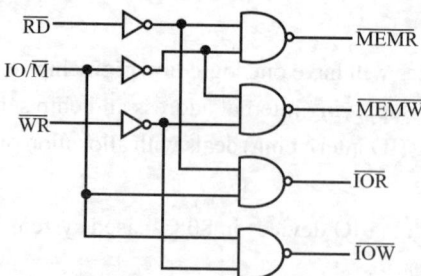

Fig. 6.25: Circuit to generate separate read and write signals for memory and IO devices in an 8085-based system.

When the devices are IO-mapped, then only IN and OUT instructions have to be used for data transfer between the device and the processor. For the IO-mapped devices a separate decoder should be used to generate the required chip select signals.

Table 6.18: Comparison of Memory Mapping and IO Mapping of IO Device in an 8085-Based System

Memory mapping of IO device	IO mapping of IO device
1. 16-bit addresses are provided for IO devices.	1. 8-bit addresses are provided for IO devices.
2. The devices are accessed by memory read or memory write cycles.	2. The devices are accessed by IO read or IO write cycle. During these cycles, the 8-bit address is available on both low order address lines and high order address lines.
3. The IO ports or peripherals can be treated like memory locations and so all instructions related to memory can be used for data transfer between the IO device and the processor.	3. Only IN and OUT instructions can be used for data transfer between the IO device and the processor.
4. In memory-mapped ports, the data can be moved from any register to the ports and vice versa.	4. In IO-mapped ports, the data transfer can take place only between the accumulator and the ports.
5. When memory mapping is used for IO devices, the full memory address space cannot be used for addressing memory. Hence memory mapping is useful only for small systems, where the memory requirement is less.	5. When IO mapping is used for IO devices, then the full memory address space can be used for addressing the memory. Hence it is suitable for systems which requires a large memory capacity.
6. In memory-mapped IO devices, a large number of IO ports can be interfaced.	6. In IO mapping, only 256 ports ($2^8 = 256$) can be interfaced.
7. For accessing memory-mapped devices, the processor executes the memory read or write cycle. During this cycle, IO/$\overline{M}$ is asserted **low**(IO/$\overline{M}$= 0)	7. For accessing the IO-mapped devices, the processor executes the IO read or write cycle. During this cycle, IO/$\overline{M}$ is asserted **high** (IO/$\overline{M}$=1).

6.6.6 IO Mapping in an 8086 Microprocessor-Based System

In memory mapping of IO devices, the ports are allotted a 20-bit address like that of memory location. Some of the chip select signals generated to select memory ICs are used for selecting the IO port devices. Hence, the processor treats the IO ports as memory locations for reading and writing (i.e., the devices which are mapped by memory mapping are accessed by executing memory read cycle or memory write cycle).

In standard IO mapping or isolated IO mapping, a separate 8-bit or 16-bit address is allotted for IO ports and the peripheral ICs. The processor differentiates the IO-mapped devices, from the memory-mapped devices in the following ways:

1. **For accessing the IO-mapped devices, the processor executes IO read or write cycle.**

2. **During IO read or write cycle, the 8-bit or 16-bit address is placed on low order address lines and the high order address lines are asserted zero.**

3. **M/$\overline{\text{IO}}$ is asserted low to indicate the IO operation (for read as well as write).**

The 8086 processor does not provide separate read ($\overline{\text{RD}}$) and write ($\overline{\text{WR}}$) signals for memory and IO devices. But it differentiates the memory and IO device accesses by M/$\overline{\text{IO}}$ signal. The three signals $\overline{\text{RD}}$, $\overline{\text{WR}}$ and M/$\overline{\text{IO}}$ can be decoded as shown in Fig. 6.26 to provide separate read and write control signals for IO devices and memory devices.

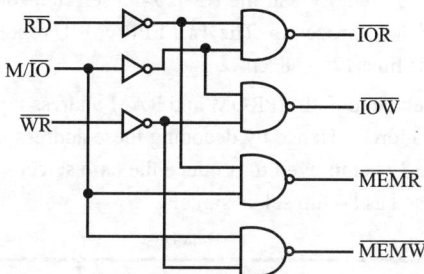

Fig. 6.26: Circuit to generate separate read and write signals for memory and IO devices in an 8086 system.

Table 6.19: Comparison of Memory Mapping and IO Mapping of IO Device in an 8086-Based System

Memory mapping of IO device	IO mapping of IO device
1. 20-bit addresses are provided for IO devices.	1. 8-bit or 16-bit addresses are provided for IO devices.
2. The IO ports or peripherals can be treated like memory locations and so all instructions related to memory can be used for data transfer between IO device and the processor.	2. Only IN and OUT instructions can be used for data transfer between IO device and the processor.
3. In memory-mapped ports, the data can be moved from any register to the ports and vice versa.	3. In IO-mapped ports, the data transfer can take place only between the accumulator and the ports.
4. When memory mapping is used for IO devices, the full memory address space cannot be used for addressing memory. Hence, memory mapping is useful only for small systems, where the memory requirement is less.	4. When IO mapping is used for IO devices, then the full memory address space can be used for addressing memory. Hence it is suitable for systems which requires large memory capacity.
5. For accessing the memory-mapped devices, the processor executes memory read or write cycle. During this cycle M/IO is asserted **high**.	5. For accessing the IO-mapped devices, the processor executes IO read or write cycle. During this cycle M/IO is asserted **low**.

When the devices are IO-mapped then only IN and OUT instructions have to be used for data transfer between the device and the processor. For IO-mapped devices, a separate decoder should be used to generate the required chip select signals.

6.7 MEMORY AND IO INTERFACE EXAMPLES

6.7.1 Examples of Memory and IO Interface in 8085-Based System

DESIGN EXAMPLE 85.1

Interface two numbers of 4 kB EPROM and one number of 8 kB RAM with 8085 processor. Explain the interface diagram and allocate binary addresses to memory ICs.

Solution

The IC 2732 is selected for EPROM memory and the IC 6264 is selected for RAM memory. Both the memory IC's have time compatibility with 8085 processor. The $4\,kB$ EPROM IC requires 12 address lines ($2^{12} = 4$ k). The 8 kB RAM IC requires 13 address lines ($2^{13} = 8$ k).

The address lines $A_0 - A_{11}$ are connected to both EPROM and RAM address input pins. The address lines A_{13}, A_{14} and A_{15} are not used for memory address. Hence by decoding these address lines we can generate chip select signals. The 3-to-8 decoder, 74LS138 is employed to produce the chip select signals for the system. The decoder has 8-output lines which can be used as 8-chip select signals.

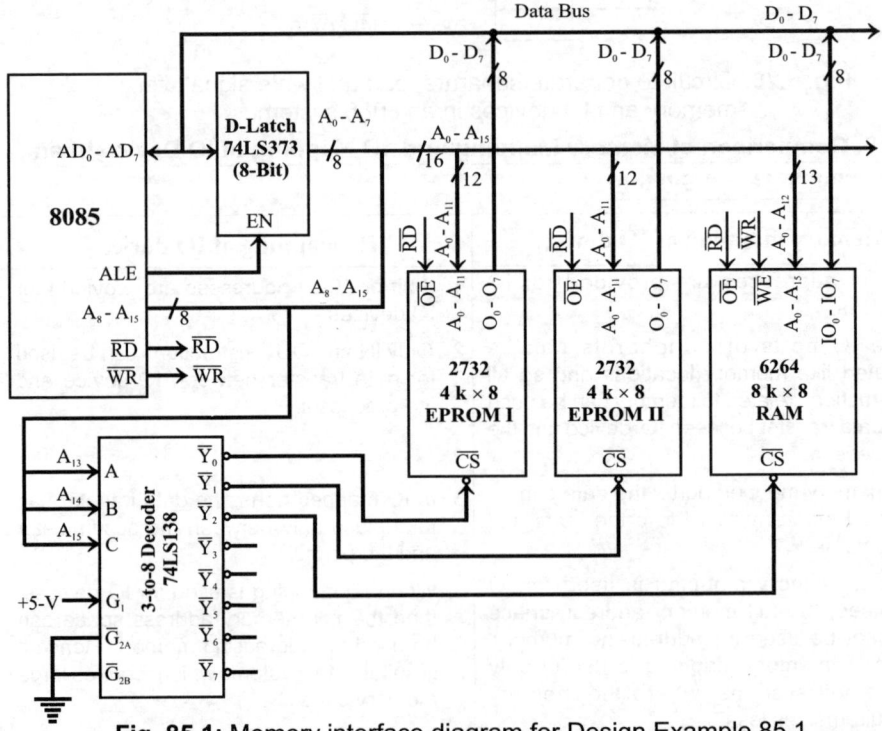

Fig. 85.1: Memory interface diagram for Design Example 85.1.

In this, three chip select signals are used for selecting memory ICs and the remaining five can be used for selecting other peripheral ICs in the system or for future expansion of the memory capacity. The interface diagram is shown in Fig. 85.1. Address allotted to memory ICs are shown in Table 85.1.

The EPROM's are mapped in the beginning of memory space. The remaining addresses can be allotted to RAM's. The EPROM memory is mapped from 0000_H to $0FFF_H$ and 2000_H to $2FFF_H$. The RAM memory is mapped from 4000_H to $5FFF_H$.

Table 85.1: Address Allocation Table for Design Example 85.1

Memory IC	Decoder input A_{15} A_{14} A_{13} A_{12}	Input to memory address pins A_{11} A_{10} A_9 A_8	A_7 A_6 A_5 A_4	A_3 A_2 A_1 A_0	Hexa address
EPROM I 2732	0 0 0 X 0 0 0 X	0 0 0 0 0 0 0 0	0 0 0 0 0 0 0 0	0 0 0 0 0 0 0 1	0000 0001
	. . . . 0 0 0 X	. . . . 1 1 1 1	. . . . 1 1 1 1	. . . . 1 1 1 1	. 0FFF
EPROM II 2732	0 0 1 X 0 0 1 X	0 0 0 0 0 0 0 0	0 0 0 0 0 0 0 0	0 0 0 0 0 0 0 1	2000 2001
	. . . . 0 0 1 X	. . . . 1 1 1 1	. . . . 1 1 1 1	. . . . 1 1 1 1	. 2FFF
RAM 6264	0 1 0 0 0 1 0 0	0 0 0 0 0 0 0 0	0 0 0 0 0 0 0 0	0 0 0 0 0 0 0 1	4000 4001
	. . . . 0 1 0 1	. . . . 1 1 1 1	. . . . 1 1 1 1	. . . . 1 1 1 1	. 5FFF

Note: X indicates the unused address line for the particular memory IC and they are considered as zero.

DESIGN EXAMPLE 85.2

Interface three numbers of 8 kB EPROM and 5 numbers of 8 kB static RAM to microprocessor 8085 to have a total memory capacity of 64 kB.

Solution

The IC 2764 is selected for EPROM memory and the IC 6264 is selected for RAM memory. Both the memory ICs have time compatibility with the 8085 processor.

The 8 kB EPROM IC requires 13 address lines ($2^{13} = 8$ k). The 8 kB RAM IC also requires 13 address lines ($2^{13} = 8$ k). The address lines $A_0 - A_{12}$ are connected to all the EPROM's and RAMs. Hence $A_0 - A_{12}$ will select the required memory location. The address lines A_{13}, A_{14} and A_{15} are not used for memory address. Hence by decoding these address lines we can generate chip select signals. The 3-to-8 decoder, 74LS138 is employed to produce the chip select signals for the system. The decoder has 8-output lines which can be used as 8-chip select signals. All the 8-chip select signals are used to select memory ICs. EPROM's are mapped at the beginning of memory space. The decoder will select a memory IC by decoding the address lines A_{13}, A_{14} and A_{15}. The address lines $A_0 - A_{12}$ will select a particular memory location in the selected IC. The interface diagram is shown in Fig. 85.2 and address allocation table is shown in Table 85.2.

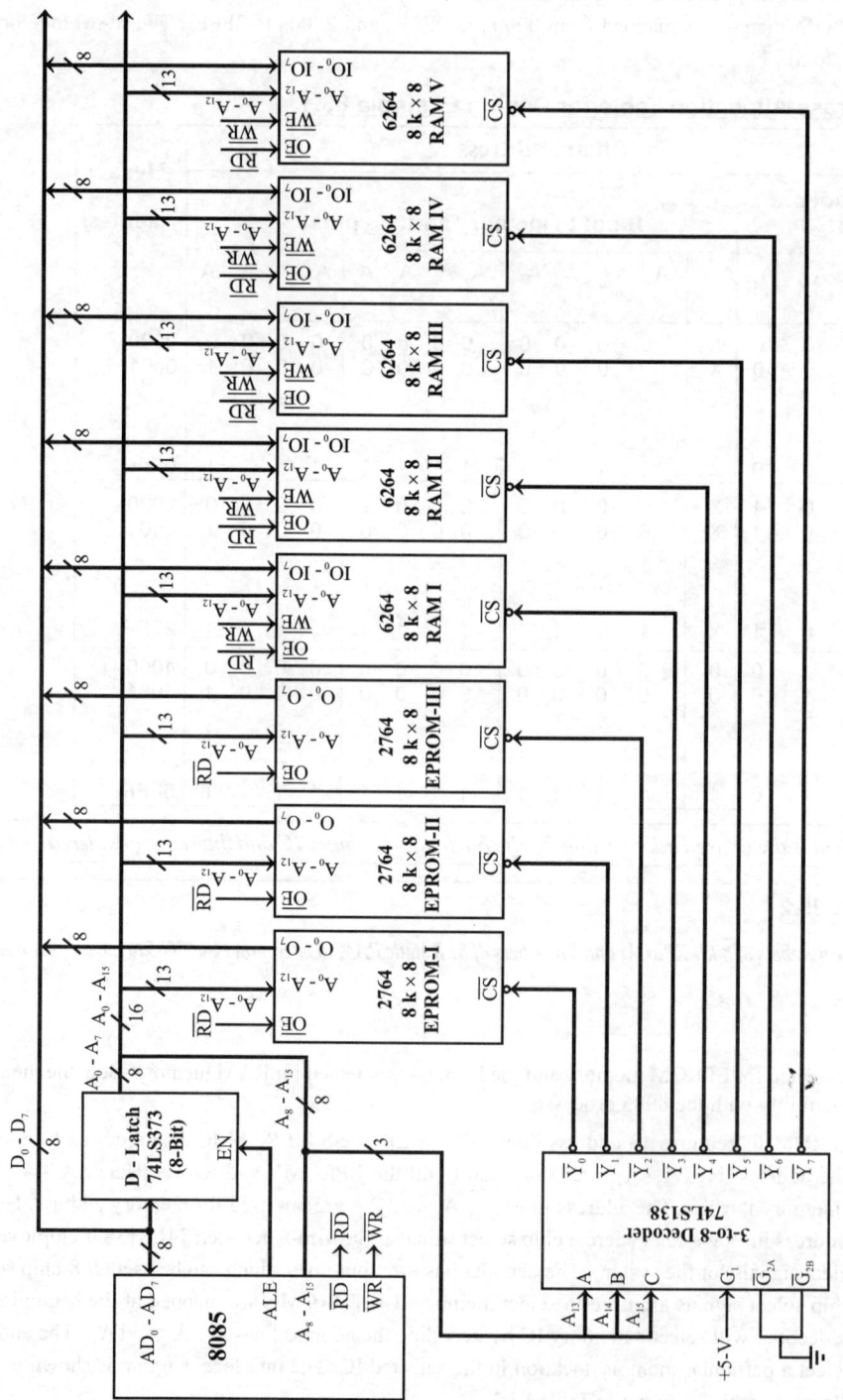

Fig. 85.2: Memory interface diagram for Design Example 85.2.

Table 85.2: Address Allocation Table for Design Example 85.2

Memory IC chip	A_{15}	A_{14}	A_{13}	A_{12}	A_{11}	A_{10}	A_9	A_8	A_7	A_6	A_5	A_4	A_3	A_2	A_1	A_0	Hexa address
Binary address	Decoder input			Input to memory address pins													
EPROM I	0	0	0	0	0	0	0	0	0	0	0	0	0	0	0	0	0000
	0	0	0	0	0	0	0	0	0	0	0	0	0	0	0	1	0001
	0	0	0	0	0	0	0	0	0	0	0	0	0	0	1	0	0002
	.	.	.	.	.	.	.	.	.	.	.	.	.	.	.	.	.
	0	0	0	1	1	1	1	1	1	1	1	1	1	1	1	1	1FFF
EPROM II	0	0	1	0	0	0	0	0	0	0	0	0	0	0	0	0	2000
	0	0	1	0	0	0	0	0	0	0	0	0	0	0	0	1	2001
	0	0	1	0	0	0	0	0	0	0	0	0	0	0	1	0	2002
	.	.	.	.	.	.	.	.	.	.	.	.	.	.	.	.	.
	0	0	1	1	1	1	1	1	1	1	1	1	1	1	1	1	3FFF
EPROM III	0	1	0	0	0	0	0	0	0	0	0	0	0	0	0	0	4000
	0	1	0	0	0	0	0	0	0	0	0	0	0	0	0	1	4001
	0	1	0	0	0	0	0	0	0	0	0	0	0	0	1	0	4002
	.	.	.	.	.	.	.	.	.	.	.	.	.	.	.	.	.
	0	1	0	1	1	1	1	1	1	1	1	1	1	1	1	1	5FFF
RAM I	0	1	1	0	0	0	0	0	0	0	0	0	0	0	0	0	6000
	0	1	1	0	0	0	0	0	0	0	0	0	0	0	0	1	6001
	0	1	1	0	0	0	0	0	0	0	0	0	0	0	1	0	6002
	.	.	.	.	.	.	.	.	.	.	.	.	.	.	.	.	.
	0	1	1	1	1	1	1	1	1	1	1	1	1	1	1	1	7FFF
RAM II	1	0	0	0	0	0	0	0	0	0	0	0	0	0	0	0	8000
	1	0	0	0	0	0	0	0	0	0	0	0	0	0	0	1	8001
	1	0	0	0	0	0	0	0	0	0	0	0	0	0	1	0	8002
	.	.	.	.	.	.	.	.	.	.	.	.	.	.	.	.	.
	1	0	0	1	1	1	1	1	1	1	1	1	1	1	1	1	9FFF
RAM III	1	0	1	0	0	0	0	0	0	0	0	0	0	0	0	0	A000
	1	0	1	0	0	0	0	0	0	0	0	0	0	0	0	1	A001
	1	0	1	0	0	0	0	0	0	0	0	0	0	0	1	0	A002
	.	.	.	.	.	.	.	.	.	.	.	.	.	.	.	.	.
	1	0	1	1	1	1	1	1	1	1	1	1	1	1	1	1	BFFF
RAM IV	1	1	0	0	0	0	0	0	0	0	0	0	0	0	0	0	C000
	1	1	0	0	0	0	0	0	0	0	0	0	0	0	0	1	C001
	1	0	0	0	0	0	0	0	0	0	0	0	0	1	0	0	C002
	.	.	.	.	.	.	.	.	.	.	.	.	.	.	.	.	.
	1	1	0	1	1	1	1	1	1	1	1	1	1	1	1	1	DFFF
RAM V	1	1	1	0	0	0	0	0	0	0	0	0	0	0	0	0	E000
	1	1	1	0	0	0	0	0	0	0	0	0	0	0	0	1	E001
	1	1	1	0	0	0	0	0	0	0	0	0	0	0	1	0	E002
	.	.	.	.	.	.	.	.	.	.	.	.	.	.	.	.	.
	1	1	1	1	1	1	1	1	1	1	1	1	1	1	1	1	FFFF

In this system (Design Example 2) the full memory capacity of 64 kB is utilized for memory. Hence the peripheral ICs and the IO ports should be IO-mapped in the system. The EPROM is mapped from 0000_H to $5FFF_H$. The RAM is mapped from 6000_H to $FFFF_H$. The EPROM capacity is 24 kB. The RAM capacity is 40 kB.

DESIGN EXAMPLE 85.3

In a microprocessor system using 8085, the memory requirement is 8 kB EPROM and 8 kB RAM. For interfacing IO devices, three numbers of 8255 are required. Select suitable memories and explain how they are interfaced to the system. Interface the 8255 by memory mapping.

Solution

The IC 2764 is selected for EPROM memory and the IC 6264 is selected for RAM memory. Both the memory IC's have time compatibility with 8085 processor.

The 8 kB EPROM, 2764 requires 13 address lines (2^{13} = 8 k). The 8 kB RAM, 6264 also requires 13 address lines (2^{13} = 8 k). The address lines A_0 to A_{12} are connected to both EPROM and RAM memory ICs. The 8255 requires four internal addresses. Let us connect A_1 of 8085 to A_0 of 8255 and A_2 of 8085 to A_1 of 8255. The 8255 is memory-mapped in the system.

> **Note:** *The internal devices of 8255 can be selected by connecting any two address lines of the processor to A_0 and A_1 of 8255.*

For the memories and 8255's we require 5 chip select signals. Hence we can use a 3-to-8 decoder 74LS138 for generating eight chip select signals by decoding the unused address lines A_{13}, A_{14} and A_{15}. The decoder enabled pins are permanently tied to appropriate levels. In the eight chip select signals, five are used for selecting memory ICs and 8255 and the remaining three can be used for future expansion. The memory/8255 interface diagram is shown in Fig. 85.3.

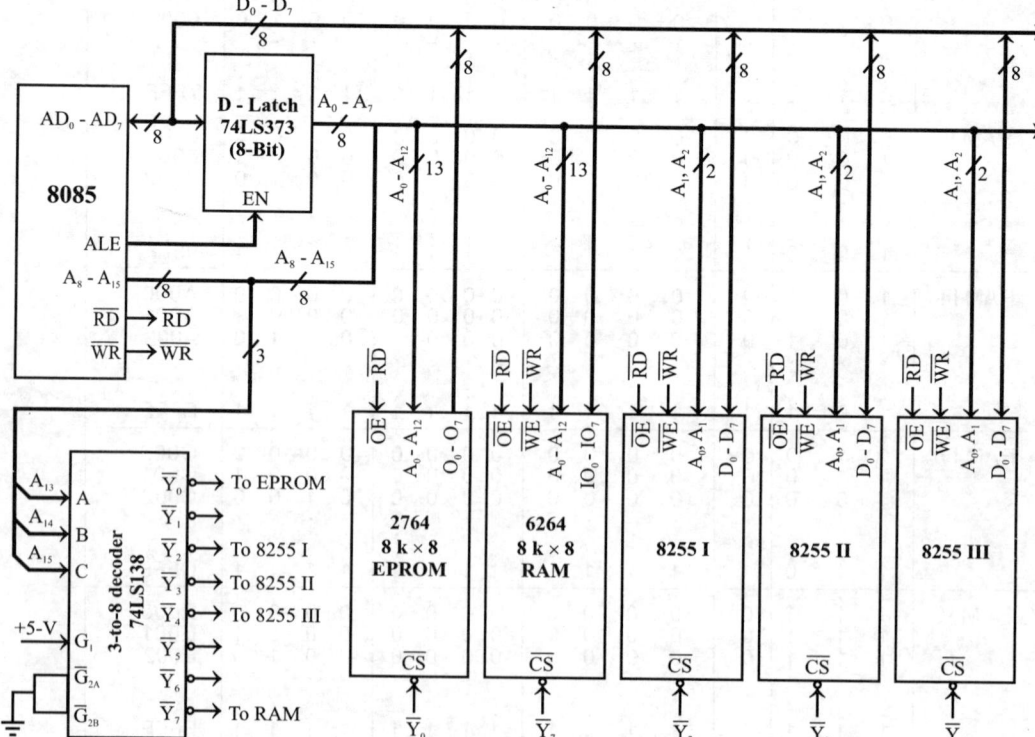

Fig. 85.3: Memory interface diagram for Design Example 85.3.

Table 85.3: Address Allocation Table for Design Example 85.3

Device	Decoder input A_{15}	A_{14}	A_{13}	A_{12}	A_{11}	A_{10}	A_9	A_8	A_7	A_6	A_5	A_4	A_3	A_2	A_1	A_0	Hexa address
	0	0	0	0	0	0	0	0	0	0	0	0	0	0	0	0	0000
2764	0	0	0	0	0	0	0	0	0	0	0	0	0	0	0	1	0001
EPROM	0	0	0	0	0	0	0	0	0	0	0	0	0	0	1	0	0002
	.	.	.	.	.	.	.	.	.	.	.	.	.	.	.	.	.
	.	.	.	.	.	.	.	.	.	.	.	.	.	.	.	.	.
	.	.	.	.	.	.	.	.	.	.	.	.	.	.	.	.	.
	0	0	0	1	1	1	1	1	1	1	1	1	1	1	1	1	1FFF
	1	1	1	0	0	0	0	0	0	0	0	0	0	0	0	0	E000
	1	1	1	0	0	0	0	0	0	0	0	0	0	0	0	1	E001
6264	1	1	1	0	0	0	0	0	0	0	0	0	0	0	1	0	E002
RAM	.	.	.	.	.	.	.	.	.	.	.	.	.	.	.	.	.
	.	.	.	.	.	.	.	.	.	.	.	.	.	.	.	.	.
	1	1	1	1	1	1	1	1	1	1	1	1	1	1	1	1	FFFF
8255 I																	
Port-A	0	1	0	X	X	X	X	X	X	X	X	X	X	0	0	X	4000
Port-B	0	1	0	X	X	X	X	X	X	X	X	X	X	0	1	X	4002
Port-C	0	1	0	X	X	X	X	X	X	X	X	X	X	1	0	X	4004
Control register	0	1	0	X	X	X	X	X	X	X	X	X	X	1	1	X	4006
8255 II																	
Port-A	0	1	1	X	X	X	X	X	X	X	X	X	X	0	0	X	6000
Port-B	0	1	1	X	X	X	X	X	X	X	X	X	X	0	1	X	6002
Port-C	0	1	1	X	X	X	X	X	X	X	X	X	X	1	0	X	6004
Control register	0	1	1	X	X	X	X	X	X	X	X	X	X	1	1	X	6006
8255 III																	
Port-A	1	0	0	X	X	X	X	X	X	X	X	X	X	0	0	X	8000
Port-B	1	0	0	X	X	X	X	X	X	X	X	X	X	0	1	X	8002
Port-C	1	0	0	X	X	X	X	X	X	X	X	X	X	1	0	X	8004
Control register	1	0	0	X	X	X	X	X	X	X	X	X	X	1	1	X	8006

Note: X indicates that the address line is not used for the particular device and considered as zero.

The EPROM is mapped at the starting of memory space. The RAM is mapped at the end of memory space. The EPROM is mapped from 0000_H to $1FFF_H$. The RAM is mapped from $E000_H$ to $FFFF_H$. The four internal devices of 8255 are control register, port-A, port-B and port-C. A 16-bit address is allotted to each internal device of 8255 as shown in Table 85.3.

DESIGN EXAMPLE 85.4

Interface 2 kB RAM and 256 × 8 ROM with 8085 processor for a memory capacity of 8 kB RAM and 1 kB ROM.

Solution

The memory requirement of 8 kB RAM can be achieved with 4 numbers of 2 kB RAM. The memory requirement of 1kB ROM can be achieved with 4 numbers of 256 × 8 ROM. ($4 \times 256 = 1024 = 1k$). The 2 kB RAM requires 11 address lines ($2^{11} = 2$ k). The 256 × 8 ROM requires 8 address lines ($2^8 = 256$).

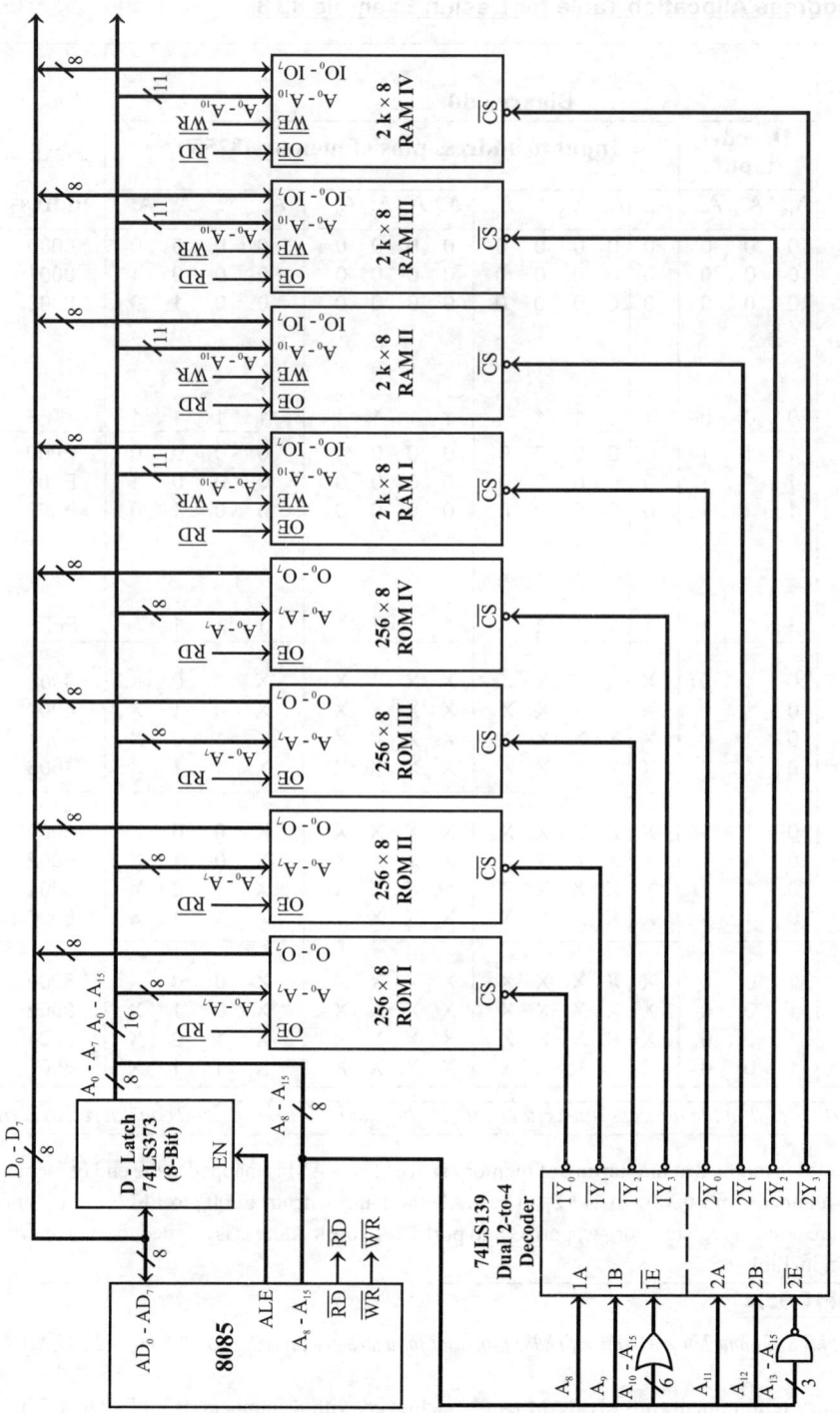

Fig. 85.4: Memory interface diagram for Design Example 85.4.

Table 85.4: Address Allocation Table for Design Example 85.4

Device	Binary address							Hexa address
	ROM decoder enable		Decoder input	Input to ROM memory address pins				
	A_{15} A_{14} A_{13} A_{12}	A_{11} A_{10}	A_9 A_8	A_7 A_6 A_5 A_4	A_3 A_2 A_1 A_0			
256 × 8 ROM I	0 0 0 0	0 0	0 0	0 0 0 0	0 0 0 0			0000
	0 0 0 0	0 0	0 0	0 0 0 0	0 0 0 1			0001
	0 0 0 0	0 0	0 0	0 0 0 0	0 0 1 0			0002
	0 0 0 0	0 0	0 0	1 1 1 1	1 1 1 1			00FF
256 × 8 ROM II	0 0 0 0	0 0	0 1	0 0 0 0	0 0 0 0			0100
	0 0 0 0	0 0	0 1	0 0 0 0	0 0 0 1			0101
	0 0 0 0	0 0	0 1	0 0 1 0	0 0 1 0			0102
	0 0 0 0	0 0	0 1	1 1 1 1	1 1 1 1			01FF
256 × 8 ROM III	0 0 0 0	0 0	1 0	0 0 0 0	0 0 0 0			0200
	0 0 0 0	0 0	1 0	0 0 0 0	0 0 0 1			0201
	0 0 0 0	0 0	1 0	0 0 0 0	0 0 1 0			0202
	0 0 0 0	0 0	1 0	1 1 1 1	1 1 1 1			02FF
256 × 8 ROM IV	0 0 0 0	0 0	1 1	0 0 0 0	0 0 0 0			0300
	0 0 0 0	0 0	1 1	0 0 0 0	0 0 0 1			0301
	0 0 0 0	0 0	1 1	0 0 0 0	0 0 1 0			0302
	0 0 0 0	0 0	1 1	1 1 1 1	1 1 1 1			03FF

Device	Binary address					Hexa address
	RAM decoder enable	Decoder input	Input to RAM memory address pins			
	A_{15} A_{14} A_{13}	A_{12} A_{11}	A_{10} A_9 A_8	A_7 A_6 A_5 A_4	A_3 A_2 A_1 A_0	
2 k × 8 RAM I	1 1 1	0 0	0 0 0	0 0 0 0	0 0 0 0	E000
	1 1 1	0 0	0 0 0	0 0 0 0	0 0 0 1	E001
	1 1 1	0 0	0 0 0	0 0 0 0	0 0 1 0	E002
	1 1 1	0 0	1 1 1	1 1 1 1	1 1 1 1	E7FF
2 k × 8 RAM II	1 1 1	0 1	0 0 0	0 0 0 0	0 0 0 0	E800
	1 1 1	0 1	0 0 0	0 0 0 0	0 0 0 1	E801
	1 1 1	0 1	0 0 0	0 0 0 0	0 0 1 0	E802
	1 1 1	0 1	1 1 1	1 1 1 1	1 1 1 1	EFFF
2 k × 8 RAM III	1 1 1	1 0	0 0 0	0 0 0 0	0 0 0 0	F000
	1 1 1	1 0	0 0 0	0 0 0 0	0 0 0 1	F001
	1 1 1	1 0	0 0 0	0 0 0 0	0 0 1 0	F002
	1 1 1	1 0	1 1 1	1 1 1 1	1 1 1 1	F7FF
2 k × 8 RAM IV	1 1 1	1 1	0 0 0	0 0 0 0	0 0 0 0	F800
	1 1 1	1 1	0 0 0	0 0 0 0	0 0 0 1	F801
	1 1 1	1 1	0 0 0	0 0 0 0	0 0 1 0	F802
	1 1 1	1 1	1 1 1	1 1 1 1	1 1 1 1	FFFF

The address lines A_0 - A_{10} are connected to RAM ICs. Hence, they will select the required memory location in that ICs. The address lines A_0 - A_7 are connected to ROM ICs. Hence, they will select the required memory location in those ICs.

Totally there are 8-memory ICs hence we require 8-chip select signals. The 8-chip select signals can be generated by using a dual 2-to-4 decoder 74LS139. One of the 2-to-4 decoder is used to generate chip select signals for ROM memory ICs and the other decoder is used to generate chip select signals for RAM memory ICs. The address lines A_8 and A_9 are used to generate chip select signals for ROM memory. The address lines A_{10} to A_{15} are logically ORed and used as enable for ROM decoder. The address lines A_{11} and A_{12} are used to generate chip select signals for RAM memory. The address lines A_{13} to A_{15} are logically NANDed and used as enable for RAM decoder.

ROM memories are mapped in the beginning of memory space. The RAM memories are mapped at the end of memory space. The ROM memories are mapped from 0000_H to $03FF_H$. The RAM memories are mapped from $E000_H$ to $FFFF_H$.

DESIGN EXAMPLE 85.5

A system requires 16 kB EPROM and 16 kB RAM. Also the system has 2 numbers of 8255, one number of 8279, one number of 8251 and one number of 8254.

(8255 - Programmable peripheral interface, 8279-Keyboard/display controller, 8251 - USART and 8254 - Timer)

Draw the Interface diagram. Allocate addresses to all the devices. The peripheral IC's should be IO-mapped.

Solution

The IO devices in the system should be mapped by standard IO mapping. Hence separate decoders can be used to generate chip select signals for memory IC's and peripheral IC's.

For 16 *kB* EPROM, we can provide 2 numbers of 2764 (8 k × 8) EPROM. For 16 *kB* RAM we can provide 2 numbers of 6264 (8 k × 8) RAM.

The 8 *kB* memory requires 13 address lines (2^{13} = 8 k). Hence the address lines A_0 - A_{12} are used for selecting the memory locations. The unused address lines A_{13}, A_{14} and A_{15} are used as input to decoder 74LS138 (3-to-8-decoder) of memory IC. The logic **low** enables of this decoder are tied to IO/M of 8085, so that this decoder is enabled for memory read/write operation. The other enable pins of decoder are tied to appropriate logic levels permanently. The 4 outputs of the decoder are used to select memory IC's and the remaining 4 are kept for future expansion.

The EPROM is mapped in the beginning of memory space from 0000_H to $3FFF_H$. The RAM is mapped at the end of memory space from $C000_H$ to $FFFF_H$.

There are five peripheral IC's to be interfaced to the system. The chip select signals for these IC's are given through another 3-to-8 decoder 74LS138 (IO decoder). The input to this decoder is A_{10}, A_{11} and A_{12}. The address lines A_{13}, A_{14} and A_{15} are logically ORed and applied to **low** enable of IO decoder. The logic **high** enable of IO decoder is tied to IO/$\overline{M}$ signal of 8085, so that this decoder is enabled for IO read/write operation.

Here, the high order address lines can be used for decoding because the processor outputs the 8-bit port address both on AD_0 to AD_7 and A_8 to A_{15}. The address lines A_0 and A_1 are used to select the internal devices of the peripheral ICs. The output of the decoder are used to select the ICs. Three outputs of the decoder will be spare for future expansion.

Note: Since the IO devices are IO-mapped in the system, 8-bit addresses have been allotted to them.

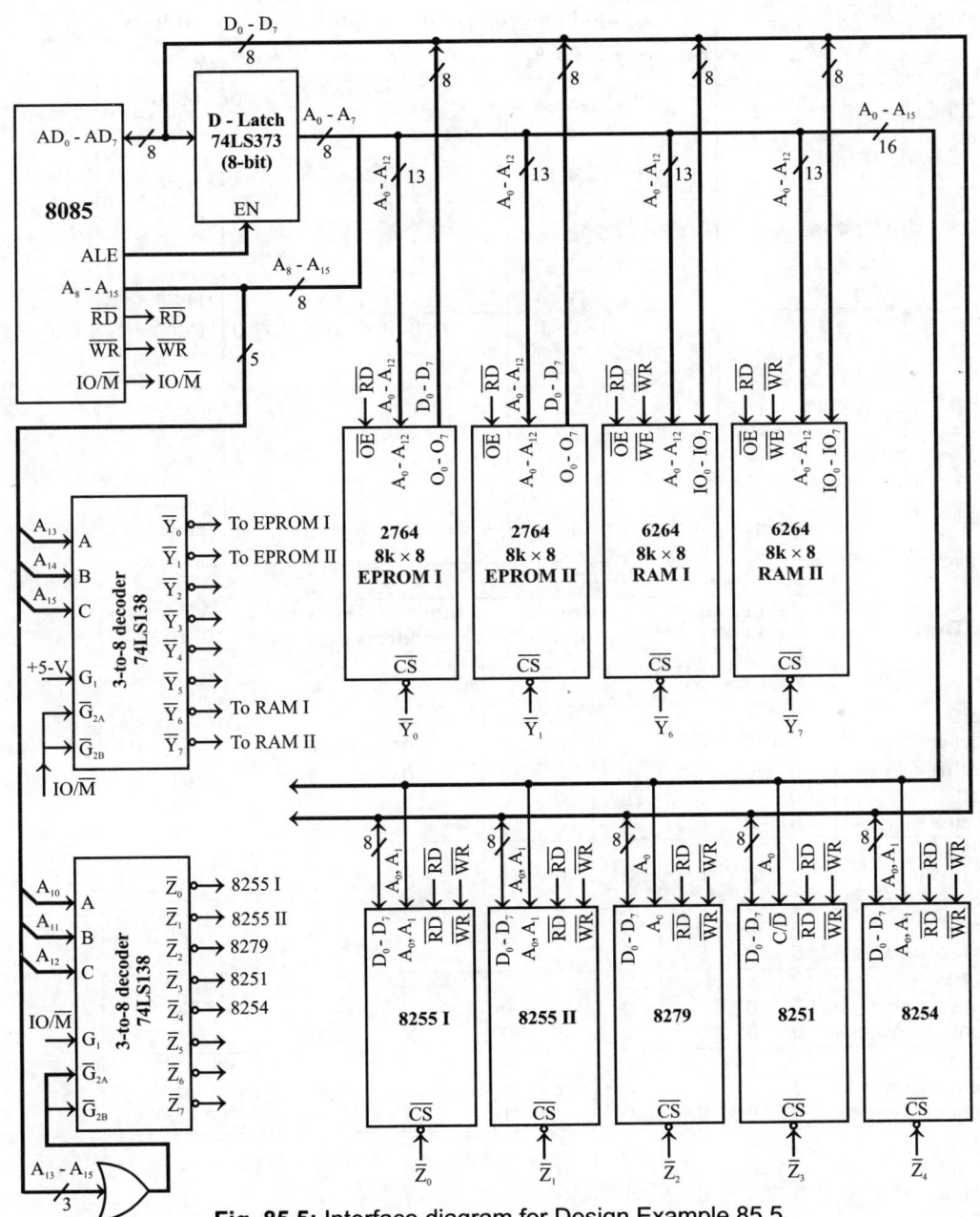

Fig. 85.5: Interface diagram for Design Example 85.5.

Table 85.5: Address Allocation Table for Design Example 85.5

Device	Input to memory decoder			Input to memory address pins				Hexa address
	A_{15}	A_{14}	A_{13}	A_{12}	$A_{11}\ A_{10}\ A_9\ A_8$	$A_7\ A_6\ A_5\ A_4$	$A_3\ A_2\ A_1\ A_0$	
2764 I	0	0	0	0	0 0 0 0	0 0 0 0	0 0 0 0	0000
8 k × 8	.	.	.	.	. . . .	. . . .	. . . .	.
	0	0	0	1	1 1 1 1	1 1 1 1	1 1 1 1	1FFF
2764 II	0	0	1	0	0 0 0 0	0 0 0 0	0 0 0 0	2000
8 k × 8	.	.	.	.	. . . .	. . . .	. . . .	.
	0	0	1	0	1 1 1 1	1 1 1 1	1 1 1 1	3FFF
6264 I	1	1	0	0	0 0 0 0	0 0 0 0	0 0 0 0	C000
8 k × 8	.	.	.	.	. . . .	. . . .	. . . .	.
	1	1	0	1	1 1 1 1	1 1 1 1	1 1 1 1	DFFF
6264 II	1	1	1	0	0 0 0 0	0 0 0 0	0 0 0 0	E000
8 k × 8	.	.	.	.	. . . .	. . . .	. . . .	.
	1	1	1	1	1 1 1 1	1 1 1 1	1 1 1 1	FFFF

Device	IO decoder enable			IO decoder input			Input to IO device address pins		Hexa address
	A_{15} A_7	A_{14} A_6	A_{13} A_5	A_{12} A_4	A_{11} A_3	A_{10} A_2	A_9 A_1	A_8 A_0	
8255 I									
Port-A	0	0	0	0	0	0	0	0	00
Port-B	0	0	0	0	0	0	0	1	01
Port-C	0	0	0	0	0	0	1	0	02
Control register	0	0	0	0	0	0	1	1	03
8255 II									
Port-A	0	0	0	0	0	1	0	0	04
Port-B	0	0	0	0	0	1	0	1	05
Port-C	0	0	0	0	0	1	1	0	06
Control register	0	0	0	0	0	1	1	1	07
8279									
Data register	0	0	0	0	1	0	X	0	08
Control register	0	0	0	0	1	0	X	1	09
8251									
Data register	0	0	0	0	1	1	X	0	0C
Control register	0	0	0	0	1	1	X	1	0D
8254									
Counter-0	0	0	0	1	0	0	0	0	10
Counter-1	0	0	0	1	0	0	0	1	11
Counter-2	0	0	0	1	0	0	1	0	12
Control register	0	0	0	1	0	0	1	1	13

Note: Don't care (X) is considered as zero.

DESIGN EXAMPLE 85.6

In a microprocessor-based system 8085, 8 kB EPROM and 8 kB RAM are needed. For interfacing IO devices two numbers of 8155 are required. Select suitable memories and explain how they are interfaced in the system. Interface the 8155 ports by IO mapping.

Solution

The IC 2764 (8 k × 8) is selected for EPROM memory and IC 6264 (8 k × 8) is selected for RAM memory. Both the memory IC's have time compatibility with 8085 processor.

The 8 *kB* memories require 13 address lines ($2^{13} = 8$ k). Hence, the address lines A_0 - A_{12} are used to select memory locations.

In addition to 6264, each one of the 8155 chip provides a static RAM capacity of 256 bytes. The RAM locations of 8155 are selected by address lines A_0-A_6.

A 3-to-8 decoder, 74LS138 is used for generating chip select signals by decoding the address lines A_{13}, A_{14} and A_{15}.

The 8155 has internal address latch and decoder to differentiate memory operation and IO operation. To utilize this facility, the control signals ALE and IO/$\overline{M}$ are connected to 8155.

The 8155 ports and memory locations can be selected from the decoder used for memory devices. It differentiates the memory and IO operation from IO/$\overline{M}$ signal. Eight bit addresses are allotted to ports of 8155 and sixteen bit addresses are allotted to RAM memory locations of 8155.

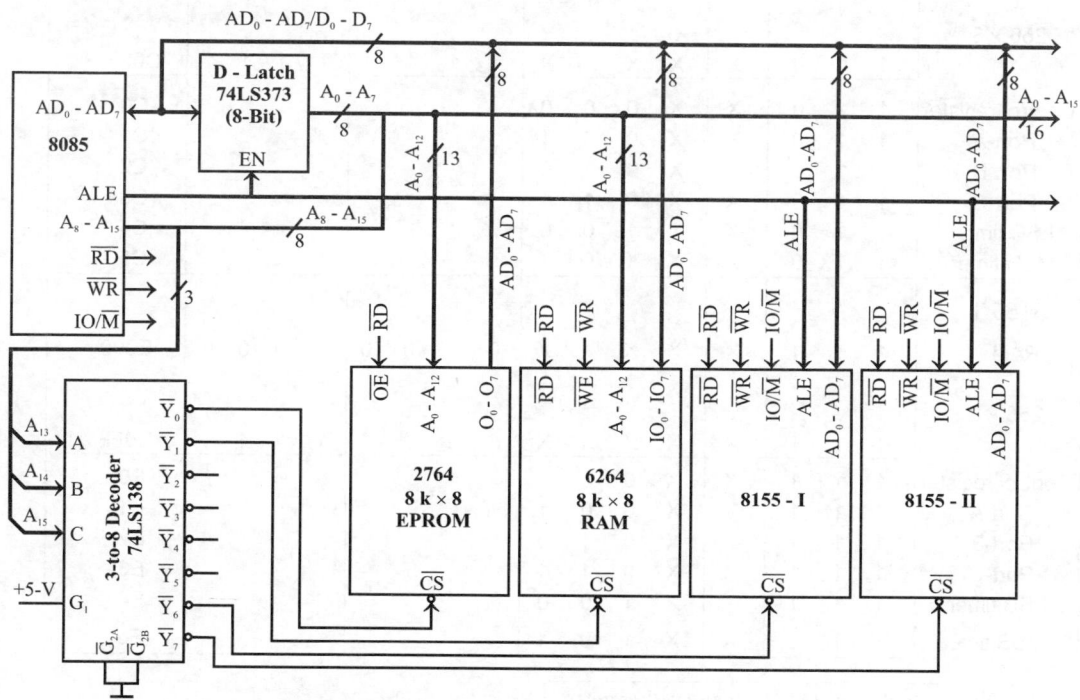

Fig. 85.6: Interface diagram for Design Example 85.6.

Table 85.6: Address Allocation Table for Design Example 85.6

Device	Decoder input			Input to address pins of memory/8155													Hexa address
	A_{15}	A_{14}	A_{13}	A_{12}	A_{11}	A_{10}	A_9	A_8	A_7	A_6	A_5	A_4	A_3	A_2	A_1	A_0	
2764 8k × 8 EPROM	0	0	0	0	0	0	0	0	0	0	0	0	0	0	0	0	0000
	0	0	0	0	0	0	0	0	0	0	0	0	0	0	0	1	0001
	.	.	.	.	.	.	.	.	.	.	.	.	.	.	.	.	.
	0	0	0	1	1	1	1	1	1	1	1	1	1	1	1	1	1FFF
6264 8k × 8 RAM	0	0	1	0	0	0	0	0	0	0	0	0	0	0	0	0	2000
	0	0	1	0	0	0	0	0	0	0	0	0	0	0	0	1	2001
	.	.	.	.	.	.	.	.	.	.	.	.	.	.	.	.	.
	0	0	1	1	1	1	1	1	1	1	1	1	1	1	1	1	3FFF
8155 I RAM 256 × 8	1	1	0	X	X	X	X	X	0	0	0	0	0	0	0	0	C000
	.	.	.	.	.	.	.	.	.	.	.	.	.	.	.	.	.
	1	1	0	X	X	X	X	X	1	1	1	1	1	1	1	1	C0FF
Control register	1	1	0	X	X	0	0	0									C0
Port-A	1	1	0	X	X	0	0	1									C1
Port-B	1	1	0	X	X	0	1	0									C2
Port-C	1	1	0	X	X	0	1	1									C3
LSB timer	1	1	0	X	X	1	0	0									C4
MSB timer	1	1	0	X	X	1	0	1									C5
8155 II RAM 256 × 8	1	1	1	X	X	X	X	X	0	0	0	0	0	0	0	0	E000
	.	.	.	.	.	.	.	.	.	.	.	.	.	.	.	.	.
	1	1	1	X	X	X	X	X	1	1	1	1	1	1	1	1	E0FF
Control register	1	1	1	X	X	0	0	0									E0
Port-A	1	1	1	X	X	0	0	1									E1
Port-B	1	1	1	X	X	0	1	0									E2
Port-C	1	1	1	X	X	0	1	1									E3
LSB timer	1	1	1	X	X	1	0	0									E4
MSB timer	1	1	1	X	X	1	0	1									E5

Note: Don't care (X) is considered as zero.

6.7.2 Examples of Memory and IO Interface in 8086-Based System

DESIGN EXAMPLE 86.1

In a microprocessor system using 8086, the memory requirement is 16 kB EPROM and 16 kB RAM. The system requires 8279 for keyboard and display interface, and 8255 for IO ports. Draw an interface diagram for memory and peripheral devices, and allot addresses for each device.

Solution

The 16 kB EPROM can be implemented in two units of 8 kB EPROM (2764). One of the 8 kB EPROM can be mapped as even bank and the other as odd bank. The address lines A_1-A_{13} are connected to each EPROM IC to select the internal locations of EPROM.

The 16 kB RAM can be implemented in two units of 8 kB RAM (6264). One of the 8 kB RAM can be mapped as even bank and the other as odd bank. The address lines A_1-A_{13} are connected to each RAM IC to select the internal locations of RAM.

Since a large amount of memory space is free, we can interface the 8279 and 8255 as memory- mapped device. In the interface diagram shown in Fig. 86.1.1, these devices are interfaced such that even addresses are allotted to them. The address line A_1 of 8086 is connected to the address line A_0 of 8279, and the address lines A_1 and A_2 of 8086 are connected to address lines A_0 and A_1 of 8255 to provide the required internal addresses.

Let us employ two 2-to-4 decoders to generate chip select signals. Both the decoders take A_{18} and A_{19} as input, and each decoder produces four decoded output signals. One of the decoders is enabled by the address line A_0 and the output of this decoder is used as chip select signals for even bank memory ICs, 8279 and 8255.

The other decoder is enabled by the control signal $\overline{BHE}$ and the output of this decoder is used as chip select signals for odd bank memory ICs.

The address lines A_{14} - A_{17} are not used for memory ICs. Similarly, the address lines A_2 - A_{17} are not used for 8279 and the address lines A_3 - A_{17} are not used for 8255.

The addresses allotted to the memory and IO devices are shown in Table 86.1. The unused address lines are denoted by "x"(don't care) in Table 1.

While framing hexa address, the don't cares are considered zero for RAM and IO devices, and they are considered one for EPROM. This will map RAM memory in the beginning of the address space and EPROM at the end of the address space, which is a necessary requirement for the 8086 system.

The don't care can be avoided by generating additional chip select signals using unused address lines and logically ORing this additional chip select signal with output of decoder and then the combined chip select signal can be used to select the memory and IO devices as shown in Fig. 86.1.2.

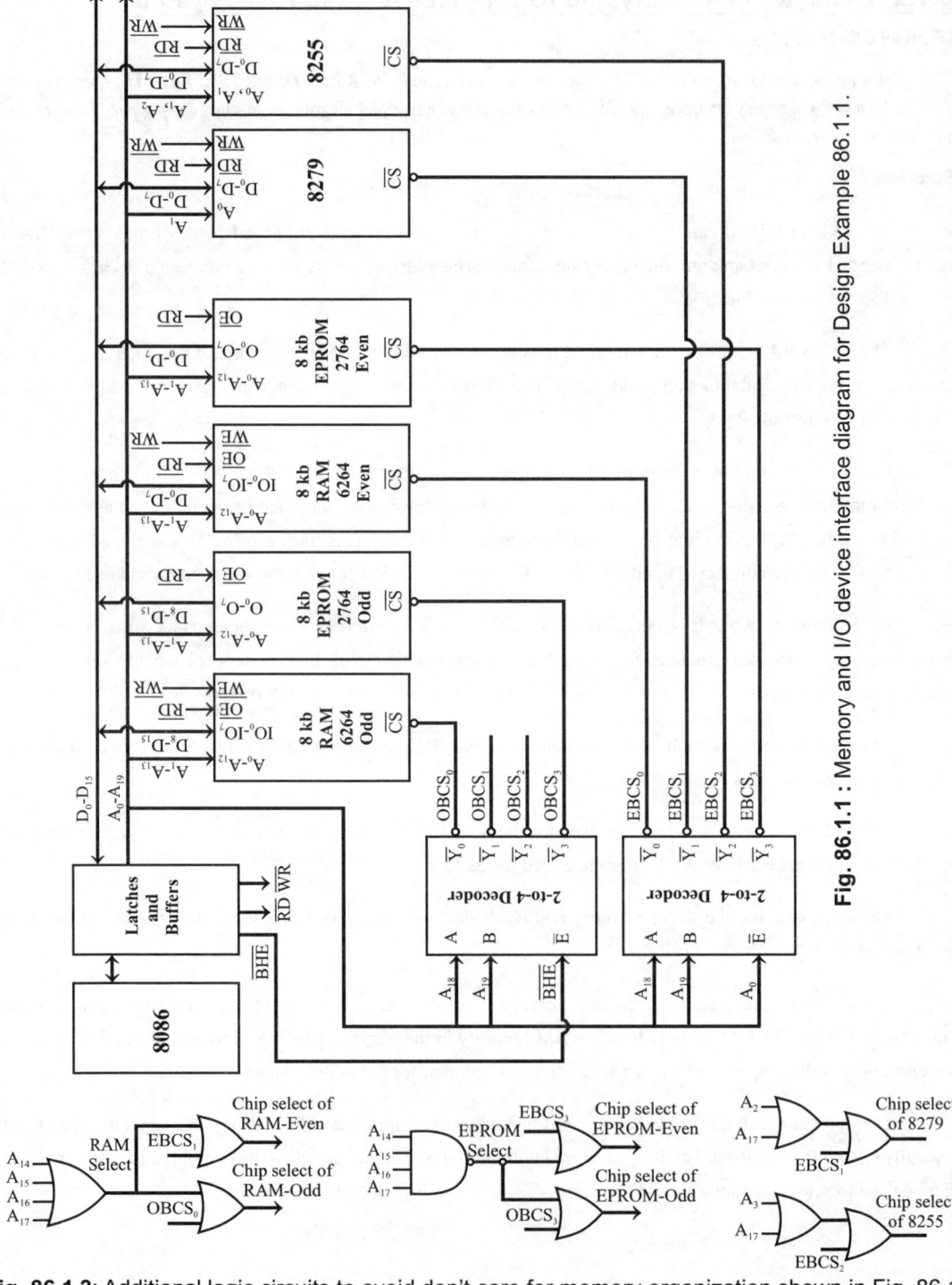

Fig. 86.1.1 : Memory and I/O device interface diagram for Design Example 86.1.1.

Fig. 86.1.2: Additional logic circuits to avoid don't care for memory organization shown in Fig. 86.1.

Table 86.1: Address Allocation Table for Design Example 86.1

Device	Binary Address									Hexa address	
	Decoder input	Unused address lines		Input to memory/ IO device address pins							
	A_{19} A_{18}	A_{17} A_{16}	A_{15} A_{14}	A_{13} A_{12}	A_{11} A_{10} A_9 A_8	A_7 A_6 A_5 A_4	A_3 A_2 A_1	A_0			
8 kB RAM Memory Even	0 0	x x	x x	0 0	0 0 0 0	0 0 0 0	0 0 0	0	0 0 0 0 0		RAM Address Range 00000_H to $03FFF_H$
	0 0	x x	x x	0 0	0 0 0 0	0 0 0 0	0 0 1	0	0 0 0 0 2		
	0 0	x x	x x	0 0	0 0 0 0	0 0 0 0	0 1 0	0	0 0 0 0 4		
	.	.	.	.	.	.	.		.		
	0 0	x x	x x	1 1	1 1 1 1	1 1 1 1	1 1 1	0	0 3 F F E		
8 kB RAM Memory Odd	0 0	x x	x x	0 0	0 0 0 0	0 0 0 0	0 0 0	1	0 0 0 0 1		
	0 0	x x	x x	0 0	0 0 0 0	0 0 0 0	0 0 1	1	0 0 0 0 3		
	0 0	x x	x x	0 0	0 0 0 0	0 0 0 0	0 1 0	1	0 0 0 0 5		
	.	.	.	.	.	.	.		.		
	0 0	x x	x x	1 1	1 1 1 1	1 1 1 1	1 1 1	1	0 3 F F F		
8 kB EPROM Memory Even	1 1	x x	x x	0 0	0 0 0 0	0 0 0 0	0 0 0	0	F C 0 0 0		EPROM Address Range $FC000_H$ to $FFFFF_H$
	1 1	x x	x x	0 0	0 0 0 0	0 0 0 0	0 0 1	0	F C 0 0 2		
	1 1	x x	x x	0 0	0 0 0 0	0 0 0 0	0 1 0	0	F C 0 0 4		
	.	.	.	.	.	.	.		.		
	1 1	x x	x x	1 1	1 1 1 1	1 1 1 1	1 1 1	0	F C 0 0 E		
8 kB EPROM Memory Odd	1 1	x x	x x	0 0	0 0 0 0	0 0 0 0	0 0 0	1	F C 0 0 1		
	1 1	x x	x x	0 0	0 0 0 0	0 0 0 0	0 0 1	1	F C 0 0 3		
	1 1	x x	x x	0 0	0 0 0 0	0 0 0 0	0 1 0	1	F C 0 0 5		
	.	.	.	.	.	.	.		.		
	1 1	x x	x x	1 1	1 1 1 1	1 1 1 1	1 1 1	1	F F F F F		
8279 Data Register Control Register	0 1	x x	x x	x x	x x x x	x x x x	x x 0	0	4 0 0 0 0		
	0 1	x x	x x	x x	x x x x	x x x x	x x 1	0	4 0 0 0 2		
8255 Port-A	1 0	x x	x x	x x	x x x x	x x x x	x 0 0	0	8 0 0 0 0		
Port-B	1 0	x x	x x	x x	x x x x	x x x x	x 0 1	0	8 0 0 0 2		
Port-C	1 0	x x	x x	x x	x x x x	x x x x	x 1 0	0	8 0 0 0 4		
Control Register	1 0	x x	x x	x x	x x x x	x x x x	x 1 1	0	8 0 0 0 6		

DESIGN EXAMPLE 86.2

Repeat Design Example 86.1 by providing IO mapping for 8279 and 8255.

Solution

The 16 kB EPROM can be implemented in two numbers of 8 kB EPROM-2764 and 16 kB RAM can be implemented in two numbers of 8 kB RAM-6264. The odd bank will consist of 8 kB EPROM and 8 kB RAM, and the even bank will consist of 8 kB EPROM and 8 kB RAM.

Table 86.2: Address Allocation Table for Design Example 86.2

Device	Binary Address								Hexa address	
	Decoder input	Unused address lines	Input address lines to memory IC							
	$A_{19} A_{18} A_{17}$	A_{16} $A_{15} A_{14}$	$A_{13} A_{12}$	$A_{11} A_{10} A_9 A_8$	$A_7 A_6 A_5 A_4$	$A_3 A_2 A_1$	A_0			
8 kB RAM Memory Even	0 0 0	x x x	0 0	0 0 0 0	0 0 0 0	0 0 0	0	0 0 0 0 0	
	0 0 0	x x x	0 0	0 0 0 0	0 0 0 0	0 0 1	0	0 0 0 0 2	RAM Address Range 00000_H to $03FFF_H$
	0 0 0	x x x	0 0	0 0 0 0	0 0 0 0	0 1 0	0	0 0 0 0 4	
	⋮	⋮	⋮	⋮	⋮	⋮		⋮	
	0 0 0	x x x	1 1	1 1 1 1	1 1 1 1	1 1 1	0	0 3 F F E	
8 kB RAM Memory Odd	0 0 0	x x x	0 0	0 0 0 0	0 0 0 0	0 0 0	1	0 0 0 0 1	
	0 0 0	x x x	0 0	0 0 0 0	0 0 0 0	0 0 1	1	0 0 0 0 3	
	0 0 0	x x x	0 0	0 0 0 0	0 0 0 0	0 1 0	1	0 0 0 0 5	
	⋮	⋮	⋮	⋮	⋮	⋮		⋮	
	0 0 0	x x x	1 1	1 1 1 1	1 1 1 1	1 1 1	1	0 3 F F F	
8 kB EPROM Memory Even	1 1 1	x x x	0 0	0 0 0 0	0 0 0 0	0 0 0	0	F C 0 0 0	
	1 1 1	x x x	0 0	0 0 0 0	0 0 0 0	0 0 1	0	F C 0 0 2	EPROM Address Range $FC000_H$ to $FFFFF_H$
	1 1 1	x x x	0 0	0 0 0 0	0 0 0 0	0 1 0	0	F C 0 0 4	
	⋮	⋮	⋮	⋮	⋮	⋮		⋮	
	1 1 1	x x x	1 1	1 1 1 1	1 1 1 1	1 1 1	0	F F F F E	
8 kB EPROM Memory Odd	1 1 1	x x x	0 0	0 0 0 0	0 0 0 0	0 0 0	1	F C 0 0 1	
	1 1 1	x x x	0 0	0 0 0 0	0 0 0 0	0 0 1	1	F C 0 0 3	
	1 1 1	x x x	0 0	0 0 0 0	0 0 0 0	0 1 0	1	F C 0 0 5	
	⋮	⋮	⋮	⋮	⋮	⋮		⋮	
	1 1 1	x x x	1 1	1 1 1 1	1 1 1 1	1 1 1	1	F F F F E	

Device	Binary address			Hexa address
	Decoder input	Input address lines to IO devices		
	A_7 A_6 A_5	A_4 A_3 A_2 A_1	A_0	
8279 Data register Control register	0 0 0	x x x 0	0	0 0
	0 0 0	x x x 1	0	0 2
8255 Port-A	0 0 1	x x 0 0	0	2 0
Port-B	0 0 1	x x 0 1	0	2 2
Port-C	0 0 1	x x 1 0	0	2 4
Control register	0 0 1	x x 1	1	0 2 6

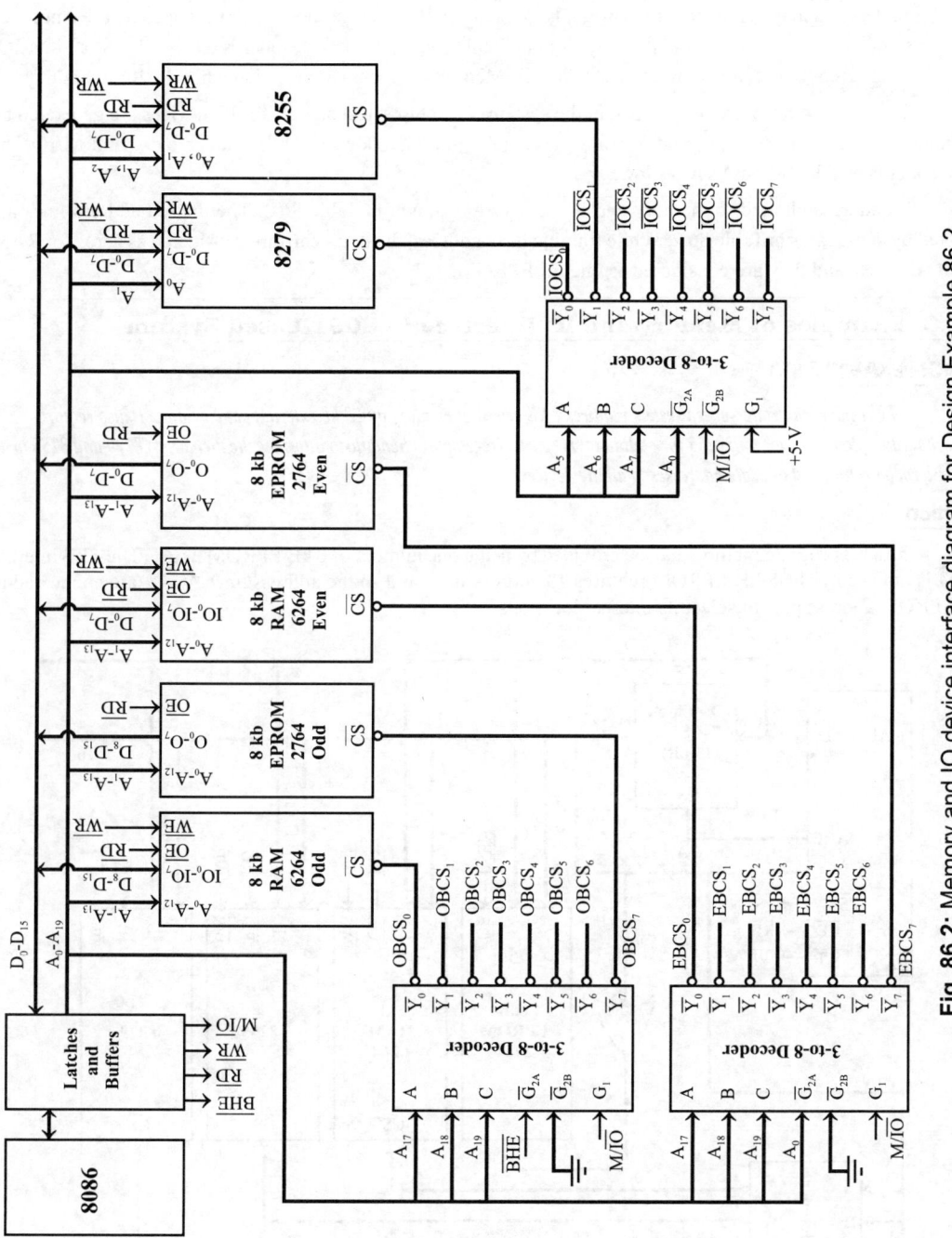

Fig. 86.2: Memory and IO device interface diagram for Design Example 86.2.

The address lines A_1 - A_{13} are used to select the internal locations of memory. The address lines A_{17}, A_{18} and A_{19} are decoded to generate chip select signals for memory ICs. Two numbers of 3-to-8 decoder are employed in the system, one for even bank and the other for odd bank. The even bank decoder is enabled by A_0 and $M/\overline{IO}$. The odd bank decoder is enabled by $\overline{BHE}$ and $M/\overline{IO}$. Memory decoders are enabled when $M/\overline{IO}$ is **high**.

The 8279 and 8255 are IO-mapped in the system. The chip select signals for IO devices are generated by decoding the address lines A_5, A_6 and A_7 using a separate 3-to-8 decoder which is enabled by A_0 and $M/\overline{IO}$ The IO decoder is enabled when $M/\overline{IO}$ is **low**.

The address allotted to memory and IO devices are shown in Table 86.2. The unused address lines are denoted by don't cares in Table 86.2. While framing hexa address, the don't care are considered as zeros for RAM and IO devices, and they are considered as one for EPROM.

6.7.3 Examples of Memory and IO Interface in 8051-Based System

DESIGN EXAMPLE 51.1

An 8051 microcontroller-based system requires 8 kB program memory and 8 kb external data memory. It also requires 8279 for keyboard/display interface and 8255 for additional IO ports. Develop a schematic to interface the memories, 8279 and 8255 to an 8051 microcontroller, and allocate addresses to all the devices.

Solution

The 8 kB program memory can be provided by using one number of 8 kB EPROM 2764. (The 4 kB internal ROM is not used.) The 8 kB EPROM requires 13 address lines and so the address lines A_0-A_{12} are connected to the EPROM address pins to select its internal locations.

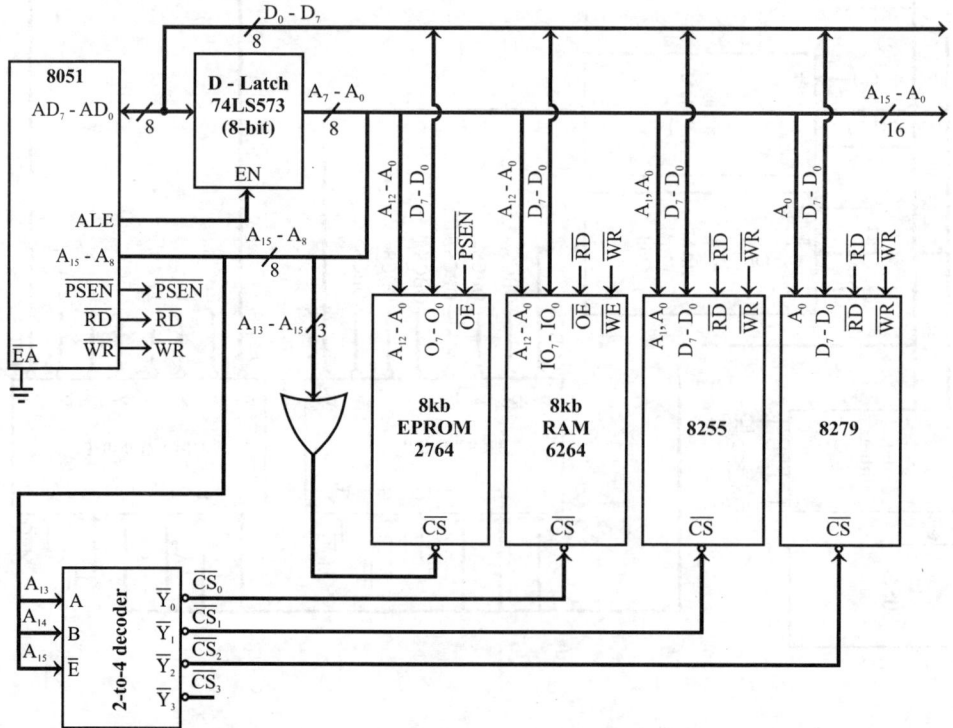

Fig. 51.1: Memory and IO interface diagram for Design Example 51.1.

The remaining address lines A_{13}, A_{14} and A_{15} are logically ORed and used as chip select signal for EPROM. The signal PSEN is used as read control signal for EPROM and so the EPROM can be accessed only as a program memory. Here, the EPROM is mapped in the first 8 kB of the program memory address space with address range 0000_H to $1FFF_H$. (Here, the remaining 56 kB program memory address space is not utilized.)

The 8051 provides a separate 64 kB external data memory address space. The RAM memory, 8279 and 8255 can be interfaced to 8051 as data memory. (The 8051 does not provide a separate IO space and so the IO devices 8279 and 8255 should be mapped only as memory-mapped IO.) The 8 kb RAM can be provided by using one number of 8 kB RAM 6264. The 8 kB RAM requires 13 address lines and so the address lines A_0-A_{12} are connected to address pins of the RAM to select its internal location.

The 8255 requires two address lines to select its internal devices port-A, port-B, port-C and the control register. Hence, the address lines A_0 and A_1 of the controller are connected to A_0 and A_1 of the 8255 respectively. The 8279 requires one address line to select its data register and control register. Hence the address line A_0 of 8051 is connected to A_0 of 8279.

Table 51.1: Address Allocation Table for Design Example 51.1

Device	A_{15} A_{14} A_{13} A_{12}	A_{11} A_{10} A_9 A_8	A_7 A_6 A_5 A_4	A_3 A_2 A_1 A_0	Hexa address	Comment
8 kB EPROM	0 0 0 0	0 0 0 0	0 0 0 0	0 0 0 0	0000	
	0 0 0 0	0 0 0 0	0 0 0 0	0 0 0 1	0001	Program
	0 0 0 0	0 0 0 0	0 0 0 0	0 0 1 0	0002	memory
						address
						space
	0 0 0 1	1 1 1 1	1 1 1 1	1 1 1 1	1FFF	
256 bytes internal RAM/SFR			0 0 0 0	0 0 0 0	00	Internal
			0 0 0 0	0 0 0 1	01	data
			0 0 0 0	0 0 1 0	02	memory
						address
			1 1 1 1	1 1 1 1	FF	space
8kB RAM	0 0 0 0	0 0 0 0	0 0 0 0	0 0 0 0	0000	External
	0 0 0 1	0 0 0 0	0 0 0 0	0 0 0 1	0001	data
	0 0 0 1	0 0 0 0	0 0 0 0	0 0 1 0	0002	memory
						address
						space
	0 0 0 1	1 1 1 1	1 1 1 1	1 1 1 1	1FFF	
8255 Port-A	0 0 1 X	X X X X	X X X X	X X 0 0	2000	External data
Port-B	0 0 1 X	X X X X	X X X X	X X 0 1	2001	memory
Port-C	0 0 1 X	X X X X	X X X X	X X 1 0	2002	address
Control register	0 0 1 X	X X X X	X X X X	X X 1 1	2003	space
8279 Data register	0 1 0 X	X X X X	X X X X	X X X 0	4000	External data
Control register	0 1 0 X	X X X X	X X X X	X X X 1	4001	memory address space

Note : The don't care "X" is considered as zero.

A 2-to-4 decoder is employed in the system to generate the chip select signals required for the RAM, 8255 and 8279. The address lines A_{13} and A_{14} are connected to the input of the decoder to generate four chip select signals. The address line A_{15} is used as logic **low** chip enable for the decoder.

Since the internal ROM is not used, the pin $\overline{EA}$ of the 8051 is permanently grounded. The 8051 provides separate read and write control signals $\overline{RD}$ and $\overline{WR}$ for reading and writing with devices interfaced as data memory. The memory and IO interface diagram is shown in Fig. 51.1. The addresses allotted to various devices are listed in Table 51.1. The 8051 has a separate 256 bytes internal data memory address space allotted to internal RAM and SFR.

DESIGN EXAMPLE 51.2

An 8051 microcontroller-based system requires a 32 kB program memory and a 16 kB data memory. The system should also utilize the internal 4 kB ROM. Also, the system should include a keyboard/display controller 8279 and an 8255 IO port device. Draw the interface diagram and allocate addresses to the memory and the IO devices.

Solution

The 8051 has 4 kB internal ROM which is mapped in the beginning of the program memory address space with the address in the range 0000_H to $0FFF_H$. Hence for a total requirement of 32 kB, we have to provide 28 kB external memory. Therefore, one number 32 kB EPROM 27256 can be interfaced to the microcontroller in which the first 4 kB will not be accessed by the controller if $\overline{EA}$ pin is tied to V_{CC} or +5-V. The 32 kB EPROM requires 15 address lines and so the address lines A_0 to A_{14} are connected to the address pins of the EPROM to select its internal locations. The address line A_{15} is used as chip select signal. The EPROM is selected for program memory addresses when $A_{15} = 0$. The signal $\overline{PSEN}$ is used as the read control signal for the EPROM and so the EPROM can be accessed only as program memory.

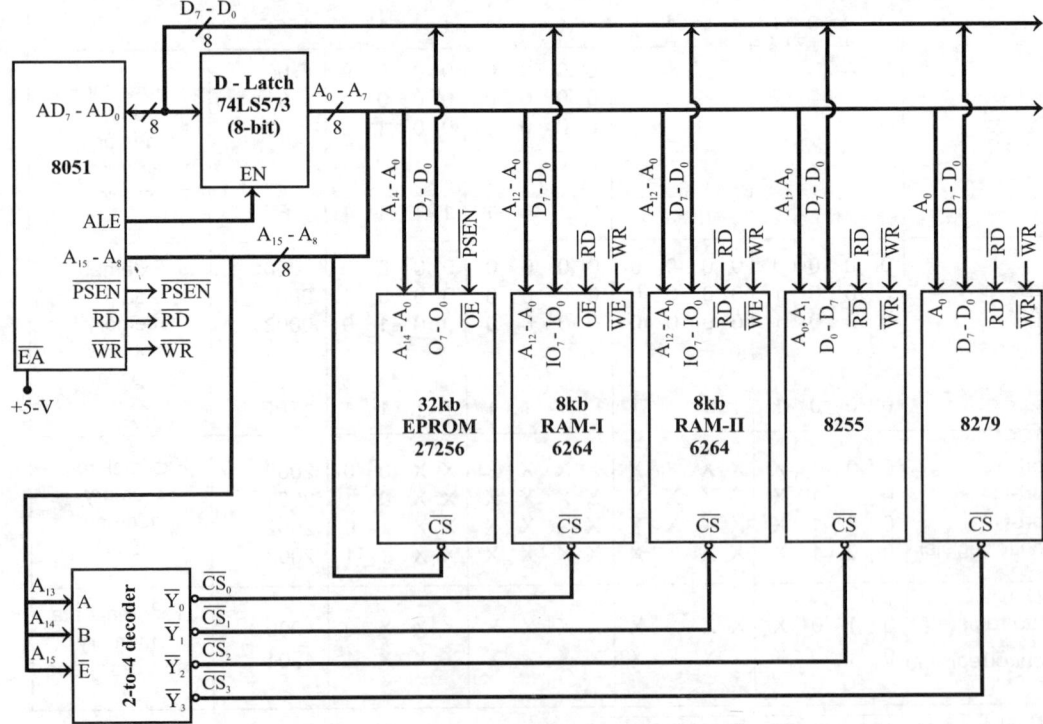

Fig. 51.2: Memory and IO interface diagram for Design Example 51.2.

Table 51.2: Address Allocation Table for Design Example 51.2

Device	$A_{15}\ A_{14}\ A_{13}\ A_{12}$	$A_{11}\ A_{10}\ A_9\ A_8$	$A_7\ A_6\ A_5\ A_4$	$A_3\ A_2\ A_1\ A_0$	Hexa address	Comment
Internal ROM	0 0 0 0 0 0 0 0 0 0 0 0 . 0 0 0 0	0 0 0 0 0 0 0 0 0 0 0 0 . 1 1 1 1	0 0 0 0 0 0 0 0 0 0 0 0 . 1 1 1 1	0 0 0 0 0 0 0 1 0 0 1 0 . 1 1 1 1	0000 0001 0002 . 0FFF	Internal program memory address space 4 kB
32 kB EPROM	0 0 0 1 0 0 0 1 0 0 0 1 . 0 1 1 1	0 0 0 0 0 0 0 0 0 0 0 0 . 1 1 1 1	0 0 0 0 0 0 0 0 0 0 0 0 . 1 1 1 1	0 0 0 0 0 0 0 1 0 0 1 0 . 1 1 1 1	1000 1001 1002 . 7FFF	External program memory address space 28 kB
256 bytes internal RAM/SFR			0 0 0 0 0 0 0 0 0 0 0 0 . 1 1 1 1	0 0 0 0 0 0 0 1 0 0 1 0 . 1 1 1 1	00 01 02 . FF	Internal data memory address space
8kB RAM-I	0 0 0 0 0 0 0 0 0 0 0 0 . 0 0 0 1	0 0 0 0 0 0 0 0 0 0 0 0 . 1 1 1 1	0 0 0 0 0 0 0 0 0 0 0 0 . 1 1 1 1	0 0 0 0 0 0 0 1 0 0 1 0 . 1 1 1 1	0000 0001 0002 . 1FFF	External data memory address space
8 kB RAM-II	0 0 1 0 0 0 1 0 0 0 1 0 . 0 0 1 1	0 0 0 0 0 0 0 0 0 0 0 0 . 1 1 1 1	0 0 0 0 0 0 0 0 0 0 0 0 . 1 1 1 1	0 0 0 0 0 0 0 1 0 0 1 0 . 1 1 1 1	2000 2001 2002 . 3FFF	External data memory address space
8255 Port-A Port-B Port-C Control register	0 1 0 X 0 1 0 X 0 1 0 X 0 1 0 X	X X X X X X X X X X X X X X X X	X X X X X X X X X X X X X X X X	X X 0 0 X X 0 1 X X 1 0 X X 1 1	4000 4001 4002 4003	External data memory address space
8279 Data register Control register	0 1 1 X 0 1 1 X	X X X X X X X X	X X X X X X X X	X X X 0 X X X 1	6000 6001	External data memory address space

Note: *The don't care "X" is considered as zero.*

The 16 kB RAM memory can be implemented by using two numbers of 8 kB RAM 6264. The RAM memories 8255 and 8279 can be interfaced to the 8051 controller as data memory. The 8 kB RAM requires 13 address lines and so the address lines A_0-A_{12} are connected to the address pins of the RAM to select its internal locations. The address lines A_0 and A_1 are connected to 8255 and the address line A_0 is connected to 8279 to select their internal devices.

A 2-to-4 decoder is employed in the system to generate the chip select signals required for RAM, 8255 and 8279. The address lines A_{13} and A_{14} are connected to the input of the decoder to generate four chip select signals. The address line A_{15} is used as logic low chip enable for the decoder. Since the internal ROM is used in the system, the pin $\overline{EA}$ should be tied to V_{CC} or +5-V. The 8051 provides separate read and write control signals $\overline{RD}$ and $\overline{WR}$ for reading and writing with devices interfaced as data memory. The memory and IO interface diagram is shown in Fig. 51.2. The addresses allotted to the various devices are listed in Table 51.2. The 8051 has separate 256 bytes internal data memory address space allotted to internal RAM and SFR.

6.8 SHORT-ANSWER QUESTIONS

Q6.1 *What is memory ?*

A memory is a storage device in a microprocessor-based system and its primary function is to store programs and data.

Q6.2 *Why are semiconductor memories used as main memory in microprocessor system ?*

Semiconductor memories have processor compatible access time for read and write operations, they are used as main memories.

Q6.3 *What are the different types of semiconductor memory ?*

The different types of semiconductor memory are RAM, PROM, EPROM, static RAM, DRAM and NVRAM.

Q6.4 *List the features of semiconductor memories.*

1. Semiconductor memories are random access memories.

2. In semiconductor memories, a read operation by the processor will not destroy the stored information.

3. The read and write time of the semiconductor memory is compatible for the microprocessor.

Q6.5 *What is meant by volatile and non-volatile memories ?*

If the information stored in the memory is lost when the power supply is switched OFF, then the memory is called volatile.

If the content of memory is preserved even after the power supply is switched OFF, then the memory is called non-volatile.

Q6.6 *List the volatile and non-volatile semiconductor memories.*

The volatile semiconductor memories are static RAM and DRAM. The non-volatile semiconductor memories are ROM, PROM, EPROM and NVRAM.

Q6.7 *What are the characteristics of* **ROM** *memory ?*

The characteristics of ROM are as follows:

1. It is non-volatile memory.

2. The contents of ROM can be read by the processor but it cannot write into it.

3. The ROM has the feature of random access.

4. The memory cell has a MOS transistor either with an open gate or a closed gate.

Q6.8   How are ROMs classified?

ROMs can be classified into the following three categories based on the method of programming.

1. **Custom programmed or Mask programmed ROM**
2. **Programmable or Field programmable ROM**
3. **Reprogrammable or Erasable - programmable ROM.**

Q6.9   List the characteristics of EPROM.

1. **The EPROM is non-volatile.**
2. **It has random-access feature.**
3. **The contents of EPROM can be erased by passing UV light and then the device can be reprogrammed.**
4. **The EPROM is read only memory and for writing into EPROM, a separate hardware set up is required.**

Q6.10 Write a short note on the memory cell of EPROM.

The memory cell of EPROM contains a MOS transistor with isolated gate. The isolated gate is located between the normal control gate and the source/drain region of transistor. The information is stored as a charge or no charge in the floating gate.

Q6.11 What is NVRAM?

The non-volatile read/write memories are called NVRAM. The various types of NVRAMs are flash memory, EEPROM and Shadow RAM.

Q6.12 List the features of static RAM.

1. **Static RAMs are read/write memories.**
2. **They are volatile and have random access feature.**
3. **The memory cell is a flip-flop constructed using 6 to 8 MOS transistors.**

Q6.13 What is DRAM ?

DRAMs are read/write semiconductor memories in which the information is stored in the form of electric charge on the gate to substrate capacitance of a MOS transistor.

Q6.14 List the characteristics of DRAM.

1. **The DRAMs are volatile and have random access feature.**
2. **They are read/write memories.**
3. **The contents of DRAM have to be refreshed periodically using refreshing circuits.**
4. **The memory cell of DRAM will have 3 to 4 MOS transistor.**

Q6.15 Compare the Static RAM with DRAM.

Static RAM	DRAM
1. Information is stored as voltage level in a flip-flop.	1. Information is stored as a charge in the gate to substrate capacitance.
2. Six to eight transistors are required to form one memory cell.	2. Three to four transistors are required to form one memory cell.
3. Packing density is low.	3. Packing density is high.
4. The contents of memory need not be refreshed.	4. The contents of memory has to be refreshed periodically.

Q6.16 What is physical memory space?

The memory locations that are directly addressed by the microprocessor is called physical memory space.

Q6.17 What is memory word size?

The size of data that can be stored in memory location is called memory word size.

Q6.18 What is meant by memory mapping?

Memory mapping is the process of interfacing memories to microprocessor and allocating addresses to each memory location.

Q6.19 What is memory access time?

Memory access time is the time taken by the processor to read or write a memory location. Read operation is the time between a valid address on the bus and the end of read control signal. Write operation is the time between a valid address on the bus and the end of write control signal.

Q6.20 What are the factors to be considered while selecting a semiconductor memory for a microprocessor system ?

The following are the factors to be considered while selecting a semiconductor memory IC:

1. Capacity and organization (Memory word size).
2. Timings of various signals.
3. Power consumption and bus loading (Current levels).
4. Physical dimensions and packaging.
5. Cost, reliability and availability.

Q6.21 What is bus contention?

If two devices drive the data bus simultaneously then it is called bus contention. It may lead to following undesirable events:

1. Damaging one or both the IC chip.
2. The high current may cause a voltage spike in the supply system leading to data loss.

Q6.22 Why is EPROM mapped at the beginning of memory space in 8085 ?

When EPROM is mapped at the beginning of memory space, then 0000_H address will be allotted to EPROM. The monitor program can be stored from 0000_H address. Whenever the processor is reset, the program counter will be cleared (i.e it will have 0000_H address) and the monitor program will be executed automatically.

Q6.23 Why is EPROM mapped at the end of memory space in an 8086 ?

The mapping of EPROM at the end of memory space will facilitate automatic execution of the monitor program upon reset. Whenever the processor is resetted, the IP is cleared and CS is initialized with $FFFF_H$, and so after a reset the processor will start executing the instructions from $FFFF0_H$ after a reset. The system designer has to permanently store the monitor/boot program starting from this address, which is possible only if EPROM is mapped at the end of memory space.

Q6.24 What is chip select and how it is generated?

Chip select is the control signal that has to be asserted TRUE to bring an IC from **high impedance** state to normal state. Generally the chip select signals are generated in a system by decoding the unused address lines with the help of decoders.

Q6.25 What are the typical control signals involved in EPROM interfacing ?

The control signals needed for EPROM are chip select and output enable.

Q6.26 What are the typical control signals involved in RAM interfacing ?

The control signals needed for RAM interfacing are chip enable, output enable and write enable.

Q6.27 What is the relation between memory capacity and address and data pins of memory IC?

If a memory IC has "m" data pins and "n" address pins, then the memory IC will have a capacity of $2^n \times m$ bits. When m = 8, the memory capacity is 2^n bytes.

Q6.28 How memory space is organized in an 8086?

In 8086, the one mega-byte (1 MB) of addressable memory space is divided into two banks: Even (or lower) memory bank and Odd (or upper) memory bank. Each bank will have an addressable space of 512 kB.

Q6.29 How are the data lines connected to memory banks in an 8086?

In an 8086-based system, the lower eight lines of the data bus, D_0 - D_7 are connected to the even bank memory ICs and the upper eight lines of data bus, D_8-D_{15} are connected to the odd bank memory ICs.

Q6.30 What are the signals involved in memory bank selection?

In 8086-based system, the even bank is selected by the address line A_0 and the odd bank is selected by the control signal $\overline{BHE}$.

Q6.31 What is the available address space in an 8051-based system ?

The 8051-based system has 64 kB program memory address space, 64 kB external data memory address space and 256 bytes internal data memory address space.

Q6.32 How is the program memory organized in an 8051 based system ?

In an 8051-based system the entire 64 kB program memory can be external or 4 kB is internal and the remaining 60 kB is external. This is decided by the logic level of the signal at $\overline{EA}$ pin. When $\overline{EA}$ pin is tied **high** $(+ V_{cc}$ or 5-V) the first 4 kB of program memory is internal and the remaining 60 kB is external. When $\overline{EA}$ pin is tied **low** (GND or 0-V) the internal ROM is ignored and the entire 64 kB is external.

Q6. 33 What is programmed IO ?

If the data transfer between an IO device and the processor is accomplished through an IO port and controlled by a program then the IO device is called programmed IO.

Q6.34 What is interrupt IO?

If the IO device initiates the data transfer through interrupt, then the IO is called interrupt driven IO.

Q6.35 What is DMA?

The direct data transfer between the IO device and the memory is called DMA.

Q6.36 What is the need for port?

IO devices are generally slow devices and their timing characteristics do not match with processor timings. Hence, the IO devices are connected to a system bus through the ports.

Q6.37 What is a port?

A port is a buffered I/C which is used to hold the data transmitted from the microprocessor to IO device or vice versa.

Q6.38 Give some examples of port devices used in a 8085 microprocessor-based system.

The various INTEL IO port devices used in 8085 microprocessor-based system are 8212, 8155,8156, 8255, 8355 and 8755.

Q6.39 Write a short note on INTEL 8255?

The INTEL 8255 is a IO port device consisting of 3 numbers of 8-bit parallel IO ports. The ports can be programmed to function either as an input port or as an output port in different operating modes. It requires 4 internal addresses and has one logic **low** chip select pin.

Q6.40 What are the different methods of interfacing IO devices to 8085 and 8086-based system?

There are two methods of interfacing IO devices to 8085 and 8086-based system. They are memory mapping of IO device and standard IO mapping.

Q6.41 Draw a simple circuit to decode the three control signals $\overline{RD}, \overline{WR}$ and $IO/\overline{M}$ and to produce separate read/write control signals for memory and IO devices.

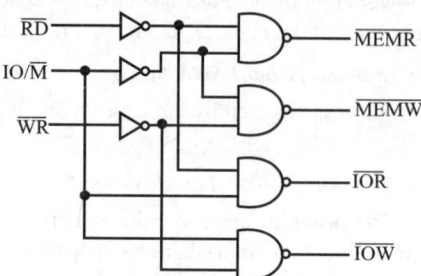

Fig. Q6.41: Circuit to generate separate read and write signals for memory and IO devices in an 8085-based system.

Q6.42 Compare the memory-mapped IO with the standard IO-mapped IO in an 8085-based system.

Memory-mapped IO	Standard IO-mapped IO
1. Sixteen bit address is allotted to an IO device.	1. Eight bit address is allotted to an IO device.
2. The devices are accessed by memory read or memory write cycle.	2. The devices are accessed by IO read or IO write cycle.
3. All instructions related to memory can be used for data.	3. Only IN and OUT instructions can be used for data transfer.
4. A large number of IO ports can be interfaced.	4. Only 256 ports can be interfaced.

Q6.43 Draw a simple circuit to decode three control signals $\overline{RD}, \overline{WR}$ and $M/\overline{IO}$ and to produce separate read/write control signals for memory and IO devices in 8086-based system.

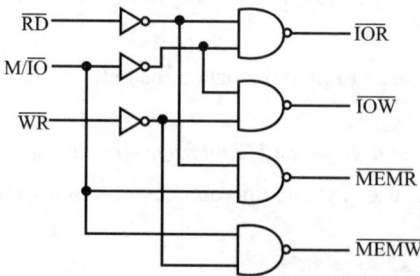

Fig. Q6.43: Circuit to generate separate read and write signals for memory and IO devices in an 8086-based system.

Q6.44 Compare the memory-mapped IO and standard IO-mapped IO in an 8086-based system.

Memory mapping of IO device	IO mapping of IO device
1. 20-bit addresses are provided for IO devices.	1. 8-bit or 16-bit addresses are provided for IO devices.
2. The IO ports or peripherals can be treated like memory locations and so all instructions related to memory can be used for data transfer between the IO device and the processor.	2. Only IN and OUT instructions can be used for data transfer between IO device and the processor.
3. In memory-mapped ports, the data can be moved from any register to the ports and vice versa.	3. In IO-mapped ports, the data transfer can take place only between the accumulator and ports.
4. When memory mapping is used for IO devices, the full memory address space cannot be used for addressing memory. Hence memory mapping is useful only for small systems, where the memory requirement is less.	4. When IO mapping is used for IO devices, then the full memory address space can be used for addressing memory. Hence it is suitable for systems which requires large memory capacity.
5. For accessing the memory-mapped devices, the processor executes memory read or write cycle. During this cycle $M/\overline{IO}$ is asserted high.	5. For accessing the IO-mapped devices, the processor executes IO read or write cycle. During this cycle $M/\overline{IO}$ is asserted low.

6.9 EXERCISES

I. Fill in the blanks with appropriate words

1. A semiconductor memory with n address lines and m data lines has a memory capacity of _____ bits.

2. The memory which retains data even after the power supply to the chip is switched OFF is called _____ memory.

3. RAM is a _____ memory.

4. The ROM which can be erased by passing electrical current or UV light is called _____.

5. The memory which requires frequent refreshing is called _____.

6. The 1024×4 bit memory chip has _____ addres lines and _____ data lines.

7. The $2^{11} \times 8$ bit memory chip has _____ number of memory locations and each location can store _____ number of bits.

8. The number of address lines needed to address a location in 8 kB memory is _____.

9. The _____ decoder is used to generate four chip select signals.

10. If the starting address of 4 kB RAM chip is mapped to 2000_H the end address will be _____.

11. _____ signal is used to select the odd memory bank in 8086.

12. The status of $A_0 =$ _____ and $\overline{BHE} =$ _____ when a word stored at address 20002_H is accessed by the 8086 based system.

13. When $A_0 = 0$ and $\overline{BHE} = 1$, the data byte from _____ memory bank of 8086 is accessed.

14. To access a data byte stored at 32001_H the $A_0 =$ _____ $\overline{BHE} =$ _____.

15. In 8086 processor the number of bus cycles required to access a word stored at odd address is _____.

16. The address range of 32 kB EPROM chip interfaced to 8085 system is from _____ to _____ if $\overline{CS} = \overline{A}_{15}$.

17. The number of 256×4 bits RAM chips required to make up 1 kB memory is _____.

18. The size of the 8051 external program memory when $\overline{EA} = 1$ is _____ bytes and mapped in the address range _____ to _____.

19. The _____ is the data transfer scheme which enables direct data transfer between memory and IO devices.

20. Only _____ and _____ instructions can be used for data transfer between the IO devices and the processor in IO mapped devices.

Answers

1. $2^n \times m$	6. 10, 4	11. $\overline{BHE}$	16. 0000_H, $7FFF_H$
2. non volatile	7. 2048, 8	12. 0, 0	17. 8
3. volatile	8. 13	13. even	18. 60k, 1000_H, $FFFF_H$
4. EPROM	9. 2 to 4	14. 1, 0	19. direct memory access (DMA)
5. dynamic RAM	10. $2FFF_H$	15. two	20. IN, OUT

II. State whether the following statements are True/False.

1. The registers of microprocessors can be accessed faster than the main memory.
2. The primary memory is slower than secondary memory.
3. The volatile memory loses its data when the power supply to the chip is switched OFF.
4. The EPROM is a non-volatile memory.
5. All the semiconductor memories are non-destructive readout memory.
6. PROMs are many time programmable by the user.
7. Static RAM can pack more bits than dynamic RAM in a given physical area.
8. The dynamic RAM (DRAM) is faster than static RAM and consumes less power.
9. The flash memory is a type of non-volatile RAM.
10. In 8086 system, EPROMs are mapped at the end of memory space.
11. In 8086 system, RAMs are mapped at the beginning of memory space.
12. The monitor program in 8086 processor based system is stored from the address 00000_H.
13. The unused address lines of microprocessor are generally used for generating chip select signals.
14. All the memory/peripheral ICs always remain in active state.
15. There are no even and odd address memory banks in 8088 based system.
16. It is always advised to store the data at even address boundary to avoid extra time taken to execute extra bus cycles.
17. The internal 4 kB EPROM of 8051 is not used when $\overline{EA} = 0$.
18. It is not possible to have a single memory bank for both data and program in 8051.
19. IO devices can be directly connected to microprocessors.
20. In memory mapped I/O technique, both memory and IO devices are treated in a same way.
21. In I/O mapped ports, the data can be moved from any registers to the ports and vice versa.
22. A separate decoder is required to generate the chip select signals for IO mapped devices.

Answers

1. True	5. True	9. True	13. True	17. True	21. False
2. False	6. False	10. True	14. False	18. False	22. True
3. True	7. False	11. True	15. True	19. False	
4. True	8. True	12. False	16. True	20. True	

III. Choose the right answer for the following questions.

1. *Which of the following is volatile memory?*

   a) RAM   b) ROM   c) PROM   d) EPROM

2. *What is the range of address (in hexadecimal) for 16 KB memory?*

   a) 0000_H to $FFFF_H$ b) 0000_H to $03FF_H$ c) 0000_H to $3FFF_H$ d) 0000 to $7FFF_H$

3. *Which of the following instruction is used for data transfer between IO device and the processor*

   a) MOV   b) IN   c) XCHG   d) all the above

4. *Which of the following pin is used to differentiate the memory access and I/O access?*

   a) $M/\overline{IO}$ b) $\overline{RD}$ c) $\overline{WR}$ d) $DT/\overline{R}$

5. *In 8085 processor, how many 1024×4 bits memory chips are required for a system memory of 4 kB?*

   a) 4   b) 8   c) 12   d) 16

6. *The starting address of a 32 kB memory chip in 8085 processor is 0000H. What is the end address?*

   a) $FFFF_H$ b) $7FFF_H$ c) $3FFF_H$ d) $1FFF_H$

7. *A memory chip has 11 address lines and 4 data lines. How many such chips are required to design a memory of 4 kB?*

   a) 4   b) 8   c) 16   d) 32

8. *In 8085 microprocessor, a 4 kB EPROM and 4 kB RAM are selected when $A_{15} = A_{14} = A_{13} = A_{12} = 0$ and $A_{15} = A_{14} = A_{13} = A_{12} = 1$ respectively. What is the unused address space?*

   a) 1000_H to $EFFF_H$ b) 2000_H to $DFFF_H$ c) 4000_H to $FFFF_H$ d) 6000_H to $CFFF_H$

9. *An 8 kB EPROM and 8 kB RAM are interfaced to 8085 microprocessor. Which of the following can be possible correct address range for this memory devices,*

   i) EPROM = 0000_H to $1FFF_H$ ii) EPROM = 0000_H to $2FFF_H$
        RAM   = 2000_H to $3FFF_H$ RAM   = 3000_H to $3FFF_H$

   iii) EPROM = 0000_H to $0FFF_H$ iv) EPROM = 0000_H to $1FFF_H$
         RAM   = 1000_H to $2FFF_H$ RAM   = $D000_H$ to $FFFF_H$

   a) i   b) iii   c) i and iv   d) ii and iv

10. *A 16 kB RAM in an 8085 system has address range 4000_H to $7FFF_H$. If the unused address lines are used to generate the chip select signals the $\overline{CS} = ?$*

    a) $\overline{CS} = \overline{A}_{15}\overline{A}_{14}$ b) $\overline{CS} = \overline{A}_{15}A_{14}$ c) $\overline{CS} = A_{15}$ d) $\overline{CS} = A_{14}$

11. *How many address lines are used to generate $\overline{CS}$ signal in memory organization of 128 kB EPROM and 384 kB RAM in 8086 system? Memory sytem is designed only by using 64 kB chips.*

    a) 4   b) 3   c) 2   d) 1

12. *How many address lines are needed to access the full memory space in 8088 system. RAM and ROM shares equal space.*

    a) 20   b) 19   c) 18   d) 17

13. *If the 64 kB even bank RAM chip in 8086 has $\overline{CS} = \overline{A}_{19}A_{18}\overline{A}_{17}$, then its memory range will be*

    a) 00000_H to $1FFFE_H$ b) 0000_H to $1FFFF_H$
    c) 40000_H to $5FFFE_H$ d) 40000_H to $5FFFF_H$

14. *In 8086 based system it is required to read the word which is stored at* 28005_H. *Which of the following conditions to be satisfied to perform the task?*

    a) i) $A_0 = \overline{BHE} = 0$ *b)* i) $A_0 = 0$; $\overline{BHE} = 1$ *c)* i) $A_0 = 1$; $\overline{BHE} = 0$ *d)* i) $A_0 = \overline{BHE} = 1$

    ii) $A_0 = \overline{BHE} = 1$ ii) $A_0 = 1$; $\overline{BHE} = 0$ ii) $A_0 = 0$; $\overline{BHE} = 1$ ii) $A_0 = \overline{BHE} = 0$

15. *In 8086 system, the memory locations* 42005_H *and* 42006_H *contain the bytes* 07_H *and* $2A_H$ *respectively. Which data bytes will be accessed through data lines* D_0-D_7 *and* D_8-D_{15} *while accessing the word using instruction MOV AX,* $[2005_H]$*?. Assume DS =* 40000_H

    a) D_0-$D_7 = 2AH$ *b)* D_0-$D_7 = 07H$ *c)* D_0-$D_7 = 05H$ *d)* D_0-$D_7 = 20H$

    D_8-$D_{15} = 07H$ D_8-$D_{15} = 2AH$ D_8-$D_{15} = 20H$ D_8-$D_{15} = 05H$

Answers							
1. a	3. b	5. b	7. a	9. c	11. b	13. c	15. a
2. c	4. a	6. b	8. a	10. c	12. b	14. c	

IV. Answer the following questions.

E6.1   An EPROM of size 262144 bit is interfaced with 8086 microprocessor. How many address lines are required to address one of the memory location?

E6.2   A semiconductor memory has 16 address lines and 8 data lines. What is the capacity of the memory?

E6.3   Show the memory interfacing of 32 kB RAM and 32 kB ROM with 8085 microprocessor.

E6.4   Implement the exercise E6.3 using 8086 microprocessor.

E6.5   How many 64 kB memory chips are required to provide complete 1 MB memory space in 8086 microprocessor based system? Explain the allocation of address lines and even and odd memory bank selection.

E6.6   Explain how the chip select ($\overline{CS}$) signals are generated to interface 256 kB RAM, 256 kB ROM and an 8255 chip with 8086.

E6.7   How many 256 kB memory chips are needed to provide a memory capacity of 512 kB RAM and 512kB ROM in 8086? How many address lines are commonly connected to all the chips and how many address lines are used for generating chip select signals?

E6.8   Consider the following decoding circuit for an 8086 processor. The address lines which are not mentioned in this circuit are used to select the memory location.

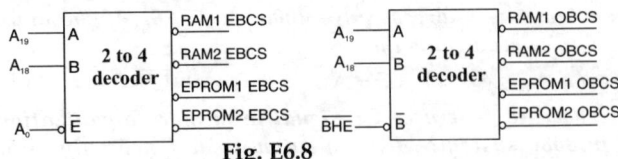

Fig. E6.8

    a. What is the memory size of the system?

    b. What is the memory size of total RAM memory?

    c. What is the memory size of total EPROM memory?

    d. What is the memory size of RAM and EPROM chip?

    e. Give the memory map

E6.9   Show the complete interface of 16 kB EPROM IC27128 and 8 kB RAM IC6264 with the 8051. Ignore the internal 4 kB ROM.

CHAPTER 7

INTERRUPTS

7.1 INTERRUPT AND ITS NEED

The microprocessors allow normal program execution to be interrupted in order to carry out a specific task/work. The processor can be interrupted in the following ways:

1. **by an external signal generated by a peripheral**

2. **by an internal signal generated by a special instruction in the program**

3. **by an internal signal generated due to an exceptional condition which occurs while executing an instruction.**

   (For example, in an 8086 processor, divide by zero is an exceptional condition which initiates a type-0 interrupt and such an interrupt is also called exception.)

In general, the process of interrupting the normal program execution to carry out a specific task/work is referred to as interrupt.

The interrupt is initiated by a signal generated by an external device or by a signal generated internal to the processor. When a microprocessor receives an interrupt signal, it stops executing the current normal program, saves the status (or content) of various registers (IP, CS and flag registers in case of 8086) in a stack and then the processor executes a subroutine/procedure in order to perform the specific task/work requested by the interrupt.

The subroutine/procedure that is executed in response to an interrupt is also called Interrupt Service Routine (ISR) or Interrupt Service Subroutine (ISS). At the end of ISR, the stored status of the registers in the stack are restored to the respective registers, and the processor resumes the normal program execution from the point (instruction) where it was interrupted.

The external interrupts are used to implement interrupt-driven data transfer schemes. The interrupts generated by special instructions are called software interrupts and they are used to implement system services/calls (or monitor services/calls). The system/monitor services are procedures developed by the system designer for various operations and stored in the memory. The user can call these services through software interrupts. The interrupts generated by exceptional conditions are used to implement error conditions in the system.

7.1.1 Interrupt Driven Data Transfer Scheme

Interrupts are useful for efficient data transfer between the processor and the peripheral. When a peripheral is ready for data transfer, it interrupts the processor by sending an appropriate signal. Upon receiving an interrupt signal, the processor suspends the current program execution, saves the status in a stack and executes an ISR to perform the data transfer between the peripheral and the processor. At the end of ISR the processor status is restored from stack and the processor resumes its normal program execution. This type of data transfer scheme is called interrupt driven data transfer scheme.

The data transfer between the processor and peripheral devices can be implemented either by polling technique or by interrupt method. In polling technique, the processor has to periodically poll or check the status/readiness of the device and can perform data transfer only when the device is ready. In polling technique the processor time is wasted, because the processor has to suspend its work and check the status of the device in predefined intervals.

Alternatively, if the device interrupts the processor to initiate a data transfer whenever it is ready then the processor time is effectively utilized because the processor need not suspend its work and check the status of the device in predefined intervals.

For example, consider the data transfer from a keyboard to the processor. Normally a keyboard has to be checked by the processor once in every 10 millisecond for a key press. Therefore, once in every 10 milliseconds the processor has to suspend its work and then check the keyboard for a valid key code. Alternatively, the keyboard can interrupt the processor, whenever a key is pressed and a valid key code is generated. In this way the processor need not waste its time to check the keyboard once in every 10 milliseconds.

7.2 CLASSIFICATION OF INTERRUPTS

In general interrupts can be classified in the following three ways:

1. **Hardware and software interrupts.**
2. **Vectored and non-vectored interrupts.**
3. **Maskable and non-maskable interrupts.**

The interrupts initiated by external hardware by sending an appropriate signal to the interrupt pin of the processor is called hardware interrupt. The 8085 processor has five interrupt pins: TRAP, RST 7.5, RST 6.5, RST 5.5 and INTR, and the interrupts initiated by applying the appropriate signal to these pins are called hardware interrupts of an 8085. The 8086 processor has two interrupt pins: INTR and NMI, and the interrupts initiated by applying the appropriate signal to these pins are called hardware interrupts of an 8086.

The software interrupts are program instructions. These instructions are inserted at desired locations in a program. While running a program, if a software interrupt instruction is encountered then the processor initiates an interrupt. The 8085 processor has 8 types of software interrupts and the 8086 processor has 256 types of software interrupts. The software interrupt instruction is INT n, where n is the type number in the range 0 to 7 in case of 8085 and 0 to 255_{10} in case of 8086.

When an interrupt signal is accepted by the processor, if the program control automatically branches to a specific address (called vector address) then the interrupt is called vectored interrupt. The automatic branching to a vector address is predefined by the manufacturer of the processors. [In these vector addresses, the Interrupt Service Routines(ISR) are stored.] In non-vectored interrupts the interrupting device should supply the address of the ISR to be executed in response to the interrupt.

All the 8085 interrupts except INTR are vectored interrupts. In 8086 all the interrupts are vectored interrupts and the vector address for an 8086 interrupt is obtained from a vector table implemented in the first 1 kB memory space.

A processor has the facility for accepting or rejecting hardware interrupts. Programming the processor to reject an interrupt is referred to as masking or disabling and programming the processor to accept an interrupt is referred to as unmasking or enabling. In 8085, the hardware interrupts RST 7.5, RST 6.5, and RST 5.5 can be masked/unmasked using an SIM instruction. Also, in 8085 all the hardware interrupts except TRAP are disabled by executing the DI instruction and they are enabled by executing the EI instruction. In 8086, the Interrupt Flag (IF) can be set to one to unmask or enable all hardware interrupts and IF is cleared to zero to mask or disable all hardware interrupts except NMI.

The interrupts whose request can be either accepted or rejected by the processor are called maskable interrupts. The interrupts whose request have to be definitely accepted (or cannot be rejected) by the processor are called non-maskable interrupts. Whenever a request is made by the non-maskable interrupt, the processor has to definitely accept that request and service that interrupt by suspending its current program and executing an ISR. In an 8085 processor, all the hardware interrupts except TRAP are maskable and the interrupt initiated through TRAP pin and all software interrupts are non-maskable. In an 8086 processor all the hardware interrupts initiated through INTR pin are maskable by clearing Interrupt Flag (IF). The interrupt initiated through NMI pin and all software interrupts are non-maskable.

7.3 SOURCES OF INTERRUPTS

7.3.1 Sources of Interrupts in 8085

The interrupt in 8085 can come from one of the following two sources:

1. One source is from an external signal applied to TRAP, RST7.5, RST6.5, RST5.5 or INTR pin of the processor. The interrupts initiated by applying appropriate signals to these pins are called hardware interrupts.

2. The second source of an interrupt is the execution of the interrupt instruction "RST n" where n can take values from 0 to 7. The interrupts initiated by "RST n" instructions are called software interrupts.

7.3.2 Sources of Interrupts in 8086

An interrupt in 8086 can come from one of the following three sources:

1. One source is from an external signal applied to NMI or INTR input pin of the processor. The interrupts initiated by applying appropriate signals to these input pins are called hardware interrupts.

2. A second source of an interrupt is execution of the interrupt instruction "INT n", where n is the type number. The interrupts initiated by "INT n" instructions are called software interrupts.

3. The third source of an interrupt is from some condition produced in the 8086 by the execution of an instruction. An example of this type of interrupt is divide by zero interrupt. Program execution will be automatically interrupted if you attempt to divide an operand by zero. Such conditional interrupts are also known as exceptions.

7.3.3 Sources of Interrupts in 8051

The 8051 microcontroller has five interrupts. In this, two interrupts are external interrupts and the remaining three are internal interrupts. The external interrupts are hardware interrupts that are initiated by applying an appropriate signals at the pins INT0 or INT1 by the external hardware. The three internal interrupts are initiated by timer-0, timer-1 and serial port. Every interrupt has a vector address to which program control is transferred and a flag. The interrupts of 8051 in the order of highest to lowest priority along with their vector address are listed in Table 7.1.

The priority of 8051 interrupts is alterable by programming the interrupt priority register. The interrupts of 8051 can be also disabled and enabled by programming the interrupt enable register. Sometimes, the reset is also considered as an interrupt with vector address 0000H.

Table 7.1: Interrupts of 8051 and their Vector Address

Interrupt	Vector address	Normal priority
External interrupt-0	0003_H	Highest
Timer-0 interrupt	$000B_H$	
External interrupt-1	0013_H	
Timer-1 interrupt	$001B_H$	
Serial port interrupt	0023_H	Lowest

7.4 INTERRUPTS OF 8085

7.4.1 Software Interrupts of 8085

Software interrupts are program instructions. When a software interrupt instruction is executed, the processor executes an Interrupt Service Routine(ISR) stored in the vector address of that software interrupt instruction.

The software interrupts of 8085 are RST0, RST1, RST2, RST3, RST4, RST5, RST6 and RST7. The software interrupts of 8085 are vectored interrupts. Software interrupts cannot be masked or be disabled. The Vector addresses of software interrupts are given in Table 7.2.

Table 7.2

Interrupt	Vector address
RST 0	0000_H
RST 1	0008_H
RST 2	0010_H
RST 3	0018_H
RST 4	0020_H
RST 5	0028_H
RST 6	0030_H
RST 7	0038_H

Software interrupt instructions are included at the appropriate (or required) place in the main program. When the processor encounters the software instruction, it pushes the content of PC (**P**rogram **C**ounter) to stack. Then, it loads the vector address in to the PC and starts executing an ISR stored in this address. The last instruction of the ISR will be RET instruction. When the RET instruction is executed, the processor POPs the content of top of stack to PC. Hence, the processor control returns to main program after servicing the interrupt. *[Execution of ISR is referred to as servicing of interrupt.]*

7.4.2 Hardware Interrupts of 8085

The hardware interrupts of 8085 are initiated by an external device by placing an appropriate signal at the interrupt pin of the processor. The processor keeps on checking the interrupt pins at the second T-state of the last machine cycle of every instruction. If the processor finds a valid interrupt signal and if the interrupt is unmasked and enabled, then the processor accepts the interrupt. The acceptance of the hardware interrupt is acknowledged by sending an INTA signal to the interrupting device.

When the interrupt is accepted, the processor saves the content of the PC (**P**rogram **C**ounter) in stack and then loads the vector address of the interrupt to the PC. (If the interrupt is non-vectored, then the interrupting device has to supply the address of ISR when it receives INTA signal.) Then the processor starts executing ISR in this address. The last instruction of ISR will be an RET instruction. When the processor executes the RET instruction, it POP the content of top of stack to PC. Thus the processor control returns to the main program after servicing the interrupt.

The hardware interrupts of 8085 are TRAP, RST 7.5, RST6.5, RST5.5 and INTR. TRAP, RST7.5, RST6.5 and RST5.5 are vectored interrupts. In vectored interrupts the address to which the program control is transferred (when the interrupt is accepted) is fixed by the manufacturer. The vector addresses of hardware interrupts are given in Table 7.3. The INTR is a non-vectored interrupt. Hence when a device interrupts through INTR, it has to supply the address of ISR after receiving interrupt acknowledge signal.

Table 7.3

Interrupt	Vector address
RST 7.5	$003C_H$
RST 6.5	0034_H
RST 5.5	$002C_H$
TRAP	0024_H

The type of signal that has to be placed on the interrupt pin of hardware interrupts of 8085 are defined by INTEL. The TRAP interrupt is edge and level sensitive. Hence, to initiate TRAP, the interrupt signal has to make a **low** to **high** transition and then it has to remain **high** until the interrupt is recognized. The RST 7.5 interrupt is edge sensitive (positive edge). In order to initiate the RST7.5, the interrupt signal has to make a **low** to **high** transition and it need not remain **high** until it is recognized. The RST6.5, RST5.5 and INTR are level sensitive interrupts. Hence, for these interrupts the interrupt signal should remain **high**, until it is recognized.

TRAP is a non-maskable interrupt and RST7.5, RST6.5 and RST5.5 are maskable interrupts, which use the SIM (**S**et **I**nterrupt **M**ask) instruction. Interrupts can be masked by moving an appropriate data (or code) to the accumulator and then executing the SIM instruction. The status of maskable interrupts can be read into the accumulator by executing the RIM instruction (RIM - **R**ead **I**nterrupt **M**ask).

All the hardware interrupts, except TRAP are disabled when the processor is reset and they can also be disabled by executing the DI instruction. (DI - **D**isable **I**nterrupt). When an interrupt is disabled, it will not be accepted by the processor (i.e., INTR, RST5.5, RST6.5 and RST7.5 are disabled by the DI instruction and upon hardware reset). In order to enable (or to allow) the disabled interrupts, the processor has to execute the EI instruction (EI - **E**nable **I**nterrupt).

7.4.3 Priorities of Interrupts of 8085

When all the interrupts are enabled, the priority sequence of hardware interrupts from highest to lowest is TRAP, RST 7.5, RST 6.5, RST 5.5 and INTR. When the 8085 processor accepts an interrupt it will disable all the hardware interrupts except TRAP. Hence in order to allow the higher priority interrupt while executing Interrupt Service Subroutine (ISR) for lower priority interrupt, enable the interrupt system in the beginning of ISR of lower priority interrupt, by executing EI instruction.

For example, if the processor accepts RST 5.5 interrupt, then it will disable RST 7.5, RST 6.5 and INTR interrupts. In order to allow the higher priority interrupt RST 7.5 and RST 6.5 while executing ISR of RST 5.5, the EI instruction should be executed in the beginning of ISR of RST 5.5.

The execution of software interrupt will not disable any hardware interrupt. Therefore while executing ISR of software interrupts, the processor will recognize or allow the hardware interrupts.

7.4.4 Enabling, Disabling and Masking of 8085 Interrupts

TRAP

The interrupt TRAP is non-maskable and it cannot be disabled by DI instruction. Also the TRAP is not disabled by system (processor) reset or after recognition of another interrupt. The only signal which can override TRAP is HOLD signal. (i.e., If the processor receives HOLD and TRAP at the same time then HOLD is recognized first and only then is TRAP recognized.)

INTR

The interrupt INTR is disabled by any one of the following operations:

1. **Executing DI instruction.**
2. **System or processor reset.**
3. **After recognition (acceptance) of an interrupt.**

The interrupt INTR can be enabled by executing EI instruction.

RST 7.5, RST 6.5 and RST 5.5

The interrupt RST 7.5, RST 6.5 and RST 5.5 are disabled by any one of the following operations.

1. **Executing DI instruction.**
2. **System or processor reset.**
3. **After recognition (acceptance) of an interrupt.**

These hardware interrupts can be enabled by executing EI instruction.

The 8085 provides additional masking facility for RST 7.5, RST 6.5 and RST 5.5 using SIM instruction. The status of these interrupts can be read by executing RIM instruction.

The masking or unmasking of RST 7.5, RST 6.5 and RST 5.5 interrupts can be performed by moving an 8-bit data to accumulator and then executing SIM instruction. The format of the 8-bit data is shown in Fig. 7.1.

The status of pending interrupts can be read from accumulator after executing RIM instruction. When RIM instruction is executed, an 8-bit data is loaded to the accumulator, which can be interpreted as shown in Fig. 7.2.

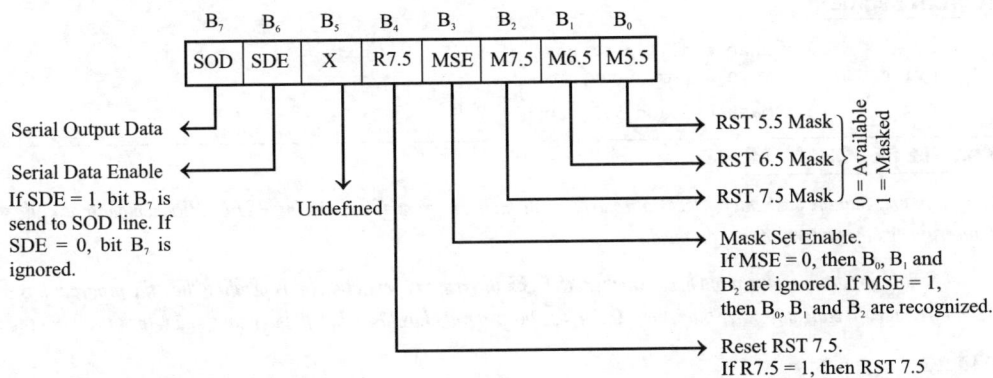

Fig. 7.1: Format of the 8-bit data to be loaded in the accumulator before executing a SIM instruction.

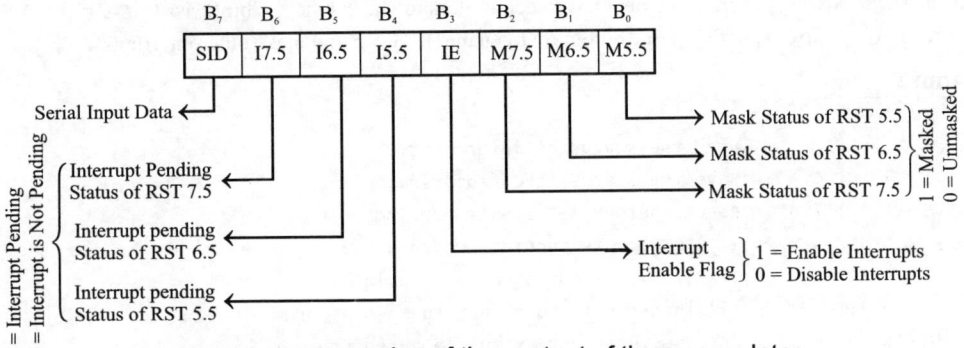

Fig. 7.2: Interpretation of the content of the accumulator after executing a RIM instruction.

EXAMPLE PROGRAM 7.1

Write a program segment to mask RST 6.5 and RST 5.5 interrupts and enable RST 7.5 interrupt.

Solution

The 8-bit data format to be loaded in the accumulator for enabling RST 7.5 and masking RST 6.5 and RST 5.5 is shown below. The data to be loaded in accumulator is $0B_H$.

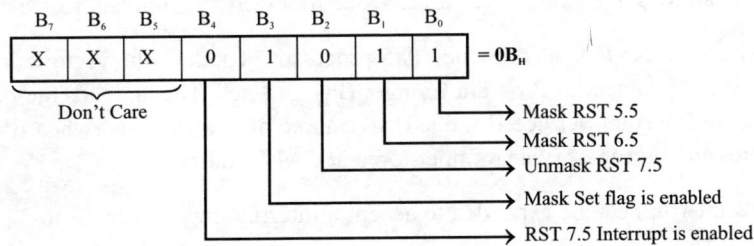

Program Segment

```
EI          ; Enable all interrupts of 8085
MVI A,0BH   ; Move 0B_H to A-register
SIM         ; Mask 6.5 and 5.5, Enable 7.5
```

EXAMPLE PROGRAM 7.2

Assume that the 8085 microprocessor returns to the main program after servicing RST 6.5. (Remember that while servicing an interrupt all other interrupts are disabled.)

Write a program segment to check whether RST 5.5 interrupt is pending. If it is pending then the program has to enable RST 5.5 without affecting any other interrupts. Otherwise the program has to enable all interrupts and return to main program.

Solution

The status of pending interrupts can be read by executing RIM instruction. This will load an 8-bit data in accumulator. If RST 5.5 is pending then the bit D_4 in accumulator will be **1** and if it is not pending then bit D_4 will be **0**. The following program segment have been written to check whether bit D_4 is **1** or **0**. If it is **1** then the program control jumps to another part of program to enable RST 5.5 and mask other interrupts.

Program Segment

```
        RIM          ;Read the status of interrupts.
        MOV C,A      ;Save the status in C-register.
        ANI 10H      ;Check whether RST 5.5 is pending.
        JNZ NEXT     ;If  RST 5.5 is pending, go to NEXT.
        EI           ;If RST 5.5 is not pending, enable
        RET          ;all interrupts and return to main program.
NEXT:   MOV A,C      ;Get the interrupt status in A-register.
        ANI FEH      ;Set D_0 = 0, for enabling RST 5.5
        ORI 08H      ;Set D_3 =1, for enabling interrupt enable flag.
        SIM          ;Enable RST 5.5
        JMP ISR55    ;Jump to Interrupt Service Routine of RST 5.5
```

7.4.5 INTR and its Expansion in 8085

The INTR is general interrupt request. An external device can interrupt the processor by placing a **high** signal on INTR pin of 8085. If the processor accepts the interrupt, then it will send an acknowledge signal $\overline{\text{INTA}}$ to the interrupting device. On receiving the acknowledge signal, the interrupting device has to place either an **RST n** opcode (or CALL opcode followed by 16-bit address) on the data bus.

On receiving the **RST n** opcode, the 8085 processor generates the vector address of **RST n** instruction. It saves the content of **P**rogram **C**ounter (PC) in stack. Then it loads the vector address in PC and executes an **I**nterrupt **S**ervice **R**outine (ISR) stored at this address. (when it receives CALL opcode it executes an interrupt service routine stored at CALL address.)

The INTR interrupt can be expanded to accept 8-interrupt inputs using 8-to-3 priority encoder as shown in Fig. 7.3.

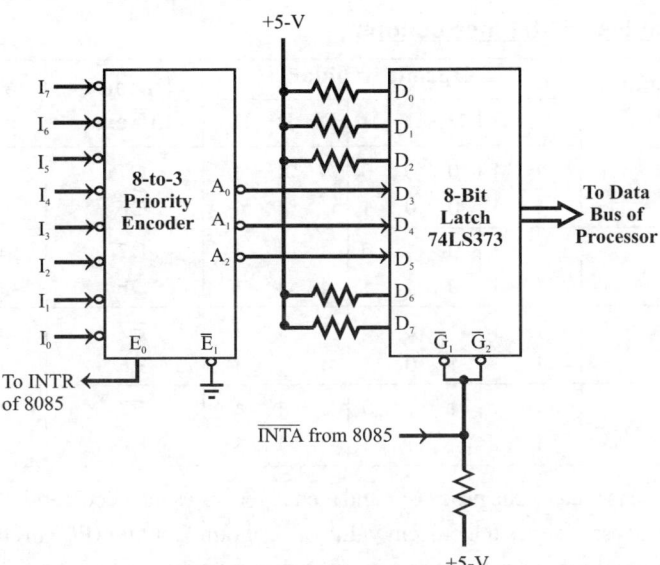

Fig. 7.3: Expanding an INTR of the 8085 using an 8-to-3 priority encoder.

The priority encoder has 8 inputs I_0 to I_7 and three outputs A_0 to A_2. It also has an output control signal, E_0. If the priority encoder receives a logic **low** at one of the inputs, for example I_n, then it asserts E_0 **high** and outputs the binary value of n on the output lines A_0, A_1 and A_2 lines (i.e., if input I_0 is **low** then output is 000; if input I_1 is **low** then output is 001 and so on). In this scheme I_7 has the highest priority and I_0 has the lowest priority.

Eight external devices can interrupt the processor through I_0 to I_7 lines, by placing a logic **low** on these pins. On receiving a valid interrupt signal the priority encoder allows the highest priority interrupt by asserting E_0 **high** and sending the corresponding binary value on A_0, A_1 and A_2 lines. The E_0 is connected to INTR of 8085 and A_0, A_1 and A_2 are connected to the inputs D_3, D_4 and D_5 of an 8-bit latch. All other inputs of the latch are tied to +5 V (logic 1) permanently.

The opcodes and vector addresses of RST n instructions are shown in Table 7.4. If we carefully look at the opcode of RST instruction, the binary bits D_3, D_4, D_5 constitutes the binary value of n in RST n instruction and all other bits are 1's. The priority encoder helps in placing the RST opcodes at the input of latch (74LS373). *[The priority encoder places the **RST n** opcode for the interrupt I_n.]*

When the processor accepts the interrupt, it sends $\overline{\text{INTA}}$ signal to the interrupting device. This signal is used to enable the latch. When the latch is enabled, the RST opcode available at the input is latched into output lines. The output of latch is connected to data bus of the processor. Hence, the opcode will be placed on the data bus.

Table 7.4: Opcodes of RST Instructions

RST instruction	Opcode in binary								Opcode in hexa	Vector address
	D_7	D_6	D_5	D_4	D_3	D_2	D_1	D_0		
RST 0	1	1	0	0	0	1	1	1	C7	0000_H
RST 1	1	1	0	0	1	1	1	1	CF	0008_H
RST 2	1	1	0	1	0	1	1	1	D7	0010_H
RST 3	1	1	0	1	1	1	1	1	DF	0018_H
RST 4	1	1	1	0	0	1	1	1	E7	0020_H
RST 5	1	1	1	0	1	1	1	1	EF	0028_H
RST 6	1	1	1	1	0	1	1	1	F7	0030_H
RST 7	1	1	1	1	1	1	1	1	FF	0038_H

This opcode is read by the processor and then it generates the vector address of the RST instruction internally. The processor saves the current value of **P**rogram **C**ounter (PC) in stack and loads the vector address in PC. Now the processor starts servicing the interrupt.

Alternatively, programmable interrupt controller, 8259 can be interfaced to 8085 processor to handle multiple interrupt request and allow one by one to the processor INTR input pin. A detailed discussion on 8259 and its interfacing with 8085 processor are presented in Section 7.7.

7.5 INTERRUPTS OF 8086

The 8086 microprocessor has 256 types of interrupts which come from any one of the three sources mentioned above. INTEL has assigned a type number to each interrupt. The type numbers are in the range of 0 to 255_{10}. The 8086 processor has dual facility of initiating these 256 interrupts. The interrupts can be initiated either by executing "INT n" instruction where n is the type number or the interrupt can be initiated by sending an appropriate signal to INTR input pin of the processor.

For the interrupts initiated by software instruction "INT n", the type number is specified by the instruction itself. When the interrupt is initiated through INTR pin then the processor runs an interrupt acknowledge cycle to get the type number (i.e., the interrupting device should supply the type number through D_0-D_7 lines when the processor requests for the same through an interrupt acknowledge cycle).

In these 256 interrupts, INTEL has defined the functions of the first five interrupts, i.e., the interrupts type-0 to type-4 are dedicated for specific functions by INTEL and they are called INTEL predefined interrupts. The next 27 interrupts, i.e., from type-5 to type-31 are reserved by INTEL for use in future microprocessors or for system calls/services. The upper 224 interrupts, i.e., from type-32 to type-255 are available for the user as hardware or software interrupts.

The interrupts of 8086 can be classified into following three groups:

1. **Intel Predefined (or dedicated) Interrupts**
2. **Software Interrupts**
3. **Hardware Interrupts**

7.5.1 INTEL Predefined (or Dedicated) Interrrupts

The INTEL predefined interrupts are as follows:

1. **Division by zero (Type-0 interrupt)**
2. **Single step (Type-1 interrupt)**
3. **Non-maskable interrupt, NMI (Type-2 interrupt)**
4. **Breakpoint interrupt (Type-3 interrupt)**
5. **Interrupt on overflow (Type-4 interrupt)**

The predefined interrupts are only defined by INTEL and INTEL has not provided any subroutine/procedure to be executed for these interrupts. To use the predefined interrupts the user/system designer has to write an **Interrupt Service Subroutine (ISS)** for each interrupt and store them in memory. The corresponding address of the ISS should be stored in an interrupt vector table. If a predefined interrupt is not used in a system then the user may assign some other functions to these interrupts.

Divide-by-zero Interrupt (type-0 interrupt)

Type-0 interrupt is implemented by INTEL as a part of the execution of the divide instruction. The 8086 will automatically do a type-0 interrupt if the result of a division operation is too large to fit in the destination register and this interrupt is non-maskable. Since the type-0 interrupt cannot be disabled in any way, we have to account for it in the programs using divide instructions. To account for this, we have to write an ISS which takes the desired action or indicate error condition when an invalid division occurs. The ISS should be stored in memory and the address of ISS is stored in the interrupt vector table.

Single-step Interrupt (type-1 interrupt)

When the **Trap/Trace Flag (TF)** is set to one, the 8086 processor will automatically generate a type-1 interrupt after execution of each instruction. The user can write an ISS for type-1 interrupt to halt the processor temporarily and return the control to the user so that after execution of each instruction, the processor status (content of register/memory) can be verified. If they are correct then we can proceed to execute the next instruction. Execution of one instruction by one instruction is known as single step and this feature will be useful to debug a program.

Non-maskable Interrupt, NMI (type-2 interrupt)

The 8086 processor will automatically generate a type-2 interrupt when it receives a **low-to-high** transition on its NMI input pin. This interrupt cannot be disabled or masked. Usually, the type-2 interrupt is used to save program data or processor status in case of system AC power failure.

The AC power failure is detected by an external hardware and whenever the AC power fails, the external hardware will send an interrupt signal to the NMI input pin of the processor. The rectifier which converts AC to DC usually has a large filter capacitor and so it can retain the DC power for at least 50 millisecond, after the AC power supply is interrupted. This 50 millisecond time will be sufficient to run an ISS by type-2 interrupt to save program data or processor status to NVRAM or RAM with battery back-up power supply.

Breakpoint Interrupt (type-3 interrupt)

Type-3 interrupt is used to implement a breakpoint function, which executes a program partly or up to the desired point and then returns the control to the user.

The breakpoint interrupt is initiated by the execution of "INT 3" instructions. To implement the breakpoint function, the system designer has to write an ISS for type-3, which takes care of displaying a message and return the control to the user whenever a type-3 interrupt is initiated.

This interrupt will be useful to debug a program by executing the program part by part. The user can insert "INT 3" instruction at the desired location and execute the program. Whenever the "INT 3" instruction is encountered, the processor halts the program execution and returns the control to the user. Now the user can verify the processor status (contents of register/memory). If they are correct then the user can proceed to execute the next part of the program.

Overflow Interrupt (type-4 interrupt)

In the 8086 processor, the Overflow Flag (OF) will be set if the signed arithmetic operation generates a result whose size is larger than the size of the destination register/memory. During such conditions, the type-4 interrupt can be used to indicate an error condition. The type-4 interrupt is initiated by the "INTO" instruction.

One way of detecting the overflow error is to put the INTO instruction immediately after the arithmetic instruction in the program. After the arithmetic operation if the overflow flag is not set then the processor will consider the "INTO" instruction as NOP (No operation). However, if the overflow flag is set then the 8086 will generate a type-4 interrrupt, which executes an ISS to indicate overflow condition.

7.5.2 Software Interrupts of 8086

The "INT n" instructions are called software interrupts. The "INT n" instruction will initiate a type-n interrupt, and the value of n is in the range of 0 to 255_{10}. Therefore, all the 256 type interrupts including the INTEL predefined and reserved interrupts can be initiated through "INT n" instruction. The software interrupts are non-maskable and have higher priority than hardware interrupts.

7.5.3 Hardware Interrupts of 8086

The interrupts initiated by applying appropriate signals to INTR and NMI pins of 8086 are called hardware interrupts. All the 256 types of interrupts including INTEL predefined and reserved interrupts can be initiated by applying a high signal to a INTR pin of 8086. When a high signal is applied to the INTR pin and the hardware interrupt is enabled/unmasked then the processor runs an interrupt acknowledge cycle to get the type number of the interrupt from the device which sends the interrupt signal. The interrupting device can send a type number in the range of 0 to 255_{10}. Therefore, all the 256 types of interrupts can be initiated through INTR pin.

The hardware interrupts initiated through INTR are maskable by clearing the Interrupt Flag (IF), i.e., the hardware interrupts are masked/disabled when IF = 0 and they are unmasked/enabled when IF = 1. The interrupts initiated through INTR have lower priority than software interrupts.

The hardware interrupt NMI is non-maskable and has higher priority than interrupts initiated through INTR. The NMI is initiated by a rising edge (or low-to-high transition) of the signal applied to the NMI pin of the processor. The processor will execute a type-2 interrupt in response to interrupt initiated through NMI pin and this type number is fixed by INTEL. The external device, interrupting the processor through NMI pin, need not supply the type number for this interrupt.

7.5.4 Priorities of Interrupts of 8086

The priorities of the interrupts of 8086 are shown in Table 7.5. The 8086 processor checks for internal interrupts before it checks for any hardware interrupt. Therefore, software interrupts has higher priority than hardware interrupts. But the processor can accept the NMI interrupt request and execute a procedure for it even in between the execution of procedure for higher priority interrupts.

For example, if the NMI is initiated by an external hardware while the processor internally generates the divide error interrupt, then the processor goes to the start of the divide error procedure and then suspends it to service NMI. Only after servicing the NMI, will the processor complete the divide error procedure.

Table 7.5: Interrupt Priority

Interrupt	Priority
Divide error, INT n, INTO	Highest
NMI	
INTR	↓
SINGLE STEP	Lowest

7.6 IMPLEMENTING INTERRUPT SCHEME IN 8086

The 8086 processor has 256 types of interrupts and these interrupts can be implemented either as hardware or software interrupts. The number of interrupts to be implemented and used in a system depends on the system designer and also on the application for which the system is designed. The choice of implementing the INTEL predefined interrupts also depends on the system designer.

Except some of the INTEL predefined interrupts, for all other interrupts, the system designer has to decide on the method of initiating the interrupts selected to implement on a system. The interrupts can be initiated either by external hardware or internally by software instruction "INT n". In a system, some interrupt types are chosen to be initiated by the hardware, some other interrupt types are chosen to be initiated by software and some of the interrupts are left unused. The unused interrupts can be implemented by the user for user-defined functions.

7.6.1 Interrupt Vector Table

For each and every interrupt decided to be implemented in the system, the system designer has to write an **Interrupt Service Subroutine** (ISS) and store them in memory. Then the system designer has to create an interrupt vector table in the first 1 kB memory space (i.e., in the memory space with address range 00000_H to $003FF_H$) of the 8086 system.

In this vector table, the 16-bit offset address and 16-bit segment base address of each ISS are stored in four consecutive memory locations. The address stored in this table are called vector addresses. For storing the vector addresses of all the 256 interrupt types, the vector table requires 1 kB (256×4 = 1 kB) memory space.

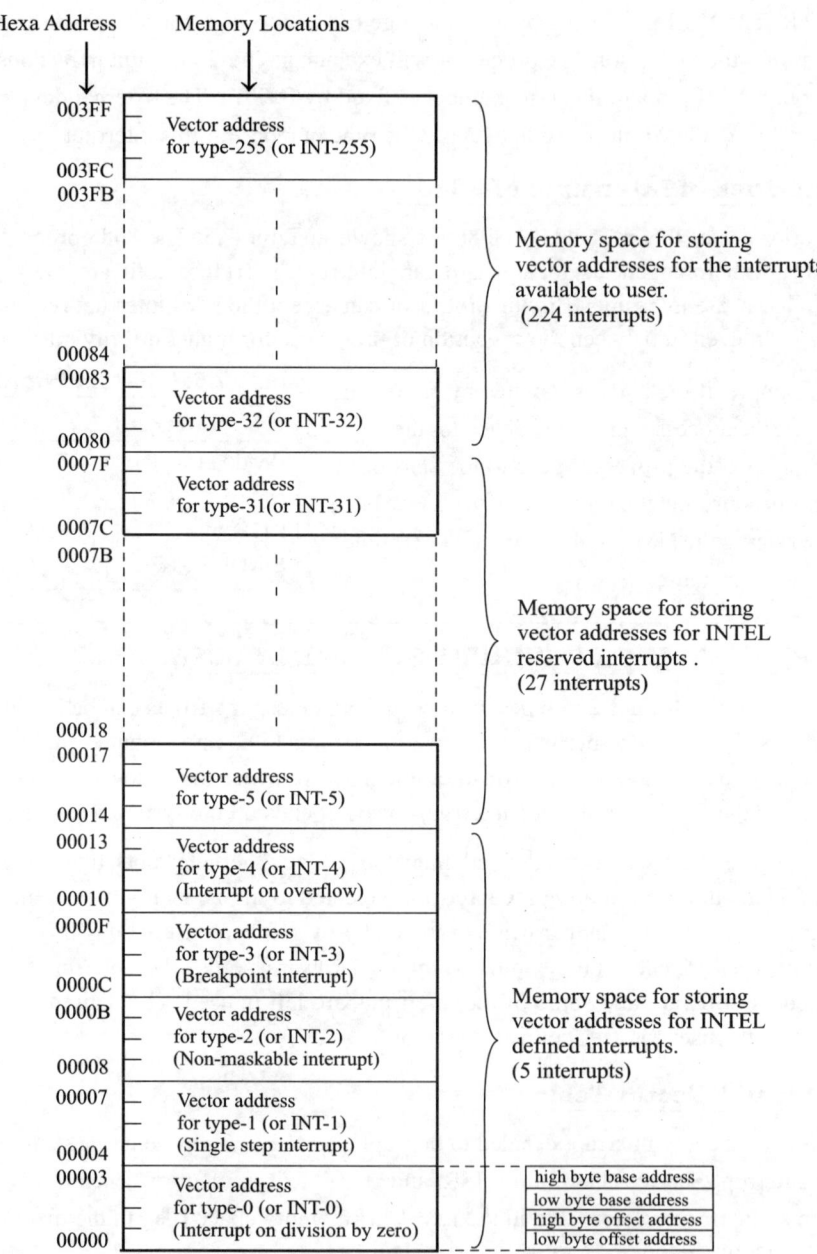

Fig. 7.4: Organization of an interrupt vector table in 8086.

The memory address for storing the vector address for an interrupt is given by multiplying the type number by four and sign extending it to 20-bit. The vector address for an interrupt is stored in four consecutive memory locations starting from this 20-bit address. The first two locations are used to store the low byte and high byte of offset address, and next two locations are used to store low byte and high byte of segment base address of ISS to be executed for an interrupt. The organization of interrupt vector table of 8086-based system is shown in Fig. 7.4.

7.6.2 Servicing an Interrupt By 8086

The 8086 processor checks for interrupt request at the end of each instruction cycle. If an interrupt request is deducted then the 8086 processor responds to the interrupt by performing the following operations:

1. The SP is decremented by two and the content of flag register is pushed to stack memory.
2. The interrupt system is disabled by clearing Interrupt Flag (IF).
3. The single-step trap flag is disabled by clearing Trap Flag (TF).
4. The stack pointer is decremented by two and the content of CS-register is pushed to stack memory.
5. Again, the stack pointer is decremented by two and the content of IP is pushed to stack memory.
6. In case of hardware interrupt through INTR, the processor runs an interrupt acknowledge cycle to get the interrupt type number. For software interrupts, the type number is specified in the instruction itself. For NMI and exceptions, the type number is defined by INTEL.
7. The processor generates a 20-bit memory address by multiplying the type number by four and sign extending it to 20-bit. This memory address is the address of the interrupt vector table, where the vector address of the Interrupt Service Subroutine (ISS) is stored by the user/system designer.
8. The first word pointed by the vector table address is loaded in IP and the next word is loaded in CS-register. Now the content of the IP is the offset address and the content of the CS-register is the segment base address of the ISS to be executed.
9. The 20-bit physical memory address of ISS is calculated by multiplying the content of CS-register by 16_{10} and adding it to the content of IP.
10. The processor executes the ISS to service the interrupt.
11. The ISS will be terminated by the IRET instruction. When this instruction is executed, the top of stack is popped to IP, CS and flag register one word by one word. After every pop operation, the SP is incremented by two.
12. Thus, at the end of ISS, the previous status of the processor is restored and so the processor will resume the execution of the normal program from the instruction where it was suspended.

7.6.3 INTR and its Expansion in 8086

The hardware interrupt INTR can be used by any external device to interrupt the processor. When an interrupt request is made through the INTR and the INTR interrupt is enabled/unmasked then the processor will run an interrupt acknowledge cycle. During this cycle, the processor asserts $\overline{\text{INTA}}$ signal twice. The first $\overline{\text{INTA}}$ signal is to inform the interrupting device about the acceptance of the interrupt.

The second time, $\overline{\text{INTA}}$ is asserted to request the interrupting device to supply the type number and to read the type number from the low order data bus. Therefore, the processor expects a type number on the low order data bus whenever it is interrupted through the INTR input pin.

A scheme for loading the type number on low order data bus is shown in Fig. 7.5. In this scheme the desired type number is applied to the input of the tristate octal buffer through switch settings and the buffer is enabled by the $\overline{\text{INTA}}$ signal. The output of the buffer is connected to low order data bus. Hence, whenever the buffer is enabled by $\overline{\text{INTA}}$ signal, the type number available on its input pins are transferred to its output lines. Thus, the type number is loaded on the low order data bus.

In the schematic shown in Fig. 7.5 the switches can be manually set to create a binary type number in the range $0000\ 0000_2$ to $1111\ 1111_2$ corresponding to 0_{10} to 255_{10}. When a switch is open, the voltage applied to the corresponding input pin of tristate buffer is +5-V and so it is logic-1. When a switch is closed, the voltage applied to the corresponding input pin of tristate buffer is zero volt and so it is logic-0. Thus, by closing/opening the switches an 8-bit binary number can be created which is the desired type number.

The scheme shown in Fig. 7.5 can be used to implement only one interrupt and so only one external device can interrupt the processor. But the INTR interrupt can be used to initiate all the 256 type interrupts. To initiate multiple interrupts through INTR, there should be some provision to supply different type numbers for various interrupts. Such a scheme is possible with programmable interrupt controller INTEL 8259.

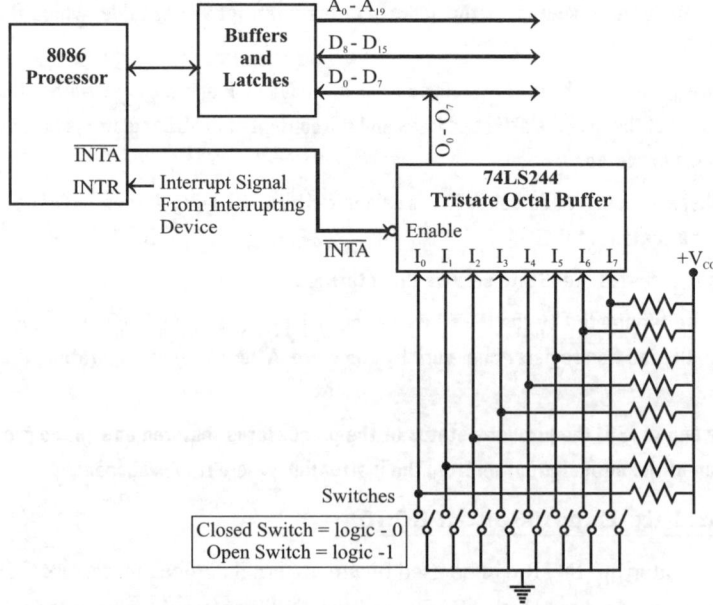

Fig. 7.5: A schematic to load any type number on low order
data bus (D_0 - D_7) in response to the $\overline{\text{INTA}}$.

The programmable interrupt controllers can be interfaced to the 8086 processor to handle multiple interrupt requests and allow one by one to the processor INTR input pin. One interrupt controller can accept up to eight interrupt requests and allow one by one to the processor. Multiple interrupt controllers can be interfaced to the processor in cascaded mode, to handle up to 64 interrupt requests.

In the cascaded mode, one master interrupt controller and a maximum of eight slave interrupt controllers can be interfaced to the processor to handle 64 interrupt requests and allow one by one to the processor INTR input pin.

A detailed discussion on the programmable interrupt controller, 8259 and its interfacing with 8086 processor are presented in the following sections.

7.7 PROGRAMMABLE INTERRUPT CONTROLLER - INTEL 8259

The 8259 is a programmable interrupt controller. It is used to expand the interrupts of an 8085 or 8086 processor. One 8259 can accept eight interrupt requests and allow one by one to the processor INTR pin. The interrupt controller can be used in cascaded mode in a system to expand the interrupts up to 64.

Features of an 8259

1. It is programmed to work with either an 8085 or 8086 processor.

2. It manages 8 interrupts according to the instructions written into its control registers.

3. In an 8085 processor-based system, it vectors an interrupt request anywhere in the memory map and the interrupt vector address is programmable.

4. The priorities of the interrupts are programmable. The different operating modes which decides the priorites are automatic rotation mode, specific rotation mode and fully nested mode.

5. The interrupts can be masked or unmasked individually.

6. The 8259 is programmed to accept either level triggered interrupt signals or edge triggered interrupt signals.

7. The 8259 provides the status of the pending interrupts, masked interrupts and interrupt being serviced.

8. The 8259s can be cascaded to accept a maximum of 64 interrupts.

7.7.1 Interfacing 8259 with an 8085 Microprocessor

The 8259 is a 28-pin IC packed in DIP. The various pins of an 8259 are shown in Fig. 7.6. It requires two internal address and they are $A_0 = 0$ or $A_0 = 1$. It can be either memory-mapped or IO-mapped in the system. The interfacing of an 8259 to 8085 is shown in Fig. 7.7. In Fig. 7.7, the 8259 is IO-mapped in the system.

The low order data bus lines D_0 - D_7 are connected to D_0 - D_7 of the 8259. The address line A_0 of the 8085 processor is connected to A_0 of the 8259 to provide the internal address. The 8259 requires one chip select signal. The chip select signal for 8259 is generated by using 3-to-8 decoder.

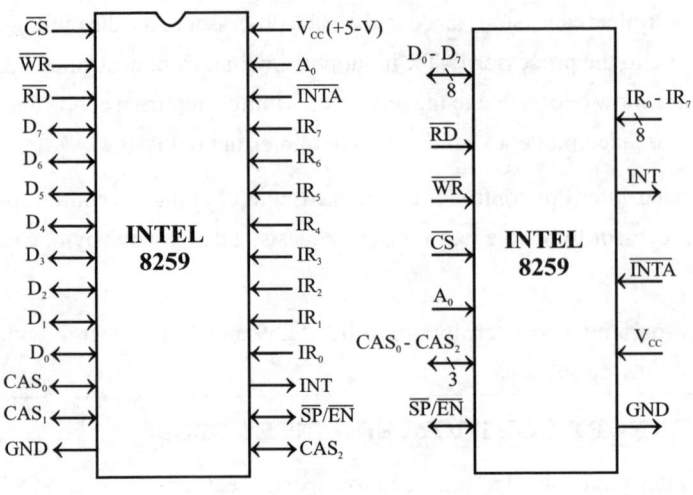

Fig. 7.6: Pin details of 8259.

Pins	Description
$D_0 - D_7$	Bidirectional datalines
$\overline{RD}$	Read control
$\overline{WR}$	Write control
A_0	Internal address
$\overline{CS}$	Chip select
$CAS_0 - CAS_2$	Cascade lines
$\overline{SP/EN}$	Slave program /Enable buffer
INT	Interrupt output
$\overline{INTA}$	Interrupt acknowledge input
$IR_0 - IR_7$	Interrupt request inputs

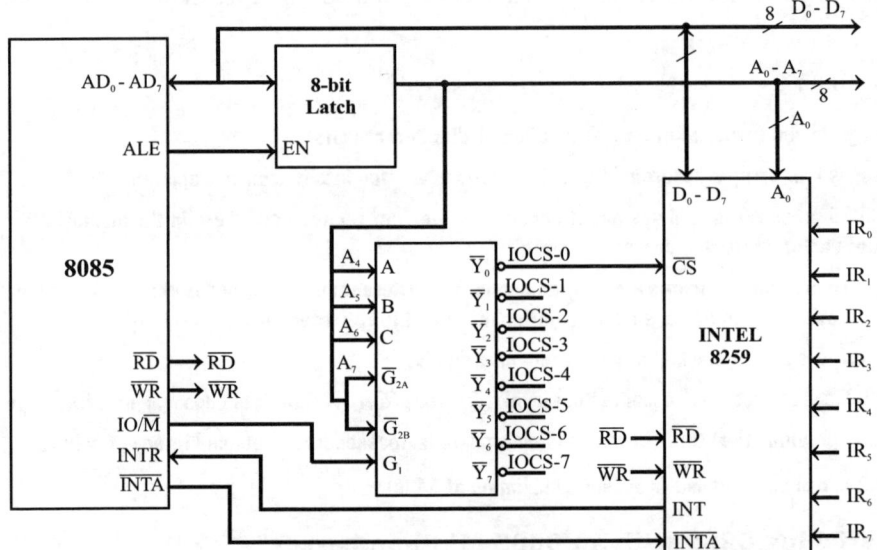

Fig. 7.7: Interfacing 8259 to 8085 microprocessor.

Table 7.6: IO Address of 8259 Interfaced to 8085 as Shown in Fig. 7.7

	Binary Address								Hexa address
	Decoder input/ enable				Input to address pin of 8259				
	A_7	A_6	A_5	A_4	A_3	A_2	A_1	A_0	
For A_0 of 8259 to be zero	0	0	0	0	x	x	x	0	00
For A_0 of 8259 to be one	0	0	0	0	x	x	x	1	01

Note : Don't care "x" is considered as zero.

The address lines A_4, A_5 and A_6 are used as input to the decoder. The control signal $IO/\overline{M}$ is used as logic **high** enables for decoder and the address line A_7 is used as logic **low** enable for decoder. The IO addresses of 8259 are shown in Table 7.6. The signals CAS_0-CAS_2 are used only in cascade operations of 8259s.

The $\overline{SP}/\overline{EN}$ pin can be used as input or output signal. In non-buffered mode, it is used as an input signal and tied to logic-1 in master 8259 and logic-0 in slave 8259. In buffered mode it is used as an output signal to disable the data buffers while the data is transferred from 8259A to the CPU.

Working of 8259 with an 8085 processor

First, the 8259 should be programmed by sending Initialization Command Word (ICW) and Operational Command Word (OCW). These command words will inform 8259 about the following,

1. **Type of interrupt signal (Level triggered/Edge triggered).**   4. **Masking of interrupts.**

2. **Type of processor (8085/8086).**   5. **Priority of interrupts.**

3. **Call address and its interval (4 or 8).**   6. **Type of end of interrupt.**

Once the 8259 is programmed, it is ready for accepting interrupt signals. When it receives an interrupt through any one of the interrupt lines IR_0-IR_7, it checks for its priority and also checks whether it is masked or not. If the previous interrupt is completed and if the current request has highest priority and is unmasked, then it is serviced.

For servicing this interrupt, the 8259 will send INT signal to the INTR pin of the 8085. In response it expects an acknowledge $\overline{INTA}$ from the processor. When the processor accepts the interrupt, it sends three $\overline{INTA}$ one by one. In response to the first, second and third $\overline{INTA}$ signals, the 8259 will supply the CALL opcode, the low byte of call address and high byte of the call address respectively. Once the processor receives the call opcode and its address, it saves the content of the Program Counter (PC) in the stack and loads the CALL address in the PC and starts executing the interrupt service routine stored in this call address.

7.7.2 Interfacing 8259 with an 8086 Microprocessor

The 8259 is a 28-pin IC packed in DIP. The various pins of 8259 are shown in Fig. 7.6. It requires two internal address and they are $A_0 = 0$ or $A_0 = 1$. It can be either memory-mapped or IO mapped in the system. The interfacing of 8259 to 8086 is shown in Fig. 7.8. In Fig. 7.8, the 8259 is IO mapped in the system. The low order data bus lines D_0 - D_7 are connected to D_0 - D_7 of 8259. The address line A_1 of the 8086 processor is connected to A_0 of 8259 to provide the internal address. The 8259 requires one chip select signal.

The chip select signal for 8259 is generated by using 3-to-8 decoder. The address lines A_5, A_6 and A_7 are used as input to decoder. The address line A_0 and control signal $M/\overline{IO}$ are used as logic low enables for the decoder. The IO addresses of 8259 are shown in Table 7.7. The signals CAS_0 - CAS_2 are used only in cascade operation of 8259s.

The $\overline{SP/EN}$ pin can be used as input or output signal. In nonbuffered mode, it is used as input signal and tied to logic-1 in master 8259 and logic-0 in slave 8259. In buffered mode, it is used as output signal to disable the data buffers while the data is transferred from 8259A to the CPU.

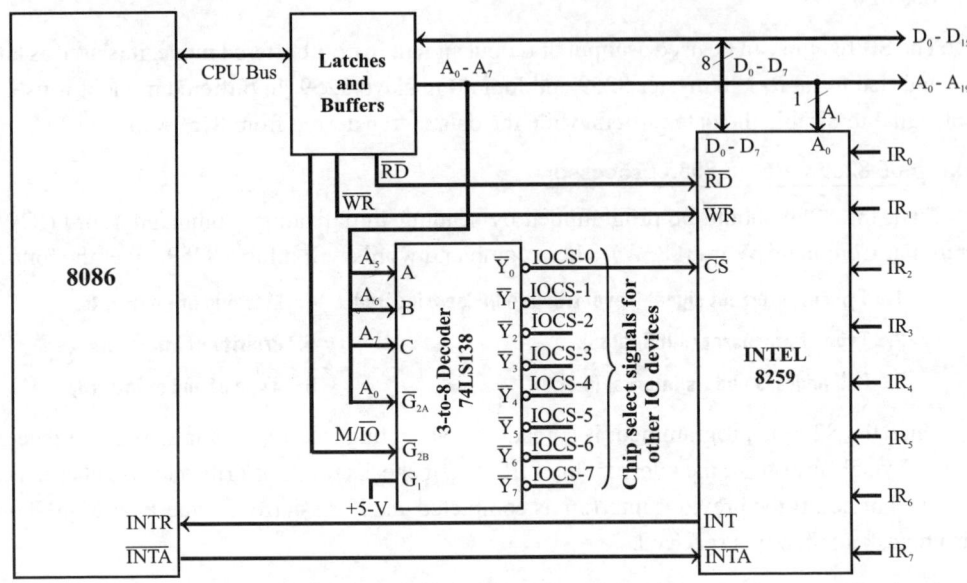

Fig. 7.8: Interfacing 8259 to 8086 microprocessor.

Table 7.7: IO Address of 8259 Interfaced to 8085 as Shown in Fig. 7.8

	Binary address								Hexa address
	Decoder input			Input to address pin of 8259				Decoder enable	
	A_7	A_6	A_5	A_4	A_3	A_2	A_1	A_0	
For A_0 of 8259 to be zero	0	0	0	x	x	x	0	0	00
For A_0 of 8259 to be one	0	0	0	x	x	x	1	0	02

Note: Don't care "x" is considered as zero.

Working of 8259 with an 8086 Processor

First the 8259 should be programmed by sending Initialization Command Word (ICW) and Operational Command Word (OCW). These command words will inform 8259 about the following:

1. **Type of interrupt signal (Level triggered/Edge trigerred).**

2. **Type of processor (8085/8086).**

3. **Call address and its interval (4 or 8).**

4. **Masking of interrupts.**

5. **Priority of interrupts.**

6. **Type of end of interrupt.**

Once 8259 is programmed, it is ready for accepting an interrupt signal. When it receives an interrupt through any one of the interrupt lines $IR_0 - IR_7$, it checks for its priority and also checks whether it is masked or not. If the previous interrupt is completed and if the current request has highest priority and unmasked then it is serviced.

For servicing this interrupt, the 8259 will send INT signal to INTR pin of 8086. In response it expects an acknowledge $\overline{INTA}$ from the processor. When the processor accepts the interrupt, it sends two $\overline{INTA}$ one by one. The first $\overline{INTA}$ is sent to 8259, to inform the acceptance of interrupt and to prepare 8259 for supplying type number. The second $\overline{INTA}$ is sent to 8259 to read the type number from the 8259.

Once the processor receives the type number, it starts processing the interrupt corresponding to this type number. The 8086 processor multiplies the type number by four and sign extends to 20-bit to generate a 20-bit vector table address.

From this vector table, the vector addresses of the interrupt type requested are read and loaded in IP and CS-register. Then the processor executes the ISR stored in this address.

7.7.3 Functional Block Diagram of 8259

The functional block diagram of an 8259 is shown in Fig. 7.9 and it shows eight functional blocks. They are Control logic, Read/Write logic, Data bus buffer, **I**nterrupt **R**equest **R**egister (IRR), **I**n-Service **R**egister (ISR), **I**nterrupt **M**ask **R**egister (IMR), **P**riority **R**esolver (PR), and Cascade buffer.

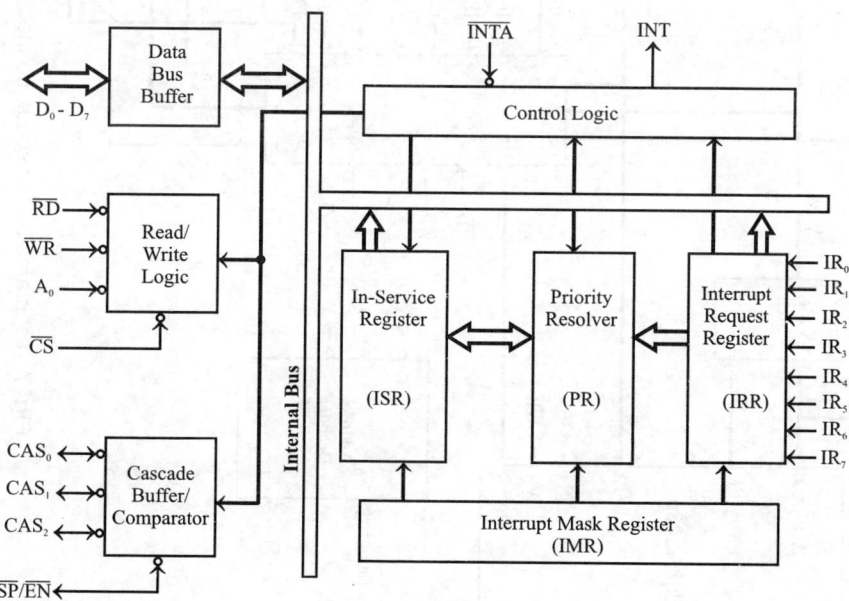

Fig. 7.9: Functional block diagram of 8259.

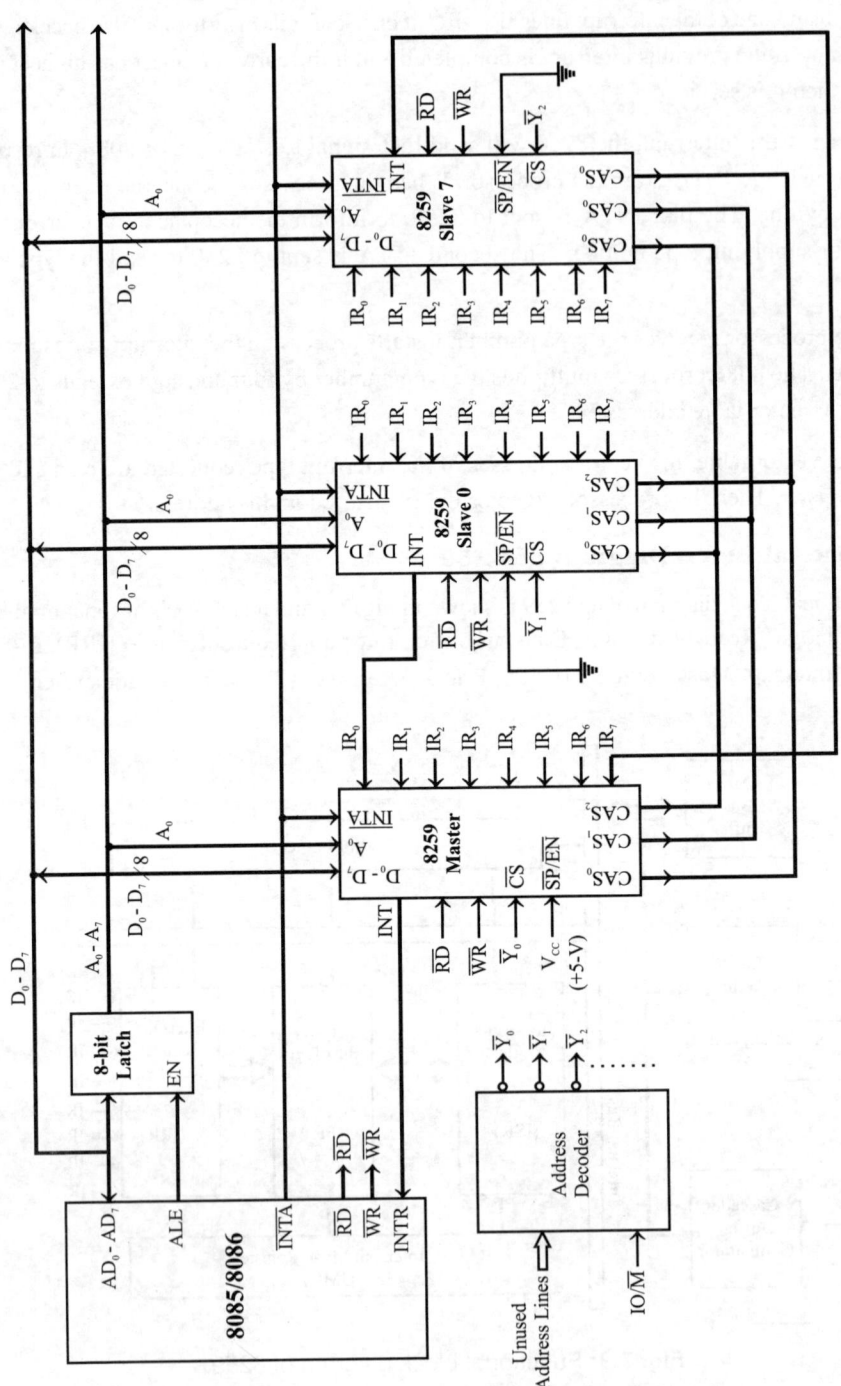

Fig. 7.10: Example of cascade connection of programmable interrupt controllers-8259.

The data bus and its buffer are used for the following activities:

1. **The processor sends the control word to the data bus buffer through D_0- D_7.**
2. **The processor reads the status word from the data bus buffer through D_0- D_7.**
3. **From the data bus buffer the 8259 call opcode and address (in case of an 8085) through D_0- D_7 to the processor.**

The processor uses the $\overline{RD}$, $\overline{WR}$ and A_0 to read or write the 8259. The 8259 is selected by the $\overline{CS}$. The IRR has eight input lines (IR_0 - IR_7) for interrupts. When these lines go **high**, the requests are stored in the IRR. It registers a request only if the interrupt is unmasked. Normally IR_0 has the highest priority and IR_7 the lowest. The priorities of an interrupt request input are also programmable.

The interrupt mask register stores the masking bits of the interrupt lines to be masked. The relevant information is sent by the processor through the OCW1. The in-service register keeps track of which interrupt input is currently being serviced. For each input that is currently being serviced, the corresponding bit will be set in the in-service register. The priority resolver examines the interrupt request, mask and in-service registers and determines whether the INT signal should be sent to the processor or not.

The cascade buffer/comparator is used to expand the interrupts of the 8259. Figure 7.10 is an example of 8259s in cascade connection. It is called the master 8259. To each interrupt request input of the master 8259 (IR_0-IR_7) one slave 8259 can be connected. The 8259s interrupting the master 8259 are called slave 8259s.

Each 8259 has its own addresses so that each 8259 can be programmed independently by sending the command words and the status bytes can be read from it independently.

The cascade pins (CAS_0, CAS_1 and CAS_2) from the master are connected to the corresponding pins of the slave. For the master, these pins function as output, and for the slave device they function as input. For the slave 8259, the $\overline{SP/EN}$ pin is tied **low** to let the device know that it is a slave.

7.7.4 Processing of Interrupts by 8259

Processing of Interrupts by 8259 in 8085

To implement interrupts, the processor interrupt should be enabled and 8259 be initialized by sending ICWs and OCWs. The ICWs are used to set up the proper conditions and specify the CALL vector addresses.

The OCWs are used to perform functions such as masking interrupts, setting up status, read operations, etc. After the 8259 is initialized, the following sequence of events occur when one or more interrupt request lines go **high**:

1. **The IRR stores the request.**

2. **The priority resolver checks three registers (IRR, IMR, ISR). The IRR is checked for interrupt request. The IMR is checked for masking bits and the ISR for the interrupt request being served. It resolves the priority and sets the INT high when appropriate.**

3. **The processor acknowledges the interrupt by sending the $\overline{INTA}$ signal.**

4. When the $\overline{INTA}$ is received, the appropriate priority bit in the ISR is set to indicate which interrupt level is being served, and the corresponding bit in the IRR is reset to indicate that the request is accepted. Then, the opcode of the CALL instruction is placed on the data bus.

5. When the processor decodes the CALL instruction, it places two more $\overline{INTA}$ signals on the data bus.

6. When the 8259 receives the second $\overline{INTA}$, it outputs the low-order byte of the CALL address on the data bus. When the third $\overline{INTA}$ signal is received, the 8259 outputs high-order byte of CALL address on the data bus. The CALL address is the vector memory location for the interrupt. (this address is programmed by sending ICW1 and ICW2 to the control register during the initialization.)

7. Once the processor reads the CALL opcode and address from the 8259 the bit corresponds to the current interrupt being serviced in the in-service register should be resetted to allow next interrupt. This is done automatically if the 8259 is programmed for Automatic End Of Interrupt (AEOI). Alternatively, the processor can sends a command word at the end of the interrupt service routine to inform 8259 about the end of interrupt.

8. After receiving the CALL opcode and address, the processor saves the content of the Program Counter (PC) in the stack and loads the call address in the PC. Thus the program control is transferred to the memory location specified by the CALL instruction.

Processing of Interrupts by 8259 in 8086

To implement interrupts, the processor interrupt should be enabled and 8259 is initialized. The 8259 is initialized by sending ICWs and OCWs. The ICWs are used to set up the proper conditions and to specify interrupt type number. The OCWs are used to perform functions such as masking interrupts, setting up status, read operations, etc. After the 8259 is initialized, the following sequence of events occur when one or more interrupt request lines go **high**.

1. The IRR stores the request.

2. The priority resolver checks three registers. The IRR is checked for interrupt request. The IMR is checked for masking bits and the ISR for the interrupt request being served. It resolves the priority and sets the INT high when appropriate.

3. The processor acknowledges the interrupt by sending two $\overline{INTA}$ signals one by one.

4. When the first $\overline{INTA}$ is received, the appropriate priority bit in the ISR is set to indicate which interrupt level is being served, and the corresponding bit in the IRR is reset to indicate that the request is accepted.

5. When the 8259 receives the second $\overline{INTA}$, it places the type number on the data bus.

6. Once the processor reads the type number from 8259, the bit corresponds to the current interrupt being serviced in the ISR should be reset to allow the next interrupt. This is done automatically if the 8259 is programmed for Automatic End Of Interrupt (AEOI). Alternatively the processor can send command word at the end of interrupt service routine to inform 8259 about the end of interrupt.

7. The 8086 processor multiplies the type number by four to generate a vector table address and from vector table, the processor reads the vector address of the interrupt type and loads in IP and CS-register. Then the processor starts executing the ISR.

7.7.5 Programming (or Initializing) 8259

The 8259 has four numbers of Initialization Command Word (ICW) and three numbers of Operational Command Word (OCW). The command words are sent to the 8259 by selecting it by $\overline{CS}$ = 0 and A_0 = 0 or 1. Certain command word are sent to the internal address, A_0 = 0 and others with A_0 = 1.

The OCW1 should be sent to the 8259 after sending the ICWs. The OCW2 can be sent at any time (either before servicing a interrupt or at the end of the interrupt service routine). The order of sending the ICWs and OCWs are shown as a flowchart in Fig. 7.11. The format of the ICWs and OCWs are shown in Figs. 7.12 and 7.13 respectively.

The ICWs are used to program the following features of an 8259.

1. **Call address interval in case of an 8085**
2. **Level or edge triggered**
3. **Cascade mode or single**
4. **Vector addresses or type number**
5. **8085 mode**
6. **Auto or normal end of interrupt**
7. **Special fully nested mode**

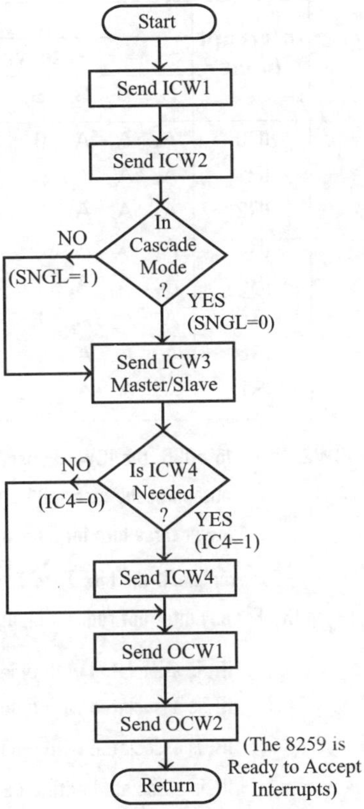

The OCWs are used to read the status of the interrupts and also to program the following features of a 8259:

1. **Masking or unmasking of individual interrupts.**
2. **Specific or non-specific end of interrupt.**
3. **Priority modes.**

A brief discussion about ICWs and OCWs are presented in the following sections.

Fig. 7.11: Sending order of the ICWs and OCWs.

7.7.6 Initialization Command Words (ICWs)

The 8259A has four ICWs and they are named as ICW1, ICW2, ICW3 and ICW4. When only one 8259 is used in the system, then we have to program the 8259 by sending ICW1, ICW2 and ICW4.

When a number of 8259s are used in the system, then we have to program each 8259 by sending all the four ICWs. The format of the ICW3 for a master and slave 8259 are different.

ICW1 :   The ICW1 programs the basic operations of 8259. In 8085 based-system, the bit ADI is used to program a call address interval of 4 or 8 and the upper three bits (B_5, B_6 and B_7) of ICW1 are used to program the upper three bits of low byte of the call address. The lower five bits of the low byte of the call address are automatically inserted by the 8259 as shown in Table 7.8. The single or cascade mode of operation is selected by programming the "SNGL" bit. The LTIM bit determines whether the interrupt request input is positive edge-triggered or level-triggered.

Table 7.8: Low Byte CALL Address

Interrupt input	Low byte call address															
	Interval = 4								Interval = 8							
	B_7	B_6	B_5	B_4	B_3	B_2	B_1	B_0	B_7	B_6	B_5	B_4	B_3	B_2	B_1	B_0
IR0	A_7	A_6	A_5	0	0	0	0	0	A_7	A_6	0	0	0	0	0	0
IR1	A_7	A_6	A_5	0	0	1	0	0	A_7	A_6	0	0	1	0	0	0
IR2	A_7	A_6	A_5	0	1	0	0	0	A_7	A_6	0	1	0	0	0	0
IR3	A_7	A_6	A_5	0	1	1	0	0	A_7	A_6	0	1	1	0	0	0
IR4	A_7	A_6	A_5	1	0	0	0	0	A_7	A_6	1	0	0	0	0	0
IR5	A_7	A_6	A_5	1	0	1	0	0	A_7	A_6	1	0	1	0	0	0
IR6	A_7	A_6	A_5	1	1	0	0	0	A_7	A_6	1	1	0	0	0	0
IR7	A_7	A_6	A_5	1	1	1	0	0	A_7	A_6	1	1	1	0	0	0

ICW2 : In 8085, the ICW2 is used to program the high byte of call address. The lower three bits of type number are automatically inserted by the 8259 and the upper five bits are programmable. The binary code inserted in the lower three bits for interrupt request IR_0 to IR_7 are 000 to 111.

For example, if the bits T_3 to T_7 are chosen as 10010, then the following interrupt type numbers are associated with IR_0 to IR_7. For any interrupt request input through IR_0-IR_7 lines, the associated interrupt type is executed by the processor.

IR_0 is associated with type-90_H interrupt (90_H = 1001 0000)
IR_1 is associated with type-91_H interrupt (91_H = 1001 0001)
IR_2 is associated with type-92_H interrupt (92_H = 1001 0010)
IR_3 is associated with type-93_H interrupt (93_H = 1001 0011)
IR_4 is associated with type-94_H interr upt (94_H = 1001 0100)
IR_5 is associated with type-95_H interrupt (95_H = 1001 0101)
IR_6 is associated with type-96_H interrupt (96_H = 1001 0110)
IR_7 is associated with type-97_H interrupt (97_H = 1001 0111)

ICW3 : The ICW3 should be sent to 8259s in cascade operations. Separate formats are provided for master and slave 8259s. In the cascade mode, slave 8259s are connected to one or more IR inputs of the master 8259 and each slave is provided with a slave ID number. The connection of slave 8259s to the IR inputs of the master are informed to the master through ICW3. For slave 8259s, the ID numbers are informed through the ICW3.

ICW4 : The ICW4 is used to inform 8259 whether it is connected to an 8085 or an 8086-based system. For an 8085- based system the right most bit is set to zero, whereas in an 8086-based system it is set to one. The AEOI bit is used to program the method of terminating the interrupt. If AEOI is set to one, then the 8259 will automatically reset the interrupt request bit in the in-service register after supplying the type number to the processor. If the AEOI bit is programmed as zero, then the processor has to send the OCW2 to terminate the interrupt.

The BUF and M/S bits are used to select the buffered or non-buffered operation of a master/slave 8259. The SFNM bit is used to nest or include the priorities of the slave IR input with the master IR input. For example if IR_4 of a master 8259 has a slave 8259 connected to it and they are programmed for SFNM operation. Now the priorities of IR_0 to IR_7 of slave 8259 will be higher than IR_5 to IR_7 of master 8259.

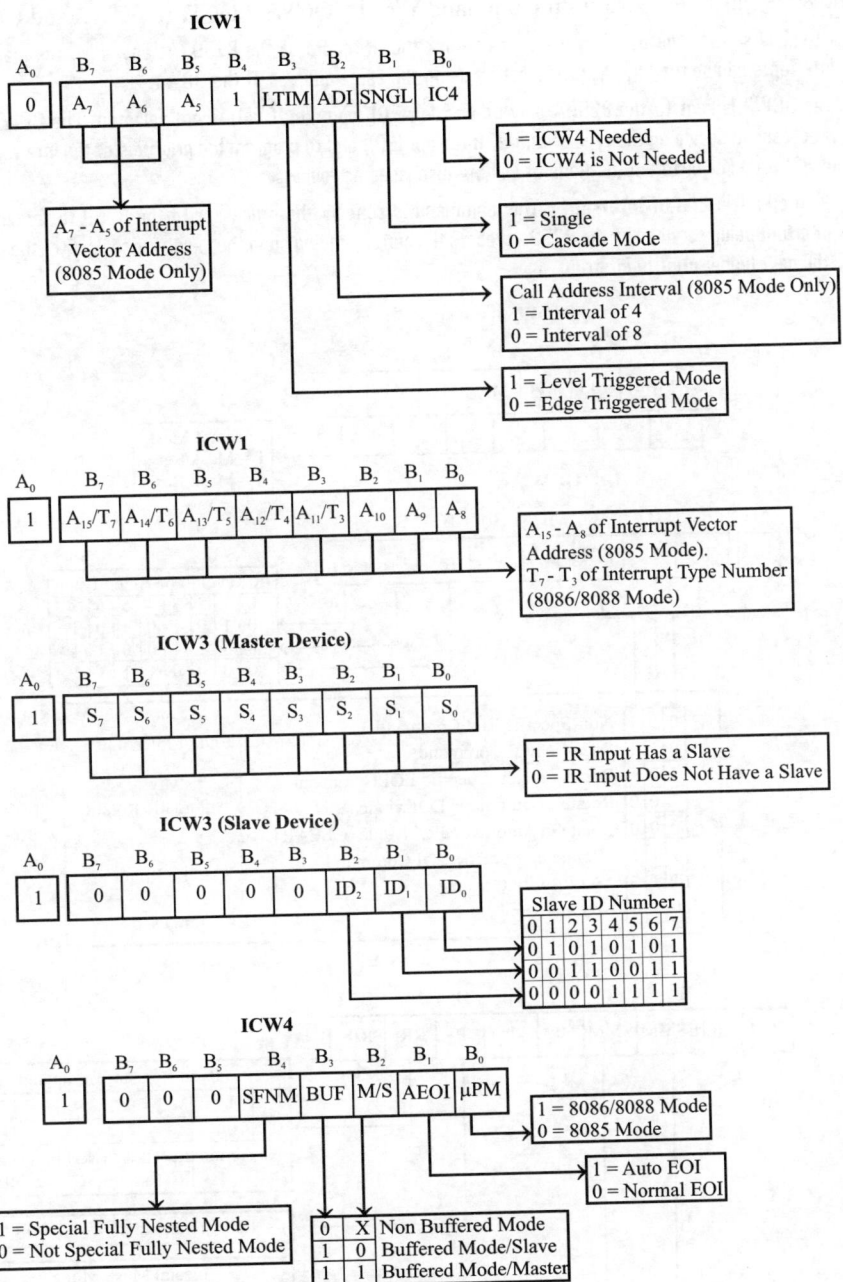

Fig. 7.12: Format of the ICWs.

7.7.7 Operation Command Words (OCWs)

The 8259 has three Operation Command Words (OCWs): OCW1, OCW2 and OCW3.

OCW1 : OCW1 is sent to the 8259 to mask or unmask the IR inputs of the 8259. At any time, the mask status of the interrupts can be read by the processor by using the same address of the OCW1.

OCW2 : The OCW2 is sent to the 8259A only when the AEOI mode (in ICW4) is not selected. The OCW2 is sent by the processor to decide on the type of End-Of-Interrupt (EOI) and to program the priorities of the interrupt (i.e., IR inputs of 8259A). The different methods of EOI are discussed as follows.

1. **Non-specific End-of-Interrupt** : This command is sent by the processor to the 8259 to terminate the current interrupt being serviced by the 8259. It resets the corresponding bit in the in-service register of the 8259 and allows the next higher priority interrupt.

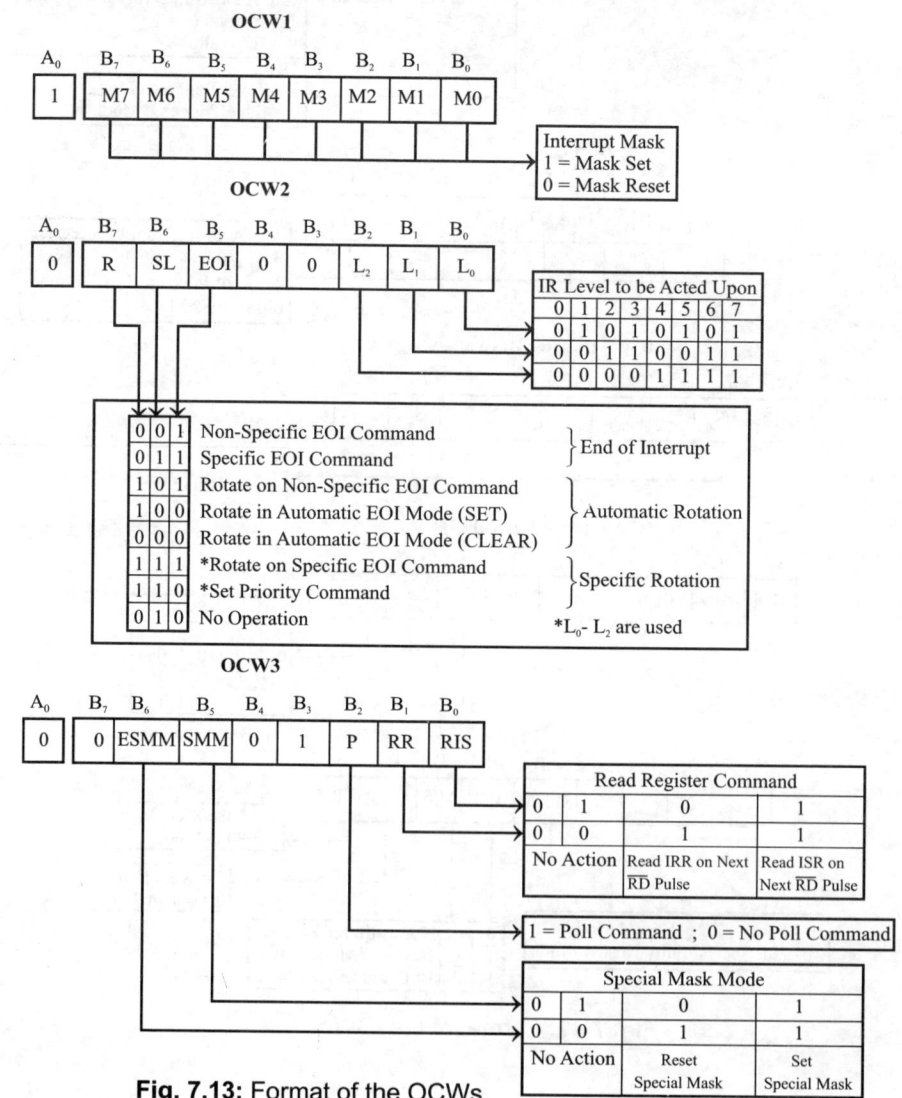

Fig. 7.13: Format of the OCWs.

2. **Specific End-of-Interrupt :** This command is sent by the processor to reset or terminate a specific interrupt request, decided by the lower 3-bits of OCW2.

3. **Rotate on Non-specific EOI :** This command will take action same as that of a non-specific EOI except that it rotates the priorities after resetting the bit in-service register. In this case, the interrupts will have rotating priority, in which the priority of the currently serviced interrupt becomes the least.

4. **Rotate on Automatic EOI :** This command is sent to the 8259 to select automatic EOI with rotating priority.

5. **Rotate on Specific EOI :** This command will take action similar to that of a specific EOI except that it rotates the priorities of the interrupts after they are serviced.

6. **Set priority :** The command is sent to set the priority of the interrupt level specified by the lower three bits of the OCW2 as the least.

OCW3 : The OCW3 is used to set any special mask mode, poll the active interrupt request and to read the in-service and interrupt request registers. In special mask mode, the mask status are negated to allow the interrupts masked by and interrupt mask register.

7.8 INTERRUPT PROGRAMING IN 8051

The interrupts are signals that are generated to interrupt the normal program execution of the microcontroller, in order to carry out a specific task/work. The specific task/work to be executed by an interrupt will be written as a program and stored in code memory as a subroutine program called the interrupt service subroutine.

While executing a program if the microcontroller encounters an interrupt, it completes the current instruction execution and saves the content of the Instruction pointer in stack and load the starting address of the interrupt service subroutine in the instruction pointer and start executing the subroutine.

At the end of the subroutine, the saved content of the instruction pointer is retrieved and loaded in the instruction pointer, so that the program execution continued from the point it was stopped.

7.8.1 Interrupt Enable (IE) Register

The IE-register is used to enable/disable the interrupts of an 8051. The interrupts are recognized (or accepted) by the controller only if they are enabled. The IE-register can be programmed to enable/disable all the five interrupts of an 8051 totally or individually.

The format of an IE-register is shown in Fig. 7.14. The EA bit of the IE-register can be programmed as zero, to disable all the five interrupts of 8051. When the EA bit is programmed as one, the interrupts are enabled provided their individual enable bits are programmed as one. (The EA bit is also called global enable.)

Each interrupt has a one-bit field to enable or disable it individually. When EA = 1, if the enable bit of a particular interrupt is programmed as one then it is enabled, and if the enable bit is programmed as zero then it is disabled.

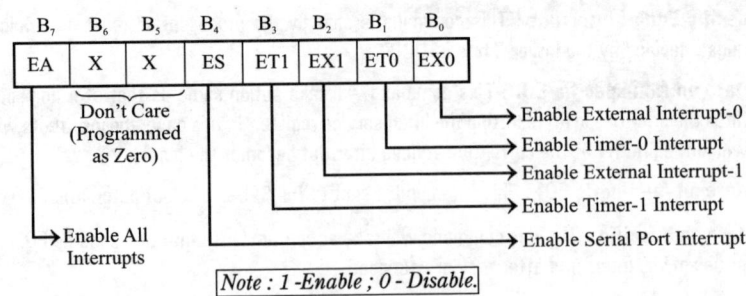

Fig. 7.14: Format of the IE register of the 8051 family of microcontrollers.

7.8.2 Interrupt Priority (IP) Register

The 8051 has five interrupts and the normal priority of these interrupts from highest to lowest are external interrupt-0, Timer-0 interrupt, External interrupt-1, Timer-1 interrupt and serial port interrupt.

The IP-register can be programmed to make the priority of any of the interrupt as highest. The format of an IP-register is shown in Fig. 7.15. The IP-register has one-bit field for the priority of each interrupt.

When the priority bit of a particular bit is programmed as one then its priority will be highest. In 8051, while servicing a lower priority interrupt a higher priority interrupt will be recognized but another lower priority interrupt will not be recognized.

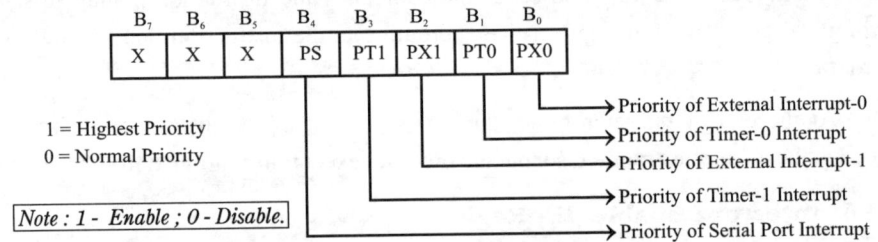

Fig. 7.15: Format of the IP-register of the 8051 family of microcontrollers.

7.8.3 Handling Interrupts of 8051

Enabling and Disabling of Interrupts

The 8051 interrupts are disabled upon power ON or after a reset. The desired interrupts has to be enabled by programming the appropriate bits in the interrupt enable register. The interrupts has a global enable which can be programmed to enable/disable all the interrupts.

Also, every interrupt has an individual enable that can be programmed to enable/disable a particular interrupt. In order to enable a particular interrupt, both global enable and particular enable bit has to be set to high. For disabling an interrupt, it is enough if the particular enable bit of an interrupt is reset to low.

Storing the Interrupt Subroutine Program in Memory

For every interrupt, a subroutine program has to be developed and stored in code memory starting from the vector address of that interrupt. Normally, very few memory locations are reserved for storing the interrupt subroutine. If the interrupt subroutine requires larger memory space then it is stored in any available code memory space. A branch instruction with starting address of interrupt subroutine as branch address is stored in vector address of the interrupt.

Initializing the Stack Pointer

The stack pointer is initialized with the data memory address 07H upon power ON or after a reset. The content of memory addressed by the stack pointer is considered occupied stack (or top of stack). Then for every push operation, the stack pointer is incremented and using the incremented address the content of memory/register is stored in the stack. Now, the default data memory space available for the stack will be 08H to 1FH, with a memory size of 24 bytes. Also, this memory space is common for register banks 1,2 and 3. If the programmer needs register banks 1,2 and 3, or a larger memory space for stack then the stack pointer has to be initialized with an address of available data memory space.

> **Note:** *The data memory address 00H to 07H is reserved for Bank 0 registers R0 to R7. The data memory spaces 20H to 2FH are reserved for bit addressable RAM.*

7.8.4 Examples of Interrupt Programming in 8051 Controller

EXAMPLE PROGRAM 7.3

This example program is developed for generation of unipolar square waveform of 1 kHz frequency using the timer-0 in mode-1 by using timer-0 interrupt, while the controller is performing continuous data transfer from port-0 to port-2. Assume that the crystal frequency of the controller is 12 MHz.

Program Analysis

In order to generate a square wave, a port pin can be set to high first and then after a time delay the port pin is reset to zero, then the process of set and reset are repeated continuously with a uniform time delay. The desired time delay can be achieved using the timer of 8051. This is achieved by the timer-0 interrupt service subroutine program.

The time period of 1 kHz square wave is $1/1 \times 10^3 = 1$ millisecond = 1000 microseconds. But in a square wave during half the period, the square wave will be high and during the next half of the period, the square wave will be low. So the time delay requires is 1000/2 = 500 microseconds.

Since the crystal frequency is 12 MHz, the timer clock will 1/12 = 1 MHz, and so the time period of the timer clock will be 1 microsecond. Therefore, the count for 500 microseconds time delay is 500. The count register in mode-1 is 16-bit and so the maximum count is, $2^{16} = 65536_{10}$. The initial count for any required delay has to be calculated by subtracting the delay count from the maximum count value.

Therefore, the initial count = $65536_{10} - 500_{10} = 65036_{10} =$ FE0CH

The byte to be loaded in TMOD register to select mode-1 operation of timer-0 is framed as shown below:

$$\text{TMOD} = \boxed{\text{Gate} \mid \text{C/}\overline{\text{T}} \mid \text{M1} \mid \text{M0} \mid \text{Gate} \mid \text{C/}\overline{\text{T}} \mid \text{M1} \mid \text{M0}} = \text{X X X X 0 0 0 1} = 01_{\text{H}}$$

The byte to be loaded in the IE register to enable timer-0 interrupt is framed as shown below:

$$\text{IE} = \boxed{\text{EA} \mid \text{X} \mid \text{X} \mid \text{ES} \mid \text{ET1} \mid \text{EX1} \mid \text{ET0} \mid \text{EX0}} = \text{1 0 0 0 0 0 1 0} = 82_{\text{H}}$$

Flowchart

Flowchart for Main Program

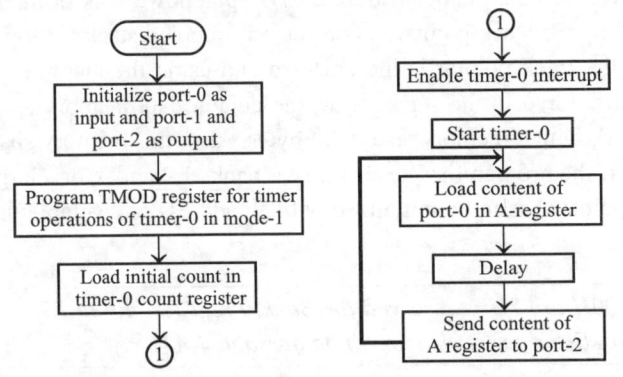

Flowchart for Interrupt Service Subroutine

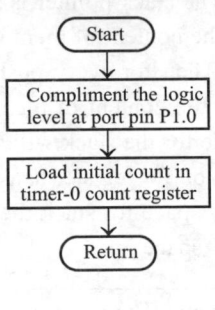

Assembly Language Program

```
;Program to generate square wave using 8051 timer-0 interrupt in mode-1

        ORG   0000H          ;Skip interrupt vector address and jump to main program
        SJMP  MAIN

;Inerrupt subroutine for timer-0

        ORG   000BH          ;Start from vector address of timer-0 interrupt
        CPL   P1.0           ;Compliment the port pin P1.0
        MOV   TL0,#0CH        ;Load low byte of count in timer-0 low order count register
        MOV   TH0,#FEH        ;Load high byte of count in timer-0 high order count register
        RETI                 ;Return from interrupt to main program

;Main Program

        ORG   0040H          ;Starting address to store main program

MAIN:   MOV   P0,#FFH        ;Initialize port-0 as input port
        MOV   P1,#00H        ;Initialize port-1 as output port
        MOV   p2,#00H        ;Initialize port-2 as output port

        MOV   TMOD,#01H       ;Program TMOD register for mode-1 operation of timer-0
        MOV   TL0,#0CH        ;Load low byte of count in timer-0 low order count register
        MOV   TH0,#FEH        ;Load high byte of count in timer-0 high order count register
        MOV   IE,#82H        ;Enable timer-0 interrupt
        SETB  TR0            ;Set timer run flag, to start timer
AGAIN:  MOV   A,P0           ;Get data from port-0
        NOP                  ;Wait for sometime
        NOP
        NOP
        MOV   P2,A           ;Send data to port-2
        SJMP  AGAIN          ;Repeat data transfer from port-0 to port-2

        END                  ;Assemby end
```

EXAMPLE PROGRAM 7.4

This example program is developed to transmit the parallel data in port-0, by serial communication to a standard PC at 9600 baud rate, by using serial port interrupt, while the controller is performing continuous data transfer from port-0 to port-2.

Program Analysis

In order to transmit serial data at 9600 baud rate, the timer-1 should be programmed for mode-2, and an initial count suitable for 9600 baud rate should be loaded in timer count register. The SCON register is programmed for mode-1 serial communication. Then the timer-1 is started by setting the timer-1 run flag. Then the data to be transmitted should be loaded in the SBUF register. At the end of serial transmission, the transmit interrupt flag is set, and interrupt service subroutine will take care of clearing this interrupt flag and allow the next data transfer.

The byte to be loaded in the TMOD register to select mode-2 operation of timer-1 is framed as shown below:

TMOD = | Gate | C/$\overline{T}$ | M1 | M0 | Gate | C/$\overline{T}$ | M1 | M0 | = 0 0 1 0 X X X X = 20_H

The byte to be loaded in the SCON register to select mode-1 serial communication is framed as shown below:

SCON = | SM0 | SM1 | SM2 | REN | TB8 | RB8 | TI | RI | = 0 1 0 1 0 0 0 0 = 50_H

The byte to be loaded in the IE register to enable serial port interrupt is shown below:

IE = | EA | X | X | ES | ET1 | EX1 | ET0 | EX0 | = 1 0 0 1 0 0 0 0 = 90_H

Flowchart

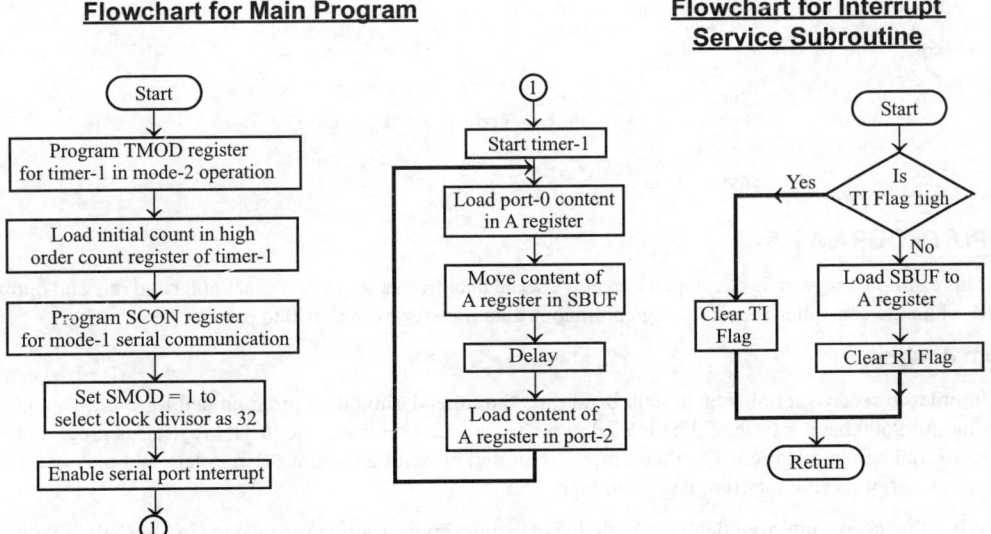

Flowchart for Main Program

Flowchart for Interrupt Service Subroutine

Assembly Language Program

```
;Program for serial data transmission in 8051 using serial port interrupt

        ORG  0000H          ;Skip interrupt vector address and jump to main program
        SJMP MAIN

;Inerrupt subroutine for serial data transmission

        ORG  0023H          ;Start from vector address of serial port interrupt
        JB   TI,CLRTI       ;If TI flag is high then clear it to allow next data transfer
        MOV  A,SBUF         ;If TI flag not high then clear receive buffer to Set RI flag
        CLR  RI             ;and then clear RI flag to return from interrupt
        RETI                ;Return from interrupt to main program
CLRTI:  CLR  TI
        RETI

;Main Program

        ORG  0040H          ;Starting address to store main program

MAIN:   MOV  P0,#FFH        ;Initialize port-0 as input port
        MOV  p2,#00H        ;Initialize port-2 as output port

        MOV  TMOD,#20H      ;Program timer-1 for mode-2 operation
        MOV  TH1,#FDH       ;Load the initial count for 9600 baud rate in timer-1 high
                             order register
        MOV  SCON,#50H      ;Program SCON reister for mode-1 serial communication
        MOV  A,PCON         ;Set SMOD=1, via A register

        SETB ACC.7          ;to choose clock dividor as 32
        MOV  PCON,A
        MOV  IE,90H         ;Enable serial port interrupt
        SETB TR1            ;Start timer-1

AGAIN:  MOV  A,P0           ;Get data from port-0
        MOV  SBUF,A         ;Transmit data serially
        NOP                 ;Wait for sometime
        NOP
        NOP
        MOV  P2,A           ;Send data to port-2
        SJMP AGAIN          ;Repeat data transfer from port-0 to port-2

        END                 ;Assemby end
```

EXAMPLE PROGRAM 7.5

This example program is developed to receive serial data from a standard PC at 9600 baud rate and output to port-1, while the controller is performing continuous data transfer from port-0 to port-2.

Program Analysis

In order to receive serial data at 9600 baud rate, the timer-1 should be programmed for mode-1, and an initial count for 9600 baud rate should be loaded in the timer count register. The SCON register is programmed for mode-1 serial communication. Then the timer-1 is started by setting the timer-1 run flag and the controller has to wait for serial receive interrupt flag to go high.

When the receive interrupt flag goes high, the interrupt service routine is called to read the SBUF register and transmit data to port-1 and then the receive interrupt flag is cleared, in order to enable serial port to receive next data.

The byte to be loaded in the TMOD register to select mode-2 operation of timer-1 is framed as shown below:

TMOD = | Gate | C/$\overline{T}$ | M1 | M0 | Gate | C/$\overline{T}$ | M1 | M0 | = 0 0 1 0 X X X X = 20_H

The byte to be loaded in the SCON register to select mode-1 serial communication is framed as shown below:

SCON = | SM0 | SM1 | SM2 | REN | TB8 | RB8 | TI | RI | = 0 1 0 1 0 0 0 0 = 50_H

The byte to be loaded in the IE register to enable serial port interrupt is shown below:

IE = | EA | X | X | ES | ET1 | EX1 | ET0 | EX0 | = 1 0 0 1 0 0 0 0 = 90_H

Flowchart

Flowchart for Main Program

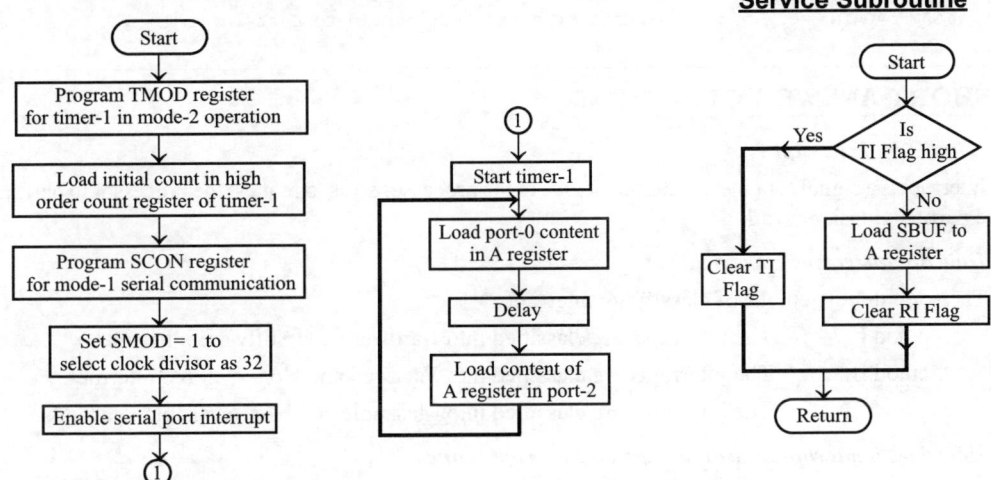

Flowchart for Interrupt Service Subroutine

Assembly Language Program

```
;Program to receive serial data in 8051 using serial port interrupt
        ORG   0000H          ;Skip interrupt vector address and jump to main program
        SJMP  MAIN
;Inerrupt subroutine to receive serial data
        ORG   0023H          ;Start from vector address of serial port interrupt
        JB    TI,CLRTI       ;If TI flag is high then clear TI flag to allow next data
                             ; transfer
        MOV   A,SBUF         ;Load received data in A register
        MOV   P1,A           ;Send data to port-1
        CLR   RI             ;and then clear RI flag to return from interrupt
        RETI                 ;Return from interrupt to main program
CLRTI:  CLR   TI
        RETI
;Main Program
        ORG   0040H          ;Starting address to store main program
```

```
MAIN:     MOV   P0,#FFH        ;Initialize port-0 as input port
          MOV   p2,#00H        ;Initialize port-2 as output port

          MOV   TMOD,#20H      ;Program timer-1 for mode-2 operation
          MOV   TH1,#FDH       ;Load the initial count for 9600 baud rate in timer-1 high
                                order register
          MOV   SCON,#50H      ;Program SCON reister for mode-1 serial communication
          MOV   A,PCON         ;Set SMOD=1, via A register
          SETB  ACC.7          ;to choose clock dividor as 32
          MOV   PCON,A
          MOV   IE,90H         ;Enable serial port interrupt
          SETB  TR1            ;Start timer-1

AGAIN:    MOV   A,P0           ;Get data from port-0
          NOP                  ;Wait for sometime
          NOP
          NOP
          MOV   P2,A           ;Send data to port-2
          SJMP  AGAIN          ;Repeat data transfer from port-0 to port-2
          END                  ;Assemby end
```

7.9 SHORT-ANSWER QUESTIONS

Q7.1   *What is an Interrupt ?*

Interrupt is a signal sent by an external device to the processor so as to request the processor to perform a particular task or work.

Q7.2   *How are interrupts classified ?*

There are three methods of classifying interrupts.

Method I   :   The interrupts are classified into Hardware and Software interrupts.

Method II   :   The interrupts are classified into Vectored and Non-vectored interrupt.

Method III   :   The interrupts are classified into Maskable and Non-maskable interrupts.

Q7.3   *How does a microprocessor service an interrupt request ?*

When the processor recognizes an interrupt, it saves the processor status in stack. Then it calls and executes an Interrupt Service Routine (ISR). At the end of ISR, it restores the processor status and the program control is transferred to the main program.

Q7.4   *What is the function of interrupt service routine?*

For each interrupt the processor has to perform a specific job. An interrupt service routine has been developed in order to perform the operations required for a device that is interrupting the processor.

Q7.5   *How are interrupts affected by system reset?*

Whenever the processor or system is reset, all the interrupts except TRAP are disabled. In order to enable the interrupts, EI instruction has to be executed after a reset.

Q7.6   *What are Software interrupts?*

Software interrupts are program instructions. These instructions are inserted at desired locations in a program. While running a program, if a software interrupt instruction is encountered then the processor executes an interrupt service routine.

Q7.7 What is Hardware interrupt?

If an interrupt is initiated in a processor by applying an appropriate signal to an interrupt pin, then the interrupt is called Hardware interrupt.

Q7.8 What is the difference between Software and hardware interrupts?

Software interrupt is initiated by the main program, but a hardware interrupt is initiated by an external device.

In 8085, the software interrupt cannot be disabled or masked but the hardware interrupt except TRAP can be disabled or masked.

Q7.9 What are vectored and non-vectored interrupt?

When an interrupt is accepted, if the processor control branches to a specific address defined by the manufacturer, then the interrupt is called vectored interrupt.

In non-vectored interrupt, there is no specific address for storing the interrupt service routine. Hence, the interrupting device should give the address of the interrupt service routine.

Q7.10 What is masking and why it is required?

Masking is preventing the interrupt from disturbing the current program execution. When the processor is performing an important job (process) and if the process should not be interrupted then all the interrupts should be masked or disabled.

In processor with multiple interrupts, the lower priority interrupt can be masked so as to prevent it from interrupting, the execution of interrupt service routine of higher priority interrupt.

Q7.11 What is vectoring?

Vectoring is the process of generating the address of interrupt service routine to be loaded in program counter.

Q7.12 List the software and hardware interrupts of 8085.

Software interrupts : RST 0, RST1, RST 2, RST 3, RST 4, RST 5, RST 6 and RST 7.

Hardware interrupts : TRAP, RST 7.5, RST 6.5, RST 5.5 and INTR.

Q7.13 What is TRAP?

TRAP is a non-maskable interrupt of 8085. It is not disabled by processor reset or after recognition of interrupt.

Q7.14 Does HOLD has higher priority than TRAP or not?

The interrupts including TRAP are recognized only if the HOLD is not valid, hence TRAP has lower priority than HOLD.

Q7.15 When does the 8085 processor accept a hardware interrupt?

The processor keeps on checking the interrupt pins at the second T-state of the last machine cycle of every instruction. If the processor finds a valid interrupt signal and if the interrupt is unmasked and enabled then the processor accepts the interrupt. The acceptance of the interrupt is acknowledged by sending an $\overline{\text{INTA}}$ signal to the interrupting device.

Q7.16 List the type of signals that have to be applied to initiate a hardware interrupt in 8085.

The TRAP is level and edge-sensitive and so the interrupt signal has to take a **low** to **high** transition and then remain **high** until it is recognized. The RST 7.5 is edge-sensitive and so the interrupt signal has to take a **low** to **high** transition and need not remain **high** until it is recognized. The RST 6.5, RST 5.5 and INTR are level-senstive and so the interrupt signal should be **high** until the interrupt is recognized.

Q7.17 What are maskable and non-maskable interrupts of 8085?

The TRAP is non-maskable interrupt. The RST 7.5, RST 6.5 and RST 5.5 are maskable interrupts. The INTR of 8085 can also be disabled by DI instruction.

Q7.18 When will the 8085 processor disable the interrupt system ?

The interrupts of 8085 except TRAP are disabled after any one of the following operations.

    1. Executing EI instruction.

    2. System or processor reset.

    3. After recognition (acceptance) of an interrupt.

Q7.19 What is the function performed by DI instruction?

The function of DI instruction is to disable the entire interrupt system.

Q7.20 What is the function performed by EI instruction?

The EI instruction can be used to enable all the interrupts after disabling.

Q7.21 How can the interrupt INTR of 8085 be expanded?

The interrupt INTR of 8085 can be expanded upto eight interrupts using 8-to-3 priority encoder. It can also be expanded to eight interrupts using one number of 8259 (Programmable interrupt controller) or upto 64 interrupts using 8259's in cascaded mode.

Q7.22 How can the hardware interrupt of 8085 be masked or unmasked?

The masking or unmasking of RST 7.5, RST 6.5 and RST 5.5 interrupts can be performed by moving an 8-bit data to accumulator and then executing SIM instruction. The format of the 8-bit data is shown in Fig. Q7.22.

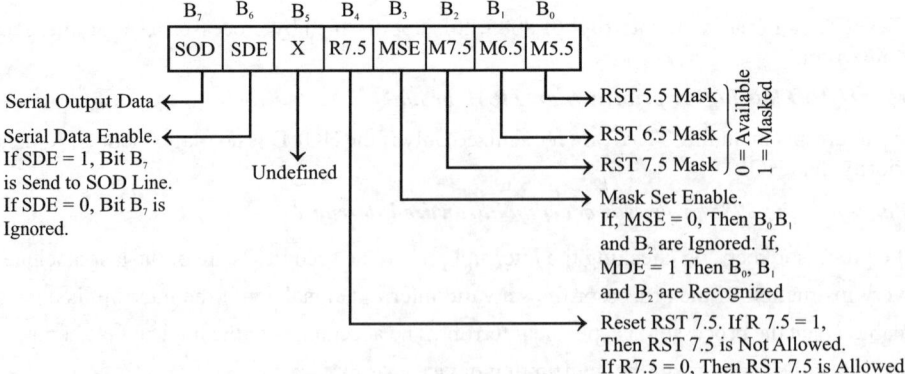

Fig. Q7.22: Format of the 8-bit data to be loaded in the accumulator before executing a SIM instruction.

Q7.23 How can the status of maskable interrupts be read in 8085 processor?

The status of hardware interrupts like interrupt request pending or not, interrupts enabled or not, and masked or unmasked can read from accumulator after executing RIM instruction. When RIM instruction is executed an 8-bit data is loaded in accumulator which can be interpreted as shown in Fig. Q7.23.

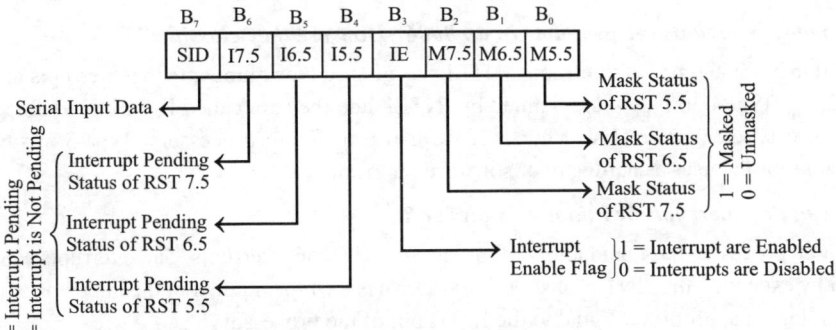

Fig. Q7.23: Interpretation of the content of accumulator after executing RIM instruction.

Q7.24 How can we check whether an 8085 interrupt is masked or not?

The masking status of an 8085 interrupt can be obtained by executing RIM instruction. When RIM instruction is executed, a 8-bit data is loaded in the accumulator. The bits B_0, B_1 and B_2 will give the masking status of RST 5.5, RST 6.5 and RST 7.5 respectively. If this bit is 1, then the corresponding interrupt is masked, otherwise it is unmasked.

Q7.25 How can we check the interrupt request pending status of 8085 interrupt?

The pending status of an 8085 interrupt can be obtained by executing RIM instruction. When the RIM instruction is executed an 8-bit data is loaded in accumulator. The bits B_4, B_5 and B_6 will give the pending status of RST 5.5, RST 6.5 and RST 7.5 respectively. If this bit is 1, then the interrupt is pending, otherwise it is not pending.

Q7.26 How are vector addresses generated for hardware interrupts of 8085?

For the hardware interrupts TRAP, RST 7.5, RST 6.5 and RST 5.5 the vector addresses are generated by the processor itself. These addresses are fixed by the manufacturer.

Q7.27 How is a vector address generated for the INTR interrupt of 8085?

For the INTR interrupt, the interrupting device has to place either RST opcode or CALL opcode followed by a 16-bit address. If RST opcode is placed, then the corresponding vector address is generated by the processor. In case of CALL opcode the given 16-bit address will be the vector address.

Q7.28 How are vector addresses generated for software interrupts of 8085?

For the software interrupts RST 0 to RST 7, the vector addresses are generated internal to the processor. These vector addresses are fixed by the manufacturer.

Q7.29 What are the sources of an 8086 interrupt?

There are three sources for interrupts in an 8086.

1. One source is from an external signal applied to the INTR or NMI pin of the processor.

2. The second source of an interrupt is execution of the interrupt instruction "INT n".

3. The third source of an interrupt is from some condition produced in the 8086 by the execution of an instruction.

Q7.30 What is exception? Give an example.

Exception is an interrupt generated due to exceptional condition (i.e., impossible situation) which occurs while executing an instruction. An example of exception is divide by zero interrupt in 8086. While executing the division instruction if the divisor is zero, then the 8086 will generate a divide by zero (type-0) interrupt.

Q7.31 How many interrupts are available in an 8086? How are they classified?

The 8086 has 256 types of interrupts. INTEL has given a type number to the interrupts in the range of 0 to 255_{10}. Type-0 to type-4 are defined by INTEL and they are called INTEL predefined interrupts. Type-5 to type-31 are reserved by INTEL for use in future processors. Type-32 to type-255 are available for the user as hardware or software interrupts.

Q7.32 How can the interrupts be initiated in an 8086?

The 8086 processor has dual facility of initiating all the 256 interrupts. The interrupts can be initiated either by executing the "INT n" instruction where n is the type number or the interrupt can be initiated by sending an appropriate signal to the INTR pin of the processor.

Q7.33 List the INTEL predefined interrupts.

The INTEL predefined interrupts are:

1. Divide by zero (Type-0 interrupt)   4. Breakpoint interrupt (Type-3 interrupt)
2. Single step (Type-1 interrupt)   5. Interrupt on overflow (Type-4 interrupt)
3. Non-maskable interrupt, NMI (Type-2 interrupt)

Q7.34 What are software and hardware interrupts of an 8086?

In 8086 the interrupts initiated by executing "INT n" instruction are called software interrupts.

The interrupts initiated by applying appropriate signals to the INTR and NMI pins of the 8086 are called hardware interrupts.

Q7.35 What are maskable and non-maskable interrupts of an 8086?

The hardware interrupts initiated by applying an appropriate signal to the INTR pin of an 8086 are maskable interrupts.

The software interrupts and the hardware interrupt NMI are non-maskable.

Q7.36 How can the interrupts be masked/unmasked in an 8086?

The maskable interrupts of 8086 can be masked by clearing the interrupt flag to zero and they can be unmasked/allowed by setting the interrupt flag to one.

Q7.37 What is a vector table? Where is it located?

The memory block consisting of vector addresses of all the 256 types of interrupts of an 8086 is called a vector table. The vector table is stored in the first 1 kB of physical memory space.

Q7.38 How is the interrupt address generated in 8086?

The 8086 will multiply the type number by four and sign extend to 20-bit to get a memory address of the vector table. The vector address for an interrupt will be available in four consecutive memory location starting from this 20-bit address. The first word in the table is the offset address of ISR (Interrupt Service Routine) and the next word is the segment base address of the ISR.

Q7.39 What is the need for an interrupt controller?

The interrupt controller is employed to expand the interrupt input. It can handle the interrupt request from various devices and allow them one by one to the processor.

Q7.40 List some of the features of INTEL 8259 (Programmable Interrupt Controller).

1. It manage eight interrupt requests   3. The priorities of interrupts are programmable
2. The interrupt vector addresses are programmable   4. The interrupt can be masked or unmasked individually

Q7.41 Write the various functional blocks of INTEL 8259?

The various functional blocks of 8259 are Control logic, Read/ Write logic, Data bus buffer, Interrupt Request Register (IRR), Interrupt Mask Register (IMR) and In-Service Register (ISR), Priority Resolver (PR) and Cascade buffer.

Q7.42 What is master and slave 8259 ?

When 8259s are connected in cascade, one 8259 will be directly interrupting the processor and it is called master 8259. To each interrupt request input of master 8259, one slave 8259 can be connected. The 8259's interrupting the master 8259 are called slave 8259.

Q7.43 How is 8259 programmed?

The 8259 is programmed by sending Initialization Command Words (ICWs) and Operational Command Words (OCWs).

Q7.44 What are the features of 8259 that are programmed using ICWs?

The ICWs are used to program the following features of an 8259:

1. Call address interval (in case of 8085)   5. 8085 or 8086 modes
2. Cascade mode or single   6. Auto or Normal end of interrupt
3. Level or Edge triggered   7. Special fully nested mode
4. Vector address (in case of 8085)
    or Type number (in case of 8086)

Q7.45 What are features of 8259 that can be programmed using OCWs?

The OCWs are used to program the following features of an 8259:

1. Masking of individual interrupts.
2. Specific or Non-specific end of interrupt.
3. Priority modes.

Q7.46 Write the format of ICW1?

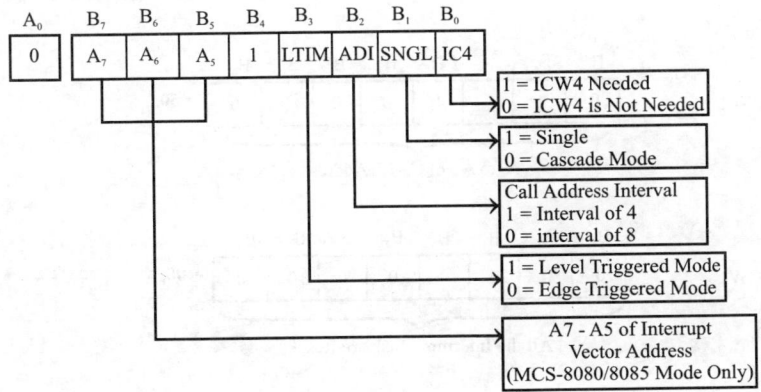

Fig. Q7.46: Format of the ICW1.

Q7.47 What is the difference in programming master 8259 and slave 8259 ?

The ICW 3 will be different for master 8259 and slave 8259. For master, the ICW3 will inform the IR input that are having slaves. For slave, the ICW3 will inform its slave ID number.

Q7.48 When is ICW4 send to 8259 ?

The ICW4 is send to 8259 to perform any one of the following features:

    1. **8085 or 8086 mode**

    2. **Special fully nested mode**

    3. **Auto or Normal end of interrupt**

    4. **Buffered or Non-buffered mode.**

Q7.49 Write a program segment to initialize a single 8259 connected to an 8085 processor.

Let us assume that 8259 is IO-mapped in the system. The 8259 can be initialized by sending ICW1, ICW2 and OCW1. Let the 8-bit address when $A_0 = 0$ be 00_H and when $A_0 = 1$ be 01_H.

```
MVI A,ICW1 ; Move ICW1 to A-register.
OUT 00H    ; Send ICW1 to 8259.
MVI A,ICW2 ; Move ICW2 to A-register.
OUT 01H    ; Send ICW2 to 8259.
MVI A,OCW1 ; Move OCW1 to A-register.
OUT 01H    ; Send OCW1 to 8259.
HLT        ; Halt program execution.
```

Q7.50 Frame the Command words ICW1, ICW2 and OCW1 for initializing a single 8259 interfaced to 8085 with the call address interval of 8 and for level triggered interrupt. Also unmask all interrupt inputs. The desired vector address is 5000_H.

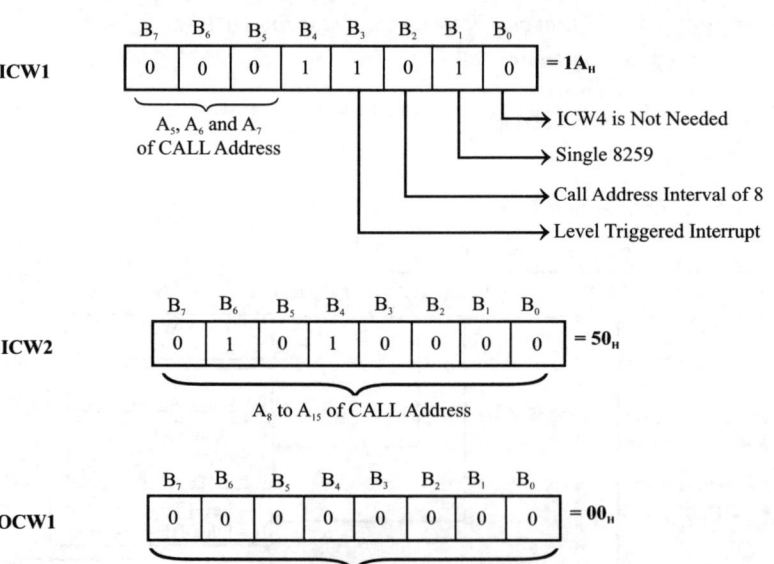

Q7.51 *Frame the command words ICW1, ICW2, ICW3 and OCW1 for initializing a single 8259 to initiate INT 40H to INT 47H in an 8086-based system. The desired features are level triggered interrupt and automatic end of interrupt.*

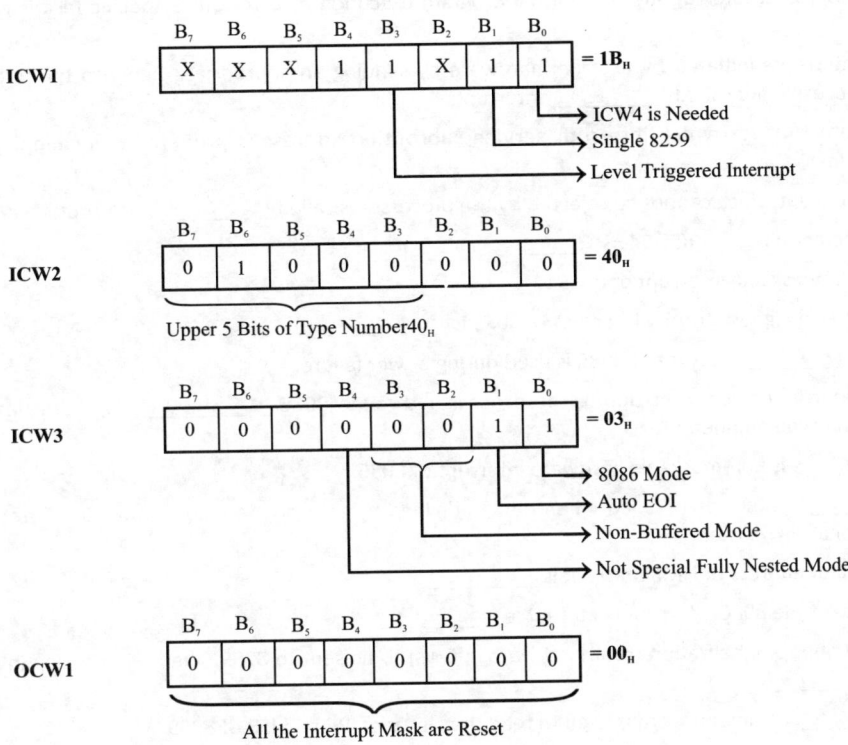

Q7.52 *Write a program segment to initialize a single 8259 connected to an 8086 processor.*

Let us assume that 8259 is IO-mapped in the system with an even address. The 8259 can be initialized by sending ICW1, ICW2, ICW4 and OCW1. Let the 8-bit address with $A_0 = 0$ be 00_H and when $A_0 = 1$ be 02_H.

```
MOV AL,ICW1      ;Move ICW1 to AL-register
OUT [00H]        ;Send ICW1 to 8259
MOV AL,ICW2      ;Move ICW2 to AL-register
OUT [02H]        ;Send ICW2 to 8259
MOV AL,ICW4      ;Move ICW4 to AL-register
OUT [02H]        ;Send ICW4 to 8259
MOV AL,OCW1      ;Move OCW1 to AL-register
OUT [02H]        ;Send OCW1 to 8259
HLT              ;Stop
```

7.10 EXERCISES

I. Fill in the blanks with appropriate words

1. The process of interrupting the normal program execution to carry out a specific task is referred to as _____.

2. The interrupts initiated by external hardware by sending an appropriate signal to the interrupt pin of the processor are called _____ interrupts.

3. The interrupt for which interrupt service subroutine address is predefined is called _____ interrupt.

4. The interrupt which cannot be rejected by the processor is called _____ interrupt.

5. The vector address of RST 5.5 is _____.

6. The non-maskable interrupt of 8085 is _____.

7. The second highest priority interrupt of 8085 is _____.

8. The _____ interrupt of 8085 is used during power failure.

9. The external device interrupting the processor through 8086 _____ pin, need not supply the interrupt type number.

10. _____ is non-maskable hardware interrupt of 8086.

11. _____ interrupt is called when data is divided by zero or the result of division is too large to fit in destination register.

12. The vector address of NMI interrupt is _____.

13. The size of the 8086 interrupt vector table is _____.

14. The number of intialization command words (ICWs) to be sent to 8259 when it is working in single mode is _____.

15. _____ command word is used to read the status of IRR and ISR of 8259

16. The _____ of 8259 is used to store the status of interrupts being serviced currently.

17. The instruction used to execute an interrupt sub-routine whose starting address is stored at an address 00006_H to 00009_H is _____.

18. An 8086 instruction used to transfer the program control back to main program and to restore the status of flag register is _____.

Answers

1. interrupt	6. TRAP	11. Divide by zero (type-0)	16. ISR(In-Service Register)
2. hardware	7. RST 7.5	12. 0000 : 0004H	17. INT 03H
3. vectored	8. TRAP	13. 1 KB	18. IRET
4. non-maskable	9. NMI	14. 3	
5. $002C_H$	10. NMI	15. OCW3	

II. State whether the following statements are True/False.

1. In polling technique, the processor has to check the readiness of the devie periodically for data transfer.

2. In interrupt driven data transfer scheme, the external device interrupts the processor only when the data is ready.

3. The Interrupt Service Subroutine (ISR) address is predefined by the processor manufacturers for non vectored interrupts.

4. The processor should compulsorily accept all interrupts any time.

5. The software interrupts cannot be masked.

6. The vector address of 8085 software interrupt RSTn is equivalent to n×8.

7. All the hardware interrupts are disabled when the 8085 processor is reset.

8. INTR interrupt of 8085 is the highest priority interrupt.

9. During 8086 reset, all the interrupts are disabled.

10. All the software interrupts of 8086 can be initiated through INTR pin.

11. In 8086, INTR interrupt cannot be masked.

12. All the 256 software interrupts of 8086 are available for the user as hardware or software interrupts.

13. NMI can also be triggered by calling type-2 software instruction

14. Divide error interrupt has the highest priority among all the interrupts of 8086.

15. The software interrupts are non-maskable and have higher priority than the hardware interrupts.

16. The memory address for storing the vector address for an 8086 interrupt is given by multiplying the type number by eight

17. In AEOI mode of 8259 the interrupt is reset immediately after receiving the second $\overline{INTA}$ pulse from 8086.

18. ICW4 command word is not necessary when 8259 is interfaced with 8086 based system.

19. ICW3 is not necessary for single mode operation of 8259.

20. Priorities of IR_0-IR_7 are fixed in 8259.

Answers

1. True	4. False	7. False	10. True	13. True	16. False	19. True
2. True	5. True	8. False	11. False	14. True	17. True	20. False
3. False	6. True	9. False	12. False	15. True	18. False	

III. Choose the right answer for the following questions.

1. *The interrupts whose request can be either accepted or rejected by the processor are called as*

    a) vectored interrupts   b) Non-vectored interrupts

    c) maskable interrupts   d) non-maskable interrupts

2. *Which of the following is not a hardware interrupt of 8085?*

    a) TRAP   b) RST 7.5   c) INTR   d) RST n

3. *The vector address of the 8085 software interrupt RST 3 is*

    a) 0008_H b) $000C_H$ c) 0010_H d) 0018_H

4. *Which of the following signal on 8085 TRAP line initiates the interrupt?*

    a)   b) 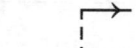   c)    d) none

5. *Which of the following 8085 interrupt is not a level sensitive interrupt?*

    a) RST 7.5   b) RST 6.5   c) RST 5.5   d) INTR

6. **The 8085 receives both DMA request and an interrupt on TRAP pin. Which of the following statement is true?**

   a) It recognizes TRAP first and then DMA request

   b) It recognizes DMA request first and then TRAP

   c) It recognizes both TRAP and DMA request simultaneously.

   d) It rejects both requests and execute the next instruction in the main program.

7. **In the 8085 microprocessor, the RST 5 instruction transfers the program control to the following location.**

   a) 0010_H b) 0018_H c) 0020_H *d)* 0028_H

8. **The following interrupt of 8086 is non-maskable.**

   a) NMI   b) all software interrupt   c) both a and b   *d)* neither a nor b

9. **Which of the following is a hardware interrupt in 8086?**

   a) RST 5.5   b) RST 4.5   c) INTR   *d)* INT

10. **How many software interrupts are there in 8086 system?**

    a) 100   b) 32   c) 64   *d)* 256

11. **What is the vector address of the interrupt INT 03H**

    a) 0000:0002H   b) 0000 : 000CH   c) 0000 : 0008H   *d)* 0000 : 0000H

12. **Which of the following has lowest priority?**

    a) NMI   b) INTR   c) single step   *d)* INT0

13. **Which of the following IC is programmable interrupt controller?**

    a) 8255   b) 8253   c) 8279   *d)* 8259

14. **How many interrupts can be accepted in cascaded mode in 8259?**

    a) 8   b) 16   c) 32   *d)* 64

15. **Which of the following register of 8259 is used to store the interrupt requests?**

    a) ISR   b) IMR   c) IRR   *d)* PR

16. **Which of the following command word is used to mask/unmask individual interrupt?**

    a) ICW1   b) ICW2   c) OCW1   *d)* none of the three

17. **Which interrupt has the highest priority by default in 8259?**

    a) IR7   b) IR0   c) IR1   *d)* none of the three

18. **What is the call address interval for 8086 vectored interrupts?**

    a) 8   b) 4   c) 2   *d)* 16

Answers

1. c	4. c	7. d	10. d	13. d	16. c
2. d	5. a	8. c	11. b	14. d	17. b
3. d	6. b	9. c	12. c	15. c	18. b

IV. Answer the following questions.

E7.1   Write an 8085 ALP to enable all the interrupts of 8085.

E7.2   The 8085 processor is executing the following ISR of RST 6.5. Rearrange the program to allow the higher priority interrupt RST 7.5 to interrupt the processor while execution of RST 6.5 ISR.

```
;ISR of RST 6.5

        MVI   A,00H
Back:   INR   A
        CPI   A,0FFH
        JNZ   BACK
        RET
```

E7.3   What will be content of A and B registers after executing the following program in 8086 processor?

```
2000:   XRA   A           ;ISR of RST 03H
2001:   ADI   02H         0018:   XRI   0FFH
2003:   MVI   B,82H       001A:   MOV   B,A
2005:   ADD   B           001B:   RET
2006:   RST   03H
2008:   DCR   A
2009:   ADD   B
200A:   HLT
```

E7.4   Identify the RST instruction generated by the following 8 to 3 priority encoder.

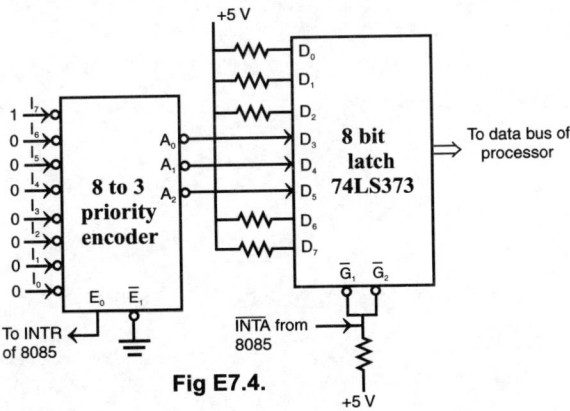

Fig E7.4.

E7.5   What will happen after executing the following program in 8086 processor?

```
        MOV   AX,8123H
        MOV   BL,00H
        DIV   BL
```

E7.6   Write an 8086 assembly language program to store 55_H in memory location 50001_H if divide by 0 interrupt generated in E7.5.

E7.7   What addresses in the 8086 interrupt vector table are used for type-1 or single step interrupt?

E7.8   How do you calculate the vector address of the interrupt INT 80H.

E7.9 What will be the content of 'AL' and BL registers after executing the following program?

```
MOV   AL,03H
ADD   AL,20H
MOV   BL,02H
INT   03H
SUB   AL,BL
INC   BL
```

E7.10 What will be the content of memory locations 1000_H and 1001_H after executing the following program? Identify and explain the error if any.

```
        MOV   CL,00H                        ;Delay Subroutine
        MOV   AL,90H                  PROCEDURE DELAY
        MOV   BL,81H                  MOV  DL,FFH
        ADD   AL,BL          YY:      DEC  DL
        CALL  DELAY                   JNZ  YY
        JNC   XY                      CLC
        INC   CL                      RET
XY:     MOV   [1000H],AL              ENDP DELAY
        MOV   [1001H],CL
```

E7.11 Write an operational command word to mask the interrupt on line 3 in 8259.

E7.12 Write an intialization command word to interface 8259A with 8085 in cascaded level triggered mode with call address interval of 8. Assume interrupt vector address 8800H.

E7.13 Write an operational command word to set IR5 as bottom priority level, with rotate on specific EOI command mode.

CHAPTER 8

ASSEMBLY LANGUAGE PROGRAMMING

8.1 INTRODUCTION TO ASSEMBLY LANGUAGE PROGRAMMING

Programs are a set of instructions or commands needed to perform a specific task by a programmable device such as a microprocessor. The programs needed for a programmable device can be developed at three different levels and they are as follows:

1. **Machine level programming**
2. **Assembly level programming**
3. **High level programming**

8.1.1 Machine Level Programming

In machine level programming, instructions are written using binary codes which uses only two symbols '0' and '1'. The manufacturer of microprocessors will give a set of instructions for each microprocessor in binary codes, i.e., one binary code will represent one operation performed by the microprocessor. The language in which the instructions are represented by binary codes is called machine language. A microprocessor can understand and execute the machine language programs directly.

The binary instructions of one microprocessor will not be same as that of another microprocessor. Therefore, the machine language programs developed for one microprocessor cannot be used for another microprocessor i.e., the machine level programs are machine dependent. Moreover, it is highly tedious for a programmer to write programs in the machine language.

8.2.2 Assembly Level Programming

In assembly level programming, instructions are written using mnemonics. A mnemonic comprises of a few letters of the English language which represent the operation performed by the instruction. For example, the mnemonic for the instruction which performs **addition** operation is **ADD**. The manufacturer of the microprocessors will provide a set of instructions in the form of a mnemonic for each microprocessor. Also, for each mnemonic a binary code will be specified by the manufacturer. If the program is developed using binary codes then it is called machine level programming and if the program is developed using mnemonics then it is called assembly level programming.

The language in which the instructions are represented by mnemonics is called assembly language. Microprocessors cannot execute the assembly language programs directly. The assembly language programs have to be converted to machine language for execution. This conversion is performed using a software tool called assembler.

The mnemonics of one microprocessor will not be same as that of another microprocessor. Therefore, the assembly language programs developed for one microprocessor cannot be used for another microprocessor directly i.e. the assembly language programs are machine dependent. But certain manufacturers provide upward compatability for the same family of microprocessors. (i.e., the program developed for a the lower version of a microprocessor of a family can be run on the higher version without modifications.) For example, consider the INTEL 80x86 family of microprocessors. The program developed for 8086 microprocessor can be run on 80186, 80286, 80386 or 80486 microprocessor-based system without any modifications.

8.1.3 High Level Programming

In high level programming the instructions will be in the form of statements written using symbols, English words and phrases. Each high level language will have its own vocabulary of words, symbols, phrases and sentences. Examples of high level languages are BASIC, C, C++, etc. The programs written in high level languages are easy to understand and machine independent. So they are known as portable programs. A high level language program has to be converted into machine language programs in order to be executed by the microprocessor. This conversion is performed by a software tool called compiler.

8.2 FLOWCHART

Flowchart is a graphical representation of the operation flow of a program. It is also the graphical form of an algorithm. Flowcharts can be a valuable aid in visualizing programs. The various symbols used for drawing flowcharts are shown in Fig. 8.1. The operations represented by various symbols of flowchart are explained in Table 8.1. A sample flowchart is shown in Fig. 8.2.

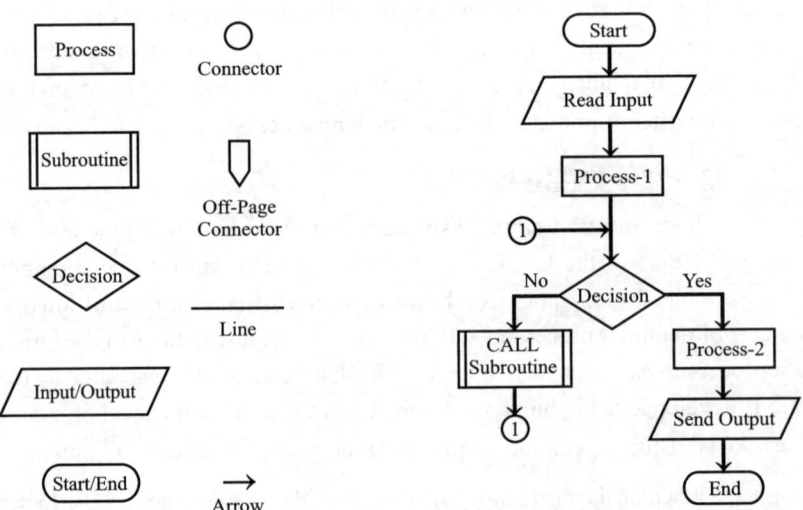

Fig. 8.1: Symbols used in a flowchart.   **Fig. 8.2:** A sample flowchart.

Table 8.1: Operations Represented by the Symbols used in Flowchart

Symbol	Operation
Racetrack shape box	A racetrack shaped symbol is used to indicate the beginning (start) or end of a program.
Parallelogram	A parallelogram is used to represent input or output operation.
Rectangular box	A rectangular box is used to represent simple operations other than input and output operations.
A rectangular box with double lines on vertical sides	A rectangular box with double lines on vertical sides is used to represent a subroutine or procedure.
Diamond shaped box	A diamond shaped box is used to represent a decision point or cross road in the programs
Small circle	A small circle is used as a connector to show the connections between various parts of a flowchart within a page. Identical numbers are entered inside the circles that represent the same connecting points.
Five-sided box	A five-sided box symbol is used as an off-page connector to show the connections between various sections of a flowchart in different pages. Identical numbers are entered inside the boxes that represent the same connecting point.
Line	Lines are drawn between boxes and diamonds to indicate the program flow.
Arrow	Arrows are placed on the lines to indicate the direction of program flow.

8.3 ASSEMBLY LANGUAGE PROGRAM DEVELOPMENT TOOLS

Development system is used by system designers to design and test the software and hardware of a microprocessor-based system before going for practical implementation (or fabrication). The microprocessor development system consists of a set of hardware and software tools. The hardware of a development system usually contain a standard PC (**P**ersonal **C**omputer), printer and an emulator. The software tools are also called program development tools and they are editor, assembler, library builder, linker, debugger and simulator. These software tools can run on a PC in order to write, assemble, debug, modify and test the assembly language programs.

8.3.1 Editor (Text Editor)

Editor is a software tool which, when run on a PC, allows the user to type/enter and modify the assembly language program. The editor provides a set of commands for insertion, deletion and modification of letters, characters, statements, etc. The main function of an editor is to help the user to construct the assembly language program in the right format. The program created using editor is known as source program and it is usually saved with the file extension .ASM For example, if a program for addition is developed using editor then it can be saved as ADDITION.ASM. Some examples of editors are NE (Norton Editor), EDIT (DOS Editor), etc.

8.3.2 Assembler

The assembler is a software tool which, when run on a PC, converts the assembly language program to a machine language program. Several types of assemblers are available and they are one-pass assembler, two-pass assembler, macro assembler, cross assembler, resident assembler and meta assembler.

In one-pass assembler the source code is processed only once and we can use only backward reference. In a one-pass assembler as the source code is processed, any labels encountered are given an address and stored in a table. Whenever a label in encountered, the assembler may look backward to find the address of the label. If the label is not yet defined then it issues an error message (because the assembler will not look forward). Since only one pass is used to translate the source code, a one-pass assembler is very fast, but because of the forward reference problem, the one-pass assembler is not used often.

Most of the popularly used assemblers are the two-pass assemblers. In a two-pass assembler, the first pass is made through source code for the purpose of assigning an address to all the labels and to store this information in a symbol table. The second pass is made to actually translate the source code into machine code.

The input for the assembler is the source program which is saved with file extension .ASM. The assembler usually generates two output files called object file and list file. The object file consist of relocatable machine codes of the program and it is saved with file extension .OBJ. The list file contains the assembly language statements, the binary codes for each instruction and address of each instruction. The list file is saved with file extension .LST.

The list file also indicates any syntax errors in the source program. The assembler will not identify the logical errors in the source program. In order to correct the errors indicated on the list file, the user have to use the editor again. The corrected source program is saved again and then reassembled. Usually, it may take several times through edit-assemble loop to eliminate the syntax errors from the source program.

Some examples of assemblers are TASM (Borland's Turbo Assembler), MASM (Microsoft's Macro Assembler), ASM86 (INTEL'S 8086 Assembler), ASM85 (INTEL'S 8085 Assembler), etc.

Advantages of the Assembler

1. The assembler translates mnemonics into binary code with speed and accuracy, thus eliminating human errors in looking up the codes.
2. The assembler assigns appropriate values to the variables used in a program. This feature offers flexibility in specifying jump locations.

3. It is easy to insert or delete instructions in a program and reassemble the entire program quickly with new memory locations and modified addresses for jump locations. This avoids rewriting the program manually.

4. The assembler checks syntax errors, such as wrong labels, opcodes, expressions, etc., and provides error messages. However, it cannot check logic errors in a program.

5. The assembler can reserve memory locations for data or results.

6. The assembler provides list file for documentation.

8.3.3 Library Builder

The library builder is used to create library files which are a collection of procedures of frequently used functions. Actually a library file is a collection of assembled object files. While developing a software for a particular application, the programmers can link the library files in their programs. When the library file is linked with a program, only the procedure required by the program are copied from library file and added to the program.

The input to library builder is a set of assembled object files of program modules/procedures. The library builder combines the program modules/procedures into a single file known as library file and it is saved with file extension ".LIB". Some examples of library builder are microsoft's LIB, Borlands TLIB, etc.

8.3.4 Linker

The linker is a software tool which is used to combine relocatable object files of program modules and library functions into a single executable file.

While developing program for a particular application it is much more efficient to develop the program in modules. The entire task of the program can be divided into smaller tasks and procedures for each task can be developed individually. These procedures are called program modules. For certain tasks we can use library files if they are available. Each module can be individually assembled, tested and debugged. Then the object files of program modules and the library files can be linked to get an executable file.

The linker also generates a link map file which contains the address information about the linked files. Some examples of linkers are microsoft's linker LINK, Borland's Turbo linker TLINK, etc.

8.3.5 Debugger

Debugger is a software tool that allows the execution of a program in a single step or break-point mode under the control of user. The process of locating and correcting the errors in a program using a debugger is known as debugging.

The debugger allows the designer to load the object code program into the memory of the PC, execute the program and troubleshoot or debug it. The debugger allows the designer to look at the contents of registers and memory locations after running the program. It allows the system designer to change the contents of registers and memory locations and return the program.

Some debuggers allow the user to stop execution after each instruction so that the memory/register content can be checked or altered. A debugger also allows the user to set a breakpoint at any point in user program. When the user runs the program, the PC will execute instructions up to this breakpoint and stop. The user can then examine register and memory contents to see whether the results are correct upto that point. If the results are correct, the user can move the breakpoint to a later point in the program. If the results are not correct, the user can check the program up to that point to find out why they are not correct.

Debugger tools can help the user to isolate a problem in the program. Once the problem/errors are identified, the algorithm can be modified. Then the user can use the editor to correct the source program, reassemble the corrected source program, relink and run the program again.

8.3.6 Simulator

The simulator is a program which can run on the development system (Personal computer) to simulate the operations of the newly designed system. Some of the operations that can be simulated are as follows:

1. Execute a program and display result.
2. Single step execution of a program.
3. Break-point execution of a program.
4. Display the contents of register/memory.

Simulator usually shows the content of registers and memory locations on the screen of the computer and allows the system designer to perform all the operations listed above, with the added advantage of watching the data change as the program operates. This feature saves considerable time because the register/memory contents do not have to be displayed using separate commands. The visual representation also gives the programmer a better feel for what is taking place in the program.

The simulators do not have the ability to perform actual IO or internal hardware operations such as timing or data transmission and reception.

8.3.7 Emulator

An emulator is a mixture of hardware and software. It is usually used to test and debug the hardware and software of a newly designed microprocessor-based system. The emulator has a multicore cable which connects the PC of the development system and the newly designed hardware of the microprocessor system. A connector/plug at one end of the cable is plugged into new hardware in place of its microprocessor. The other end of cable is connected to parallel port of PC. Through this connection the software of the emulator allows the designer to download the object code program into RAM in the system being tested and run it.

Like a debugger, an emulator allows the system designer to load and run programs, examine and change the contents of registers, examine and change the contents of memory locations and insert breakpoints in the program.

The emulator also takes a snapshot of the content of registers, activity on the address and data bus and the state of the flags as each instruction executes. Also, the emulator stores this trace data. The user can have a printout of the trace data to see the results that the program produced on a step-by-step basis. Another powerful feature of an emulator is the ability to use either development system memory or the memory on the hardware under test for the program that is being debugged.

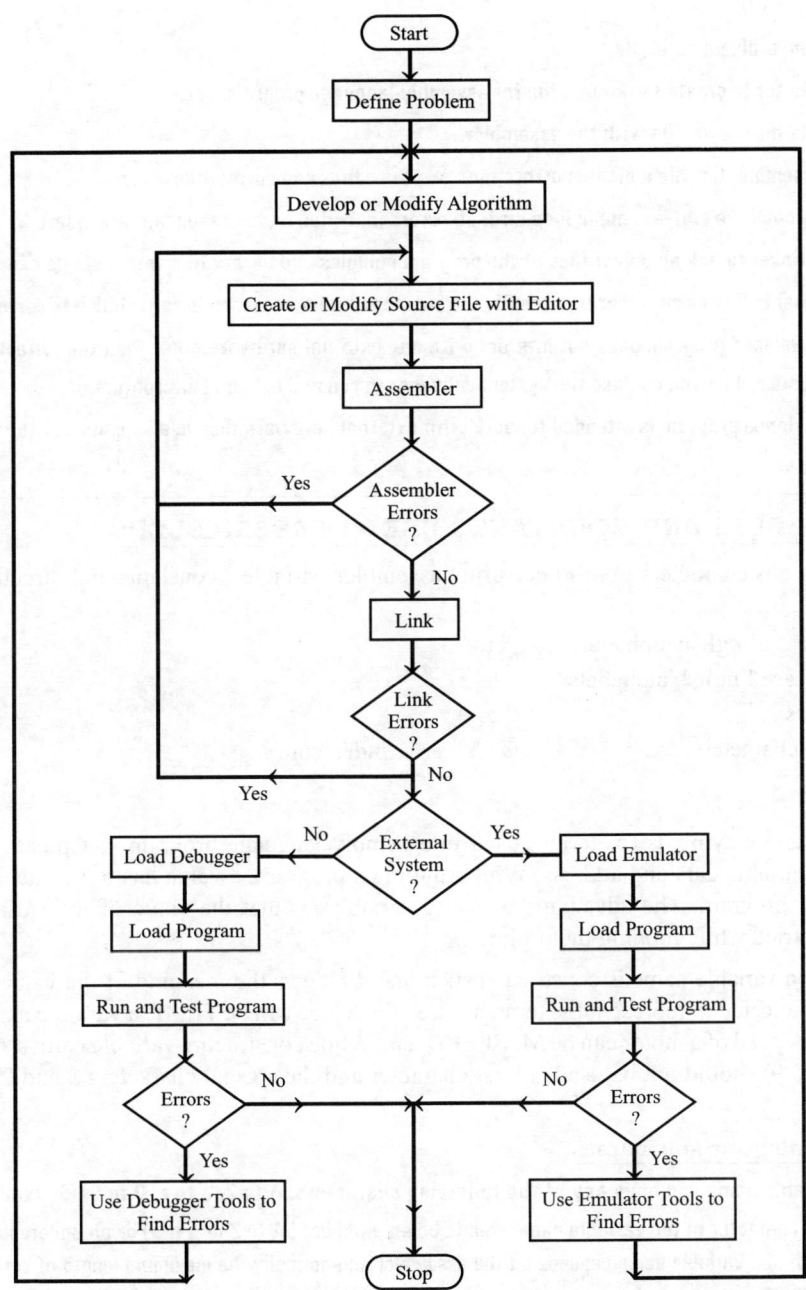

Fig. 8.3: Development process of an assembly language program.

Summary of the Use of Program Development Tools

The various steps in the development of an assembly language program are as follows, and also as a flowchart in Fig. 8.3.

1. Define the problem carefully.

2. Use an editor to create the source file for assembly-language program.

3. Assemble the source file with the assembler.

4. If the assembler list file indicates errors then use the editor and correct the errors.

5. Cycle through the edit-assemble loop until all errors indicated by the assembler are cleared.

6. Use the linker to link all object files of the program modules and library files into a single executable file.

7. If the linker indicates any error then modify the source program, reassemble and relink it to correct the errors.

8. If the developed program does not interact with any external hardware other than that directly connected to the system then you can use the system debugger to run and debug your program.

9. If the designed program is intended to work with external hardware then use an emulator to run and debug the program.

8.4 VARIABLES AND CONSTANTS USED IN ASSEMBLERS

The various characters used to construct assembler variables, constants and directives are the following :

Upper case English alphabets	:	A to Z
Lower case English alphabets	:	a to z
Numbers	:	0 to 9
Special characters	:	@, $, ?, _ (Underscore)

Variables

Variables are symbols (or terms) used in assembly language program statements in order to represent the variable data and address. While running a program, a value has to be attached to each variable in the program. The advantage of using variables is that the value of the variable can be dynamically varied while running the program.

Usually a variable name is constructed such that it reflects the meaning of the value it holds. A variable name selected to represent the temperature of a device can be TEMP, a variable name selected to represent the speed of a motor can be M_SPEED, etc. While constructing variable names, the numeric characters (0 to 9) should not be used as first character and the special characters $ and ? should not be used.

Rules for framing variable names

1. The variable name can have any of the following characters. A to Z, a to z, 0 to 9, @, _(underscore).

2. The first character in the variable name should be an alphabet (A to Z or a to z) or an underscore.

3. The length of a variable name depends on the assembler and normally the maximum length of variable name is 32 characters.

4. Variable names are case insensitive. Therefore, the assembler does not distinguish between the upper and lower case letters/alphabets.

Constants

The decimal, binary or hexadecimal number used to represent the data or address in assembly language program statement are called constants or numerical constants.

When constants are used to represent the address/data then their values are fixed and cannot be changed while running a program. The binary, hexadecimal and decimal constants can be differentiated by placing a specific alphabet at the end of the constant.

A valid binary constant/number is framed using numeric characters 0 and 1 and the alphabet B is placed at the end.

A valid decimal (BCD) constant/number is framed using numeric characters 0 to 9 and the alphabet D is placed at the end. However, a constant/number which does not end with any alphabet is also treated as a decimal constant.

A valid hexadecimal constant/number is framed using numeric characters 0 to 9 and alphabets A to F and the alphabet H is placed at the end. A zero should be placed/inserted at the beginning of a hexadecimal number if the first digit is an alphabet character from A to F, otherwise the assembler will consider the constant starting with A to F as a variable.

Examples of valid constant

1011	-	Decimal (BCD) constant
1060D	-	Decimal constant
1101B	-	Binary constant
92ACH	-	Hexadecimal constant
0E2H	-	Hexadecimal constant

Examples of invalid constant

1131B	-	The character 3 should not be used in binary constant.
0E2	-	The character H at the end of hexadecimal number is missing.
C42AH	-	Zero is not inserted in the beginning of hexadecimal number and so it is treated as a variable.
1A65D	-	The character A should not be used in decimal constant.

8.5 ASSEMBLER DIRECTIVES

The assembler directives are the instructions to the assembler regarding the program being assembled. They are also called pseudo instructions or pseudo opcodes.

The assembler directives are used to specify start and end of a program, attach value to variables, allocate storage locations for input/output data, to define start and end of segments, procedures, macros, etc.

The assembler directives control the generation of machine code and organization of the program. But no machine codes are generated for assembler directives. Some of the assembler directives that can be used for 8085/8086/8051 assembly language program development are listed in Table 8.2. The assembler directives that can be used only for 8086 assembly language program development are listed in Table 8.3. A brief discusion about some of the assembler directives are presented in the following sections:

Table 8.2: Assembler Directives of 8085/8051/8086 Assemblers

S.No	Assembler directives	Functions
1.	DB	Define Byte. Used to define byte type variable.
2.	DW	Define word. Used to define 16-bit variable.
3.	END	Indicates the end of the program.
4.	ENDM	End of macro. Indicates the end of a macro sequence.
5.	EQU	Equate. Used to equate numeric value or constant to a variable.
6.	MACRO	Defines the name, parameters and start of a macro.
7.	ORG	Origin. Used to assign the starting address for a program.

Table 8.3: Assembler Directives of an 8086 Assembler

Assembler Directive	Function
ASSUME	Indicates the name of each segment to the assembler.
BYTE	Indicates a byte sized operand.
DD	Define double word. Used to define 32-bit variable.
DQ	Define quad word. Used to define 64-bit variable.
DT	Define ten bytes. Used to define ten bytes of a variable.
DUP	Duplicate. Generate duplicates of characters or numbers.
DWORD	Double word. Indicates a double word sized operand.
ENDP	End of procedure. Indicates the end of a procedure.
ENDS	End of segment. Indicates the end of a memory segment.
EVEN	Informs the assembler to align the data array starting from an even address.
FAR	Used to declare the procedure as far which assigns a far address.
NEAR	Used to declare a procedure as near which assigns a near address.
OFFSET	Specifies an offset address.
PROC	Procedure. Defines the beginning of a procedure.
PTR	Pointer. It is used to indicate the type of memory access (BYTE/ WORD/ DWORD).
PUBLIC	Used to declare variables as common to various program modules.
SEGMENT	Defines the start of a memory segment.
STACK	Indicates that a segment is a stack segment.
SHORT	Used to assign one byte displacement to jump instructions.
THIS	Used with EQU directive to set a label to a byte, word or double word.
WORD	Indicates a word sized operand.

DB (DEFINE BYTE)

The directive DB is used to define a byte type variable. It reserves specific amount of memory to variables and stores the values specified in the statement as initial values in the allotted memory locations. The range of value that can be stored in a byte type variable is 0 to 255_{10} (00_H to FF_H) for unsigned value and -128_{10} to 127_{10} for signed value (00_H to $7F_H$ for positive values and 80_H to FF_H for negative values).

The general form of the statement to define the byte variable is,

variable DB value/values

Examples:	
AREA DB 45	One memory location is reserved for the variable AREA and 45_{10} is stored as initial value in that memory location.
LIST DB 7FH, 42H, 35H	Three consecutive memory locations are reserved for the variable LIST and $7F_H$, 42_H, and 35_H are stored as initial value in the reserved memory location.

DW (DEFINE WORD)

The directive DW is used to define a word type (16-bit) variable. It reserves two consecutive memory locations to each variable and stores the 16-bit values specified in the statement as the initial value in the allotted memory locations. The range of values that can be stored in word type variable is 0 to 65535_{10} (0000_H to $FFFF_H$) for unsigned value, and -32768 to $+32767$ for signed value (0000_H to $7FFF_H$ for positive value and 8000_H to $FFFF_H$ for negative value).

The general form of the statement to define the word type variable is,

variable DW value/values

Examples:	
WEIGHT DW 1250	Two consecutive memory locations are reserved for the variable WEIGHT and initialized with value 1250_{10}.
ALIST DW 6512H, 0F251H, 0CDE2H	Six consecutive memory locations are reserved for the variable ALIST and each 16-bit data specified in the instruction is stored in two consecutive memory locations.

SEGMENT AND ENDS (END OF SEGMENT)

The directive SEGMENT is used to indicate the begirning of a code/data/stack segment. The directive ENDS is used to indicate the end of a code/data/stack segment. The directives SEGMENT and ENDS must enclose the program or data defining segment. The general form of writing a program or data defining segment is given below:

segnam   SEGMENT

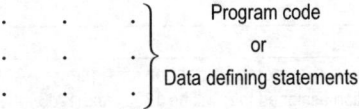

segnam   ENDS

where "segnam" is the user defined name of the segment and it can be any valid assembler variable.

Examples:

_DATA SEGMENT · · · ⎫ · · · ⎬ Data defining statements · · · ⎭ _DATA ENDS	The _DATA is the name of the data segment enclosed by the directive ENDS.
_CODE SEGMENT · · · ⎫ · · · ⎬ Program codes · · · ⎭ · · · _CODE ENDS	The _CODE is the name of the program segment enclosed by the directives SEGMENT and ENDS.

ASSUME

The directive ASSUME informs the assembler the name of the program/data segment that should be used for a specified segment. The general form of a statement using ASSUME directive is given below:

ASSUME segreg : segnam, ..., segreg : segnam

where, "segreg" is the segment register.

"segnam" is user defined name of the segment.

The segment register can be any of the CS, SS, DS and ES registers and segment name can be any valid assembler variable. In a single statement, logical segments can be assigned to one or all the segment registers.

Examples:

ASSUME CS : _CODE	The directive ASSUME informs the assembler that the instruction of the program are stored in the user defined logical segment _CODE.
ASSUME DS : _DATA	The directive ASSUME informs the assembler that the data of the program are stored in the user defined logical segment _DATA.
ASSUME CS : ACODE, DS: ADATA	The directive ASSUME informs the assembler that the instructions of the program are stored in the segment ACODE and data are stored in the segment ADATA.

ORG, END AND EQU

The directive ORG (Origin) is used to assign the starting address for a program. The directive END is used to terminate a program. The statements after the directive END will be ignored by the assembler.

The directive EQU (Equate) is used to attach a value to a variable.

Examples:

ORG 1000H	This directive informs the assembler that the statements following ORG 1000H should be stored in memory starting with address 1000_H.
PORT1 EQU 0F2H	The value of variable PORT1 is $F2_H$.
LOOP EQU 10FEH	The value of variable LOOP is $10FE_H$.

_SDATA SEGMENT
    ORG 1200H
    A DB 4CH
    EVEN
    B DW 1052H
 	
_ SDATA ENDS
	In this data segment, the effective address of the memory location assigned to A will be 1200_H and the effective address of memory location assigned to B will be 1202_H and 1203_H.

PROC, FAR, NEAR and ENDP

The directives PROC, FAR, NEAR and ENDP are used to define a procedure/subroutine. The directive PROC indicates the beginning of a procedure and the directive ENDP indicates the end of a procedure. The FAR or NEAR, are type specifier which is used by the assembler to differentiate intrasegment call (call within segment/near call) and intersegment call (call from another segment/far call).

The general form of writing a procedure is given below:

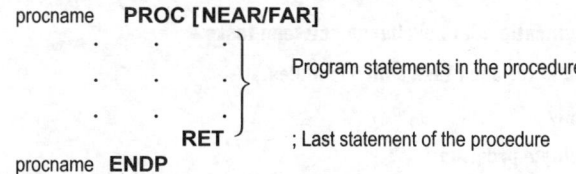

procname   **PROC [NEAR/FAR]**

      .   .   .   ⎫
      .   .   .   ⎬ Program statements in the procedure
      .   .   .   ⎭

            RET   ; Last statement of the procedure
procname **ENDP**

where "procname" is the user defined name of the procedure.

The procedure name can be any valid assembler variable. The type specifier NEAR/FAR is optional and if it is discarded then the assembler assumes the procedure as near call. Also the use of a specifier helps the assembler to decide whether to code RET as near return or far return.

Examples :

ADD64   PROC NEAR
 	
   .   .   .   ⎫
   .   .   .   ⎬ Program statements in
   .   .   .   ⎭ the procedure
   RET
ADD64   ENDP
	The subroutine/procedure named ADD64 is declared as NEAR and so the assembler will code the CALL and RET instructions involved in this procedure as near call and return.

CONVERT PROC FAR
 	
   .   .   .   ⎫
   .   .   .   ⎬ Program statements in
   .   .   .   ⎭ the procedure
   RET
CONVERT ENDP
	The subroutine/procedure named CONVERT is declared as FAR and so the assembler will code the CALL and RET instructions involved in this procedure as far call and return.

SHORT

The directive SHORT is used to reserve one memory location for an 8-bit signed displacement in jump instructions.

Example :

JMP SHORT AHEAD	The directive will reserve one memory location for 8-bit displacement named AHEAD.

MACRO and ENDM

The directive MACRO is used to indicate the beginning of a macro and the directive ENDM is used to indicate the end of a macro. The directives MACRO and ENDM must enclose the definitions, declarations and program statements which are to be substituted at the invocation of a macro.

The general form of writing a macro is given below:

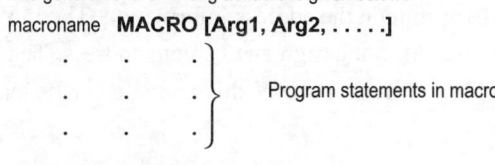

macroname **MACRO [Arg1, Arg2, ...]**

      .   .   .   ⎫
      .   .   .   ⎬ Program statements in macro
      .   .   .   ⎭

macroname **ENDM**

where, "macroname" is the user defined name of the macro.

The macroname can be any valid assembler variable.

8.6 MODULAR PROGRAMMING

The various steps in the development of an assembly language program are the following:

1. **Defining the overall work to be done by the program.**
2. **Breaking the overall program task into smaller tasks.**
3. **Determine the various communication/data exchange between tasks.**
4. **Writing assembly-language code for each task called modules.**
5. **Testing each module separately.**
6. **Combine the modules into a single program.**
7. **Testing and debugging the program.**
8. **Documenting the program.**

The concept of modular programming refers to development of program codes in modules and merging the codes of various modules into a single program code. When the program to be developed is too large to be handled by a single programmer, a team can be formed to develop the program. The overall task can be divided into a number of smaller tasks and each smaller task can be developed as a module by a team member, and the modules can be integrated by the team leader to obtain the program for the overall task.

The advantages of modular programming are given below:

1. **Modules are easier to develop.**
2. **Modules can be developed independently by different programmers.**
3. **Debugging and testing of modules can be carried out independently.**
4. **Any future modifications may be localized.**
5. **Repeated tasks can be developed as modules and stored as subroutine/macro.**
6. **Common tasks can be developed as modules and stored as library.**
7. **Documentation of modules can be made independently.**

8.6.1 Linking and Relocation

The process of combining various program modules into a single program is called **linking** and it is usually performed using a software tool called a linker. The linker will generate a link file which contains the binary codes for all the combined modules.

The linker also generates a link map file which contain the address information. The linker will assign only a relative address starting from zero and will not assign an absolute address. The linked modules with relative address is called a **relocatable program,** because this program can be loaded in any user memory space in a system for execution.

A software tool called a **locator** can be used to assign absolute or specific address to a **link file**. In order to run a program in an 8085 microprocessor-based system, the program should be mapped to specific user memory address.

In a typical assembly language development process, the source code for various modules can be developed using any text editor like EDIT, NOTEPAD, WORDPAD, etc., and have to be saved as .ASM files.

Then .ASM modules can be individually assembled using the software tool MASM assembler to generate .OBJ files then using the software tool LINK, the .OBJ files of all modules can be combined to generate a single relocatable .EXE file, and using a locator software tool called EXE2BIN, the .EXE file can be converted to .BIN file, and this .BIN file can be downloaded to the user memory space of the 8086 microprocessor-based system for execution.

8.6.2 Procedure (or) Subroutine

When a group of instructions are to be used several times to perform a same function in a program, then we can write them as a separate subprogram called procedure or subroutine. Whenever required the procedures can be called in a program using CALL instructions.

Procedures are written and assembled as separate program modules and stored in memory. When a procedure is called in the main program, the program control is transferred to the procedure and after executing the procedure the program control is transferred back to the main program. In 8085/8086 processor, the instruction CALL is used to call a procedure in the main program and the instruction RET is used to return the control to the main program.

The main advantage of using a procedure is that the machine codes for the group of instructions in the procedure has to be put in memory only once. The disadvantages of using the procedure are the need for a stack and the overhead time required to call the procedure and return to the calling program.

Handling subroutine in 8085

While executing a program, if the 8085 processor encounters a CALL the instruction, then it saves the content of the program counter in a stack and loads the subroutine address in the program counter. (The content of program counter that is saved in stack is the address of the instruction next to CALL in the main program. The subroutine address is the address given in the CALL instruction.)

When the subroutine address is loaded in the program counter, the processor starts executing the subroutine. The last instruction of the subroutine will be RET instruction and when it is executed, the processor moves the top of the stack memory to the program counter. (The top of stack memory is the address which is saved in stack before executing subroutine.) Now the program control (execution) is returned to main program.

The subroutine program may use the registers that are used by the main program. If in the main program the content of these registers are to be preserved then they have to be saved (PUSHed) in stack before calling the subroutine.

After returning from subroutine, they can be retrieved (POPed) from the stack back to the respective register. In 8085 the type of stack is LIFO(Last-In-First-Out). Hence, the order of retrieving (POPing) should be opposite to that of storing (PUSHing). For example, if the content of register pair HL is stored first followed by DE then while retrieving the DE pair should be poped first followed by HL pair.

Handling procedure in 8086

The 8086 processor has two types of call instructions: Intra-segment call or near call (call within a segment) and inter-segment call or far call (call outside a segment). A procedure can be called using near call instruction, if it is stored in the same segment where the main program is also stored.

A procedure can be called using far call instruction, if the procedure and main program are stored in different memory segments.

The procedures are terminated with RET instructions. The 8086 has two types of RET instructions: Near return and Far return. The near return instruction is used to terminate a procedure stored in the same segment. The far return instruction is used to terminate a procedure stored in a different segment.

When a procedure is called using far call instruction, the 8086 processor will push the content of the IP and CS-register in stack and the segment base address of procedure is loaded in the CS-register and the effective address of the procedure is loaded in the IP.

Now the program control is transferred to the procedure stored in another segment and so the processor will start executing the instructions of the procedure. At the end of the procedure, RET instruction is encountered. On executing the RET instruction, the top of stack (which is the previous stored value) is poped to the CS-register and the IP. Thus, the program control is returned to the main program.

When a procedure is called using near call instruction, the 8086 processor will push the content of the IP alone in stack and the effective address of procedure is loaded in the IP. Here the content of the CS-register is not altered. Now the program control is transferred to the procedure stored in same segment and so, the processor will start executing the instructions of the procedures.

At the end of the procedure, the RET instruction is encountered. On executing the RET instruction, the top of stack (which is the previous stored value) is poped to IP. Thus the program control is returned to the main program.

8.6.3 Macros

When a group of instructions are to be used several times to perform a same function in a program, and they are too small to be written as a procedure, then they can be defined as a macro. **Macro** is a small group of instructions enclosed by the assembler directives MACRO and ENDM. Macros are identified by their name and usually defined at the start of a program.

The macro is called by its name in the program. Whenever a macro is called in a program, the assembler will insert the defined group of instructions in place of the call. In other words, the macro call is like shorthand expression which tells the assembler, *"Every time you see a macro name in the program, replace it with the group of instructions defined as macro"*. Actually, the assembler generates machine codes for the group of instructions defined as macro, whenever it is called in the program. The process of replacing the macro with the instructions it represents is called expanding the macro. Hence, macros are also known as **open subroutines** because they get expanded at the point of macro invocation.

When macros are used, the generated machine codes are right-in-line with the rest of the program and so the processor does not have to go off to a procedure call and return. This results in avoiding the overhead time involved in calling and returning from a procedure. The disadvantage of using a macro is that the program may take up more memory due to insertion of the machine codes in the program at the place of macros. Hence, the macros should be used only when its body has a few program statements.

Table 8.4: Comparison of Procedure and Macro

Procedure	Macro
1. Accessed by CALL and RET mechanism during program execution.	1. Accessed during assembly with name given to macro when defined.
2. Machine code for instructions are stored in memory once.	2. Machine codes are generated for instructions in the macro each time it is called.
3. Parameters are passed in registers, memory locations or stack.	3. Parameters are passed as part of statement which calls macro.

8.6.4 Delay Routine

Delay routines are the subroutines used for maintaining the timings of various operations in a microprocessor. In control applications, certain equipment need to be ON/OFF after a specified time delay. In some applications, a certain operation has to be repeated after a specified time interval. In such cases simple time delay routines can be used to maintain the timings of the operations.

A delay routine is generally written as a subroutine (It need not be a subroutine always. It can even be a part of the main program.) In a delay routine a count (number) is loaded in a register of microprocessor. Then it is decremented by one and the zero flag is checked to verify whether the content of register is zero or not. This process is continued until the content of the register is zero. When it is zero the time delay is over and the control is transferred to the main program to carry out the desired operation.

The delay time is given by the total time taken to execute the delay routine. It can be computed by multiplying the total number of T-states required to execute the subroutine and the time for one T-state of the processor. The total of number of T-states can be computed from the knowledge of T-states required for each instruction. The time for one T-state of the processor is given by the inverse of the internal clock frequency of the processor. For example, if the 8085 microprocessor has 5 *MHz* quartz crystal then,

The internal clock frequency $= \dfrac{5}{2} = 2.5 \; MHz$

Time for one T-state $= \dfrac{1}{2.5 * 10^6} = 0.4 \; ms$

Two example delay routines that can be used in 8085 assembly language programs are presented in this section with details of timing calculations. For small time delays (< 0.5 millisecond) an 8-bit register can be used as counter, but for large time delays (< 0.5 second) 16-bit register should be used as counter. For very large time delays (>0.5 second), a delay routine can be repeatedly called in the main program. The disadvantage in delay routines is that the processor time is wasted. An alternate solution is to use a dedicated timer like 8253/8254 to produce time delays or to maintain timings of various operations.

EXAMPLE DELAY ROUTINE - 1

Write a delay routine to produce a time delay of 0.5 millisecond in 8085 processor-based system whose clock source is 6 MHz quartz crystal.

Solution

The delay required is 0.5 millisecond, hence an 8-bit register of 8085 can be used to store a count value. The count is decremented by one and the zero flag is verified. If zero flag is set then decrement operation is terminated. The delay routine is written as a subroutine as shown below:

Delay Routine

```
        MVI  D,N    ; Load the count value, N in D-register.
LOOP:   DCR  D      ; Decrement the count.
        JNZ  LOOP   ; If count is not zero go to LOOP.
        RET         ; If count is zero return to main program.
```

The following table shows the T-state required for execution of the instructions in the subroutine.

Instruction	T-state required for execution of an instruction	Number of times the instruction is executed	Total T-states	
CALL addr16	18	1	18×1	$= 18$
MVI D,N	7	1	7×1	$= 7$
DCR D	4	N times	$4 \times N$	$= 4\,N$
JNZ LOOP	10	(N−1) times	$10 \times (N-1)$	$= 10\,N - 10$
	or 7	1	7×1	$= 7$
RET	10	1	10×1	$= 10$
Total T-state required for subroutine				$= 14\,N + 32$

Calculation to find the count value, N

External clock frequency $= 6$ MHz

Internal clock frequency $= \dfrac{\text{External clock}}{2} = \dfrac{6}{2} = 3$ MHz

Time period of one T-state $= \dfrac{1}{\text{Internal clock frequency}} = \dfrac{1}{3 \times 10^6} = 0.33\ \mu s$

$\left.\begin{array}{l}\text{Number of T-states}\\ \text{required for 0.5 ms}\end{array}\right\} = \dfrac{\text{Required time delay}}{\text{Time for one T-state}} = \dfrac{0.5 \times 10^{-3}}{0.33} = 1500.15 = 1500_{10}$

On equating the total T-states required for the subroutine and the number of T-states for the required time delay, the count value, N can be calculated.

$\therefore\ 14N + 32 = 1500_{10}$

$$N = \dfrac{1500 - 32}{40} = 104.857_{10} = 105_{10} = 69_H$$

$\therefore$ Count value, $N = 69_H$

If the above delay routine is called by a program and executed with count value of 69_H then the delay produced will be 0.5 millisecond.

Note: *The register used in the delay routine is D-register. Also the execution of delay routine will alter the flags. Hence, if the contents of these registers are to be preserved, the main program has to save them in the stack before calling the delay routine.*

EXAMPLE DELAY ROUTINE - 2

Write a delay routine to produce a time delay of 0.5 second in 8085 processor-based system whose internal clock frequency is 3 MHz.

Solution

The delay required is large, hence a 16-bit register can be used for storing the count value. The count is decremented one by one until it is zero. After each decrement operation we have to verify whether the content of register pair is zero or not. This can be performed by logically ORing the content of low order and high order register and then checking the zero flag. (Because the 16-bit increment/decrement instruction will not modify any flag.) The delay routine is written as a subroutine as shown below:

Delay Routine

```
        LXI  D,N    ; Load the count value, N in DE-register pair.
LOOP:   DCX  D      ; Decrement the count.
        MOV  A,E    ; Logically OR the content of
        ORA  D      ; E-register with D-register.
        JNZ  LOOP   ; If count is not zero, go to LOOP.

        RET         ; If count is zero, return to main program.
```

The following table shows the T-states required for execution of the instructions in the subroutine.

Instructions	T-state required for the execution of an instruction	Number of times the instruction is executed	Total T-states	
CALL addr16	18	1	18×1	$= 18$
LXI D,N	10	1	10×1	$= 10$
DCX D	6	N times	$6 \times N$	$= 6 N$
MOV A,E	4	N times	$4 \times N$	$= 4 N$
ORA D	4	N times	$4 \times N$	$= 4 N$
JNZ LOOP	10	(N−1) times	$10 \times (N-1)$	$= 10 N - 10$
	or 7	1	7×1	$= 7$
RET	10	1	10×1	$= 10$
		Total T states required for subroutine		$= 24 N + 35$

Calculation to find the count value, N

Internal Clock frequency $\quad = 3 \text{ MHz}$

Time period of one T-state $\quad = \dfrac{1}{\text{Internal clock frequency}} = \dfrac{1}{3 \times 10^6} = 0.3333 \ \mu s$

$\left. \begin{array}{l} \text{Number of T states required} \\ \text{for 0.5 second} \end{array} \right\} = \dfrac{\text{Required time delay}}{\text{Time for one T-state}} = \dfrac{0.5 \text{ sec}}{0.3333 \times 10^{-6}}$

$$= 1500150.015_{10} = 1500150_{10}$$

On equating the total T-states required for the subroutine and the number of T-states for the required time delay, the count value, N can be calculated.

$$\therefore 24N + 35 = 1500150_{10}$$

$$N = \frac{1500150 - 35}{24} = 62504.79_{10} \approx 62505_{10} = F429_H$$

$\therefore$ Count value, $N = F429_H$

If the above delay routine is called by a program and executed with count value of $F429_H$ then the delay produced will be 0.5 second.

> **Note:** *The registers used in the delay routine are A, D and E. Also the execution of delay routine will alter the flags. Hence if the contents of these register are to be preserved, then the main program has to save them in stack before calling the delay routine.*

8.6.5 List and Array

List

List is a linked data structure used in programming techniques. The linked data structure will have a number of components linked in a particular fashion. Each component will consist of a string data and a pointer to the next component. The basic idea of a linked data structure is that each component within the structure includes a pointer indicating where the next component can be found. Therefore, the relative order of the components can be changed by altering the pointers. In addition, individual components can be easily added or deleted, again by altering the pointer. As a result, a linked data structure is not confined to some maximum number of components, but whenever required the data structure can be expanded or contracted in size.

The different types of linked data structures are linear linked lists, linked lists with multiple pointers, circular linked lists and trees.

Array

An array is a series of data of the same type stored in successive memory locations. Each value in the array is referred to as an element of the array. In programming techniques, array is created when we want to perform some operation on a series of data items.

8.7 STACK

The stack is a portion of RAM memory defined by the user for temporary storage and retrieval of data while executing a program. The microprocessor will have a dedicated internal register called **Stack Pointer (SP)** to hold the address of the stack. Also, the processor will have a facility to automatically decrement/increment the content of SP after every write/read operation into stack.

The user can initialize or create a stack by loading a RAM address in the **Stack Pointer (SP)**. Once an address is loaded in SP, the RAM memory locations below the address pointed by SP are reserved for stack. Typically 25 to 100 RAM memory locations are sufficient for stack. The user should take care that the reserved RAM memory locations for stack are not used for any other purpose.

The user has to create/implement a stack whenever the program consists of PUSH, POP, RST n, CALL and RET instructions. Also, the stack is needed whenever the system uses interrupt facility.

In a program, when the number of available registers are not sufficient for storing intermediate result and data, then some of intermediate result and data can be stored in a stack using PUSH instruction and retrieved whenever required using POP instruction.

The CALL instruction and the interrupts store the return address (content of program counter) in stack before executing the subroutine. Usually the subroutines are terminated with RET instruction. When RET instruction is executed, the top of stack is poped to program counter and the program control returns to the main program after the execution of subroutine.

8.7.1 Stack in an 8085 Microprocessor

In an 8085 processor, the stack is created by loading a 16-bit address in the stack pointer. Upon reset, the stack pointer is cleared to zero.

In an 8085 processor, for every write operation into stack, the SP is automatically decremented by two and for every read operation from stack, the SP is automatically incremented by two. Hence, data can be stored only in lower addresses from the address pointed by SP. Therefore, we can say that the SP holds the address of the top of stack. All the RAM addresses higher than that pointed by the SP can be considered as occupied stack and all the RAM addresses lower than that pointed by the SP can be considered as empty stack as shown in Fig. 8.4. However, in practice only few memory locations are needed for stack.

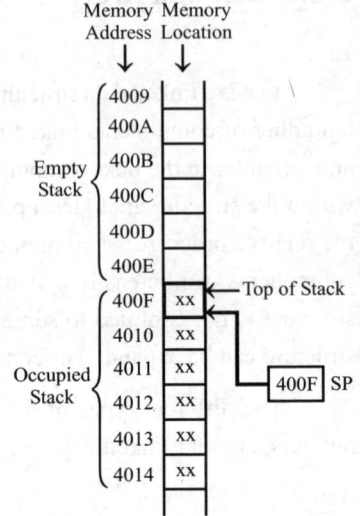

Fig. 8.4: Example of stack in 8085.

In 8085 processor the content of register pairs can be stored in stack using PUSH instruction and the stored information can be retrieved back to register pair using POP instruction.

When a number of register pairs have to be stored and retrieved in the stack, the order of retrieval should be reverse of that of the order of storage.

For example, let BC pair be pushed to stack first and DE pair next. When the stored information has to be retrieved to appropriate registers, the top of stack should be poped to the DE pair first and then to the BC pair next.

The storage and retrieval in stack are in reverse order, because the SP is decremented for every write operation into stack and SP is incremented for every read operation from stack.Therefore, the stack in 8085 is called **L**ast-**I**n-**F**irst-**O**ut (LIFO) stack, i.e., the last stored information can be read first.

8.7.2 Stack in an 8086 Microprocessor

In an 8086 microprocessor-based system, the stack is created by loading a 16-bit base address in a Stack Segment (SS) register and a 16-bit offset address in the Stack **P**ointer (SP). The 20-bit physical address of the stack is computed by multiplying the contents of SS-register by 16_{10} and then adding the contents of SP to this product. Here the content of SP is the offset address of the stack. Upon reset, the SS-register and SP are cleared to zero.

For every write operation into stack, the SP is automatically decremented by two and for every read operation from stack, the SP is automatically incremented by two. The contents of the SS-register will not be altered while reading or writing into the stack.

In an 8086 processor, the content of the register can be stored in the stack using the PUSH instruction and the stored information can be retrieved back to the register using the POP instruction. When a number of registers have to be stored and retrieved in the stack, the order of retrieval should be reverse that of the order of the storage. For example, let BX be pushed to the stack first and DX next. When the stored information has to be retrieved to appropriate registers then the top of the stack should be popped to DX first and then to BX next. The storage and retrieval in the stack are in reverse order, because the SP is decremented for every write operation into the stack and SP is incremented for every read operation form the stack. Therefore, the stack in an 8086 is called the **Last-In-First-Out** (LIFO) stack, i.e., the last stored information can be read first. A typical example of stack in 8086 is shown in Fig. 8.5.

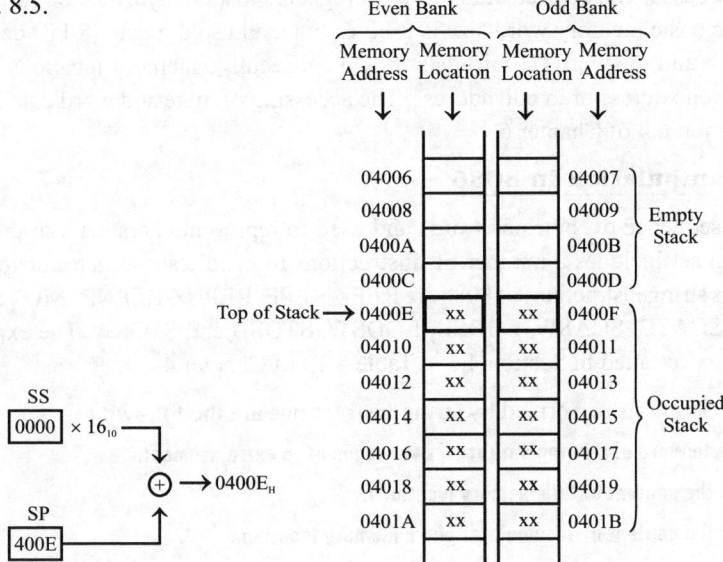

Fig. 8.5: Example of stack in 8086.

8.7.3 Stack in an 8051 Microcontroller

The stack in an 8051 is also a LIFO stack similar to that in an 8085/8086, but in an 8051 the SP is only 8 bits wide and so it can hold only an 8-bit address.

Hence the stack in an 8051 can reside anywhere in the internal RAM. The stack is initialized by loading an 8-bit address in the SP. Upon reset the stack is initialized with 07_H.

In an 8051 the stack can be accessed in bytes, whereas in an 8085/8086 the stack can be accessed only in words. For every write operation into a stack in the 8051, the SP is automatically decremented by one.

For every read operation from the stack, the SP is automatically incremented by one. An example of stack in an 8051 is shown in Fig. 8.6.

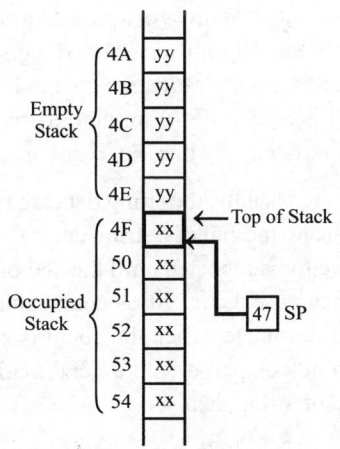

Fig. 8.6 : Example of stack in 8051.

8.8 BYTE AND STRING MANIPULATION IN 8086

8.8.1 Byte Manipulation in 8086

The 8086 processor is basically a 16-bit processor and so the basic data size is 16-bit, but it is designed to handle both byte data (8-bit data) and word data (16-bit data). The general-purpose registers of the 8086 microprocessor can be used either as 8-bit registers to store byte data or 16-bit registers to store word data. The basic memory word size is 8-bit or byte, and so the byte (8-bit) data is stored in one memory location and the word (16-bit) data in two consecutive memory locations. The data can be stored either in even address or in odd address. The accessing of byte and word data from memory are discussed in Section 6.4 of Chapter 6.

8.8.2 String Manipulation in 8086

A string is a sequence of bytes or words and used to represent character variables. The 8086 processor instruction set includes a number of instructions to handle string data for read, write and compare. The various string instructions of 8086 are REPZ/REPE, REPNZ/REPNE, MOVSB, MOVSW, CMPSB, CMPSW, SCASB, SCASW, LODSB, LODSW, STOSB and STOSW. The explanations for these instructions are presented in Section 4.6.4, Table 4.13 of Chapter 4.

The various operations performed by string instructions are the following:

1. Copy a byte/word of a string data from data segment to extra segment.

2. Compare the content of two memory locations.

3. Compare the content of accumulator and a memory location .

4. Load a byte/word of a string data from memory to accumulator.

5. Store a byte/word of a string data from accumulator to memory.

The string instructions have two default or dedicated index registers SI and DI for storing offset or effective address of string data and the direction flag is dedicated to increment/decrement the content of SI and DI registers. The SI register holds the address of the source operand and the DI register holds the address of destination operand. The direction flag is set (DF=1) for auto decrement, and the direction flag is reset (DF=0) for auto increment. For byte operand, the contents of SI and DI are incremented/ decremented by one. For word operand, the contents of SI and DI are incremented/decremented by two.

Usually, the string data are framed for a specified length and terminated with a fixed character to indicate the end of a string data. Therefore, while manipulating string data in a program, the operations have to be repeated until the end of the string and requires a looping operation. For looping operation, we can use the length as count, and decrement the count after manipulating each byte/word of a string and terminate when the count is zero. Alternatively, after accessing each byte/word of string data, it can be compared with an end marker, and the operation can be repeated until the end marker or the end of string data.

The 8086 processor instruction set includes the REP instruction that can be used as prefix in string instructions to perform this loop operation and the count for number of operations (i.e., length of string) should be loaded in CX register. The explanation for REP instructions are presented in Section 4.6.4, Table 4.13 of Chapter 4.

8.9 INTERRUPTS OF PERSONAL COMPUTERS

The 8086 assembly language program can be executed in any **P**ersonal **C**omputer (PC) based on a 80x86/pentium processor or its compatibility. While executing the programs in PC, the IO devices of the PC like keyboard, monitor, printer, etc., can be used as interactive IO devices for input data to the program and outputs the result of the program. These devices can be accessed by the programmer through the predefined interrupts of personal computer.

In the personal computers based on 80x86/pentium processor, specific interrupt type number are assigned to various activities. The interrupts predefined in personal computers can be broadly classified into the following three groups.

    1. Interrupts generated from peripherals or exceptions.

    2. Interrupts for services (system calls) through software interrupts.

    3. Interrupts used to store pointers to the device parameters.

The interrupts of each group along with function assigned are listed in Table 8.5 to Table 8.7. In personal computers, the BIOS and OS programs will initialize the vector table for the interrupts listed from Table 8.5 to Table 8.7. These interrupt vector tables should not be modified by the programmer. The interrupts which are not mentioned in the Tables 8.5 to 8.7, are not predefined in personal computers and so the undefined interrupts can be used by the programmer for any specific task/function.

Table 8.5: Hardware or Exception Interrupts of PC

Interrupt number	Function assigned
INT 00H	Division by zero
INT 01H	Single-step
INT 02H	Nonmaskable
INT 03H	Breakpoint
INT 04H	Overflow
INT 05H	Print screen
INT 06H	Reserved
INT 07H	Reserved
INT 08H	Timer
INT 09H	Keyboard
INT 0AH to INT 0DH	Hardware Interrupts
INT 0EH	Diskette
INT 0FH	Hardware Interrupt

Table 8.6: Software Interrupts in PC for Implementing System Calls

Interrupt number	Function assigned
INT 10H to INT 17H	BIOS Interrupts
INT 18H	ROM - BASIC
INT 19H	Bootstrap
INT 1AH	Time IO
INT 1BH	Keyboard Break
INT 1CH	User timer Interrupt
INT 20H to INT 2FH	DOS Interrupts
INT 67H	Expanded Memory Functions

Table 8.7: Interrupts used in PC to Store Pointers to Device Parameters

Interrupt number	Function assigned
INT 1DH	Video Parameters
INT 1EH	Diskette Parameters
INT 1FH	Graphics Characters
INT 41H	Hard Disk-0 Parameters
INT 46H	Hard Disk-1 Parameters
INT 44H	EGA Graphic Characters
INT 4AH	User Alarm Address
INT 50H	CMOS Timer Interrupt

Some of the DOS and BIOS interrupts are explained in the following sections. For detailed discussion on the interrupts of PC, readers are advised to refer the IBM PC technical reference manual and DOS reference manual.

DOS Interrupts

The DOS (**D**isk **O**perating **S**ystem) provides a large number of procedures to access devices, files, memory and process control services. These procedures can be called in any user program using software interrupts "INT n" instruction. The various DOS interrupts are listed in Table 8.8.

The DOS interrupt INT 21H provides a large number of services. A function code has been allotted to each service provided by INT 21H. The function code should be loaded in AH-register before calling INT 21H to avail the service provided by the function.

Table 8.8: DOS Interrupts

Interrupt type	Service provided by the interrupt
INT 20H	Program Terminate
INT 21H	DOS services (DOS system call)
INT 22H	Terminate Address
INT 23H	Control Break Address
INT 24H	Critical Error Handler Address
INT 25H	Absolute Disk Read
INT 26H	Absolute Disk Write
INT 27H	Terminate and Stay Resident (TSR)
INT 28H	DOS time slice
INT 2EH	Perform DOS Command
INT 2FH	Multiplex Interrupts

The various services provided by the INT 21H are classified depending on the function performed by them and they are listed in Appendix-VII.

The following steps are involved in accessing DOS services:

1. **Load a DOS function number in AH-register. If there is a subfunction, then its code is loaded in AL register.**

2. **Load the other registers as indicated in the DOS service formats.**

3. **Prepare buffers, ASCIIZ (ASCII string terminated by zero) and control blocks, if necessary.**

4. **Set the location of Disk Transfer Area, if necessary.**

5. **Invoke DOS service INT 21H.**

6. **The DOS service will return the required parameters in the specified registers.**

> *Note : All values entered in the register are preserved by the DOS service call except when information is returned in a register.*

BIOS Interrupts

In personal computers, the basic interface between the hardware and software is provided by a program stored in ROM called BIOS program. (BIOS-**B**asic **I**nput **O**utput control **S**ystem). The BIOS program consists of a large number of procedures to access various hardwares in a PC. These procedures can be called in any user program using software interrupts "INT n" instruction. Even the DOS uses BIOS interrupts to control the hardware. The various BIOS interrupts are listed in Table 8.9.

Table 8.9: BIOS Interrupts

Interrupt type	Service name
INT 10H	Video services
INT 11H	Machine configuration
INT 12H	Usable RAM Memory size
INT 13H	Disk IO
INT 14H	Serial port IO (RS 232C)
INT 15H	AT services
INT 16H	Keyboard IO
INT 17H	Printer IO

Each BIOS interrupt provides a large number of services. A function code has been allotted to each service provided by the BIOS interrupts. The function code should be loaded in the AH-register before calling the BIOS interrupt to avail the service provided by the function. The various functions performed by BIOS interrupts are listed in Appendix-VII.

The following steps are involved in accessing the BIOS services:

1. **Load a BIOS function number in the AH-register. If there is a subfunction, then its code is loaded in AL-register.**

2. **Load the other register as indicated in the BIOS service formats.**

3. **Prepare buffers, ASCIIZ (ASCII string terminated by zero) and control blocks, if necessary.**

4. **Invoke BIOS call.**

5. **The BIOS service will return the required parameters in the specified register.**

> *Note: All values entered in the register are preserved except when information is returned in a register.*

Explanation of DOS and BIOS Interrupts

The explanation provided for DOS and BIOS interrupts consists of following three sections :

1. Operation : This section explains the dedicated operation performed by the interrupt with specified function code.

2. Expects : This section explains the register and parameter settings required before accessing the service.

3. Returns : This section explains the status of a service call and return parameters after the response of the service.

INT 10H, FUNCTION CODE 02H : SET CURSOR POSITION

Operation : The INT 10H with function code 02H is used to set the position of the cursor on the monitor using text coordinates (row and column).

Expects : AH = 02H

BH = Video page (must be zero in graphics mode)

DH = Row (y–coordinate)

DL = Column (x–coordinate)

Returns : None

INT 10H, FUNCTION CODE 03H : READ CURSOR POSITION

Operation : The INT 10H with function code 03H is used to read the current position of cursor on the monitor in text coordinates.

Expects : AH = 03H

BH = Video page

Returns : DH = Current row (y-coordinate)

DL = Current column (x-coordinate)

CH = Starting line for cursor

CL = Ending line for cursor

INT 10H, FUNCTION CODE 06H : INITIALIZE/SCROLL RECTANGULAR WINDOW UP

Operation : The INT 10H with function code 06H is used to initialize a specified rectangular window on the monitor or scrolls the contents of a window up by a specified number of lines.

Expects : AH = 06H

AL = Number of lines to scroll up

(If AL = zero, entire window is cleared or blanked)

BH = Blanked area attributes

CH = y-coordinate, upper left corner of window

CL = x-coordinate, upper left corner of window

DH = y-coordinate, lower right corner of window

DL = x-coordinate, lower right corner of window

Returns : None

INT 10H, FUNCTION CODE 07H: INITIALIZE/SCROLL RECTANGULAR WINDOW DOWN

Operation   : The INT 10H with function code 07H is used to initialize a specified rectangular window or scrolls the contents of a window down by a specified number of lines.

Expects   :   AH = 07H
                AL   = Number of lines to scroll down
                      (If AL = zero, entire window is cleared or blanked)
                BH = Blanked area attributes
                CH = y-coordinate, upper left corner of window
                CL = x-coordinate, upper left corner of window
                DH = y-coordinate, lower right corner of window
                DL = x-coordinate, lower right corner of window

Returns   : None

INT 10H, FUNCTION CODE 09H : WRITE CHARACTER AND ATTRIBUTE AT CURSOR

Operation   : The INT 10H with function code 09H is used to write a specified ASCII character and its attribute to the monitor at the current cursor position.

Expects   : AH = 09H
                AL = ASCII character code
                BH = Video page
                BL = Attribute (in text mode) or colour (in graphics mode)
                CX = Count of character to write (replication factor).

Returns   : None

INT 10H, FUNCTION CODE 0AH : WRITE CHARACTER ONLY AT CURSOR

Operation   : The INT 10H with function code 0AH is used to write an ASCII character to the monitor at current cursor position. The character uses the attribute of the previous character displayed at the same position.

Expects   : AH = 0AH
                AL = ASCII character code
                BH = Video page
                BL = Colour (graphics mode)
                CX = Count of character to write (replication factor)

Returns   : None

INT 16H, FUNCTION CODE 00H : READ KEYBOARD CHARACTER

Operation   : The INT 16H with function code 00H is used to read a character from the keyboard. It also returns the keyboard scan code.

Expects   : AH = 00H

Returns   : AH = Keyboard scan code
                AL = ASCII character code

INT 17H, FUNCTION CODE 00H : WRITE TO PRINTER

Operation   : The INT 17H with function code 00H is used to send a character to the specified parallel port to which a printer is connected. It also returns the current status of the printer.

Expects   : AH = 00H
                AL = Character to be written
                DX = Port number ("0" for LPT1, "1" for LPT2 and "2" for LPT3)

Returns   : AH = Printer status as shown below

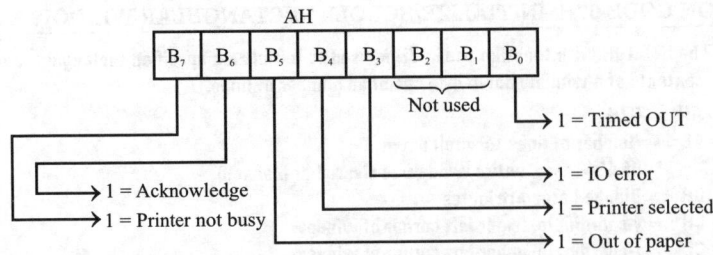

INT 21H, FUNCTION CODE 01H : READ CHARACTER FROM STANDARD INPUT DEVICE

Operation   :   The INT 21H with function code 01H reads a character from the standard input device (keyboard) and echoes (send) the character to the standard output device (monitor). It waits for the character if no character is available on the input device.

Expects   :   AH = 01H

Returns   :   AL = ASCII code of the input key.

INT 21H, FUNCTION CODE 02H : WRITE CHARACTER TO STANDARD OUTPUT DEVICE

Operation   :   The INT 21H with function code 02H writes a character to the standard output device (monitor).

Expects   :   AH = 02H

                 DL = ASCII character code

Returns   :   None

INT 21H, FUNCTION CODE 05H : SEND A CHARACTER TO PRINTER

Operation   :   The INT 21H with function code 05H send a character to printer. The default parallel port is LPT1, unless explicitly redirected by the DOS.

Expects   :   AH = 05H

                 DL = ASCII code of the output character

Returns   :   None

INT 21H, FUNCTION CODE 08H : READ CHARACTER WITHOUT ECHO

Operation   :   The INT 21H with function code 08H reads a character from the standard input device (keyboard) without echo to the standard output device (monitor). It waits for the character, if no character is available on the input device.

Expects   :   AH = 08H

Returns   :   AL = ASCII code of the input key.

INT 21H, FUNCTION CODE 09H : DISPLAY STRING

Operation   :   The INT 21H with function code 09H writes a string terminated with $ to the standard output device(monitor).

Expects   :   AH = 09

                 DS = Segment address of the string terminated by the symbol $.

                 DX = Offset address of the string terminated by the symbol $.

Returns   :   None

INT 21H, FUNCTION CODE 4CH : TERMINATE WITH RETURN CODE-EXIT(n)

Operation	:	The INT 21H with function code 4CH is used to terminate the program with return code. The return code "0" is generally considered as program terminating with successful execution. DOS sets the error level to the return code.
Expects	:	AH = 4CH
		AL = return code
Returns	:	None

8.10 ASSEMBLY LANGUAGE FORMAT

The assembly language program consists of a sequence of instructions written using assembly language instructions of a particular processor. Every line of the assembly language program is called a statement and each statement is divided into four parts called fields.

The four fields of a statement are label, operation code or opcode, operand and comment.

Label field is not compulsory for all the statements and necessary only when the instruction is referred for jump and call operations. Every statement should have an opcode and operand that specify the operation performed by the instruction or statement. The comment in the statement is optional and written for human or user understanding.

The four fields are separated by delimiters which specify the end or boundary of the field. Also, any number of blank spaces can be given in between fields which is also known as free-field format.

The commonly used delimiters are colon, space, comma and semicolon. The usage of these delimiters are as follows:

1. Colon is used at the end of label field
2. Space is used between opcode and operand
3. Comma is used between two operands
4. Semicolon is used at the start of a comment.

A simple example program (Example Program - 85.1) with parts of statements placed under various fields is given below.

Label	Opcode	Operand	Comment
START:	LDA	4200H	;Get 1st data in A and save in B.
	MOV	B,A	
	LDA	4201H	;Get 2nd data in A-register.
	MVI	C,00H	;Clear C-register to account for carry.
	ADD	B	;Get the sum in A-register.
	JNC	AHEAD	;If CF=0, go to AHEAD.
	INR	C	;If CF=1, increment C-register.
AHEAD:	STA	4202H	;Store the sum in memory.
	MOV	A,C	
	STA	4203H	;Store the carry in memory.
STOP:	HLT		;Halt program execution.

8.11   HAND CODING OF ASSEMBLY LANGUAGE PROGRAMS IN 8086

The 8086 assembly language programs should be converted to machine codes (binary codes) for execution. This can be acheived by two methods.

In one method, the software development tools like editor, assembler and linker are used to generate the machine codes of the program. Using an editor, the assembly language program is typed and saved as ".asm" file. Using an assembler, it is converted to machine code and saved as ".obj" file. Using a linker, the machine codes are mapped to the memory of the target hardware and saved as ".exe" file. The ".exe" file is the machine language program which can be run on an 8086 system or its compatible.

In another method, the machine-language code of each instruction is obtained manually by referring to the machine-code templates of 8086 provided by INTEL. This method is referred to as hand coding of an assembly language program.

The template of 8086 instructions are listed in Appendix VI. The templates of each instruction will have a fixed binary code called opcode, programmable fields (like mod, reg, segreg, r/m) and one bit special indicators (like w, d, sw, v or z). The various choice of binary codes for programmable fields and one-bit special indicators are listed in Tables A-1 to A-5 in Appendix VI:

While hand coding 8086 instructions, the following hints will be useful.

1. **The term "mem" in the operand field of instructions refers to operand in memory (or memory operand) and it can be specified in 24 different ways as shown in Table A-5 of Appendix VI. The different methods of specifying memory operands differ in the way of calculating effective and physical address of memory. These calculations are shown in Table 4.4 of Chapter 4.**

2. **The term "disp8" in the operand field of jump instructions (for both conditional and unconditional) refer to the number of memory locations to be jumped forward or backward. For forward jump the disp8 is a positive integer and for backward jump the disp8 is a negative integer. Therefore, for backward jump, the disp8 should be expressed in 2's complement form.**

The hand coding of Example Program 86.19 in Section 8.14 is shown in Table 8.10.

After hand coding of the instructions, they should be stored in memory locations. In the Example Program 86.19, the origin of the program effective address is specified as 1000_H. Therefore, the instructions are stored in the memory starting from the address 1000_H.

The machine codes of the instructions in Hexa along with the address of each instruction are listed in Table 8.11. In this table, the address in the address column refers to the starting address of each instruction.

Table 8.10: Handing of Example Program 86.19

Instruction	Template	Binary code	Hex code
MOV SI, 1100H	1011 w reg | l.b.data | h.b.data w = 1, reg = 110, l.b.data = 00_H, h.b.data = 11_H	1011 1110 | 0000 0000 | 0001 0001	BE 00 11
MOV DL, [SI]	1000 10dw | mod reg r/m d = 1, w = 0, mod = 00, reg = 010, r/m = 100	1000 1010 | 0001 0100	8A 14
MOV DI, 1200H	1011 w reg | l.b.data | h.b.data w = 1, reg = 111, l.b.data = 00_H, h.b.data = 12_H	1011 1111 | 0000 0000 | 0001 0010	BF 00 12
MOV BL, 01H	1011 w reg | l.b.data w = 0, reg = 011, l.b.data = 01_H	1011 0011 | 0000 0001	B3 01
MOV [DI], BL	1000 10dw | mod reg r/m d = 0, w = 0, mod = 00, reg = 011, r/m = 101	1000 1000 | 0001 1101	88 1D
INC DI	0100 0 reg reg = 111	0100 0111	47
INC BL	1111 111w | mod 000 r/m w = 0, mod = 11, r/m = 011	1111 1110 | 1100 0011	FE C3
MOV CL, 02H	1011 w reg | l.b.data w = 0, reg = 001, l.b.data = 02_H	1011 0001 | 0000 0010	B1 02
CMP BL, CL	0011 10dw | mod reg r/m d = 1, w = 0, mod = 11, reg = 011, r/m = 001	0011 1010 | 1101 1001	3A D9
JZ STORE	0111 0100 | disp8 disp8 = STORE	0111 0100 | STORE	74 <u>STORE</u>
MOV AH, 00H	1011 w reg | l.b.data w = 0, reg = 100, l.b.data = 00_H	1011 0100 | 0000 0000	B4 00

Table 8.10 continued..

Instruction	Template	Binary code	Hex code
MOV AL, BL	1000 10dw | mod reg r/m d = 1, w = 0, mod = 11, reg = 000, r/m = 011	1000 1010 | 1100 0011	8A C3
DIV CL	1111 011w | mod 110 r/m w = 0, mod = 11, r/m = 001	1111 0110 | 1111 0001	F6 F1
CMP AH, 00H	1000 00sw | mod 111 r/m | l.b.data sw = 00, mod = 11, r/m = 100, l.b.data = 00$_H$	1000 0000 | 1111 1100 | 0000 0000	80 FC 00
JZ NEXT	0111 0100 | disp8 disp8 = NEXT	0111 0100 | NEXT	74 NEXT
INC CL	1111 111w | mod 000 r/m w = 0, mod = 11, r/m = 001	1111 1110 | 1100 0001	FE C1
JMP REPEAT	1110 1011 | disp8 disp8 = REPEAT	1110 1011 | REPEAT	EB REPEAT
MOV [DI], BL	1000 10dw | mod reg r/m d = 0, w = 0, mod = 00, reg = 011, r/m = 101	1000 1000 | 0001 1101	88 1D
INC DI	0100 0reg reg = 111	0100 0111	47
INC BL	1111 111 w | mod 000 r/m w = 0, mod = 11, r/m = 011	1111 1110 | 1100 0011	FE C3
CMP BL, DL	0011 10dw | mod reg r/m d = 1, w = 0, mod = 11, reg = 011, r/m = 001	0011 1010 | 1101 1001	3A D9
JNZ GENERAT	0111 0101 | disp8 disp8 = GENERAT	0111 0101 | GENERAT	75 GENERAT
HLT	1111 0100	1111 0100	F4

Table 8.11: Machine Codes of Instructions in Hexa

Instruction	Effective address in Hex	Hex code
MOV SI, 1100H	1000	BE 00 11
MOV DL, [SI]	1003	BA 14
MOV DI, 1200H	1005	BF 00 12
MOV BL, 01H	1008	B3 01
MOV [DI], BL	100A	88 1D
INC DI	100C	47
INC BL	100D	FE C3
GENERAT : MOV CL, 02H	100F	B1 02
REPEAT : CMP BL, CL	1011	3A D9
JZ STORE	1013	74 0F
MOV AH, 00H	1015	B4 00
MOV AL, BL	1017	8A C3
DIV CL	1019	F6 F1
CMP AH, 00H	101B	80 FC 00
JZ NEXT	101E	74 07
INC CL	1020	FE C1
JMP REPEAT	1022	EB ED
STORE : MOV [DI], BL	1024	88 1D
INC DI	1026	47
NEXT : INC BL	1027	FE C3
CMP BL, DL	1029	3A DA
JNZ GENERAT	102B	75 E2
HLT	102D	F4

The machine codes (or binary codes) for the disp8 in the jump instructions are determined as explained in the following paragraph.

The instruction "JZ STORE" is a forward jump. In this instruction, the program control should be transferred to the instruction labelled "STORE" if zero flag is one. The instruction labelled "STORE" is stored in memory after 15_{10} (or $0F_H$) memory locations from "JZ STORE". Hence, the machine code for "STORE" is $0F_H$.

The instruction "JZ NEXT" is a forward jump. In this instruction, the program control should be transferred to the instruction labelled NEXT if the zero flag is one. The instruction labelled "NEXT" is stored in the memory after 07_{10} (or 07_H) memory locations from "JZ NEXT". Hence, the machine code for "NEXT" is 07_H.

The instruction "JMP REPEAT" is a backward jump. In this instruction, the program control is unconditionally transferred to the instruction labelled REPEAT, which is stored at the memory address 1011_H. After executing, the "JMP REPEAT" instruction, the content of IP (**I**nstruction **P**ointer) will be the address of the next instruction, which is 1024_H.

The difference between these two addresses gives the disp8, which is the number of locations to be jumped backward. Here, $1024_H - 1011_H = 13_H = 19_{10}$. Since JMP REPEAT is a backward jump, we have to express 13_H in 2's complement. The 2's complement of 13_H is ED_H. Therefore, the machine code for REPEAT is ED_H.

The instruction "JNZ GENERAT" is a backward jump. In this instruction, the program control is transferred to the instruction labelled GENERAT if the zero flag is zero. The instruction labelled GENERAT is stored at memory address $100F_H$.

After executing "JNZ GENERAT" instruction, the content of IP (**I**nstruction **P**ointer) will be the address of the next instruction, which is $102D_H$. The difference between these two addresses gives the disp8, which is the number of locations to be jumped backward. Here, $102D_H - 100F_H = 1E_H = 30_{10}$. Since "JNZ GENERAT" is a backward jump, we have to express $1E_H$ in 2's complement. The 2's complement of $1E_H$ is $E2_H$. Therefore, the machine code for GENERAT is $E2_H$.

8.12 EXAMPLES OF 8085 ASSEMBLY LANGUAGE PROGRAMS

EXAMPLE PROGRAM - 85.1 : 8-Bit Addition

Write an assembly language program to add two numbers of 8-bit data stored in memory locations 4200_H and 4201_H and store the result in 4202_H and 4203_H.

Problem Analysis

In order to perform addition in 8085, one of the data should be in accumulator and another data can be in any one of the general purpose register or in the memory.

After addition, the sum is stored in the accumulator. The sum of two 8-bit data can be either 8-bits (sum only) or 9-bits (sum and carry). The accumulator can accommodate only the sum and if there is a carry, the 8085 will indicate by setting the carry flag. Hence, one of the registers is used to account for carry.

In method-1, direct addressing is used to address the data. But in method-2, register indirect addressing is used to the address data. Here, HL-register is used to hold the address of the data and it is called pointer.

Algorithm (Method-1)

1.   Load the first data from memory to accumulator and move it to B-register.
2.   Load the second data from memory to accumulator.
3.   Clear C-register.
4.   Add the content of B-register to accumulator.
5.   Check for carry. If carry = 1, go to step 6 or if carry = 0, go to step 7.
6.   Increment the C-register.
7.   Store the sum in memory.
8.   Move the carry to accumulator and store in memory.
9.   Stop.

Flowchart for Example Program 85.1

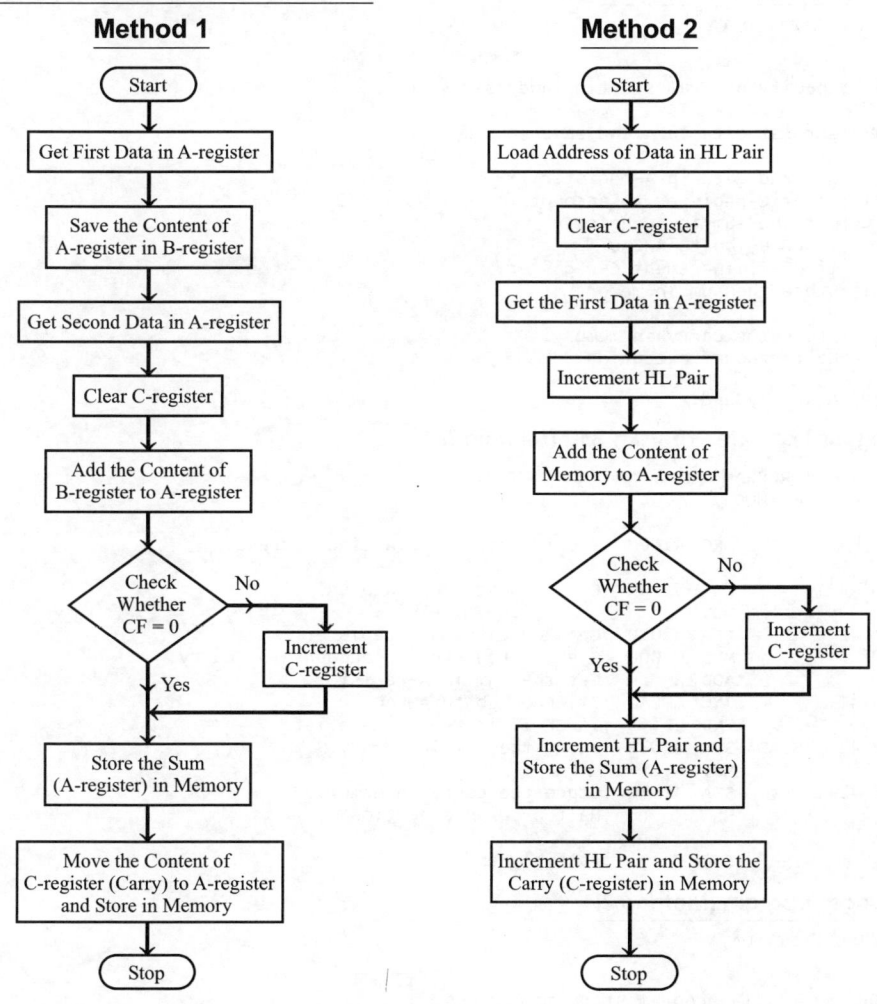

Method 1	Method 2
Start	Start
Get First Data in A-register	Load Address of Data in HL Pair
Save the Content of A-register in B-register	Clear C-register
Get Second Data in A-register	Get the First Data in A-register
Clear C-register	Increment HL Pair
Add the Content of B-register to A-register	Add the Content of Memory to A-register
Check Whether CF = 0 — No → Increment C-register	Check Whether CF = 0 — No → Increment C-register
Yes	Yes
Store the Sum (A-register) in Memory	Increment HL Pair and Store the Sum (A-register) in Memory
Move the Content of C-register (Carry) to A-register and Store in Memory	Increment HL Pair and Store the Carry (C-register) in Memory
Stop	Stop

Algorithm (Method-2)

1. Load the address of the data memory in HL pair (i.e., set HL pair as pointer for data).

2. Clear C-register.

3. Move the first data from memory to accumulator.

4. Increment the pointer (HL pair).

5. Add the content of memory addressed by HL with accumulator.

6. Check for carry. If carry = 1, go to step 7 or if carry = 0, go to step 8.

7. Increment the C-register.

8. Increment the pointer and store the sum.

9. Increment the pointer and store the carry.

10. Stop.

Assembly Language Program (Method 1)

```
;PROGRAM TO ADD TWO 8-BIT DATA
;METHOD-1

        ORG  4100H  ;specify program starting address.

        LDA  4200H  ;Get 1st data in A and save in B.
        MOV  B,A
        LDA  4201H  ;Get 2nd data in A-register.
        MVI  C,00H  ;Clear C-register to account for carry.
        ADD  B      ;Get the sum in A-register.
        JNC  AHEAD  ;If CF=0, go to AHEAD.
        INR  C      ;If CF=1, increment C-register.
AHEAD:  STA  4202H  ;Store the sum in memory.
        MOV  A,C
        STA  4203H  ;Store the carry in memory.
        HLT         ;Halt program execution.

        END         ;Assembly end.
```

Assembler Listing for Example Program 85.1 (Method 1)

```
 1                      ;PROGRAM TO ADD TWO 8-BIT DATA
 2                      ;METHOD-1
 3   0000
 4   4100                      ORG  4100H  ;specify program starting address.
 5
 6   4100  3A 00 42            LDA  4200H  ;Get 1st data in A and save in B.
 7   4103  47                  MOV  B,A
 8   4104  3A 01 42            LDA  4201H  ;Get 2nd data in A-register.
 9   4107  0E 00               MVI  C,00H  ;Clear C-register to account for carry.
10   4109  80                  ADD  B      ;Get the sum in A-register.
11   410A  D2 0E 41.           JNC  AHEAD  ;If CF=0, go to AHEAD.
12   410D  0C                  INR  C      ;If CF=1, increment C-register.
13   410E  32 02 42  AHEAD: STA  4202H  ;Store the sum in memory.
14   4111  79                  MOV  A,C
15   4112  32 03 42            STA  4203H  ;Store the carry in memory.
16   4115  76                  HLT         ;Halt program execution.
17
18   4116                      END         ;Assembly end.
```

Assembly Language Program (Method 2)

```
;PROGRAM TO ADD TWO 8-BIT DATA
;METHOD-2

        ORG  4100H   ;specify program starting address.

        LXI  H,4200H ;Set pointer for data.
```

```
          MVI  C,00H    ;Clear C-register to account for carry.
          MOV  A,M      ;Get 1st data in A-register.
          INX  H        ;Add 2nd data which is available
          ADD  M        ;in memory to A. Sum in A-register.
          JNC  AHEAD    ;If CF=0, go to AHEAD.
          INR  C        ;If CF=1, increment C-register.
AHEAD:    INX  H
          MOV  M,A      ;Save the sum in memory.
          INX  H
          MOV  M,C      ;Save the carry in memory.
          HLT           ;Halt program execution.

          END           ;Assembly end.
```

Assembler Listing for Example Program 85.1 (Method 2)

```
 1                              ;PROGRAM TO ADD TWO 8-BIT DATA
 2                              ;METHOD-2
 3
 4   4100                       ORG  4100H    ;specify program starting address.
 5
 6   4100  21  00  42           LXI  H,4200H  ;Set pointer for data.
 7   4103  0E  00               MVI  C,00H    ;Clear C-register to account for carry.
 8   4105  7E                   MOV  A,M      ;Get 1st data in A-register.
 9   4106  23                   INX  H        ;Add 2nd data which is available
10   4107  86                   ADD  M        ;in memory to A. Sum in A-register.
11   4108  D2  0C  41           JNC  AHEAD    ;If CF=0, go to AHEAD.
12   410B  0C                   INR  C        ;If CF=1, increment C-register.
13   410C  23            AHEAD: INX  H
14   410D  77                   MOV  M, A     ;Save the sum in memory.
15   410E  23                   INX  H
16   410F  71                   MOV  M,C      ;Save the carry in memory.
17   4110  76                   HLT           ;Halt program execution.
18
19   4111                       END           ;Assembly end.
```

Sample Data

Input Data :   Data-1 = E2$_H$

Data-2 = 45$_H$

Output data :   Sum   = 27$_H$

Carry = 01$_H$

Memory address	content
4200	E2
4201	45
4202	27
4203	01

EXAMPLE PROGRAM -85. 2 :   16-Bit Addition

Write an assembly language program to add two numbers of 16-bit data stored in memory locations from 4200$_H$ to 4203$_H$. The data are stored such that the low byte first and then the high byte is stored. Store the result from 4204$_H$ to 4206$_H$.

Problem Analysis

The 16-bit addition can be performed in 8085 either in terms of 8-bit addition or by using DAD instruction. In addition using DAD instruction, one of the data should be in HL pair and another data can be in another register pair. After addition the sum is stored in HL pair. If there is a carry in addition then that is indicated by setting a carry flag. Hence, one of the registers is used to account for carry.

Algorithm

1. Load the first data in HL-register pair.
2. Move the first data to DE- register pair.
3. Load the second data in HL-register pair.
4. Clear A-register for carry.

5.   **Add the content of DE pair to HL pair.**
6.   **Check for carry. If carry = 1, go to step 7 or If carry = 0, go to step 8.**
7.   **Increment A-register to account for carry.**
8.   **Store the sum and carry in memory.**
9.   **Stop.**

Flowchart for Example Program 85.2

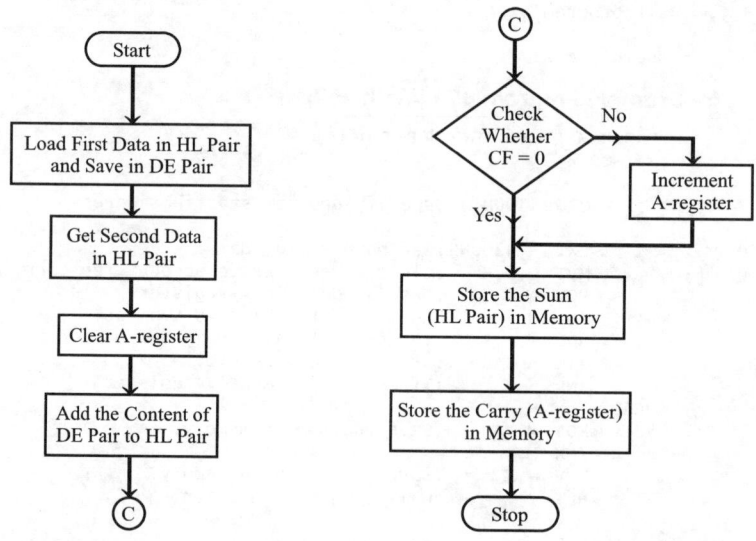

Assembly Language Program

```
;PROGRAM TO ADD TWO 16-BIT DATA

        ORG   4100H  ;specify program starting address.

        LHLD  4200H  ;Get 1st data in HL pair.
        XCHG         ;Save 1st data in DE pair.
        LHLD  4202H  ;Get 2nd data in HL pair.
        XRA   A      ;Clear A-register for carry.
        DAD   D      ;Get the sum in HL pair.
        JNC   AHEAD  ;If CF=0, go to AHEAD.
        INR   A      ;If CF=1, increment A-register.
AHEAD:  SHLD  4204H  ;Store the sum in memory.
        STA   4206H  ;Store the carry in memory.
        HLT          ;Halt program execution.

        END          ;Assembly end.
```

Assembler Listing for Example Program 85.2

```
1                        ;PROGRAM TO ADD TWO 16-BIT DATA
2
3 4100                   ORG   4100H  ;specify program starting address.
4
5 4100   2A 00 42        LHLD  4200H  ;Get 1st data in HL pair.
6 4103   EB              XCHG         ;Save 1st data in DE pair.
7 4104   2A 02 42        LHLD  4202H  ;Get 2nd data in HL pair.
8 4107   AF              XRA   A      ;Clear A-register for carry.
9 4108   19              DAD   D      ;Get the sum in HL pair.
```

```
10  4109  D2 0D 41          JNC   AHEAD  ;If CF=0, go to AHEAD.
11  410C  3C                INR   A      ;If CF=1, increment A-register.
12  410D  22 04 42  AHEAD:  SHLD  4204H  ;Store the sum in memory.
13  4110  32 06 42          STA   4206H  ;Store the carry in memory.
14  4113  76                HLT          ;Halt program execution.
15
16  4114                    END          ;Assembly end.
```

Sample Data

Input Data :   Data-1 = $C254_H$

                Data-2 = $8A92_H$

Output Data :   Sum   = $4CE6_H$

                Carry = 01

Memory address	Content
4200	54
4201	C2
4202	92
4203	8A

Memory address	Content
4204	E6
4205	4C
4206	01

EXAMPLE PROGRAM - 85.3 : 8-Bit Subtraction

Write an assembly language program to subtract two numbers of 8-bit data stored in memory locations 4200_H and 4201_H. Store the magnitude of the result in 4202_H. If the result is positive store 00 in 4203_H or if the result is negative store 01 in 4203_H.

Problem Analysis

In order to perform subtraction in 8085, one of the data should be in accumulator and another data can be in any one of the general purpose register or in the memory. After subtraction the result is stored in the accumulator. The 8085 perform 2's complement subtraction and then complement the carry. Therefore, if the result is negative then the carry flag is set and the accumulator will have 2's complement of the result. One of the register is used to account for sign of the result. In order to get the magnitude of the result again take 2's complement of the result.

Flowchart for Example Program 85.3

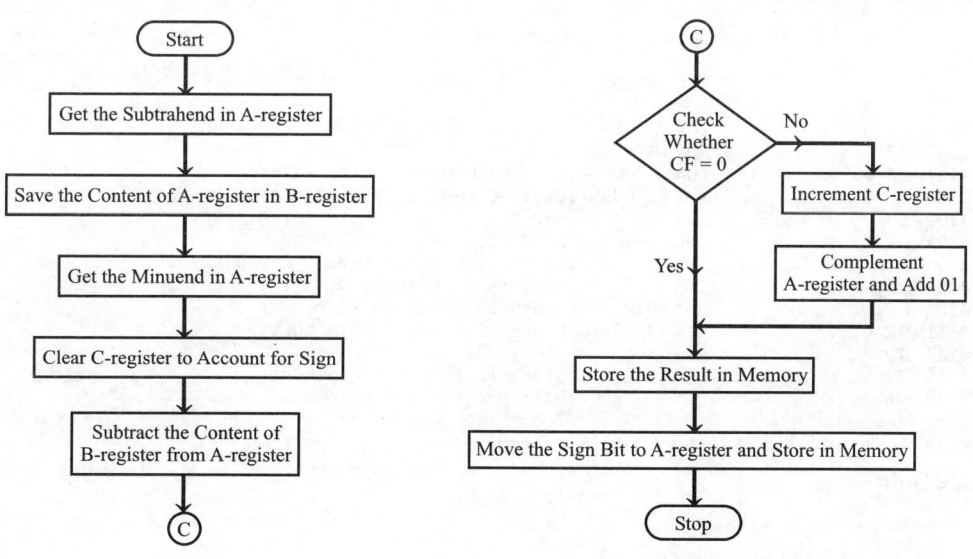

Algorithm

1. Load the subtrahend (the data to be subtracted) from memory to accumulator and move it to B-register.
2. Load the minuend from memory to accumulator.
3. Clear C-register to account for sign of the result.
4. Subtract the content of B-register (subtrahend) from the content of accumulator (minuend).
5. Check for carry. If carry = 1, go to step 6 or if carry = 0, go to step 7.
6. Increment C-register. Complement the accumulator and add 01_H.
7. Store the difference (accumulator) in memory.
8. Move the content of C-register (sign bit) to accumulator and store in memory.
9. Stop.

Assembly Language Program

```
;PROGRAM TO SUBTRACT TWO 8-BIT DATA


        ORG  4100H  ;specify program starting address.

        LDA  4201H  ;Get the subtrahend in B-register.
        MOV  B,A
        LDA  4200H  ;Get the minuend in A-register.
        MVI  C,00H  ;Clear C-register to account for sign.
        SUB  B      ;Get the difference in A-register.
        JNC  AHEAD  ;If CF=0, then go to AHEAD.
        INR  C      ;If CF=1, then increment C-register.
        CMA         ;Get 2's complement of difference
        ADI  01H    ;(result) in A-register.
AHEAD:  STA  4202H  ;Store the result in memory.
        MOV  A,C
        STA  4203H  ;Store the sign bit in memory.
        HLT         ;Halt program execution.

        END         ;Assembly end.
```

Assembler Listing for Example Program 85.3

```
1                           ;PROGRAM TO SUBTRACT TWO 8-BIT DATA
2
3   4100                        ORG  4100H  ;specify program starting address.
4
5   4100  3A 01 42              LDA  4201H  ;Get the subtrahend in B-register.
6   4103  47                    MOV  B,A
7   4104  3A 00 42              LDA  4200H  ;Get the minuend in A-register.
8   4107  0E 00                 MVI  C,00H  ;Clear C-register to account for sign.
9   4109  90                    SUB  B      ;Get the difference in A-register.
10  410A  D2 11 41              JNC  AHEAD  ;If CF=0, then go to AHEAD.
11  410D  0C                    INR  C      ;If CF=1, then increment C-register.
12  410E  2F                    CMA         ;Get 2's complement of difference
13  410F  C6 01                 ADI  01H    ;(result) in A-register.
14  4111  32 02 42   AHEAD:     STA  4202H  ;Store the result in memory.
15  4114  79                    MOV  A,C
16  4115  32 03 42              STA  4203H  ;Store the sign bit in memory.
17  4118  76                    HLT         ;Halt program execution.
18
19  4119                        END         ;Assembly end.
```

Sample Data

```
    Input Data  :  Minuend    = 5E_H
                   Subtrahend = 34_H

    Output Data :  Difference = 2A_H
                   Sign bit   = 00_H
```

EXAMPLE PROGRAM - 85.4:   16-Bit Subtraction

Write an assembly language program to subtract two numbers of 16-bit data stored in memory locations from 4200$_H$ to 4203$_H$. The data are stored such that the low byte is stored first and then the high byte is stored. Store the result in 4204$_H$ and 4205$_H$.

Problem Analysis

The 16-bit subtraction is performed in terms of 8-bit subtraction. First low bytes of the data are subtracted and the result is stored in memory. Then high bytes of the data are subtracted along with borrow (carry) in the previous subtraction and the result is stored in memory.

Algorithm

1. Load the low byte of subtrahend (the data to be subtracted) in accumulator from memory and move it to B-register.
2. Load the low byte of minuend in accumulator from memory.
3. Subtract the content of B-register (subtrahend) from the content of accumulator (minuend).
4. Store the low byte of result in memory.
5. Load the high byte of subtrahend in accumulator from memory and move it to B-register.
6. Load the high byte of minuend in accumulator from memory.
7. Subtract the content of B-register and the carry (borrow) from the content of accumulator.
8. Store the high byte of the result in memory.
9. Stop.

Flowchart for Example Program 85.4

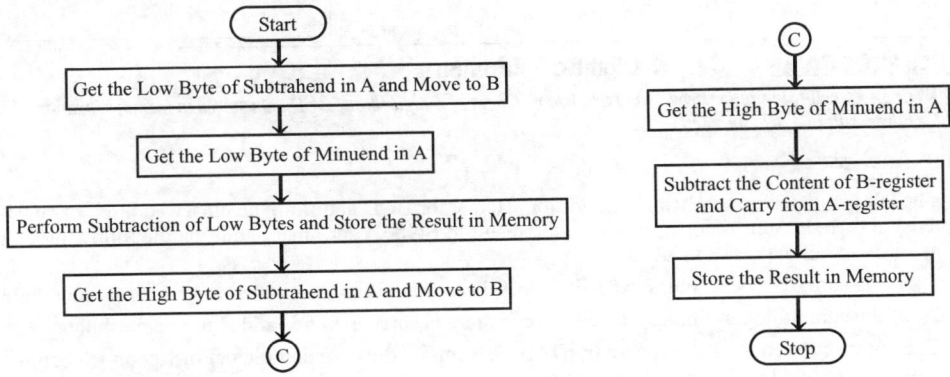

Assembly Language Program

```
;PROGRAM TO SUBTRACT TWO 16-BIT DATA

        ORG  4100H   ;specify program starting address.

        LDA  4202H
        MOV  B,A     ;Get low byte of subtrahend in B-register.
        LDA  4200H   ;Get low byte of minuend in A-register.
        SUB  B       ;Get difference of low bytes in A-register.
        STA  4204H   ;Store the result in memory.
        LDA  4203H
        MOV  B,A     ;Get high byte of subtrahend in B-register.
        LDA  4201H   ;Get high byte of minuend in A-register.
        SBB  B       ;Get difference of high bytes in A-register.
        STA  4205H   ;Store the result.
        HLT          ;Halt program execution.

        END          ;Assembly end.
```

Assembler Listing for Example Program 85. 4

```
1                              ;PROGRAM TO SUBTRACT TWO 16-BIT DATA
2
3     4100                     ORG 4100H ;specify program starting address.
4
5     4100    3A 02 42   LDA 4202H
6     4103    47         MOV B,A    ;Get low byte of subtrahend in B-register.
7     4104    3A 00 42   LDA 4200H  ;Get low byte of minuend in A-register.
8     4107    90         SUB B      ;Get difference of low bytes in A-register.
9     4108    32 04 42   STA 4204H  ;Store the result in memory.
10    410B    3A 03 42   LDA 4203H
11    410E    47         MOV B,A    ;Get high byte of subtrahend in B-register.
12    410F    3A 01 42   LDA 4201H  ;Get high byte of minuend in A-register.
13    4112    98         SBB B      ;Get difference of high bytes in A-register.
14    4113    32 05 42   STA 4205H  ;Store the result.
15    4116    76         HLT        ;Halt program execution.
16
17    4117               END        ;Assembly end.
```

Sample Data

Memory address	Content
4200	AB
4201	B2
4202	2C
4203	92
4204	7F
4205	20

Input Data : Minuend   = B2AB$_H$

Subtrahend = 922C$_H$

Output Data : Difference = 207F$_H$

EXAMPLE PROGRAM - 85.5 : 2-Digit BCD Addition

Write an assembly language program to add two numbers of 2-digit (8-bit) BCD data stored in memory locations 4200$_H$ and 4201$_H$. Store the result in 4202$_H$ and 4203$_H$.

Problem Analysis

The 8085 will perform only binary addition. Hence for BCD addition, the binary addition of BCD data is performed and then the sum is corrected to get the result in BCD. After binary addition the following correction should be made to get the result in BCD.

1. If the sum of lower nibbles exceeds **9** or if there is auxiliary carry then **6** is added to lower nibble.
2. If the sum of upper nibbles exceeds **9** or if there is carry then **6** is added to upper nibble.

The above correction is taken care by DAA (**D**ecimal **A**djust **A**ccumulator) instruction. Therefore after binary addition, execute DAA instruction to do the above correction in the sum.

Flowchart for Example Program 85.5

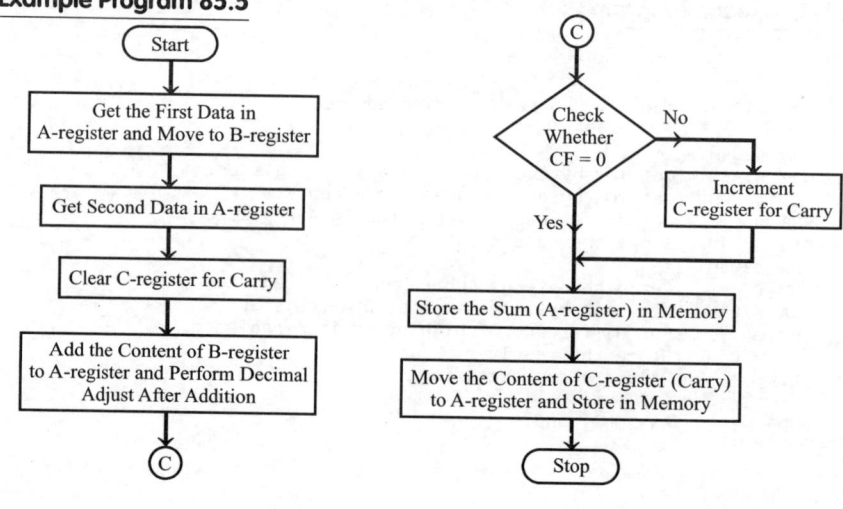

Algorithm

1. Load the first data in accumulator and move it to B-register.
2. Load the second data in accumulator.
3. Clear C-register for storing carry.
4. Add the content of B-register to accumulator.
5. Execute DAA instruction.
6. Check for carry. If carry = 1, go to step 7 or if carry = 0, go to step 8.
7. Increment C-register to account for carry.
8. Store the sum (content of accumulator) in memory.
9. Move the carry (content of C-register) to accumulator and store in memory.
10. Stop.

Assembly Language Program

```
;PROGRAM TO ADD TWO 2-DIGIT BCD DATA

        ORG  4100H   ;specify program starting address.

        LDA  4200H
        MOV  B,A     ;Get 1st data in B-register.
        LDA  4201H   ;Get 2nd data in A-register.
        MVI  C,00H   ;Clear C-register for accounting carry.
        ADD  B
        DAA          ;Get the sum of BCD data in A-register.
        JNC  AHEAD   ;If CF=0, go to AHEAD.
        INR  C       ;If CF=1, increment C-register.
AHEAD:  STA  4202H   ;Store the sum in memory.
        MOV  A,C
        STA  4203H   ;Store the carry in memory.
        HLT          ;Halt program execution.

        END          ;Assembly end.
```

Assembler Listing for Example Program 85.5

```
1                              ;PROGRAM TO ADD TWO 2-DIGIT BCD DATA
2
3    4100                      ORG  4100H   ;specify program starting address.
4
5    4100   3A 00 42           LDA  4200H
6    4103   47                 MOV  B,A     ;Get 1st data in B-register.
7    4104   3A 01 42           LDA  4201H   ;Get 2nd data in A-register.
8    4107   0E 00              MVI  C,00H   ;Clear C-register for accounting carry.
9    4109   80                 ADD  B
10   410A   27                 DAA          ;Get the sum of BCD data in A-register.
11   410B   D2 0F 41           JNC  AHEAD   ;If CF=0, go to AHEAD.
12   410E   0C                 INR  C       ;If CF=1, increment C-register.
13   410F   32 02 42   AHEAD:  STA  4202H   ;Store the sum in memory.
14   4112   79                 MOV  A,C
15   4113   32 03 42           STA  4203H   ;Store the carry in memory.
16   4116   76                 HLT          ;Halt program execution.
17
18   4117                      END          ;Assembly end.
```

Sample Data

Memory address	content
4200	72
4201	99
4202	71
4203	01

Input Data : Minuend   = $B2AB_H$

Subtrahend = $922C_H$

Output Data : Difference = $20\ F_H$

EXAMPLE PROGRAM - 85.6 : 4-Digit BCD Addition

Write an assembly language program to add two numbers of 4-digit BCD data stored in memory locations from 4200$_H$ to 4203$_H$ and store the result from 4204$_H$ to 4206$_H$.

Problem Analysis

The 4-digit BCD addition is performed in terms of 2-digit BCD addition. First lower order two digits are added and the sum is stored in the memory. Then the higher order two digits are added along with previous carry and the sum and final carry are stored in the memory.

Algorithm

1. Load the low order two digits of first data in accumulator and move it to B-register.
2. Load the low order two digits of second data in accumulator.
3. Clear C-register for storing carry.
4. Add the content of B-register to accumulator.
5. Execute DAA instruction.
6. Store the low order two digits of the result in memory.
7. Load the high order two digits of first data in accumulator and move it to B-register.
8. Load the high order two digits of second data in accumulator.
9. Add the content of B-register and carry (from previous addition) to accumulator.
10. Execute DAA instruction.
11. Check for carry. If carry = 1, go to step 12 or if carry = 0, go to step 13.
12. Increment C-register to account for final carry.
13. Store the high order two digits of the result in memory.
14. Move the carry (content of C-register) to accumulator and store in memory.
15. Stop.

Flowchart for Example Program 85.6

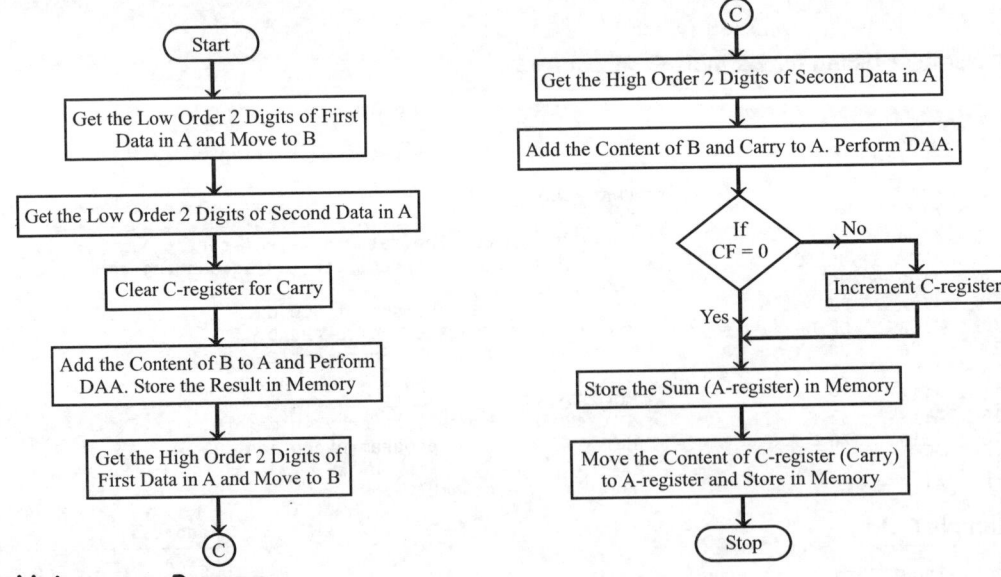

Assembly Language Program

```
;PROGRAM TO ADD TWO 4-DIGIT BCD DATA

        ORG 4100H  ;specify program starting address.

        LDA 4200H
        MOV B,A    ;Get low order 2 digits of 1st data in B.
```

```
         LDA  4202H  ;Get low order 2 digits of 2nd data in A.
         MVI  C,00H  ;Clear C-register to account for carry.
         ADD  B
         DAA         ;Get the sum of low order two digits in A
         STA  4204H  ;and store it in memory.
         LDA  4201H
         MOV  B,A    ;Get high order 2 digits of 1st data in B.
         LDA  4203H  ;Get high order 2 digits of 2nd data in A.
         ADC  B
         DAA         ;Get the sum of high order two digits in A
         STA  4205H  ;and store the same in memory.
         JNC  AHEAD
         INR  C      ;If CF=1, increment C-register.
AHEAD:   MOV  A,C
         STA  4206H  ;Store the carry in memory.
         HLT         ;Halt program execution.

         END         ;Assembly end.
```

Assembler Listing for Example Program 85.6

```
 1                        ;PROGRAM TO ADD TWO 4-DIGIT BCD DATA
 2
 3   4100                 ORG  4100H  ;specify program starting address.
 4
 5   4100  3A 00 42       LDA  4200H
 6   4103  47             MOV  B,A    ;Get low order 2 digits of 1st data in B.
 7   4104  3A 02 42       LDA  4202H  ;Get low order 2 digits of 2nd data in A.
 8   4107  0E 00          MVI  C,00H  ;Clear C-register to account for carry.
 9   4109  80             ADD  B
10   410A  27             DAA         ;Get the sum of low order 2 digits in A
11   410B  32 04 42       STA  4204H  ;and store it in memory.
12   410E  3A 01 42       LDA  4201H
13   4111  47             MOV  B,A    ;Get high order 2 digits of 1st data in B.
14   4112  3A 03 42       LDA  4203H  ;Get high order 2 digits of 2nd data in A.
15   4115  88             ADC  B
16   4116  27             DAA         ;Get the sum of high order 2 digits in A
17   4117  32 05 42       STA  4205H  ;and store the same in memory.
18   411A  D2 1E 41       JNC  AHEAD
19   411D  0C             INR  C      ;If CF=1, increment C-register.
20   411E  79      AHEAD: MOV  A,C
21   411F  32 06 42       STA  4206H  ;Store the carry in memory.
22   4122  76             HLT         ;Halt program execution.
23
24   4123                 END         ;Assembly end.
```

Sample Data

```
    Input Data  : Data-1 = 8067₁₀
                  Data-2 = 2892₁₀
    Output Data : Sum    = 0959₁₀
                  Carry  = 01₁₀
```

Memory address	Content	Memory address	Content
4200	67	4204	59
4201	80	4205	09
4202	92	4206	01
4203	28		

EXAMPLE PROGRAM - 85.7 : 2-Digit BCD Subtraction

Write an assembly language program to subtract two numbers of 2-digit BCD data stored in memory locations 4200_H and 4201_H and store the result in 4202_H.

Problem Analysis

The 8085 will perform only binary subtraction. Hence for BCD subtraction, 10's complement subtraction is performed. First the 10's complement of the subtrahend is obtained and then added to the minuend. The DAA instruction is executed to get the result in BCD.

Algorithm

1. Load the subtrahend in accumulator and move it to B-register.
2. Move 99 to accumulator and subtract the content of B-register from accumulator.
3. Increment the accumulator.
4. Move the content of accumulator to B-register.
5. Load the minuend in accumulator.
6. Add the content of B-register to accumulator.
7. Execute DAA instruction.
8. Store the result in memory.
9. Stop.

Flowchart for Example Program 85.7

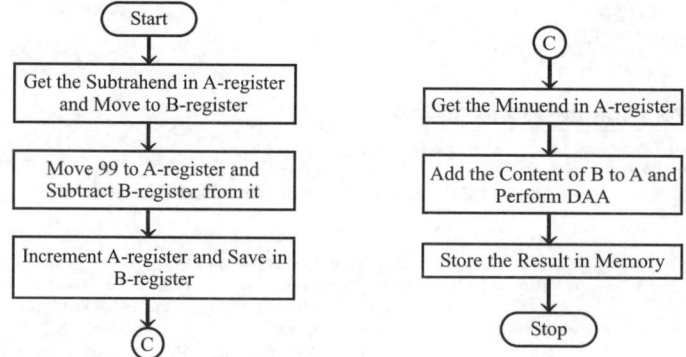

Assembly Language Program

```
;PROGRAM TO SUBTRACT TWO BCD (2-DIGIT) DATA

        ORG 4100H   ;specify program starting address.

        LDA 4201H
        MOV B,A     ;Get the subtrahend in B-register.
        MVI A,99H   ;Get 10's complement of
        SUB B       ;subtrahend in A.
        INR A
        MOV B,A     ;Save 10's complement in B.
        LDA 4200H   ;Get the minuend in A-register.
        ADD B       ;Get BCD sum of minuend and 10's
        DAA         ;complement of subtrahend. This sum
                    ;is the difference between BCD data.
        STA 4202H   ;Store the result in memory.
        HLT         ;Halt program execution.

        END         ;Assembly end.
```

Assembler Listing for Example Program 85.7

```
1                            ;PROGRAM TO SUBTRACT TWO BCD (2-DIGIT) DATA
2
3   4100                ORG 4100H   ;specify program starting address.
4
5   4100  3A 01 42      LDA 4201H
6   4103  47            MOV B,A     ;Get the subtrahend in B-register.
7   4104  3E 99         MVI A, 99H  ;Get 10's complement of
8   4106  90            SUB B       ;subtrahend in A.
9   4107  3C            INR A
10  4108  47            MOV B,A     ;Save 10's complement in B.
11  4109  3A 00 42      LDA 4200H   ;Get the minuend in A-register.
12  410C  80            ADD B       ;Get BCD sum of minuend and 10's
13  410D  27            DAA         ;complement of subtrahend. This sum
```

```
14                                  ;is the difference between BCD data.
15   410E  32 02 42   STA  4202H    ;Store the result in memory.
16   4111  76         HLT           ;Halt program execution.
17
18   4112             END           ;Assembly end.
```

Sample Data

Input Data : Minuend   = 95_{10}

Subtrahend = 32_{10}

Output Data: Difference= 63_{10}

Memory address	Content
4200	95
4201	32
4202	63

EXAMPLE PROGRAM - 85.8 :   8-Bit Multiplication

Write an assembly language program to multiply two numbers of 8-bit data stored in memory locations 4200_H and 4201_H and store the product in 4202_H and 4203_H.

Problem Analysis

In 8085, the multiplication is performed as repeated additions. The initial value of sum is assumed as zero. One of the data is used as count (N) for the number of additions to be performed. Another data is added to the sum N times, where N is the count. The result of the product of two 8-bit data will be 16 bits. Hence, another register is used to account for the overflow.

Flowchart for Example Program 85.8

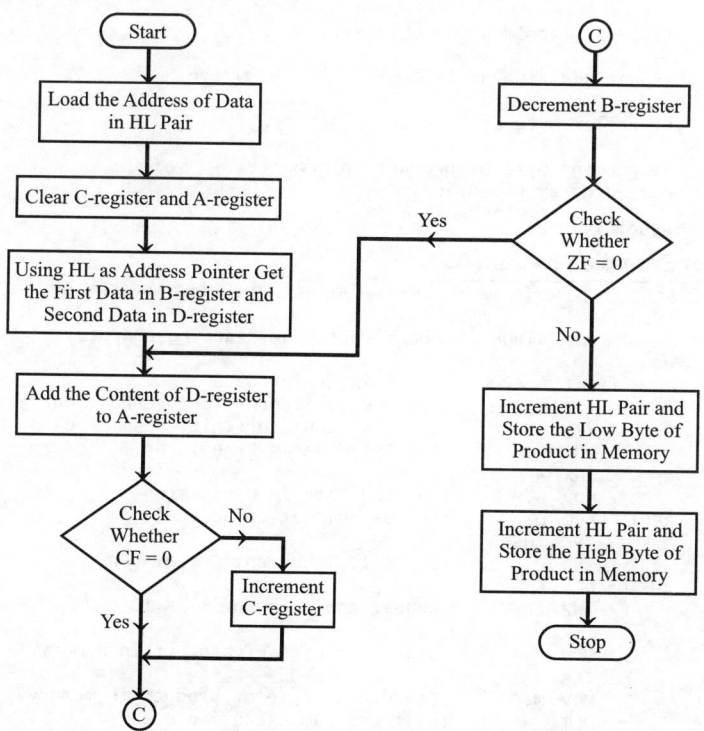

Algorithm

1. Load the address of the first data in HL pair (pointer).
2. Clear C-register for overflow (carry).
3. Clear the accumulator.
4. Move the first data to B-register (count).
5. Increment the pointer.
6. Move the second data to D-register (multiplicand).
7. Add the content of D-register to accumulator.
8. Check for carry. If carry = 1, go to step 9 or If carry = 0, go to step 10.
9. Increment C-register.
10. Decrement B-register (count).
11. Check whether count has reached zero. If ZF = 0 repeat steps 7 through 11, or if ZF = 1 go to next step.
12. Increment the pointer and store low byte of the product in memory.
13. Increment the pointer and store high byte of the product in memory.
14. Stop.

Assembly Language Program

```
;PROGRAM TO MULTIPLY TWO NUMBERS OF 8-BIT DATA

        ORG  4100H   ;specify program starting address.

        LXI  H,4200H ;Set pointer for data.
        MVI  C,00H   ;Clear C to account for overflow (Carry).
        XRA  A       ;Clear accumulator(Initial sum = 0).
        MOV  B,M     ;Get 1st data in B-register.
        INX  H
        MOV  D,M     ;Get 2nd data in D-register.
REPT:   ADD  D       ;Add D-register to accumulator.
        JNC  AHEAD
        INR  C       ;If CF=1, increment C-register.
AHEAD:  DCR  B
        JNZ  REPT    ;Repeat addition until ZF=1.
        INX  H
        MOV  M,A     ;Store low byte of product in memory.
        INX  H
        MOV  M,C     ;Store high byte of product in memory.
        HLT          ;Halt program execution.

        END          ;Assembly end.
```

Assembler Listing for Example Program 85.8

```
 1                                      ;PROGRAM TO MULTIPLY TWO NUMBERS OF 8-BIT DATA
 2
 3   4100                      ORG  4100H   ;specify program starting address.
 4
 5   4100  21 00 42            LXI  H,4200H ;Set pointer for data.
 6   4103  0E 00               MVI  C,00H   ;Clear C to account for overflow (Carry).
 7   4105  AF                  XRA  A       ;Clear accumulator(Initial sum = 0).
 8   4106  46                  MOV  B,M     ;Get 1st data in B-register.
 9   4107  23                  INX  H
10   4108  56                  MOV  D,M     ;Get 2nd data in D-register.
11   4109  82          REPT:   ADD  D       ;Add D-register to accumulator.
12   410A  D2 0E 41            JNC  AHEAD
13   410D  0C                  INR  C       ;If CF=1, increment C-register.
14   410E  05          AHEAD:  DCR  B
15   410F  C2 09 41            JNZ  REPT    ;Repeat addition until ZF=1.
16   4112  23                  INX  H
17   4113  77                  MOV  M,A     ;Store low byte of product in memory.
18   4114  23                  INX  H
19   4115  71                  MOV  M,C     ;Store high byte of product in memory.
20   4116  76                  HLT          ;Halt program execution.
21
22   4117                      END          ;Assembly end.
```

Sample Data

			Memory address	Content
Input Data :	Data-1	= C7$_H$	4200	C7
	Data-2	= 4A$_H$	4201	4A
			4202	86
Output Data :	Product	= 3986$_H$	4203	39

EXAMPLE PROGRAM - 85.9 : 16-Bit Multiplication

Write an assembly language program to multiply two numbers of 16-bit data stored in memory locations from 4200$_H$ to 4203$_H$. Store the product in memory locations from 4204$_H$ to 4207$_H$.

Problem Analysis

The 16-bit multiplication is performed as repeated 16-bit additions. The initial sum is assumed as zero. One of the data is stored in SP (Stack Pointer) and another data is stored in DE pair. The content of DE pair is used as count for number of additions. The content of SP is added to the sum N times, where N is the count. The maximum size of product will be 32-bit. Hence BC pair is used to account for overflow. In 16-bit decrement no flags are affected. Hence, to check zero of the count (DE pair), move E-register to A-register and logically ORed with D-register.

Flowchart for Example Program 85.9

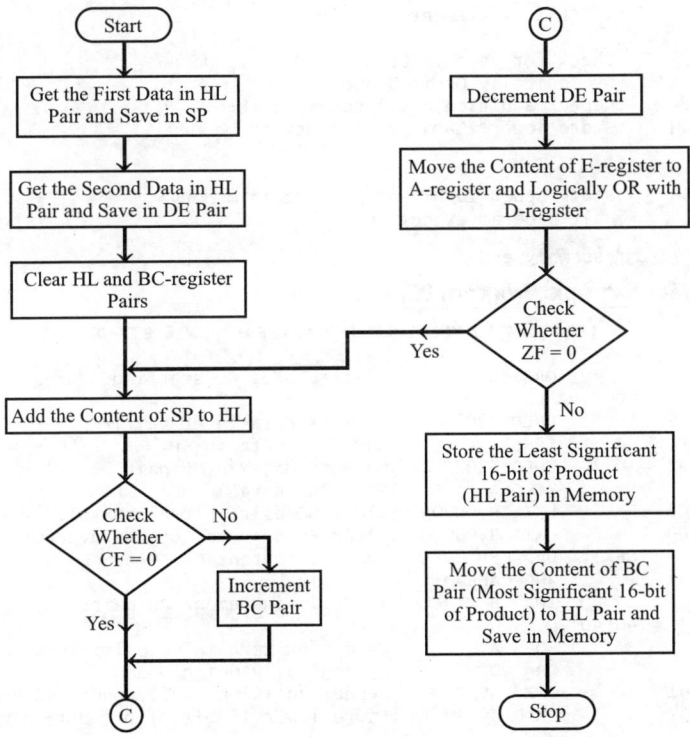

Algorithm

1. Load the first data in HL pair and move to SP.
2. Load the second data in HL and move to DE (count).
3. Clear HL pair (Initial sum).
4. Clear BC pair for overflow (carry).
5. Add the content of SP to HL.
6. Check for carry. If carry=1, go to step 7 or If carry=0, go to step 8.

7. Increment BC pair.
8. Decrement the count.
9. Check whether count has reached zero.
10. To check for zero of the count, move the content of E-register to A-register and logically OR with D-register.
11. Check the zero flag. If ZF=0, repeat steps 5 through 11 or If ZF=1, go to next step.
12. Store the content of HL in memory. (Least significant 16 bits of the product).
13. Move the content of C to L and B to H and store HL in memory. (Most significant 16 bits of the product).
14. Stop.

Assembly Language Program

```
;PROGRAM TO MULTIPLY TWO NUMBERS OF 16-BIT DATA

         ORG    4100H        ;specify program starting address.

         LHLD   4200H        ;Get 1st data in HL pair.
         SPHL                ;Save 1st data in SP.
         LHLD   4202H        ;Get 2nd data in HL pair.
         XCHG                ;Save 2nd data in DE pair.
         LXI    H,0000H      ;Clear HL pair(initial sum=0).
         LXI    B,0000H      ;Clear BC pair to account overflow.
NEXT:    DAD    SP           ;Add the content of SP to sum(HL).
         JNC    AHEAD
         INX    B            ;If CF=1, increment BC pair.
AHEAD:   DCX    D
         MOV    A,E          ;Check for zero in DE pair. This is done
         ORA    D            ;by logically OR in D and E.
         JNZ    NEXT         ;Repeat addition until count is zero.
         SHLD   4204H        ;Store lower 16-bit of product in memory.
         MOV    L,C
         MOV    H,B
         SHLD   4206H        ;Store upper 16-bit of product in memory.
         HLT                 ;Halt program execution.

         END                 ;Assembly end.
```

Assembler Listing for Example Program 85.9

```
 1                          ;PROGRAM TO MULTIPLY TWO NUMBERS OF 16-BIT DATA
 2
 3  4100                            ORG    4100H        ;specify program starting address.
 4
 5  4100   2A 00 42                 LHLD   4200H        ;Get 1st data in HL pair.
 6  4103   F9                       SPHL                ;Save 1st data in SP.
 7  4104   2A 02 42                 LHLD   4202H        ;Get 2nd data in HL pair.
 8  4107   EB                       XCHG                ;Save 2nd data in DE pair.
 9  4108   21 00 00                 LXI    H,0000H      ;Clear HL pair(initial sum=0).
10  410B   01 00 00                 LXI    B,0000H      ;Clear BC pair to account overflow.
11  410E   39            NEXT:      DAD    SP           ;Add the content of SP to sum(HL).
12  410F   D2 13 41                 JNC    AHEAD
13  4112   03                       INX    B            ;If CF=1, increment BC pair.
14  4113   1B            AHEAD:     DCX    D
15  4114   7B                       MOV    A,E          ;Check for zero in DE pair. This is done
16  4115   B2                       ORA    D            ;by logically ORing D and E.
17  4116   C2 0E 41                 JNZ    NEXT         ;Repeat addition until count is zero.
18  4119   22 04 42                 SHLD   4204H        ;Store lower 16-bit of product in memory.
19  411C   69                       MOV    L,C
20  411D   60                       MOV    H,B
21  411E   22 06 42                 SHLD   4206H        ;Store upper 16-bit of product in memory.
22  4121   76                       HLT                 ;Halt program execution.
23
24  4122                            END                 ;Assembly end.
```

Sample Data

Input Data   :   Data-1   =   5A 24$_H$

Data-2   =   47C2$_H$

Output Data   :   Product   =   19444B48$_H$

Memory address	Content
4200	24
4201	5A
4202	C2
4203	47

Memory address	Content
4204	48
4205	4B
4206	44
4207	19

EXAMPLE PROGRAM - 85.10 :   8-Bit Division

Write an assembly language program to divide two numbers of 8-bit data stored in memory locations 4200$_H$ and 4201$_H$. Store the quotient in 4202$_H$ and the remainder in 4203$_H$.

Problem Analysis

The division in 8085 is performed as repeated subtraction. The dividend is stored in A-register and divisor in B-register. The initial value of quotient is assumed as zero. Subtraction should be performed only when dividend is greater than divisor. So repeated subtraction is performed until dividend is lesser than the divisor. For each subtraction, the quotient is incremented by one. Then store the quotient and remainder in the memory.

Algorithm

1.   Load the divisor in accumulator and move it to B-register.
2.   Load the dividend in accumulator.
3.   Clear C-register to account for quotient.
4.   Check whether divisor is less than dividend. If divisor is less than dividend, go to step 8, otherwise go to next step.
5.   Subtract the content of B-register from accumulator.
6.   Increment the content of C-register (quotient).
7.   Go to step 4.
8.   Store the content of accumulator (remainder) in memory.
9.   Move the content of C-register (quotient) to accumulator and store in memory.
10.   Stop.

Flowchart for Example Program 85.10

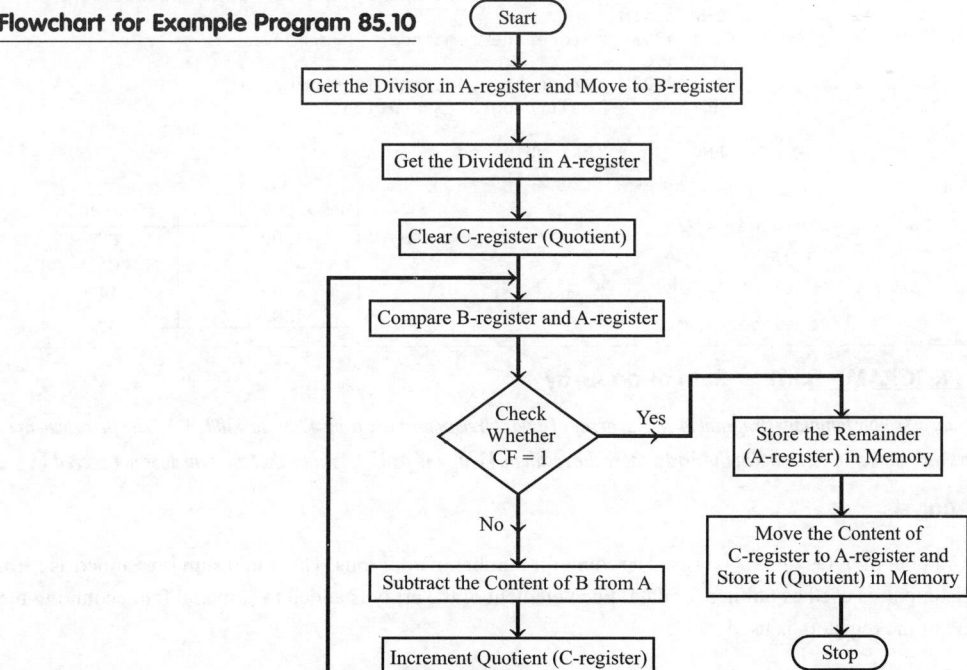

Assembly Language Program

```
;PROGRAM TO DIVIDE TWO NUMBERS OF 8-BIT DATA

            ORG  4100H   ;specify program starting address.

            LDA  4201H
            MOV  B,A      ;Get the divisor in B-register.
            LDA  4200H    ;Get the dividend in A-register.
            MVI  C,00H    ;Clear C-register for quotient.
AGAIN:      CMP  B
            JC   STORE    ;If divisor is less than dividend go to store.
            SUB  B        ;Subtract divisor from dividend.
            INR  C        ;Increment quotient by one for each subtraction.
            JMP  AGAIN
STORE:      STA  4203H    ;Store the remainder in memory.
            MOV  A,C
            STA  4202H    ;Store the quotient in memory.
            HLT           ;Halt program execution.

            END           ;Assembly end.
```

Assembler Listing for Example Program 85.10

```
1                              ;PROGRAM TO DIVIDE TWO NUMBERS OF 8-BIT DATA
2
3    4100                      ORG 4100H ;specify program starting address.
4
5    4100  3A 01 42            LDA  4201H
6    4103  47                  MOV  B,A      ;Get the divisor in B-register.
7    4104  3A 00 42            LDA  4200H    ;Get the dividend in A-register.
8    4107  0E 00               MVI  C,00H    ;Clear C-register for quotient.
9    4109  B8          AGAIN:  CMP  B
10   410A  DA 12 41            JC   STORE    ;If divisor is less than dividend go to store.
11   410D  90                  SUB  B        ;Subtract divisor from dividend.
12   410E  0C                  INR  C        ;Increment quotient by one for each subtraction.
13   410F  C3 09 41            JMP  AGAIN
14   4112  32 03 42    STORE:  STA  4203H    ;Store the remainder in memory.
15   4115  79                  MOV  A,C
16   4116  32 02 42            STA  4202H    ;Store the quotient in memory.
17   4119  76                  HLT           ;Halt program execution.
18
19   411A                      END           ;Assembly end.
```

Sample Data

				Memory address	Content
Input Data :	Dividend	= C9$_H$		4200	C9
	Divisor	= 0A$_H$		4201	0A
Output Data :	Quotient	= 14$_H$		4202	14
	Remainder	= 01$_H$		4203	01

EXAMPLE PROGRAM - 85.11 : Sum of an Array

Write an assembly language program to add an array of data stored in memory from 4200$_H$ to 4200$_H$ + N. The first element of the array, gives the number of elements in the array. Store the result in 4300$_H$ and 4301$_H$. Assume that the sum does not exceed 16-bit.

Problem Analysis

The number of bytes (data) N, is used as count for number of additions. The initial sum is assumed as zero. The HL register pair is used as pointer for data. Each element of the array is added to sum and for accounting the overflow one of the registers is used.

Flowchart for Example Program 85.11

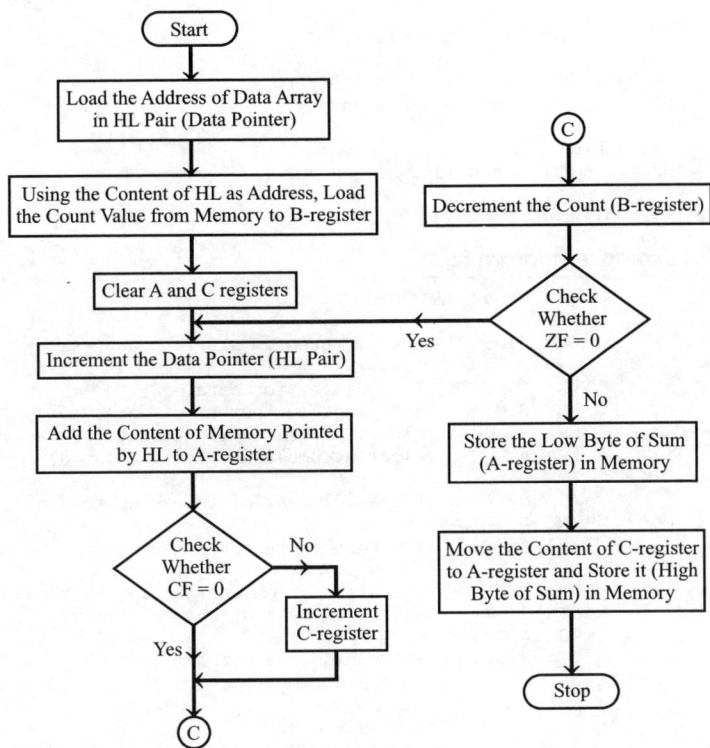

Algorithm

1. Load the address of the first element of the array in HL pair (pointer).
2. Move the count to B-register.
3. Clear C-register for carry.
4. Clear accumulator for sum.
5. Increment the pointer (HL pair).
6. Add the content of memory addressed by HL to accumulator.
7. Check for carry. If carry = 1, go to step 8, or If carry = 0, go to step 8.
8. Increment C-register.
9. Decrement the count.
10. Check for zero of the count. If ZF = 0 go to step 5 or If ZF = 1, go to next step.
11. Store the content of accumulator (low byte of sum).
12. Move the content of C-register (high byte of sum) to accumulator. Store the content of accumulator in memory.
13. Stop.

Assembly Language Program

```
; PROGRAM TO ADD AN ARRAY OF DATA

        ORG  4100H   ;specify program starting address.

        LXI  H,4200H ;Set pointer for data.
        MOV  B,M     ;Set count for number of data.
        MVI  C,00H   ;Clear C-register to account for carry.
        XRA  A       ;Clear accumulator. Initial sum=0.
```

```
REPT:     INX H
          ADD M         ;Add an element of the array to sum.
          JNC AHEAD
          INR C         ;If CF=1, increment C-register.
AHEAD:    DCR B
          JNZ REPT      ;Repeat addition until count is zero.
          STA 4300H     ;Store low byte of sum in memory.
          MOV A,C
          STA 4301H     ;Store high byte of sum in memory.
          HLT           ;Halt program execution.

          END           ;Assembly end.
```

Assembler Listing for Example Program 85.11

```
 1                                ;PROGRAM TO ADD AN ARRAY OF DATA
 2
 3    4100                        ORG 4100H    ;specify program starting address.
 4
 5    4100   21 00 42             LXI H,4200H  ;Set pointer for data.
 6    4103   46                   MOV B,M      ;Set count for number of data.
 7    4104   0E 00                MVI C,00H    ;Clear C-register to account for carry.
 8    4106   AF                   XRA A        ;Clear accumulator. Initial sum=0.
 9    4107   23          REPT:    INX H
10    4108   86                   ADD M        ;Add an element of the array to sum.
11    4109   D2 0D 41             JNC AHEAD
12    410C   0C                   INR C        ;If CF=1, increment C-register.
13    410D   05          AHEAD:   DCR B
14    410E   C2 07 41             JNZ REPT     ;Repeat addition until count is zero.
15    4111   32 00 43             STA 4300H    ;Store low byte of sum in memory.
16    4114   79                   MOV A,C
17    4115   32 01 43             STA 4301H    ;Store high byte of sum in memory.
18    4118   76                   HLT          ;Halt program execution.
19
20    4119   END                               ;Assembly end.
```

Sample Data

Input Data   : Count = 07_H

Array = $C2_H$

45_H

$B3_H$

$F4_H$

$7C_H$

ED_H

16_H

Output Data   : Sum = $042D_H$

Memory address	Content	
4200	07	Count
4201	C2	
4202	45	
4203	B3	
4204	F4	Array
4205	7C	
4206	ED	
4207	16	
4300	2D	Sum
4301	04	

EXAMPLE PROGRAM - 85.12 :   Search for Smallest Data in an Array

Write an assembly language program to search the smallest data in an array of N data stored in memory locations from 4200_H *to* $(4200_H + N)$. *The first element of the array gives the number of data in the array. Store the smallest data in* 4300_H.

Problem Analysis

The HL register pair is used as pointer for the array. One of the general purpose register is used as count. A data in the array is moved to A-register and compared with next data. After each comparison, the smallest data is brought to accumulator. The comparisons are carried N–1 times. After N–1 comparisons, the smallest data will be in A-register and store it in memory.

Algorithm

1. Load the address of the first element of the array in HL register pair (pointer).
2. Move the count to B-register.
3. Increment the pointer.
4. Get the first data in accumulator.
5. Decrement the count.
6. Increment the pointer.
7. Compare the content of memory addressed by HL pair with that of accumulator.
8. If carry = 1, go to step 10 or If carry = 0, go to step 8.
9. Move the content of memory addressed HL to accumulator.
10. Decrement the count.
11. Check for zero of the count. If ZF = 0, go to step 6, or If ZF = 1 go to next step.
12. Store the smallest data in memory.
13. Stop.

Flowchart for Example Program 85.12

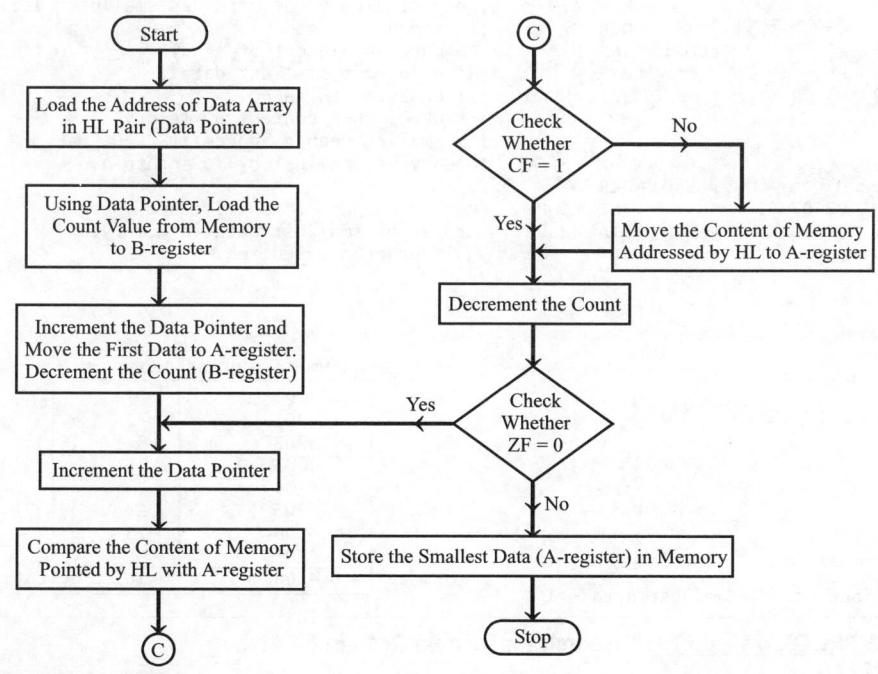

Assembly Language Program

```
;PROGRAM TO SEARCH SMALLEST DATA IN AN ARRAY

        ORG  4100H     ;specify program starting address.

        LXI  H,4200H   ;Set pointer for array.
        MOV  B,M       ;Set count for number of elements in array.
        INX  H
        MOV  A,M       ;Set 1st element of array as smallest data.
        DCR  B         ;Decrement the count.
```

```
LOOP:    INX  H          ;Compare an element of array
         CMP  M          ;with current smallest data.
         JC   AHEAD      ;If CF=1, go to AHEAD.
         MOV  A,M        ;If CF=0, then content of memory
                         ;is smaller than A, hence if CF=0, make
                         ;memory as smallest by moving to A.
AHEAD:   DCR  B
         JNZ  LOOP       ;Repeat comparison until count is zero.
         STA  4300H      ;Store the smallest data in memory.
         HLT             ;Halt program execution.

         END             ;Assembly end.
```

Assembler Listing for Example Program 85.12

```
 1                               ;PROGRAM TO SEARCH SMALLEST DATA IN AN ARRAY
 2
 3   4100                        ORG  4100H   ;specify program starting address.
 4
 5   4100   21 00 42             LXI  H,4200H ;Set pointer for array.
 6   4103   46                   MOV  B,M     ;Set count for number of elements in array.
 7   4104   23                   INX  H
 8   4105   7E                   MOV  A,M     ;Set 1st element of array as smallest data.
 9   4106   05                   DCR  B       ;Decrement the count.
10   4107   23          LOOP:    INX  H       ;Compare an element of array
11   4108   BE                   CMP  M       ;with current smallest data..
12   4109   DA 0D 41             JC   AHEAD   ;If CF=1, go to AHEAD.
13   410C   7E                   MOV  A,M     ;If CF=0, then content of memory
14                                            ;is smaller than A, hence if CF=0, make
15                                            ;memory as smallest by moving to A.
16   410D   05          AHEAD:   DCR  B
17   410E   C2 07 41             JNZ  LOOP    ;Repeat comparison until count is zero.
18   4111   32 00 43             STA  4300H   ;Store the smallest data in memory.
19   4114   76                   HLT          ;Halt program execution.
20
21   4115                        END          ;Assembly end.
```

Sample Data

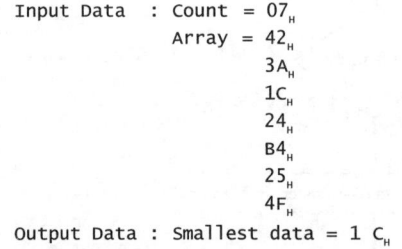

Input Data : Count = 07_H
                     Array = 42_H
                             $3A_H$
                             $1C_H$
                             24_H
                             $B4_H$
                             25_H
                             $4F_H$
Output Data : Smallest data = $1C_H$

Memory address	Content	
4200	07	Count
4201	42	
4202	3A	
4203	1C	
4204	24	Array
4205	B4	
4206	25	
4207	4F	
4300	1C	Smallest data

EXAMPLE PROGRAM - 85.13 :   Search for Largest Data in an Array

Write an assembly language program to search the largest data in an array of N data stored in memory locations from 4200_H to $4200_H + N$. The first element of the array is the number of data (N) in the array. Store the largest data in 4300_H.

Problem Analysis

The HL register pair is used as pointer for the array. One of the general purpose register is used as count. A data in the array is moved to A-register and compared with next data. After each comparison, the largest data is brought to A-register. The comparisons are performed $N-1$ times. After $N-1$ comparisons the largest data will be in A-register and store it in memory.

Algorithm

1. Load the address of the first element of the array in HL register pair (pointer).
2. Move the count to B-register.
3. Increment the pointer.
4. Get the first data in accumulator.
5. Decrement the count.
6. Increment the pointer.
7. Compare the content of memory addressed by HL pair with that of accumulator.
8. If carry = 0, go to step 10 or If carry = 1, go to step 8.
9. Move the content of memory addressed HL to accumulator.
10. Decrement the count.
11. Check for zero of the count. If ZF = 0, go to step 6, or If ZF = 1 go to next step.
12. Store the largest data in memory.
13. Stop.

Flowchart for Example Program 85.13

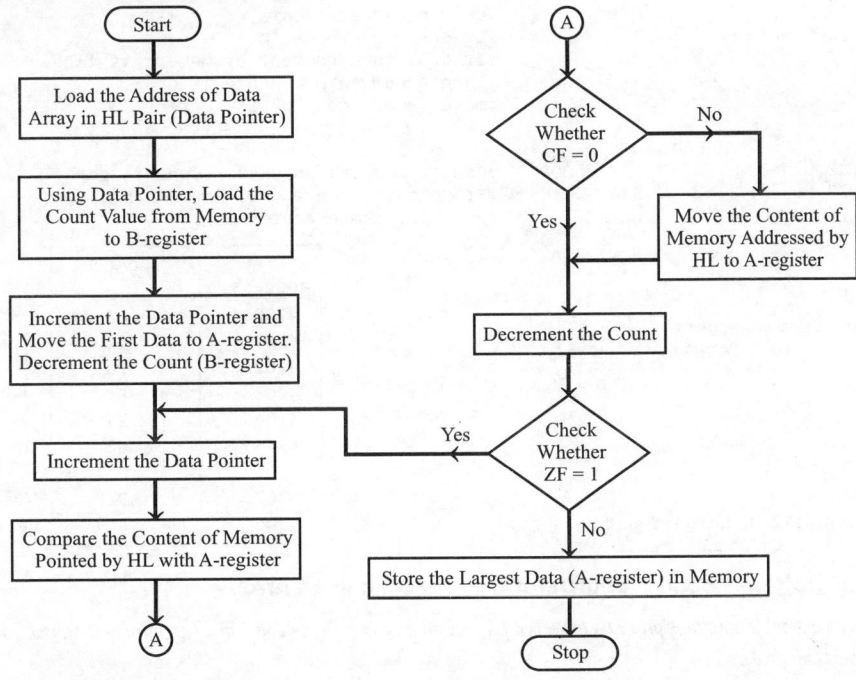

Assembly Language Program

```
;PROGRAM TO SEARCH LARGEST DATA IN AN ARRAY

        ORG   4100H    ;specify program starting address.

        LXI   H,4200H  ;Set pointer for array.
        MOV   B,M      ;Set count for number of elements in array.
        INX   H
        MOV   A,M      ;Set 1st element of array as largest data.
        DCR   B        ;Decrement the count.
LOOP:   INX   H        ;Compare an element of array with
        CMP   M        ;current largest data.
        JNC   AHEAD    ;If CF=0, go to AHEAD.
        MOV   A,M      ;If CF=1,then content of memory is larger
                       ;than accumulator.Hence if CF=1,
```

```
                       ;make memory content as current
                       ;largest by moving it to A-register.
AHEAD:  DCR  B
        JNZ  LOOP      ;Repeat comparison until count is zero.
        STA  4300H     ;Store the largest data in memory.
        HLT            ;Halt program execution.

        END            ;Assembly end.
```

Assembler Listing for Example Program 85.13

```
 1                              ;PROGRAM TO SEARCH LARGEST DATA IN AN ARRAY
 2
 3   4100                       ORG  4100H    ;specify program starting address.
 4
 5   4100  21 00 42             LXI  H,4200H  ;Set pointer for array.
 6   4103  46                   MOV  B,M      ;Set count for number of elements in array.
 7   4104  23                   INX  H
 8   4105  7E                   MOV  A,M      ;Set 1st element of array as largest data.
 9   4106  05                   DCR  B        ;Decrement the count.
10   4107  23           LOOP:   INX  H        ;Compare an element of array with
11   4108  BE                   CMP  M        ;current largest data.
12   4109  D2 0D 41             JNC  AHEAD    ;If CF=0, go to AHEAD.
13   410C  7E                   MOV  A,M      ;If CF=1,then content of memory is larger
14                                            ;than accumulator. Hence if CF=1,
15                                            ;make memory content as current
16                                            ;largest by moving it toA-register.
17   410D  05           AHEAD:  DCR  B
18   410E  C2 07 41             JNZ  LOOP     ;Repeat comparison until count is zero.
19   4111  32 00 43             STA  4300H    ;Store the largest data in memory.
20   4114  76                   HLT           ;Halt program execution.
21
22   4115                       END           ;Assembly end.
```

Sample Data

Memory address	Content	
4200	07	Count
4201	62	
4202	7D	
4203	FC	
4204	24	Array
4205	C2	
4206	0F	
4207	92	
4300	FC	Laragest data

```
Input Data  : Count  = 07ₕ
              Array  = 62ₕ
                       7Dₕ
                       FCₕ
                       24ₕ
                       C2ₕ
                       0Fₕ
                       92ₕ

Output Data : Largest data = FCₕ
```

Input Data : Count = 07_H

Array = 62_H, $7D_H$, FC_H, 24_H, $C2_H$, $0F_H$, 92_H

Output Data : Largest data = FC_H

EXAMPLE PROGRAM - 85.14 : Search for a Given Data in an Array

Write an assembly language program to search for a given data (stored in 4250_H) in an array of data stored from 4200_H. The end of the array is marked by 20_H.

If the data is available store FF_H in 4251_H. Store the position of the data and its address in 4252_H, 4253_H and 4254_H respectively. If the data is not available store 00_H in memory locations from 4251_H to 4254_H.

Problem Analysis

The HL pair is used as pointer for given data. B-register is used as pointer for position of the data. C-register is used to record the availability of given data. The given data is moved to A-register and compared with each element of the array one-by-one. If the data is available, terminate the comparison and store the position, address and FF_H (for availability) in memory.

Algorithm

1. Load the address of the data array in HL register pair.
2. Load the given data in accumulator.
3. Clear B-register.
4. Increment B-register.
5. Compare the content of memory addressed by HL pair with that of accumulator.
6. If ZF = 0, go to next step or if ZF = 1, go to step 8.
7. Check for end of array by comparing the data with 20_H. If ZF = 0, go to step 4, or If ZF = 1, go to next step.
8. Clear B, C, H, L registers and jump to step 10.
9. Move FF_H to C-register.
10. Store the content of H and L registers in memory.
11. Move C to L and B to H and store HL in memory.
12. Stop.

Flowchart for Example Program 85.14

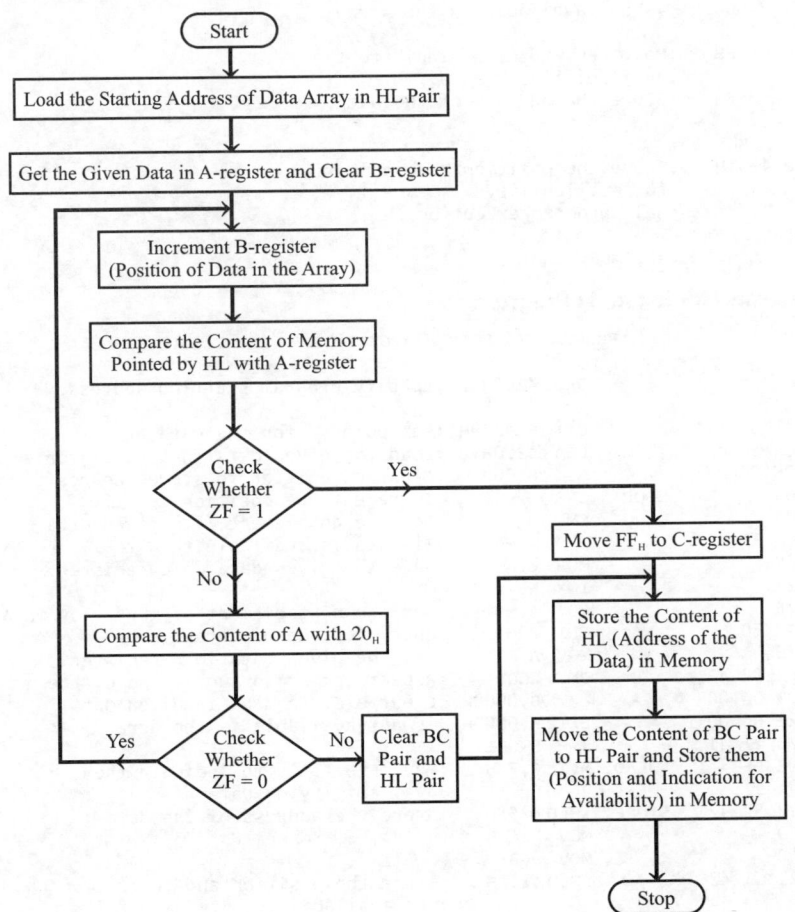

Assembly Language Program

```
;PROGRAM TO SEARCH A GIVEN DATA IN AN ARRAY

        ORG   4100H    ;specify program starting address.

        LXI   H,4200H  ;Set pointer for the data array.
        LDA   4250H    ;Load the given data in accumulator.
        MVI   B,00H    ;Clear B-register to store the position.
LOOP:   INR   B        ;Increment the position count.
        CMP   M        ;Compare an element with given data.
        JZ    AHEAD    ;If data is available,then ZF=1.
        MOV   C,A      ;Save the given data in C-register.
        INX   H
        MOV   A,M      ;Get the next element of the array in A and
        CPI   20H      ;check for end of array.
        MOV   A,C      ;Get the given data in A-register.
        JNZ   LOOP     ;Repeat comparison until end of the array.
        LXI   B,0000H  ;Clear B,C,H and L if given data
        LXI   H,0000H  ;is not available in the array.
        JMP   STORE
AHEAD:  MVI   C,FFH    ;Move FFH  to C, to indicate the
                       ;availability of data.
STORE:  SHLD  4253H    ;Store the address of the data.
        MOV   L,C
        MOV   H,B
        SHLD  4251H    ;Store the position and indication
                       ;for availability.
        HLT            ;Halt program execution.

        END            ;Assembly end.
```

Assembler Listing for Example Program 85.14

```
1                                 ;PROGRAM TO SEARCH A GIVEN DATA IN AN ARRAY
2
3    4100                         ORG 4100H    ;specify program starting address.
4
5    4100  21 00 42               LXI H,4200H  ;Set pointer for the data array.
6    4103  3A 50 42               LDA 4250H    ;Load the given data in accumulator.
7    4106  06 00                  MVI B,00H    ;Clear B-register to store the position.
8    4108  04           LOOP:     INR B        ;Increment the position count.
9    4109  BE                     CMP M        ;Compare an element with given data.
10   410A  CA 1F 41               JZ  AHEAD    ;If data is available,then ZF=1.
11   410D  4F                     MOV C,A      ;Save the given data in C-register.
12   410E  23                     INX H
13   410F  7E                     MOV A,M      ;Get the next element of the array in A and
14   4110  FE 20                  CPI 20H      ;check for end of array.
15   4112  79                     MOV A,C      ;Get the given data in A-register.
16   4113  C2 08 41               JNZ LOOP     ;Repeat comparison until end of the array.
17   4116  01 00 00               LXI B,0000H  ;Clear B,C,H and L  if given data
18   4119  21 00 00               LXI H,0000H  ;is not available in the array.
19   411C  C3 21 41               JMP STORE
20   411F  0E FF        AHEAD:    MVI C,FFH    ;Move FFH to C, to indicate the
21                                             ;availability of data.
22   4121  22 53 42     STORE:    SHLD 4253H   ;Store the address of the data.
23   4124  69                     MOV L,C
24   4125  60                     MOV H,B
25   4126  22 51 42               SHLD 4251H   ;Store the position and indication
26                                             ;for availability.
27   4129  76                     HLT          ;Halt program execution.
28
29   412A                         END          ;Assembly end.
```

Sample Data

Input Data :

Array = 45_H
72_H
CA_H
$2F_H$
$C2_H$
$D1_H$
$4F_H$
20_H

Given Data = $2F_H$

Memory address	Content
4200	45
4201	72
4202	CA
4203	2F
4204	C2
4205	D1
4206	4F
4207	20

Output Data :

Availability = FF_H
Position = 04_H
Address = 4203_H

Memory address	Content
4250	2F
4251	FF
4252	04
4253	03
4254	42

EXAMPLE PROGRAM - 85.15 : Sorting an Array in Ascending Order

Write an assembly language program to sort an array of data in ascending order. The array is stored in memory starting from 4200_H*. The first element of the array gives the count value for the number of elements in the array.*

Problem Analysis

The algorithm for bubble sorting is given below. In bubble sorting of N-data, N−1 comparisons are carried by taking two consecutive data at a time. After each comparison, the data is rearranged such that smallest among the two is in the first memory location and the largest in the next memory location. (Here the data is rearranged within the two memory locations whose contents are compared). When we perform N−1 comparisons as mentioned above for N−1, times then the array consisting of N-data will be sorted in the ascending order.

Flowchart for Example Program 85.15

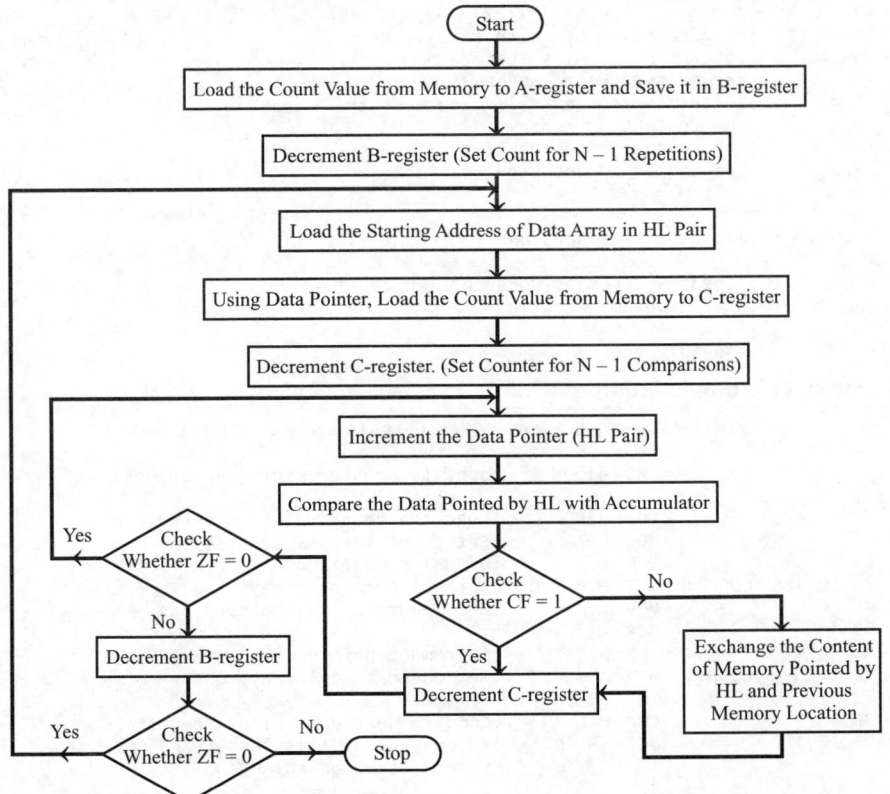

Algorithm

1. Load the count value from memory to A-register and save it in B-register.
2. Decrement B-register (B is counter for N−1 repetitions).
3. Set HL pair as data array address pointer.
4. Set C-register as counter for N−1 comparisons.
5. Load a data of the array in accumulator using the data address pointer.
6. Increment the HL pair (data address pointer).
7. Compare the data pointed by HL with accumulator.
8. If carry flag is set (If the content of accumulator is smaller than memory) then go to step 10, otherwise go to next step.
9. Exchange the content of memory pointed by HL and the accumulator.
10. Decrement C-register. If zero flag is reset go to step 6 otherwise go to next step.
11. Decrement B-register. If zero flag is reset go the step 3 otherwise go to next step.
12. Stop.

Assembly Language Program

```
;PROGRAM TO SORT AN ARRAY OF DATA IN ASCENDING ORDER

        ORG  4100H      ;specify program starting address.

        LDA  4200H      ;Load the count value in A-register.
        MOV  B,A        ;Set count for N-1 repetitions
        DCR  B          ;of N-1 comparisons.
LOOP2:  LXI  H,4200H    ;Set pointer for array.
        MOV  C,M        ;Set count for N-1 comparisons.
        DCR  C
        INX  H          ;Increment pointer.
LOOP1:  MOV  A,M        ;Get one data of array in A.
        INX  H
        CMP  M          ;Compare next data with A-register.
        JC   AHEAD      ;If content of A is less than
                        ;memory then go to AHEAD.
        MOV  D,M        ;If the content of A is greater than
        MOV  M,A        ;the content of memory,
        DCX  H          ;then exchange the content of memory
        MOV  M,D        ;pointed by HL and previous location.
        INX  H
AHEAD:  DCR  C
        JNZ  LOOP1      ;Repeat comparisons until C count is zero.
        DCR  B
        JNZ  LOOP2      ;Repeat N-1 comparisons until B count is zero.
        HLT             ;Halt program execution.

        END             ;Assembly end.
```

Assembler Listing for Example Program 85.15

```
1                               ;PROGRAM TO SORT AN ARRAY OF DATA IN ASCENDING ORDER
2
3   4100                            ORG  4100H      ;specify program starting address.
4
5   4100    3A 00 42                LDA  4200H      ;Load the count value in A-register.
6   4103    47                      MOV  B,A        ;Set count for N-1 repetitions
7   4104    05                      DCR  B          ;of N-1 comparisons.
8   4105    21 00 42    LOOP2:      LXI  H,4200H    ;Set pointer for array.
9   4108    4E                      MOV  C,M        ;Set count for N-1 comparisons.
10  4109    0D                      DCR  C
11  410A    23                      INX  H          ;Increment pointer.
12  410B    7E          LOOP1:      MOV  A,M        ;Get one data of array in A.
13  410C    23                      INX  H
14  410D    BE                      CMP  M          ;Compare next data with A-register.
15  410E    DA 16 41                JC   AHEAD      ;If content of A is less than
16                                                  ;memory  then go to AHEAD.
```

```
17   4111   56                    MOV  D,M      ;If the content of A  is greater than
18   4112   77                    MOV  M,A      ;the content of memory,
19   4113   2B                    DCX  H        ;then exchange the content of memory
20   4114   72                    MOV  M,D      ;pointed by HL and previous location.
21   4115   23                    INX  H
22   4116   0D        AHEAD:      DCR  C
23   4117   C2 0B 41              JNZ  LOOP1    ;Repeat comparisons until C count is zero.
24   411A   05                    DCR  B
25   411B   C2 05 41              JNZ  LOOP2    ;Repeat N-1 comparisons until B count is zero.
26   411E   76                    HLT           ;Halt program execution.
27
28   411F                         END           ;Assembly end.
```

Sample Data

Input Data: 07
AB
92
84
4F
69
F2
34

Memory address	Content
4200	07
4201	AB
4202	92
4203	84
4204	4F
4205	69
4206	F2
4207	34
(Before sorting)	

Output data: 07
34
69
84
92
AB
F2

Memory address	Content
4200	07
4201	34
4202	4F
4203	69
4204	84
4205	92
4206	AB
4207	F2
(After sorting)	

EXAMPLE PROGRAM - 85.16 : Sorting an Array in Descending Order

Write an assembly language program to sort an array of data in descending order. The array is stored in the memory location starting from 4200_H. The first element of the array gives the count value for the number of elements in the array.

Problem Analysis

The algorithm for bubble sorting is given below. In bubble sorting of N-data, N–1 comparisons are carried by taking two consecutive data at a time. After each comparison, the data is rearranged such that largest among the two is in first memory location and the smallest in the next memory location. (Here the data is rearranged within the two memory locations whose contents are compared). When we perform N –1 comparisons as mentioned above for N–1 times, then the array consisting of N-data will be sorted in descending order.

Algorithm

1. Load the count value from memory to A-register and save it in B-register.
2. Decrement B-register (B is counter for N–1 repetitions).
3. Set HL pair as data array address pointer.
4. Set C-register as counter for N–1 comparisons.
5. Load a data of the array in accumulator using the data address pointer.
6. Increment the HL pair (data address pointer).
7. Compare the data pointed by HL with accumulator.
8. If carry flag is reset (If the content of accumulator is larger than memory) then go to step 10, otherwise go to next step.
9. Exchange the content of memory pointed by HL and the accumulator.
10. Decrement C-register. If zero flag is reset go to step 6 otherwise go to next step.
11. Decrement B-register. If zero flag is reset go the step 3 otherwise go to next step.
12. Stop.

Flowchart for Example Program 85.16

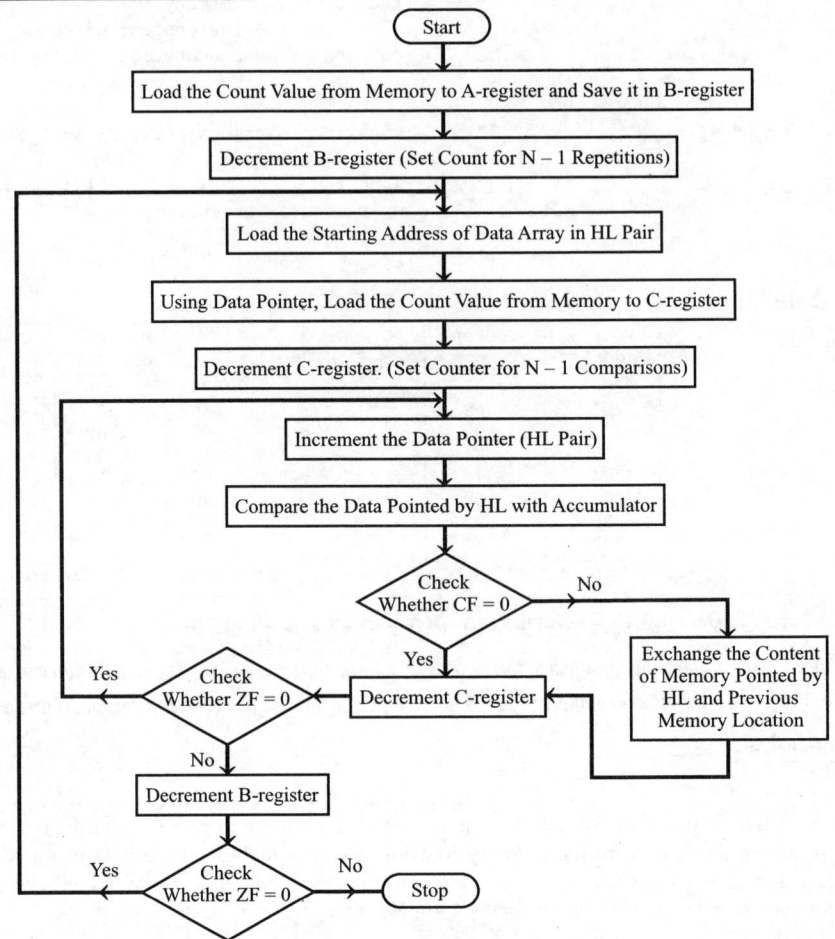

Assembly Language Program

```
;PROGRAM TO SORT AN ARRAY OF DATA IN DESCENDING ORDER

        ORG  4100H    ;specify program starting address.

        LDA  4200H    ;Load the count value in A-register.
        MOV  B,A      ;Set count for N-1 repetitions
        DCR  B        ;of N-1 comparisons.
LOOP2:  LXI  H,4200H  ;Set pointer for array.
        MOV  C,M      ;Set count for N-1 comparisons.
        DCR  C
        INX  H        ;Increment the pointer.
LOOP1:  MOV  A,M      ;Get one data of array in A.
        INX  H        ;Compare the next data of array with
        CMP  M        ;the content of A-register.
        JNC  AHEAD    ;If content of A is greater than content
                      ;of memory addressed by HL pair,
                      ;then go to AHEAD.
```

```
          MOV  D,M      ;If the content of A is less than content
          MOV  M,A      ;of memory addressed by HL pair,
          DCX  H        ;then exchange content of memory pointed
          MOV  M,D      ;by HL and previous memory location.
          INX  H
AHEAD:    DCR  C
          JNZ  LOOP1    ;Repeat comparisons until C count is zero.
          DCR  B
          JNZ  LOOP2    ;Repeat N-1 comparisons until B count is zero.
          HLT           ;Halt program execution.

          END           ;Assembly end.
```

Assembler Listing for Example Program 85.16

```
 1                              ;PROGRAM TO SORT AN ARRAY OF DATA IN DESCENDING ORDER
 2
 3   4100                       ORG  4100H    ;specify program starting address.
 4
 5   4100  3A 00 42             LDA  4200H    ;Load the count value in A-register.
 6   4103  47                   MOV  B,A      ;Set counter for N-1 repetitions
 7   4104  05                   DCR  B        ;of N-1 comparisons.
 8   4105  21 00 42   LOOP2:    LXI  H,4200H  ;Set pointer for array.
 9   4108  4E                   MOV  C,M      ;Set count for N-1 comparisons.
10   4109  0D                   DCR  C
11   410A  23                   INX  H        ;Increment the pointer.
12   410B  7E         LOOP1:    MOV  A,M      ;Get one data of array in A.
13   410C  23                   INX  H        ;Compare the next data of array with
14   410D  BE                   CMP  M        ;the content of A-register.
15   410E  D2 16 41             JNC  AHEAD    ;If content of A is greater than content
16                                            ;of memory addressed by HL pair,
17                                            ;then go to AHEAD.
18   4111  56                   MOV  D,M      ;If the content of A is less than content
19   4112  77                   MOV  M,A      ;of memory addressed by HL pair,
20   4113  2B                   DCX  H        ;then exchange content of memory pointed
21   4114  72                   MOV  M,D      ;by HL and previous memory location.
22   4115  23                   INX  H
23   4116  0D         AHEAD:    DCR  C
24   4117  C2 0B 41             JNZ  LOOP1    ;Repeat comparisons until C count is zero.
25   411A  05                   DCR  B
26   411B  C2 05 41             JNZ  LOOP2    ;Repeat N-1 comparison until B count is zero.
27   411E  76                   HLT           ;Halt program execution.
28
29   411F                       END           ;Assembly end.
```

Sample data

Input Data:	Memory address	Content	Output Data:		Memory address	Content
07	4200	07	07		4200	07
C4	4201	C4	F4		4201	F4
84	4202	84	E2		4202	E2
9A	4203	9A	C4		4203	C4
7B	4204	7B	B2		4204	B2
E2	4205	E2	9A		4205	9A
F4	4206	F4	84		4206	84
B2	4207	B2	7B		4207	7B
	(Before sorting)				(After sorting)	

EXAMPLE PROGRAM - 85.17 :   BCD to 7-Segment LED Code

Write an assembly language program to find the 7-segment LED code for a 2-digit BCD data, by using the look up table. The BCD data is stored in 4200_H. Store the 7-segment code in 4201_H and 4202_H.

Problem Analysis

The 7-segment LED codes for decimal digit 0 to 9 are determined and stored in memory locations from 5000_H to 5009_H respectively. The look-up table is created such that the low order address is same as that of decimal digit. Hence, by this method the high order address is fixed (50) and the low order address is the decimal digit itself.

In order to find the 7-segment code, the BCD data is split into lower nibble and upper nibble. The code is determined by taking each nibble as low order address of the **look up table.**

Flowchart for Example Program 85.17

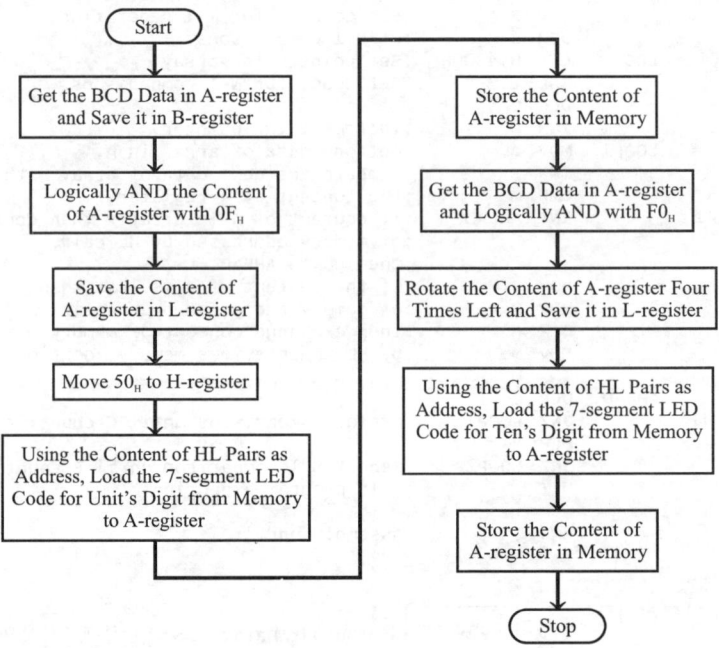

Algorithm

1. Load the BCD data in A-register and save in B-register.
2. Logically AND A-register with $0F_H$ to mask upper nibble (ten's digit).
3. Move A-register to L-register and move 50_H to H-register.
4. Get the LED code for lower nibble (unit's digit) in A-register and store in memory.
5. Move the BCD data from B-register to A-register and mask the lower nibble (unit's digit).
6. Rotate the upper nibble to lower nibble position.
7. Move A-register to L-register.
8. Get the LED code for ten's digit in A-register and store in memory.
9. Stop.

Assembly Language Program

```
;PROGRAM TO FIND THE 7-SEGMENT LED CODE FOR A BCD DATA

        ORG  4100H  ;specify program starting address.

        LDA  4200H  ;Get BCD data in A and save in B.
        MOV  B,A
        ANI  0FH    ;Mask the upper nibble (ten's digit).
        MOV  L,A    ;Get memory address of LED code
        MVI  H,50H  ;for unit's digit, in HL pair.
        MOV  A,M    ;Get LED code for unit's digit in A
        STA  4201H  ;and store in memory.
        MOV  A,B    ;Get the BCD data in A-register and
        ANI  F0H    ;mask the lower nibble (unit's digit).
        RLC         ;Rotate upper nibble to
        RLC         ;lower nibble position.
        RLC
        RLC
        MOV  L,A    ;Get memory address of LED code
                    ;for ten's digit in HL pair.
        MOV  A,M    ;Get LED code for ten's digit in A
        STA  4202H  ;and store in memory.
        HLT         ;Halt program execution.

        END         ;Assembly end.
```

Assembler Listing for Example Program 85.17

```
 1                      ;PROGRAM TO FIND THE 7-SEGMENT LED CODE FOR A BCD DATA
 2
 3   4100                     ORG  4100H  ;specify program starting address.
 4
 5   4100   3A 00 42          LDA  4200H  ;Get BCD data in A and save in B.
 6   4103   47                MOV  B,A
 7   4104   E6 0F             ANI  0FH    ;Mask the upper nibble (ten's digit).
 8   4106   6F                MOV  L,A    ;Get memory address of LED code
 9   4107   26 50             MVI  H,50H  ;for unit's digit, in HL pair.
10   4109   7E                MOV  A,M    ;Get LED code for unit's digit in A
11   410A   32 01 42          STA  4201H  ;and store in memory.
12   410D   78                MOV  A,B    ;Get the BCD data in A-register and
13   410E   E6 F0             ANI  F0H    ;mask the lower nibble (unit's digit).
14   4110   07                RLC         ;Rotate upper nibble to
15   4111   07                RLC         ;lower nibble position.
16   4112   07                RLC
17   4113   07                RLC
18   4114   6F                MOV  L,A    ;Get memory address of LED code
19                                        ;for ten's digit in HL pair.
20   4115   7E                MOV  A,M    ;Get LED code for ten's digit in A
21   4116   32 02 42          STA  4202H  ;and store in memory.
22   4119   76                HLT         ;Halt program execution.
23
24   411A                     END         ;Assembly end.
```

Sample Data 1 :

Loop-up table for common cathode 7-segment LED

```
Input Data  :  45₁₀
Output Data :  6D_H
               66_H
```

Memory address	Content
5000	3F
5001	06
5002	5B
5003	4F
5004	66

Memory address	Content
5005	6D
5006	7D
5007	07
5008	7F
5009	6F

Memory address	Content
4200	45
4201	6D
4202	66

Sample Data 2 :

Loop-up table for common anode 7-segment LED

Memory address	Content
5000	C0
5001	F9
5002	A4
5003	B0
5004	99

Memory address	Content
5005	92
5006	82
5007	F8
5008	80
5009	90

Input Data   : 45_{10}
Output Data : 92_H
                     99_H

Memory address	Content
4200	45
4201	92
4202	99

EXAMPLE PROGRAM - 85.18 :   Square Root of 8-Bit Binary Number

Write an assembly language program to find the square root of an 8-bit binary number. The binary number is stored in memory location 4200_H and store the square root in 4201_H.

Problem Analysis

Square root can be computed by an iterative technique. First an initial value is assumed. Here the initial value of square root is taken as half the value of given number. The new value of square root is computed by using an expression, XNEW = (X + Y/X)/2 where X is the initial value of square root and Y is the given number. Then XNEW is compared with initial value. If they are not equal then the above process is repeated until X is equal to XNEW after taking XNEW as initial value, (i.e., X ← XNEW).

Algorithm

1. Load the given data (Y) in A-register.
2. Save the content of A-register in B-register.
3. Move 02_H (divisor) to C-register.
4. Call DIV subroutine to get initial value of square root (X) in D-register.
5. Save the content of D-register (initial value X) in E-register.
6. Move the given data (Y) from B-register to A-register.
7. Move the initial value (X) from D-register to C-register.
8. Call DIV subroutine to get Y/X in D-register.
9. Move the Y/X available in D-register to A-register.
10. Add the value of X in E-register to A-register to get X+Y/X in A-register.
11. Move 02_H to C-register.
12. Call DIV subroutine to get new value of square root (XNEW) in D-register.
13. Compare X and XNEW.
14. If ZF = 1, go to next step. If ZF = 0, go to step 5.
15. Store the value of square root (A-register) in memory.
16. Stop.

Algorithm for Subroutine DIV

1. Clear D-register.
2. Subtract the content of C-register (divisor) from the content of A-register (dividend).
3. Increment quotient (D-register).
4. Compare A-register and C-register.
5. If CF = 1, go to next step. If CF = 0 go to step 2.
6. Return to main program.

Flowchart for Example Program 85.18

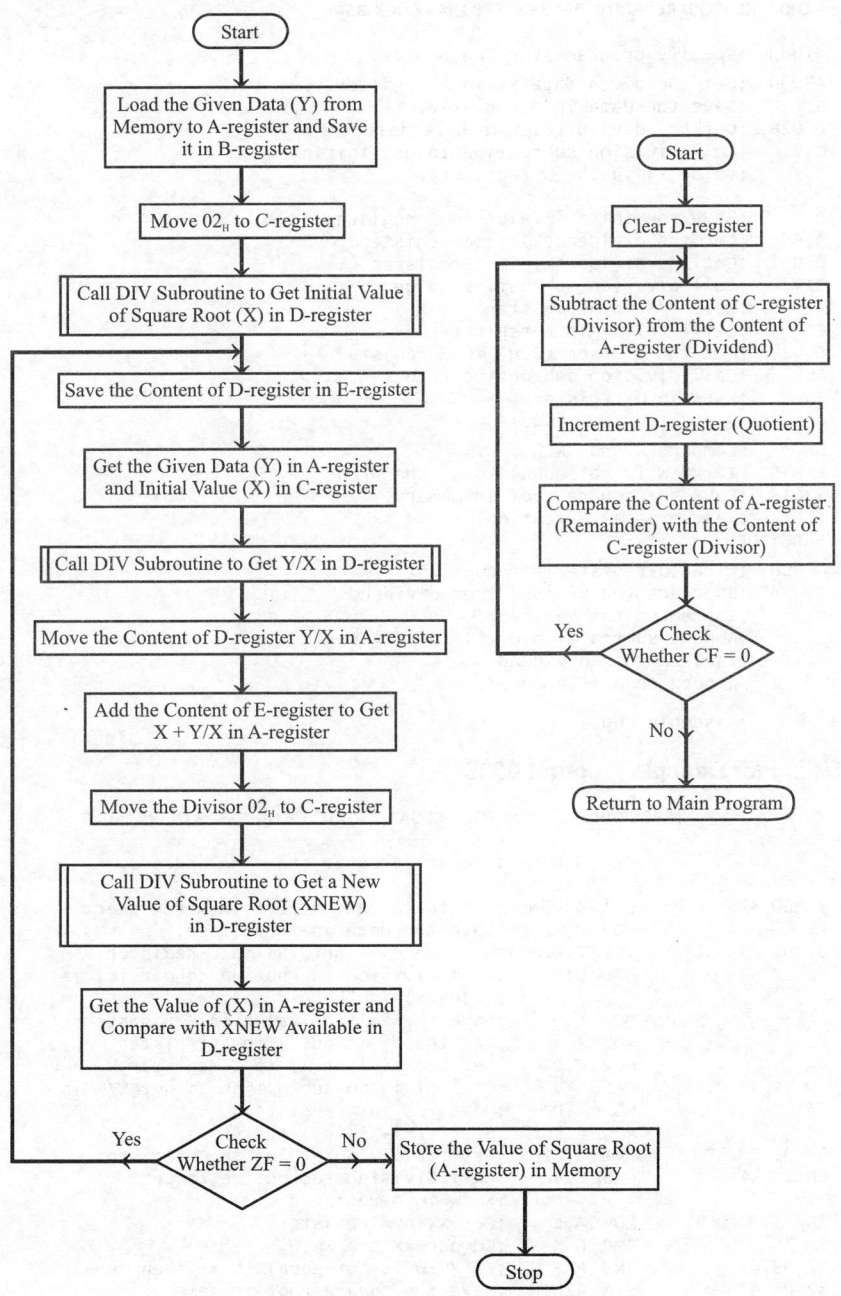

Assembly Language Program

```
;PROGRAM TO FIND THE SQUARE ROOT OF 8-BIT BINARY NUMBER

        ORG   4100H   ;specify program starting address.
        LDA   4200H   ;Get the given data(Y) in A-register.
        MOV   B,A     ;Save the data in B-register.
        MVI   C,02H   ;Get the divisor(02ₕ) in C-register.
        CALL  DIV     ;Call division subroutine to get initial
                      ;value(X) in the D-register.

REP:    MOV   E,D     ;Save the initial value in E-register.
        MOV   A,B     ;Get the dividend(Y) in A-register.
        MOV   C,D     ;Get the divisor(X) in C-register.
        CALL  DIV     ;Call division subroutine to get Y/X in D.
        MOV   A,D     ;Move Y/X in A-register.
        ADD   E       ;Get((Y/X)+X) in A-register.
        MVI   C,02H   ;Get the divisor (02H) in C-register.
        CALL  DIV     ;Call division subroutine to get
                      ;XNEW in D-register.

        MOV   A,E     ;Get X in A-register.
        CMP   D       ;Compare X and XNEW.
        JNZ   REP     ;If XNEW is not equal to X, then repeat.
        STA   4201H   ;Save the square root in memory.
        HLT           ;Halt program execution.
;DIVISION SUBROUTINE
DIV:    MVI   D,00H   ;Clear D-register for quotient.
NEXT:   SUB   C       ;Subtract the divisor from dividend.
        INR   D       ;Increment the quotient.
        CMP   C       ;Repeat subtraction until the divisor
        JNC   NEXT    ;is less than dividend.
        RET           ;Return to main program.

        END           ;Assembly end.
```

Assembler Listing for Example Program 85.18

```
 1                              ;PROGRAM TO FIND THE SQUARE ROOT OF 8-BIT BINARY NUMBER
 2
 3    4100                          ORG   4100H   ;specify program starting address.
 4
 5    4100   3A 00 42               LDA   4200H   ;Get the given data(Y) in A-register.
 6    4103   47                     MOV   B,A     ;Save the data in B-register.
 7    4104   0E 02                  MVI   C,02H   ;Get the divisor(02H) in C-register.
 8    4106   CD 1F 41               CALL  DIV     ;Call division subroutine to get initial
 9                                                ;value(X) in the D-register.
10    4109   5A          REP:  MOV   E,D     ;Save the initial value in E-register.
11    410A   78                     MOV   A,B     ;Get the dividend(Y) in A-register.
12    410B   4A                     MOV   C,D     ;Get the divisor(X) in C-register.
13    410C   CD 1F 41               CALL  DIV     ;Call division subroutine to get Y/X in D.
14    410F   7A                     MOV   A,D     ;Move Y/X in A-register.
15    4110   83                     ADD   E       ;Get((Y/X)+X) in A-register.
16    4111   0E 02                  MVI   C,02H   ;Get the divisor (02H) in C-register.
17    4113   CD 1F 41               CALL  DIV     ;Call division subroutine to get
18                                                ;XNEW in D-register
19    4116   7B                     MOV   A,E     ;Get X in A-register.
20    4117   BA                     CMP   D       ;Compare X and XNEW.
21    4118   C2 09 41               JNZ   REP     ;If XNEW is not equal to X, then repeat.
22    411B   32 01 42               STA   4201H   ;Save the square root in memory.
23    411E   76                     HLT           ;Halt program execution.
24
25
26                              ;DIVISION SUBROUTINE
27
```

```
28   411F   16 00      DIV:  MVI D,00H   ;Clear D-register for quotient.
29   4121   91         NEXT: SUB C       ;Subtract the divisor from dividend.
30   4122   14               INR D       ;Increment the quotient.
31   4123   B9               CMP C       ;Repeat subtraction until the divisor
32   4124   D2 21 41         JNC NEXT    ;is less than dividend.
33   4127   C9               RET         ;Return to main program.
34
35   4128                    END         ;Assembly end.
```

Sample Data

Input Data : 64_H

Output Data : $0A_H$

Memory address	Content
4200	64
4201	0A

EXAMPLE PROGRAM - 85.19 : Binary to ASCII Conversion

Write an assembly language program to convert an 8-bit binary (2-digit hexa) to ASCII code. The binary data is stored in 4200_H *and store the ASCII code in* 4201_H *and* 4202_H.

Problem Analysis

Each Hexa digit (4-bit binary) is represented by an 8-bit ASCII. The Hexa digit 0 through 9 are represented by 30_H to 39_H in ASCII. Hence, for Hexa 0 to 9, if we add 30_H, we will get the corresponding ASCII.

The Hexa digit A through F are represented by 41_H to 46_H in ASCII. Hence, for Hexa digit A to F if we add 37_H we will get the corresponding ASCII.

In the following algorithm the given 8-bit data is split into two nibbles. The ASCII code for each nibble is found by calling a subroutine, which takes care of adding 30_H to the nibble if it is less than $0A_H$, or adding 37_H if the nibble is greater than 09_H.

Algorithm

1. Load the given data in A-register and move to B-register.
2. Mask the upper nibble of the binary (hexa) data in A-register.
3. Call subroutine ACODE to get ASCII code of the lower nibble and store in memory.
4. Move B-register to A-register and mask the lower nibble.
5. Rotate the upper nibble to lower nibble position.
6. Call subroutine ACODE to get the ASCII code of upper nibble and store in memory.
7. Stop.

Algorithm for Subroutine Code

1. Compare the content of A-register with $0A_H$.
2. If CF = 1, go to step 4. If CF = 0, go to next step.
3. Add 07_H to A-register.
4. Add 30_H to A-register.
5. Return to main program.

Flowchart for Example Program 85.19

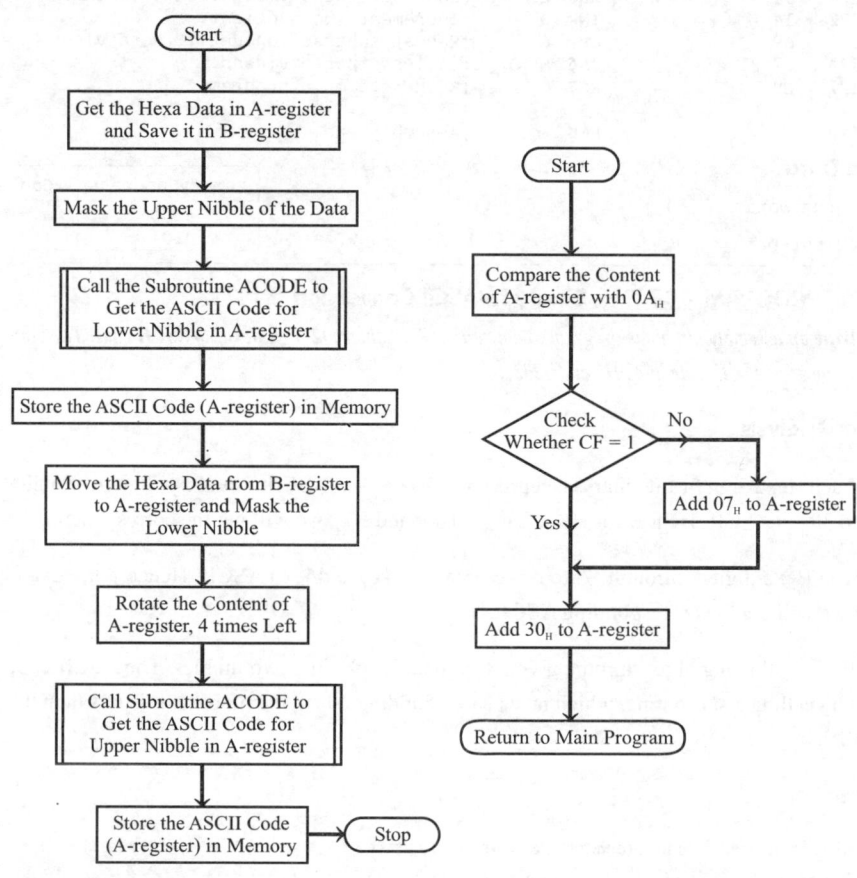

Assembly Language Program

```
;PROGRAM TO CONVERT 8-BIT BINARY TO ASCII CODE

        ORG 4100H ;specify program starting address.

        LDA 4200H ;Get binary data in A.
        MOV B,A   ;Save the binary data in B-register.
        ANI 0FH   ;Mask the upper nibble.
        CALL ACODE ;Call subroutine to get ASCII code for
        STA 4201H ;lower nibble in A and store in memory.
        MOV A,B   ;Get data in A-register.
        ANI F0H   ;Mask the lower nibble.
        RLC       ;Rotate upper nibble to
        RLC       ;lower nibble position.
        RLC
        RLC
        CALL ACODE ;Call subroutine to get ASCII code for
        STA 4202H ;upper nibble in A and store in memory.
        HLT       ;Halt program execution.
;SUBROUTINE ACODE
ACODE:  CPI 0AH   ;If the content of A is less than 0AH,
        JC  SKIP  ;then add 30H to A  otherwise
```

```
        ADI   07H    ;add 37H to A-register.
SKIP:   ADI   30H
        RET          ;Return to main program.
        END          ;Assembly end.
```

Assembler Listing for Example program 85.19

```
1                             ;PROGRAM TO CONVERT 8-BIT BINARY TO ASCII CODE
2
3    4100                     ORG   4100H ;specify program starting address.
4
5    4100  3A 00 42           LDA   4200H ;Get binary data in A.
6    4103  47                 MOV   B,A   ;Save the binary data in B-register.
7    4104  E6 0F              ANI   0FH   ;Mask the upper nibble.
8    4106  CD 1A 41           CALL  ACODE ;Call subroutine to get ASCII code for
9    4109  32 01 42           STA   4201H ;lower nibble in A and store in memory.
10   410C  78                 MOV   A,B   ;Get data in A-register.
11   410D  E6 F0              ANI   F0H   ;Mask the lower nibble.
12   410F  07                 RLC         ;Rotate upper nibble to
13   4110  07                 RLC         ;lower nibble position.
14   4111  07                 RLC
15   4112  07                 RLC
16   4113  CD 1A 41           CALL  ACODE ;Call subroutine  to get ASCII code for
17   4116  32 02 42           STA   4202H ;upper nibble in A and store in memory.
18   4119  76                 HLT         ;Halt program execution.
19
20
21                           ;SUBROUTINE ACODE
22   411A
23   411A  FE 09       ACODE: CPI   0AH   ;If the content of A is less than 0AH,
24   411C  DA 21 41           JC    SKIP  ;then add 30H to A otherwise
25   411F  C6 07              ADI   07H   ;add 37H to A-register.
26   4121  C6 30       SKIP:  ADI   30H
27   4123  C9                 RET         ;Return to main program.
28
29   4124                     END         ;Assembly end.
```

Sample Data

Input Data   : $E4_H$

Output Data : 34 (ASCII code for 4)

45 (ASCII code for E)

Memory address	Content
4200	E4
4201	34
4202	45

EXAMPLE PROGRAM - 85.20 : ASCII to Binary Conversion

Write an assembly language program to convert an array of ASCII codes to corresponding binary (hexa) value. The ASCII array is stored starting from 4200ₕ. The first element of the array gives the number of elements in the array.

Problem Analysis

The hexa digit 0 through 9 are represented by 30_H to 39_H in ASCII. Hence, for ASCII code 30_H to 39_H if we subtract 30_H then we will get the corresponding binary (hexa) value. The hexa digit A through F are represented by 41_H to 46_H in ASCII. Hence for ASCII code 41_H to 46_H we have to subtract 37_H to get corresponding binary (hexa) value. In the following algorithm, a subroutine has been written to subtract either 30_H or 37_H from the given data.

Flowchart for Example Program 85.20

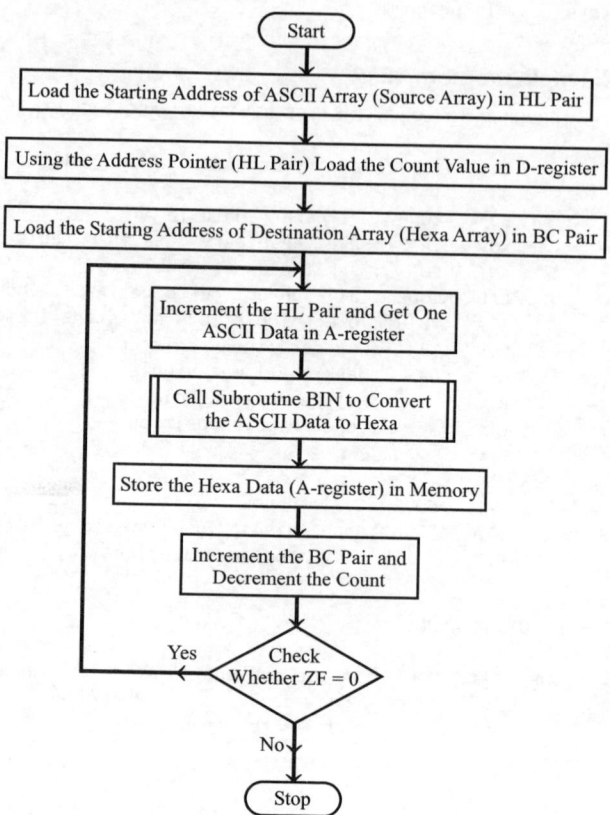

Flowchart for Subroutine BIN

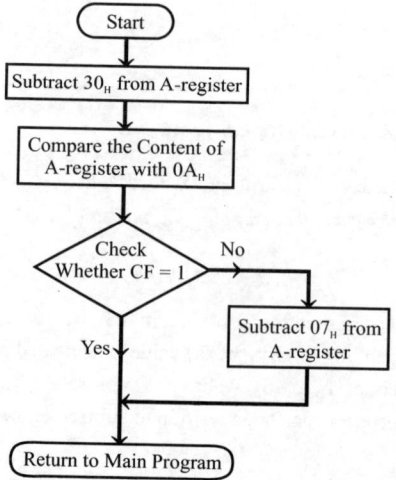

Algorithm

1. Set HL pair as pointer for ASCII array.
2. Set D-register as count for number of data in the array.
3. Set BC pair as pointer for binary (hexa) array.
4. Increment HL pair and move a data of ASCII array to A-register.
5. Call subroutine BIN to find the binary (hexa) value.
6. The binary (hexa) value available in A-register is stored in memory.
7. Increment BC pair.
8. Decrement D-register. If ZF = 0, then go to step 4. If ZF = 1, then stop.

Algorithm for Subroutine BIN

1. Subtract 30_H from A-register.
2. Compare the content of A-register with $0A_H$.
3. If CF = 1, go to step 5. If CF = 0, go to next step.
4. Subtract 07_H from A-register.
5. Return to main program.

Assembly Language Program

```
;PROGRAM TO CONVERT ASCII CODE TO BINARY VALUE

        ORG  4100H    ;specify program starting address.

        LXI  H,4200H  ;Set pointer for ASCII array.
        MOV  D,M      ;Set count for number of data.
        LXI  B,4300H  ;Set pointer for binary(hexa) array.
LOOP:   INX  H
        MOV  A,M      ;Get an ASCII data in A-register.
        CALL BIN      ;Call subroutine to get binary
        STAX B        ;value in A and store in memory.
        INX  B        ;Increment the binary array pointer.
        DCR  D
        JNZ  LOOP     ;Repeat conversion until count is zero.
        HLT           ;Halt program execution.


;SUBROUTINE BIN
BIN:    SUI  30H      ;Subtract 30H from the data.
        CPI  0AH
        RC            ;If CF=1, Return to main program.
        SUI  07H      ;If data is greater than 0AH, then subtract
        RET           ;07H and return to main program.

        END           ;Assembly end.
```

Assembler Listing for Example Program 85.20

```
1                         ;PROGRAM TO CONVERT ASCII CODE TO BINARY VALUE
2
3    4100                 ORG  4100H    ;specify program starting address.
4
5    4100  21 00 42       LXI  H,4200H  ;Set pointer for ASCII array.
6    4103  56             MOV  D,M      ;Set count for number of data.
7    4104  01 00 43       LXI  B,4300H  ;Set pointer for binary(hexa) array.
8    4107  23       LOOP: INX  H
9    4108  7E             MOV  A,M      ;Get an ASCII data in A-register.
10   4109  CD 13 41       CALL BIN      ;Call subroutine to get binary
11   410C  02             STAX B        ;value in A and store in memory.
12   410D  03             INX  B        ;Increment the binary array pointer.
13   410E  15             DCR  D
14   410F  C2 07 41       JNZ  LOOP     ;Repeat conversion until count is zero.
15   4112  76             HLT           ;Halt program execution.
```

```
16    4113
17
18                        ;SUBROUTINE BIN
19
20    4113  D6 30    BIN:   SUI  30H    ;Subtract 30H from the data.
21    4115  FE 0A           CPI  0AH
22    4117  D8              RC           ;If CF = 1, Return to main program.
23    4118  D6 07           SUI  07H     ;If data is greater than 0AH then subtract
24    411A  C9              RET          ;07H and return to main program.
25
26    411B                  END          ;Assembly end.
```

Sample Data

Input Data : Count : 07 ASCII Array: 31 42 35 46 43 39 38		
Memory address	Content	
4200	07	
4201	31	
4202	42	
4203	35	
4204	46	
4205	43	
4206	39	
4207	38	

Output Data : Binary array= 01 0B 05 0F 0C 09 08		
Memory address	Content	
4300	01	
4301	0B	
4302	05	
4303	0F	
4304	0C	
4305	09	
4306	08	

EXAMPLE PROGRAM - 85.21 : BCD to Binary Conversion

Write an assembly language program to convert a two-digit BCD (8-bit) data to binary data. The BCD data is stored in 4200_H and store the binary data in 4201_H.

Problem Analysis

The 2-digit BCD data will have units digit and tens digit. When the tens digit (upper nibble) is multiplied by $0A_H$ and the product is added to units digit (lower nibble), the result will be in binary, because the microprocessor performs binary arithmetic.

Algorithm

1. Get the BCD data in A-register and save in stack.
2. Mask the lower nibble (units) of the BCD data in A-register.
3. Rotate the upper nibble to lower nibble position and save in B-register.
4. Clear the accumulator.
5. Move $0A_H$ to C-register.
6. Add B-register to A-register.
7. Decrement C-register. If ZF = 0 go to step 6. If ZF = 1, go to next step.
8. Save the product in B-register.
9. Get the BCD data from stack in A-register and mask the upper nibble (tens).
10. Add the units (A-register) to product (B-register).
11. Store the binary value (A-register).
12. Stop.

Flowchart for Example Program 85.21

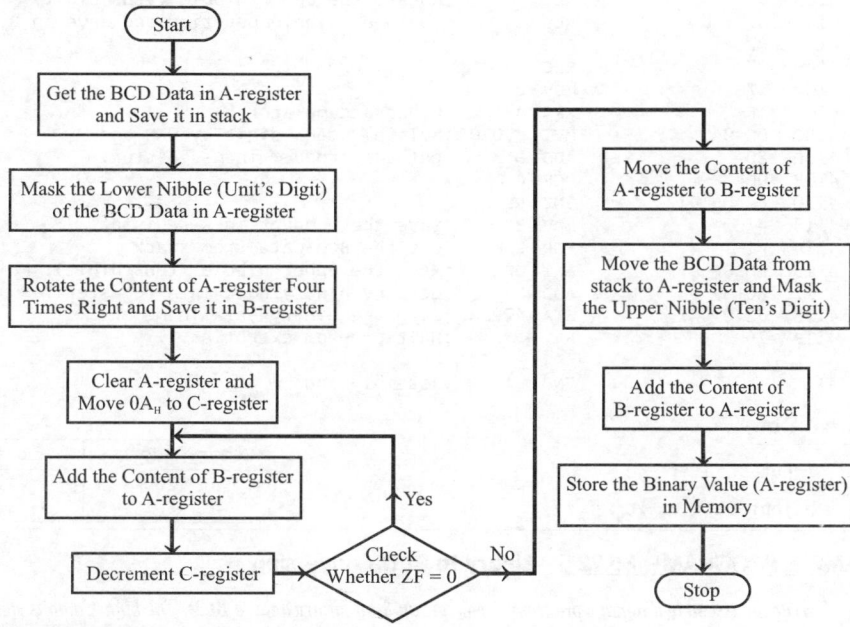

Assembly Language Program

```
;PROGRAM TO CONVERT 2-DIGIT BCD TO BINARY NUMBER

        ORG  4100H   ;specify program starting address.

        LDA  4200H   ;Get the data in A-register,
        PUSH PSW     ;save data in Stack.
        ANI  F0H     ;Mask the lower nibble (units digit).
        RLC          ;Rotate the upper nibble (tens digit)
        RLC          ;to lower nibble position and save in B.
        RLC
        RLC
        MOV  B,A
        XRA  A       ;Clear accumulator.
        MVI  C,0AH   ;Multiply tens digit by 0AH and
REP:    ADD  B       ;get the product in A-register.
        DCR  C
        JNZ  REP
        MOV  B,A     ;Save the product in B-register.
        POP  PSW     ;Get the BCD data from stack.
        ANI  0FH     ;Mask the upper nibble (tens digit).
        ADD  B       ;Get the binary data in A-register.
        STA  4201H   ;Save the binary data in memory.
        HLT          ;Halt program execution.

        END          ;Assembly end.
```

Assembler Listing for Example Program 85.21

```
1                         ;PROGRAM TO CONVERT 2-DIGIT BCD TO BINARY NUMBER
2
3   4100                  ORG  4100H   ;specify program starting address.
4
5   4100   3A 00 42       LDA  4200H   ;Get the data in A-register,
6   4103   F5             PUSH PSW     ;save data in stack.
```

7	4104	E6 F0		ANI F0H	;Mask the lower nibble (units digit).
8	4106	07		RLC	;Rotate the upper nibble (tens digit)
9	4107	07		RLC	;to lower nibble position and save in B.
10	4108	07		RLC	
11	4109	07		RLC	
12	410A	47		MOV B,A	
13	410B	AF		XRA A	;Clear accumulator.
14	410C	0E 0A		MVI C,0AH	;Multiply tens digit by 0AH and
15	410E	80	REP:	ADD B	;get the product in A-register.
16	410F	0D		DCR C	
17	4110	C2 0E 41		JNZ REP	
18	4113	47		MOV B,A	;Save the product in B-register.
19	4114	F1		POP PSW	;Get the BCD data from stack.
20	4115	E6 0F		ANI 0FH	;Mask the upper nibble (tens digit).
21	4117	80		ADD B	;Get the binary data in A-register.
22	4118	32 01 42		STA 4201H	;Save the binary data in memory.
23	411B	76		HLT	;Halt program execution.
24					
25	411C			END	;Assembly end.

Sample Data

Input Data   : 45_{10}

Output Data : $2D_H$

Memory address	Content
4200	45
4201	2D

EXAMPLE PROGRAM - 85.22 :   Binary to BCD Conversion

Write an assembly language program to convert an 8-bit binary data to BCD. The binary data is stored in 4200_H. Store the hundred's digit in 4251_H. Store the ten's and unit's digits in 4250_H.

Problem Analysis

The maximum value of 8-bit binary is $FF_H = 256_{10}$. Hence the maximum size of the data will have hundreds, tens and units. The algorithm given below uses two counters to count hundreds and tens. Initially counters are cleared. First let us subtract all hundreds from the binary data. For each subtraction, hundred's register is incremented by one. Then, let us subtract all tens. For each subtraction, ten's register is incremented by one. The remaining will be units. The tens and units are combined to form 2-digit BCD (8-bit binary).

Algorithm

1. Clear D and E registers to account for hundreds and tens.

2. Load the binary data in A-register.

3. Compare A-register with 64_H. If carry flag is set, go to step 7 otherwise go to next step.

4. Subtract 64_H from A-register.

5. Increment E-register (Hundred's register).

6. Go to step 3.

7. Compare the A-register with $0A_H$. If carry flag is set, go to step 11, otherwise go to next step.

8. Subtract $0A_H$ from A-register.

9. Increment D-register (ten's register).

10. Go to step 7.

11. Combine the units and tens to form 8-bit result.

12. Save the units, tens and hundreds in memory.

Flowchart for Example Program 85.22

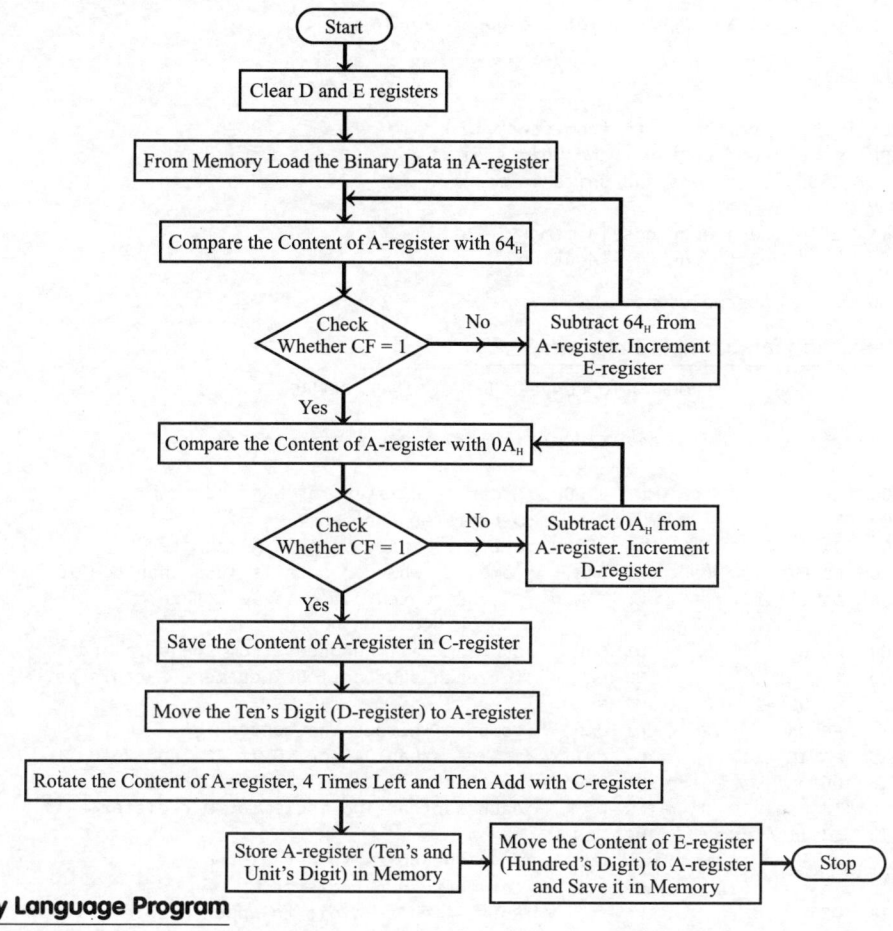

Assembly Language Program

```
;PROGRAM TO CONVERT 8-BIT BINARY NUMBER TO BCD

        ORG  4100H  ;specify program starting address.

        MVI  E,00H  ;Clear E-register for hundreds and
        MOV  D,E    ;D-register for tens.
        LDA  4200H  ;Get the binary data in A-register.
HUND:   CPI  64H    ;Compare, whether data is less than 64H(100).
        JC   TEN     ;If the content of A is less than
                     ;100 or 64H then go to TEN.
        SUI  64H    ;Subtract all hundreds from the data and
        INR  E       ;for each subtraction increment E-register.
        JMP  HUND
TEN:    CPI  0AH    ;Compare whether the content of A
        JC   UNIT    ;is less than 0AH or 10.If CF=1 go to UNIT.
        SUI  0AH    ;Subtract all tens from the data and for
        INR  D       ;each subtraction increment D-register.
        JMP  TEN
```

```
UNIT:   PUSH PSW    ;Save the units in stack.
        MOV A,D     ;Get tens in A-register.
        RLC         ;Rotate ten's digit to upper nibble position.
        RLC
        RLC
        RLC
        POP B       ;get the units from stack in B.
        ADD B       ;Combine ten's and unit's digits.
        STA 4250H   ;Save tens and units in memory.
        MOV A,E
        STA 4251H   ;Save hundreds in memory.
        HLT         ;Halt program execution.

        END         ;Assembly end.
```

Assembler Listing for Example Program 85.22

```
1                           ;PROGRAM TO CONVERT 8-BIT BINARY NUMBER TO BCD
2
3    4100                   ORG 4100H  ;specify program starting address.
4
5    4100  1E 00            MVI E,00H  ;Clear E-register for hundreds and
6    4102  53               MOV D,E    ;D-register for tens.
7    4103  3A 00 42         LDA 4200H  ;Get the binary data in A-register.
8    4106  FE 64     HUND:  CPI 64H    ;Compare, whether data is less than 64H(100).
9    4108  DA 11 41         JC  TEN    ;If the content of A is less than
10                                     ;100 or 64H then go to TEN.
11   410B  D6 64            SUI 64H    ;Subtract all hundreds from the data and
12   410D  1C               INR E      ;for each subtraction increment E-register.
13   410E  C3 06 41         JMP HUND
14   4111  FE 0A     TEN:   CPI 0AH    ;Compare whether the content of A
15   4113  DA 1C 41         JC  UNIT   ;is less than 0AH or 10.If CF=1 go to UNIT.
16   4116  D6 0A            SUI 0AH    ;Subtract all tens from the data and for
17   4118  14               INR D      ;each subtraction increment D-register.
18   4119  C3 11 41         JMP TEN
19   411C  4F        UNIT:  MOV C,A    ;Save the units in C-register.
20   411D  7A               MOV A,D    ;Get tens in A-register.
21   411E  07               RLC        ;Rotate ten's digit to upper nibble position.
22   411F  07               RLC
23   4120  07               RLC
24   4121  07               RLC
25   4122  c1               POP B      ; get the units from stack in B.
26   4123  80               ADD B      ;Combine ten's and unit's digits.
27   4124  32 50 42         STA 4250H  ;Save tens and units in memory.
28   4127  7B               MOV A,E
29   4128  32 51 42         STA        4251H ;Save hundreds in memory.
30   412B  76               HLT        ;Halt program execution.
31
32   412C                   END        ;Assembly end.
```

Sample Data

Input Data : $B9_H$

Output Data : 0185_{10}

Memory address	Content	
4200	B9	Binary data
4250	85	BCD data
4251	01	

8.13 EXAMPLES OF 8051 ASSEMBLY LANGUAGE PROGRAMS

EXAMPLE PROGRAM 51.1: 8-Bit Addition

Write an assembly language program to add two numbers of 8-bit data stored in memory 2400_H and 2401_H and store the result in 2402_H and 2403_H.

Problem Analysis

In order to perform addition in 8051, one of the data should be in the accumulator and the other data can be in any SFR/internal RAM or can be an immediate data. After addition, the sum is stored in the accumulator. The sum of a two 8-bit data can be either 8 bits (sum only) or 9 bits (sum and carry). The accumulator can accommodate only the sum and if there is carry, the 8051 will indicate by setting the carry flag. Hence, one of the internal registers/RAM location is used to account for carry.

Algorithm

1.   Set DPTR as pointer for data (load address of data in DPTR).
2.   Move first data from external memory to accumulator and save it in R1-register.
3.   Increment DPTR.
4.   Move second data from external memory to accumulator.
5.   Clear R0-register to account for carry.
6.   Add the content of R1-register to accumulator.
7.   Check for carry. If carry is not set go to step 8, otherwise go to next step.
8.   Increment R0-register.
9.   Increment DPTR and save the sum(accumulator) in external memory.
10.   Increment DPTR, move carry to accumulator and save it in external memory.
11.   Stop.

Flowchart for Example Program 51.1

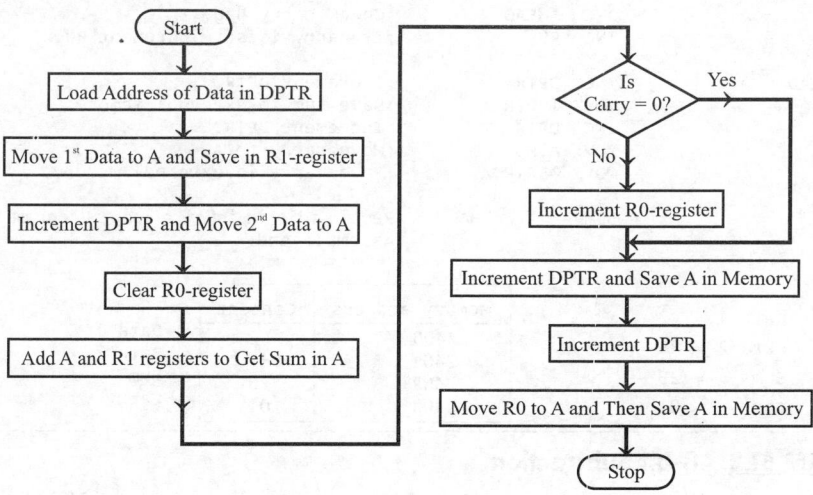

Assembly Language Program

```
;PROGRAM TO ADD TWO 8-BIT DATA
        ORG   2100H              ;specify program starting address.
        MOV   DPTR,#2400H        ;Load address of 1st data in DPTR.
        MOVX  A,@DPTR            ;Move the 1st data to A.
        MOV   R1,A               ;Save the first data in R1.
        INC   DPTR               ;Increment DPTR to point 2nd data.
        MOVX  A,@DPTR            ;Load 2nd data in A.
        MOV   R0,#00H            ;Clear R0 to account for carry.
        ADD   A,R1               ;Get sum of data in A.
        JNC   AHEAD              ;Check carry flag.
        INC   R0                 ;If carry is set increment R0.
AHEAD:  INC   DPTR               ;Increment DPTR.
        MOVX  @DPTR,A            ;Save sum in external memory.
        INC   DPTR               ;Increment DPTR.
        MOV   A,R0               ;Move carry to A.
        MOVX  @DPTR,A            ;Save carry in external memory.
HALT:   SJMP  HALT              ;Remain idle in infinite loop. Program end.
        END                     ;Assembly end.
```

Assembler Listing for Example Program 51.1

```
 1                                               ;PROGRAM TO ADD TWO 8-BIT DATA
 2
 3   2100                   ORG   2100H           ;specify program starting address.
 4   2100   90 24 00        MOV   DPTR,#2400H     ;Load address of 1st data in DPTR.
 5   2103   E0                    MOVX  A,@DPTR   ;Move the 1st data to A.
 6   2104   F9                    MOV   R1,A      ;Save the first data in R1.
 7   2105   A3                    INC   DPTR      ;Increment DPTR to point 2nd data.
 8   2106   E0                    MOVX  A,@DPTR   ;Load 2nd data in A.
 9   2107   78 00                 MOV   R0,#00H   ;Clear R0 to account for carry.
10   2109   29                    ADD   A,R1      ;Get sum of data in A.
11
12   210A   50 01                 JNC   AHEAD     ;Check carry flag.
13   210C   08                    INC   R0        ;If carry is set increment R0.
14
15   210D   A3    AHEAD:          INC   DPTR      ;Increment DPTR.
16   210E   F0                    MOVX  @DPTR,A   ;Save sum in external memory.
17   210F   A3                    INC   DPTR      ;Increment DPTR.
18   2110   E8                    MOV   A,R0      ;Move carry to A.
19   2111   F0                    MOVX  @DPTR,A   ;Save carry in external memory.
20
21   2112   80 FE HALT:           SJMP       HALT ;Remain idle in infinite loop. Program end.
22   2114         END                             ;Assembly end.
```

Sample Data

```
Input Data :   Data-1 = F2_H
               Data-2 = 34_H
Output Data:   Sum    = 26_H
               Carry  = 01_H
```

Memory address	Content	
2400	F2	← Data 1
2401	34	← Data 2
2402	26	← Sum
2403	01	← Carry

EXAMPLE PROGRAM 51.2: 8-bit Subtraction

Write an assembly language program to subtract two numbers of 8-bit data stored in memory 2400$_H$ and 2401$_H$. Store the magnitude of the result in 2402$_H$. If the result is positive store 00 in 2403$_H$ or if the result is negative store 01 in 2403$_H$.

Problem Analysis

In order to perform subtraction in an 8051, one of the data should be in the accumulator and the other data can be in any one of the internal memory/registers or can be an immediate data. The controller stores the result in the accumulator after subtraction. The 8051 perform 2's complement subtraction and then complements the carry. Therefore, if the result is negative then the carry flag is set and the accumulator will have 2's complement of the result. In order to get the magnitude of the result, again take 2's complement of the result. One of the registers is used to account for sign of the result. The 8051 will consider previous carry while performing subtraction and so the carry should be cleared before performing the subtraction.

Flowchart for Example Program 51.2

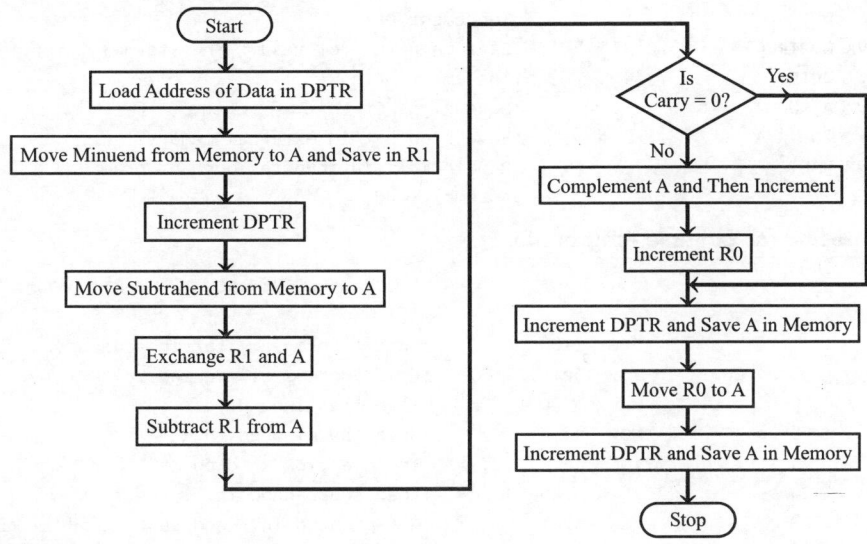

Algorithm

1. Set DPTR as pointer for data (Load address of data in DPTR).
2. Move the minuend from external memory to accumulator (A) and save in R1-register.
3. Increment DPTR and move the subtrahend from external memory to accumulator (A).
4. Exchange the contents of R1 and A, such that minuend is in A and subtrahend is in B.
5. Clear R0-register to account for sign.
6. Clear carry flag.
7. Subtract the content of R1 from A.
8. Check carry flag. If carry flag is not set go to step 10, otherwise go to next step.
9. Complement the content of A and increment by one to get 2's complement of result in A. Also increment R0 by one to indicate negative result.
10. Increment DPTR and save the content of A (which is magnitude of result) in external memory.
11. Increment DPTR, move R0 (sign bit) to A and then save sign bit in external memory.
12. Stop.

Assembly Language Program

```
;PROGRAM TO SUBTRACT TWO 8-BIT DATA
        ORG   2100H              ;specify program starting address.
        MOV   DPTR,#2400H        ;Load address of minuend in DPTR.
        MOVX  A,@DPTR            ;Move the minuend to A.
        MOV   R1,A               ;Save the minuend in R1.
        INC   DPTR               ;Increment DPTR to point subtrahend.
        MOVX  A,@DPTR            ;Load subtrahend in A.
        XCH   A,R1               ;Get minuend in A and subtrahend in R1.
```

```
        MOV  R0,#00H              ;Clear R0 to account for sign.
        CLR  C                    ;Clear carry.
        SUBB A,R1                 ;Subtract R1 from A.
        JNC  AHEAD                ;Check carry flag,If carry is set then,
        CPL  A                    ;get 2's complement of result in A.
        INC  A
        INC  R0                   ;Set R0 as one to indicate negative result.

AHEAD:  INC  DPTR                 ;Increment DPTR.
        MOVX @DPTR,A              ;Save magnitude of result in external memory.
        INC  DPTR                 ;Increment DPTR.
        MOV  A,R0                 ;Move sign bit to A.
        MOVX @DPTR,A              ;Save sign bit in external memory.
HALT:   SJMP HALT                 ;Remain idle in infinite loop. program end.
        END                       ;Assembly end.
```

Assembler Listing for Example Program 51.2

```
1                                            ;PROGRAM TO SUBTRACT TWO 8-BIT DATA
2
3    2100                    ORG   2100H     ;specify program starting address.
4    2100   90 24 00    MOV   DPTR,#2400H    ;Load address of minuend in DPTR.
5    2103   E0          MOVX  A,@DPTR        ;Move the minuend to A.
6    2104   F9          MOV   R1,A           ;Save the minuend in R1.
7    2105   A3          INC   DPTR           ;Increment DPTR to point subtrahend.
8    2106   E0          MOVX  A,@DPTR        ;Load subtrahend in A.
9    2107   C9          XCH   A,R1           ;Get minuend in A and subtrahend in R1.
10   2108   78 00       MOV   R0,#00H        ;Clear R0 to account for sign.
11   210A   C3          CLR   C              ;Clear carry.
12   210B   99          SUBB  A,R1           ;Subtract R1 from A.
13
14   210C   50 03       JNC   AHEAD          ;Check carry flag,If carry is set then,.
15   210E   F4          CPL   A              ;get 2's complement of result in A.
16   210F   04          INC   A
17   2110   08          INC   R0             ;Set R0 as one to indicate negative result.
18   2111
19   2111   A3     AHEAD:INC   DPTR          ;Increment DPTR.
20   2112   F0          MOVX  @DPTR,A        ;Save magnitude of result in external memory.
21   2113   A3          INC   DPTR           ;Increment DPTR.
22   2114   E8          MOV   A,R0           ;Move sign bit to A.
23   2115   F0          MOVX  @DPTR,A        ;Save sign bit in external memory.
24
25   2116   80 FE  HALT: SJMP HALT           ;Remain idle in infinite loop. program end.
26   2118              END                   ;Assembly end.
```

Sample Data

Input Data : Minuend   = $4C_H$
               Subtrahend = $F7_H$

Output Data : Difference = AB_H
               Sign bit   = 01_H

Memory address	Content
2400	4C
2401	F7
2402	AB
2403	01

EXAMPLE PROGRAM 51.3: 8-bit Multiplication

Write an assembly language program to multiply two numbers of 8-bit data stored in the memory 2400$_H$ and 2401$_H$ and store the product in 2402$_H$ and 2403$_H$.

Problem Analysis

In order to perform multiplication in 8051, the two 8-bit data should be stored in A and B registers, then multiplication can be performed by using "MUL AB" instruction. After multiplication, the 16-bit product will be in A and B-register such that the low byte is in A and the high byte is in B.

Algorithm

1. Load address of data in DPTR.
2. Move first data from external memory to A and save in B.
3. Increment DPTR and move second data from external memory to A.
4. Perform multiplication to get the product in A and B.
5. Increment DPTR and save A (which is low byte of product) in memory.
6. Increment DPTR, move B (which is high byte of product) to A and save it in memory.
7. Stop.

Flowchart for Example Program 51.3

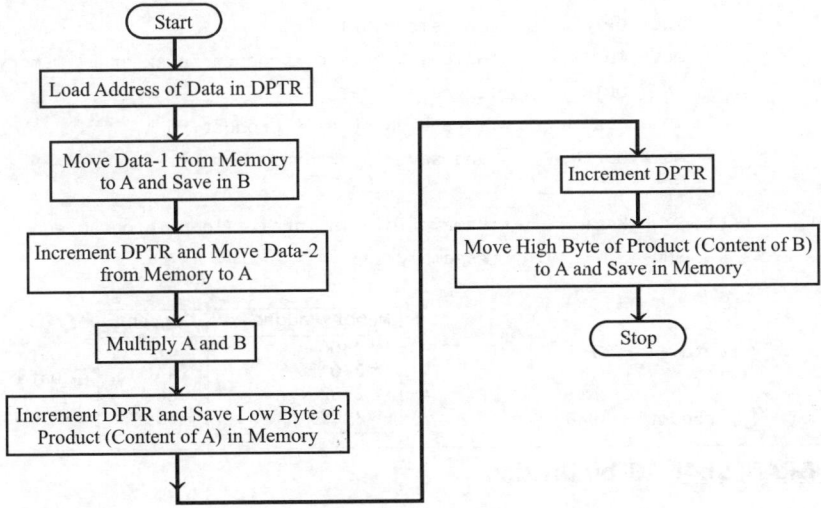

Assembly Language Program

```
;PROGRAM TO MULTIPLY TWO 8-BIT DATA
        ORG2100H          ;specify program starting address.
        MOV  DPTR,#2400H  ;Load address of 1st data in DPTR.
        MOVX A,@DPTR      ;Move the 1st data to A.
        MOV  B,A          ;Save the first data in B.
        INC  DPTR         ;Increment DPTR to point 2nd data.
        MOVX A,@DPTR      ;Load 2nd data in A.
        MUL  AB           ;Get the product in A and B.
```

```
        INC   DPTR          ;Increment DPTR.
        MOVX @DPTR,A         ;Save low byte of product in external memory.
        INC  DPTR           ;Increment DPTR.
        MOV   A,B            ;Move high byte of product to A,
        MOVX @DPTR,A         ;and save in external memory.

HALT:  SJMP HALT            ;Remain idle in infinite loop. program end.
        END                 ;Assembly end.
```

Assembler Listing for Example Program 51.3

```
1                                                   ;PROGRAM TO MULTIPLY TWO 8-BIT DATA
2
3    2100                       ORG   2100H         ;specify program starting address.
4    2100    90 24 00          MOV   DPTR,#2400H    ;Load address of 1st data in DPTR.
5    2103    E0               MOVX A,@DPTR          ;Move the 1st data to A.
6    2104    F5 F0            MOV   B,A             ;Save the first data in B.
7    2106    A3               INC   DPTR            ;Increment DPTR to point 2nd data.
8    2107    E0               MOVX A,@DPTR          ;Load 2nd data in A.
9    2108    A4               MUL   AB              ;Get the product in A and B.
10
11   2109    A3               INC   DPTR            ;Increment DPTR.
12   210A    F0               MOVX @DPTR,A          ;Save low byte of product in external memory.
13   210B    A3               INC   DPTR            ;Increment DPTR.
14   210C    E5 F0            MOV   A,B             ;Move high byte of product to A,
15   210E    F0               MOVX @DPTR,A          ;and save in external memory.
16
17   210F    80 FE      HALT: SJMP HALT             ;Remain idle in infinite loop. program end.
18   211                      END                   ;Assembly end.
```

Sample Data

		Memory address	Content	
Input Data	: Data -1 = C7$_H$	2400	C7	← Data 1
	Data -2 = 4A$_H$	2401	4A	← Data 2
		2402	86	} Product
Output Data	: Product = 3986$_H$	2403	39	

EXAMPLE PROGRAM 51.4:   8-bit Division

Write an assembly language program to divide the 8-bit data stored in the memory location 2400$_H$ by the 8-bit data in 2401$_H$. Store the quotient in 2402$_H$ and remainder in 2403$_H$.

Problem Analysis

In order to perform division in 8051, the dividend should be stored in A and the divisor should be stored in B. Then the content of A can be divided by B using the instruction "DIV AB". After division, the quotient will be in A and the remainder will be in B.

Algorithm

1.   **Load address of data in DPTR.**
2.   **Move the dividend from external memory to A and save it in R0-register.**

3. Increment DPTR and move the divisor from external memory to A and save it in B-register.
4. Move the dividend from R0 to A.
5. Perform division to get quotient in A and remainder in B.
6. Increment DPTR and save quotient (content of A)in memory.
7. Increment DPTR.
8. Move the remainder (content of B) to A and save in memory.
9. Stop.

Flowchart for Example Program 51.4

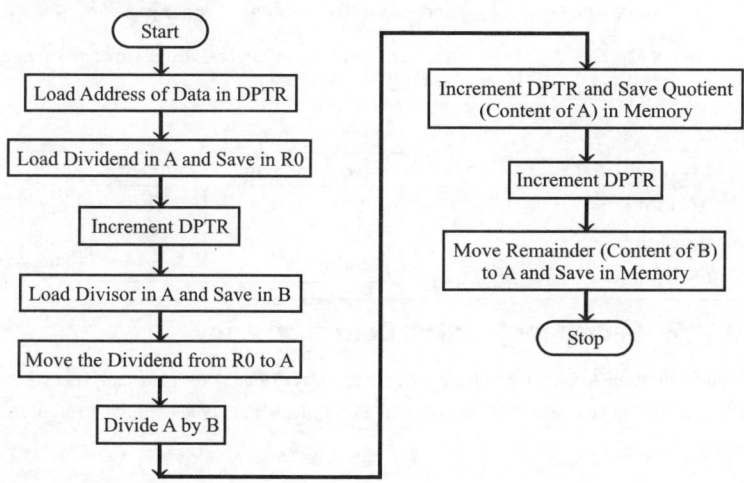

Assembly Language Program

```
;PROGRAM TO DIVIDE TWO 8-BIT DATA
                ORG      2100H           ;specify program starting address.

                MOV      DPTR,#2400H     ;Load address of dividend in DPTR.
                MOVX     A,@DPTR         ;Load the dividend in A.
                MOV      R0,A            ;Save the dividend in R0.
                INC      DPTR            ;Let DPTR point to divisor.
                MOVX     A,@DPTR         ;Load the divisor in A.
                MOV      B,A             ;Move the divisor to B.
                MOV      A,R0            ;Move the dividend to A.
                DIV      AB              ;Divide the content of A by B.

                INC      DPTR            ;Increment DPTR.
                MOVX     @DPTR,A         ;Save quotient in external memory.
                INC      DPTR            ;Increment DPTR.
                MOV      A,B             ;Move remainder to A,
                MOVX     @DPTR,A         ;and save in external memory.

HALT:           SJMP     HALT            ;Remain idle in infinite loop. program end.
                END                      ;Assembly end.
```

Assembler Listing for Example Program 51.4

```
1                                       ;PROGRAM TO DIVIDE TWO 8-BIT DATA
2
3  2100                        ORG  2100H      ;specify program starting address.

4  2100    90 24 00            MOV  DPTR,#2400H ;Load address of dividend in DPTR.
5  2103    E0                  MOVX A,@DPTR     ;Load the dividend in A.
6  2104    F8                  MOV  R0,A        ;Save the dividend in R0.
```

7	2105	A3		INC	DPTR	;Let DPTR point to divisor.
8	2106	E0		MOVX	A,@DPTR	;Load the divisor in A.
9	2107	F5 F0		MOV	B,A	;Move the divisor to B.
10	2109	E8		MOV	A,R0	;Move the dividend to A.
11	210A	84		DIV	AB	;Divide the content of A by B.
12						
13	210B	A3		INC	DPTR	;Increment DPTR.
14	210C	F0		MOVX	@DPTR,A	;Save quotient in external memory.
15	210D	A3		INC	DPTR	;Increment DPTR.
16	210E	E5 F0		MOV	A,B	;Move remainder to A,
17	2110	F0		MOVX	@DPTR,A	;and save in external memory.
18						
19	2111	80 FE	HALT:	SJMP	HALT	;Remain idle in infinite loop. program end.
20	2113			END		;Assembly end.

Sample Data

Input Data	:	Dividend	= 64_H
		Divisor	= 07_H
Output Data	:	Quotient	= $0E_H$
		Remainder	= 02_H

Memory address	Content	
2400	64	← Divider
2401	07	← Divisor
2402	0E	← quotient
2403	02	← Remainder

EXAMPLE PROGRAM 51.5: Search for Smallest Data in an Array

Write an assembly language program to search the smallest data in an array of data stored in memory. Let the array be stored in memory starting from 2400$_H$, with the first element of the array as count for the number of data in the array. Store the smallest data in memory location 2500$_H$.

Problem Analysis

The DPTR is used as the pointer for the array. One of the register of the registers bank is used as counter and another register is used to store the current smallest data. Initially, the first data of the array is considered as the current smallest. The smallest data is searched by performing subtraction of a data of the array with the current smallest. The condition of the carry flag after subtraction is used to determine the smaller among the two and the smallest among the two is moved to the register reserved to store the current smallest data. The comparison by subtraction is performed N – 1 times (where N is the count for the number of data in the array). After N – 1 comparisons, the smallest data in the array will be in the register reserved for current smallest data, which can be stored in the memory.

Algorithm

1. Set DPTR as pointer for data array.
2. Load the count value, N in A and save in R0.
3. Decrement R0 to set count for N–1 comparisons.
4. Increment DPTR.
5. Load the first data of array in A and save it as current smallest in R4-register.
6. Increment DPTR.
7. Get a data of the array in A-register and save it in R2-register.
8. Clear carry flag and subtract the current smallest in R4 from A.
9. Check carry flag. If carry is not set then go to step 11, otherwise go to next step.
10. If carry is set, then the content of R2 is smaller than R4 and so, move R2 to R4 via A.
11. Decrement R0 and check whether it is zero. If R0 is not zero then go to step 6, otherwise go to next step.
12. Load the address of the memory where smallest data to be stored in DPTR.
13. Move the smallest data from R4 to A and save in memory pointed by DPTR.
14. Stop.

Flowchart for Example Program 51.5

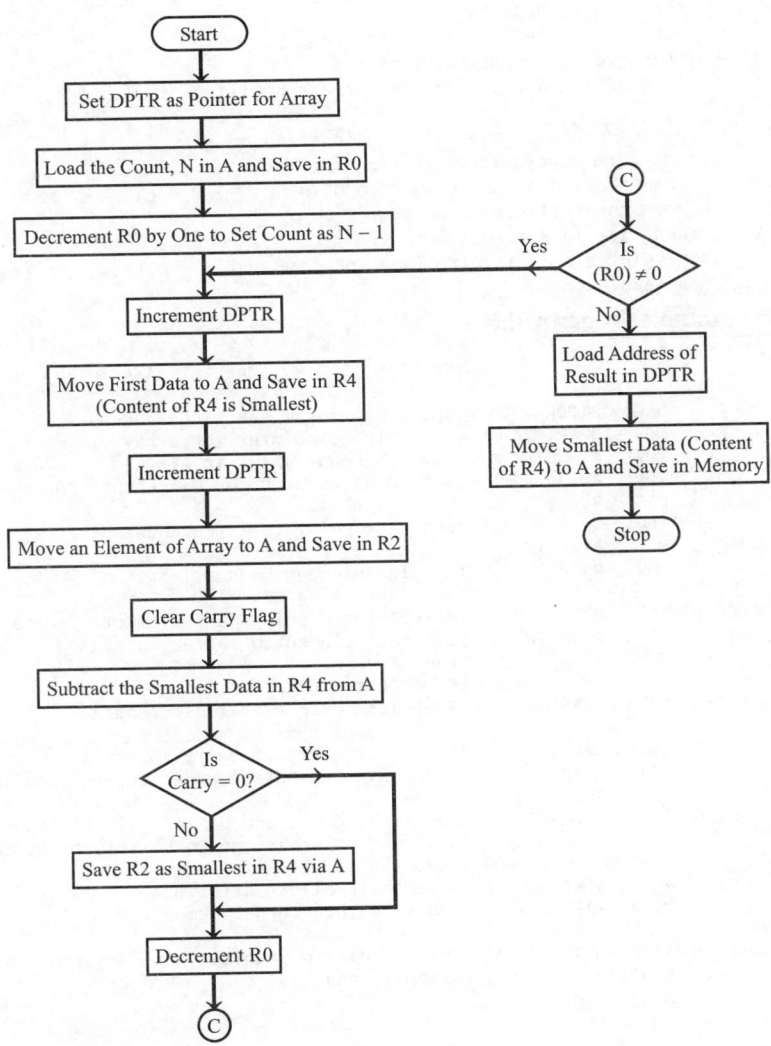

Assembly Language Program

```
;PROGRAM TO FIND SMALLEST DATA IN AN ARRAY
      ORG   2100H        ;specify program starting address.
      MOV   DPTR,#2400H  ;Set DPTR as pointer for array.
      MOVX  A,@DPTR      ;Get the count value in A.
      MOV   R0,A         ;Set R0 as counter for N-1 comparisons.
      DEC   R0
      INC   DPTR         ;Let DPTR point to 1st element of array.
      MOVX  A,@DPTR
      MOV   R4,A         ;Let 1st element be smallest,
                         ;and save it in R4.
```

```
AGAIN: INC  DPTR        ;Make DPTR to point next element of array.
       MOVX A,@DPTR     ;Get next element of array in A,
       MOV  R2,A        ;and save in R2.
       CLR  C           ;Clear carry flag.
       SUBB A,R4        ;Subtract current smallest from A.
       JNC  AHEAD       ;Check for carry,If carry is set.
       MOV  A,R2        ;then save content of R2 as current smallest.
       MOV  R4,A
AHEAD: DJNZ R0,AGAIN    ;Decrement count and go to again if count is
                        ;not zero,otherwise go to next instruction.
       MOV  DPTR,#2500H ;Load the address of result in DPTR.
       MOV  A,R4        ;Move the smallest data to A,
       MOVX @DPTR,A     ;and save in external memory.
HALT:  SJMP HALT        ;Remain idle in infinite loop. program end.
       END              ;Assembly end.
```

Assembler Listing for Example Program 51.5

```
 1                                      ;PROGRAM TO FIND SMALLEST DATA IN AN ARRAY
 2
 3  2100                  ORG   2100H   ;specify program starting address.
 4  2100  90 24 00        MOV   DPTR,#2400H ;Set DPTR as pointer for array.
 5  2103  E0             MOVX  A,@DPTR  ;Get the count value in A.
 6  2104  F8              MOV   R0,A     ;Set R0 as counter for N-1 comparisons.
 7  2105  18              DEC   R0
 8  2106  A3              INC   DPTR     ;Let DPTR point to 1st element of array.
 9  2107  E0             MOVX  A,@DPTR
10  2108  FC              MOV   R4,A     ;Let 1st element be smallest,
11                                       ;and save it in R4.
12  2109  A3      AGAIN: INC   DPTR     ;Make DPTR to point next element of array.
13  210A  E0             MOVX  A,@DPTR  ;Get next element of array in A,
14  210B  FA              MOV   R2,A     ;and save in R2.
15  210C  C3              CLR   C        ;Clear carry flag.
16  210D  9C              SUBB  A,R4     ;Subtract current smallest from A.
17
18  210E  50 02           JNC   AHEAD    ;Check for carry,If carry is set.
19  2110  EA              MOV   A,R2     ;then save content of R2 as current smallest.
20  2111  FC              MOV   R4,A
21  2112  D8 F5   AHEAD: DJNZ  R0,AGAIN ;Decrement count and go to again if count is
22                                       ;not zero,otherwise go to next instruction.
23  2114  90 25 00        MOV   DPTR,#2500H ;Load the address of result in DPTR.
24  2117  EC              MOV   A,R4     ;Move the smallest data to A,
25  2118  F0             MOVX  @DPTR,A  ;and save in external memory.
26
27  2119  80 FE   HALT:  SJMP  HALT     ;Remain idle in infinite loop. program end.
28  211B                  END            ;Assembly end.
```

Sample Data

Input Data :   Count = 06_H

Array = $7F_H$
$1C_H$
42_H
57_H
13_H
FE_H

Output Data: 13

Memory address	Content	
2400	06	Count
2401	7F	⎫
2402	1C	⎪
2403	42	⎬ Array
2404	57	⎪
2405	13	⎪
2406	FE	⎭
2500	13	Smallest data

EXAMPLE PROGRAM 51.6:   Search for Largest Data in an Array

Write an assembly language program to search the largest data in an array of data stored in the memory. Let the array be stored in the memory starting from 2400$_H$, with the first element of the array as count for the number of data in the array. Store the largest data in memory location 2500$_H$.

Problem Analysis

The DPTR is used as a pointer for the array. One of the registers of register bank is used as the counter and another register is used to store current largest data. Initially, the first data of the array is considered as current largest. The largest data is searched by performing subtraction of a data of the array with current largest. The condition of the carry flag after subtraction is used to determine the larger among the two and the largest among the two is moved to the register reserved to store current largest data. The comparison by subtraction is performed N – 1 times (where N is count for the number of data in the array). After N – 1 comparisons, the largest data in the array will be in the register reserved for the current largest data, which can be stored in the memory.

Flowchart for Example Program 51.6

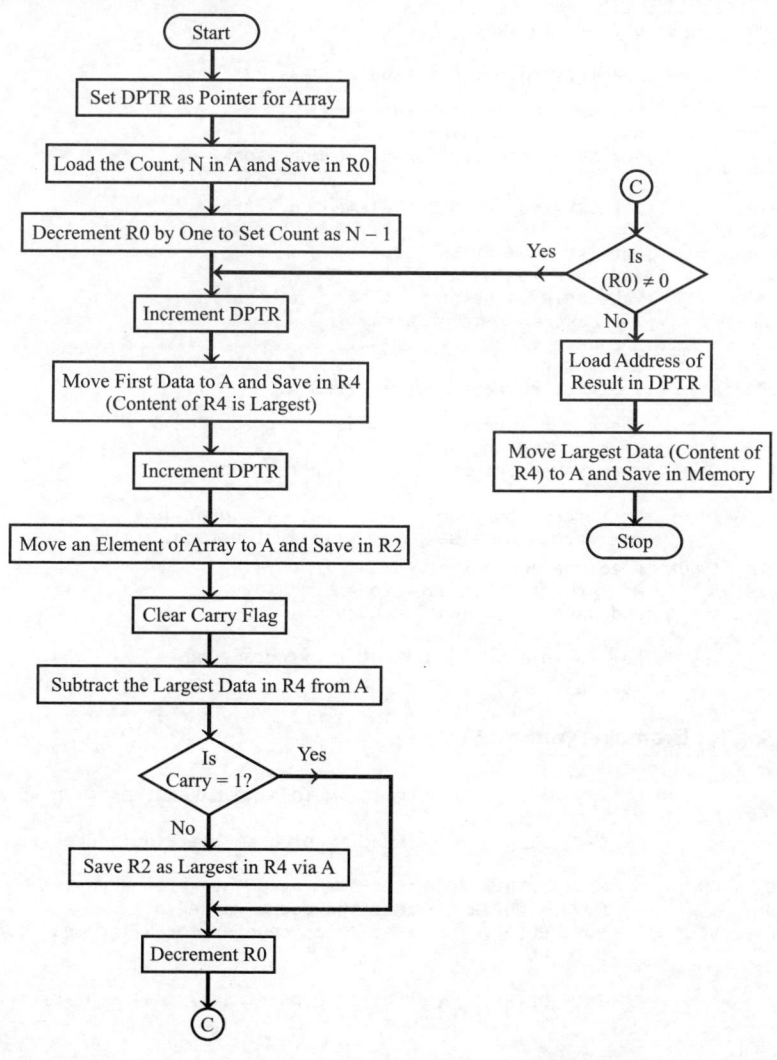

Algorithm

1. Set DPTR as pointer for data array.
2. Load the count value, N in A and save in R0.
3. Decrement R0 to set count for N−1 comparisons.
4. Increment DPTR.
5. Load the first data of array in A and save it as current largest in R4-register.
6. Increment DPTR.
7. Get a data of the array in A-register and save it in R2-register.
8. Clear carry flag and subtract the current largest in R4 from A.
9. Check carry flag. If carry is set then go to step 11, otherwise go to next step.
10. If carry is not set, then the content of R2 is larger than R4 and so, move R2 to R4 via A.
11. Decrement R0 and check whether it is zero. If R0 is not zero then go to step 6, otherwise go to next step.
12. Load the address of the memory where largest data to be stored in DPTR.
13. Move the largest data from R4 to A and save in memory pointed by DPTR.
14. Stop.

Assembly Language Program

```
;PROGRAM TO FIND LARGEST DATA IN AN ARRAY

        ORG   2100H        ;Specify program starting address.

        MOV   DPTR,#2400H  ;Set DPTR as pointer for array.
        MOVX  A,@DPTR       ;Get the count value in A.
        MOV   R0,A          ;Set R0 as counter for N-1 comparisons.
        DEC   R0

        INC   DPTR          ;Let DPTR point to 1st element of array.
        MOVX  A,@DPTR
        MOV   R4,A          ;Let 1st element be largest,
                            ;and save it in R4.
AGAIN:  INC   DPTR          ;Make DPTR to point next element of array.
        MOVX  A,@DPTR       ;Get next element of array in A,
        MOV   R2,A          ;and save in R2.
        CLR   C             ;Clear carry flag.
        SUBB  A,R4          ;Subtract current largest from A.

        JC    AHEAD         ;Check for carry,If carry is set go to AHEAD.
        MOV   A,R2          ;If carry is not set,
        MOV   R4,A          ;then save content of R2 as current largest.

AHEAD:  DJNZ  R0,AGAIN      ;Decrement count and go to again if count is
                            ;not zero,otherwise go to next instruction.
        MOV   DPTR,#2500H   ;Load the address of result in DPTR.
        MOV   A,R4          ;Move the largest data to A,
        MOVX  @DPTR,A       ;and save in external memory.

HALT:   SJMP  HALT         ;Remain idle in infinite loop. Program end.
        END                 ;Assembly end.
```

Assembler Listing for Example Program 51.6

```
1                                        ;PROGRAM TO FIND LARGEST DATA IN AN ARRAY
2
3    2100                       ORG   2100H      ;Specify program starting address.
4
5    2100   90 24 00           MOV   DPTR,#2400H ;Set DPTR as pointer for array.
6    2103   E0                 MOVX  A,@DPTR     ;Get the count value in A
7    2104   F8                 MOV   R0,A        ;Set R0 as counter for N-1 comparisons.
8    2105   18                 DEC   R0
9
10   2106   A3                 INC   DPTR        ;Let DPTR point to 1st element of array.
```

11	2107	E0		MOVX A,@DPTR	
12	2108	FC		MOV R4,A	;Let 1st element be largest,
13					;and save it in R4.
14	2109	A3	AGAIN:	INC DPTR	;Make DPTR to point next element of array.
15	210A	E0		MOVX A,@DPTR	;Get next element of array in A,
16	210B	FA		MOV R2,A	;and save in R2.
17	210C	C3		CLR C	;Clear carry flag.
18	210D	9C		SUBB A,R4	;Subtract current largest from A.
19					
20	210E	40 02		JC AHEAD	;Check for carry,If carry is set go to AHEAD.
21	2110	EA		MOV A,R2	;If carry is not set,
22	2111	FC		MOV R4,A	;then save content of R2 as current largest.
23					
24	2112	D8 F5	AHEAD:	DJNZ R0,AGAIN	;Decrement count and go to again if count is
25					;not zero,otherwise go to next instruction.
26					
27	2114	90 25 00		MOV DPTR,#2500H	;Load the address of result in DPTR.
28	2117	EC		MOV A,R4	;Move the largest data to A,
29	2118	F0		MOVX @DPTR,A	;and save in external memory.
30					
31	2119	80 FE	HALT:	SJMP HALT	;Remain idle in infinite loop. Program end.
32					
33	211B			END	;Assembly end.

Sample Data

Input Data : Count = 06$_H$
            Array = 7F$_H$
                    1C$_H$
                    42$_H$
                    57$_H$
                    13$_H$
                    FE$_H$

Output Data : FE$_H$

Memory address	Content	
2400	06	Count
2401	7F	⎫
2402	1C	⎪
2403	42	⎬ Array
2404	57	⎪
2405	13	⎪
2406	FE	⎭
2500	FE	Largest data

EXAMPLE PROGRAM 51.7: Sorting an Array in Ascending Order

Write an assembly language program to sort an array of data in ascending order. The array is stored in memory starting from 2400$_H$. The first element of the array gives the count value for the number of elements in the array.

Problem Analysis

The algorithm for bubble sorting is given below. In bubble sorting of N-data, N–1 comparisons are carried by taking two consecutive data at a time. After each comparison, the data are rearranged such that the smallest among the two is in the first memory location and the largest in the next memory location. (Here the data are rearranged within the two memory locations whose contents are compared.) When we perform N–1 comparisons as mentioned above for N–1, times then the array consisting of N-data will be sorted in the ascending order.

Algorithm

1. Load address of data array in DPTR, and using DPTR load the count value in R2 via A.
2. Decrement R2-register (R2 is counter for N–1 repetitions).
3. Set DPTR as data array address pointer.
4. Set R1-register as counter for N–1 comparisons.
5. Increment DPTR, and using DPTR load two consecutive data of the array in R3 and A.
6. Compare the content of A and R3 by performing subtraction.
7. If carry flag is not set (If the content of A is greater than R3) then go to step 9, otherwise go to next step.
8. If carry flag is set (If the content of A is less than R3), then exchange the content of memory pointed by DPTR and previous memory location.

9. Decrement R1-register. If zero flag =0, go to step 5 otherwise go to next step.
10. Decrement R2-register. If zero flag =0, go to step 3 otherwise go to next step.
11. Stop.

Flowchart for Example Program 51.7

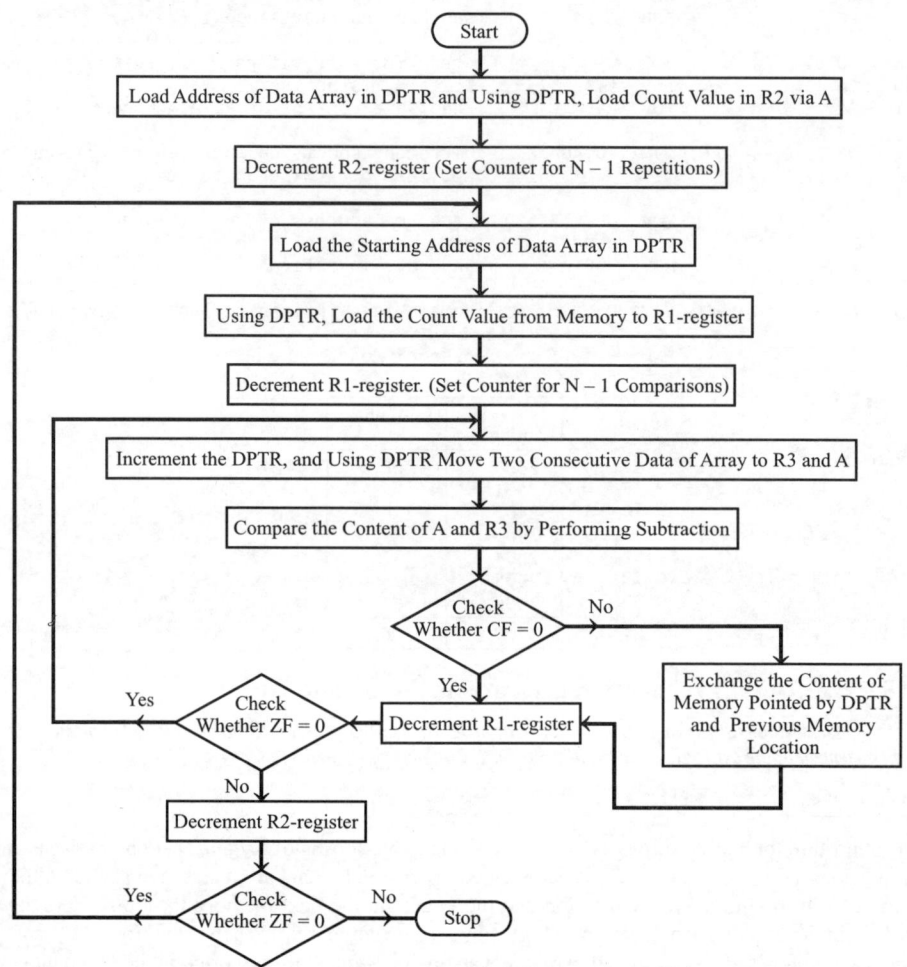

Assembly Language Program

```
;PROGRAM TO SORT AN ARRAY OF DATA IN ASCENDING ORDER

        ORG   2100H          ;Specify program starting address.

        MOV   DPTR,#2400H    ;Load the count value in A-register.
        MOVX  A,@DPTR
        MOV   R2,A            ;Set counter for (N-1) repetitions
        DEC   R2             ;of (N-1) comparisons.
LOOP2:  MOV   DPTR,#2400H    ;Set pointer for array.
        MOVX  A,@DPTR        ;Set count for (N-1) comparisons.
        MOV   R1,A
```

```
        DEC   R1

        INC   DPTR      ;Increment pointer.
LOOP1:  MOVX  A,@DPTR   ;Get two consecutive data of array in
        MOV   R3,A      ;R3 and A-register.
        INC   DPTR
        MOVX  A,@DPTR

        CLR   C         ;Compare A and R3-register.
        SUBB  A,R3      ;If content of A is greater than
        JNC   AHEAD     ;R3 then go to AHEAD.

        MOVX  A,@DPTR   ;If the content of A is less than
        DEC   DPTR      ;the content of R3-register.
        MOVX  @DPTR,A   ;then exchange the content of memory
        INC   DPTR      ;pointed by DPTR and previous location.
        MOV   A,R3
        MOVX  @DPTR,A

AHEAD:  DJNZ  R1,LOOP1  ;Repeat comparisons until R1 count is zero.

        DJNZ  R2,LOOP2  ;Repeat until R2 count is zero.

HALT:   SJMP  HALT      ;Remain idle in infinite loop. Program end.

        END             ;Assembly end.
```

Assembler Listing for Example Program 51.7

```
1                              ;PROGRAM TO SORT AN ARRAY OF DATA IN ASCENDING ORDER
2
3    2100                      ORG   2100H       ;Specify program starting address.
4
5    2100  90 24 00            MOV   DPTR,#2400H ;Load the count value in A-register.
6    2103  E0                  MOVX  A,@DPTR
7    2104  FA                  MOV   R2,A        ;Set counter for (N-1) repetitions
8    2105  1A                  DEC   R2          ;of (N-1) comparisons.
9
10   2106  90 24 00  LOOP2:    MOV   DPTR,#2400H ;Set pointer for array.
11   2109  E0                  MOVX  A,@DPTR     ;Set count for (N-1) comparisons.
12   210A  F9                  MOV   R1,A
13   210B  19                  DEC   R1
14
15   210C  A3                  INC   DPTR        ;Increment pointer.
16   210D  E0        LOOP1:    MOVX  A,@DPTR     ;Get two consecutive data of array in
17   210E  FB                  MOV   R3,A        ;R3 and A-register.
18   210F  A3                  INC   DPTR
19   2110  E0                  MOVX  A,@DPTR
20   2111  C3                  CLR   C           ;Compare A and R3-register.
21   2112  9B                  SUBB  A,R3        ;If content of A is greater than
22   2113  50 07               JNC   AHEAD       ;R3 then go to AHEAD.
23
24   2115  E0                  MOVX  A,@DPTR     ;If the content of A is less than
25   2116  15 82               DEC   DPTR        ;the content of R3-register.
26   2118  F0                  MOVX  @DPTR,A     ;then exchange the content of memory
27   2119  A3                  INC   DPTR        ;pointed by DPTR and previous location.
28   211A  EB                  MOV   A,R3
29   211B  F0                  MOVX  @DPTR,A
30
31   211C  D9 EF     AHEAD:    DJNZ  R1,LOOP1    ;Repeat comparisons until R1 count is zero.
32   211E
33   211E  DA E6               DJNZ  R2,LOOP2    ;Repeat until R2 count is zero.
34
35   2120  80 FE     HALT:     SJMP  HALT        ;Remain idle in infinite loop. Program end.
36
37   2122                      END               ;Assembly end.
```

Sample Data

Memory address	Content		Memory address	Content	
		Input Data: 07	2400	07	Output data : 07

Input Data:

	Memory address	Content
07	2400	07
AB	2401	AB
92	2402	92
84	2403	84
4F	2404	4F
69	2405	69
F2	2406	F2
34	2407	34

(Before sorting)

Output data :

	Memory address	Content	
07	2400	07	← Count
34	2401	34	⎫
4F	2402	4F	⎪
69	2403	69	Array in
84	2404	84	ascending
92	2405	92	order
AB	2406	AB	⎪
F2	2407	F2	⎭

(After sorting)

EXAMPLE PROGRAM 51.8: Sorting an Array in Descending Order

Write an assembly language program to sort an array of data in descending order. The array is stored in memory starting from 2400$_H$. The first element of the array gives the count value for the number of elements in the array.

Problem Analysis

The algorithm for bubble sorting is given below. In bubble sorting of N-data, N−1 comparisons are carried by taking two consecutive data at a time. After each comparison, the data are rearranged such that the largest among the two is in the first memory location and the smallest in the next memory location. (Here the data are rearranged within the two memory locations whose contents are compared.) When we perform N −1 comparisons as mentioned above for N−1 times, then the array consisting of N-data will be sorted in the descending order.

Algorithm

1. Load address of data array in DPTR, and using DPTR load the count value in R2 via A.

2. Decrement R2-register (R2 is counter for N−1 repetitions).

3. Set DPTR as data array address pointer.

4. Set R1-register as counter for N−1 comparisons.

5. Increment DPTR, and using DPTR load two consecutive data of the array in R3 and A.

6. Compare the content of A and R3 by performing subtraction.

7. If carry flag is set (If the content of A is smaller than R3) then go to step 9, otherwise go to next step.

8. If carry flag is not set (If the content of A is larger than R3), then exchange the content of memory pointed by DPTR and next memory location.

9. Decrement R1-register. If zero flag =0, go to step 5 otherwise go to next step.

10. Decrement R2-register. If zero flag =0, go the step 3 otherwise go to next step.

11. Stop.

Flowchart for Example Program 51.8

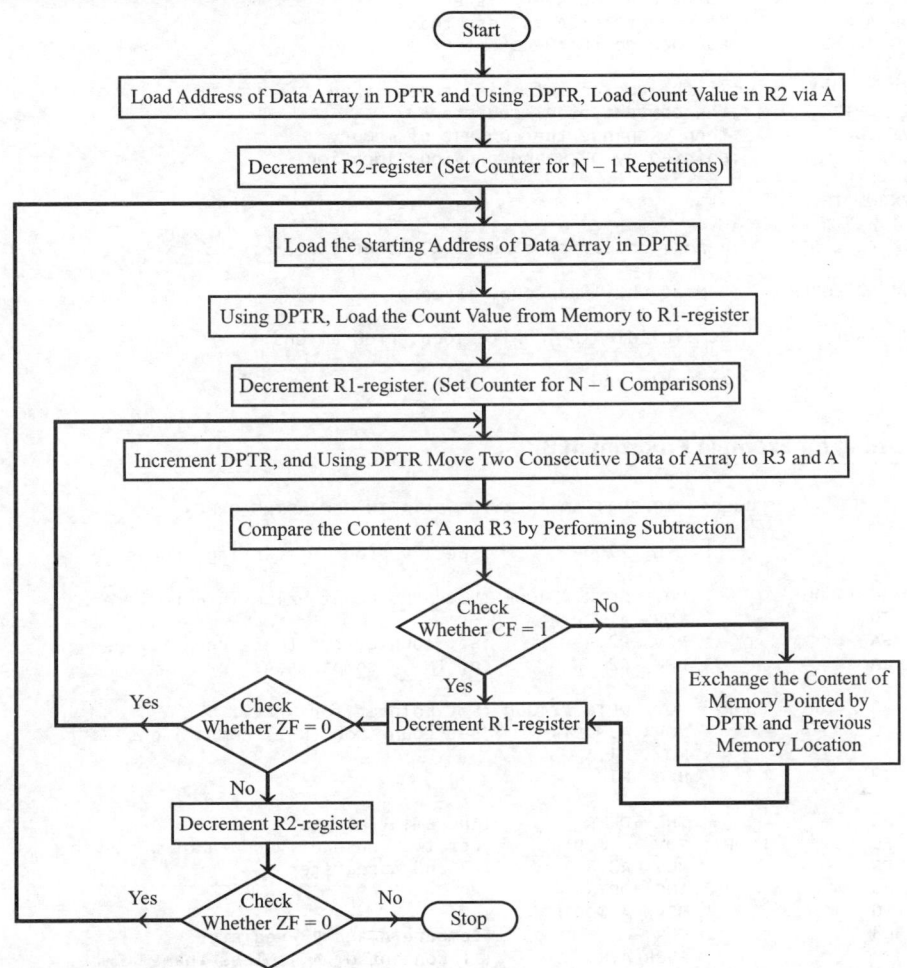

Assembly Language Program

```
;PROGRAM TO SORT AN ARRAY OF DATA IN DESCENDING ORDER

        ORG   2100H         ;Specify program starting address.

        MOV   DPTR,#2400H   ;Load the count value in A-register.
        MOVX  A,@DPTR
        MOV   R2,A           ;Set counter for (N-1) repetitions
        DEC   R2             ;of (N-1) comparisons.

LOOP2:  MOV   DPTR,#2400H   ;Set pointer for array.
        MOVX  A,@DPTR        ;Set count for (N-1) comparisons.
        MOV   R1,A
        DEC   R1

        INC   DPTR           ;Increment pointer.

LOOP1:  MOVX  A,@DPTR        ;Get two consecutive data of array in
        MOV   R3,A           ;R3 and A-register.
        INC   DPTR
```

```
          MOVX  A,@DPTR
          CLR   C              ;Compare A and R3-register.
          SUBB  A,R3           ;If content of A is less than
          JC    AHEAD          ;R3 then go to AHEAD.

          MOVX  A,@DPTR        ;If the content of A is greater than
          DEC   DPTR           ;the content of R3-register
          MOVX  @DPTR,A        ;then exchange the content of memory
          INC   DPTR           ;pointed by DPTR and previous location.
          MOV   A,R3
          MOVX  @DPTR,A

AHEAD:    DJNZ  R1,LOOP1       ;Repeat comparisons until R1 count is zero.

          DJNZ  R2,LOOP2       ;Repeat until R2 count is zero.

HALT:     SJMP  HALT           ;Remain idle in infinite loop. Program end.

          END                  ;Assembly end.
```

Assembler Listing for Example Program 51.8

```
1                          ;PROGRAM TO SORT AN ARRAY OF DATA IN DECENDING ORDER
2
3    2100                      ORG   2100H          ;Specify program starting address.
4
5    2100   90 24 00           MOV   DPTR,#2400H ;Load the count value in A-register.
6    2103   E0                 MOVX  A,@DPTR
7    2104   FA                 MOV   R2,A            ;Set counter for (N-1) repetitions
8    2105   1A                 DEC   R2              ;of (N-1) comparisons.
9
10   2106   90 24 00    LOOP2: MOV   DPTR,#2400H ;Set pointer for array.
11   2109   E0                 MOVX  A,@DPTR         ;Set count for (N-1) comparisons.
12   210A   F9                 MOV   R1,A
13   210B   19                 DEC   R1
14
15   210C   A3                 INC   DPTR           ;Increment pointer.
16   210D   E0          LOOP1: MOVX  A,@DPTR        ;Get two consecutive data of array in
17   210E   FB                 MOV   R3,A           ;R3 and A-register.
18   210F   A3                 INC   DPTR
19   2110   E0                 MOVX  A,@DPTR
20   2111   C3                 CLR   C              ;Compare A and R3-register.
21   2112   9B                 SUBB  A,R3           ;If content of A is less than
22   2113   40 07              JC    AHEAD          ;R3 then go to AHEAD.
23
24   2115   E0                 MOVX  A,@DPTR        ;If the content of A is greater than
25   2116   15 82              DEC   DPTR           ;the content of R3-register.
26   2118   F0                 MOVX  @DPTR,A        ;then exchange the content of memory
27   2119   A3                 INC   DPTR           ;pointed by DPTR and previous location.
28   211A   EB                 MOV   A,R3
29   211B   F0                 MOVX  @DPTR,A
30
31   211C   D9 EF       AHEAD: DJNZ  R1,LOOP1       ;Repeat comparisons until R1 count is zero.
32
33   211E   DA E6              DJNZ  R2,LOOP2       ;Repeat until R2 count is zero.
34
35   2120   80 FE       HALT:  SJMP  HALT           ;Remain idle in infinite loop. Program end.
36
37   2122                      END                  ;Assembly end.
```

Sample Data

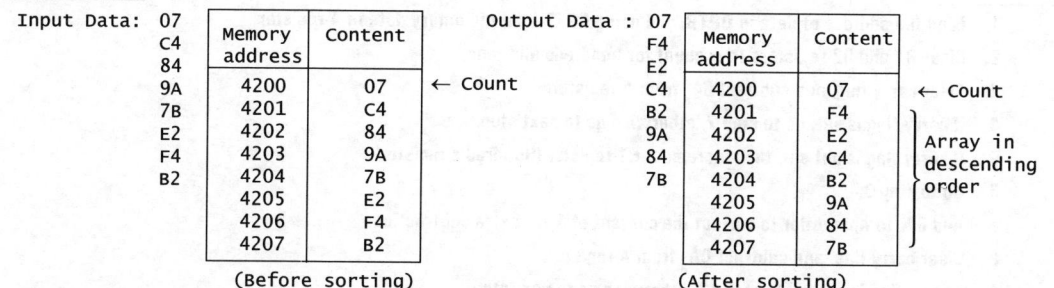

Input Data: 07
C4
84
9A
7B
E2
F4
B2

Memory address	Content
4200	07 ← Count
4201	C4
4202	84
4203	9A
4204	7B
4205	E2
4206	F4
4207	B2

(Before sorting)

Output Data : 07
F4
E2
C4
B2
9A
84
7B

Memory address	Content
4200	07 ← Count
4201	F4
4202	E2
4203	C4
4204	B2
4205	9A
4206	84
4207	7B

Array in descending order

(After sorting)

EXAMPLE PROGRAM 51.9: Binary to BCD Conversion (or) Hexadecimal to BCD conversion

Write an assembly language program to convert an 8-bit binary data (2-digit hexa) to BCD. The binary data is stored in 2400_H. Store the ten's and unit's digits in 2401_H. Store the hundred's digit in 2402_H.

Problem Analysis

The maximum value of 8-bit binary is $FF_H = 256_{10}$. Hence, the maximum size of the data will have hundreds, tens and units. The algorithm given below uses two counters to count hundreds and tens. Initially the counters are cleared. First, let us subtract all hundreds from the binary data. For each subtraction, the hundred's register is incremented by one. Then let us subtract all tens. For each subtraction ten's register is incremented by one. The remaining will be units. The tens and units are combined to form 2-digit BCD (8-bit binary representation of BCD).

Flowchart for Example Program 51.9

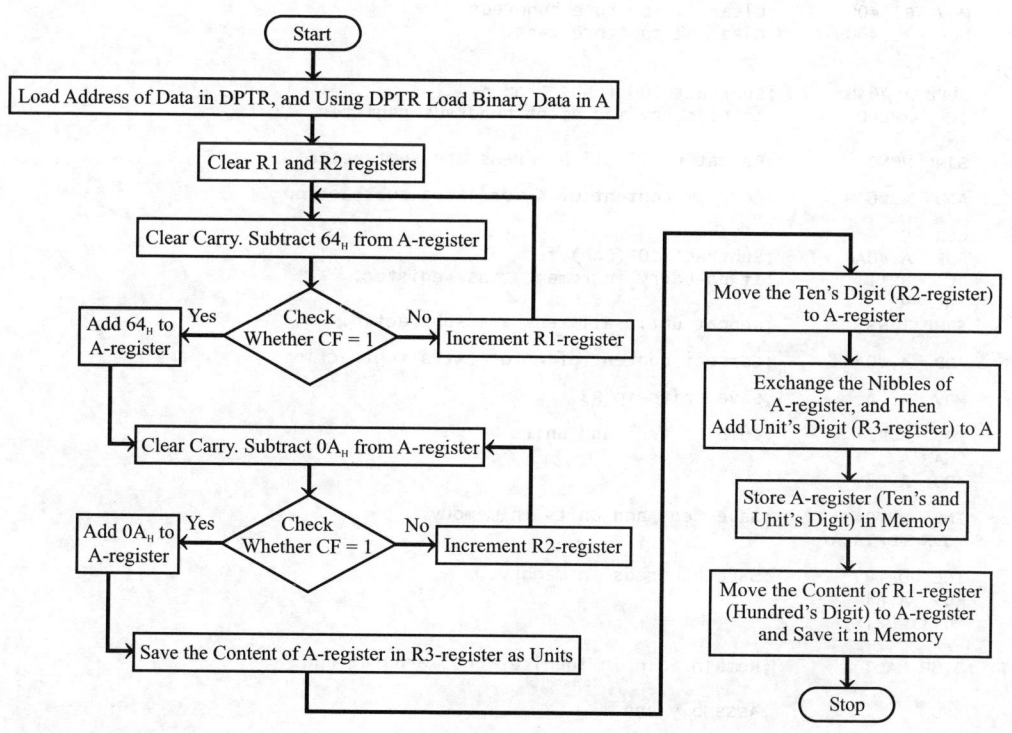

Algorithm

1. Load the address of data in DPTR, and using DPTR load the binary data in A-register.
2. Clear R1 and R2 registers to account for hundreds and tens.
3. Clear carry flag, and subtract 64_H from A-register.
4. If carry flag is set, go to step 7, otherwise go to next step.
5. If carry flag is not set, then increment R1-register (Hundred's register).
6. Go to step 3.
7. Add 64_H to A, in order to correct the content of A for extra subtraction.
8. Clear carry flag, and subtract $0A_H$ from A-register.
9. If carry flag is set, go to step 12, otherwise go to next step.
10. If carry flag is not set, then increment R2-register (Ten's register).
11. Go to step 8.
12. Add $0A_H$ to A, in order to correct the content of A for extra subtraction.
13. Save the content of A as units in R3-register.
14. Combine the units and tens to form 8-bit result.
15. Save the units, tens and hundreds in memory.

Assembly Language Program

```
;PROGRAM TO CONVERT A BINARY DATA TO BCD DATA

        ORG   2100H        ;Specify program starting address.

        MOV   DPTR,#2400H  ;Load the data address in DPTR.
        MOVX  A,@DPTR       ;Get the binary data in A-register.

        MOV   R1,#00H       ;Clear R1 to store hundreds.
        MOV   R2,#00H       ;Clear R2 to store tens.

HUND:   CLR   C
        SUBB  A,#64H        ;Subtract 100 (64h) from A.
        JC    AHEAD         ;If no carry increment hundreds register.
        INC   R1
        SJMP  HUND          ;Repeat until all hundreds are subtracted.

AHEAD:  ADD   A,#64H        ;Correct content of A for extra subtraction.

TENS:   CLR   C
        SUBB  A,#0AH        ;Subtract 10 (0Ah) from A.
        JC    UNIT          ;If no carry increment tens register.
        INC   R2
        SJMP  TENS          ;Repeat until all tens are subtracted.

UNIT:   ADD   A,#0AH        ;Correct content of A for extra subtraction.

        MOV   R3,A          ;Save units in R3.

        MOV   A,R2          ;Combine tens and units.
        SWAP  A
        ADD   A,R3

        INC   DPTR          ;Save tens and units in memory.
        MOVX  @DPTR,A

        INC   DPTR          ;Save hundreds in memory.
        MOV   A,R1
        MOVX  @DPTR,A

HALT:   SJMP  HALT          ;Remain idle in infinite loop. Program end.

        END                 ;Assembly end.
```

Assembler Listing for Example Program 51.9

```
1                              ;PROGRAM TO CONVERT A BINARY DATA TO BCD DATA
2
3    2100                      ORG   2100H        ;Specify program starting address.
4
5    2100    90 24 00          MOV   DPTR,#2400H  ;Load the data address in DPTR.
6    2103    E0                MOVX  A,@DPTR      ;Get the binary data in A-register.
7
8    2104    79 00             MOV   R1,#00H      ;Clear R1 to store hundreds.
9    2106    7A 00             MOV   R2,#00H      ;Clear R2 to store tens.
10
11   210A    C3       HUND:    CLR   C
12   210B    94 64             SUBB  A,#64H       ;Subtract 100 (64h) from A.
13   210D    40 03             JC    AHEAD        ;If no carry, increment hundreds register.
14   210F    09                INC   R1
15   2110    80 F8             SJMP  HUND         ;Repeat until all hundreds are subtracted.
16
17   2112    24 64    AHEAD:   ADD   A,#64H       ;Correct content of A for extra subtraction.
18
19   2114    C3       TENS:    CLR   C
20   2115    94 0A             SUBB  A,#0AH       ;Subtract 10 (0Ah) from A.
21   2117    40 03             JC    UNIT         ;If no carry, increment tens register.
22   2119    0A                INC   R2
23   211A    80 F8             SJMP  TENS         ;Repeat until all tens are subtracted.
24
25   211C    24 0A    UNIT:    ADD   A,#0AH       ;Correct content of A for extra subtraction.
26
27   211E    FB                MOV   R3,A         ;Save units in R3.
28
29   211F    EA                MOV   A,R2         ;Combine tens and units.
30   2120    C4                SWAP  A
31   2121    2B                ADD   A,R3
32
33   2122    A3                INC   DPTR         ;Save tens and units in memory.
34   2123    F0                MOVX  @DPTR,A
35
36   2124    A3                INC   DPTR         ;Save hundreds in memory.
37   2125    E9                MOV   A,R1
38   2126    F0                MOVX  @DPTR,A
39
40   2127    80 FE    HALT:    SJMP  HALT         ;Remain idle in infinite loop. Program end.
41
42   2129                      END                ;Assembly end.
```

Sample Data

Input Data : B9$_H$

Output Data : 0185$_{10}$

Memory address	Content	
2400	B9	Binary data
2401	85	} BCD data
2402	01	

EXAMPLE PROGRAM 51.10: 2-Digit Hexa to ASCII Conversion or 8-bit Binary to ASCII Conversion

Write an assembly language program to convert a 2-digit hexa (8-bit binary) to ASCII code. The binary data is stored in 2400$_H$ and store the ASCII code in 2401$_H$ and 2402$_H$.

Problem Analysis

Each hexa digit (4-bit binary) is represented by an 8-bit ASCII. The hexa digit **0** through **9** are represented by 30_H to 39_H in ASCII. Hence for hexa **0** to **9**, if we add 30_H, we will get the corresponding ASCII. The hexa digit **A** through **F** are represented by 41_H to 46_H in ASCII. Hence for hexa digit **A** to **F** if we add 37_H we will get the corresponding ASCII.

In the following algorithm, the given 8-bit data is split into two nibbles. The ASCII code for each nibble is found by calling a subroutine, which takes care of adding 30_H to the nibble if it is less than $0A_H$, or adding 37_H if the nibble is greater than 09_H.

Flowchart for Example Program 51.10

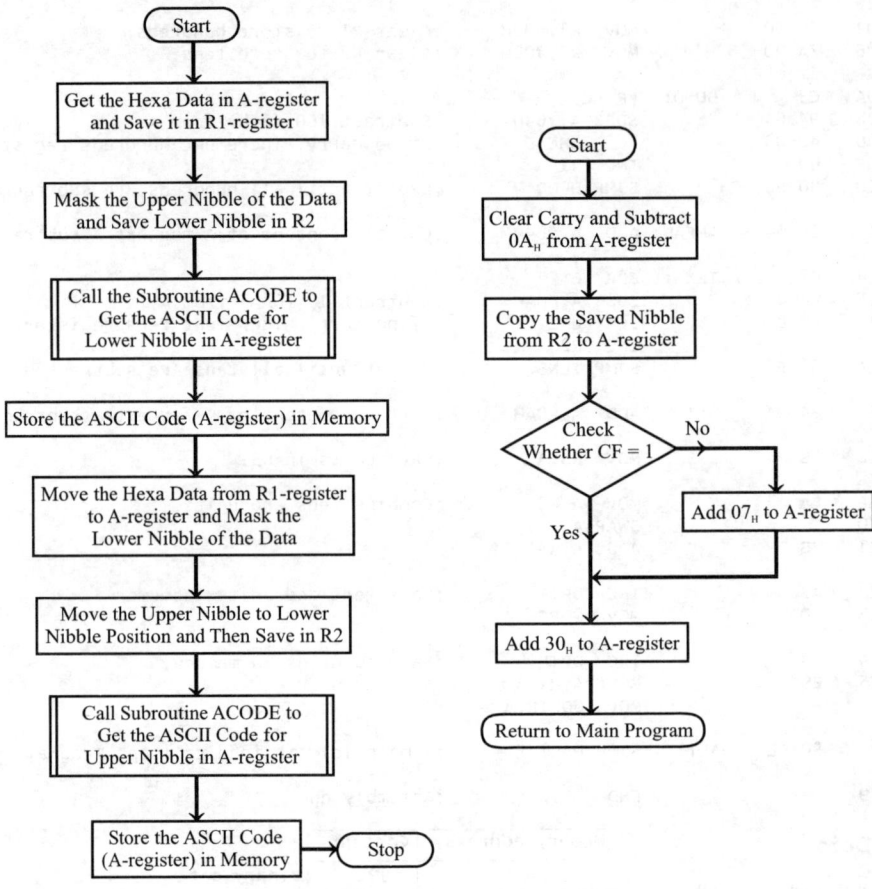

Algorithm

1. Load the given data in A-register and move to R1-register.
2. Mask the upper nibble of the binary (hexa) data in A-register and save the lower nibble in R2-register.
3. Call subroutine acode to get ASCII code of the lower nibble and store in memory.
4. Move the content of R1-register to A-register and mask the lower nibble.
5. Move the upper nibble to lower nibble position and save in R2-register.
6. Call subroutine acode to get the ASCII code of upper nibble and store in memory.
7. Stop.

Algorithm for Subroutine ACODE

1. Clear carry flag, and subtract $0A_H$ from the content of A-register. Move R2 to A.
2. If CF = 1, go to step 4. If CF = 0, go to next step.
3. Add 07_H to A-register.
4. Add 30_H to A-register.
5. Return to main program.

Assembly Language Program

```
;PROGRAM TO CONVERT 2-DIGIT HEXA TO ASCII CODE

        ORG     2100H       ;Specify program starting address.

        MOV     DPTR,#2400H
        MOVX    A,@DPTR      ;Get hexa data in A.
        MOV     R1,A         ;Save the hexa data in R1-register.

        ANL     A,#0FH       ;Mask the upper nibble.
        MOV     R2,A         ;Save lower nibble in R2.
        LCALL   ACODE        ;Call subroutine ACODE to get ASCII code.

        INC     DPTR         ;Increment DPTR.
        MOVX    @DPTR,A      ;save ASCII code of lower nibble in memory.

        MOV     A,R1         ;Get binary data in A.
        ANL     A,#F0H       ;Mask the lower nibble.
        SWAP    A            ;Move upper nibble to lower nibble position.
        MOV     R2,A         ;Save upper nibble in R2.
        LCALL   ACODE        ;Call subroutine ACODE to get ASCII code.

        INC     DPTR         ;Increment DPTR.
        MOVX    @DPTR,A      ;Save ASCII code of upper nibble in memory.

HALT:   SJMP    HALT         ;Remain idle in infinite loop. Program end.

;SUBROUTINE ACODE

ACODE:CLR     C             ;Clear carry flag.
        SUBB    A,#0AH       ;Check whether content A is less than 0AH.
        MOV     A,R2         ;If content of A-register is less than 0AH,
        JC      SKIP         ;then add 30H to A-register,
        ADD     A,#07H       ;otherwise add 37H to A-register.
SKIP:   ADD     A,#30H
        RET                  ;Return to main program.

        END                  ;Assembly end.
```

Assembler Listing for Example Program 51.10

```
1                       ;PROGRAM TO CONVERT 2-DIGIT HEXA TO ASCII CODE
2
3     2100              ORG     2100H        ;Specify program starting address.
4
5     2100  90 24 00    MOV     DPTR,#2400H
6     2103  E0          MOVX    A,@DPTR      ;Get hexa data in A.
7     2104  F9          MOV     R1,A         ;Save the hexa data in R1-register.
8
9     2105  54 0F       ANL     A,#0FH       ;Mask the upper nibble.
10    2107  FA          MOV     R2,A         ;Save lower nibble in R2.
11    2108  12 21 19    LCALL   ACODE        ;Call subroutine ACODE to get ASCII code.
12
13    210B  A3          INC     DPTR         ;Increment DPTR.
14    210C  F0          MOVX    @DPTR,A      ;Save ASCII code of lower nibble in memory.
15
```

16	210D	E9		MOV	A,R1	;Get binary data in A.
17	210E	54 F0		ANL	A,#F0H	;Mask the lower nibble.
18	210O	C4		SWAP	A	;Move upper nibble to lower nibble position.
19	2111	FA		MOV	R2,A	;Save upper nibble in R2.
20	2112	12 21 19		LCALL	ACODE	;Call subroutine ACODE to get ASCII code.
21						
22	2115	A3		INC	DPTR	;Increment DPTR.
23	2116	F0		MOVX	@DPTR,A	;Save ASCII code of upper nibble in memory.
24						
25	2117	80 FE	HALT:	SJMP	HALT	;Remain idle in infinite loop. Program end.
26						
27			;SUBROUTINE ACODE			
28						
29	2119	C3	ACODE:	CLR	C	;Clear carry flag.
30	211A	94 0A		SUBB	A,#0AH	;Check whether content A is less than 0AH.
31	211C	EA		MOV	A,R2	;If content of A-register is less than 0AH,
32	211D	40 02		JC	SKIP	;then add 30H to A-register,
33	211F	24 07		ADD	A,#07H	;otherwise add 37H to A-register.
34	2121	24 30	SKIP:	ADD	A,#30H	
35	2123	22		RET		;Return to main program.
36						
37	2124			END		;Assembly end.

Sample Data

			Memory address	Content	
Input Data	:	E4$_H$			←Hexadecimal
Output Data	:	34 (ASCII code for 4)	2400	E4	
			2401	34	} ASCII
		45 (ASCII code for E)	2402	45	

Example Program 51.11: GCD of Two 8-Bit Data

Write an assembly language program to determine the GCD of two 8-bit data.

Problem Analysis

First divide the smaller data by the larger data and check for remainder. If remainder is zero then the smaller data is the GCD.

If the remainder is not zero then take the remainder as the divisor and the previous divisor as the dividend and repeat the division until the remainder is zero. When the remainder is zero, we can store the divisor as GCD.

Before performing division we can even check whether the dividend and divisor are equal. If they are equal then we can directly store the divisor as GCD without performing division.

Algorithm

1. Set DPTR as pointer for input data.
2. Get one data in A-register and save in R1-register.
3. Increment DPTR and get another data in A-register.
4. Check whether the content of A and R1 are equal using compare instruction.
5. If zero flag is set then go to step 13, otherwise go to next step.
6. Save A in R2. Check whether A is greater than R1 using subtract instruction. Then restore the content of A from R2.
7. If carry flag is not set then go to step 10, otherwise go to next step.
8. Exchange the content of A and R1, so that the larger among the two is dividend and smaller is the divisor.
9. Copy R1 to B, and divide A-register by B-register.
10. Move remainder in B to A. Check whether remainder is zero using compare instruction.
11. If zero flag is set then go to next step, otherwise go to step 4.
12. Move the content of R1 to A.
13. Increment DPTR and store the content of A-register as GCD in memory.
14. Stop.

Flowchart for Example Program 51.11

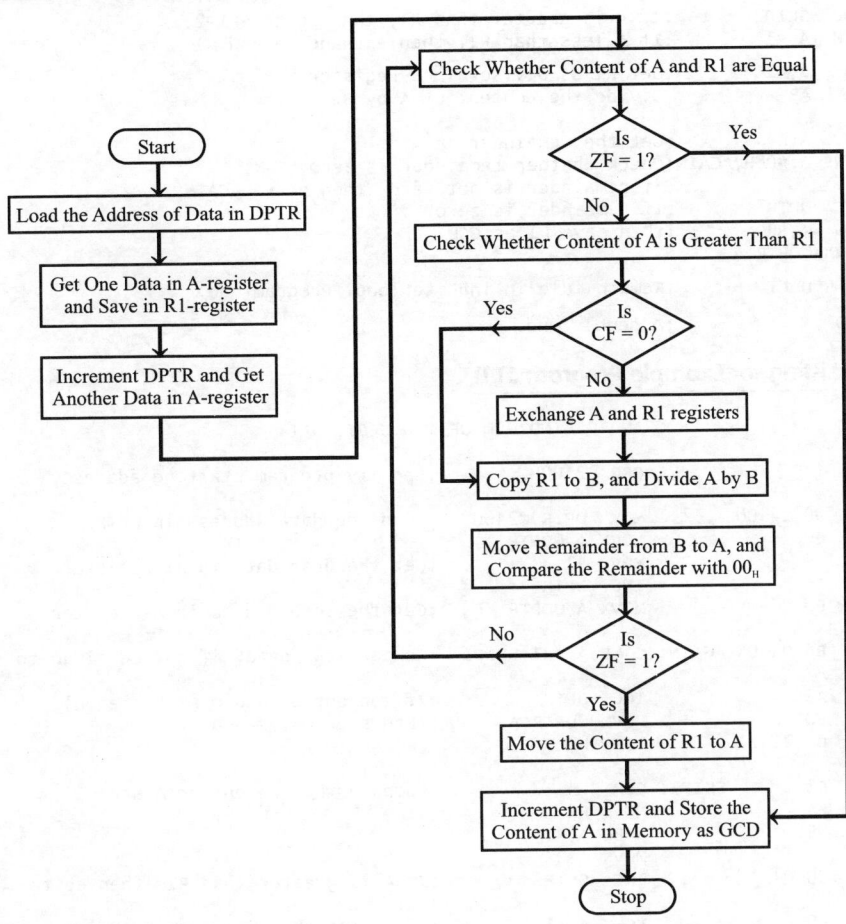

Assembly Language Program

```
;PROGRAM TO FIND GCD OF TWO 8-BIT DATA

        ORG   2100H       ;Specify program starting address.

        MOV   DPTR,#2400H ;Load the data address in DPTR.
        MOVX  A,@DPTR
        MOV   R1,A        ;Get the first data in R1-register.
        INC   DPTR
        MOVX  A,@DPTR     ;Get the second data in A-register.

AGAIN:  CJNE  A,#R1,SKIP1 ;Compare two data, if not equal go to SKIP1.

        INC   DPTR        ;If content of A and R1 are equal,
        MOVX  @DPTR,A     ;then store A as GCD.
        LJMP  HALT

SKIP1:  CLR   C           ;Compare the content of A and R1.
        MOV   R2,A
```

```
            SUBB  A,R1
            MOV   A,R2
            JNC   SKIP2      ;If A is greater than R1, then go to SKIP2.
            XCH   A,R1       ;If A less than R1, then exchange A and R1.
SKIP2:      MOV   B,R1       ;Set R1 as divisor in B-register.
            DIV   AB         ;Divide the content of A by B.

            MOV   A,B        ;Get the remainder in A.
            CJNE  A,#00H,AGAIN;Check whether remainder is zero,
                             ;if remainder is not zero, then go to AGAIN.
            INC   DPTR       ;If remainder is zero,
            MOV   A,R1       ;then save R1 as GCD.
            MOVX  @DPTR,A
HALT:       SJMP  HALT       ;Remain idle in infinite loop. Program end.

            END              ;Assembly end.
```

Assembler Listing for Example Program 51.11

```
1                           ;PROGRAM TO FIND GCD OF TWO 8-BIT DATA
2
3    2100                   ORG   2100H        ;Specify program starting address.
4
5    2100  90 24 00         MOV   DPTR,#2400H  ;Load the data address in DPTR.
7    2103  E0               MOVX  A,@DPTR
8    2104  F9               MOV   R1,A         ;Get the first data in R1-register.
9    2105  A3               INC   DPTR
10   2106  E0               MOVX  A,@DPTR      ;Get the second data in A-register.
11
12   2107  B4 01 05  AGAIN: CJNE  A,#R1,SKIP1  ;Compare two data, if not equal go to SKIP1.
13
14   210A  A3               INC   DPTR         ;If content of A and R1 are equal,
15   210B  F0               MOVX  @DPTR,A      ;then store A as GCD.
16   210C  02 21 21         LJMP  HALT
17
18   210F  C3        SKIP1: CLR   C            ;Compare the content of A and R1.
19   2110  FA               MOV   R2,A
20   2111  99               SUBB  A,R1
21   2112  EA               MOV   A,R2
22   2113  50 01            JNC   SKIP2        ;If A is greater than R1, then go to SKIP2.
23
24   2115  C9               XCH   A,R1         ;If A less than R1, then exchange A and R1.
25
26   2116  89 F0     SKIP2: MOV   B,R1         ;Set R1 as divisor in B-register.
27   2118  84               DIV   AB           ;Divide the content of A by B.
28
29   2119  E5 F0            MOV   A,B          ;Get the remainder in A.
30   211B  B4 00 E9         CJNE  A,#00H,AGAIN ;Check whether remainder is zero,
31                                             ;if remainder is not zero, then go to AGAIN.
32   211E  A3               INC   DPTR         ;If remainder is zero,
33   211F  E9               MOV   A,R1         ;then save R1 as GCD.
34   2120  F0               MOVX  @DPTR,A
35
36   2121  80 FE     HALT:  SJMP  HALT         ;Remain idle in infinite loop. Program end.
37
38   2123                   END                ;Assembly end.
```

Sample Data

Input Data : Data1 = 0C$_H$ Output Data : 04$_H$

Data2 = 04$_H$

Memory address	Content	
2400	0C	← Data 1
2401	04	← Data 2
2402	04	← GCD

Example Program 51.12: LCM of Two 8-Bit Data

Write an assembly language program to determine the LCM of the two 8-bit data.

Problem Analysis

First determine the GCD of two data. Then determine the product of the two data. Here, it is assumed that the product does not exceed 8 bits. When the product is divided by GCD, the quotient will be the LCM of the two data. (For the GCD of two data please refer to example program 11.)

Algorithm

1. Set DPTR as pointer for input data.
2. Get one data in A-register and save in R1 and R4 registers.
3. Increment DPTR and get another data in A-register and save in R5-register.
4. Call subroutine GCD to get the GCD in A-register.
5. Save GCD in R3-register.
6. Copy two input data from R4 and R5 into A and B registers, and get the product of two data in A-register.
7. Copy GCD from R3 to B. Divide the product in A by GCD in B. The quotient is LCM.
8. Increment DPTR, and save the LCM in memory.
9. Stop.

> **Note:** The algorithm for subroutine GCD can be obtained from example program 11.

Flowchart for Example Program 51.12

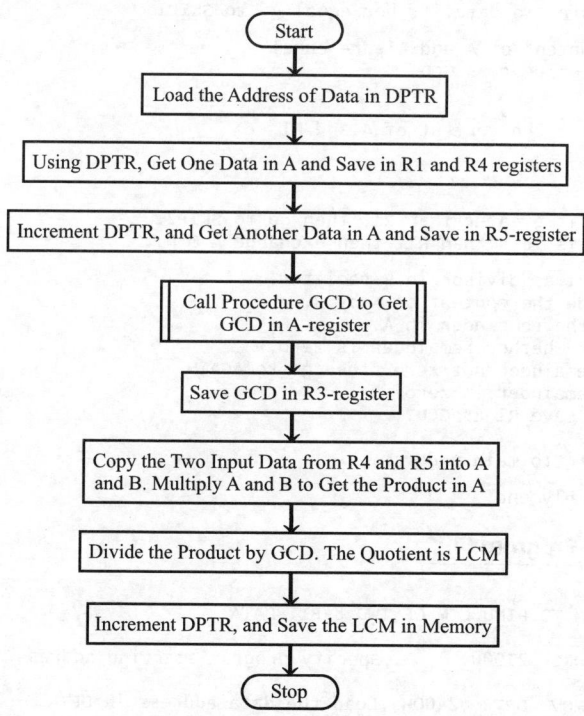

> **Note:** The flowchart for procedure GCD can be obtained from example program 11.

Assembly Language Program

```
;PROGRAM TO FIND LCM OF TWO 8-BIT DATA

        ORG    2100H       ;Specify program starting address.
        MOV  DPTR,#2400H   ;Load the data address in DPTR.
        MOVX A,@DPTR
        MOV  R1,A          ;Get the first data in R1-register.
        MOV  R4,A          ;Save the first data in R4-register.
        INC  DPTR
        MOVX A,@DPTR       ;Get the second data in A-register.
        MOV  R5,A          ;Save the second data in R5.

        LCALLGCD
        MOV  R3,A

        MOV  B,R4          ;Get the product of two data in A.
        MOV  A,R5
        MUL  AB
        MOV  B,R3          ;Set R3 (GCD) as divisor in B-register.
        DIV  AB            ;Divide product by GCD, the quotient is LCM.

        INC  DPTR          ;Store LCM in memory.
        MOVX @DPTR,A
HALT:   SJMP HALT          ;Remain idle in infinite loop. Program end.

;SUBROUTINE TO FIND GCD OF TWO 8-BIT DATA

GCD:    CJNE A,#R1,SKIP1   ;Compare two data, if not equal go to SKIP1.

        INC  DPTR          ;If content of A and R1 are equal,
        MOVX @DPTR,A       ;then store A as GCD.
        LJMP RET1

SKIP1:  CLR  C             ;Compare the content of A and R1.
        MOV  R2,A
        SUBB A,R1
        MOV  A,R2
        JNC  SKIP2         ;If A is greater than R1, then go to SKIP2.
        XCH  A,R1          ;If A is less than R1, then exchange A & R1.

SKIP2:  MOV  B,R1          ;Set R1 as divisor in B-register.
        DIV  AB            ;Divide the content of A by B.
        MOV  A,B           ;Get the remainder in A.
        CJNE A,#00H,GCD    ;Check whether remainder is zero,
                           ;if remainder not zero, then go to AGAIN.
        INC  DPTR          ;If remainder is zero,
        MOV  A,R1          ;then save R1 as GCD.
        MOVX @DPTR,A
RET1:   RET                ;Return to main program.

        END                ;Assembly end.
```

Assembler Listing for Example Program 51.12

```
1                        ;PROGRAM TO FIND LCM OF TWO 8-BIT DATA
2
3   2100                     ORG  2100H         ;Specify program starting address.
4
5   2100   90 24 00          MOV  DPTR,#2400H   ;Load the data address in DPTR.
6   2103   E0                MOVX A,@DPTR
7   2104   F9                MOV  R1,A          ;Get the first data in R1-register.
```

8	2105	FC	MOV R4,A	;Save the first data in R4-register.
9	2106	A3	INC DPTR	
10	2107	E0	MOVX A,@DPTR	;Get the second data in A-register.
11	2108	FD	MOV R5,A	;Save the second data in R5.
12				
13	2109	12 21 18	LCALL GCD	
14	210C	FB	MOV R3,A	
15				
16	210D	8C F0	MOV B,R4	;Get the product of two data in A.
17	210F	ED	MOV A,R5	
18	2110	A4	MUL AB	
19	2111	8B F0	MOV B,R3	;Set R3 (GCD) as divisor in B-register.
20	2113	84	DIV AB	;Divide product by GCD, the quotient is LCM.
21				
22	2114	A3	INC DPTR	;Store LCM in memory.
23	2115	F0	MOVX @DPTR,A	
24				
25	2116	80 FE	HALT: SJMP HALT	;Remain idle in infinite loop. Program end.
26				
27			;SUBROUTINE TO FIND GCD OF TWO 8-BIT DATA	
28				
29	2118	B4 01 05	GCD: CJNE A,#R1,SKIP1	;Compare two data, if not equal go to SKIP1.
30				
31	211B	A3	INC DPTR	;If content of A and R1 are equal,
32	211C	F0	MOVX @DPTR,A	;then store A as GCD.
33	211D	02 21 32	LJMP RET1	
34				
35	2120	C3	SKIP1: CLR C	;Compare the content of A and R1.
36	2121	FA	MOV R2,A	
37	2122	99	SUBB A,R1	
38	2123	EA	MOV A,R2	
39	2124	50 01	JNC SKIP2	;If A is greater than R1, then go to SKIP2.
40	2126	C9	XCH A,R1	;If A is less than R1, then exchange A & R1.
41				
42	2127	89 F0	SKIP2: MOV B,R1	;Set R1 as divisor in B-register.
43	2129	84	DIV AB	;Divide the content of A by B.
44	212A	E5 F0	MOV A,B	;Get the remainder in A.
45	212C	B4 00 E9	CJNE A,#00H,GCD	;Check whether remainder is zero,
46				;if remainder not zero, then go to AGAIN.
47	212F	A3	INC DPTR	;If remainder is zero,
48	2130	E9	MOV A,R1	;then save R1 as GCD.
49	2131	F0	MOVX @DPTR,A	
50	2132	22	RET1: RET	;Return to main program.
51				
52	2133		END	;Assembly end.

Sample Data

Input Data :	Output Data :
Data1 = 0C$_H$	GCD = 04$_H$
Data2 = 04$_H$	LCM = 0C$_H$

Memory address	Content	
2400	0C	} Input data
2401	04	
2402	04	← GCD
2403	0C	← LCM

8.14 EXAMPLES OF 8086 ASSEMBLY LANGUAGE PROGRAMS

Note : 1. *The example programs 86.1 to 86.26 given in this book can be run on any 8086 microprocessor trainer kit. Since the initialization of segment registers and stack are taken care by the monitor program in the trainer kits, those initializations are not included in the example programs 86.1 to 86.26.*

2. *The example programs 86.27 to 86.30 given in this book can be run on INTEL microprocessor-based PC (Personal Computer) or its compatibility. In these programs the PC keyboard is used as input device and the monitor is used as the output device. The DOS and BIOS interrupts are used to access these devices.*

EXAMPLE PROGRAM 86.1: 16-bit Addition

Write an assembly language program to add two numbers of a 16-bit data.

Problem Analysis

To perform addition in 8086, one of the data should be stored in a register and another data can be stored in the register/memory. After addition, the sum will be available in the destination register/memory. The sum of two 16-bit data can be either 16 bits (sum only) or 17 bits (sum and carry). The destination register/memory can accommodate only the sum and if there is a carry, the 8086 will indicate by setting carry flag. Hence, one of the registers is used to account for carry. The program for addition in 8086 has been presented by three methods. In Method I, immediate addressing is used for input data and direct addressing is used for output data. In Method II, direct addressing is used for input and output data. In Method III, indexed addressing is employed.

Algorithm (Method I)

1. Load the first data in AX-register.

2. Load the second data in BX-register.

3. Clear CL-register.

4. Add the two data and get the sum in AX-register.

5. Store the sum in memory.

6. Check for carry. If carry flag is set then go to next step, otherwise go to step 8.

7. Increment CL-register.

8. Store the carry in memory.

9. Stop.

Flowchart (Method I)

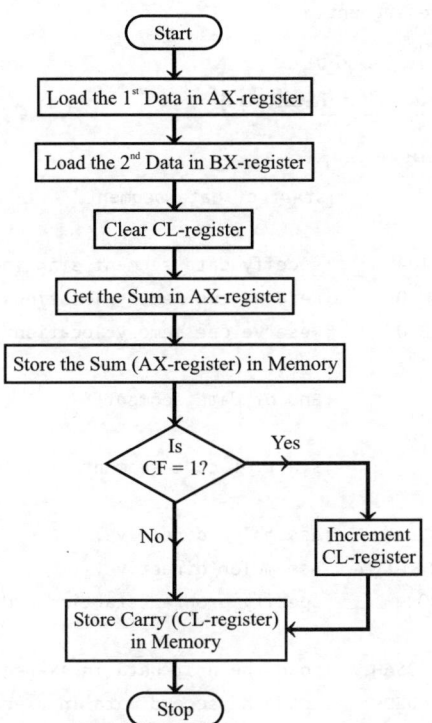

Assembly Language Program (Method I)

```
;PROGRAM TO ADD TWO 16-BIT DATA (METHOD-1)

DATA SEGMENT                 ;start of data segment.

        ORG      1104H       ;specify data segment starting address.
        SUM      DW 0        ;Reserve two memory locations for sum.
        CARRY    DB 0        ;Reserve one memory location for carry.

DATA ENDS                    ;End of data segment.

CODE SEGMENT                 ;Start of code segment.

        ASSUME CS:CODE       ;Assembler directive.
        ASSUME DS:DATA       ;Assembler directive.
        ORG 1000H            ;specify program starting address.

        MOV AX,205AH         ;Load the first data in AX-register.
        MOV BX,40EDH         ;Load the second data in BX-register.
        MOV CL,00H           ;Clear the CL-register for carry.
        ADD AX,BX            ;Add the two data, sum will be in AX.
        MOV SUM,AX           ;Store the sum in memory location (1104H).
        JNC AHEAD            ;Check the status of carry flag.
        INC CL               ;If carry flag is set,increment CL by one.
AHEAD:  MOV CARRY,CL         ;Store the carry in memory location (1106H).
```

```
        HLT                   ;Halt program execution.

CODE ENDS                     ;End of code segment.
END                           ;Assembly end.
```

Assembler Listing for Example Program 86.1 (Method I)

```
;PROGRAM TO ADD TWO 16-BIT DATA (METHOD-1)

0000                   DATA SEGMENT           ;start of data segment.

1104                        ORG    1104H      ;specify data segment starting address.
1104   0000                 SUM    DW 0       ;Reserve two memory locations for sum.
1106   00                   CARRY  DB 0       ;Reserve one memory location for carry.

1107                   DATA ENDS              ;End of data segment.

0000                   CODE SEGMENT           ;start of code segment.

                            ASSUME CS:CODE    ;Assembler directive.
                            ASSUME DS:DATA    ;Assembler directive.
1000                        ORG 1000H         ;specify program starting address.

1000   B8 205A              MOV AX,205AH      ;Load the first data in AX-register.
1003   BB 40ED              MOV BX,40EDH      ;Load the second data in BX-register.
1006   B1 00                MOV CL,00H        ;Clear the CL-register for carry.
1008   03 C3                ADD AX,BX         ;Add the two data, sum will be in AX.
100A   A3 1104 R            MOV SUM,AX        ;Store the sum in memory location (1104H).
100D   73 02                JNC AHEAD         ;Check the status of carry flag.
100F   FE  C1               INC CL            ;If carry flag is set,increment CL by one.
1011   88  0E 1106 R AHEAD: MOV CARRY,CL      ;Store the carry in memory location (1106H).
1015   F4                   HLT               ;Halt program execution.

1016                   CODE ENDS              ;End of code segment.
                       END                    ;Assembly end.
```

Sample Data

Input Data : $205A_H$　　Output Data : SUM = 6147_H

　　　　　　$40ED_H$　　　　　　　　　CARRY = 00_H

Memory address	Content
1104	47
1105	61
1106	00

Algorithm (Method II)

1. **Get the first data in AX-register.**
2. **Clear CL register.**
3. **Add the second data to AX-register and get the sum in AX-register.**
4. **Store the sum in memory.**

5. Check for carry. If carry flag is set then go to next step, otherwise go to step 7.
6. Increment CL-register.
7. Store the carry in memory.
8. Stop.

Flowchart (Method II)

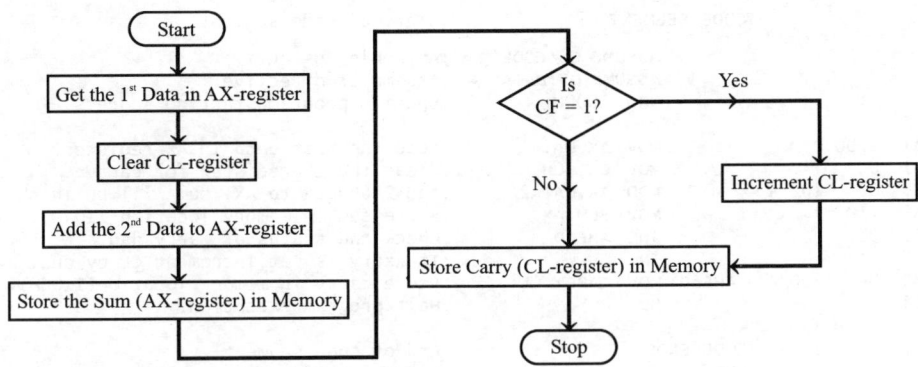

Assembly Language Program (Method II)

```
;PROGRAM TO ADD TWO 16-BIT DATA (METHOD-2)

DATA SEGMENT                ;Start of data segment.

        ORG    1100H        ;Specify data segment starting address.
        DATA1 DW  0         ;Reserve two memory locations for DATA1.
        DATA2 DW  0         ;Reserve two memory locations for DATA2.
        SUM   DW  0         ;Reserve two memory locations for sum.
        CARRY DB  0         ;Reserve one memory location for carry.

DATA ENDS                   ;End of data segment.
CODE SEGMENT                ;Start of code segment.

        ASSUME CS:CODE      ;Assembler directive.
        ASSUME DS:DATA      ;Assembler directive.
        ORG 1000H           ;specify program starting address.
        MOV AX,DATA1        ;Load the first data in AX-register.
        MOV CL,00H          ;Clear the CL-register for carry.
        ADD AX,DATA2        ;Add 2nd data to AX, sum will be in AX.
        MOV SUM,AX          ;Store sum in memory location (1104H).
        JNC AHEAD           ;Check the status of carry flag.

        INC CL              ;If carry is set,increment CL by one.
AHEAD:  MOV CARRY,CL        ;Store carry in memory location (1106H).

        HLT                 ;Halt program execution.
CODE ENDS                   ;End of code segment.
END                         ;Assembly end.
```

Assembler Listing for Example Program 86.1 (Method II)

```
;PROGRAM TO ADD TWO 16-BIT DATA (METHOD-2)

0000              DATA SEGMENT              ;Start of data segment.
```

```
1100                          ORG    1100H         ;Specify data segmentstarting address.
1100   0000                   DATA1  DW 0          ;Reserve two memory locations for DATA1.
1102   0000                   DATA2  DW 0          ;Reserve two memory locations for DATA2.


1104   0000                   SUM    DW 0          ;Reserve two memory locations for sum.
1106   00                     CARRY  DB 0          ;Reserve one memory location for carry.

1107                  DATA ENDS                    ;End of data segment.
0000                  CODE SEGMENT                 ;Start of code segment.

                             ASSUME CS:CODE        ;Assembler directive.
                             ASSUME DS:DATA        ;Assembler directive.
1000                         ORG 1000H             ;Specify program starting address.

1000   A1  1100 R            MOV AX,DATA1          ;Load the first data in AX-register.
1003   B1  00                MOV CL,00H            ;Clear the CL-register for carry.
1005   03  06   1102 R       ADD AX,DATA2          ;Add 2nd data to AX, sum will be in AX.
1009   A3  1104 R            MOV SUM,AX            ;Store sum in memory location(1104H).
100C   73  02                JNC AHEAD             ;Check the status of carry flag.
100E   FE  C1                INC CL                ;If carry is set,increment CL by one.
1010   88  0E   1106 R  AHEAD: MOV CARRY,CL        ;Store carry in memory location(1106H).
1014   F4                    HLT                   ;Halt program execution.

1015                  CODE ENDS                    ;End of code segment.
                      END                          ;Assembly end.
```

Sample Data

Input Data : Data1 = F048$_H$ Output Data : Sum   = 009A$_H$
              Data2 = 1052$_H$ Carry = 01$_H$

Memory address	Content
1100	48
1101	F0
1102	52
1103	10

Memory address	Content
1104	9A
1105	00
1106	01

Algorithm (Method III)

1. Set SI-register as pointer for data.
2. Get the first data in AX-register.
3. Get the second data in BX-register.
4. Clear CL-register.
5. Get the sum in AX-register.
6. Store the sum in memory.
7. Check for carry. If carry flag is set then go to next step, otherwise go to step 9.
8. Increment CL-register.
9. Store the carry in memory.
10. Stop.

Flowchart (Method III)

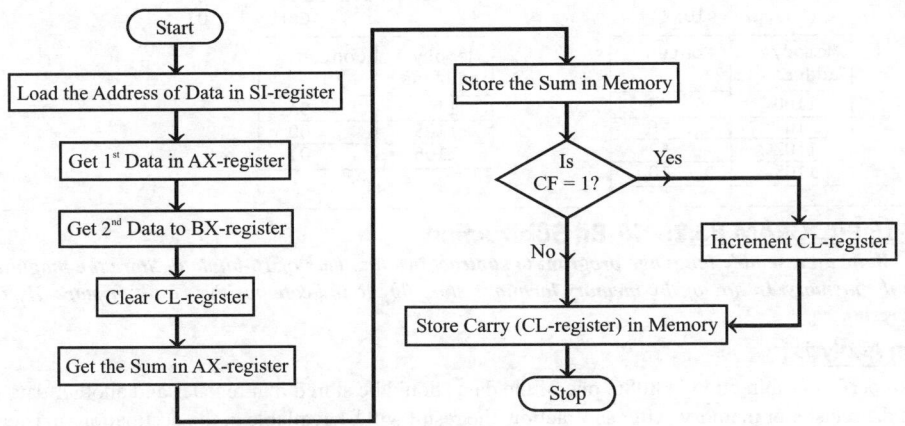

Assembly Language Program (Method III)

```
;PROGRAM TO ADD TWO 16-BIT DATA (METHOD-3)

CODE SEGMENT                    ;Start of code segment.

        ASSUME CS:CODE          ;Assembler directive.
        ORG 1000H               ;specify program starting address.

        MOV  SI,1100H           ;Set SI-register as pointer for data.
        MOV  AX,[SI]            ;Get the first data in AX-register.
        MOV  BX,[SI+2]          ;Get the second data in BX-register.
        MOV  CL,00H             ;Clear the CL-register for carry.
        ADD  AX,BX              ;Add the two data, sum will be in AX-register.
        MOV  [SI+4],AX          ;Store the sum in memory location (1104H).
        JNC  AHEAD              ;Check the status of carry flag.
        INC  CL                 ;If carry flag is set,increment CL by one.
AHEAD:  MOV  [SI+6],CL          ;Store carry in memory location (1106H).
        HLT                     ;Halt program execution.
CODE ENDS                       ;End of code segment.
END                             ;Assembly end.
```

Assembler Listing for Example Program 86.1 (Method III)

```
;PROGRAM TO ADD TWO 16-BIT DATA (METHOD-3)

0000                CODE SEGMENT            ;Start of code segment.

                    ASSUME CS:CODE          ;Assembler directive.
1000                ORG 1000H               ;specify program starting address.

1000  BE 1100       MOV  SI,1100H           ;Set SI-register as pointer for data.
1003  8B 04         MOV  AX,[SI]            ;Get the first data in AX-register.
1005  8B 5C 02      MOV  BX,[SI+2]          ;Get the second data in BX-register.
1008  B1 00         MOV  CL,00H             ;Clear the CL-register for carry.
100A  03 C3         ADD  AX,BX              ;Add the two data, sum will be in AX-register.
100C  89 44 04      MOV  [SI+4],AX          ;Store the sum in memory location (1104H).
100F  73 02         JNC  AHEAD              ;Check the status of carry flag.
1011  FE C1         INC  CL                 ;If carry flag is set,increment CL by one.
1013  88 4C 06 AHEAD: MOV [SI+6],CL         ;Store carry in memory location (1106H).
1016  F4            HLT                     ;Halt program execution.
1017            CODE ENDS                   ;End of code segment.
                END                         ;Assembly end.
```

Sample Data

Input Data : Data1 = F048$_H$
                Data2 = 1052$_H$

Output Data : Sum = 009A$_H$
                  Carry = 01$_H$

Memory address	Content
1100	48
1101	F0
1102	52
1103	10

Memory address	Content
1104	9A
1105	00
1106	01

EXAMPLE PROGRAM 86.2: 16-Bit Subtraction

Write an assembly language program to subtract two numbers of 16-bit data. Store the magnitude of the result in the memory. In one of the memory locations store 00$_H$ to indicate positive result or store 01$_H$ to indicate negative result.

Problem Analysis

To perform subtraction in 8086, one of the data should be stored in a register and another data should be stored in the register or memory. After subtraction, the result will be available in the destination register/memory. The 8086 will perform 2's complement subtraction and then complement the carry. Therefore, if the result is negative then the carry flag is set and the destination register/memory will have 2's complement of the result. Hence, one of the registers is used to account for sign of the result. To get the magnitude of the result, again take 2's complement of the result.

Flowchart for Example Program 86.2

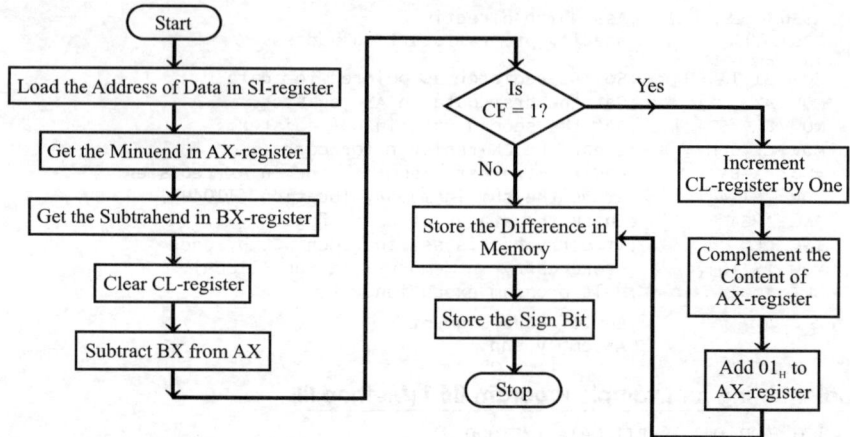

Algorithm

1. Set SI-register as pointer for data.
2. Get the minuend in AX-register.
3. Get the subtrahend in BX-register.
4. Clear CL-register to account for sign.
5. Subtract the content of BX from AX, the difference will be in AX.
6. Check for carry. If carry flag is set then go to next step, otherwise go to step 9.
7. Increment CL-register by one.
8. Take 2's complement of the difference in AX-register (For this, complement AX and add one.)
9. Store the magnitude of difference in memory.
10. Store the sign bit in memory.
11. Stop.

Assembly Language Program

```
;PROGRAM TO SUBTRACT TWO 16-BIT DATA
CODE SEGMENT              ;Start of code segment.
        ASSUME CS:CODE   ;Assembler directive.
        ORG 1000H        ;specify program starting address.
        MOV SI,1100H     ;Load the address of data in SI-register.
        MOV AX,[SI]      ;Get the minuend in AX-register.
        MOV BX,[SI+2]    ;Get the subtrahend in BX-register.
        MOV CL,00H       ;Clear the CL-register to account for sign.
        SUB AX,BX        ;Get the difference in AX-register.
        JNC STORE        ;Check the status of carry flag.
        INC CL           ;If carry flag is set,increment CL by one,
        NOT AX           ;then take 2's complement of difference.
        ADD AX,0001H
STORE:  MOV [SI+4],AX    ;Store difference in memory location (1104H).
        MOV [SI+6],CL    ;Store sign bit in memory location (1106H).
        HLT              ;Halt program execution.

CODE ENDS                ;End of code segment.
END                      ;Assembly end.
```

Assembler Listing for Example Program 86.2

```
;PROGRAM TO SUBTRACT TWO 16-BIT DATA
0000                CODE SEGMENT          ;Start of code segment.

                    ASSUME CS:CODE ;Assembler directive.
1000                ORG 1000H      ;specify program starting address.

1000  BE 1100       MOV  SI,1100H  ;Load the address of data in SI-register.
1003  8B 04         MOV  AX,[SI]   ;Get the minuend in AX-register.
1005  8B 5C 02      MOV  BX,[SI+2] ;Get the subtrahend in BX-register.
1008  B1 00         MOV  CL,00H    ;Clear the CL-register to account for sign.
100A  2B C3         SUB  AX,BX     ;Get the difference in AX-register.
100C  73 07         JNC  STORE     ;Check the status of carry flag.
100E  FE C1         INC  CL        ;If carry flag is set,increment CL by one,
1010  F7 D0         NOT  AX        ;then take 2's complement of the difference.
1012  05 0001       ADD  AX,0001H

1015  89 44 04  STORE: MOV [SI+4],AX ;Store difference in memory location (1104H).
1018  88 4C 06         MOV [SI+6],CL ;Store sign bit in memory location (1106H).
101B  F4               HLT           ;Halt program execution.

101C                CODE ENDS      ;End of code segment.
                    END            ;Assembly end.
```

Sample Data

Input Data   : Minuend   = 840C$_H$
               Subtrahend = B2CA$_H$

Output Data : Difference = 2EBE$_H$
               Sign Bit   = 01$_H$

Memory address	Content	
1100	0 C	} Minuend
1101	84	
1102	CA	} Subtrahend
1103	B 2	
1104	B E	} Difference
1105	2 E	
1106	01	← Sign bit

EXAMPLE PROGRAM 86.3:   Multibyte Addition

Write an assembly language program to add two numbers of multibyte data.

Problem Analysis

In the 8086 processor, the multibyte data can be added either byte by byte or word by word. The number of bytes in the data can be used as a count for the number of additions. One of the registers is used to account for the final carry.

To perform addition, we require three address pointers: Two pointers for input data and one pointer for output data.

Flowchart for Example Program 86.3

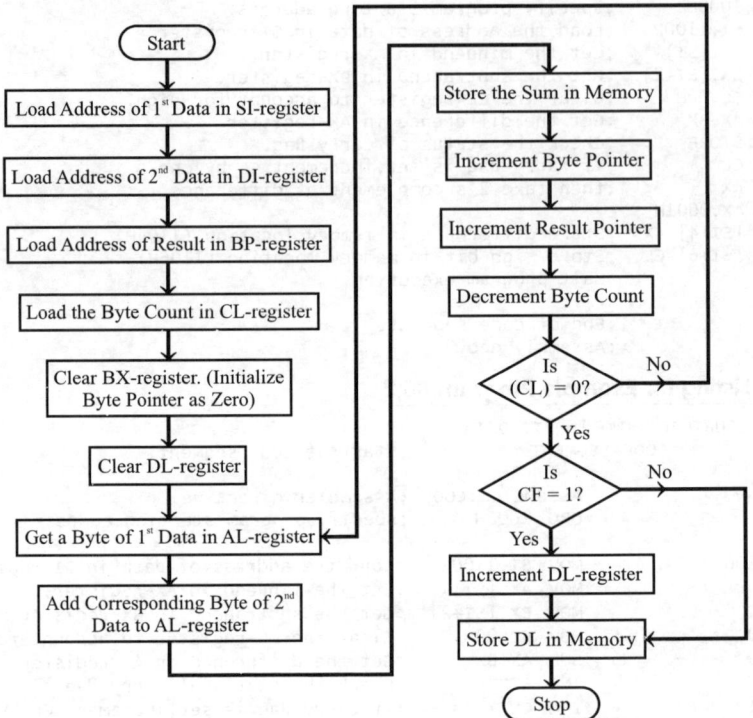

Algorithm

1. Load the starting address of 1st data in SI-register.
2. Load the starting address of 2nd data in DI-register.
3. Load the starting address of result in BP-register.
4. Load the byte count in CL-register.
5. Let BX-register be byte pointer. Initialize byte pointer as zero.
6. Clear DL-register to account for final carry.
7. Clear carry flag (i.e., initial carry is zero).
8. Load a byte of 1st data in AL-register.
9. Add the corresponding byte of 2nd data in memory to AL-register along with previous carry.
10. Store the sum in memory.
11. Increment the byte pointer (BX) and result pointer (BP).
12. Decrement the byte count (CL).
13. If byte count (CL) is zero, go to next step, otherwise go to step 8.
14. Check for carry. If carry flag is set then go to next step, otherwise go to step 16.
15. Increment DL-register.
16. Store final carry in memory.
17. Stop.

Assembly Language Program

```
;PROGRAM TO ADD TWO MULTIBYTE DATA

CODE SEGMENT                    ;Start of code segment.

        ASSUME CS:CODE  ;Assembler directive.
        ORG 1000H       ;specify program starting address.

        MOV  SI,1100H   ;Set SI-register as pointer for 1st data.
        MOV  DI,1201H   ;Set DI-register as pointer for 2nd data.
        MOV  BP,1301H   ;Set BP-register as pointer for result.
        MOV  CL,[SI]    ;Load the count for number of bytes in CL.
        INC  SI         ;Set SI to point to 1st byte of 1st data.
        MOV  BX,00H     ;Initialize byte pointer as zero.
        MOV  DL,00H     ;Initialize final carry as zero.
        CLC             ;Clear carry flag.

REPEAT: MOV  AL,[SI+BX] ;Get a byte of 1st data in AL-register.
        ADC  AL,[DI+BX] ;Add the corresponding byte of 2nd data to AL.
        MOV  [BP],AL    ;Store sum of corresponding bytes in memory.
        INC  BX         ;Increment the byte pointer.
        INC  BP         ;Increment the result pointer.
        LOOP REPEAT     ;Repeat addition until byte count is zero.

        JNC  AHEAD      ;Check for final carry.
        INC  DL         ;If carry flag is set then increment DL.
AHEAD:  MOV  [BP],DL    ;Store the final carry in memory.
        HLT             ;Halt program execution.

CODE ENDS                       ;End of code segment.
END                             ;Assembly end.
```

Assembler Listing for Example Program 86.3

```
;PROGRAM TO ADD TWO MULTIBYTE DATA

0000                    CODE SEGMENT            ;Start of code segment.

                        ASSUME CS:CODE  ;Assembler directive.
1000                    ORG 1000H       ;specify program starting address.

1000  BE 1100           MOV  SI,1100H   ;Set  SI-register as pointer for 1st data.
1003  BF 1201           MOV  DI,1201H   ;Set  DI-register as pointer for 2nd data.
1006  BD 1301           MOV  BP,1301H   ;Set  BP-register as pointer for result.
1009  8A 0C             MOV  CL,[SI]    ;Load the count for number of bytes in CL.
100B  46                INC  SI         ;Set SI to point to 1st byte of 1st data.
100C  BB 0000           MOV  BX,00H     ;Initialize byte pointer as zero.
100F  B2 00             MOV  DL,00H     ;Initialize final carry as zero.
1011  F8                CLC             ;Clear carry flag.

1012  8A 00     REPEAT: MOV  AL,[SI+BX] ;Get a byte of 1st data in AL-register.
1014  12 01             ADC  AL,[DI+BX] ;Add corresponding byte of 2nd da      AL.
1016  88 46 00          MOV  [BP],AL    ;Store sum of corresponding bytes in memory.
1019  43                INC  BX         ;Increment the byte pointer.
101A  45                INC  BP         ;Increment the result pointer.
101B  E2 F5             LOOP REPEAT     ;Repeat addition until byte count is zero.

101D  73 02             JNC  AHEAD      ;Check for final carry.
101F  FE C2             INC  DL         ;If carry flag is set then increment DL.
1021  88 56 00  AHEAD:  MOV  [BP],DL    ;Store the final carry in memory.
1024  F4                HLT             ;Halt program execution.

1025                    CODE ENDS               ;End of code segment.
                        END                     ;Assembly end.
```

Sample Data

1ˢᵗ Data : F5C2647217ₕ

Memory address	Content	
1100	05	← Count
1101	17	
1102	72	
1103	64	Data 1
1104	C2	
1105	F5	

2ⁿᵈ Data : C265750712ₕ

Memory address	Content	
1200	05	← Count
1201	12	
1202	07	
1203	75	Data 2
1204	65	
1205	C2	

Output Data : 01B827D97929ₕ

Memory address	Content	
1301	29	
1302	79	
1303	D9	Sum
1304	27	
1305	B8	
1306	01	← Carry

EXAMPLE PROGRAM 86.4:   Multibyte Subtraction

Write an assembly language program to subtract two numbers of multibyte data.

Problem Analysis

In the 8086 processor, the multibyte data can be subtracted either byte by byte or word by word. The number of bytes in the data can be used as count for number of subtractions. One of the registers is used to account for the final borrow.

To perform subtraction we require three pointers: Two pointers for input data and one pointer for output data.

Algorithm

1. Load the starting address of minuend in SI-register.
2. Load the starting address of subtrahend in DI-register.
3. Load the starting address of result in BP-register.
4. Load the byte count in CL-register.
5. Let BX-register be byte pointer. Initialize byte pointer as zero.
6. Clear DL-register to account for final borrow.
7. Clear carry flag, i.e., initial borrow is zero.
8. Load a byte of minuend in AL-register.
9. Subtract the corresponding byte of subtrahend in memory from AL-register along with previous borrow.
10. Store the difference in memory.
11. Increment the byte pointer (BX) and result pointer (BP).
12. Decrement the byte count (CL).
13. If byte count (CL) is zero then go to next step, otherwise go to step 8.
14. Check for carry flag, if carry flag is set then go to next step, otherwise go to step 16.
15. Increment DL-register.
16. Store final borrow in memory.
17. Stop.

Flowchart

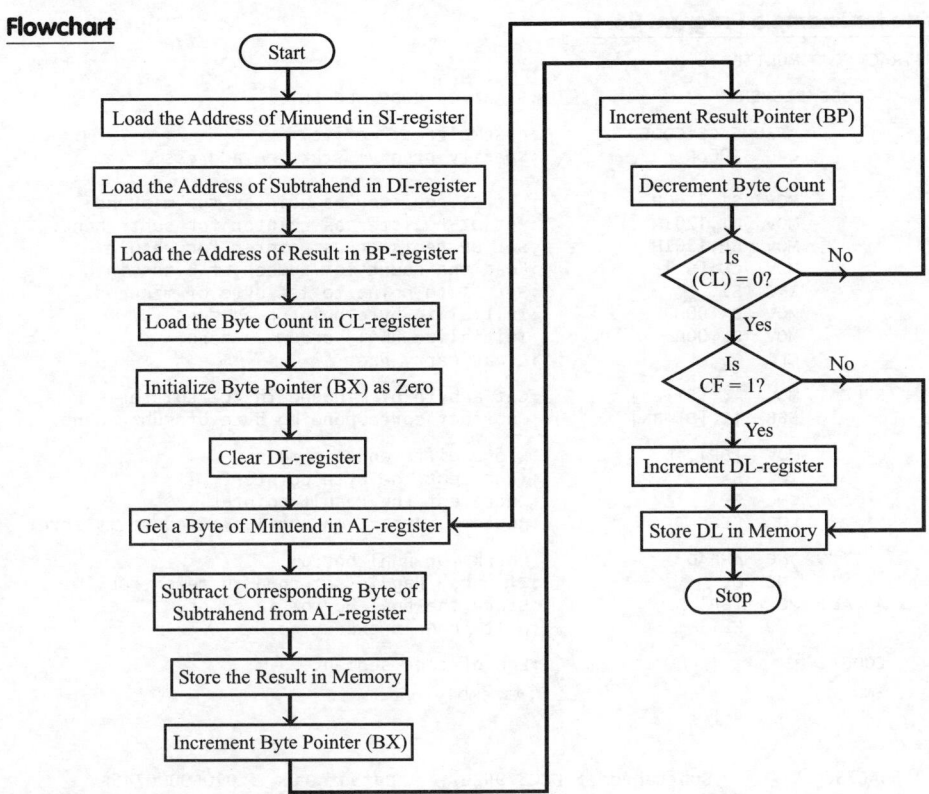

Assembly Language Program

```
;PROGRAM TO SUBTRACT TWO MULTIBYTE DATA

CODE SEGMENT                    ;Start of code segment.

        ASSUME CS:CODE          ;Assembler directive.
        ORG  1000H              ;specify program starting address.

        MOV  SI,1100H           ;Set SI-register as pointer for minuend.
        MOV  DI,1201H           ;Set DI-register as pointer for subtrahend.
        MOV  BP,1301H           ;Set BP-register as pointer for result.
        MOV  CL,[SI]            ;Load the count for number of bytes in CL.
        INC  SI                 ;Set SI to point to 1st byte of minuend.
        MOV  BX,00H             ;Initialize byte pointer as zero.
        MOV  DL,00H             ;Initialize final borrow as zero.
        CLC                     ;Clear carry flag.

REPEAT: MOV  AL,[SI+BX]         ;Get a byte of minuend in AL-register.
        SBB  AL,[DI+BX]         ;Subtract corresponding byte of subtrahend.
        MOV  [BP],AL            ;Store difference in memory.
        INC  BX                 ;Increment the byte pointer.
        INC  BP                 ;Increment the result pointer.
        LOOP REPEAT             ;Repeat subtraction until byte count is zero.

        JNC  AHEAD              ;Check for final borrow.
        INC  DL                 ;If carry flag is set then increment DL.

AHEAD:  MOV  [BP],DL            ;Store the final borrow in memory.
        HLT                     ;Halt program execution.
CODE ENDS                       ;End of code segment.
END                             ;Assembly end.
```

Assembler Listing for Example Program 86.4

```
;PROGRAM TO SUBTRACT TWO MULTIBYTE DATA

0000                    CODE SEGMENT              ;START OF CODE SEGMENT.
                        ASSUME CS:CODE            ;Assembler directive.
1000                    ORG  1000H                ;Specify program starting address.

1000  BE 1100           MOV  SI,1100H             ;Set SI-register as pointer for minuend.
1003  BF 1201           MOV  DI,1201H             ;Set DI-register as pointer for subtrahend.
1006  BD 1301           MOV  BP,1301H             ;Set BP-register as pointer for result.
1009  8A 0C             MOV  CL,[SI]              ;Load the count for number of bytes in CL.
100B  46                INC  SI                   ;Set SI to point to 1st byte of minuend.
100C  BB 0000           MOV  BX,00H               ;Initialize byte pointer as zero.
100F  B2 00             MOV  DL,00H               ;Initialize final borrow as zero.
1011  F8                CLC                       ;Clear carry flag.

1012  8A 00      REPEAT: MOV  AL,[SI+BX]          ;Get a byte of minuend in AL-register.
1014  1A 01            SBB  AL,[DI+BX]           ;Subtract corresponding byte of subtrahend.

1016  88 46 00          MOV  [BP],AL              ;Store difference in memory.
1019  43                INC  BX                   ;Increment the byte pointer.
101A  45                INC  BP                   ;Increment the result pointer.
101B  E2 F5             LOOP REPEAT               ;Repeat subtraction until byte count is zero.

101D  73 02             JNC  AHEAD                ;Check for final borrow.
101F  FE C2             INC  DL                   ;If carry flag is set then increment DL.
1021  88 56 00   AHEAD: MOV  [BP],DL              ;Store the final borrow in memory.
1024  F4                HLT                       ;Halt program execution.

1025                    CODE ENDS                 ;End of code segment.
                        END                       ;Assembly end.
```

Sample Data

Minuend : D2564A6756ₕ   Subtrahend : F2C579F2E7ₕ   Output Data : 01DF90D0746F

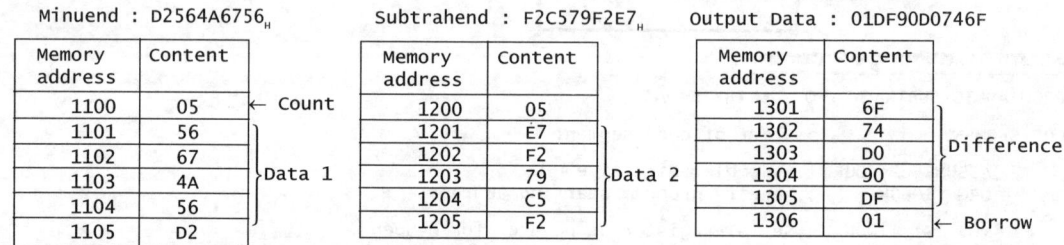

Memory address	Content	
1100	05	← Count
1101	56	⎫
1102	67	⎬ Data 1
1103	4A	
1104	56	
1105	D2	⎭

Memory address	Content	
1200	05	
1201	E7	⎫
1202	F2	⎬ Data 2
1203	79	
1204	C5	
1205	F2	⎭

Memory address	Content	
1301	6F	
1302	74	⎫
1303	D0	⎬ Difference
1304	90	
1305	DF	⎭
1306	01	← Borrow

EXAMPLE PROGRAM 86.5:   Sum of an Array

Write an assembly language program to determine the sum of elements in an array.

Problem Analysis

Declare the content of one of the registers as sum and take initial value of sum as zero. The sum of elements of array can be obtained by adding the elements of array one by one (i.e., byte by byte) to sum. The number of bytes in the array can be used as count for the number of additions to be performed. The carry in each addition can be separately added in a register and saved as high byte of sum.

Algorithm

1. Load the address of the array in SI-register.
2. Load the address of the result in DI-register.
3. Load the count value in CL-register.
4. Let the content of AX be sum and keep initial sum as zero.
5. Add a byte of array to sum.
6. Check for carry. If carry flag is set then go to next step, otherwise go to step 8.
7. Increment high byte of sum (AH-register).
8. Increment the array pointer.
9. Decrement the count (CL-register).
10. If count (CL) is zero then go to next step, otherwise go to step 5.
11. Store the 16-bit sum in memory.
12. Stop.

Flowchart for Example Program 86.5

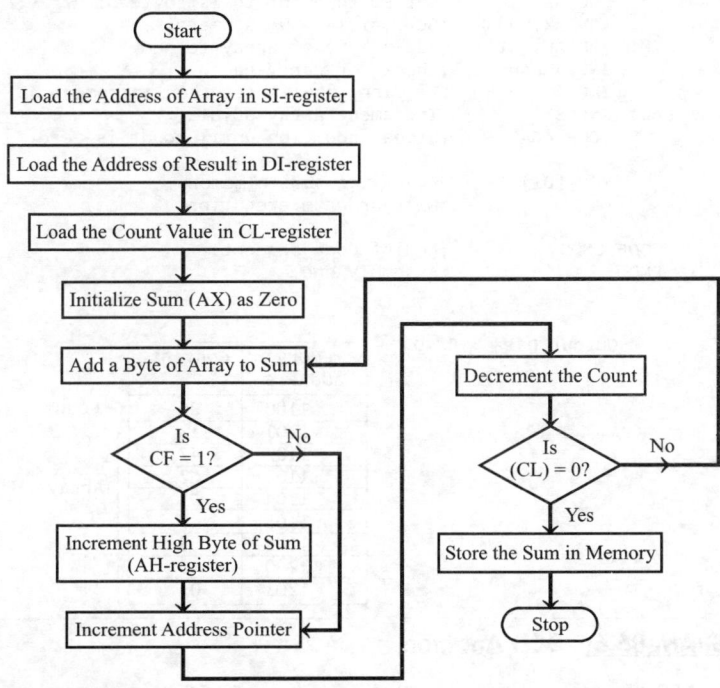

Assembly Language Program

```
;PROGRAM TO FIND THE SUM OF THE ELEMENTS IN AN ARRAY
CODE SEGMENT            ;Start of code segment.
      ASSUME CS:CODE    ;Assembler directive.
      ORG  1000H        ;specify program starting address.

      MOV  SI,1100H     ;Set SI-register as pointer for array.
      MOV  DI,1200H     ;Set DI-register as pointer for result.
      MOV  CL,[SI]      ;Set CL as count for number of bytes in array.
      INC  SI           ;Set SI to point to 1st byte of array.
      MOV  AX,0000H     ;Set initial sum as zero.
AGAIN: ADD  AL,[SI]     ;Add a byte of array to sum.
      JNC      AHEAD    ;Check for carry flag.
      INC  AH           ;If carry flag is set then increment AH.
```

```
AHEAD: INC SI                    ;Increment array pointer.
       LOOP AGAIN                ;Repeat addition until count is zero.
       MOV [DI],AX               ;Store the sum in memory.
       HLT                       ;Halt program execution.

CODE ENDS                        ;End of code segment.
END                              ;Assembly end.
```

Assembler Listing for Example Program 86.5

```
;PROGRAM TO FIND THE SUM OF THE ELEMENTS IN AN ARRAY
0000                 CODE SEGMENT             ;Start of code segment.

                     ASSUME CS:CODE;Assembler directive.
1000                 ORG 1000H               ;specify program starting address.

1000  BE 1100        MOV SI,1100H   ;Set SI-register as pointer for array.
1003  BF 1200        MOV DI,1200H   ;Set DI-register as pointer for result.
1006  8A 0C          MOV CL,[SI]    ;Set CL as count for number of bytes in array.
1008  46             INC SI         ;Set SI to point to 1st byte of array.
1009  B8 0000        MOV AX,0000H   ;Set initial sum as zero.
100C  02 04   AGAIN: ADD AL,[SI]    ;Add a byte of array to sum.
100E  73 02          JNC AHEAD      ;Check for carry flag.
1010  FE C4          INC AH         ;If carry flag is set then increment AH.
1012  46      AHEAD: INC SI         ;Increment array pointer.
1013  E2 F7          LOOP AGAIN     ;Repeat addition until count is zero.

1015  89 05          MOV [DI],AX    ;Store the sum in memory.
1017  F4             HLT            ;Halt program execution.

1018                 CODE ENDS      ;End of code segment.
                     END            ;Assembly end.
```

Sample Data

Input Data :06   Output Data : 02AD$_H$

Memory address	Content	
1100	06	←Count
1101	12	
1102	47	
1103	C2	
1104	F5	Array
1105	47	
1106	56	
1200	AD	
1201	02	Sum

Input Data :
```
06
12
47
C2
F5
47
56
```

EXAMPLE PROGRAM 86.6: BCD Addition

Write an assembly language program to add two numbers of BCD data.

Problem Analysis

The 8086 processor will perform only binary addition. Hence, for BCD addition, the binary addition of BCD data is performed and then the sum is corrected to get the result in BCD. After the binary addition the following correction should be made to get the result in BCD:

1. If the sum of lower nibble exceeds 9 or if there is auxiliary carry then 6 is added to lower nibble.

2. If the sum of upper nibble exceeds 9 or if there is carry then 6 is added to upper nibble.

The above correction is taken care of by the DAA (**D**ecimal **A**djust **A**ccumulator) instruction. Therefore, after binary addition, execute the DAA instruction to do the above correction in the sum.

Algorithm

1. Load the address of data in SI-register.
2. Clear CL-register to account for carry.
3. Load the first data in AX-register and second data in BX-register.
4. Perform binary addition of low byte of data to get the sum in AL-register.
5. Adjust the sum of low bytes to BCD.
6. Save the sum of low bytes in DL-register.
7. Get the high byte of first data in AL-register.
8. Add the high byte of second data and previous carry to AL-register. Now the sum of high bytes will be in AL-register.
9. Adjust the sum of high bytes to BCD.
10. Save the sum of high bytes in DH-register.
11. Check for carry. If carry flag is set then go to next step, otherwise go to step 13.
12. Increment CL-register.
13. Save the sum (DX-register) in memory.
14. Save the carry (CL-register) in memory.
15. Stop.

Flowchart for Example Program 86.6

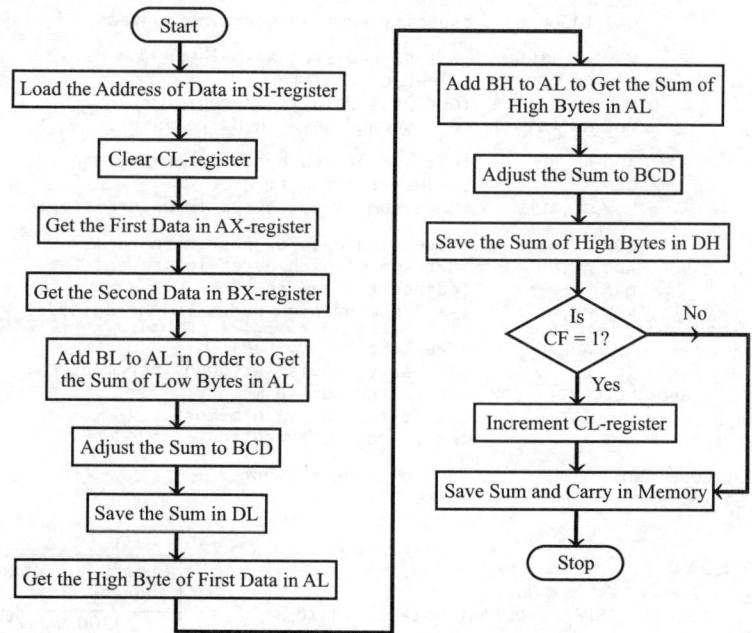

Assembly Language Program

```
;PROGRAM TO ADD TWO BCD DATA

CODE SEGMENT              ;Start of code segment.

        ASSUME CS:CODE   ;Assembler directive.
        ORG 1000H        ;specify program starting address.
        MOV SI,1100H     ;Set SI-register as pointer for data.
        MOV CL,00H       ;Clear CL-register.
        MOV AX,[SI]      ;Get first data in AX-register.
        MOV BX,[SI+2]    ;Get second data in BX-register.
```

```
          ADD  AL,BL        ;Get sum of low bytes in AL-register.
          DAA               ;Adjust the sum to BCD.
          MOV  DL,AL        ;Save sum of low bytes in DL-register.

          MOV  AL,AH        ;Move high byte of first data to AL.
          ADC  AL,BH        ;Get sum of high bytes in AL-register.
          DAA               ;Adjust the sum to BCD.
          MOV  DH,AL        ;Save sum of high bytes in DH-register.

          JNC  AHEAD        ;Check for carry flag.
          INC  CL           ;If carry flag is set then increment CL.
AHEAD:    MOV  [SI+4],DX    ;Store the sum in memory.
          MOV  [SI+6],CL    ;Store the carry in memory.
          HLT               ;Halt program execution.

CODE ENDS                   ;End of code segment.
END                         ;Assembly end.
```

Assembler Listing for Example Program 86.6

```
;PROGRAM TO ADD TWO BCD DATA
0000              CODE SEGMENT              ;Start of code segment.

                  ASSUME CS:CODE ;Assembler directive.
1000              ORG 1000H                 ;specify program starting address.

1000  BE 1100     MOV SI,1100H     ;Set SI-register as pointer for data.
1003  B1 00       MOV CL,00H       ;Clear CL-register.
1005  8B 04       MOV AX,[SI]      ;Get first data in AX-register.
1007  8B 5C 02    MOV BX,[SI+2]    ;Get second data in BX-register.

100A  02 C3       ADD AL,BL        ;Get sum of low bytes in AL-register.
100C  27          DAA              ;Adjust the sum to BCD.
100D  8A D0       MOV DL,AL        ;Save sum of low bytes in DL-register.

100F  8A C4       MOV AL,AH        ;Move high byte of first data to AL.
1011  12 C7       ADC AL,BH        ;Get sum of high bytes in AL-register.
1013  27          DAA              ;Adjust the sum to BCD.
1014  8A F0       MOV DH,AL        ;Save sum of high bytes in DH-register.

1016  73 02       JNC AHEAD        ;Check for carry flag.
1018  FE C1       INC CL           ;If carry flag is set then increment CL.
101A  89 54 04  AHEAD: MOV [SI+4],DX ;Store the sum in memory.
101D  88 4C 06    MOV [SI+6],CL    ;Store the carry in memory.
1020  F4          HLT              ;Halt program execution.

1021              CODE ENDS                 ;End of code segment.
                  END                       ;Assembly end.
```

Sample Data

Input Data : Data1 = 4578_{10} Output Data : 013176_{10}

Data2 = 8598_{10}

Memory address	Content	
1100	78	} Data 1
1101	45	
1102	98	} Data 2
1103	85	
1104	76	} Sum
1105	31	
1106	01	←Carry

EXAMPLE PROGRAM 86.7: BCD Subtraction

Write an assembly language program to subtract two numbers of BCD data.

Problem Analysis

The 8086 processor will perform only binary subtraction. Hence, for BCD subtraction, the binary subtraction of BCD data is performed and then the difference is corrected to get the result in BCD. After binary subtraction, the following correction should be made to get the result in BCD.

1. If the difference of lower nibble exceeds 9 or if there is auxiliary carry then 6 is subtracted from lower nibble.

2. If the difference of upper nibble exceeds 9 or if there is carry then 6 is subtracted from upper nibble.

The above correction is taken care by the DAS (**D**ecimal **A**djust after **S**ubtraction) instruction. Therefore, after binary subtraction, execute DAS to do the above correction in the difference.

Flowchart for Example Program 86.7

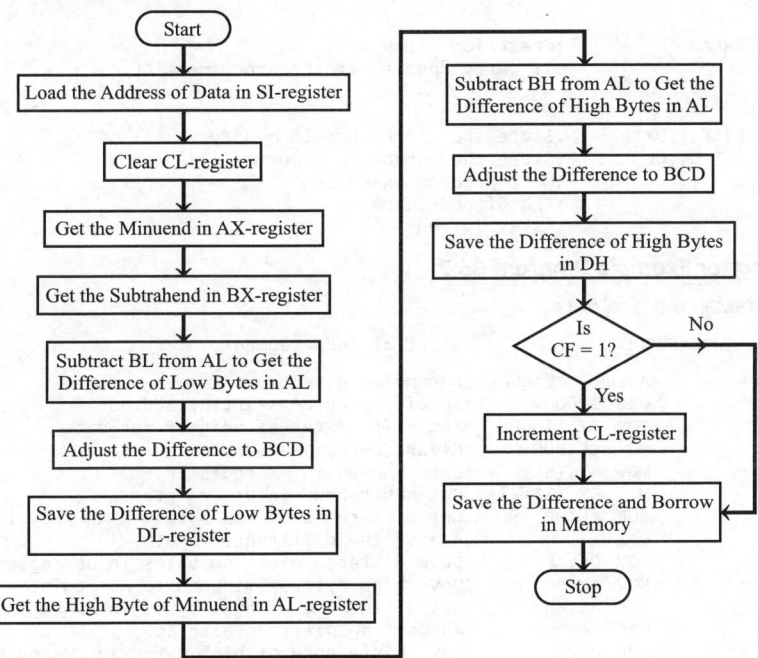

Algorithm

1.   Load the address of data in SI-register.
2.   Clear CL-register to account for borrow.
3.   Load the minuend in AX-register.
4.   Get the subtrahend in BX-register.
5.   Subtract BL from AL to get the difference of low bytes in AL.
6.   Adjust the difference of low bytes to BCD and then save it in DL-register.
7.   Get the high byte of minuend in AL-register.
8.   Subtract BH and the previous borrow from AL to get the difference of high bytes in AL-register.
9.   Adjust the difference of high bytes to BCD and save it in DH-register.
10.   Check for carry. If carry flag is set then go to next step, otherwise go to step 12.
12.   Increment CL-register.
13.   Save the difference (DX-register) and the borrow (CL-register) in memory.
14.   Stop.

Assembly Language Program

```
;PROGRAM TO SUBTRACT TWO BCD DATA

CODE SEGMENT                                      ;Start of code segment.
            ASSUME CS:CODE                        ;Assembler directive.
            ORG 1000H                             ;specify program starting address.
            MOV SI,1100H                          ;Set SI-register as pointer for data.
            MOV CL,00H                            ;Clear CL-register.
            MOV AX,[SI]                           ;Get minuend in AX-register.
            MOV BX,[SI+2]                         ;Get subtrahend in BX-register.
            SUB AL,BL                             ;Get difference of low bytes in AL-register.
            DAS                                   ;Adjust the difference to BCD.
            MOV DL,AL                             ;Save difference of low bytes in DL-register.
            MOV AL,AH                             ;Move high byte of minuend to AL-register.
            SBB AL,BH                             ;Get difference of high bytes in AL-register.
            DAS                                   ;Adjust the difference to BCD.
            MOV DH,AL                             ;Save difference of high bytes in DH-register.


            JNC AHEAD                             ;Check for carry flag.
            INC CL                                ;If carry flag is set then increment CL.

AHEAD:      MOV [SI+4],DX                         ;Store the difference in memory.
            MOV [SI+6],CL                         ;Store the borrow in memory.
            HLT                                   ;Halt program execution.
CODE ENDS                                         ;End of code segment.
END                                              ;Assembly end.
```

Assembler Listing for Example Program 86.7

```
;PROGRAM TO SUBTRACT TWO BCD DATA

0000                CODE SEGMENT                 ;Start of code segment.

                    ASSUME CS:CODE  ;Assembler directive.
1000                ORG 1000H                    ;specify program starting address.
1000   BE 1100      MOV SI,1100H                 ;Set SI-register as pointer for data.
1003   B1 00        MOV CL,00H                   ;Clear CL-register.
1005   8B 04        MOV AX,[SI]                  ;Get minuend in AX-register.
1007   8B 5C 02     MOV BX,[SI+2]                ;Get subtrahend in BX-register.
100A   2A C3        SUB AL,BL                    ;Get difference of low bytes in AL-register.
100C   2F           DAS                          ;Adjust the difference to BCD.
100D   8A D0        MOV DL,AL                    ;Save difference of low bytes in DL-register.
100F   8A C4        MOV AL,AH                    ;Move high byte of minuend to AL-register.
1011   1A C7        SBB AL,BH                    ;Get difference of high bytes in AL-register.
1013   2F           DAS                          ;Adjust the difference to BCD.
1014   8A F0        MOV DH,AL                    ;Save difference of high bytes in DH-register.
1016   73 02        JNC AHEAD                    ;Check for carry flag.
1018   FE C1        INC CL                       ;If carry flag is set then increment CL.
101A   89 54 04   AHEAD:  MOV [SI+4],DX          ;Store the difference in memory.
101D   88 4C 06     MOV [SI+6],CL               ;Store the borrow in memory.
1020   F4           HLT                          ;Halt program execution.
1021                CODE ENDS                    ;End of code segment.
                    END                          ;Assembly end.
```

Sample Data

Memory address	Content	
1100	72	} Minuend
1101	95	
1102	93	} Subtrahend
1103	47	
1104	79	} Difference
1105	47	
1106	00	← Borrow

Input Data   : Minuend $= 9572_{10}$
                 Subtrahend $= 4793_{10}$

Output Data   : 004779_{10}

EXAMPLE PROGRAM 86.8: Multiplication

Write an assembly language program to multiply two numbers of 16-bit data.

Problem Analysis

To perform multiplication in the 8086 processor, one of the data should be stored in AX-register and another data can be stored in the register/memory. After multiplication, the product will be in AX and DX registers.

Algorithm

1. Load the address of data in SI-register.
2. Get the first data in AX-register.
3. Get the second data in BX-register.
4. Multiply the content of AX and BX. The product will be in AX and DX.
5. Save the product (AX and DX) in memory.
6. Stop.

Flowchart for Example Program 86.8

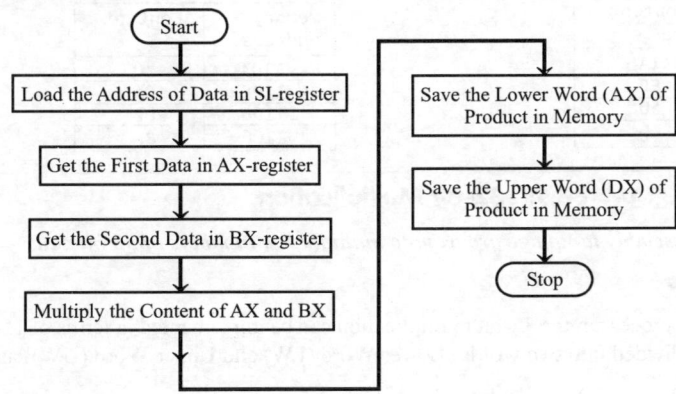

Assembly Language Program

```
;PROGRAM TO MULTIPLY TWO 16-BIT DATA

CODE SEGMENT             ;Start of code segment.

        ASSUME CS:CODE   ;Assembler directive.
        ORG 1000H        ;specify program starting address.

        MOV SI,1100H     ;Set SI as pointer for data.
        MOV AX,[SI]      ;Get the 1st data in AX-register.
        MOV BX,[SI+2]    ;Get the 2nd data in BX-register.
        MUL BX           ;Multiply AX and BX.
                         ;The product will be in AX and DX registers.
        MOV [SI+4],AX    ;Save the lower 16 bits of product in memory.
        MOV [SI+6],DX    ;Save the upper 16 bits of product in memory.
        HLT              ;Halt program execution.

CODE ENDS                ;End of code segment.
END                      ;Assembly end.
```

Assembler Listing for Example Program 86.8

```
;PROGRAM TO MULTIPLY TWO 16-BIT DATA
0000                 CODE SEGMENT        ;Start of code segment.

                     ASSUME CS:CODE      ;Assembler directive.
1000                 ORG  1000H          ;specify program starting address.

1000   BE 1100       MOV  SI,1100H       ;Set SI as pointer for data.
1003   8B 04         MOV  AX,[SI]        ;Get the 1st data in AX-register.
1005   8B 5C 02      MOV  BX,[SI+2]      ;Get the 2nd data in BX-register.
1008   F7 E3         MUL  BX             ;Multiply AX and BX.
                                         ;The product will be in AX and DX registers.
100A   89 44 04      MOV  [SI+4],AX      ;Save the lower 16 bits of product in memory.
100D   89 54 06      MOV  [SI+6],DX      ;Save the upper 16 bits of product in memory.
1010   F4            HLT                 ;Halt program execution.

1011                 CODE ENDS           ;End of code segment.
                     END                 ;Assembly end.
```

Sample Data

Input Data : Data1 = EF1A$_H$
              Data2 = CD50$_H$

Output Data : BFC28A20$_H$

Memory address	Content
1100	1A
1101	EF
1102	50
1103	CD

Memory address	Content
1104	20
1105	8A
1106	C2
1106	BF

EXAMPLE PROGRAM 86.9: 32-Bit Multiplication

Write an assembly language program to multiply two numbers of 32-bit data.

Problem Analysis

In the 8086 processor, the 32-bit multiplication can be implemented in terms of 16-bit multiplication. The given data can be divided into two words (**Lower Word (LW)** and **Upper Word (UW)**) as shown below:

Data1 (32-bit) $\rightarrow$ D1$_{UW}$ (16-bit), D1$_{LW}$ (16-bit)
Data2 (32-bit) $\rightarrow$ D2$_{UW}$ (16-bit), D2$_{LW}$ (16-bit)

Then perform the following four multiplications. Each multiplication will give a 32-bit result which can be divided into **Lower Word (LW)** and **Upper Word (UW)** as shown below:

Product 1 (P1)   : D1$_{LW}$ $\times$ D2$_{LW}$ = P1$_{UW}$, P1$_{LW}$
Product 2 (P2)   : D1$_{UW}$ $\times$ D2$_{LW}$ = P2$_{UW}$, P2$_{LW}$
Product 3 (P3)   : D1$_{LW}$ $\times$ D2$_{UW}$ = P3$_{UW}$, P3$_{LW}$
Product 4 (P4)   : D1$_{UW}$ $\times$ D2$_{UW}$ = P4$_{UW}$, P4$_{LW}$

The result of the above four multiplications can be added to get the final result as shown below: The final product will have a size of four words and they are denoted as P$_{W1}$, P$_{W2}$, P$_{W3}$ and P$_{W4}$.

$$
\begin{array}{cccc}
 & & D1_{UW} & D1_{LW} \\
 & \times & D2_{UW} & D2_{LW} \\
\hline
 & & P1_{UW} & P1_{LW} \\
 & P2_{UW} & P2_{LW} & \\
 & P3_{UW} & P3_{LW} & \\
P4_{UW} & P4_{LW} & & \\
\hline
P_{W4} & P_{W3} & P_{W2} & P_{W1} \\
\hline
\end{array}
$$

Flowchart for Example Program 86.9

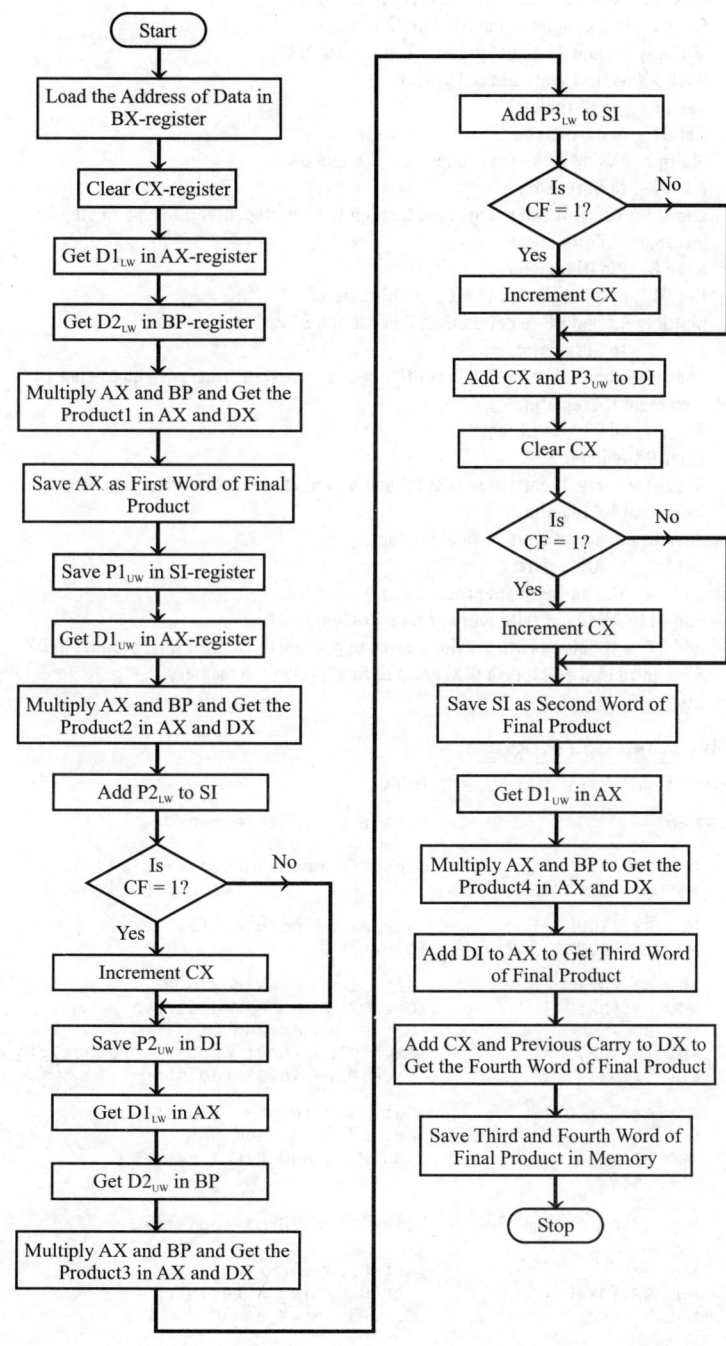

Algorithm

1. Load the address of data in BX-register.
2. Clear CX-register to account for carry in additions.
3. Get $D1_{LW}$ in AX-register and $D2_{LW}$ in BP-register.
4. Multiply AX and BP to get product 1 in AX and DX.
5. Save AX as first word of final product.
6. Save $P1_{UW}$ in SI-register.
7. Get $D1_{UW}$ in AX-register.
8. Multiply AX and BP to get product 2 in AX and DX.
9. Add $P2_{LW}$ to SI-register.
10. Check for carry. If carry flag is set then go to next step, otherwise go to step 12.
11. Increment CX-register.
12. Save $P2_{UW}$ in DI-register.
13. Get $D1_{LW}$ in AX-register and $D2_{UW}$ in BP-register.
14. Multiply AX and BP to get product 3 in AX and DX.
15. Add $P3_{LW}$ to SI-register.
16. Check for carry. If carry flag is set then go to next step, otherwise go to step 18.
17. Increment CX-register.
18. Add CX and $P3_{UW}$ to DI-register.
19. Clear CX-register.
20. Check for carry. If carry flag is set then go to next step, otherwise go to step 22.
21. Increment CX-register.
22. Save SI as second word of final product.
23. Get $D1_{UW}$ in AX-register.
24. Multiply AX and BP to get product 4 in AX and DX.
25. Add DI to AX to get third word of final product in AX.
26. Add CX to DX along with previous carry to get fourth word of final product in DX.
27. Save third (AX) and fourth (DX) word of final product in memory.
28. Stop.

Assembly Language Program

```
;PROGRAM TO MULTIPLY TWO 32-BIT DATA

CODE SEGMENT                      ;Start of code segment.

        ORG 1000H                 ;specify program starting address.
        ASSUME CS:CODE                      ;Assembler directive.

        MOV BX,1100H              ;Set BX as pointer for data.
        MOV CX,0000H             ;Clear CX.

P1:     MOV AX,[BX]              ;Get D1LW in AX-register.
        MOV BP,[BX+04]           ;Get D2LW in BP-register.
        MUL BP                    ;Get P1 in AX and DX.
        MOV [BX+08],AX           ;Save P1LW (first word of product) in memory.
        MOV SI,DX                ;Save P1UW in SI-register.

P2:     MOV AX,[BX+02]           ;Get D1UW in AX-register.
        MUL BP                    ;Get P2 in AX and DX.
        ADD SI,AX                ;Add P1UW and P2LW.
        JNC SKIP1
        INC CX
SKIP1:  MOV DI,DX                ;Save P2UW in DI-register.

P3:     MOV AX,[BX]              ;Get D1LW in AX-register.
        MOV BP,[BX+06]           ;Get D2UW in BP-register.
        MUL BP                    ;Get P3 in AX and DX.

        ADD  SI,  AX             ;Get sum of P1UW, P2LW and P3LW in SI.
```

```
              JNC  SKIP2
              INC  CX
SKIP2:  ADD  DX,CX        ;Get sum of P2UW and P3UW in DI-register.
              MOV  CX,0000H
              ADC  DI,DX
              JNC  SKIP3
              INC  CX
SKIP3:  MOV  [BX+0AH],SI ;Save second word of the product in memory.

P4:     MOV  AX,[BX+02]  ;Get D1UW in AX-register.
              MUL  BP           ;Get P4 in AX and DX.
              ADD  AX,DI        ;Get sum of P2UW, P3UW and P4LW in AX.
              ADC  DX,CX
              MOV  [BX+0CH],AX ;Save third word of the product in memory.
              MOV  [BX+0EH],DX ;Save fourth word of the product in memory.
              HLT               ;Halt program execution.

CODE ENDS                    ;End of code segment.
END                          ;Assembly end.
```

Assembler Listing for Example Program 86.9

```
;PROGRAM TO MULTIPLY TWO 32-BIT DATA
0000                    CODE SEGMENT              ;Start of code segment.

1000                    ORG  1000H               ;specify program starting address.
                        ASSUME CS:CODE           ;Assembler directive.

1000  BB 1100           MOV  BX,1100H            ;Set BX as pointer for data.
1003  B9 0000           MOV  CX,0000H            ;Clear CX.

1006  8B 07      P1:    MOV  AX,[BX]             ;Get D1LW in AX-register.
1008  8B 6F 04          MOV  BP,[BX+04]          ;Get D2LW in BP-register.
100B  F7 E5             MUL  BP                  ;Get P1 in AX and DX.
100D  89 47 08          MOV  [BX+08],AX          ;Save P1LW (first word of product) in memory.
1010  8B F2             MOV  SI,DX               ;Save P1UW in SI-register.

1012  8B 47 02   P2:    MOV  AX,[BX+02]          ;Get D1UW in AX-register.
1015  F7 E5             MUL  BP                  ;Get P2 in AX and DX.
1017  03 F0             ADD  SI,AX               ;Add P1UW and P2LW.
1019  73 01             JNC  SKIP1
101B  41                INC  CX
101C  8B FA     SKIP1:  MOV  DI,DX               ;Save P2UW in DI-register.

101E  8B 07      P3:    MOV  AX,[BX]             ;Get D1LW in AX-register.
1020  8B 6F 06          MOV  BP,[BX+06]          ;Get D2UW in BP-register.
1023  F7 E5             MUL  BP                  ;Get P3 in AX and DX.
1025  03 F0             ADD  SI,AX               ;Get sum of P1UW, P2LW and P3LW in SI.
1027  73 01             JNC  SKIP2
1029  41                INC  CX
102A  03 D1     SKIP2:  ADD  DX,CX               ;Get sum of P2UW and P3UW in DI-register.
102C  B9 0000           MOV  CX,0000H
102F  13 FA             ADC  DI,DX
1031  73 01             JNC  SKIP3
1033  41                INC  CX
1034  89 77 0A  SKIP3:  MOV[BX+0AH],SI           ;Save second word of the product in memory.

1037  8B 47 02   P4:    MOV  AX,[BX+02]          ;Get D1UW in AX-register.
103A  F7 E5             MULn BP                  ;Get P4 in AX and DX.
103C  03 C7             ADDAX,DI                 ;Get sum of P2UW, P3UW and P4LW in AX.

103E  13 D1             ADC  DX,CX
1040  89 47 0C          MOV  [BX+0CH],AX         ;Save third word of the product in memory.

1043  89 57 0E          MOV  [BX+0EH],DX         ;Save fourth word of the product in memory.
1046  F4                HLT                      ;Halt program execution.

1047                    CODE ENDS                ;End of code segment.
                        END                      ;Assembly end.
```

Sample Data

Input Data : Data1 : 42107F6C$_H$
               Data2 : 1052C26F$_H$

Output Data : 0436 636C B64F 17D4$_H$

Memory address	Content
1100	6C
1101	7F
1102	10
1103	42
1104	6F
1105	C2
1106	52
1107	10

Memory address	Content
1108	D4
1109	17
110A	4F
110B	B6
110C	6C
110D	63
110E	36
110F	04

EXAMPLE PROGRAM 86.10: Division

Write an assembly language program to divide 32-bit data by 16-bit data.

Problem Analysis

To perform division in the 8086 processor, the 32-bit dividend should be stored in AX and DX registers (The lower word in AX and upper word in DX). The 16-bit divisor can be stored in the register/memory. After division, the quotient will be in AX-register and the remainder will be in DX-register.

Algorithm

1. Load the address of data in SI-register.
2. Get the lower word of dividend in AX-register.
3. Get the upper word of dividend in DX-register.
4. Get the divisor in BX-register.
5. Perform division to get quotient in AX and remainder in DX.
6. Save the quotient (AX) and the remainder (DX) in memory.
7. Stop.

Flowchart for Example Program 86.10

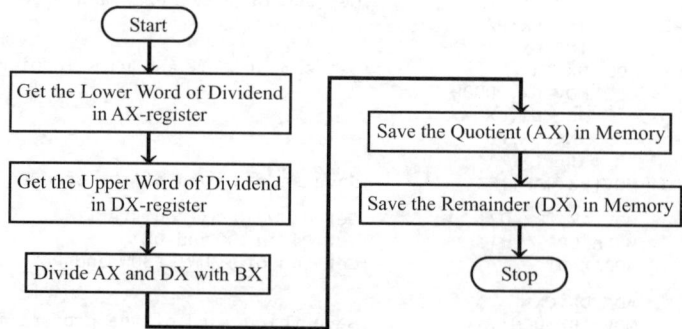

Assembly Language Program

```
;PROGRAM TO DIVIDE 32-BIT DATA BY 16-BIT DATA

CODE SEGMENT              ;Start of code segment.

        ASSUME CS:CODE  ;Assembler directive.
        ORG 1000H       ;specify program starting address.

        MOV SI,1100H    ;Set SI as pointer for data.
        MOV AX,[SI]     ;Get the lower 16-bit of dividend in AX-register.
        MOV DX,[SI+2]   ;Get the upper 16-bit of dividend in DX-register.
        MOV BX,[SI+4]   ;Get the divisor in BX-register.
        DIV BX          ;Divide the content of AX and DX with content of BX.
                        ;The quotient will be in AX-register.
                        ;The remainder will be in DX-register.
        MOV [SI+6],AX   ;Save the quotient in memory.
        MOV [SI+8],DX   ;Save the remainder in memory.
        HLT             ;Halt program execution.

CODE ENDS                ;End of code segment.
END                      ;Assembly end.
```

Assembler Listing for Example Program 86.10

```
;PROGRAM TO DIVIDE 32-BIT DATA BY 16-BIT DATA

0000              CODE SEGMENT      ;Start of code segment.

                  ASSUME CS:CODE  ;Assembler directive.
1000              ORG    1000H     ;specify program starting address.

1000   BE 1100    MOV SI,1100H    ;Set SI as pointer for data.
1003   8B 04      MOV AX,[SI]     ;Get the lower 16-bit of dividend in AX-register.
1005   8B 54 02   MOV DX,[SI+2]   ;Get the upper 16-bit of dividend in DX-register.
1008   8B 5C 04   MOV BX,[SI+4]   ;Get the divisor in BX-register.
100B   F7 F3      DIV BX          ;Divide the content of AX and DX with content of BX.
                                  ;The quotient will be in AX-register.
                                  ;The rezmainder will be in DX-register.
100D   89 44 06   MOV [SI+6],AX   ;Save the quotient in memory.
1010   89 54 08   MOV [SI+8],DX   ;Save the remainder in memory.
1013   F4         HLT             ;Halt program execution.

1014              CODE ENDS       ;End of code segment.
                  END             ;Assembly end.
```

Sample Data

Input Data : Dividend = 71C2580A$_H$ Output Data : Quotient = 75EE$_H$
            Divisor = F6F2$_H$ Remainder = 290E$_H$

Memory address	Content	
1100	0A	}
1101	58	
1102	C2	Dividend
1103	71	
1104	F2	}
1105	F6	Divisor

Memory address	Content	
1106	EE	}
1107	75	Quotient
1108	0E	}
1109	29	Remainder

EXAMPLE PROGRAM 11:   Search for a Given Data

Write an assembly language program to search a given data in an array. Also determine the position and address of the data in the array.

Problem Analysis

The given data is stored in a register, and then it is compared with each element of the array. The comparison is terminated once the data is found or after comparing all elements of the array. One register can be used to keep track of the position of the element being compared. One of the index registers can be used to hold the address of the element being compared.

If the data is found then store FF_H in a memory location to show availability, and store the position and address in consecutive memory locations. If data is not available then store zero in all these locations. The array is terminated with character 20_H.

Flowchart for Example Program 86.11

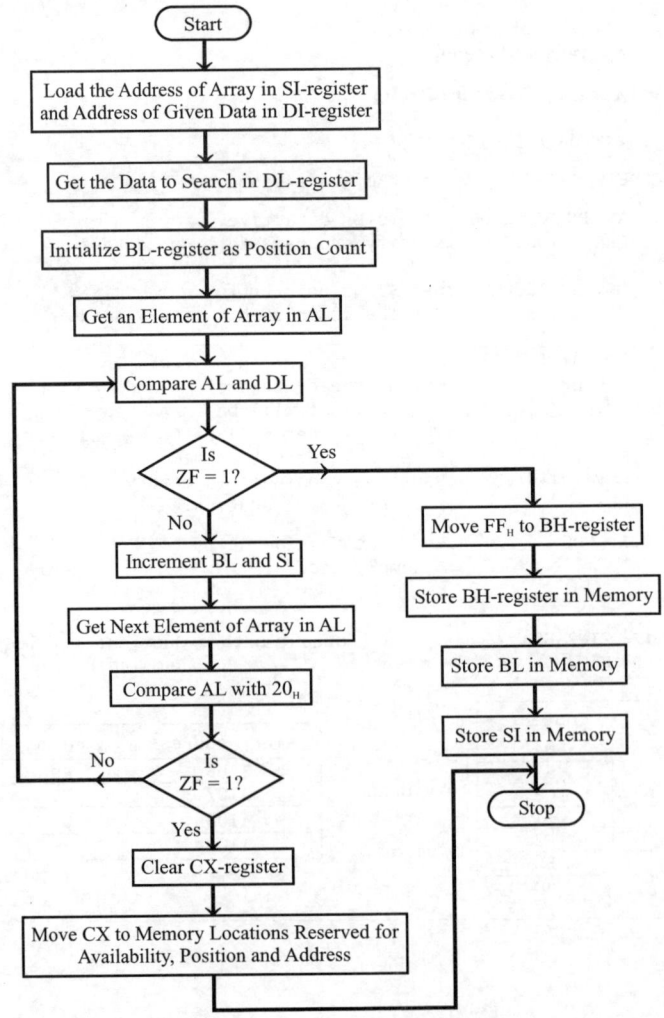

Algorithm

1. Set SI-register as pointer for the array.
2. Set DI-register as pointer for given data and result.
3. Get the data to search in DL-register.
4. Let BL-register keep track of position. Initialize the position count as one.
5. Get an element of array in AL.
6. Compare an element of array (AL) with given data (DL).
7. Check for zero flag. If zero flag is set then go to step 14, otherwise go to next step.
8. Increment the array pointer (SI) and position count (BL).
9. Get next element of array in AL-register.
10. Compare AL with end marker (20_H).
11. Check zero flag. If zero flag is not set then go to step 6, otherwise go to next step.
12. Clear CX-register and store CX-register in four consecutive locations in memory after the given data.
13. Jump to end (step 17).
14. Move FF_H to BH-register and store it in memory.
15. Store the position count (BL) in memory.
16. Store the address (SI) in memory.
17. Stop.

Assembly Language Program

```
;PROGRAM TO SEARCH A GIVEN DATA IN AN ARRAY

CODE SEGMENT              ;Start of code segment.

        ASSUME CS:CODE    ;Assembler directive.
        ORG 1000H         ;specify program starting address.

START:  MOV SI,1100H      ;Set SI-register as pointer for array.
        MOV DI,1200H      ;Load the address of data to search in DI-register.
        MOV DL,[DI]       ;Get the data to search in DL-register.
        MOV BL,01H        ;Set BL-register as position count.
        MOV AL,[SI]       ;Get first element of array in AL-register.

AGAIN:  CMP AL,DL         ;Compare an element of array with data to search.
        JZ AVAIL          ;If data are equal then jump to AVAIL.
        INC SI            ;If data are not equal, increment address pointer.
        INC BL            ;Increment position count.
        MOV AL,[SI]       ;Get the next element of array in AL-register.
        CMP AL,20H        ;Check for end of array,
        JNZ AGAIN         ;if not end, repeat search,otherwise go to NOTAVA.

NOTAVA: MOV CX,0000H      ;If search data is not found, then store zero.
        MOV [DI+1],CX
        MOV [DI+3],CX
        JMP OVER

AVAIL:  MOV BH,0FFH
        MOV [DI+1],BH     ;Store FFH to indicate availability of data.
        MOV [DI+2],BL     ;Store the position of data.
        MOV [DI+3],SI     ;Store the address of data.
OVER:   HLT               ;Halt program execution.

CODE ENDS                 ;End of code segment.
END                       ;Assembly end.
```

Assembler Listing for Example Program 86.11

```
;PROGRAM TO SEARCH A GIVEN DATA IN AN ARRAY

0000                    CODE SEGMENT            ;Start of code segment.

                        ASSUME CS:CODE ;Assembler directive.
1000                    ORG 1000H              ;specify program starting address.
1000   BE 1100   START: MOV SI,1100H           ;Set SI-register as pointer for array.
1003   BF 1200          MOV DI,1200H           ;Load the address of data to search in DI-register.
1006   8A 15            MOV DL,[DI]            ;Get the data to search in DL-register.
1008   B3 01            MOV BL,01H             ;Set BL register as position count.
100A   8A 04            MOV AL,[SI]            ;Get first element of array in AL-register.

100C   3A C2     AGAIN: CMP AL,DL              ;Compare an element of array with data to search.
100E   74 15            JZ AVAIL              ;If data are equal then jump to AVAIL.
1010   46               INC SI                ;If data are not equal,increment address pointer.
1011   FE C3            INC BL                ;Increment position count.
1013   8A 04            MOV AL,[SI]            ;Get the next element of array in AL-register.
1015   3C 20            CMP AL,20H             ;Check for end of array,
1017   75 F3            JNZ AGAIN             ;if not end, repeat search, otherwise go to NOTAVA.

1019   B9 0000   NOTAVA: MOV CX,0000H         ;If search data is not found, then store zero.
101C   89 4D 01          MOV [DI+1],CX
101F   89 4D 03          MOV [DI+3],CX
1022   EB 0C 90          JMP OVER

1025   B7 FF     AVAIL : MOV BH,0FFH
1027   88 7D 01          MOV [DI+1],BH         ;Store FFH to indicate availability of data.
102A   88 5D 02          MOV [DI+2],BL         ;Store the position of data.
102D   89 75 03          MOV [DI+3],SI         ;Store the address of data.
1030   F4        OVER :  HLT                   ;Halt program execution.

1031                     CODE ENDS             ;End of code segment.
                         END                   ;Assembly end.
```

Sample Data

Input Data :	Memory address	Content
Array = 1F$_H$	1100	1F
AC$_H$	1101	AC
D0$_H$	1102	D0
89$_H$	1103	89
72$_H$	1104	72
20$_H$	1105	20
Given Data = 89H	1200	89

Output Data :	Memory address	Content
Availability = FF$_H$		
Position = 04$_H$	1201	FF
Address = 1103$_H$	1202	04
	1203	03
	1204	11

EXAMPLE PROGRAM 86.12: Search for the Smallest Data

Write an assembly language program to search the smallest data in an array.

Problem Analysis

Let the size of the array be N bytes. Let us reserve AL-register to store the smallest data. The first byte of the array is assumed to be the smallest and it is saved in AL-register. Then each byte of the array is compared with AL. After each comparison, the smaller among the two is brought to AL-register. Therefore, after the N − 1 comparison, the AL-register will have the smallest data.

Algorithm

1. Load the starting address of the array in SI-register.
2. Load the address of the result in DI-register.
3. Load the number of bytes in the array in CL-register.
4. Increment the array pointer (SI-register).
5. Get the first byte of the array in AL-register.
6. Decrement the byte count (CL-register).
7. Increment the array pointer (SI-register).
8. Get next byte of the array in BL-register.
9. Compare current smallest (AL) and next byte (BL) of the array.
10. Check carry flag. If carry flag is set then go to step 12, otherwise go to next step.
11. Move BL to AL.
12. Decrement the byte count (CL-register).
13. Check zero flag. If zero flag is reset then go to step 7, otherwise go to next step.
14. Save the smallest data in memory pointed by DI.
15. Stop

Flowchart for Example Program 86.12

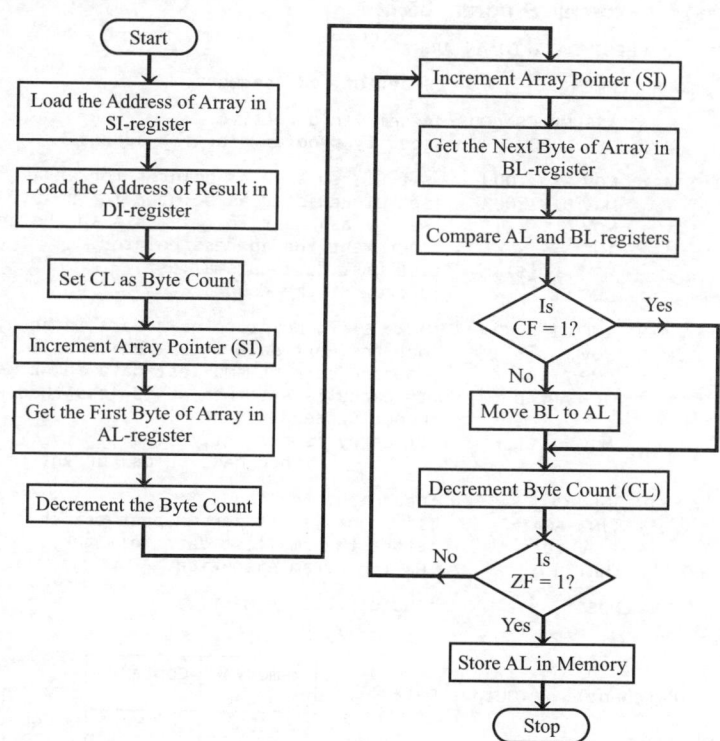

Assembly Language Program

```
;PROGRAM TO FIND SMALLEST DATA IN AN ARRAY

CODE SEGMENT            ;Start of code segment.
     ASSUME CS:CODE ;Assembler directive.
     ORG 1000H        ;specify program starting address.
```

```
START:   MOV  SI,1100H    ;Set SI-register as pointer for array.
         MOV  DI,1200H    ;Set DI-register as pointer for result.
         MOV  CL,[SI]     ;Set CL as count for elements in the array.
         INC  SI          ;Increment the address pointer.
         MOV  AL,[SI]     ;Set first data as smallest.
         DEC  CL          ;Decrement the count.

AGAIN:   INC  SI          ;Make SI to point to next data in array.
         MOV  BL,[SI]     ;Get the next data in BL-register.
         CMP  AL,BL       ;Compare current smallest data in AL with BL.
         JC   AHEAD       ;If carry is set then  AL is less than BL,
                          ;hence proceed to AHEAD.
         MOV  AL,BL       ;If carry is not set,
                          ;then make BL as current smallest.

AHEAD:   DEC  CL          ;Decrement the count.
         JNZ  AGAIN       ;If count is not zero repeat search.
         MOV  [DI],AL     ;Store the smallest data in memory.
         HLT              ;Halt program execution.

CODE ENDS                 ;End of code segment.
END                       ;Assembly end.
```

Assembler Listing for Example Program 86.12

```
;PROGRAM TO FIND SMALLEST DATA IN AN ARRAY

0000              CODE SEGMENT       ;Start of code segment.

                  ASSUME CS:CODE  ;Assembler directive.
1000              ORG 1000H          ;specify program starting address.

1000  BE 1100  START: MOV SI,1100H   ;Set SI-register as pointer for array.
1003  BF 1200         MOV DI,1200H   ;Set DI-register as pointer for result.
1006  8A 0C           MOV CL,[SI]    ;Set CL as count for elements in the array.
1008  46              INC SI         ;Increment the address pointer.
1009  8A 04           MOV AL,[SI]    ;Set first data as smallest.
100B  FE C9           DEC CL         ;Decrement the count.

100D  46       AGAIN: INC SI         ;Make SI to point to next data in array.
100E  8A 1C           MOV BL,[SI]    ;Get the next data in BL-register.
1010  3A C3           CMP AL,BL      ;Compare current smallest data in AL with BL.
1012  72 02           JC  AHEAD      ;If carry is set then AL is less than BL,
                                     ;hence proceed to AHEAD.
1014  8A C3           MOV AL,BL      ;If carry is not set,
                                          ;then make BL as current smallest.

1016  FE C9    AHEAD: DEC CL         ;Decrement the count.
1018  75 F3       ,   JNZ AGAIN      ;If count is not zero repeat search.
101A  88 05           MOV [DI],AL    ;Store the smallest data in memory.
101C  F4              HLT            ;Halt program execution.

101D             CODE ENDS           ;End of code segment.
                 END                 ;Assembly end.
```

Sample Data

Input Data : 06_H(count)   Output Data : $2D_H$
                $4E_H$
                $2D_H$
                30_H
                98_H
                AC_H
                FE_H

Memory address	Content	
1100	06	← Count
1101	4E	⎫
1102	2D	
1103	30	
1104	98	⎬ Array
1105	AC	
1106	FE	⎭
1200	2D	← Smallest data

EXAMPLE PROGRAM 86.13: Search for Largest Data

Write an assembly language program to search the largest data in an array.

Problem Analysis

Let the size of the array be N bytes. Let us reserve AL-register to store the largest data. The first byte of the array is assumed to be the largest and it is saved in AL-register. Then each byte of the array is compared with AL. After each comparison, the larger among the two is brought to AL-register. Therefore, after N – 1 comparisons, the AL-register will have the largest data.

Flowchart for Example Program 86.13

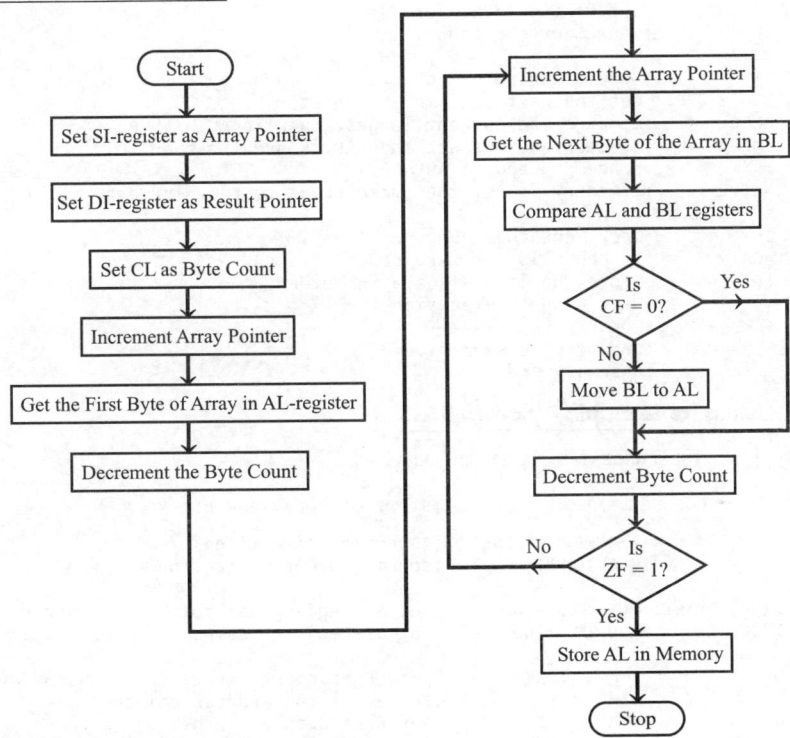

Algorithm

1. Load the starting address of the array in SI-register.
2. Load the address of result in DI-register.
3. Load the number of bytes in the array in CL-register.
4. Increment the array pointer (SI-register).
5. Get the first byte of the array in AL-register.
6. Decrement the byte count (CL-register).
7. Increment the array pointer (SI-register).
8. Get next byte of the array in BL-register.
9. Compare current largest (AL) and next byte (BL) of the array.
10. Check carry flag. If carry flag is reset then go to step 12, otherwise go to next step.
11. Move BL to AL.
12. Decrement byte count (CL-register).
13. Check zero flag. If zero flag is reset then go to step 7, otherwise go to next step.
14. Store the largest data in memory pointed by DI.
15. Stop.

Assembly Language Program

```
;PROGRAM TO FIND THE LARGEST DATA IN AN ARRAY

CODE SEGMENT              ;Start of code segment.

        ASSUME CS:CODE ;Assembler directive.
        ORG 1000H        ;specify program starting address.

START: MOV SI,1100H      ;Set SI-register as pointer for array.
       MOV DI,1200H      ;Set DI-register as pointer for result.

       MOV CL,[SI]       ;Set CL as count for elements in the array.
       INC SI            ;Increment the address pointer.
       MOV AL,[SI]       ;Set first data as largest.
       DEC CL            ;Decrement the count.

AGAIN: INC SI            ;Make SI to point to next data in array.
       MOV BL,[SI]       ;Get the next data in BL-register.
       CMP AL,BL         ;Compare the current largest data in AL with BL.
       JNC AHEAD         ;If carry is not set then AL is greater than BL,
                         ;hence proceed to AHEAD.
       MOV AL,BL         ;If carry is set then make BL as current largest.

AHEAD: DEC CL            ;Decrement the count.
       JNZ AGAIN         ;If count is not zero repeat search.
       MOV [DI],AL       ;Store the largest data in memory.
       HLT               ;Halt program execution.

CODE ENDS                ;End of code segment.
END                      ;Assembly end.
```

Assembler Listing for Example Program 86.13

```
;PROGRAM TO FIND THE LARGEST DATA IN AN ARRAY

0000              CODE SEGMENT                ;Start of code segment.

                       ASSUME CS:CODE         ;Assembler directive.
1000                   ORG 1000H              ;specify program starting address.

1000   BE 1100   START: MOV SI,1100H          ;Set SI-register as pointer for array.
1003   BF 1200          MOV DI,1200H          ;Set DI-register as pointer for result.

1006   8A 0C            MOV CL,[SI]            ;Set CL as count for elements in the array.
1008   46               INC SI                ;Increment the address pointer.
1009   8A 04            MOV AL,[SI]            ;Set first data as largest.
100B   FE C9            DEC CL                 ;Decrement the count.

100D   46        AGAIN: INC SI                ;Make SI to point to next data in array.
100E   8A 1C            MOV BL,[SI]            ;Get the next data in BL-register.
1010   3A C3            CMP AL,BL              ;Compare the current largest data in AL with BL.
1012   73 02            JNC AHEAD              ;If carry is not set then AL is greater than BL,
                                               ;hence proceed to AHEAD.
1014   8A C3            MOV AL,BL              ;If carry is set then make BL as current largest.
1016   FE C9     AHEAD: DEC CL                 ;Decrement the count.
1018   75 F3            JNZ AGAIN              ;If count is not zero repeat search.
101A   88 05            MOV [DI],AL            ;Store the largest data in memory.
101C   F4               HLT                   ;Halt program execution.

101D              CODE ENDS                    ;End of code segment.
                  END                          ;Assembly end.
```

Sample Data

Input Data :   06 (count)   Output Data :FE$_H$
4E$_H$
2D$_H$
30$_H$
98$_H$
AC$_H$
FE$_H$

Memory address	Content	
1100	06	← Count
1101	4E	⎫
1102	2D	
1103	30	
1104	98	⎬ Array
1105	AC	
1106	FE	⎭
1200	FE	← Largest data

EXAMPLE PROGRAM 86.14:   Sorting an Array in Ascending Order

Write an assembly language program to sort an array of data in ascending order.

Problem Analysis

The array can be sorted in ascending order by bubble sorting. In bubble sorting of N-data, N – 1 comparisons are performed by taking two consecutive data at a time. After each comparison the two data can be rearranged in the ascending order in the same memory locations, i.e., smaller first and larger next. When the above N – 1 comparisons are performed N – 1 times, the array will be sorted in ascending order in the same locations.

Algorithm

1.   Set SI-register as pointer for array.

2.   Set CL-register as count for N – 1 repetitions.

3.   Initialize array pointer.

4.   Set CH as count for N – 1 comparisons.

5.   Increment the array pointer.

6.   Get an element of array in AL-register.

7.   Increment the array pointer.

8.   Compare the next element of the array with AL.

9.   Check carry flag. If carry flag is set then go to step 12, otherwise go to next step.

10.   Exchange the content of memory pointed by SI and the content of previous memory location. (For this, exchange AL and memory pointed by SI, and then exchange AL and memory pointed by SI–1.)

11.   Decrement the count for comparisons (CH-register).

12.   Check zero flag. If zero flag is reset then go to step 6, otherwise go to next step.

13.   Decrement the count for repetitions (CL-register).

14.   Check zero flag. If zero flag is reset then go to step 3, otherwise go to next step.

15.   Stop.

Flowchart for Example Program 86.14

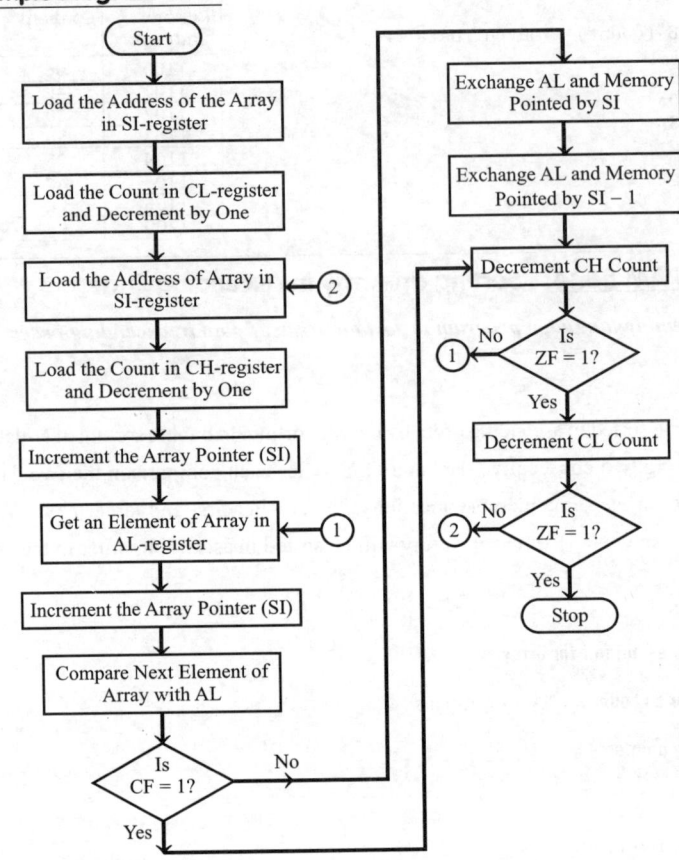

Assembly Language Program

```
;PROGRAM TO SORT AN ARRAY OF DATA IN ASCENDING ORDER

CODE SEGMENT            ;Start of code segment.

        ASSUME CS:CODE ;Assembler directive.
        ORG  1000H     ;specify program starting address.

START:  MOV  SI,1100H  ;Set SI-register as pointer for array.
        MOV  CL,[SI]   ;Set CL as count for N-1 repetitions.
        DEC  CL

REPEAT: MOV  SI,1100H  ;Initialize pointer.
        MOV  CH,[SI]   ;Set CH as count for N-1 comparisons.
        DEC  CH
        INC  SI        ;Increment the pointer.

REPCOM: MOV  AL,[SI]   ;Get an element of array in AL-register.
        INC  SI
        CMP  AL,[SI]   ;Compare with next element of array in memory.
        JC   AHEAD     ;If AL is lesser than memory, then go to AHEAD.
        XCHG AL,[SI]   ;If AL is less than memory then,
        XCHG AL,[SI-1] ;exchange the content of memory pointed by SI,
                       ;and the previous memory location.
```

```
AHEAD:  DEC  CH          ;Decrement the count for comparisons.
        JNZ  REPCOM      ;Repeat comparison until CH count is zero.
        DEC  CL          ;Decrement the count for repetitions.
        JNZ  REPEAT      ;Repeat N-1 comparisons until CL count is zero.
        HLT              ;Halt program execution.

CODE ENDS                ;End of code segment.
END                      ;Assembly end.
```

Assembler Listing for Example Program 86.14

```
;PROGRAM TO SORT AN ARRAY OF DATA IN ASCENDING ORDER

0000                 CODE SEGMENT          ;Start of code segment.

                     ASSUME CS: CODE  ;Assembler directive.
1000                 ORG 1000H             ;specify program starting address.

1000  BE 1100  START: MOV SI,1100H         ;Set SI-register as pointer for array.
1003  8A 0C           MOV CL,[SI]          ;Set CL as count for N-1 repetitions.
1005  FE C9           DEC CL

1007  BE 1100  REPEAT: MOV SI,1100H        ;Initialize pointer.
100A  8A 2C           MOV CH,[SI]          ;Set CH as count for N-1 comparisons.
100C  FE CD           DEC CH
100E  46              INC SI               ;Increment the pointer.

100F  8A 04    REPCOM: MOV AL,[SI]         ;Get an element of array in AL-register.
1011  46              INC SI
1012  3A 04           CMP AL,[SI]          ;Compare with next element of array in memory.
1014  72 05           JC  AHEAD            ;If AL is lesser than memory, then go to AHEAD.

1016  86 04           XCHG AL,[SI]         ;If AL is less than memory then,
1018  86 44 FF        XCHG AL,[SI-1]       ;exchange the content of memory pointed by SI,
                                           ;and the previous memory location.

101B  FE CD    AHEAD:  DEC CH              ;Decrement the count for comparisons.
101D  75 F0           JNZ REPCOM           ;Repeat comparison until CH count is zero.
101F  FE C9           DEC CL               ;Decrement the count for repetitions.
1021  75 E4           JNZ REPEAT           ;Repeat N-1 comparisons until CL count is zero.
1023  F4              HLT                  ;Halt program execution.

1024                 CODE ENDS             ;End of code segment.
                     END                   ;Assembly end.
```

Sample Data

	Before Execution				After Execution		
Input Data : 07	Memory address	Content		Output Data :11	Memory address	Content	
AA				22			
77	1100	07	← Count	44	1101	11	
FF	1101	AA		77	1102	22	
22	1102	77		AA	1103	44	Array in
11	1103	FF	Input Array	BB	1104	77	ascending
44	1104	22		FF	1105	AA	order
BB	1105	11			1106	BB	
	1106	44			1107	FF	
	1107	BB					

EXAMPLE PROGRAM 86.15:   Sorting an Array in Descending Order

Write an assembly language program to sort an array of data in descending order.

Problem Analysis

The array can be sorted in descending order by bubble sorting. In bubble sorting of N-data, N − 1 comparisons are performed by taking two consecutive data at a time. After each comparison, the two data can be rearranged in the descending order in the same memory locations, i.e., larger first and smaller next. When the above N − 1 comparisons are performed N − 1 times, the array will be sorted in descending order in the same locations.

Flowchart for Example Program 86.15

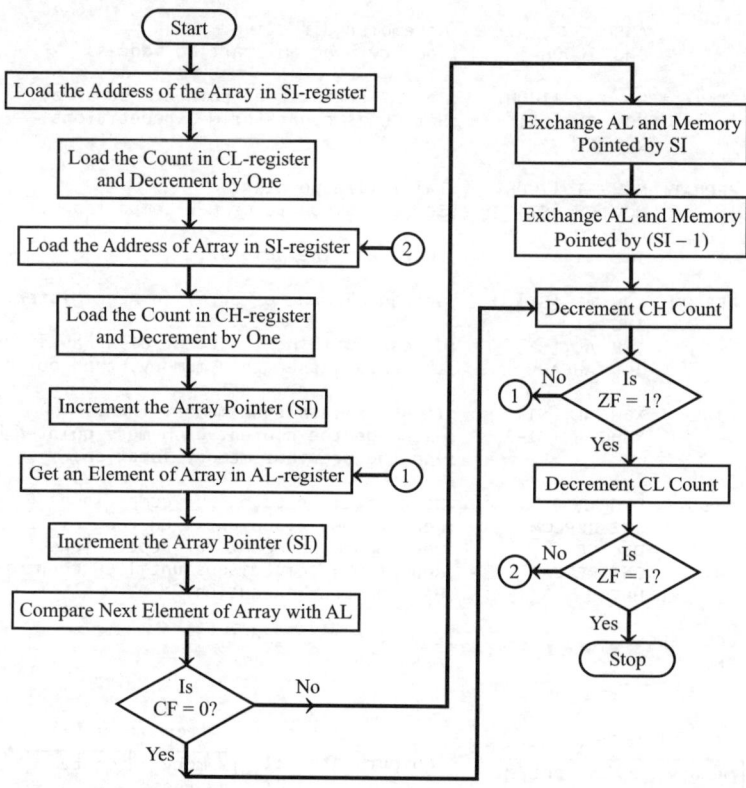

Algorithm

1.   Set SI-register as pointer for array.
2.   Set CL-register as count for N − 1 repetitions.
3.   Initialize array pointer.
4.   Set CH as count for N − 1 comparisons.
5.   Increment the array pointer.
6.   Get an element of array in AL-register.
7.   Increment the array pointer.
8.   Compare the next element of the array with AL.
9.   Check carry flag. If carry flag is reset then go to step 12, otherwise go to next step.
10.   Exchange the content of memory pointed by SI and the content of previous memory location. (For this, exchange AL and memory pointed by SI, and then exchange AL and memory pointed by SI − 1.)
11.   Decrement the count for comparisons (CH-register).

12. **Check zero flag. If zero flag is reset then go to step 6, otherwise go to next step.**
13. **Decrement the count for repetitions (CL-register).**
14. **Check zero flag. If zero flag is reset then go to step 3, otherwise go to next step.**
15. **Stop.**

Assembly Language Program

```
;PROGRAM TO SORT AN ARRAY OF DATA IN DESCENDING ORDER

CODE SEGMENT                 ;Start of code segment.

        ASSUME CS:CODE       ;Assembler directive.
        ORG    1000H         ;specify program starting address.

START:  MOV  SI,1100H        ;Set SI-register as pointer for array.
        MOV  CL,[SI]         ;Set CL as count for N-1 repetitions.
        DEC  CL

REPEAT: MOV  SI,1100H        ;Initialize pointer.
        MOV  CH,[SI]         ;Set CH as count for N-1 comparisons.
        DEC  CH
        INC  SI              ;Increment the pointer.

REPCOM: MOV  AL,[SI]         ;Get an element of array in AL-register.
        INC  SI
        CMP  AL,[SI]         ;Compare with next element of the array in memory.
        JNC  AHEAD           ;If AL is greater than memory, then go to AHEAD.

        XCHG AL,[SI]         ;If AL is less than memory then,
        XCHG AL,[SI-1]       ;exchange the content of memory pointed by SI,
                             ;and the previous memory location.

AHEAD:  DEC  CH              ;Decrement the count for comparisons.
        JNZ  REPCOM          ;Repeat comparison until CH count is zero.
        DEC  CL              ;Decrement the count for repetitions.
        JNZ  REPEAT          ;Repeat N-1 comparisons until CL count is zero.
        HLT                  ;Halt program execution.

CODE ENDS                    ;End of code segment.
END                          ;Assembly end.
```

Assembler Listing for Example Program 86.15

```
;PROGRAM TO SORT AN ARRAY OF DATA IN DESCENDING ORDER

0000              CODE SEGMENT              ;Start of code segment.

                  ASSUME CS:CODE  ;Assembler directive.
1000              ORG    1000H    ;specify program starting address.

1000  BE 1100  START:  MOV  SI,1100H   ;Set SI-register as pointer for array.
1003  8A 0C            MOV  CL,[SI]    ;Set CL as count for N-1 repetitions.
1005  FE C9            DEC  CL

1007  BE 1100  REPEAT: MOV  SI,1100H   ;Initialize pointer.
100A  8A 2C            MOV  CH,[SI]    ;Set CH as count for N-1 comparisons.
100C  FE CD            DEC  CH
100E  46               INC  SI         ;Increment the pointer.

100F  8A 04    REPCOM: MOV  AL,[SI]    ;Get an element of array in AL-register.

1011  46               INC  SI
1012  3A 04            CMP  AL,[SI]    ;Compare with next element of the array in memory.
1014  73 05            JNC  AHEAD      ;If AL is greater than memory, then go to AHEAD.

1016  86 04            XCHG AL,[SI]    ;If AL is less than memory then,
1018  86 44 FF         XCHG AL,[SI-1]  ;exchange the content of memory pointed by SI,
                                       ;and the previous memory location.
```

```
101B   FE CD    AHEAD:   DEC   CH          ;Decrement the count for comparisons.
101D   75 F0             JNZ   REPCOM      ;Repeat comparison until CH count is zero.
101F   FE C9             DEC   CL          ;Decrement the count for repetitions.
1021   75 E4             JNZ   REPEAT      ;Repeat N-1 comparisons until CL count is zero.
1023   F4                HLT               ;Halt program execution.

1024                     CODE ENDS         ;End of code segment.
                         END               ;Assembly end.
```

Sample Data

		Before Execution					After Execution		

Input Data : 07
              AA
              77
              FF
              22
              11
              44
              BB

Memory address	Content	
1100	07	← Count
1101	AA	⎫
1102	77	
1103	FF	Input
1104	22	Array
1105	11	
1106	44	
1107	BB	⎭

Output Data : FF
               BB
               AA
               77
               44
               22
               11

Memory address	Content	
1101	FF	⎫
1102	BB	Array
1103	AA	in
1104	77	descending
1105	44	order
1106	22	
1107	11	⎭

EXAMPLE PROGRAM 86.16:   GCD of Two 16-Bit Data

Write an assembly language program to determine the GCD of two 16-bit data.

Problem Analysis

First divide the larger data by the smaller data and check for the remainder. If the remainder is zero, then smaller data is the GCD.

If the remainder is not zero then take the remainder as the divisor and the previous divisor as the dividend and repeat division until the remainder is zero. When the remainder is zero, we can store the divisor as GCD. Before performing division, we can even check whether the dividend and divisor are equal. If they are equal then we can directly store the divisor as GCD without performing division.

Algorithm

1.   Set BX as pointer for input data.
2.   Set DI as pointer for result.
3.   Get one data in AX-register.
4.   Get another data in CX-register.
5.   Compare the two data (AX and CX).
6.   Check zero flag. If zero flag is set then go to step 14, otherwise go to next step.
7.   Check carry flag. If carry flag is reset then go to step 9, otherwise go to next step.
8.   Exchange the content of AX and CX, so that the larger among the two is dividend and smaller is the divisor.
9.   Clear DX-register.
10. Divide AX-register by CX-register.
11. Compare DX-register (Remainder) with 0000_H.
12. Check zero flag. If zero flag is set then go to step 14, otherwise go to next step.
13. Move the remainder (DX-register) to AX and go to step 5.
14. Save the content of CX-register as GCD in memory.
15. Stop.

Flowchart for Example Program 86.16

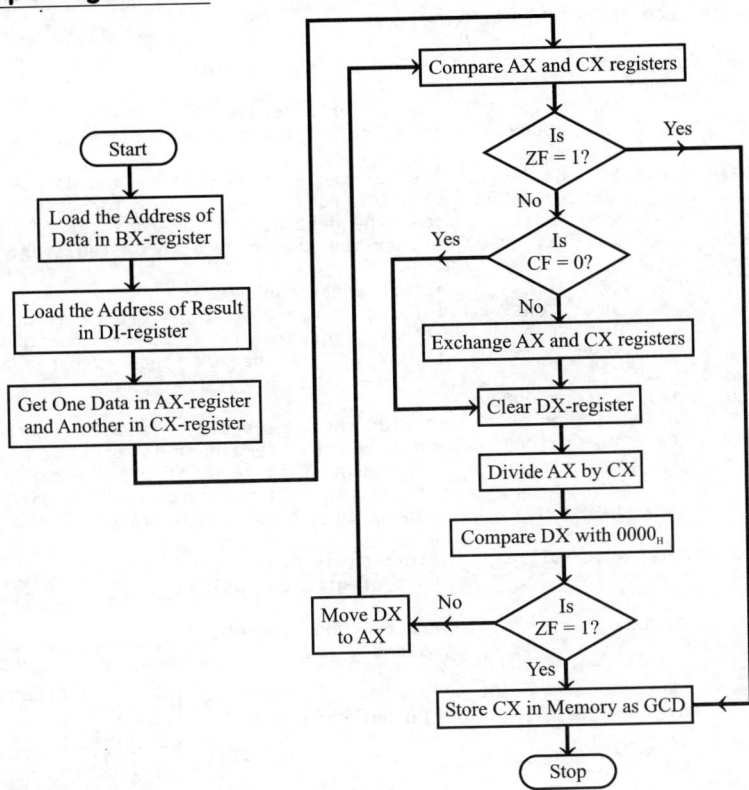

Assembly Language Program

```
; PROGRAM TO FIND GCD OF TWO 16-BIT DATA
CODE SEGMENT                ;start of code segment.

        ASSUME CS:CODE      ;Assembler directive.
        ORG   1000H         ;specify program starting address.

        MOV   BX,1100H      ;Set BX-register as pointer for data.
        MOV   DI,1200H      ;Set DI-register as pointer for result.
        MOV   AX,[BX]       ;Get the first data in AX-register.
        MOV   CX,[BX+02]    ;Get the second data in CX-register.

RPT:    CMP   AX,CX         ;Compare the two data.
        JE    STORE         ;If the data are equal, store CX as GCD.
        JNC   SKIP          ;If AX is greater than CX, then go to SKIP.
        XCHG  AX,CX         ;If AX is less than CX, then exchange AX and CX.

SKIP:   MOV   DX,0000H
        DIV   CX            ;Divide the two data.
        CMP   DX,0000H      ;Check whether remainder is zero.
        JE    STORE         ;If remainder is zero, th
        MOV   AX,DX         ;If remainder is not zero, move remainder to AX.
        JMP   RPT           ;Repeat comparison and division.
STORE:  MOV   [DI],CX       ;Store CX as GCD.
        HLT                 ;Halt program execution.
CODE ENDS                   ;End of code segment.
END                         ;Assembly end.
```

Assembler Listing for Example Program 86.16

```
;PROGRAM TO FIND GCD OF TWO 16-BIT DATA

0000                  CODE SEGMENT          ;Start of code segment.

                      ASSUME CS:CODE        ;Assembler directive.
1000                  ORG   1000H           ;specify program starting address.

1000  BB 1100         MOV   BX,1100H        ;Set BX-register as pointer for data.
1003  BF 1200         MOV   DI,1200H        ;Set DI-register as pointer for result.
1006  8B 07           MOV   AX,[BX]         ;Get the first data in AX-register.
1008  8B 4F 02        MOV   CX,[BX+02]      ;Get the second data in CX-register.

100B  3B C1    RPT:   CMP   AX,CX           ;Compare the two data.
100D  74 11           JE    STORE           ;If the data are equal, store CX as GCD.
100F  73 01           JNC   SKIP            ;If AX is greater than CX, then go to SKIP.
1011  91              XCHG  AX,CX           ;If AX is less than CX, then exchange AX and CX.

1012  BA 0000   SKIP: MOV   DX,0000H
1015  F7 F1           DIV   CX              ;Divide the two data.
1017  83 FA 00        CMP   DX,0000H        ;Check whether remainder is zero.
101A  74 04           JE    STORE           ;If remainder is zero, then store CX as GCD.
101C  8B C2           MOV   AX,DX           ;If remainder is not zero, move remainder to AX.
101E  EB EB           JMP   RPT             ;Repeat comparison and division.

1020  89 0D    STORE: MOV   [DI],CX         ;Store CX as GCD.
1022  F4              HLT                   ;Halt program execution.

1023                  CODE ENDS             ;End of code segment.
                      END                   ;Assembly end.
```

Sample Data

			Memory address	Content
Input Data	: Data1 = 358E$_H$	Output Data : 0005$_H$	1100	8E
	Data2 = 01BD$_H$		1101	35
			1102	BD
			1103	01
			1200	05
			1201	00

EXAMPLE PROGRAM 86.17:   LCM of Two 16-Bit Data

Write an assembly language program to determine the LCM of two 16-bit data.

Problem Analysis

First determine the GCD of two data. Then determine the product of two data. Here it is assumed that the product does not exceed 32 bits. When the product is divided by GCD, the quotient will be the LCM of the two data. (For the GCD of two data, please refer to Example Program 16.)

Algorithm

1.   Set BX as pointer for input data.
2.   Set DI as pointer for result.
3.   Get first data in AX-register and second data in CX-register.
4.   Call procedure GCD to get the GCD in SI-register.
5.   Again get first data in AX-register and second data in CX-register.
6.   Determine the product of two data. The product will be in AX and DX registers.
7.   Divide the product (AX and DX registers) by GCD (SI-register).
8.   Save the quotient (AX-register) as LCM in memory.
9.   Stop.

Note: *The algorithm for procedure GCD can be obtained from Example Program 16.*

Flowchart for Example Program 86.17

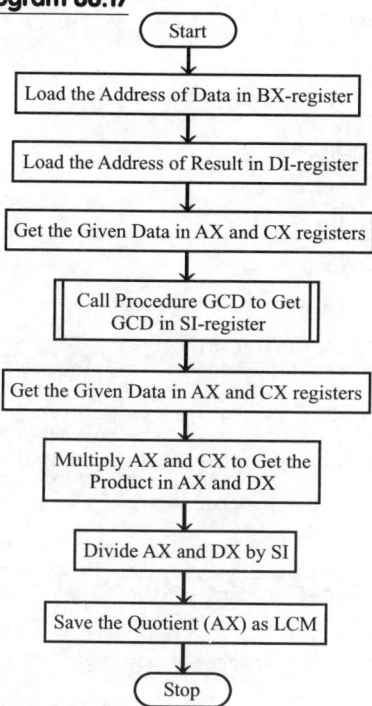

Start

Load the Address of Data in BX-register

Load the Address of Result in DI-register

Get the Given Data in AX and CX registers

Call Procedure GCD to Get GCD in SI-register

Get the Given Data in AX and CX registers

Multiply AX and CX to Get the Product in AX and DX

Divide AX and DX by SI

Save the Quotient (AX) as LCM

Stop

Note: *The flowchart for procedure GCD can be obtained from Example Program 16*

Assembly Language Program

```
;PROGRAM TO FIND LCM OF TWO 16-BIT DATA
CODE SEGMENT              ;Start of code segment.

       ASSUME CS:CODE ;Assembler directive.
       ORG 1000H         ;specify program starting address.

       MOV BX,1100H      ;Set BX-register as pointer for data.
       MOV DI,1200H      ;Set DI-register as pointer for result .
       MOV AX,[BX]       ;Get the first data in AX-register.
       MOV CX,[BX+02] ;Get the second data in CX-register.
       CALL GCD          ;Call procedure GCD.

       MOV AX,[BX]       ;Get the first data in AX-register.
       MOV CX,[BX+02] ;Get the second data in CX-register.
       MUL CX            ;Get product of two numbers in AX and DX.
       DIV SI            ;Divide the product with GCD.
       MOV [DI],AX       ;Save the quotient as LCM.
       HLT               ;Halt program execution.
GCD PROC NEAR

RPT:   CMP AX,CX         ;Compare the two data.
       JE  SAVE          ;If the data are equal, store CX as GCD.
       JNC SKIP          ;If AX is greater than CX, then go to SKIP.
       XCHG AX,CX        ;If AX is less than CX, then exchange AX and CX.
SKIP:  MOV DX,0000H
       DIV CX            ;Divide the two data.
       CMP DX,0000H      ;Check whether remainder is zero.
       JE  SAVE          ;If remainder is zero, then store CX as GCD.
```

```
MOV    AX,DX              ;If remainder is not zero, move remainder to AX.
       JMP RPT            ;Repeat comparison and division.

SAVE:  MOV SI,CX          ;Store CX as GCD.
       RET

GCD    ENDP              ;Assembler directive.
CODE   ENDS              ;End of code segment.
END                      ;Assembly end.
```

Assembler Listing for Example Program 86.17

```
;PROGRAM TO FIND LCM OF TWO 16-BIT DATA

0000                     CODE SEGMENT      ;Start of code segment.

                         ASSUME CS:CODE    ;Assembler directive.
1000                     ORG 1000H         ;specify program starting address.

1000  BB 1100 R          MOV  BX,1100H     ;Set BX-register as pointer for data.
1003  BF 1200 R          MOV  DI,1200H     ;Set DI-register as pointer for result.
1006  8B 07              MOV  AX,[BX]      ;Get the first data in AX-register.
1008  8B 4F 02           MOV  CX,[BX+02]   ;Get the second data in CX-register.
100B  E8 000C R          CALL GCD          ;Call procedure GCD.

100E  8B 07              MOV  AX,[BX]      ;Get the first data in AX-register.
1010  8B 4F 02           MOV  CX,[BX+02]   ;Get the second data in CX-register.
1013  F7 E1              MUL  CX           ;Get product of two numbers in AX and DX.
1015  F7 F6              DIV  SI           ;Divide the product with GCD.
1017  89 05              MOV  [DI],AX      ;Save the quotient as LCM.
1019  F4                 HLT               ;Halt program execution.

101A             GCD PROC NEAR

101A  3B C1     RPT:    CMP  AX,CX         ;Compare the two data.
101C  74 11             JE   SAVE          ;If the data are equal, store CX as GCD.
101E  73 01             JNC  SKIP          ;If AX is greater than CX, then go to SKIP.
1020  91                XCHG AX,CX         ;If AX is less than CX, then exchange AX and CX.
1021  BA 0000   SKIP:   MOV  DX,0000H
1024  F7 F1             DIV  CX            ;Divide the two data.
1026  83 FA 00          CMP  DX,0000H      ;Check whether remainder is zero.
1029  74 04             JE   SAVE          ;If remainder is zero, then store CX as GCD.
102B  8B C2             MOV  AX,DX         ;If remainder is not zero, move remainder to AX.
102D  EB EB             JMP  RPT           ;Repeat comparison and division.

102F  8B F1     SAVE:   MOV SI,CX          ;Store CX as GCD.
1031  C3                RET

1032             GCD    ENDP              ;Assembler directive.
1032             CODE ENDS               ;End of code segment.
                 END                      ;Assembly end.
```

Sample Data

Input Data :   Output Data : 077A$_H$
Data1 = 0042$_H$

Data2 = 003A$_H$

Memory address	Content	
1100	42	} Data 1
1101	00	
1102	3A	} Data 2
1103	00	
1200	7A	} LCM
1201	07	

EXAMPLE PROGRAM 86.18: Factorial of 8-Bit Data

Write an assembly language program to determine the factorial of 8-bit data.

Problem Analysis

The factorial can be calculated by repeated multiplication. In the first multiplication, the given data is taken as multiplicand and data − 1 (data minus one) is taken as multiplier. In each subsequent multiplication, the previous product is taken as the multiplicand and the previous multiplier is decremented by one and used as the current multiplier. The multiplications are repeated until the multiplier becomes zero. The final product after data − 1 (data minus one) multiplications will be the factorial of the data.

In this example, it is assumed that the product/factorial does not exceed 32 bits. The given data is converted to 16-bit data by taking the high byte as zero and in each multiplication, 32-bit by 16-bit multiplication is performed. The logic of 32-bit by 16-bit multiplication is as follows:

The multiplicand can be divided into two words: **L**ower **W**ord (LW) and **U**pper **W**ord (UW) as shown below.

Multiplicand (32-bit) $\rightarrow$ MD_{UW} (16-bit), MD_{LW} (16-bit)

Let the 16-bit multiplier be MR. Then perform the following two multiplications. Each multiplication will give a 32-bit result which can be divided into **L**ower **W**ord (LW) and **U**pper **W**ord (UW) as shown below:

Product 1 (P1)   : $MD_{LW} \times MR = P1_{UW}, P1_{LW}$

Product 2 (P2)   : $MD_{UW} \times MR = P2_{UW}, P2_{LW}$

The result of the above two multiplications can be added to get the final result as shown below. The final product will have a size of three words and they are denoted as P_{W1}, P_{W2} and P_{W3}. Since we restrict the product to 32-bit, the third word PW_3 is discarded.

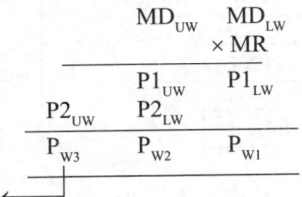

Algorithm

    Discard $\leftarrow$

1.   Set SI as pointer for data.
2.   Get the data in AL-register and clear AH-register to convert the data to 16-bit.
3.   Clear BP-register to keep initial value of second word of final product as zero.
4.   Compare AX-register with 01_H.
5.   Check zero flag. If zero flag is set then go to step 19, otherwise go to next step.
6.   Set CX-register as count for data−1 (data minus one) multiplications.
7.   Move AX-register to BX-register, so that the initial mutiplier in BX is the given data.
8.   Decrement the multiplier (BX-register).
9.   Multiply AX and BX to get the product1 in AX and DX.
10.   Save the product1 in stack.
11.   Load the second word of previous product in BP to AX-register.
12.   Multiply AX and BX to get the product 2 in AX and DX.
13.   Get the upper word of product1 in DX.
14.   Add AX and DX to get the second word of final product in AX.
15.   Move AX to BP to save the second word of final product in BP.
16.   Get the first word of final product in AX-register.
17.   Decrement multiplication count (CX-register).
18.   If content of CX-register is not zero, then go to step 8, otherwise go to next step.
19.   Store AX and BP in memory.
20.   Stop.

Flowchart of Example Program 86.18

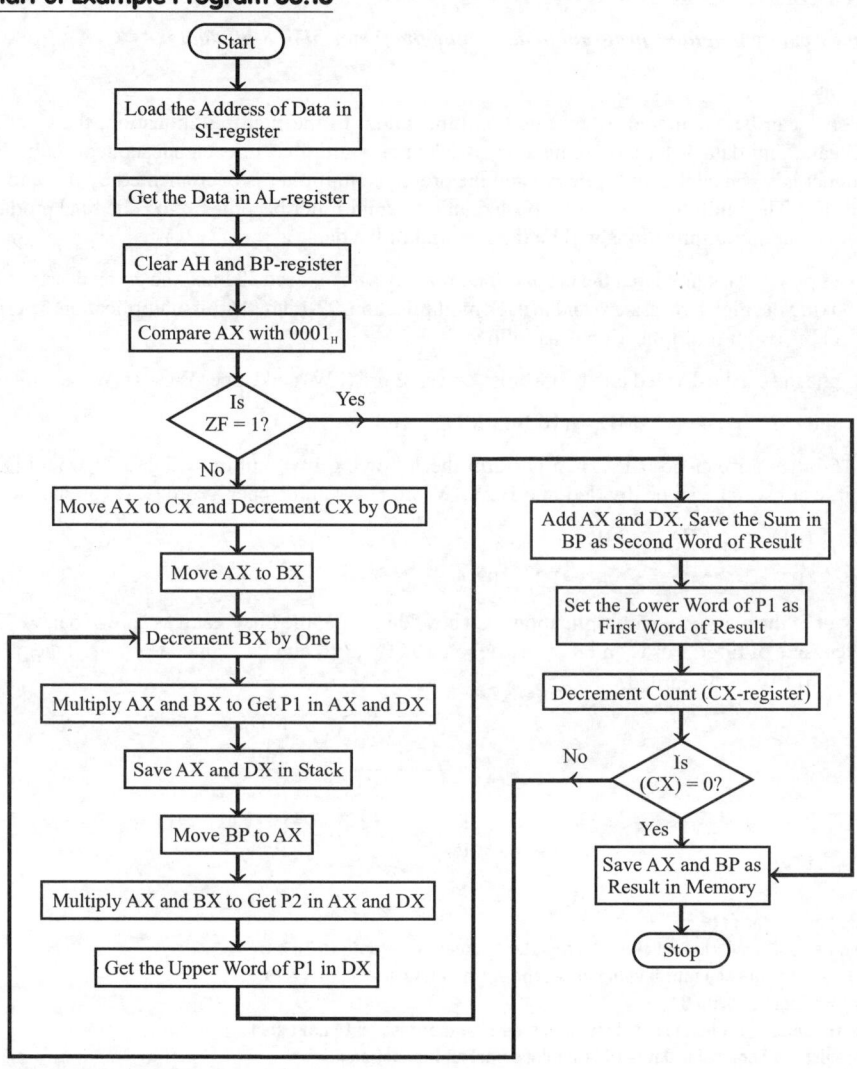

Assembly Language Program

```
;PROGRAM TO FIND FACTORIAL OF 8-BIT DATA

CODE SEGMENT              ;Start of code segment.

        ASSUME CS:CODE   ;Assembler directive.
        ORG  1000H       ;specify program starting address.

        MOV  SI,1100H    ;Set SI as pointer for data.
        MOV  AL,[SI]     ;Get the data in AL.
        MOV  AH,00H      ;Clear AH-register.
        MOV  BP,0000H    ;Initialize upper word of the result as zero.
        CMP  AX,0001H    ;Check whether data is 01.
        JNG  STORE       ;If data is 01, then store 01 as factorial.

        MOV  CX,AX       ;Set CX as count for number of multiplications.
        DEC  CX          ;Decrement the count.
```

```
          MOV  BX,AX       ;Set the data as multiplier.
REPEAT: DEC  BX            ;Decrement the multiplier.
        MUL  BX            ;Get the product1(P1) in AX and DX registers.
        PUSH AX            ;Save lower word of product 1 in stack.
        PUSH DX            ;Save upper word of product 1 in stack.
        MOV  AX,BP
        MUL  BX            ;Get the product 2(P2) in AX and DX registers.
        POP  DX            ;Get the upper word of product 1 in DX-register.
        ADD  AX,DX         ;Get sum of lower word of P1 and
                           ;upper word of P2 in AX.
        MOV  BP,AX         ;Set the sum as second word of the result.
        POP  AX            ;Set lower word of P1 as first word of result.
        LOOP REPEAT        ;Repeat multiplication until count is zero.

STORE:  MOV  [SI+1],AX     ;Store the lower word of the result in memory.
        MOV  [SI+3],BP     ;Store the upper word of the result in memory.
        HLT                ;Halt program execution.

CODE ENDS                  ;End of code segment.
END                        ;Assembly end.
```

Assembler Listing for Example Program 86.18

```
;PROGRAM TO FIND FACTORIAL OF 8-BIT DATA

0000              CODE SEGMENT              ;Start of code segment.

                  ASSUME CS:CODE            ;Assembler directive.
1000              ORG 1000H                 ;specify program starting address.

1000  BE 1100     MOV  SI,1100H             ;Set SI as pointer for data.
1003  8A 04       MOV  AL,[SI]              ;Get the data in AL.
1005  B4 00       MOV  AH,00H               ;Clear AH-register.
1007  BD 0000     MOV  BP,0000H             ;Initialize upper word of the result as zero.
100A  3D 0001     CMP  AX,0001H             ;Check whether data is 01.
100D  7E 16       JNG  STORE                ;If data is 01, then store 01 as factorial.

100F  8B C8       MOV  CX,AX                ;Set CX as count for number of multiplications.
1011  49          DEC  CX                   ;Decrement the count.

1012  8B D8       MOV  BX,AX                ;Set the data as multiplier.
1014  4B  REPEAT: DEC  BX                   ;Decrement the multiplier.
1015  F7 E3       MUL  BX                   ;Get the product 1(P1) in AX and DX registers.
1017  50          PUSH AX                   ;Save lower word of product 1 in stack.
1018  52          PUSH DX                   ;Save upper word of product 1 in stack.
1019  8B C5       MOV  AX,BP
101B  F7 E3       MUL  BX                   ;Get the product 2(P2) in AX and DX registers.
101D  5A          POP  DX                   ;Get the upper word of product 1 in DX-register.
101E  03 C2       ADD  AX,DX                ;Get sum of lower word of P1 and
                                            ;     upper word of P2 in AX.
1020  8B E8       MOV  BP,AX                ;Set the sum as second word of the result.
1022  58          POP  AX                   ;Set lower word of P1 as first word of result.
1023  E2 EF       LOOP REPEAT               ;Repeat multiplication until count is zero.

1025  89 44 01  STORE: MOV  [SI+1],AX       ;Store the lower word of the result in memory.
1028  89 6C 03         MOV  [SI+3],BP        ;Store the upper word of the result in memory.
102B  F4                HLT                  ;Halt program execution.

102C              CODE ENDS                 ;End of code segment.
                  END                       ;Assembly end.
```

Sample Data

Input Data : 0B$_H$ Output Data : 02611500$_H$

Memory address	Data	
1100	0B	← Data
1101	00	⎫
1102	15	⎬ Factorial
1103	61	
1104	02	⎭

EXAMPLE PROGRAM 86.19: Generation of Prime Numbers

Write an assembly language program to generate all possible prime numbers less than the given data.

Problem Analysis

A number is prime if it is divisible only by one and the same number, and it should not be divisible by any other number. Hence to check whether a number is prime or not, we can divide the number by all possible integers less than the given number and verify the remainder. (The initial divisor is 02.) If the remainder is zero in any of the division then the number is not prime. If the remainder is nonzero in all the divisions then we can say the number is prime.

Algorithm

1. Set SI-register as pointer for data.
2. Load the given data in DL-register.
3. Set DI as pointer for result.
4. Initialize the number to checked as 01_H in BL-register.
5. Save 01_H as first prime number.
6. Increment the result pointer (DI).
7. Increment the number (BL-register) to be checked.
8. Load the initial divisor 02_H in CL-register.
9. Compare BL and CL registers.
10. Check zero flag. If zero flag is set then go to step 16, otherwise go to next step.
11. Clear AH-register and load the number to be checked in AL-register.
12. Divide AX by CL-register.
13. Compare the remainder (AH-register) with 00_H.
14. Check zero flag. If zero flag is set then go to step 18, otherwise go to next step.
15. Increment the divisor (CL-register) and go to step 9.
16. Save the prime number.
17. Increment the result pointer (DL-register).
18. Increment the number to be checked (BL-register)
19. Compare DL and BL registers.
20. Check zero flag. If zero flag is reset then go to step 8, otherwise stop.

Flowchart of Example Program 86.19

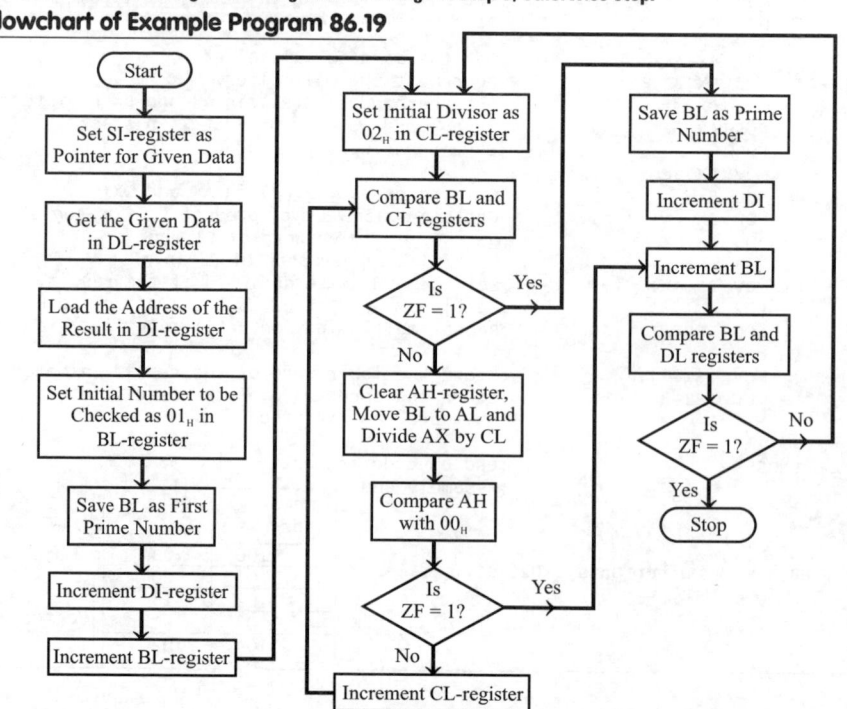

Assembly Language Program

```
;PROGRAM TO GENERATE PRIME NUMBERS
CODE SEGMENT                    ;Start of code segment.

          ASSUME CS:CODE        ;Assembler directive.
          ORG 1000H             ;specify program starting address.

          MOV SI,1100H          ;Set SI-register as pointer for end data N.
          MOV DL,[SI]           ;Get the data N in DL-register.
          MOV DI,1200H          ;Set DI as pointer for storing prime numbers.
          MOV BL,01H            ;Initialize the number to be checked as 01.
          MOV [DI],BL           ;Save the first prime number.
          INC DI                ;Increment address pointer.
          INC BL                ;Increment the number to be checked.

GENERAT:  MOV CL,02H            ;Set initial divisor as 02.
REPEAT:   CMP BL,CL             ;If BL=CL, jump to store.
          JZ  STORE

          MOV AH,00H            ;Clear AH register.
          MOV AL,BL             ;Set the number to be checked as dividend.
          DIV CL                ;Check whether divisible by any other number.
          CMP AH,00H            ;Check whether the remainder is zero.
          JZ  NEXT              ;If remainder is zero, verify next number.
          INC CL                ;If remainder is non-zero then,
          JMP REPEAT            ;increment the divisor and jump to REPEAT.

STORE:    MOV [DI],BL           ;Save the prime number.
          INC DI                ;Increment address pointer.
NEXT:     INC BL                ;Increment the number to be checked.
          CMP BL,DL             ;Check whether number to be checked is N.
          JNZ GENERAT           ;If number to be checked is not equal to N,
          HLT                   ;then continue generation,otherwise stop.

CODE ENDS                       ;End of code segment.
END                             ;Assembly end.
```

Assembler Listing for Example Program 86.19

```
;PROGRAM TO GENERATE PRIME NUMBERS

0000              CODE SEGMENT          ;Start of code segment.

                  ASSUME CS:CODE ;Assembler directive.
1000              ORG 1000H             ;specify program starting address.

1000  BE 1100     MOV SI,1100H          ;Set SI-register as pointer for end data N.
1003  8A 14       MOV DL,[SI]           ;Get the data N in DL-register.
1005  BF 1200     MOV DI,1200H          ;Set DI as pointer for storing prime numbers.
1008  B3 01       MOV BL,01H            ;Initialize the number to be checked as 01
100A  88 1D       MOV [DI],BL           ;Save the first prime number.
100C  47          INC DI                ;Increment address pointer.
100D  FE C3       INC BL                ;Increment the number to be checked.

100F  B1 02   GENERAT: MOV CL,02H       ;Set initial divisor as 02.
1011  3A D9   REPEAT:  CMP BL,CL        ;If BL=CL, jump to store.
1013  74 0F            JZ  STORE

1015  B4 00            MOV AH,00H       ;Clear AH-register.
1017  8A C3            MOV AL,BL        ;Set the number to be checked as dividend.
1019  F6 F1            DIV CL           ;Check whether divisible by any other number.
101B  80 FC 00         CMP AH,00H       ;Check whether the remainder is zero.
101E  74 07            JZ  NEXT         ;If remainder is zero, verify next number.
1020  FE C1            INC CL           ;If remainder is non-zero then,
1022  EB ED            JMP REPEAT       ;increment the divisor and jump to REPEAT.

1024  88 1D   STORE:   MOV [DI],BL      ;Save the prime number.
1026  47               INC DI           ;Increment address pointer.
```

```
1027  FE C3    NEXT:    INC  BL           ;Increment the number to be checked.
1029  3A DA             CMP  BL,DL         ;Check whether number to be checked is N.
102B  75 E2             JNZ  GENERAT       ;If number to be checked is not equal to N,
102D  F4                HLT                ;then continue generation, otherwise stop.

102E           CODE ENDS                   ;End of code segment.
               END                         ;Assembly end.
```

Sample Data

Input Data : $0C_H$ Output Data : 01_H
                                    02_H
                                    03_H
                                    05_H
                                    07_H
                                    $0B_H$

Memory address	Content
1100	0C
1200	01
1201	02
1202	03

Memory address	Content
1203	05
1204	07
1205	0B

EXAMPLE PROGRAM 86.20:   Generation of Fibonacci Series

Write an assembly language program to generate the Fibonacci series.

Problem Analysis

The first and second terms of the fibonacci series are 00_H and 01_H. The third element is given by the sum of the first and second elements. The fourth element is given by the sum of the second and third elements, and so on. In general, an element of the fibonacci series is given by the sum of the immediate two previous element.

Algorithm

1. Set SI-register as pointer for Fibonacci series.

2. Set CL-register as count for number of elements to be generated.

3. Increment the pointer (SI).

4. Initialize the first element of Fibonacci series as 00_H in AL-register.

5. Store first element in memory.

6. Increment the pointer (SI).

7. Increment AL to get second element (01_H) of Fibonacci series in AL-register.

8. Store the second element in memory.

9. Decrement the count (CL-register) by 02.

10. Decrement the pointer (SI).

11. Get the element prior to last generated element in AL.

12. Increment the pointer (SI).

13. Get the last generated element in BL.

14. Add the previous two elements (AL and BL) to get the next element in AL.

15. Increment the pointer.

16. Store the next element (AL) of the Fibonacci series in memory.

17. Decrement the count (CL-register).

18. If the content of CL is not zero then go to step 10, otherwise stop.

Flowchart for Example Program 86.20

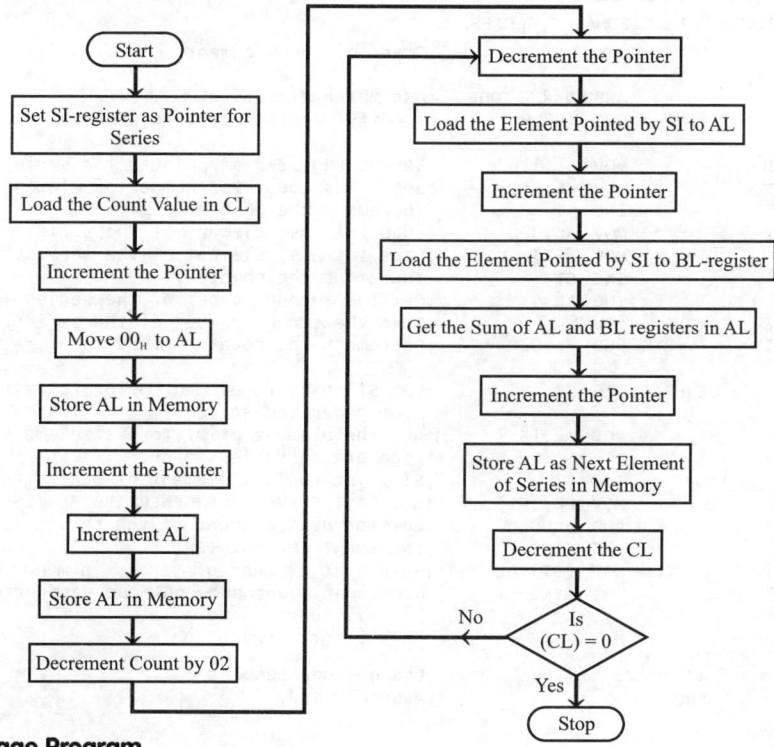

Assembly Language Program

```
;PROGRAM TO GENERATE FIBONACCI SERIES
CODE SEGMENT                ;Start of code segment.

        ASSUME CS:CODE ;Assembler directive.
        ORG 1000H      ;specify program starting address.
        MOV  SI,1100H  ;Set SI-register as pointer for series.
        MOV  CL,[SI]   ;Set CL as count for number of elements of series.
        INC  SI        ;Increment the pointer.
        MOV  AL,00H    ;Load the first element of the series in AL.
        MOV  [SI],AL   ;Save the first element of the series.
        INC  SI        ;Increment the pointer.
        INC  AL        ;Get the second number of the series in AL.
        MOV  [SI],AL   ;Save the second number of the series.
        SUB  CL,02H    ;Decrement the count by two.
GENERAT: DEC SI        ;Let SI point to element prior to last element
                              ;of generated series.
        MOV  AL,[SI]   ;Get the element prior to last element of
                              ;generated series in AL.
        INC  SI        ;SI point to last element of generated series.
        MOV  BL,[SI]   ;Get last element of generated series in BL.
        ADD  AL,BL     ;Get the next element of the series in AL.
        INC  SI        ;Increment the pointer.
        MOV  [SI],AL   ;Save next element of series in memory.
        LOOP GENERAT   ;Decrement count and continue generation
                              ;until count is zero.
        HLT            ;Halt program execution.

CODE ENDS             ;End of code segment.
END                   ;Assembly end.
```

Assembler Listing for Example Program 86.20

```
;PROGRAM TO GENERATE FIBONACCI SERIES
0000            CODE SEGMENT          ;Start of code segment.

                ASSUME CS:CODE        ;Assembler directive.
1000            ORG 1000H             ;specify program starting address.

1000  BE 1100   MOV SI,1100H          ;Set SI-register as pointer for series.
1003  8A 0C     MOV CL,[SI]           ;Set CL as count for number of elements of series.
1005  46        INC SI                ;Increment the pointer.
1006  B0 00     MOV AL,00H            ;Load the first element of the series in AL.
1008  88 04     MOV [SI],AL           ;Save the first element of the series.
100A  46        INC SI                ;Increment the pointer.
100B  FE C0     INC AL                ;Get the second number of the series in AL.
100D  88 04     MOV [SI],AL           ;Save the second number of the series.
100F  80 E9 02  SUB CL,02H            ;Decrement the count by two.

1012  4E  GENERAT: DEC SI             ;Let SI point to element prior to last element
                                         ;of generated series.
1013  8A 04     MOV AL,[SI]           ;Get the element prior to last element of
                                         ;generated series in AL.
1015  46        INC SI                ;SI point to last element of generated series.
1016  8A 1C     MOV BL,[SI]           ;GET LAST ELEMENT OF GENERATED SERIES IN BL.
1018  02 C3     ADD AL,BL             ;Get the next element of the series in AL.
101A  46        INC SI                ;INCREMENT THE POINTER.
101B  88 04     MOV [SI],AL           ;Save next element of series in memory.
101D  E2 F3     LOOP GENERAT          ;Decrement count and continue generation
                                         ;until count is zero.

101F  F4        HLT                   ;Halt program execution.

1020            CODE ENDS             ;End of code segment.
                END                   ;Assembly end.
```

Sample Data

Input Data : 08_H Output Data : 00_H
                                    01_H
                                    01_H
                                    02_H
                                    03_H
                                    05_H
                                    08_H
                                    $0D_H$

Memory address	Content	Memory address	Content
1100	08	1105	03
1101	00	1106	05
1102	01	1107	08
1103	01	1108	0D
1104	02		

EXAMPLE PROGRAM 86.21:   Matrix Addition

Write an assembly language program to add two numbers of 3×3 matrices.

Problem Analysis

While storing the matrices in the memory, the first-row elements are stored first, followed by second-row elements and then third-row elements. For addition operation, the matrices can be addressed as an array with number of elements in the matrices as the count value.

The two input matrices can be stored in different memory areas in the same order as mentioned above. The base registers BX and BP can be used to hold the base address of the input matrices and SI-register can be used as pointer for the elements in the matrices. Another index register DI can be used as pointer to store the sum matrix.

Flowchart of Example Program 86.21

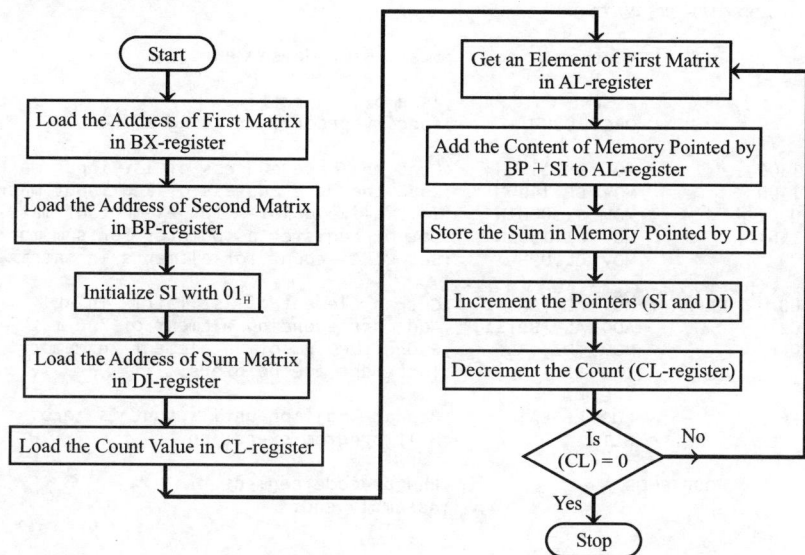

Algorithm

1. **Load the base address of first input matrix in BX-register.**
2. **Load the base address of second input matrix in BP-register.**
3. **Set SI-register as index (or pointer) for elements of the matrix.**
4. **Set DI-register as pointer for sum matrix.**
5. **Load the count value in CL-register.**
6. **Get an element of first matrix in AL-register.**
7. **Add the corresponding element of second matrix to AL-register.**
8. **Store the sum in memory.**
9. **Increment the SI and DI registers.**
10. **Repeat steps 6 to 9 until the count value in CL-register is zero.**
11. **Stop.**

Assembly Language Program

```
;PROGRAM TO ADD TWO 3X3 MATRIX
CODE SEGMENT                 ;Start of code segment.

        ASSUME CS:CODE   ;Assembler directive.
        ORG 1000H        ;specify program starting address.

        MOV BX,1300H     ;Load the base address of 1st input matrix in BX.
        MOV BP,1400H     ;Load the base address of 2nd input matrix in BP.
        MOV SI,0001H     ;Initialize pointer for element of matrix.
        MOV DI,1501H     ;Set DI-register as pointer for sum matrix.
        MOV CL,09H       ;Set CL as count for elements in matrix.

REPEAT: MOV AL,[BX+SI]   ;Get an element of 1st matrix in AL.
        ADD AL,[BP+SI]   ;Add corresponding element of 2nd matrix to AL.
        MOV [DI],AL      ;Store the sum of an element in memory.
        INC SI           ;Increment the pointers.
        INC DI
        LOOP REPEAT      ;Repeat addition until count is zero.
        HLT              ;Halt program execution.

CODE ENDS                ;End of code segment.
END                      ;Assembly end.
```

Assembler Listing for Example Program 86.21

```
;PROGRAM TO ADD TWO 3X3 MATRIX

0000                    CODE SEGMENT          ;Start of code segment.

                        ASSUME CS:CODE        ;Assembler directive.
1000                    ORG  1000H            ;specify program starting address.

1000  BB 1300           MOV  BX,1300H         ;Load the base address of 1st input matrix in BX.
1003  BD 1400           MOV  BP,1400H         ;Load the base address of 2nd input matrix in BP.
1006  BE 0001           MOV  SI,0001H         ;Initialize pointer for element of matrix.
1009  BF 1501           MOV  DI,1501H         ;Set DI-register as pointer for sum matrix.
100C  B1 09             MOV  CL,09H           ;Set CL as count for elements in matrix.

100E  8A 00     REPEAT: MOV  AL,[BX+SI]       ;Get an element of 1st matrix in AL.
1010  02 02             ADD  AL,[BP+SI]       ;Add corresponding element of 2nd matrix to AL.
1012  88 05             MOV  [DI],AL          ;Store the sum of an element in memory.
1014  46                INC  SI               ;Increment the pointers.
1015  47                INC  DI
1016  E2 F6             LOOP REPEAT           ;Repeat addition until count is zero.
1018  F4                HLT                   ;Halt program execution.

1019                    CODE ENDS             ;End of code segment.
                        END                   ;Assembly end.
```

Sample Data

MATRIX 1				MATRIX 2				SUM MATRIX		
[01 02 03]	Memory address	Content	[F0 E1 D2]	Memory address	Content	[F1 E3 D5]	Memory address	Content		
[04 05 06]			[C3 B4 A5]			[C7 B9 AB]				
[07 08 09]	1301	01	[96 87 78]	1401	F0	[9D 8F 81]	1501	F1		
	1302	02		1402	E1		1502	E3		
	1303	03		1403	D2		1503	D5		
	1304	04		1404	C3		1504	C7		
	1305	05		1405	B4		1505	B9		
	1306	06		1406	A5		1506	AB		
	1307	07		1407	96		1507	9D		
	1308	08		1408	87		1508	8F		
	1309	09		1409	78		1509	81		

EXAMPLE PROGRAM 86.22: Matrix Multiplication

Write an assembly language program to multiply two numbers of 3×3 matrices.

Problem Analysis

While storing the matrices in memory, the first-row elements are stored first, followed by second-row elements and then third-row elements. For multiplication operation, the matrices should be addressed as two dimensional array. Here the two-dimensional arrays are addressed by using pointers and counters.

The SI and DI registers are used as address pointers for two input matrices. The CL and CH registers are used as row and column counts, respectively. The BP-register is used as pointer for storing the elements of product matrix.

Algorithm

1.   Load the address of first input matrix in SI-register.

2.   Load the address of second input matrix in DI-register.

3.   Load the address of product matrix in BP-register.

4.   Load the row count in CL-register.

5.   Load the column count in CH-register.

6.   Copy the column count in BL-register (Let it be second column count).

7.   Initialize sum as zero in DL-register.

8.   Get the column count in DH-register.

9.   Get a row element of first matrix in AL-register.

10. Multiply a column element of second matrix with AL, the product will be in AL. (Because it is assumed that the product does not exceed 8-bit.)

11. Add the product (AL) to sum (DL).

12. Increment SI to point to next element of same row in first input matrix.

13. Increment DI by 03 to point to next element of same column in second input matrix.

14. Decrement the column count (DH-register).

15. Check zero flag. If zero flag is reset then go to step 9, otherwise go to next step.

16. Store an element of product matrix (DL) in memory.

17. Increment the product matrix pointer (BP).

18. Subtract 03_H from SI to point to the first element of same row.

19. Subtract 09_H from DI to point to first element of next row.

20. Decrement BL-register (second column count).

21. Check zero flag. If zero flag is reset then go to step 7, otherwise go to next step.

22. Add 03_H to SI to point to first element of next row in first matrix.

23. Load the starting address of second matrix in DI-register.

24. Decrement the row count.

25. Check zero flag. If zero is reset then go to step 6, otherwise stop.

Flowchart of Example Program 86.22

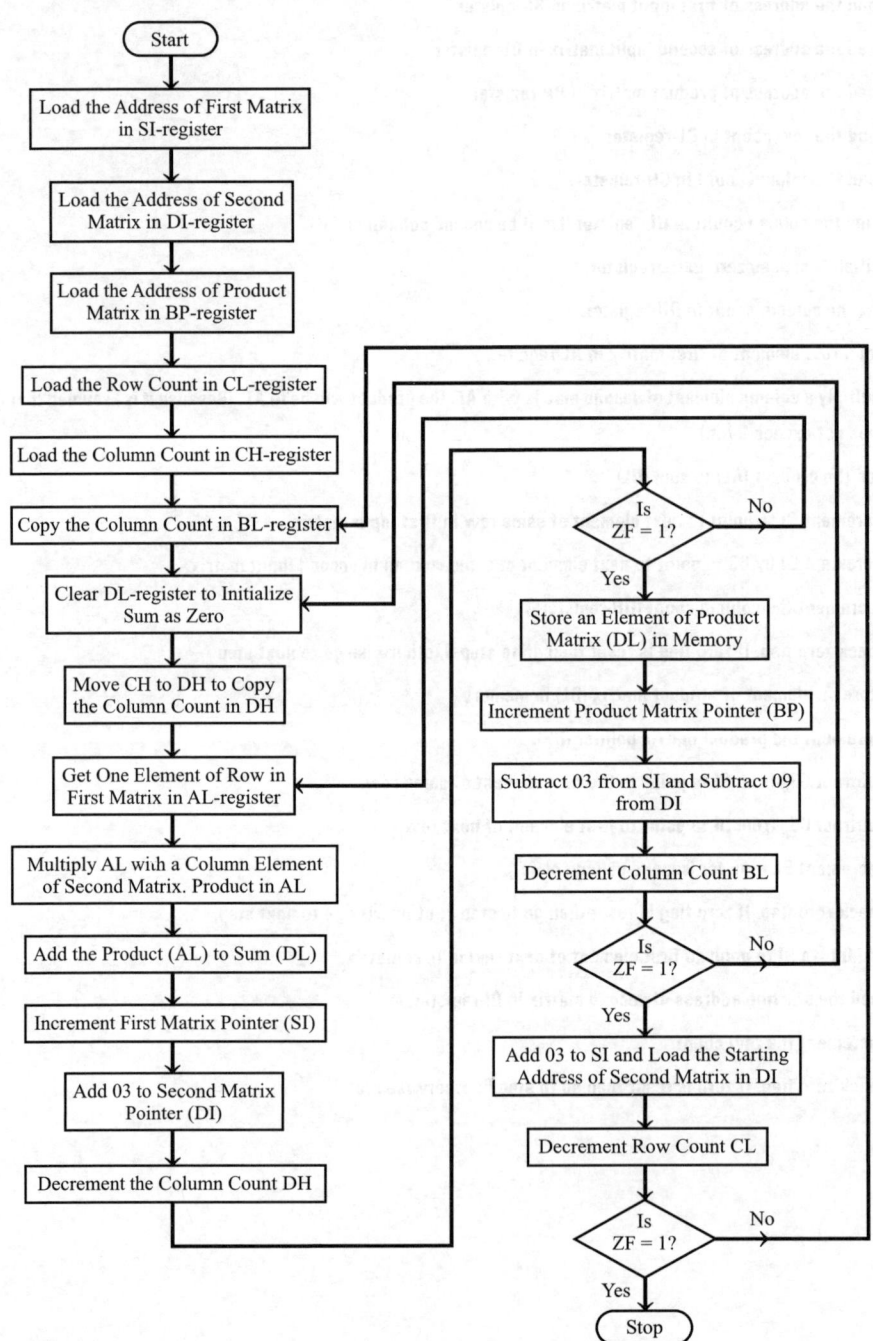

Assembly Language Program

```
;PROGRAM TO MULTIPLY TWO 3X3 MATRIX

CODE SEGMENT                    ;Start of code segment.

        ASSUME CS:CODE ;Assembler directive.
        ORG 1000H      ;specify program starting address.

        MOV SI,1301H   ;Set SI as pointer for first input matrix.
        MOV DI,1401H   ;Set DI as pointer for second input matrix.
        MOV BP,1501H   ;Set BP as pointer for product matrix.
        MOV CL,03H     ;Set CL as count for elements in a row.
        MOV CH,03H     ;Set CH as count for elements in a column.

REPEAT3: MOV BL,CH     ;Copy the column count in BL-register.

REPEAT2: MOV DL,00H    ;Initialize sum as zero.
         MOV DH,CH     ;Get the column count in DH.

REPEAT1: MOV AL,[SI]   ;Get one element of the row in AL-register.
         MUL [DI]      ;Get product of row and column element in AL.
         ADD DL,AL     ;Add the product to sum.
         INC SI        ;Increment the first input matrix pointer.
         ADD DI,03H    ;Let DI point to next element of
                       ;same column of 2nd matrix.
         DEC DH        ;Decrement the column count.
         JNZ REPEAT1   ;Repeat multiplication and addition until
                       ;DH count is zero.

         MOV [BP],DL   ;Store an element of product matrix in memory.
         INC BP        ;Increment the product matrix pointer.
         SUB SI,03H    ;Make SI to point to first element of the row.
         SUB DI,09H
         INC DI
         DEC BL        ;Decrement the column count.
         JNZ REPEAT2   ;Repeat multiplication and addition of a row
                       ;in 1st matrix with next column of 2nd matrix.
         ADD SI,03H    ;Let SI point to first element of
                       ;next row of 1st matrix.
         MOV DI,1401H  ;Make DI to point to first element of 2nd matrix.
         DEC CL        ;Decrement the row count.
         JNZ REPEAT3   ;Repeat multiplication and addition of next row
                       ;in 1st matrix with all column of 2nd matrix.
         HLT           ;Halt program execution.

CODE ENDS                       ;End of code segment.
END                             ;Assembly end.
```

Assembler Listing for Example Program 86.22

```
;PROGRAM TO MULTIPLY TWO 3X3 MATRIX

0000               CODE SEGMENT          ;Start of code segment.

                   ASSUME CS:CODE ;Assembler directive.
1000               ORG 1000H      ;specify program starting address.

1000  BE 1301      MOV SI,1301H   ;Set SI as pointer for first input matrix.
1003  BF 1401      MOV DI,1401H   ;Set DI as pointer for second input matrix.
1006  BD 1501      MOV BP,1501H   ;Set BP as pointer for product matrix.
1009  B1 03        MOV CL,03H     ;Set CL as count for elements in a row.
100B  B5 03        MOV CH,03H     ;Set CH as count for elements in a column.
100D  8A DD   REPEAT3: MOV BL,CH  ;Copy the column count in BL-register.

100F  B2 00   REPEAT2: MOV DL,00H ;Initialize sum as zero.
1011  8A F5        MOV DH,CH      ;Get the column count in DH.
```

```
1013   8A 04      REPEAT1: MOV AL,[SI]     ;Get one element of the row in AL-register.
1015   F6 25           MUL [DI]            ;Get product of row and column element in AL.
1017   02 D0           ADD DL,AL           ;Add the product to sum.
1019   46              INC SI              ;Increment the first input matrix pointer.
101A   83 C7 03        ADD DI,03H          ;Let DI point to next element of
                                           ;same column of 2nd matrix.
101D   FE CE           DEC DH              ;Decrement the column count.
101F   75 F2           JNZ REPEAT1         ;Repeat multiplication and addition until
                                           ;DH count is zero.

1021   88 56 00        MOV [BP],DL         ;Store an element of product matrix in memory.
1024   45              INC BP              ;Increment the product matrix pointer.
1025   83 EE 03        SUB SI,03H          ;Make SI to point to first element of the row.
1028   83 EF 09        SUB DI,09H
102B   47              INC DI
102C   FE CB           DEC BL              ;Decrement the column count.
102E   75 DF           JNZ REPEAT2         ;Repeat multiplication and addition of a row
                                           ;in 1st matrix with next column of 2nd matrix.
1030   83 C6 03        ADD SI,03H          ;Let SI point to first element of
                                           ;next row of 1st matrix.
1033   BF 1401         MOV DI,1401H        ;Make DI to point to first element of 2nd matrix.
1036   FE C9           DEC CL              ;Decrement the row count.
1038   75 D3           JNZ REPEAT3         ;Repeat multiplication and addition of next row
                                           ;in 1st matrix with all column of 2nd matrix.

103A   F4              HLT                 ;Halt program execution.

103B           CODE ENDS                   ;End of code segment.
                END
```

Sample Data

MATRIX 1					MATRIX 2					PRODUCT MATRIX		

MATRIX 1	Memory address	Content	MATRIX 2	Memory address	Content	PRODUCT MATRIX	Memory address	Content
[01 01 01]			[04 04 04]			[0F 0F 0F]		
[02 02 02]			[05 05 05]			[1E 1E 1E]		
[03 03 03]	1301	01	[06 06 06]	1401	04	[2D 2D 2D]	1501	0F
	1302	01		1402	04		1502	0F
	1303	01		1403	04		1503	0F
	1304	02		1404	05		1504	1E
	1305	02		1405	05		1505	1E
	1306	02		1406	05		1506	1E
	1307	03		1407	06		1507	2D
	1308	03		1408	06		1508	2D
	1309	03		1409	06		1509	2D

Example Program 86.23:   BCD to Binary Conversion

Write an assembly language program to convert a BCD data (2-digit/8-bit) to binary.

Problem Analysis

The 2-digit BCD data will have units digit and tens digit. When the tens digit (upper nibble) is multiplied by $0A_H$ and the product is added to units digit (lower nibble), the result will be in binary, because the microprocessor will perform binary arithmetic.

Algorithm

1.   Load the address of BCD data in BX-register.
2.   Get the BCD data in AL-register.
3.   Copy the BCD data in DL-register.
4.   Logically AND DL with $0F_H$ to mask upper nibble and get units digit in DL.
5.   Logically AND AL with $F0_H$ to mask lower nibble.
6.   Move the count value for rotation in CL-register.
7.   Rotate the content of AL to move the upper nibble to lower nibble position.
8.   Move $0A_H$ to DH-register.

9. Multiply AL with DH-register. The product will be in AL-register.
10. Add the units digit in DL-register to product in AL-register.
11. Save the binary data (AL) in memory.
12. Stop.

Flowchart of Example Program 86.23

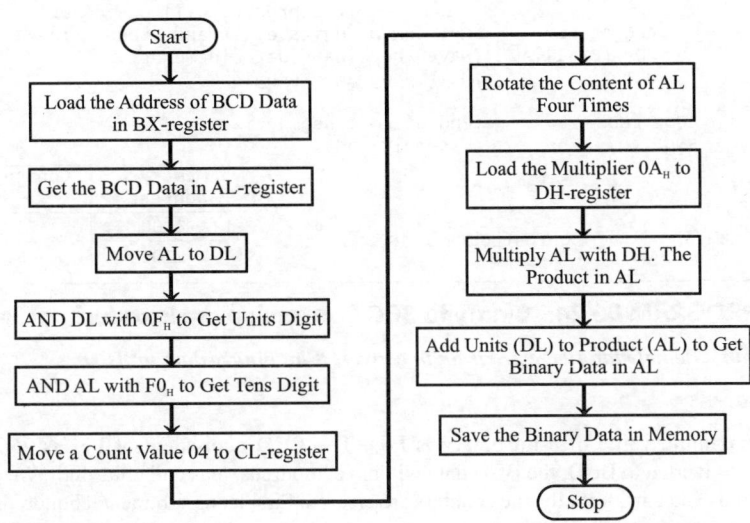

Assembly Language Program

```
;PROGRAM TO CONVERT A BCD DATA TO BINARY DATA
CODE SEGMENT              ;Start of code segment.

        ASSUME CS:CODE   ;Assembler directive.
        ORG 1000H        ;specify program starting address.
        MOV BX,1100H     ;Load the address of the data in BX-register.
        MOV AL,[BX]      ;Get the BCD data in AL-register.
        MOV DL,AL        ;Copy the data in DL-register.
        AND DL,0FH       ;Mask upper nibble (tens digit).
        AND AL,0F0H      ;Mask lower nibble (units digit).
        MOV CL,4         ;Rotate the upper nibble to lower nibble position.
        ROR AL,CL
        MOV DH,0AH       ;Set multiplier as 0AH.
        MUL DH           ;Multiply tens digit by 0AH,
                         ;the product will be in AL.
        ADD AL,DL        ;Get sum of units digit and product in AL.
        MOV [BX+1],AL    ;Save the binary data in memory.
        HLT              ;Halt program execution.

CODE ENDS                ;End of code segment.
END                      ;Assembly end.
```

Assembler Listing for Example Program 86.23

```
;PROGRAM TO CONVERT A BCD DATA TO BINARY DATA

0000            CODE SEGMENT        ;Start of code segment.

                ASSUME CS:CODE  ;Assembler directive.
1000            ORG 1000H       ;specify program starting address.
1000  BB 1100   MOV BX,1100H    ;Load the address of the data in BX-register.
```

```
1003   8A 07        MOV AL,[BX]      ;Get the BCD data in AL-register.
1005   8A D0        MOV DL,AL        ;Copy the data in DL-register.
1007   80 E2 0F     AND DL,0FH       ;Mask upper nibble (tens digit).
100A   24 F0        AND AL,0F0H      ;Mask lower nibble (units digit).
100C   B1 04        MOV CL,4         ;Rotate the upper nibble to lower nibble position.
100E   D2 C8        ROR AL,CL
1010   B6 0A        MOV DH,0AH       ;Set multiplier as 0AH.
1012   F6 E6        MUL DH           ;Multiply tens digit by 0AH,
                                         ;the product will be in AL.
1014   02 C2        ADD AL,DL        ;Get sum of units digit and product in AL.
1016   88 47 01     MOV [BX+1],AL    ;Save the binary data in memory.
1019   F4           HLT              ;Halt program execution.

101A                CODE ENDS        ;End of code segment.
                    END              ;Assembly end.
```

Sample Data

Input Data : 75_{10} Output Data : $4B_H$

Memory address	Content	
1100	75	← BCD
1101	4B	← Binary

EXAMPLE PROGRAM 86.24: Binary to BCD Conversion or Hexadecimal to BCD Conversion

Write an assembly language program to convert 8-bit binary data to BCD.

Problem Analysis

The maximum value of 8-bit binary is FF_H. The BCD equivalent of FF_H is 256_{10}. Hence, when an 8-bit binary is converted to BCD, the BCD data will have hundreds, tens and units digit. We can use two counters to count hundreds and tens. Initially, the counters are cleared. First let us subtract all hundreds from the given data and for each subtraction hundreds counter is incremented by one. Then we can subtract all tens from the given data and for each subtraction tens counter is incremented by one. The remaining will be units. The tens and units can be combined as 2-digit BCD and stored in the memory. The hundreds can be separately stored in the memory.

Algorithm

1. Load the address of data in BX-register.
2. Get the binary data in AL-register.
3. Clear DX-register for storing hundreds and tens.
4. Compare AL with 64_H (100_{10}).
5. Check carry flag. If carry flag is set then go to step 9, otherwise go to next step.
6. Subtract 64_H (100_{10}) from AL-register.
7. Increment hundreds register (DL).
8. Go to step 4.
9. Compare AL with $0A_H$ (10_{10}).
10. Check carry flag. If carry flag is set then go to step 14, otherwise go to next step.
11. Subtract $0A_H$ (10_{10}) from AL-register.
12. Increment tens register (DH).
13. Go to step 9.
14. Move the count value 04_H for rotation in CL-register.
15. Rotate the content of DH four times.
16. Add DH to AL to combine tens and units as 2-digit BCD.
17. Save AL and DL in memory.
18. Stop

Flowchart of Example Program 86.24

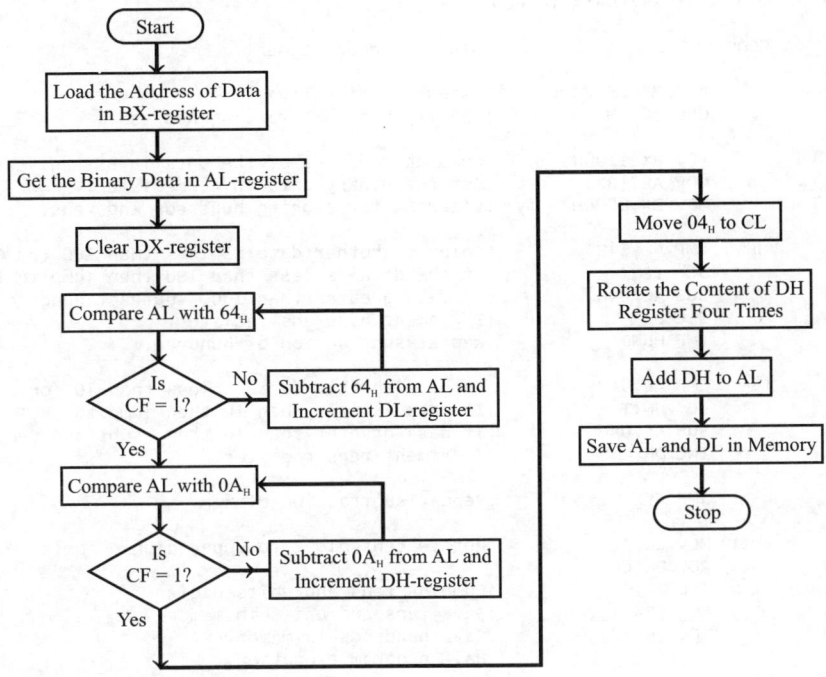

Assembly Language Program

```
;PROGRAM TO CONVERT A BINARY DATA TO BCD DATA

CODE SEGMENT          ;Start of code segment.

      ASSUME CS:CODE ;Assembler directive.
      ORG 1000H      ;specify program starting address.

      MOV BX,1100H   ;Load the address of the data in BX-register.
      MOV AL,[BX]    ;Get the binary data in AL-register.
      MOV DX,0000H   ;Clear DX for storing hundreds and tens.

HUND: CMP AL,64H     ;Compare whether data is less than 100 (or 64H).
      JC  TEN        ;If the data is less than 100 then jump to TEN.
      SUB AL,64H     ;If data greater than 100, subtract  hundred.
      INC DL         ;Increment hundreds register.
      JMP HUND       ;Repeat subtraction of hundred.
TEN:  CMP AL,0AH     ;Compare whether data is less than 10 (or 0AH).
      JC  UNIT       ;If data is less than 10 then jump to Unit.
      SUB AL,0AH     ;If data greater than 10 then, subtract ten.
      INC DH         ;Increment tens register.
      JMP TEN        ;Repeat subtraction of ten.

UNIT: MOV CL,4       ;Rotate tens digit to upper nibble position.
      ROL DH,CL
      ADD AL,DH      ;Combine tens and units digit.
      MOV [BX+1],AL  ;Save tens and units in memory.
      MOV [BX+2],DL  ;Save hundreds in memory.
      HLT            ;Halt program execution.
CODE ENDS            ;End of code segment.
END                  ;Assembly end.
```

Assembler Listing for Example Program 86.24

```
;PROGRAM TO CONVERT A BINARY DATA TO BCD DATA

0000                  CODE SEGMENT              ;Start of code segment.

                      ASSUME CS:CODE           ;Assembler directive.
1000                  ORG 1000H                ;specify program starting address.

1000 BB 1100          MOV BX,1100H             ;Load the address of the data in BX-register.
1003 8A 07            MOV AL,[BX]              ;Get the binary data in AL-register.
1005 BA 0000          MOV DX,0000H             ;Clear DX for storing hundreds and tens.

1008 3C 64     HUND:  CMP AL,64H               ;Compare whether data is less than 100 (or 64H).
100A 72 06            JC  TEN                  ;If the data is less than 100 then jump to TEN.
100C 2C 64            SUB AL,64H               ;If data greater than 100, subtract hundred.
100E FE C2            INC DL                   ;Increment hundreds register.
1010 EB F6            JMP HUND                 ;Repeat subtraction of hundred.

1012 3C 0A     TEN:   CMP AL,0AH               ;Compare whether data is less than 10 (or 0AH).
1014 72 06            JC  UNIT                 ;If data is less than 10 then jump to UNIT.
1016 2C 0A            SUB AL,0AH               ;If data greater than 10 then, subtract ten.
1018 FE C6            INC DH                   ;Increment tens register.

101A EB F6            JMP TEN                  ;Repeat subtraction of ten.

101C B1 04     UNIT:  MOV CL,4                 ;Rotate tens digit to upper nibble position.
101E D2 C6            ROL DH,CL
1020 02 C6            ADD AL,DH                ;Combine tens and units digit.
1022 88 47 01         MOV [BX+1],AL            ;Save tens and units in memory.
1025 88 57 02         MOV [BX+2],DL            ;Save hundreds in memory.
1028 F4               HLT                      ;Halt program execution.

1029                  CODE ENDS                ;End of code segment.
                      END                      ;Assembly end.
```

Sample Data

Input Data : E4$_H$ Output Data : 0228$_{10}$

Memory address	Content	
1100	E4	← Binary
1101	28	} BCD
1102	02	

EXAMPLE PROGRAM 86.25: Binary to ASCII Conversion or Hexadecimal to ASCII Conversion

Write an assembly language program to convert an array of 8-bit binary data to ASCII code.

Problem Analysis

The 8-bit binary can be represented by 2-digit hexa. Each hexa digit can be converted to an 8-bit ASCII code (i.e., each nibble of binary data can be converted to an 8-bit ASCII code). The hexa digit 0 through 9 are represented by 30 to 39 in ASCII and the hexa digit A through F are represented by 41 to 46 in ASCII. Therefore, the 8-bit binary data is split into two nibbles: lower nibble and upper nibble. Then check each nibble whether it is less than $0A_H$ or not, if it is less than $0A_H$ then add 30_H to convert to ASCII or of it is greater than/equal to $0A_H$ then add 37_H to convert to ASCII.

Flowchart of Example Program 86.25

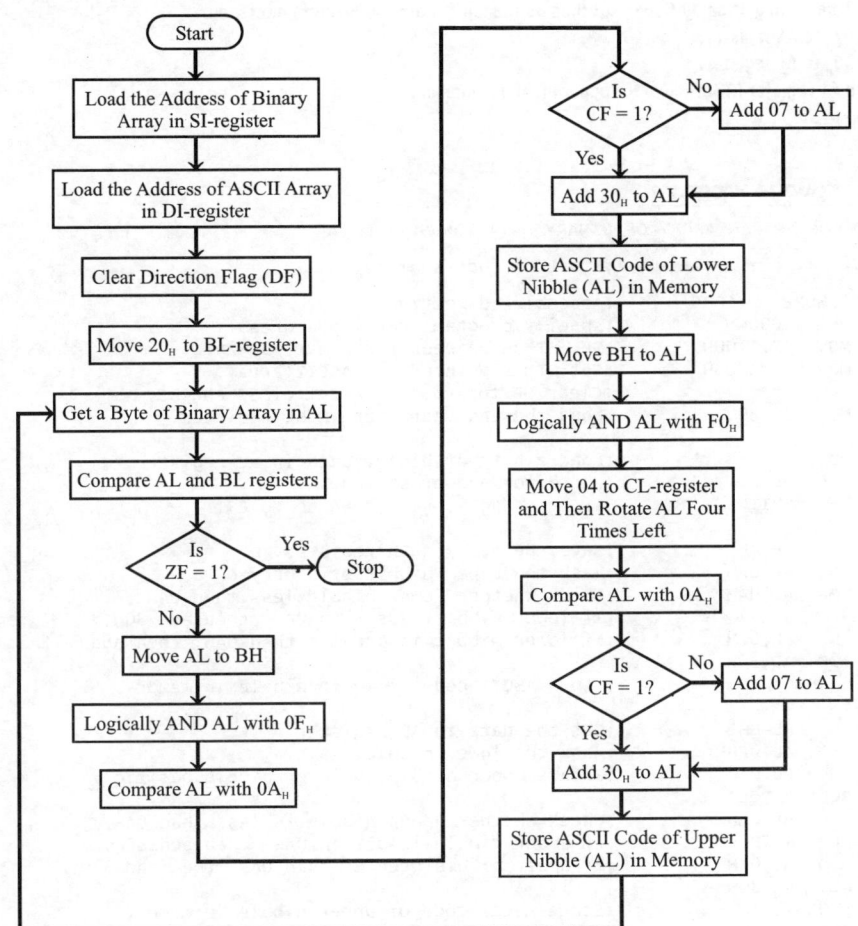

Algorithm

1. Set SI as pointer for binary array.
2. Set DI as pointer for ASCII array.
3. Clear Direction Flag (DF) for autoincremming of pointers.
4. Move the end character 20_H to BL-register.
5. Load a byte of binary array in AL-register.
6. Compare AL and BL.
7. Check zero flag. If zero flag is set then go to step 24, otherwise go to next step.
8. Save the byte in BH-register.
9. Logically AND AL with $0F_H$ to mask the upper nibble.
10. Compare AL with $0A_H$.
11. Check carry flag. If carry flag is set then go to step 13, otherwise go to next step.
12. Add 07_H to AL-register.
13. Add 30_H to AL-register.
14. Store AL-register (ASCII code for lower nibble) in memory.
15. Move the saved byte in BH register to AL-register.
16. Logically AND AL-register with $F0_H$ to mask the lower nibble.

17. Rotate the upper nibble in AL-register to lower nibble position.
18. Compare AL with 0A$_H$.
19. Check carry flag. If carry flag is set then go to step 21, otherwise go to next step.
20. Add 07$_H$ to AL-register.
21. Add 30$_H$ to AL-register.
22. Store AL-register (ASCII code for upper nibble) in memory.
23. Jump to step 5.
24. Stop.

Assembly Language Program

```
;PROGRAM TO CONVERT AN ARRAY OF BINARY DATA TO ASCII DATA

CODE SEGMENT                              ;Start of code segment.

            ASSUME CS:CODE               ;Assembler directive.
            ORG   1000H                  ;specify program starting address.
            MOV   SI,1100H               ;Set SI as pointer for binary array.
            MOV   DI,1400H               ;Set DI as pointer for ASCII array.
            CLD                          ;Clear DF for autoincrement of SI and DI.
            MOV   BL,20H                 ;Load the end character in BL-register.

NEXT:       LODSB                        ;Load a byte of binary data in AL-register.
            CMP   AL,BL                  ;Check for end of string.
            JE    EXIT                   ;If zero flag is set, then go to EXIT.

            MOV   BH,AL                  ;Save the byte in BH-register.
            AND   AL,0FH                 ;Mask the upper nibble of binary data.
            CMP   AL,0AH                 ;Check whether lower nibble less than 0AH.
            JL    SKIP1                  ;If lower nibble less than 0AH, then add 30H.
            ADD   AL,07H                 ;If lower nibble is greater than 0AH, then add 37H.
SKIP1:      ADD   AL,30H
            STOSB                        ;Store ASCII code of lower nibble in memory.

            MOV   AL,BH                  ;Get the data in AL-register.
            AND   AL,0F0H                ;Mask the lower nibble.
            MOV   CL,04H                 ;Rotate upper nibble to lower nibble position.
            ROL   AL,CL
            CMP   AL,0AH                 ;Check whether upper nibble is less than 0AH.
            JL    SKIP2                  ;If upper nibble less than 0AH, then add 30H.
            ADD   AL,07H                 ;If upper nibble greater than 0AH, then add 37H.
SKIP2:      ADD   AL,30H
            STOSB                        ;Store ASCII code of upper nibble in memory.
            JMP   NEXT                   ;Jump to NEXT to convert next byte.
EXIT:       HLT                          ;Halt program execution.

CODE ENDS                                ;End of code segment.
END                                      ;Assembly end.
```

Assembler Listing for Example Program 86.25

```
;PROGRAM TO CONVERT AN ARRAY OF BINARY DATA TO ASCII DATA

0000                CODE SEGMENT              ;Start of code segment.

                    ASSUME CS:CODE           ;Assembler directive.
1000                ORG  1000H               ;specify program starting address.
1000   BE 1100      MOV SI,1100H             ;Set SI as pointer for binary array.
1003   BF 1400      MOV DI,1400H             ;Set DI as pointer for ASCII array.
1006   FC           CLD                      ;Clear DF for autoincrement of SI and DI.
1007   B3 20        MOV BL,20H               ;Load the end character in BL-register.

1009   AC     NEXT: LODSB                    ;Load a byte of binary data in AL-register.
100A   3A C3        CMP AL,BL                ;Check for end of string.
```

```
100C   74 20            JE   EXIT         ;If zero flag is set, then go to EXIT.

100E   8A F8            MOV  BH,AL        ;Save the byte in BH-register.
1010   24 0F            AND  AL,0FH       ;Mask the upper nibble of binary data.
1012   3C 0A            CMP  AL,0AH       ;Check whether lower nibble less than 0AH.
1014   7C 02            JL   SKIP1        ;If lower nibble less than 0AH, then add 30H.
1016   04 07            ADD  AL,07H       ;If lower nibble greater than 0AH, then add 37H.
1018   04 30    SKIP1:  ADD  AL,30H
101A   AA               STOSB            ;Store ASCII code of lower nibble in memory.

101B   8A C7            MOV  AL,BH        ;Get the data in AL-register.
101D   24 F0            AND  AL,0F0H      ;Mask the lower nibble.
101F   B1 04            MOV  CL,04H       ;Rotate upper nibble to lower nibble position.
1021   D2 C0            ROL  AL,CL
1023   3C 0A            CMP  AL,0AH        ;Check whether upper nibble is less than 0AH.
1025   7C 02            JL   SKIP2        ;If upper nibble less than 0AH, then add 30H.
1027   04 07            ADD  AL,07H       ;If upper nibble greater than 0AH, then add 37H.
1029   04 30    SKIP2:  ADD  AL,30H
102B   AA               STOSB            ;Store ASCII code of upper nibble in memory.
102C   EB DB            JMP  NEXT         ;Jump to NEXT to convert next byte.
102E   F4       EXIT:   HLT              ;Halt program execution.

102F            CODE ENDS                ;End of code segment.
                END                      ;Assembly
```

Sample Data

Input Data: 4E$_H$
           15$_H$
           87$_H$
           C0$_H$
           20$_H$

Memory address	Content
1100	4E
1101	15
1102	87
1103	C0
1104	20

Output Data: 45
           34
           35
           31
           37
           38
           30
           43

Memory address	Content
1400	45
1401	34
1402	35
1403	31
1404	37
1405	38
1406	30
1407	43

Binary	ASCII
4E	34 45
15	31 35
87	38 37
c0	43 30

EXAMPLE PROGRAM 86.26: ASCII to Binary Conversion or ASCII to Hexadecimal Conversion

Write an assembly language program to convert an array of ASCII character to binary array.

Problem Analysis

The hexa digit 0 through 9 are represented by 30$_H$ to 39$_H$ in ASCII. Hence, if the ASCII code is in the range 30$_H$ to 39$_H$, then we can subtract 30$_H$ to get the binary value. The hexa digit A through F are represented by 41$_H$ to 46$_H$ in ASCII. Hence, if the ASCII code is in the range 41$_H$ to 46$_H$ then we can subtract 37$_H$ to get the binary value.

Flowchart of Example Program 86.26

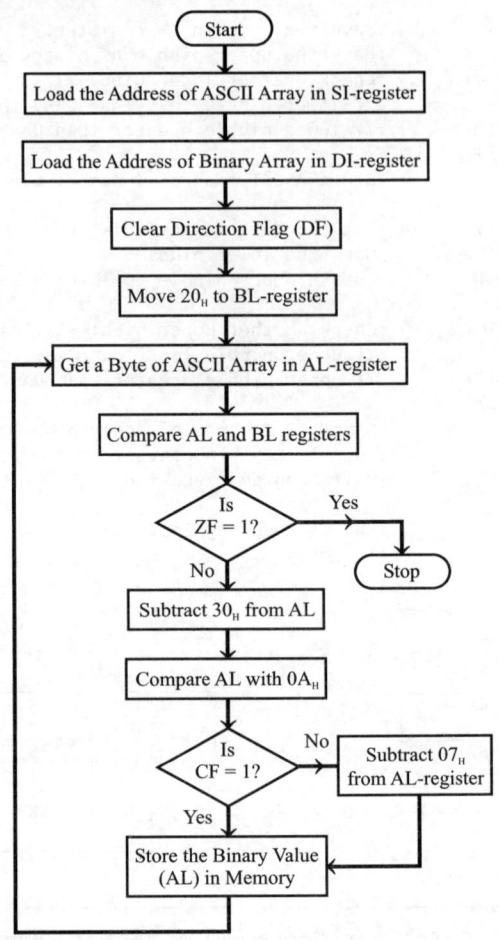

Algorithm

1.   Set SI as pointer for ASCII array.
2.   Set DI as pointer for binary array.
3.   Clear direction flag (DF) for autoincrement of pointers.
4.   Move the end character 20_H to BL.
5.   Get a byte of ASCII array in AL-register.
6.   Compare AL and BL registers.
7.   Check zero flag. If zero flag is set then go to step 14, otherwise go to next step.
8.   Subtract 30_H from AL-register.
9.   Compare AL with $0A_H$.
10. Check carry flag. If carry flag is set then go to step 12, otherwise go to next step.
11. Subtract 07_H from AL-register.
12. Store the binary value (AL) in memory.
13. Go to step 5.
14. Stop.

Assembly Language Program

```
;PROGRAM TO CONVERT AN ARRAY OF ASCII DATA TO BINARY DATA

CODE SEGMENT                ;Start of code segment.

        ASSUME CS:CODE ;Assembler directive.
        ORG 1000H       ;specify program starting address.

        MOV SI,1100H    ;Set SI as pointer for ASCII array.
        MOV DI,1400H    ;Set DI as pointer for binary array.
        CLD             ;Clear DF for autoincrement of SI and DI.
        MOV BL,20H      ;Load the end character in BL-register.

NEXT:   LODSB           ;Load a byte of ASCII array in AL-register.
        CMP AL,BL       ;Check for end of string.
        JE  EXIT        ;If zero flag is set, then go to EXIT.
        SUB AL,30H      ;If zero flag is not set,
                        ;then subtract 30H from AL.

        CMP AL,0AH      ;Check whether AL is greater than 0AH.
        JC  STORE       ;If AL is less than 0AH, then go to STORE.
        SUB AL,07H      ;If AL is greater than or equal to 0AH,
                        ;then subtract 07H from AL.
STORE:  STOSB           ;Store the binary value in memory.
        JMP NEXT        ;Jump to convert next byte.

EXIT:   HLT             ;Halt program execution.

CODE ENDS               ;End of code segment.
END                     ;Assembly end.
```

Assembler Listing for Example Program 86.26

```
PROGRAM TO CONVERT AN ARRAY OF ASCII DATA TO BINARY DATA

0000            CODE SEGMENT            ;Start of code segment.

                ASSUME CS:CODE          ;Assembler directive.
1000            ORG  1000H              ;specify program starting address.
1000  BE 1100   MOV  SI,1100H           ;Set SI as pointer for ASCII array.
1003  BF 1400   MOV  DI,1400H           ;Set DI as pointer for binary array.
1006  FC        CLD                     ;Clear DF for autoincrement of SI and DI.
1007  B3 20     MOV  BL,20H             ;Load the end character in BL-register.

1009  AC        NEXT: LODSB             ;Load a byte of ASCII array in AL-register.
100A  3A C3           CMP  AL,BL        ;Check for end of string.
100C  74 0B           JE   EXIT         ;If zero flag is set, then go to EXIT.
100E  2C 30           SUB  AL,30H       ;If zero flag is not set,
                                         ;then subtract 30H from AL.

1010  3C 0A           CMP  AL,0AH       ;Check whether AL is greater than 0AH.
1012  72 02           JC   STORE        ;If AL is less than 0AH, then go to STORE.

1014  2C 07           SUB  AL,07H       ;If AL is greater than or equal to 0AH,
                                         ;then subtract 07H from AL.
1016  AA        STORE: STOSB            ;Store the binary value in memory.
1017  EB F0           JMP NEXT          ;Jump to convert next byte.

1019  F4        EXIT: HLT               ;Halt program execution.

101A            CODE ENDS               ;End of code segment.
                END                     ;Assembly end.
```

Sample Data

Input Data :	42	Memory address	Content
	37	1100	42
	46	1101	37
	39	1102	46
	38	1103	39
	20	1104	38
		1105	20

Output Data :	0B	Memory address	Content
	07	1400	0B
	0F	1401	07
	09	1402	0F
	08	1403	09
		1404	08

ASCII	Binary
42	0B
37	07
46	0F
39	09
38	08

EXAMPLE PROGRAM 86.27:   Program to Display the ASCII code of the key pressed

Write an 8086 assembly language program to read the PC (Personal Computer) keyboard and display the ASCII code of the key pressed in the PC monitor.

Problem Analysis

This assembly language program can be developed using BIOS and DOS interrupts. The BIOS interrupt INT 16H with function code 00_H can be used to read the ASCII code of the key pressed. Using the received ASCII value and the DOS interrupt INT 21H with function code 02_H the symbol/character representing the key which is being pressed can be displayed.

In order to display the ASCII value of the key, the ASCII value has to be considered as hexcode and then the ASCII value of each nibble of the hexcode has to be determined. A separate procedure can be written to convert the hexcode to ASCII. (The logic of converting hex to ASCII has been discussed in example program 25.) The ASCII values of lower and upper nibbles can be stored as ASCII string and then displayed on the PC monitor using the DOS interrupt INT 21H with function code 09_H.

Apart from displaying the key symbol and ASCII value, some message can also be displayed for the user. Since the program involves display of codes/messages a number of times, a macro can be written for display of ASCII string on PC the monitor.

One of the PC keys can be used to terminate the program. Here the ESC key is used to terminate the program. When the ESC key is pressed, the DOS interrupt INT 21H with function code $4C_H$ can be initiated to terminate the program and return the control to the command prompt.

Flowchart of Example Program 86.27

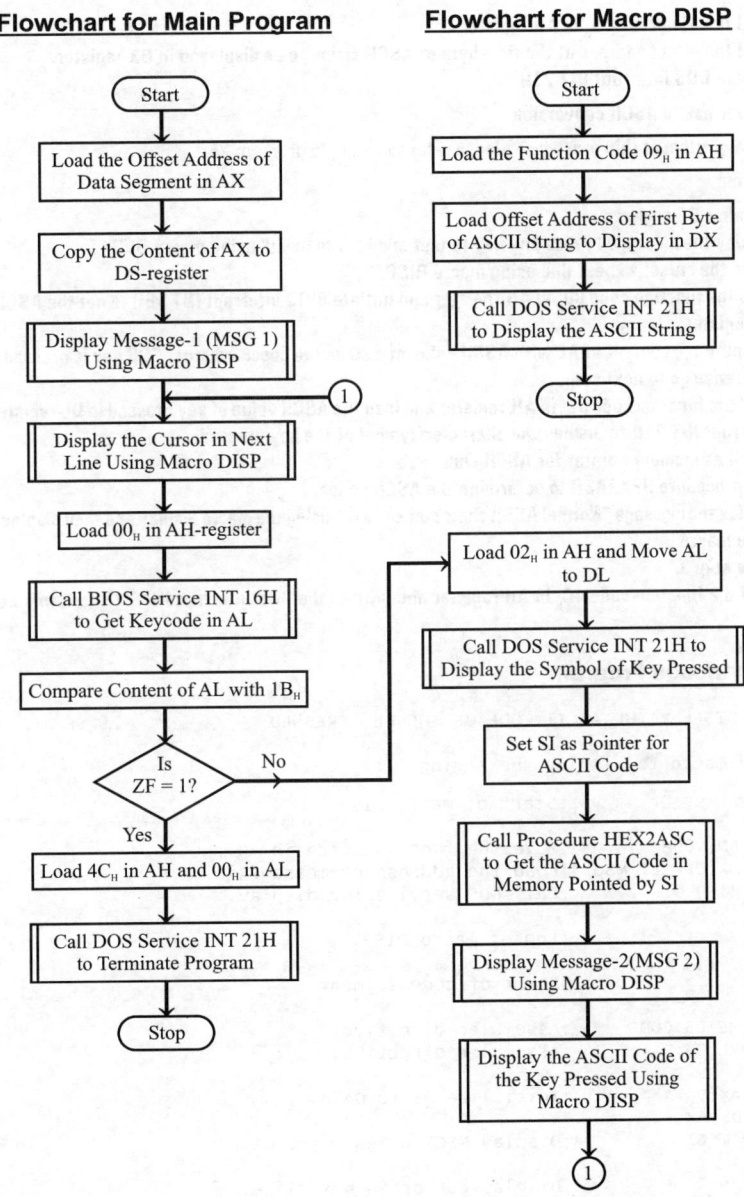

Flowchart for Main Program

- Start
- Load the Offset Address of Data Segment in AX
- Copy the Content of AX to DS-register
- Display Message-1 (MSG 1) Using Macro DISP
- ① → Display the Cursor in Next Line Using Macro DISP
- Load 00$_H$ in AH-register
- Call BIOS Service INT 16H to Get Keycode in AL
- Compare Content of AL with 1B$_H$
- Is ZF = 1? — No →
- Yes
- Load 4C$_H$ in AH and 00$_H$ in AL
- Call DOS Service INT 21H to Terminate Program
- Stop

Flowchart for Macro DISP

- Start
- Load the Function Code 09$_H$ in AH
- Load Offset Address of First Byte of ASCII String to Display in DX
- Call DOS Service INT 21H to Display the ASCII String
- Stop

- Load 02$_H$ in AH and Move AL to DL
- Call DOS Service INT 21H to Display the Symbol of Key Pressed
- Set SI as Pointer for ASCII Code
- Call Procedure HEX2ASC to Get the ASCII Code in Memory Pointed by SI
- Display Message-2(MSG 2) Using Macro DISP
- Display the ASCII Code of the Key Pressed Using Macro DISP
- ①

Note: For flowchart of the procedure HEX2ASC please refer to Example program 25.

Algorithm

i) **Macro DISP to display ASCII string**

    1.   Load function code 09_H in AH-register.

    2.   Load the offset address of the first byte of ASCII string to be displayed in DX-register.

    3.   Initiate DOS interrupt INT 21H.

ii) **Procedure for hex to ASCII conversion**

    For algorithm of this procedure please refer to Example program 25.

iii) **Main program**

    1.   Initialize DS-register.

    2.   Display the message "Press any key to test and ESC to exist" using macro DISP.

    3.   Move the cursor to next line using macro DISP.

    4.   Load the function code 00_H in AH-register and initiate BIOS interrupt INT 16H to get the ASCII value of key pressed in AL-register.

    5.   Compare the content of AL with ASCII value of ESC key to check whether ESC key is pressed. If ZF = 1, go to step 11, otherwise go to next step.

    6.   Load the function code 02_H in AH-register and load the ASCII value of key pressed in DL-register and then initiated DOS interrupt INT 21H to display the character/symbol of the key pressed.

    7.   Set SI as memory pointer for ASCII code.

    8.   Call procedure HEX2ASC to determine the ASCII codes.

    9.   Display the message "Normal ASCII character code = " using the macro display and then display the ASCII code using the same macro.

    10.   Go to step 3.

    11.   Load the function code $4C_H$ in AH-register and initiate the DOS interrupt INT 21H to terminate the program.

    12.   Stop.

Assembly Language Program

```
;PROGRAM TO DISPLAY THE ASCII CODE OF THE KEY PRESSED

;user defined macro to display the string

DISP MACRO MSG              ;Start of macro DISP.

        MOV AH, 09H         ;Move the function code in AH.
        MOV DX,OFFSET MSG   ;Load the address of message in DX.
        INT 21H             ;Call DOS service for display.

ENDM                        ;End of macro DISP.

CODE SEGMENT                ;Start of code segment.

        ASSUME CS:CODE      ;Assembler directive.
        ASSUME DS:DATA      ;Assembler directive.

        MOV AX,DATA         ;Initialize DS to DATA.
        MOV DS,AX
        DISP MSG1           ;Display MSG1 using macro DISP.

RDKEY:  DISP NL             ;Display cursor in next line.
        MOV AH,00H          ;Move the function code in AH.
        INT 16H             ;Call BIOS service to read key code.

        CMP AL,ESC          ;Check whether ESC key is pressed.
        JE  OVER            ;If yes, jump to OVER, otherwise write
                            ;character to standard output device.
        MOV AH,02H          ;Move the function code in AH.
        MOV DL,AL           ;Load hexcode of key pressed in DL.
        INT 21H             ;Call DOS service for display.
```

```
                MOV  SI,OFFSET ASCI  ;Get address for saving ASCII code.
                CALL HEX2ASC        ;Call procedure hex to ASCII conversion.
                DISP MSG2           ;Display MSG2 using macro DISP.
                DISP ASCI           ;Display ASCII code using macro DISP.
                JMP RDKEY           ;Jump to RDKEY to read next key code.

OVER:           MOV AH,4CH          ;Move the function code in AH.
                MOV AL,00H
                INT 21H             ;Return to command prompt.

HEX2ASC PROC NEAR                   ;Start of procedure HEX2ASC.

                MOV BL,AL           ;Save the key code in BL-register.
                AND AL,0F0H         ;Mask the lower nibble of key code.
                MOV CL,04H          ;Rotate upper nibble to
                ROL AL,CL           ;lower nibble position.
                MOV DL,0AH          ;Check whether upper nibble is
                CMP AL,DL           ;less than 0AH.
                JL SKIP1            ;If upper nibble is less than 0AH,
                ADD AL,07H          ;then add 30H or if upper nibble is
SKIP1:          ADD AL,30H          ;greater than 0AH, then add 37H.
                MOV [SI],AL         ;Store ASCII code of upper nibble.

                MOV AL,BL           ;Get the key code in AL-register.
                AND AL,0FH          ;Mask the upper nibble.
                CMP AL,DL           ;Check whether lower nibble is less than 0AH.
                JL SKIP2            ;If lower nibble is less than 0AH,
                ADD AL,07H          ;then add 30H or if lower nibble is
SKIP2:          ADD AL,30H          ;greater than 0AH, then add 37H.
                MOV [SI+1],AL       ;Store ASCII code of lower nibble.
                MOV AL,'$'          ;Append end of string.
                MOV [SI+2],AL
                RET                 ;Return to main program.

HEX2ASC ENDP                        ;End of procedure HEX2ASC.

CODE ENDS                           ;End of code segment.

DATA SEGMENT                        ;start of data segment.


                CR   EQU 0DH        ;ASCII for carriage return.
                LF   EQU 0AH        ;ASCII for line feed.
                ESC EQU 1BH         ;ASCII for escape.
                ASCI DB 8 DUP(0)    ;Assembler directive.

                MSG1 DB   'Press any key to test and esc to exit ','$'.
                MSG2 DB   ', Normal ASCII character code = ','$'.
                NL   DB   CR,LF,'$'.

DATA ENDS                           ;End of data segment.
END                                 ;Assembly end.
```

Assembler Listing for Example Program 86. 27

```
;PROGRAM TO DISPLAY THE ASCII CODE OF THE KEY PRESSED

;user defined macro to display the string

DISP MACRO MSG                      ;Start of macro DISP.

                MOV AH, 09H         ;Move the function code in AH.
                MOV DX,OFFSET MSG   ;Load the address of message in DX.
                INT 21H             ;Call DOS service for display.
```

```
                        ENDM                    ;End of macro DISP.

0000                    CODE SEGMENT            ;Start of code segment.

                        ASSUME CS:CODE          ;Assembler directive.
                        ASSUME DS:DATA          ;Assembler directive.

0000  B8 - R            MOV  AX,DATA            ;Initialize DS to DATA.
0003  8E D8             MOV  DS,AX
                        DISPMSG1                ;Display MSG1 using macro DISP.
0005  B4 09      +      MOV  AH, 09H            ;Move the function code in AH.
0007  BA 0008 R +       MOV  DX,OFFSET MSG1     ;Load the address of message in DX.
000A  CD 21      +      INT  21H                ;Call DOS service for display.

000C            RDKEY:  DISP NL                 ;Display cursor in next line.
000C  B4 09      +      MOV  AH, 09H            ;Move the function code in AH.
000E  BA 0050 R +       MOV  DX,OFFSET NL       ;Load the address of message in DX.
0011  CD 21      +      INT  21H                ;Call DOS service for display.
0013  B4 00             MOV  AH,00H             ;Move the function code in AH.
0015  CD 16             INT  16H                ;Call BIOS service to read key code.

0017  3C 1B 90 90       CMP  AL,ESC             ;Check whether ESC key is pressed.
001B  74 1C             JE   OVER               ;If yes, jump to OVER, otherwise write
                                                ;character to standard output device.
001D  B4 02             MOV  AH,02H             ;Move the function code in AH.
001F  8A D0             MOV  DL,AL              ;Load hexcode of key pressed in DL.
0021  CD 21             INT  21H                ;Call DOS service for display.

0023  BE 0000 R         MOV  SI,OFFSET ASCI     ;Get address for saving ASCII code.
0026  E8 003F R         CALL HEX2ASC            ;Call procedure hex to ASCII conversion.
                        DISP MSG2               ;Display MSG2 using macro DISP.
0029  B4 09      +      MOV  AH, 09H            ;Move the function code in AH.
002B  BA 002F R +       MOV  DX,OFFSET MSG2     ;Load the address of message in DX.
002E  CD 21      +      INT  21H                ;Call DOS service for display.
                        DISP ASCI               ;Display ASCII code using macro DISP.
0030  B4 09      +      MOV  AH,09H             ;Move the function code in AH.
0032  BA 0000 R +       MOV  DX,OFFSET ASCI     ;Load the address of message in DX.
0035  CD 21      +      INT  21H                ;Call DOS service for display.
0037  EB D3             JMP  RDKEY              ;Jump to RDKEY to read next key code.

0039  B4 4C      OVER:  MOV  AH,4CH             ;Move the function code in AH.
003B  B0 00             MOV  AL,00H
003D  CD 21             INT  21H                ;Return to command prompt.

003F                    HEX2ASCPROC  NEAR       ;Start of procedure HEX2ASC.

003F  8A D8             MOV  BL,AL              ;Save the key code in BL-register.
0041  24 F0             AND  AL,0F0H            ;Mask the lower nibble of key code.
0043  B1 04             MOV  CL,04H             ;Rotate upper nibble to
0045  D2 C0             ROL  AL,CL              ;lower nibble position.
0047  B2 0A             MOV  DL,0AH             ;Check whether upper nibble is
0049  3A C2             CMP  AL,DL              ;less than 0AH.
004B  7C 02             JL   SKIP1              ;If upper nibble is less than 0AH,
004D  04 07             ADD  AL,07H             ;then add 30H or if upper nibble is
004F  04 30      SKIP1: ADD  AL,30H             ;greater than 0AH, then add 37H.
0051  88 04             MOV  [SI],AL            ;Store ASCII code of upper nibble.

0053  8A C3             MOV  AL,BL              ;Get the key code in AL-register.
0055  24 0F             AND  AL,0FH             ;Mask the upper nibble.
0057  3A C2             CMP  AL,DL              ;Check lower nibble is less than 0AH.
0059  7C 02             JL   SKIP2              ;If lower nibble is less than 0AH,
005B  04 07             ADD  AL,07H             ;then add 30H or if lower nibble is
005D  04 30      SKIP2: ADD  AL,30H             ;greater than 0AH, then add 37H.
005F  88 44 01          MOV  [SI+1],AL          ;Store ASCII code of lower nibble.
```

```
0062  B0 24           MOV    AL,'$'       ;Append end of string.
0064  88 44 02        MOV    [SI+2],AL
0067  C3              RET                  ;Return to main program.

0068                  HEX2ASC ENDP         ;End of procedure HEX2ASC.

0068                  CODE ENDS            ;End of code segment.

0000                  DATA SEGMENT         ;start of data segment.

= 000D               CR      EQU 0DH      ;ASCII for carriage return.
= 000A               LF      EQU 0AH      ;ASCII for line feed.
= 001B               ESC     EQU 1BH      ;ASCII for escape.
0000  08  [          ASCI    DB 8 DUP(0)  ;Assembler directive.
           00
         ]
0008  50 72 65 73 73 20   MSG1   DB  'Press any key to test and esc to exit ','$'.
      61 6E 79 20 6B 65
      79 20 74 6F 20 74
      65 73 74 20 61 6E
      64 20 65 73 63 20
      74 6F 20 65 78 69
      74 20 24
002F  2C 20 4E 6F 72 6D   MSG2   DB  ', Normal ASCII character code = ','$'.
      61 6C 20 41 53 43
      49 49 20 63 68 61
      72 61 63 74 65 72
      20 63 6F 64 65 20
      3D 20 24
0050  0D 0A 24            NL DB CR,LF,'$'

0053                  DATA ENDS            ;End of data segment.
                      END                  ;Assembly end.
```

EXAMPLE PROGRAM 86.28 : Program to Find the Length of a String

Write an 8086 assembly language program to determine the length of ASCII string. Use PC keyboard to input the string and display the length on PC monitor.

Problem Analysis

The ASCII string is input through PC keyboard and so DOS interrupt INT 21H with function code 01_H can be used to read the key code. On counting the number of input characters the length can be determined. The carriage return (Enter key) can be used to terminate the input string. The input string can also be displayed on PC monitor.

The count value representing the length of string will be in hex. In order to display the count value in PC monitor it has to be converted to decimal and ASCII value of each decimal digit has to be determined and stored in memory as ASCII string. Then the DOS interrupt INT 21H with function code 09_H can be used to display the length in decimal.

A separate procedure can be written to convert the count value in hex to decimal and then to ASCII. In this procedure the hex value is divided by ten ($0A_H$). Now the quotient is ten's digit and remainder is unit's digit. (Here it is assumed that the count does not exceeds 99_{10}.) Then the ASCII value of zero can be added to ten's digit and unit's digit to get their respective ASCII values.

Apart from displaying the input string and its length some message can also be displayed for the user. Since the program involves display of messages a number of times, a macro can be written for display of ASCII string on PC monitor.

After displaying the length of string the DOS interrupt INT 21H with function code $4C_H$ can be initiated to terminate the program execution and return the control to the command prompt.

Flowchat of Example Program 86.28

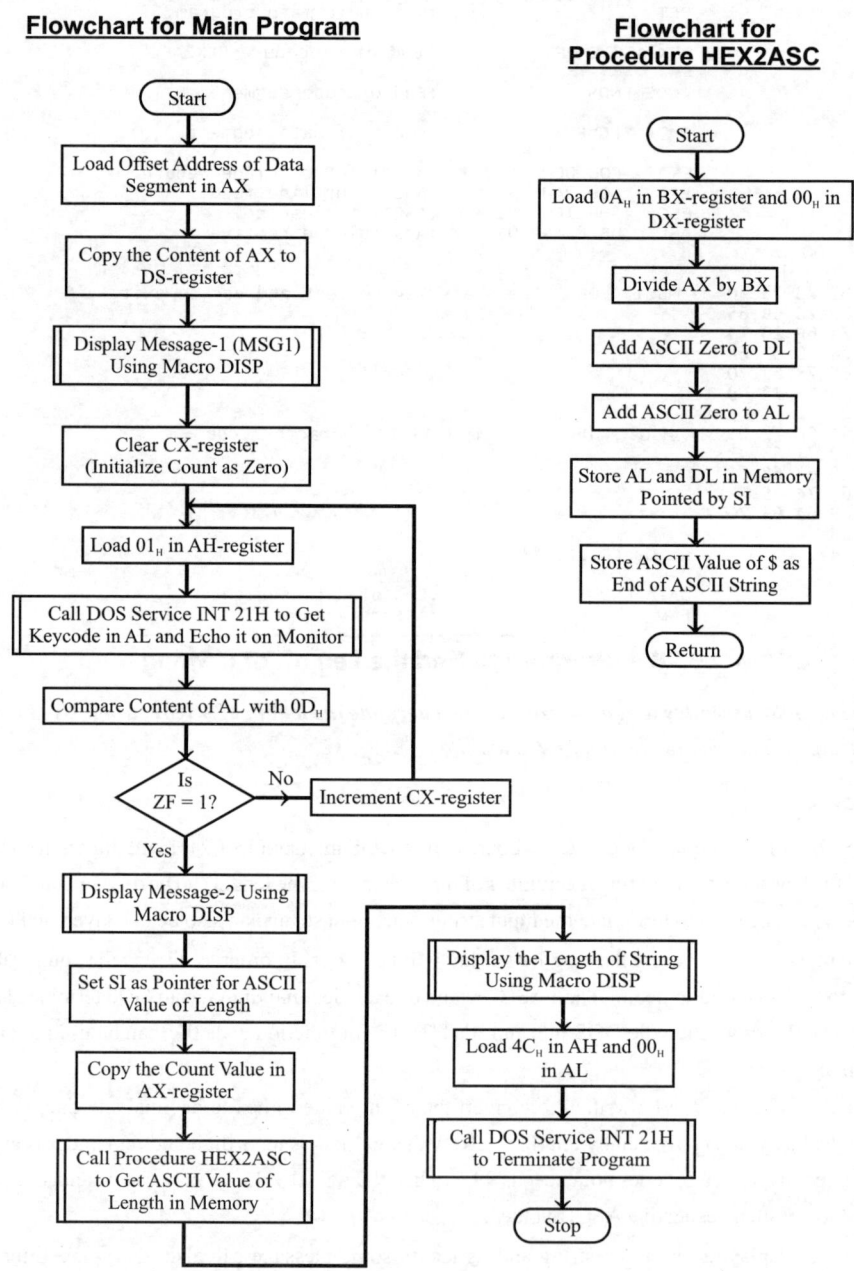

Flowchart for Main Program

Flowchart for Procedure HEX2ASC

Algorithm

i) Main program

1. Initialize DS-register
2. Display the message "ENTER THE STRING.INPUT = " using macro DISP.
3. Let CX-register be used as counter. Initialize count as zero.
4. Load the function code 01_H in AH-register and initiate DOS interrupt INT 21H to get the ASCII value of key pressed in AL and echo (display) the key symbol/character in the PC monitor.
5. Compare the content of AL with ASCII value of ENTER key to check whether ENTER key is pressed. If ZF =1, go to step 8, otherwise go to next step.
6. Increment the counter (CX-register).
7. Go to step 4.
8. Display the message " THE LENGTH OF STRING = " using the macro DISP.
9. Set SI as memory pointer for ASCII value of length.
10. Copy the count value in AX-register.
11. Call procedure HEX2ASC to get the ASCII value of count in memory pointed by SI.
12. Display the length of string using the macro DISP.
13. Load the function code $4C_H$ in AH-register and initiate the DOS interrupt INT 21H to terminate the program.
14. Stop.

ii) Procedure HEX2ASC

1. Load the divisor ($0A_H$) in BX-register.
2. Divide the hex value of length in AX with $0A_H$ to get units in DX and tens in AX.
3. Add the ASCII value of zero to DL-register to get ASCII value of units.
4. Add the ASCII value of zero to AL-register to get ASCII value of tens.
5. Store the ASCII values in memory pointed by SI.
6. Return.

> *Note:* *Refer to Example program 86.27 for algorithm of macro DISP.*

Assembly Language Program

```
;PROGRAM TO FIND THE LENGTH OF STRING
;User defined macro to display message

DISP MACRO MSG                  ;Start of macro.

        MOV  AH, 09H            ;Move function code in AH.
        MOV  DX,OFFSET MSG      ;Get address of string to display in DX.
        INT  21H               ;Call DOS service for display.

ENDM                            ;End of macro.

DATA SEGMENT                    ;Start of data segment.

        CR   EQU 0DH            ;ASCII for carriage return.
        LF   EQU 0AH            ;ASCII for line feed.
        LEN  DB 04 DUP(0)       ;Assembler directive.
        MSG1 DB  'ENTER THE STRING.INPUT = ','$'.
        MSG2 DB CR,LF,  'THE LENGTH OF STRING = ','$'.

DATA ENDS                       ;End of data segment.

CODE SEGMENT                    ;Start of code segment.

        ASSUME CS:CODE          ;Assembler directive.
        ASSUME DS:DATA          ;Assembler directive.

        MOV  AX,DATA            ;Initialize DS to data.
        MOV  DS,AX
        DISP MSG1               ;Display MSG1 using macro DISP.
        MOV  CX,00H             ;Initialize count to zero.
```

```
RDKEY:   MOV  AH,01H            ;Move function code in AH.
         INT  21H              ;Call DOS service to read keycode.

         CMP  AL,CR            ;Compare keycode with carriage return.
         JE   AHEAD            ;If key is carriage return, go to AHEAD.

         INC  CX               ;Increment the counter.
         JMP  RDKEY            ;Jump to RDKEY to get next keycode.

AHEAD:   DISP MSG2             ;Display MSG2 using macro disp.
         MOV  SI,OFFSET LEN    ;Get the address for saving the length.
         MOV  AX,CX            ;Get the count in AX-register.
         CALL HEX2ASC          ;Call procedure HEX2ASC.
         DISP LEN              ;Display length of string using macro.

         MOV  AH,4CH           ;Move function code in AH.
         MOV  AL,00H
         INT  21H              ;Return to command prompt.

HEX2ASC PROC NEAR              ;Start of procedure HEX2ASC.

         MOV  BX,0AH           ;Load the divisor in BX-register.
         MOV  DX,00            ;Clear DX-register.
         DIV  BX               ;Divide the count value by 0AH to get
                                    ;unit digit in DL and ten's digit in AL.
         ADD  DL,'0'           ;Convert the unit digit to ASCII.
         ADD  AL,'0'           ;Convert the ten's digit to ASCII.
         MOV  [SI],AL          ;Store the ASCII values in memory.
         MOV  [SI+1],DL
         MOV  AL,'$'           ;Append end of string.
         MOV  [SI+2],AL
         RET

HEX2ASC ENDP                   ;End of procedure HEX2ASC.

CODE ENDS                      ;End of code segment.
END                            ;Assembly end.
```

Assembler Listing for Example Program 86.28

```
;PROGRAM TO FIND THE LENGTH OF STRING
;User defined macro to display message

                    DISP MACRO MSG            ;Start of macro.

                    MOV AH, 09H               ;Move function code in AH.
                    MOV DX,OFFSET MSG         ;Get address of string to display in DX.
                    INT 21H                   ;Call DOS service for display.
                    ENDM                      ;End of macro.

0000                DATA SEGMENT              ;start of data segment.

= 000D              CR EQU 0DH                ;ASCII for carriage return.
= 000A              LF EQU 0AH                ;ASCII for line feed.
0000    04  [       LEN DB 04 DUP(0)          ;Assembler directive.
            00
         ]
0004 45 4E 54 45 52 20   MSG1 DB  'ENTER THE STRING.INPUT = ',','$'.
     54 48 45 20 53 54
     52 49 4E 47 2E 49
     4E 50 55 54 20 3D
     20 24

001E 0D 0A 54 48 45 20   MSG2 DB CR,LF,  'THE LENGTH OF STRING = ',','$'.
     4C 45 4E 47 54 48
```

```
            20 4F 46 20 53 54
            52 49 4E 47 20 3D
            20 24

0038                            DATA ENDS                    ;End of data segment.

0000                            CODE SEGMENT                 ;Start of code segment.

                                ASSUME CS:CODE               ;Assembler directive.
                                ASSUME DS:DATA               ;Assembler directive.

0000  B8 -    R                 MOV   AX,DATA                ;Initialize DS to data.
0003  8E D8                     MOV   DS,AX
                                DISP MSG1                    ;Display MSG1 using macro DISP.
0005  B4 09      +             MOV   AH, 09H                 ;Move function code in AH.
0007  BA 0004 R  +             MOV   DX,OFFSET MSG1          ;Get address of string to display in DX.
000A  CD 21      +             INT   21H                     ;Call DOS service for display.
000C  B9 0000                  MOV   CX,00H                  ;Initialize count to zero.

000F  B4 01          RDKEY:    MOV   AH,01H                  ;Move function code in AH.
0011  CD 21                    INT   21H                     ;Call DOS service to read keycode.

0013  3C 0D                    CMP   AL,CR                   ;Compare keycode with carriage return.
0015  74 03                    JE    AHEAD                   ;If key is carriage return, go to AHEAD.

0017  41                       INC   CX                      ;Increment the counter.
0018  EB F5                    JMP   RDKEY                   ;Jump to RDKEY to get next keycode.

001A           AHEAD:          DISP MSG2                    ;Display MSG2 using macro disp.
001A  B4 09      +             MOV   AH, 09H                 ;Move function code in AH.
001C  BA 001E R  +             MOV   DX,OFFSET MSG2          ;Get address of string to display in DX.
001F  CD 21      +             INT   21H                     ;Call DOS service for display.
0021  BE 0000 R                MOV   SI,OFFSET LEN           ;Get the address for saving the length.
0024  8B C1                    MOV   AX,CX                   ;Get the count in AX-register.
0026  E8 0036 R                CALL  HEX2ASC                 ;Call procedure HEX2ASC.
                                DISP LEN                    ;Display length of string using macro.
0029  B4 09      +             MOV   AH, 09H                 ;Move function code in AH.
002B  BA 0000 R  +             MOV   DX,OFFSET LEN           ;Get address of string to display in DX.
002E  CD 21      +             INT   21H                     ;Call DOS service for display.

0030  B4 4C                    MOV   AH,4CH                  ;Move function code in AH.
0032  B0 00                    MOV   AL,00H
0034  CD 21                    INT   21H                     ;Return to command prompt.

0036                            HEX2ASC  PROC NEAR           ;Start of procedure HEX2ASC.

0036  BB 000A                  MOV   BX,0AH                  ;Load the divisor in BX-register.
0039  BA 0000                  MOV   DX,00                   ;Clear DX-register.
003C  F7 F3                    DIV   BX                      ;Divide the count value by 0AH to get
                                                             ;unit digit in DL and ten's digit in AL.
003E  80 C2 30                 ADD   DL,'0'                  ;Convert the unit's digit to ASCII.
0041  04 30                    ADD   AL,'0'                  ;Convert the ten's digit to ASCII.
0043  88 04                    MOV   [SI],AL                 ;Store the ASCII values in memory.
0045  88 54 01                 MOV   [SI+1],DL
0048  B0 24                    MOV   AL,'$'                  ;Append end of string.
004A  88 44 02                 MOV   [SI+2],AL
004D  C3                       RET
004E                            HEX2ASC ENDP                 ;End of procedure HEX2ASC.

004E                            CODE ENDS                    ;End of code segment
                                END                          ;Assembly end.
```

EXAMPLE PROGRAM 86.29 : Program to Find Palindrome

Write an 8086 assembly language program to verify whether an input string is a palindrome or not. Use PC keyboard to input the string and display the result on the PC monitor.

Problem Analysis

The input ASCII string can be read using DOS interrupt INT 21H with function code 01_H and stored in memory. Then the input string can be arranged in the reverse order to get reversed string and store the reversed string in another memory location.

Compare the input string and reversed string byte by byte to verify whether it is a palindrome. If the input string is byte by byte same as reversed string then it is a palindrome, otherwise it is not a palindrome.

For the convenience of the user, the input string and reversed string can also be displayed on the PC monitor. The result can be displayed as a message on the PC monitor. Since the program involves display of messages a number of times, a macro can be written for display of ASCII string on the PC monitor. After displaying the result the DOS interrupt INT 21H with function code $4C_H$ can be initiated to terminate the program execution and return the control to the command prompt.

Algorithm

1. Initialize DS-register.
2. Set SI as pointer for input string and DI as pointer for reversed string.
3. Display the message "ENTER THE STRING. INPUT STRING = " using the macro DISP.
4. Let CX-register be used as counter for number of bytes in the input string. Initialize count as zero.
5. Load the function code 01_H in AH-register and initiate DOS interrupt INT 21H to get the ASCII value of key pressed in AL and echo (display) the key symbol/character in the PC monitor.
6. Compare the content of AL with ASCII value of ENTER key to check whether ENTER key is pressed. If ZF = 1, go to step 10, otherwise go to next step.
7. Save the content of AL-register in memory pointed by SI and increment SI-register.
8. Increment the counter (CX-register).
9. Go to step 5.
10. Save the count value in BX-register.
11. Decrement SI register and copy a byte of input string in AL-register.
12. Copy the content of AL-register to the memory pointed by DI and increment DI-register.
13. Decrement CX-register. Check whether content of CX is zero and if it is true, go to next step, otherwise go to step 11.
14. Display the message "REVERSE OF STRING = " using the macro DISP.
15. Display the reversed string using the macro DISP.
16. Again set SI as pointer for input string and DI as pointer for reversed string.
17. Load the count value from BX-register to CX-register.
18. Get a byte of input string in AL and compare with corresponding byte of reverse string in memory.
19. Check zero flag. If ZF = 0, then go to step 24, otherwise go to next step.
20. Increment SI and DI-register.
21. Decrement CX-register. Check whether content of CX is zero and if it is true, go to next step, otherwise go to step 18.
22. Display the message "INPUT STRING IS A PALINDROME" using macro DISP.
23. Go to step 25.
24. Display the message "INPUT STRING IS NOT A PALINDROME" using macro DISP.
25. Load function code $4C_H$ in AH-register and initiate the DOS interrupt INT 21H to terminate the program.
26. Stop.

Note: *For algorithm and flowchart of macro DISP refer to Example program 27.*

Flowchart of Example Program 86.29

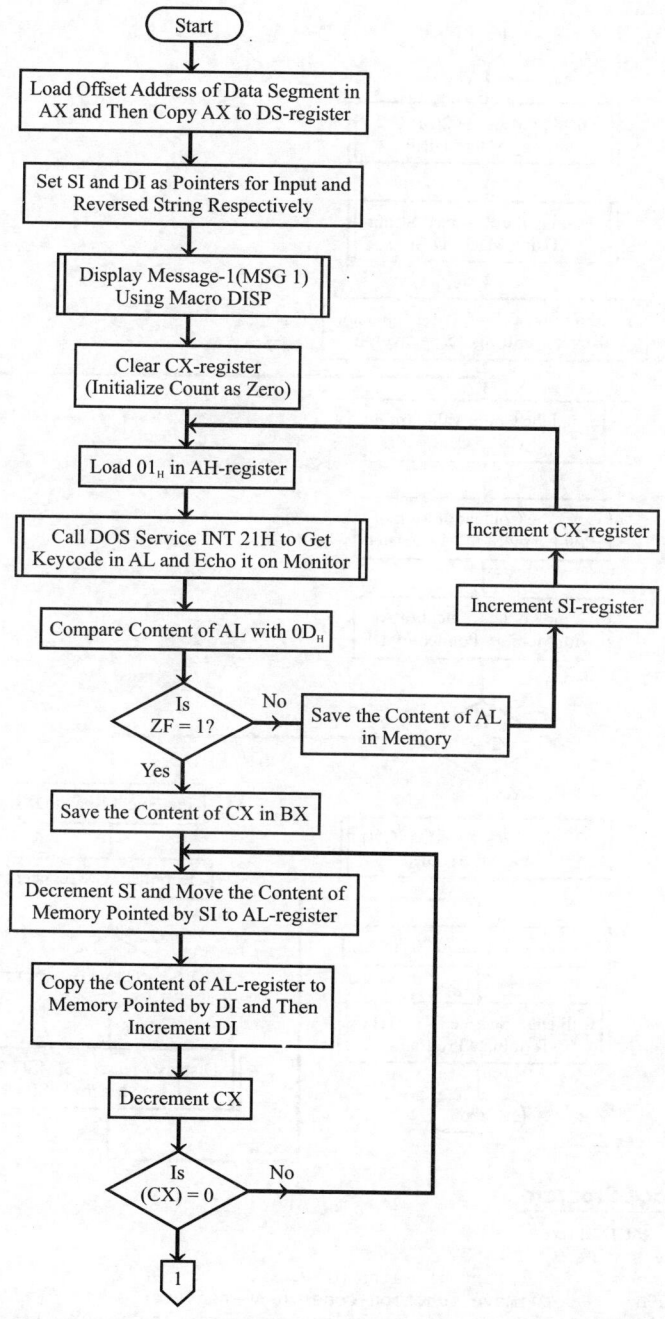

Flowchart continued..

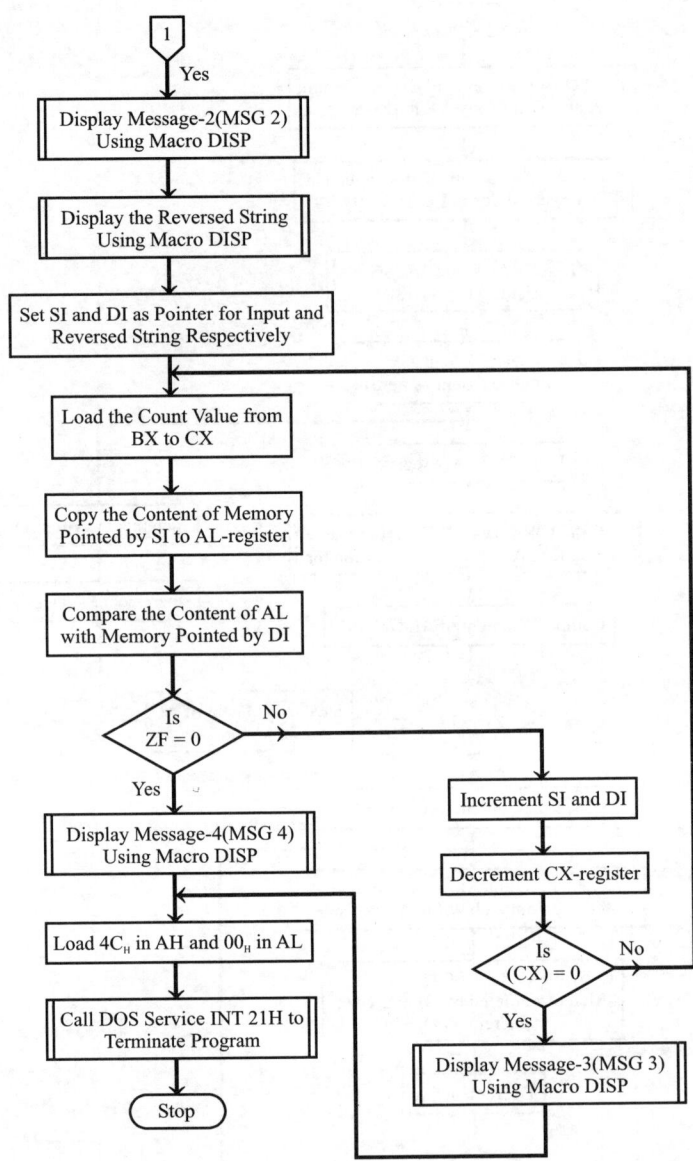

Assembly Language Program

```
;PROGRAM TO FIND PALINDROME
;Macro to display string
DISP MACRO MSG              ;Start of macro DISP.
        MOV AH,09H         ;Move function code to AH-register.
        MOV DX,OFFSET MSG  ;Move address of string to display in DX.
        INT 21H            ;Call DOS service for display.
ENDM                       ;End of macro DISP.
;Define code segment
CODE SEGMENT               ;Start of code segment.
```

```
        ASSUME CS:CODE        ;Assembler directive.
        ASSUME DS:DATA        ;Assembler directive.
        MOV AX,DATA           ;Initialize DS to DATA.
        MOV DS,AX
        MOV SI,OFFSET INS     ;Load address of input string in SI.
        MOV DI,OFFSET RES     ;Load address of reverse string in DI.
        DISP MSG1             ;Display MSG1 using macro DISP.
        MOV CX,00H            ;Initialize the count.
RDCHAR: MOV AH,01H            ;Move the function code in AH.
        INT 21H               ;Call DOS service to read keyboard.
        CMP AL,CR             ;Check for carriage return.
        JE  AHEAD             ;If key entered is carriage return,
                              ;then go to AHEAD.
        MOV [SI],AL           ;Save character in memory pointed by SI.
        INC SI                ;Increment the pointer.
        INC CX                ;Increment the counter.
        JMP RDCHAR            ;Jump to RDCHAR to get next character.
AHEAD:  MOV BX,CX             ;Save the count in BX.
REVERS: DEC SI                ;Copy characters from memory pointed
        MOV AL,[SI]           ;by SI to memory pointed by DI.
        MOV [DI],AL           ;in reverse order.
        INC DI
        LOOP REVERS
        MOV AL,'$'            ;Put end of string.
        MOV [DI],AL
        DISP MSG2             ;Display the MSG2 using macro DISP.
        DISP RES              ;Display reversed string using macro DISP.
        MOV SI,OFFSET INS     ;Get address of input string in SI.
        MOV DI,OFFSET RES     ;Get address of reversed string in DI.
        MOV CX,BX             ;Initialize the counter.
CHECK:  MOV AL,[SI]           ;Get a character of input string in AL.
        CMP AL,[DI]           ;Compare with character of reversed string.
        JNE FALSE             ;If not equal jump to FALSE.
        INC SI                ;Increment the pointers.
        INC DI
        LOOP CHECK            ;Repeat comparison until counter expires.
        DISP MSG3             ;Display MSG3 using macro DISP.
        JMP EXIT
FALSE:  DISP MSG4             ;Display MSG4 using macro DISP.
EXIT:   MOV AH,4CH            ;Move function code to AH-register.
        MOV AL,00H
        INT 21H               ;Return to command prompt.
CODE ENDS                     ;End of code segment.
;Define data segment
DATA SEGMENT                  ;start of data segment.
        CR  EQU 0DH           ;ASCII code for carriage return.
        LF  EQU 0AH           ;ASCII code for line feed.
        INS DB 40 DUP(0)      ;Reserve 40 locations for input string.
        RES DB 40 DUP(0)      ;Reserve 40 locations for reverse string.
        MSG1 DB 'ENTER THE STRING.INPUT STRING= ','$'.
        MSG2 DB CR,LF,'REVERSE OF STRING = ','$'.
        MSG3 DB CR,LF,'INPUT STRING IS A PALINDROME','$'.
        MSG4 DB CR,LF,'INPUT STRING IS NOT A PALINDROME','$'.
DATA ENDS                     ;End of data segment.
END                           ;Assembly end.
```

Assembler Listing for Example Program 86.29

```
;PROGRAM TO FIND PALINDROME
;Macro to display string

DISP MACRO MSG                               ;Start of macro DISP.

                MOV AH,09H                   ;Move function code to AH-register.
```

```
                             MOV   DX,OFFSET MSG      ;Move address of string to display in DX.
                             INT   21H                ;Call DOS service for display.
                             ENDM                     ;End of macro DISP.
                 ;Define code segment
0000                  CODE SEGMENT                    ;Start of code segment.
                      ASSUME CS:CODE                  ;Assembler directive.
                      ASSUME DS:DATA                  ;Assembler directive.
0000 B8 —  R                 MOV   AX,DATA            ;Initialize DS to DATA.
0003 8E D8                   MOV   DS,AX
0005 BE 0000 R               MOV   SI,OFFSET INS      ;Load address of input string in SI.
0008 BF 0028 R               MOV   DI,OFFSET RES      ;Load address of reverse string in DI.
                             DISP MSG1                ;Display MSG1 using macro DISP.
000B B4 09     +             MOV   AH,09H             ;Move function code to AH-register.
000D BA 0050 R +             MOV   DX,OFFSET MSG1     ;Move address of string to display in DX.
0010 CD 21     +             INT   21H                ;Call DOS service for display.
0012 B9 0000                 MOV   CX,00H             ;Initialize the count.
0015 B4 01       RDCHAR:     MOV   AH,01H             ;Move the function code in AH.
0017 CD 21                   INT   21H                ;Call DOS service to read keyboard.
0019 3C 0D 90 90             CMP   AL,CR              ;Check for carriage return.
001D 74 06                   JE    AHEAD              ;If key entered is carriage return,
                                                      ;then go to AHEAD.
001F 88 04                   MOV   [SI],AL            ;Save character in memory pointed by SI.
0021 46                      INC   SI                 ;Increment the pointer.
0022 41                      INC   CX                 ;Increment the counter.
0023 EB F0                   JMP   RDCHAR             ;Jump to RDCHAR to get next character.
0025 8B D9       AHEAD:      MOV   BX,CX              ;Save the count in BX.
0027 4E          REVERS:     DEC   SI                 ;Copy characters from memory pointed
0028 8A 04                   MOV   AL,[SI]            ;by SI to memory pointed by DI
002A 88 05                   MOV   [DI],AL            ;in reverse order.
002C 47                      INC   DI
002D E2 F8                   LOOP REVERS
002F B0 24                   MOV   AL,'$'             ;Put end of string.
0031 88 05                   MOV   [DI],AL
                             DISP MSG2                ;Display the MSG2 using macro DISP.
0033 B4 09     +             MOV   AH,09H             ;Move function code to AH-register.
0035 BA 0063 R +             MOV   DX,OFFSET MSG2     ;Move address of string to display in DX.
0038 CD 21     +             INT   21H                ;Call DOS service for display.
                             DISP RES                 ;Display reversed string using macro DISP.
003A B4 09     +             MOV   AH,09H             ;Move function code to AH-register.
003C BA 0028 R +             MOV   DX,OFFSET RES      ;Move address of string to display in DX.
003F CD 21     +             INT   21H                ;Call DOS service for display.
0041 BE 0000 R               MOV   SI,OFFSET INS      ;Get address of input string in SI.
0044 BF 0028 R               MOV   DI,OFFSET RES      ;Get address of reversed string in DI.
0047 8B CB                   MOV   CX,BX              ;Initialize the counter.
0049 8A 04       CHECK:      MOV   AL,[SI]            ;Get a character of input string in AL.
004B 3A 05                   CMP   AL,[DI]            ;Compare with character of reversed string.
004D 75 0E                         JNE   FALSE        ;If not equal jump to FALSE.
004F 46                      INC   SI                 ;Increment the pointers.
0050 47                      INC   DI
0051 E2 F6                   LOOP CHECK               ;Repeat comparison until counter expires.
                                    DISP MSG3                              ;Display
MSG3 using macro DISP.
0053 B4 09     +             MOV   AH,09H             ;Move function code to AH-register.
0055 BA 007A R +             MOV   DX,OFFSET MSG3     ;Move address of string to display in DX.
0058 CD 21     +             INT   21H                ;Call DOS service for display.
005A EB 08 90                JMP   EXIT
005D             FALSE:      DISP MSG4                ;Display MSG4 using macro DISP.
005D B4 09     +             MOV   AH,09H             ;Move function code to AH-register.
005F BA 0099 R +             MOV   DX,OFFSET MSG4     ;Move address of string to display in DX.
0062 CD 21     +             INT   21H                ;Call DOS service for display.
0064 B4 4C       EXIT:       MOV   AH,4CH             ;Move function code to AH-register.
0066 B0 00                   MOV   AL,00H
0068 CD 21                   INT   21H                ;Return to command prompt.
006A             CODE ENDS                            ;End of code segment.
```

```
                                              ;Define data segment
0000                    DATA SEGMENT          ;start of data segment.
= 000D                       CR  EQU 0DH      ;ASCII code for carriage return.
= 000A                       LF  EQU 0AH      ;ASCII code for line feed.
0000 28 [                    INS DB 40 DUP(0) ;Reserve 40 locations for input string.
         00
      ]
0028 28 [                    RES DB 40 DUP(0) ;Reserve 40 locations for reverse string.
         00
      ]
0050 45 4E 54 45 52 20           MSG1 DB'ENTER THE STRING.INPUT STRING= ','$'.
     54 48 45 20 53 54
     52 49 4E 47 2E 49
     4E 50 55 54 20 53
     54 52 49 4E 47 3D
     20 24
0070 0D 0A 52 45 56 45           MSG2 DB CR,LF,'REVERSE OF STRING = ','$'.
     52 53 45 20 4F 46
     20 53 54 52 49 4E
     47 20 3D 20 24
0087 0D 0A 49 4E 50 55           MSG3 DB CR,LF,'INPUT STRING IS A PALINDROME','$'.
     54 20 53 54 52 49
     4E 47 20 49 53 20
     41 20 50 41 4C 49
     4E 44 52 4F 4D 45
     24
00A6 0D 0A 49 4E 50 55           MSG4 DB CR,LF,'INPUT STRING IS NOT A PALINDROME','$'.
     54 20 53 54 52 49
     4E 47 20 49 53 20
     4E 4F 54 20 41 20
     50 41 4C 49 4E 44
     52 4F 4D 45 24
00C9                    DATA ENDS             ;End of data segment.
                       END                    ;Assembly end.
```

EXAMPLE PROGRAM 86.30 : Program to Verify Password

Write an 8086 assembly language program to read a password through the PC keyboard and validate the user. Use the PC monitor to display the validity of the password.

Problem Analysis

The actual password can be stored in the memory as an ASCII array. The user entered password through the PC keyboard can be read by using BIOS service INT 16H with function code 00_H. A-register can be used as the counter to count the number of characters received and the received characters can be stored in the memory. The carriage return/enter key can be used to terminate the password.

First the count value is checked with the number of characters in the stored password. If it is not equal then an invalid password message can be displayed on the PC screen and request the user either to enter the correct password or to enter the ESC key to exit.

If the count is equal to the number of characters in the stored password, then the user entered password is checked with the stored password byte by byte. If the passwords are equal then an acceptance message can be displayed and the control can be returned to DOS prompt. If they are not equal then an invalid message can be displayed and request the user either to enter the correct password or to enter ESC key to exit.

In this program separate procedures has been written to clear the PC monitor screen and position the cursor at the desired location using the BIOS services.

Algorithm

1. Call procedure CLRS to clear the PC monitor screen.
2. Call procedure POS to position the cursor.
3. Initialize DS-register.
4. Load 09_H in AH and 0450_H in DX, and then call DOS service INT 21H to display the message "ENTER THE PASSWORD :".
5. Set SI as memory pointer to store user entered password.
6. Let CX be counter for number of characters in user entered password. Initialize CX as zero.
7. Load 00_H in AH and call BIOS service INT 16H to get the keycode in AL-register.
6. Compare AL with $0D_H$ (ASCII value of ENTER key).
9. If ZF = 1, go to step 13, otherwise go to next step.
10. Load 02_H in AH and ASCII value of "*" in DL, and then call DOS service INT 21H to display the symbol "*".
11. Increment the pointer(SI) and count(CL).
12. Go to step 7.
13. Compare the count (CX-register) with 07_H (Here the stored password has 7 characters).
14. If ZF = 0, then go to step 20, otherwise go to next step.
15. Set SI as pointer for user entered password and DI as pointer for stored password.
16. Load a character of entered password in AL and compare with corresponding character stored password.
17. If ZF = 0, then go to step 25, otherwise go to next step.
16. Increment the pointer (SI and DI).
19. Decrement CX-register and check whether CX is zero.
20. If content of CX is zero then go to next step, otherwise go to step 16.
21. Call procedure CLRS to clear the PC monitor screen.
22. Call procedure POS to position the cursor.
23. Load 09_H in AH-register and 0600_H in DX-register and then call DOS service INT 21H to display the message "ENTRY ACCEPTED".
24. Go to step 32.
25. Call procedure CLRS to clear the PC monitor screen.
26. Call procedure POS to position the cursor.
27. Load 09_H in AH and 0550_H in DX, and then call DOS service INT 21H to display the message "INCORRECT PASSWORD".
26. Load 09_H in AH and 0580_H in DX, and then call DOS service INT 21H to display the message "PRESS ANY KEY TO TRY AGAIN OR PRESS ESC TO EXIT."
29. Load 00_H in AH and call BIOS service INT 16H to get the keycode in AL.
30. Compare AL with $1B_H$ (ASCII value of ESC key).
31. If ZF = 0, then go to step 1, otherwise go to next step.
32. Load $4C_H$ in AH-register and call DOS service INT 21H to terminate the program and return the control to DOS prompt.
33. Stop.

Algorithm for Procedure CLRS

1. Load the function code 07_H in AH-register.
2. Load the number of lines to scroll down in AL-register.
3. Load the blanked area attribute (07_H) in BL-register.
4. Load the x and y coordinates of upper left corner in CL and CH registers.
5. Load the x and y coordinates of lower right corner in DL and DH.
6. Call BIOS video service INT 10H to clear the screen.
7. Return.

Algorithm for Procedure POS

1. Load the function code 02$_H$ in AH-register.
2. Load the video page (00$_H$) in BH-register.
3. Load the x and y coordinates of video page in DL and DH.
4. Call BIOS video service INT 10H to position the cursor.
5. Return.

Flowchart of Example Program 86.30

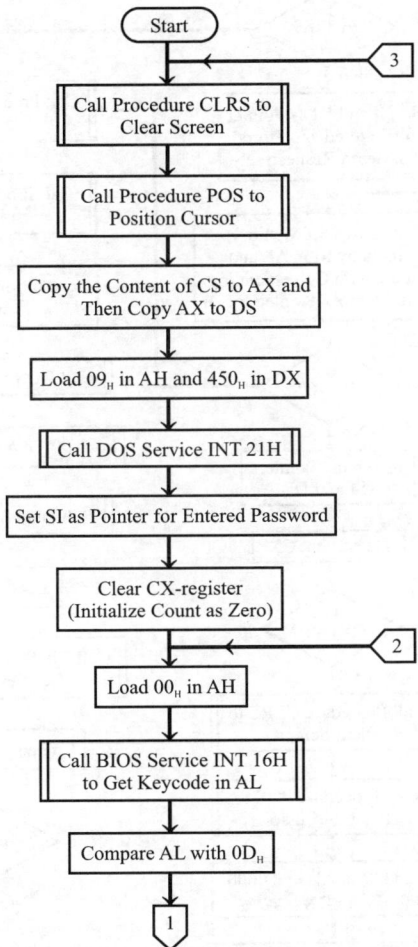

Flowchart continued..

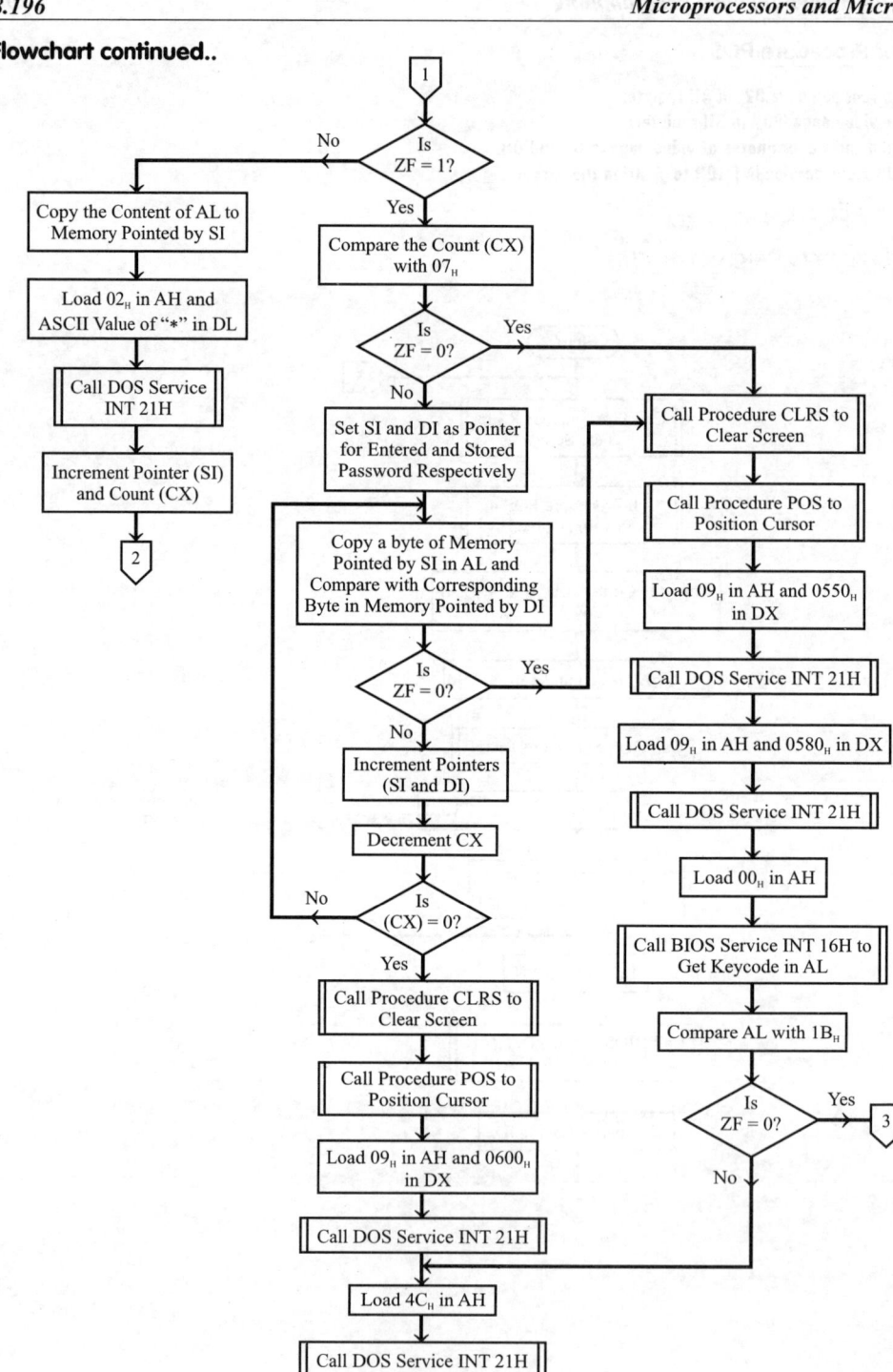

Assembly Language Program

```
;PROGRAM TO READ THE PASSWORD AND VALIDATE THE USER

CODE SEGMENT                    ;Start of code segment.

        ASSUME CS:CODE          ;Assembler directive.
        ASSUME DS:CODE          ;Assembler directive.
        CR   EQU ODH            ;Assign ASCII value for carriage return.

AGAIN:  CALL CLRS               ;Clear the screen.
        CALL POS                ;Position the cursor.

        MOV AX,CS               ;Initialize DS to code segment.
        MOV DS,AX

        MOV AH,09H              ;Load the function code in AH.
        MOV DX,450H             ;Load address of string to display in DX.
        INT 21H                 ;Call DOS service for display.

        MOV SI,400H             ;Initialize the pointer.
        MOV CX,0000H            ;Initialize count in CX-register as zero.

L1:     MOV AH,00H              ;Load the function code in AH.
        INT 16H                 ;Call BIOS service to get keyboard character in AL.
        CMP AL,CR               ;Compare the keycode with carriage return.
        JE  CHECK               ;If equal go to CHECK.
        MOV [SI],AL             ;Store the keyboard character in memory.
        MOV AH,02H              ;Load the function code in AH.
        MOV DL,'*'              ;Load the ASCII value of '*' in DL.
        INT 21H                 ;Call DOS service for display.
        INC SI                  ;Increment pointer.
        INC CX                  ;Increment the count.
        JMP L1                  ;Jump to L1 to get next character.

CHECK:  CMP CX,0007H            ;Compare count with 07H.
        JNE REPEAT              ;If count is not equal to 07H,
                                ;then go to REPEAT.
        MOV SI,400H             ;Initialize the pointers.
        MOV DI,500H
L2:     MOV AL,[SI]             ;Get a character of entered password in AL.
        CMP AL,[DI]             ;Compare entered character with
                                ;corresponding character of stored password.
        JNE REPEAT              ;If not equal jump to REPEAT.
        INC SI                  ;Increment the pointers.
        INC DI
        LOOP L2                 ;Repeat comparison until counter expires.

        CALL CLRS               ;Clear the screen.
        CALL POS                ;Position the cursor.
        MOV AH,09H              ;Load the function code in AH.
        MOV DX,600H             ;Load address of string to display in DX.
        INT 21H                 ;Call DOS service for display.
        JMP EXIT                ;Go to exit.

REPEAT: CALL CLRS               ;Clear the screen.
        CALL POS                ;Position the cursor.

        MOV AH,09H              ;Load the function code in AH.
        MOV DX,550H             ;Load address of string to display in DX.
        INT 21H                 ;Call DOS service for display.

        MOV AH,09H              ;Load the function code in AH.
        MOV DX,580H             ;Load address of string to display in DX.
        INT 21H                 ;Call DOS service for display.
```

```
          MOV    AH,00H          ;Load the function code in AH.
          INT    16H             ;Call BIOS service to get keyboard
                                 ;Character in AL.
          CMP    AL,1BH          ;Check for esc key.
          JNE    AGAIN           ;If key pressed is not esc, then go to AGAIN.
          JMP    EXIT            ;If key pressed is esc, then go to EXIT.

POS PROC NEAR                    ;Start of procedure to position the cursor.

          MOV    AH,02H          ;Load the function code in AH.
          MOV    BH,00H          ;Load the video page in BH.
          MOV    DX,0000H        ;Load the x and y coordinates in DL and DH.
          INT    10H             ;Call BIOS service for cursor positioning.
          RET                    ;Return to main program.

POS ENDP                         ;End of procedure to position the cursor.

CLRS PROC NEAR                   ;Start of procedure to clear the screen.

          MOV    AH,07H          ;Load the function code in AH.
          MOV    AL,00H          ;Load number of lines to scroll down in AL.
          MOV    BH,07H          ;Load the blanked area attribute in BL.
          MOV    CX,0000H        ;Load the x and y coordinates of
                                 ;upper left corner in CL and CH.
          MOV    DX,184FH        ;Load the x and y coordinates
                                 ;of lower right corner in DL and DH.
          INT    10H             ;Call BIOS video service for clearing screen.
          RET                    ;Return to main program.

CLRS ENDP                        ;End of procedure to clear the screen.
          ORG 450H
          DB 'ENTER THE PASSWORD : ','$'.
          ORG 500H
          DB 'WELCOME','$'.
          ORG 550H
          DB 'INCORRECT PASSWORD.  ','$'.
          ORG 580H
          DB 'PRESS ANY KEY TO TRY AGAIN OR PRESS ESC TO EXIT ','$'.
          ORG 600H
          DB 'ENTRY ACCEPTED','$'.
EXIT:     MOV  AH,4CH            ;Load the function code in AH.
          INT  21H              ;Call DOS service to return to command prompt.

CODE ENDS                        ;End of code segment.
END                              ;Assembly end.
```

Assembler Listing for Example Program 86.30

```
;PROGRAM TO READ THE PASSWORD AND VALIDATE THE USER

0000                    CODE SEGMENT         ;Start of code segment.

                        ASSUME CS:CODE       ;Assembler directive.
                        ASSUME DS:CODE       ;Assembler directive.
=000D                   CR  EQU 0DH          ;Assign ASCII value for carriage return.

0000  E8 0079 R  AGAIN: CALL CLRS            ;Clear the screen.
0003  E8 006F R         CALL  POS            ;Position the cursor.

0006  8C C8            MOV AX,CS             ;Initialize DS to code segment.
0008  8E D8            MOV DS,AX

000A  B4 09            MOV AH,09H            ;Load the function code in AH.
000C  BA 0450          MOV DX,450H           ;Load address of string to display in DX.
000F  CD 21            INT 21H               ;Call DOS service for display.
```

```
0011  BE 0400           MOV SI,400H        ;Initialize the pointer.
0014  B9 0000           MOV CX,0000H       ;Initialize count in CX-register as zero.

0017  B4 00      L1:    MOV AH,00H         ;Load the function code in AH.
0019  CD 16             INT 16H            ;Call BIOS service to get keyboard character in
AL.
001B  3C 0D             CMP AL,CR          ;Compare the keycode with carriage return.
001D  74 0C             JE  CHECK          ;If equal go to CHECK.
001F  88 04             MOV [SI],AL        ;Store the keyboard character in memory.
0021  B4 02             MOV AH,02H         ;Load the function code in AH.
0023  B2 2A             MOV DL,'*'         ;Load the ASCII value of '*' in DL.
0025  CD 21             INT 21H            ;Call DOS service for display.
0027  46                INC SI             ;Increment pointer.
0028  41                INC CX             ;Increment the count.
0029  EB EC             JMP L1             ;Jump to L1 to get next character.

002B  83 F9 07   CHECK: CMP CX,0007H       ;Compare count with 07H.
002E  75 20             JNE REPEAT         ;If count is not equal to 07H,
                                           ;then go to REPEAT.
0030  BE 0400           MOV SI,400H        ;Initialize the pointers.
0033  BF 0500           MOV DI,500H
0036  8A 04      L2:    MOV AL,[SI]        ;Get a character of entered password in AL.
0038  3A 05             CMP AL,[DI]        ;Compare entered character with
                                           ;corresponding character of stored password.
003A  75 14             JNE REPEAT         ;If not equal jump to REPEAT.
003C  46                INC SI             ;Increment the pointers.
003D  47                INC DI
003E  E2 F6             LOOP L2            ;Repeat comparison until counter expires.

0040  E8 0079 R         CALL CLRS          ;Clear the screen.
0043  E8 006F R         CALL POS           ;Position the cursor.
0046  B4 09             MOV AH,09H         ;Load the function code in AH.
0048  BA 0600           MOV DX,600H        ;Load address of string to display in DX.
004B  CD 21             INT 21H            ;Call DOS service for display.
004D  E9 060F R         JMP EXIT           ;Go to exit.

0050  E8 0079 R  REPEAT: CALL CLRS         ;Clear the screen.
0053  E8 006F R         CALL POS           ;Position the cursor.

0056  B4 09             MOV AH,09H         ;Load the function code in AH.
0058  BA 0550           MOV DX,550H        ;Load address of string to display in DX.
005B  CD 21             INT 21H            ;Call DOS service for display.

005D  B4 09             MOV AH,09H         ;Load the function code in AH.
005F  BA 0580           MOV DX,580H        ;Load address of string to display in DX.
0062  CD 21             INT 21H            ;Call DOS service for display.

0064  B4 00             MOV AH,00H         ;Load the function code in AH.
0066  CD 16             INT 16H            ;Call BIOS service to get keyboard
                                           ;character in AL.
0068  3C 1B             CMP AL,1BH         ;Check for esc key.
006A  75 94             JNE AGAIN          ;If key pressed is not esc, then go to AGAIN.
006C  E9 060F R         JMP EXIT           ;If key pressed is esc,then go to EXIT.
006F             POS PROC NEAR             ;Start of procedure to position the cursor.

006F  B4 02             MOV AH,02H         ;Load the function code in AH.
0071  B7 00             MOV BH,00H         ;Load the video page in BH.
0073  BA 0000           MOV DX,0000H       ;Load the x and y coordinates in DL and DH.
0076  CD 10             INT 10H            ;Call BIOS service for cursor positioning.
0078  C3                RET                ;Return to main program.
0079             POS ENDP                  ;End of procedure to position the cursor.
```

```
0079                    CLRS PROC NEAR              ;Start of procedure to clear the screen.

0079  B4 07                 MOV   AH,07H            ;Load the function code in AH.
007B  B0 00                 MOV   AL,00H            ;Load number of lines to scroll down in AL.
007D  B7 07                 MOV   BH,07H            ;Load the blanked area attribute in BL.
007F  B9 0000               MOV   CX,0000H          ;Load the x and y coordinates of
                                                    ;upper left corner in CL and CH.
0082  BA 184F               MOV   DX,184FH          ;Load the x and y coordinates
                                                    ;of lower right corner in DL and DH.
0085  CD 10                 INT   10H               ;Call BIOS video service for clearing screen.
0087  C3                    RET                     ;Return to main program.
0088                    CLRS ENDP                   ;End of procedure to clear the screen.

0450                        ORG 450H
0450  45 4E 54 45 52 20   DB 'ENTER THE PASSWORD : ','$'.
      54 48 45 20 50 41
      53 53 57 4F 52 44
      20 3A 20 24

0500                        ORG 500H
0500  57 45 4C 43 4F 4D   DB 'WELCOME','$'.
      45 24

0550                        ORG 550H
0550  49 4E 43 4F 52 52   DB 'INCORRECT PASSWORD.  ','$'.
      45 43 54 20 50 41
      53 53 57 4F 52 44
      2E 20 20 24

0580                        ORG 580H
0580  50 52 45 53 53 2C   DB 'PRESS ANY KEY TO TRY AGAIN OR PRESS ESC TO EXIT ','$'.
      41 4E 59 20 4B 45
      59 20 54 4F 20 54
      52 59 20 41 47 41
      49 4E 20 4F 52 20
      50 52 45 53 53 20
      45 53 43 20 54 4F
      20 45 58 49 54 20
      24

0600                        ORG 600H
0600  45 4E 54 52 59 20   DB 'ENTRY ACCEPTED','$'.
      41 43 43 45 50 54
      45 44 24

060F  B4 4C        EXIT:  MOV   AH,4CH              ;Load the function code in AH.
0611  CD 21               INT   21H                 ;Call DOS service to return to command
prompt.

0613                  CODE ENDS                     ;End of code segment.
                      END                           ;Assembly end.
```

8.15 SHORT-ANSWER QUESTIONS

Q8.1 *What is software and hardware?*

Software is a set of instructions or commands needed for performing a specific task by a programmable device or a computing machine.

The hardware refers to the components or devices used to form computing machine in which the software can be run and tested. Without software the hardware is an idle machine.

Q8.2 *What is a machine language?*

The language that can be understood by a programmable machine is called machine language. The machine language program are developed using 1's and 0's.

Q8.3 *What is an assembly language?*

The language in which the mnemonics (short-hand form of instructions) are used to write a program is called assembly language. The mnemonics are given by the manufacturers of microprocessor.

Q8.4 *What are machine language and assembly language programs?*

The software developed using 1's and 0's are called machine language programs. The software developed using mnemonics are called assembly language programs.

Q8.5 *What is the drawback in the machine language and assembly language programs?*

The machine language and assembly language programs are machine dependent. The programs developed using these languages for a particular machine cannot be directly run on another machine (But, after conversion using suitable conversion software, it can be run on another machine.)

Q8.6 *What is meant by a program?*

A program is a set of instructions written to perform a certain task.

Q8.7 *What is assembler, interpreter and compiler?*

(a) **Assembler :** It is a software that converts assembly language program codes to machine language codes.

(b) **Compiler :** It is a software that converts the programs written in high level language to machine language.

(c) **Interpreter** : It is similar to a compiler but it converts the instructions one by one.

Q8.8 *What is the need for a assembler?*

An assembler is used to translate assembly language programs to machine language programs (i.e., in the executable format). Without the assembler it is very difficult to convert very large assembly language programs to machine codes.

Q8.9 *What are the advantages of an assembler?*

The advantages of an assembler are:

1. The assembler translates mnemonics into binary code with speed and accuracy.
2. It allows the programmer to use variables in the program.
3. It is easy to alter the program and reassemble.
4. The assembler identifies the syntax errors.
5. The assembler can reserve memory locations for data or result.
6. The assembler provides list file for documentation.

Q8.10 *What are assembler directives or pseudo instructions?*

A assembler directives are the instructions to the assembler regarding the program being assembled. They are also called pseudo instructions or pseudo opcodes.

The assembler directives will give information like start and end of a program, values of variables used in the program, storage locations for input and output data, etc.

Q8.11 List the assembler directives of a typical 8085 assembler.

The assembler directives of a typical 8085 assembler are the following:

Assembler directive	Function
DB	Define Byte. Used to define byte type variable.
DW	Define word. Used to define 16-bit variable.
END	Indicates the end of the program.
ENDM	End of macro. Indicates the end of a macro sequence.
EQU	Equate. Used to equate numeric value or constant to a variable.
MACRO	Defines the name, parameters and start of a macro.
ORG	Origin. Used to assign the starting address for a program.

Q8.12 What is a macro and when is it used?

A macro is a group of instructions written within brackets and identified by a name. A macro is written when a repeated group of instructions is too short or not appropriate to be written as a subroutine.

Q8.13 What is the meaning of expanding the macro?

While assembling a program, the assembler replaces the instructions represented by a macro in the place where macro is called. This is called expanding the macro.

Q8.14 What is the disadvantage in a macro?

The disadvantage in a macro is that, if it is expanded or used a number of times in a program then the program may occupy more memory.

Q8.15 What is a subroutine (or procedure)?

A subroutine (or procedure) is a group of instructions written separately from the main program to perform a function that occurs repeatedly in the main program.

Q8.16 What are the advantages of a subroutine?

   1. **Modular programming : The various tasks in a program can be developed as separate modules and called in the main program.**

   2. **Reduction in the amount of work and program development time.**

   3. **Reduces memory requirement for program storage.**

Q8.17 What is a flowchart?

A flowchart is graphical representation of the operation flow of a program. It is the graphical (pictorial) form of an algorithm.

Q8.18 List the symbols used for drawing a flowchart.

The following are the symbols used for drawing flowchart:

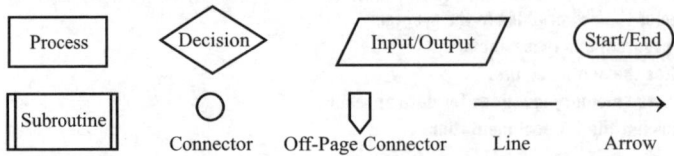

Fig. Q8.18: Symbols used in flowcharts.

Q8.19 What is a development system? What are its components?

A development system is a system used by the microprocessor-based system designer to design and test the software and hardware aspects of a new system under development.

The components of a development system are a microcomputer with standard accessories, emulator and program development tools like editor, assembler, linker, locator, debugger, simulator, etc.

Q8.20 *Write a short note on assembly language program development tools.*

The program development tools for an assembly language program are editor, assembler, linker, locator, debugger and simulator. These tools are softwares that can be run on the development system in order to write, assemble, debug, modify and test the assembly language programs.

Q8.21 *What is an Editor?*

An editor is a program which when run on a microcomputer system, allows the user to type and modify the assembly language program statements. The main function of an editor is to help the user to construct the assembly language program in the right format and save as a file.

Q8.22 *What is a one-pass assembler?*

A one-pass assembler is an assembler in which the source codes are processed only once. A one- pass assembler is very fast and in one-pass assembler only backward reference may be used.

Q8.23 *What is a two-pass assembler?*

The two-pass assembler is an assembler in which the source codes are processed two times. In the first pass, the assembler assigns addresses to all the labels and attach values to all the variables used in the program. In the second pass it converts the source code into machine code.

Q8.24 *What is the drawback of a one-pass assembler?*

The drawback of a one-pass assembler is that the program cannot have forward reference, because, a one-pass assembler issues an error message if it encounters a label or variable that is defined at a later part of a program.

Q8.25 *What is linker and locator?*

i) A linker is a program used to join together several object files into one large object file.

ii) A locator is a program used to assign specific addresses to the object codes to be loaded into memory.

Q8.26 *What is debugging?*

The process of locating and correcting an error using a debugger is known as debugging.

Q8.27 *What is a debugger?*

A debugger is a software used to locate and troubleshoot errors in a program.

Q8.28 *What is a simulator?*

A simulator is a program which can be run on the development system to simulate the operations of the newly designed system. Some operations that can be simulated are given below:

1. **Execute a program and display the result.**
2. **Break-point execution of a program.**
3. **Single step execution of a program.**
4. **Display the content of register/memory.**

Q8.29 *What is an emulator?*

An emulator is a system that can be used to test the hardware and software of a newly developed microprocessor-based system.

Q8.30 *What is the difference between an emulator and a simulator?*

A simulator can be used to run and check the software of a newly developed microprocessor -based system but an emulator can be used to run and check both the hardware and software of a newly developed microprocessor-based system.

Q8.31 How to access DOS services?

The steps involved in accessing DOS services are:

1. Load a DOS function number in AH-register. If there is a sub-function then its code is loaded in the AL-register.
2. Load the other registers as indicated in the DOS service formats.
3. Prepare buffers, ASCIIZ (ASCII string terminated by zero) and control blocks if necessary.
4. Set the location of a Disk Transfer Area if necessary.
5. Invoke DOS service INT 21H.
6. The DOS service will return the required parameters in the specified registers.

Q8.32 How to access BIOS services?

The steps involved in accessing the BIOS services are :

1. Load a BIOS function number in the AH-register. If there is a sub-function, then its code is loaded in the AL-register.
2. Load the other register as indicated in the BIOS service formats.
3. Prepare Buffers, ASCIIZ (ASCII string terminated by zero) and control blocks if necessary.
4. Invoke the BIOS call.
5. The BIOS service will return the required parameters in the specified register.

Q8.33 Write a subroutine to output the content of flag register to LED's connected to the port of a 8085 microprocessor-based system.

Subroutine to display the content of flag register

```
PUSH     PSW    ;Push the A-register and flag register to stack.
POP      B      ;POP the top of stack to BC pair.
MOV      A,C    ;Get the content of flag register in A-register.
OUT      PORT   ;Output the content of A-register to PORT.
RET             ;Return to main program.
```

> *Note:* *The A-register and Flag register are together called PSW (Program Status Word). In this, the A-register is high order register and Flag register is low order register.*

Q8.34 Write a simple program to find the smallest among the two data stored in memory.

(Assume that data are stored in 4200_H and 4201_H. Store the result in 5000_H)

```
        LDA  4200H ;Get first data in A-register.
        MOV  B,A   ;Save first data in B-register.
        LDA  4201H ;Get second data in A-register.
        CMP  B     ;Compare the two data.
        JC   AHEAD ;If CF = 1, go to AHEAD.
        MOV  A,B   ;If CF = 0, move B-register to A-register.
AHEAD:  STA  5000H ;Store the smallest data in memory
        HLT        ;Stop
```

Q8.35 Write a simple program to multiply an 8-bit data stored at 4200_H by 02_H and store the result at 4300_H and 4301_H.

```
        MVI B,00H ;Clear B-register.
        XRA A     ;Clear A-register and carry
        LDA 4200H ;Get the data in A-register.
        RAL       ;Multiply the content of A by 02.
        JNC AHEAD ;If CF = 0, go to AHEAD.
        INR B     ;If CF = 1, increment B-register.
AHEAD:  STA 4300H ;Store the product in memory.
        MOV A,B
        STA 4301H ;Store the carry in memory.
        HLT       ;Halt program execution.
```

Q8.36 *Write a simple program to divide an 8-bit data stored at 4200$_H$ by 02$_H$ and store the result at 4300$_H$ and 4301$_H$.*

```
        MVI  B,00H   ;Clear B-register.
        XRA  A       ;Clear A-register and carry
        LDA  4200H   ;Get the data in A-register.
        RAR          ;Divide the content of A by 02.
        JNC  AHEAD   ;If CF = 0, go to AHEAD.
        INR  B       ;If CF = 1, increment B-register.
AHEAD:  STA  4300H   ;Store the quotient in memory.
        MOV  A,B
        STA  4301H   ;Store the remainder in memory.
        HLT          ;Halt program execution.
```

Q8.37 *Write a simple program to split a hexa data into two nibbles and store in memory.*

```
        LXI  H,4200H ;Set pointer for data array.
        MOV  B,M     ;Get the data in B-register.
        MOV  A,B     ;Copy the data to A-register.
        ANI  0FH     ;Mask the upper nibble.
        INX  H
        MOV  M,A     ;Store the lower nibble in memory.
        MOV  A,B     ;Get the data in A-register
        ANI  F0H     ;Mask the lower nibble.
        RRC          ;Bring the upper nibble to lower nibble position.
        RRC
        RRC
        RRC
        INX  H
        MOV  M,A     ;Store the upper nibble in memory.
        HLT          ;Halt program execution.
```

Q8.38 *Explain the mathematical functions performed by the following instructions.*

```
        MVI  A,07H
        RLC
        MOV  B,A
        RLC
        RLC
        ADD  B
```

The operations performed by each of the above given mathematical instruction are as follows:

1. An 8-bit data 07$_H$ is moved to A-register.

2. The content of A-register is multiplied by 02.

3. The content of A-register (07 × 02 = 0E$_H$) is copied to B-register.

4. The content of A-register is multiplied by 02.

5. The content of A-register is multiplied by 02.

6. The content of B-register is added to B-register.

The result of the above operations is that the content of A-register is multiplied by 0A$_H$. Therefore, after executing the above instructions, the content of A-register will be 46$_H$.

Q8.39 *Write a subroutine to clear a flag register and an accumulator.*

Subroutine to clear flag register

```
LXI  SP,4200H  ;Initialize stack.
LXI  B,0000H   ;Clear BC register pair.
PUSH B         ;Push the content of BC pair to stack.
POP  PSW       ;Pop the top of stack to A-register and flag register (PSW).
RET            ;Return to main program.
```

> **Note:** *The A-register and Flag register are together called PSW (Program Status Word). The A-register is a high order register and the flag register is a low order register.*

Q8.40 *Write a subroutine program to exchange the content of BC pair and DE pair?*

Subroutine to exchange BC pair and DE pair

```
LXI   SP,4200H  ;Initialize stack.
PUSH  B         ;Store the content of BC pair in stack.
PUSH  D         ;Store the content of DE pair in stack.
POP   B         ;Move the content of DE pair stored in stack to BC pair.
POP   D         ;Move the content of BC pair stored in stack to DE pair.
RET             ;Return to main program.
```

8.16 EXERCISES

I. Fill in the blanks with appropriate words

1. The language in which instructions are represented by binary codes is called _____ language.

2. The _____ software tool is used to convert assembly language into machine language.

3. The _____ is used to create library files which are collections of procedures of frequently used functions.

4. An _____ is a mixture of hardware and software.

5. The _____ assembler directive is used to define byte type variable.

6. The assembler directive which is used to indicate the end of the procedure is _____

7. The assembler directive _____ is used to declare variables as common to various program modules.

8. Calling an 8086 subroutine within the same segment is defined as _____.

9. The _____ register of 8085 has to be preserved in stack whenver a subroutine is called.

10. The _____ are also called as open subroutines.

11. In 8086, the _____ register holds the offset address of the stack.

12. The _____ instruction of 8085 is equivalent to the following set of instructions,

```
MOV  A,L
STA  3000H
MOV  A,H
STA  3001H
```

13. The 8085 instruction _____ is used to add two 16-bit numbers which are available in HL and BC register pairs.

Answers

1. machine	4. emulator	7. PUBLIC	10. macros	13. DAD B
2. assembler	5. DB	8. intrasegment cell	11. stack pointer(SP)	
3. library builder	6. ENDP	9. program counter(PC)	12. SHLD 3000H	

II. State whether the following statements are True/False.

1. Any programming language has to be converted into machine language finally.

2. The compiler software converts high level language into assembly level language.

3. The simulators do not have the ability to perform timing and data transmission and reception.

4. Assembler directives are pseudo-instructions.

5. No machine codes are generated for assembler directives.

6. The assembler directive DT is used to define ten words of a variable.

7. The 8086 statement X DB 32H reserves 2 memory locations.

8. Both CS register and IP register have to be pushed into stack memory while calling a 8086 near call procedure.

9. There are no near and far procedures in 8085.

10. Subroutines are stored in memory only once.

11. Macros are copied in the program everytime they are called in.

12. The stack pointer(SP) is automatically incremented or decremented by the processor after every stack read/write operation.

13. The SP register of 8085 is incremented by one after read operation from stack.

14. The content of stack segment(SS) register will not be altered while reading or writing into the stack of 8086.

15. The stack memory of 8051 can be located anywhere in external RAM of 8051.

16. Upon reset, the SP of 8085 and 8086 are cleared to zero.

17. Upon reset, the SP of 8051 are cleared to zero.

18. The 8051 SP register is incremented by one after every write operation.

19. The 8051 SP register is decremented by one after every read operation.

Answers

1. True	4. True	7. False	10. True	13. False	16. True	19. True
2. False	5. True	8. False	11. True	14. True	17. False	
3. True	6. False	9. True	12. True	15. False	18. True	

III. Choose the right answer for the following questions.

1. *The assembly language programs are written using _____.*
   a) binary code   *b)* Mnemonics   *c)* english words   *d)* all the three

2. *Which of the following is a high level language?*
   a) c   b) c++   c) JAVA   d) all the three

3. *Which of the following is used to combine relocatable object files and library functions into a single executable file?*
   a) library builder   b) linker   c) debugger   d) simulator

4. *Which of the following tools is used to test and run the programs of the newly designed system?*
   a) linker   b) debugger   c) simulator   d) builder

5. *Identify the valid constants of assembler*
   a) A2H   b) F2   c) 112B   d) 8BC2H

6. The _____ assembler directive of 8085 is used to define a 8-bit variable.

    a) DD   b) DB   c) DT   d) DW

7. Which of the following 8085 assembler directive is used to indicate the end of a program?

    a) END   b) ENDP   c) ENDS   d) none

8. The 8086 statement ARRAY DB 25 DUP(0) assigns _____ number of memory locations with _____ as initial value in all locations.

    a) 25 and 00   b) 25H and 00H   c) 50 and 00   d) 50H and 00H

9. What will be the content of SP register of 8085 after a writing operation into stack? SP = 4218_H before write operation.

    a) 4219_H b) 4217_H c) $421A_H$ d) 4216_H

10. Which of the following processor/controller uses an 8-bit SP register

    a) 8085   b) 8086   c) 8051   d) all the three

11. Identify the operations performed by the following 8085 program.

```
LXI   H,2000H
MVI   M,28H
MOV   A,M
XRI   0FFH
INR   A
MOV   M,A
```

    a) Inverts the content of 2000_H b) complements the content of 2000_H

    c) change the sign of content of 2000_H d) Increment content of 2000_H by one

12. what will be the content of SP register and DE register pair after executing the following 8085 instruction?

```
2000: LXI   SP,3020H
2003: CALL  2006
2006: NOP
2007: POP   H
```

    a) SP = 3020_H ; HL = 2003_H b) SP = 3022_H ; HL = 2006_H

    c) SP = 3020_H ; HL = 2006_H d) SP = 3020_H ; HL = 2003_H

13. Consider the following 8085 main program and subroutine. What will be the value of the accumulator after executing the line 4,5 and 6?

```
1.    MVI    A,27H        SUBROUTINE: INR  A
2.    ANI    0FH                      ORI  00H
3.    PUSH   A                        RET
4.    CALL   SUBROUTINE
5.    ADD    A
6.    POP    A
7.    HLT
```

    a) 08_H, 10_H and 07_H respectvely   b) 07_H, 08_H and 10_H respectively

    c) 10_H, 07_H and 08_H respectively   d) 08_H, 07_H and 10_H respectively

14. The following program searches for the byte 05_H in an array of size 6 bytes which starts at 2000_H and stores 01_H if the data byte is found else 00_H in memory location 2100_H. A part of the program is given below:

```
          LXI   H,2000H
          MOV   B,06H
CONTINUE: MOV   A,M
```

```
        CPI  05H
        JZ   FOUND
        ─ ─ ─ ─
        ─ ─ ─ ─
        ─ ─ ─ ─
        MVI  A,00H
        JMP  HERE
FOUND:  MVI  A,01H
HERE:   STA  2100H
        HLT
```

The sequence of instructions to complete the program would be

a) INX H
   DCR B
   JNZ CONTINUE

b) INX H
   DCR B
   JNZ HERE

c) INX H
   DCR B
   JZ CONTINUE

d) INX H
   DCR B
   JNZ FOUND

Answers

1. b   3. b   5. d   7. a   9. c   11. c   13. a

2. d   4. c   6. b   8. a   10. c   12. c   14. a

IV. Answer the following questions

E8.1 Write an 8085 assembly language program without using MOV instruction to load the accumulator with content of memory location 3600_H.

E8.2 Write an 8085 assembly language program to transfer 08 bytes of data from memory location which starts at 2050_H to another memory location which starts at 2100_H.

E8.3 Write an 8085 assembly language program to exchange 08 bytes of data from memory location which starts at 2050_H to another memory location which starts at 2100_H.

E8.4 Write an 8085 assembly language program to square the given data stored at 3000H and store the results at 3001_H.

E8.5 Write an 8085 ALP to calculate the average room temperature. Use 5 different readings noted down at 5 different instances throughout the day. Assume that five readings are stored in memory 2500_H onwards

E8.6 It is desired to multiply the numbers $0A_H$ and $0B_H$ and store the result in the accumulator. The numbers are available in registers B and C respectively. A part of 8085 program for theis purpose is given below.

```
        MVI  A,00H
LOOP:   ─ ─ ─ ─
        ─ ─ ─ ─
        HLT
        END
```

Complete the program by writing the missing sequence of instructions

E8.7 The question i) and ii) are linked questions. Consider an 8085 microprocessor based system.

i) The following program starts at location 0100_H.

```
        LXI  SP,00FFH
        LXI  H,0701H
        MVI  A,20H
        SUB  M
```

What will be the content of accumulator when the program counter reaches 0109_H?

ii) If in addition the following code exists from 0109_H onwards

```
        ORI  40H
        ADD  M
```

What will be the result in the accumulator after the last instruction is executed?

E8.8 Consider the following sequence of instructions for an 8085 microprocessor based system.

Memory Address	Instructions	
FF00	MVI	A,FFH
FF02	INR	A
FF03	JC	FF0CH
FF06	ORI	A8H
FF08	JM	FF15H
FF0B	XRA	A
FF0C	OUT	PORT1
FF0E	HLT	
FF0F	NOP	
FF10	XRI	FFH
FF12	OUT	PORT2
FF14	HLT	
FF15	MVI	A,FFH
FF17	ADI	02H
FF19	RAL	
FF1A	JZ	FF23H
FF1D	JC	FF12H
FF20	JNC	FF12H
FF23	CMA	
FF24	OUT	PORT3
FF26	HLT	

a) If the program execution begins at location FF00$_H$, write down the sequence of instruction which are actually executed till a HLT instruction. (Assume all flags are initially RESET).

b) Which of the three ports (port-1, port-2, and port-3) will be loaded with data. What is the bit pattern of the data?

E8.9 The program and machine code for an 8085 microprocessor are given below.

```
3E   MVI   A,C3
C3
00   NOP
80   ADD   B
3D   DEC   A
C2   JNZ   800A
0A
80
C3   JMP   800C
0C
80
D3   OUT   10
10
76   HLT
```

The starting address of the above program is 7FFF$_H$. What would happen if it is executed from 8000$_H$?

E8.10 An 8085 assembly language program is given below

```
        MVI   C,03H
        LXI   H,2000H
        MOV   A,M
        DCR   C
LOOP1:  INX   H
        MOV   B,M

        CMP   B
        JNC   LOOP2

        MOV   A,B
LOOP2:  DCR   C

        JNZ   LOOP1

        STA   2100H

        HLT
```

The contents of memory locations are

$2000_H = 18_H$; $\quad$ $2001_H = 10_H$; $\quad$ $2002_H = 2B_H$

a) What does the above program do?

b) At the end of the program what will be

   i) the content of the registers A, B, C, H and L?

   ii) the condition of the carry and zero flags.

   iii) the content of the memory locations 2000_H, 2001_H, 2002_H and 2100_H

E8.11 Write an assembly language program without using any arithemetic instruction to store hexadecimal $5D_H$ in the flag register of 8085 microprocessor. Data in other registers of the processor must not alter upon executing this program.

E8.12 An 8085 assembly language program is given below.

```
Line 1:   MVI   A,B5H
Line 2:   MVI   B,0EH
Line 3:   XRI   69H
Line 4:   ADD   B
Line 5:   ANI   9FH
Line 6:   CPI   9FH
Line 7:   STA   3010H
Line 8:   HLT
```

a) What will be the content of the accumulator just after execution of the ADD instruction in line 4?

b) What will be the status of the cy and z flag after executing line 7 of the program.

E8.13 Following is the segment of a 8085 assembly language program.

```
          LXI    SP,EFFFH
          CALL   3000H
              .
              .
              .
              .
3000H:    LXI    H,3CF4H
          PUSH   PSW
          SPHL
          POP    PSW
          RET
```

What will be the contents of SP on completion of RET execution

E8.14 Write an 8085 ALP to find out how many positive integers and negative integers are there in a six byte array which starts at 2000_H.

E8.15 The following program is written to find whether the given signed numbers located at 2100_H is positive or negative and stores 00_H if the data is positive and 01_H is it is negative at memory loation 2101_H. Identify the logical error if any.

```
              LDA    2100H
              RRC    A
              JNC    POSITIVE
              MVI    A,01H
              JMP    HALT
POSITIVE:     MVI    A,00H
HALT:         HLT
```

E8.16 In 8085 microprocessor the following program is executed. What will be the accumulator content at the end of program?

```
          MVI   A,04H
          MVI   B,04H
HERE:     ADD   B
          DCR   B
          JNZ   HERE
          ADI   A,06H
          HLT
```

E8.17 Give the content of 8085 register A and the memory locations 2000_H and 2100_H after executing the following program?

```
LXI   D,2000H
LXI   H,2100H
XRA   A
INR   A
MOV   M,A
XCHG
MOV   M,A
MOV   A,M
DCR   A
XCHG
DCR   M
```

E8.18 What will be the content of SP, A and B registers after executing the lines 6, 7, 8 and 9 in the following program? Assume content of 2000_H is 08_H.

```
1. MVI   C,00H
2. LDA   2000H            2800H:  MVI   A,099H
3. ADI   A,02H                    MVI   B,03H
4. MOV   B,A                      SUB   B
5. INR   B                        RET
6. LXI   SP,2500H
7. PUSH  B
8. CALL  2800H
9. POP   B
```

E8.19 Write an 8051 assembly language program to reset all the seven registers (R_0-R_7) of register bank-3.

E8.20 Identify the operations performed by the following program

```
MOV   A,#00H            MOV   @R0,A
MOV   R0,08H            INC   R0
MOV   R1,#05H           DJNZ  R1,BACK
BACK: ADD    A,#02H     HALT: SJMP HALT
```

E8.21 Write an 8051 program to exchange the contents of internal RAM 12_H and external data memory 0012_H.

E8.22 Write an 8051 program to exchange the contents of stack memory 50_H and the content of port-2.

E8.23 Write an 8051 program to perform the following simple task.

(i)   Move a block of data (6 bytes) from internal RAM memory locations(60_H to 65_H) to internal RAM (70_H to 75_H)

(ii) Exchange the contents of 60_H to 65_H and 70_H to 75_H

(iii) Copy the content of 60_H to 65_H at external data memory which starts at 2500_H.

E8.24 Write an 8051 assembly language program to add the data 12_H and 35_H. Also store the status of all flags of 8051.

E8.25 Write an 8051 program to add two 16-bit numbers.

E8.26 Write an 8051 program to add two BCD numbers.

E8.27 Identify the operation performed by the following 8051 program. Assume content of memory location $2000_H = 48_H$

```
MOV    DPTR,#2000H
MOVX   A,@DPTR
CPL    A
INC    A
INC    DPTR
MOVX   @DPTR,A
HALT:  SJMP   HALT
```

E8.28 Write an 8051 program to calculate the distance (in km) travelled by a car when it travels at a speed of 50 kmph and reaches the destination in 2 hours.

E8.29 Write an 8051 program to count the number of 100's in the given data which is stored at 2300_H and store the result in 2301_H.

E8.30 Write an 8051 program to convert a packed BCD number into an unpacked BCD number. Assume the packed BCD number is at internal memory 20_H. Store the result in external memory location 3100_H and 3101_H.

E8.31 Write an 8051 program to find the square of a number.

E8.32 Write an 8051 program to evaluate the expression, $y = 4x + 6$, where x is an 8-bit hexadecimal number stored at 2500_H. Store the result at 3000_H and 3001_H.

E8.33 Write an 8051 program to perform the following tasks

a) Read the status of port-1 and if it is FF_H clear port-2

b) Set only 1,3,5 and 7th bit of port-1 to 1.

c) Reset only 1,3,5 and 7th bit of port-1 to 0.

d) Swap the lower nibble of port-0 and port-1.

e) To invert only the lower nibble of port-1

E8.34 Given an array of 6 data which is stored from 2501_H. Search for the data 25_H in the array, and if it is found store the value FF_H else 00_H in memory location 3000_H.

E8.35 Write an 8051 program to count the number of ones and zeros in the given data stored at 2000_H. store the number of zeros at 2001_H and the number of ones at 2002_H.

E8.36 Write an 8051 program to display 00_H at location 2001_H if the data available at memory location 2000_H is positive number else display 01_H.

E8.37 Check whether the data at location 2000H is even or odd number. Display 00_H if even else 01_H at memory location 2001_H. Use 8051 microcontroller to program.

E8.38 Write an 8051 ALP to check whether the two numbers stored at memory locations 2000_H and 2001_H are same or not. If same display 00_H else 01_H at memory location 2002_H. Use logical instructions to compare.

E8.39 Write an 8051 ALP to convert a positive number into negative number. Assume positive data is stored at 3000_H.

E8.40 Write an 8086 assembly language program to complement the highest nibble of each byte in an array of size 30 which is located in a data segment with base address 2000_H and an offset of 0100_H. Store the result in the same segment at an offset address 0200_H.

E8.41 Write an 8086 ALP to multiply the contents of AL and Bx register.

E8.42   In the following 8086 program what is the data in AX after execution of fourth instruction and where is the program control transfered after executing fifth instruction?

```
            CLC
            MOV     AL,80H
            ADD     AL,AL
            RCL     AL,1
            JZ      Loop
            REP     AL,1
   LOOP:    MOV     [0200H],AL
```

E8.43   Write an 8086 ALP to find and count the prime numbers in an array of size 50. The array is stored in data segment with base address 2500H and at an offset of 0010H. Store the result from 2500 : 0100$_H$.

E8.44   Write an 8086 ALP to find the average of 10 words which are stored from memory address 3000$_H$: 0025$_H$.

E8.45   Write an 8086 ALP to swap all the elements of an 8 byte array which starts at 2000 : 0010 with another 8-byte array which starts at 2000 : 0020$_H$.

E8.46   Write an 8086 ALP to check whether an input byte available at 2000 : 0001$_H$ is a valid BCD number or not. If valid store 01$_H$ else 00H at 2000 : 0002$_H$.

E8.47   Write an 8086 program to mask the higher nibble of AL.

E8.48   Write an 8086 program to swap the nibbles of AL register.

E8.49   Write an 8086 ALP to convert the BCD numbers 0-9 into corresponding ASCII code 30-39$_H$. Use XLAT instruction to read the ASCII code for a given BCD number from the table stored at memory locations starting from 3000 : 0000$_H$

E8.50   Write an 8086 ALP to add two BCD words which are stored 2000 : 0010 and 2000 : 0020 respectively. Store the result at 2000 : 0030.

E8.51   What will be the contents of A register after the execution of following program?

```
            MOV     AL,0A3H
            CBW
            CWD
```

E8.52   Write an 8086 ALP to reverse the word stored at 3000 : 0010H. Store the result at 3000 : 0020.

E8.53   Write an 8086 ALP to calculate the average of 5 words stored consecutively from 2000 : 0001$_H$. Store the result in 2000 : 0030$_H$.

E8.54   Write an 8086 ALP to find the factorial of a byte stored at 2000 : 0001$_H$. Store the result at 2000 : 0002$_H$.

E8.55   Write an 8086 ALP to move a block of byte array from 2000 : 0001 to 2000 : 0101. Initialize the array size at 2000 : 0000.

E8.56   Write an 8086 ALP to exchange two blocks of byte arrays which are stored at 2000 : 0001 and 2000 : 0101. Initialize the array size at 2000 : 0000.

E8.57   Write an 8086 ALP to scan for a given character in a string array.

E8.58   Write an 8086 ALP to compare two strings in a string array.

E8.59   Write an 8086 ALP to reverse a string in a string array.

E8.60   Write an 8086 ALP to check whether the given string is palindrome or not. Use macros.

CHAPTER 9

PERIPHERAL DEVICES AND INTERFACING

9.1 PROGRAMMABLE PERIPHERAL DEVICES

Programmable peripheral devices are designed to perform various input/output functions and specific routine activities. Every programmable device will have one or more control registers. The programmable devices can be set up to perform specific functions by writing control words into the control registers. The control word is an instruction which informs the peripheral about various functions it has to perform. The format of the control word will be specified by the manufacturer of the peripheral devices.

INTEL have developed a number of peripheral devices that can be used with 8085/8086/8088 based systems. Some of the peripheral devices developed by INTEL for 8085/8086/8088 based system are Parallel peripheral interface-8255, Serial communication interface-8251, Keyboard/Display controller-8279 and Programmable Timer 8254 and DMA controller-8237. A brief discussion about these devices and their interfacing with an 8086 processor are presented in this chapter.

The parallel peripheral interface-8255 is used to interface a slow IO device to the fast processor and to achieve an efficient data transfer between them. The USART is used to provide serial communication between processor and another system. The 8279 is used to relieve the processor from time-consuming routine activities like keyboard scanning and display refreshing. The programmable timers are used to maintain various timings and to initiate time-based activities. The DMA controllers are used to achieve very fast data transfer between memory and IO devices by bypassing the processor.

9.2 PARALLEL DATA COMMUNICATION INTERFACE

In microprocessor-based systems, digital informations can be transmitted from one system to another system either by parallel or serial data transfer scheme.

In parallel data transfer, a group of bits (e.g., 8 bits) are transmitted from one device to another at any one time. To achieve the parallel data transfer scheme, a group of data lines will be connecting the processor and peripheral devices. Normally, in microprocessor-based systems, the parallel data transfer schemes are adopted to transfer data between various devices inside the system.

Basically, the microprocessor-based system has been fabricated on a PCB (Printed Circuit Board) in which a bus is formed with the required number of data lines and the bus connects all the devices in the system. The data transmitted over the bus in a PCB are highly reliable. In a well-designed board, there will not be any loss of data and the data will not be corrupted.

When data has to be transmitted over longer distances (i.e., greater than 0.5 m), we require high current signals to drive the data for a longer distance. In such cases, data are transmitted bit by bit through a single data line.

9.2.1 Parallel Data Transfer Schemes

The data transfer schemes refer to the method of data transfer between the processor and peripheral devices. In a typical microcomputer, data transfer takes place between any two devices: microprocessor and memory, microprocessor and IO devices, memory and IO devices. For effective data transfer between these devices, the timing parameters of the devices should be matched. But most of the devices have incompatible timings. For example, an IO device may be slower than the processor due to which it cannot send data to the processor at the expected time.

The semiconductor memories are available with compatible timings. Moreover, slow memories can be interfaced using additional hardware to introduce wait states in machine cycles. The microprocessor system designer often faces difficulties while interfacing IO devices and magnetic memories (like floppy or hard disk) to achieve efficient data transfer to or from microprocessor. Several data transfer schemes have been developed to solve the interfacing problems with IO devices.

The data transfer schemes have been broadly classified into the following two categories:

1. **Programmed data transfer.**

2. **Direct Memory Access (DMA) data transfer.**

In programmed data transfer, a memory resident routine (subroutine) requests the device for data transfer to or from one of the processor register.

Programmed data transfer scheme is used when relatively small amount of data are to be transferred. In these schemes, usually one-byte or word of data is transferred at a time. Examples of devices using programmed data transfer are ADC, DAC, Hex-keyboard, 7-segment LEDs, etc.

The programmed data transfer scheme can be further classified into the following three types:

1. **Synchronous data transfer scheme.**

2. **Asynchronous data transfer scheme.**

3. **Interrupt driven data transfer scheme.**

In DMA data transfer, the processor is forced to **HOLD** state (**high impedance** state) by an IO device until the data transfer between the device and the memory is complete. The processor does not execute any instructions during the **HOLD** period.

The DMA data transfer is used for a large block of data transfer between the IO device and memory. Typical examples of devices using DMA are CRT controller, floppy disk, hard disk, high speed line printer, etc.

The different types of DMA data transfer schemes are:

a) **Cycle stealing DMA or Single transfer mode DMA.**

b) **Block or Burst mode DMA.**

c) **Demand transfer mode DMA.**

Figure 9.1 shows the various types of data transfer scheme. All the data transfer schemes discussed above require both software and hardware for their implementation. Within a microcomputer, more than one scheme can be used for interfacing different IO devices. However, some of these schemes require specific hardware features in the microprocessor for implementing the scheme.

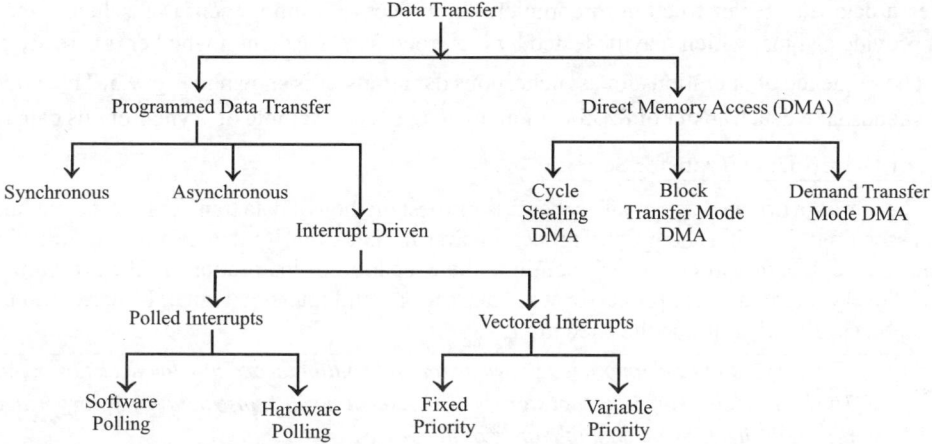

Fig. 9.1: Types of data transfer schemes.

Synchronous Data Transfer Scheme

The synchronous data transfer scheme is the simplest of all data transfer schemes. In this scheme, the processor does not check the readiness of the device. The IO device or peripheral should have matched timing parameters. Whenever data is to be obtained from the device or transferred to the device, the user program can issue a suitable instruction for the device. At the end of the execution of this instruction, the transfer would have been completed.

The synchronous data transfer scheme can also be implemented with small delay (if the delay is tolerable) after the request has been made. The sequence of operations for synchronous data transfer scheme is shown in Fig. 9.2. The mode-0 input/output in 8255 is an example of synchronous data transfer.

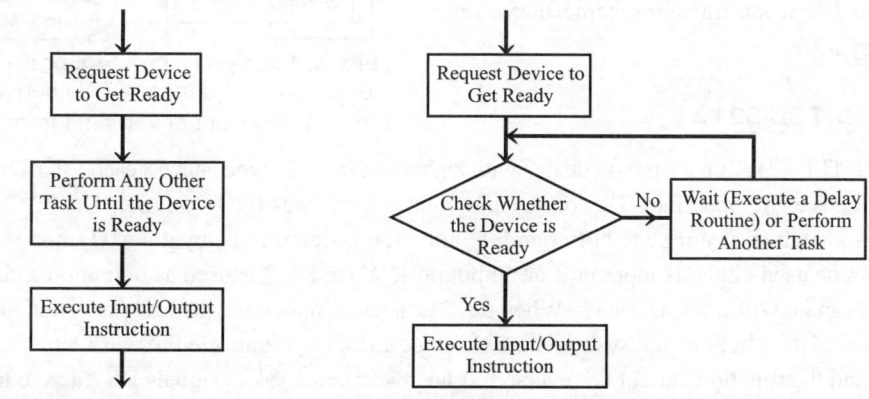

Fig. 9.2: Synchronous data transfer scheme.   **Fig. 9.3:** Asynchronous data transfer scheme.

Asynchronous Data Transfer Scheme

The asynchronous data transfer scheme is employed when the speed of the processor and IO device do not match. In this scheme, the processor sends a request to the device for read/write operation. Then the processor keeps on polling the status of the device. Once the device is ready, the processor executes a data transfer instruction to complete the process. To implement this scheme, the device should provide a signal which may be tested by the processor to ascertain whether it is ready or not.

The sequence of operations for asynchronous data transfer is shown in Fig. 9.3. The mode-1 and mode-2 handshake data transfer of 8255 without interrupt is an example of asynchronous data transfer.

Interrupt Driven Data Transfer Scheme

The interrupt driven data transfer scheme is the best method of data transfer for efficient utilization of processor time. In this scheme, the processor first initiates the IO device for data transfer. After initiating the device, the processor will continue the execution of instructions in the program. Also at the end of every instruction the processor will check for a valid interrupt signal. If there is no interrupt then the processor will continue the execution.

> *Note: The user/system designer need not write any subroutine/procedure to check for an interrupt. The logic of checking interrupt signals while executing each instruction is incorporated in the processor itself by the manufacturer of the processor.*

When the IO device is ready, it will interrupt the processor. On receiving an interrupt signal the processor will complete the current instruction execution and save the processor status in stack. Then the processor call an **Interrupt Service Routine** (ISR) to service the interrupting device. At the end of ISR, the processor status is retrieved from stack and the processor starts executing its main program. The sequence of operations for an interrupt driven data transfer scheme is shown in Fig. 9.4. (For detailed discussion on interrupt driven data transfer scheme, please refer to Chapter-7.)

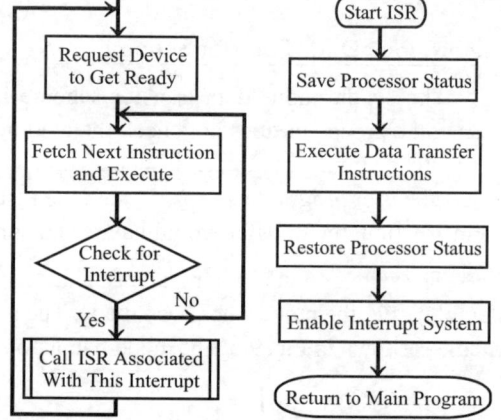

Fig. a: Main program execution sequence.   **Fig. b:** ISR execution sequence.

Fig. 9.4: Interrupt driven data transfer scheme.

9.2.2 INTEL 8212

INTEL 8212 is a 24-pin IO device with eight number of D-type latches, each followed by a tristate buffer. It has eight input lines (DI_1 to DI_8) and eight output lines (DO_1 to DO_8). The 8212 can be used either as a latch or as a tristate buffer and the function is determined by pin MD (mode). The INTEL 8212 can be used either as input port or output port. When 8212 is used as output port, the MD pin is tied **high** and it will work as a latch. When 8212 is used as input port, the MD pin is tied **low** and it will work as a tristate buffer. In a system the 8212 is permanently connected to work either as input or as output and the function cannot be reversed. It has two device select signals $\overline{DS}_1$ (active **low**) and DS_2 (active **high**) and three control pins $\overline{CLR}$ (clear), STB (strobe) and $\overline{INT}$ (Interrupt). The pin description of 8212 and its internal block diagram are shown in Fig. 9.5.

Note: *In microprocessor-based systems, the input port should be a tristate buffer and the output port should be a latch.*

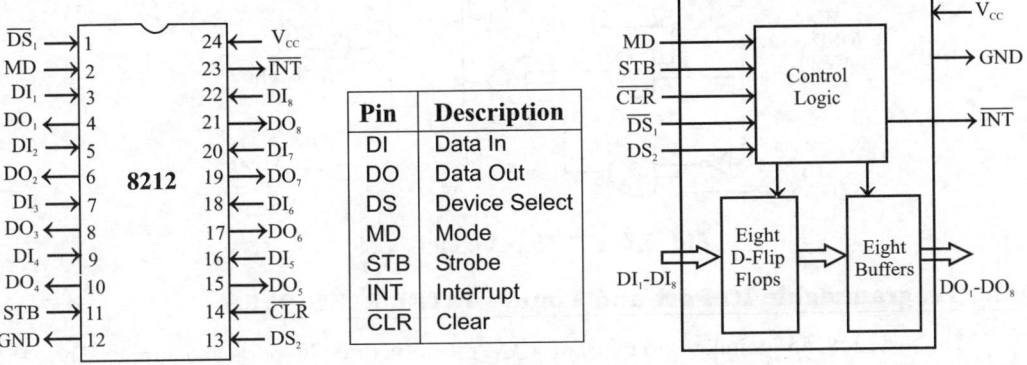

Fig. 9.5: 8212 pin description and internal (functional) block diagram.

The INTEL 8212 can be used for simple data transfer or data transfer with handshake signals. The strobe and interrupt signals are used for handshake data transfer. For simple data transfer, the STB is permanently tied **high** and $\overline{INT}$ is not connected (not used) in the system.

The output logic of 8212 is shown in simplified form in the Fig. 9.6. The input lines DI_1 to DI_8 are connected to data bus of microprocessor system and the output lines DO_1 to DO_8 are connected to the output device. In the output mode, MD, STB and $\overline{CLR}$ are **high**. When MD is **high**, the output of gate G_2 is **high**, which enables the tristate buffer. The D-flip-flop functions as a latch. Now, the output of gate G_4 is **low**, which makes the STB signal non-functional. When the the device is selected by chip select circuit by making $\overline{DS}_1 = 0$ and $DS_2 = 1$, the output signals of gates G_1, G_3 and G_5 goes **high** and the clock signal of the flip-flop goes **high**. The data on pins DI flow to the output of the flip-flops and are latched when the clock pulse goes **low**.

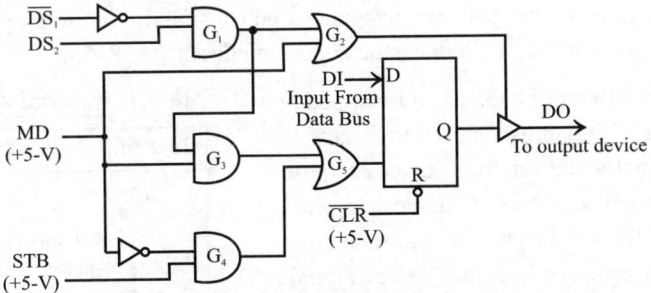

Fig. 9.6: 8212 Output control logic.

The INTEL 8212 functions as an input device when the mode signal is **low**. Figure 9.7 shows the simplified logic of the 8212 in the input mode. The input lines DI_1 to DI_8 are connected to input device and the output lines DO_1 to DO_8 are connected to data bus of microprocessor system. When the mode pin is **low**, all tristate buffers are disabled until the device is selected. However, when the STB is **high**, the output of G_4 and G_5 goes **high** and external data can be loaded into the flip-flops even if the 8212 is not selected. When the microprocessor selects the 8212, the tristate buffers are enabled and the data flow from output of D-latch (Q) to the data bus.

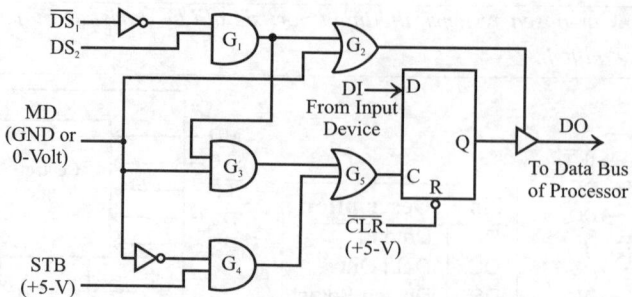

Fig. 9.7: 8212 Input control logic.

9.2.3 Programmable IO Port and Timer - INTEL 8155/8156

The INTEL 8155 includes 256 bytes of RAM memory, three IO ports and a timer. The 8156 is identical with the 8155 except that the 8156 requires active high **Chip Enable (CE)**.

Functionally 8155 can be viewed as two independent chips, one having static RAM and the other having IO ports and a timer. The IO section of 8155 includes 2 numbers of 8-bit parallel IO ports called port-A and port-B, one number of 6-bit port called port-C and a programmable timer.

All the ports can be configured as simple input or output ports. Ports A and B can be programmed in the handshake mode. In the handshake mode, each port uses three signals as handshake signals and the port-C pins are used for handshake signals. When some of the port-C pins are used for handshake signals, the remaining pins can be used as simple input or output lines.

The timer has a 14-bit counter which can be programmed to work in four operating modes. The internal block diagram of 8155 and its internal decoding logic are shown in Fig. 9.8.

The control logic of the 8155 is specifically designed to eliminate the need for external demultiplexing of AD_0 - AD_7 and generating separate control signals for memory and IO. The ALE, $IO/\overline{M}$, $\overline{RD}$ and $\overline{WR}$ signals from the 8085 can be connected directly to 8155.

The ports and the timer of 8155 are IO-mapped in the system. Hence an 8-bit address is used to select the internal devices. Actually the internal devices require a 3-bit address to select any one of the five internal devices as shown in Table 9.1. The remaining address lines are decoded to produce the chip select signal. (For interfacing of 8155 with 8085, please refer to Chapter-6, Design Example-85.6.)

Ports are programmed to work as input or output port in simple or handshake modes. The timer is also

Table 9.1: Internal Address of 8155

Internal device	Internal address		
	A_2	A_1	A_0
Control register/ Status register	0	0	0
Port-A	0	0	1
Port-B	0	1	0
Port-C	0	1	1
LSB timer	1	0	0
MSB timer	1	0	1

programmed to work in any one of the four operating modes. The programming of the ports and the timer is accomplished by writing a control word in the control register. The control word is framed in the specified format as shown in Fig. 9.9 and then loaded in the control register. Each bit of control word defines a function as given in Table 9.2.

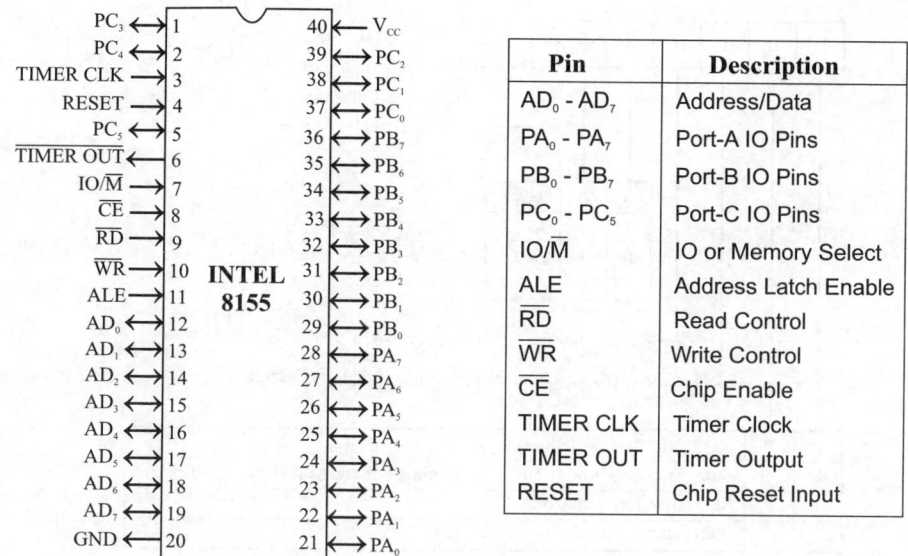

Fig. a: Pin description of 8155.

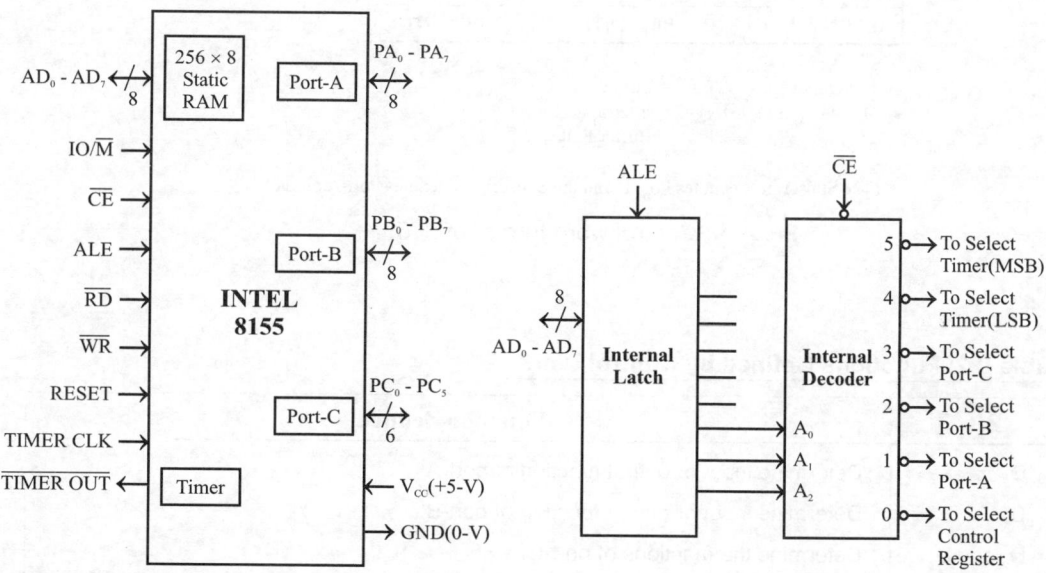

Fig. b: Internal block diagram of 8155.   **Fig. c:** Internal decoding logic of 8155.

Fig. 9.8: Internal block diagram of 8155 and its internal decoding logic.

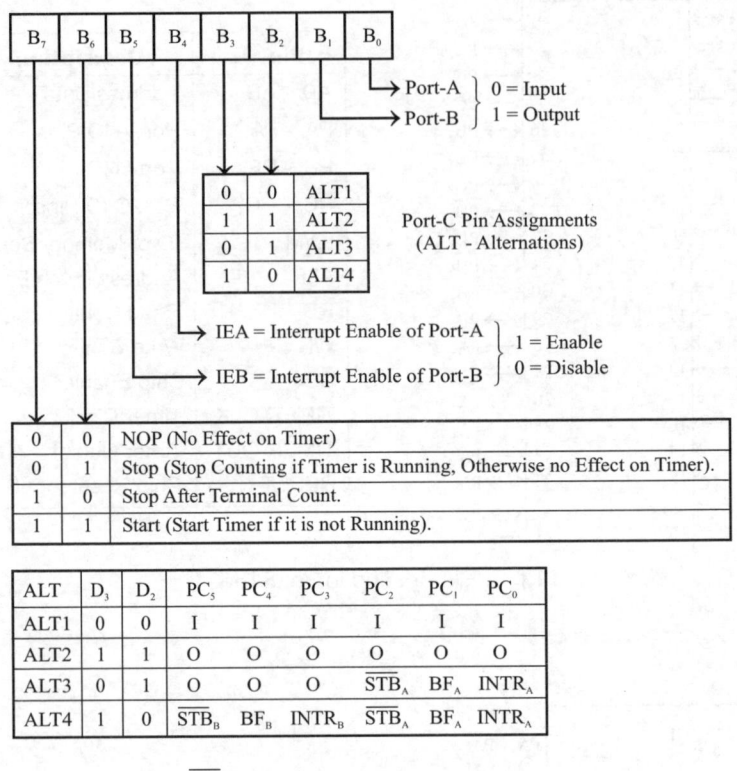

| B$_7$ | B$_6$ | B$_5$ | B$_4$ | B$_3$ | B$_2$ | B$_1$ | B$_0$ |

→ Port-A   0 = Input
→ Port-B   1 = Output

0	0	ALT1
1	1	ALT2
0	1	ALT3
1	0	ALT4

Port-C Pin Assignments
(ALT - Alternations)

→ IEA = Interrupt Enable of Port-A   1 = Enable
→ IEB = Interrupt Enable of Port-B   0 = Disable

0	0	NOP (No Effect on Timer)
0	1	Stop (Stop Counting if Timer is Running, Otherwise no Effect on Timer).
1	0	Stop After Terminal Count.
1	1	Start (Start Timer if it is not Running).

ALT	D$_3$	D$_2$	PC$_5$	PC$_4$	PC$_3$	PC$_2$	PC$_1$	PC$_0$
ALT1	0	0	I	I	I	I	I	I
ALT2	1	1	O	O	O	O	O	O
ALT3	0	1	O	O	O	$\overline{STB}_A$	BF$_A$	INTR$_A$
ALT4	1	0	$\overline{STB}_B$	BF$_B$	INTR$_B$	$\overline{STB}_A$	BF$_A$	INTR$_A$

I = Input $\overline{STB}$ = Strobe
O = Output   INTR = Interrupt Request
              BF   = Buffer Full

(The Subscript B denotes Port-B and the Subscript A denotes Port-A Signal).

Fig. 9.9: Control word format of 8155.

Table 9.2: Functions Defined by Control Word

Bit	Function defined
D$_0$	Determine input or output function of port-A.
D$_1$	Determine input or output function of port-B.
D$_2$ and D$_3$	Determine the functions of port-C.
D$_4$ and D$_5$	Used to enable or disable the internal flip-flop of 8155. If this internal flip-flop is enabled, then it generates an interrupt signal during handshake mode of IO.
D$_6$ and D$_7$	Used for timer control.

The timer section of the 8155 has two 8-bit registers as shown in Fig. 9.10 and in this register a 14-bit count value and a two bit mode code should be loaded. The timer requires an input clock signal and for every clock pulse, the timer decrements the count value by one. An appropriate control word starts the counter which decrements the count at each clock pulse. The timer output will vary according to the mode specified. The timer can be stopped either during counting or at the end of the count. In addition, the actual count at a given moment can be obtained by reading the status register.

The timer output normally remains **high** (while the timer is not running). When the timer starts running the output changes according to the mode of operation. The timer output for different operating modes are shown in Fig. 9.11.

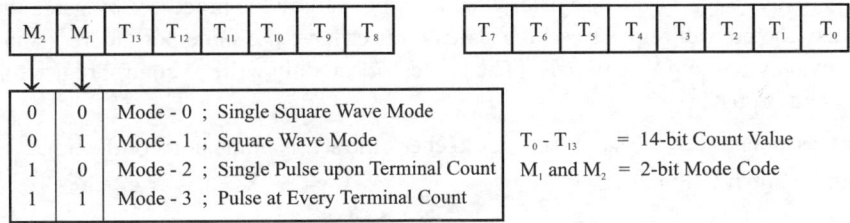

M_2	M_1		
0	0	Mode - 0 ;	Single Square Wave Mode
0	1	Mode - 1 ;	Square Wave Mode
1	0	Mode - 2 ;	Single Pulse upon Terminal Count
1	1	Mode - 3 ;	Pulse at Every Terminal Count

$T_0 - T_{13}$ = 14-bit Count Value
M_1 and M_2 = 2-bit Mode Code

Fig. 9.10: Timer count value format.

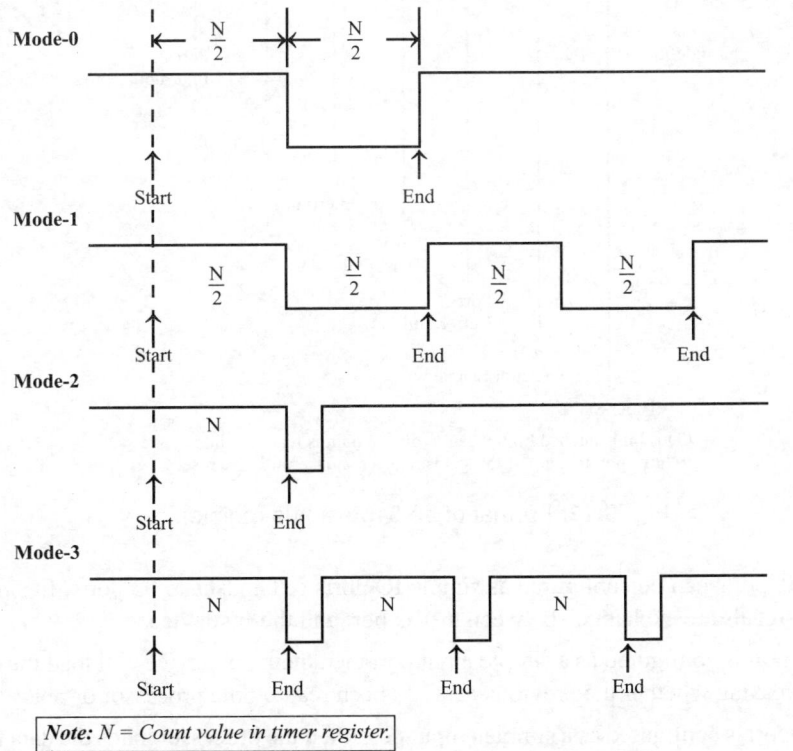

Note: N = Count value in timer register.

Fig. 9.11: Output waveforms of the timer in 8155.

Mode-0 : In this mode the timer output remains **high** for half the count and goes **low** for the remaining count, thus providing a single square wave.

Mode-1 : In this mode, the timer output is a square wave. The initial timer count is automatically reloaded at the end of each count. The timer output remains **high** for half the count and remains **low** for the other count.

Mode-2 : In this mode a single clock pulse is provided at the end of the count. The width of the pulse is equal to the time period of the input clock pulse.

Mode-3 : This is similar to mode-2, except that the initial count is reloaded to provide a continuous waveform.

The INTEL 8155 has a status register which can be read by the processor by using the control register address. The control register and the status register have the same port address and they are differentiated by only $\overline{RD}$ and $\overline{WR}$ signals. (The processor can only write to control register and it can only read the status register.)

The processor can read the status register to check the status of the ports or the timer. The status register format is shown in Fig. 9.12.

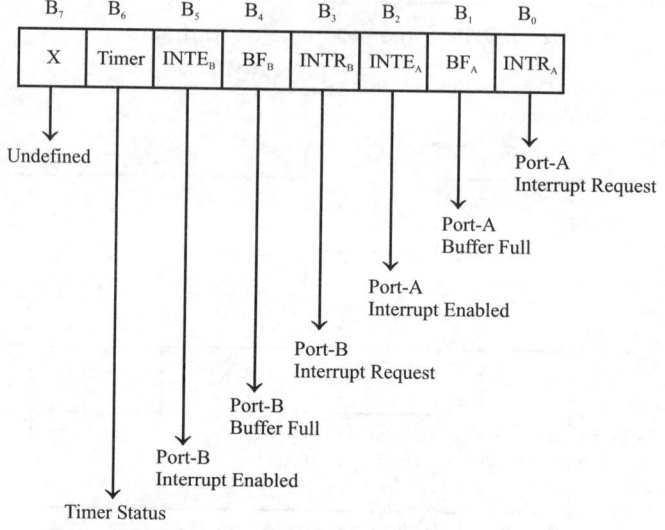

(This bit is latched **high** when terminal count is reached and is reset to **low** upon reading of the status register and by hardware reset.)

Fig. 9.12: Format of an 8155 status register.

The 8155 ports can be configured as simple IO ports or handshake IO ports. In simple IO port no handshake signals are exchanged between the IO port and the IO device.

When a port is configured as a simple input port then the input device will load the data into the port without checking whether the previous data has been read by the processor or not.

When a port is configured as a simple output port, then the processor loads the data to the output port without checking whether the previous data is accepted by the output device or not.

When a port is programmed as a handshake input port then handshake signals are used to transfer the data from input device to the port. When the port receives a data, it interrupts the processor for executing a subroutine for reading the data from the port and storing in appropriate place.

Alternatively, the processor can check the status register of 8155 to know any data is available on the port, if a data is available on the port then the processor executes a subroutine to read the data and store it in appropriate place.

When a port is programmed as a handshake output port then the handshake signals are used to transfer the data from the port to output device. The processor first loads a data to the port. When the data is accepted by the output device the port will send an interrupt signal to the processor.

Now the processor can load the next data to the port. Alternatively the processor can check the status register of 8155 before loading the next data.

8155 Handshake Input Port

The signals used for data transfer between input device and microprocessor using port-A of 8155 as handshake input port are shown in Fig. 9.13.

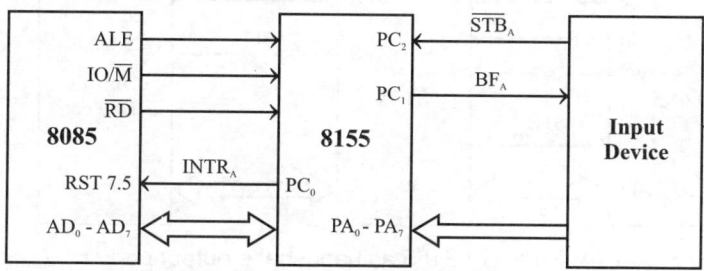

Fig. 9.13: Port-A of 8155 as handshake input port.

1. The input device checks the Buffer Full (BF_A) signal. If BF_A is low then it places the data on port lines and asserts strobe ($\overline{STB}_A$) low, to inform the port.

2. At the falling edge of $\overline{STB}_A$, the BF signal is asserted high to inform the input device that the port is full and it has to wait.

3. The input device asserts $\overline{STB}_A$ signal as high after a predefined time and when it is asserted high, the 8155 generates an interrupt signal to the processor ($INTR_A$).

4. On receiving an interrupt request the processor executes a subroutine to read the data from the port.

5. The data is read by the processor using $\overline{RD}$ signal. At the rising edge of RD the BF_A and $INTR_A$ signal are asserted low, and now the input device can send the next data to the port.

> **Note:** *Instead of interrupting the processor, the system can be designed to have data transfer by polling technique. In this method the processor polls the status register at regular intervals.*

8155 Handshake Output Port

The signals used for data transfer between the output device and microprocessor using port-A of 8155 as handshake output port are shown in Fig. 9.14.

1. When the port is empty the processor writes a byte into the port.

2. For writing a data to the port, the processor asserts $\overline{WR}$ signal as low and then high. At the falling edge of $\overline{WR}$ the $INTR_A$ is reset (asserted low) and at the rising edge of $\overline{WR}$ the BF_A is asserted high.

3. The BF_A signal informs the output device that the data is ready for it. If the output device accepts the data byte, then it asserts $\overline{STB}_A$ low and then high.

4. When strobe is low, the BF_A is reset to low and at the rising edge of the strobe the $INTR_A$ goes high to interrupt the processor.

5. When the processor is interrupted, it executes an interrupt service routine to load the next data in the output port.

> *Note:* *The data transfer between the port and the processor can also be achieved by status check technique instead of using interrupt.*

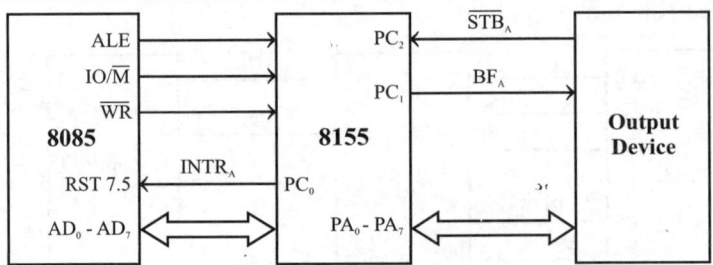

Fig. 9.14: Port-A of 8155 as handshake output port.

9.2.4 Programmable IO Port and Memory - INTEL 8355

The INTEL 8355 is a ROM and IO port chip that can be used in the 8085A and 8088 microprocessor systems. The ROM portion has 2048 (2 k) locations with a word size of 8 bits. It has a maximum access time of 400 ns so that the device can be used without wait states in the 8085A CPU. The 8355-2 has 300 ns access time for compatibility with the 8085A-2 and full speed 5 MHz 8088 microprocessors. The internal block diagram and the pin description of 8355 are shown in Fig. 9.15.

The 8355 is a 40-pin IC available in DIP. It has multiplexed address and data lines. It has an internal address latch to demultiplex the address and data lines using the signal ALE. The IO portion consists of two general purpose IO ports. Each IO port has eight port lines and each IO port line is individually programmable as input or output.

The ports are programmed as input or output port by loading an 8-bit word in the **Data Direction Register (DDR)** of the concerned port. Each line of the port can be individually used as input or output line. A **zero** loaded in the DDR register makes the corresponding line of port as input line. A **one** loaded in the DDR register makes the corresponding line of port as output line.

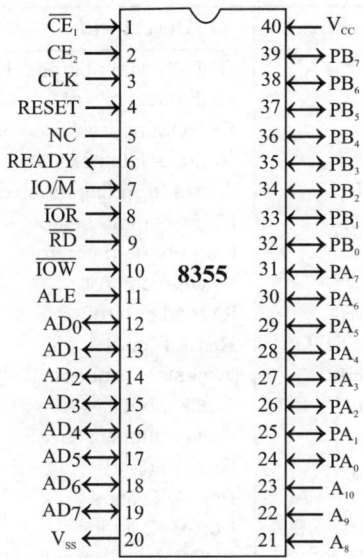

Fig. a: Pin description.

Fig. 9.15: Pin description and internal block diagram of 8355.

Fig. b: Internal block diagram of 8355.

Table 9.3

Internal address		Device selected
A_1	A_0	
0	0	Port-A
0	1	Port-B
1	0	DDR A
1	1	DDR B

Pin	Description
$AD_0 - AD_7$	Bidirectional address/data
$A_8 - A_{10}$	High order bits of ROM address
ALE	Address latch enable
$\overline{CE}_1$	Active **low** chip enable
CE_2	Active **high** chip enable
$IO/\overline{M}$	IO or memory select
$\overline{RD}$	Memory read control
$\overline{IOW}$	IO write control
$\overline{IOR}$	IO read control
RESET	RESET input
READY	Wait state request
CLK	Clock input
V_{CC}	Power supply (+5-V)
V_{SS}	Ground (0-V)
$PA_0 - PA_7$	Port-A IO lines
$PB_0 - PB_7$	Port-B IO lines

The ports and data direction registers are selected by two bit internal address, A_0 and A_1 as shown in Table 9.3. When the device is reset by applying a **high** signal at RESET pin, the DDRs are cleared and the ports are initialized to input mode.

9.2.5 Programmable IO Port and Memory - INTEL 8755

The INTEL 8755 is an EPROM and IO chip that can be used in the 8085A and 8088 microprocessor systems. The EPROM portion has 2048 (2 k) locations with a word size of 8 bits. It has a maximum access time of 450 ns so that the device can be used without wait states in an 8085A CPU. The pin description and internal block diagram of 8755 are shown in Fig. 9.16. The 8255 is a 40-pin IC available in DIP. It has multiplexed address and data lines.

It has an internal latch to demultiplex the address and data lines using the signal ALE. The IO portion consists of two general purpose IO ports. Each IO port has eight port lines and each IO port line is individually programmable as input or output.

The ports are programmed as input or output port by loading an 8-bit word in the **D**ata **D**irection **R**egister (DDR) of the concerned port. Each line of the port can be individually used as input or output lines. A **zero** loaded in the DDR register makes the corresponding line of the port as input line. A **one** loaded in the DDR register makes the corresponding line of the port as output line.

Pin	Description
AD_0 - AD_7	Bidirectional address/data
A_8 - A_{10}	High order bits of EPROM address
ALE	Address latch enable
CE_2	Active **high** chip enable
IO/$\overline{M}$	IO or memory select
$\overline{RD}$	Memory read control
$\overline{IOW}$	IO write control
$\overline{IOR}$	IO read control
RESET	RESET input
READY	Wait state request
CLK	Clock input
V_{CC}	Power supply (+5-V)
V_{SS}	Ground (0-V)
PA_0 - PA_7	Port-A IO lines
PB_0 - PB_7	Port-B IO lines
PROG/$\overline{CE}_1$	EPROM programming / Active **low** chip enable

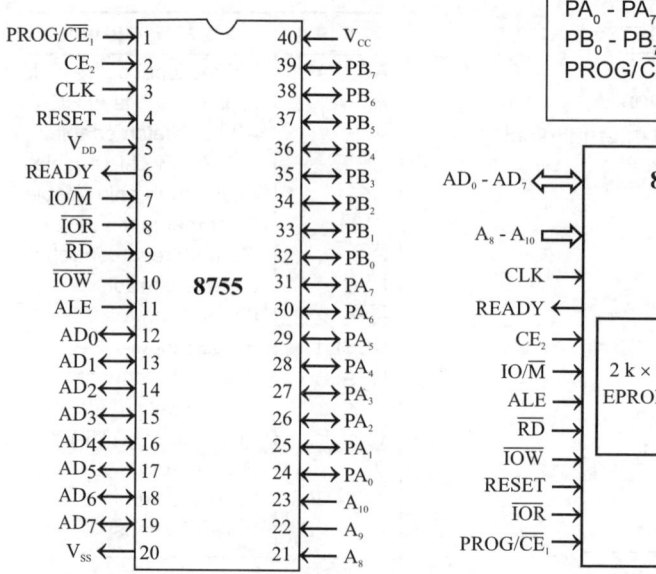

Fig. a: Pin description.　　　　**Fig. b:** Internal block diagram of 8755.

Fig. 9.16: Pin description and block diagram of 8755.

The ports and data direction registers are selected by two bit internal address, A_0 and A_1 as shown in Table 9.4. When the device is reset by applying a **high** signal at RESET pin, the DDRs are cleared and the ports are initialized in the input mode. The pin PROG/CE_1 is used to program the EPROM.

Table 9.4:

Internal address		Device selected
A_1	A_0	
0	0	Port-A
0	1	Port-B
1	0	DDR A
1	1	DDR B

9.3 PROGRAMMABLE PERIPHERAL INTERFACE - INTEL 8255

The INTEL 8255 is a device used to implement parallel data transfer between processor and slow peripheral devices like ADC, DAC, keyboard, 7-segment display, LCD, etc.

The 8255 has three ports: Port-A, Port-B and Port-C. The ports A and B are 8-bit parallel ports. Port-A can be programmed to work in any one of the three operating modes as input or output port. The three operating modes are:

Mode - 0 → Simple IO port

Mode - 1 → Handshake IO port

Mode - 2 → Bidirectional IO port

The port-B can be programmed to work either in mode-0 or mode-1 as input or output port. The port-C pins (8 pins) have different assignments depending on the mode of ports A and B. If ports A and B are programmed in mode-0, then the port-C can perform any one of the following function:

1. **As 8-bit parallel port in mode-0 for input or output.**

2. **As two numbers of 4-bit parallel port in mode-0 for input or output.**

3. **The individual pins of port-C can be set or reset for various control applications.**

If port-A is programmed in mode-1/mode-2 and port-2 is programmed in mode-1 then some of the pins of port-C are used for handshake signals and the remaining pins can be used as input/output lines or individually set/reset for control applications.

<u>IO Modes of 8255</u>

<u>Mode-0 :</u>   In this mode, all the three ports can be programmed either as the input or the output port. In mode-0, the outputs are latched and the inputs are not latched. The ports do not have handshake or interrupt capability.

The ports in mode-0 can be used to interface DIP switches, hexa-keypad, LEDs and 7-segment LEDs to the processor.

<u>Mode-1 :</u>   In this mode, only ports A and B can be programmed either as the input or output port. In mode-1, handshake signals are exchanged between the processor and peripherals prior to data transfer. The port-C pins are used for handshake signals. Input and output data are latched. Interrupt driven data transfer scheme is possible.

<u>Mode-2 :</u>   In this mode, the port will be a bidirectional port. (i.e., the processor can perform both read and write operations with an IO device connected to a port in mode-2.) Only port-A can be programmed to work in mode-2.

Five pins of port-C are used for handshake signals. This mode is used primarily in applications such as data transfer between two computers or floppy disk controller interface.

9.3.1 Pins, Signals and Internal Block Diagram of 8255

The pin description of 8255 is shown in Fig. 9.17. It has 40 pins and requires a single +5V supply. The internal block diagram of 8255 is shown in Fig. 9.18.

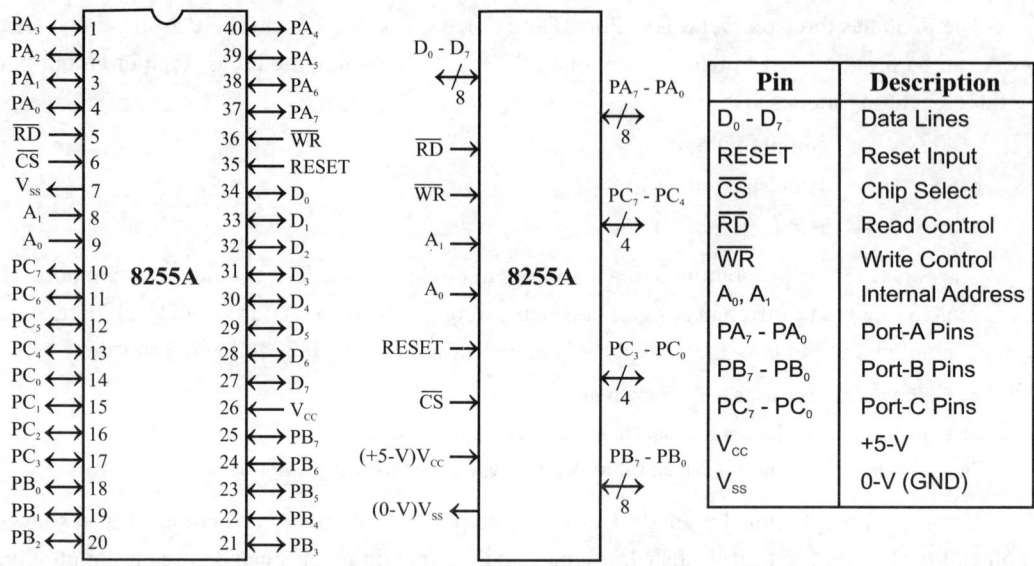

Fig. 9.17: Pin description of 8255.

The ports are grouped as Group A and Group B. The group A has port-A, port-C upper and its control circuit. The group B comprises port-B, port-C lower and its control circuit. The read/write control logic requires six control signals. These signals are:

$\overline{RD}$ (Read)   :   This control signal enables the read operation. When this signal is **low**, the microprocessor reads data from a selected IO port of the 8255A.

$\overline{WR}$ (Write)   :   This control signal enables the write operation. When this signal goes **low**, the microprocessor writes into a selected IO port or the control register.

$\overline{RESET}$:   This is an active **high** signal. It clears the control register and sets all ports in the input mode.

$\overline{CS}$, A_0 and A_1 :   These are device select signals. The address lines A_0 and A_1 of 8255 can be connected to any two address lines of the processor to provide internal addresses. The A_0 and A_1 selects any one of the 4 internal devices as shown in Table 9.5. The 8255 will remain in **high impedance** state if the signal input to $\overline{CS}$ is **high** and the device can be brought to normal logic by making the signal input to $\overline{CS}$ as logic **low**.

Table 9.5:

Internal address		Device selected
A_1	A_0	
0	0	Port-A
0	1	Port-B
1	0	Port-C
1	1	Control Register

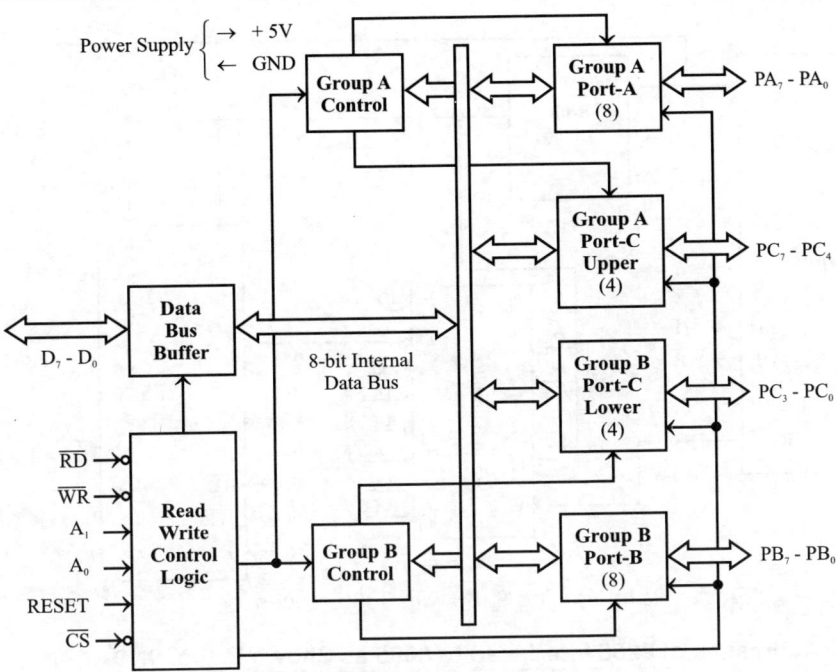

Fig. 9.18: Internal block diagram of 8255.

9.3.2 Interfacing of 8255 with 8085 Processor

A simple schematic for interfacing the 8255 with an 8085 processor is shown in Fig. 9.19. The 8255 can be either memory-mapped or IO-mapped in the system. In the schematic shown in Fig. 9.19, the 8255 is IO-mapped in the system.

The chip select signals for IO-mapped devices are generated by using a 3-to-8 decoder. The address lines A_4, A_5 and A_6 are decoded to generate eight chip select signals (IOCS-0 to IOCS-7) and in this, the chip select IOCS-1 is used to select 8255. The address line A_7 and the control signal $IO/\overline{M}$ are used as enable for the decoder.

The address line A_0 of 8085 is connected to A_0 of 8255 and A_1 of 8085 is connected to A_1 of 8255 to provide the internal addresses. The IO addresses allotted to the internal devices of 8255 are listed in Table 9.6. The data lines D_0 - D_7 are connected to D_0 - D_7 of the processor to achieve parallel data transfer.

In the schematic shown in Fig. 9.19, the interrupt scheme is not included and so the data transfer can be performed only by checking the status of 8255 and not by interrupt method.

For interrupt driven data transfer scheme, the interrupt controller 8259 has to be interfaced to the system and the interrupts of port-A (PC_3) and port-B (PC_0) should be connected to two IR inputs of 8259.

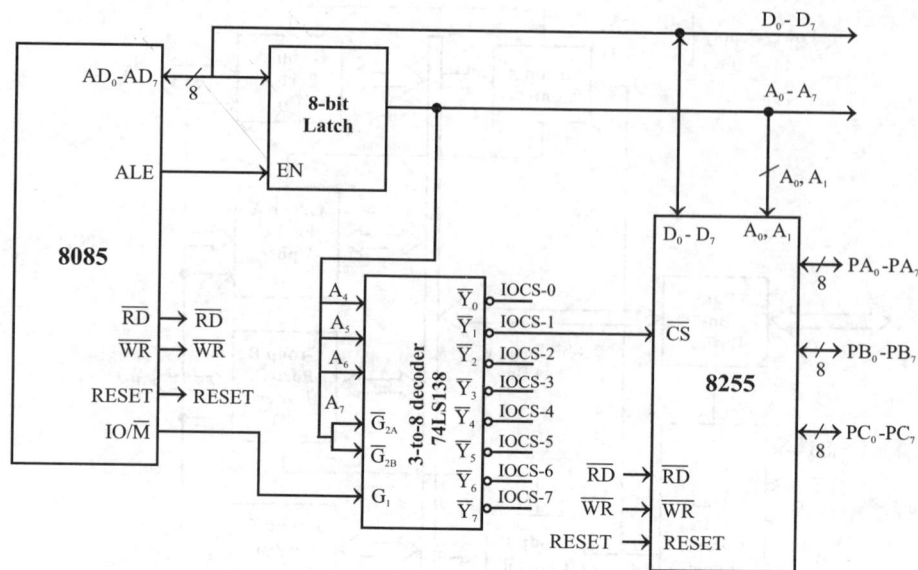

Fig. 9.19: Interfacing 8255 with 8085 processor.

Table 9.6: IO Addresses of 8255 Interfaced to 8085 as Shown in Fig. 9.19

Internal device	Binary address								Hexa address
	Decoder input and enable				Input to address pins of 8255				
	A_7	A_6	A_5	A_4	A_3	A_2	A_1	A_0	
Port-A	0	0	0	1	x	x	0	0	10
Port-B	0	0	0	1	x	x	0	1	11
Port-C	0	0	0	1	x	x	1	0	12
Control Register	0	0	0	1	x	x	1	1	13

Note: Don't care "x" is considered as zero.

9.3.3 Interfacing of 8255 with 8086 Processor

A simple schematic for interfacing the 8255 with the 8086 processor is shown in Fig. 9.20. The 8255 can be either memory-mapped or IO-mapped in the system. In the schematic shown in Fig. 9.20, the 8255 is IO-mapped in the system with even addresses. The chip select signals for IO-mapped devices are generated by using a 3-to-8 decoder.

The address lines A_5, A_6 and A_7 are decoded to generate eight chip select signals (IOCS-0 to IOCS-7) and in this, the chip select IOCS-1 is used to select 8255. The address line A_0 and the control signal $M/\overline{IO}$ are used as enable for decoder. The address line A_1 of 8086 is connected to A_0 of 8255 and A_2 of 8086 is connected to A_1 of 8255 to provide the internal addresses. The IO addresses allotted to the internal devices of 8255 are listed in Table 9.7. The data lines D_0 - D_7 are connected to D_0 - D_7 of the processor to achieve parallel data transfer.

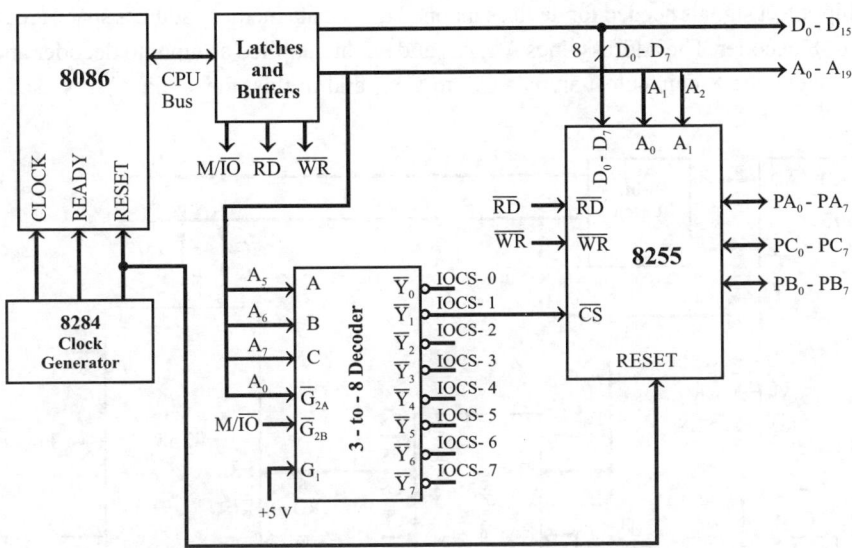

Fig. 9.20: Interfacing of 8255 with the 8086 processor.

Table 9.7: IO Addresses of 8255 Interfaced to 8086 as shown in Fig. 9.20

Internal device	Binary address								Hexa address
	Decoder input			Input to address pins of 8255				Decoder enable	
	A_7	A_6	A_5	A_4	A_3	A_2	A_1	A_0	
Port-A	0	0	1	x	x	0	0	0	20
Port-B	0	0	1	x	x	0	1	0	22
Port-C	0	0	1	x	x	1	0	0	24
Control Register	0	0	1	x	x	1	1	0	26

Note: *Don't care "x" is considered as zero.*

In the schematic shown in Fig. 9.20, the interrupt scheme is not included and so the data transfer can be performed only by checking the status of 8255 and not by interrupt method. For interrupt-driven data transfer scheme, the interrupt controller 8259 has to be interfaced to system and the interrupts of port-A (PC_3) and port-B (PC_0) should be connected to two IR inputs of 8259.

9.3.4 Interfacing of 8255 with 8051 Microcontroller

The INTEL 8255 can be interfaced to an 8051 microcontroller for additional port requirement. A simple schematic for interfacing the 8255 with 8051 microcontroller is shown in Fig. 9.21. The 8255 can be mapped only as a memory device in an 8051 microcontroller. Moreover, both the read and write operation is possible only if the devices are mapped in the data memory address space. (If the devices are mapped in the program memory address space, then only read operation is possible.)

The chip select signals needed for devices mapped in the data memory address space are generated by using a 3-to-8 decoder. The address lines A_{13}, A_{14} and A_{15} are applied as input to decoder and so they are decoded to generate 8-chip select signals $\overline{CS}_0$ to $\overline{CS}_7$, and in this, the signal $\overline{CS}_1$ is used to select the 8255.

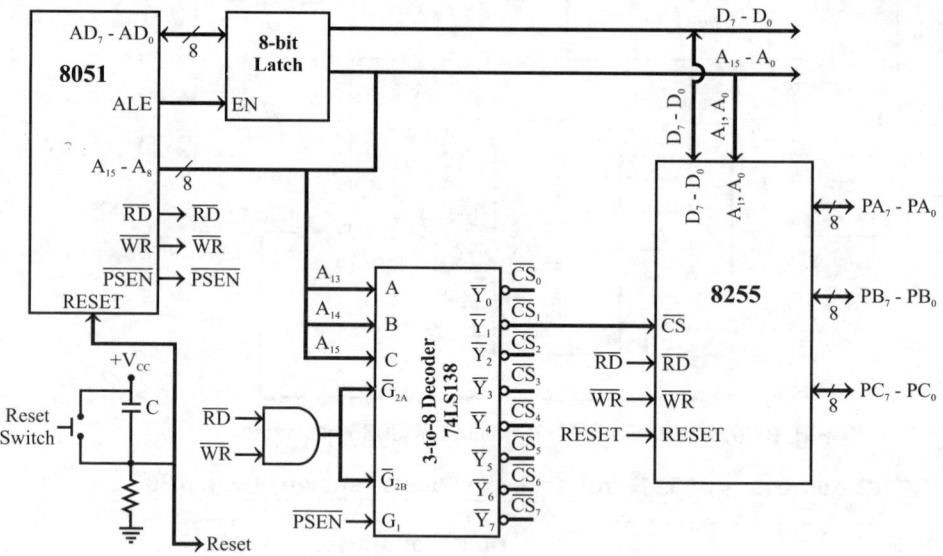

Fig. 9.21: Interfacing of 8255 with 8051 Microcontroller.

Table 9.8: Address Allotted to 8255 Interfaced to 8051 as Shown in Fig. 9.21

Internal device	Binary address																	Hexa address	Comment
	Decoder input			Input to address pins of 8255															
	A_{15}	A_{14}	A_{13}	A_{12}	A_{11}	A_{10}	A_9	A_8	A_7	A_6	A_5	A_4	A_3	A_2	A_1	A_0		
Port-A	0	0	1	x	x	x	x	x	x	x	x	x	x	x	0	0	2000	External
Port-B	0	0	1	x	x	x	x	x	x	x	x	x	x	x	0	1	2001	data
Port-C	0	0	1	x	x	x	x	x	x	x	x	x	x	x	1	0	2002	memory
Control Register	0	0	1	x	x	x	x	x	x	x	x	x	x	x	1	1	2003	address space

Note: *Don't care "x" is considered as zero.*

The control signals $\overline{RD}$ and $\overline{WR}$ are logically ANDed and applied as logic **low** enable for the decoder. The $\overline{PSEN}$ is connected to logic **high** enable of the decoder. $\overline{PSEN}$ will be asserted **low** while accessing program memory and so, the data memory decoder will be disabled. When program memory is not accessed, $\overline{PSEN}$ will be **high** and so the data memory decoder will be enabled.

The address line A_0 of 8031 is connected to A_0 of 8255 and the address line A_1 of 8031 is connected to A_1 of 8255 to provide internal addresses. The addresses allotted to the internal devices of 8255 are listed in Table 9.8. The data lines D_7 - D_0 are connected to D_7 - D_0 of the processor to achieve parallel data transfer.

9.3.5 Programming (or Initializing) 8255

The 8255 has two control words: IO **M**ode **S**et control **W**ord (MSW) and **B**it **S**et/**R**eset (BSR) control word. The MSW is used to specify IO functions and BSR word is used to set/reset individual pins of port-C. Both the control words are written in the same control register.

The control register differentiates them by the value of bit B_7. The BSR control word does not affect the functions of ports A and B.

Bit B_7 of the control register specifies either the IO function or the bit set/reset function. If $B_7 = 1$, then the bits B_6- B_0 determine the IO functions in various modes. If bit $B_7 = 0$, then the bits B_6- B_0 determine the pin of port-C to be set or reset.

The 8255 ports are programmed (or initialized) by writing a control word in the control register. For setting IO functions and mode of operation, the IO mode set control word is sent to the control register. For setting/resetting a pin of port-C, the bit set/reset control word is sent to the control register.

The format of the IO mode set control word is shown in Fig. 9.22 and the format of bit set/reset control word is shown in Fig. 9.23. The various functions (assignments) of port-C pins during the different operating modes of ports A and B are listed in Table 9.9.

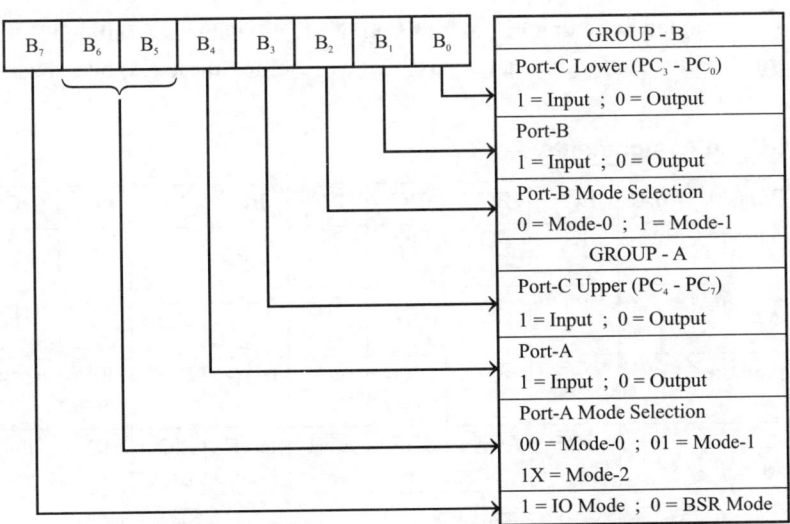

Fig. 9.22: Format of IO mode set control word of 8255.

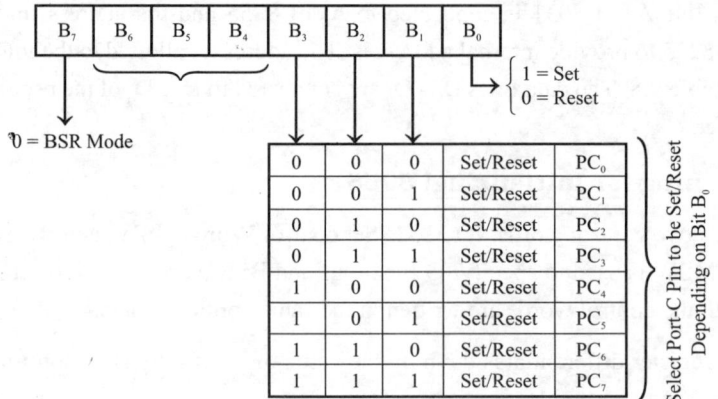

Fig. 9.23: Format of Bit Set/Reset control word of 8255.

In handshake mode (i.e., in mode-1 and mode-2), the data transfer between the processor and the port can be implemented either by the interrupt method or by checking the status of the 8255 ports. In an interrupt-driven data transfer scheme, when the port is ready it interrupts the 8086 processor through the NMI/INTR pin for a read/write operation. In status check technique, the 8086 processor can check the status of ports A and B by reading port-C. When the port is ready for data transfer, the processor executes a read/write cycle.

The 8255 has two internal flip-flops as interrupt enables ($INTE_A$ and $INTE_B$) for port-A and port-B interrupt signals. In interrupt driven data transfer scheme, the 8255 generates an interrupt signal, only if these flip-flops are enabled by using BSR control word.

The $INTE_A$ is enabled by setting PC_4 to **high** and $INTE_B$ is enabled by setting PC_2 to **high** using BSR control word. The interrupt signal can be disabled by resetting these two bits to zero using BSR control word.

Table 9.9: Port-C Pin Assignments

Functions of Ports A and B	PC_7	PC_6	PC_5	PC_4	PC_3	PC_2	PC_1	PC_0
Ports A and B in mode-0 Input/Output	IO	IO	IO	IO	IO	IO	IO	IO
Ports A and B in mode-1 Input ports	IO	IO	IBF_A	$\overline{STB}_A$	$INTR_A$	$\overline{STB}_B$	IBF_B	$INTR_B$
Ports A and B in mode-1 Output ports	$\overline{OBF}_A$	$\overline{ACK}_A$	IO	IO	$INTR_A$	$\overline{ACK}_B$	$\overline{OBF}_B$	$INTR_B$
Port-A in mode-2 Port-B in mode-0	$\overline{OBF}_B$	$\overline{ACK}_A$	IBF_A	$\overline{STB}_A$	$INTR_A$	IO	IO	IO

IO	-	Input /Output line	$\overline{OBF}$	- Output Buffer Full
$\overline{STB}$	-	Strobe	$\overline{ACK}$	- Acknowledge
IBF	-	Input Buffer Full	The subscript A denotes port-A signal.	
INTR	-	Interrupt Request	The subscript B denotes port-B signal.	

When port-A and port-B are programmed in handshake mode (i.e., in mode-1 and mode-2), the port-C can be read to know the readiness of the ports for data transfer. The format of the status word read from port-C is shown in Fig. 9.24.

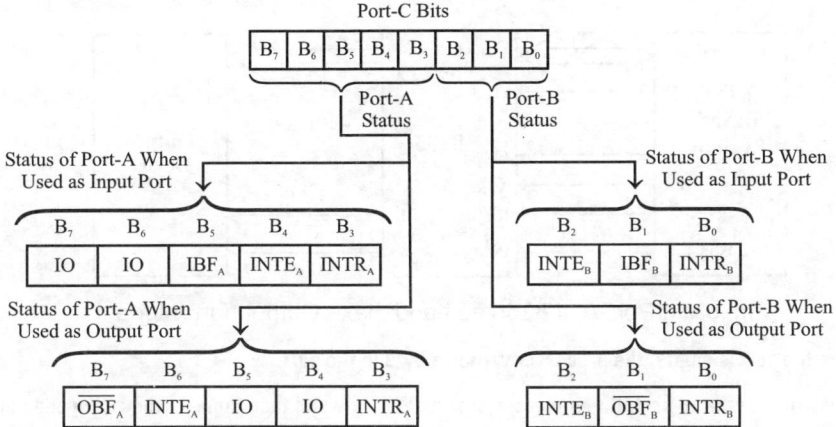

Fig. 9.24: Format of status word of 8255 for handshake input and output operation.

8255 Handshake Input Port (Mode-1)

The signals used for data transfer between input device and 8085 microprocessor using port-A of 8255 as handshake input port (Mode-1) are shown in Fig. 9.25.

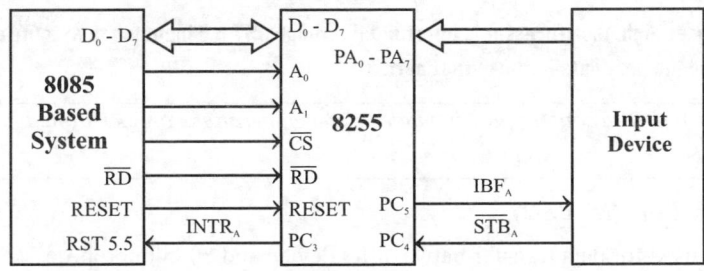

Fig. 9.25: Port-A of 8255 as handshake input port (Mode-1).

1. The input device checks the IBF_A signal. If it is low, then the input device places the data on the port lines PA_0-PA_7 and asserts $\overline{STB}_A$ low and after a delay time $\overline{STB}_A$ is asserted high.

2. When $\overline{STB}_A$ is low, the 8255 asserts IBF signal high and at the rising edge of $\overline{STB}_A$ the data is latched to the port and $INTR_A$ is set high.

3. When $INTR_A$ goes high, the processor is interrupted through the RST 5.5 input pin to execute a subroutine for reading the data from the port. For a read operation, the processor asserts $\overline{RD}$ low and then high.

4. When $\overline{RD}$ is low, the $INTR_A$ is reset (asserted low) by 8255 and at the rising edge of $\overline{RD}$, the IBF is asserted low and the input device can send the next data.

> **Note:** For port-B as input port in mode-1, same operations are performed, but for handshake signals PC_0, PC_1 and PC_2 are used.

8255 Handshake Output Port (Mode-1)

The signals used for data transfer between output device and an 8085 microprocessor using port-A of 8255 as the handshake output port (Mode-1) are shown in Fig. 9.26.

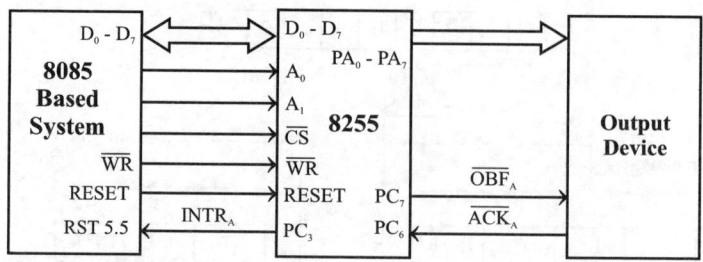

Fig. 9.26: Port-A of 8255 as handshake Output port (Mode-1).

1. When the port is empty, the processor writes a byte in the port.

2. For writing a data to port, the processor asserts $\overline{WR}$ low and then high. At the rising edge of $\overline{WR}$, both the $INTR_A$ and $\overline{OBF}_A$ are asserted low by the 8255.

3. The $\overline{OBF}_A$ signal informs the output device that the data is ready. If the output device accepts the data then it sends an acknowledgement signal by asserting $\overline{ACK}_A$ low and then high.

4. When $\overline{ACK}_A$ is low, the $\overline{OBF}_A$ is asserted high by the 8255. When $\overline{ACK}_A$ is high, the $INTR_A$ is set (asserted high), to interrupt the processor.

5. When $INTR_A$ goes high, the processor is interrupted through RST 5.5 input pin to execute an interrupt service routine to load the next data in the output port.

> *Note:* For port-B as output port in mode-1, same operations are performed, but for handshake signals PC_0, PC_1 and PC_2 are used.

8255 Bidirectional Port (Mode-2)

The signals used for data transfer between IO device and 8085 microprocessor using port-A of 8255 as a bidirectional port (Mode-2) are shown in Fig. 9.27.

> *Note: Only port-A can work in mode-2.*

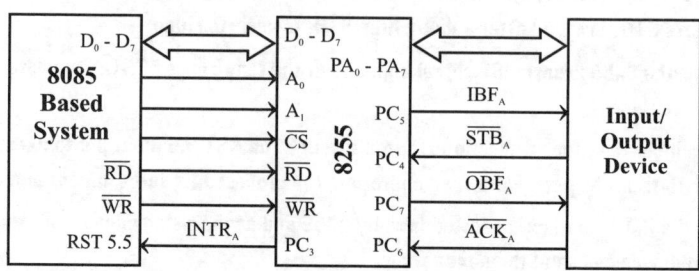

Fig. 9.27: Port-A of 8255 as bidirectional port (Mode-2).

In mode-2, the port can be used either as an input port or as an output port. At any one time, the processor will perform either read or write operation. In mode-2, the read operation can be followed by write, or write operation can be followed by read.

The signals involved and the operations performed for read operation are similar to mode-1 input port. The signals involved and the operations performed for write operation are similar to mode-1 output port.

9.4 DMA DATA TRANSFER SCHEME

In **D**irect **M**emory **A**ccess (DMA) data transfer, the processor is forced to **HOLD** state (**high impedance** state) by an IO device until the data transfer between the device and the memory is complete. The processor does not execute any instructions during the **HOLD** period.

The DMA data transfer is used for a large block of data transfer between the IO device and memory. Typical examples of devices using DMA are CRT controller, floppy disk, hard disk, high-speed line printer, etc.

The different types of DMA data transfer schemes are

1. **Cycle stealing DMA or Single transfer mode DMA.**

2. **Block or Burst mode DMA.**

3. **Demand transfer mode DMA.**

Normally the data transfer from memory to IO device or IO device to memory can be acheived only through the microprocessor. When data has to be transferred from memory to IO device, first the processor sends address and control signals to memory to read the data from memory. Then the processor sends address and control signals to IO device to write data to IO device.

Similarly, when the data has to be transferred from IO device to memory, first the processor sends address and control signals to IO device to read data from IO device. Then the processor sends address and control signals to memory device to write data to memory.

In the data transfer method described above, the data cannot be directly transferred between memory and IO device, even though they are connected to a common bus. The above process is inevitable because the processor cannot simultaneously select two devices.

Hence a scheme called **D**irect **M**emory **A**ccess (DMA) has been developed in which the IO device can access the memory directly for data transfer. The DMA data transfer will be useful to transfer large amounts of data between the memory and IO device in a short time.

For direct data transfer between IO device and memory a dedicated hardware device called **D**irect **M**emory **A**ccess controller (DMA controller) is used.

A DMA controller temporarily borrow the address bus, data bus and control bus from the microprocessor, and transfers the data bytes directly from the IO ports to a series of memory locations or vice versa. Some DMA controllers can also perform memory-to-memory transfer.

9.4.1 A Microcomputer System with a DMA Controller

The simplified diagram of a microcomputer system with a DMA controller is shown in Fig. 9.28. In the system shown in Fig. 9.28, the DMA controller has one channel, which serves for one IO device. In an actual DMA controller we may have more than one channel and each channel may service an IO device independently. Each channel contains an address register, a control register and a count register. For simplicity let us consider a one-channel DMA controller.

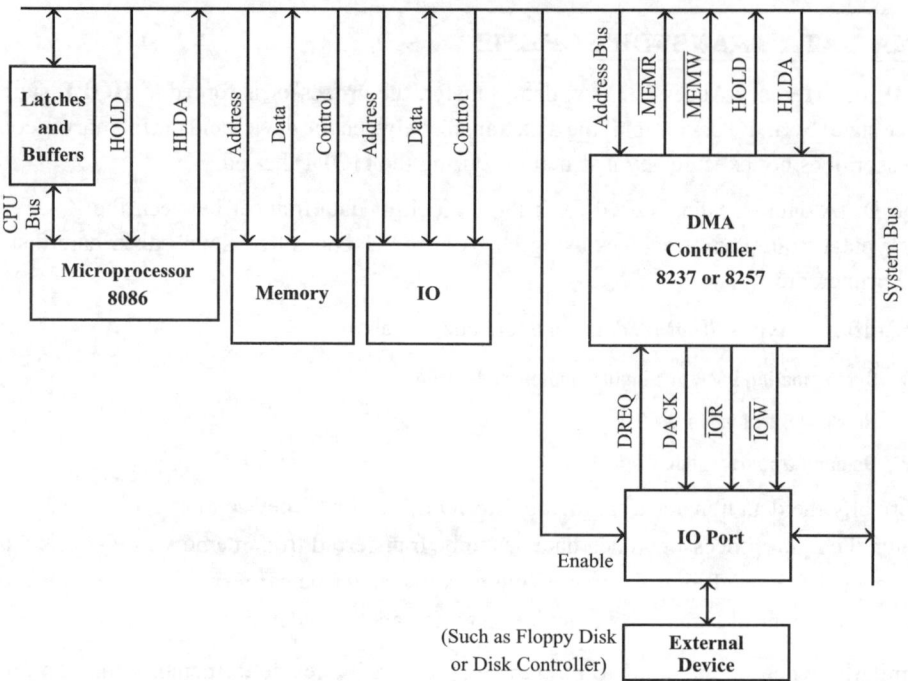

Fig. 9.28: Block diagram of a microcomputer system with DMA controller.

The DMA controller can work as a slave or as a master. In the slave mode, the microprocessor loads the address register with the starting address of the memory, loads the count register with the number of bytes to be transferred and loads the control register with the control information.

For performing DMA operation the processor has to initialize or program the IO device and DMA controller. Consider an example of transferring a bulk data from floppy to memory by DMA. In this case the processor initializes both DMA controller and floppy controller, so that DMA controller is informed about the address, type of DMA and number of bytes to be transferred, and the floppy controller is informed to go for a DMA.

When the IO device needs a DMA transfer it sends a **DMA req**uest signal (DREQ) to the DMA controller. When the DMA controller receives a DMA request, it sends a HOLD request to the processor. At the end of the current instruction execution, the processor relieves the bus by asserting all its data, address and control pins to **high impedance** state. Then the processor sends an acknowledge (HLDA) signal to the DMA controller.

When the controller receives an acknowledge signal it takes control of the system bus and begins to work as a master. The DMA controller sends a **DMA ack**nowledge signal (DACK) to IO device. The DACK signal will inform the device to get ready for DMA transfer.

For a read operation, the DMA controller outputs the memory address on the address bus and asserts $\overline{MEMR}$ and $\overline{IOW}$ signals. The DMA read refers to reading memory locations. Hence, for a read operation, the memory outputs the data on the data bus and this data will be written into IO port.

For a write operation, the DMA controller outputs the memory address on the address bus and asserts $\overline{MEMW}$ and $\overline{IOR}$ signals. The DMA write refers to writing data to the memory. Hence, for a write operation, the IO device outputs the data on the data bus and this data will be written into memory. When the data transfer is complete the DMA controller unasserts its HOLD request signal to the processor and the processor takes control of the system bus.

The DMA transfer may be performed to transfer a byte at a time or in blocks. In cycle stealing DMA or single transfer mode, the DMA controller will perform one-byte transfer in between instruction cycles. In burst mode or block transfer mode, the DMA controller will transfer a block of data.

9.5 DMA CONTROLLER - INTEL 8237

The DMA controller-8237 has been developed for 8085/8086/8088 microprocessor-based system. It is a device dedicated to perform high speed data transfer between memory and IO device. The 8237 has four channels and so it can be used to provide DMA to four IO devices. When more than four devices require DMA, then a number of 8237 can be connected in cascade to increase the DMA channels.

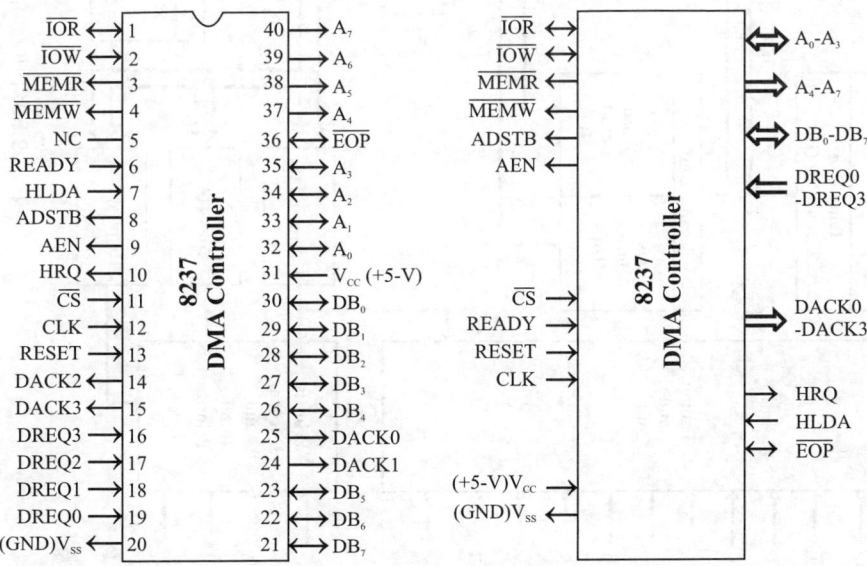

Fig. 9.29: Pin configuration of 8237.

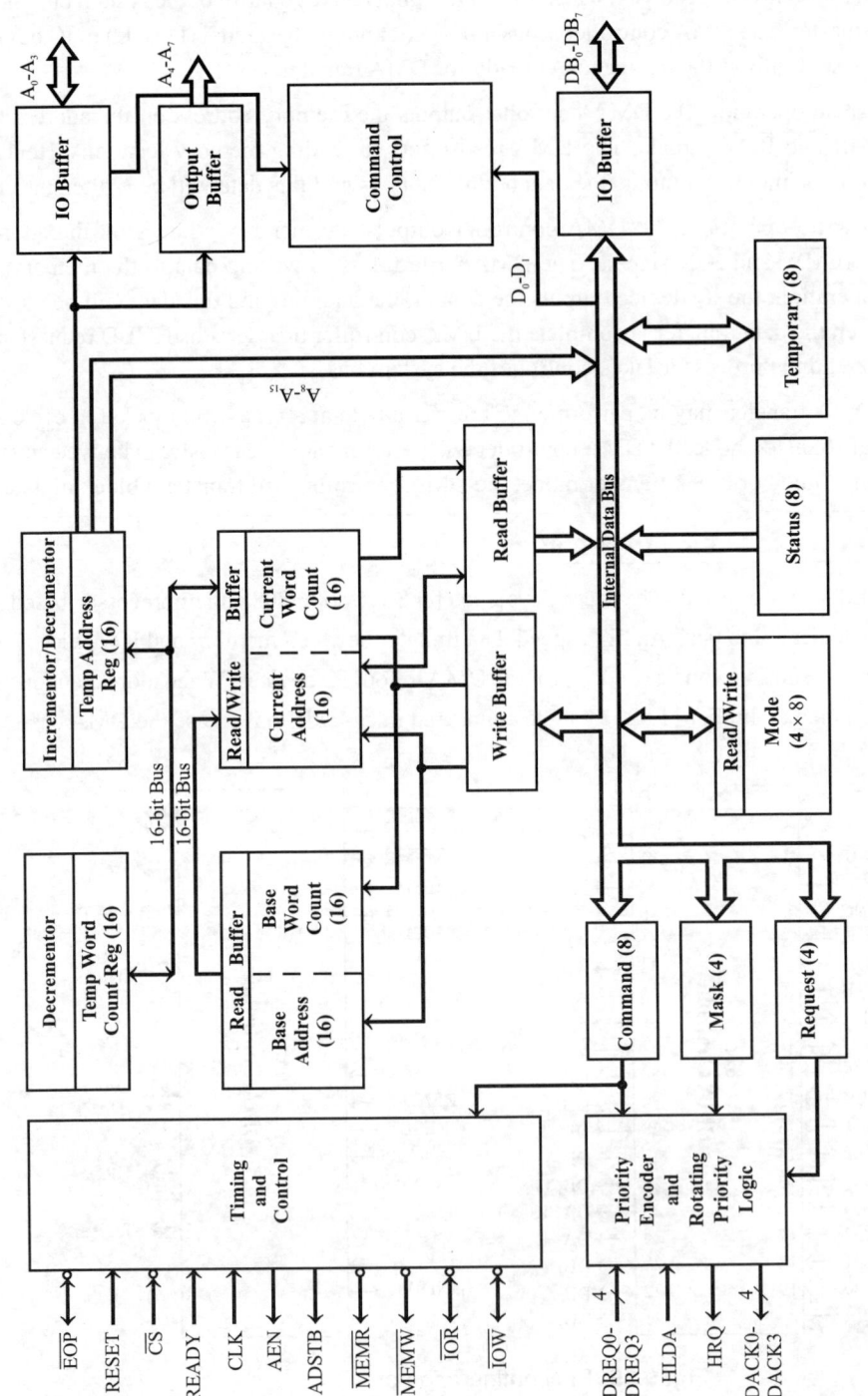

Fig. 9.30: Functional block diagram of an 8237.

For each DMA channel, a set of registers has been dedicated to store the memory address and the count value for the number of bytes to be read/write by DMA. These registers are base address, current address, base word count, current word count and mode registers. Apart from these dedicated registers, the 8237 has temporary registers, status, command, mask and request registers.

9.5.1 Pins, Signals and Functional Block Diagram of 8237

The 8237 is a 40-pin IC and available in a **D**ual **I**n-**L**ine **P**ackage (DIP). The pin configuration of 8237 is shown in Fig. 6.27. A brief description about the pins and signals of 8237 are listed in Table 9.10. The functional block diagram of 8237 is shown in Fig. 9.30.

Features of 8237

1. It has four independent DMA channels to service four IO devices.

2. Number of channels can be increased by cascading any number of 8237.

3. Each channel can be independently programmable to transfer up to 64 kB of data by DMA.

4. Each channel can independently perform read transfer, write transfer and verify transfer.

5. Channel-0 and channel-1 are used to perform memory-to-memory transfer.

6. Each channel can be independently programmable to perform demand transfer DMA, single-transfer DMA and block-transfer DMA.

Table 9.10: Pin Description of 8237

Pin	Description
CLK	Clock input to 8237. Maximum clock frequency is 5 MHz. In the 8086/8088 system, the processor clock is inverted and applied to the CLK of 8237.
$\overline{CS}$	Logic **low** chip select signal. It is input signal to select the 8237 during the programming mode.
RESET	Reset input to 8237. Connected to system reset, when the RESET signal goes **high** the command, status, request and temporary registers are cleared. It also clears the first-last flip-flop and sets the mask register.
READY	Ready input signal and it is tied to V_{cc} for normal timings. When READY input is tied **low**, the 8237 enters a wait state. This is used to get extra time in the DMA machine cycles to transfer data between slow memory and IO devices.
HRQ	Hold request output signal. It is the hold request signal sent by the 8237 to the processor HOLD pin, to make a request for bus to perform DMA transfer.
HLDA	Hold acknowledge input signal. It is the hold acknowledge signal to be sent by the processor to inform the acceptance of hold request.
DREQ3 to DREQ0	DMA request inputs (four channel inputs). Used by IO devices to request for DMA transfer.
DACK3 to DACK0	DMA acknowledge output signals. These are output signals from 8237 to the IO devices inform the acceptance of DMA request. These outputs are programmable as either active **high** or active **low** signals.
DB_7 to DB_0	Data bus lines. These pins are used for data transfer between the processor and the DMA during the programming mode. During DMA mode, these lines are used as multiplexed high order address and data lines.

Table 9.10 continued...

Pin	Description
$\overline{IOR}$	Bidirectional IO read control signal. It is an input control signal for reading the DMA controller during the programming mode and the output control signal for reading the IO device during DMA (memory) write cycle.
$\overline{IOW}$	Bidirectional IO write control signal. It is an input control signal for writing the DMA controller during the programming mode and the output control signal for writing the IO device during DMA (memory) read cycle.
$\overline{EOP}$	End of process. It is a bidirectional active low signal. It is used either as input to terminate a DMA process or as output to inform the end of the DMA transfer to the processor. This output can be used as interrupt to terminate the DMA.
A_3 to A_0	Four bidirectional address lines. Used as input address during the programming mode, to select internal registers. During DMA mode, the low order four bits of memory address are output by 8237 on these lines.
A_7 to A_4	Four unidirectional address lines. Used to output the memory address bits A_7 to A_4 during DMA mode.
AEN	Address enable output signal. It is used to enable the address latch connected to the DB_7 - DB_0 pins of 8237. It is also used to disable any buffers in the system connected to the processor.
ADSTB	Address strobe output signal. It is used to latch the high-byte memory address issued through DB_7 to DB_0 lines by 8237 during the DMA mode into an external latch.
$\overline{MEMR}$	Memory read control signal. It is an output control signal issued during DMA read operation.
$\overline{MEMW}$	Memory write control signal. It is an output control signal issued during DMA write operation.

The various internal registers of 8237 are listed in Table 9.11. The processor can read or write into these registers. But with certain registers the processor can perform only read operation and with certain registers the processor can perform only write operation. The internal registers are selected by a 4-bit address supplied through A_0 - A_3 lines of 8237. The addresses of the internal registers and the operations (read/write) that can be performed on these registers are listed in Table 9.12.

Table 9.11: Internal Registers of 8237

Name of the register	Size of register in bits	Number of registers available
Base address register	16	4
Base word count register	16	4
Current address register	16	4
Current word count register	16	4
Temporary address register	16	1
Temporary word count register	16	1
Status register	8	1
Command register	8	1
Temporary register	8	1
Mode register	8	4
Mask register	4	1
Request register	3	1

Table 9.12: Address of Internal Registers of 8237

Name of the register	Operation performed	Binary address			
		A_3	A_2	A_1	A_0
Channel-0 Base and Current address	Write	0	0	0	0
Channel-0 Current address	Read	0	0	0	0
Channel-0 Base and Current word count	Write	0	0	0	1
Channel-0 Current word count	Read	0	0	0	1
Channel-1 Base and Current address	Write	0	0	1	0
Channel-1 Current address	Read	0	0	1	0
Channel-1 Base and Current word count	Write	0	0	1	1
Channel-1 Current word count	Read	0	0	1	1
Channel-2 Base and Current address	Write	0	1	0	0
Channel-2 Current address	Read	0	1	0	0
Channel-2 Base and Current word count	Write	0	1	0	1
Channel-2 Current word count	Read	0	1	0	1
Channel-3 Base and Current address	Write	0	1	1	0
Channel-3 Current address	Read	0	1	1	0
Channel-3 Base and Current word count	Write	0	1	1	1
Channel-3 Current word count	Read	0	1	1	1
Command register	Write	1	0	0	0
Status register	Read	1	0	0	0
Request register	Write	1	0	0	1
Write single mask register bit	Write	1	0	1	0
Mode register	Write	1	0	1	1
Clear byte pointer flip-flop	Write	1	1	0	0
Temporary register	Read	1	1	0	1
Master clear	Write	1	1	0	1
Clear mask register	Write	1	1	1	0
Write all mask register bits	Write	1	1	1	1

The 16-bit internal registers of 8237 are read/write through an 8-bit data bus. The 8237 has an internal first-last flip-flop which has to be cleared to zero for reading/writing low byte first and then high byte. The first last flip-flop can be set to one for reading/writing high byte first and then low byte. (However, the 8237 does not have the facility to directly set the first-last flip-flop, but it has the facility to directly clear the first-last flip-flop.) After each read or write operation, the state of flip-flop automatically toggles.

9.5.2 Internal Registers of 8237

Current Address (CA) Register

It is used to hold the 16-bit memory address of the next memory location to be accessed by DMA. The 8237 outputs the content of the CA-register as the memory address and increments/decrements it by one. Each channel has its own CA-register. Initially, the starting address of memory is loaded in CA-register from base address register.

Current Word Count (CWC) Register

It holds the count value of the number of bytes to be transferred by the DMA. Initially, the count value is loaded to the CWC register from the base count register. After each byte transfer by the DMA, the count value is decremented by one. Therefore, at any one time it holds the count value for the number of bytes (pending) to be transferred by DMA.

Base Address (BA) Register

It is used to hold the starting address of the memory block to be accessed by the DMA. During the start of the DMA process, the content of BA-register is loaded in CA-register. If auto initialization is enabled in the mode register then the content of BA-register is reloaded in the CA-register at the end of the DMA process.

Base Word Count (BWC) Register

It is used to hold the count value for the number of bytes to be transferred by DMA. During the start of DMA process, the content of BWC register is loaded in CWC register. If auto initialization is enabled in mode register then the content of BWC register is reloaded in CWC register at the end of DMA process.

Command Register

The command register is used to program the following features of 8237:

1. Enable/Disable memory-to-memory transfer.
2. Enable/Disable the DMA controller.
3. Normal/Compressed timing.
4. Fixed/Rotating priority.
5. Type of (active low/high) DMA request and acknowledge signal.

The format of the control word to be loaded in the command register to program the above features is shown in Fig. 9.31. During memory-to-memory DMA transfer, the channel-0 registers are used to hold the source address and the channel-1 registers are used to hold the destination address. The data transfer takes place via the temporary register in 8237. The number of bytes transferred is determined by the channel-1 count register.

The bit B_2 is used to turn ON/OFF the entire controller by the software. The bit B_3 is used to program the normal/compressed timing. In normal timing, the time taken to perform one DMA transfer will be four clock periods. In compressed timing, the time taken to perform one DMA transfer will be two clock periods.

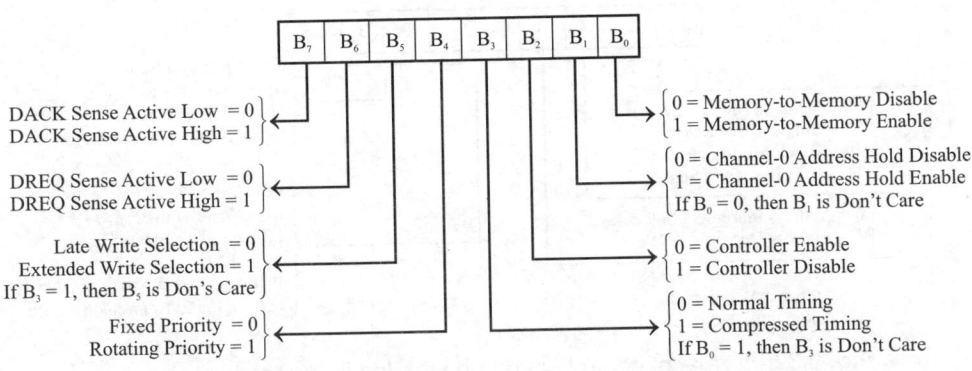

Fig. 9.31: Format of control word to be loaded in command register.

The bit B_4 is used to select fixed/rotating priority for DMA channels. In fixed priority, the channel-0 has the highest priority and channel-3 has the lowest priority. In rotating priority, after servicing a channel its priority is made as the lowest. For example, if DMA request is made to channel-2 and there is no DMA request in other channels. Now after servicing channel-2 in the rotating priority scheme, the priorities of the channels from the highest to the lowest will be channel-0, channel-1, channel-3 and channel-2. Alternately, if the 8237 is programmed for fixed priority then for the same situation after servicing channel-2, the priorities of the DMA channels from highest to lowest will be channel-0, channel-1, channel-2 and channel-3.

The bit B_5 is used to extend the timing of the write pulse when the IO devices require wider write pulse. This is possible only in normal timing. The bit B_6 and B_7 are used to program the polarities (logic **low/high**) of the DMA request input and DMA acknowledge output.

Mode Register

Each channel has its own mode register and it is used to program the following features of each channel of 8237:

1. **Read/Write/Verify transfer**
2. **Demand/Single/Block transfer mode**
3. **Single/Cascaded operation of 8237**
4. **Enable/Disable auto initialization**

The format of the control word to be loaded in the mode register is shown in Fig. 9.32. The control word of all the four mode registers are sent to the same internal address, but 8237 identifies the control word of a channel from the bits B_0 and B_1. The bits B_2 and B_3 are used to program the read/write/verify transfer. In read transfer, the data is transferred from memory to the IO device. In write transfer, the data is transferred from the IO device to the memory. Verification operations generate the DMA addresses without generating the DMA memory and IO control signals.

The bit B_4 is used to enable/disable auto initialization of DMA channels. When it is enabled, the memory address and count value from base registers are loaded in the current registers after completion of DMA process, which are used to repeat the DMA process between the IO device and same block of memory.

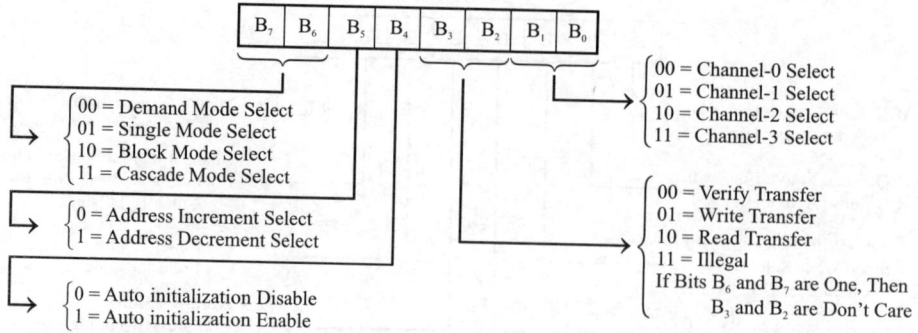

Fig. 9.32: Format of control word to be loaded in the mode register.

The bits B_6 and B_7 are used to program various modes of operations like demand transfer mode, single transfer mode, block transfer mode and cascade mode. In demand transfer mode, the DMA transfer is performed until an external signal is applied to the $\overline{EOP}$ pin of 8237 or until the DREQ input becomes inactive.

In single transfer mode, the 8237 releases the bus to the processor by deactivating the HOLD signal after transfer of each byte by the DMA. If the DREQ pin is held active then 8237 will make a request for the DMA to the processor through the HOLD pin again after a small delay. This will allow the processor to execute one instruction and the 8237 to perform one DMA transfer alternatively.

In block transfer mode, the 8237 will transfer an entire block of data specified by the count register and then release the bus to the processor by deactivating the HOLD signal. In cascaded operation, the **Hold Request** pin (HRQ) of one 8237 will be connected to the HOLD pin of the processor and to each DREQ pin of this 8237, the HRQ pin of another 8237 can be connected. This connection can be extended until we get the required number of DMA channels.

When DMA request is made to a channel by another 8237 then this channel cannot perform read/write/verify transfer.

Request Register

It is used to request a DMA transfer via software. The format of the control word to be loaded in the request register is shown in Fig. 9.33. The bit B_0 and B_1 selects the channel in which DMA transfer is required and the bit B_2 is used to set/reset the DMA request.

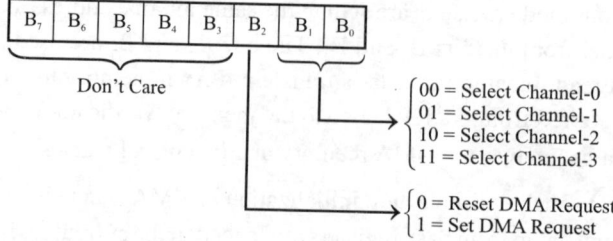

Fig. 9.33: Format of control word to be loaded in a request register.

Mask Register

This register is used to mask (or disallow) the DMA request made through channels and to unmask (or enable) the DMA request made through channels. Please remember that after a RESET all the channels are masked and so after a RESET, the channels have to be unmasked by sending a control word to the mask register.

The mask register has two internal addresses. One address is used to set/reset the single mask bit (i.e., to mask/unmask one channel at a time) and the other address is used to set/reset all the mask bits (i.e., to mask/unmask all the channels). The format of the two control words for mask register are shown in Fig. 9.34.

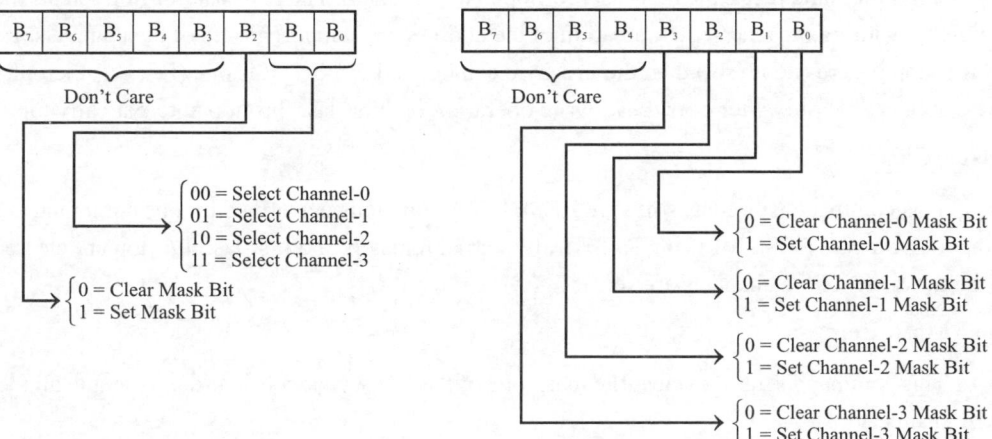

Fig. a: Format of the control word to mask/unmask one channel.

Fig. b: Format of the control word to mask/unmask all channel.

Fig. 9.34: Format of the control word to be loaded in the mask register.

Status Register

The status register can be read to know whether the channels have reached their **T**erminal **C**ount (TC) or not and also to know whether the DMA request on the DREQ pins are active or not. The format of status register is shown in Fig. 9.35.

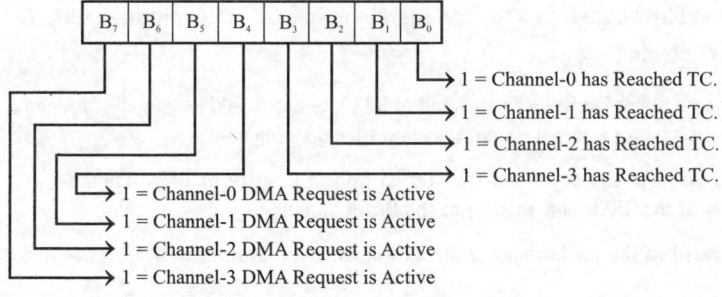

Fig. 9.35: Format of a status register.

Software Commands of 8237

The 8237 has three software commands to control its operation and they are Clear first-last flip-flop, Master clear and Clear mask register. These software commands can be enabled by executing a write operation to the internal address allotted to these commands. (Please refer to Table 7.6 for internal addresses of these commands.) We need not worry about the data sent to these ports during write operation because the 8237 will ignore the data. The function of the software commands are given in the following subsections.

Clear First-Last Flip-Flop

This command resets the first-last flip-flop in 8237 to zero. The first-last flip-flop selects the low or high byte during the read/write operation of the address and count registers of the channels. If first-last flip-flop is zero (i.e., reset) then the low byte can be read/write. If it is one (i.e., set) then the high byte can be read/write. After every read/write operation the first-last flip-flop automatically toggles.

Master Clear

This command is used as software RESET. The functions performed by this command is same as that of hardware RESET. During RESET all internal registers and first-last flip-flop are cleared and all the mask bits of the channels are set.

Clear Mask Register

This command is used to clear the mask bits of the DMA channels in order to enable all the four DMA channels.

9.5.3 Programming (or initializing) 8237

The 8237 can work as a slave or as a temporary master in a microprocessor system. Normally the 8237 is interfaced to a system as a slave device. During the DMA operation, it works as a temporary master. For proper DMA operation, the 8237 has to be programmed when it is working as a slave. The programming of 8237 refers to sending software commands and various controlwords to 8237, in order to inform the types of DMA, memory address, count value, etc., for each channel. At the start of programming, all DMA channels have to be disabled and then they are enabled at the end of programming. Also the first-last flip-flop has to be cleared before sending 16-bit address/count value to 8237 in order to load low byte first and then high byte in address/count registers. The various steps in programming 8237 are as follows:

1. First send a "master clear" software command to 8237, which mask/disable all DMA channels, clears first-last flip-flop and clears all internal register, except the mask register.

2. Send a control word to the command register to inform priority of DMA channels, normal/compressed timings, polarity of the DREQ and polarity of the DACK signals.

3. Write a mode word to the mode register of each channel to inform the DMA mode and the type of DMA transfer.

4. Send a "clear first-last flip-flop" software command to reset it to zero.

5. After ensuring that the first-last flip-flop is zero, write the 16-bit address in the address register of each channel, by sending the low byte first and then high byte.

6. Then write the 16-bit count value in the count register of each channel, by sending the low byte first and then the high byte. It is sufficient if the first-last flip-flop is cleared at the beginning of sending a series of 16-bit address/count values, because after each write operation it automatically toggles to keep track of low byte and high byte.

7. Finally, send "the clear mask register" software command to enable all DMA channels. Now 8237 is ready to perform the DMA process.

9.5.4 Interfacing 8237 with 8085 Processor

A simple schematic for interfacing the 8237 with 8085 processor is shown in Fig. 9.36. The 8237 can be either memory-mapped or IO-mapped in the system. In the schematic shown in Fig. 9.36, the 8237 is IO-mapped in the system.

The chip select signals for IO-mapped devices are generated by using a 3-to-8 decoder. The address lines A_4, A_5 and A_6 are decoded to generate eight chip select signals (IOCS-0 to IOCS-7) and in this, the chip select signal IOCS-6 is used to select 8237. The address line A_7 and the control signal IO/$\overline{M}$ are used as enable for decoder. The IO addresses of the internal register of 8237 are listed in Table 9.13.

The DB_0 - DB_7 lines of 8237 are connected to data bus lines D_0-D_7 for data transfer with processor during programming mode. These lines (DB_0 - DB_7) are also used by 8237 to supply the memory address A_8 - A_{15} during the DMA mode. The 8237 also supplies two control signals ADSTB and AEN to latch the address supplied by it during DMA mode on external latches.

In the schematic shown in Fig. 9.36, two 8-bit latches are provided to hold the 16-bit memory address during DMA mode. During DMA mode, the AEN signal is also used to disable the buffers and latches used for address, data and control signals of the processor.

The 8237 provides separate read and write control signals for memory and IO devices during DMA. Therefore, the $\overline{RD}$, $\overline{WR}$ and IO/$\overline{M}$ of the 8085 processor are decoded by a suitable logic circuit to generate separate read and write control signals for memory and IO devices. (Please refer to Chapter-6, Fig. 6.21 for the logic circuit to generate separate read and write signals for memory and IO devices.)

The output clock of the 8085 processor should be inverted and supplied to the 8237 clock input for proper operation. The HRQ output of the 8237 is connected to HOLD input of the 8085 in order to make a HOLD request to the processor.

The HLDA output of 8085 is connected to HLDA input of 8237, in order to receive the acknowledge signal from the processor once the HOLD request is accepted. The RESET OUT of the 8085 processor is connected to RESET of the 8237 processor.

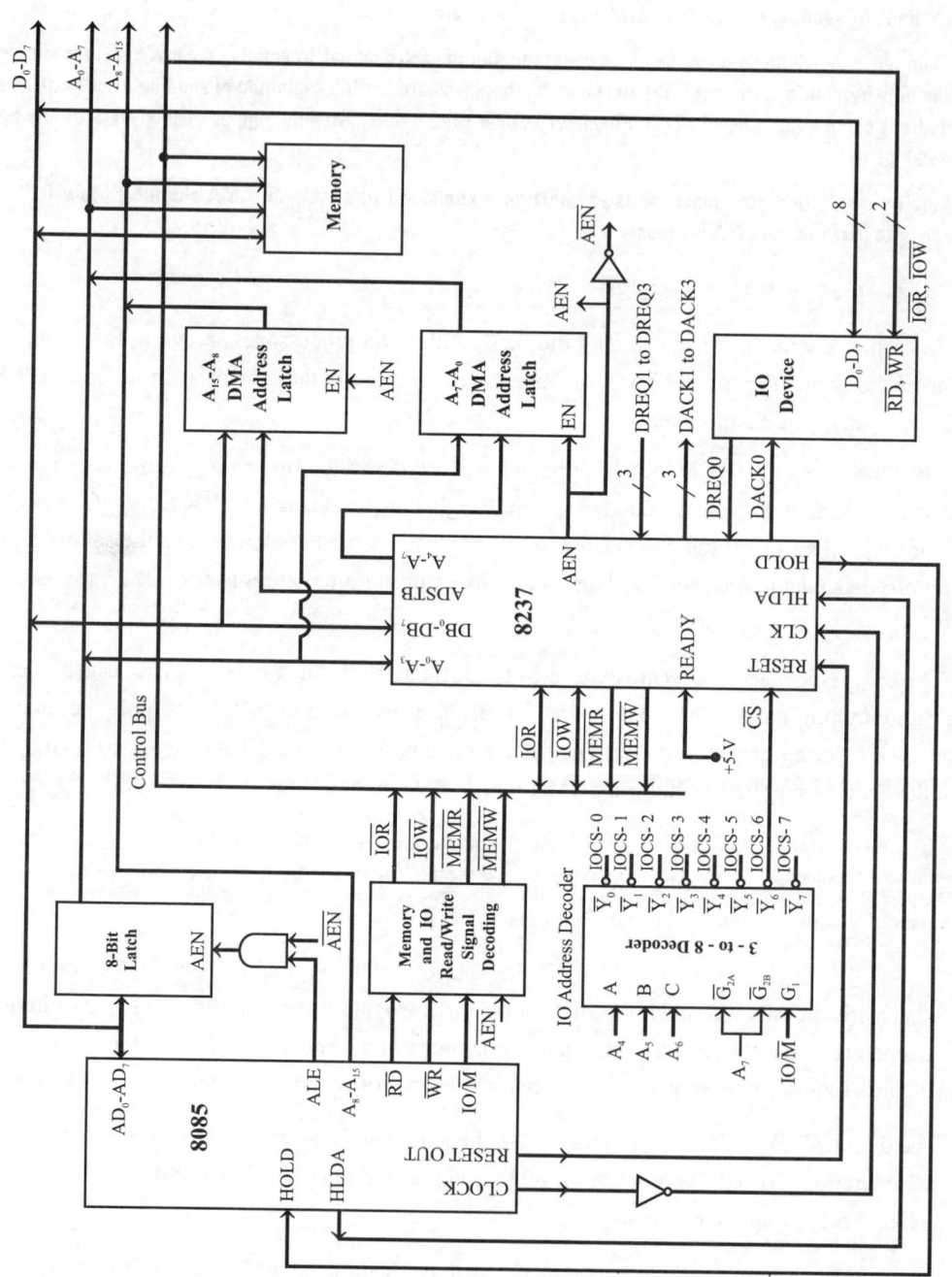

Fig. 9.36: Interfacing of 8237 with 8085 processor.

Table 9.13: IO Addresses of 8237 Interfaced to 8085 as Shown in Fig. 9.36

Name of the internal register of 8237	Binary address		Hexa address
	Decoder input and enable A_7 A_6 A_5 A_4	Input to address pins of 8237 A_3 A_2 A_1 A_0	
Channel-0 Base and Current address register	0 1 1 0	0 0 0 0	60
Channel-0 Base and Current word count register	0 1 1 0	0 0 0 1	61
Channel-1 Base and Current address register	0 1 1 0	0 0 1 0	62
Channel-1 Base and Current word count register	0 1 1 0	0 0 1 1	63
Channel-2 Base and Current address register	0 1 1 0	0 1 0 0	64
Channel-2 Base and Current word count register	0 1 1 0	0 1 0 1	65
Channel-3 Base and Current address register	0 1 1 0	0 1 1 0	66
Channel-3 Base and Current word count register	0 1 1 0	0 1 1 1	67
Status/Command register	0 1 1 0	1 0 0 0	68
Request register	0 1 1 0	1 0 0 1	69
Write single mask register bit	0 1 1 0	1 0 1 0	6A
Mode register	0 1 1 0	1 0 1 1	6B
Clear first-last flip-flop	0 1 1 0	1 1 0 0	6C
Temporary register/Master clear	0 1 1 0	1 1 0 1	6D
Clear mask register	0 1 1 0	1 1 1 0	6E
Write all mask register bits	0 1 1 0	1 1 1 1	6F

9.5.5 Interfacing 8237 with 8086 Processor

A simple schematic for interfacing the 8237 with 8086 processor is shown in Fig. 9.37. The 8237 can be either memory-mapped or IO-mapped in the system. In the schematic shown in Fig. 9.37, the 8237 is IO-mapped in the system with even addresses.

The chip select signals for IO-mapped devices are generated by using a 3-to-8 decoder. The address lines A_5, A_6 and A_7 are decoded to generate eight chip select signals (IOCS-0 to IOCS-7) and in this the chip select signal IOCS-6 is used to select 8237. The address line A_0 and the control signal M/IO are used as enables for the decoder.

The address lines A_1 - A_4 of 8086 are connected to A_0 - A_3 of 8237 through a latch enabled by ALE to provide the internal addresses. The intermediate latch is necessary because the address lines A_0 - A_3 of 8237 should be connected to A_1 to A_4 of the address bus during the programming mode and they are connected to A_0 - A_3 of the address bus during the DMA mode.

The IO addresses of the internal register of 8237 are listed in Table 9.14.

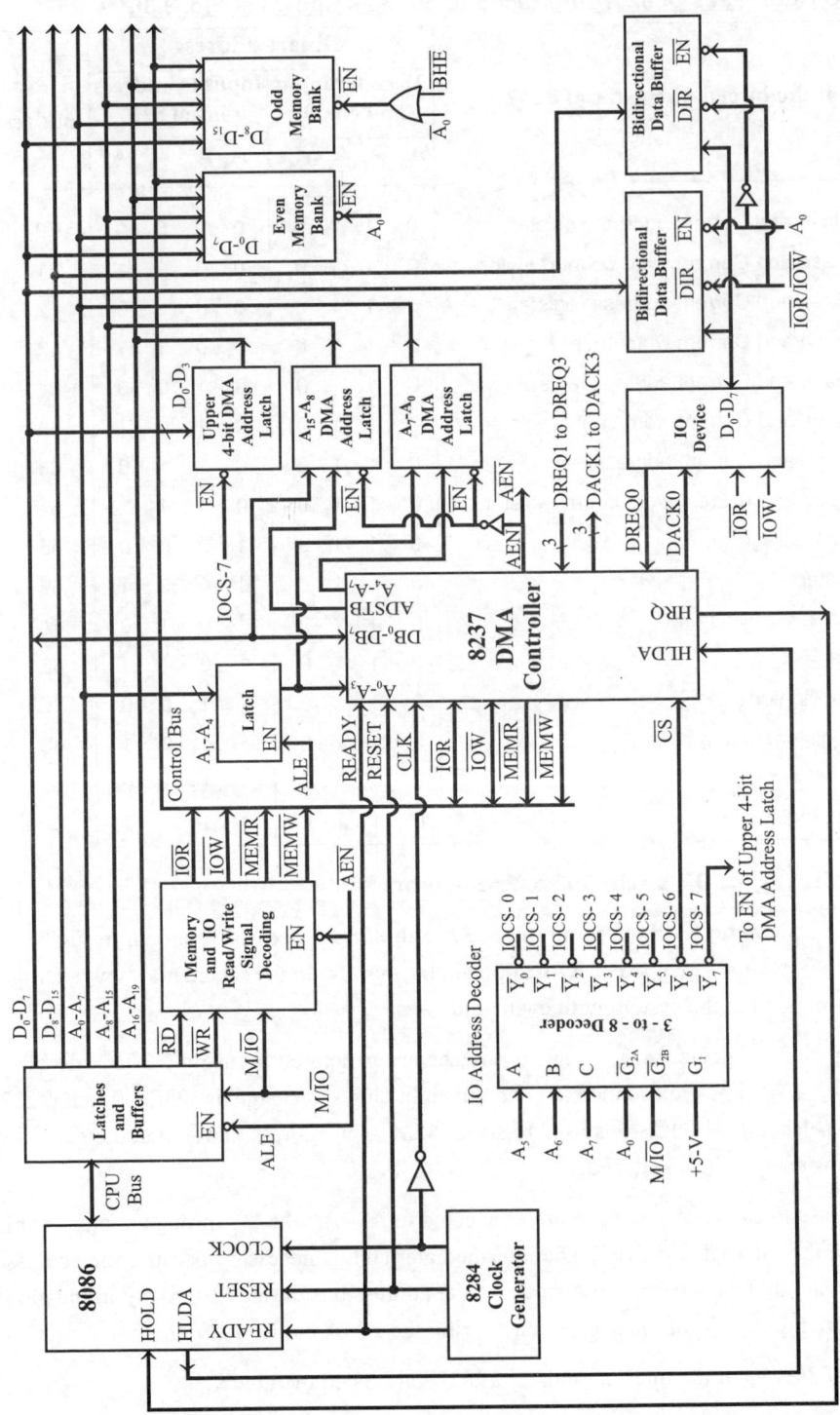

Fig. 9.37: Interfacing of 8237 with the 8086 processor.

The 8237 supplies only lower 16 bits (A_0-A_{15}) of memory address during the DMA mode. The upper 4 bits of the memory address (A_{16}-A_{19}) should be generated by some means. For this a latch is provided in the system to hold the upper 4-bits of the DMA memory address.

This latch is enabled by the chip select signal IOCS-7 and the IO address of this latch is also listed in Table 9.14. The processor can load the upper four bits of the DMA memory address in this latch while initializing the DMA controller.

The DB_0- DB_7 lines of 8237 are connected to the data bus lines D_0 - D_7 for data transfer with the processor during the programming mode. These lines (DB_0 - DB_7) are also used by 8237 to supply the memory address A_8 - A_{15} during the DMA mode.

The 8237 also supplies two control signals ADSTB and AEN to latch the address supplied by it during the DMA mode on the external latches.In the schematic shown in Fig. 9.37, two 8-bit latches are provided to hold the lower 16-bit memory address during DMA mode.

During the DMA mode, the AEN signal is also used to disable the buffers and latches used for address, data and control signals of the processor.

The 8237 provides separate read and write control signals for memory and IO devices during DMA.Therefore, $\overline{RD}$, $\overline{WR}$ and $M/\overline{IO}$ of the 8086 processor are decoded by a suitable logic circuit to generate separate read and write control signals for memory and IO devices. (Please refer Chapter- 6, Fig. 6.22 for the logic circuit to generate separate read and write signals for memory and IO devices.)

The RESET, READY and clock signal for 8237 are provided by 8284 clock generator. The clock used for the processor should be inverted and supplied to 8237 for proper operation. The HRQ output of 8237 is connected to the HOLD input of 8086 in order to make a HOLD request to the processor.

The HLDA output of 8086 is connected to HLDA input of 8237 in order to receive the acknowledge signal from the processor once the HOLD request is accepted.

The IO device has to read/write data from/to both the even memory bank and the odd memory bank. In order to provide this facility, two bidirectional data buffers are provided in the system. These buffers act as switches to connect the D_0 - D_7 of the IO device to D_0 - D_7 of the system bus for even addresses and to D_8 - D_{15} of the system bus for odd addresses. The address A_0 is used to enable these buffers.

The buffer which is enabled by A_0 is used for read/write data from/to the even memory bank via the low order data bus. The buffer which is enabled by $\overline{A_0}$ is used for read/write data from/to the odd memory bank via the high order data bus. The direction control for the bidirectional buffer can be provided by using either $\overline{IOR}$ or $\overline{IOW}$ signal.

Table 9.14: IO Addresses of 8237 Interfaced to 8086 as Shown in Fig. 9.37

Name of the internal register of 8237	Binary address			Hexa address
	Decoder input	Input to address pins of 8237	Decoder enable	
	A_7 A_6 A_5	A_4 A_3 A_2 A_1	A_0	
Channel-0 Base and Current address register	1 1 0	0 0 0 0	0	C0
Channel-0 Base and Current word count register	1 1 0	0 0 0 1	0	C2
Channel-1 Base and Current address register	1 1 0	0 0 1 0	0	C4
Channel-1 Base and Current word count register	1 1 0	0 0 1 1	0	C6
Channel-2 Base and Current address register	1 1 0	0 1 0 0	0	C8
Channel-2 Base and Current word count register	1 1 0	0 1 0 1	0	CA
Channel-3 Base and Current address register	1 1 0	0 1 1 0	0	CC
Channel-3 Base and Current word count register	1 1 0	0 1 1 1	0	CE
Status/Command register	1 1 0	1 0 0 0	0	D0
Request register	1 1 0	1 0 0 1	0	D2
Write single mask register bit	1 1 0	1 0 1 0	0	D4
Mode register	1 1 0	1 0 1 1	0	D6
Clear first-last flip-flop	1 1 0	1 1 0 0	0	D8
Temporary register/Master clear	1 1 0	1 1 0 1	0	DA
Clear mask register	1 1 0	1 1 1 0	0	DC
Write all mask register bits	1 1 0	1 1 1 1	0	DE

Name of the external device	Binary address		Hexa address
	Decoder input	Unused address lines	
	A_7 A_6 A_5	A_4 A_3 A_2 A_1 A_0	
Upper 4-bit DMA address latch	1 1 1	x x x x x	E0

Note: Here don't care "x" is considered zero.

DMA Operation in 8085/8086 using 8237

After programming the 8237 in the slave mode, it will be ready to perform DMA. Once the 8237 is programmed it keeps on checking DMA request input from IO devices. When the 8237 detects a valid DMA request then it performs the following activity:

1. When the 8237 receives a DMA request from a peripheral it sends a hold request to the processor (provided the channel should be enabled and there should not be any pending higher priority DMA request).

2. When the 8085/8086 processor receives a hold request, it will complete the current instruction execution and drive all its tristate (address, data and control) pins to high impedance state. Then, the 8085/8086 sends an acknowledge signal to 8237.

3. On receiving an acknowledge from 8085/8086, the 8237 will send an acknowledge to the peripheral device which requested DMA.

4. The 8237 asserts AEN high, which enables DMA memory address latches and disables the processor address latch.

5. Then the 8237 outputs the low byte address on A_0 - A_7 lines and high byte address on DB_0 to DB_7 lines. Also, the control signal ADSTB is asserted high to latch this address into external latches.

6. Also the DMA controller asserts appropriate read and write control signals to perform DMA transfer.

7. In block transfer mode, after performing one byte transfer the steps 4, 5 and 6 are repeated again and again until the count is zero. In demand transfer mode the steps 4, 5 and 6 are repeated until an external end of process signal is applied or till the DMA request is deactivated. In single transfer mode the 8237 deactivate the hold request to the processor after one byte transfer by DMA.

9.6 DMA CONTROLLER - INTEL 8257

The DMA controller-8257 has been developed for 8085/8086/8088 microprocessor-based systems. It is a device dedicated to perform a high-speed data transfer between memory and IO device. The 8257 has four channels and so it can be used to provide DMA to four IO devices. It cannot be connected in cascade like the 8237 and it has less features than the 8237.

For each DMA channel, an address register and a count register has been dedicated to store the memory address and the count value for the number of bytes to be read/write by DMA respectively. Apart from these dedicated registers, the 8257 has mode set and status registers.

9.6.1 Pins, Signals and Functional Block Diagram of 8257

The 8237 is a 40-pin IC and available in **D**ual **I**n-line **P**ackage (DIP). The pin configuration of the 8257 is shown in Fig. 9.38. A brief description about the pins and signals of the 8257 are listed in Table 9.15.

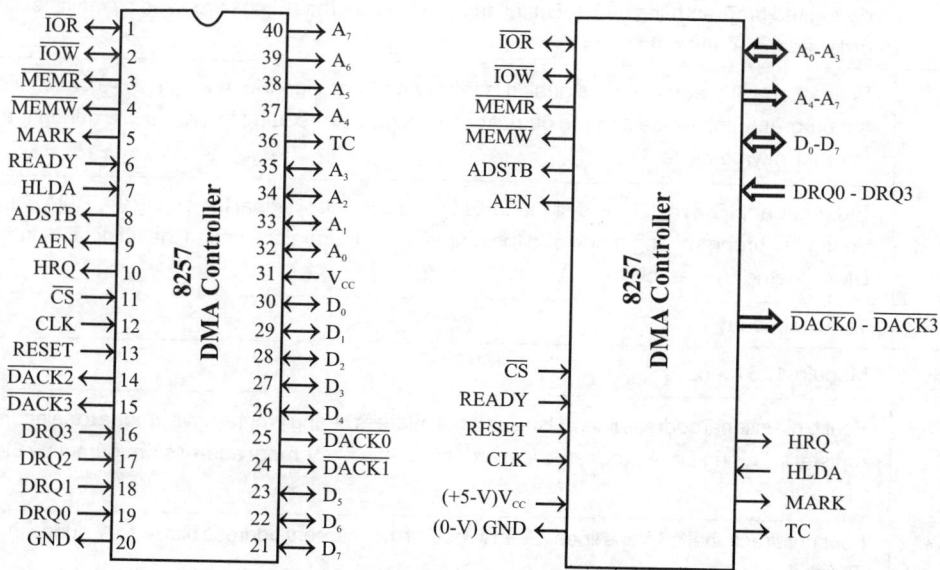

Fig. 9.38: Pin configuration of an 8257.

Table 9.15: Pin Description of 8257

Pin	Description
CLK	Clock input to 8257. The maximum clock frequency is 5 MHz. In an 8086/8088 system, the processor clock signal is inverted and applied to the CLK of 8257.
$\overline{CS}$	Logic **low** chip select signal. It is the input signal to select 8257 during the programming mode.
RESET	Reset input to 8257. Connected to the system reset, when the RESET signal goes **high** all the internal registers are cleared.
READY	Ready input signal and it is tied to V_{cc} for normal timings. When READY input is tied **low**, the 8257 enters a wait state. This is used to get extra time in DMA machine cycles to transfer data between slow memory and IO devices.
HRQ	Hold request output signal. It is the hold request signal sent by 8257 to the processor HOLD pin to make a request for bus to perform DMA transfer.
HLDA	Hold acknowledge input signal. It is the hold acknowledge signal to be sent by the processor to inform the acceptance of a hold request.
DREQ3 to DREQ0	DMA request inputs (four channel inputs). Used by IO devices to request for DMA transfer.
DACK3 to DACK0	DMA acknowledge output signals. These are active **low** output signals from 8257 to the IO devices to inform the acceptance of a DMA request.
D_0 - D_7	Data bus lines. These pins are used for data transfer between the processor and the DMA controller during the programming mode. During the DMA mode, these lines are used as multiplexed high order address and data lines.
$\overline{IOR}$	Bidirectional IO read control signal. It is the input control signal for reading DMA controller during the programming mode and the output control signal for reading the IO device during the DMA (memory) write cycle.
$\overline{IOW}$	Bidirectional IO write control signal. It is the input control signal for writing the DMA controller during the programming mode and the output control signal for writing the IO device during the DMA (memory) read cycle.
TC	Terminal count.
MARK	Modulo-128 mark.
A_3 to A_0	Four bidirectional address lines. Used as input address during programming mode to select internal registers. During DMA mode, the low order four bits of memory address are output by 8257 on these lines.
A_7 to A_4	Four unidirectional address lines. Used to output the memory address bits A_7 to A_3 during the DMA mode.

Table 9.15 continued...

Pin	Description
AEN	Address enable output signal. It is used to enable the address latch connected to D_7 - D_0 pins of 8257. It is also used to disable any buffers in the system connected to the processor.
ADSTB	Address strobe output signal. It is used to latch the high byte memory address issued through D_7 - D_0 lines by 8257 during the DMA mode into an external latch.
$\overline{MEMR}$	Memory read control signal. It is an output control signal issued during the DMA read operation.
$\overline{MEMW}$	Memory write control signal. It is an output control signal issued during the DMA write operation.

Features of 8257

1. It has four independent DMA channels to service four IO devices.

2. Each channel is independently programmable to transfer up to 64 kB of data by DMA.

3. Each channel can independently perform read transfer, write transfer and verify transfer.

9.6.2 Functional Block Diagram of 8257

The functional block diagram of 8257 is shown in Fig. 9.39. The functional blocks of 8257 are the data bus buffer, read/write logic, control logic and four numbers of the DMA channels.

Each channel has two programmable 16-bit registers. One register is used to program the starting address of the memory location for DMA data transfer and another register is used to program a 14-bit count value and a 2-bit code for the type of DMA transfer (Read/Write/Verify transfer).

The address in the address register is automatically incremented after every read/write/verify transfer. The format of the count register is shown in Fig. 9.40(a).

In read transfer, the data is transferred from the memory to the IO device. In write transfer, the data is transferred from the IO device to memory. Verification operations generate the DMA addresses without generating the DMA memory and IO control signals.

Apart from the address and count registers of each channel, the 8257 has a mode set register and status register. The mode set register is used to program various features of 8257 and the status register can be read to know the terminal count status of the channels.

The registers of 8257 are selected for read/write operation during the slave/programming mode by sending a 4-bit address to 8257 through A_0 - A_3 lines. The internal addresses of the registers of 8257 are listed in Table 9.16.

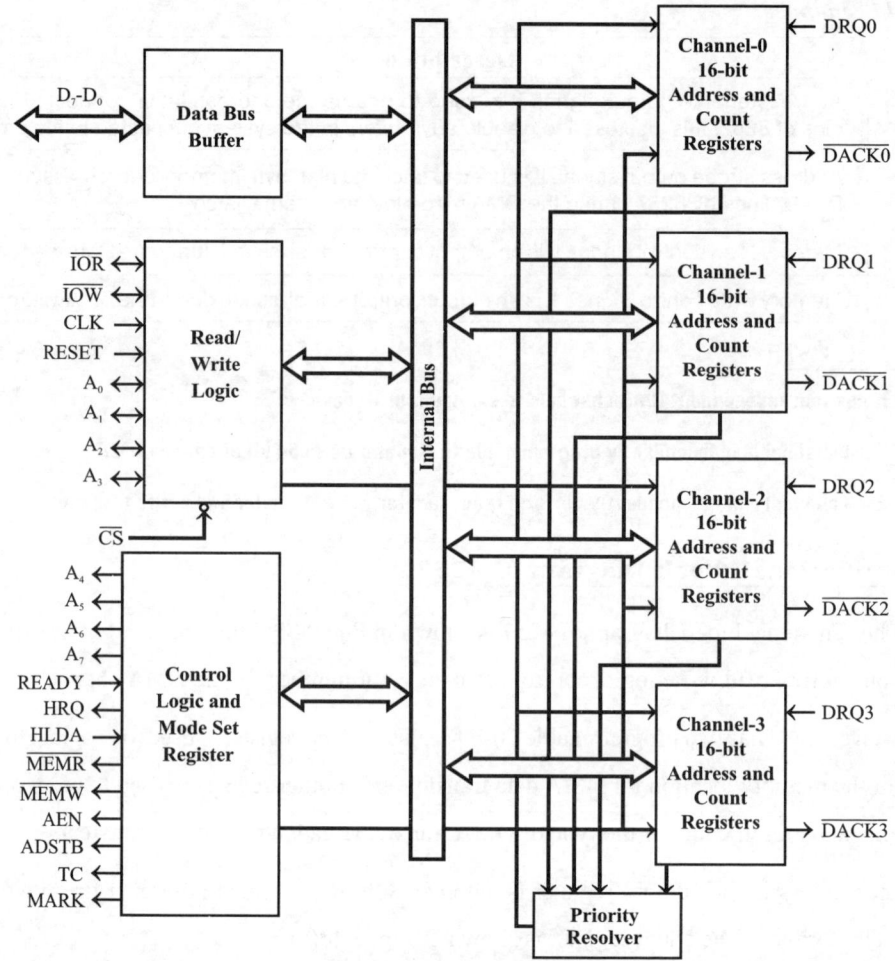

Fig. 9.39: Functional block diagram of DMA controller 8257.

Table 9.16: Internal Address of 8257 Registers

Register	Address			
	A_3	A_2	A_1	A_0
Channel-0 DMA address register	0	0	0	0
Channel-0 Count register	0	0	0	1
Channel-1 DMA address register	0	0	1	0
Channel-1 Count register	0	0	1	1
Channel-2 DMA address register	0	1	0	0
Channel-2 Count register	0	1	0	1
Channel-3 DMA address register	0	1	1	0
Channel-3 Count register	0	1	1	1
Mode set register (Write only)	1	0	0	0
Status register (Read only)	1	0	0	0

While programming the 16-bit register, the low byte has to be sent first and then the high byte. Internally, the loading of low byte and high byte into 16-bit register are taken care by a first-last flip-flop.

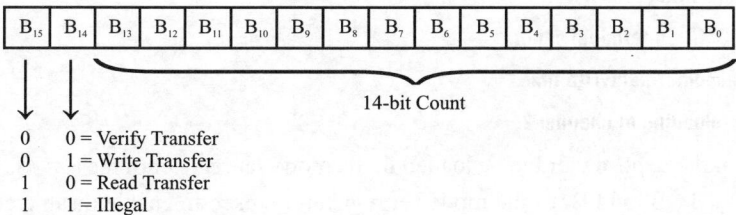

Fig. a: Format of count to be loaded in the count register of 8257.

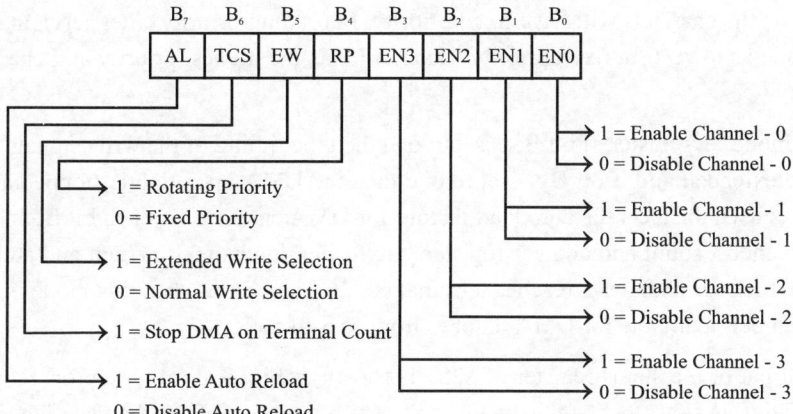

Fig. b: Format of control word to be loaded in mode set register of 8257.

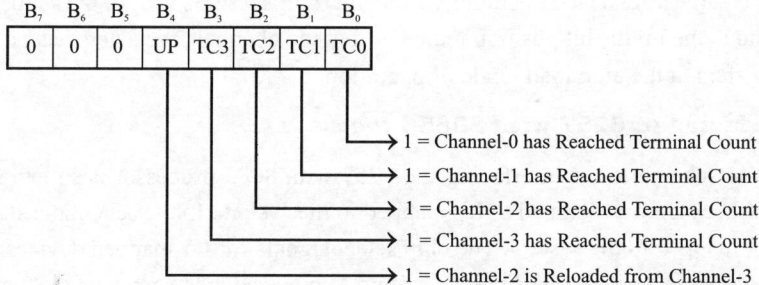

Fig. c: Status register of 8257.

Fig. 9.40: Format of registers of 8257.

Mode Register

The mode set register is used to program the following features of 8257:

1. **Enable/disable a channel.**

2. **Fixed/rotating priority.**

3. **Stop DMA on terminal count.**

4. **Extended/normal write time.**

5. **Auto reloading of channel-2.**

The format of the control word to be loaded in the mode set register of the 8257 is shown in Fig. 9.40(b). The bits B_0, B_1, B_2 and B_3 of the mode set register are used to enable/disable channel-0, 1, 2 and 3, respectively. A one in these bit position will enable a particular channel and a zero will disable it.

In the mode set register, if bit B_4 is set to one then the channels will have rotating priority and if it is zero then the channels will have fixed priority. In rotating priority, after servicing a channel its priority is made as lowest. In fixed priority, channel-0 has the highest priority and channel-2 has the lowest priority.

In the mode set register, if bit B_5 is set to one then the timing of the write signals ($\overline{\text{MEMW}}$ and $\overline{\text{IOW}}$) will be extended and if bit B_6 is set to one then the DMA operation is stopped at the terminal count. Bit B_7 is used to select the auto load feature for DMA channel-2. When bit B_7 is set to one, the content of channel-3 count and address registers are loaded in channel-2 count and address registers respectively whenever channel-2 reaches terminal count. Therefore, when this mode is activated, the number of channels available for DMA reduces from four to three.

The format of the status register of 8257 is shown in Fig. 9.40(c). The processor can read the status of 8257 during slave mode to know the terminal count status of the channels. The bits B_0, B_1, B_2 and B_3 of the status register indicate the terminal count status of channels-0, 1, 2 and 3, respectively.

A one in these bit positions indicate that the particular channel has reached terminal count. These status bits are cleared after a read operation by the microprocessor. Bit B_4 of the status register is called update flag and a one in this bit position indicates that the channel-2 registers has been reloaded from channel-3 registers in the auto load mode of operation.

9.6.3 Interfacing of 8257 with 8085 Processor

A simple schematic for interfacing the 8257 with 8085 processor is shown in Fig. 9.41. The 8257 can be either memory-mapped or IO-mapped in the system. In the schematic shown in Fig. 9.41, the 8257 is IO-mapped in the system. The chip select signals for IO-mapped devices are generated by using a 3-to-8 decoder. The address lines A_4, A_5 and A_6 are decoded to generate eight chip select signals (IOCS-0 to IOCS-7) and in this the chip select signal IOCS-6 is used to select 8257. The address line A_7 and the control signal IO/$\overline{\text{M}}$ are used as enable for decoder. The IO addresses of the internal registers of 8257 are listed in Table 9.17.

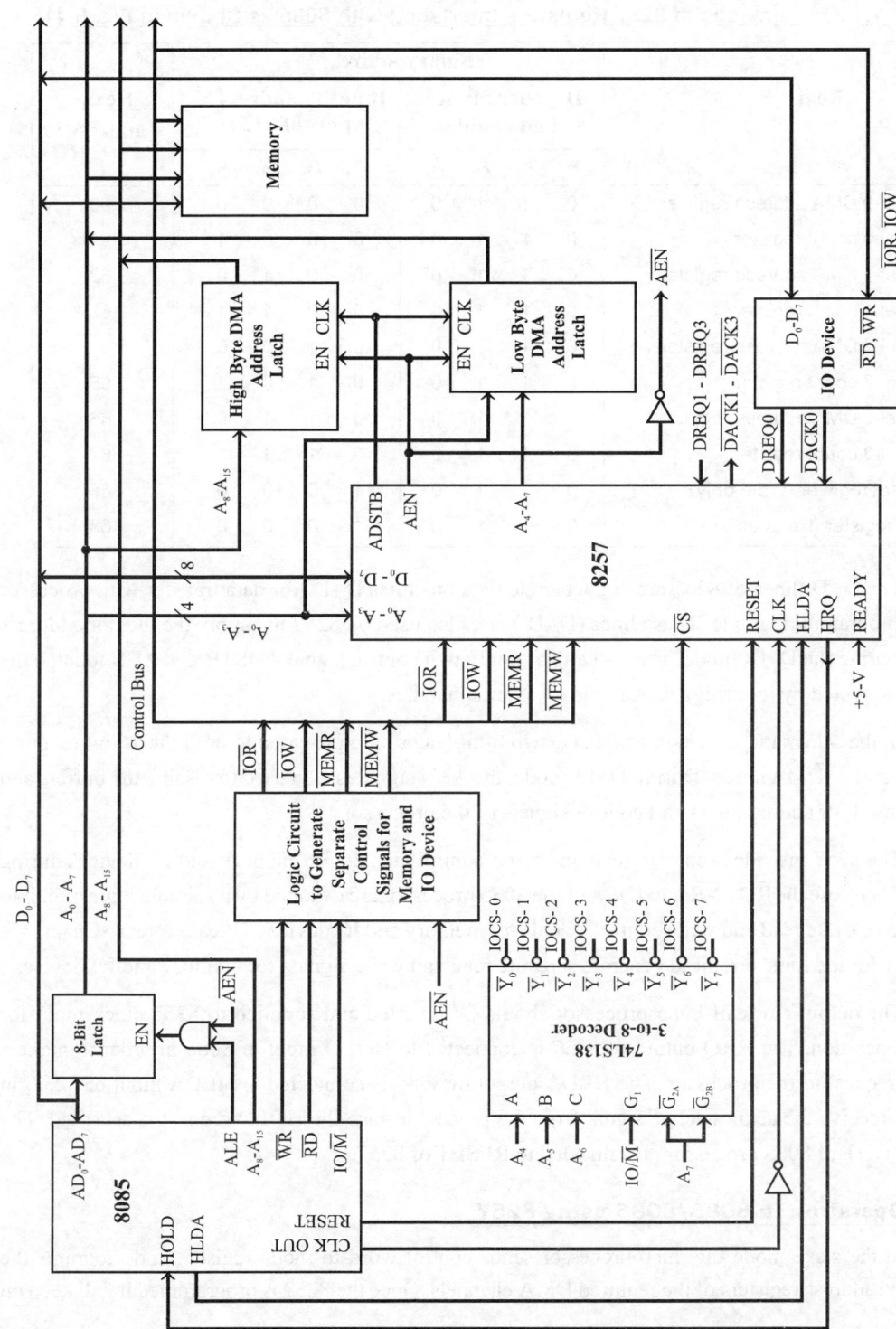

Fig. 9.41: Interfacing DMA controller 8257 with 8085 microprocessor.

Table 9.17: IO Addresses of 8257 Registers Interfaced with 8085 as Shown in Fig. 9.41

Register	Binary address								Hexa address
	Decoder input and enable				Input to address pins of 8257				
	A_7	A_6	A_5	A_4	A_3	A_2	A_1	A_0	
Channel-0 DMA address register	0	1	1	0	0	0	0	0	60
Channel-0 count register	0	1	1	0	0	0	0	1	61
Channel-1 DMA address register	0	1	1	0	0	0	1	0	62
Channel-1 count register	0	1	1	0	0	0	1	1	63
Channel-2 DMA address register	0	1	1	0	0	1	0	0	64
Channel-2 count register	0	1	1	0	0	1	0	1	65
Channel-3 DMA address register	0	1	1	0	0	1	1	0	66
Channel-3 count register	0	1	1	0	0	1	1	1	67
Mode set register (Write only)	0	1	1	0	1	0	0	0	68
Status register (Read only)	0	1	1	0	1	0	0	0	68

The D_0-D_7 lines of 8257 are connected to data bus lines D_0-D_7 for data transfer with processor during programming mode. These lines (D_0-D_7) are also used by 8257 to supply the memory address A_8-A_{15} during the DMA mode. The 8257 also supply two control signal ADSTB and AEN to latch the address supplied by it during DMA mode on external latches.

In the schematic shown in Fig. 9.41, two 8-bit latches are provided to hold the 16-bit memory address during DMA mode. During DMA mode, the AEN signal is also used to disable the buffers and latches used for address, data and control signals of the processor.

The 8257 provides separate read and write control signals for memory and IO devices during DMA. Therefore the $\overline{RD}$, $\overline{WR}$ and IO/$\overline{M}$ of the 8085 processor are decoded by a suitable logic circuit to generate separate read and write control signals for memory and IO devices. (Please refer to Chapter-6, Fig. 6.21 for the logic circuit to generate separate read and write signals for memory and IO devices.)

The output clock of 8085 processor should be inverted and supplied to 8257 clock input for proper operation. The HRQ output of 8257 is connected to HOLD input of 8085 in order to make a HOLD request to the processor. The HLDA output of 8085 is connected to HLDA input of 8257, in order to receive the acknowledge signal from the processor once the HOLD request is accepted. The RESET OUT of 8085 processor is connected to RESET of 8257.

DMA Operation in 8085/8086 using 8257

In the slave mode the microprocessor sends control word to mode register, and programs the count and address registers of the required DMA channels. Once the 8257 is programmed it will keep on checking DMA request input from IO devices. Whenever a DMA request is made by an IO device the DMA operation is performed and the various steps of DMA operation are as follows :

1. When a peripheral device requires a DMA, it will assert DRQ signal high.

2. When the DRQ of a channel is asserted high and if the channel is enabled then the 8257 will assert HRQ (HOLD Request) as high.

3. When the 8085/8086 processor receives a high signal on its HOLD pin, it will complete the current instruction execution and then drive all its tristate (address, data and control) pins to high impedance state and send an acknowledge signal to 8257 by asserting HLDA signal as high.

4. When the 8257 receive an acknowledge signal from 8086, the 8257 will send an acknowledge signal to the peripheral which requested DMA, by asserting $\overline{DACK}$ signal as low.

5. The 8257 asserts AEN high, which enable the DMA memory address latches and disables the processor address latch. Then the 8257 outputs low byte DMA address on A_0-A_7 lines and high byte DMA address on D_0 - D_7 lines. Also the ADSTB signal is asserted high to latch this address into external latches. Once the address is output on the address lines the content of address register is incremented by one and the count register is decremented by one.

6. Also the 8257 asserts appropriate read and write control signal to perform DMA transfer from peripheral to memory.

7. After performing one byte transfer the steps 5 and 6 are repeated again and again, until the terminal count (i.e., until the count reaches zero).

9.7 SERIAL DATA COMMUNICATION INTERFACE

9.7.1 Serial Data Communication

The fastest way of transmitting data within a microcomputer is parallel data transfer. For transferring data over long distances, however, parallel data transmission requires too many wires. Therefore, for long distance transmission, data is usually converted from parallel form to serial form so that it can be sent on a single wire or pair of wires. Serial data received from a distant source is converted to parallel form so that it can be easily transferred on the microcomputer buses.

Three terms often encountered in literature on communication systems are simplex, half-duplex and full-duplex. A simplex data line can transmit data only in one direction. Data from sensors to the processor and commercial radio stations are examples of simplex transmission.

Half-duplex transmission means that communication can take place in either direction between two systems, but can occur only in one direction at a time. An example of a half-duplex transmission is a two-way radio system, where one user always listens while the other talks because the receiver circuitry is turned off during transmit.

The term full-duplex means that each system can send and receive data at the same time. A normal phone conversation is an example of a full-duplex operation.

Serial data can be sent synchronously or asynchronously. In synchronous transmission, data are transmitted in blocks at a constant rate. The start and end of a block are identified with specific bytes or bit patterns. In asynchronous transmission, data is transmitted one by one. Each data has a bit which identifies its start and 1 or 2 bits which identifies its end. Since each data is individually identified, data can be sent at any time. Figure 9.42 shows the bit format often used for transmitting asynchronous serial data.

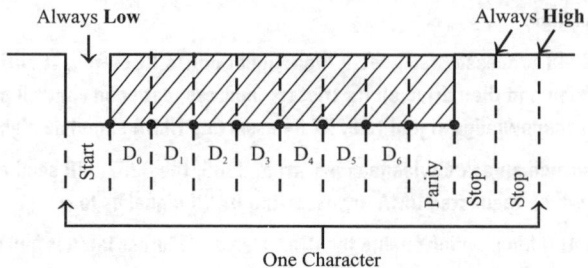

Fig. 9.42: Bit format used for sending asynchronous serial data.

When no data is being sent, the signal line will be at constant **high** or marking state. The beginning of a data character is indicated by the line going **low** for 1-bit time. This bit is called a start bit. The data bits are then sent out on the line one after the other. Note that the least significant bit is sent out first. Depending on the system, the data word may consist of 5, 6, 7 or 8 bits.

Following the data bits is a parity bit, which is used to check for errors in received data. Some systems do not insert or look for a parity bit. After the data bits and the parity bit, the signal line is returned **high** for at least 1-bit time to identify the end of the character. This **always-high** bit is referred to as a stop bit. Some systems may use 2 stop bits.

The term baud rate is used to indicate the rate at which serial data is being transferred. **Baud rate** is defined as $\dfrac{1}{\text{(The time for a bit cell)}}$. In some systems, one-bit cell has one data bit, then the baud rate and bits/second are same. In other cases, 2 to 4 actual data bits are encoded within one transmitted bit time, so data bits per second and baud do not correspond. Commonly used baud rates are 110, 300, 1200, 2400, 4800, 9600 and 19,200 bauds.

In order to interface a microcomputer with serial data lines, the data must be converted to and from serial form. A parallel-in-serial-out shift register and a serial-in-parallel-out shift register can be used to do this.

In some cases of serial data transfer, handshake signals are needed to make sure that a transmitter does not send data faster than it can be read in by the receiving system. The programmable devices INTEL 8251A, National JNS8250, etc., can be interfaced to microprocessors to perform such functions.

A device such as INTEL 8251A which can be programmed to do either asynchronous or synchronous communication is often called USART (**U**niversal **S**ynchronous **A**synchronous **R**eceiver **T**ransmitter). A device such as the National INS8250 which can only do asynchronous communication is often referred to as a **U**niversal **A**synchronous **R**eceiver **T**ransmitter (UART).

Once the data is converted to serial form, it must be in some way sent from the transmitting UART to the receiving UART. There are several ways in which serial data is commonly sent. One method is to use a current to represent a "**1**" in the signal line and no current to represent a "**0**". Another approach is to add line drivers at the output of the UART to produce a sturdy voltage signal. The range of each of these methods, however is limited to a few thousand feet.

For sending serial data over long distances, the standard telephone system is a convenient path, because the wiring and connections are already in place. Standard phone lines, often referred to as switched lines because any two points can be connected together through a series of switches and have a bandwidth of about 300 to 3000 Hz. But digital signals require very large bandwidth (typically 5 MHz). Therefore, for several reasons, digital signals cannot be sent directly over standard phone lines.

The solution to this problem is to convert the digital signals to audio-frequency tones, which are in the frequency range that the phone lines can transmit. The device used to do this conversion and to convert transmitted tones back to digital information is called a MODEM. The term is a contraction of **mo**dulator-**dem**odulator.

Modems and other equipment used to send serial data over long distances are known as data communication equipment or DCE. The terminals and computers that are sending or receiving the serial data are referred to as data terminal equipment or DTE.

9.7.2 RS-232C Serial Data Standard

In serial IO, data can be transmitted as either current or voltage. Several standards have been developed for serial communication. When data are transmitted as voltage, the commonly used standard is known as RS-232C. It was developed by **E**lectronics **I**ndustries **A**ssociation (EIA), USA and adopted by IEEE. This standard, proposes a maximum of 25 signals for the bus used for serial data transfer. The 25 signals of RS-232C are listed in Table 9.18. In practice, the first 9 signals are sufficient for most of the serial data transmission scheme and so the RS-232C bus signals are terminated on a D-type 9-pin connector. (When all the 25 signals are used, then the RS-232C serial bus is terminated on a 25-pin connector.)

The voltage levels for all RS-232C signals are,

Logic **low** = –3 V to –15 V under load (–25 V on no-load)

Logic **high** = +3 V to +15 V under load (+25 V on no-load)

Commonly used voltage levels are,

+12 V (logic **high**) and –12 V(logic **low**).

The RS-232C signal levels are not compatible with TTL logic levels. Hence, for interfacing TTL devices, level converters or RS-232C line drivers are employed. The popularly used level converters are:

MC1488   -   TTL to RS-232C level converter.

MC1489   -   RS-232C to TTL level converter.

MAX 232   -   Bidirectional level converter.

(Max 232 is equivalent to a combination of MC1488 and MC1489 in single IC.)

The signal level conversion using the above converters are shown in Fig. 9.43.

Table 9.18: RS-232C Pin Names and Signal Description

Pin number	Common name	RS-232C name	Description	Signal direction on DCE
1	-	AA	Protective ground	-
2	TxD	BA	Transmitted data	IN
3	RxD	BB	Received data	OUT
4	$\overline{RTS}$	CA	Request to send	IN
5	$\overline{CTS}$	CB	Clear to send	OUT
6	$\overline{DSR}$	CC	Data Set ready	OUT
7	GND	AB	Signal ground (Common return)	-
8	$\overline{CD}$	CF	Received line signal detector	OUT
9		-	Reserved for data set testing	-
10		-	Reserved for data set testing	-
11		-	Unassigned	-
12		SCF	Secondary received line signal detector	OUT
13		SCB	Secondary clear to send	OUT
14		SBA	Secondary transmitted data	IN
15		DB	Transmission signal element timing (DCE source)	OUT
16		SBF	Secondary received data	OUT
17		DD	Receiver signal element timing (DCE source)	OUT
18		-	Unassigned	-
19		SCA	Secondary request to send	IN
20	$\overline{DTR}$	CD	Data terminal ready	IN
21		CG	Signal quality detector	OUT
22		CE	Ring indicator	OUT
23		CH/CI	Data signal rate selector (DTE/DCE Source)	IN/OUT
24		DA	Transmit signal element timing (DTE source)	IN
25		-	Unassigned	-

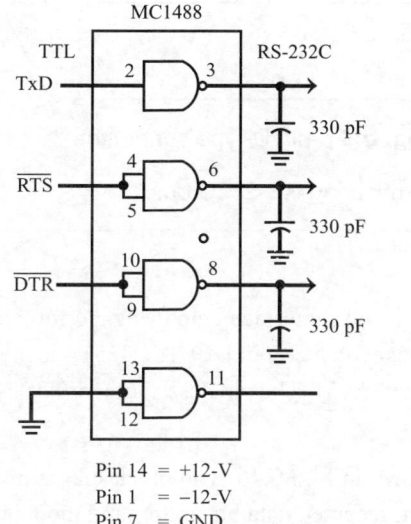

Pin 14 = +12-V
Pin 1 = −12-V
Pin 7 = GND

Fig. a: TTL to RS-232C Signal Conversion.

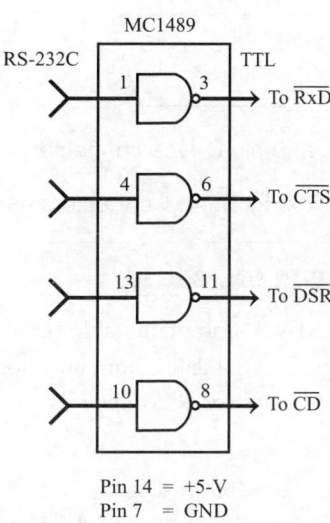

Pin 14 = +5-V
Pin 7 = GND

Fig. b: RS-232C to TTL Signal Conversion.

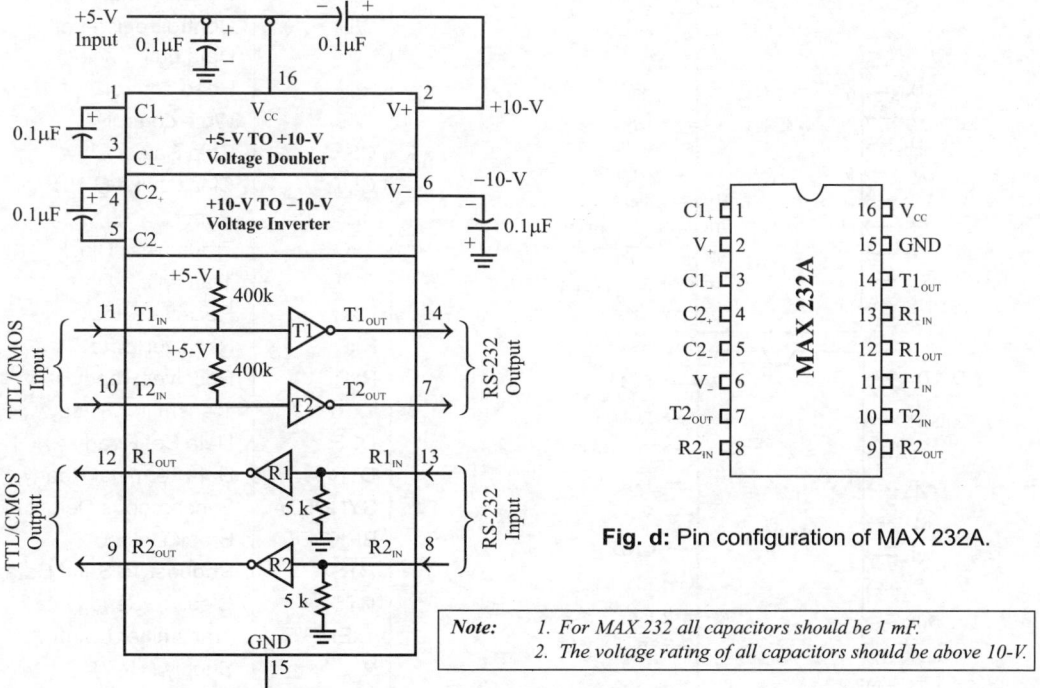

Fig. d: Pin configuration of MAX 232A.

| *Note:* | 1. For MAX 232 all capacitors should be 1 mF. |
| | 2. The voltage rating of all capacitors should be above 10-V. |

Fig. c: Typical circuit connection of MAX 232A.

Fig. 9.43: TTL to RS-232C and RS-232C to TTL signal conversion
and circuit connection of MAX 232.

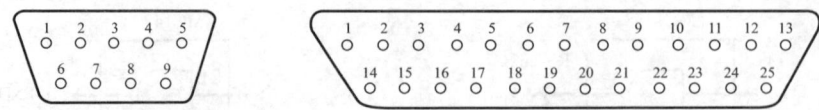

Fig. a: 9-pin D-type connector.   **Fig. b:** 25-pin D-type connector.

Fig. 9.44: Connectors used for terminating RS-232C bus.

9.8 USART-INTEL 8251A

The 8251A is a programmable serial communication interface chip designed for synchronous and asynchronous serial data communication. It is packed in a 28-pin DIP. The 8251A is the enhanced version of its predecessor, 8251 and it is compatible with 8251. The pin description of 8251A is shown in Fig. 9.45.

The functional block diagram of 8251A is shown in Fig. 9.46. The block diagram shows five sections, they are read/write control logic, transmitter, receiver, data bus buffer and modem control.

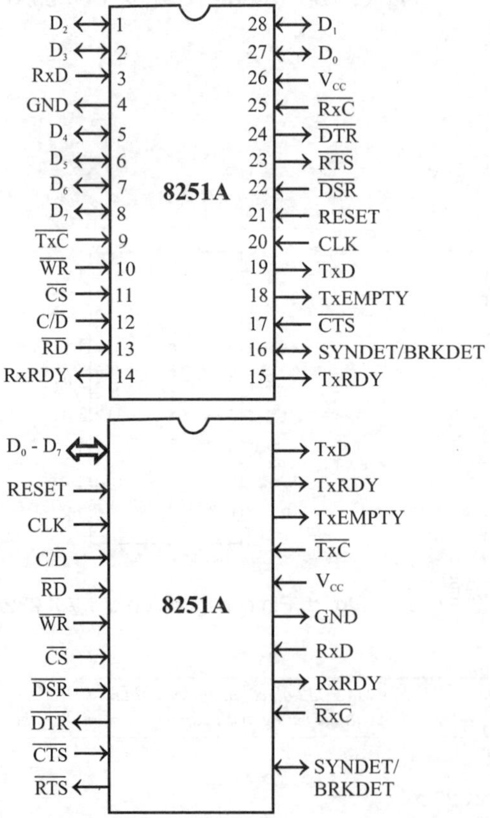

Pin	Description
D_0-D_7	Parallel data
C/$\overline{D}$	Control register or Data buffer select
$\overline{RD}$	Read control
$\overline{WR}$	Write control
$\overline{CS}$	Chip Select
CLK	Clock pulse (TTL)
RESET	Reset
$\overline{TxC}$	Transmitter Clock
TxD	Transmitter Data
$\overline{RxC}$	Receiver Clock
RxD	Receiver Data
RxRDY	Receiver Ready
TxRDY	Transmitter Ready
$\overline{DSR}$	Data Set Ready
$\overline{DTR}$	Data Terminal Ready
SYNDET/ BRKDET	Synchronous Detect / Break Detect
$\overline{RTS}$	Request To Send Data
$\overline{CTS}$	Clear To Send Data
TxEMPTY	Transmitter Empty
V_{cc}	Supply (+5-V)
GND	Ground (0-V)

Fig. 9.45: Pin description of 8251A.

Read/Write Control Logic

The Read/Write control logic interfaces the 8251A with the CPU, determines the functions of the 8251A according to the control word written into its control register and monitors the data flow. This section has three registers and they are control register, status register and data buffer.

The signals $\overline{RD}, \overline{WR}, C/\overline{D}$, and $\overline{CS}$ are used for read/write operations with these registers. When $C/\overline{D}$ is **high**, the control register is selected for writing control word or reading status word. When $C/\overline{D}$ is **low**, the data buffer is selected for read/write operation.

A **high** on the reset input forces 8251A into the idle mode. The clock input is necessary for 8251A for communication with CPU and this clock does not control either the serial transmission or the reception rate.

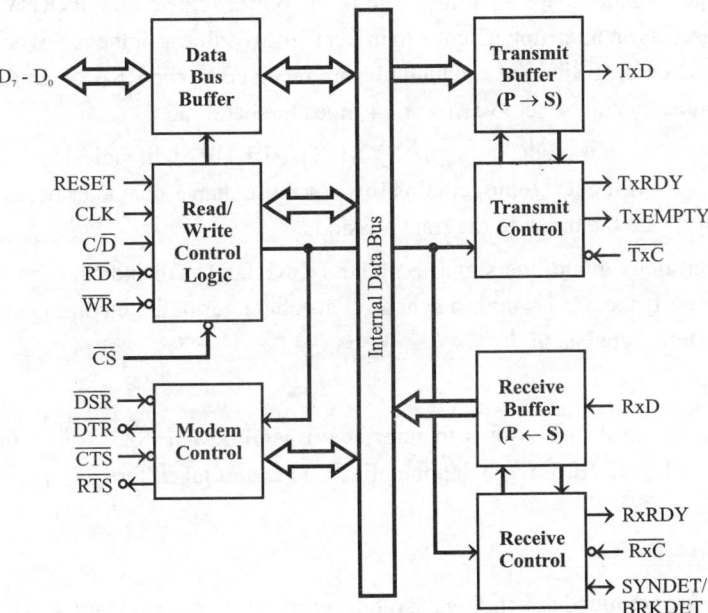

Fig. 9.46: The functional block diagram of 8251A - USART.

Transmitter Section

The transmitter section accepts parallel data from the CPU and converts them into serial data. The transmitter section is double buffered, i.e., it has a buffer register to hold an 8-bit parallel data and another register called output register to convert the previous data into a stream of serial bits.

The processor loads a data into the buffer register. When the output register is empty, the data is transferred from the buffer to the output register. Now, the processor can again load another data in the buffer register. If the buffer register is empty then TxRDY is asserted **high** and if the output register is empty then TxEMPTY is asserted **high**. These signals can also be used as interrupt or status for data transmission.

The clock signal, $\overline{TxC}$ controls the rate at which the bits are transmitted by the USART. The clock frequency can be 1, 16 or 64 times the baud rate.

Receiver Section

The receiver section accepts serial data and converts them into parallel data. The receiver section is double buffered, i.e., it has an input register to receive the serial data and convert it to parallel, and a buffer register to hold the previous converted data.

Normally, RxD line is **high**, when the RxD line goes **low**, the control logic assumes it as a START bit, waits for half a bit time and samples the line again. If the line is still **low** then the input register accepts the following bits, forms a character and loads it into the buffer register. The CPU reads the parallel data from the buffer register.

When the input register loads a parallel data to the buffer register, the RxRDY line goes **high**. This signal can be used as an interrupt or status to indicate the readiness of the receiver section to CPU. The clock signal $\overline{RxC}$ controls the rate at which bits are received by the USART. In the asynchronous mode, the clock frequency can be set to 1, 16 or 64 times the baud rate.

During the asynchronous mode, the signal SYNDET/BRKDET will indicate the intentional break in the data transmission. If the RxD line remains **low** for more than 2 character times then this signal is asserted **high** to indicate the break in the transmission.

During synchronous mode, the signal SYNDET/BRKDET will indicate the reception of the synchronous character. If the 8251A finds a synchronous character in the incoming string of data bits then it asserts SYNDET signal as **high**.

MODEM Control

The MODEM control unit allows to interface a MODEM to 8251A and to establish data communication through MODEM over telephone lines. This unit takes care of handshake signals for MODEM interface.

Programming the 8251A

The 8251A is programmed by sending the mode word and command word. First, reset the 8251A and then send a mode word to control register address. Next, the command word is sent to the same address. The CPU can check the readiness of the 8251A for data transfer by reading the status register. The format of control and status words are shown in Fig. 9.47.

The mode word informs 8251 about the baud rate, character length, parity and stop bits. The command word can be sent to enable the data transmission and/or reception. The information regarding the readiness of transmitter/receiver and the transmission errors can be obtained from the status word.

If 8251A is programmed for a baud rate factor of 64x through mode word then the baud rate is clock frequency divided by 64. If the baud rate factor is 16x then the baud rate is clock frequency divided by 16. If the baud rate factor is 1x then the baud rate is given by clock frequency.

9.8.1 Interfacing of 8251A with 8085

A simple schematic for interfacing the 8251A with 8085 processor is shown in Fig. 9.48. The 8251A can be either memory-mapped or IO-mapped in the system.

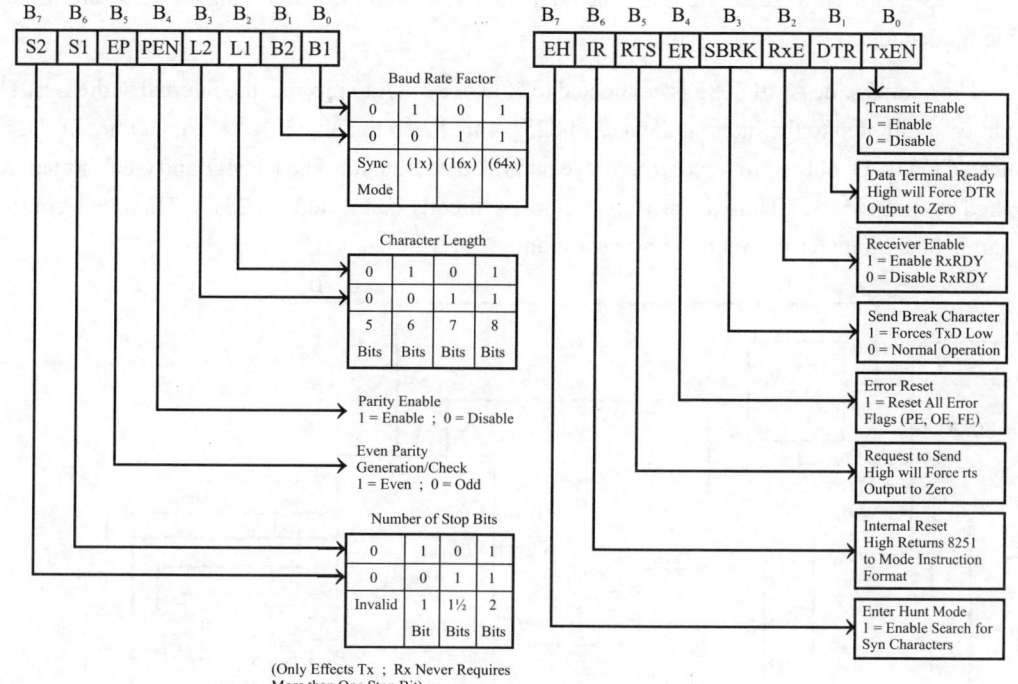

Fig.a: Mode word.   **Fig. b:** Command word.

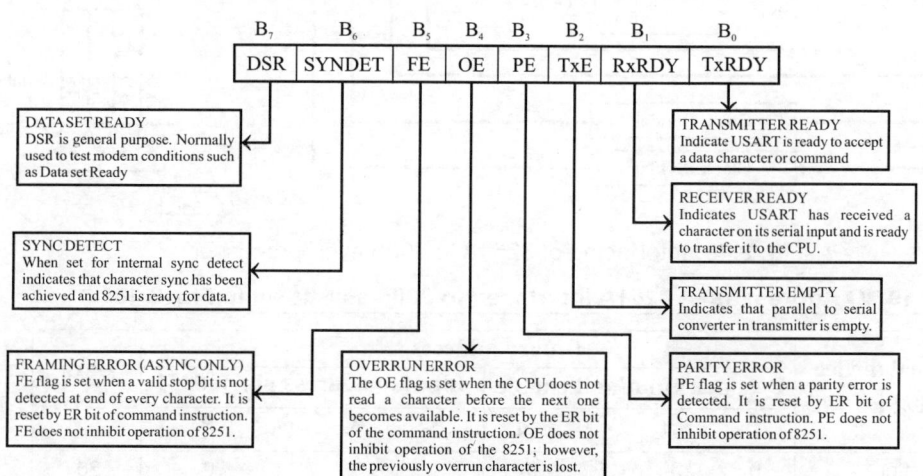

Fig. c: Status word.

Fig. 9.47: Format of 8251A mode, command and status words.

In the schematic shown in Fig. 9.48, the 8251A is IO-mapped in the system. The chip select signals for IO-mapped devices are generated by using a 3-to-8 decoder. The address lines A_4, A_5 and A_6 are decoded to generate eight chip select signals (IOCS-0 to IOCS-7) and in this, the chip select signal IOCS-2 is used to select 8251A. The address line A_7 and the control signal $IO/\overline{M}$ are used as enable for decoder.

The address line A_0 of 8085 is connected to $C/\overline{D}$ of 8251A to provide the internal addresses. The IO addresses allotted to the internal devices of 8251A are listed in Table 9.18. The data lines D_0-D_7 are connected to D_0-D_7 of the processor to achieve parallel data transfer. The RESET and clock signals are supplied by the processor. Here the processor clock is directly connected to 8251A. This clock controls the parallel data transfer between the processor and 8251A.

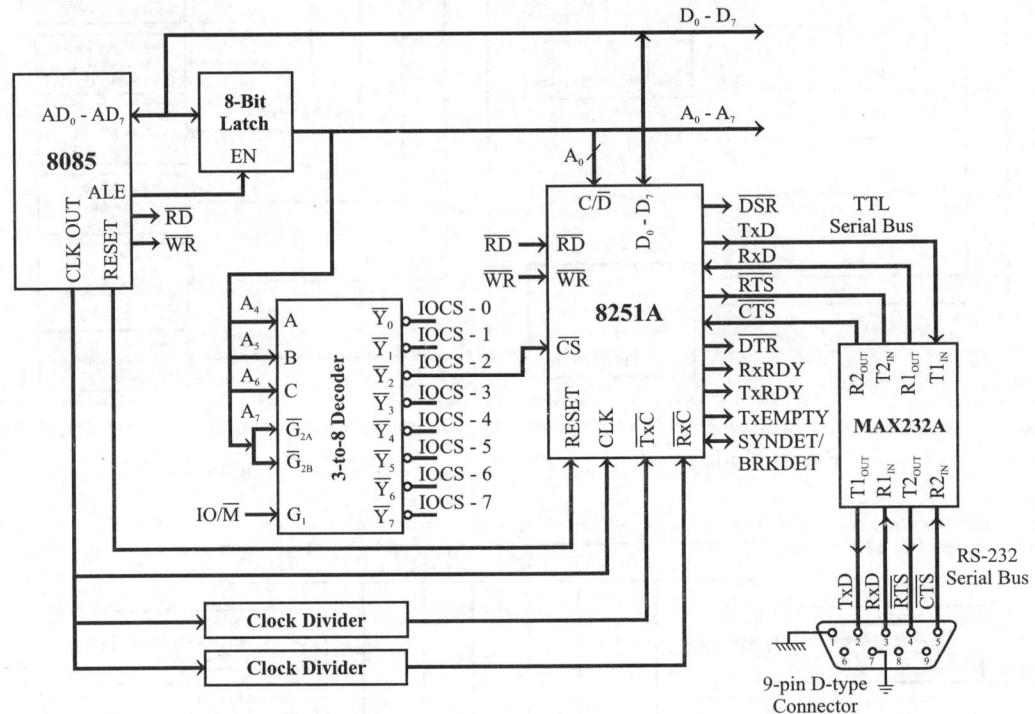

Fig. 9.48: Interfacing of 8251A to 8085 microprocessor.

Table 9.19: IO Addresses of 8251A Interfaced to 8085 as Shown in Fig. 9.48

Internal device of 8251A	Binary address								Hexa addresss
	Decoder input and enable				Input to address pin of 8251				
	A_7	A_6	A_5	A_4	A_3	A_2	A_1	A_0	
Data buffer	0	0	1	0	x	x	x	0	20
Control register	0	0	1	0	x	x	x	1	21

Note: The don't care "x" is considered as zero.

The output clock signal of 8085, is divided by suitable clock dividers and then used as clock for serial transmission and reception ($\overline{\text{TxC}}$ and $\overline{\text{RxC}}$). In 8251A the transmission and reception baud rates can be different or same. Usually a programmable timer, 8254 (which is discussed in Section 9.5) is used to divide the processor output clock and supply to $\overline{\text{TxC}}$ and $\overline{\text{RxC}}$ at the required rate.

The TTL logic levels of the serial data lines (RxD and TxD) and the control signals necessary for serial transmission and reception are converted to RS232 logic levels using MAX232 and then terminated on a standard 9-pin D-type connector. The device which requires serial communication with processor can be connected to this 9-pin D-type connector using 9-core cable.

The signals TxEMPTY, TxRDY and RxRDY can be used as interrupt signals to initiate interrupt driven data transfer scheme between processor and 8251A.

9.8.2 Interfacing of 8251A with 8086

A simple schematic for interfacing the 8251A with 8086 processor is shown in Fig. 9.49. The 8251A can be either memory-mapped or IO-mapped in the system. In the schematic shown in Fig. 9.49, the 8251A is IO-mapped in the system, with even addresses.

The chip select signals for IO-mapped devices are generated by using a 3-to-8 decoder. The address lines A_5, A_6 and A_7 are decoded to generate eight chip select signals (IOCS-0 to IOCS-7) and in this, the chip select signal IOCS-2 is used to select 8251A. The address line A_0 and the control signal $M/\overline{IO}$ are used as enable for the decoder.

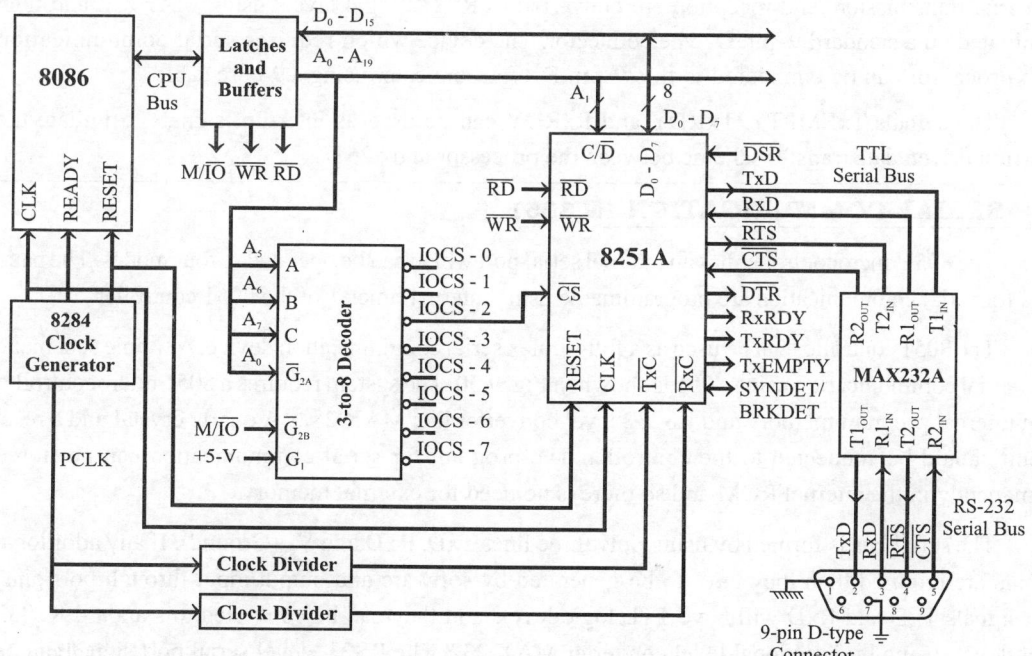

Fig. 9.49: Interfacing of 8251A to 8086 microprocessor.

Table 9.20: IO Addresses of 8251A Interfaced to 8086 as Shown in Fig. 9.49

Internal device of 8251A	Binary address								Hexa address
	Decoder input			Input to address pin of 8251				Decoder enable	
	A_7	A_6	A_5	A_4	A_3	A_2	A_1	A_0	
Data buffer	0	1	0	x	x	x	0	0	40
Control register	0	1	0	x	x	x	1	0	42

Note: The don't care "x" is considered as zero.

The address line A_1 of 8086 is connected to $C/\overline{D}$ of 8251A to provide the internal addresses. The IO addresses allotted to the internal devices of 8251A are listed in Table 9.20. The data lines D_0 - D_7 are connected to D_0 - D_7 of the processor to achieve parallel data transfer. The RESET and clock signals are supplied by 8284 clock generator. Here the processor clock is directly connected to 8251A.This clock controls the parallel data transfer between the processor and 8251A.

The **Peripheral Clock** (PCLK) supplied by 8284, is divided by suitable clock dividers and then used as clock for serial transmission and reception ($\overline{TxC}$ and $\overline{RxC}$). In 8251A, the transmission and reception baud rates can be different or same. Usually a programmable timer, 8254 (which is discussed in Section 9.5) is used to divide the PCLK, and supply to $\overline{TxC}$ and $\overline{RxC}$ at the required rate.

The TTL logic levels of the serial data lines (RxD and TxD) and the control signals necessary for serial transmission and reception are converted to RS232 logic levels using MAX232 and then terminated on a standard 9-pin D-type connector. The device which requires serial communication with processor can be connected to this 9-pin D-type connector using a 9-core cable.

The signals TxEMPTY, TxRDY and RxRDY can be used as interrupt signals to initiate the interrupt driven data transfer scheme between the processor and 8251A.

9.9 SERIAL COMMUNICATION IN 8051

The 8051 microcontroller has an internal serial port which can be operated in four modes. The baud rates for serial communication are programmable using internal timer-1 of the 8051 controller.

The 8051 controller can be used as a full-duplex serial communication device. A simple schematic for a serial communication using 8051 is shown in Fig. 9.50. The system requires a 8051 microcontroller with internal program memory and RS232 level converter like MAX 232. A quartz crystal and a reset circuit should be connected to the controller. The program for serial communication can be stored permanently in the internal ROM and so there is no need for external memory.

The serial bus is formed by using only three lines TxD, RxD and V_{ss}(Ground). If any additional signals are required then they have to be generated by software and output/input through port pins. The signals TxD and RxD will have TTL logic levels and they can be converted to standard RS232 logic levels using bi-directional level converter MAX 232. The RS232 level serial port signal can be terminated on a standard 9-pin D-type connector, so that any standard serial device can be connected to the 8051 controller for serial communication.

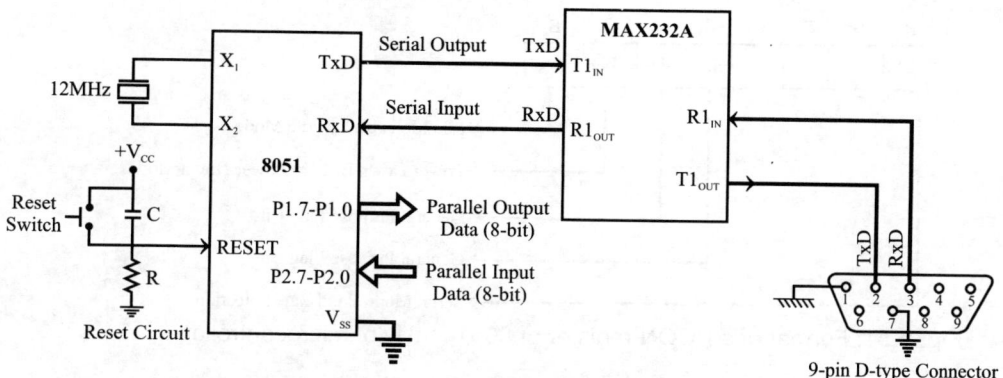

Fig. 9.50: Serial communication using 8051 microcontroller.

The parallel data transfer can be achieved through ports. In the schematic shown in Fig. 9.50, the parallel input data is received through port-2 and the parallel output data is sent through port-1. For example, the controller can receive parallel data from an ADC and convert it to serial and transmit via the serial port to another serial device.

Also, the controller can receive a serial data from another serial device via the serial port and convert to parallel, and then outputs through port-1 to a parallel device like DAC. The 8051 controller supports full-duplex communication and so the transmission and reception can be performed simultaneously.

9.9.1 Serial Data Buffer (SBUF) Register

The SBUF register is used to hold the parallel data during transmission and reception. During serial reception, the serial data is received via RxD pin and converted to parallel data and stored in the receive buffer. During serial transmission, the parallel data is stored in the transmit buffer and then converted to serial data to transmit via TxD pin.

The transmit and receive buffers are assigned the same internal address 99_H but the transmit buffer can be accessed only for write operation and the receive buffer can be accessed only for read operation. When data is written to SBUF, it goes to the transmit buffer and when data is read from SBUF it comes from the receive buffer.

9.9.2 Power Control Register (PCON)

The PCON register is used for power control and baud rate selection. The format of the PCON register is shown in Fig. 9.51.

The SMOD bit is used to decide the baud rate in serial port operating modes 1, 2 or 3. In mode-2, if SMOD = 0 then the baud rate is 1/64 of the oscillator frequency and if SMOD = 1 then the baud rate is 1/32 of oscillator frequency. (In 8051, the oscillator frequency and microcontroller internal frequency are same). In modes 1 and 3, the baud rate depends on the SMOD and timer-1 overflow rate.

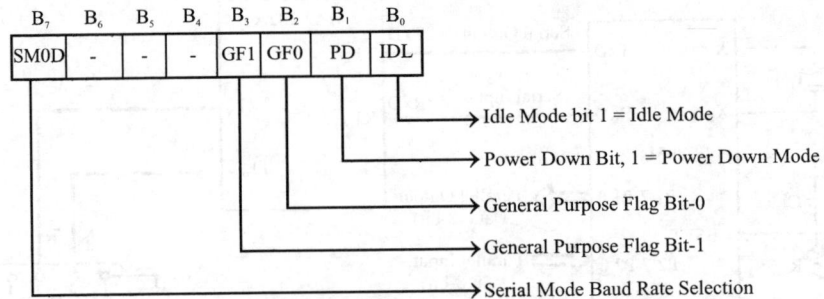

Fig. 9.51: Format of a PCON register of 8051 family of microcontrollers.

The baud rate in mode 1 or 3 $= \dfrac{2^{SMOD}}{32} \times$ (Timer-1 overflow rate).

When timer-1 is configured for auto reload mode then,

$$\text{Timer-1 overflow rate} = \dfrac{\text{Oscillator frequency}}{12 \times [256 - (TH1)^2]}$$

where, TH1 = Reload count value (8-bit) in higher order timer-1 count register.

The function of other bits of the PCON register are expalined in Chapter 2, Section 2.2.2.

9.9.3 Serial Port Control Register (SCON)

The format of a SCON register is shown in Fig. 9.52. The SCON register consists of mode selection bits, the 9th data bit (bit-B_8) for transmit and receive, and the serial port interrupt bits TI and RI. The bits SM0 and SM1 are used to select any one of the four operating modes for serial transmission and reception. The four modes of a serial port are mode-0, mode-1, mode-2 and mode-3.

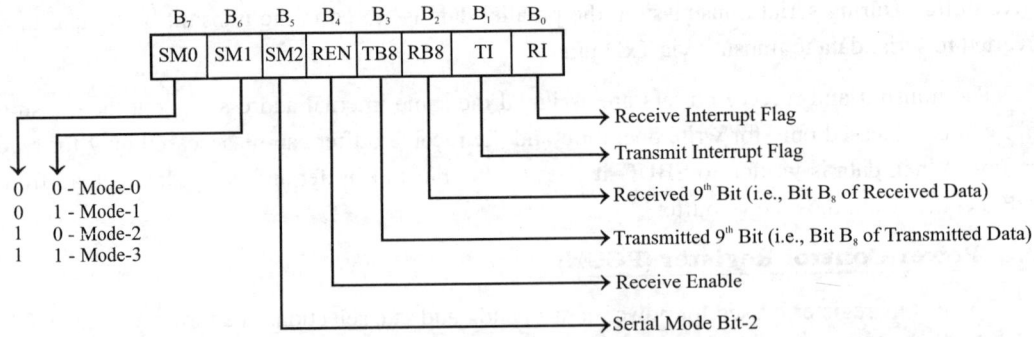

Fig. 9.52: Format of an SCON register of the 8051 family of microcontrollers.

Serial Port Mode-0

In this mode, the serial port functions as a half-duplex serial port with a fixed baud rate. The 8-bit serial data is received and transmitted through the RxD pin and the controller outputs the shift clock through the TxD pin during reception and transmission. The baud rate is fixed at 1/12 of the oscillator frequency.

Serial Port Mode-1

In this mode, the serial port functions as a full-duplex serial port with variable baud rates. In this mode, one data consists of 10 bits which includes one start bit, eight data bit and one stop bit. During reception, the stop bit is stored as RB8 in the SCON register. The baud rate in mode-1 depends on the value of the SMOD bit in the PCON register and the timer-1 overflow rate.

Serial Port Mode-2

In this mode, the serial port functions as a full duplex serial port with a baud rate of either 1/32 or 1/64 of the oscillator frequency. In this mode, one data consists of 11 bits which includes one start bit, eight data bits a programmable 9th data bit and one stop bit.

During transmission, the TB8 of the SCON register is added as the 9th data bit and during reception, the 9th data bit is stored as RB8 in the SCON register. (This 9th data bit can be used as the parity bit.) The baud rate depends on the value of the SMOD bit in the PCON register.

Serial Port Mode-3

Mode-3 is same as mode-2, except the baud rate. In mode-3, the baud rate is variable. The baud rate depends on the value of the SMOD bit in the PCON register and the timer-1 overflow rate.

The serial mode bit-2 (SM2) has no effect in mode-0 and when programmed for mode-0, the SM2 should be equal to zero. In mode-1, SM2 is used to check a valid stop bit during reception. In mode-1, if SM2 = 0 then **R**eceive **I**nterrupt (RI) is activated only when a valid stop bit is received.

In mode-2 and mode-3, the SM2 bit is used to enable multiprocessor communication. In multiprocessor communication, the serial port of a number of microcontrollers can be connected to a common serial bus. One controller will act as a master and all other controllers will act as slaves. A unique 8-bit address is assigned to each slave and the SM2 bit in all the slaves is set to 1.

When the SM2 bit is one, the slaves will consider the received byte as address and when the SM2 bit is zero, the slaves will consider the received byte as data. For communication with a slave, the master will first send an address byte and then a data byte.

The master initiates communication with a slave by sending the address of the slave on the bus. The address byte will be received by all the slaves. Since SM2 = 1, initially in all the slaves, the received byte will be considered as the address and the slaves will verify whether the received address matches with the assigned address.

The slave whose assigned address matches with the received address will clear its SM2 bit. Now, the SM2 bit of only one of the slaves will be zero.

Next, the master will send a data byte which is also received by all the slaves, but the data byte is accepted by the slave whose SM2 = 0 and so the receive interrupt is activated only in one of the slaves whose SM2 = 0. After reading the received data from the SBUF register, the SM2 bit of the slave should be set to one again to receive the next data.

The REN bit of the SCON register can be used to enable or disable the serial reception. When REN is set to one, the serial reception is enabled and when REN is cleared to zero, the serial reception is disabled.

The bits TI and RI of the SCON register are a transmit interrupt flag and a receive interrupt flag, respectively. They are also called serial data interrupt flags. The controller will set the TI bit during the transmission of stop bit of a data character in modes 1 to 3 and during the transmission of the 8th bit of a data character in mode-0.

Similarly, the controller will set the RI bit during the reception of stop bit of a data character in modes 1 to 3 and during the reception of the 8th bit of a data character is mode-0.

These two flags are logically ORed internally and used as an internal interrupt signal (called serial port interrupt) to interrupt the current program being executed by the controller. On receiving the serial port interrupt, the controller has to suspend the current program execution and begins to execute a subroutine program to check the value of bits TI and RI of the SCON register.

If TI is one then the controller can understand that the previous character has been transmitted and so the controller can execute another subroutine to clear TI flag and load the next data in the SBUF register.

If RI is one, the controller can understand that a character has been received and so the controller can execute another subroutine to clear the RI flag and read the data from the SBUF register.

9.9.4 Baud Rate in 8051

In the 8051 microcontroller, the baud rate is programmable in serial communication modes 1 and 3, and the baud rate is fixed in modes 0 and 2. In the 8051 microcontroller, the timer-1 is dedicated to generate the required baud rate clock for serial communication modes 1 and 3.

The timer-1 overflow rate or output frequency is divided by 32 or 64 and used as baud rate clock (or transmit and receive clock) in serial communication modes 1 and 3.

The crystal frequency of 11.0592 MHz is used for serial communication with standard PC (Personal Computer) in order to generate the clock for right baud rate for serial communication. The standard baud rates for serial communication with PC are 110, 150, 300, 600, 1200, 2400, 4800, 9600 and 19200. When the crystal frequency is 11.0592 MHz, the timer input clock frequency will be 11.0592/12 = 0.9216 MHz = 921.6 kHz.

The SMOD bit in the PCON register can be programmed to divide the timer overflow rate or clock frequency by either 64 or 32, to generate a timer clock frequency for right baud rate.

The various timer clock frequency, and the initial count for timer overflow to achieve right baud rate clock for serial transmission and reception are listed in Table 9.21.

Table 9.21: Baud Rate Frequency and Timer Initial Count for crystal frequency of 11.059 MHz

Timer overflow rate	Initial count	SMOD	Baud rate frequency
$\dfrac{921.6 \times 10^3}{3}$	$256 - 3 = 253 = FD_H$	0	$\dfrac{921.6 \times 10^3}{3 \times 64} = 4800\,Hz$
		1	$\dfrac{921.6 \times 10^3}{3 \times 32} = 9600\,Hz$
$\dfrac{921.6 \times 10^3}{6}$	$256 - 6 = 250 = FA_H$	0	$\dfrac{921.6 \times 10^3}{6 \times 64} = 2400\,Hz$
		1	$\dfrac{921.6 \times 10^3}{6 \times 32} = 4800\,Hz$
$\dfrac{921.6 \times 10^3}{12}$	$256 - 12 = 244 = F4_H$	0	$\dfrac{921.6 \times 10^3}{12 \times 64} = 1200\,Hz$
		1	$\dfrac{921.6 \times 10^3}{12 \times 32} = 2400\,Hz$
$\dfrac{921.6 \times 10^3}{24}$	$256 - 24 = 232 = E8_H$	0	$\dfrac{921.6 \times 10^3}{24 \times 64} = 600\,Hz$
		1	$\dfrac{921.6 \times 10^3}{24 \times 32} = 1200\,Hz$

9.9.5 Examples of Serial Port Programming in 8051 Controller

EXAMPLE PROGRAM 9.1

This example program is developed to receive serial data from a standard PC at 9600 baud rate and store as parallel data in internal RAM of microcontroller.

Problem Analysis

In order to receive serial data at 9600 baud rate, the timer-1 should be programmed for mode-1, and an initial count for 9600 baud rate should be loaded in the timer count register. The SCON register is programmed for mode-1 serial communication. Then the timer-1 is started by setting the timer 1 run flag and the controller has to wait for serial receive interrupt flag to go high.

When a serial data is received and converted to parallel data and loaded in the SBUF register, the receive interrupt flag will go high. Now the controller has to read the SBUF data and load in memory and clear the receive interrupt flag, in order to enable serial port to receive next data. Here, the data is stored in the memory location 7FH, and memory will have the last received data.

The byte to be loaded in the TMOD register to select mode-2 operation of timer-1 is framed as follows.

TMOD = | Gate | C/$\overline{T}$ | M1 | M0 | Gate | C/$\overline{T}$ | M1 | M0 | = 0 0 1 0 X X X X = 20_H

The byte to be loaded in the SCON register to select mode-1 serial communication is framed as follows.

SCON = | SM0 | SM1 | SM2 | REN | TB8 | RB8 | TI | RI | = 0 1 0 1 0 0 0 0 = 50_H

Flowchart

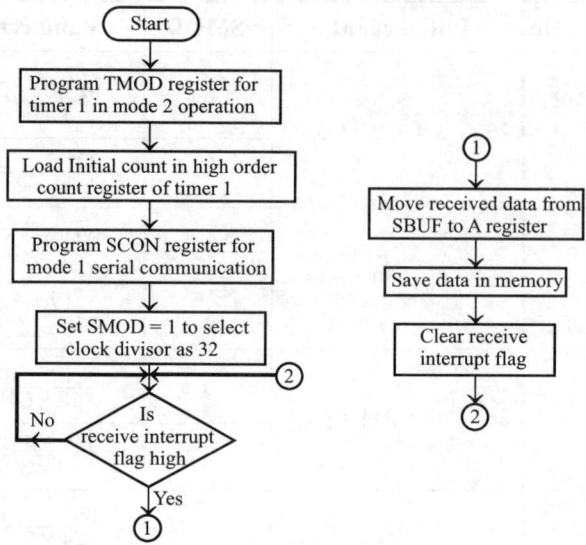

Assembly Language Program

```
;Program to receive serial data in 8051

        MOV   TMOD,#20H    ;Program timer-1 for mode-2 operation
        MOV   TH1,#FDH     ;Load initial count for 9600 baud rate in timer-1 high order register
        MOV   SCON,#50H    ;Program SCON reister for mode-1 serial communication
        MOV   A,PCON       ;Set SMOD = 1, via A register
        SETB  ACC.7        ;to choose clock divisor as 32
        MOV   PCON, A
        SETB  TR1          ;Start timer-1
WAIT:   JNB   RI,WAIT      ;Wait for receive interrupt
        MOV   A,SBUF       ;Save serial data in A register
        MOV   7FH,A        ;Save data in internal RAM
        CLR   RI           ;Clear receive interrupt flag to receive next data
        SJMP  WAIT         ;Wait for next receive interrupt

        END                ;Assemby end
```

EXAMPLE PROGRAM 9.2

This example program is developed to transmit the word 'HELO' by serial communication to a standard PC at 9600 baud rate.

Problem Analysis

In order to transmit serial data at 9600 baud rate, the timer-1 should be programmed for mode-1, and an initial count suitable for 9600 baud rate should be loaded in timer count register. The SCON register is programmed for mode-1 serial communication.

Then the timer 1 is started by setting the timer-1 run flag. The ASCII value of first character to be transmitted should be loaded in SBUF register. At the end of serial transmission, the transmit interrupt flag is set. Now the controller can clear this transmit interrupt flag and load the next character in SBUF, and this process is repeated until transmission of all the characters one by one.

The byte to be loaded in the TMOD register to select mode-2 operation of timer-1 is framed as follows.

TMOD = | Gate | C/$\overline{T}$ | M1 | M0 | Gate | C/$\overline{T}$ | M1 | M0 | = 0 0 1 0 X X X X = 20_H

The byte to be loaded in the SCON register to select mode-1 serial communication is framed as follows.

SCON = | SM0 | SM1 | SM2 | REN | TB8 | RB8 | TI | RI | = 0 1 0 1 0 0 0 0 = 50_H

Flowchart

Flowchart For MainProgram

Flowchart For Subroutine STRF

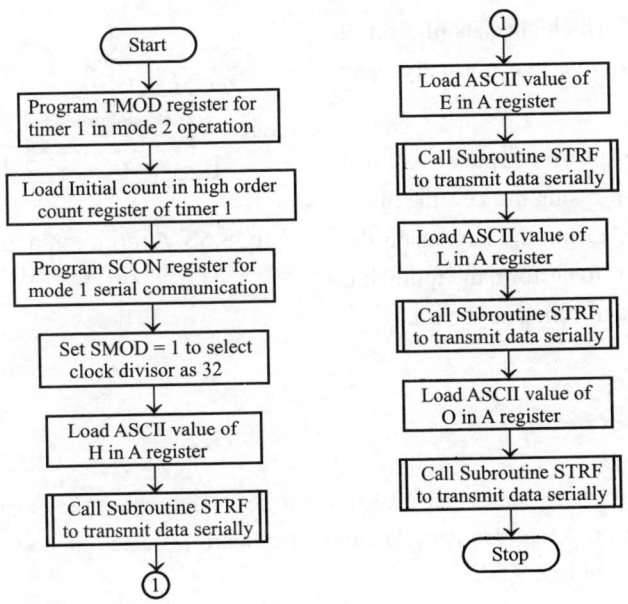

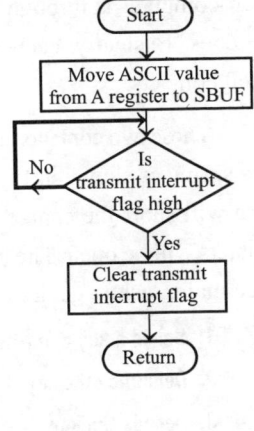

Assembly Language Program

```
;Program for serial data transmission in 8051
        MOV   TMOD,#20H    ;Program timer-1 for mode-2 operation
        MOV   TH1,#FDH     ;Load initial count for 9600 baud rate in timer-1 high order register
        MOV   SCON,#50H    ;Program SCON reister for mode 1 serial communication
        MOV   A,PCON       ;Set SMOD = 1, via A register
        SETB  ACC.7        ;to choose clock divisor as 32
        MOV   PCON, A
        SETB  TR1          ;Start timer-1

        MOV   A,#'H'       ;Load ASCII value of H in A register
        ACALL STRF         ;Transfer the content of A register serially

        MOV   A,#'E'       ;Load ASCII value of E in A register
        ACALL STRF         ;Transfer the content of A register serially

        MOV   A,#'L'       ;Load ASCII value of L in A register
        ACALL STRF         ;Transfer the content of A register serially
```

```
            MOV  A,#'0'      ;Load ASCII value of 0 in A register
            ACALL STRF       ;Transfer the content of A register serially
NEXT:       SJMP NEXT        ;Wait in infinite loop for next task
STRF:       MOV  SBUF,A      ;Load data to be transmitted serially in SBUF
WAIT:       JNB  TI,WAIT     ;Wait for end of serial transmission
            CLR  TI          ;Clear transmit interrupt flag
            RET

            END              ;Assembly end
```

9.10 KEYBOARD AND DISPLAY INTERFACE

9.10.1 Keyboard Interface using 8255 Ports and 8085/8086 Processor

A common method of entering programs into a microcomputer is through a keyboard which consists of a set of switches. Basically, each switch will have two normally open metal contacts.

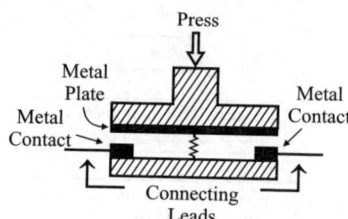

Fig. 9.53: A representation of keyboard switch.

These two contacts can be shorted by a metal plate supported by a spring as shown in Fig. 9.53. On pressing the key, the metal plate will short the contacts and on releasing the key, again the contacts will be open. The processor has to perform the following three major task to get a meaningful data from a keyboard.

1. **Sense a key actuation**

2. **Debounce the key**

3. **Decode the key**

The three major tasks mentioned above can be performed by the software, when a keyboard is connected through ports to 8085/8086 processor. Consider a simple keyboard in which the keys are arranged in rows and columns as shown in Fig. 9.54.

The rows are connected to port-A lines of 8255. The columns are connected to port-B lines, of the same chip. The rows and columns are normally tied **high**. At the intersection of a row and column, a key is placed such that pressing a key will short the row and the column.

A key actuation is sensed by sending a **low** to all the rows through port-A. Pressing a key will short the row and column to which it is connected, and so the column to which the key is connected will be pulled **low**.

Therefore, the columns are read through port-B to see whether any of the normally **high** columns are pulled **low** by a key actuation. If they are, then rows can be checked individually to determine the row in which the key is down. For checking each row, the scan code of the type shown in Table 9.22 are output to port-A one by one. This process of sensing a key actuation is called keyboard scanning.

Table 9.22: Scan Code for Keyboard Scanning

PA_3	PA_2	PA_1	PA_0
1	1	1	0
1	1	0	1
1	0	1	1
0	1	1	1

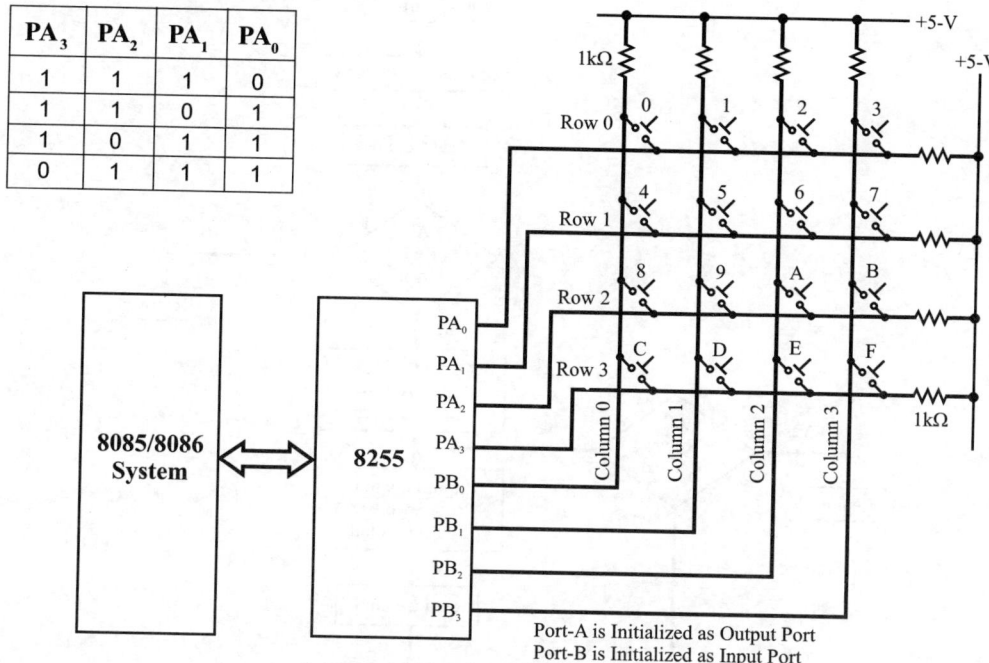

Port-A is Initialized as Output Port
Port-B is Initialized as Input Port

Fig. 9.54: Keyboard interfacing using ports.

A key press has to be accepted only after debouncing. Normally, the key bounces for 10 to 20 milliseconds when it is pressed and released. The bouncing time depends on the type of key. When this bounce occurs, it may appear to the microcomputer that the same key has been actuated several times instead of just once. This problem can be eliminated by scanning the row in which the key press is deducted after 10 to 20 milliseconds and then verifying to see if the same key is still down. If it is then the key actuation is valid. This process is called **key debouncing**.

After debouncing, the code for the key has to be generated. Each key can be individually identified by the port-A output value (row code) and port-B input value (column code). The next step is to translate the row and column code into a more popular code such as a hexadecimal or an ASCII. This can easily be accomplished by a program. The flowchart for the keyboard scanning when the keyboard is interfaced using ports is shown in Fig. 9.55.

In keyboard interfacing, there are two methods of handling multiple key presses and they are two-key lockout and N-key rollover. The two-key lockout takes into account only one key pressed. An additional key pressed and released does not generate any codes. The system is simple to implement and more often used. However, it might slow down the typing since each key must be fully released before the next one is pressed down. On the other hand, the N-key rollover will detect all the keys pressed in the order of entry and generate a corresponding keycode.

The disadvantage in keyboard interfacing using ports is that most of the processor time is utilized (or wasted) in keyboard scanning and debouncing.

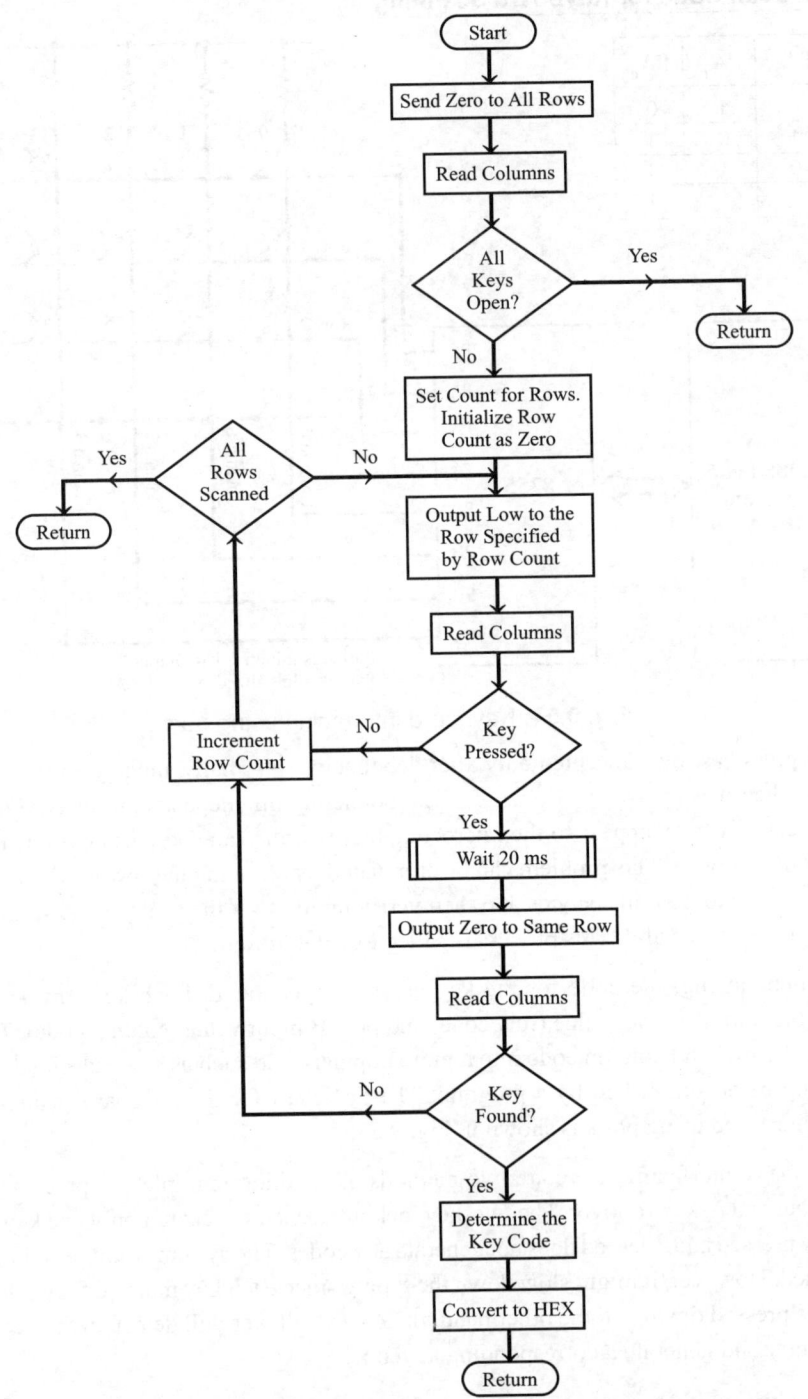

Fig. 9.55: Flowchart for keyboard scanning subroutine.

EXAMPLE PROGRAM 9.3

This example program is developed for the hex keyboard interface using 8255 ports and 8085 microprocessor based system shown in Fig. 9.54. The hex keys are interfaced to microprocessor via IO ports of 8255 and the IO ports are IO mapped in the system with 8-bit IO port addresses. The scan codes and the corresponding hex keys are listed in the following table. The scan codes are stored in system memory and SI and DI registers are used as address pointers for scan codes.

The program has been developed to wait for a key press. When a key is pressed the program starts scanning. The A-register is initialized with the hex code 00H. The scanning is performed by output a row scan code to port-A and reading column data through port-B and check with column scan code. After every column scan the hex value in A-register is incremented. When a column data, matches with column scan code, the scanning process stopped, and the hex key value in A-register is stored in memory location 1400_H. Then the program wait for key debouncing and start scanning next key press. The content of memory at 1400_H will be the hex value of last pressed key.

Table 1: Scan Code and Corresponding Hex Key

Row scan code									Column scan code									Hex key
PA_7	PA_6	PA_5	PA_4	PA_3	PA_2	PA_1	PA_0	Hex	PB_7	PB_6	PB_5	PB_4	PB_3	PB_2	PB_1	PB_0	Hex	
*	*	*	*	1	1	1	0	$0E_H$	*	*	*	*	1	1	1	0	0EH	0
									*	*	*	*	1	1	0	1	0DH	1
									*	*	*	*	1	0	1	1	0BH	2
									*	*	*	*	0	1	1	1	07H	3
*	*	*	*	1	1	0	1	$0D_H$	*	*	*	*	1	1	1	0	0EH	4
									*	*	*	*	1	1	0	1	0DH	5
									*	*	*	*	1	0	1	1	0BH	6
									*	*	*	*	0	1	1	1	07H	7
*	*	*	*	1	0	1	1	$0B_H$	*	*	*	*	1	1	1	0	0EH	8
									*	*	*	*	1	1	0	1	0DH	9
									*	*	*	*	1	0	1	1	0BH	A
									*	*	*	*	0	1	1	1	07H	B
*	*	*	*	0	1	1	1	07_H	*	*	*	*	1	1	1	0	0EH	C
									*	*	*	*	1	1	0	1	0DH	D
									*	*	*	*	1	0	1	1	0BH	E
									*	*	*	*	0	1	1	1	07H	F

Flowchart

Flowchart for Main Program

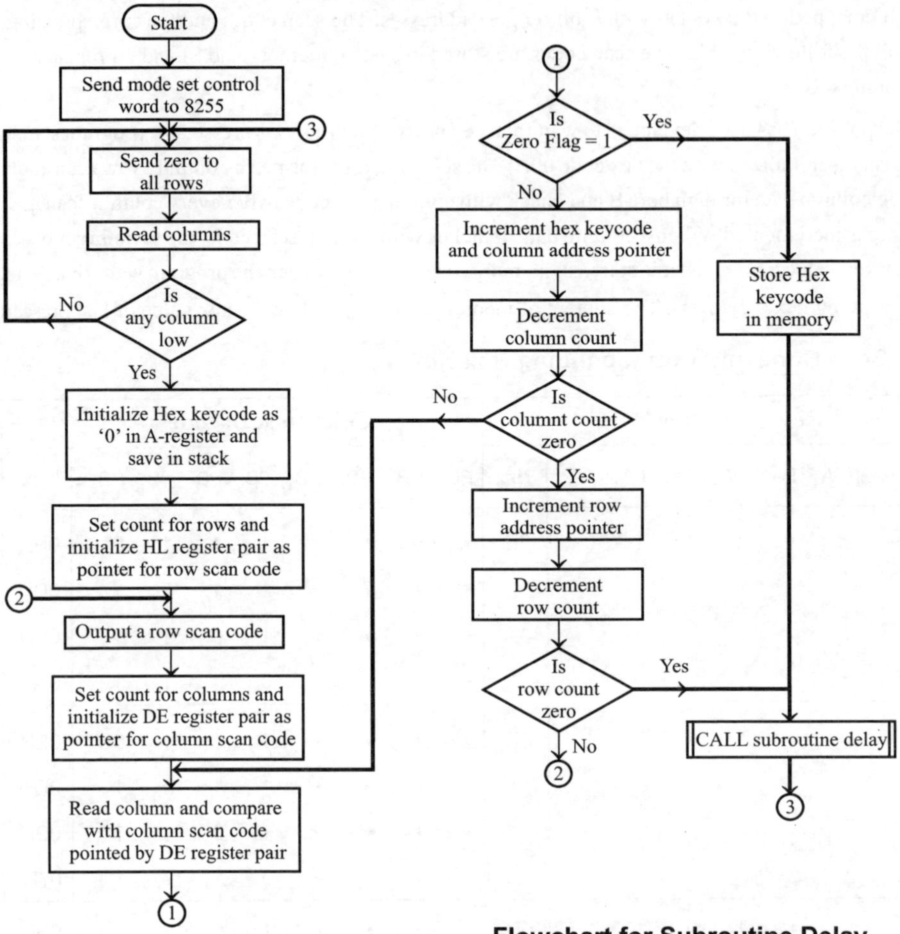

(Start)

↓

Send mode set control
word to 8255

③ →

Send zero to
all rows

↓

Read columns

↓

Is
any column
low

No →

Yes ↓

Initialize Hex keycode as
'0' in A-register and
save in stack

↓

Set count for rows and
initialize HL register pair as
pointer for row scan code

② →

Output a row scan code

↓

Set count for columns and
initialize DE register pair as
pointer for column scan code

↓

Read column and compare
with column scan code
pointed by DE register pair

↓

①

---

①

↓

Is
Zero Flag = 1   Yes →

No ↓

Increment hex keycode
and column address pointer

↓

Decrement
column count

↓

No ←   Is
columnt count
zero

Yes ↓

Increment row
address pointer

↓

Decrement
row count

↓

Is
row count
zero   Yes →

No ↓

②

Store Hex
keycode
in memory

CALL subroutine delay

↓

③

Flowchart for Subroutine Delay

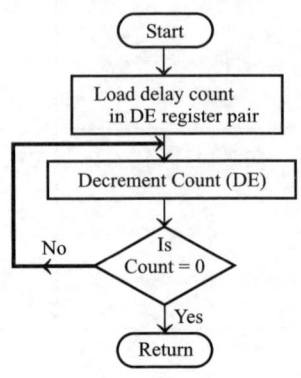

(Start)

↓

Load delay count
in DE register pair

↓

Decrement Count (DE)

↓

Is
Count = 0

No ←

Yes ↓

(Return)

Assembly Language Program

```
;Program to scan hex keyboard and store the hex value of key pressed in memory
PORTA  EQU  100H          ;Address of port-A of 8255
PORTB  EQU  11H           ;Address of port-B of 8255
CNREG  EQU  13H           ;Address of 8255 control register
MSCW   EQU  82H           ;Mode set Control word of 8255-Port-A input and Port-B output

       ORG  1000H         ;Assembler directive
       MVI  A,MSCW        ;Load control word in A-register to set 8255 ports in mode-0
       OUT  CNREG         ;Send mode set control word to control register
       LXI  SP, 1500H     ;Initialize stack
WAIT:  MVI  A,00H         ;Send zero to all rows
       OUT  PORTA

       IN   PORTB         ;Read columns. If any keypress then one column will low
       ANI  0FH           ;Mask upper nibble
       CPI  0FH           ;If no keypress, then ZF is set, go to WAIT
       JZ   WAIT          ;If any keypress, then ZF is reset, start scanning
AGAIN: MVI  A,00H         ;Set initial keycode in A-register as zero
       PUSH PSW           ;Save keycode in stack
       LXI  H,RSCAN       ;Initialize HL register pair as address pointer of row scan code
       MVI  C,04H         ;Load C-register with row count
L2:    MOV  A,M           ;Get row scan code in A-register
       OUT  PORTA         ;Send a scan code to rows
       LXI  D,CSCAN       ;Initialise DE register pair as address pointer of column scan code
       MVI  B,04H         ;Load B-register with column count

L1:    IN   PORTB         ;Read the columns
       ANI  0F            ;Mask upper nibble
       XCHG               ;Exchange address pointers
       CMP  A,M           ;Check column data with column scan code
       XCHG               ;Exchange address pointers
       JZ   STORE         ;If equal store keycode

       POP  PSW           ;Load previous keycode in A-register from stack
       INR  A             ;Increment A-register to get next keycode
       PUSH PSW           ;Save keycode in stack
       INX  D             ;Increment column pointer
       DCR  B             ;Decrement column count
       JNZ  L1            ;Jump on non-zero to scan next column
       INX  H             ;Increment row pointer
       DCR  C             ;Decrement row counter
       JNZ  L2            ;Jump on non-zero to L2
       JMP  SKIP

STORE: POP  PSW           ;Retrieve keycode from stack
       STA  1400H         ;Save hex code in memory
SKIP:  CALL DELAY         ;Wait for key debouncing
       JMP  WAIT          ;Start scanning next keypress

;DELAY ROUTINE
DELAY: LXI  D,0F429H      ;Load DE register pair with delay count
REPT:  DCX  D             ;Decrement DE register pair
       MOV  A,E           ;Check whether both D and E register
       CMP  D             ;contents are zeros
       JNZ  REPT          ;Jump on non-zero to REPT
RET                       ;If zero, return to main program

RSCAN: DB   0EH,0DH,0BH,07H
CSCAN: DB   0EH,0DH,0BH,07H

END                      ;Assembler directive
```

EXAMPLE PROGRAM 9.4

This example program is developed for the hex keyboard interface using 8255 ports and 8086 microprocessor based system shown in Fig. 9.54. The hex keys are interfaced to microprocessor via IO ports of 8255 and the IO ports are IO mapped in the system with 16-bit IO addresses. The scan codes and the corresponding hex keys are listed in the following table. The scan codes are stored in system memory and SI and DI registers are used as address pointers for scan codes.

The program has been developed to wait for a key press. When a key is pressed the program starts scanning. The AH register is initialized with the hex code 00_H. The scanning is performed by output a row scan code to port-A and reading column data through port-B and check with column scan code. After every column scan the hex value in AH register is incremented. When a column data, matches with column scan code, the scanning process stopped, and the hex key value in AH is stored in memory location 01400H. Then the program wait for key debouncing and start scanning next keypress. The content of memory at 01400H will be the hex value of last pressed key.

The scan code and corresponding Hex key are same as that of Table-1 in Example Program 9.3.

Flowchart

Flowchart for Main Program

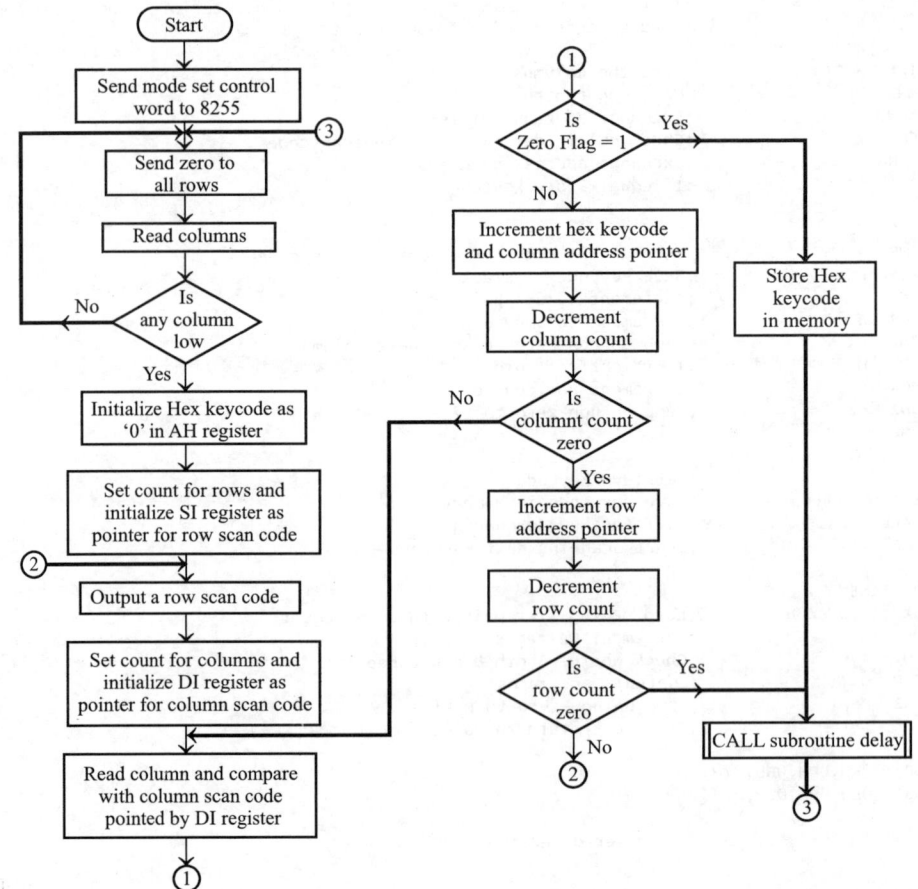

Flowchart for Subroutine Delay

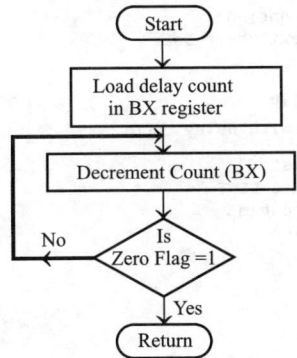

Assembly Language Program

```
;Program to scan hex keyboard and store the hex value of key pressed in memory
PORTA EQU 0000H          ;Address of port-A of 8255
PORTB EQU 0002H          ;Address of port-B of 8255
CNREG EQU 0006H          ;Address of 8255 control register
MSCW  EQU 82H            ;Mode set Control word of 8255

CODE SEGMENT             ;Assembler directive
ASSUME CS:CODE           ;Assembler directive
      ORG 1000H          ;Assembler directive
      MOV DX,CNREG        ;Get control register address in DX register
      MOV AL,MSCW         ;Set 8255 ports in mode-0
      OUT DX,AL           ;Send mode set control word to controlregister

WAIT: MOV DX,PORTA        ;Get port-A address in DX register
      MOV AL,00H          ;Send zero to all rows
      OUT DX,AL
      MOV DX,PORTB        ;Read columns
      IN  AL,DX           ;If any keypress then one column will low
      AND AL,0FH          ;Mask upper nibble
      CMP AL,0FH          ;If no keypress, then ZF is set, go to WAIT
      JZ  WAIT            ;If any keypress, then ZF is reset, start scanning

AGAIN: MOV AH,00H         ;Set initial keycode in AH as zero
       MOV SI,OFFSET RSCAN ;Initialize SI register with offset address of row scan code
       MOV CH,04H         ;Load CH register with row count

L2:   MOV AL,[SI]         ;Get row scan code in AL register
      MOV DX,PORTA        ;Load port-A address in DX register
      OUT DX,AL           ;Send a scan code to rows
      MOV DI,OFFSET CSCAN ;Initialise DI register with offset address of column scan code
      MOV CL,04H          ;Load CL register with column count
      MOV DX,PORTB        ;Load port-B address in DX register

L1:   IN  AL,DX           ;Read the columns
      AND AL,0FH          ;Mask upper nibble
      CMP AL,[DI]         ;Check column data with column scan code
      JZ  STORE           ;If equal store keycode
      INC AH              ;Increment AH register to get next keycode
      INC DI              ;Increment column pointer
      DEC CL              ;Decrement column count
      JNZ L1              ;Jump on non-zero to scan next column
      INC SI              ;Increment row pointer
      DEC CH              ;Decrement row counter
      JNZ L2              ;Jump on non-zero to L2
      JMP SKIP
```

```
STORE: MOV DI,1400H          ;Load offset address in DI register
       MOV [DI],AL           ;Save hex code in memory
SKIP:  CALL DELAY            ;Wait for key debouncing
       JMP WAIT              ;Start scanning next keypress

;DELAY ROUTINE
DELAY PROC NEAR              ;Assembler directive

      MOV BX,7FFFH           ;Load BX register with delay count

REPT: DEC BX                 ;Decrement BX register
      JNZ REPT               ;Jump on non-zero to REPT
      RET                    ;Return to main program
DELAY ENDP                   ;Assembler directive

RSCAN: DB 0EH,0DH,0BH,07H
CSCAN: DB 0EH,0DH,0BH,07H
CODE ENDS                    ;Assembler directive
END                          ;Assembler directive
```

9.10.2 Keyboard Interface using Ports of 8051 Controller

The general concepts of keyboard interface using ports are discussed in Section 9.10.1. The interfacing of hex keyboard with the ports of a 8051 microcontroller is shown in Fig. 9.56. Here, the 4×4 matrix keyboard is formed using the lower 4 lines of port-0 and upper 4 lines of port-2. Normally, all the port pins are pulled high by pullup resistors. A hex key is placed at the intersection of a row and a column.

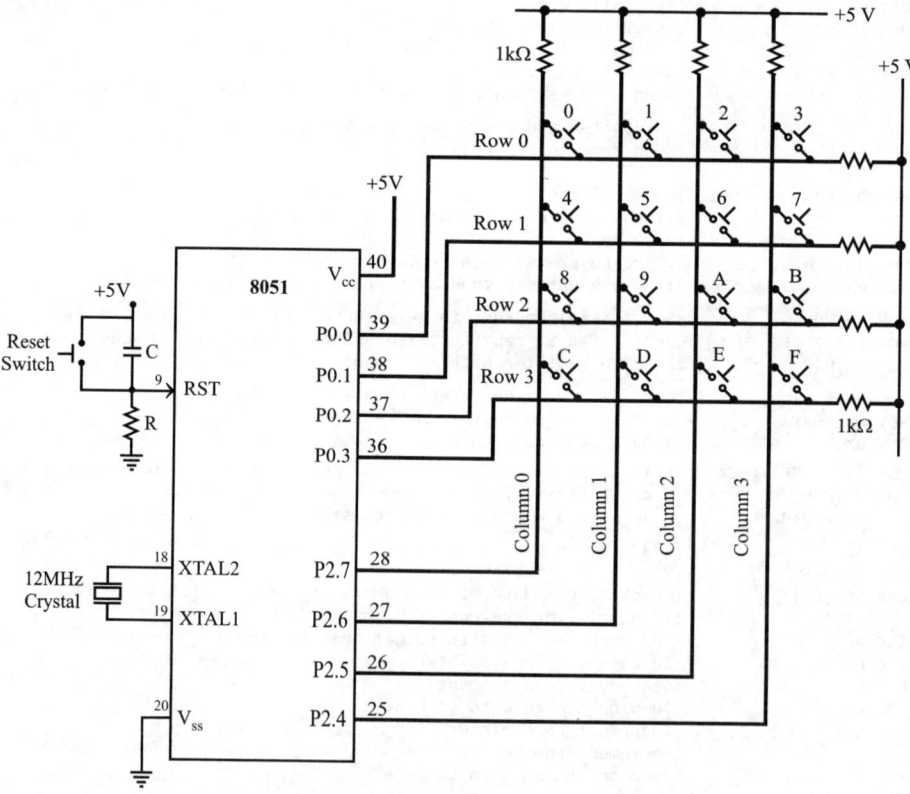

Fig. 9.56: Hex keyboard interfacing to 8051 microcontroller.

A keypress is sensed by sending a low to a row and reading the columns. A key pressed in the row that made low, will make the corresponding column also low, and so a keypress can be deducted from the data read from columns.

EXAMPLE PROGRAM 9.5

This example program is developed for the hex keyboard interface shown in Fig. 9.56. The scan codes and the corresponding hex keys are listed in the following table. The row scan codes are stored in 8051 internal RAM starting from address 60H and the column scan codes starting from address 70H.

The program has been developed to wait for a keypress. When a key is pressed, the program starts scanning. The R0 register is initialized with the hex code 00H. The scanning is performed by sending a row scan code to port-0 and reading column data through port-2 and check with column scan code. After every column scan, the hex value in the R0 register is incremented. When a column data matches with the column scan code, the scanning process stopped, and the hex key value in R0 is stored in the 8051 microcontroller internal memory at the address 7FH. Then the program waits for key debouncing and start scanning next key press. The content of memory at 7FH will be the hex value of the last pressed key.

Table 1: Scan Code and Corresponding Hex key

Row Scan Code									Column Scan Code									Hex Key
P0.7	P0.6	P0.5	P0.4	P0.3	P0.2	P0.1	P0.0	Hex	P2.7	P2.6	P2.5	P2.4	P2.3	P2.2	P2.1	P2.0	Hex	
*	*	*	*	1	1	1	0	$0E_H$	0	1	1	1	*	*	*	*	70_H	0
									1	0	1	1	*	*	*	*	$B0_H$	1
									1	1	0	1	*	*	*	*	$D0_H$	2
									1	1	1	0	*	*	*	*	$E0_H$	3
*	*	*	*	1	1	0	1	$0D_H$	0	1	1	1	*	*	*	*	70_H	4
									1	0	1	1	*	*	*	*	$B0_H$	5
									1	1	0	1	*	*	*	*	$D0_H$	6
									1	1	1	0	*	*	*	*	$E0_H$	7
*	*	*	*	1	0	1	1	$0B_H$	0	1	1	1	*	*	*	*	70_H	8
									1	0	1	1	*	*	*	*	$B0_H$	9
									1	1	0	1	*	*	*	*	$D0_H$	A
									1	1	1	0	*	*	*	*	$E0_H$	B
*	*	*	*	0	1	1	1	07_H	0	1	1	1	*	*	*	*	70_H	C
									1	0	1	1	*	*	*	*	$B0_H$	D
									1	1	0	1	*	*	*	*	$D0_H$	E
									1	1	1	0	*	*	*	*	$E0_H$	F

Flowchart

Flowchart for Main Program

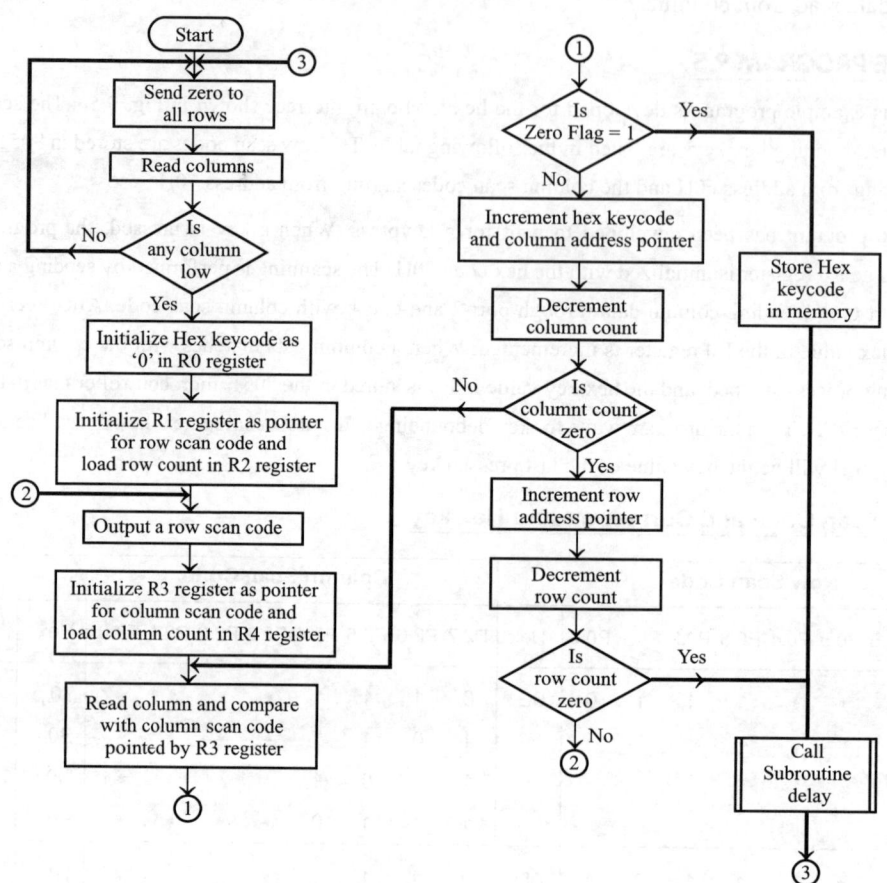

Flowchart for Subroutine Delay

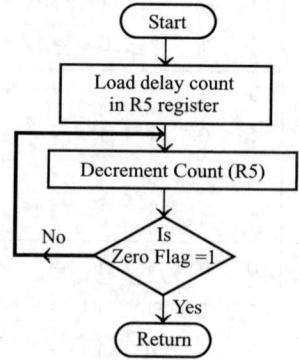

Assembly Language Program

```
;Program to scan hex keyboard and store the hex value of key pressed in memory

WAIT:   MOV  P0,#00H      ;Send zero to all rows
        MOV  A,P2         ;Read columns
        ANL  A,#F0H       ;Mask lower nibble
        CJNE A,#F0H,SCAN  ;If keypress deducted, start scanning
        SJMP WAIT         ;Wait for keypress

SCAN    MOV  R0,#00H      ;Set initial keycode in R0 as zero

        MOV  R1,#60H      ;Set R1 register as pointer for row scan code
        MOV  R2,#04H      ;Load R2 register with row count

L2:     MOV  P0,@R1       ;Send a row scan code to port-0

        MOV  R3,#70H      ;Set R3 register as pointer for column scan code
        MOV  R4,#04H      ;Load R4 register with column count

L1:     MOV  A,P2         ;Read column data
        ANL  A,#F0H       ;Mask lower nibble

        CJNE A,03H,AHEAD  ;Compare column data with column scan code
                          ;if not equal jump to AHEAD(Here 03H is address of R3 register)

STORE:  MOV  7FH, R0      ;Store hex key code in internal RAM
        SJMP SKIP

AHEAD:  INC  R0           ;Get next hex key code in R0

        INC  R3           ;Increment column scan code pointer
        DJNZ R4,L1        ;Repeat until column count is zero

        INC  R1           ;Increment row scan code pointer
        DJNZ R2,L2        ;Repeat until row count is zero

SKIP:   ACALL DELAY       ;Wait for key debouncing
        SJMP WAIT         ;Start scanning next keypress


;Delay Subroutine

DELAY:  MOV  R5,#FFH
AGAIN:  DJNZ R5,AGAIN
        RET

        ORG  60H                  ;Store row scan codes in memory
RSCAN:  DB   0EH,0DH,0BH,07H      ;Row scan code

        ORG  70H                  ;Store column scan codes in memory
CSCAN:  DB   70H,B0H,D0H,E0H      ;Column scan code

        END                       ;Assemby end
```

9.10.3 Seven-Segment Display Interface Using 8255 Ports and 8085/8086 Porcessor

The 7-segment LEDs are the most popular display devices used for single board microcomputers (microprocessor trainer kits). The 7-segment LEDs can be either common anode type or common cathode type.

Each 7-segment LED will have seven **L**ight **E**mitting **D**iodes (LEDs) arranged in the form of small rectangular segments and another LED as a dot point in a single package. In common cathode type, all the cathode terminals of LEDs are internally shorted and one/two pins are provided for external connection. The anode of the LEDs are terminated on separate pins for external connection. The pin configuration and the internal connection of a common cathode 7-segment LED are shown in Fig. 9.57.

In common anode type, all the anode terminals of LEDs are internally shorted and one/two pins are provided for external connection. The cathode of LEDs are terminated on separate pins for external connection. The pin configuration and the internal connection of the common anode 7-segment LED are shown in Fig. 9.58.

In a 7-segment LED, a segment will glow or emit light when it is forward biased. Therefore, a segment can be made to glow by applying a **high** (logic-1/+5 V) to the anode and a **low**(logic-0/0 V) to cathode. The alphabetic/numeric characters can be displayed on the 7-segment LED by forward biasing the appropriate segments.

In a common cathode 7-segment LED, the common point is tied to logic-0. To display a character, logic-1 is applied to the anode of segments which has to emit light and logic-0 is applied to anode of segments which should not emit light. The binary and hex codes for displaying the decimal digits 0 to 9 in the common cathode 7-segment LED are listed in Table 9.23.

In a common anode 7-segment LED, the common point is tied to logic-1. To display a character, logic-0 is applied to the cathode of segments which has to emit light and logic-1 is applied to the cathode of segments which should not emit light. The binary and hex codes for displaying the decimal digits 0 to 9 in the common anode 7-segment LED are listed in Table 9.24.

The display codes for LEDs can be generated by using the BCD to 7-segment decoder, IC 7447. When a BCD code is sent to the input of the 7447, it outputs **low** on the segments required to display the number represented by the BCD code. A simple schematic is shown in Fig. 9.59, to interface a common anode 7-segment LED to 8085/8086 system using a port device. This circuit connection is referred to as static display because current is being passed through the display at all times.

A typical microprocessor system normally requires 6 to 8 numbers of 7-segment LEDs. The current requirement of each 7-segment LED is 140 mA to 200 mA. Hence, the total current requirement for 6 units of 7-segment LEDs will be 1200 mA. Also each 7-segment LED requires a 7447 decoder and 4 lines of a port. The current required by the decoder and the LED displays might be several times the current required by the rest of the circuit in the microprocessor system.

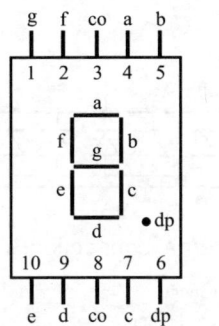

Fig. a: Pin configuration.

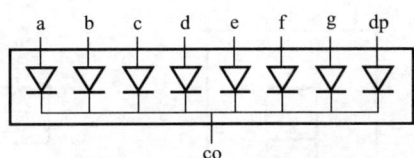

Fig. b: Internal connection.

co - common cathode
dp - anode of dot point
a, b, c, d, e, f, g - anodes of segments

Fig. 9.57: Common cathode 7-segment LED.

Table 9.23: 7-Segment Display Code for Common Cathode LED

BCD digit	Binary code								Hexa code
	dp	g	f	e	d	c	b	a	
0	0	0	1	1	1	1	1	1	3F
1	0	0	0	0	0	1	1	0	06
2	0	1	0	1	1	0	1	1	5B
3	0	1	0	0	1	1	1	1	4F
4	0	1	1	0	0	1	1	0	66
5	0	1	1	0	1	1	0	1	6D
6	0	1	1	1	1	1	0	1	7D
7	0	0	0	0	0	1	1	1	07
8	0	1	1	1	1	1	1	1	7F
9	0	1	1	0	1	1	1	1	6F

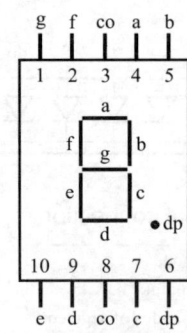

Fig. a: Pin configuration.

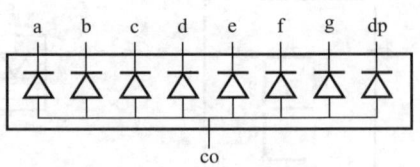

Fig. b: Internal connection.

co - common anode
dp - cathode of dot point
a, b, c, d, e, f, g - cathodes of segments

Fig. 9.58: Common anode 7-segment LED.

Table 9.24: 7-Segment Display Code for Common Anode LED

BCD digit	Binary code								Hexa code
	dp	g	f	e	d	c	b	a	
0	1	1	0	0	0	0	0	0	C0
1	1	1	1	1	1	0	0	1	F9
2	1	0	1	0	0	1	0	0	A4
3	1	0	1	1	0	0	0	0	B0
4	1	0	0	1	1	0	0	1	99
5	1	0	0	1	0	0	1	0	92
6	1	0	0	0	0	0	1	0	82
7	1	1	1	1	1	0	0	0	F8
8	1	0	0	0	0	0	0	0	80
9	1	0	0	1	0	0	0	0	90

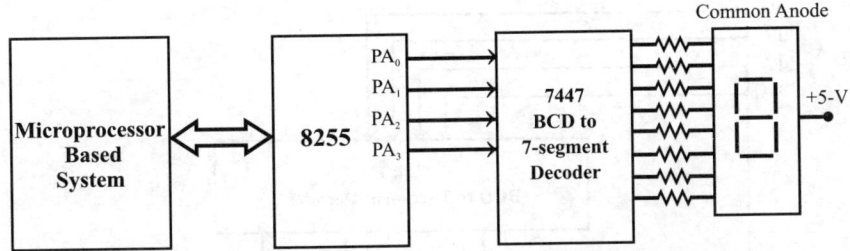

Fig. 9.59: 7-Segment LED display using port.

The heavy current requirement in the static display can be reduced drastically by using multiplexed display scheme. In multiplexed display, only one 7-segment display is made to glow at a time. Each 7-segment LED is turned ON at definite intervals. Due to persistence of vision the display appears to be continuous to a human eye. (Actually LEDs are turned ON and OFF.)

> **Note:** *A human eye can retain an image for 125 milliseconds.*

The advantages of a multiplexed display are as follows:

1. **Only one 7447 is needed for all the 7-segment LEDs.**
2. **In a current requirement of one-7-segment LED, 6 to 8 LEDs can be interfaced.**

Figure 9.60 shows a multiplexed display of 6 units of 7-segment common anode LEDs. The segment pins (cathodes) of 7-segment LEDs are connected to a common bus. The output of the decoder (7447) is connected to this common bus. The BCD code for the character to be displayed is sent to 7447 through port-A lines. The common anode of each 7-segment LED has a driver transistor (PNP type). A driver transistor can be turned ON by sending **low** to the base of the transistor through port-B lines.

The trick of multiplexed display is that the segment information is sent out to all of the digits on the common bus, but only one display digit is turned on at a time. The PNP transistors in series with the common-anode of each digit acts as an ON and OFF switch for that digit.

The BCD code for digit-1 is the first output from port-A to the 7447. The 7447 outputs the corresponding 7-segment code on the segment bus lines. The transistor connected to digit-1 is then turned ON by outputting a **low** on the corresponding bit of port-B (remember, a **low** turns ON a PNP transistor). All the rest bits of port-B should be **high** to make sure no other digits are turned ON. After a few milliseconds, digit-1 is turned OFF by outputting all **high** to port-B.

Next, the BCD code for digit-2 is output to the 7447 on port-A and a data to turn ON the driver transistor of digit-2 is output on port-B. After a few milliseconds, digit-2 is turned OFF and the process is repeated for digit-3. This process is continued until all of the digits have had a turn. Then digit-1 and the following digits are turned ON again in turn. This process is also called display refreshing.

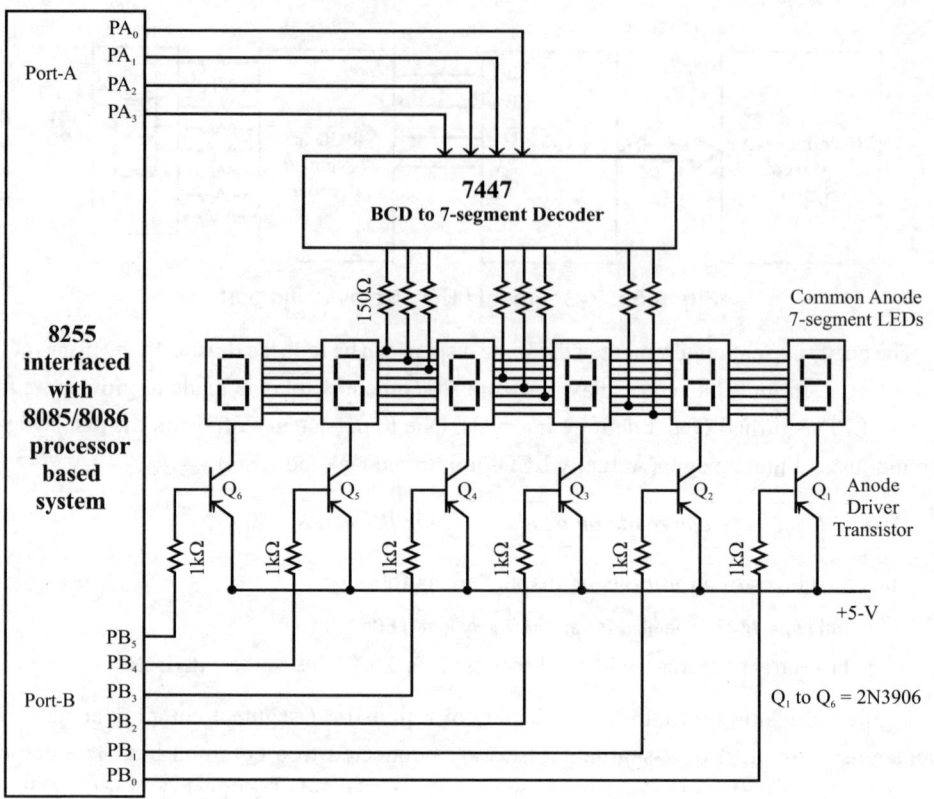

Fig. 9.60: A schematic diagram of a multiplexed display using ports.

With 6 digits and 5 ms per digit, we can get back to digit-1 every 25 ms or about 40 times a second. This refresh rate is fast enough so that all the digits appear to be turned ON all the time.

Refresh rates of 40 to 200 times a second are acceptable. A flowchart for the operational flow in a multiplexed display is shown in Fig. 9.61.

The greatest advantages of multiplexing the displays are that only one 7447 is required and only one digit is ON at a time. Hence, it results in large saving of power and parts.

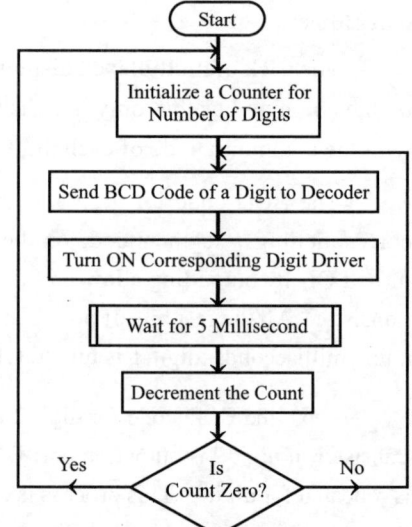

Fig. 9.61: Flowchart for multiplexed display.

EXAMPLE PROGRAM 9.6

This program is developed for multiplexed display for the 7-segment LED interface with 8085 processor based system shown in Fig. 9.60 to display the numeric digits 0,1,2,3,4 and 5. The flowchart for program logic is shown below.

Flowchart

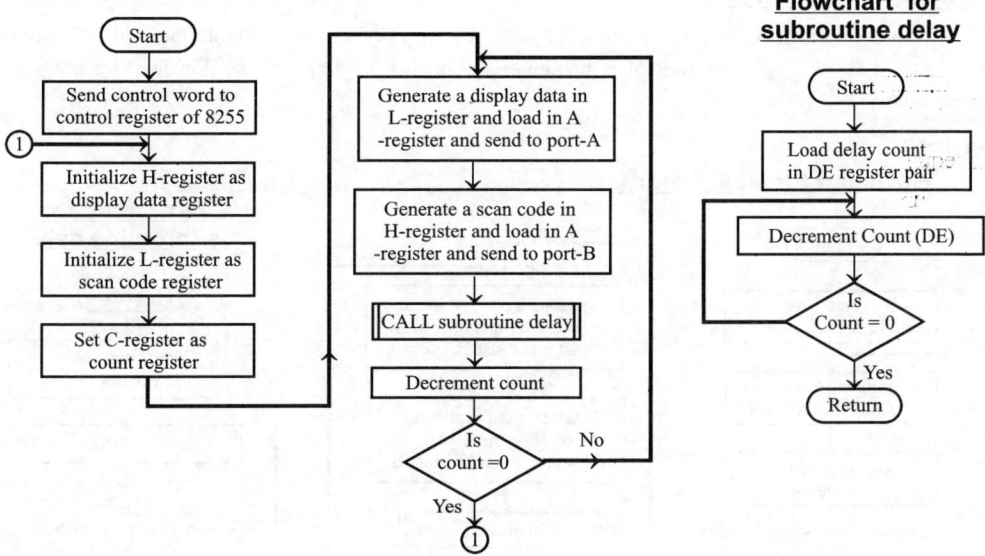

Flowchart for Main Program

Flowchart for subroutine delay

Assembly Language Program

```
;PROGRAM FOR MULTIPLEXED LED DISPLAY
PORTA    EQU 10H        ;Port-A address-8255 is IO mapped with 8-bit port addresses
PORTB    EQU 11H        ;Port-B address
PORTC    EQU 12H        ;Port-C address
CNTPORT  EQU 13H        ;Control Port address
CNTWORD  EQU 80H        ;Control Word to initialize ports in mode-0 output

         ORG 1000H      ;Assembler directive

         MVI A,CNTWORD  ;Load control word in A-register to set 8255 ports in mode-0 output
         OUT CNTPORT    ;Send control word to control register

START:   MVI H,0FFH     ;Initialize H-register to generate display data
         MVI L,80H      ;Initialize L-register to generate scan code
         MVI C,06H      ;Load count value in C-register

AGAIN:   INR H          ;Get display data in H-register
         MOV A,H        ;Load display data in A-register
         OUT PORTA      ;Send the display data to port-A
         MOV A, L       ;Get previous scan code in A-register
         RLC            ;Rotate left Get current scan code in A-register
         MOV L,A        ;Save scan code in L-register
         OUT PORTB      ;Send scan code to port-B
         CALL DELAY     ;Call delay subroutine
         DCR C          ;Repeat until the count in C-register is zero
         JNZ AGAIN
         JMP START      ;Jump to START
```

```
;DELAY ROUTINE
DELAY: LXI D, 0F429H      ;Load DE register pair with delay count
REPT:  DCX D              ;Decrement DE register pair
       MOV A,E            ;Check whether both D and E register
       CMP D              ;contents are zeros
       JNZ REPT           ;Jump on non-zero to REPT
       RET                ;If zero, return to main program

END                      ;Assembler directive
```

EXAMPLE PROGRAM 9.7

This program is developed for multiplexed display for the 7-segment LED interface with 8086 processor based system shown in Fig. 9.60 to display the numeric digits 0,1,2,3,4 and 5. The flowchart for program logic is shown below.

Flowchart

Flowchart for Main Program

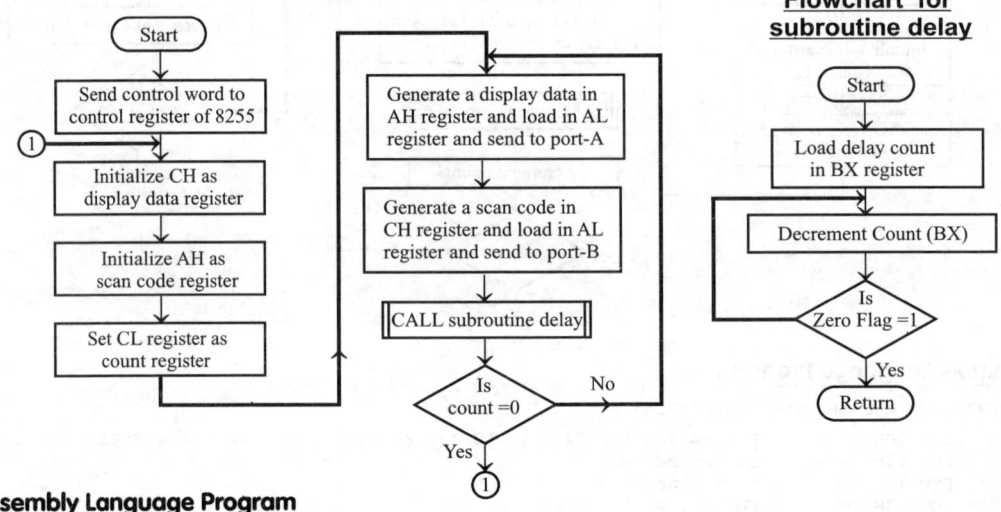

```
;PROGRAM FOR MULTIPLEXED LED DISPLAY

PORTA    EQU 0000H        ;Port-A address
PORTB    EQU 0002H        ;Port-B address
PORTC    EQU 0004H        ;Port-C address
CNTPORT  EQU 0006H        ;Control Port address
CNTWORD  EQU 80H          ;Control Word to initialize ports in mode-0 output

CODE SEGMENT              ;Assembler directive
        ASSUME CS:CODE    ;Assembler directive
        ORG 1000H         ;Assembler directive
        MOV DX,CNTPORT    ;Get control register address in DX register
        MOV AL,CNTWORD    ;Set 8255 ports in mode-0 output
        OUT DX,AL         ;Send control word to control register
START:  MOV AH,0FFH       ;Initialize AH register to generate display data
        MOV CH,80H        ;Initialize CH register to generate scan code
        MOV CL,06H        ;Load count value in CL register
AGAIN:  MOV DX,PORTA      ;Get port-A addressin DX register
        INC AH            ;Get display data in AH register
        MOV AL,AH         ;Load display data in AL register
        OUT DX,AL         ;Send the display data to port-A
        MOV DX,PORTB      ;Get port-B addressin DX register
```

```
          ROL CH,1            ;Get scan code in CH register
          MOV AL,CH           ;Load scan code in AL register
          OUT DX,AL           ;Send scan code to port-B
          CALL DELAY          ;Call delay subroutine
          LOOP AGAIN          ;Repeat until the count in CL register is zero
          JMP START           ;Jump to START

;DELAY ROUTINE
DELAY  PROC NEAR              ;Assembler directive
          MOV BX,0F429H       ;Load BX register with delay count
REPT:   DEC BX                ;Decrement BX register
          JNZ REPT            ;Jump on non-zero to REPT
          RET                 ;Return to main program
DELAY  ENDP                   ;Assembler directive

CODE ENDS                     ;Assembler directive
END                           ;Assembler directive
```

> **Note:** *In the assembly-language program, it is assumed that 8255 ports are IO mapped in the system with 16-bit IO port addresses.*

9.10.4 Latches and Buffers as IO Devices

The Latches and Buffers can be used as IO ports. Basically in programmable IO ports of 8155/8255/8355/8755 the ports are made of latches and buffers. Latches can be used as output ports and buffers can be used as input ports. Examples of 8-bit latches are 74LS373, 74LS273, 74LS573, 74LS574, etc. Examples of 8-bit buffers are 74LS245, 74LS244, 74LS240, INTEL 8286, etc.

The 8-bit latch can be used as an output device to interface LEDs or seven segment LEDs as shown in Fig. 9.62. A Latch is selected by a chip select signal. If can be mapped in the system either by IO mapping or memory mapping. The processor has to send an address to select the latch. A decoder in the system produces Chip Select signal (CS) which enables the latch. Then the processor loads the display code on the data bus. The latch will hold the display code, until it is loaded with another display code.

In Fig. 9.62, a segment is turned ON if a one is sent to anode through the latch. The cathode is permanently tied to ground.

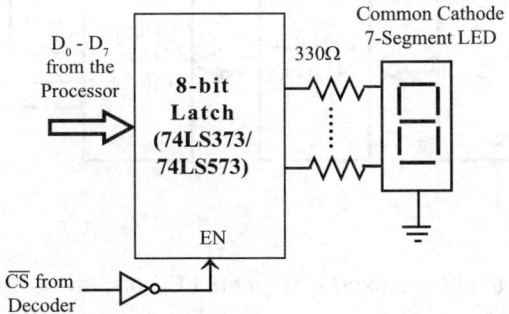

Fig. 9.62: An 8-bit latch as output port.

Using latches and decoders multiplexed display is also possible as shown in Fig. 9.63. In this scheme two latches are used. One for turning ON the segment driver transistor and another for turning ON the digit driver transistor. The segments (Anodes in case of common cathode LEDs) are connected to a common bus. Driver transistors are provided to satisfy the current requirement of LEDs. The segment transistors can be turned ON by sending appropriate code through Latch-1.

The common point (Cathode in case of common cathode LEDs) of each 7-segment LED is connected to a driver transistor (digit driver). The digit driver transistors can be turned ON, one at a time by the decoder.

The processor selects Latch-1 and sends the display code (7-segment code) to it. Then the processor selects Latch-2 and send an appropriate binary count to turn on a particular digit. The input to the decoder 74LS138 is a binary count and the output of the decoder will turn ON only one digit transistor. After a delay time (typically 3 to 10 ms) all the segments are turned OFF by sending appropriate code to segment drivers. Then the display code for next digit is sent to segment drivers and the corresponding digit driver is turned ON, by sending an appropriate count to decoder.

The disadvantage in using ports and latches as output devices is that a considerable processor time is consumed for display refreshing.

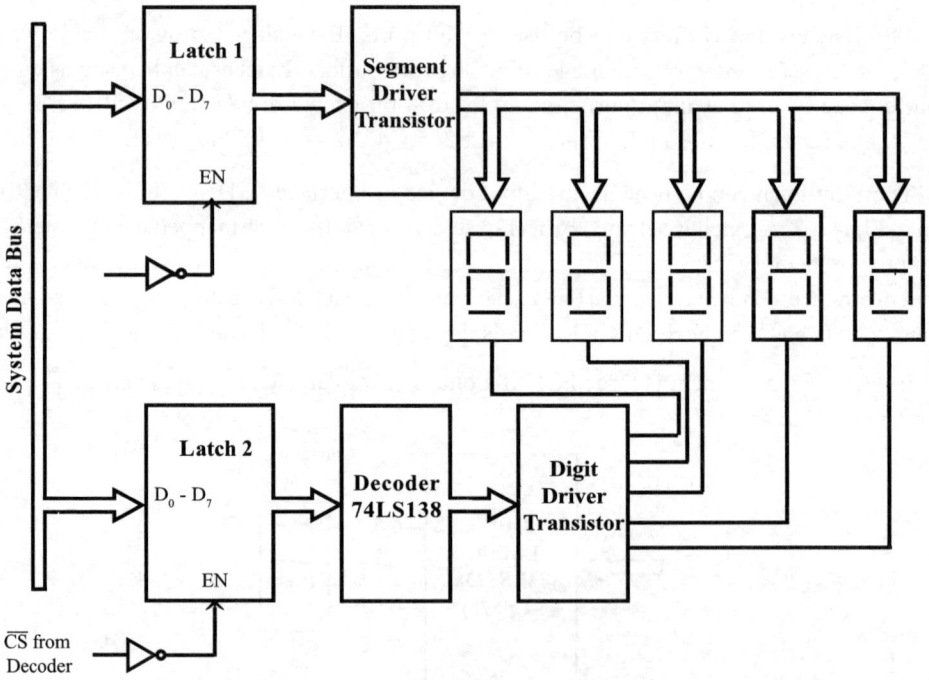

Fig. 9.63: Multiplexed 7-segment LED display using latches.

Similarly an input device such as a keyboard can also be interfaced using buffers. The interfacing of keyboard using buffers will be similar to that of keyboard interface using ports of 8255 discussed in Section 9.10.1. The disadvantage in using buffers for keyboard interfacing is that most of the processor time is consumed for keyboard scanning and debouncing.

9.11 KEYBOARD/DISPLAY CONTROLLER - INTEL 8279

The INTEL 8279 is a dedicated controller specially developed for interfacing keyboard and display devices to 8085/8086/8088 microprocessor based system. It relieves the processor from the time-consuming task like keyboard scanning and display refreshing.

The important features of 8279 are,

1. **Simultaneous keyboard and display operations**
2. **2-key lockout or N-key rollover with contact debounce**
3. **Scanned keyboard mode**
4. **Scanned sensor mode**
5. **Strobed input entry mode**
6. **8-character keyboard FIFO**
7. **16-character display**
8. **Right or left entry 16-byte display RAM**
9. **Mode programmable from CPU**
10. **Programmable scan timing**
11. **Interrupt output on key entry**

The 8279 provides an interface for a maximum of 64-contact key matrix (arranged as 8×8 matrix array of key switches).

The keyboard entries are debounced and stored in the internal FIFO RAM. It generates an interrupt signal for each key entry, to inform the processor to read the keycode from FIFO.

The 8279 provides a multiplexed interface for 7-segment LEDs and other popular display devices. It consist of 16×8 display RAM which can also be organized into dual 16×4 RAM. The CPU have to load the display codes in this RAM.

Once the data is loaded, the 8279 takes care of display and refreshing. A maximum of 16 numbers of 7-segment LEDs can be interfaced using 8279.

9.11.1 Pins, Signals and Functional Block Diagram of 8279

The 8279 is a 40-pin IC available in DIP (**D**ual **I**n-line **P**ackage). The pin configuration of 8279 is shown in Fig. 9.64. The 8279 has two internal addresses decided by the logic level of A_0.

If A_0 is **low** then the processor can read or write to the data register of 8279. If A_0 is **high** then the processor can write to control register or read status register. The 8279 can be either IO-mapped or memory-mapped in the system.

The functional block diagram of 8279 is shown in Fig. 9.65. The four major sections of 8279 are keyboard, scan, display and CPU interface.

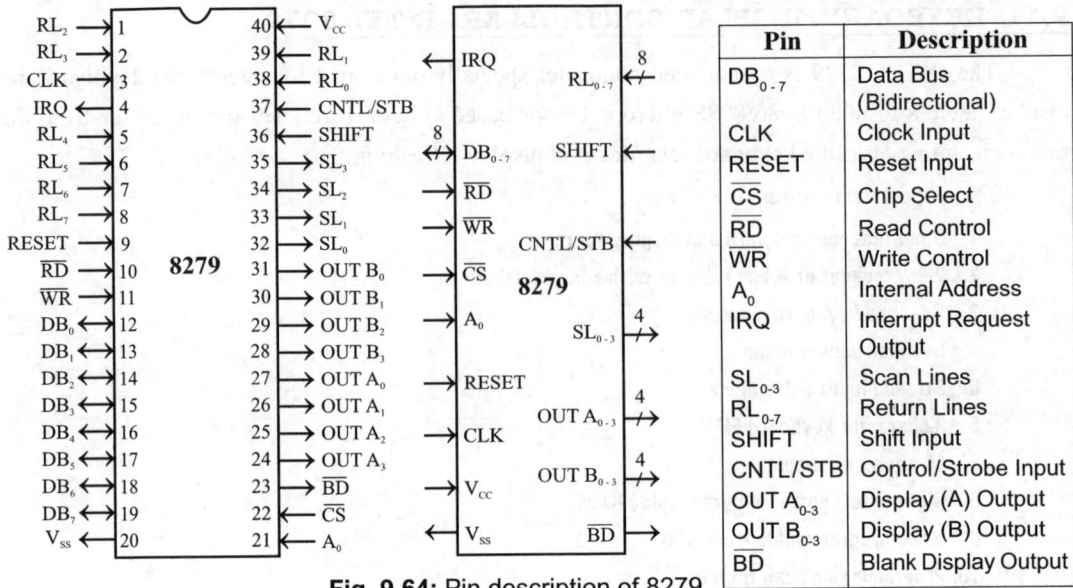

Fig. 9.64: Pin description of 8279.

Fig. 9.65: Block diagram of 8279.

Keyboard Section

The keyboard section consists of eight return lines RL_0 - RL_7 that can be used to form the columns of a keyboard matrix. It has two additional inputs: shift and control/strobe. The keys are automatically debounced. The two operating modes of keyboard section are 2-key lockout and *N*-key rollover. In the 2-key lockout mode, if two keys are pressed simultaneously, only the first key is recognized. In the *N*-key rollover mode simultaneous keys are recognized and their codes are stored in the FIFO.

The keyboard section also has an 8×8 FIFO (**F**irst-**I**n-**F**irst-**O**ut) RAM. The FIFO can store eight keycodes in the scan keyboard mode. The status of the shift key and control key are also stored along with keycode. The 8279 generates an interrupt signal when there is an entry in the FIFO. The format of the keycode entry in FIFO for the scan keyboard mode is shown in Fig. 9.66.

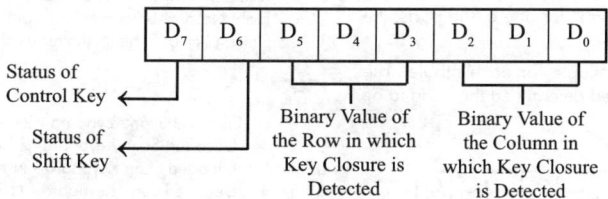

Fig. 9.66: Keycode entry in FIFO for scan keyboard mode.

In a sensor matrix mode, the condition (i.e., open/close status) of 64 switches is stored in FIFO RAM. If the condition of any of the switches changes then the 8279 asserts IRQ as **high** to interrupt the processor.

Display Section

The display section has eight output lines divided into two groups A_0 - A_3 and B_0 - B_3. The output lines can be used either as a single group of eight lines or as two groups of four lines, in conjunction with the scan lines for a multiplexed display.

The output lines are connected to the anodes through a driver transistor in case of common cathode 7-segment LEDs. The cathodes are connected to scan lines through driver transistors. The display can be blanked by BD line. The display section consists of 16×8 display RAM. The CPU can read from or write into any location of the display RAM.

Scan Section

The scan section has a scan counter and four scan lines, SL_0 to SL_3. In a decoded scan mode, the output of the scan lines will be similar to a 2-to-4 decoder. In encoded scan mode, the output of the scan lines will be binary count, and so an external decoder should be used to convert the binary count to the decoded output. The scan lines are common for keyboard and display. The scan lines are used to form the rows of a matrix keyboard and are also connected to the digit drivers of a multiplexed display to turn ON/OFF.

Write Display RAM

Code : | 1 | 0 | 0 | AI | A | A | A | A |

The CPU sets up the 8279 for a write to the Display RAM by first writing this command. After writing the command with $A_0=1$, all subsequent writes with $A_0=0$ will be to the Display RAM. The addressing and Auto increment functions are identical to those for the Read Display RAM.

Display Write Inhibit/Blanking

				A	B	A	B
Code : | 1 | 0 | 1 | X | IW | IW | BL | BL |

The IW Bits can be used to mask nibble A and nibble B in application requiring separate 4-bit display ports. By setting the IW flag (IW=1) for one of the ports, the port becomes masked.

The BL flags are available for each nibble. The last Clear command issued determined the code to be used as a blank.

Clear

Code : | 1 | 1 | 0 | C_D | C_D | C_D | C_F | C_A |

The CD bits are available in this command to clear all rows of the Display RAM to a selectable blanking code as follows.

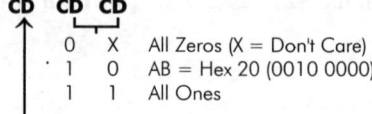

CD	CD	CD	
0	X		All Zeros (X = Don't Care)
1	0		AB = Hex 20 (0010 0000)
1	1		All Ones

└─ Enable clear display if this bit is 1

If the C_F bit is asserted ($C_F = 1$), the FIFO status is cleared and the interrupt output line is reset.

C_A, the clear all bit, has the combined effect of C_D and C_F; it uses the C_D clearing code on the Display RAM and also clears FIFO status. Furthermore, it resynchronizes the internal timing chain.

Read FIFO/Sensor RAM

Code : | 0 | 1 | 0 | AI | X | A | A | A |

X = Don't care

The CPU sets up the 8279 for a read of the FIFO/Sensor RAM by first writing this command. In the Scan keyboard Mode, the Auto-Increment flag (AI) and the Ram address bits (AAA) are irrelevant.

In the Sensor Matrix Mode, the RAM address bits AAA select one of the 8 rows of the Sensor RAM. If the AI flag is set (AI = 1), each successive read will be from the subsequent row of the sensor RAM.

End Interrupt/Error Mode Set

Code : | 1 | 1 | 1 | E | X | X | X | X |

X = Don't care

For the sensor matrix modes this command lowers the IRQ line and enables further writing into RAM. For the N-Key rollover mode , if the E bit is programmed to **1** the chip will operate in the special Error mode.

Keyboard/Display Mode Set

Code : | 0 | 0 | 0 | D | D | K | K | K |

DD

0	0	Eight No.of 8-bit character display	-Left entry
0	1	Sixteen No.of 8-bit character display	-Left entry
1	0	Eight No.of 8-bit character display	-Right entry
1	1	Sixteen No.of 8-bit character display	-Right entry

KKK

0	0	0	Encoded Scan Keyboard - 2-Key Lockout
0	0	1	Decoded Scan Keyboard- 2-Key Lockout
0	1	0	Encoded Scan Keyboard - N-Key Rollover
0	1	1	Decoded Scan Keyboard - N-Key Rollover
1	0	0	Encoded Scan Sensor Matrix
1	0	1	Decoded Scan Sensor Matrix
1	1	0	Strobed Input, Encoded Display Scan
1	1	1	Strobed Input, Decoded Display Scan.

Program Clock

Code : | 0 | 0 | 1 | P | P | P | P | P |

All timing and multiplexing signals for the 8279 are generated by an internal prescaler. This prescaler divides the external clock (pin 3) by a programmable integer. Bits PPPPP determine the value of this integer which ranges from 2 to 31. Choosing a divisor that yields 100 kHz will give the specified scan and debounce times.

Read Display RAM

Code : | 0 | 1 | 1 | AI | A | A | A | A |

The CPU sets up the 8279 for a read of the Display RAM by first writing this command. The address bits AAAA select one of the 16 rows of the Display RAM. If the AI flag is set (AI = 1), this row address will be incremented after each read or write to the Display RAM.

Fig. 9.67: 8279 Command word formats.

CPU Interface Section

The CPU interface section takes care of the data transfer between 8279 and the processor. This section has eight bidirectional data lines DB_0 - DB_7 for data transfer between 8279 and CPU. It requires two internal address $A_0 = 0$ or 1, for selecting either data buffer or control register of 8279. The control signals $\overline{WR}$, $\overline{RD}$, $\overline{CS}$ and A_0 are used for read/write to 8279. It has an interrupt request line IRQ for interrupt driven data transfer with the processor.

The 8279 requires an internal clock frequency of 100 kHz. This can be obtained by dividing the input clock by an internal prescaler. The prescaler can take a value from 2 to 31, which is programmable. The RESET signal sets the 8279 in 16-character display with two-key lockout keyboard mode. Also the reset will set the clock prescaler to 31.

Programming the 8279

The 8279 can be programmed to perform various functions through eight command words. The formats of the command words and a brief explanation are presented in Fig. 9.67 .

9.11.2 Keyboard and Display Interface using 8279

In a microprocessor-based system, when the keyboard and the 7-segment LED display are interfaced using ports or latches then the processor has to carryout the following tasks:

1. Keyboard scanning
2. Key debouncing
3. Keycode generation
4. Sending display code to LED
5. Display refreshing

The above functions has to be performed continuously in specified time intervals. Hence most of the processor time will be utilized for the above task. To overcome this problem, the dedicated keyboard/display controller such as INTEL 8279 can be employed in the system. The 8279 provides a hardware solution for keyboard and display interfacing in microprocessor-based system.

When 8279 is employed, the processor task is to program the 8279 by sending the control words and load the display code in display RAM of 8279. Once the 8279 is programmed it takes care of keyboard scanning, debouncing, keycode generation and display refreshing. Whenever 8279 detects a key press, it informs the processor through interrupt so that the processor can read the keycode from FIFO of 8279.

9.11.3 Interfacing of 8279 with 8085 Processor

A typical hexa keyboard and 7-segment LED display interfacing circuit using 8279 is shown in Fig. 6.60. The circuit can be used in an 8085 microprocessor system and consists of 16 numbers of hexa-keys and 6 numbers of 7-segment LEDs. The 7-segment LEDs can be used to display a six digit alphanumeric character.

The 8279 can be either memory-mapped or IO-mapped in the system. In the circuit of Fig. 6.60, the 8279 is IO-mapped. The address line A_0 of the system is used as A_0 of 8279. The clock signal for 8279 is obtained by dividing the output clock signal of 8085 by a clock divider circuit.

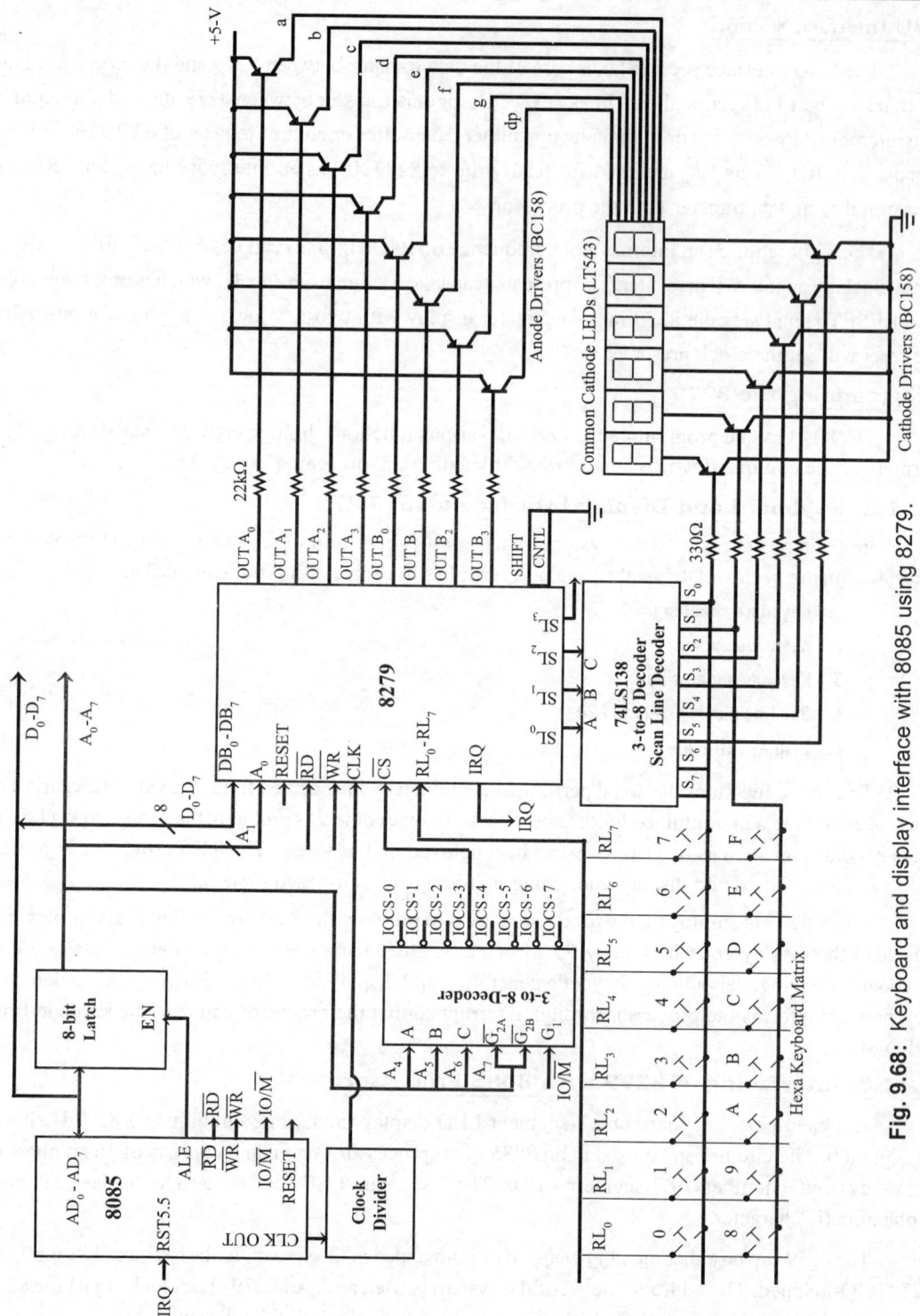

Fig. 9.68: Keyboard and display interface with 8085 using 8279.

The chip select signal $\overline{CS}$ is obtained from the IO address decoder of the 8085 system. The chip select signals for IO-mapped devices are generated by using a 3-to-8 decoder. The address lines A_4, A_5 and A_6 are used as input to the decoder. The address line A_7 and the control signal $IO/\overline{M}$ are used as enable for the decoder. The chip select signal IOCS-3 is used to select 8279. The IO address of the internal devices of 8279 are shown in Table 9.25.

Table 9.25: IO Addresses of 8279 Interfaced to 8085 as Shown in Fig. 9.68

Internal device	Binary address								Hexa address
	Decoder input			Input to address pin of 8279				Decoder enable	
	A_7	A_6	A_5	A_4	A_3	A_2	A_1	A_0	
Data register	0	1	1	x	x	x	0	0	60
Control register	0	1	1	x	x	x	1	0	62

Note: Don't care "x" is considered as zero.

The circuit has 6 numbers of 7-segment LEDs and so the 8279 has to be programmed in encoded scan. (Because in decoded scan, only 4 numbers of 7-segment LEDs can be interfaced.) In encoded scan, the output of the scan lines will be binary count. Therefore an external 3-to-8 decoder is used to decode the scan lines SL_0, SL_1 and SL_2 of 8279 to produce eight scan lines S_0 to S_7. The decoded scan lines S_0 and S_1 are common for keyboard and display. The decoded scan lines S_2 to S_5 are used only for display and the decoded scan lines S_6 and S_7 are not used in the system.

The common cathode LEDs, LT543 has been used in the circuit shown in Fig. 9.68. The corresponding segments of the anodes are connected to the common line to form a bus and this bus can be called a segment bus. (i.e., segment "a" of all 7-segment LEDs are connected to a common line, similarly segment "b" is connected and so on.)

Anode and cathode drivers are provided to take care of the current requirement of LEDs. The pnp transistors, BC158 are used as driver transistors. The anode drivers are called segment drivers and cathode drivers are called digit drivers.

The 8279 outputs the display code for one digit through its output lines (OUT A_0 to OUT A_3 and OUT B_0 to OUT B_3) and sends a scan code through SL_0 - SL_3. The display code is inverted by segment drivers and sent to the segment bus.

The scan code is decoded by the decoder and turns ON the corresponding digit driver. Now one digit of the display character is displayed. After a small interval (10 milliseconds, typical), the display is turned OFF (i.e., display is blanked) and the above process is repeated for the next digit. Thus multiplexed display is performed by 8279.

> *Note:* *Since the anode drivers invert the display code, the complement of the data required to turn ON a common cathode LED should be loaded in display RAM of 8279.*

The keyboard matrix is formed using the return lines, RL_0 to RL_7 of 8279 as columns and decoded scan lines S_0 and S_1 as rows. A hexa key is placed at the crossing point of each row and column. A key press will short the row and column.

Normally the column and row line will be **high**. (i.e., the 8279 will tie the return line as **high** and the decoder will tie the scan line as **high.**) During scanning the 8279 will output the binary count on SL_0 to SL_3, which is decoded by the decoder to make a row as zero. When a row is zero, the 8279 reads the columns. If there is a key press then the corresponding column will be zero.

If 8279 detects a key press then it waits for debounce time and again reads the columns to generate the keycode. In encoded scan keyboard mode, the 8279 stores an 8-bit code for each valid key press. The keycode consists of the binary value of the column and row in which the key is found and the status of shift and control key.

The format of the code entered in FIFO RAM is shown in Fig. 9.66. After a scan time, the next row is made zero and the above process is repeated and so on. Thus 8279 continuously scans the keyboard.

9.11.4 Interfacing of 8279 with an 8086 Processor

A typical Hexa keyboard and 7-segment LED display interfacing circuit using 8279 for 8086-based system is shown in Fig. 9.69. The system consist of 16 numbers of hexa-keys and 8 numbers of 7-segment LEDs. The 7-segment LEDs can be used to display eight digit alphanumeric characters.

The 8279 can be either memory-mapped or IO-mapped in the system. In the circuit of Fig. 9.69, the 8279 is IO-mapped. The address line A_1 of the system is used as A_0 of 8279. The clock signal for 8279 is obtained by dividing the PCLK (peripheral clock) of the 8284 by a clock divider circuit.

The chip select signal $\overline{CS}$, is obtained from the IO address decoder of the 8086 system. The chip select signals for IO mapped devices are generated using a 3-to-8 decoder. The address lines A_5, A_6 and A_7 are used as inputs to the decoder.

The address line A_0 and the control signal $M/\overline{IO}$ are used as enable for the decoder. The chip select signal IOCS-3 is used to select 8279. The IO address of the internal devices of 8279 are shown in Table 9.26. The working of the system shown in Fig. 9.69 is similar to that of the system shown in Fig. 9.68.

Table 9.26: IO Addresses of 8279 Interface to 8086 as Shown in Fig. 9.69

Internal device	Binary address								Hexa address
	Decoder input			Input to address pin of 8279				Decoder enable	
	A_7	A_6	A_5	A_4	A_3	A_2	A_1	A_0	
Data register	0	1	1	x	x	x	0	0	60
Control register	0	1	1	x	x	x	1	0	62

9.11.5 Interfacing of 8279 with an 8051 Microcontroller

The 8279 can be interfaced to an 8051 controller to provide the keyboard and 7-segment LED display. A simple schematic for interfacing 8279 with 8051 microcontroller is shown in Fig. 9.70.

In this scheme the display drivers, scan line decoders and keyboard matrix are same as that in Fig. 9.68. The working of this keyboard and display interface is also similar to the interface shown in Fig. 9.68.

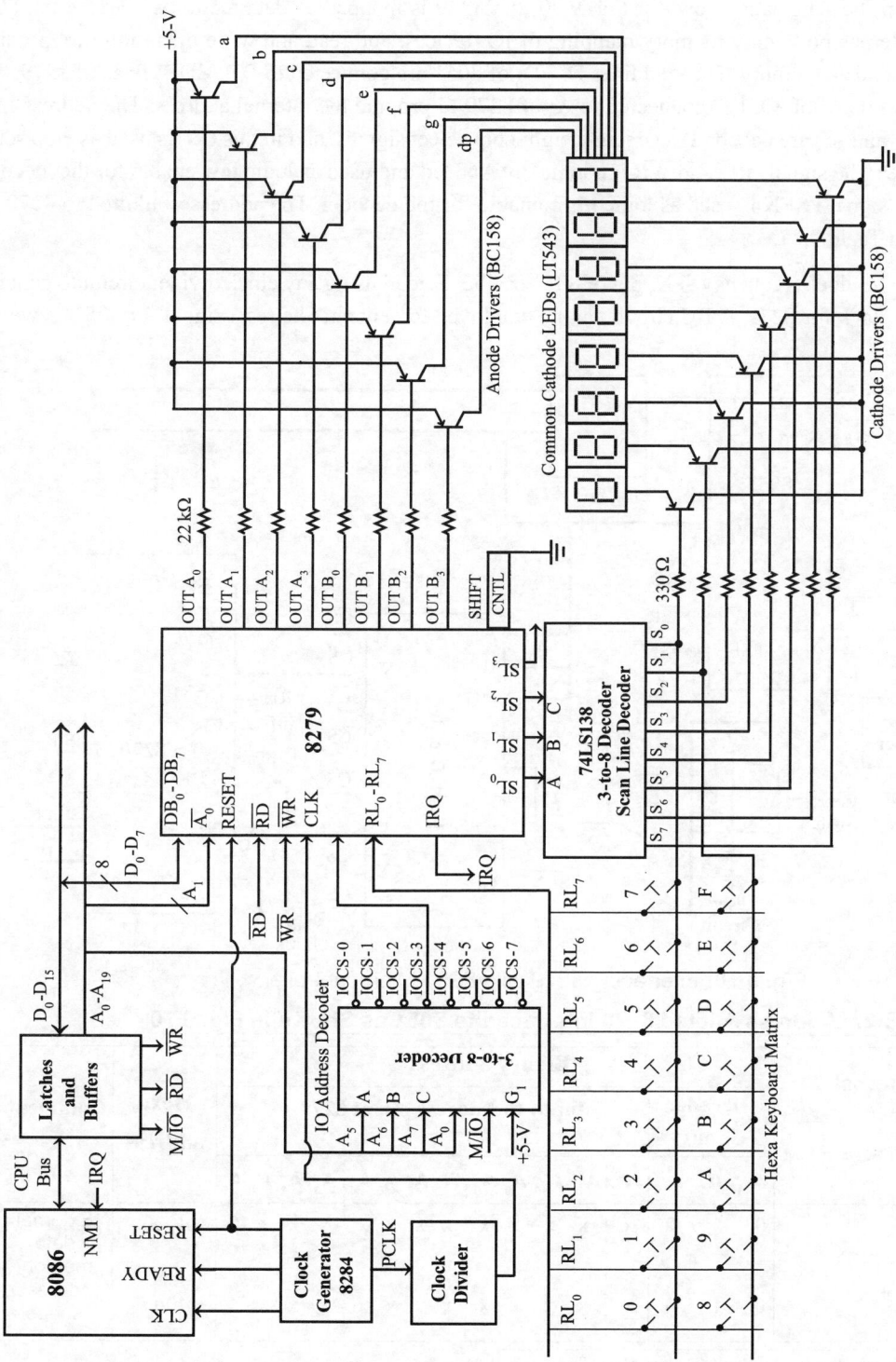

Fig. 9.69: Keyboard and display interface with 8086 using 8279.

In the schematic shown in Fig. 9.70, the 8279 is mapped as data memory, because the 8051 controller supports only memory mapping of IO devices, and read and write operations are possible only with data memory. The data lines D_7 - D_0 of 8051 are connected to DB_7 - DB_0 lines of 8279. The address line A_0 of 8051 is connected to A_0 of 8279 to provide the internal address. The address lines A_{13}, A_{14} and A_{15} are decoded to generate eight chip select signals, and in this $\overline{CS_2}$ is used as chip select for 8279. The signals $\overline{RD}$ and $\overline{WR}$ are logically ANDed and used as logic **low** enable for the decoder and the signal $\overline{PSEN}$ is used as logic **high** enable for the decoder. The addresses allotted to 8279 are listed in Table 9.27.

The clock frequency at X_2 pin of 8051 can be divided using any clock divider circuit to generate 3 MHz clock for 8279. A RC circuit can be employed to generate the reset signal for 8051 as well as for 8279.

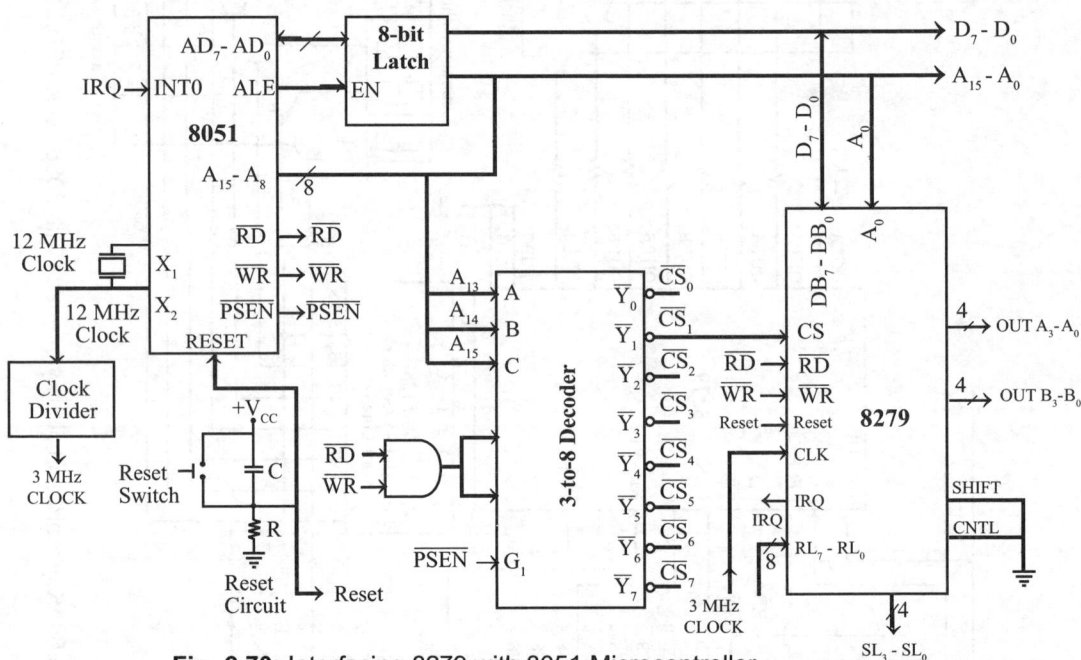

Fig. 9.70: Interfacing 8279 with 8051 Microcontroller.

Table 9.27: Address Alloted 8279 Interfaced to 8051 as Shown in Fig. 9.70

Internal device	Binary address		Hexa address	Comment
	Decoder input	Input to Address pin of 8279		
	A_{15} A_{14} A_{13}	A_{12} A_{11} A_{10} A_9 A_8 A_7 A_6 A_5 A_4 A_3 A_2 A_1 A_0		
Data Register	0 1 0	x x x x x x x x x x x x x 0	4000	External data memory address space
Control Register	0 1 0	x x x x x x x x x x x x x 1	4001	

Note: Don't care "x" is considered as zero.

9.12 PROGRAMMABLE TIMER - INTEL 8254

When the processor has to perform time-based activities, there are two methods to maintain the timings of the operations. In one method, the processor can execute a delay subroutine. In this method, the delay subroutine will load a count value in one of the register of the processor and start decrementing the count value.

After every decrement operation, the zero flag is checked to verify whether the count has reached zero or not. If the count has reached zero, the delay subroutine is terminated. Now the desired time will be elapsed and the processor can perform the desired time-based task. In this method, the time is estimated in terms of processor clock periods needed to execute the delay subroutine.

In the second method, an external timer can maintain the timings and interrupt the processor at periodic intervals. In the first method, the processor time is wasted by simply decrementing a register. But in the second method, the processor time can be efficiently utilized, because the processor can perform other tasks in between timer interrupts.

One of the programmable external timer device is the 8254 developed by INTEL. The INTEL 8254 timer has three independent counters. In each counter, a count value can be incremented loaded and decremented by applying a clock signal. At the end of count, each counter will generate an output which can be used as interrupt to processor to initiate the time-based activity. Some of the applications of programmable timers are given below:

1. The timer can interrupt a time-sharing operating system at specified intervals so that it can switch programs.

2. The timer can send timing signals at periodic intervals to IO devices. (For example, start of conversion signal to ADC.)

3. The timer can be used as baud rate generator. (For example, the timer can be used as a clock divider to divide the processor clock to the desired frequency for TxC and RxC of USART-8251A.)

4. The timer can be used to measure the time between the external events.

5. The timer can be used as an external event counter to count repetitive external operations and inform the count value to the processor.

6. The timer can be used to initiate an activity through the interrupt after a programmed number of external events have occured.

9.12.1 Pins, Signals and Functional Block Diagram of 8254

The 8254 is a 24-pin IC packed in DIP and requires a single +5 V supply. The pin configuration of 8254 is shown in Fig. 9.71. The functional block diagram of 8254 is shown in Fig. 9.72.

The 8254 has three independent 16-bit counters which can be programmed to work in any one of the possible six modes. Each counter has a clock input, gate input and counter output.

To operate a counter, a count value has to be loaded in count register, gate should be tied **high** and a clock signal should be applied through clock input.

The counter counts by decrementing the count value by one in each cycle of clock signal and generates an output depending on the mode of operation. The maximum input clock frequency for 8254 is 10 MHz.

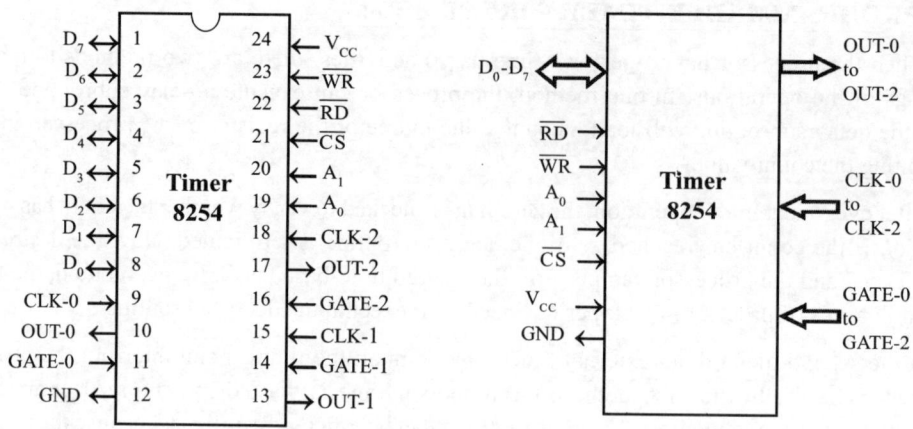

Pin	Description
D_0 - D_7	Bidirectional data lines
$\overline{CS}$	Chip select
$\overline{RD}$	Read control
WR	Write control
A_0, A_1	Internal address
CLK-0 to CLK-2	Clock input to counters
GATE-0 to GATE-2	Gate control input to counters
OUT-0 to OUT-2	Output of counters

Fig. 9.71: Pin configuration of an 8254 timer.

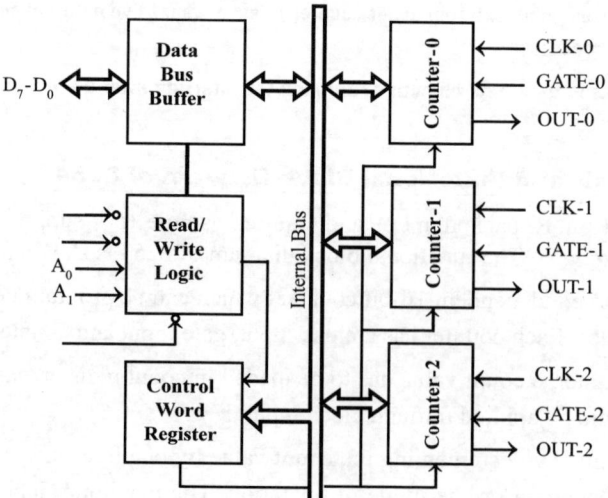

Fig. 9.72: Functional block diagram of an 8254 timer.

Note:	*Another timer released by INTEL is the 8253 which is a low clock version of the 8254. The maximum input clock frequency to the 8253 is 2.6 MHz. The 8253 and 8254 are pin to pin compatible and functionally same except the clock frequency.*

The 8254 has eight data lines which can be used for communication with the processor. The control words and count values are written into the 8254 registers through the data bus buffer. The $\overline{CS}$ is used to select the chip. The address lines A_0 and A_1 are used to select any one of the four internal devices as shown in Table 9.28. The control signals $\overline{RD}$ and $\overline{WR}$ are used by the processor to perform read/write operation. The processor can read the count value in the count register with/without stopping the counter at any time.

Table 9.28: Internal Addresses of 8254

Internal address		Device selected
A_1	A_0	
0	0	Counter-0
0	1	Counter-1
1	0	Counter-2
1	1	Control Register

9.12.2 Interfacing of 8254 with an 8085 Processor

A simple schematic for interfacing the 8254 with an 8085 processor is shown in Fig. 9.73. The 8254 can be either memory-mapped or IO-mapped in the system. In the schematic shown in Fig. 9.73, the 8254 is IO-mapped in the system. The chip select signals for IO-mapped devices are generated by using a 3-to-8 decoder. The address lines A_4, A_5 and A_6 are decoded to generate eight chip select signals (IOCS-0 to IOCS-7) and in this, the chip select IOCS-5 is used to select the 8254. The address line A_7 and the control signal IO/$\overline{M}$ are used to enable the decoder.

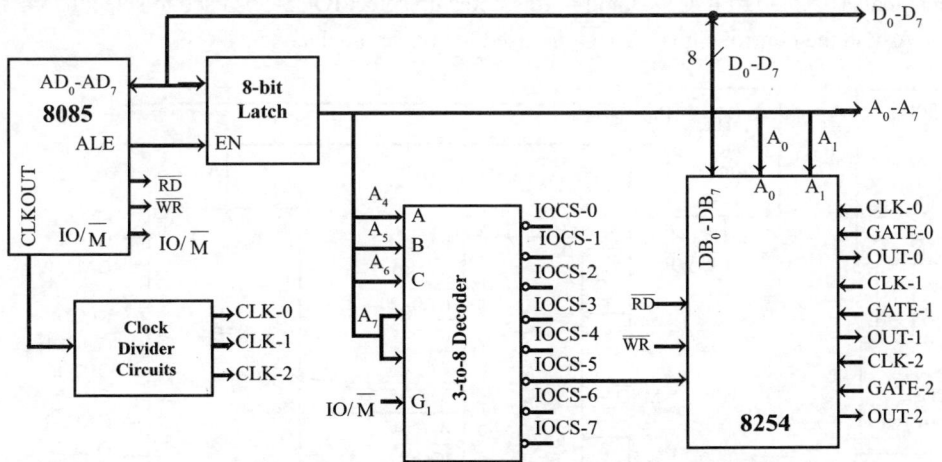

Fig. 9.73: Interfacing of 8254 with 8085 processor.

The address lines A_0 and A_1 of the 8085 are connected to A_0 and A_1 of the 8254 to provide the internal addresses. The IO addresses allotted to the internal devices of 8254 are listed in Table 9.29. The data lines D_0-D_7, $\overline{RD}$ and $\overline{WR}$ signals of the 8254 are connected to the D_0-D_7, $\overline{RD}$ and $\overline{WR}$ of the processor respectively to achieve parallel data transfer.

Table 9.29: IO Addresses of 8254 Interfaced to 8085 as Shown in Fig. 9.73

Internal device	Binary address								Hexa address
	Decoder input and enable				Input to address pins of 8254				
	A_7	A_6	A_5	A_4	A_3	A_2	A_1	A_0	
Counter - 0	0	1	0	1	x	x	0	0	50
Counter - 1	0	1	0	1	x	x	0	1	51
Counter - 2	0	1	0	1	x	x	1	0	52
Control Register	0	1	0	1	x	x	1	1	53

Note: Don't care "x" is considered as zero.

The clock signals required for the counters can be obtained either from the processor clock output or from an external clock source. The clock signal from a 8085 can also be divided to lower values by using clock divider circuits and then applied to clock input of counters.

9.12.3 Interfacing of 8254 with an 8086 Processor

A simple schematic for interfacing the 8254 with the 8086 processor is shown in Fig. 9.74. The 8254 can be either memory-mapped or IO-mapped in the system. In the schematic shown in Fig. 9.74, the 8254 is IO-mapped in the system with even addresses. The chip select signals for IO-mapped devices are generated by using a 3-to-8 decoder. The address lines A_5, A_6 and A_7 are decoded to generate eight chip select signals (IOCS-0 to IOCS-7) and in this, the chip select IOCS-5 is used to select 8254. The address line A_0 and the control signal $M/\overline{IO}$ are used to enable the decoder.

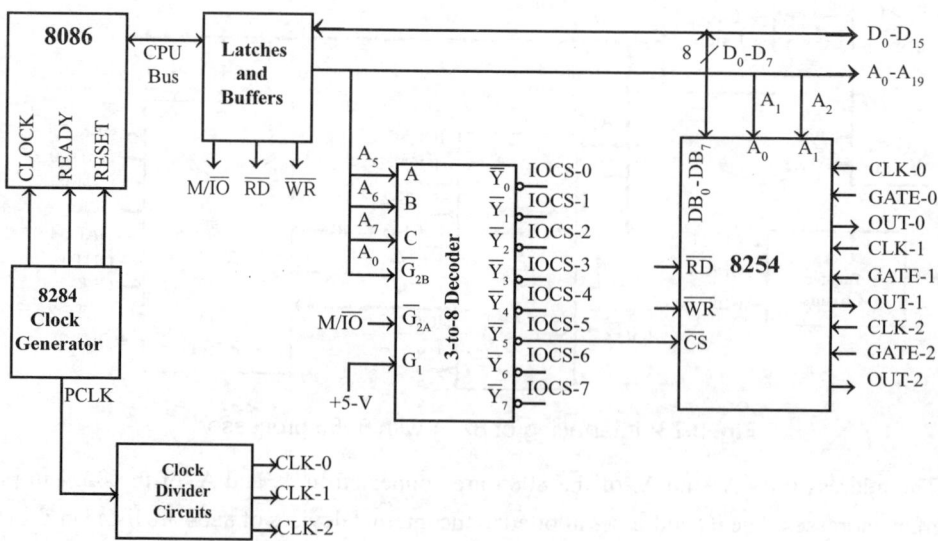

Fig. 9.74: Interfacing of 8254 with 8086 processor.

The address lines A_1 and A_2 of the 8086 are connected to A_0 and A_1 of 8254 to provide the internal addresses. The IO addresses allotted to the internal devices of 8254 are listed in Table 9.30. The data lines $D_0 - D_7$ are connected to $D_0 - D_7$ of the processor, and $\overline{RD}$ and $\overline{WR}$ signals of 8254 are connected to the $\overline{RD}$ and $\overline{WR}$ of the processor respectively to acheive parallel data transfer.

The clock signals required for the counters can be obtained either from the processor clock or from the **Peripherals clock** (PCLK) supplied by the clock generator 8284. The clock signals from 8284 can also be divided to lower values by using clock divider circuits and then applied to the clock input of counters.

Table 9.30: IO Addresses of 8254 Interfaced to 8086 as Shown in Fig. 9.74

| Internal device | Binary address | | | | | | | | Hexa address |
| | Decoder input | | | Input to address pins of 8254 | | | | Decoder enable | |
	A_7	A_6	A_5	A_4	A_3	A_2	A_1	A_0	
Counter-0	1	0	1	x	x	0	0	0	A0
Counter-1	1	0	1	x	x	0	1	0	A2
Counter-2	1	0	1	x	x	1	0	0	A4
Control Register	1	0	1	x	x	1	1	0	A6

Note: Don't care "x" is considered as zero.

9.12.4 Programming 8254

Each counter of 8254 can be individually programmed by writing a control word followed by the count value. The format of the control word is shown in Fig. 9.75.

The bit B_0 (BCD) of control word is used to select BCD or binary count and the bits B_1 to B_3 (M0, M1 and M2) are used to select the mode of operation for the counter specified by bits B_6 and B_7 of control word. Please remember that for each counter separate control word has to be sent to the same control register address. The 8254 identifies the control word for a particular counter from bits B_6 and B_7 of the control word.

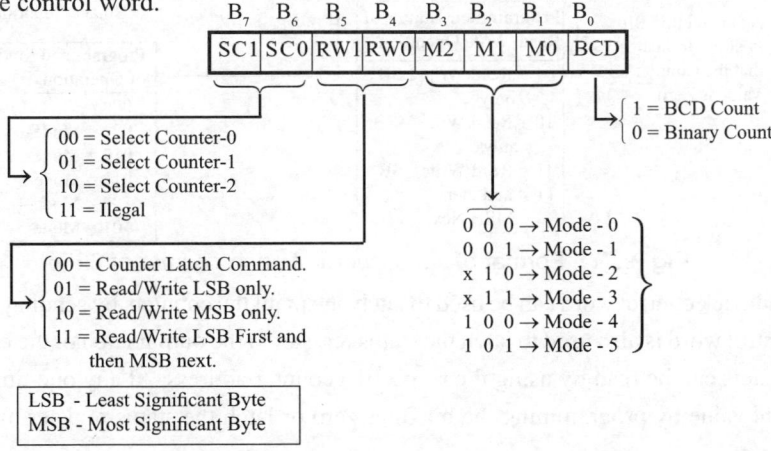

Fig. 9.75: Format of control word for timer 8254.

The bits B_4 and B_5 are used for read/write command. These bits are programmed for reading/writing the 16-bit count value in the proper order. If the count value is read without stopping the counter, then the count value may change between reading the LSB and MSB. To avoid this, the counter latch command can be used to latch the count value to an internal latch available at the output of each counter before the read operation.

Alternatively, a separate read-back control word is available for latching the count value in the 8254. (This control word is not available in 8253.) The format of read-back control word of 8254 is shown in Fig. 9.76. This control word has to be send to the same control register address before the read operation to latch the count value. The control register identifies this control word from the value of bits B_6 and B_7.

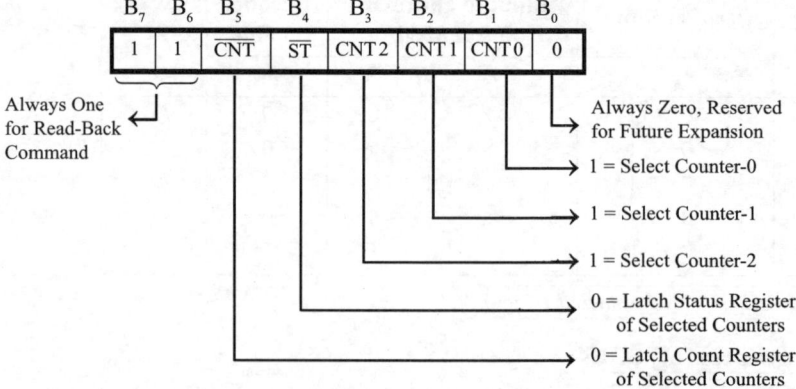

Fig. 9.76: Format of read-back control word of 8254.

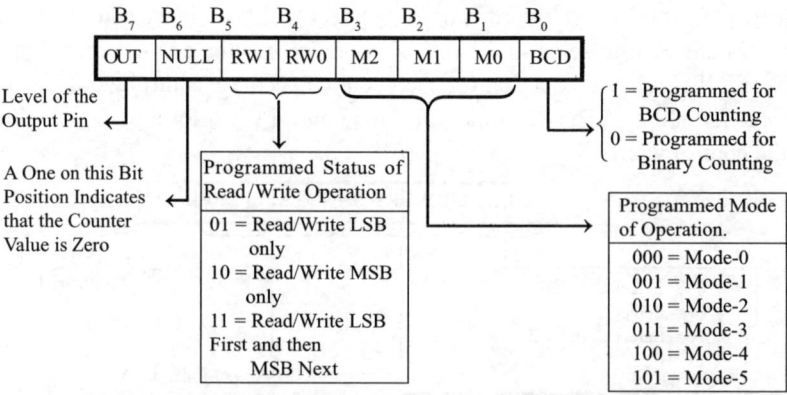

Fig. 9.77: Format of status word of each counter of 8254.

The read-back control word can be used to latch one or all the counters by sending a single control word. This control word is also used to latch the status register to the output latch of the counters, so that the status registers can be read by using the respective counter address. At any one time we can latch either the count value by programming the bit B_5 as zero or latch the status register by programming the bit B_4 as zero.

The format of the status register of each counter is shown in Fig. 9.77. The status word of a counter can be read to check the programmed status of the counter and also to verify whether the count value has reached terminal count i.e., zero or not.

9.12.5 Operating Modes of 8254

The 8254 has six modes of operation. Each counter of 8254 can be independently programmed to work in one of the possible six operating modes. The six modes are:

Mode - 0 → Interrupt on terminal count.

Mode - 1 → Hardware retriggerable one shot.

Mode - 2 → Rate generator or Timed interrupt generator.

Mode - 3 → Square wave mode.

Mode - 4 → Software triggered strobe.

Mode - 5 → Hardware triggered strobe.

The initialization procedure for each mode is almost same, but the output of each mode will be different. To initialize a counter, the following steps are necessary:

1. **Write a control word into the control register.**

2. **Write a count value in the count register.**

The writing of count value depends on the control word. There may be three possible choice. These are:

i) **If the control word is framed for writing LSB only, then write LSB alone.**

ii) **If the control word is framed for writing MSB only, then write MSB alone.**

iii) **If the control word is framed for writing LSB first and MSB next, then write LSB first and write MSBnext.**

> *Note:* LSB - Least Significant Byte (Low order byte).
>
> MSB - Most Significant Byte (High order byte).

In all the modes the GATE signal acts as a control signal to start, stop or maintain the counting process. In modes 0, 2, 3 and 4 once the count value is loaded in the counter, the timer starts decrementing the count value if the GATE is **high**.

Whenever the GATE signal goes **low**, the counter stops counting and will resume counting only when the GATE is made **high** again.

In modes 1 and 5 the GATE act as a triggering pulse. In these modes, the count value is loaded in the counter and it starts the decrementing process only when the GATE signal makes a low-to-high transition (i.e., the count process is initiated only on the rising edge of the GATE signal.) In modes 1 and 5 the GATE signal need not remain **high** (after initiation), to maintain the counting process.

A brief description about each mode of operation is presented here. In the following discussions it is assumed that the counter is initialized for binary count, by writing only the LSB of the count.

Mode-0 : Interrupt on terminal count

In mode-0 operation when a count value is loaded in a counter it starts decrementing the count value by one for each input clock pulse (provided the GATE is **high**) and asserts the output as **high** when the count value is zero (i.e., on terminal count). This low-to-high transition of the counter output can be used as an interrupt to the processor to initiate any activity. In mode-0 the 8254 will count as long as the GATE is **high**. Whenever the GATE signal goes **low** the counter stops counting and will resume counting only when the GATE is made **high** again.

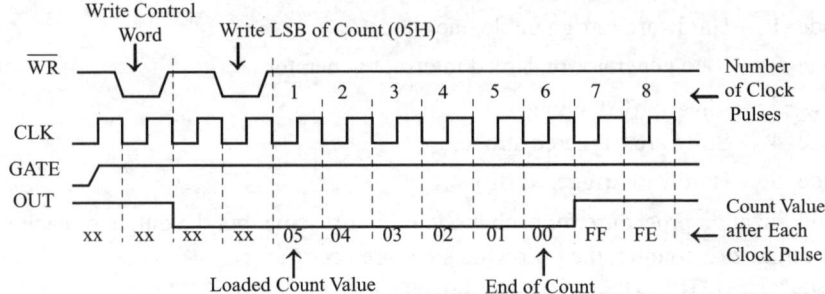

Fig. a: Timing diagram of Mode-0 with GATE always **high**.

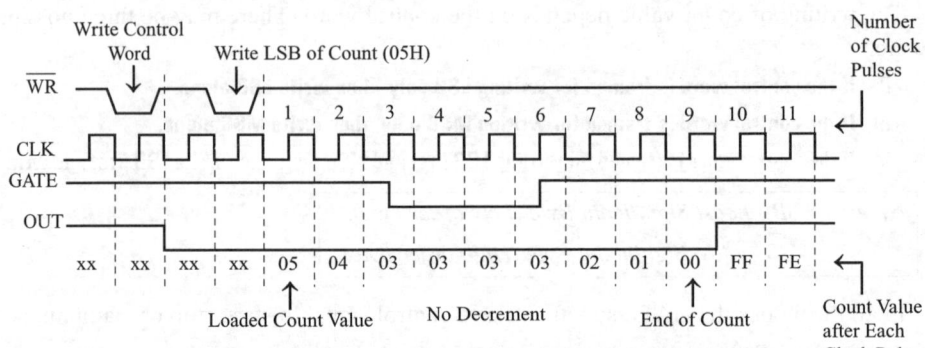

Fig. b: Example diagram of mode-0 when the GATE is made **low** for small duration before the terminal count.

Note :   *"xx" represents undefined count value.*

Fig. 9.78: Timing diagram of mode-0 of 8254.

The timing diagram for mode-0 operation is shown in Fig. 9.78. In the timing diagram of Fig. 9.78(a) initially the counter output remains **high**, and it is assumed that the GATE is always **high**. The processor writes the control word and count value using write control signal ($\overline{\text{WR}}$). Once the control word is written into the control register, the output goes **low**. After the write operation of count value by the processor, the 8254 requires one clock pulse to load the count value in the respective count register. Therefore in the first clock pulse after $\overline{\text{WR}}$ goes **high**, the 8254 loads the count value in the count register and in each subsequent clock pulse, the count value is decremented by one. When the count value becomes zero, the output of the counter is asserted **high**.

Figure 9.78 shows the timing diagram for a count value of 05H initially loaded in the count register. Here the output goes **high** after 6 (5 + 1 = 6) clock pulse. In general, if a count value of N is loaded in the count register then the output goes **high** after N + 1 clock pulses. Please remember that the counter continues to decrement the count value even after zero (00 → FF ; FF → FE and so on) as the long as the GATE is **high** and the clock signal is supplied. The output of the counter remains **high** until a new count or command is sent to the counter.

In the timing diagram shown in Fig. 9.78 (b), the GATE is made **low** for a small period before the terminal count value. It is observed that, in this period, the count value is not decremented and previous value is maintained as such. The counter resumes operation only when the GATE is made **high** again.

Mode-1 : Hardware Retriggerable One Shot

In mode-1, the counter functions as a retriggerable monostable multivibrator (one shot). In this mode, the output will be **high** once the control word is sent to the control register. The GATE acts as a trigger pulse to start the count process. When a low-to-high transition of GATE signal occurs, the count value is loaded in the counter and the count is decremented by one for each clock pulse. When the count value is loaded in the counter the OUTPUT goes **low** and it becomes **high** when the count value is zero. Therefore mode-1 produces a logic **low** pulse output whose width is equal to the duration of the count.

The timing diagram of mode-1 operation is shown in Fig. 9.79. The processor writes the control word and count value using $\overline{WR}$ control signal. Initially the output is assumed to be **high**. Even if it is **low**, it is asserted **high**, once the control word is written into the control register. Initially the GATE can be **high** or **low**.

If the GATE is **low** then it is made **high** to initiate the count process. If it is **high** then it is made **low** and after a small delay it is made again **high**, because the count process is initiated only after a low-to-high transition of GATE. After the trigger pulse (i.e., low-to-high transition) the gate can remain either in **high** state or in **low** state.

The first clock pulse after a low-to-high transition of gate is used to load the count value in the counter and for each subsequent clock the count value is decremented by one. Once the count value is loaded in the counter the output is asserted **low** and at the end of the count, when the count value is zero, the output is asserted **high**. In the timing diagram shown in Fig. 9.79(a), a count value of 05H is loaded and so the output remains **low** for 5 clock periods.

In general, if a count value of N is loaded in the counter then the output will remain **low** for N clock periods. Therefore the output low pulse width will be N times the clock period.

In the timing diagram of Fig. 9.79(b), the GATE is retriggered before the end of the count. In this case, the original count value is reloaded again in the clock pulse after gate retriggering and the count value is decremented by one in each subsequent clock pulse.

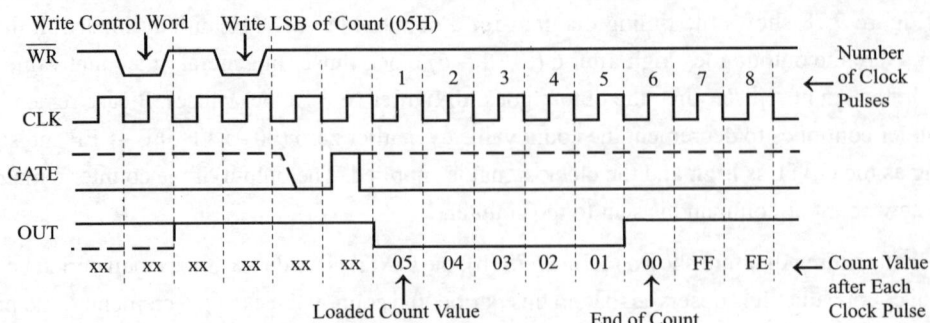

Fig. a: Timing diagram of mode-1.

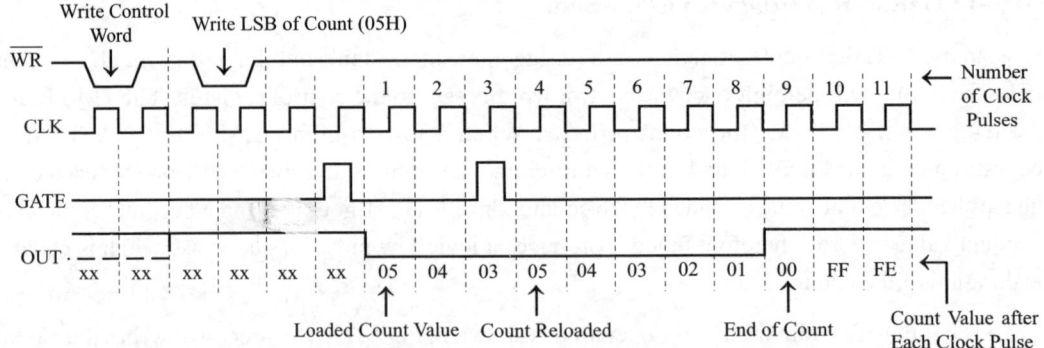

Fig. b: Timing diagram of mode-1 with GATE retriggering before end of count.

Note: "xx" represents undefined count value.

Fig. 9.79: Timing diagram of mode-1 of 8254.

Mode-2 : Rate Generator or Timed Interrupt Generator

The mode-2 is used to generate a periodic low pulse of width equal to one clock period. If a count value of N is loaded in the counter then the output will go **low** once in N clock periods. Therefore the frequency of low pulse generated will be equal to the input clock frequency divided by N. For mode-2 operation GATE should be always **high**.

The timing diagram of mode-2 operation is shown in Fig. 9.809. The processor writes the control word and count value using $\overline{WR}$ control signal. Initially the output is assumed to be **high**. Even if it is **low**, it is asserted **high**, once the control word is written into the control register. The GATE input is permanently tied to logic **high**. In the first clock pulse after the $\overline{WR}$ signal goes **high**, the count value is loaded in the counter and the count value is decremented by one for each subsequent clock pulse.

Initially the output is **high**. When the count reaches one, the output is asserted **low**. In the next clock pulse, the output is asserted **high** and the original count value is reloaded. In the subsequent clock pulses the count value is decremented. The above process is repeated again and again until a next command by the processor. In the timing diagram shown in Fig. 9.80, a count value of 03 H is loaded in the counter. In a total period of 3 clock periods, the output goes **low** for one clock period. If the gate is made **low** at any time during the count process, the counter will stop the operation and resumes the counting only when the gate is made **high** again.

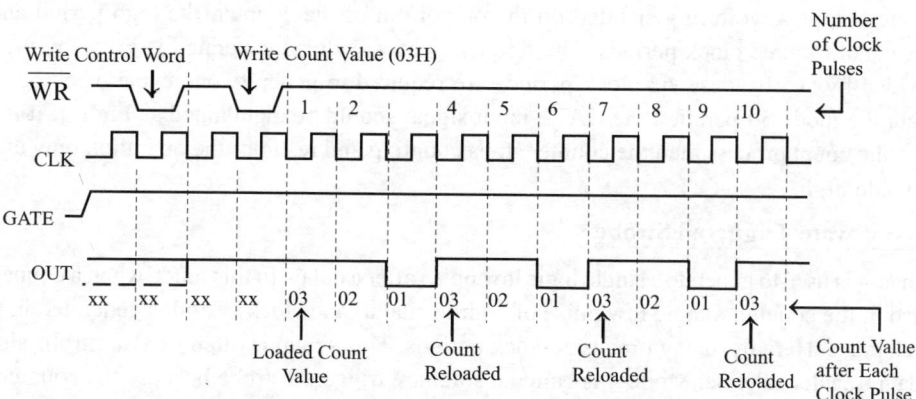

Fig. 9.80: Timing diagram of mode-2 of 8254.

Mode-3 : Square Wave Mode

In mode-3, the counter generates a square wave at the output pin. The frequency of the square wave will be given by the frequency of the input clock signal divided by the count value loaded in the count register. If the count value N is an even number, then the output will be alternatively **high** for N/2 clock periods and **low** for N/2 clock periods. If the count value is an odd number, then the output will be alternatively **high** for $\frac{N+1}{2}$ clock periods and **low** for clock periods (i.e., when the count value is odd, then the output **high** period will be more than low period by one clock period.) The timing diagram of mode-3 is shown in Fig. 9.81.

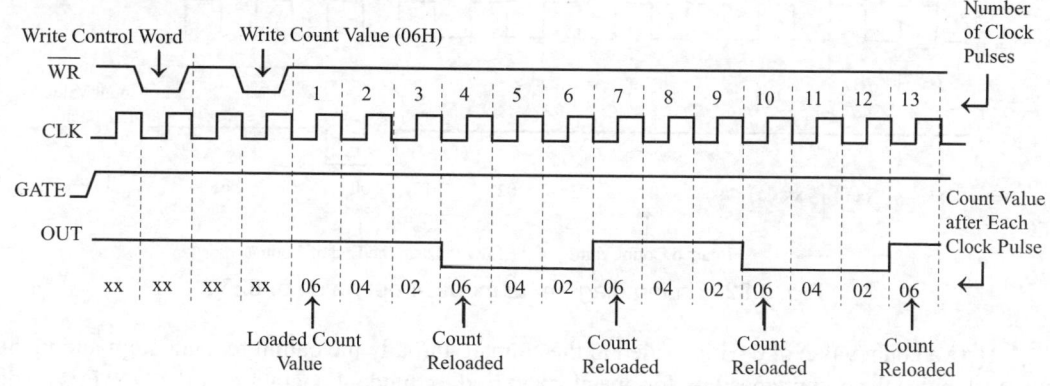

Fig. 9.81: Timing diagram of mode-3 of 8254.

In the timing diagram shown in Fig. 9.81 a count value of 06H is loaded in the counter. The count value is loaded in the counter in the first clock pulse after the $\overline{\text{WR}}$ signal goes **high**. Then for each subsequent clock pulse the count is decremented by two. When the count value reaches two then in the next clock pulse, the output is asserted **low** and original/initial count is reloaded in the counter and for each subsequent clock pulse the count is decremented by two. When the count value reaches two then in the next clock pulse, the output is asserted **high** and the original/initial count is reloaded and the above process is repeated again and again.

In the output waveform generated on the output pin of the counter, the high period and low period are equal to three clock periods. The frequency of waveform generated is given by the clock signal divided by six, because six clock periods are required to generate one cycle of output wave. Throughout the mode-3 operation the GATE input signal should be maintained as **high**. If it is made **low** during the count process then the counter stops counting and resumes the operation only after the GATE is made **high**.

Mode-4 : Software Triggered Strobe

Mode-4 is used to generate a single logic **low** pulse after a delay. In this mode when a count value, N is loaded in the counter, a logic **low** pulse of width equal to one clock period is generated in the (N + 1)th clock pulse. Here the delay time is N clock periods. This signal is often used as strobe signal in parallel data transfer scheme. Mode-4 is called a software triggered strobe because the counter starts its operation once the count value is written into the count register by a software instruction. However the GATE input signal should remain **high** throughout the mode-4 operation.

The timing diagram of the mode-4 operation is shown in Fig. 9.82. The GATE is permanently tied to **high**. The processor writes the control word and count value using the write control signal. In the first clock pulse after $\overline{WR}$ signal goes **high**, the count value is loaded in the counter and in each subsequent clock pulses the count value is decremented by one. When the count value reaches zero, the output is asserted **low** for one clock period and then it is made **high**.

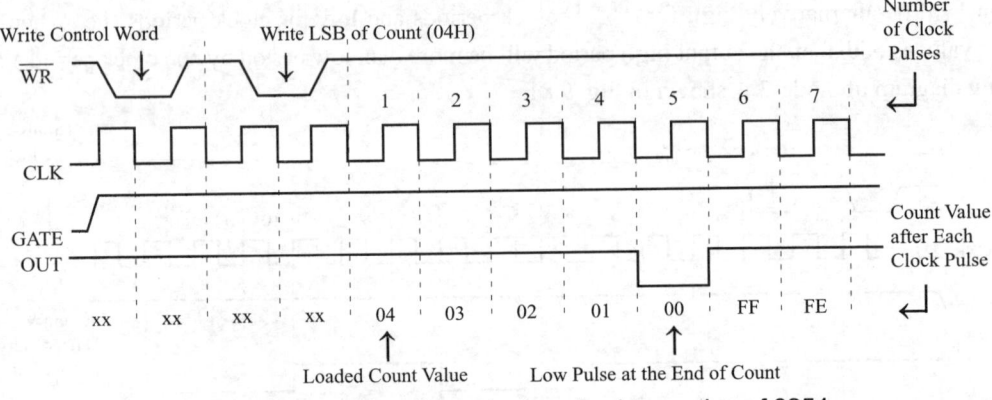

Fig. 9.82: Timing diagram of mode-4 operation of 8254.

Here a count value of 04H is loaded in the counter. Initially the output remains **high** and in the fifth clock pulse the output goes **low** for one clock period. In mode-4 operation, if the GATE is made **low** during count process then the counter stops counting and resumes the operation only when the GATE is made **high**.

Mode-5 : Hardware Triggered Strobe

The mode-5 is same as that of mode-4, except that the counter is initiated by a low-to-high transition of the GATE signal. In mode-4, the counter will start decrementing the count value immediately after the write operation of count value by the processor. But in mode-5, the counter will wait for a low-to-high transition of GATE signal after the write operation of count value by the processor.

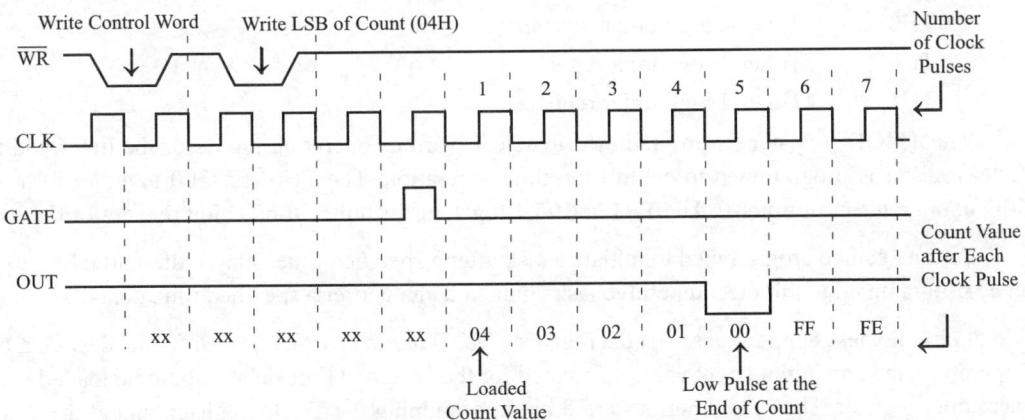

Fig. 9.83: Timing diagram of mode-5 operation of 8254.

The timing diagram of a mode-5 operation is shown in Fig. 9.83. In the first clock pulse after a low-to-high transition of GATE, the count value is loaded in the counter and for each subsequent clock pulse the count value is decremented by one. When the count value reaches zero, the output is asserted **low** for one clock period and then it is made **high**. Here a count value of 04 H is loaded in the counter. Initially the output remains **high** and the counter wait for a low-to-high transition of the GATE signal. In the fifth clock pulse after a low-to-high trasition of GATE signal, the output goes **low** for one clock period.

In mode-5 operation, if the gate signal makes another low-to-high transition (i.e., retriggered) before the end of count then the original count value is reloaded in the clock pulse after gate retriggering and count value is decremented by one in each subsequent clock pulse.

9.13 PROGRAMMING 8051 TIMERS

The 8051 has two internal 16-bit timers/counters that can be programmed to work independently. They are called timer-0/counter-0 and timer-1/counter-1. In the counter mode of operation, the timer will count the high-to-low transition of the signal applied at the corresponding timer pin (port-3 pin, P3.2 for timer-0 and P3.3 for timer-1), by incrementing the content of the timer register associated with the timer by one for every high-to-low transition. (The signal applied at the port pin will act as a clock for incrementing the content of the timer register.)

In the timer mode of operation, the internal timer clock will increment the content of the associated timer register for every clock pulse. The timer clock is internally derived by dividing the crystal frequency by 12. Therefore, the timer clock is an independent clock, and the frequency of the timer clock will be 1/12 of the system clock frequency.

The various special function registers associated with internal timers/counters of 8051 are,

TMOD : Timer/Counter mode control register

TCON : Timer/Counter control register

TL0   : Timer-0 low order register

TH0   :   Timer-0 high order register

TL1   :   Timer-1 low order register

TH1   :   Timer-1 high order register

The TMOD register is programmed to select various operating modes of the timer and the TCON register is programmed to control the timer operation. The TH0 and TL0 together form the 16-bit count register of timer-0. The TH1 and TL1 together form the 16-bit count register of timer-1.

A timer can be programmed to initiate a task after a specified time delay. Alternatively, a timer can be programmed to initiate a repetitive task again and again after a specified time delay.

For both these applications, first the register TMOD has to be programmed for the desired mode of operation, then an initial count value calculated for the specified time delay should be loaded in the timer count register. Then the timer is started by programming the TCON register, and at the end of time delay, an interrupt is generated which can be used to initiate the specified task.

9.13.1 Timer Mode Control (TMOD) Register

The TMOD register is used to select the operating mode and the timer/counter operation of the timers. The format of a TMOD register is shown in Fig. 9.84. The lower 4-bits of the TMOD register are used to control timer-0 and the upper four bits are used to control timer-1.

The two timers can be independently programmed to operate in various modes. The TMOD register has two separate 2-bit fields, M0 and M1, to program the operating mode of the timers. The operating modes of the timers are mode-0, mode-1, mode-2 and mode-3. In all these operating modes, the oscillator clock is divided by 12 and applied as the input clock to the timer.

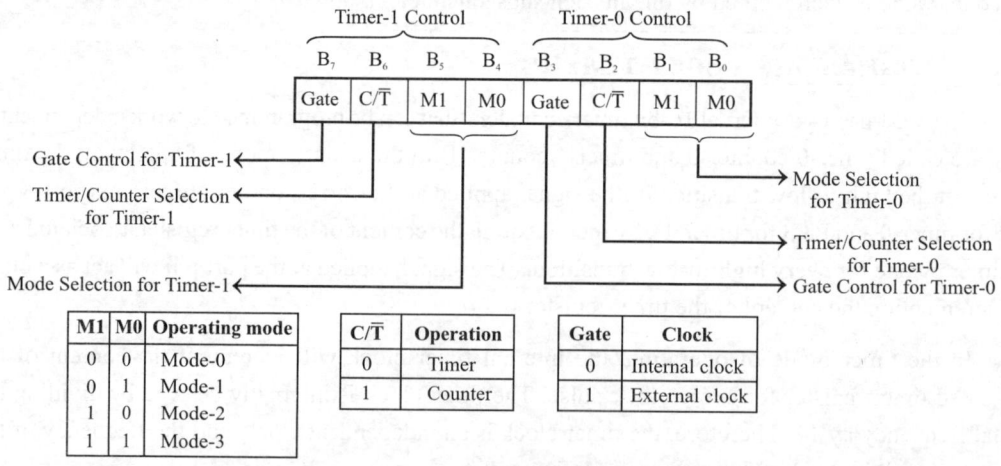

Fig. 9.84: Format of the TMOD register of the 8051 family of microcontrollers.

Timer Port Mode-0

In mode-0, the timer register is configured as a 13-bit register. For timer-1, the 8 bits of TH1 and lower 5 bits of TL1 are used to form the 13-bit register. For timer-0, the 8-bit of TH0 and lower 5 bits of TL0 are used to form the 13-bit register. (The upper three bits of the TL registers are ignored). For every clock input to the timer, the 13-bit timer register is incremented by one. When the timer count rolls over from all 1s to all 0s (i.e., 11111 1111 1111 to 00000 0000 0000), the timer interrupt flag in the TCON register is set to one.

Timer Port Mode-1

Mode-1 is same as mode-0 except the size of the timer register. In mode-1, the TH and TL registers are cascaded to form a 16-bit timer register.

Timer Port Mode-2

In mode-2, the timers function as 8-bit timers with automatic reload feature. The TL register will function as an 8-bit timer count register and the TH-register will hold an initial count value. When the timer is started, the initial value in TH is loaded to TL and for each clock input to the timer, the 8-bit timer count register is incremented by one. When the timer count rolls over from all 1s to all 0s (i.e., 1111 1111 to 0000 0000), the timer interrupt flag in the TCON register is set to one and the content of the TH register is reloaded in the TL register and the count process starts again from this initial value.

Timer Port Mode-3

In mode-3, the timer-0 is configured as two separate 8-bit timers and timer-1 is stopped. In mode-3, the TL0 will function as an 8-bit timer controlled by standard timer-0 control bits and the TH0 will function as an 8-bit timer controlled by timer-1 control bits. While timer-0 is programmed in mode-3, timer-1 can be programmed in mode-0, 1 or 2 and can be used for applications that do not require an interrupt.

The $C/\overline{T}$ bit of the TMOD register is used to program the counter or timer operation of the timer. When the $C/\overline{T}$ bit is set to one, the timer will function as an event counter. The $C/\overline{T}$ bit is programmed to zero for timer operation.

Normally, the TR (**T**imer **R**un) bit in the TCON register is used to control the clock input to the timer. In order to allow the clock input (*Note: the timer will run only if the clock input is allowed*), the TR bit should be set to one. In addition, the GATE bit in the TMOD register will facilitate the external signal applied to the $\overline{INT}$ pin to act as an additional control signal to allow or disallow the clock input to the timer. When GATE = 1, the clock input to the timer is allowed only if the signal at the $\overline{INT}$ pin is **high** (as well as TR should be set to one.) When GATE = 0, the signal at the $\overline{INT}$ pin is ignored (but the TR should be set to one.)

9.13.2 TIMER CONTROL (TCON) REGISTER

The TCON register consists of timer overflow flags, timer run control bits, external interrupt flags and external interrupt-type control bits. The format of a TCON register is shown in Fig. 9.85.

The timers in an 8051 microcontroller are upcounters and keep on incrementing as long as the clock is applied. Therefore, when the clock is applied after reaching the maximum value (i.e., the content of the counter is all 1s), the content of the counter will become zero (i.e., all 0s). This condition is called timer overflow and it is also the end of timing which a program wants to maintain by using the timer. The TCON register has a 1-bit flag, TF for each timer to indicate the timer overflow or end of timing. Whenever the timer/counter overflows, the TF flag is set to one.

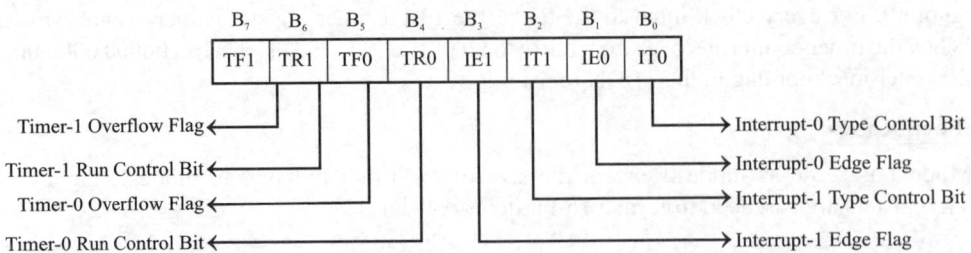

Fig. 9.85: Format of a TCON register of the 8051 family of microcontrollers.

The TF flag is also used as an interrupt signal to initiate the execution of a subroutine. When the controller vectors to subroutine, the TF flag is cleared.

The TR bit is used to start/stop the timer/counter. When the TR bit is set to one, the timer/counter will start counting and continue the counting as long as the TR bit is one. The timer/counter will stop counting when the TR bit is cleared to zero.

When a valid external interrupt signal is detected, the IE flag is set to one. When the controller accepts the external interrupt and starts processing it, the IE flag is cleared to zero. The IT bit is used to program the type of external interrupt signal to be recognized by the controller. The IT bit is programmed as one to recognize the falling edge triggered external interrupt and it is programmed as zero to recognize logic **low**-level external interrupt.

9.13.3 Calculating the Time Delay for 8051 Timers

Case 1: Crystal Frequency is 12 MHz

When the crystal frequency is 12 MHz, the timer clock frequency will be 1 MHz. Therefore, the time period of 1 clock is 1/1 MHz = 1 microsecond. Therefore, when the timer clock is a 1 MHz clock, the timer count gets incremented once in every 1 microsecond. Hence, the timer count value for a specified delay in microseconds will be same as the value of delay in microsecond itself. For example, for 100 microseconds delay, the count value is 100. In 8051 timers, there are three choices for the size of count value, and they are 8-bit, 13-bit and 16-bit.

When the 8-bit count register is selected, the maximum count value is $2^8 = 256_{10}$ = FFH, and maximum possible time delay is 256 microseconds.

When the 13-bit count register is selected, the maximum count value is $2^{13} = 8192_{10}$ = 2000H, and maximum possible time delay is 8192 microseconds = 8.192 milliseconds.

When the 16-bit count register is selected, the maximum count value is $2^{16} = 65536_{10}$ = FFFFH, and maximum possible time delay is 65536 microseconds = 65.536 milliseconds.

The 8051 counters are up-counters, and overflow or end of count occurs only after reaching maximum value, therefore the initial count cannot be 0 for any desired time delay. The initial count for any required delay has to be calculated by subtracting the delay count from maximum value. For example, when a 10 millisecond delay is required while using a 16-bit count register, the initial count will be 65536 – 10000 = 55536 = D8F0H.

Hence, in order to get a time delay of 10 milliseconds, if the initial value D8F0H is loaded in the 16-bit count register, and the timer operation is started, then the overflow occurs when the count reaches FFFFH and the time delay achieved will be 10 milliseconds.

Case-2: Crystal Frequency is 11.0592 MHz

The crystal frequency of 11.0592 MHz is used when serial communication with standard PC (Personal Computer) is employed in order to generate the clock for right baud rate for serial communication with standard PC. When the crystal frequency is 11.0592 MHz, the timer clock frequency will be $11.0592/12 = 0.9216$ MHz. Therefore, the time period of 1 clock is $1/0.9216$ MHz $= 1.085$ microseconds. Now, the timer can increment the count by one in 1.085 microseconds. The count value for a specified delay in microseconds will be microseconds delay value divided by 1.085. For example, for 100 microseconds delay, the count value is $100/1.085 = 92$. In 8051 timers, there are three choices for the size of count value, and they are 8-bit, 13-bit and 16-bit.

When an 8-bit count register is selected, the maximum count value is $2^8 = 256_{10} =$ FFH, and maximum possible time delay is $256 \times 1.085 = 277$ microseconds.

When a 13-bit count register is selected, the maximum count value is $2^{13} = 8192_{10} = 2000$H, and maximum possible time delay is $8192 \times 1.085 = 8888$ microseconds $= 8.888$ milliseconds.

When the 16-bit count register is selected, the maximum count value is, $2^{16} = 65536_{10} =$ FFFFH, and maximum possible time delay is $65536 \times 1.085 = 71106$ microseconds $= 71.106$ milliseconds.

9.13.4 Examples of Timer Programming in an 8051 Controller

EXAMPLE PROGRAM 9.8

This example program is developed for the generation of unipolar square waveform of 1 kHz frequency using the timer-0 of 8051 in mode-1. Assume that the crystal frequency of the controller is 12 MHz.

Problem Analysis

In order to generate a square wave, a port pin can be set to high first and then after a time delay, the port pin is reset to zero; then the process of set and reset are repeated continuously with a uniform time delay. The desired time delay can be achieved using the timer of 8051.

The time period of 1 kHz square wave is $1/1*10^3 = 1$ millisecond $= 1000$ microseconds. But in a square wave during half the period, the square wave will be high and during the next half of the period, the square wave will be low. So the time delay required is $1000/2 = 500$ micro-seconds.

Since the crystal frequency is 12 MHz, the timer clock will be $1/12 = 1$ MHz, and so the time period of the timer clock will be 1 microsecond. Therefore, the count for 500 microseconds time delay is 500. The count register in mode-1 is 16-bit and so the maximum count is $2^{16} = 65536_{10}$. The initial count for any required delay has to be calculated by subtracting the delay count from the maximum count value.

Therefore, the initial count $= 65536_{10} - 500_{10} = 65036_{10} =$ FE0CH

The byte to be loaded in the TMOD register to select mode-1 operation of timer-0 is framed as follows:

TMOD =	Gate	C/$\overline{T}$	M1	M0	Gate	C/$\overline{T}$	M1	M0	= X X X X 0 0 0 1 = 01_H

Flowchart

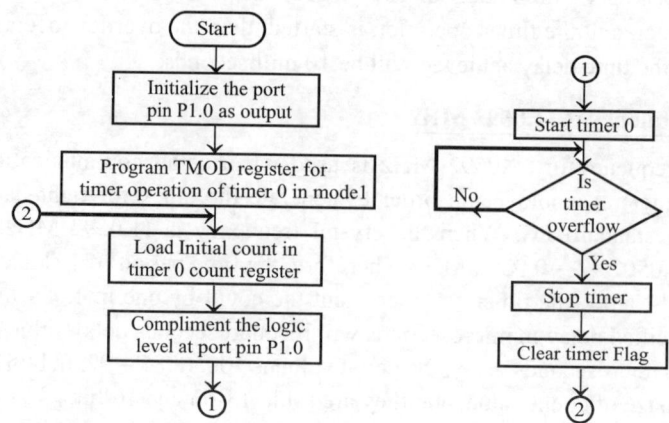

Assembly Language Program

```
;Program to generate square wave using 8051 timer in mode-1

        CLR   P1.0          ;Initialize port pin P1.0 as output
        MOV   TMOD,#01H      ;Program TMOD register for mode-1 operation of timer-0

AGAIN:  MOV   TL0,#0CH       ;Load low byte of count in timer-0 low order count register
        MOV   TH0,#FEH       ;Load high byte of count in timer-0 high order count register

        CPL   P1.0          ;Compliment the port pin P1.0
        SETB  TR0           ;Set timer run flag, to start timer

WAIT:   JNB   TF0,WAIT      ;Wait for timer overflow
        CLR   TR0           ;Clear timer run flag, to stop timer
        CLR   TF0           ;Clear timer flag

        SJMP  AGAIN         ;Repeat generation of next cycle

        END                 ;Assemby end
```

EXAMPLE PROGRAM 9.9

This example program is developed for the generation of unipolar square waveform of 1 kHz frequency using the timer-0 of 8051 in mode-0. Assume that the crystal frequency of the controller is 12 MHz.

Problem Analysis

The program logic is same as Example Program 8.

The count register in mode-0 is 13-bit and so the maximum count is $2^{13} = 8192_{10}$. The initial count for any required delay has to be calculated by subtracting the delay count from the maximum count value.

Therefore, the initial count = $8192_{10} - 500_{10} = 7692_{10} = 1E0CH$

The 13-bit initial count has to be divided into upper 8 bits and lower 5 bits are as follows and the upper 8 bits are loaded in timer high order register and the lower 5 bits are loaded in timer low-order register.

1E0CH = 1 1110 0000 1100 = 11110000 01100

11110000 = 1111 0000 = F0H (Initial count for high order timer register)

01100 = 0 1100 = 0000 1100 = 0CH (Initial count for low order timer register)

The byte to be loaded in the TMOD register to select mode-0 operation of timer-0 is framed as follows.

TMOD = | Gate | C/$\overline{T}$ | M1 | M0 | Gate | C/$\overline{T}$ | M1 | M0 | = X X X X 0 0 0 0 = 00_H

Flowchart

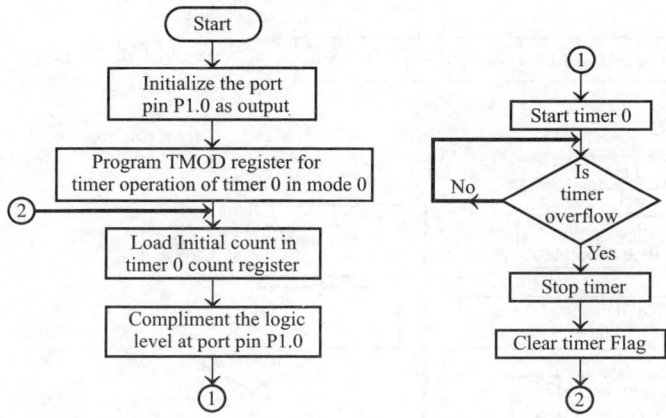

Assembly Language Program

```
;Program to generate square wave using 8051 timer in mode-0

        CLR   P1.0           ;Initialize port pin P1.0 as output
        MOV   TMOD,#00H      ;Program TMOD register for mode-0 operation of timer-0

AGAIN:  MOV   TL0,#0CH       ;Load low byte of count in timer-0 low order count register
        MOV   TH0,#F0H       ;Load high byte of count in timer-0 high order count register

        CPL   P1.0           ;Compliment the port pin P1.0
        SETB  TR0            ;Set timer run flag, to start timer

WAIT:   JNB   TF0,WAIT       ;Wait for timer overflow
        CLR   TR0            ;Clear timer run flag, to Stop timer
        CLR   TF0            ;Clear timer flag

        SJMP  AGAIN          ;Repeat generation of next cycle
        END                  ;Assemby end
```

EXAMPLE PROGRAM 9.10

This example program is developed for the generation of unipolar square waveform of 2 kHz frequency using the timer-0 of 8051 in mode-2. Assume that the crystal frequency of the controller is 12 MHz.

Problem Analysis

The program logic is same as Example Program 51.8. Here we have chosen the frequency as 2 kHz to reduce the time delay count.

Now, time delay count = $(1/2*10^3)/2 = 250$

The count register in mode-2 is 8-bit and so the maximum count is $2^8 = 256_{10}$. The initial count for any required delay has to be calculated by subtracting the delay count from maximum count value.

Therefore, the initial count = $256_{10} - 250_{10} = 6_{10} = 06H$

In mode-2, the 8-bit count value is loaded in high order count register only once. The controller will copy the value of the count in low order register and start the count operation. At the end of timer operation, the count value is reloaded from high order register to low order register and again restart the count operation. Therefore, the count register need not be loaded with the count value for every cycle of wave generation.

The byte to be loaded in the TMOD register to select mode-2 operation of timer-0 is framed as follows.

$$\text{TMOD} = \boxed{\text{Gate} \mid \text{C/}\overline{\text{T}} \mid \text{M1} \mid \text{M0} \mid \text{Gate} \mid \text{C/}\overline{\text{T}} \mid \text{M1} \mid \text{M0}} = \text{X X X X 0 0 1 0} = 02_\text{H}$$

Flowchart

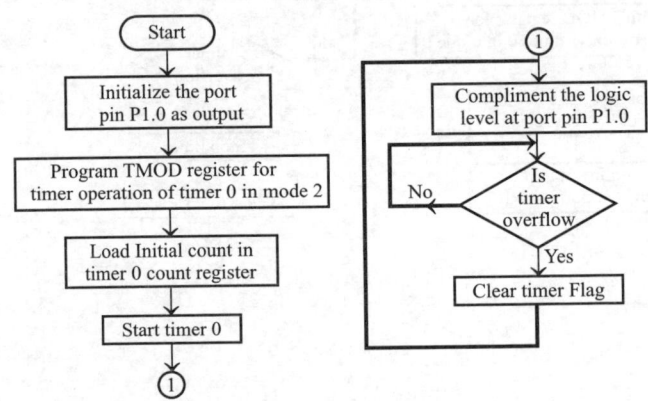

Assembly Language Program

```
;Program to generate square wave using 8051 timer in mode-2

        CLR   P1.0        ;Initialize port pin P1.0 as output
        MOV   TMOD,#02H   ;Program TMOD register for mode-2 operation of timer-0

        MOV   TH0,#06H    ;Load initial count in timer-0 high order count register
        SETB  TR0         ;Set timer run flag, to start timer

AGAIN:  CPL   P1.0        ;Compliment the port pin P1.0

WAIT:   JNB   TF0,WAIT    ;Wait for timer overflow
        CLR   TF0         ;Clear timer flag

        SJMP  AGAIN       ;Repeat generation of next cycle

        END               ;Assemby end
```

9.14 DAC INTERFACE

In many applications, the microprocessor has to produce analog signals for controlling certain analog devices. Basically, the microprocessor system can produce only digital signals. In order to convert the digital signal to analog signal, a **Digital-to-Analog** Converter (DAC) has to be employed.

The DAC will accept a digital (binary) input and convert to analog voltage or current. Every DAC will have "n" input lines and an analog output. The DAC requires a reference analog voltage (V_{ref}) or current (I_{ref}) source. The smallest possible analog value that can be represented by the n-bit binary code is called resolution. The resolution of DAC with n-bit binary input is $\frac{1}{2^n}$ of the reference analog value.

Every analog output will be a multiple of the resolution. In some converters, the input reference analog signal will be multiplied or divided by a constant to get full scale value. In this case the resolution will be $\frac{1}{2^n}$ of the full scale value. For example, consider an 8-bit DAC with reference analog voltage of 5 volts. Now the resolution of the DAC is $(1/2^8) \times 5$ volts. The 8-bit digital input can take, $2^8 = 256$ different values. The analog values for all possible digital inputs are as shown in Table 9.86.

The maximum input digital signal will have an analog value which is equal to reference analog value minus resolution. The digital-to-analog converters can be broadly classified into three categories, and they are current output, voltage output and multiplying type DAC. The current output DAC provides an analog current as output signal. In voltage output DAC, the analog current signal is internally converted to voltage signal.

Table 9.31: Analog value for Digital Input

Digital input	Analog output
0000 0000	$\frac{0}{2^8} \times 5$ Volts
0000 0001	$\frac{1}{2^8} \times 5$ Volts
0000 0010	$\frac{2}{2^8} \times 5$ Volts
0000 0011	$\frac{3}{2^8} \times 5$ Volts
.	.
.	.
.	.
1111 1111	$\frac{255}{2^8} \times 5$ Volts

In multiplying type DAC, the output is given by the product of the input signal and the reference source and the product is linear over a broad range. Basically, there is not much difference between these three types and any DAC can be viewed as multiplying DAC. The basic components of a DAC are resistive network with appropriate values, switches, a reference source and a current to voltage converter as shown in Fig. 9.86.

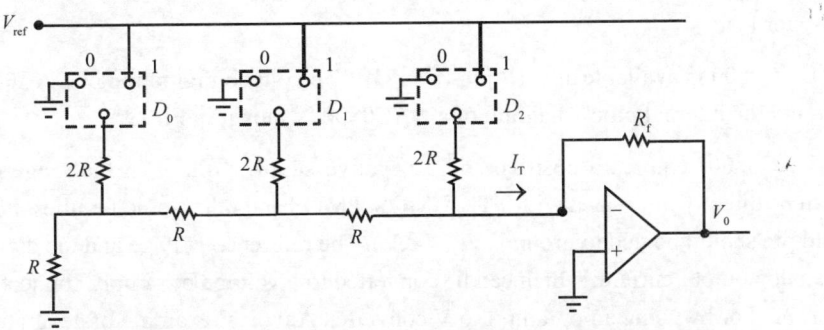

Fig. 9.86: A typical R/2R ladder resistive network as DAC.

The switches in the circuit of Fig. 9.86 can be transistors which connect the resistance either to ground or V_{ref}. The resistors are connected in such a way that for any possible binary input, the total current I_T is in binary proportion. The operational amplifier converts the current I_T to a voltage signal V_0, which can be calculated from the following equation.

$$V_0 = V_{ref} \frac{R_f}{R}\left(\frac{D_2}{2^1} + \frac{D_1}{2^2} + \frac{D_0}{2^3}\right)$$

The circuit of Fig. 9.86 can be modified as 8-bit DAC by increasing the number of R/2R ladder. For an 8-bit DAC the output voltage is given by,

$$V_0 = V_{ref} \frac{R_f}{R}\left(\frac{D_7}{2^1} + \frac{D_6}{2^2} + \frac{D_5}{2^3} + \frac{D_4}{2^4} + \frac{D_3}{2^5} + \frac{D_2}{2^6} + \frac{D_1}{2^7} + \frac{D_0}{2^8}\right)$$

The time required for converting the digital signal to analog signal is called **conversion time**. It depends on the response time of the switching transistors and the output amplifier. If the DAC is interfaced to the microprocessor, then the digital data (signal) should remain at the input of DAC, until the conversion is complete. Hence, to hold the data a latch is provided at the input of DAC.

The Digital-to-Analog converters compatible to the microprocessors are available with or without internal latch and I to V converting amplifier. The AD558 of the Analog Device is an example of an 8-bit DAC with an internal latch and I to V converting amplifier. The output of AD558 is an analog voltage signal.

The AD558 can be directly interfaced to 8086 microprocessor bus and it requires only two control signals: **Chip Select** ($\overline{CS}$) and **Chip Enable** ($\overline{CE}$). [No handshake signals are necessary for interfacing a DAC. The time between loading two digital data to the DAC is controlled by software time delay.]

The DAC0800 of the National Semiconductor Corporation is an example of an 8-bit DAC without internal latch and I to V converting amplifier. The DAC0800 can be interfaced to the microprocessor using either a port device or a latch.

9.14.1 DAC0800

The DAC0800 is an 8-bit, high speed, current output DAC with a typical settling time (conversion time) of 100 ns. It produces complementary current output which can be converted to voltage by using a simple resistor load.

The DAC0800 is available as a 16-pin IC in DIP. The pin configuration of DAC0800 is shown in Fig. 9.87 and the internal block diagram of a DAC0800 is shown in Fig. 9.88.

The DAC0800 requires a positive and a negative supply voltage in the range of ± 5 V to ± 18 V. It can be directly interfaced with TTL, CMOS, PMOS and other logic families. For TTL input, the threshold pin should be tied to ground ($V_{LC} = 0$ V). The reference voltage and the digital input will decide the analog output current, which can be converted to a voltage by simply connecting a resistor to output terminal or by using an op-amp I to V converter. A typical example of generating a positive voltage output using DAC0800 is shown in Fig. 9.89.

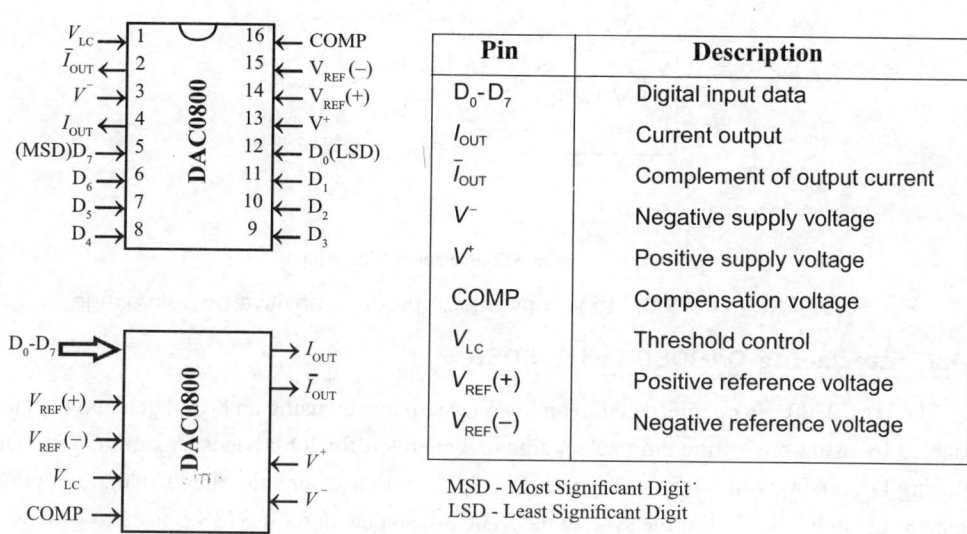

Pin	Description
D_0-D_7	Digital input data
I_{OUT}	Current output
$\bar{I}_{OUT}$	Complement of output current
V^-	Negative supply voltage
V^+	Positive supply voltage
COMP	Compensation voltage
V_{LC}	Threshold control
$V_{REF}(+)$	Positive reference voltage
$V_{REF}(-)$	Negative reference voltage

MSD - Most Significant Digit
LSD - Least Significant Digit

Fig. 9.87: Pin description of DAC0800.

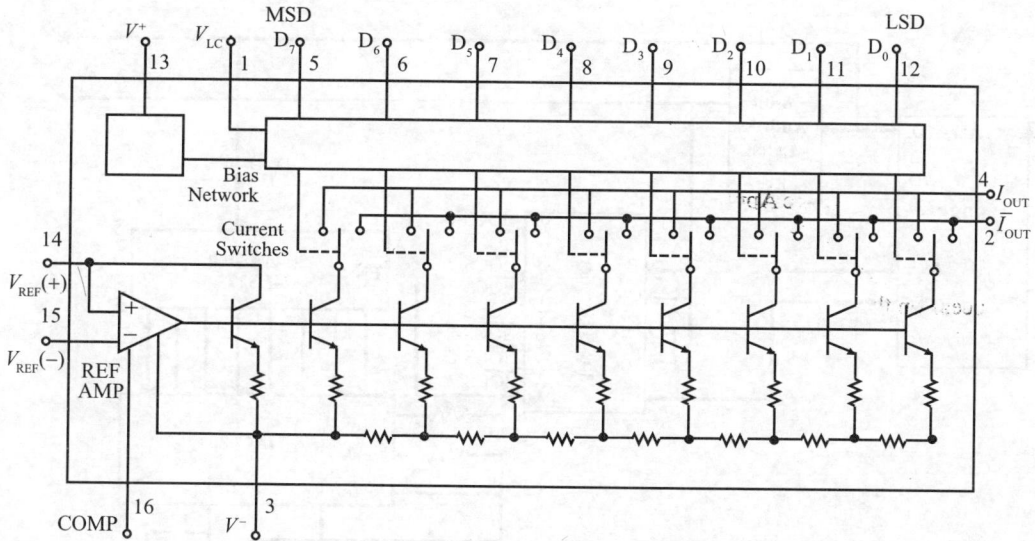

Fig. 9.88: Block diagram of DAC0800.

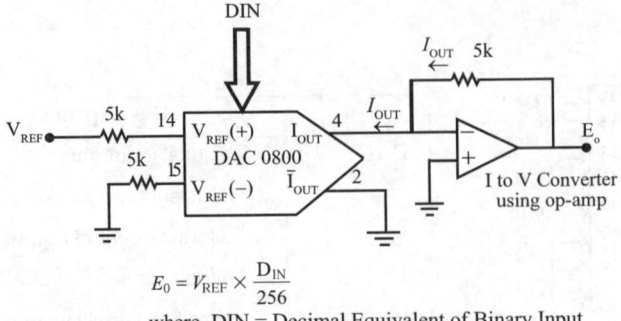

$$E_0 = V_{REF} \times \frac{D_{IN}}{256}$$

where, DIN = Decimal Equivalent of Binary Input

Fig. 9.89: DAC 0800 with I to V converter to produce positive output voltage.

9.14.2 Interfacing DAC0800 with 8085

The DAC0800 can be interfaced to an 8085 system bus by using an 8-bit latch and the latch can be enabled by using one of the chip select signals generated for IO devices. A simple schematic for interfacing DAC0800 with 8085 is shown in Fig. 9.90. In this schematic, the DAC0800 is interfaced using an 8-bit latch 74LS273 to the system bus. The 3-to-8 decoder 74LS138 is used to generate chip select signals for IO devices. The address lines A_4, A_5 and A_6 are used as input to decoder. The address line A_7 and the control signal IO/$\overline{M}$ are used as enable for the decoder. The decoder will generate eight chip select signals and in this, the signal IOCS-7 is used as enable for latch of the DAC. The IO address of the DAC is shown in Table 9.32.

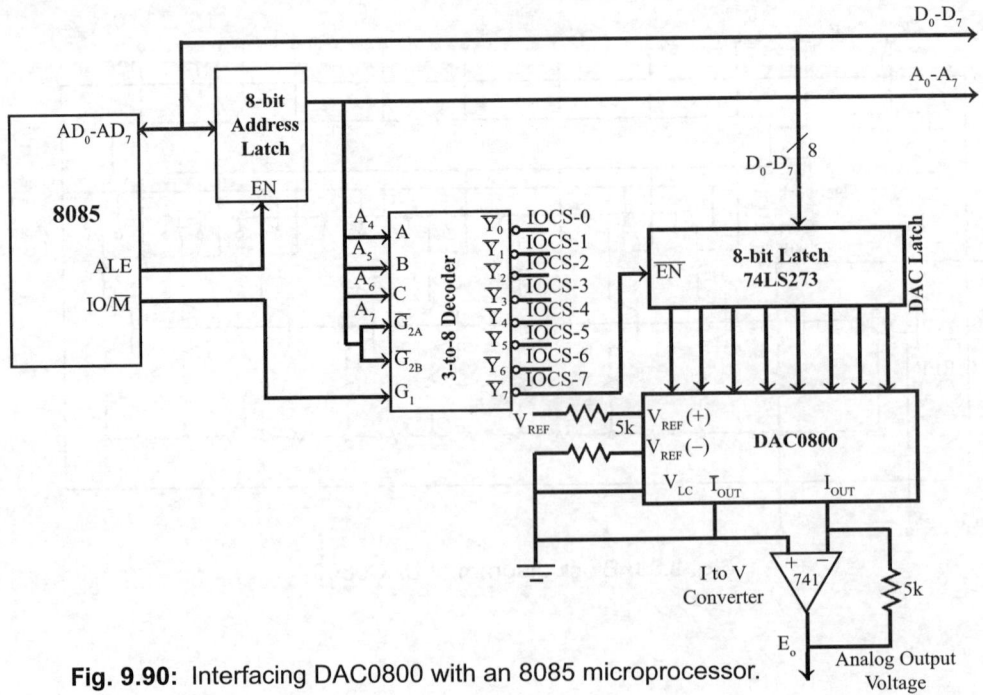

Fig. 9.90: Interfacing DAC0800 with an 8085 microprocessor.

In order to convert a digital data to analog value, the processor has to load the data to latch. The latch will hold the previous data until the next data is loaded. The DAC will take definite time to convert the data. The software should take care of loading successive data only after the conversion time. The DAC 0800 produces a current output, which is converted to voltage output using a I to V converter.

Table 9.32: IO Address of DAC Latch Shown in Fig. 9.90

Device	Binary address								Hexa address
	Decoder Input and enable				Unused address lines				
	A_7	A_6	A_5	A_4	A_3	A_2	A_1	A_0	
DAC Latch 74LS273	0	1	1	1	x	x	x	x	70

EXAMPLE PROGRAM 9.11

This example program is developed for generation of square waveform using the DAC interfaced to the 8085 system bus as shown in Fig. 9.90. The DAC is interfaced to the system bus with IO address 70_H. In order to generate a square wave, first the digital data corresponding to negative maximum is send to DAC and then after a time delay, the digital data corresponding to positive maximum is send to DAC and again after a time delay, the process is repeated continuously. The frequency of the square wave is decided by the amount of time delay introduced between two voltage levels.

Assembly Language Program

```
;Program to generate square wave using DAC interfaced to 8085 system bus

AGAIN:   MVI  A,00H      ;Load digital data to generate negative maximum
         OUT  70H        ;Send digital data to DAC
         CALL DELAY      ;Maintain output of DAC at negative maximum for half the time period
         MVI  A,0FFH     ;Load digital data to generate positive maximum
         CALL DELAY      ;Maintain output of DAC at positive maximum for half the time period
         JMP  AGAIN

DELAY:   LXI  B,0FFFH    ;Load count value in BC pair (Assume count as 0FFFH)
LOOP:    DCX  B          ;Decrement count
         MOV  A,C        ;Get C in A
         CMP  B          ;Check for count zero
         JNZ  LOOP       ;Decrement count until zero
         RET             ;Return to main program
```

9.14.3 Interfacing DAC0800 with 8086

The DAC0800 can be interfaced to the 8086 system bus by using an 8-bit latch and the latch can be enabled by using one of the chip select signals generated for IO devices. A simple schematic for interfacing the DAC0800 with 8086 is shown in Fig. 9.91.

In this schematic, the DAC0800 is interfaced using an 8-bit latch 74LS273 to the system bus and the latch is IO-mapped in the system with an even address. The 3-to-8 decoder 74LS138 is used to generate chip select signals for IO devices.

The address lines A_5, A_6 and A_7 are used as inputs to the decoder. The address line A_0 and the control signal $M/\overline{IO}$ are used as enable for decoder. The decoder will generate eight chip select signals and in this the signal IOCS-7 is used as enable for latch of DAC. The IO address of the DAC is shown in Table 9.33.

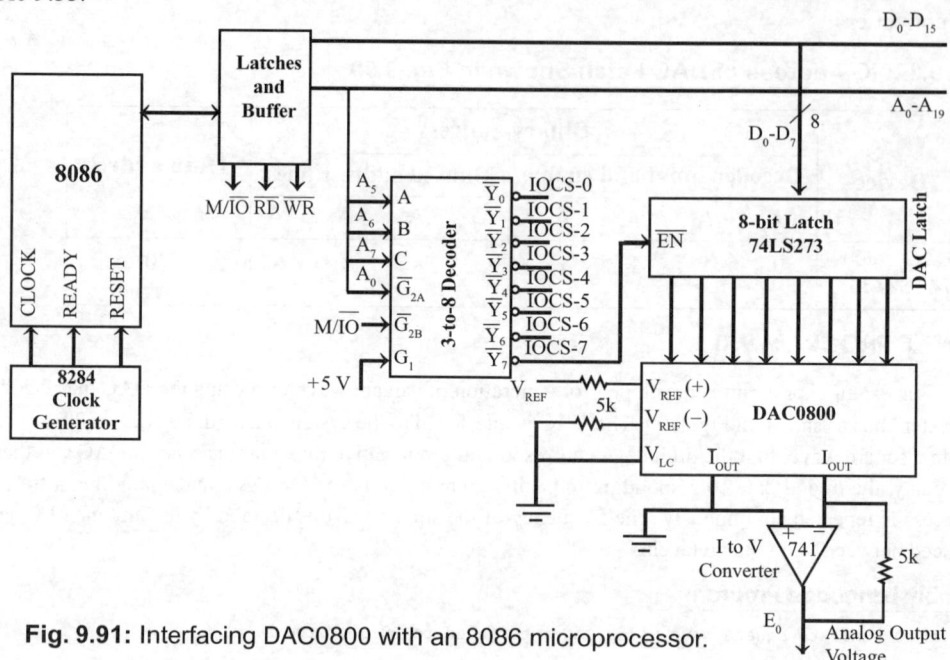

Fig. 9.91: Interfacing DAC0800 with an 8086 microprocessor.

Table 9.33: IO Address of DAC Latch as Shown in Fig. 9.91

Device	Binary address								Hexa address
	Decoder input			Unused address lines				Decoder enable	
	A_7	A_6	A_5	A_4	A_3	A_2	A_1	A_0	
DAC Latch 74LS273	0	1	1	x	x	x	x	0	60

9.14.4 Interfacing DAC0800 with 8051

In simple systems, when the ports are free the DAC0800 can be directly interfaced to an 8-bit port of any 8051 controller, as shown in Fig. 9.92. In this system, the controller can be programmed to work as a signal generator for various applications and the program can be permanently stored in the internal program memory of the controller. Since the 8x5x ports are internally provided with latch there is no need for external latch to interface DAC0800.

The DAC0800 can also be interfaced to an 8051 microcontroller as memory-mapped IO, as shown in Fig. 9.93. In this case an 8-bit latch such as 74LS273 is interfaced to the system bus and mapped in the data memory address space with 16-bit address. The DAC0800 is connected to output lines of the latch. The controller will load the digital data to the latch and it will hold the data on its output lines. The next data will be loaded to the latch only when previous data has been converted to analog value. The loading of consecutive data to the latch of DAC is controlled by software time delay.

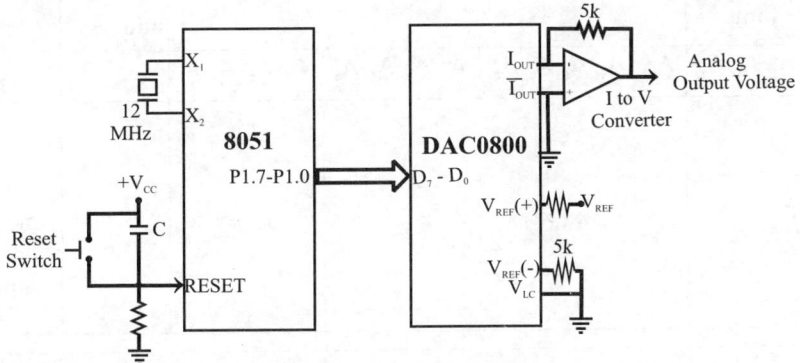

Fig. 9.92: Interfacing DAC0800 to a port of 8051 microcontroller.

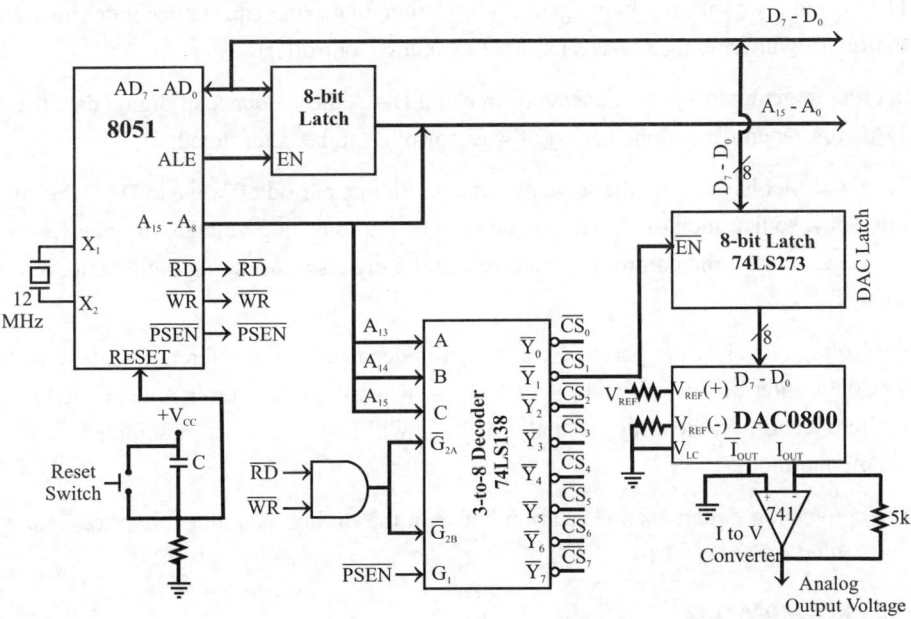

Fig. 9.93: Interfacing DAC0800 to 8051 as memory-mapped IO.

The address lines A_{13}, A_{14} and A_{15} are decoded to generate 8 chip select signals and in this the signal CS_3 is used as logic **low** enable for the DAC latch 74LS273. The signals $\overline{RD}$ and $\overline{WR}$ are logically ANDed and used as logic **low** enable for the decoder and the signal PSEN is used as logic **high** enable for the decoder. The address allotted to DAC latch is shown in Table 9.34.

Table 9.34: Address Allocation to DAC Latch Interfaced to Controller as Shown in Fig. 9.93

Device	Binary address					Hexa address	Comment
	Decoder input	Unused address lines					
	$A_{15} A_{14} A_{13}$	A_{12}	$A_{11} A_{10} A_9 A_8$	$A_7 A_6 A_5 A_4$	$A_3 A_2 A_1 A_0$		
DAC External Latch 74LS273	0 1 1	x	x x x x	x x x x	x x x x	6000	data memory address space

9.14.5 Programming the DAC Interfaced (DAC0800) with 8051

The periodic waveforms like square, ramp, triangular, sine can be generated using the 8051 microcontroller by interfacing a DAC (Digital to Analog Controller).

In order to generate a periodic waveform using DAC, the sequence of digital data that has to be sent to DAC for generation of one period of waveform has to be determined.

The controller has to send the sequence of data for one period of wave to DAC one by one with or without delay, so that the DAC converts the digital data to analog voltage. In order to generate the waveform continuously, the controller has to repeat the process of sending digital data of one period continuously.

The voltage level of the periodic waveform is decided by the reference voltage of the DAC. In the DAC0800 interface discussed in Section 9.14, when the positive reference is tied to +5 V and negative reference to 0 V, the range of analog voltage will be 0 to 5 volts in unipolar mode and −5 V to +5 V in bipolar mode.

The conversion equations and relation between the digital data and converted analog voltage are also presented in Section 9.14.

EXAMPLE PROGRAM 9.12

This example program is developed for generation of a square waveform using the DAC directly interfaced to port-1 of the 8051 controller as shown in Fig. 9.92. In order to generate a square wave, first the digital data corresponding to negative maximum is send to DAC and then after a time delay, the digital data corresponding to positive maximum is send to DAC and again after a time delay, the process is repeated continuously. The frequency of the square wave is decided by the amount of time delay introduced between two voltage levels.

Flowchart

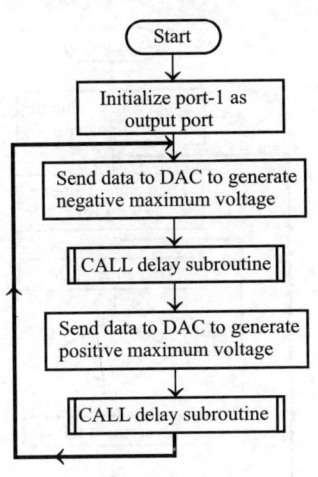

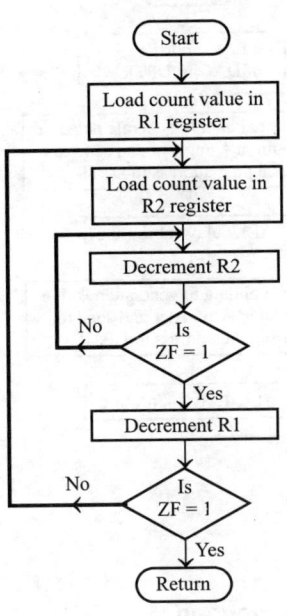

Assembly Language Program

```
;Program to generate square wave using DAC directly interfaced to port of 8051
;Program to generate square wave using DAC directly interfaced to port of 8051

        MOV     P1,#00H    ;Initialize port-1 as output port

AGAIN:  MOV     P1,#00H    ;Send digital data to DAC to generate negative maximum
        ACALL   DELAY      ;Maintain output of DAC at negative maximum for half the time period

        MOV     P1,#FFH    ;Send digital data to DAC to generate positive maximum
        ACALL   DELAY      ;Maintain output of DAC at positive maximum for half the time period

        SJMP    AGAIN

DELAY:  MOV     R1,#FFH    ;Load count value in R1 register
        MOV     R2,#FFH    ;Load count value in R2 register
LOOP1:  DJNZ    R2,LOOP1   ;Decrement count in R2 one by one until zero
LOOP2:  DJNZ    R1,LOOP2   ;Decrement count in R1 ono by one until zero
        RET

        END
```

EXAMPLE PROGRAM 9.13

This example program is developed for generation of square waveforms using the DAC interfaced to the 8051 system bus as shown in Fig. 9.93. The DAC is interfaced to the system bus with data memory address 6000H. The logic of square-wave generation is same as Example Program 9.12.

Flowchart

Flowchart for Main Program

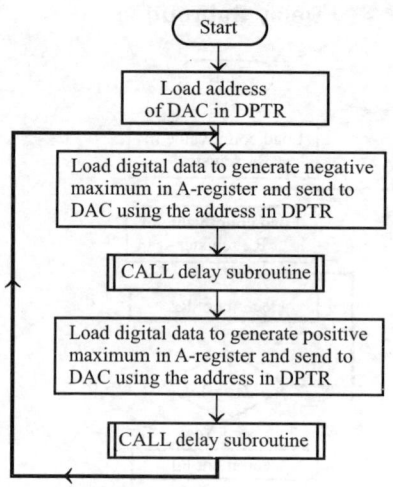

Flowchart for Delay Subroutine

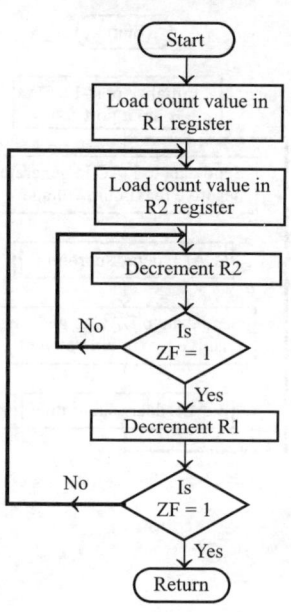

Assembly Language Program

```
;Program to generate square wave using DAC interfaced to 8051 system bus

AGAIN:  MOV    DPTR,#6000H ;Load address of DAC in DPTR
        MOV    A,#00H      ;Load digital data to generate negative maximum
        MOVX   @DPTR,A     ;Send digital data to DAC
        ACALL  DELAY       ;Maintain output of DAC at negative maximum for half the time period

        MOV    A,#FFH      ;Load digital data to generate positive maximum
        MOVX   @DPTR,A     ;Send digital data to DAC
        ACALL  DELAY       ;Maintain output of DAC at positive maximum for half the time period

        SJMP   AGAIN

DELAY:  MOV    R1,#FFH     ;Load count value in R1 register
        MOV    R2,#FFH     ;Load count value in R2 register
LOOP1:  DJNZ   R2,LOOP1    ;Decrement count in R2 one by one until zero
LOOP2:  DJNZ   R1,LOOP2    ;Decrement count in R1 one by one until zero
        RET

        END
```

EXAMPLE PROGRAM 9.14

This example program is developed for generation of ramp waveforms using the DAC directly interfaced to port-1 of the 8051 controller as shown in Fig. 9.92. In order to generate one period of a ramp wave, the digital data is send continuously from minimum to maximum (00H to FFH) one by one with or without delay. Then the process is repeated continuously to generate the ramp waveform. The time delay and the total time taken by the program to generate the all possible 256 values of digital data from 00H to FFH will decide the frequency of the ramp waveform.

Flowchart

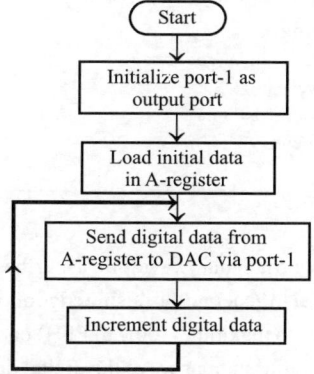

Assembly Language Program

```
;Program to generate ramp wave using the DAC directly interfaced to port of 8051

        MOV   P1,00H      ;Initialize port-1 as output port

        MOV   A,#00H      ;Load initial digital data in A register
CONTIN: MOV   P1,A        ;Send digital data to DAC

        INC   A           ;Increment digital data one by one
        SJMP  CONTIN      ;and send to DAC continuously

        END
```

EXAMPLE PROGRAM 9.15

This example program is developed for generation of ramp waveform using the DAC interfaced to the 8051 system bus as shown in Fig. 9.93. The DAC is interfaced to the system bus with data memory address 6000H. The logic of ramp wave generation is same as Example Program 9.14.

Flowchart

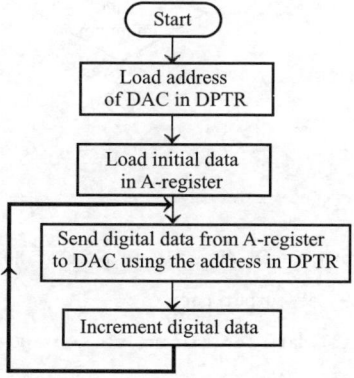

Assembly Language Program

```
;Program to generate ramp wave using the DAC interfaced to 8051 system bus
        MOV    DPTR,#6000H   ;Load address of DAC in DPTR
        MOV    A,#00H        ;Load initial digital data in A register
CONTIN: MOVX   @DPTR,A       ;Send digital data to DAC
        INC    A             ;Increment digital data one by one
        SJMP   CONTIN        ;and send to DAC continuously
        END
```

EXAMPLE PROGRAM 9.16

This example program is developed for generation of triangular waveform using the DAC directly interfaced to port-1 of the 8051 controller as shown in Fig. 9.92. In order to generate the raising edge of one period of a triangular wave, the digital data is send continuously from minimum to maximum (00H to FFH) one by one with or without delay, and then to generate the falling edge of one period of triangular wave the digital data is send continuously from maximum to minimum (FFH to 00H) one by one with or without delay. Then the process is repeated continuously to generate the triangular waveform. The time delay and the total time taken by the program to generate the all possible 256 values of digital data from 00H to FFH for raising edge and the all possible values of digital data from FFH to 00H for falling edge will decide the frequency of the triangular waveform.

Flowchart

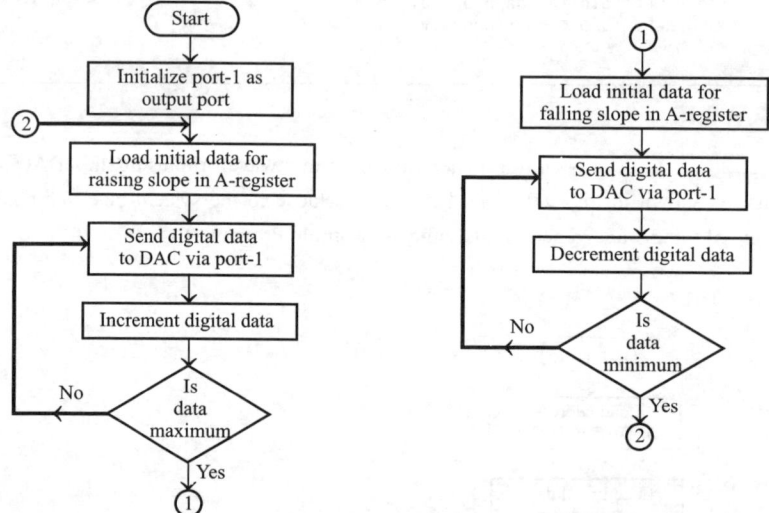

Assembly Language Program

```
;Program to generate triangular wave using DAC interfaced directly to port of 8051
        MOV    P1,#00H          ;Initialize port-1 as output port
AGAIN:  MOV    A,#00H           ;Set initial digital data for raising slope as 00H
RAISE:  MOV    P1,A             ;Send digital data to DAC
        INC    A                ;Increment the digital data
        CJNE   A,#FFH,RAISE     ;Repeat until the data reaches upper limit
        MOV    A,#FFH           ;Set initial digital data for falling slope as FFH
```

```
FALL:   MOV     P1,A            ;Send digital data to DAC
        DEC     A               ;Decrement the digital data
        CJNE    A,#00H,FALL     ;Repeat until the data reaches lower limit

        JMP     AGAIN           ;Repeat generation of next cycle
        END
```

EXAMPLE PROGRAM 9.17

This example program is developed for generation of triangular waveform using the DAC interfaced to the 8051 system bus as shown in Fig. 9.93. The DAC is interfaced to the system bus with data memory address 6000H. The logic of the triangular wave generation is same as Example Program 9.16.

Flowchart

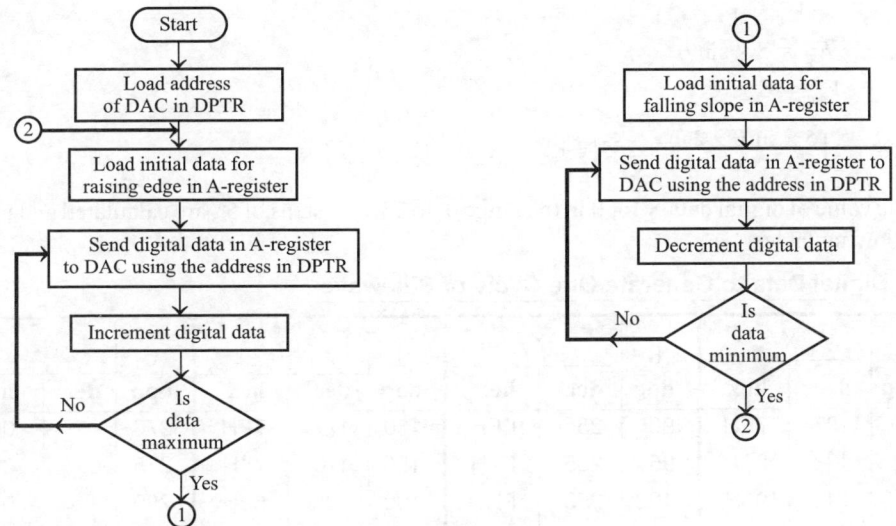

Assembly Language Program

```
;Program to generate triangular wave using the DAC interfaced to 8051 system bus

AGAIN:  MOV     DPTR,#6000H     ;Load address of DAC in DPTR
        MOV     A,#00H          ;Set initial digital data for raising slope as 00H
RAISE:  MOVX    @DPTR,A         ;Send digital data to DAC
        INC     A               ;Increment the digital data
        CJNE    A,#FFH,RAISE    ;Repeat until the data reaches upper limit

        MOV     A,#FFH          ;Set initial digital data for falling slope as FFH
FALL:   MOVX    @DPTR,A         ;Send digital data to DAC
        DEC     A               ;Decrement the digital data
        CJNE    A,#00H,FALL     ;Repeat until the data reaches lower limit

        JMP     AGAIN           ;Repeat generation of next cycle
        END
```

EXAMPLE PROGRAM 9.18

This example program is developed for generation of sinusoidal waveform using the DAC directly interfaced to port-1 of 8051 controller as shown in Fig. 9.92. In order to generate of one period of a sine wave, the digital data has to be calculated at equal time intervals of a period and stored as a data base. The controller program has to take digital data from this data base one by one and sent to DAC with or without delay. Then the process is repeated continuously to generate the sine waveform. The time delay and the total time taken by the program to send the digital values of one period will decide the frequency of the sine waveform. Here, one period of the sine wave is divided into 72 equal intervals and the digital data for each interval is calculated and tabulated in the following table.

Calculation of Digital Data For One Period of Sine Wave

In bipolar mode DAC the analog voltage, E_o for a digital data X is given by,

$$E_o = V_{REF} \times \left(\frac{-255 + 2X}{256} \right)$$

From the above equation, the equation of digital data, X can be obtained as,

$$X = \frac{1}{2} \left[\frac{E_o \times 256}{V_{REF}} + 255 \right]$$

Here, $E_o = 5 \times \sin \theta$

        $V_{REF} = 5$

$$\therefore X = \frac{1}{2} \left[\frac{5 \times \sin \theta \times 256}{5} + 255 \right] = 128 \times \sin \theta + 127.5$$

The value of digital data X for θ in the range 0° to 355°, in steps of 5°, are calculated and tabulated in the following table.

Table 1: Digital Data to Generate One Cycle of Sinewave

θ deg	X dec	X hex	θ deg	X dec	X hex	θ deg	X dec	X hex	θ deg	X dec	X hex
0	127	7FH	90	255	FFH	180	127	7FH	270	0	00H
5	138	8AH	95	255	FFH	185	116	74H	275	0	00H
10	149	95H	100	253	FDH	190	105	69H	280	1	01H
15	160	A0H	105	251	FBH	195	94	5EH	285	3	03H
20	171	ABH	110	247	F7H	200	83	53H	290	7	07H
25	181	B5H	115	243	F3H	205	73	49H	295	11	0BH
30	191	BFH	120	238	EEH	210	63	3FH	300	16	10H
35	200	C8H	125	232	E8H	215	54	36H	305	22	16H
40	209	D1H	130	225	E1H	220	45	2DH	310	29	1DH
45	218	DAH	135	218	DAH	225	36	24H	315	36	24H
50	225	E1H	140	209	D1H	230	29	1DH	320	45	2DH
55	232	E8H	145	200	C8H	235	22	16H	325	54	36H
60	238	EEH	150	191	BFH	240	16	10H	330	63	3FH
65	243	F3H	155	181	B5H	245	11	0BH	335	73	49H
70	247	F7H	160	171	ABH	250	7	07H	340	83	53H
75	251	FBH	165	160	A0H	255	3	03H	345	94	5EH
80	253	FDH	170	149	95H	260	1	01H	350	105	69H
85	255	FFH	175	138	8AH	265	0	00H	355	116	74H

Flowchart

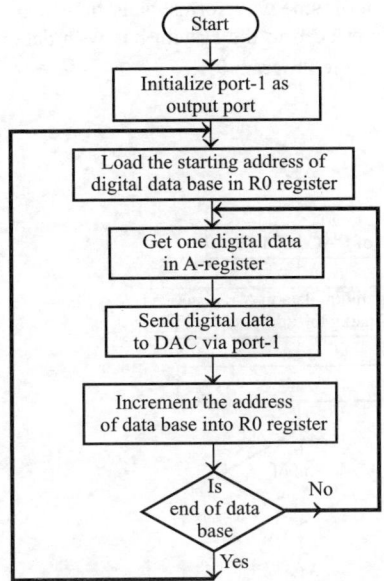

Assembly Language Program

```
;Program to generate sinewave using DAC directly interfaced to port of 8051

        MOV     P1,#00H       ;Initialize port-1 as output port

AGAIN   MOV     R0,#10H       ;Load starting address of table in R0 register

CONTIN: MOV     A,@R0         ;Get the digital data in A register
        MOV     P1,A          ;Send digital data to DAC
        INC     R0            ;Increment the address
        CJNE    R0,#58H,CONTIN ;Continue output of digital data one by one, until end of table

        SJMP    AGAIN         ;Repeat generation of next cycle

;Digital data to generate one cycle of sinewave

        ORG 10H

TABLE:  DB  7FH,8AH,95H,A0H,ABH,B5H,BFH,C8H
        DB  D1H,DAH,E1H,E8H,EEH,F3H,F7H,FBH
        DB  FDH,FFH,FFH,FFH,FDH,FBH,F7H,F3H
        DB  EEH,E8H,E1H,DAH,D1H,C8H,BFH,B5H
        DB  ABH,A0H,95H,8AH,7FH,74H,69H,5EH
        DB  53H,49H,3FH,36H,2DH,24H,1DH,16H
        DB  10H,0BH,07H,03H,01H,00H,00H,00H
        DB  01H,03H,07H,0BH,10H,16H,1DH,24H
        DB  2DH,36H,3FH,49H,53H,5EH,69H,74H

        END
```

EXAMPLE PROGRAM 9.19

This example program is developed for generation of sine waveforms using the DAC interfaced to the 8051 system bus as shown in Fig. 9.93. The DAC is interfaced to the system bus with data memory address 6000H. The logic of sine-wave generation is same as Example Program 9.18.

Flowchart

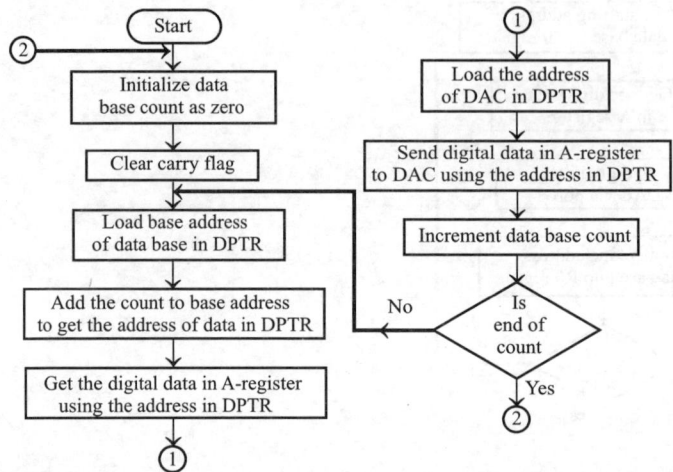

Assembly Language Program

```
;Program to generate sinewave using DAC interfaced to 8051 system bus

AGAIN:   MOV   R1,#00H         ;Initialize count as zero
         CLR   C               ;Clear carry flag

CONTIN:  MOV   DPTR,#TABLE      ;Load base address of table in DPTR
         MOV   A,R1            ;Get the count in A register
         ADDC  A,DPL           ;Add the count to low byte of DPTR
         MOV   DPL,A           ;to get the address of digital data in DPTR

         MOVX  A,@DPTR         ;Get the digital data in A register
         MOV   DPTR,#6000H     ;Load the address of DAC in DPTR
         MOVX  @DPTR,A         ;Send digital data to DAC

         INC   R1              ;Increment count
         MOV   A,R1            ;Get the count in A register
         CJNE  A,#48H,CONTIN   ;Continue output of digital data one by one, until end of count

         SJMP  AGAIN           ;Repeat generation of next cycle

;Digital data to generate one cycle of sinewave
TABLE:   DB  7FH,8AH,95H,A0H,ABH,B5H,BFH,C8H
         DB  D1H,DAH,E1H,E8H,EEH,F3H,F7H,FBH
         DB  FDH,FFH,FFH,FFH,FDH,FBH,F7H,F3H
         DB  EEH,E8H,E1H,DAH,D1H,C8H,BFH,B5H
         DB  ABH,A0H,95H,8AH,7FH,74H,69H,5EH
         DB  53H,49H,3FH,36H,2DH,24H,1DH,16H
         DB  10H,0BH,07H,03H,01H,00H,00H,00H
         DB  01H,03H,07H,0BH,10H,16H,1DH,24H
         DB  2DH,36H,3FH,49H,53H,5EH,69H,74H

         END
```

9.15 ADC INTERFACE

In many applications, an analog device has to be interfaced to the digital system. But the digital devices cannot accept the analog signals directly and so the analog signals are converted to equivalent digital signals (data) using an **A**nalog-to-**D**igital **C**onverter (ADC).

The **A**nalog-to-**D**igital (A/D) conversion is the reverse process of **D**igital to **A**nalog (D/A) conversion. The A/D conversion is also called quantization, in which the analog signal is represented by an equivalent binary data.

The analog signals vary continuously and are defined for any interval of time. The digital signals (or data) can take only finite values and defined only for discrete instant of time. If the digital data is represented by an n-bit binary then it can have 2^n different values. In A/D conversion the given analog signal has to be divided into steps of 2^n values, and each step is represented by one of the 2^n values.

The analog-to-digital converters can be classified into two groups based on the technique involved for conversion. The first group includes successive approximation, counter and flash-type converters. The technique involved in these devices is that the given analog signal is compared with internally generated analog signal.

The second group includes integrator converters and voltage to frequency converters. In the devices of the second group, the given analog signal is converted to time or frequency and the new parameters (time or frequency) is compared with the known values to produce digital signal.

The trade-off between the two techniques is based on Accuracy vs Speed. The successive approximation and the flash type are faster but generally less accurate than the integrator and the voltage-to-frequency type converters. Also, the flash type is costlier. The successive approximation type converters are used for high speed conversion and the integrating type converters are used for high accuracy.

The resolution of the converter is the minimum analog value that can be represented by the digital data. If the ADC gives n-bit digital output and the full scale analog input is X volts, then the resolution is $\frac{1}{2^n} \times X$ volts. In an ADC, another critical parameter is conversion time. The conversion time is defind as the total time required to convert an analog signal into its digital equivalent. It depends on the conversion technique and the propagation delay in various circuits.

Successive Approximation ADC

A successive approximation ADC consists of D/A converter, successive approximation register and comparator. Figure 9.94 shows the functional blocks of a typical successive approximation A/D converter. The conversion process is initiated by a **S**tart **O**f **C**onversion (SOC) signal from the processor to the ADC. On receiving the SOC, the control unit of the ADC will give a start command to the successive approximation register and it starts generating digital signal by the successive approximation method. The generated digital data is converted to analog signal by the D/A converter and then compared with the given analog signal.

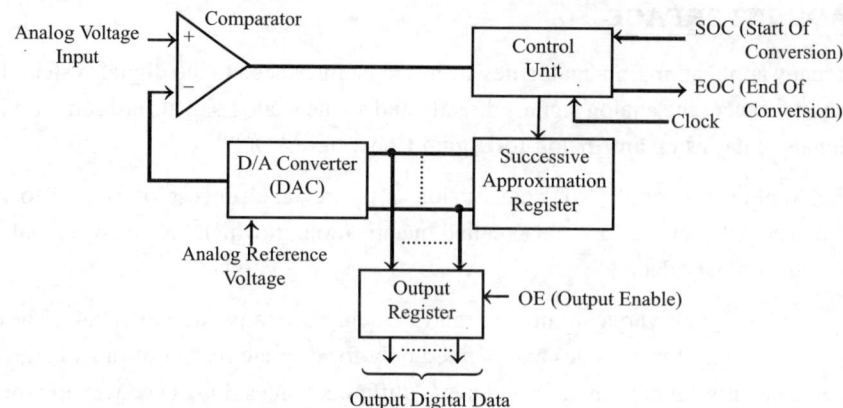

Fig. 9.94: Successive approximation A/D converter.

When the analog signals are equal, the comparator output informs the control unit to stop generation of digital signal. The digital data available at this instant is given as output through output register. Also, the control unit generates a signal to indicate the **E**nd **O**f **C**onversion (EOC) process to the processor.

Successive Approximation Method of Conversion

In this method, the MSD (**M**ost **S**ignificant **D**igit) is first set to **"1"** and all other digits are reset to **"0"**. The analog signal generated for this digital data is compared with the given analog signal. (Initially, the comparator output will be **high**.

After comparison the output of the comparator remains in **high** state if the given analog signal is higher than the generated analog signal. Otherwise, if the given signal is less than the generated signal then the output of the comparator changes from **high** to **low** state.) If the output state of the comparator changes then the MSD is reset to **"0"** otherwise it is retained as **"1"**.

Then the above process is repeated by setting the next higher order bit to **"1"**. The process is continued for each bit starting from MSD to LSD. (During a process, the higher order bits are the bits determined in earlier steps and the lower order bits are reset to "0".) After one complete cycle through MSD to LSD, the data available on the successive approximation register will be the digital equivalent of the given analog signal.

9.15.1 ADC0809

The ADC0809 is an 8-bit successive approximation type ADC with inbuilt 8-channel multiplexer. The ADC0809 is suitable for interface with 8086 microprocessor. The ADC0809 is available as a 28-pin IC in DIP (**D**ual **I**n-line **P**ackage). The ADC0809 has a total unadjusted error of ±1 LSD (**L**east **S**ignificant **D**igit). The ADC0808 is also same as ADC0809 except the error. The total unadjusted error in ADC0808 is $\pm\frac{1}{2}$ LSD. The pin configuration of ADC0809/ADC0808 is shown in Fig. 9.95.

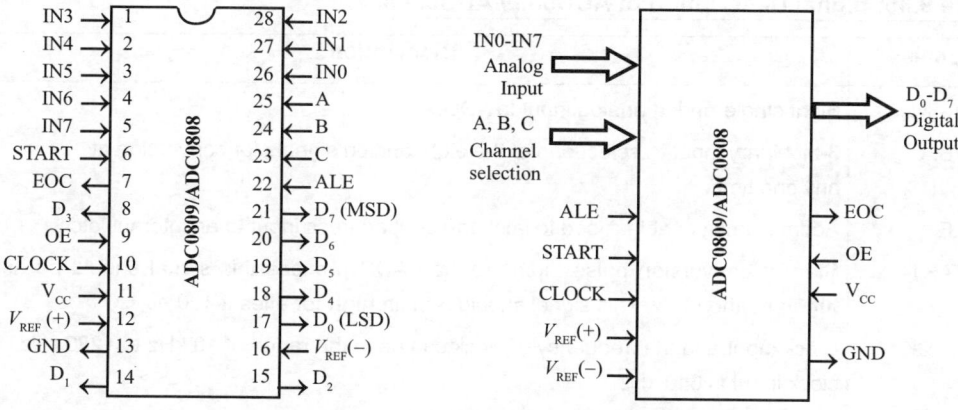

LSD = Least Significant Digit, MSD = Most Significant Digit

Fig. 9.95: Pin configuration of ADC0809/ADC0808.

The internal block diagram of ADC0809/ADC0808 is shown in Fig. 9.96. The various functional blocks of ADC are 8-channel multiplexer, comparator, 256R resistor ladder, switch tree, successive approximation register, output buffer, address latch and decoder.

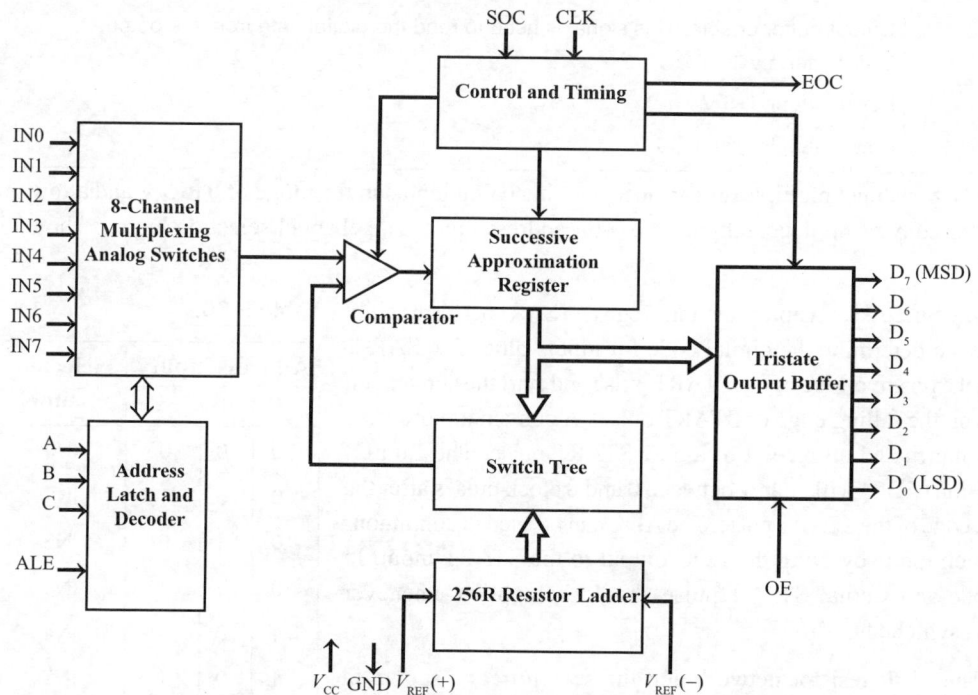

Fig. 9.96: Functional block diagram of ADC0809/ADC0808.

Table 9.35: Signal Description of ADC0809/ADC0808

Signals	Description
IN0-IN7	Eight single-ended analog input to ADC.
A, B, C	3-bit binary input to select one of the eight analog signals for conversion at any one time.
ALE	Address latch enable. Used to latch the 3-bit address input to an internal latch.
START	Start of conversion pulse input. To start ADC process this signal should be asserted **high** and then **low**. This signal should remain **high** for atleast 100 ns.
CLOCK	Clock input and the frequency of clock can be in the range of 10 kHz to 1280 kHz. Typical clock input is 640 kHz.
$V_{REF}(+), V_{REF}(-)$	Reference voltage input. The positive reference voltage can be less than or equal to V_{cc} and the negative reference voltage can be greater than or equal to ground.
D_0-D_7	The 8-bit digital output. The reference voltages will decide the mapping of the analog input to the digital data.
EOC	End of conversion. This signal is asserted **high** by the ADC to indicate the end of conversion process and it can be used as interrupt signal to processor.
OE	Output buffer enable. This signal is used to read the digital data from the output buffer after a valid EOC.
V_{cc}	Power supply, +5 V
GND	Power supply ground, 0 V

The 8-channel multiplexer can accept eight analog inputs in the range of 0 to 5 V and allow one by one for conversion depending on the 3-bit address input. The channel selection logic is shown in Table 9.36.

The Successive Approximation Register (SAR) performs eight iterations to determine the digital code for input value. The SAR is reset on the positive edge of the START pulse and start the conversion process on the falling edge of START pulse. A conversion processss will be interrupted on receipt of a new START pulse. The **End Of Conversion** (EOC) will go **low** between 0 and 8 clock pulses after the positive edge of the START pulse. The ADC can be used in continuous conversion mode by tying the EOC output to the START input. In this mode an external START pulse should be applied whenever power is switched ON.

Table 9.36:

Address input			Selected
C	**B**	**A**	channel
0	0	0	IN0
0	0	1	IN1
0	1	0	IN2
0	1	1	IN3
1	0	0	IN4
1	0	1	IN5
1	1	0	IN6
1	1	1	IN7

The 256R resistor network and the switch tree is shown in Fig. 9.97. The 256R ladder network has been provided instead of conventional R/2R ladder because of its inherent monotonicity, which guarantees no missing digital codes. Also, the 256R resistor network does not cause load variations on the reference voltage.

The comparator in the ADC0809/ ADC0808 is a chopper-stabilized comparator. It converts the DC input signal into an AC signal and amplifies the AC signal using a high gain AC amplifier. Then it converts AC signal to DC signal. This technique limits the drift component of the amplifier, because the drift is a DC component and it is not amplified/passed by the AC amplifier. This makes the ADC extremely insensitive to temperature, long term drift and input offset errors.

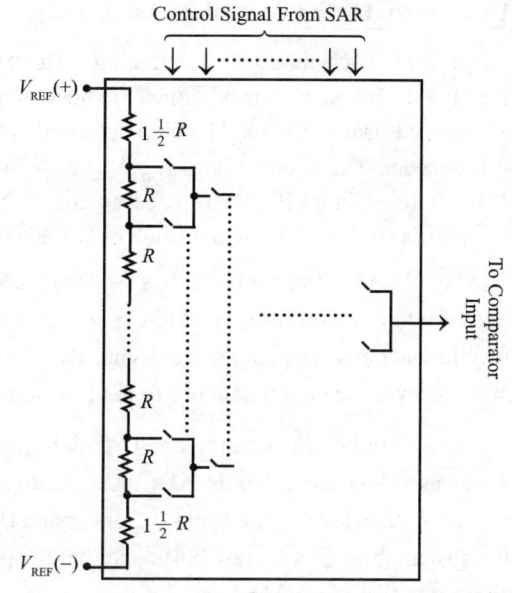

In ADC conversion process, the input analog value is quantized and each quantized analog value will have a unique binary equivalent. The quantization step in ADC0809/ADC0808 is given by,

$$Q_{step} = \frac{V_{REF}}{2^8} = \frac{V_{REF}(+) - V_{REF}(-)}{256_{10}}$$

Fig. 9.97: 256R resistor network and switch tree.

The digital data corresponding to an analog input (V_{in}) is given by,

$$\text{Digital data} = \frac{V_{in}}{Q_{step}} \pm \text{Absolute Accuracy} = \left(\frac{V_{in}}{Q_{step}} - 1\right)_{10}$$

EXAMPLE 1

Let, $V_{REF}(+) = 3.84$ V, $V_{REF}(-) = 0$ V

$$\therefore Q_{step} = \frac{V_{REF}(+) - V_{REF}(-)}{256_{10}} = \frac{3.84}{2.56} = 0.015 \text{ V} = 15 \text{ mV}$$

Let the input analog voltage be 2.56 V. Now the digital data corresponding to 2.56 V is given by,

$$\text{Digital data} = \frac{V_{in}}{Q_{step}} - 1 = \frac{2.56}{0.015} - 1 = 169_{10} = A9_H = 1010\ 1001_2$$

EXAMPLE 2

Let $V_{REF}(+) = 5$ V, $V_{REF}(-) = 0$ V

$$\therefore Q_{step} = \frac{V_{REF}(+) - V_{REF}(-)}{256_{10}} = \frac{5}{256} = 0.01953125$$

Let the input analog voltage be 1.25 V. Now the digital data corresponding to 1.25 V is given by,

$$\text{Digital data} = \frac{V_{in}}{Q_{step}} - 1 = \frac{1.25}{0.01953125} - 1 = 63_{10} = 3F_H = 0011\ 1111_2$$

9.15.2 Interfacing ADC0809 with 8085

A simple schematic for interfacing ADC0809/ADC0808 with 8085 microprocessor is shown in Fig. 9.98. The ADC can be either memory-mapped or IO-mapped in the system. Here the ADC is IO-mapped in the system. The chip select signals for IO-mapped devices are generated by using a 3-to-8 decoder. The address lines A_4, A_5 and A_6 are used as input to decoder. The address line A_7 and the control signal $IO/\overline{M}$ are used as enable for the decoder. The decoder generates eight chip select signals (IOCS-0 to IOCS-7), and out of this three chip select signals are used for ADC interface.

The chip select signal IOCS-6 is used to give Start Of Conversion (SOC) signal to ADC along with a channel address. The chip select IOCS-5 is used to enable the tristate buffer provided for interfacing EOC with data bus. The chip select signal IOCS-7 is inverted and used to enable the output buffer of ADC whenever the digital data has to read from the ADC.

The output clock signal of an 8085 microprocessor is divided by suitable clock divider circuits and used as a clock signal for the ADC. A separate voltage source has to be provided to give an accurate reference voltage levels. The End Of Conversion (EOC) signal of ADC is connected to the bus line D_0 of the system through a tristate buffer, so that the processor can check for a valid EOC before reading the output buffer of the ADC.

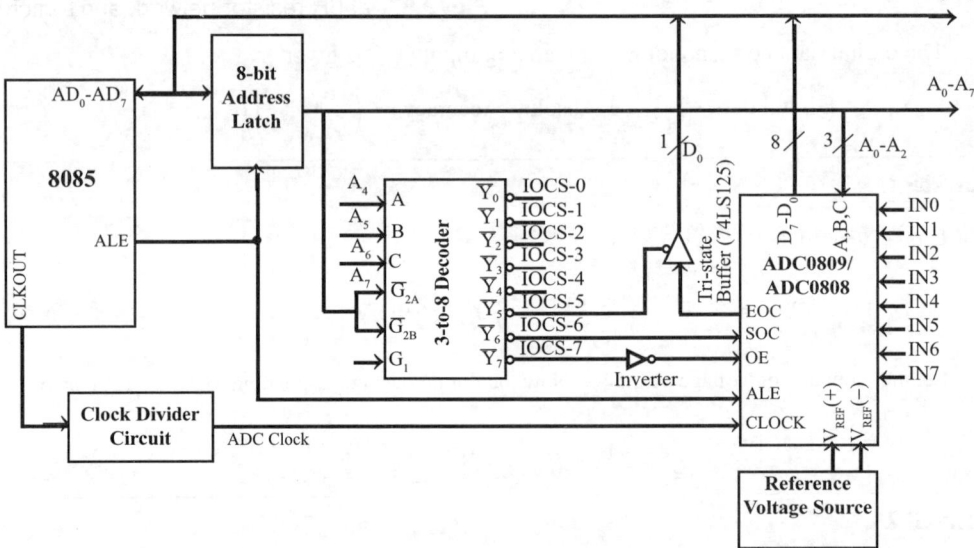

Fig. 9.98: Interfacing ADC0809/ADC0808 with 8085 microprocessor.

The working of ADC 0809 with 8085 will be as follows:

1. First the processor selects a channel by sending an address and SOC pulse is asserted high and low.
2. Once address of the channel and SOC pulse are applied, the ADC will start converting the signal at the selected channel.
3. Then the processor keeps on polling the status of the EOC to verify whether it is set to one. (when the conversion is completed by ADC0809 the EOC is set to one.)
4. When the processor finds a valid EOC, then it will read the digital value from the output buffer of ADC.

Table 9.37: IO Address of ADC0809/ADC0808 Interfaced to 8085 as Shown in Fig. 9.98

Operation performed	Binary address								Hexa address
	Decoder input/enable				Address input to ADC				
	A_7	A_6	A_5	A_4	A_3	A_2	A_1	A_0	
SOC channel-0	0	1	1	0	x	0	0	0	60
SOC channel-1	0	1	1	0	x	0	0	1	61
SOC channel-2	0	1	1	0	x	0	1	0	62
SOC channel-3	0	1	1	0	x	0	1	1	63
SOC channel-4	0	1	1	0	x	1	0	0	64
SOC channel-5	0	1	1	0	x	1	0	1	65
SOC channel-6	0	1	1	0	x	1	1	0	66
SOC channel-7	0	1	1	0	x	1	1	1	67
Read EOC	0	1	0	1	x	x	x	x	50
Read ADC output	0	1	1	1	x	x	x	x	70

9.15.3 Interfacing ADC0809 with 8086

A simple schematic for interfacing ADC0809/ADC0808 with the 8086 microprocessor is shown in Fig. 9.99. The ADC can be either memory-mapped or IO-mapped in the system. Here the ADC is IO-mapped in the system with even address. The chip select signals for IO-mapped devices are generated by using a 3-to-8 decoder. The address lines A_5, A_6 and A_7 are used as input to decoder. The address line A_0 and the control signal $M/\overline{IO}$ are used as enable for the decoder. The decoder generates eight chip select signals (IOCS-0 to IOCS-7), and these three chip select signals are used for the ADC interface. The chip select signal IOCS-6 is inverted and used to give the **S**tart **O**f **C**onversion (SOC) signal to the ADC along with a channel address.

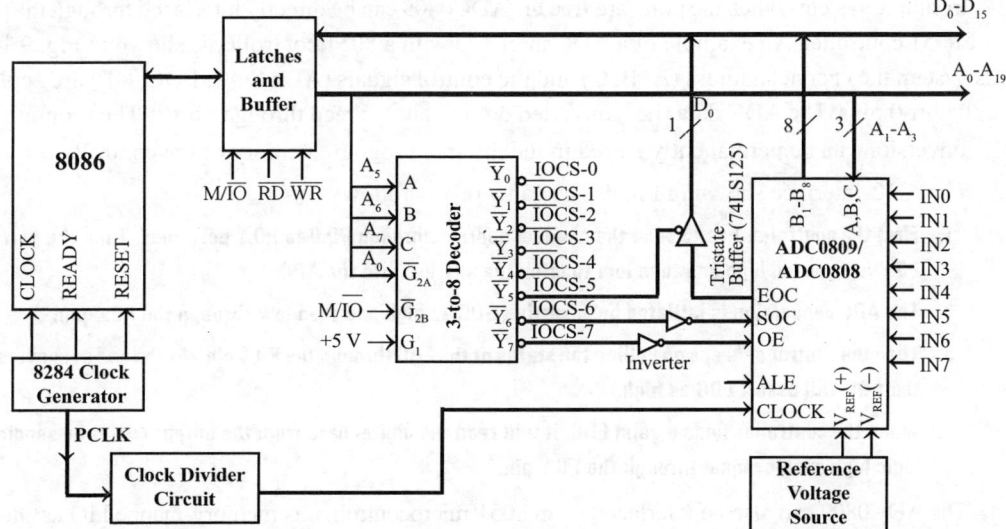

Fig. 9.99: Interfacing ADC0809/ADC0808 with the 8086 microprocessor.

Table 9.38: IO Address of ADC0809/ADC0808 Interfaced to 8086 as Shown in Fig. 9.99

Operation performed	Binary address								Hexa address
	Decoder input			Address input to ADC				Decoder enable	
	A_7	A_6	A_5	A_4	A_3	A_2	A_1	A_0	
SOC channel-0	1	1	0	x	0	0	0	0	C0
SOC channel-1	1	1	0	x	0	0	1	0	C2
SOC channel-2	1	1	0	x	0	1	0	0	C4
SOC channel-3	1	1	0	x	0	1	1	0	C6
SOC channel-4	1	1	0	x	1	0	0	0	C8
SOC channel-5	1	1	0	x	1	0	1	0	CA
SOC channel-6	1	1	0	x	1	1	0	0	CC
SOC channel-7	1	1	0	x	1	1	1	0	CE
Read EOC	1	0	1	x	x	x	x	0	A0
Read ADC Output	1	1	1	x	x	x	x	0	E0

The chip select IOCS-5 is used to enable the tristate buffer provided for interfacing EOC with the data bus. The chip select signal IOCS-7 is inverted and used to enable the output buffer of the ADC whenever the digital data has to be read from the ADC. The **Peripheral Clock Signal (PCLK)** of the 8284 is divided by a suitable clock divider circuit and used as a clock signal for the ADC. A separate voltage source has to be provided to give an accurate reference voltage levels. The **End Of** Conversion (EOC) signal of the ADC is connected to the bus line D_0 of the system through a tristate buffer, so that the processor can check for a valid EOC before reading the output buffer of ADC. The working of an ADC0809 with an 8086 is similar to the working of an ADC0809 with a processor.

9.15.4 Interfacing ADC0809 with 8051

In simple systems when the ports are free the ADC0809 can be directly interfaced through the port pins of 8051 controller. An example of a ADC interface with a 8051 controller is shown in Fig. 9.100. In this system the channel address (A, B, C) and the control signals (ALE, SOC, EOC, OE) are applied through port-0 pins. The ADC data (i.e, converted digital data) is read through port 1. The program for ADC conversion can be permanently stored in the internal program memory of the controller.

The ADC interface shown in Fig. 9.100 can work as follows:

1. First the controller has to send the channel address through P0.0 to P0.2 port lines. Then the port pin P0.3 is asserted high and then low to latch the address into the ADC.

2. The ADC conversion is initiated by asserting SOC as high and then low through the P0.4 pin.

3. Then the controller keeps on polling the status of the EOC through the P0.5 pin. (At the end of conversion the ADC will assert EOC as high.)

4. When the controller finds a valid EOC, it will read the digital data from the output buffer by sending a logic high enable signal through the P0.6 pin.

The ADC0809 can also be interfaced to an 8051 microcontroller as memory-mapped IO as shown in Fig. 9.101. The address lines A_0, A_1 and A_2 are used to select the desired channel for conversion. The signals SOC and OE are generated using a decoder. The signal EOC is read by the controller through a tristate buffer. The clock signal at X_2 pin is divided by a suitable clock divider and used as an ADC clock. A separate source is provided for reference voltage. The working of this system is similar to that shown in Fig. 9.101. The addresses allotted to initiate various operations are listed in Table 9.39.

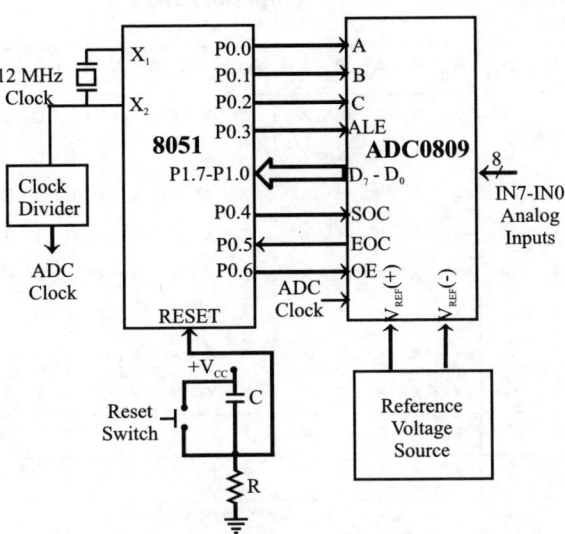

Fig. 9.100: Interfacing ADC0809 through port pins of an 8051 Microcontroller.

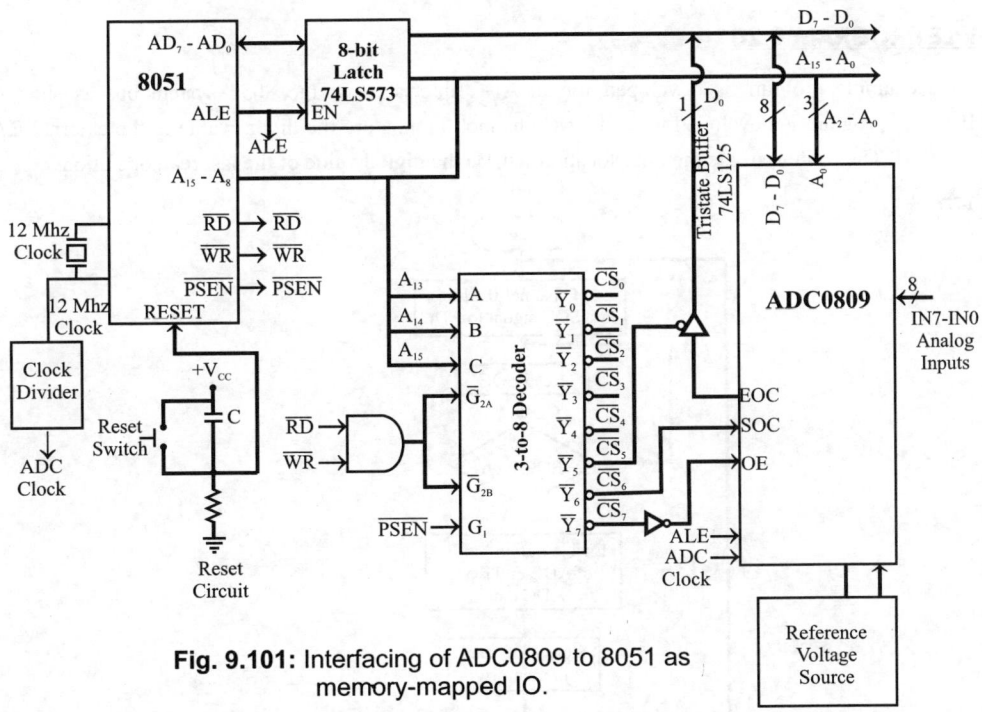

Fig. 9.101: Interfacing of ADC0809 to 8051 as memory-mapped IO.

Table 9.39: Addresss Alloted to ADC0809 Interfaced to 8051 as Shown in Fig. 9.101

Operation performed	Decoder Input A_{15} A_{14} A_{13}			Address input to ADC A_{12} A_{11} A_{10} A_9 A_8 A_7 A_6 A_5 A_4 A_3 A_2 A_1 A_0															Hexa address	Comment
SOC channel-0	1	1	0	x	x	x	x	x	x	x	x	x	x	0	0	0	C000	
SOC channel-1	1	1	0	x	x	x	x	x	x	x	x	x	x	0	0	1	C001	
SOC channel-2	1	1	0	x	x	x	x	x	x	x	x	x	x	0	1	0	C002	
SOC channel-3	1	1	0	x	x	x	x	x	x	x	x	x	x	0	1	1	C003	External
SOC channel-4	1	1	0	x	x	x	x	x	x	x	x	x	x	1	0	0	C004	data
SOC channel-5	1	1	0	x	x	x	x	x	x	x	x	x	x	1	0	1	C005	memory address
SOC channel-6	1	1	0	x	x	x	x	x	x	x	x	x	x	1	1	0	C006	space
SOC channel-7	1	1	0	x	x	x	x	x	x	x	x	x	x	1	1	1	C007	
Read EOC	1	0	1	x	x	x	x	x	x	x	x	x	x	x	x	x	A000	
Read ADC output	1	1	1	x	x	x	x	x	x	x	x	x	x	x	x	x	E000	

Note: Don't care "x" is considered as zero.

9.15.5 Programming the ADC Interfaced (ADC0809) with 8051

EXAMPLE PROGRAM 9.20

This example program is developed for an ADC interfaced to the 8051 system bus as shown in Fig. 9.101. The program is developed to read ADC channel-0 and store the digital value in the external RAM location 2400H. The content of the memory location will be the digital value of the last read operation.

Flowchart

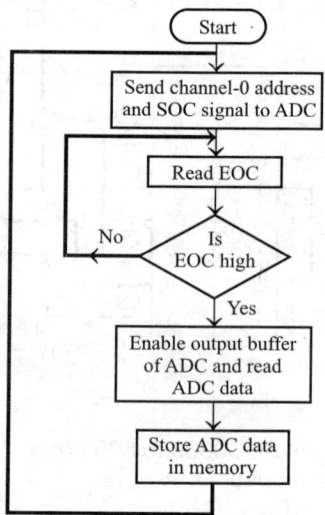

Assembly Language Program

```
;Program to read ADC interfaced to 8051 system bus

AGAIN:    MOV   DPTR,#C000H    ;Load address of ADC Channel-0 in DPTR
          MOV   A,#00          ;Move a dummy data to A
          MOVX  @DPTR,A        ;Send address and SOC to ADC

          MOV   DPTR,#A000H    ;Load address of EOC buffer in DPTR
WAIT:     MOVX  A,@DPTR        ;Get the status of EOC in A register
          RLC   A              ;Move EOC status to carry flag
          JNC   WAIT           ;Wait until EOC is high

          MOV   DPTR,#E000H    ;Load address of ADC output buffer in DPTR
          MOVX  A,@DPTR        ;Get the ADC data in A register

          MOV   DPTR,#2400H    ;Load address of data memory in DPTR
          MOVX  @DPTR,A        ;Store the ADC data in memory location 2400H
          SJMP  AGAIN          ;Go to read next ADC data

          END                  ;Assemby end
```

EXAMPLE PROGRAM 9.21

This example program is developed for ADC interfaced directly to ports of 8051 as shown in Fig. 9.100. The program is developed to read ADC channel-0 and store the digital value in the internal RAM location 7FH. The content of the memory location will be the digital value of the last read operation.

Flowchart

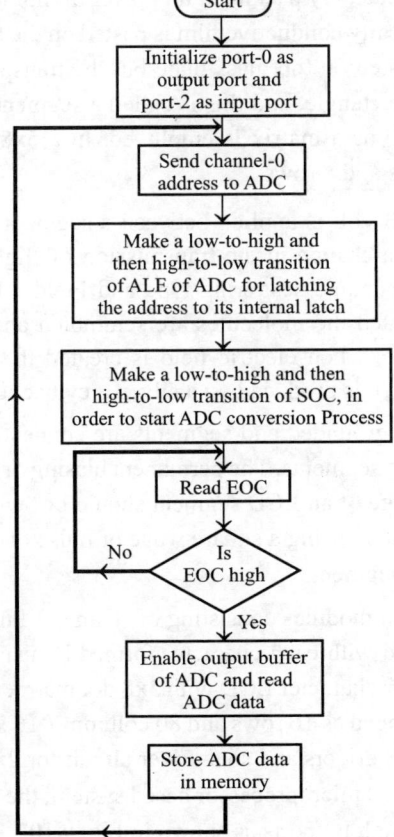

Assembly Language Program

```
;Program to read ADC interfaced directly to port of 8051

START:  MOV   P0,#00H      ;Initialize port-0 as output port
        MOV   P2,#FFH      ;Initialize port-2 as input port

AGAIN:  MOV   P0,#00H      ;Send channel-0 address to port-0
        SETB  P0.3         ;Set ALE high
        NOP                ;Delay and make ALE low
        NOP
        CLR   P0.3

        SETB  P0.4         ;Set SOC high
        NOP                ;Delay and make SOC low
        NOP

        CLR   P0.4
WAIT:   JNB   P2.7,WAIT    ;Wait until EOC is high

        SETB  P0.5         ;Set OE high, to enable ADC output buffer
        MOV   A,P1         ;Get ADC data in A register
        MOV   7FH,A        ;Store ADC data in internal RAM
        SJMP  AGAIN

        END                ;Assemby end
```

9.16 LIQUID CRYSTAL DISPLAY (LCD)

The LCD displays are created by packing a thin layer of liquid crystal fluid between two glass plates. A transparent and electrically conductive film is pasted on the bottom glass plate and it is called backplane. A similar film is pasted on top glass plate but the transparent portion of the film will be in the shape of a circle/square/rectangle, and this is called a segment. Usually, to display a character a number of segments arranged as a matrix is employed. In a 5x8 character display there will 40 segments arranged as 5 columns and 8 rows.

When a voltage of 2 to 3 volt is applied between a segment and backplane, an electric field is created, and this will create a change in the transmission of light under the segment. There are two common types of LCD: Dynamic scattering type and Field effect type. In dynamic scattering type, when electric field is created the molecules are scrambled and this produce a light on a dark background. In field effect type, when electric field is created the molecules are polarized which absorbs light and so a dark pattern is produced on a silver-grey background.

Usually the backplane is grounded and segments are connected to a positive supply of 2 to 3 volts and this is called biasing the segments. The permanent biasing of an LCD segment will destroy the crystals and so the biasing voltage of an LCD segment should be switched ON and OFF periodically. Usually, the segments are biased by using a square wave or pulse of frequency 50 to 150 Hz. This is also called refreshing the LCD segment.

The LCDs are available in modules consisting of 1-line, 2-lines or 4-lines of characters with 16 or 20 characters per line, and with each character formed by using 40 segments arranged as 5x8 dot matrix pattern. In a 2-line 16-character LCD with 5x8 dot matrix pattern for a character, there will 2x16x5x8=1280 segments, arranged as 16 rows and 80 columns (16 x 80 = 1280). The LCD segments should be biased through CMOS drivers and so a driver circuit for 2x16 LCD will consists of 16 row drivers and 80 column drivers. In a microprocessor based system, the row and column drivers of LCD segments can be interfaced through IO ports as shown in Fig. 9.103.

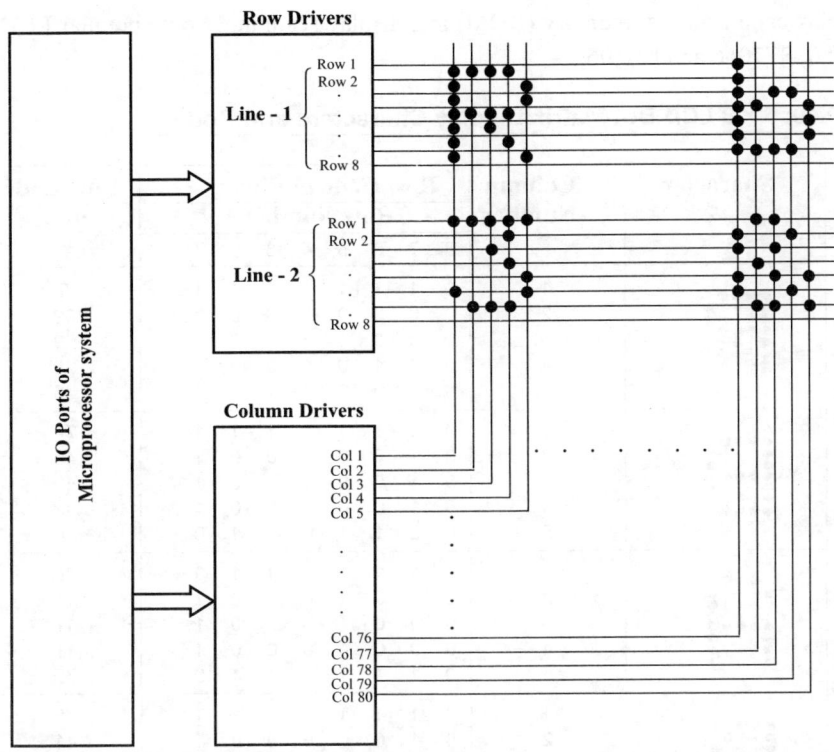

Fig. 9.102: 2 ×16 LCD segments connected to row and column drivers.

In order to display a character on a standard 5x8 dot matrix LCD, the segments to be turned ON has to be decided, and a display code is formed such that the segments to be turned ON is marked 1 and the segments to be turned OFF is marked 0. Some sample character display and their display codes are shown in Table 9.40.

For a 5 × 8 dot matrix LCD, the display code will have 5 row codes corresponding to 5 columns. In order to display a character, the row code corresponding to the first column is sent to row drivers and the first column is tuned ON through its column driver and all other columns will be OFF. Then the row code corresponding to second column is sent to row drivers and the second column is tuned ON through its column driver and all other columns will be OFF, and this is repeated for consecutive columns until all columns had a turn and then the process again starts with the first column. This is called **display scanning**.

The two major tasks in interfacing LCD displays are formation of display codes and scanning or refreshing LCD segments. In order to take care of these two tasks, dedicated LCD controllers that can be interfaced to microprocessors and microcontrollers have been developed. The LCD controller will have a dedicated memory to store the display character in a permanent semiconductor memory (ROM) and will have all necessary circuits to drive and refresh/scan the LCD segments. When LCD controllers are employed the task of microprocessor is to simply send the message or numbers to be displayed as ASCII characters to the controller, and the controller will take care of determining the display code from its memory (ROM) and display scanning. Some popular LCD controllers are HD44780, ST7066 and KS0066.

Table 9.40: 5 × 8 LCD Dot Matrix Display Characters and Codes

Character	Column Number	Row Code in Binary Row Number								Row Code in Hex
		8	7	6	5	4	3	2	1	
	1	0	1	1	1	1	1	1	1	7F
	2	0	0	0	0	1	0	0	1	09
	3	0	0	0	1	1	0	0	1	19
	4	0	0	1	0	1	0	0	1	29
	5	0	1	0	0	0	1	1	0	46
	1	0	1	1	1	1	1	1	1	7F
	2	0	1	0	0	1	0	0	1	49
	3	0	1	0	0	1	0	0	1	49
	4	0	1	0	0	1	0	0	1	49
	5	0	0	1	1	0	1	1	0	36
	1	0	1	1	1	1	1	1	0	7E
	2	0	0	0	1	0	0	0	1	11
	3	0	0	0	1	0	0	0	1	11
	4	0	0	0	1	0	0	0	1	11
	5	0	1	1	1	1	1	1	0	7E
	1	0	1	1	1	1	1	1	1	7F
	2	0	1	0	0	0	0	0	1	41
	3	0	1	0	0	0	0	0	1	41
	4	0	0	1	0	0	0	1	0	22
	5	0	0	0	1	1	1	0	0	1C
	1	0	1	1	1	1	1	1	1	7F
	2	0	1	0	0	1	0	0	1	49
	3	0	1	0	0	1	0	0	1	49
	4	0	1	0	0	1	0	0	1	49
	5	0	1	0	0	0	0	0	1	41
	1	0	1	1	1	1	1	0	0	7C
	2	0	0	0	0	1	0	0	0	08
	3	0	0	0	0	0	1	0	0	04
	4	0	0	0	0	0	1	0	0	04
	5	0	0	0	0	1	0	0	0	08
	1	0	1	1	1	1	1	1	1	7F
	2	0	1	0	0	1	0	0	0	48
	3	0	1	0	0	0	1	0	0	44
	4	0	1	0	0	0	1	0	0	44
	5	0	0	1	1	1	0	0	0	38
	1	0	0	1	0	0	0	0	0	20
	2	0	1	0	1	0	1	0	0	54
	3	0	1	0	1	0	1	0	0	54
	4	0	1	0	1	0	1	0	0	54
	5	0	1	1	1	1	0	0	0	78

Table 9.40 continued...

Character	Column Number	Row Code in Binary Row Number								Row Code in Hex
		8	7	6	5	4	3	2	1	
	1	0	0	1	1	1	0	0	0	38
	2	0	1	0	0	0	1	0	0	44
	3	0	1	0	0	0	1	0	0	44
	4	0	1	0	0	1	0	0	0	48
	5	0	1	1	1	1	1	1	1	7F
	1	0	0	1	1	1	0	0	0	38
	2	0	1	0	1	0	1	0	0	54
	3	0	1	0	1	0	1	0	0	54
	4	0	1	0	1	0	1	0	0	54
	5	0	0	0	1	1	0	0	0	18
	1	0	0	0	0	0	0	0	0	00
	2	0	1	0	0	0	0	1	0	42
	3	0	1	1	1	1	1	1	1	7F
	4	0	1	0	0	0	0	0	0	40
	5	0	0	0	0	0	0	0	0	00
	1	0	1	0	0	0	0	1	0	42
	2	0	1	1	0	0	0	0	1	61
	3	0	1	0	1	0	0	0	1	51
	4	0	1	0	0	1	0	0	1	49
	5	0	1	0	0	0	1	1	0	46
	1	0	0	1	0	0	0	0	1	21
	2	0	1	0	0	0	0	0	1	41
	3	0	1	0	0	0	1	0	1	45
	4	0	1	0	0	1	0	1	1	4B
	5	0	0	1	1	0	0	0	1	31
	1	0	0	0	1	1	0	0	0	18
	2	0	0	0	1	0	1	0	0	14
	3	0	0	0	1	0	0	1	0	12
	4	0	1	1	1	1	1	1	1	7F
	5	0	0	0	1	0	0	0	0	10
	1	0	0	1	0	0	1	1	1	27
	2	0	1	0	0	0	1	0	1	45
	3	0	1	0	0	0	1	0	1	45
	4	0	1	0	0	0	1	0	1	45
	5	0	0	1	1	1	0	0	1	39
	1	0	0	0	1	0	1	0	0	14
	2	0	0	0	0	1	0	0	0	08
	3	0	0	1	1	1	1	1	0	3E
	4	0	0	0	0	1	0	0	0	08
	5	0	0	0	1	0	1	0	0	14
	1	0	0	0	1	0	1	0	0	14
	2	0	1	1	1	1	1	1	1	7F
	3	0	0	0	1	0	1	0	0	14
	4	0	1	1	1	1	1	1	1	7F
	5	0	0	0	1	0	1	0	0	14
	1	0	0	0	0	0	0	0	0	00
	2	0	0	0	1	1	1	0	0	1C
	3	0	0	1	0	0	0	1	0	22
	4	0	1	0	0	0	0	0	1	41
	5	0	0	0	0	0	0	0	0	00

Table 9.40 continued...

Character	Colum Number	Row Code in Binary								Row Code in Hex
		Row Number								
		8	7	6	5	4	3	2	1	
(character 1)	1	0	0	0	0	0	0	0	0	00
	2	0	1	0	0	0	0	0	1	41
	3	0	0	1	0	0	0	1	0	22
	4	0	0	0	1	1	1	0	0	1C
	5	0	0	0	0	0	0	0	0	00
(character 2)	1	0	0	1	1	0	1	1	0	36
	2	0	1	0	0	1	0	0	1	49
	3	0	1	0	1	0	1	0	1	55
	4	0	0	1	0	0	0	1	0	22
	5	0	1	0	1	0	0	0	0	50

LCD Controller-HD44780

The HD44780 is HITACHI make common LCD controller and is available in widespread formats 1 x 8, 2 x 16, 2 x 20, 4 x 20, etc.

The HD44780 dot-matrix liquid crystal display controller and the driver is designed to display alphanumeric characters and symbols that are custom programmed in its internal permanent ROM.

It is designed to drive a dot-matrix liquid crystal display under the control of a 4- or 8-bit microprocessor/microcontroller. Since all the functions such as display RAM, character generator, and liquid crystal driver required for driving a dot-matrix liquid crystal display are internally provided on one chip, a minimal microprocessor system can be interfaced with this HD44780 controller/driver. A single HD44780 can display up to one/two lines of 8-characters and with extended driver it can drive 16-characters per line.

9.16.1 Internal Architecture of LCD Controller-HD44780

The simplified functional blocks of LCD Controller HD44780U is shown in Fig. 9.103. The various functional blocks of LCD Controller are as follows:

1. Instruction Register (IR)
2. Data Register (DR)
3. Busy Flag (BF)
4. Address Counter (AC)
5. Display Data RAM (DDRAM)
6. Character Generator ROM (CGROM)
7. Character Generator RAM (CGRAM)
8. Timing Generation Circuit
9. Liquid Crystal Display Driver Circuit
10. Cursor/Blink Control Circuit

Instruction Register (IR) and Data Register (DR)

The LCD Controller HD44780U has two 8-bit registers, an Instruction Register (IR) and a Data Register (DR). The IR is used to hold the instructions and address temporarily during read/write operation.

The DR is used to hold the display data temporarily during a read/write operation. The microprocessor can perform a read/write operation to IR after setting the signal RS = 0.

The instruction codes, such as display clear and cursor shift, and address information for display data RAM (DDRAM) and character generator RAM (CGRAM), all are written to the controller through IR. Also, the status of busy flag and count value of AC can be read through IR.

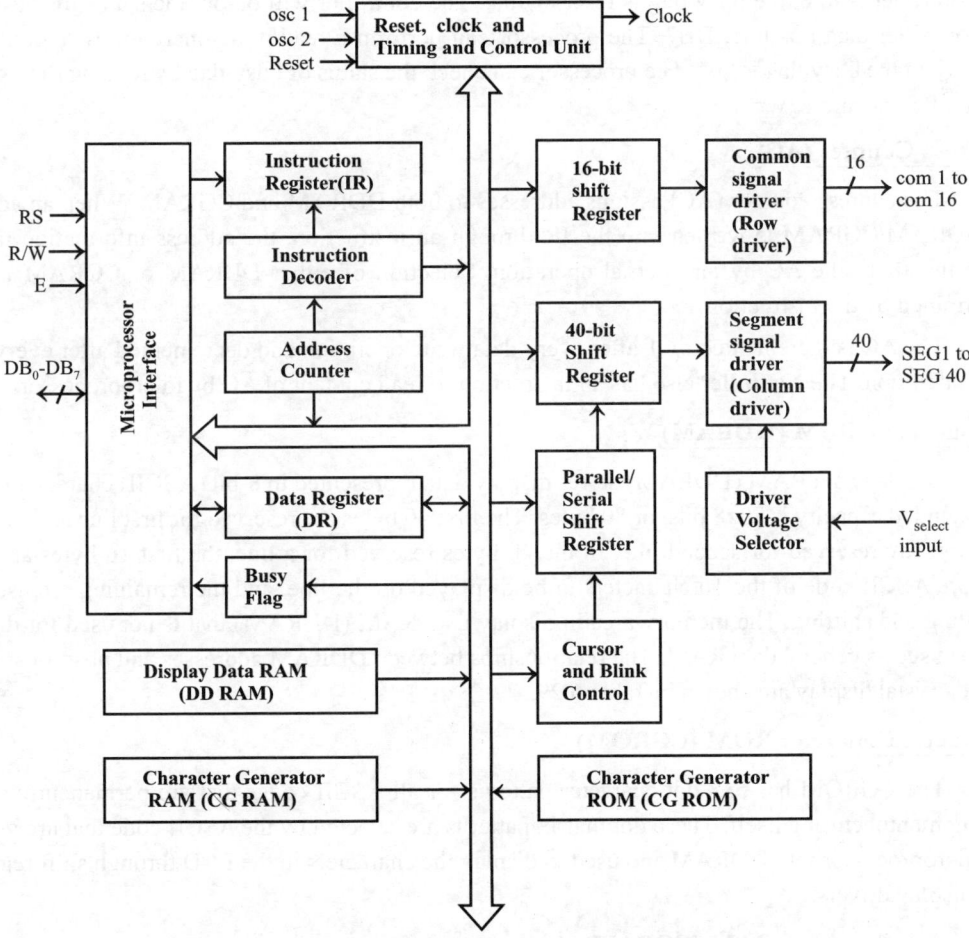

Fig. 9.103: Block Diagram of LCD Controller Hd44780.

The microprocessor can perform a read/write operation to DR after setting the signal RS = 1. In order to display a character the address should be written to IR by selecting RS = 0, and then the ACSII code should be write to DR by selecting the signal RS = 1. Data written into the DR from the microprocessor is automatically written into DDRAM or CGRAM by an internal operation.

In order to read a stored value in DDRAM or CGRAM, the address should be sent to IR after setting RS = 0. When address information is written into the IR, data is read and then stored into the DR from DDRAM or CGRAM by an internal operation. Then the processor can read the data from DR after setting RS = 1.

Busy Flag (BF)

The busy flag is used to inform the processor that the controller is performing an internal work or idle. When the controller is performing an internal operation, the busy flag is set to "1", and when the controller is idle, the busy flag is reset to "0",. The controller will output the status of busy flag on controller data bus line, DB7. The processor can perform a read/write operation with controller only when the busy flag is "0". The processor can check the status of busy flag by reading the IR after setting RS = 0 and R/$\overline{\text{W}}$= 1.

Address Counter (AC)

The address counter (AC) assigns addresses to both DDRAM and CGRAM. When an address of DDRAM/CGRAM is written into the IR through an instruction, the address information is sent from the IR to the AC by an internal operation. Selection of either DDRAM or CGRAM is also determined by the instruction.

The AC is incremented by 1 after every data write operation and decremented after every data read operation. The controller also has an instruction to read content of AC by microprocessor.

Display Data RAM (DDRAM)

Display data RAM (DDRAM) stores display data represented in 8-bit (ASCII) character codes. Its extended capacity is 80x8 bits, or 80 bytes. The first 40 bytes are reserved for first line and the next 40 bytes are reserved for second line. In the 40 bytes reserved for a line, the first 16 bytes are used to store ASCII code of the 16 characters to be displayed on the line, and the remaining are used for scrolling and shifting. The memory area in display data RAM (DDRAM) that is not used for display can be used as general data RAM. The relationships between DDRAM addresses and positions on the liquid crystal display are shown in Table 6.29.

Character Generator ROM (CGROM)

The CGROM has 5x8 dot character patterns for all ASCII codes that are permanently stored during manufacturing itself. These dot matrix patterns are selected by the ASCII code that are written by microprocessor into DDRAM and used to display the characters in the LCD through shift registers and display drivers.

Character Generator RAM (CGRAM)

The character generator RAM, can be used to develop and store new dot matrix character patterns. For 5x8 dots, eight character patterns can be developed and stored by user in CGRAM.

Timing Generation Circuit

The timing generation circuit generates separate timing signals for the internal operation and external read/write with microprocessor in order to avoid interference with each other. For example, when writing data to DDRAM, there will be no undesirable interferences, such as flickering.

Liquid Crystal Display Driver Circuit

The liquid crystal display driver circuit consists of 16 common signal drivers (for rows) and 40 segment signal drivers (for columns). The drivers are interfaced to internal bus via shift registers in order to perform the display scanning by sending display data to a row and turning ON a column.

Cursor/Blink Control Circuit

The cursor/blink control circuit generates the cursor or character blinking. The cursor or the blinking will appear under the digit whose address is currently available in address counter. For example, when the address counter is C2H, the cursor position is displayed at DDRAM address C2H.

Initializing by Internal Reset Circuit

An internal reset circuit automatically initializes the LCD controller when the power is turned ON. The following instructions are executed during the initialization. The busy flag (BF) is kept in the busy state until the initialization ends (BF = 1). The busy state lasts for 10 milliseconds after VCC rises to 4.5 V.

1. Display clear: To clear the display and bring the cursor to home.

2. Function set: Initialized for 8-bit interface, 1-line display and 5x8 dot character pattern.

3. Display ON/OFF control: Initialized for Display OFF, Cursor OFF and Blinking OFF.

4. Entry mode set: Initialized for address increment with no shift.

9.16.2 Pins and Signals of HD44780 LCD Display Module

The 2x16, HD44780 LCD display module layout is shown in Fig. 9.104. The LCD display module includes 32 numbers of 5x8 LCD arranged in two lines with 16-character per line. It has 16 connecting pads/pins for interfacing to microprocessor. The pin description is listed in Table 9.42. Each character position is identified by its row number (1 or 2), and its column number (from 1 to 16). The address for line-1 characters are 80H to 8FH, and address for line-2 characters are C0H to CFH. The address format is shown in Table 9.41.

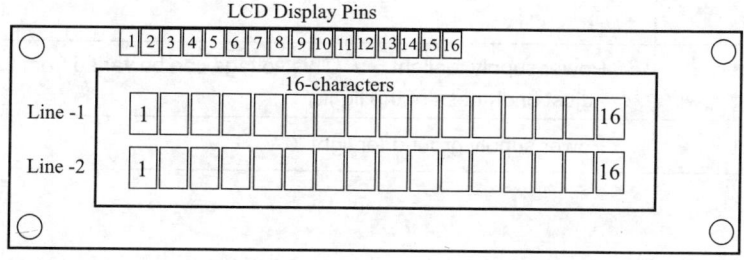

Fig. 9.104: LCD Layout.

Table 9.41: DDRAM Hexa Address of 2 × 16 LCD

Line Numbers	Character position															
	1	2	3	4	5	6	7	8	9	10	11	12	13	14	15	16
Line - 1	80	81	82	83	84	85	86	87	88	89	8A	8B	8C	8D	8E	8F
Line - 2	C0	C1	C2	C3	C4	C5	C6	C7	C8	C9	CA	CB	CC	CD	CE	CF

Table 9.42: LCD Module Pin Description

Pin No	Symbol	Description
1	Gnd	Power supply ground, 0 V
2	V_{cc}	Power supply, +5 V
3	Contrast	Contrast adjustment, 0 to 5 V
4	RS	Register select control signal RS = 0 ; Select Instruction(or command) Register RS = 1 ; Select Data Register
5	R/$\overline{W}$	Read/Write control signal R/$\overline{W}$ = 0 ; Write to Instruction/Data Register R/$\overline{W}$ = 1 ; Read from Instruction/Data Register
6	E	Enable LCD module. Enable signal has to make a high-to-low transition for latching data into the LCD module after a write operation and to acknowledge the data read after a read operation.
7	DB0	
8	DB1	4-bit low order data bus
9	DB2	
10	DB3	
11	DB4	
12	DB5	4-bit high order data bus
13	DB6	
14	DB7	
15	Light +	Power supply for light 5 V. (This voltage can be varied from 0 to 5V to adjust brightness of the light.)
16	Light -	Power supply ground for light, 0 V.

9.16.3 Interfacing HD44780 LCD Display Module to Microprocessor

The LCD module has eight data pins and three control pins for interface with microprocessor. The LCD module can be interfaced with microprocessor for either 4-bit data or 8-bit data. When interfaced for 4-bit data, the number of IO lines required for interfacing will be very less. The interfacing of LCD module with the 8085 microprocessor is shown in Fig. 9.105. In this interface, IO ports of 8255 are used to interface LCD for 8-bit operation.

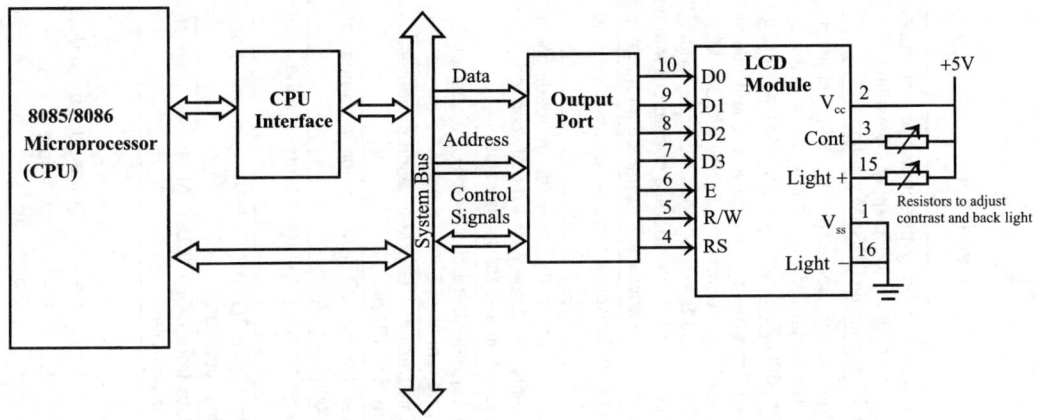

Fig. 9.105: Interfacing LCD module to 8085/8086 Microprocessor.

Instruction set of LCD module

The LCD module has an instruction set consisting 11 instructions listed in Table 9.43. The LCD controller has instructions/command for selecting various features of the controller. The features that are programmable are display format (5x8 or 5x10), data length (4-bit or 8-bit), cursor blinking, right/ left shift. The LCD controller has instructions to set DDRAM/CGRAM address, to read/write display data in DDRAM and to read the address in address count register and busy flag.

Programming the LCD Module

First, the LCD has to be initialized by sending some command words, then the AC is initialized with DDRAM address. The command words and address should be sent to IR register after setting the signal RS = 0. Then the ACSII value of the character to be displayed is written into DR register after setting the control signal RS = 1. The LCD commands and ASCII values of characters are 8-bit/byte. If LCD is interfaced for 4-bit operation, then the LCD commands/ASCII values has to send nibble by nibble, upper nibble first and lower nibble next. The LCD requires a time delay between consecutive read or write operation. When the LCD is ready to receive a byte, the busy flag will be reset to low. When a byte is send, the LCD will set the busy flag high, and then cleared to low after a predefined time.

Normally, the processor read/write operation will be very fast and so the consecutive write can be made either by checking busy flag or after a delay of 50 microseconds.

Table 9.43: Instruction Set of LCD Module

S.No.	Instructions	Control Bits RS	Control Bits R/W̄	DB7	DB6	DB5	DB4	DB3	DB2	DB1	DB0	Description
1.	Clear Display	0	0	0	0	0	0	0	0	0	1	Clear display and bring the cursor to home(Line-1, first character). The controller clear the display by writing display code for blank character (20 H) in all DDRAM locations.
2.	Cursor At Home	0	0	0	0	0	0	0	0	1	*	Bring the cursor home (Line-1, first character) without clearing display. The controller will set the cursor home by setting the address counter to 00H. The content of DDRAM is not altered.
3.	Entry Mode Set	0	0	0	0	0	0	0	1	I/D	S	Increment/Decrement the address count along with left/right shift after a read/write operation of DDRAM/CGRAM. When I/D = 1 and S = 1, address increment with left shift. When I/D = 0 and S = 1, address decrement with right shift. When S = 0; only address increment/decrement without shift
4.	Display ON/OFF control	0	0	0	0	0	0	1	D	C	B	Switch ON/OFF display, cursor and cursor blinking 1= ON and 0 = OFF.
5.	Cursor/Display Shift	0	0	0	0	0	1	S/C	R/L	*	*	Shift the cursor or display without reading/writing data. Useful to search/verify the display and modify/convert. The content of address counter is not altered for display shift
6.	Function Set	0	0	0	0	1	DL	N	F	*	*	Set the data length (8-bit or 4-bit), number of display lines(2-lines or 1- line) and font size(5 × 10 dots or 5 × 8 dots)
7.	CGRAM Address set	0	0	0	1	6-Bit CGRAM Address						Set the address of CGRAM location in address counter for read/write operation in CGRAM
8.	DDRAM Address set	0	0	1	7-Bit DDRAM Address							Set the address of DDRAM location in address counter for read/write operation in DDRAM
9.	Busy Flag/ Address Read	0	1	BF	7-Bit AC Register value							Readbusy flag and the current address in the address counter.
10.	CGRAM/DDRAM Data write	1	0	8-Bit Write Data								Write data into CGRAM/DDRAM location whose address is pointer by address counter.
11.	CGRAM/DDRAM Data Read	1	1	8-Bit Read Data								Read data from CGRAM/DDRAM location whose address is pointed by address counter.

I/D : 1 = Increment ; 0 = Decrement
S : 1 = Shift ; 0 = No shift
S/C : 1 = Display Shift : 0 = Cursor move
R/L : 1 = Shift to right ; 0 = Shift to left
DL : 1 = 8-bits ; 0 = 4-bits
N : 1 = 2-lines ; 0 = 1-line
F : 1 = 5 × 10 dots ; 0 = 5 × 8 dots
BF : 1 = Busy ; 0 = Idle
* : Don't care

EXAMPLE PROGRAM 9.22

This example program is developed to display 8085 in the LCD interface shown in Fig. 9.105. It is assumed that 8255 IO ports are IO mapped in the system and the details of LCD pins that are connected to 8255 IO port lines are shown in the following table.

Table 1: Details of LCD Pins Connected to 8255 IO Port Lines

Port-A	PA_7	PA_6	PA_5	PA_4	PA_3	PA_2	PA_1	PA_0
LCD	DB_7	DB_6	DB_5	DB_4	DB_3	DB_2	DB_1	DB_0
Port-B	PB_7	PB_6	PB_5	PB_4	PB_3	PB_2	PB_1	PB_0
LCD	*	*	*	*	*	RS	R/$\overline{W}$	E

The LCD is initialized for a 5 × 8 dot matrix pattern, 8-bit interface and address increment. The formation of command words are listed in the following table.

Table 2: Command Word/Control Signals for LCD Interface Program

Command/ Port-B	Feature Selected	Binary code								Hexa code	Variable name in the program
Function set	8-bit interface 2-line display and 5×8 dot pattern	0	0	1	DL	N	F	*	*		FUNSET
	DL = 1, N = 1, F = 0	0	0	1	1	1	0	0	0	38_H	
Display ON/OFF control	Display ON, cursor OFF and No blinking	0	0	0	0	1	D	C	B		ONOFCNT
	D = 1, C = 0, B = 0	0	0	0	0	1	1	0	0	$0C_H$	
Entry mode set	Address increment with no shift	0	0	0	0	0	1	I/D	S		ENTRCNT
	I/D = 1, S = 0	0	0	0	0	0	1	1	0	06_H	
Display clear	Clear display RAM and bring cursor home	0	0	0	0	0	0	0	1	01_H	CLEAR
Port-B	Write IR of LCD with high enable	*	*	*	*	*	RS	R/$\overline{W}$	E		IRENHIG
	RS = 0, R/$\overline{W}$ = 0, E = 1	0	0	0	0	0	0	0	1	01_H	
Port-B	Write IR of LCD with low enable	*	*	*	*	*	RS	R/$\overline{W}$	E		IRENLOW
	RS = 0, R/$\overline{W}$ = 0, E = 0	0	0	0	0	0	0	0	0	00_H	
Port-B	Write DR of LCD with high enable	*	*	*	*	*	RS	R/$\overline{W}$	E		DRENHIG
	RS = 1, R/$\overline{W}$ = 0, E = 1	0	0	0	0	0	1	0	1	05_H	
Port-B	Write DR of LCD with low enable	*	*	*	*	*	RS	R/$\overline{W}$	E		DRENLOW
	RS = 1, R/$\overline{W}$ = 0, E = 0	0	0	0	0	0	1	0	0	04_H	

Flowchart

Flowchart for Main Program

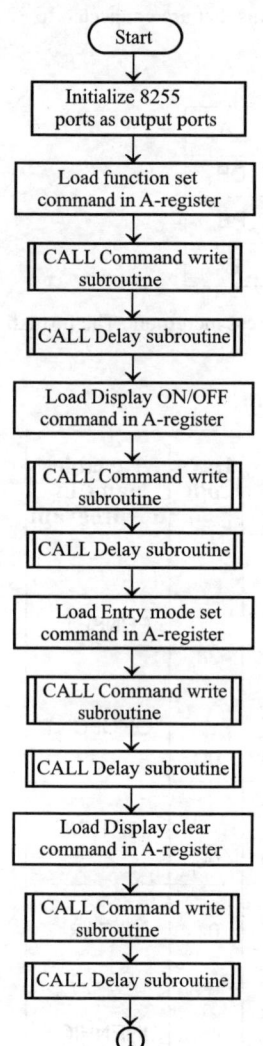

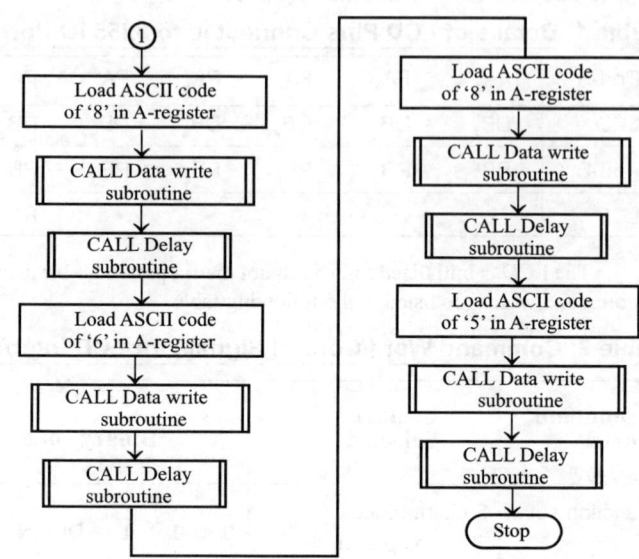

Flowchart for Command Write Subroutine

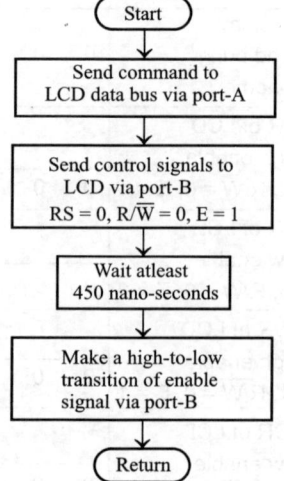

Flowchart for Data Write Subroutine

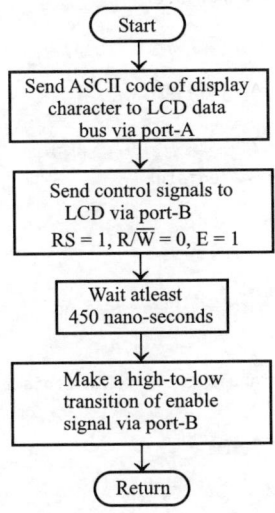

Flowchart for Data Write Subroutine

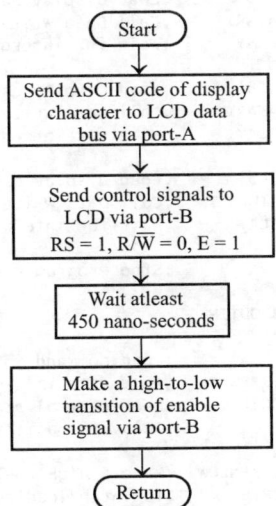

Assembly Language Program

```
;Program for LCD display

PORTA    EQU   10H        ;Address of port-A of 8255
PORTB    EQU   11H        ;Address of port-B of 8255
CNTPORT  EQU   13H        ;Address of 8255 control register
CNTWORD  EQU   80H        ;Mode set control word of 8255
FUNSET   EQU   38H        ;Function Set Command of LCD Module
ONOFCNT  EQU   0CH        ;Display ON/OFF command of LCD Module
ENTRCNT  EQU   06H        ;Entry Mode Set command of LCD Module
CLEAR    EQU   01H        ;Clear command of LCD Module
IRENHIG  EQU   01H        ;Control Signals for LCD IR Register write with High enable
IRENLOW  EQU   00H        ;Control Signals for LCD IR Register write with Low enable
DRENHIG  EQU   05H        ;Control Signals for LCD DR Register write with High enable
DRENLOW  EQU   04H        ;Control Signals for LCD DR Register write with Low enable

         ORG   1000H      ;Assembler directive
         MVI   A,CNTWORD  ;Get control word in A-register to set 8255 ports in mode-0 output
         OUT   CNTPORT    ;Send control word to control register
         MVI   A,FUNSET   ;Load Function Set Command in A-register
         CALL  CMDWRIT    ;Call Command Write subroutine
         CALL  DELAY      ;Wait for internal operations to complete

         MVI   A,ONOFCNT  ;Load Display ON/OFF Command in A-register
         CALL  CMDWRIT    ;Call Command Write subroutine
         CALL  DELAY      ;Wait for internal operations to complete

         MVI   A,ENTRCNT  ;Load Entry Mode Set Command in A- register
         CALL  CMDWRIT    ;Call Command Write subroutine
         CALL  DELAY      ;Wait for internal operations to complete

         MVI   A,CLEAR    ;Load LCD Clear Command in A-register
         CALL  CMDWRIT    ;Call Command Write subroutine
         CALL  DELAY      ;Wait for internal operations to complete
```

```
            MVI   A,'8'         ;Load Display data (ASCII value of 8) in A-register
            CALL  DATWRIT       ;Call Data Write subroutine
            CALL  DELAY         ;Wait for internal operations to complete

            MVI   A,'0'         ;Load Display data (ASCII value of 0)  in A-register
            CALL  DATWRIT       ;Call Data Write subroutine
            CALL  DELAY         ;Wait for internal operations to complete

            MVI   A,'8'         ;Load Display data (ASCII value of 8) in A-register
            CALL  DATWRIT       ;Call Data Write subroutine
            CALL  DELAY         ;Wait for internal operations to complete

            MVI   A,'5'         ;Load Display data (ASCII value of 5)  in A-register
            CALL  DATWRIT       ;Call Data Write subroutine
            CALL  DELAY         ;Wait for internal operations to complete

            HLT                 ;Stop Program execution

;COMMAND WRITE SUBROUTINE

CMDWRIT:    OUT   PORTA         ;Load command in data bus of LCD
            MVI   A,IRENHIG     ;Load data in A-register to set control signals to write IR register of LCD
            OUT   PORTB         ;Send control signals to LCD
            NOP                 ;Delay for a minimum of 450 nano-seconds
            NOP
            MVI   A,IRENLOW     ;Make a high-to-low transition of Enable
            OUT   PORTB         ;to latch/load command to IR register
            RET                 ;Return to main program

;DATA WRITE SUBROUTINE
DATWRIT:    OUT   PORTA         ;Load display data in data bus of LCD
            MVI   A,DRENHIG     ;Load data in A-register to set control signals to write LCD
            OUT   PORTB         ;Send control signals to LCD
            NOP                 ;Delay for a minimum of 450 nano-seconds
            NOP
            MVI   A,DRENLOW     ;Set a high-to-low transition of Enable
            OUT   PORTB         ;to latch/load display data to DR register
            RET                 ;Return to main program

;DELAY ROUTINE
DELAY:      LXI   D,0F429H      ;Load DE register pair with delay count
REPT:       DCX   D             ;Decrement DE register pair
            MOV   A,E           ;Check whether both D and E register
            CMP   D             ;contents are zeros
            JNZ   REPT          ;Jump on non-zero to REPT
            RET                 ;If zero, return to main program

END                            ;Assembler directive
```

EXAMPLE PROGRAM 9.23

This example program is developed to display 86 in the LCD interface shown in Fig 9.105. It is assumed that 8255 IO ports are IO mapped in the system and the details of LCD pins that are connected to 8255 IO port lines are same as that of Table 1 in Example Program 9.22.

The LCD is intialized for a 5×8 dot matrix pattern, 8-bit interface and address increment. The formation of command words are same as that of Table 2 in Example Program 9.22.

Flowchart

Flowchart for Main Program

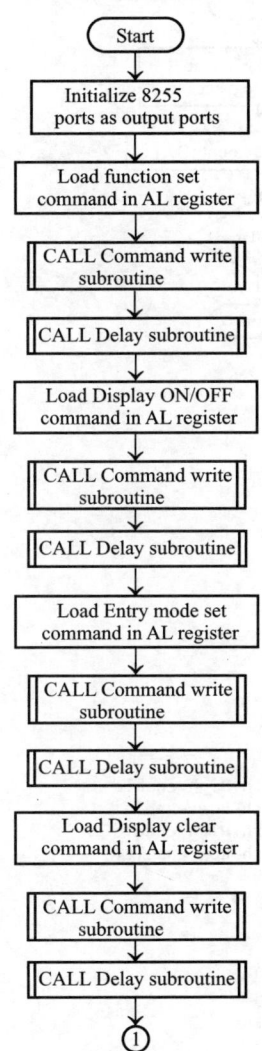

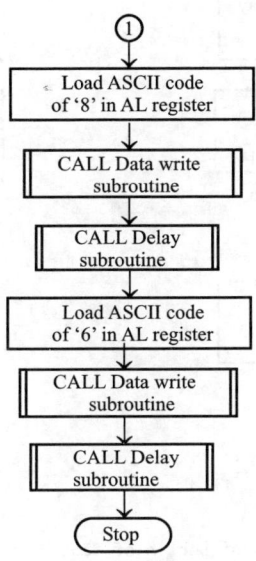

Flowchart for Command Write Subroutine

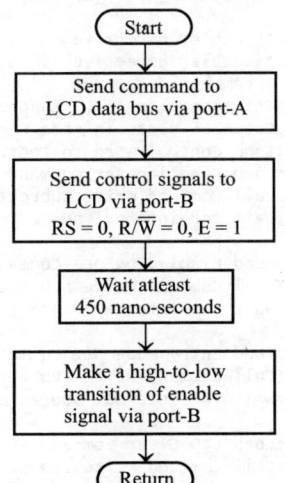

Flowchart for Data Write Subroutine

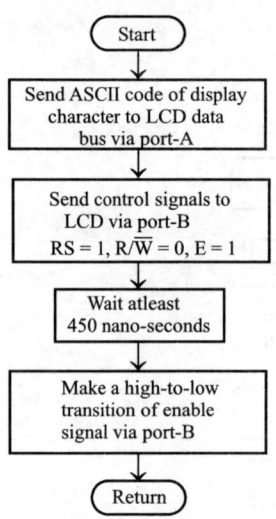

Flowchart for Delay Subroutine

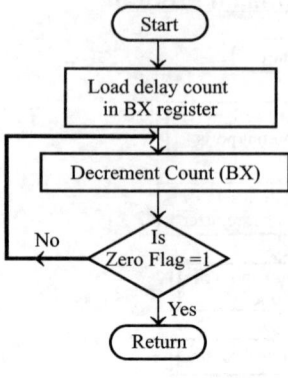

Assembly Language Program

```
;Program for LCD display

PORTA EQU 0000H        ;Address of port-A of 8255
PORTB EQU 0002H        ;Address of port-B of 8255
CNTPORT EQU 0006H      ;Address of 8255 control register
CNTWORD EQU 80H        ;Mode set control word of 8255
FUNSET  EQU 38H        ;Function Set Command of LCD Module
ONOFCNT EQU 0CH        ;Display ON/OFF command of LCD Module
ENTRCNT EQU 06H        ;Entry Mode Set command of LCD Module
CLEAR   EQU 01H        ;Clear command of LCD Module
IRENHIG EQU 01H        ;Control Signals for LCD IR Register write with High enable
IRENLOW EQU 00H        ;Control Signals for LCD IR Register write with Low enable
DRENHIG EQU 05H        ;Control Signals for LCD DR Register write with High enable
DRENLOW EQU 04H        ;Control Signals for LCD DR Register write with Low enable

CODE SEGMENT           ;Assembler directive
ASSUME CS:CODE         ;Assembler directive
        ORG 1000H      ;Assembler directive
        MOV DX,CNTPORT ;Get control register address in DX register
        MOV AL,CNTWORD ;Set 8255 ports in mode-0 output
        OUT DX,AL      ;Send control word to control register
        MOV AL,FUNSET  ;Load Function Set Command in AL register
        CALL CMDWRIT   ;Call Command Write subroutine
        CALL DELAY     ;Wait for internal operations to complete

        MOV AL,ONOFCNT ;Load Display ON/OFF Command in AL register
        CALL CMDWRIT   ;Call Command Write subroutine
        CALL DELAY     ;Wait for internal operations to complete

        MOV AL,ENTRCNT ;Load Entry Mode Set Command in AL register
        CALL CMDWRIT   ;Call Command Write subroutine
        CALL DELAY     ;Wait for internal operations to complete

        MOV AL,CLEAR   ;Load LCD Clear Command in AL register
        CALL CMDWRIT   ;Call Command Write subroutine
        CALL DELAY     ;Wait for internal operations to complete
```

```
        MOV AL,'8'        ;Load Display data (ASCII value of 8) in AL register
        CALL DATWRIT      ;Call Data Write subroutine
        CALL DELAY        ;Wait for internal operations to complete

        MOV AL,'6'        ;Load Display data (ASCII value of 6)  in AL register
        CALL DATWRIT      ;Call Data Write subroutine
        CALL DELAY        ;Wait for internal operations to complete
        HLT               ;Stop Program execution

;COMMAND WRITE SUBROUTINE
CMDWRIT PROC NEAR         ;Assembler directive
        MOV DX,PORTA      ;Get port-A address in DX register
        OUT DX,AL         ;Load command in data bus of LCD
        MOV DX,PORTB      ;Get port-B address in DX register
        MOV AL,IRENHIG    ;Set control signals to write IR register of LCD
        OUT DX,AL         ;Send control signals to LCD
        NOP               ;Delay for a minimum of 450 nano-seconds
        NOP
        MOV AL,IRENLOW    ;Make a high-to-low transition of Enable
        OUT DX,AL         ;to latch/load command to IR register
        RET               ;Return to main program
CMDWRIT ENDP             ;Assembler directive

;DATA WRITE SUBROUTINE

DATWRIT PROC NEAR        ;Assembler directive
        MOV DX,PORTA      ;Get port-A address in DX register
        OUT DX,AL         ;Load display data in data bus of LCD
        MOV DX,PORTB      ;Get port-B address in DX register
        MOV AL,DRENHIG    ;Set control signals to write LCD
        OUT DX,AL         ;Send control signals to LCD
        NOP               ;Delay for a minimum of 450 nano-seconds
        NOP
        MOV AL,DRENLOW    ;Set a high-to-low transition of Enable
        OUT DX,AL         ;to latch/load display data to DR register
        RET               ;Return to main program
DATWRIT ENDP             ;Assembler directive

;DELAY ROUTINE
DELAY PROC NEAR          ;Assembler directive
        MOV BX,0F429H     ;Load BX register with delay count
REPT:   DEC BX            ;Decrement BX register
        JNZ REPT          ;Jump on non-zero to REPT
        RET               ;Return to main program
DELAY ENDP               ;Assembler directive

CODE ENDS                ;Assembler directive
END                      ;Assembler directive
```

9.16.4 LCD Interfacing with the 8051 Microcontroller

The details of 2×16 LCD module are discussed in Section 9.16. The LCD module has eight data pins and three control pins for interface with the microcontroller. The LCD module can be interfaced with the microcontroller for either 4-bit or 8-bit data. When interfaced for 4-bit data, the number of IO lines required for interfacing will be very less.

The interfacing of the LCD module with the 8051 microcontroller is shown in Fig. 9.106. In this interface, the LCD data bus lines DB0-DB7 are connected to Port-0 of 8051. Three pins of Port-2 (P2.7,P2.6 and P2.5) are used to provide control signals E, R/$\overline{\text{W}}$ and RS for the LCD module.

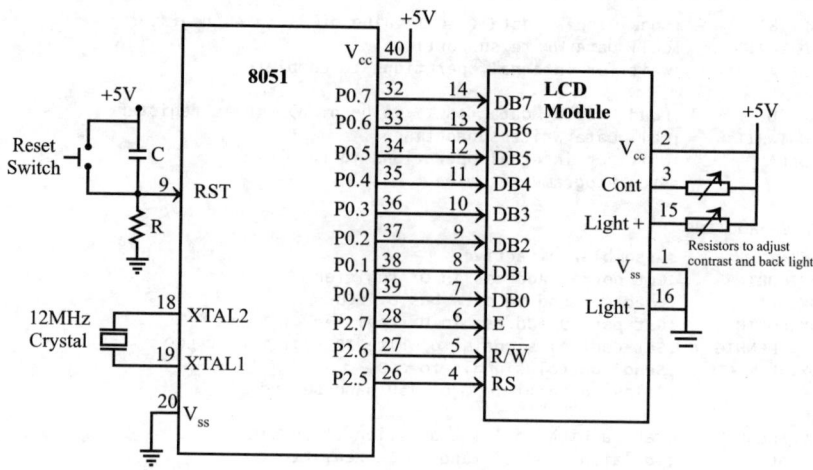

Fig. 9.106: Interfacing LCD Display module to 8051 Microcontroller.

The instruction set of the LCD module are presented in Table 9.43. The brief discussion presented in Section 9.16, about instruction set of LCD module and programming the LCD module are also applicable for programming LCD modules interfaced with the 8051 microcontroller.

EXAMPLE PROGRAM 9.24

This example program is developed to display 51 in the LCD interface shown in Fig. 9.106. The LCD pins are directly connected to port pins of the 8051 microcontroller as shown in the following table.

Table 1: Details of LCD Pins Connected to 8255 IO Port Lines

Port-0	P0.7	P0.6	P0.5	P0.4	P0.3	P0.2	P0.1	P0.0
LCD	DB7	DB6	DB5	DB4	DB3	DB2	DB1	DB0
Port-2	P2.7	P2.6	P2.5	P2.4	P2.3	P2.2	P2.1	P2.0
LCD	E	R/W̄	RS	*	*	*	*	*

The LCD is intialized for 5×8 dot matrix pattern, 8-bit interface and address increment. The formation of command words are listed in the following table.

Table 2: Command Word/Control Signals for LCD Interface Program

Command/ Port-2	Feature Selected	Binary code								Hexa code
Function set	8-bit interface 2-line display and	0	0	1	DL	N	F	*	*	
	5×8 dot pattern DL = 1, N = 1, F = 0	0	0	1	1	1	0	0	0	38_H
Display ON/OFF control	Display ON, cursor OFF and No blinking	0	0	0	0	1	D	C	B	
	D = 1, C = 0, B = 0	0	0	0	0	1	1	0	0	$0C_H$

Table 2 continued....

Command/ Port-2	Feature Selected	Binary code								Hexa code
Entry mode set	Address increment with no shift I/D = 1, S = 0	0	0	0	0	0	1	I/D	S	
		0	0	0	0	0	1	1	0	06$_H$
Display clear	Clear display RAM and bring cursor home	0	0	0	0	0	0	0	1	
		E	R/$\overline{W}$	RS	*	*	*	*	*	01$_H$
Port-2	Write IR of LCD with high enable RS = 0, R/$\overline{W}$ = 0, E = 1	1	0	0	0	0	0	0	1	
										80$_H$
Port-2	Write IR of LCD with low enable RS = 0, R/$\overline{W}$ = 0, E = 0	E	R/$\overline{W}$	RS	*	*	*	*	*	
		0	0	0	0	0	0	0	0	00$_H$
Port-2	Write DR of LCD with high enable RS = 1, R/$\overline{W}$ = 0, E = 1	E	R/$\overline{W}$	RS	*	*	*	*	*	
		1	0	1	0	0	0	0	0	A0$_H$
Port-2	Write DR of LCD with low enable RS = 1, R/$\overline{W}$ = 0, E = 0	E	R/$\overline{W}$	RS	*	*	*	*	*	
		0	0	1	0	0	0	0	0	20$_H$

Flowchart

Flowchart for Main Program

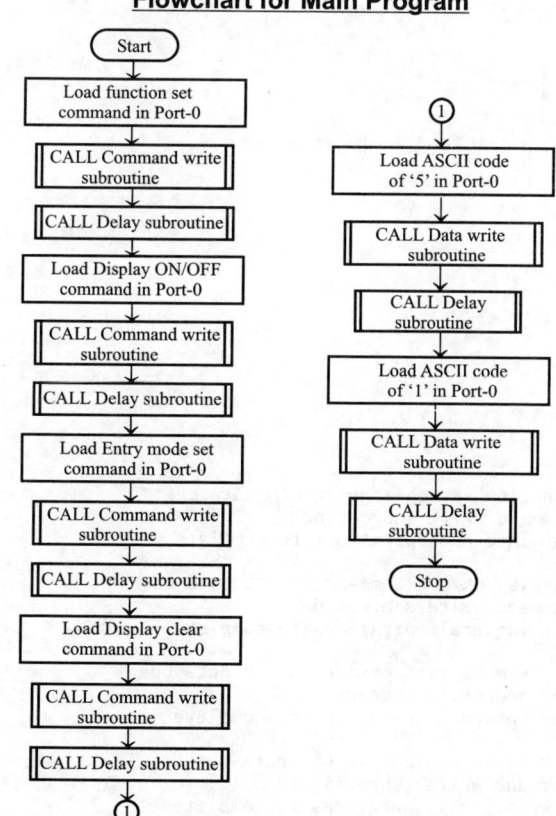

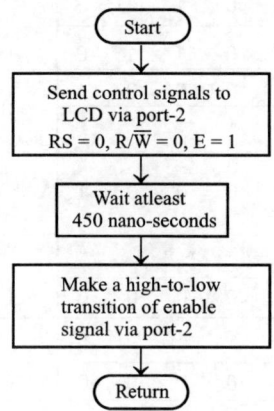

Flowchart for command write subroutine

Flowchart for data write subroutine

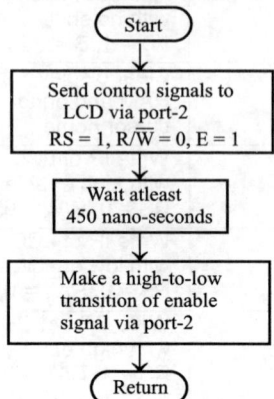

Flowchart for delay subroutine

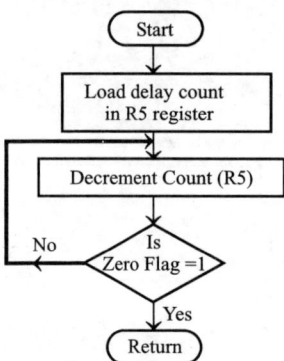

Assembly Language Program

```
;Program for LCD display

        MOV  P0,#38H      ;Load function set command in LCD data bus
        ACALL CMDWRITE    ;Call command write subroutine
        ACALL DELAY       ;Wait for internal operations to complete

        MOV  P0,#0CH      ;Load display ON/OFF command in LCD data bus
        ACALL CMDWRIT     ;Call command write subroutine
        ACALL DELAY       ;Wait for internal operations to complete

        MOV  P0,#06H      ;Load entry mode set command in LCD data bus
        ACALL CMDWRIT     ;Call command write subroutine
        ACALL DELAY       ;Wait for internal operations to complete

        MOV  P0,#01H      ;Load LCD clear command in LCD data bus
        ACALL CMDWRIT     ;Call command write subroutine
        ACALL DELAY       ;Wait for internal operations to complete
```

```
        MOV   PO,'5'        ;Load display data (ASCII value of 5) in LCD data bus
        ACALL DATWRIT       ;Call data write subroutine
        ACALL DELAY         ;Wait for internal operations to complete

        MOV   PO,'1'        ;Load display data (ASCII value of 1) in LCD data bus
        ACALL DATWRIT       ;Call data write subroutine
        ACALL DELAY         ;Wait for internal operations to complete

HALT:   SJMP HALT           ;Remain in infinite loop until reset, program end

;Command write subroutine

CMDWRIT: MOV P2,#80H        ;Set control signal to write IR of LCD
         NOP                ;Wait for a minimum of 450 nano-seconds
         NOP
         MOV P2,#00H        ;Make a high-to-low transition of enable
         RET                ;Return to main program

;Command data subroutine

DATWRIT: MOV P2,#A0H        ;Set control signal to write DR of LCD
         NOP                ;Wait for a minimum of 450 nano-seconds
         NOP
         MOV P2,#20H        ;Make a high-to-low transition of enable
         RET                ;Return to main program

;Delay Subroutine

DELAY:   MOV R5,#FFH
AGAIN:   DJNZ R5,AGAIN
         RET
         END                ;Assemby end
```

9.17 SHORT-ANSWER QUESTIONS

Q9.1 *What is a programmable peripheral device?*

If the functions performed by a peripheral device can be altered or changed by a program instruction then the peripheral device is called programmable device. Usually the programmable devices will have control registers. The device can be programmed by sending control word in the prescribed format to the control register.

Q9.2 *What is data transfer scheme and what are its types ?*

The data transfer scheme refers to the method of data transfer between the processor and peripheral devices.

The different types of data transfer schemes are shown below :

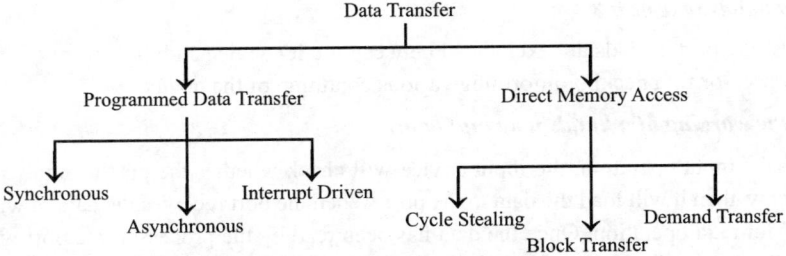

Fig. Q9.2: Types of data transfer scheme.

Q9.3   What is synchronous data transfer scheme?

In synchronous data transfer scheme, the processor does not check the readiness of the device after a command has been issued for read/write operation. In this scheme the processor will request the device to get ready and then read/write to the device immediately after the request. In some synchronous schemes a small delay is allowed after the request.

Q9.4   What is asynchronous data transfer scheme?

In asynchronous data transfer scheme, first the processor sends a request to the device for read/write operation. Then the processor keeps on polling the status of the device. Once the device is ready, the processor executes a data transfer instruction to complete the process.

Q9.5   What are the operating modes of an 8212 ?

The 8212 can be hardwired to work either as a latch or tristate buffer. If mode (MD) pin is tied **high,** then it will work as a latch and so it can be used as an output port. If mode (MD) pin is tied **low,** then it work as tristate buffer and so it can be used as an input port.

Q9.6   What are the various internal devices of INTEL 8155?

The INTEL 8155 is an IC consisting of static RAM, IO ports and a timer. The internal devices of 8155 are 256 bytes of static RAM, three numbers of programmable IO ports and a 14-bit programmable timer.

Q9.7   What are the internal devices of an 8255 ?

The internal devices of an 8255 are port-A, port-B and port-C. The ports can be programmed for either input or output function in different operating modes.

Q9.8   What are the operating modes of port-A of an 8255?

The port-A of an 8255 can be programmed to work in any one of the following operating modes as input or output port:

      Mode-0 : Simple IO port

      Mode-1 : Handshake IO port

      Mode-2 : Bidirectional IO port

Q9.9   What are the functions performed by port-C of an 8255?

1.   **The port-C pins are used for handshake signals.**
2.   **Port-C can be used as an 8-bit parallel IO port in mode-0.**
3.   **It can be used as two numbers of 4-bit parallel port in mode-0.**
4.   **The individual pins of port-C can be set or reset for various control applications.**

Q9.10 What is a handshake port ?

In a handshake port, signals are exchanged between the IO device and the port or between the port and the processor for checking/informing various condition of the device.

Q9.11 Explain the working of a handshake input port.

In handshake input operation, the input device will check whether the port is empty or not. If the port is empty, then it will load the data to the port. When the port receives the data, it will inform the processor for read operation. Once the data has been read by the processor, the port will signal the input device that it is empty. Now the input device can load another data to the port and the above process is repeated.

Q9.12 Explain the working of a handshake output port.

In handshake output operation, the processor will load a data to the port. When the port receives the data, it will inform the output device to collect the data. Once the output device accepts the data, the port will inform the processor that it is empty. Now the processor can load another data to the port and the above process is repeated.

Q9.13 How is DMA initiated?

When the IO device needs a DMA transfer, it will send a DMA request signal to the DMA controller. The DMA controller inturn sends a HOLD request to the processor. When the processor receives a HOLD request, it will drive its tristate pins to **high impedance** state at the end of current instruction execution and sends an acknowledge signal to the DMA controller. Now the DMA controller will perform DMA transfer.

Q9.14 What are the different types of DMA?

The different types of DMA data transfer are cycle stealing (or single transfer) DMA, Block transfer (or Burst mode) DMA and Demand transfer DMA.

Q9.15 What is cycle stealing DMA?

In cycle stealing DMA (or single transfer mode) the DMA controller will perform one DMA transfer in between the instruction cycles (i.e., In this mode the execution of one processor instruction and one DMA data transfer will take place alternatively).

Q9.16 What are block and demand transfer modes DMA?

In block transfer mode, the DMA controller will transfer a block of data and relieve the bus to the processor. After sometime another block of data is transferred by the DMA and so on.

In demand transfer mode the DMA controller will complete the entire data transfer at a stretch and then relieve the bus to the processor.

Q9.17 What are the programmable registers of 8237?

The programmable registers of 8237 are Base address registers, Base word count registers, Command register, Request register, Mode registers end Mask register.

Q9.18 What is the first-last flip-flop?

The 8237 has an internal flip-flop called first-last flip-flop which takes care of reading/writing 16-bit information through 8-data lines.

The first-last flip-flop selects the low or high byte during read/write operation of the address and count registers of channels. If first-last flip-flop is zero (i.e. reset),then the low byte can be read/write. If it is one (i.e., set), then the high byte can be read/write. After every read/write operation the first-last flip-flop automatically toggles.

Q9.19 What is RS-232C standard ?

RS-232C is a serial bus consisting of a maximum of 25 signals. These bus signals are standardized by EIA (**E**lectronics **I**ndustries **A**ssociation), USA and adopted by IEEE. Usually the first 9 signals are sufficient for most of the serial data transmission. The RS-232C serial bus is usually terminated using either a 9-pin connector or a 25-pin connector.

Q9.20 What is baud rate ?

The baud rate is the rate at which the serial data is transmitted. Baud rate is defined as $\dfrac{1}{(\text{The time for a bit cell})}$. In some systems one bit cell has one data bit, then the baud rate and bits per second are same.

Q9.21 What is the bit format used for sending asynchronous serial data?

In asynchronous transmission, each data character has a bit which identifies its start and 1 or 2 bits which identify its end. A typical bit format is shown in Fig. Q9.21.

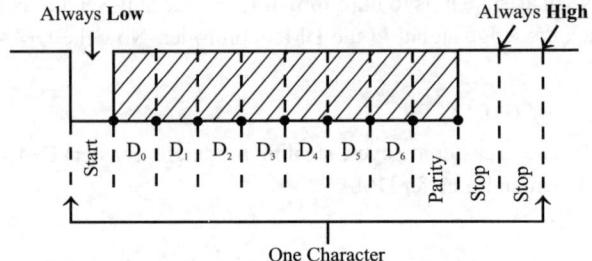

Fig. Q9.21: Bit format used for sending asynchronous serial data.

Q9.22 What voltage levels are used in RS-232C serial communication standard?

The voltage levels for all RS-232C signals are as follows.

Logic low = -3-V to -15-V under load (-25-V on no load)
Logic high = $+3$-V to $+15$-V under load ($+25$-V on no load)

Commonly used voltage levels are $+12$-V (logic **high**) and -12-V (logic **low**).

Q9.23 How is the RS-232C serial bus interfaced to TTL logic device ?

The RS-232C signal voltage levels are not compatible with TTL logic levels. Hence, for interfacing TTL devices to RS-232C serial bus, level converters are used. The popularly used level converters are MC 1488 and MC 1489 or MAX 232.

Q9.24 What is USART ?

The device which can be programmed to perform synchronous or asynchronous serial communication is called USART (**U**niversal **S**ynchronous **A**synchronous **R**eceiver **T**ransmitter). INTEL 8251A is an example of USART.

Q9.25 What are the functions performed by INTEL 8251A?

The INTEL 8251A is used for converting parallel data to serial or vice versa. The data transmission or reception can be either asynchronous or synchronous. The 8251A can be used to interface MODEM and establish serial communication through MODEM over telephone lines.

Q9.26 What are the control words of 8251A and what are its functions?

The control words of 8251A are Mode word and Command word. The mode word informs 8251 about the baud rate, character length, parity and stop bits. The command word can be sent to enable the data transmission and/or reception.

Q9.27 What is the information that can be obtained from the status word of 8251?

The status word can be read by the CPU to check the readiness of the transmitter or receiver and to check the character synchronization in synchronous reception. It also provides information regarding various errors in the data received. The various error conditions that can be checked from the status word are parity error, overrun error and framing error.

Q9.28 What are the tasks involved in keyboard interface ?

The tasks involved in keyboard interfacing are sensing a key actuation, debouncing the key and generating keycodes (Decoding the key). These tasks are performed by software if the keyboard is interfaced through ports and they are performed by hardware if the keyboard is interfaced through 8279.

Q9.29 What is debouncing ?

When a key is pressed it bounces for a short time. If a key code is generated immediately after sensing a key actuation, then the processor will generate the same keycode a number of times. (A key typically bounces for 10 to 20 millisecond.) Hence, the processor has to wait for the key bounces to settle before reading the keycode. This process is called keyboard debouncing.

Q9.30 What is scanning in keyboard and what is scan time?

The process of sending a zero to each row of a keyboard matrix and reading the columns for key actuation is called scanning. The scan time is the time taken by the device/processor to scan all the rows one by one starting from first row and coming back to the first row again.

Q9.31 What is the disadvantage in keyboard interfacing using ports?

The disadvantage in keyboard interfacing using ports is that most of the processor time is utilized in keyboard scanning and debouncing. As a result the computational speed/efficiency of the processor will be reduced.

Q9.32 What is multiplexed display? What is its advantage?

The process of switching ON the display devices one by one for a specified time interval is called multiplexed display. In microprocessor-based systems, six to eight 7-segment LEDs are interfaced to provide multiplexed display. At any one time only one 7-segment LED is made to glow at a time. After a few milliseconds, the next 7-segment LED is made to glow and so on. Due to persistence of vision, it will appear as if the LEDs are glowing continuously. The advantage in multiplexed display is that the power requirement of the display devices are reduced to a very large extent.

Q9.33 What is scanning in display and what is the scan time?

In display devices, the process of sending display codes to 7-segment LEDs to display the LEDs one by one is called scanning (or multiplexed display). The scan time is the time taken to display all the 7-segment LEDs one by one, starting from first LED and coming back to the first LED again.

Q9.34 What is the disadvantage in 7-segment LED interfacing using ports?

The disadvantage in using ports for 7-segment LED interfacing is that most of the processor time is utilized for display refreshing.

Q9.35 What is the advantage in using INTEL 8279 for keyboard and display interfacing?

When 8279 is used for keyboard and display interfacing, it takes care of all the tasks involved in keyboard scanning and display refreshing. Hence the processor is relieved from the task of keyboard scanning, debouncing, keycode generation and display refreshing. So, the processor time can be more efficiently used for computing.

Q9.36 List the functions performed by 8279.

The functions performed by 8279 are:

1. Keyboard scanning
2. Informing the key entry to CPU
3. Display refreshing
4. Key debouncing
5. Storing display codes
6. Keycode generation
7. Output display codes to LEDs

Q9.37 What is the maximum number of keycodes that can be generated by 8279?

In scanned keyboard mode the maximum size of keyboard matrix array that can be interfaced to 8279 is 8 x 8, which consists of 64 keys. In addition, the 8279 has two control keys called shift and control. For each key press, an 8-bit code is generated and stored in FIFO (keyboard RAM of 8279). The keycode consists of row and column number of the key in binary along with the status of shift and control key. Hence with 64 contact keys, shift and control key, a maximum of 256 keycodes can be generated by 8279.

Q9.38 What are the programmable display features of 8279 ?

The 8279 can be used for interfacing LEDs or 7 segment LEDs. In decoded scan, 4 numbers of 7-segment LEDs can be interfaced and in encoded scan, a maximum of 16 numbers of 7-segment LEDs can be interfaced. The 8279 can be programmed for left entry or right entry.

Q9.39 What are the different scan modes of of 8279?

The different scan modes of 8279 are decoded scan and encoded scan. In decoded scan mode, the output of scan lines will be similar to a 2-to-4 decoder. In encoded scan mode, the output of scan lines will be binary count. So, an external decoder should be used to convert the binary count to decoded output.

Q9.40 What is the difference in programming the 8279 for encoded scan and decoded scan?

If the 8279 is programmed for decoded scan then the output of scan lines will be decoded output and if it is programmed for encoded scan then the output of scan lines will be binary count. In encoded mode, an external decoder should be used to decode the scan lines.

Q9.41 How is a keyboard matrix is formed in a keyboard interface using 8279?

The return lines, RL_0 to RL_7 of 8279 are used to form the columns of keyboard matrix. In decoded scan the scan lines SL_0 to SL_3 of 8279 are used to form the rows of keyboard matrix. In encoded scan mode, the scan line SL_0 to SL_3 are connected to input of a decoder and the output lines of decoder are used as rows of keyboard matrix.

Q9.42 What are the operating modes of a timer 8254?

The 8254 timer has six operating modes. These are:

1. Mode-0 → Interrupt on terminal count
2. Mode-1 → Hardware retriggerable one shot
3. Mode-2 → Rate generator or Timed interrupt generator
4. Mode-3 → Square wave mode
5. Mode-4 → Software triggered strobe
6. Mode-5 → Hardware triggered strobe

Q9.43 What is the function of the GATE signal in timer 8254?

In timer 8254, the GATE signal acts as a control signal to start, stop or maintain the counting process. In modes 0, 2, 3 and 4, the GATE signal should remain **high** to start and maintain the counting process. In modes 1 and 5, the GATE signal has to make a low-to-high transition to start the counting process and need not remain **high** to maintain the counting process.

Q9.44 What will be the frequency of the square wave generated by a 8254 timer in mode-3?

The frequency of the generated square wave is given by the frequency of input clock signal divided by the count value loaded in the count register. If the count value N is an even number then the square wave will be alternatively **high** and **low** for N/2 clock periods. If the count value N is an odd number then the **high** time of square wave will be $\frac{N+1}{2}$ clock periods and **low** time will be $\frac{N-1}{2}$ clock periods.

Q9.45 What is resolution in DAC?

The resolution in DAC is the smallest possible analog value that can be generated by the n-bit binary input. If the reference voltage in n-bit DAC is V_{REF}, then the resolution is $(1/2^n) \times V_{REF}$ volts.

Q9.46 What are the internal devices of a typical DAC?

The internal devices of a DAC are R/2R resistive network, an internal latch and current to voltage converting amplifier.

Q9.47 What is settling or conversion time in DAC?

The time taken by the DAC to convert a given digital data to corresponding analog signal is called conversion time.

Q9.48 What are the different types of ADC?

The different types of ADC are successive-approximation ADC, counter type ADC, flash type ADC, integrator converters and voltage-to-frequency converters.

Q9.49 What is resolution and conversion time in ADC?

The resolution in ADC is the minimum analog value that can be represented by the digital data. If the ADC gives n-bit digital output and the analog reference voltage is V_{REF}, then the resolution is $(1/2^n) \times V_{REF}$ Volts. The conversion time in ADC is defined as the total time required to convert an analog signal into its digital equivalent.

9.18 EXERCISES

I. Fill in the blanks with appropriate words

1. In _____ data transfer scheme, the processor is forced to high impedance state by an IO device till the data transfer is complete.

2. The _____ data transfer scheme is employed when the speeds of the processor and IO device do not match.

3. The timer of INTEL 8155/8156 has a _____ bit counter.

4. In Intel 8255, the port whose individual pins can be set or reset is _____.

5. The internal register which is used to store the command words of 8255 is _____ register.

6. The strobed input/output mode of 8255 is _____.

7. The mode of 8255 which is used for bidirectional data transfer between processor and pheripherals is _____ .

8. The _____ control word of Intel 8255 is used to set/reset individual pins of Port-C.

9. The control word which is used to initialize all the ports of 8255 as input ports in mode-0 is _____ .

10. The _____ signals of 8255 are used to provide synchronization between transmitter and receiver.

11. The _____ of each port of 8355 is used to initialize the ports as input or output.

12. The _____ controller temporarily borrows the buses from microprocessor for direct data transfer between memory and IO device.

13. The _____ and _____ signals of microprocessor are employed in DMA data transfer.

14. The _____ register of Intel 8237 is used to hold the starting address of the memory block to be accessed by DMA.

15. The transmission mode in which both transmission and reception are possible but only one at a time is called _____ mode.

16. The normal phone conversation is an example of _____ duplex mode operation.

17. In _____ transmission, data are transmitted in blocks at a constant rate.

18. The baud rate is defined as _____ .

19. The device which can perform only asynchronous serial communication is _____ .

20. The commonly used standard for serial I/O when data is transmitted as voltage or current is _____ .

21. The Intel _____ is a programmable serial communication interface chip designed for synchronous and asynchronous serial data communication.

22. In common _____ type LED, all the anode terminals of LEDs are internally shorted.

23. The method by which current requirement for LEDs can be reduced is called _____ .

24. The 8254 has _____ numbers of independent 16-bit counters.

25. The maximum input clock frequency of 8254 is _____ .

26. The time taken by the DAC to convert a given digital data to a corresponding analog signal is called _____ time.

27. _____ signal of ADC0809 is used to indicate the end of conversion process and can be used as interrupt signal to the processor.

Answers

1. DMA	8. Bit Set/Reset(BSR)	15. half-duplex	22. anode
2. asynchronous	9. 9BH	16. full	23. multiplexed display
3. 14	10. handshaking	17.synchronous	24. three
4. port-C	11.Data Direction Register(DDR)	18. $1/$Time for a bit cell	25. 2.6 MHz
5. control word	12. DMA	19. UART	26. conversion
6. mode-1	13.HOLD, HLDA	20. RS232C	27. End of Conversion (EOC)
7. mode-2	14. Base Address(BA)	21. 8251A	

II. State whether the following statements are True/False.

1. Data over longer distance are generally transmitted serially.

2. The IO device or peripherals and microprocessors should have matched timing parameters.

3. When the INTEL 8212 is used as an output port, it will work as a tri-state buffer.

4. The Intel 8212 is permanently connected to work either as input or as output.

5. All the three IO ports of INTEL 8155 are 8 bit long.

6. The timer of INTEL 8155 is an increment counter.

7. The port-B of Intel 8255 can work either in mode-0 or mode-1.

8. Port C of 8255 can function independently as either input or output port.

9. In BSR mode, all the ports can be used to set and reset individual pins.

10. The mode-2 of Intel 8255 is generally used for data transfer between two computers.

11. In Intel 8255, only port-A can work in mode-2.

12. The Intel 8755 has RAM and IO(Input/Output) ports.

13. The DMA controller 8237 has four DMA channels.

14. The first-last flip flop of Intel 8237 has to be cleared to zero for reading/writing high byte first and then low byte.

15. The 8237 does not have the facility to directly set the first last flip flop.

16. The Intel 8257 cannot be connected in cascade like the 8237.

17. The bit rate and baud rate for serial data communication are same.

18. The standard telephone systems are used for serial data transfer over long distances.

19. Digital signals require less bandwidth than analog signals.

20. All the 25 signals of RS-232C are mandatory for serial communication.

21. keyboard interfacing using ports consumes more time than keyboard interfacing using hardware.

22. In common cathode LED, all the cathods of LEDs are connected to logic 1.

23. Maximum of eight 7-segment LEDs can only be interfaced with 8086 using 8279.

24. More than one key pressed can be recognized by 8279 in N-key rollover mode.

25. In 8279 sensor matrix mode, the condition of all 64 switches are stored in FIFO RAM.

26. The input clock frequency of 8253 and 8254 are same.

27. ADC0809 uses successive approximation method to convert analog to digital.

28. The ADC conversion is also called as quantization.

Answers

1. True	5. False	9. False	13. True	17. False	21. True	25. True
2. True	6. False	10. True	14. False	18. True	22. False	26. False
3. False	7. True	11. True	15. True	19. False	23. False	27. True
4. True	8. True	12. False	16. True	20. False	24. True	28. True

III. Choose the right answer for the following questions.

1. Which of the following devices use programmed data transfer scheme?

    a) hex-keyboard   *b)* CRT controller   *c)* floppy disk   *d)* hard disk

2. Which of the following data transfer scheme is used for large block of data transfer?

    a) synchronous   *b)* asynchronous   *c)* interrupt driven   *d)* DMA

3. The Intel 8155 does not include the following

    a) 256 bytes RAM   *b)* 128 byte EPROM   *c)* timer   *d)* three I/O ports

4. Which of the following mode of Intel 8155 timer produces a single pulse upon terminal count?

    a) mode-0   *b)* mode-1   *c)* mode-2   *d)* mode-3

5. The Programmable Peripheral Interface(PPI) is also known as

    a) Serial IO port   *b)* Parallel IO port   *c)* Serial Input Port   *d)* Parallel Output port

6. Which of the following 8255 port can work in all the three modes?

    a) port-A   *b)* port-B   *c)* port-C   *d)* all the three

7. Which of the following port is used to generate handshake signals in mode-1 and mode-2 of 8255?

    a) Port-A   *b)* Port-B   *c)* Port-C lower   *d)* Port-C upper

8. Which of the following are Group-B port of 8255?

    a) port-A   *b)* port-B   *c)* port-C lower   *d)* b and c

9. Which of the following control word of 8255 is used to set 4th pin of port-C?

    a) 18_H *b)* 19_H *c)* 08_H *d)* 09_H

10. Which of the following is ROM and IO port chip of Intel?

    a) 8212   *b)* 8155   *c)* 8255   *d)* 8355

11. Which of the following register of Intel 8237 is used to hold the address of the next memory location to be accessed by DMA?

    a) CA register   *b)* CWC regiser   *c)* BA register   *d)* BWC register

12. Which of the following is a simplex device?

    a) sensors   *b)* FM radio   *c)* pager   *d)* all the three

13. Which of the following is a half-duplex device?

    a) walkie-talkie   *b)* telephone line   *c)* pager   *d)* television

14. A system which has 1000 bits/sec bit rate encodes 1 bit within one transmitted bit time. What is the baud rate?

    a) 500 baud   *b)* 1000 baud   *c)* 2000 baud   *d)* 4000 baud

15. *The following device is used to perform both synchronous and asynchronous communication*

    a) USART   *b)* UART   *c)* RS232   *d)* none of the above

16. *Which of the following device is used to convert digital signal into audio frequency tones and vice versa?*

    a) USART   *b)* UART   *c)* RS232   *d)* MODEM

17. *The voltage levels for logic low for RS232C signals is*

    a) -5 V to -15 V   *b)* -5 V to +5 V   *c)* -3 V to -15 V   *d)* -3 V to -10 V

18. *Which of the following is used to convert parallel data into serial data?*

    a) USART   *b)* UART   *c)* DMA   *d)* a and b

19. *The process of rechecking the keypress in a row of hex keyboard after 10 to 20 ms is called as* _____

    a) key actuation   *b)* key debouncing   *c)* key bouncing   *d)* key decoding

20. *Which of the following chip is used to interface keyboard and display with microprocessor?*

    a) 8255   *b)* 8254   *c)* 8259   *d)* 8279

21. *Which of the following method consider only one key pressed and rejects all other simultaneous keypress in keyboard?*

    a) two-key lockout   *b)* N-key lockout   *c)* two-key roll over *d)* N-key rollover

22. *Which of the following mode of 8279 is used when 4-digit 7 segment LEDs are required?*

    a) encoded scan keyboard   *b)* Decoded scan keyboard

    c) Encoded scan sensor matrix   *d)* none of the above

23. *The* _____ *Intel chip is used to provide timings and to interrupt the processor at periodic interval*

    a) 8254   *b)* 8259   *c)* 8279   *d)* none of the above

24. *Which of the following operating mode of 8254 is used to generate square wave?*

    a) mode-0   *b)* mode-1   *c)* mode-2   *d)* mode-3

25. *The minimum analog value that can be represented by the digital data is*

    a) average   *b)* resolution   *c)* percentage   *d)* none of the above

Answers

1. a	4. c	7. d	10. d	13. a	16. d	19. b	22. b	25. b
2. d	5. b	8. d	11. a	14. b	17. c	20. d	23. a	
3. b	6. a	9. d	12. d	15. a	18. d	21. a	24. d	

IV. Answer the following questions.

E9.1 Write an 8086 program to read the status of 8 switches connected to 8255 port-A and switch status to be displayed on 8 LEDs connected to port-B. Let the address of port-A and port-B and control register are 80_H, 82_H and 86_H respectively.

E9.2 Write an 8086 program to blink an LED connected to pin PC.0 of port-C of 8255 interfaced with 8086.

E9.3 Write an 8086 program to switch ON all the eight LEDs one by one which are connected to Port-C of 8255.

E9.4 Interface 8254 with 8086 and write a program to load counter-0 with 1234_H and enable its output OUT 0 upon termination of count.

E9.5 Repeat exercise E9.4 to count in BCD mode

E9.6 Write a program to generate a square wave of period 10ms using counter 2 of 8254 operating at 2 MHz clock frequency.

E9.7 Write a program to generate a 'strobe' pulse at the output of counter 1 of 8255 after 1 ms delay. Take 8254 clock frequency = 2 MHz.

E9.8 Find the command word to initialize 8279 in 2 key lockout encoded scan keyboard mode and to display sixteen digits in left entry mode and write an ALP in 8086.

E9.9 Write a program to display all 'ones' as clear command only in left most four LEDs. Use N-key rollover mode.

E9.10 If the clock signal to 8279 is 1.5 MHz, calculate the prescalar value to obtain the operating frequency of 100 kHZ.

CHAPTER 10

INTEL 80x86 FAMILY OF PROCESSORS

10.1 INTRODUCTION

The microprocessors are used as CPU in personal computers. The IBM Corporation, USA, designed the first Personal Computer(PC) using INTEL 8088 as CPU. Ever since the introduction of the PC, the applications and usage of personal computers has increased day by day. This leads to great improvement and advancement in a PC, which in turn demands improvement in microprocessors.

INTEL has realized the demand for improvement in microprocessors and so keeps on updating the features of microprocessors year after year. Every year, INTEL releases new processors with improved features. (Please refer to Appendix I for complete list of processors released by INTEL.) The improvements in microprocessors are related to the following:

1. **Increased word length and memory space**
2. **Increased internal performance and clock rating**
3. **Increased external communications and error detection**
4. **Improved instruction set and support to software**
5. **Improved trouble shooting aids/provision**

Initially, INTEL had retained the 8086 architecture as base architecture and improved the other features of the processor and released the 80x86 family of processors. The 80x86 family includes 80186, 80286, 80386 and 80486 processors. Then INTEL switched to superscalar architecture and released the Pentium family of processors. The Pentium family of processors includes Pentium, Pentium Pro, Pentium II, Pentium III and Pentium 4 processors. The data and address bus size, addressable memory space and internal clock ratings of 80x86 and Pentium family of processors are listed in Table 10.1. The major new features of the 80x86 and Pentium family of processors are listed in Table 10.2.

Table 10.1: INTEL 80 X86 and Pentium Family of Processors

INTEL processor	Internal data bus	External data bus	Address bus	Physical memory space	Virtual memory space	Internal clock rating
8086	16-bit	16-bit	20-bit	1 MB	-	5/8/10 MHz
8088	16-bit	8-bit	20-bit	1 MB	-	5/8 MHz
80186	16-bit	16-bit	20-bit	1 MB	-	10/12 MHz
80286	16-bit	16-bit	24-bit	16 MB	1 GB	6/10/12 MHz
80386 DX	32-bit	32-bit	32-bit	4 GB	64 TB	16/20/25/33 MHz
80486 DX	32-bit	32-bit	32-bit	4 GB	64 TB	25/33/50 MHz
Pentium	32-bit	64-bit	32-bit	4 GB	64 TB	60 - 200 MHz
Pentium pro	32-bit	64-bit	36-bit	64 GB	64 TB	150/166/180/200 MHz
Pentium II	32-bit	64-bit	36-bit	64 GB	64 TB	233 - 450 MHz
Pentium III	32-bit	64-bit	36-bit	64 GB	64 TB	450 MHz - 1 GHz
Pentium 4	32-bit	64-bit	36-bit	64 GB	64 TB	1.4 - 3.3 GHz

Table 10.2: New Features of 80X86 and Pentium Family of Processors

INTEL processor	Major new features
8086/8088	Pipelined architecture, Instruction queue, Coprocessor support, Segmented memory.
80286	Instruction pre-decode, Multitasking, Memory protection, Virtual memory, Auto-shutdown.
80386DX	Instruction pipelining, Break-point instruction, Built-in self test, 32-bit internal registers.
80486DX	Cache memory, Internal floating-point unit, Restartable instruction, Data-bus parity.
Pentium	Superscalar architecture, Dual-processor configuration, Internal error detection, Dynamic branch prediction, Performance monitoring, Power management, Functional redundancy check, Machine check, Address bus parity.
Pentium Pro	Out-of-order execution, Speculative execution, Register renaming, Secondary cache, ECC (**E**rror **C**hecking and **C**orrecting codes), DIB (**D**ual **I**ndependent **B**us).
Pentium II	MMX (**M**ulti-**M**edia **E**xtension), System management bus, Integrated thermal diode.
Pentium III	Processor serial number, Improved MMX, SIMD (**S**ingle **I**nstruction **M**ultiple **D**ata) extensions.
Pentium 4	Rapid execution engine, Hyperpipelined and **H**yper **T**hreading (HT) technology, Advanced dynamic execution, 400/533/800 MHz system bus.

10.2 INTEL 80186 MICROPROCESSOR

The 80186 microprocessor was released in 1982. The 80186 is a 16-bit microprocessor and it consists of 15 to 20 of the most common microprocessor-based system components on a single chip. It is actually an integration of BIU (**B**us **I**nterface **U**nit) and EU (**E**xecution **U**nit) of the 8086 processor with the following functional units on a single chip.

1. **Clock generator**
2. **Programmable interrupt controller**
3. **Programmable timers (3 independent 16-bit timers)**
4. **Programmable DMA controller (2 independent DMA channels)**
5. **Programmable memory and peripheral chip selection logic**
6. **Programmable wait-state generator**
7. **Local bus controller**

The 80186 is object-code compatible with the 8086 and provides twice the performance of the standard 8086. The instruction set of 80186 includes all the instructions of 8086 with additional 10 new instructions.

The 80186 uses a 20-bit address to access memory and, hence, it can directly address up to one megabyte (2^{20} = 1 Mega) of memory space. The memory organization in 80186 and the IO addressing are similar to that of 8086.

The 80186 is available with maximum internal clock frequency of 6, 8, 10 and 12 MHz.

10.2.1 Pins and Signals of 80186

The 80186 is a 68-pin IC available in three different packages : PLCC (**P**lastic **L**eaded **C**hip **C**arrier), LCC (Ceramic **L**eadless **C**hip **C**arrier) and PGA (Ceramic **P**in **G**rid **A**rray). The pin configuration of the PLCC package of 80186 is shown in Fig. 10.1. The name of the pins and signals of 80186 processor are listed in Table 10.3.

Table 10.3: Pin Description of 80186

Pin name	Description	Type
AD_{15}-AD_0	Multiplexed address/data	Bidirectional, Tristate
A_{19}/S_6-A_{16}/S_3	Multiplexed address/status signals	Output, Tristate
$\overline{BHE}/S_7$	Bus high enable or status signal S_7	Output, Tristate
ALE/QS_0	Address latch enable or Queue status 0	Output
$\overline{WR}/QS_1$	Write control signal or Queue status 1	Output, Tristate
$\overline{RD}/\overline{QSMD}$	Read control signal or Queue status mode	Bidirectional
$\overline{S_0}, \overline{S_1}, \overline{S_2}$	Bus cycle status indicators	Output, Tristate
$DT/\overline{R}$	Data transmit/receive	Output, Tristate
$\overline{DEN}$	Data enable	Output, Tristate
ARDY	Asynchronous ready signal	Input
SRDY	Synchronous ready signal	Input
$\overline{RES}$	System reset	Input
RESET	Peripheral reset	Output
X_1, X_2	Crystal connection	Input
CLKOUT	Peripheral clock	Output
$\overline{LOCK}$	Bus priority lock control	Output, Tristate
HOLD	Hold request	Input
HLDA	Hold acknowledge	Output
$\overline{TEST}$	Wait on test control	Input
TMR IN 0	Timer 0 input clock or control signal	Input
TMR IN 1	Timer 1 input clock or control signal	Input
TMR OUT 0	Timer 0 output signal	Output
TMR OUT 1	Timer 1 output signal	Output
DRQ 0	Channel 0 DMA request	Input
DRQ 1	Channel 1 DMA request	Input
NMI	Nonmaskable interrupt	Input
INT0, INT1	Maskable interrupt request	Input
INT2, $\overline{INTA0}$	Maskable interrupt request or Acknowledge of INT0	Bidirectional

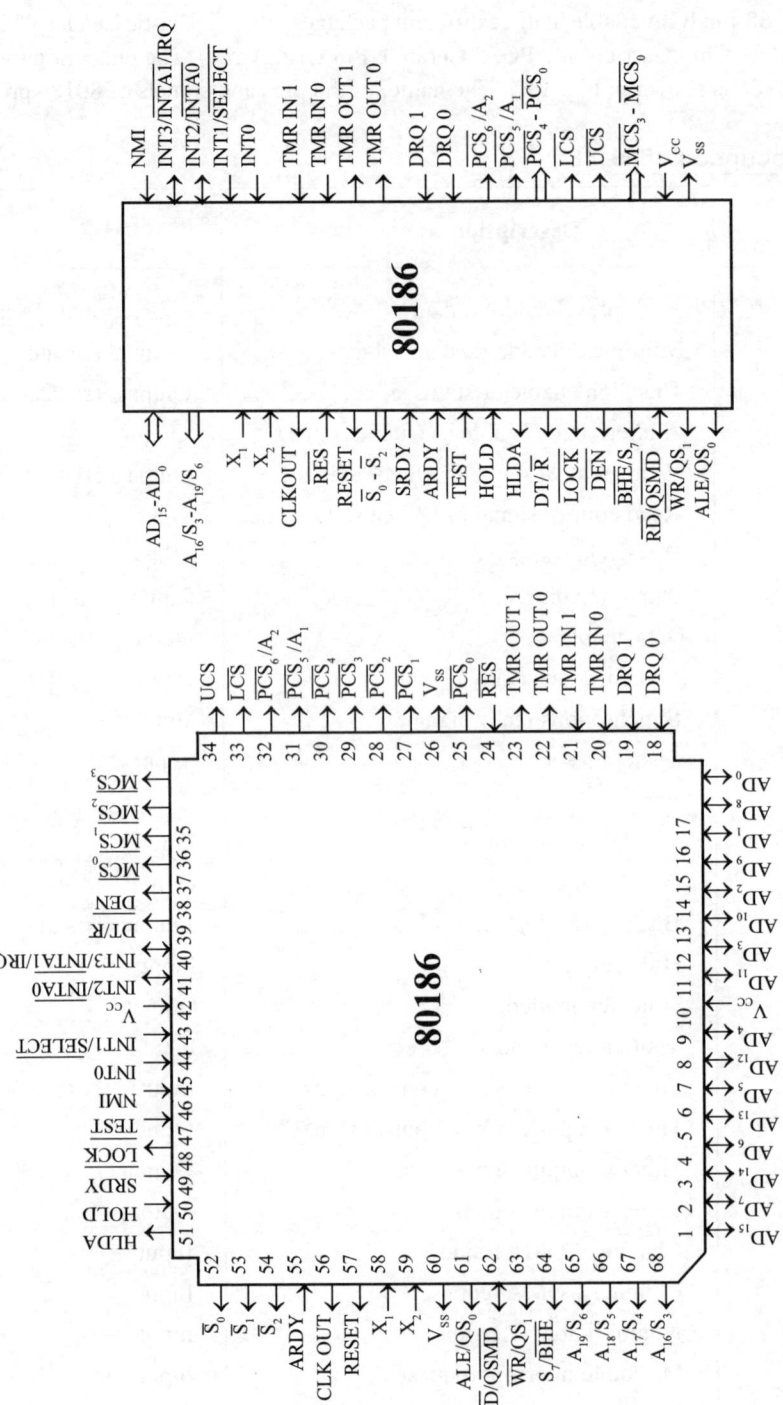

Fig.10.1: Pin configuration of 80186.

Table 10.3 continued...

Pin name	Description	Type
INT3, $\overline{\text{INTA1}}$	Maskable interrupt request or	
	Acknowledge of INT1	Bidirectional
$\overline{\text{UCS}}$	Upper memory chip select signal	Output
$\overline{\text{LCS}}$	Lower memory chip select signal	Output
$\overline{\text{MCS}}_0$ to $\overline{\text{MCS}}_3$	Mid-range memory chip select signals	Output
$\overline{\text{PCS}}_0$ to $\overline{\text{PCS}}_4$	Peripheral chip select signals	Output
$\overline{\text{PCS}}_5/\text{A}_1$	Peripheral chip select 5 or	
	Latched address A_1	Output
$\overline{\text{PCS}}_6/\text{A}_2$	Peripheral chip select 6 or	
	Latched address A_2	Output
V_{CC}	Power supply, +5V	Input
V_{SS}	Power supply ground, 0V	Output

In an 80186 processor, the lower sixteen lines of address are multiplexed with the data and the upper four lines of address are multiplexed with the status signals. During the first clock period of a bus cycle, the entire 20-bit address is available on these lines. During all other clock periods of a bus cycle, the data and status signals will be available on these lines. The signal ALE is used to demultiplex address and data/ status lines using external latches.

During the processor initiated bus cycle, the status signal S_6 is asserted low and during DMA cycle, it is asserted high. The status signals S_3, S_4 and S_5 always remain at logic low.

The 80186 outputs a low on $\overline{\text{BHE}}$ pin during read, write and interrupt acknowledge cycles when the data is to be transferred to the high order data bus. The output signal $\overline{\text{BHE}}$ on the first T-state of a bus cycle is maintained as status signal S7 during all other T-states of the bus cycle. The $\overline{\text{BHE}}$ can be used in conjunction with address bit A_0 (AD_0) to select memory banks. The status of $\overline{\text{BHE}}$ and A_0 during word/ byte transfer from even/odd memory bank are shown in Table 10.4.

Table 10.4: Status of $\overline{\text{BHE}}$ and A_0 During Memory Access

$\overline{\text{BHE}}$	A_0	Function
0	0	Word transfer
0	1	Byte transfer on upper half of data bus (D_{15} - D_8)
1	0	Byte transfer on lower half of data bus (D_7 - D_0)
1	1	Reserved

The queue status are output through ALE and $\overline{\text{WR}}$ pins during the queue status mode. This mode is selected by the coprocessor 80187 by sending a logic low signal to $\overline{\text{RD}}$ pin. The queue status can be used to track the internal status of the queue. The output on QS_0 and QS_1 can be interpreted as shown in Table 10.5.

Table 10.5: Queue Status

QS_1	QS_0	Queue operation
0	0	No queue operation
0	1	First opcode byte fetched from the queue
1	0	Empty the queue
1	1	Subsequent byte fetched from the queue

The status signals $\overline{S}_0, \overline{S}_1$ and $\overline{S}_2$ will provide information regarding the nature of the bus cycle performed by the processor. The output on $\overline{S}_0, \overline{S}_1$ and $\overline{S}_2$ during various bus cycles are listed in Table 10.6.

Table10.6: Bus Status Signals

S_2	S_1	S_0	Bus cycle
0	0	0	Input acknowledge
0	0	1	Read IO
0	1	0	Write IO
0	1	1	Halt
1	0	0	Instruction fetch
1	0	1	Read data from memory
1	1	0	Write data to memory
1	1	1	Passive (No bus cycle)

The signal $\overline{DEN}$ is used to enable the external data bus buffers/transceivers and DT/$\overline{R}$ is used for direction control of data bus buffers.

The ARDY and SRDY are input signals to the processor, used by the memory or IO devices to get extra time for data transfer or to introduce wait states in the bus cycles. The ARDY pin accepts a rising edge that is asynchronous to CLKOUT and it is active high. The SRDY pin accepts an active high input synchronized to CLKOUT. When one of the ready input is used, the other should be tied to logic low. When both ready inputs are not used, they should be tied to logic high.

The $\overline{RES}$ is the input reset signal to be applied to bring the processor to a known state. The $\overline{RES}$ pin should be held low for atleast 50 milliseconds after power is switched-ON. Whenever $\overline{RES}$ goes low, the processor generates a logic high output RESET signal and this signal can be used to reset the peripheral devices in the system. When the processor is reset, the DS, SS, ES, IP and flag register are cleared, Code Segment (CS) register is initialized to $FFFF_H$ and queue is emptied. After reset, the processor will start fetching instructions from the 20-bit physical address $FFFF0_H$.

The pins X_1 and X_2 are provided to connect an external quartz crystal. The frequency of the quartz crystal should be double that of internal clock frequency or processor clock frequency. The processor clock signal is also given out through CLKOUT pin for used by the peripheral devices.

The $\overline{LOCK}$ is an active low output signal activated by the LOCK prefix instruction and remains active until the completion of the instruction prefixed by LOCK, in order to prevent other bus masters from gaining control of the system bus. The $\overline{TEST}$ input is tested by the WAIT instruction. The processor will enter a wait state after execution of the WAIT instruction, and it will resume execution only when $\overline{TEST}$ is made low by an external hardware.

The $\overline{LCS}$ is a programmable memory chip select signal to select a memory block of size 1kB to 256 kB with starting address 00000_H. The $\overline{UCS}$ is a programmable memory chip select signal to select a memory block of size 1 kB to 256 kB ending with the address $FFFFF_H$. The $\overline{MCS}_0$ to $\overline{MCS}_3$ are programmable to select four middle memory blocks of size 8 kB to 512 kB.

The $\overline{PCS}_0$ to $\overline{PCS}_6$ are programmable peripheral chip select signals to select seven IO devices. The $\overline{PCS}_5$ can also be programmed to provide internally latched address bit A_1 and $\overline{PCS}_6$ can be programmed to provide internally latched address bit A_2.

10.2.2 Architecture of 80186

The architecture (functional block diagram) of the 80186 microprocessor is shown in Fig. 10.2. The various functional blocks of the 80186 are **B**us Interface Unit (BIU), **E**xecution Unit (EU), Clock generator, Programmable interrupt controller, Programmable timers, DMA controller and Chip select unit.

The BIU of the 80186 is identical to the BIU of the 8086 processor. The BIU consists of a dedicated adder to generate a 20-bit address, 16-bit segment registers CS, DS, SS and ES, logic circuit to produce bus control signals and 6-byte instruction queue. The 20-bit physical address is generated by multiplying the content of one of the segment registers by 16_{10} and then adding it to a 16-bit offset. The EU of 80186 is identical to the EU of the 8086 processor. The EU consists of 16-bit ALU, flag register, and general purpose registers AX, BX, CX, DX, SP, BP, SI and DI. The functions of the general-purpose registers and the flags of 80186 processor are same as that of the 8086 processor.

The on-chip (internal) clock generator consists of a crystal oscillator, a divide-by-two counter, synchronous and asynchronous ready inputs and reset circuitry. An external quartz crystal of frequency double that of the processor clock should be connected through X_1 and X_2 pins to the oscillator of the clock generator.

The clock generated by the oscillator is divided by two and used as a processor clock. The clock signal is also given out through the CLKOUT pin for use by peripheral devices. The clock generator also provides the internal timing for synchronizing the ready input signals.

The timer unit consists of three programmable 16-bit timers. The timer-0 and timer-1 can be used to count external events, provide timings for external events, generate waveforms, etc., depending on the mode of operation. The timer-0 and timer-1 can be driven either by a processor clock or by an external clock. The timer-2 can be used only for internal timing operations. It can be used as a clock source for other timers, as a watchdog timer or as a DMA request source. [A watchdog timer is a timer which can internally interrupt the processor after a programmed time interval.]

The DMA controller unit consists of two independent DMA channels. DMA data transfer can be performed between memory spaces or between IO spaces or between memory and IO space. The data can be transferred either in bytes or in words to/from even/odd addresses. Each data transfer consumes two bus cycles, one to read data and the other to write. Each DMA channel has a 20-bit source pointer, a 20-bit destination pointer, a 16-bit count register and a 16-bit control register. The pointers are used to hold the source and destination addresses. The count register is used to program the number of bytes/words to be transferred by DMA. The control register is programmed for the type of DMA and various other functions. The DMA channels can be programmed such that one channel can have higher priority over the other.

The interrupt controller unit arbitrates all internal and external interrupts. The 80186 has five external (hardware) interrupt inputs and they are INT0, INT1, INT2, INT3 and NMI. The external hardware interrupts can be expanded by connecting INTEL 8259 (external Programmable Interrupt Controller) to INT0 and INT1 inputs.

The internal interrupts of 80186 includes INTEL predefined interrupts, software interrupts and interrupts from internal timers and DMA channels. The INTEL predefined interrupts of 80186 includes the five INTEL predefined interrupts of 8086 (Divide error, Single step, NMI, Break point and Interrupt on overflow) and in addition has three predefined interrupts : Array BOUNDS Interrupt, Unused opcode interrupt and ESC opcode interrupt.

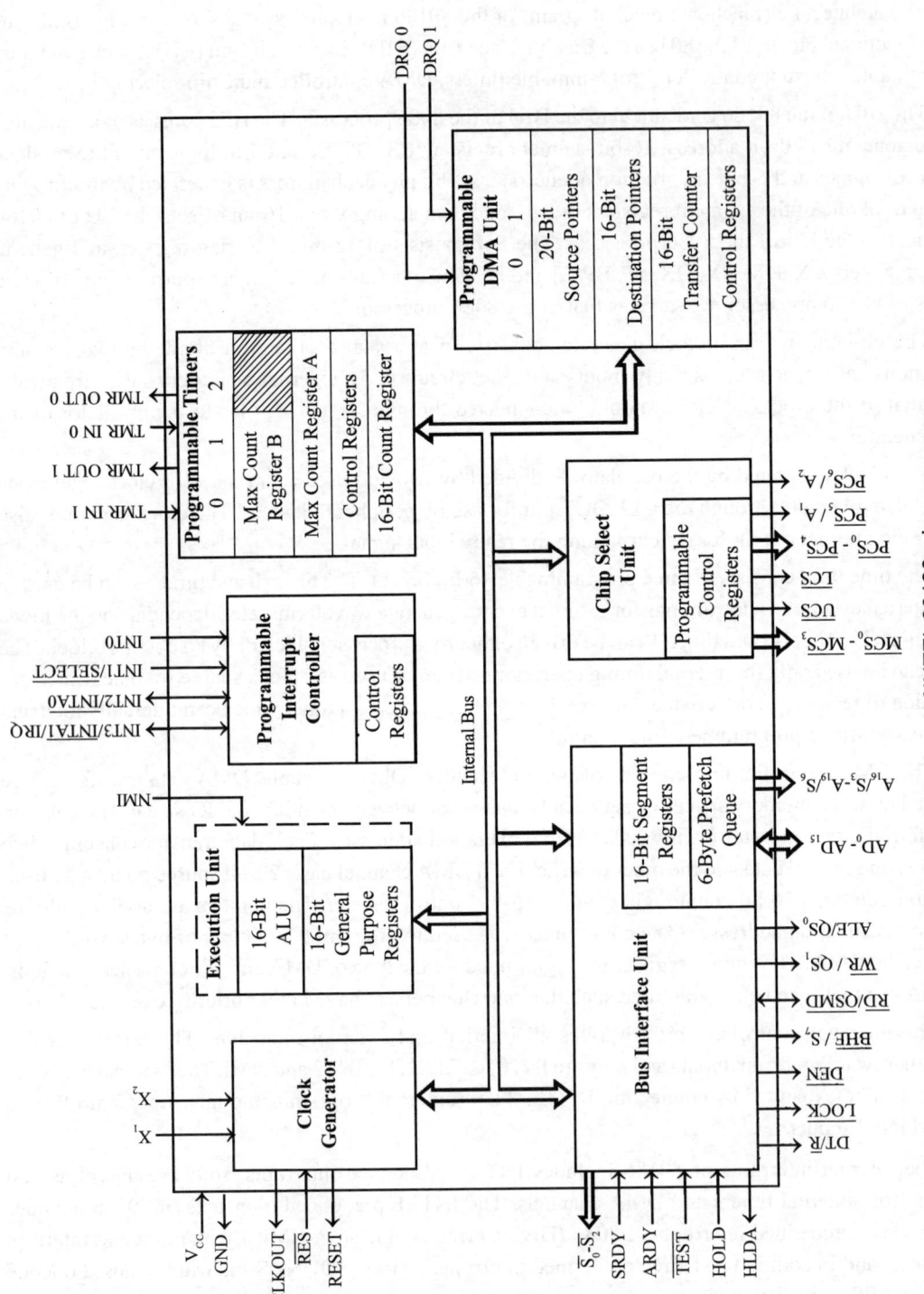

Fig. 10.2: Functional block diagram of INTEL 80186 microprocessor.

The array BOUNDS interrupt occurs if the boundary of an index register is outside the values set up in the memory. The unused opcode interrupt occurs whenever the processor executes any undefined opcode. The ESC opcode interrupt occurs if ESC opcodes are executed.

Each interrupt of 80186 has been allotted a type number and a vector address like that of 8086. The interrupt vector table of 80186 occupies the first 1kB memory space like that of 8086. The type number, vector address and priority of the internal and external interrupts of 80186 are listed in Table 10.7.

Table 10.7: Type Number and Priorities of Interrupts of 80186

Interrupt	Type number	Vector address	Priority level
Divide error	0	00000 - 00003	1
Single step	1	00004 - 00007	1A
NMI	2	00008 - 0000B	1
Break-point	3	0000C - 0000F	1
Interrupt on overflow	4	00010 - 00013	1
Array BOUNDS	5	00014 - 00017	1
Unused opcode	6	00018 - 0001B	1
ESC opcode	7	0001C - 0001F	1
Timer-0	8	00020 - 00023	2A
Timer-1	18	00048 - 0004B	2B
Timer-2	19	0004C - 0004F	2C
Reserved	9	00024 - 00027	3
DMA 0	10	00028 - 0002B	4
DMA 1	11	0002C - 0002F	5
INT0	12	00030 - 00033	6
INT1	13	00034 - 00037	7
INT2	14	00038 - 0003B	8
INT3	15	0003C - 0003F	9
80187	16	00040 - 00043	1

Note : Interrupt priority level-1 is the highest and level-9 is the lowest. Some interrupts have the same priority.

The chip select unit generates the chip select signals for memories and peripherals. This unit provides 6 memory chip select outputs, namely $\overline{UCS}$, $\overline{LCS}$, $\overline{MCS_0}$, $\overline{MCS_1}$, $\overline{MCS_2}$ and $\overline{MCS_3}$. The $\overline{UCS}$ is used to select the upper/top memory space of size 1 kB to 256 kB ending with address $FFFFF_H$. The $\overline{LCS}$ is used to select the lower/bottom memory space of size 1 kB to 256 kB starting with address 00000_H. The $\overline{MCS_0}$ to $\overline{MCS_3}$ can be used to select four address spaces of size 8 kB to 512 kB within 1MB address space, excluding the address space defined by $\overline{UCS}$ and $\overline{LCS}$.

The chip select unit provides seven peripheral chip select signals. Each peripheral chip select signal addresses a 128-byte block of IO address space. The programmable IO address space starts at a base IO (or memory) address programmed by the user. The seven consecutive blocks of 128 bytes starting from this base address will be the IO address space addressed by the seven peripheral chip select signals respectively.

The chip select signals are active for all memory and IO cycles in their programmed areas, whether they are generated by the BIU or the DMA unit.

The 80186 is completely object-code compatible with 8086. The instruction set of 80186 consists of the instructions of 8086 and of 10 new instructions, which are as follows:

1.	ENTER	-	Enter a procedure
2.	LEAVE	-	Leave a procedure
3.	BOUND	-	Check if an array index in a register is in range of array
4.	INS	-	Input string byte or string word
5.	OUTS	-	Output string byte or string word
6.	PUSHA	-	Push all registers to stack
7.	POPA	-	Pop all registers from stack
8.	PUSH imm	-	Push immediate (imm) data to stack
9.	IMUL reg,sou,imm	-	Multiply the immediate (imm) data and source (sou) data, and store the result in register(reg)
10.	SHIFT des,imm	-	Shift the destination (des) register/memory contents specified immediate (imm) number of times

10.3 INTEL 80286 MICROPROCESSOR

The INTEL 80286 is a 16-bit microprocessor with on-chip memory protection capabilities. The 80286 is an integration of 8086 and the memory management unit on a single chip. It is primarily designed for multiuser/multitasking systems. The 80826 is used as a CPU in IBM's personal computers PC/AT and its clones.

The 80826 has two operating modes: real address mode and protected virtual address mode. In the real address mode, the 80286 can address upto 1 MB (Megabyte) of physical memory address space like 8086. In protected virtual address mode, it can address up to 16 MB of physical memory address space and 1GB (Gigabyte) of virtual memory address space.

The instruction set of 80286 includes the instructions of 8086 and 80186, and has some extra instructions to support the operating system and memory management. In the real address mode, the 80286 is object-code compatible with 8086. In the protected virtual address mode, it is source code compatible with 8086 and the software may require some modifications to incorporate the virtual address features. The performance of 80286 is five times faster than that of a standard 8086.

The 80286 is available with maximum internal clock frequency ratings of 4, 6 and 8 MHz.

10.3.1 Pins and Signals of 80286

The 80286 is a 68-pin IC available in ceramic leadless flat package. The pin configuration of 80286 is shown in Fig. 10.3. The pins and signals of 80286 are listed in Table 10.8. The 80286 has nonmultiplexed address and data bus. It has 16 pins (D_0 - D_{15}) for data and 24 pins (A_0 - A_{23}) for address.

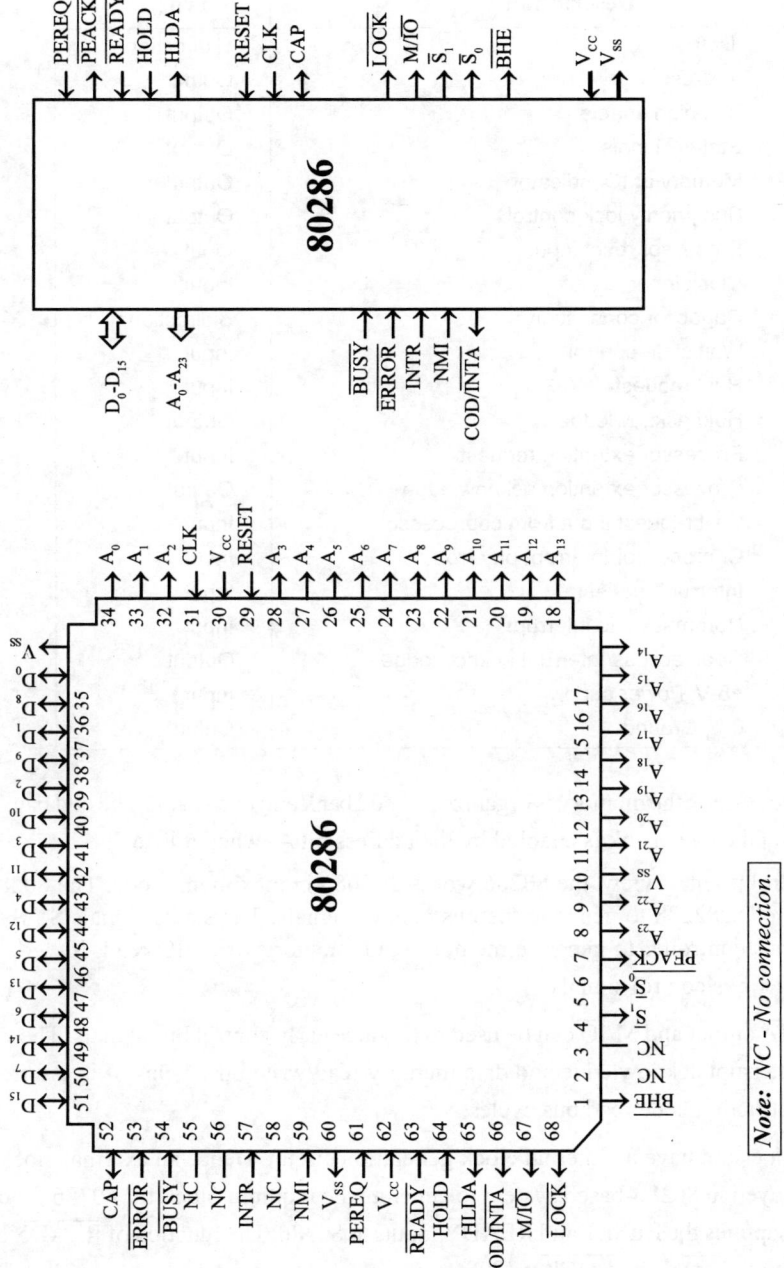

Fig. 10.3: Pin configuration of 80286.

Note: NC - No connection.

Table 10.8: Pins and Signals of 80286

Pin	Description	Type
D_{15}-D_0	Data	Bidirectional
A_{23}-A_0	Address	Output
$\overline{BHE}$	Bus high enable	Output
$\overline{S}_0$, $\overline{S}_1$	Status signals	Output
M/$\overline{IO}$	Memory or IO indicator	Output
$\overline{LOCK}$	Bus priority lock control	Output
RESET	Processor reset input	Input
CLK	Clock input	Input
CAP	Capacitor connection	Bidirectional
$\overline{READY}$	Wait state control	Input
HOLD	Hold request	Input
HLDA	Hold acknowledge	Output
PEREQ	Processor extention request	Input
PEACK	Processor extention acknowledge	Output
$\overline{BUSY}$	Wait request input from coprocessor	Input
$\overline{ERROR}$	Coprocessor interrupt on error	Input
INTR	Interrupt request	Input
NMI	Non-maskable interrupt	Input
COD/$\overline{INTA}$	Code access/Interrupt acknowledge	Output
V_{cc}	+5-V, Power supply	Input
V_{ss}	0-V, Ground	Output

In 80286-based system, the memory is organized as odd bank and even bank. The odd bank is enabled by the signal $\overline{BHE}$ and the even bank is enabled by the address bit A_0 when it is **low**.

From the control point of view, the 80286 works as 8086 in maximum mode. The 80286 requires an external bus controller 82288 to generate the bus control signals. The status signals S_0, S_1 and M/$\overline{IO}$ are decoded by the bus controller to generate memory read, memory write, IO read, IO write, interrupt acknowledge and other bus control signals.

The COD/$\overline{INTA}$ output and M/$\overline{IO}$ can be used to produce early control bus signals. The COD/$\overline{INTA}$ is asserted **low** for interrupt acknowledge and data memory read/write bus cycles. It is asserted **high** for IO read/write and instruction/code read bus cycles.

The 80286 does not have an internal clock generation circuit. Hence an external clock generator 8284 should be employed in 80286-based system to generate the required clock for 80286 processor. The clock generator also supplies the RESET and $\overline{READY}$ inputs to 80286. The function of $\overline{READY}$ and $\overline{LOCK}$ of 80286 are similar to $\overline{READY}$ and $\overline{LOCK}$ of 8086.

The 80286 has a logic **high** reset to bring the processor to a known state. After a reset, the 80286 processor will work in real address mode and start executing the program stored at FFFFF0$_H$. The reset will also initialize the internal registers as follows:

Flag register : 0002_H CS-register : $F000_H$

Machine Status Word : $FFF0_H$ DS-register : 0000_H

Instruction Pointer : $FFF0_H$ SS-register : 0000_H

                                         ES-register : 0000_H

The CAP pin has been provided to connect an external filter capacitor for the negative bias voltage generator located internally. The negative bias voltage is required for the substrate of MOS devices in 80286 in order to work at maximum speed. An electrolyte capacitor of rating 0.047 µF, ±20%, 25-V should be connected to a CAP pin. The positive end of capacitor is grounded and the negative end is connected to the CAP pin.

The HOLD input is used by the DMA controller to request the bus to perform DMA transfer. The HLDA signal is the acknowledge signal sent by 80286 to the DMA controller to inform that the bus has been released for performing DMA.

The pins PEREQ, $\overline{PEACK}$, $\overline{BUSY}$ and $\overline{ERROR}$ are used for interfacing coprocessor 80287 with 80286. The coprocessor (80287) will assert PEREQ as **high** whenever it requires a memory data transfer. When the processor (80286) starts performing the data transfer it will send an acknowledge to the coprocessor (80287) by asserting $\overline{PEACK}$ as **low**.

When the coprocessor executes an instruction, the 80286 will wait (by executing wait instruction) until the BUSY signal is asserted **low** by the coprocessor. Whenever the coprocessor finds some error during processing it will interrupt the processor by asserting $\overline{ERROR}$ as **low**. In response the processor will perform a type 16_H interrrupt call, which will automatically execute a procedure written for the desired response to error condition.

10.3.2   Architecture of 80286

The architecture (functional block diagram) of 80286 is shown in Fig. 10.4. The 80286 has four separate processing units: **B**us Unit (BU), **I**nstruction Unit (IU), **E**xecution Unit (EU) and **A**ddress Unit (AU).

The **B**us Unit (BU) performs all the memory and IO read/write operations. Whenever the bus is free, the BU will fetch instruction bytes and put them in 6-byte prefetch queue.

When a jump/call instruction is encountered, the BU will dump/clear the queue and starts filling it from the jump/call address. The BU also controls transfer of data to and from the coprocessor 80287.

The **I**nstruction Unit (IU) will read the prefetched instructions from the BU and decode them, and then put them in the instruction queue. The instruction queue can accommodate up to three decoded instructions. This additional pipelining in 80286 increases the speed of processing of 80286, when compared to an 8086.

The **E**xecution Unit (EU) will read the decoded instructions from the IU and execute them sequentially. It directs the BU to fetch memory or IO operands whenever needed.

The EU has a 16-bit ALU, a set of registers identical to that of 8086 and in addition it has a 16-bit **M**achine **S**tatus **W**ord (MSW). The MSW is used to switch from real address mode to protected virtual address mode after a reset.

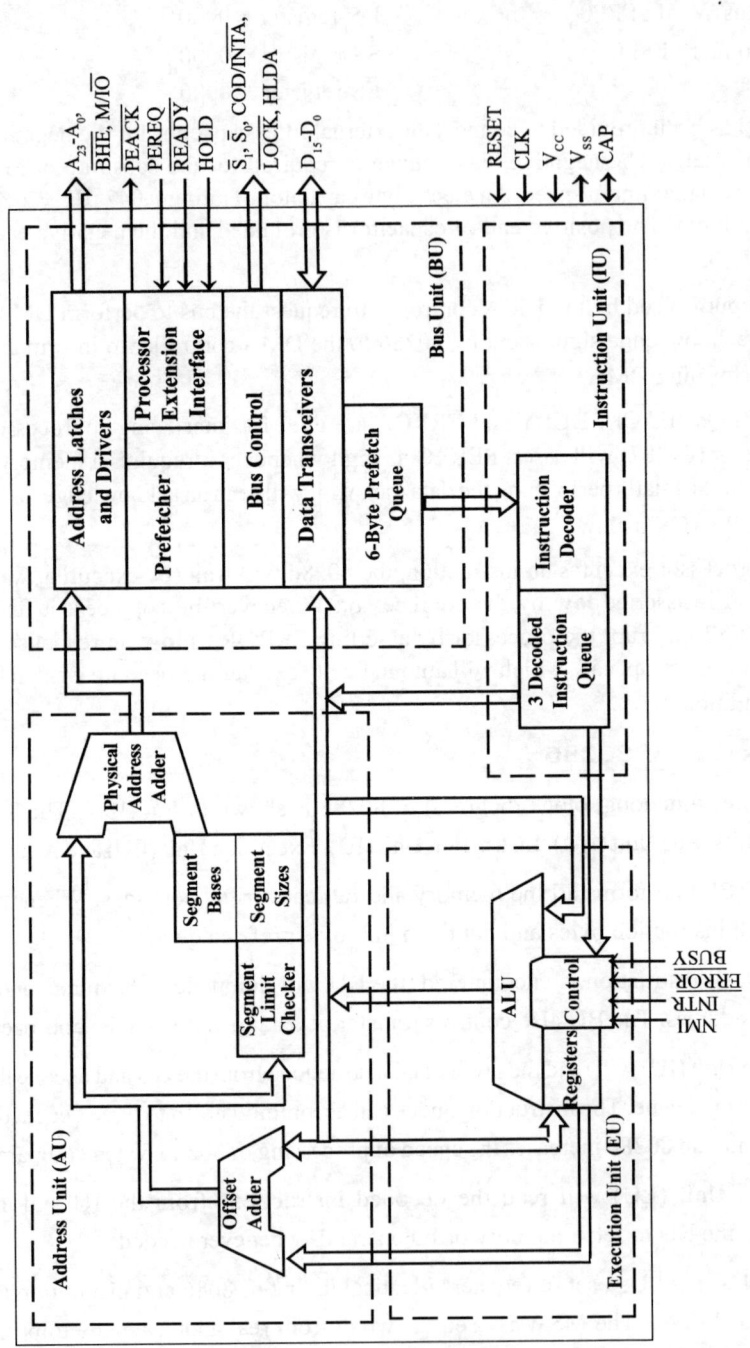

Fig. 10.4: Functional block diagram of an INTEL 80286 microprocessor.

The general-purpose registers available in the EU of 80286 are AX, BX, CX, DX, SP, BP, SI and DI. The functions of these registers are same as that of 8086. Also, the EU has a 16-bit **I**nstruction **P**ointer (IP) and a 16-bit flag register. The format of the flag register of 80286 is shown in Fig. 10.5.

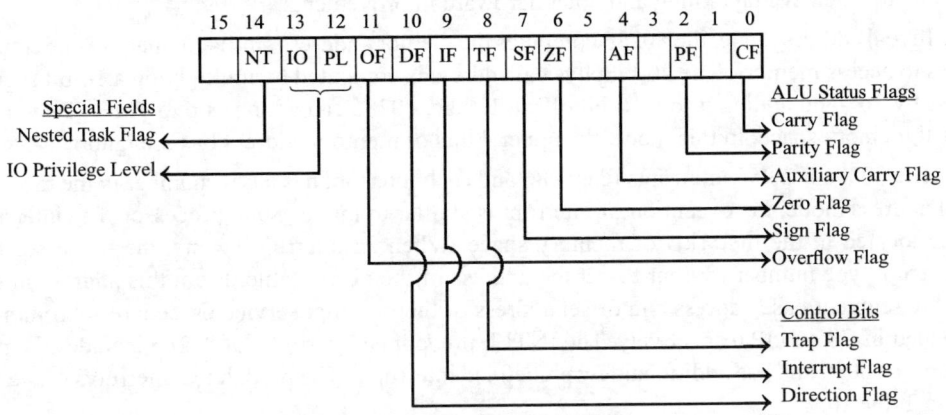

Fig. 10.5: Flag register of 80286.

The **A**ddress **U**nit (AU) takes care of computing physical address of memory or IO and sends the address to the BU. The 80286 can operate either in the real address mode or protected virtual address mode. After a reset, the 80286 processor will work in the real address mode. In the real address mode, the address computation is similar to that of 8086 and the processor uses only the lower 20 bits of the address. Therefore, the maximum physical address space in this mode is only 1MB. The AU has four numbers of 16-bit segment registers like 8086 and they are **C**ode **S**egment (CS), **D**ata **S**egment (DS), **S**tack **S**egment (SS) and **E**xtra **S**egment (ES) registers. The program/code address is computed by multiplying the content of the CS by 16_{10} and then adding it to 16-bit offset available in IP. The data address is computed by multiplying the content of the DS/ES by 16_{10} and then adding it to a 16-bit offset specified by the instruction. The stack address is computed by multiplying the content of SS by 16_{10} and then adding to a 16-bit offset specified by the instruction.

The processor can be switched from real address mode to **P**rotected **V**irtual **A**ddress **M**ode (PVAM) by loading an appropriate word in MSW register. In PVAM, the address unit functions as a complete **M**emory **M**anagement **U**nit (MMU). In PVAM, the 80286 uses all the 24 address lines and can access up to 16 MB (2^{24} = 16 Mega) of physical memory address space and 1GB of virtual address space.

In PVAM mode, each memory location is represented by a 32-bit virtual address (or logical address). The MMU translates the 32-bit virtual address to a 24-bit physical address. The 32-bit virtual address has a 16-bit selector and 16-bit offset. The 16-bit selector is used to fetch a descriptor from a descriptor table. The descriptor contains a 24-bit segment base address which is added to the 16-bit offset to get a 24-bit physical address.

> **Note:** *The 80286 has a 13-bit index for descriptor and allows two descriptor tables (Global Descriptor Table [GDT] and Local Descriptor Table [LDT]). Therefore, the processor allows $2^{13} \times 2 = 2^{14} = 16\,k$ descriptors. Each descriptor can define a segment of 1kB size to 64 kB. Hence, total virtual space is $16\,k \times 64\,kB = 2^{14} \times 2^{16}$ bytes $= 2^{10} \times 2^{20}$ bytes $= 1024 \times 1MB = 1GB$.*

10.3.3 Real Address Mode of 80286

After a reset, the 80286 processor will work in real address mode. The working of 80286 in real address mode is similar to that of 8086 in maximum mode. In real address mode, the 80286 will directly execute the 8086 machine code programs with only minor modifications. In 80286, the execution will be faster due to extensive pipelining and other hardware improvements.

In real address mode, the 80286 computes the memory address similar to that of 8086. It uses 20-bit address to access memory. The 20-bit physical address is computed by multiplying a 16-bit segment base address by 16_{10} and adding it to a 16-bit offset. Hence, 80286 can address up to 1MB ($2^{20} = 1$ Mega) of physical memory space. In this mode, the upper 4 bits of memory address bus are ignored.

The 80286 has 256 interrupts like 8086 and each interrupt has a type number in the range 0 to 255. In real address mode, the execution of interrupt is similar to that of 8086 processor. The interrupt vector table is located in the first 1kB of memory space. When an interrupt occurs, the processor multiplies the interrupt type number by four to get the address of the vector table. From this address in the vector table, the segment base address and offset address of the interrupt service procedure/subroutine are read and loaded in CS and IP respectively. The INTEL predefined interrupts of 80286 includes the predefined interrupts of 8086 and has additional few predefined interrupts. The predefined interrupts of the 80286 are listed in Table 10.9

The real address mode is mainly used to initialize peripheral devices, load the main part of the operating system from disk into memory, enable interrupts and enter the protected virtual address mode.

Table 10.9: Predefined Interrupts of 80286

Interrupt	Type number
Divide error	0
Single step	1
NMI	2
Breakpoint	3
Interrupt on overflow	4
Array bounds	5
Invalid opcode	6
Processor extension not available	7
Interrupt table limit too small	8
Processor extension segment overrun	9
Invalid task state segment	10
Segment not present	11
Stack segment overrun or not present	12
Segment overrun	13
Processor extension error	16

10.3.4 Protected Virtual Address Mode of 80286

After a reset, the 80286 processor will work in real address mode and it can be made to work in **Protected Virtual Address Mode** (PVAM) by loading an appropriate word in the **Machine Status Word** (MSW) register. The format of the MSW is shown in Fig. 10.6. In order to enter PVAM, frame a word such that the most significant bit (PE) is one (i.e., to enter PVAM, PE bit is set to one) and load this word in a register/memory, and then execute the instruction LMSW (Load **MSW**). Once the processor enters PVAM, the only way to get back to the real address mode is by resetting the system. In PVAM, the **Address**

Unit (AU) functions as a **Memory Management Unit (MMU)**. In PVAM, 24-bit address is used to access memory and this address is computed by adding a 24-bit segment base address and a 16-bit offset. The segment base address is obtained from a descriptor, which is stored in the descriptor table.

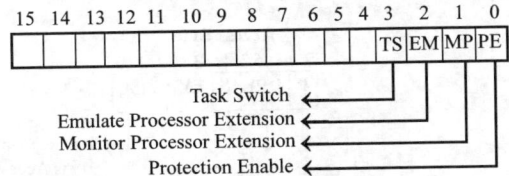

Fig. 10.6: Format of the Machine Status Word (MSW).

Memory Management in PVAM

In PVAM, the 80286 allows the user to create memory segments of length/size 1kB to 64 kB for each task/program. A size or limit is given to each segment when it is created. The 80286 allows 64 k memory segments and all these segments are not available in the physical memory space at the same time. Hence, these segments are called logical/virtual segments. The segments currently being used by a task/program are kept in physical memory. The segments which are not currently used will be in secondary memory like hard disk and they are brought to physical memory whenever needed.

In PVAM, the user has to provide a 32-bit virtual address for memory. The lower word (lower 16-bit) of virtual address is offset address and the upper word (upper 16-bit) is called selector, which is loaded in segment register. The 16-bit selector is used to fetch an 8-byte descriptor from the descriptor table. The descriptor contains the 24-bit segment base address which is added to the 16-bit offset to get a 24-bit physical address. Thus, address calculation in PVAM involves selector, descriptor and descriptor table.

Selector

When a program is assembled for execution on an 80286 processor in PVAM, each segment is assigned a unique 16-bit selector. The format of the selector is shown in Fig. 10.7.

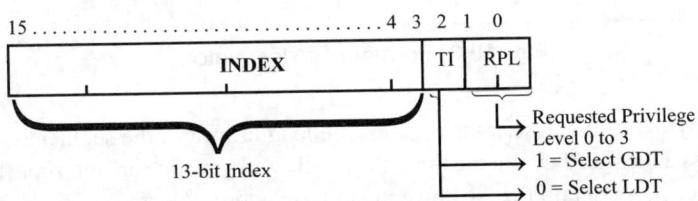

Fig. 10.7: Format of the selector.

The 2-bit RPL (**R**equested **P**rivilege **L**evel) field is used by operating system for implementing the 80286's protection features and is not used in address calculation. The 1-bit TI field is used to select one of the two descriptor tables. When TI field is one, **G**lobal **D**escriptor **T**able (GDT) is selected and when TI field is zero, **L**ocal **D**escriptor **T**able (LDT) is selected. The 13-bit index is used to create $2^{13} = 8192_{10}$ = 8 k descriptors in each table. The 13-bit index is multiplied by 8 to get the address of the descriptor in a descriptor table.

Descriptor

When a program is assembled for execution on an 80286 processor in PVAM, a unique descriptor is produced for each segment. The length of the descriptor is 8-byte. The format of descriptor is shown in Fig. 10.8.

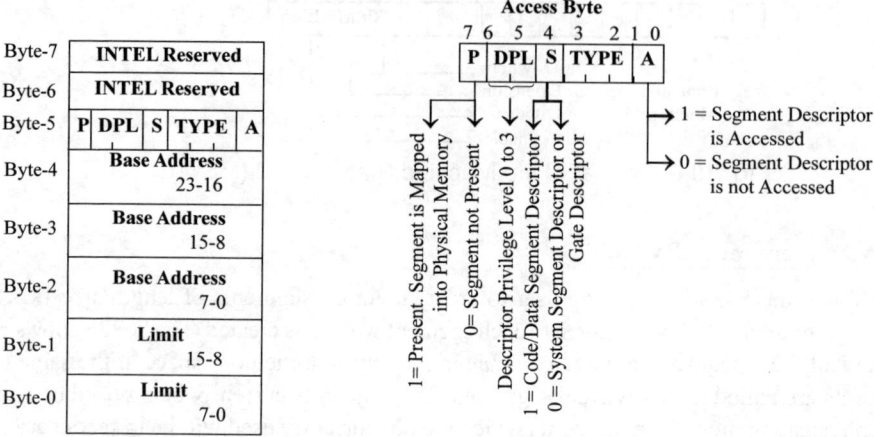

Fig. 10.8: Format of a descriptor.

The byte-0 and byte-1 of the descriptor contains the length/limit of the segment in bytes. When an attempt is made to access a location beyond the limit, the MMU will generate an interrupt. The byte-2, byte-3 and byte-4 of the descriptor contain the 24-bit segment base address. Byte-5 of the descriptor is called the access byte and it contains information regarding privilege level, access right and the type of segment. Byte-6 and byte-7 are reserved for future expansion and should be filled with zero for compatibility with a higher version of 80x86 family.

Descriptor table

In 80286-based system, the descriptors are stored in descriptor tables. There are two types of descriptor tables: **G**lobal **D**escriptor **T**able (GDT) and **L**ocal **D**escriptor **T**able (LDT). The descriptor tables are created by system software and stored in memory. The processor can read the descriptors from the tables in memory, whenever needed.

The GDT contains the descriptors for the operating system segments and descriptors for segments which need to be accessed by all user tasks. The LDT contains descriptors for each task. A four level protection scheme can be used to protect the operating system descriptors in the GDT from unauthorized access by user tasks.

Address Translation in 80286

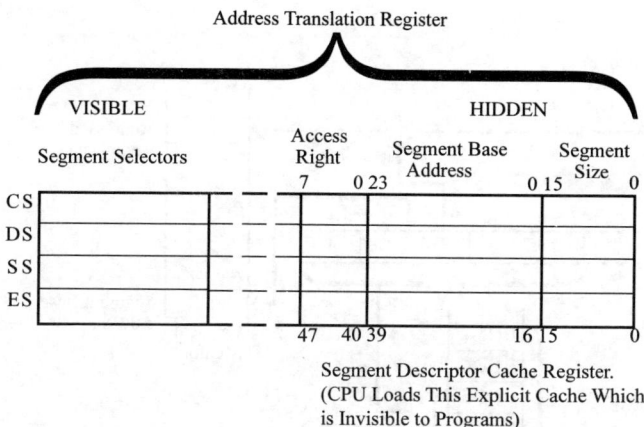

Fig. 10.9: Address translation register.

The 80286 has four numbers of segment descriptor cache registers which are invisible to the programmer. The size of these register is 48 bits (so that it can hold the 6-byte descriptor).

When the 80286 is operating in PVAM, descriptors must be copied to the processor from tables in memory and loaded in the invisible segment descriptor cache registers. The descriptors are then used for producing and checking physical addresses.

The 80286 also has internal registers for storing base addresses of descriptor tables and they are GDTR and LDTR. The 80286 keeps the base addresses and limits for the descriptor tables currently in use in these internal registers.

The **G**lobal **D**escriptor **T**able **R**egister (GDTR) contains the 24-bit base address and limit for the table containing the global address space description. This register is initialized with LGDT (Load **GDT**) instruction when the system is booted.

The **L**ocal **D**escriptor **T**able **R**egister (LDTR) in the 80286 contains the base address and limit of the local descriptor table for the task currently being executed. The LLDT(Load **LDT**) instruction is used to load this register when the system is booted.

Each address in the PVAM is represented by a 32-bit virtual address. The virtual address consists of a 16-bit selector and 16-bit offset. To access a segment, the selector for that segment is loaded into the visible part of the appropriate segment register in the 80286. The 80286 then automatically multiplies the index value of the selector by 8 and adds the result to a descriptor table base address in its GDTR or LDTR. The 80286 then fetches the segment descriptor from the resultant address in a descriptor table. Byte-0 to byte-5 of the descriptor is loaded into an invisible segment descriptor cache register. The descriptor has a 24-bit base address. This base address is added directly to the offset part of the virtual address to produce the physical address of the desired byte or word. The conversion of virtual address to physical address is diagramatically shown in Fig. 10.10.

In a system which always runs the same program, the physical base addresses in descriptors are fixed by program development tools when the program is built. In a general purpose system which runs the same program, the physical base addresses in the descriptors may be changed by the operating system when a program is loaded into the memory to be run.

This is done so that the program can be loaded into available memory without disturbing the programs or tasks already present.

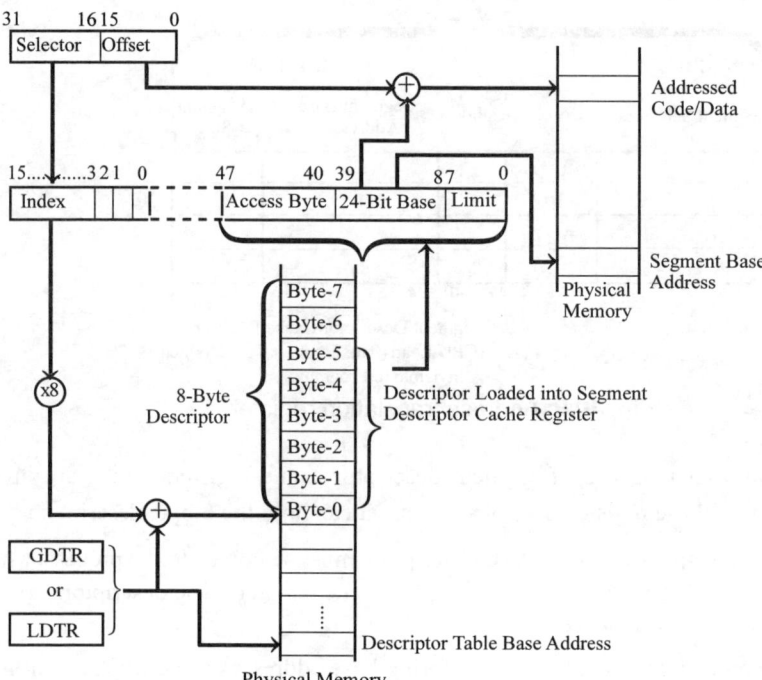

Fig. 10.10: Translation of 32-bit virtual address to 24-bit physical address in 80286.

10.4 INTEL 80386 MICROPROCESSOR

The 80386 is a 32-bit microprocessor and it is an improved version of 80286 with software compatibility with 8086, 80186 and 80286. The major improvements in 80386 over 80286 are the following:

1. **The processor registers and ALU are 32-bit wide and the instruction set is extended to support 32-bit addresses and data.**

2. **The main memory and the data path to memory can be 32-bit wide, so instructions and data read/write operations will be two times faster.**

3. **The maximum size of physical memory is extended from 16 MB (2^{24} bytes) to 4GB (2^{32} bytes).**

4. **Since 80386 runs at higher clock frequency, faster execution speed is obtained and most instructions take fewer clock cycles to execute.**

5. **The on-chip memory management supports paging.**

The 80386 is available in two versions: 80386DX and 80386SX. The internal architecture of both the versions of 80386 are same, but they differ only in external address and data bus. The 80386DX has separate external 32-bit data bus and address bus.

The 80386SX has external 16-bit data bus and 24-bit address bus. The 80386DX is called the full version of 80386 and 80386SX is called the reduced bus version of 80386. The 80386SX was developed after the 80386DX for applications that did not require an external 32-bit bus and at the same time had the advantage of internal 32-bit computation.

The 80386SX can address up to 16 MB (2^{24} = 16 M) of physical memory space and memory is organized as two banks of 8 MB. The 80386DX can address up to 4 GB (2^{32} = 4G) of physical memory space and memory is organized as four banks of 1GB. Both 80386SX and 80386DX have virtual address space of 64 Tera bytes (2^{46} = 64T).

The 80386 can work in three modes : real address mode, protected virtual address mode and virtual 8086 mode. The 80386 processor will enter real address mode after a hardware reset and in this mode it works as fast as an 8086 processor with a few additional new instructions. The real address mode is mainly used for initialization and enters into the **P**rotected **V**irtual **A**ddress **M**ode (PVAM).

In PVAM, the 80386 works as a 32-bit processor and all instructions and features of 80386 are available in this mode. While working in PVAM, the processor can switch to **V**irtual 80**86** (V86) mode to run 8086 applications and then return to PVAM. In V86 mode, the processor can run 8086 applications with protection features of 80386.

The 80386 is available with maximum clock speed rating of 12.5, 16, 20, 25 or 33 MHz.

10.4.1 Pins and Signals of 80386

The 80386DX is a 132-pin IC available in **P**in **G**rid **A**rray (PGA) package. The pin configuration of an 80386DX is shown in Fig. 10.11. The 80386SX is a 100-pin IC available in plastic quad flatpack package. The pin configuration of 80386SX is shown in Fig. 10.12. The signals of 80386DX are listed in Table 10.10.

The 80386DX has 32 pins for data transfer from/to memory or IO. It can access byte/word/double word from memory or IO in one bus cycle.

The 80386DX has 30 pins for addressing 1 GB (2^{30} = 1G) of physical address space and four byte/ bank enable signals $\overline{BE_0}$ to $\overline{BE_3}$ to enable four memory banks, each of size 1GB. The signals $\overline{BE_0}$ to $\overline{BE_3}$ are generated internally by decoding the address bits A_0 and A_1.

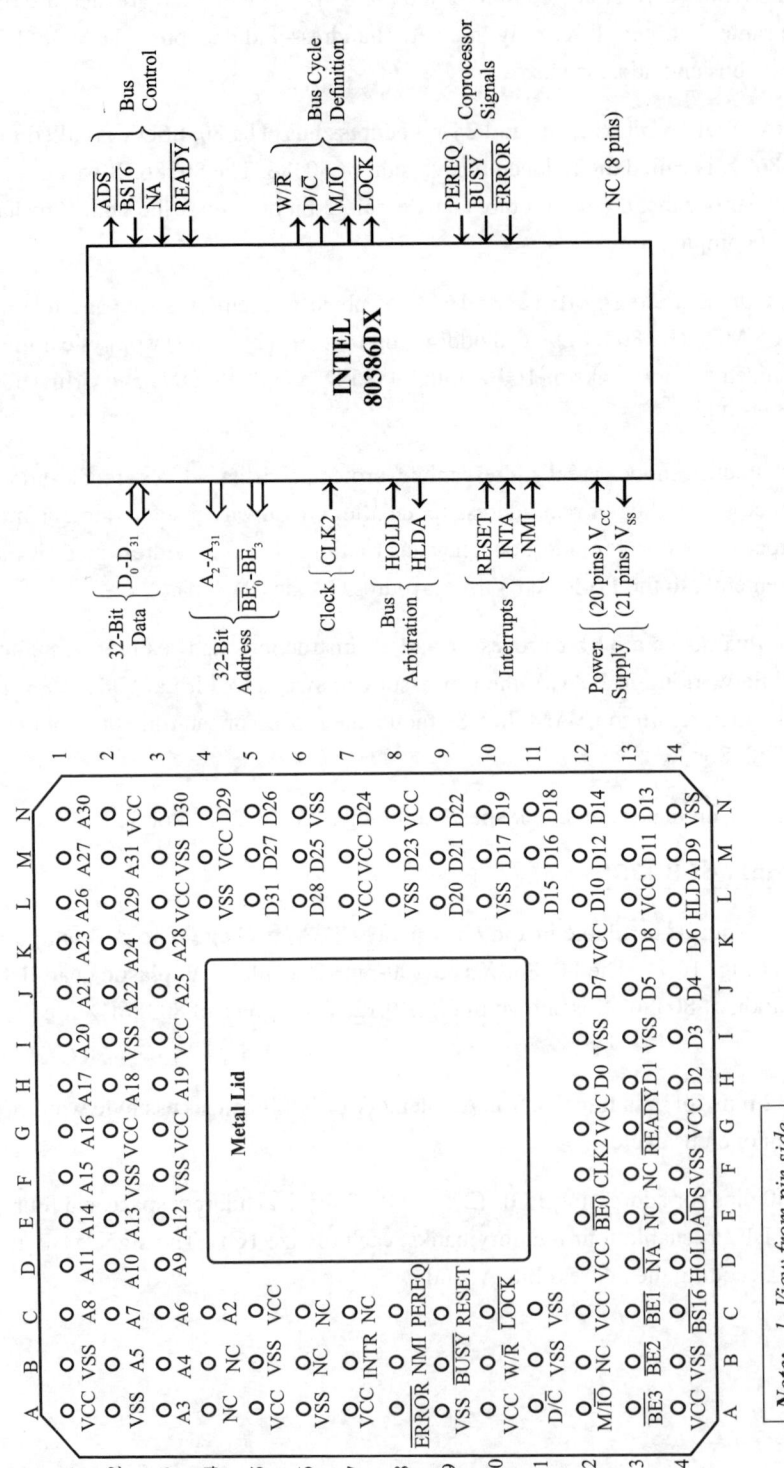

Fig. 10.11: Pin description of INTEL 80386DX microprocessor.

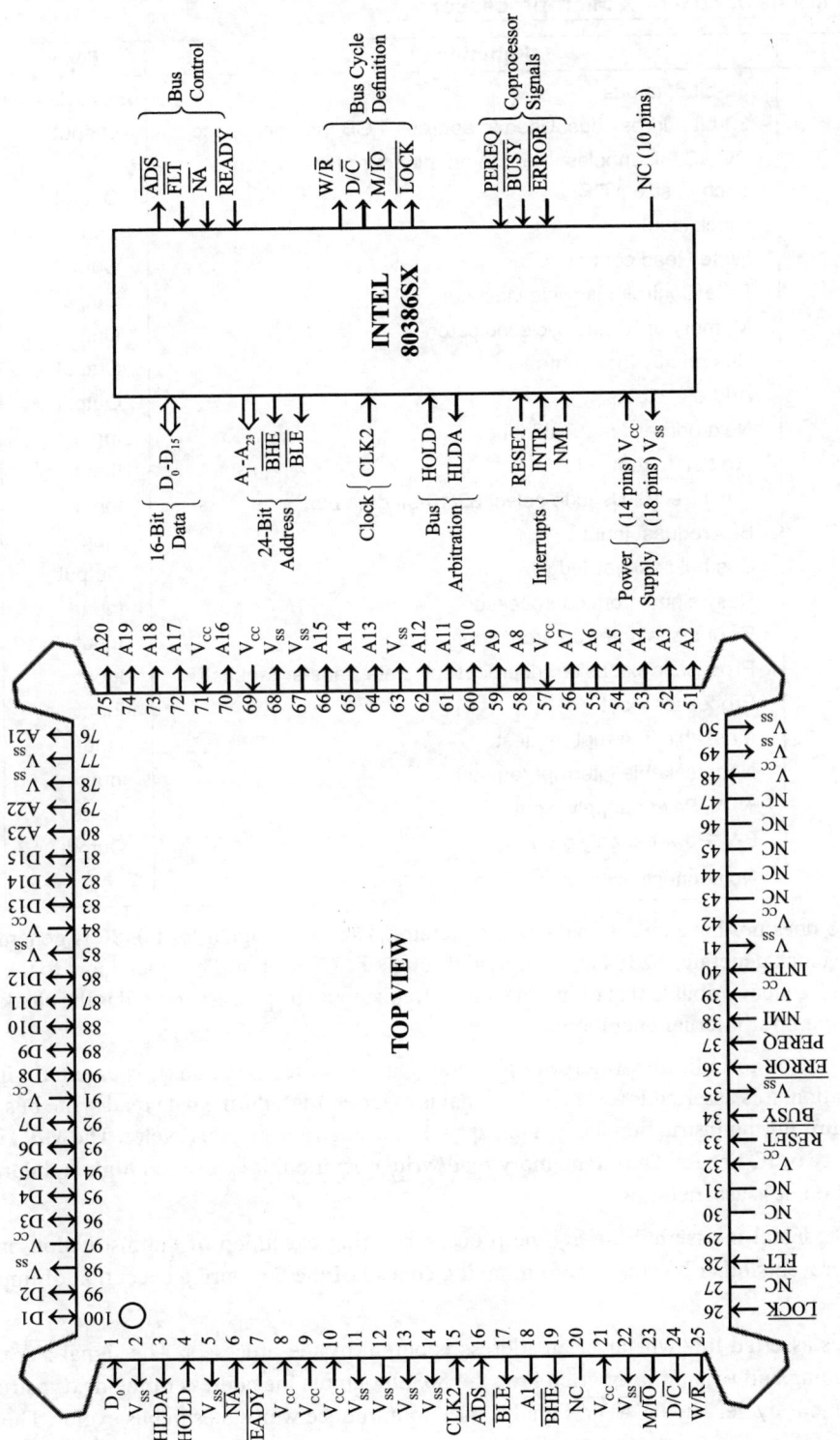

Fig. 10.12: Pin description of INTEL 80386SX microprocessor.

Table 10.10: Signals of 80386DX Microprpcessor

Signal	Function	Type
D_{31}-D_0	32-bit data bus	Bidirectional
A_{31}-A_2	30-bit address bus. Used to address 1 GB memory space	Output
BE_3-BE_0	Byte/Bank enables. Used to enable four memory banks each of size 1 GB	Output
CLK2	Clock input	Input
W/$\overline{R}$	Write/Read control	Output
D/$\overline{C}$	Data/Control bus cycle indicator	Output
M/$\overline{IO}$	Memory or IO bus cycle indicator	Output
$\overline{LOCK}$	Bus priority lock control	Output
$\overline{ADS}$	Address status	Output
$\overline{NA}$	Next address request input	Input
$\overline{READY}$	Wait state request	Input
$\overline{BS16}$	Bus size 16. Used to select 32/16-bit data bus	Input
HOLD	Bus request input	Input
HLDA	Bus hold acknowledge	Output
$\overline{BUSY}$	Busy signal from coprocessor	Input
$\overline{ERROR}$	Error signal/Interrupt from coprocessor	Input
PEREQ	Processor extension (coprocessor) data transfer request	Input
RESET	Processor reset	Input
INTR	Maskable interrupt request	Input
NMI	Nonmaskable interrupt request	Input
V_{cc}	+5-V, Power supply input	Input
V_{ss}	0-V, Power supply ground	Output
NC	No connection	-

The 80386 does not have an internal clock generator. The clock signal for 80386 is generated by using an external clock generator 82384 and supplied through the CLK2 pin. The clock input to the CLK2 pin should have a frequency double that of internal clock frequency. The processor divides the clock signal by two and uses them for internal operations.

The W/$\overline{R}$ is used to indicate write/read operation. During write operation, it is asserted **high** and during read operation, it is asserted **low**. The D/$\overline{C}$ signal is asserted **high** during data read/write bus cycles and it is asserted **low** during instruction fetch, interrupt acknowledge and halt bus cycles. The M/$\overline{IO}$ is used to indicate memory or IO access. During memory read/write operation, it is asserted **high** and during IO read/write operation, it is asserted **low**.

The $\overline{LOCK}$ signal is asserted **low** by the processor during execution of the instruction prefixed by LOCK. This prevents other bus masters from taking control of the bus during execution of important instructions.

The $\overline{ADS}$ is asserted **low** whenever an address is output by the processor. The signal $\overline{NA}$ is used for address pipelining and when it is asserted **low**, the 80386 outputs the address of the next instruction/ data in the current bus cycle. The $\overline{READY}$ signal is used to introduce wait states in bus cycles. The $\overline{BS16}$

is used to interface a 16-bit bus to a 32-bit bus. It is tied to logic **low** for 16-bit external data bus and tied to logic **high** for 32-bit external data bus.

The HOLD and HLDA signals are used for DMA data transfer. The signals $\overline{BUSY}$, $\overline{ERROR}$ and PEREQ are used for 80387 coprocessor interface. The functions of these pins are similar to that of 80286.

The 80386 has logic **high** RESET input. Whenever the processor is reset, it is initialized to execute instructions from the memory location FFFF FFF0$_H$ in the real mode. The 80386 has one maskable interrupt request input INTR, which can be expanded using interrupt controller 8259 and one non-maskable interrupt request input, NMI.

10.4.2 Architecture of 80386

The architecture of 80386 is shown in Fig. 10.13. It has a highly pipelined architecture with six functional units operating in parallel. In 80386 processor fetching, decoding, execution, memory management (address computation) and bus access for several instructions are performed simultaneously. The six functional units of 80386 are:

1. **Bus interface unit**
2. **Instruction prefetch unit**
3. **Instruction decode unit**

4. **Execution unit**
5. **Segmentation unit**
6. **Paging unit**

The bus interface unit generates signals for memory and IO interface. This unit generates the control signals for the current bus cycle based on internal requests for fetching instructions from the instruction prefetch unit and transferring data from the execution unit. The physical address of memory and IO are output through the bus interface unit and also the data transfer to the external world takes place through this unit.

Whenever the bus is free, the instructions are fetched and stored in the 16-byte code queue in the instruction prefetch unit. The instruction decode unit reads the instructions from the prefetch unit and decodes them. The decoded instructions are then stored in the queue in decode unit. The queue in the decode unit can accommodate three decoded instructions. The queues in the instruction prefetch and decode units are FIFO queues.

The execution unit reads the decoded instructions from the decode unit and processes them. The execution unit consists of a control unit, data unit and a protection test unit. The control unit contains the microcode stored in the ROM for the instructions, flag register and generates internal control signals. The data unit includes an ALU, a 64-bit barrel shifter and eight general purpose registers. The data unit performs the operations requested by the control unit. The barrel shifter is used for fast shift, rotate, multiply and divide operations. The protection test unit checks for segmentation violations under the control of the microcode.

The segmentation and paging units can be considered as MMU (**M**emory **M**anagement **U**nit) of 80386. In real address mode, the paging is disabled and the segmentation unit computes the address similar to that of 8086 or 80286 in real mode. In **P**rotected **V**irtual **A**ddress **M**ode (PVAM) the segmentation unit computes a 32-bit linear address using an offset, selector and descriptor (similar to that of 80286). In PVAM mode, each memory location is represented by a 48-bit virtual address. The MMU translates the 48-bit virtual address to the 32-bit physical address. The 48-bit virtual address has a 16-bit selector and 32-bit offset. The 16-bit selector is used to fetch a descriptor from a descriptor table. The descriptor contains a 32-bit segment base address which is added to a 32-bit offset to get a 32-bit linear address. If the paging unit is not enabled, then this linear address is used as physical address. If paging is enabled, then the paging unit will translate the 32-bit linear address to a 32-bit physical address. The paging provides an additional memory management mechanism to handle very large segments.

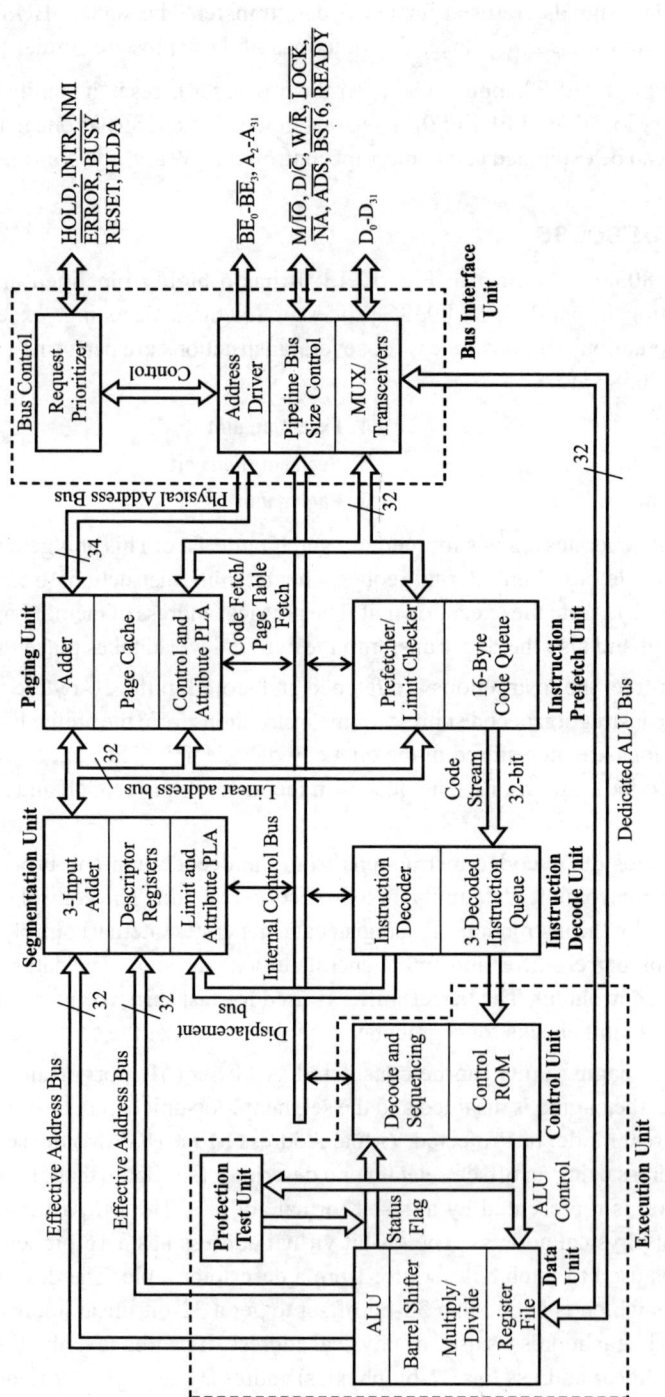

Fig. 10.13: Architecture of an 80386 microprocessor.

> *Note: The 80386 has 13-bit index for descriptor and allows two descriptor table (Global Descriptor Table [GDT] and Local Descriptor Table [LDT]). Therefore, the processor allows $2^{13} \times 2 = 2^{14} = 16\,k$ descriptors. Each descriptor can define a segment of maximum size 4GB. Hence, total virtual address space is $16k \times 4GB = 64TB(terabyte)$.*

10.4.3 Registers of 80386 Microprocessor

The 80386 processor has 32 internal registers and they can be classified into following seven categories:

1. **General purpose registers**
2. **Segment registers**
3. **Instruction pointer and flag register**
4. **Control registers**
5. **System address and segment registers**
6. **Debug registers**
7. **Test registers**

Most of the 80386 registers are 32-bit registers. The registers of 80386 are a superset of 8086, 80186 and 80286 registers and so all the registers of 8086, 80186 and 80286 are contained within the 32-bit registers of 80386. The general purpose registers, segment registers, instruction pointer and flag register are called base architecture registers. Figure 10.14 shows the base architecture registers of 80386.

General Purpose Registers

The 80386 has eight numbers of 32-bit general purpose registers and they are EAX, EBX, ECX, EDX, ESI, EDI, EBP and ESP. The least significant 16 bits of these registers can be accessed separately with their 16-bit names AX, BX, CX, DX, SI, DI, BP and SP.

Each of the 16-bit registers AX, BX, CX and DX can be accessed as two numbers of 8-bit registers. The lower 8 bits of these registers can be accessed with the name AL, BL, CL and DL. The higher 8 bits of these registers can be accessed with the name AH, BH, CH and DH.

Segment Register

The 80386 has six segment registers to address six segments of memory at any given time. The segment registers of 80386 includes the four segment registers (CS, SS, DS and ES) of 80286 and has two additional data segment registers FS and GS registers.

The four data segment registers DS, ES, FS and GS registers can be used to access four separate data areas and allow programs to access different types of data structures. In real address mode, the segment registers will hold the segment base address like that of 8086.

In PVAM, the segment registers will hold the selectors like that of 80286. The selector in CS and SS registers indicates the current code and stack segment respectively. The selectors in DS, ES, FS and GS registers indicate the current data segments.

Each segment register has a program invisible 64-bit register called segment descriptor register. These registers are used to hold the 8-byte descriptor of the current memory segment in PVAM.

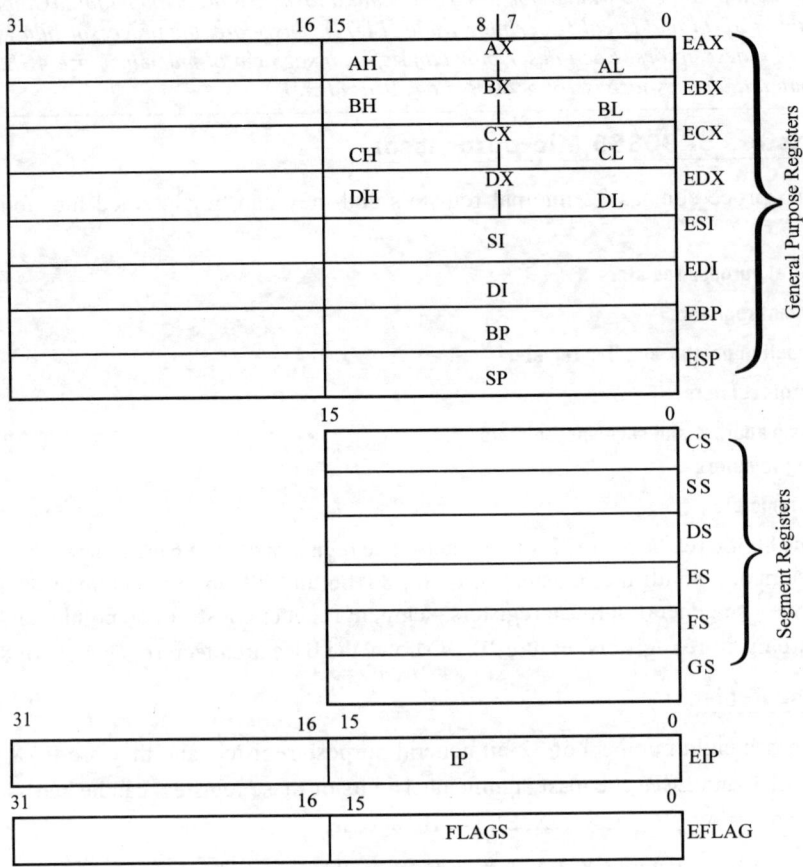

Fig. 10.14: Base architecture registers of an 80386.

Instruction Pointer and Flag Register

The 80386 has a 32-bit instruction pointer and it is named as EIP. It is used to hold a 32-bit offset. The lower 16 bits of EIP is called 16-bit instruction pointer, IP, which is used in real address mode.

The 80386 has a 32-bit flag register and it is called EFLAG. The format of flag register of 80386 is shown in Fig. 10.15. The flags of 80386 can be classified into three groups: status flags, control flags and system flags. The status flags are CF, AF, PF, ZF, SF and OF.

The control flags are TF, DF and IF. (The status and control flags are same as that of 8086.) The system flags are ID, VIP, VIF, AC, VM, RF, NT and IOPL. The system flags control IO access, maskable interrupt, debugging, task switching and operating mode switching.

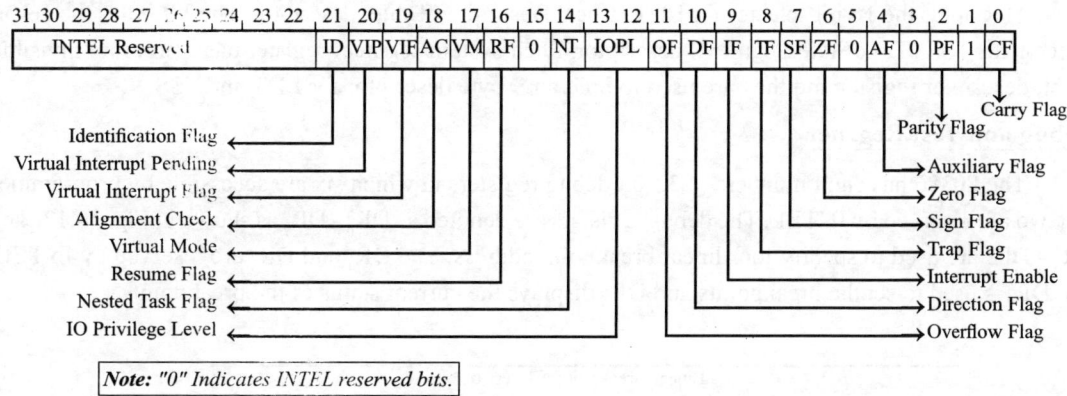

Note: "0" Indicates INTEL reserved bits.

Fig. 10.15: Flag register of 80386 microprocessor.

Control Registers

The 80386 has four numbers of 32-bit control registers CR0, CR1, CR2 and CR3. The control registers of 80386 are shown in Fig. 10.16. The lower 16-bit of CR0 is the **Machine Status Word** (MSW) similar to MSW of 80286. The most significant bit of CR0 is used to enable/disable paging. The CR1 is reserved for future expansion. The CR2 holds the 32-bit linear address of the last page accessed before a page fault interrupt. The CR3 holds the 32-bit base address of the page directory. Since the page directory table is always page aligned (4 kB - aligned), the lower 12 bits of CR3 are undefined.

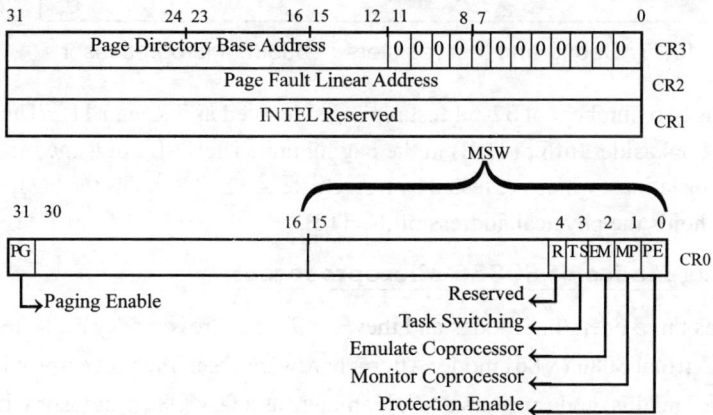

Fig. 10.16: Control registers of 80386 microprocessor.

System Address and Segment Registers

The 80386 has two numbers of 32-bit system address registers and two numbers of 16-bit system segment registers. The system address registers are **G**lobal **D**escriptor **T**able **R**egister (GDTR) and **I**nterrupt **D**escriptor **T**able **R**egister (IDTR). The GDTR holds the 32-bit base address and 16-bit limit of GDT (**G**lobal **D**escriptor **T**able). The IDTR holds the 32-bit base address and 16-bit limit of IDT (**I**nterrupt **D**escriptor **T**able).

The system segment registers are Local Descriptor Table Register (LDTR) and Task Register (TR). The LDTR holds the 16-bit selector for LDT (Local Descriptor Table) descriptor. The TR holds the 16-bit selector for TSS (Task State Segment) descriptor. Each system segment register has a program invisible 64-bit descriptor register and they are used to hold an 8-byte descriptor for LDT and TSS.

Debug and Test Register

The 80386 has eight numbers of 32-bit debug registers in which six are accessible by programmer and two are reserved by INTEL. The debug registers are denoted as DR_0 - DR_7 as shown in Fig. 10.17. The DR_0 - DR_3 are used to specify four linear breakpoint address. The DR_4 and DR_5 are reserved by INTEL. The DR_7 is used to set the breakpoints and DR_6 displays the current status of the breakpoints.

31	0	
Linear Breakpoint Address 0		DR_0
Linear Breakpoint Address 1		DR_1
Linear Breakpoint Address 2		DR_2
Linear Breakpoint Address 3		DR_3
INTEL Reserved		DR_4
INTEL Reserved		DR_5
Breakpoint Status		DR_6
Breakpoint Control		DR_7

31	0	
Test Control		TR_6
Test Status		TR_7

Fig. 10.17: Debug and test registers of 80386 microprocessor.

The 80386 has two numbers of 32-bit test registers denoted as TR_6 and TR_7. The test registers will test the Translation Lookaside Buffer (TLB) in the paging unit. The TLB holds the most commonly used page table address translations, which are tested by test registers. The TR_6 holds the tag field (linear address) of the TLB and TR_7 holds the physical address of the TLB.

10.4.4 Operating Modes of 80386 Microprocessor

The 80386 has three operating modes and they are Real address mode, Protected Virtual Address Mode (PVAM) and Virtual 8086 (V86) mode. After a hardware reset, the processor will start working in the real address mode and this mode appears to programmers as a fast 8086 processor with a few additional new instructions.

The main purpose of real address mode is to initialize the system for protected virtual address mode of operation. In PVAM, the 80386 works as a normal 32-bit processor, and all the instructions and features are available in this mode. The V86 mode allows the programmer to execute 8086 applications without disturbing the protection mechanism of PVAM. The 80386 processor can switch from PVAM to V86 mode to execute 8086 applications and then can return to PVAM.

Real Address Mode

The 80386 processor will enter the real address mode when it is reset. This mode is similar to 8086 but allows access to the 32-bit registers. In real address mode, the default operand size is 16 bits and so override prefixes should be employed to use 32-bit registers and addressing modes.

In real address mode, paging is disabled and the processor can access 1 MB of physical memory space like 8086 using a 20-bit address. Therefore, the address lines A_2 - A_{19} alone are active. The 20-bit physical address is computed by multiplying the segment register by 16_{10} and adding to an offset.

> **Note:** *Multiplying by 16_{10} is equivalent to shifting four times left.*

In real address mode, the size of memory segments are 64 kB and the segments can be overlapped like 8086. Since segment size is 64kB, when 32-bit effective address or offset is employed it should be less than 0000 FFFF$_H$. In real address mode, two memory areas are reserved.

One for system initialization and the other for interrupt pointer table. The memory address 00000$_H$ to 003FF$_H$ are reserved for interrupt pointer table and the memory address FFFF0$_H$ to FFFFF$_H$ are reserved for system initialization.

All the instructions of 80386, except a few can be executed in real address mode. The primary purpose of real address mode is to set up the 80386 processor for protected virtual address mode of operation.

Protected Virtual Address Mode

The processor can switch from the real address mode to PVAM by setting the PE bit (**Protection Enable bit**) in the control register CR_0. In PVAM, the processor can run all the 8086 and 80286 programs with sophisticated memory management and hardware assisted protection mechanism.

In PVAM, the processor has a very large address space. It has 4 GB of physical memory address space and 64 TB of virtual memory address space. The address lines A_2 - A_{31} along with four bank select signals BE_0 - BE_3 are used to address 4 GB of physical memory.

In PVAM, each physical address is represented by a 48-bit virtual address. The 80386 supports two method of converting a virtual address to physical address. In one method, the paging is disabled and the physical address is computed using the selector, descriptor and offset like 80286. In another method, the paging is enabled and a linear address is computed using the selector, descriptor and offset like 80286, and then the linear address is converted to physical address by the paging unit.

The physical address computation in 80386, when paging unit is disabled is shown in Fig. 10.18. The virtual address (pointer) can be 48-bit or 32-bit. The 48-bit pointer has a 16-bit selector and 32-bit offset. The 32-bit pointer has 16-bit selector and a 16-bit offset.

The selector is used to fetch an 8-byte segment descriptor from the descriptor table. The descriptor has a 32-bit segment base address, segment length /limit, protection level, privilege level, the default operand size and segment type. The sum of segment base address and offset gives the 32-bit physical address. The 80386 allows a maximum size of 4 GB to each segment.

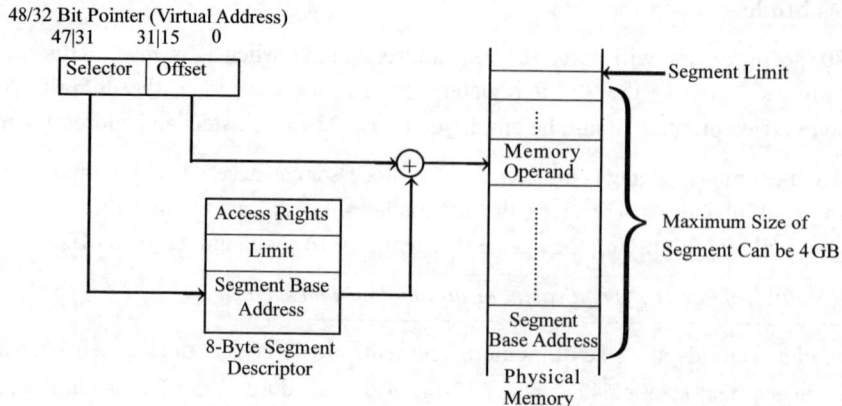

Fig. 10.18: Address computation in PVAM of 80386 when paging is disabled.

The physical address computation in 80386 when paging is enabled is shown in Fig. 10.19. When paging is enabled the 80386 processor provides additional memory management mechanism which is available only in PVAM. The paging provides a method for managing very large memory segments defined in PVAM. The paging divides each segment into equal sized pages. When paging is enabled, the segmentation unit will generate a 32-bit linear address and supply it to the paging unit, which translates this linear address to a 32-bit physical address.

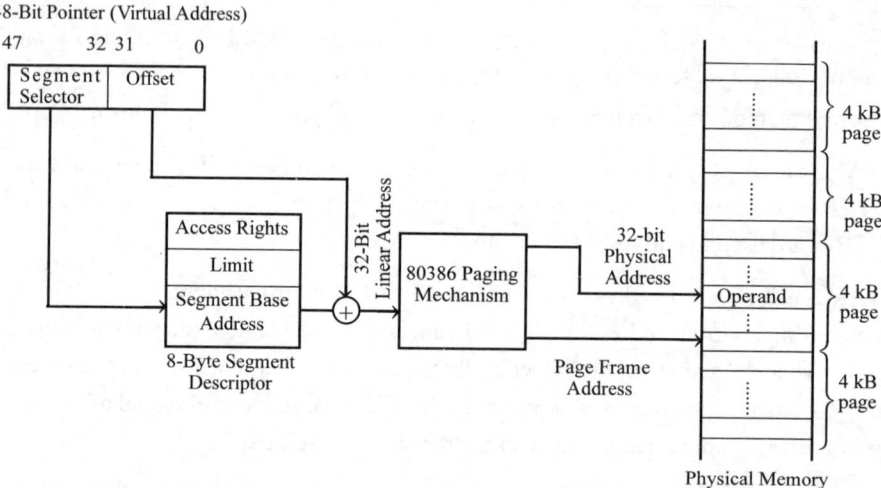

Fig. 10.19: Address computation in PVAM of 80386 when paging is enabled.

When paging in enabled the segmentation unit computes a 32-bit linear address using a selector, descriptor and offset. (This computation is similar to the computation of physical address when paging is disabled.) In fact, the output of segmentation is used as a physical address when paging is disabled and when paging is enabled the output of segmentation unit is fed as input to the paging unit.

In paging mechanism, the memory segments can be organized as pages of size 4 kB. The 80386 paging mechanism allows 2^{20} pages ($1024 \times 1024 = 2^{10} \times 2^{10}$) of size 4kB. The paging mechanism involves three elements and they are page directory, page table and page. The paging mechanism allows one page directory of size 4kB and the page directory can define 1024 (1k) page tables. Four bytes (32 bits) of the page directory are used to store information about a page table. Each page table can define 1024 pages of size 4kB. The size of each page table is also 4kB and four bytes (32 bits) of page table are used to store information about a page.

The paging mechanism of 80386 is shown in Fig. 10.20. The segmentation unit will supply a 32-bit linear address to paging unit. The upper 10 bits (A_{22}-A_{31}) of linear address is the index for page directory, the middle 10 bits (A_{12}-A_{21}) of linear address is the index for the page table and the lower 12 bits (A_0-A_{11}) of linear address is the lower 12 bits of physical address.

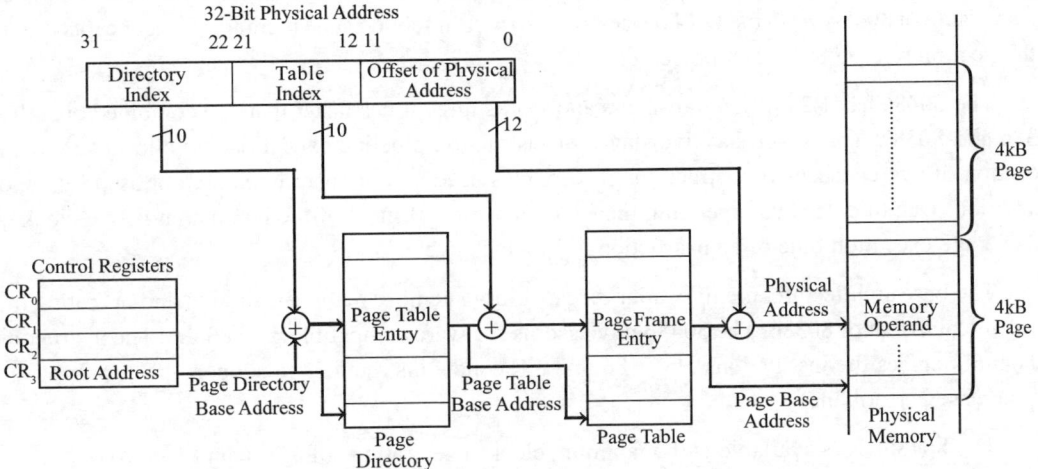

Fig. 10.20: Paging mechanism of 80386 microprocessor.

The control register CR_3 holds the base/root address of the page directory. The root address is added to the 10-bit page directory index (given by the linear address) to get the address page directory entry of the page table. The page directory entry has the base address of the page table, which is added to the 10-bit page table index (given by the linear address) to get the address of the page table entry of the page. The page table entry has the upper 20 bits of the page frame address, which is concatenated with lower 12 bits of the linear address to form the physical address.

Virtual 8086 Mode

The processor can switch from PVAM to virtual 8086 mode by setting the VM bit in the EFLAG register to logic 1. While working in privilege level 0, the processor can also enter the virtual 8086 mode by executing the IRET instruction. The processor can return to PVAM from the virtual 8086 mode only on receipt of an interrupt or exception.

The virtual 8086 mode permits the execution of 8086 applications with all protection features of 80386. In the virtual 8086 mode, the segment registers are used similar to that of the real mode. In this mode, the processor computes 20-bit address by shifting the segment register left by four times and adding to the offset. The 20-bit address can be used to access 1 MB of physical memory space.

In virtual 8086 mode, paging can be enabled to run multiple virtual mode tasks, to provide protection and operating system isolation. When paging is enabled, the 20-bit linear address can be divided into 256 pages and each page can be located anywhere in the 4 kB physical address space of 80386.

10.5 INTEL 80486 MICROPROCESSOR

The INTEL 80486 is a 32-bit processor with higher performance than 80386. It is integration of the improved 80386 processor, 80387 coprocessor and 8 kB RAM memory (called cache memory) on a single chip. The INTEL 80486 family of processors includes 80486SX, 80486DX, SX2, DX2, Write-back enhanced DX2, DX4 and Write-back enhanced DX4 processors.

The base architecture for the entire family of 80486 processor is same except minor differences. The 80486SX and SX2 processors does not have an internal coprocessor unit. The DX2 and DX4 are double clock version of 80486. Also the DX4 processor has 16 kB internal cache memory. The concepts discussed in this section refer to 80486DX processor.

The 80486 has 1.2 million transistors and works three times faster than the combined operation of 80386 and 80387. The 80486 has five stages of instruction pipeline execution and allows simultaneous execution of two consecutive instructions if resources used by one instruction are not used by the other instruction. Due to extensive pipelining the execution time of most of the instruction is one clock cycle and average execution time of an instruction is 1.6 clock cycle.

The base architecture, memory address capability, memory management unit and operating modes are identical to that of 80386. The 80486 processor is software compatible with 80386. The instruction set of 80486 includes the instructions of 80386 and a few new instructions to support the new applications and increase performance.

The 80486DX is available with maximum clock speed ratings of 33, 66 and 100 MHz.

10.5.1 Pins and Signals of 80486

The 80486DX is a 168-pin IC and available in PGA package. The pin configuration of 80486DX is shown in Fig. 10.21. The functions of pins of 80486 are listed in Table 10.11.

The 80486 has 32 data pins and so we can form a 32-bit data bus. The 80486 has dynamic bus size feature, which allows 8-bit and 16-bit devices to be interfaced with the processor through a 32-bit data bus. When the signal BS8 is asserted **low** the processor selects an 8-bit data bus and when BS16 is asserted **low** the processor selects a 16-bit data bus.

The pins DP_0 to DP_3 are used for parity bits of the data bytes of the data bus. One pin is used for each byte of data bus. During write operation, the processor generates even parity bits and output on these lines. During read operation, the external device has to supply even parity bits through these lines. When parity is not employed these pins should be tied to V_{CC} through a pull up resistor. During read operation, whenever the processor detects a parity error, it asserts PCHK signal as **low**.

In 80486, the memory is organized as 4 banks, each of size 1 GB. The address lines A_2-A_{31} are used to select the memory locations and the bank/byte enable signals BE_0 - BE_3 are used to select memory banks. The bank select signals are generated internally by decoding the address lines A_0 and A_1.

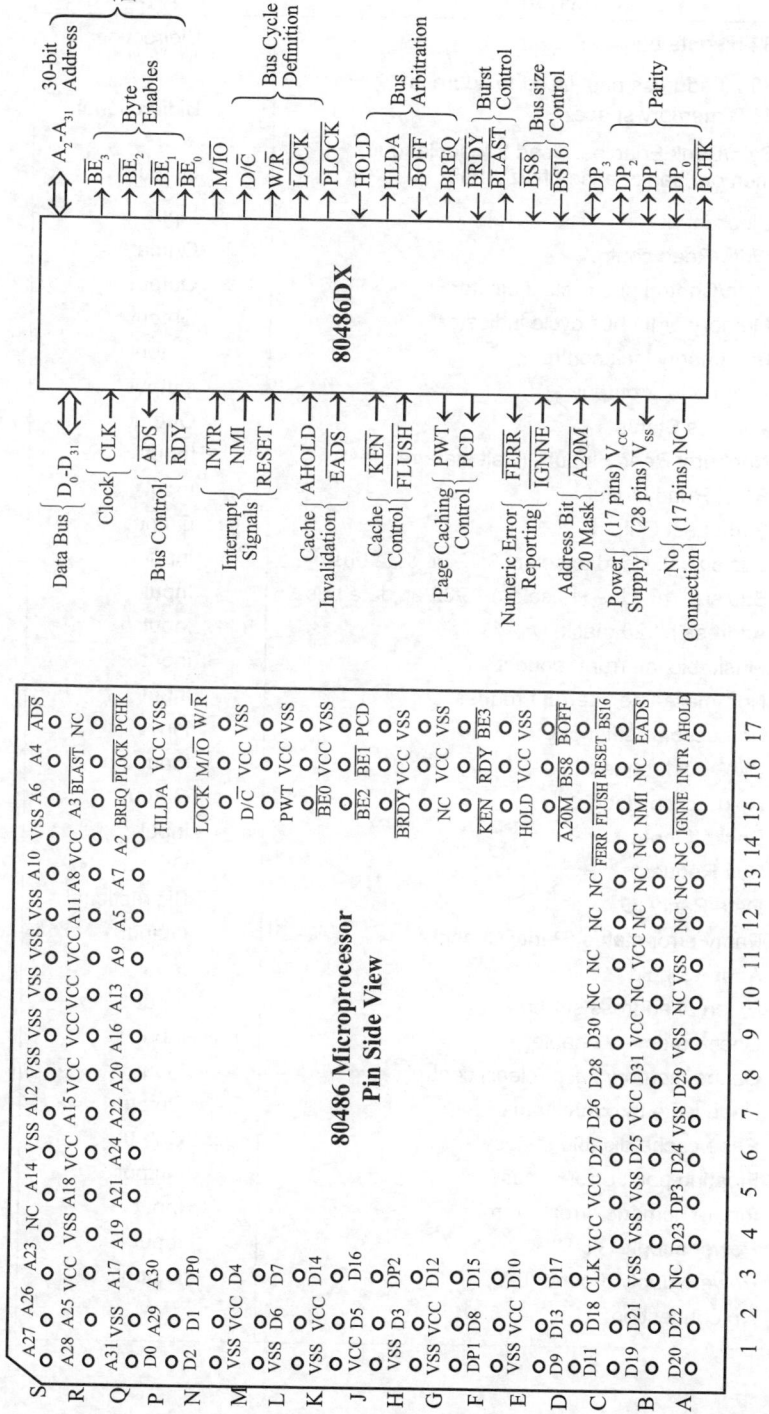

Fig. 10.21: Pin configuration of an 80486DX.

Table 10.11: Signals of 80486DX Microprocessor

Signal	Function	Type
D_{31}-D_0	32-bit data bus	Bidirectional
A_{31}-A_2	30-bit address bus. Used to address 1GB memory space	Bidirectional
BE_3-BE_0	Byte/Bank Enables. Used to enable four memory banks each of size 1GB	Output
CLK	Clock input	Input
W/$\overline{R}$	Write/Read control	Output
D/$\overline{C}$	Data/Control bus cycle indicator	Output
M/$\overline{IO}$	Memory or IO bus cycle indicator	Output
$\overline{LOCK}$	Bus priority lock control	Output
$\overline{PLOCK}$	Pseudo-lock output	Output
$\overline{ADS}$	Address Status	Output
$\overline{RDY}$	Nonburst Ready Input - (wait state request)	Input
$\overline{BRDY}$	Burst Ready Input	Input
$\overline{BLAST}$	Burst Last Output	Output
$\overline{BS8}$	Bus size 8. Used to select 8/32 bit data bus	Input
$\overline{BS16}$	Bus size 16. Used to select 16/32 bit data bus	Input
$\overline{A20M}$	Address bit 20 mask	Input
INTR	Maskable interrupt request	Input
NMI	Non-maskable interrupt request	Input
RESET	Processor reset input	Input
HOLD	DMA hold request	Input
HLDA	Hold Acknowledge	Output
$\overline{BOFF}$	Backoff Input	Input
BREQ	Bus Request	Output
DP_3-DP_0	Data Parity IO	Bidirectional
$\overline{PCHK}$	Parity error status (Parity Check)	Output
AHOLD	Address hold	Input
$\overline{EADS}$	External address strobe	Input
$\overline{KEN}$	Cache memory enable	Input
$\overline{FLUSH}$	Cache memory flush (clear) control	Input
PWT	Page write through status	Output
PCD	Page cache disable status	Output
$\overline{FERR}$	Floating point error status	Output
$\overline{IGNNE}$	Ignore numeric error control	Input
V_{cc}	Power supply (+5-V)	Input
V_{ss}	Power supply ground (0-V)	Output
NC	No connection	–

The signal $\overline{\text{A20M}}$ input can be used to mask the physical address bit 20 while working in the real address mode. When $\overline{\text{A20M}}$ is asserted **low,** the processor address space is wraparound into 1MB memory space (from FFFFF$_\text{H}$ to 00000$_\text{H}$) as that of 8086 processor.

The processor asserts $\overline{\text{ADS}}$ as **low** in the first clock of a bus cycle to indicate the start of a bus cycle and it is inactive in the subsequent clock of a bus cycle. The signal $\overline{\text{RDY}}$ is used to introduce wait states in nonburst bus cycles and the signal $\overline{\text{BRDY}}$ is used to introduce wait states in burst bus cycles. For normal bus timings, these signals should be tied **low**. The $\overline{\text{BLAST}}$ signal is asserted **low** by the processor to indicate the completion of a burst bus cycle.

The 80486 processor does not have an internal clock generator. The clock required for 80486 should be generated externally and supplied through CLK pin. The 80486 processor has active **high** reset. The RESET input is used to bring the processor to a known state. The reset will clear all segment registers except the CS-register and after a reset all general purpose registers except EDX will be in an undefined state. The reset will load F000$_\text{H}$ in CS-register, 0FFF0$_\text{H}$ in EIP and a component identifier is loaded in DX-register.

The 80486 has two hardware interrupt pins INTR and NMI. The INTR is a maskable interrupt request. The INTR is active **high** and it is not provided with an internal pull-down resistor. The NMI is nonmaskable interrupt request and when asserted, it executes a type-2 interrupt. The NMI is rising edge sensitive and it is not provided with an internal pull-down resistor.

The signals $\text{M}/\overline{\text{IO}}$, $\text{D}/\overline{\text{C}}$ and $\text{W}/\overline{\text{R}}$ are primary bus cycle definition signals and these signals are asserted after a valid $\overline{\text{ADS}}$. For memory read/write cycle $\text{M}/\overline{\text{IO}}$ is asserted **high.** For IO read/write cycle, $\text{M}/\overline{\text{IO}}$ is asserted **low**. For memory or IO data bus cycles (read/write), the $\text{D}/\overline{\text{C}}$ is asserted **high**. For interrupt acknowledge, code access and halt bus cycles the $\text{D}/\overline{\text{C}}$ is asserted **low**. The $\text{W}/\overline{\text{R}}$ differentiates between read and write bus cycles.

The $\overline{\text{LOCK}}$ signal is activated during execution of instruction prefixed by lock prefix . The $\overline{\text{PLOCK}}$ is asserted **low** during memory read/write operations that involves operands greater than 32 bits. The PLOCK is asserted **low** during floating point unit read/write (which involve 64 bits operand), segment table descriptor read (which involve 64 bits operand) and cache line fills (which involve 128 bits operand). When $\overline{\text{LOCK}}$ or $\overline{\text{PLOCK}}$ is asserted **low**, the other bus masters cannot take control of the system bus.

The HOLD input forces the processor to **high impedance** state after execution of current bus cycle and asserts HLDA to acknowledge the device which requested the HOLD. The HOLD and HLDA are mainly used for DMA data transfer. The $\overline{\text{BOFF}}$ input will force the processor to **high impedance** state immediately in the next clock. The $\overline{\text{BOFF}}$ is similar to HOLD but differs in two ways : First, the processor will not complete the current bus cycle before releasing the bus. Second, the processor does not assert HLDA.

The processor asserts BREQ whenever a bus cycle is pending internally. External logic can use BREQ signal to arbitrate among multiple processors. The AHOLD and $\overline{\text{EADS}}$ inputs are used during cache invalidation cycles. When AHOLD is asserted, the processor will stop driving its address bus in the next clock and so an external device can supply an address to the processor through the address bus. The $\overline{\text{EADS}}$ input is asserted **low** to indicate that a valid external address has been driven into the address bus.

The $\overline{\text{KEN}}$ is a cache enable pin and is used to determine whether the data being returned by the current cycle is cacheable. The $\overline{\text{FLUSH}}$ input forces the processor to flush its entire internal cache and this signal needs to be asserted **low** for one clock period to clear the cache memory.

The PWT and PCD output signals correspond to two user attribute bits in the page table entry. The PCD output reflects the state of the PCD attribute bit in the page table entry or the page directory entry. The PWT output reflects the state of the PWT attribute bit in the page table entry or page directory entry.

The processor asserts $\overline{\text{FERR}}$ pin as **low** whenever an unmasked floating point error is encountered. The assertion of $\overline{\text{IGNNE}}$ input signal informs the processor to ignore the floating point errors and continue executing noncontrol floating-point instructions.

10.5.2 Architecture of 80486

The INTEL 80486 is a 32-bit processor with on-chip memory management, floating point and cache memory units. The architecture (or functional block diagram) of 80486 processor is shown in Fig. 10.22. The various functional units of 80486 processor are:

1. **Data processing unit consisting of ALU, barrel shifter and an array of registers.**
2. **Bus interface unit consisting of drivers and various control logic unit.**
3. **32-byte instruction prefetch queue.**
4. **Instruction decode unit.**
5. **Floating point unit.**
6. **Memory management unit consisting of segmentation and paging units.**
7. **8 kB cache memory unit.**

The data processing unit consists of a 32-bit ALU, a 64-bit barrel shifter and eight 32-bit general purpose registers. The functions of ALU and barrel shifter are same as that of 80386 processor. The general purpose registers of 80486 are also same as that of 80386 processor. The general purpose registers are EAX, EBX, ECX, EDX, EBP, EDI, ESI and ESP. Part of these 32-bit registers can be accessed as 16 or 8-bit registers like that of 80386. The Instruction Pointer and Flag register of 80486 are identical to that of 80386 processor.

The bus interface unit consists of drivers for address and data bus and various control logic units, which includes bus cycle control, burst bus cycle control, bus size control and cache control logic units. The control signals necessary for memory, IO and interrupt bus cycles are generated by this unit. It also takes care of managing the control signals for cache memory control.

The bus interface unit also has a parity generation and control unit. The parity unit generates a parity bit (for even parity) for each byte of data during write operation, and output on DP_0 - DP_3 lines. These parity bits can be stored in the memory along with data when parity is employed in the system. During read operation the parity unit checks for even parity and if it finds an error then it generates a parity check error signal.

The 80486 processor has five stage instruction pipeline execution, which includes prefetch, first decode, second decode, execute and write back. Due to five stage pipeline, several instructions will be in the pipeline at a time. Hence, the 80486 processor can execute two instructions simultaneously if execution of one instruction does not depend on the other instruction. While decoding jump instructions, the processor automatically prefetch the instructions from the jump destination, which improves the processor performance greatly.

The floating point unit consists of eight numbers of 80-bit data registers, three numbers of 16-bit registers called status register, control register and tag word register and two numbers of 48-bit pointer registers called instruction pointer and data pointer. The floating point unit supports 32/64/80 bits floating point data types, 16/32/64 bits (signed) integer data types and 80-bit packed BCD data types.

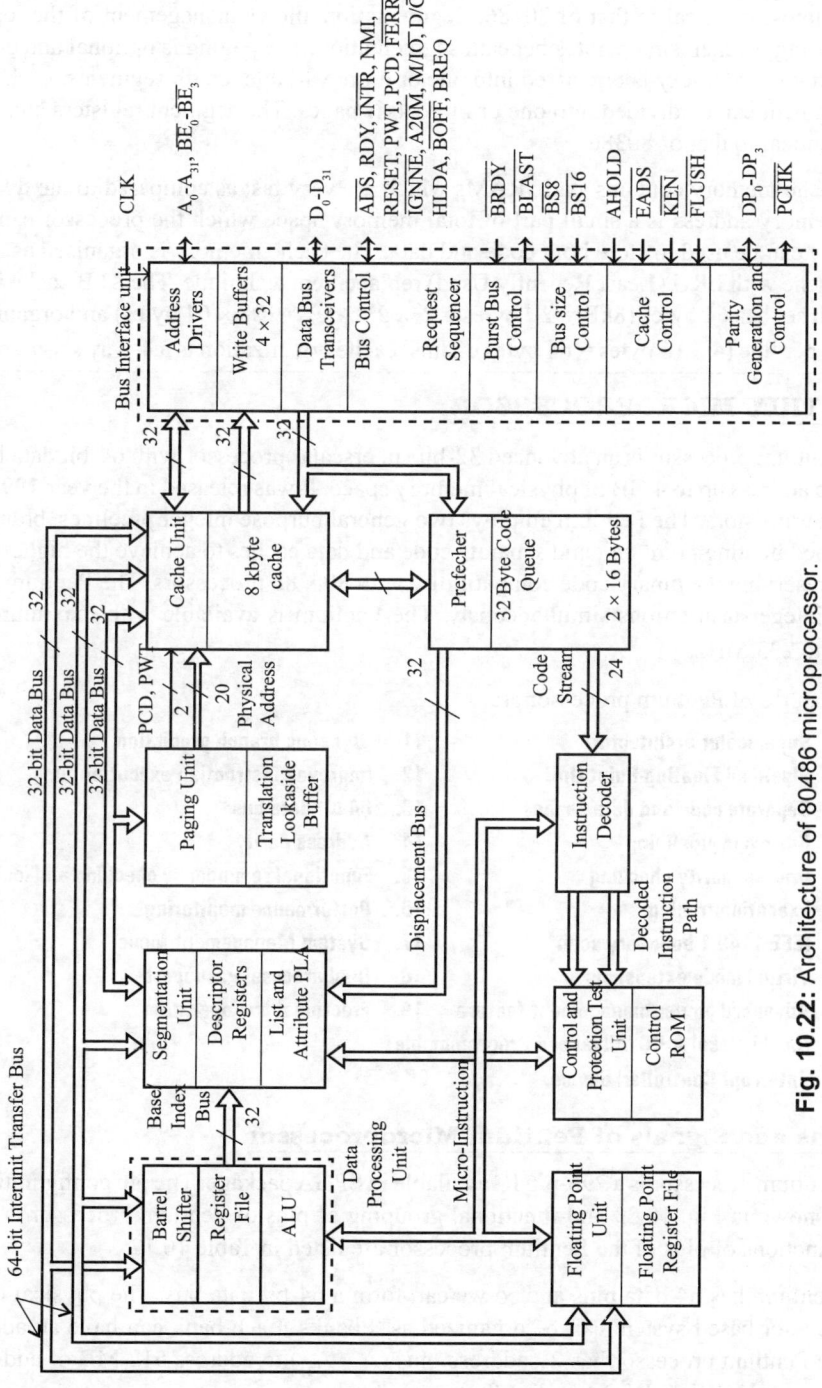

Fig. 10.22: Architecture of 80486 microprocessor.

The **Memory Management Unit (MMU)** consists of a segmentation unit and paging unit. The MMU of 80486 is almost identical to that of 80386. Segmentation allows management of the logical address space. The paging mechanism operates beneath segmentation. The paging is optional and can be disabled by system software. Memory is organized into one or more variable length segments, each up to 4 GB in size. Each segment can be divided into one or more 4kB pages. The segment registers and descriptors of 80486 are identical to that of 80386.

The cache memory contains static RAMs which are very fast as compared to the dynamic RAMs. The cache memory address is a small part of total memory space which the processor can address. The cache memory can be used to store both code and data. The cache memory is organised as a four-way set associative cache with LRU (**L**east **R**ecently **U**sed) replacement technique. The 8 kB cache is divided into 128 sets. Each set has 64 bytes ($8kB = 2^{13}$ bytes $= 2^7 \times 2^6$ bytes $= 128 \times 64$ bytes) and organized as 4 lines with 16 bytes per line (4×16 bytes $= 64$ bytes). Thus, cache organization is a 4-way set-associative cache.

10.6 PENTIUM MICROPROCESSOR

The Pentium processor is an advanced 32-bit superscalar processor with 64-bit data bus and 32-bit address bus to address up to 4 GB of physical memory space. It was released in the year **1993** and consists of **3.1** million transistors. The Pentium employs two general purpose integer pipelines, branch prediction, highly pipelined floating point unit and separate code and data caches to achieve the highest performance level while preserving the binary code compatibility with 80 x 86 processors. The Pentium processor can execute two integer instructions simultaneously. The Pentium is available with maximum clock speed ratings of 60 to 233 MHz.

The features of Pentium processor are:

1.	Superscalar architecture	11.	Dynamic branch prediction
2.	Pipelined Floating-Point Unit	12.	Improved instruction execution time
3.	Separate code and data caches	13.	64-bit data bus
4.	Bus cycle pipelining	14.	Address parity
5.	Internal parity checking	15.	Functional redundancy checking and lock-step operation
6.	Execution tracing	16.	Performance monitoring
7.	IEEE 1149.1 boundary scan	17.	System Management Mode
8.	Virtual mode extensions	18.	Dual processing support
9.	Advanced power management feature	19.	Fractional bus operation
10.	On-chip local APIC (Advanced Programmable Interrupt Controller) device.		

10.6.1 Pins and Signals of Pentium Microprocessor

The Pentium processor is a 296-pin IC available in SPGA package. The pin configuration of Pentium processor is shown in Fig. 10.23 and functional grouping of pins of Pentium processor is shown in Fig. 10.24. The functions of pins of the Pentium processor are listed in Table 10.12.

The Pentium has 64 data pins and so we can form a 64-bit data bus. The physical memory in the Pentium processor-based system can be organized as 8 banks. Each bank can have an address space of 512 MB. The Pentium processor has 29 address pins ($A_3 - A_{31}$) to address 512 MB of address space and 8-bank select signals ($\overline{BE}_0$ to $\overline{BE}_7$) to select 8-memory banks.

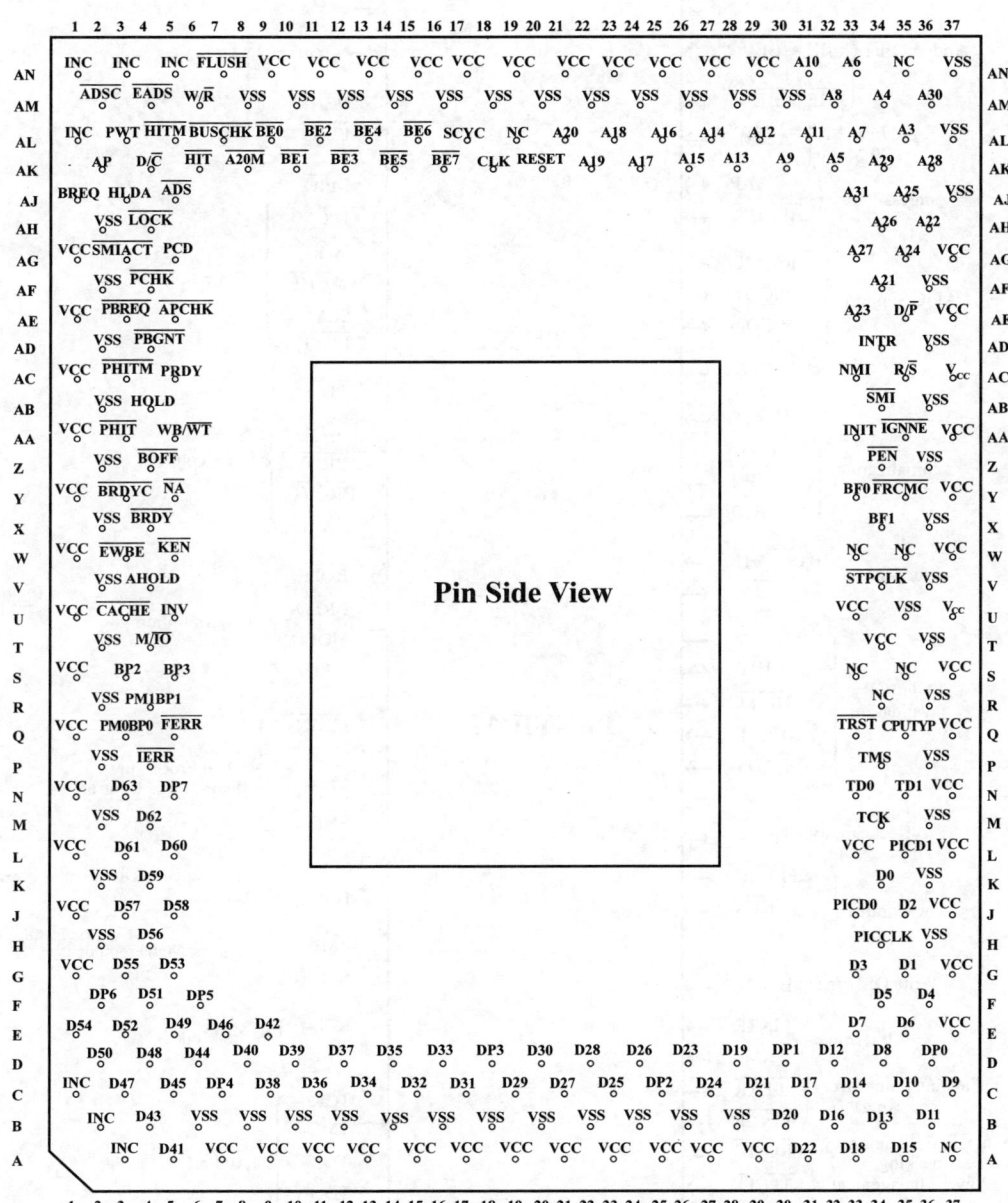

Fig. 10.23: Pin configuration of Pentium processor.

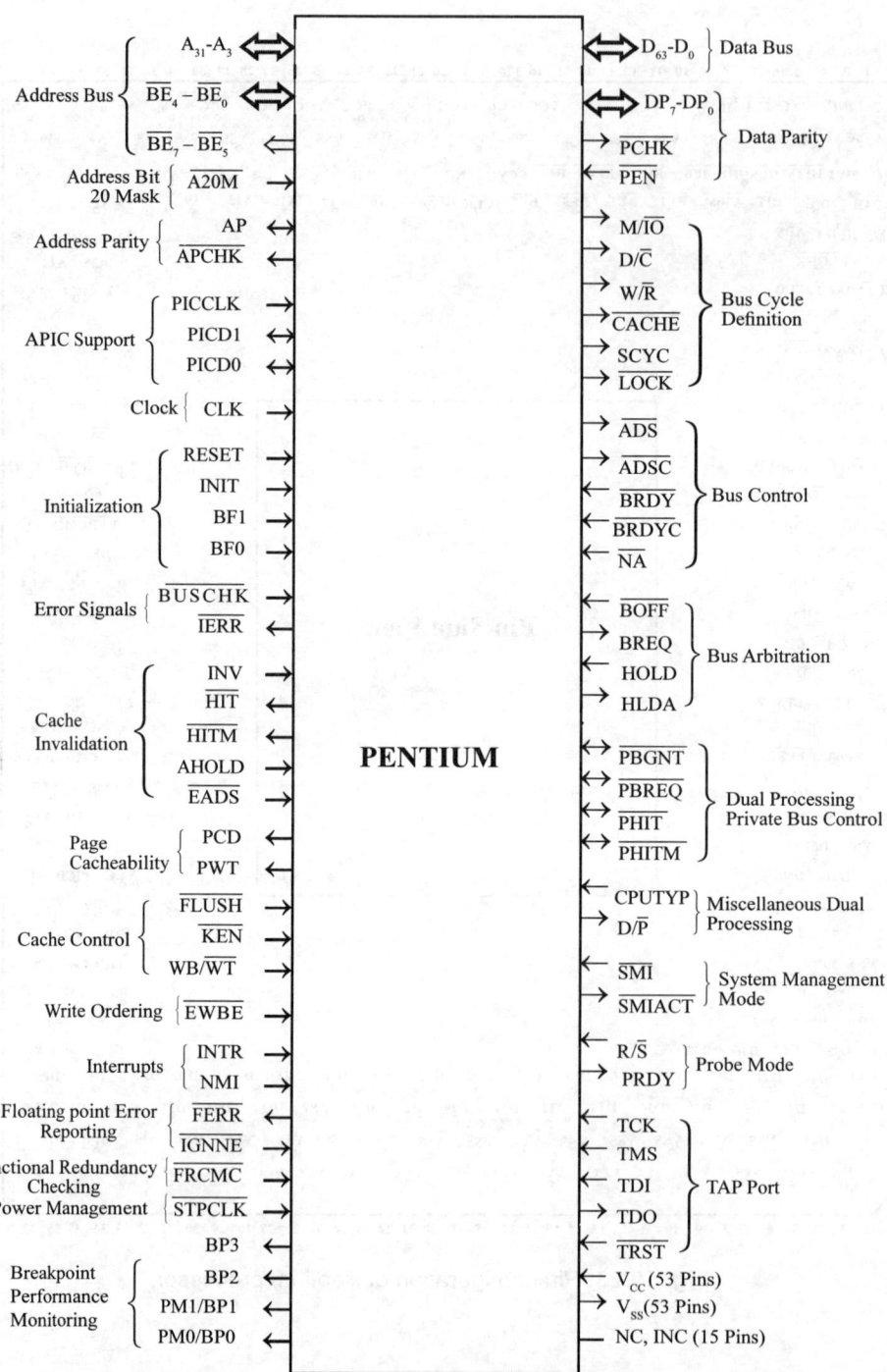

Fig. 10.24 : Functional grouping of the pins of a Pentium processor.

The pins DP_0 - DP_7 are used for parity of the data bytes of the data bus and the functions of these pins are similar to that of data parity pins of 80486. The pins $W/\overline{R}$, $D/\overline{C}$, $M/\overline{IO}$, $\overline{LOCK}$, $\overline{BRDY}$, $\overline{A20M}$, INTR, NMI, HOLD, HLDA, $\overline{BOFF}$, $\overline{BREQ}$, $\overline{PCHK}$, AHOLD, $\overline{EADS}$, $\overline{KEN}$, $\overline{FLUSH}$, PWT, PCD, $\overline{FERR}$ and $\overline{IGNNE}$ are common to 80486 and Pentium processors. The functions of these pins in Pentium processor are similar to that in 80486 processor.

The pin CPUTYP is used to distinguish the primary processor and dual processor. For primary processor CPUTYP is tied to ground (V_{SS}). For dual processor it is tied to V_{CC}. A brief description about the functions of each pin of Pentium processor are provided in Table 10.12. For detailed explanations the readers are advised to refer INTEL Pentium data sheet.

Table 10.12: Signals of Pentium Microprocessor

Signal	Function	Type
$\overline{A20M}$	Address 20 mask. Used to emulate the 1MB address wrap around of 8086.	Input
A_{31} - A_3	29-bit address bus. Used to address 512 MB memory space.	Bidirectional
$\overline{ADS}$	Address strobe. Indicates that a new valid bus cycle is currently being driven by the processor.	Output
$\overline{ADSC}$	Additional address strobe.	Output
AHOLD	Address hold. Used to get the address bus for an inquire cycle.	Input
AP	Address parity. Address parity pin for the address lines.	Bidirectional
$\overline{APCHK}$	Address parity check. The status of the address parity check is driven on this output.	Output
$\overline{BE_7}$ - $\overline{BE_0}$	Byte /Bank enables. Used to enable eight memory banks each of size 512 MB.	Output
BF_1-BF_0	Bus-to-core frequency ratio. Used to configure processor bus-to-core frequency ratio.	Input
$\overline{BOFF}$	Backoff input. This input is used to force the processor off the bus in the next clock.	Input
BP_3-BP_0	Breakpoint signals. These signals externally indicates a breakpoint match.	Output
$\overline{BRDY}$	Burst ready. Transfer complete indication.	Input

Table 10.12 continued...

Signal	Function	Type
$\overline{BRDYC}$	Additional burst ready input	Input
BREQ	Bus request. Indicates externally when a bus cycle is pending internally.	Output
$\overline{BUSCHK}$	Bus check. Allows the system to signal an unsuccessful completion of a bus cycle.	Input
$\overline{CACHE}$	Cacheability. External indication of internal cacheability.	Output
CLK	Clock. Fundamental timing source for processor.	Input
CPUTYP	Processor type definition pin. Used to configure as primary/dual processor.	Input
$D/\overline{C}$	Data/Code access. Distinguishes a data access from a code access.	Output
D_{63}- D_0	64-data lines. Forms the 64-bit data bus.	Bidirectional
DP_7- DP_0	Data parity. These are parity pins for the data bytes.	Bidirectional
$D/\overline{P}$	Dual/Primary processor indicator.	Output
$\overline{EADS}$	External address strobe. This input signals the processor to run an inquire cycle with the address on the bus.	Input
$\overline{EWBE}$	External write buffer empty. Provides the option of strong write ordering to the memory system.	Input
$\overline{FERR}$	Floating point error. The floating point error output is driven active when an unmasked floating point error occurs.	Output
$\overline{FLUSH}$	Cache flush. Writes all modified lines in the data cache back and flushes the code and data caches.	Input
$\overline{FRCMC}$	**F**unctional **R**edundancy **C**hecking **M**aster/Checker configuration. Determines whether the processor is configured as a master or checker.	Input
$\overline{HIT}$	Inquire cycle hit/miss.	Output
$\overline{HITM}$	Inquire cycle hit/miss to a modified line.	Output
HLDA	Bus hold acknowledge. External indication that the processor outputs are floated.	Output
HOLD	Bus hold request. Used for DMA transfer.	Input

Table 10.12 continued...

Signal	Function	Type
$\overline{\text{IERR}}$	Internal or functional redundancy check error. Alerts the system of internal parity errors and functional redundancy errors.	Output
$\overline{\text{IGNNE}}$	Ignore numeric exception. Determines whether or not numeric exceptions should be ignored	Input
INIT	Initialization. Forces the processor to begin execution in a known state without flushing the caches or affecting the floating point state.	Input
INTR	Non-maskable external interrupt request.	Input
INV	Invalidation request. Determines the final state of a cache line as a result of an inquire hit.	Input
$\overline{\text{KEN}}$	Cache enable. Indicates to the processor whether or not the system can support a cache line fill for the current cycle	Input
$\overline{\text{LOCK}}$	Bus lock. Indicates to system that the current bus cycle should not be interrupted.	Output
M/$\overline{\text{IO}}$	Memory or IO indicator. Distinguishes a memory access from an IO access.	Output
$\overline{\text{NA}}$	Next address. Indicates that external memory is prepared for a pipelined cycle.	Input
NMI	Non-maskable interrupt request.	Input
$\overline{\text{PBGNT}}$	Dual processor bus grant. Indicates to the LRM (**L**east **R**ecent **M**aster) processor that it will become the MRM (**M**ost **R**ecent **M**aster) processor in the next cycle.	Bidirectional
$\overline{\text{PBREQ}}$	Dual processor bus request. Indicates to MRM processor that LRM processor requires the bus.	Bidirectional
PCD	Page cacheability disable. Externally reflects the cacheability paging attribute bit in control register.	Output
$\overline{\text{PCHK}}$	Data parity check. Indicates the result of a parity check on a data read.	Output
$\overline{\text{PHIT}}$	Private inquire cycle hit/miss indication.	Bidirectional
$\overline{\text{PHITM}}$	Private inquire cycle hit/miss to a modified line indication.	Bidirectional
PICCLK	Processor interrupt controller clock. This pin drives the clock for APIC serial data bus operation.	Input
PICD1 - PICD0	Process interrupt controller data. These are the data pins for the 3-wire APIC bus.	Bidirectional

Table 10.12 continued...

Signal	Function	Type
$\overline{PEN}$	Parity enable. This signal determines whether a machine check exception has to be taken for data parity error.	Input
PM1-PM0	Performance Monitoring. Externally indicates the status of the performance monitor counter.	Output
PRDY	Probe ready. For use with INTEL debug port.	Output
PWT	Page write through status. Externally reflects the write through paging attribute bit in control register.	Output
R/$\overline{S}$	Run/Stop. For use with the INTEL debug port.	Input
RESET	Processor reset. Forces the processor to begin execution at a known state.	Input
SCYC	Split cycle indication. Indicates that a misaligned locked transfer is on the bus.	Output
$\overline{SMI}$	System management interrupt request.	Input
$\overline{SMIACT}$	System management interrupt active. Indicates that the processor is operating in SMM.	Output
STPCLK	Stop clock request. Used to stop the internal processor clock and consume less power.	Input
TCK	Test clock input. Provides boundary scan clocking function.	Input
TDI	**T**est **D**ata **I**nput. Input pin to receive serial data and instructions.	Input
TDO	**T**est **D**ata **O**utput. Output serial test data and instructions.	Output
TMS	**T**est **M**ode **S**elect. Controls TAP controller state transitions.	Input
$\overline{TRST}$	Test reset. Allows the TAP controller to be asynchronously initialized.	Input
V_{cc}	Supply to processor (3.3-V).	Input
V_{ss}	Power supply ground.	Output
W/$\overline{R}$	**W**rite/**R**ead control. Distinguishes a write cycle from a read cycle.	Output
WB/$\overline{WT}$	**W**rite **B**ack/**W**rite **T**hrough. This pin allows a cache line to be defined as write back or write through on a line by line basis.	Input

10.6.2 Architecture of Pentium Microprocessor

The Pentium processor has superscalar architecture which allows parallel execution of two instructions. The various functional blocks of Pentium processor are shown in Fig. 8.25. The functional units of Pentium processor are Bus unit, Paging unit, Prefetch buffer, Instruction decode, Control ROM, Execution unit with two integer pipelines (U-pipe and V-pipe), **B**ranch **T**arget **B**uffer (BTB), Code cache, Data cache, **F**loating **P**oint **U**nit (FPU), **D**ual **P**rocessing (DP) logic, and **A**dvanced **P**rogrammable **I**nterrupt **C**ontroller (APIC).

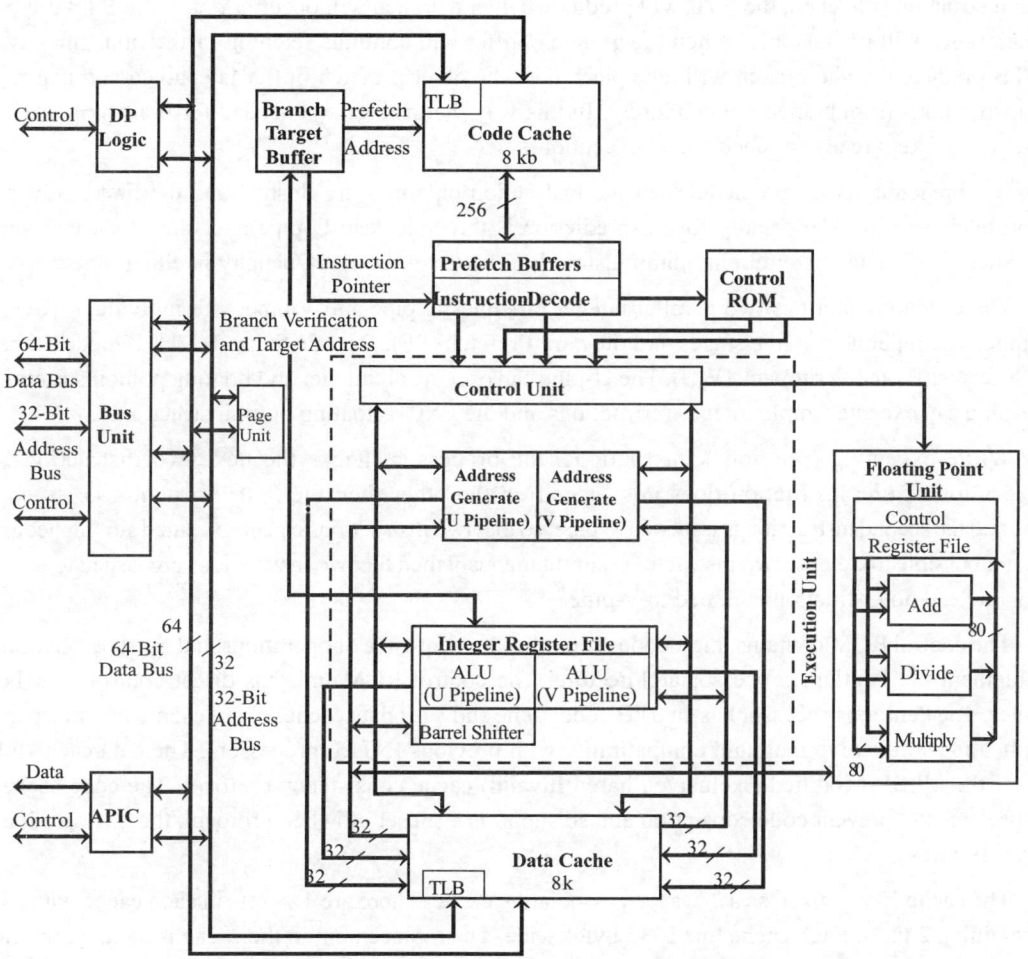

Fig. 10.25: Architecture of a Pentium processor.

The bus unit takes care of the issuing control signals and fetching the code/data from the external memory and IO devices. The Pentium processor has 64-bit external data bus and supports burst read and burst writeback cycles. In addition, bus cycle pipelining has been added to allow two bus cycles to be in progress simultaneously. The paging unit contains optional extensions to the architecture which allow 2 MB and 4 MB page sizes.

The code cache, branch target buffer and prefetch buffers are responsible for getting instructions into the execution unit. Instructions are fetched from the code cache or from the external memory. Branch addresses are remembered by the branch target buffer. The code cache TLB (**T**ranslation **L**ookaside **B**uffer) translates linear addresses to physical addresses used by the code cache.

The Pentium processor has two independent pairs of 32-byte prefetch buffers which operate in conjunction with BTB (**B**ranch **T**arget **B**uffer). Only one prefetch buffer will be active at any one time. One prefetch buffer will fetch instructions sequentially until a branch instruction is encountered. When a branch instruction is fetched, the BTB will predict whether a branch will occur or not. If the BTB predicts that the branch will not take place then the prefetch buffer will continue fetching instructions linearly. If the BTB predicts that the branch will take place then the other prefetch buffer is enabled and begins to fetch instructions from branch target address. In this way the instructions needed for branching are also prefetched and kept ready for decode and execution.

If a branching is mispredicted then the instruction pipelines are flushed and the first buffer will start prefetch activity. The penalty for misprediction is three clock in U-pipe and four clock in V-pipe. Mispredicted calls and unconditional jump instructions have a three clock penalty in either pipe.

The execution unit has two parallel integer pipelines U-pipe and V-pipe with individual ALU for each pipe. The pipeline has five stages and they are **P**refetch (PF), **D**ecode **S**tage-1(D1), **D**ecode **S**tage-2 (D2), **E**xecute (E) and **W**rite**b**ack (WB). The U-pipe can execute all integer and floating point instructions. The V-pipe can execute simple integer instructions and the FXCH floating point instruction.

While executing (previous) instructions, the processor checks the next two instructions. If the execution of one instruction does not depend on the other then the first instruction is issued to U-pipe and the second instruction is issued to V-pipe, so that two instructions can be executed simultaneously. If it is not possible to execute two instructions simultaneously then the two instructions are issued to U-pipe one by one and no instruction is issued to V-pipe.

The control ROM contains microcodes to control the sequence of operations that must be performed to implement the Pentium processor architecture. The control ROM unit has direct control over both pipelines. The Pentium processor has an 8 kB code cache and 8 kB data cache. These caches are transparent to application software to maintain compatibility with previous INTEL processors. The data cache fully supports the MESI (**M**odified/**E**xclusive/**S**hared/**I**nvalid) cache consistency protocol. The code cache is write protected to prevent code corruption and so supports a subset of MESI protocol, the S (**S**hared) and I (**I**nvalid) states.

The cache is organized as a 2-way set associative cache. There are 128 sets in each cache with each set containing 2 lines. Each cache line is 32 bytes wide. The replacement in the cache is handled by LRU (**L**east **R**ecently **U**sed) mechanism.

The code cache is connected to prefetch buffer by a 256-bit bus. Hence, in one clock cycle the instruction cache can provide up to 32 bytes ($32 \times 8 = 256$) of raw opcode to prefetch buffer. The data cache has two ports and each port is connected to pipelines by a 32-bit. Hence the data cache can provide data for two data references simultaneously. The caches are accessed with physical address and each cache has its own TLB (**T**ranslation **L**ookaside **B**uffer) to translate linear addresses to physical address.

The Pentium processor has a pipelined **F**loating **P**oint **U**nit (FPU) working independently. The FPU of Pentium processor is up to ten times faster than the FPU of 80486 processor for common operations including add, multiply and load.

The Pentium processor supports clock control. When the clock to the processor is stopped, power dissipation is virtually eliminated and this makes the Pentium processor a good choice for energy efficient designs. The Pentium processor supports fractional bus operation. This allows the processor core to operate at high frequencies, while communicating with the external bus at lower frequencies.

The Pentium processor contains an on-chip **A**dvanced **P**rogrammable **I**nterrupt **C**ontroller (APIC), which supports multiprocessor interrupt management, multiple IO subsystem support, 8259A compatiblity and inter-processor interrupt support.

The Pentium processor has in-built logic for dual processing mode of operation. In dual processor mode, two identical Pentium processors can be interfaced to a single system bus. The dual processor pair appears to the system bus as a single, unified processor. Multiprocessor operating systems properly schedule computing tasks between the two processors. This scheduling of task is transparent to software applications and the end user. Through a private bus, the two processors arbitrate for the external bus and maintains cache coherency. In dual processing mode, both the processor should have the same bus to core frequency ratio.

10.7 ADVANCED PENTIUM MICROPROCESSORS

INTEL has released a number of advanced microprocessors after pentium with advanced features. (Please refer to Appendix-III for processors released by INTEL.) In the pentium series, the INTEL has released Pentium Pro, Pentium II, Pentium III and Pentium 4 processors. The Pentium Pro, Pentium II and Pentium III have a common architecture and the architecture of these processors are known as INTEL's P6 microarchitecture. The architecture of Pentium 4 is known as NetBurst Microarchitecture. The higher clock version of pentium 4 incorporates **H**yper **T**hreading (HT) technology. In all these advanced pentium processors, the basic data size and memory addressing capability has been retained, but the processors are provided with enhanced features and advanced/sophisticated method of instruction execution. The salient features of the advanced pentium processors are presented in the following sections:

10.7.1 Pentium PRO

The Pentium pro is a 32-bit processor with 64-bit data bus and 36-bit address bus to address up to 64 GB of physical memory space. It was released in the year 1995 and consists of 5.5 million transistors. It is a 387 pin IC and available in PGA (**P**in **G**rid **A**rray) package. It is available with maximum internal clock ratings of 150/166/180/200 MHz.

The features of pentium pro processor are:

1. **Three-way superscalar architecture.**
2. **Five parallel execution units and 12-stage super pipeline.**
3. **Dual cavity PGA ceramic packages with a CPU die and a secondary cache die.**
4. **Out of order execution and speculative execution.**
5. **DIB (Dual Independent Bus) architecture.**

6. **Register renaming**
7. **Error checking and correcting codes.**
8. **Improved power management with two extra modes (Stop Grant and Auto HALT modes).**
9. **Internal micro-ops similar to RISC like instructions.**
10. **Transactional IO bus.**
11. **Scalable up to four processors.**
12. **Fault analysis/recovery.**
13. **Integrated level two (secondary) cache of 256 k/512k/1MB.**
14. **Internal thermal protection.**
15. **Automatic selection of power supply voltage.**

10.7.2 Pentium II

The Pentium II was released in 1997 and consists of 7.5 million transistors. It is actually a pentium pro processor with on-chip MMX (**M**ulti **M**edia **E**xtension). It is also a 32-bit processor with 64-bit data bus and 36-bit address bus to address up to 64 GB of physical memory space. It is available with maximum internal clock ratings of 233 MHz to 450 MHz and in SEC (**S**ingle **E**dge **C**onnector) catridge packaging or as a boxed processor along with fan/heatsink.

The features of pentium II processor are:

1. **Supports the INTEL architecture with dynamic execution.**
2. **Integrated primary (L1) 16 kB instruction cache and 16 kB write back data cache.**
3. **Integrated 256 kB second level (L2) cache.**
4. **Fully compatible with previous microprocessors.**
5. **Supports MMX technology.**
6. **Quick start and Deep sleep modes provide extremely low power dissipation.**
7. **Low power GTL + processor system bus interface (GTL : Gunning Transceiver Logic)**
8. **Integrated math co-processor (Floating point unit compatible with IEEE std 754).**
9. **Integrated thermal diode for measuring processor temperature.**

10.7.3 Pentium III

The Pentium III was released in 1999 and consists of 9.5 million transistors. The higher clock version of pentium III consists of 28 million transistors. The Pentium III is a 32-bit processor with a 64-bit data bus and a 36-bit address bus to address up to 64 GB of physical memory space. It is available with maximum internal ratings of 500 MHz to 1 GHz.

In the IC form, it is available as 370-pin IC in PGA (**P**in **G**rid **A**rray) package. The Pentium III is an advanced version of pentium II with improved MMX technology and processor serial number. The INTEL has incorporated a processor serial number in Pentium III which supports the concept of processor identification. Each pentium III processor has a 96-bit processor number accessible by software (of various applications) to identify a system. Some of the applications that may utilize processor serial number are membership authentication, data backup/restore protection, removable storage data protection, managed access to files, etc.

The features of pentium III processor are:

1. **Dynamic execution microarchitecture.**
2. **Optimized for 32-bit applications running on advanced 32-bit operating systems.**
3. **Fully compatible with previous microprocessors.**
4. **Integrated high performance 16 kB instruction and 16 kB data, nonblocking level one cache.**
5. **Integrated 512 kB full speed level two cache allows for low latency on read/store operation.**
6. **256-bit cache data bus provides extremely high throughput on read/store operation.**
7. **Eight-way cache associativity provides improved cache hit rate on read/store operations.**
8. **Error correcting code for system bus data.**
9. **Data prefetch logic**
10. **Internet streaming SIMD (Single Instruction Multiple Data) Extensions for enhanced Video, Sound and 3D performance.**
11. **System management mode and multiple low-power states.**
12. **Flip Chip Pin Grid Array (FC - PGA2) packaging technology which offers improved handling protection and socketability.**
13. **Intel processor serial number.**

10.7.4 Pentium 4

The Pentium 4 processor was released in 2000 and consists of 42 million transistors. It is available with maximum internal clock ratings of 1.4 GHz to 2.8 GHz. The Pentium 4 processor with HT (**H**yper **T**hreading) Technology was released in the year 2002 and consists of 55 million transistors. It is available with maximum internal clock ratings of 2.4 GHz to 3.3 GHz. It is available as 478-pin IC in PGA (**P**in **G**rid **A**rray) package.

The features of Pentium 4 processor are:

1. **INTEL NetBurst microarchitecture**
2. **Hyper Threading (HT) technology.**
3. **Hyperpipelined technology which supports advanced dynamic execution and very deep out-of-order execution.**
4. **Rapid execution engine-ALUs run at twice the processor core frequency.**
5. **System bus frequency at 400/533/800 MHz.**
6. **Binary compatible with applications running on previous members of INTEL processors.**
7. **8 kB level 1 data cache.**
8. **Level 1 execution trace cache stores 12 k micro-ops and removes decoder latency from main execution loops.**
9. **512 kB advanced transfer cache with 8-way associativity and error correcting code.**
10. **144 streaming SIMD Extensions 2 (SSE2) instructions. (SIMD : Single Instruction Multiple Data).**
11. **System management mode and mulitple low power states.**

The features of INTEL NetBurst microarchitecture are hyper pipelined technology, a rapid execution engine, 400/533/800 MHz system bus, execution trace cache, advanced dynamic execution, advanced transfer cache, enhanced floating point and multimedia unit and streaming SIMD Extensions 2 (SSE2).

The hyper pipelined technology doubles the pipeline depth in the Pentium-4 processor with 512 kB L2 cache, allowing the processor to reach much higher core frequencies. The rapid execution engine allows the two integer ALUs in the processor to run at twice the core frequency, which allows many integer instructions to execute in 1/2 clock tick.

The **Hyper Threading** (HT) technology allows a single physical Pentium-4 processor to function as two logical processors. Each logical processor has its own architecture state, own set of general purpose registers and control registers to provide increased system performance in multitasking environments.

10.8 SHORT-ANSWER QUESTIONS

Q10.1   List some of the advanced processors of the INTEL family.

    1. INTEL 80186   4. INTEL 80486

    2. INTEL 80286   5. Pentium processors

    3. INTEL 80386   6. Pentium PRO

Q10.2   What are the major features of the Pentium processor ?

The features of Pentium processor are,

    1. Superscalar architecture   11. Dynamic branch prediction

    2. Pipelined floating-point unit   12. Improved instruction execution time

    3. Separate code and data caches   13. 64-bit data bus

    4. Bus-cycle pipelining   14. Address parity

    5. Internal parity checking   15. Functional redundancy checking and lock-step operation

    6. Execution tracing   16. Performnce monitoring

    7. IEEE 1149.1 boundary scan   17. System management mode

    8. Virtual mode extensions   18. Dual processing support

    9. Advanced power management feature   19. Fractional bus operation

    10.on-chip local APIC (Advanced Programmable Interrupt Control) device

Q10.3   Give some examples of 32-bit processors.

    1. Pentium   4. Pentium III

    2. Pentium PRO   5. Pentium 4

    3. Pentium II

Q10.4   List the INTEL 80×86 family of processors.

    1. INTEL 8086   4. INTEL 80286

    2. INTEL 8088   5. INTEL 80386 DX

    3. INTEL 80186   6. INTEL 80486 DX

Q10.5   List the INTEL pentium family of processors.

    1. Pentium   4. Pentium III

    2. Pentium PRO   5. Pentium 4

    3. Pentium II

Q10.6   List the on chip functional units of 80186 microprocessor.

    1. Clock generator.
    2. Programmable interrupt controller.
    3. Programmable timers (3 independent 16-bit timers).
    4. Programmable DMA controller (2 independent DMA channels).
    5. Programmable memory and peripheral chip selection logic.
    6. Programmable wait-state generator.
    7. Local bus controller.

Q10.7   What is the significance of 80186 signals $\overline{LCS}$, $\overline{UCS}$ and $\overline{MCS}$?

The $\overline{LCS}$ is a programmable memory chip select signal to select a memory block of size 1 kB to 256 kB with starting address 00000H. The $\overline{UCS}$ is a programmable memory chip select signal to select a memory block of size 1 kB to 256 kB ending with the address $FFFFF_H$. The $\overline{MCS}_0$ to $\overline{MCS}_3$ are programmable to select four middle memory blocks of size 8 kB to 512 kB.

Q10.8   What are the internal interrupts of 80186 ?

The internal interrupts of 80186 includes INTEL predefined interrupts, software interrupts and interrupts from internal timers and DMA channels.

Q10.9   Write short notes on two operating modes of 80286 processor.

The 80826 has two operating modes: real address mode and protected virtual address mode. In the real address mode, the 80286 can address upto 1 MB (Megabyte) of physical memory address space like 8086. In protected virtual address mode, it can address up to 16 MB of physical memory address space and 1GB (Gigabyte) of virtual memory address space.

Q10.10   Name the four processing units of 80286 processor.

    1. Bus unit (BU)   3. Execution unit (EU)

    2. Instruction unit (IU)   4. Address unit (AU)

Q10.11   What is meant by real address mode of 80286 ?

In real address mode, the 80286 computes the memory address similar to that of 8086. It uses 20-bit address to access memory. The 20-bit physical address is computed by multiplying a 16-bit segment base address by 16_{10} and adding it to a 16-bit offset. Hence, 80286 can address up to 1MB (2^{20} = 1 Mega) of physical memory space. In this mode, the upper 4 bits of memory address bus are ignored.

The real address mode is mainly used to initialize peripheral devices, load the main part of the operating system from disk into memory, enable interrupts and enter the protected virtual address mode.

Q10.12   What is global descriptor table ?

The **G**lobal **D**escriptor **T**able **R**egister (GDTR) contains the 24-bit base address and limit for the table containing the global address space description. This register is initialized with LGDT (Load GDT) instruction when the system is booted.

Q10.13 What is local descriptor table ?

The **L**ocal **D**escriptor **T**able **R**egister (LDTR) in the 80286 contains the base address and limit of the local descriptor table for the task currently being executed. The LLDT(**L**oad **LDT**) instruction is used to load this register when the system is booted.

Q10.14 List the functional units of 80386.

1. **Bus interface unit**   4. **Execution unit**
2. **Instruction prefetch unit**   5. **Segmentation unit**
3. **Instruction decode unit**   6. **Paging unit**

Q10.15 Write short notes on bus interface unit of 80386.

The bus interface unit generates signals for memory and IO interface. This unit generates the control signals for the current bus cycle based on internal requests for fetching instructions from the instruction prefetch unit and transferring data from the execution unit. The physical address of memory and IO are output through the bus interface unit and also the data transfer to the external world takes place through this unit.

Q10.16 Write short notes on execution unit of 80386.

The execution unit reads the decoded instructions from the decode unit and processes them. The execution unit consists of a control unit, data unit and a protection test unit. The control unit contains the microcode stored in the ROM for the instructions, flag register and generates internal control signals. The data unit includes an ALU, a 64-bit barrel shifter and eight general purpose registers. The data unit performs the operations requested by the control unit. The barrel shifter is used for fast shift, rotate, multiply and divide operations. The protection test unit checks for segmentation violations under the control of the microcode.

Q10.17 Write short notes on control register of 80386.

The 80386 has four numbers of 32-bit control registers CR0, CR1, CR2 and CR3.The lower 16-bit of CR0 is the Machine Status Word (MSW) similar to MSW of 80286. The most significant bit of CR0 is used to enable/disable paging. The CR1 is reserved for future expansion. The CR2 holds the 32-bit linear address of the last page accessed before a page fault interrupt. The CR3 holds the 32-bit base address of the page directory. Since the page directory table is always page aligned (4 kB - aligned), the lower 12 bits of CR3 are undefined.

Q10.18 What is debug register of 80386 ?

The 80386 has eight numbers of 32-bit debug registers in which six are accessible by programmer and two are reserved by INTEL. The debug registers are denoted as DR_0 - DR_7. The DR_0 - DR_3 are used to specify four linear breakpoint address. The DR_4 and DR_5 are reserved by INTEL. The DR_7 is used to set the breakpoints and DR_6 displays the current status of the breakpoints.

Q10.19 Write short notes on paging mechanism of 80386.

In paging mechanism of 80386 the segmentation unit will supply a 32-bit linear address to paging unit. The upper 10 bits (A_{22}-A_{31}) of linear address is the index for page directory, the middle 10 bits (A_{12}-A_{21}) of linear address is the index for the page table and the lower 12 bits (A_0-A_{11}) of linear address is the lower 12 bits of physical address.

Q10.20 Write short notes on floating point unit of 80486 ?

The floating point unit consists of eight numbers of 80-bit data registers, three numbers of 16-bit registers called status register, control register and tag word register and two numbers of 48-bit pointer registers called instruction pointer and data pointer. The floating point unit supports 32/64/80 bits floating point data types, 16/32/64 bits (signed) integer data types and 80-bit packed BCD data types.

Q10.21 Write short notes on cache-memory of 80486?

The cache memory address is a small part of total memory space which the processor can address. The cache memory can be used to store both code and data. The cache memory is organised as a four-way set associative cache with LRU (**L**east **R**ecently **U**sed) replacement technique. The 8 kB cache is divided into 128 sets. Each set has 64 bytes (8kB = 2^{13} bytes = $2^7 \times 2^6$ bytes = 128×64 bytes) and organized as 4 lines with 16 bytes per line (4×16 bytes = 64 bytes). Thus, cache organization is a 4-way set-associative cache.

Q10.22 Discuss about pipelined floating point unit of pentium processor.

The Pentium processor has a pipelined Floating Point Unit (FPU) working independently. The FPU of Pentium processor is up to ten times faster than the FPU of 80486 processor for common operations including add, multiply and load.

Q10.23 What is hyper pipelined technology ?

The hyper pipelined technology doubles the pipeline depth in the Pentium-4 processor with 512 kB L2 cache, allowing the processor to reach much higher core frequencies. The rapid execution engine allows the two integer ALUs in the processor to run at twice the core frequency, which allows many integer instructions to execute in 1/2 clock tick.

Q10.24 What is hyper threading technology ?

The Hyper Threading (HT) technology allows a single physical Pentium-4 processor to function as two logical processors. Each logical processor has its own architecture state, own set of general purpose registers and control registers to provide increased system performance in multitasking environments.

Q10.25 Write short notes on pentium dual processing mode of operations.

The Pentium processor has in-built logic for dual processing mode of operation. In dual processor mode, two identical Pentium processors can be interfaced to a single system bus. The dual processor pair appears to the system bus as a single, unified processor. Multiprocessor operating systems properly schedule computing tasks between the two processors. This scheduling of task is transparent to software applications and the end user. Through a private bus, the two processors arbitrate for the external bus and maintains cache coherency. In dual processing mode, both the processor should have the same bus to core frequency ratio.

10.9 EXERCISES

I. Fill in the blanks with appropriate words

1. The _____ architecture is used in pentium family of processors.

2. The _____ and _____ signals of microprocessor 80186 are used to introduce wait states in the bus cycles.

3. The 80186 has _____ number of external (hardware) interrupts.

4. The _____ interrupt occurs whenever the boundary of an 80186 index register is outside the values set up in the memory.

5. When 80186 executes any undefined opcode, the _____ interrupt is triggered.

6. The 80286 processor can address upto _____ MB of physical memory address space in virtual address mode.

7. The _____ and _____ signals of 80286 are used to produce early control bus signals.

8. The _____ interrupt call is generated by 80286 whenever it encounters an error in its coprocessor.

9. The _____ of 80286 decodes the prefetched instructions and stores in instruction queue.

10. In 80286, _____ the register is used to switch from real address mode to protected virtual address mode (PVAM).

11. In 80286 protected virtual address mode, the sizes of physical memory is _____ and the virtual memory is _____.

12. The length of the 80286 descriptor is _____ byte.

13. The 80386SX propcessor can address upto _____ size of physical memory.

14. The _____ register of 80386 is used for fast shift, rotate, multiply and divide operations.

15. The _____ and _____ units of 80386 is considered as memory management unit(MMU)

16. The _____ registers are used to hold the selector in 80386 PVAM mode.

17. By setting the _____ bit on EFLAG register, the 80386 enters into virtual 8086 mode.

18. The _____ processor is actually a pentium pro processor with on-chip multi media extension.

19. The length of pentium-III processor identification number is _____ bit.

20. The technology which allows a single physical processor to function as two logical processor is called _____.

Answers

1. superscalar	6. 16	11. 16 MB, 1 GB	16. segment
2. ARDY, SRDY	7. COD/$\overline{\text{INTA}}$, M/$\overline{\text{IO}}$	12. 8	17. VM
3. five	8. type 16_H	13. 16 MB	18. Pentium II
4. array BOUNDS	9. instruction unit	14. barrel shifter	19. 96
5. unused opcode	10. machine status word(MSW)	15. segmentation, paging	20. hyper threading (HT)

II. State whether the following statements are True/False.

1. The 80186 has an internal clock generator.

2. The 80186 is object code compatible with 8086.

3. The SRDY signal of 80186 is an asynchronous signal.

4. The Bus Interface Unit (BIU) and Execution unit (EU) of 80186 is same as that of 8086.

5. The 80186 contains the same set of general purpose registers and flag resgisters as that of 8086 processor.

6. It is not possible to assign same priority to more than one interrupts of 80186.

7. The 80286 has 16 MB of virtual memory space.

8. The 80286 is source code compatible with 8086 in protected virtual address mode.

9. The 80286 has an internal clock generator.

10. 80286 works similar to 8086 in maximum mode.

11. The 80286 does not require any external bus controller to generate the bus control signals.

12. The COD/$\overline{\text{INTA}}$ signal of 80286 is asserted high for IO read/write and instruction/code read bus cycles.

13. The instruction queue of 80286 can accomodate upto six decoded instructions.

14. The default operating mode of 80286 is real address mode.

15. The maximum physical address space of 80286 in real address mode is 16 MB.

16. The 80386DX processor is the full version of 80386.

17. The instructioin prefetch unit of 80386 can store 16-byte code in it.

18. The paging is disabled in the real address mode of 80386.

19. The segment registers of 80386 are 32-bit long.

20. The real address mode is generally used to initialize the hardwares of 80×86 family of processor and to enter into protected virtual address mode.

21. It is possible for 80386 processor to switch from PVAM mode to virtual 8086 mode and vice versa without processor reset.

22. paging is not supported by 80386 in its virtual 8086 mode.

23. The 80486 processor is a 16-bit processor.

24. The 80486SX and 80486SX2 processors do not have an internal coprocessor unit.

25. The 80486 processor has 8 KB onchip cache memory.

26. The 80486 processor can execute two instructions simultaneously.

27. The 80486 processor supports parity check while reading and writing data into memory.

28. The cache-memory of 80486 can store only the data

29. The pentium processor has separate code and data cache memory.

30. The pentium processor supports dual processors.

Answers

1. True	6. False	11. False	16. True	21. True	26. True
2. True	7. False	12. True	17. True	22. False	27. True
3. False	8. True	13. False	18. True	23. False	28. False
4. True	9. False	14. True	19. False	24. True	29. True
5. True	10. True	15. False	20. True	25. True	30. True

III. Choose the right answer for the following questions.

1. *Which of the following microprocessor supports virtual memory space?*

    a) 8086　　　　*b)* 8088　　　　*c)* 80186　　　　*d)* 80286

2. *Which of the following are on-chip functional units of 80186?*

    a) clock generator　　　　　　*b)* programmable timer
    c) programmable DMA controller　　*d)* all the three

3. *Which of the following 80186 pin generates a signal to reset the processor?*

    a) $\overline{RES}$　　　　*b)* RST　　　　*c)* RESET　　　　*d)* none

4. *The crysal frequency of 80186 is 16 MHZ. What is the internal clock frequency of 80186?*

    a) 4 MHz　　　　*b)* 8 MHz　　　　*c)* 16 MHz　　　　*d)* 32 MHz

5. *Which of the following 80186 unit can be used as a watch dog timer?*

    a) Timer-0　　　　*b)* Timer-1　　　　*c)* Timer-2　　　　*d)* all the three

6. Which of the following is not an INTEL predefined interrupts of 80186?

   a) INT1   b) divide error   c) NMI   d) Breakpoint

7. The following processor is a 16-bit processor with on-chip memory protection capabailities

   a) 8086   b) 8088   c) 80186   d) 80286

8. Which of the following signal of 80286 is used to interrupt the 80286 whenever the coprocessor 80287 finds error in the process?

   a) $\overline{BUSY}$ b) PEREQ   c) $\overline{PEACK}$ d) $\overline{ERROR}$

9. Which of the following 80286 instruction is executed to switch from real address mode to PVAM mode?

   a) LGDT   b) LLDT   c) LMSW   d) all the three

10. Which of the following statements are correct for 80286 processor?

    i) The 80286 can be switched from real address mode to PVAM mode and vice versa without reseting the processor

    ii) The unused 80286 segments can be placed in hard disk

    iii) The virtual memory space of 80286 is 1 GB

    a) (i) & (ii)   b) (ii) & (iii)   c) (i) & (iii)   d) (iii)

11. Which of the following is called reduced bus version of 80386?

    a) 80386DX   b) 80386EX   c) 80386RX   d) 80386SX

12. What is the physical memory capacity of 80386DX?

    a) 1 MB   b) 16MB   c) 1GB   d) 4 GB

13. The following operating mode of 80386 allows the processor to run 8086 applications with protection features of 80386.

    a) real address mode   b) protected virtual address mode

    c) virtual 8086 mode   d) all the three

14. Which of the following fields are used to calculate the 32-bit linear address of 80386 in PVAM mode?

    a) offset   b) selector   c) descriptor   d) all the three

15. The size of the physical address is _____ bit and virtual address is _____ bit in 80386 PVAM mode.

    a) 48 bit and 32 bit   b) 32 bit and 48 bit   c) 16 bit and 24 bit   d) 24 bit and 16 bit

16. Which of the following is the data segment register of 80386?

    a) DS   b) ES   c) FS   d) all the three

17. Which of the following control register holds the machine status word of 80386?

    a) $CR0_{(0-15)}$ b) $CR0_{(16-31)}$ c) $CR1_{(0-15)}$ d) $CR1_{(16-31)}$

18. Which of the following processor is not the double clock version of 80486?

    a) 80486SX2   b) 80486DX2   c) 80486DX4   d) all the three

19. The following 80486 processor has 16 KB internal cache memory.

   a) 80486SX2   *b)* 80486DX   *c)* 80486DX2   *d)* 80486DX4

20. Which of the following unit is not present in 80386 but in 80486?

   a) instruction prefetch unit   *b)* floating-point unit

   c) bus interface unit   *d)* segmentation and paging unit

21. Which of the following processor has superscalar architecture?

   a) 80286   *b)* 80386   *c)* 80486   *d)* pentium processor

22. The _____ processor has processor serial number

   a) pentium   *b)* pentium pro   *c)* pentium II   *d)* pentium III

Answers							
1. d	4. b	7. d	10. b	13. c	16. d	19. d	22. d
2. d	5. c	8. d	11. d	14. d	17. a	20. b	
3. a	6. a	9. c	12. d	15. b	18. a	21. d	

IV. Answer the following questions

E10.1 What are the uses of $\overline{PCS}_0$ through $\overline{PCS}_6$ signals of 80186?

E10.2 Write a short on on-chip clock generator of 80186.

E10.3 Write a short on timers of 80186.

E10.4 Explain in brief about DMA operation in 80186.

E10.5 What are the external interrupts of 80186?

E10.6 What is meant by array BOUNDS interrupts?

E10.7 How does the vector address of 80186 interrupt is calculated? Explain with an example.

E10.8 Write down the priority level of all the 80186 interrups.

E10.9 How are the control bus signals generated in 80286?

E10.10 What happens to segment registers, flag register and instruction pointer when 80286 is reset?

E10.11 Name and explain the 80286 pins used for interfacing 80286 with coprocessor 80287.

E10.12 Write short notes on Bus Unit of 80286 processor.

E10.13 Write short notes on Instruction unit of 80286 processor.

E10.14 Write short notes on Execution unit of 80286 processor.

E10.15 Write short notes on address unit of 80286 processor.

E10.16 Draw and explain the flag register of 80286?

E10.17 What is meant by protected virtual address mode (PVAM) of 80286?

E10.18 Compare real address and protected virtual address mode(PVAM) of 80286.

E10.19 Write short notes on memory management unit of 80286.

E10.20 The virtual address space of 80286 is 1 GB. Justify.

E10.21 What is MSW register?

E10.22 What is the use of selector field in 80286?

E10.23 How is the physical address generated in 80286 PVAM mode?

E10.24 List the features of 80386 microprocessor.

E10.25 Compare 80386 DX and 80386SX processors.

E10.26 Briefly explain the operating modes of 80386.

E10.27 Differentiate real address mode and virtual 8086 mode of 80386.

E10.28 Write short notes on instruction prefetch unit of 80386.

E10.29 Write short notes on instruction decode unit of 80386.

E10.30 What does the segmentation unit of 80386 do?

E10.31 What is the need to have Paging unit in 80386?

E10.32 How is the physical address calculated in real address mode of 80286?

E10.33 How is the physical address calculated in PVAM mode of 80286?

E10.34 What is the size of virtual address space in 80386 PVAM mode? Justify your answer.

E10.35 Name and explain the general purpose registers of 80386.

E10.36 List and explain the segment registers of 80386.

E10.37 What is the size of Instruction Pointer (IP) of 80386? What is the use of it?

E10.38 Draw the format and explain the flag register of 80386.

E10.39 What are the uses of system address register and segment registers of 80386?

E10.40 Explain the real address mode of 80386.

E10.41 Explain the PVAM mode of 80386.

E10.42 What is virtual 8086 mode of 80386? Explain.

E10.43 List the features of 80486 processor.

E10.44 How does 80486 implement parity checking?

E10.45 List the features of pentium processor.

E10.46 What is superscalar architecture? Discuss its uses.

E10.47 Explain the branch prediction option of pentium processor.

E10.48 Discuss about parallel integer pipelining of execution unit in pentium processor.

E10.49 Write short notes on cache memory of pentium processor.

E10.50 Discuss about clock control feature of pentium processor.

E10.51 Name any two processor which support multimedia extension facility.

CHAPTER 11

MICROPROCESSOR AND MICROCONTROLLER-BASED SYSTEM DESIGN

11.1 DESIGNING A MICROPROCESSOR-BASED SYSTEM

Desiging of a microcomputer system starts with specifications. The specification of the system includes the following:

1. Input device
2. Output device
3. Memory requirement
4. System clock frequency
5. Peripheral devices required
6. Type of CPU (Microprocessor)
7. Applications or Nature of work

Input Devices

The popular input device in a single board microcomputer system (microprocessor trainer kit) is the Hex-keyboard. Other forms of input devices are DIP switches, ADC interfaced through port and floppy disk interfaced through the floppy disk controller - INTEL 8272. The Hex-keyboard is normally interfaced to the 8085/8086 system using INTEL 8279 keyboard and display controller. A maximum of 64 keys can be interfaced using 8279. Along with shift and control, 256 key-codes can be generated using 8279.

Output Devices

The popular output device used in single board microcomputer (microprocessor trainer kit) is the 7-segment LED. The seven segment LEDs are interfaced to the 8085 processor using INTEL 8279 keyboard and display controller. The 8279 is a dedicated controller which takes care of keyboard scanning and display refreshing. A maximum of 16 number of 7-segment LEDs can be interfaced using one 8279 in an 8085/8086-based system as multiplexed display.

Other output devices are LCD (**L**iquid **C**rystal **D**isplay), printer, floppy disk and CRT terminal. The LCD and printer can be interfaced using ports. Special dedicated controllers are required for interfacing floppy disk and CRT terminal. The INTEL 8272 or INTEL 82072 floppy disk controller and INTEL 8275 CRT controller are popularly used with 8085/8086/8088 systems.

Memory Requirement

The memory requirement of the system is splitted between EPROM and RAM. The memory capacity of EPROM and RAM are estimated based on the applications and work to be performed by the processor. Most of the microprocessors use memory with word size of 1-byte. Hence the memory capacity of the system is specified in kilo bytes.

The popular EPROM used in the 8085/8086-based system are 2708 (1k × 8), 2716 (2 k × 8), 2732 (4 k × 8), 2764 (8 k × 8) and 27256 (32 k × 8). The popular static RAM used in the 8085-based system are 6208 (1k × 8), 6216 (2 k × 8), 6232 (4 k × 8), 6264 (8 k × 8), and 62256 (32 k × 8). The memories are chosen with compatible access time, i.e., the access time of memories should be less than the read time and write time of the processor.

The total memory requirement of the system is implemented by using more than one memory IC. But the processor, at any one time can communicate with (or access) only one memory IC. To select a memory IC, chip select signals has to be generated using decoders. The input to the decoders are the unused address lines. Also, to each memory location, specific addresses should be allotted. [These techniques are discussed in memory interfacing.]

The EPROMs are mapped at the beginning of memory space in an 8085-based system and at the end of the memory space in 8086-based system in order to store the monitor program in the EPROM and to execute the monitor program upon power-on-reset. [Every system will be resetted when power supply is switched ON.]

In 8085-based system, the interrupt vector addresses belong to EPROM locations. Normally, a jump instruction with an address of RAM location is stored in the vector locations. Hence, the user can store the interrupt subroutines in these jump addresses.

In an 8086-based system the interrupt vector addresses belongs to RAM locations in the beginning of address space. The 8086 processor has 256 types of interrupts. For each interrupt, four locations are reserved in the first 1 kilo-byte address space. In these locations the offset and base address of the subroutine program to be executed in response to interrupts are stored.

Apart from allocating addresses to memory devices, the peripherals and IO devices should also be allotted specific addresses. The peripherals and IO devices can be either memory-mapped or IO-mapped in the system. If the memory requirement of the system is very large and in future if memory expansion is required, then the peripherals and IO devices are IO-mapped in the system. If memory requirement of the system is less, then the peripherals and IO devices are memory mapped in the system.

System Clock Frequency

The microprocessor and the peripheral devices require a clock signal for synchronizing various internal operations or devices. An oscillator is needed for generating the clock signal. The oscillator consists of an amplifier and a feedback network. The feedback network has R, L, C or Quartz crystal.

In 8085 processor, the oscillator circuit (except the Quartz crystal and L-C component) is fabricated in the processor itself. Hence, it is necessary to connect a quartz crystal external to the processor. The oscillator generates a clock whose frequency is double to that of internal clock. The processor divides the generated clock by **two** for internal operations.

The 8086 microprocessor does not have an internal clock circuit. Hence the clock has to be supplied from an external device. The INTEL 8284 clock generator can be employed to generate the clock required for 8086. The 8284 has an internal oscillator circuit. An external quartz crystal has to be connected to the 8284 to generate the clock signal. The frequency of quartz crystal should be thrice the internal clock frequency of 8086. The 8284 generates the clock at crystal frequency and divides the clock by three and then supply to 8086.

For each system a maximum clock speed is specified. Driving a system at the maximum clock is advantageous because the execution time will be minimum if the clock is maximum. When the system is driven at maximum clock, then the peripherals chosen should have speed compatibility with the processor.

Peripheral Devices

The peripheral devices required for a system depends on its applications. Some of the peripheral devices that can be interfaced to the 8085-based system are:

1. Programmable Interval Timer - INTEL 8253/8254
2. USART - INTEL 8251
3. Programmable Peripheral Interface - INTEL 8255
4. Keyboard /Display Controller - INTEL 8279
5. Programmable Interrupt Controller - INTEL 8259
6. DMA Controller - INTEL 8237/8257
7. ADC
8. DAC

When the system has to monitor an analog signal from a sensor, then an ADC can be interfaced using 8255 ports. If the processor has to control an analog device, then it has to convert the digital signal to analog signal using DAC. When the system requires a large number of interrupt inputs, the interrupt structure of the system has to be expanded by using interrupt controller 8259. One 8259 supports 8-interrupt requests.

The USART-8251 can be used for serial data communication and the programmable timer-8253/8254 can be employed for various timing operations.

Type of CPU

The CPU of the system is a microprocessor. The microprocessor is chosen based on clock speed, instruction execution time, memory capacity, size of data and address, addressing modes, the operations it can perform and the number of additional devices required to form a system.

Application or Nature of Work

The specifications of the microprocessor itself depends on the applications for the proposed system and the nature of work it is going to perform. The input device, output device, memory requirement, peripheral requirement and the choice of CPU depend on the nature of work to be performed by the system.

11.2 DESIGNING A MICROCONTROLLER-BASED SYSTEM

The microcontrollers have internal memory and internal IO ports, and so a minimum system can be formed by connecting a reset circuit and quartz crystal of required frequency. The input devices and output devices can be directly connected to IO ports.

The 8051 microcontroller-based minimum system can be formed by connecting a 12 MHz quartz crystal to X1 and X2 pins, an RC reset circuit and a 5 volt power supply as shown in Fig. 11.1. The system will have 32 IO lines organized as four 8-bit IO ports. The IO lines can also be used as individual IO lines. The devices like push-button keys, 7-segment LED display, LCD, ADC, DAC, etc., can be directly interfaced to IO lines.

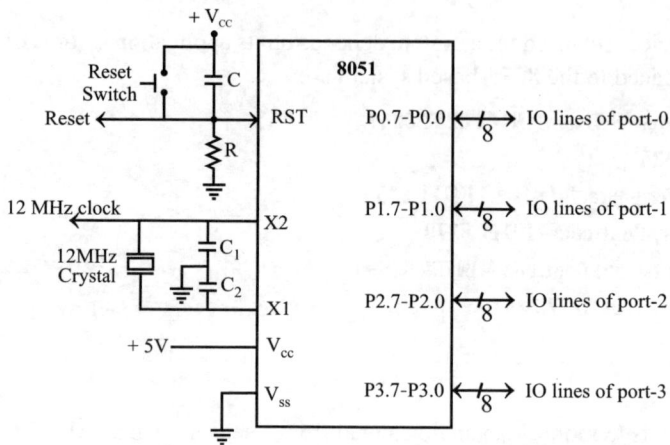

Fig. 11.1: The 8051 microcontroller-based minimum system.

The 8051 microcontroller has 4 kB internal ROM for program storage and 128 bytes internal RAM for user data storage. The 8051 supports optional external program memory and data memory. Usually, if the controller internal memory is not sufficient then external memory of required capacity is interfaced to the system.

For interfacing external memory, an external system bus has to be formed as shown in Fig. 11.2. The address and data bus is formed using port-0 and port-2. The address bus is 16-bit wide and the data bus is 8-bit wide. The port-0 lines are multiplexed low-byte address and data lines, and the port-2 lines are high-byte address lines.

The multiplexed address and data lines of port-0 are demultiplexed by using an 8-bit external latch. The input of the latch is connected to port-0 lines and the output of the latch forms the low-byte address bus. The 8051 controller provides a signal called ALE (Address Latch Enable) which is used as enable/clock signal for the external latch.

In the beginning of every machine cycle, the address is output on port-0 and port-2 lines, and then the ALE is asserted high and then low. The ALE acts as a clock/enable for the external latch, and so the address at the input of the latch (which is connected to port-0), will be transferred to the output of the latch, and so the port-0 lines are free to carry data. When external memory access is employed, the signal EA is tied low.

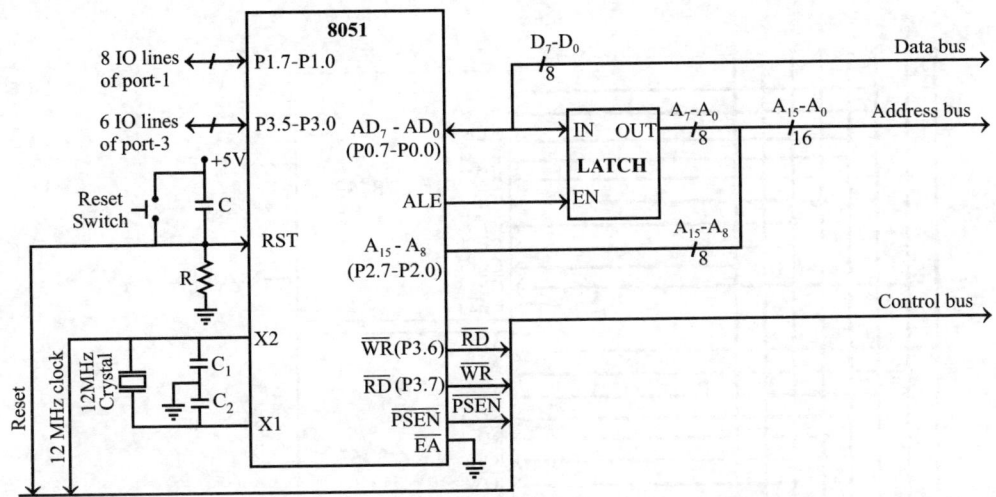

Fig. 11.2: External bus structure of the 8051 microcontroller-based system.

The control bus is formed using port-3 pins P3.6 and P3.7 (through which the controller sends $\overline{RD}$ and $\overline{WR}$), and other signals $\overline{PSEN}$, Reset and Clock signals. The $\overline{RD}$ and $\overline{WR}$ are used for read and write operations of the external data memory.

The $\overline{PSEN}$ is used for read operation of the external program memory. When external bus is formed for external memory access, the port-0, port-2 and port-3 pins P3.6 and P3.7 cannot be used as IO lines. Therefore, in the 32 IO lines of 8051, 18 IO lines are used to form the system bus, and the remaining 14 lines can be used as IO lines.

Besides accessing external memory, the external bus can be used to interface the peripheral devices like ADC, DAC, etc., as memory-mapped IO, mapped in the data memory address space. The devices that have matched read and write timing as that of the microcontroller can be directly interfaced to the 8051 system bus. The slow devices have to be interfaced to system bus via buffers/latches/8255 ports. When other devices are interfaced in the system bus, some of the memory addresses will be used to select peripheral devices, and so full data memory space cannot be used for memory.

11.3 8085-BASED MINIMUM SYSTEM

A minimum system is one which is formed using minimum number of IC chips. The minimum system in 8085 is formed using 8155, 8355, and 8755. In this, the 8085 is the CPU and the 8155, 8355, 8755 are memory and port devices. The 8155, 8355 and 8755 are called **P**rogrammable **P**eripheral **I**nterface (PPI).

The 8155 has ports and 256×8 static RAM, the 8355 has ports and $2 \text{ k} \times 8$ ROM and the 8755 has ports and $2 \text{ k} \times 8$ EPROM. The port and memory requirement of the system are provided by PPI. The input and output devices like keyboard and display LEDs are interfaced via ports of PPI to the system.

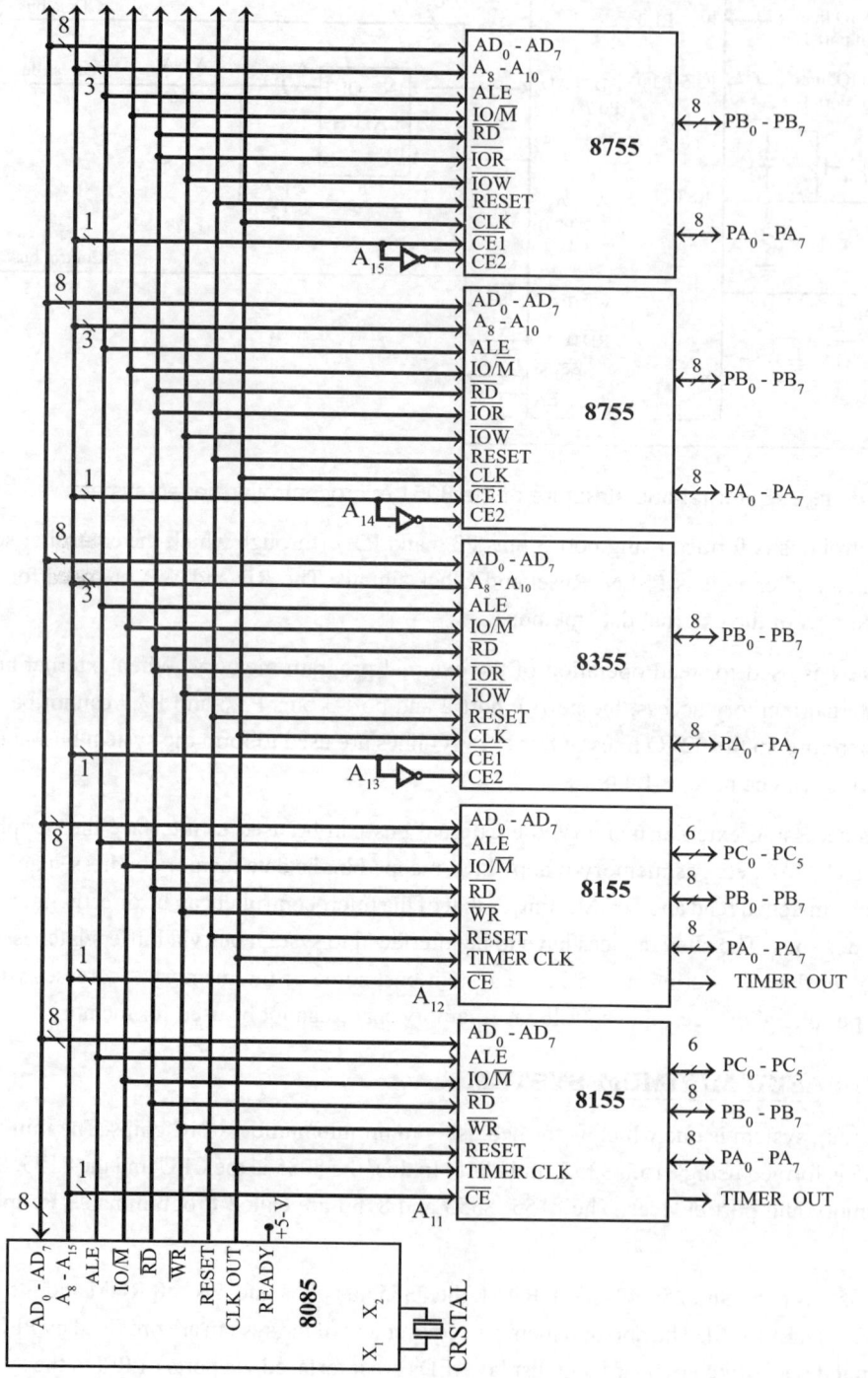

Fig. 11.3 : 8085-based minimum system.

In minimum system, the monitor program and system program are usually stored in permanent memories like ROM and EPROM of 8355 and 8755. For stack operations we require RAM memory. The RAM memory requirement is provided by 8155 .

The PPIs (8155, 8355 and 8755) have internal address latches, hence there is no need to demultiplex the AD_0-AD_7 lines of the processor. The demultiplexing is done by these devices, internally. These devices are mapped in the system by linear address selection method. In this method, the unused address lines are directly connected to chip select pins of the peripheral ICs.

The memories of the PPI are interfaced by memory mapping and the ports of the PPIs are mapped by IO mapping. The PPIs accept IO/$\overline{M}$ signal and differentiate between memory and port addresses. A 16-bit address is allotted to memory locations and 8-bit IO address is allotted to ports. An example of minimum system is shown in Fig. 11.3.

11.4 SUPPORT CHIPS NEEDED FOR 8086 MICROPROCESSOR-BASED SYSTEM

11.4.1 Clock Generator - INTEL 8284A

The clock generator 8284A is specially designed for 8086/8088 microprocessor. It generates the clock signal, READY signal and RESET signal required for 8086/8088 microprocessor. It also generates a TTL level peripheral clock signal which can be used for other peripheral devices in the system.

The INTEL 8284A is an 18-pin IC packed in DIP. The pin configuration of 8284A is shown in Fig. 11.4. The function of various pins are listed in Table 11.1.

A typical connection of 8284A with 8086 is shown in Fig. 11.5. The clock generator is used for generating clock signal and reset signal. It is assumed that the processor does not require wait states in bus cycles and so READY of 8086 is permanently tied to logic **high** (V_{CC}).

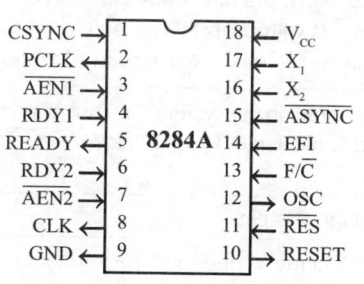

Pins	Description
CSYNC	Clock synchronization
PCLK	Peripheral clock
$\overline{AEN1}$, $\overline{AEN2}$	Address enable
RDY1, RDY2	Bus ready (Transfer complete)
READY	Ready signal to 8086/8088
CLK	Clock for 8086/8088
GND	Power supply/ground
RESET	Reset signal to 8086
$\overline{RES}$	Reset input to 8284A
OSC	Oscillator output
F/$\overline{C}$	Frequency/crystal selection
EFI	External frequency input
$\overline{ASYNC}$	Ready synchronization
X_1, X_2	Pins for crystal connection
V_{CC}	Power supply, +5-V

Fig. 11.4: Pin configuration of 8284A.

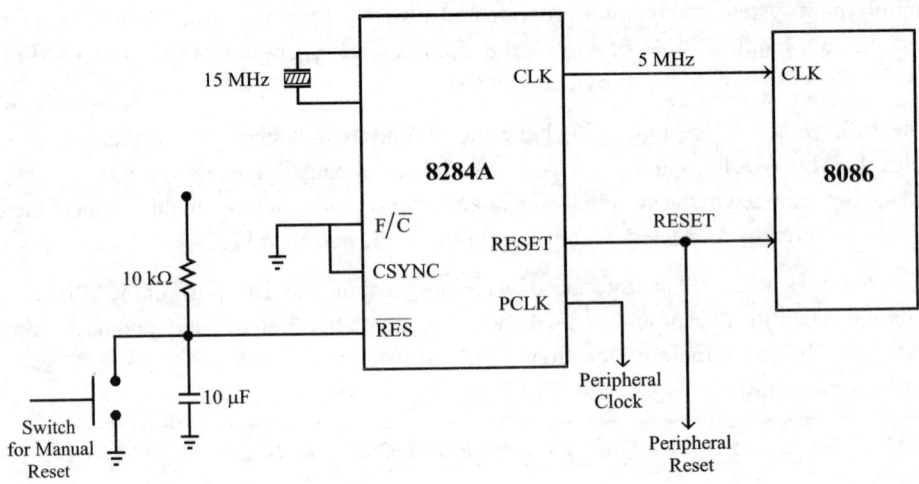

Fig. 11.5: A typical connection of 8284A with 8086.

Table 11.1: Pins and Signals of 8284A

Pin name	Function
CSYNC	When EFI is employed this clock synchronization input signal is used to synchronize the clocks of the various processors in a multimaster system. When crystal is connected to X_1 and X_2, this pin is grounded.
PCLK	It is a clock signal for peripheral devices in the system with a frequency of one-sixth the crystal frequency or EFI signal and has 50% duty cycle.
$\overline{AEN1}$, $\overline{AEN2}$	Address enables are provided to qualify (i.e., to allow) the bus ready signals RDY1 and RDY2 respectively. In a two-master system $\overline{AEN1}$ will qualify RDY1 and $\overline{AEN2}$ will qualify RDY2. In a single-master system $\overline{AEN1}$ and $\overline{AEN2}$ are tied to logic **low** (Ground).
RDY1, RDY2	These are active **high** signals from the devices located on the system data bus of each master to indicate that the data has been received or available. The RDY along with $\overline{AEN}$ are used to generate the READY signal for the 8086.
READY	The READY signal for the 8086 is supplied through this pin.
CLK	Clock signal for 8086. The frequency of this clock signal is one-third the frequency of crystal/EFI and has 33% duty cycle.
RESET	The reset signal for 8086 and other peripheral devices are supplied through this pin.
$\overline{RES}$	The reset input for 8284 and usually a RC network is connected to this pin to provide power-on reset.
OSC	A clock signal with same frequency as that of crystal/EFI. It can be used as EFI for other 8284s in a multimaster system.
$F/\overline{C}$	This is used to select crystal or EFI for clock generation. When EFI is used this pin is tied to logic **high** (V_{cc}). When a crystal is connected, this pin is tied to logic **low** (Ground).

Table 11.1 continued...

Pin name	Function
EFI	This pin is used to supply external frequency input to 8284 when F/C is tied **high**. This external signal should be a square wave with frequency three times the frequency required for the CLK output.
ASYNC̄	This signal is used to select one or two stages of synchronization for the RDY1 and RDY2 inputs. When this input is tied **low,** two stages of synchronization is provided and when this input is left open or tied **high,** one stage of synchronization is provided.
X_1, X_2	Pins for connecting quartz crystal. The frequency of crystal should be three times the desired processor clock frequency. When EFI is employed, X_1 should be tied to V_{CC} / GND and X_2 should be left open.
V_{CC}	Power supply input, +5-V
GND	Power supply ground

11.4.2 Bus Controller - INTEL 8288

The Bus controller-INTEL 8288 is specially designed for 8086/8088 microprocessor for use in maximum mode to generate bus control signals. It reads the status signals from the processor and generates bus control signal for memory and other IO devices.

The INTEL 8288 is a 20-pin IC packed in DIP. The pin configuration and internal block diagram of 8288 is shown in Fig. 11.6.

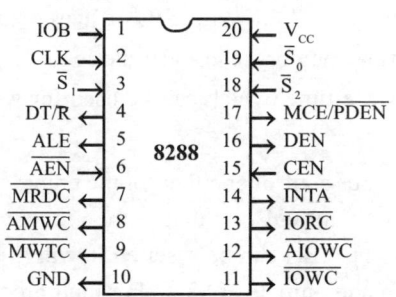

Fig. a: Pin configuration of INTEL 8288.

Pin	Description
AEN̄	Address enable
AIOWC̄	Advanced IO write control signal
ALE	Address latch enable
AMWC̄	Advanced memory write control signal
CEN	Command enable
CLK	Clock input
DEN	Data enable
DT/R̄	Data transmit/receive
INTA	Interrupt acknowledge
IOB	IO bus mode
IORC̄	IO read control signal
IOWC̄	IO write control signal
MCE/PDEN	Master cascade enable/ Peripheral data enable
MRDC̄	Memory read control signal
MWTC̄	Memory write control signal
$\bar{S}_0, \bar{S}_1, \bar{S}_2$	Input status signals
V_{CC}	Power supply, +5-V
GND	Ground, 0-V

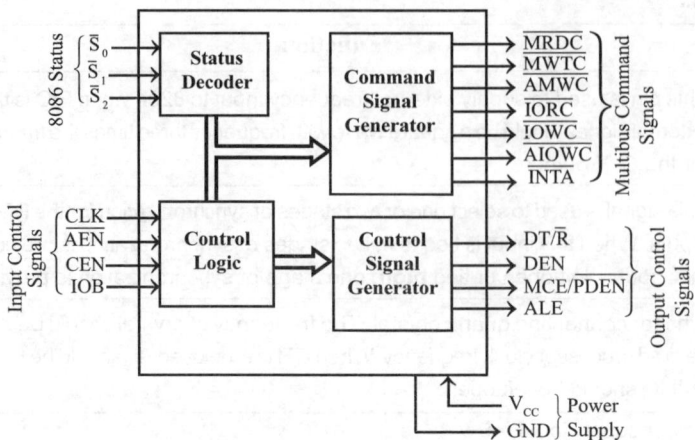

Fig. b: Internal block diagram of INTEL 8288.

Fig. 11.6: Bus controller-INTEL 8288.

The pins $\overline{S}_0$, $\overline{S}_1$, $\overline{S}_2$ are provided to receive the corresponding status bits from the 8086/8088 processor. The processor clock is directly connected to the CLK input of the 8288 in order to synchronize the activity of the bus controller with that of the processor. The signal output on ALE, DT/$\overline{R}$ and DEN are similar to that of 8086/8088 minimum mode signals for various bus cycles in order to enable the address latch and data transceivers. The bus controller issues the appropriate interrupt acknowledge signal through the $\overline{INTA}$ pin when the status signals are all zero (i.e., when $\overline{S}_0 = \overline{S}_1 = \overline{S}_2 = 0$).

The $\overline{IORC}$ and $\overline{IOWC}$ can be used as normal IO read and write control signals respectively. The $\overline{AIOWC}$ will provide extended IO write time, which can be used for writing IO devices requiring higher write time. The $\overline{MRDC}$ and $\overline{MWTC}$ can be used as normal memory read and write control signals respectively. The $\overline{AMWC}$ will provide extended memory write time, which can be used for writing memory devices requiring higher write time.

The $\overline{AEN}$, IOB and CEN are provided to configure the bus controller either for the uniprocessor or the multiprocessor system. In a uniprocessor system, $\overline{AEN}$ and IOB are tied to the ground (0-V) and CEN is tied to V_{cc}(+5-V). In a multiprocessor system, the $\overline{AEN}$ can be asserted **low/high** (by a bus arbiter such as INTEL 8289) to enable/disable the command outputs of 8288. For multiprocessor system IOB is tied to V_{cc}(+5-V).

The signal output on MCE/$\overline{PDEN}$ depends on the mode, which is determined by the signal applied to IOB. In a uniprocessor mode, IOB is grounded and so the output of MCE/$\overline{PDEN}$ will be **high** and it is used to control the cascaded 8259s (cascaded interrupt controllers). In multiprocessor mode, IOB is tied to V_{cc} (+5-V) and in this mode the MCE/$\overline{PDEN}$ is asserted **low** during the IO read/write operation and this signal is used to enable IO bus data transceivers.

> **Note:** *In multiprocessor mode when an IO transfer is made, $\overline{PDEN}$ is active and DEN is inactive. For a memory transfer DEN is active and $\overline{PDEN}$ is inactive.*

11.4.3 Coprocessor - INTEL 8087

The coprocessors has been specially designed to take care of mathematical calculations involving integer and floating point data. The coprocessor is also called math coprocessor or **Numeric Data Processor (NDP)**. A coprocessor is designed to work in parallel with a microprocessor. The INTEL has developed 80×87 series of coprocessors for 80×86 family of microprocessors. For example, INTEL has developed 8087 coprocessor for 8086 processor, 80287 coprocessor for 80286 processor and so on. From 80486DX onwards INTEL has started integrating the coprocessor with the microprocessor and started fabricating microprocessors with an on-chip coprocessor.

The coprocessor-INTEL 8087 has been developed to work with 8086 system in maximum mode. The 8087 coprocessor is a 40-pin IC packed in DIP. The pin configuration of 8087 is shown in Fig. 11.7.

The 8087 has 16 multiplexed address/data pins and 4 multiplexed address/status pins. Hence, it can have a 16-bit external data bus and a 20-bit external address bus like 8086. The processor clock, reset and ready signals are applied as clock, reset and ready signals for coprocessor. The BUSY output is used as $\overline{TEST}$ input of the processor.

One of the Request/Grant ($\overline{RQ/GT}$) signal can be connected to the corresponding pin of the processor to get the control of the bus from the processor. During an internal error condition, the coprocessor asserts the INT output as **high**, which can be used to interrupt the processor to take appropriate action. The pins $\overline{S}_1$, $\overline{S}_2$ and $\overline{S}_0$ are used to receive the bus status signal from the processor and the pins QS_1 and QS_0 are used to receive the queue status from the processor. When used with 8086, the $\overline{BHE}$ signal is asserted to enable the upper memory bank.

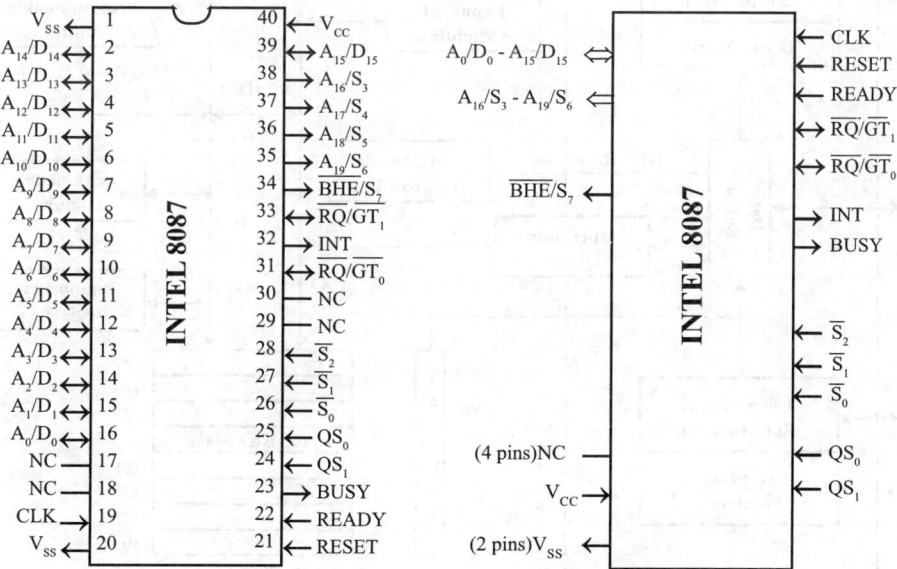

Fig. 11.7: Pin configuration of INTEL 8087 coprocessor.

Architecture of 8087

The internal architecture of 8087 is shown in Fig. 11.8. The 8087 has two funtional units: Control Unit (CU) and Numeric Execution Unit (NEU).

The control unit consists of data buffer, shared operands queue, addressing and bus tracking unit, exception pointers and control and status word registers.

The numeric execution unit consists of eight numbers of 80-bit register stack, microcode control unit, exponent module, programmable shifter, arithmetic module, temporary registers and shared operands queue.

The NEU is responsible for the execution of the coprocessor instructions under the control of the coprocessor CU. The CU transfers the numeric instructions to the microcode control unit of NEU.

The 8087 internally operates on all numbers in the 80-bit temporary real format. The 8087 has eight 80-bit registers which are used as LIFO (Last-In-First-Out) stack. This register stack holds the operands on which the coprocessor instruction operates.

The 8087 has a 3-bit stack pointer to point to the current top of the stack. The stack pointer can hold binary values 000_2 to 111_2 to represent the eight registers of the stack. The stack operates as a circular stack of fixed size (8 elements) on the basis of LIFO access. Upon reset, the stack pointer is initialized with 000_2.

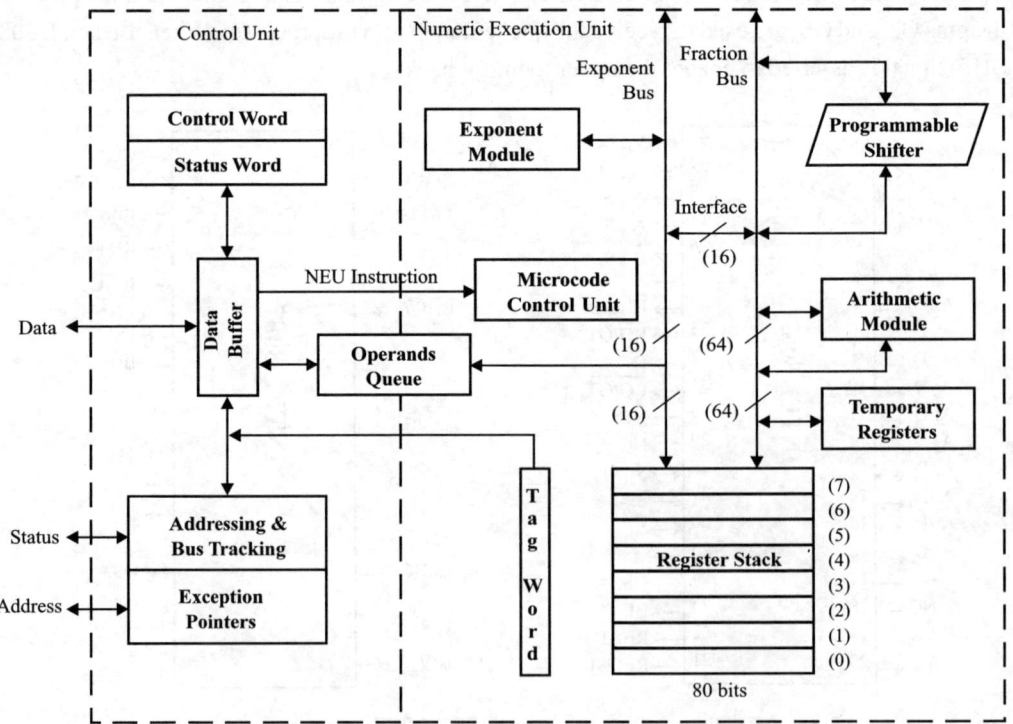

Fig. 11.8: Architecture of coprocessor-INTEL 8087.

The coprocessor works with seven types of numeric data, which are divided into the following three classes:

1. **Word integer (16-bit)** ⎫
2. **Short integer (32-bit)** ⎬ Binary integers
3. **Long integer (64-bit)** ⎭

4. **BCD (80-bit)** ⎬ Packed decimal number

5. **Short real (32-bit)** ⎫
6. **Long real (64-bit)** ⎬ Real numbers
7. **Temporary real (80-bit)** ⎭

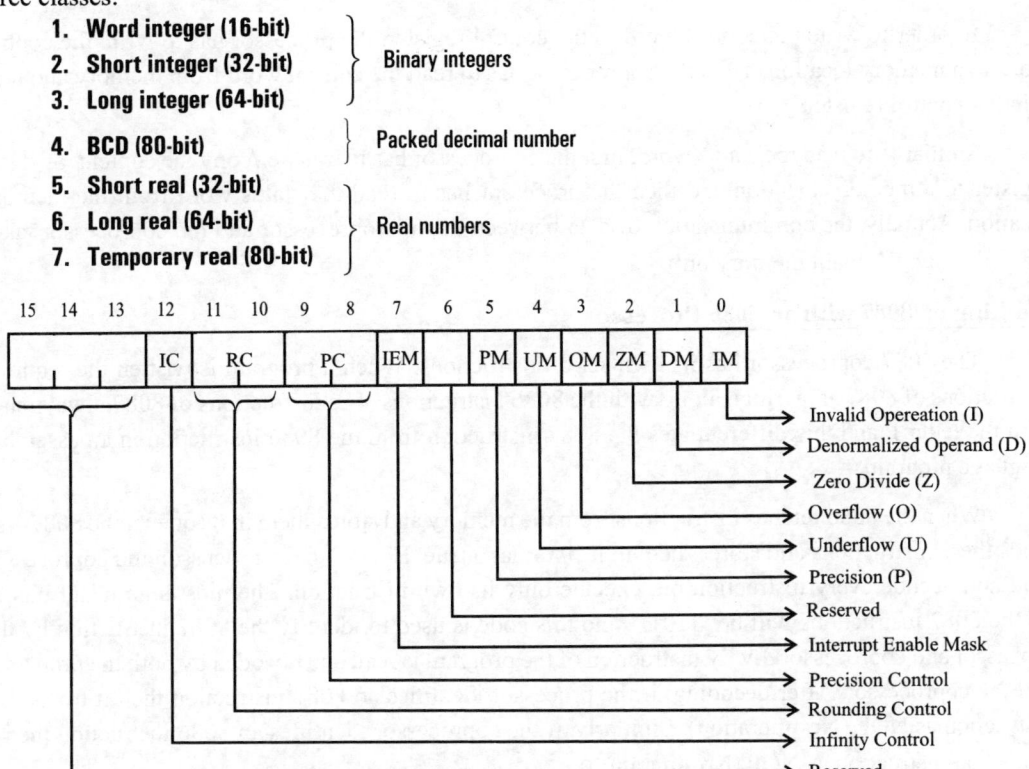

Fig. a: Format of the control word.

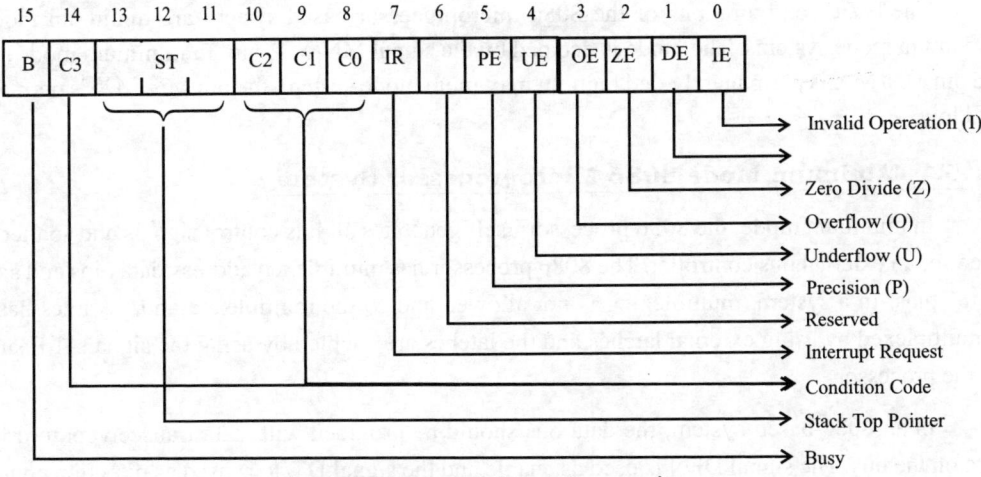

Fig. b: Format of the status word.

Fig. 11.9: Control and Status word format of coprocessor 8087

The 8087 has a 16-bit control word and a 16-bit status word. The format of the control word and the status word are shown in Fig. 11.9.

In order to write the control word to the control register, the processor has to write the control word to a memory location. Then the coprocessor has to read the control word from memory and load it in the control register.

Similarly, to read the status word, first the coprocessor has to transfer/copy the content of status register to a memory location and then the processor has to read the status word from the memory location. Actually, the communication of data between the microprocessor and the coprocessor takes place through the main memory only.

Working of 8087 with an 8086 Processor

The 8087 coprocessor has its own set of instructions. When a program is written the required instructions of 8087 are written along with the 8086 instructions. The instructions of 8087 always start with the letter F and this differentiates the 8087 instruction from the 8086 instruction in an assembly language program.

When the 8086 fetches instructions from the memory and stores them in its queue, the 8087 also reads these instructions and stores them in its internal queue. Here both the processor and coprocessor read and decode every instruction but execute only its own instruction. The most significant bits of all the 8087 instructions will be "11011" and this code is used to identify the 8087 instruction by the processor and coprocessor. Every instruction of the program is read and decoded by both the processor and the coprocessor. After decoding, if the processor identifies an 8087 instruction then it treats that instruction as NOP (**No-op**eration). Similarly, if the coprocessor identifies an 8086 instruction then it treats that instruction as NOP (**No-op**eration).

11.5 BASIC CONFIGURATION OF 8086 MICROPROCESSOR-BASED SYSTEM

The basic configurations of the 8086 microprocessor-based system are minimum mode and maximum mode systems. The mode is decided by the signal $MN/\overline{MX}$ pin. In minimum mode system, the pin $MN/\overline{MX}$ is permanently tied high. In maximum mode system, the pin $MN/\overline{MX}$ is permanently tied low.

11.5.1 Minimum Mode 8086 Microprocessor System

In minimum mode, the 8086 processor itself generates all bus control signals and so there is no need for an external bus controller. The 8086 processor has multiplexed address/data pins and address/ status pins. In a system, multiplexing is not allowed and so the multiplexed address lines has to be demultiplexed by using external latches and the latches are enabled by using the signal ALE supplied by the processor.

In an 8086-based system, the data bus should be provided with data transceivers to drive the data on the bus. The signal DEN is used as enable and the signal $DT/\overline{R}$ is used as direction control for data transceivers.

The formation of an address bus and data bus in the 8086-based minimum mode system is shown in Fig. 11.10. For minimum mode of operation, the $MN/\overline{MX}$ is tied to V_{cc} (+5-V). The clock generator of the 8284 is used to generate the clock, reset and ready signals for the processor. A quartz crystal of 15 MHz frequency is connected to the X_1 and X_2 pins of 8284 so that the clock frequency supplied to the 8086 processor will be 5 MHz. An RC circuit is connected to the reset input of 8284 to provide power-ON reset. A switch is also connected across the capacitor to provide manual reset.

In the system shown in Fig. 11.10, three numbers of 8-bit latch 74LS573 are used as address latches and two numbers of 8-bit bidirectional buffer 74LS245 are used as data transceivers. The interfacing of memory and IO (or peripheral) devices are discussed in Chapter 9.

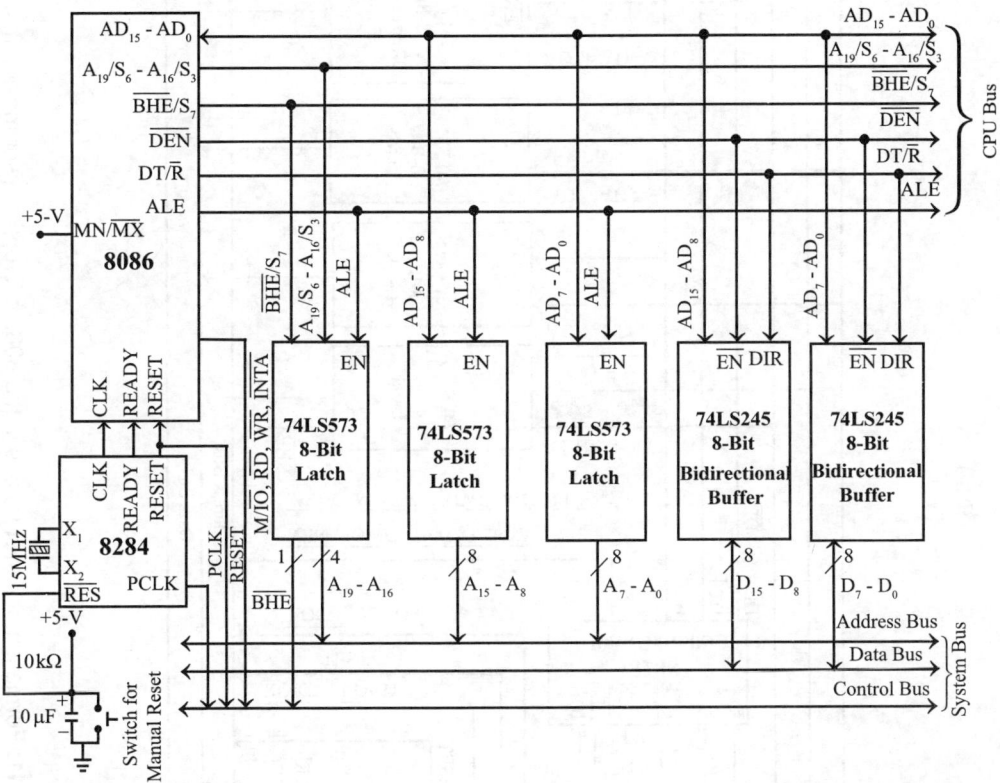

Fig. 11.10: Formation of a system bus in the 8086 microprocessor minimum mode system.

11.5.2 Maximum Mode 8086 Microprocessor System

In a maximum mode 8086-based system, an external bus controller 8288 has to be employed to generate the bus control signals. The 8288 can be configured for uniprocessor or multiprocessor modes of operation using the signals, $\overline{AEN}$, IOB and CEN. The formation of an address bus and data bus in 8086-based maximum mode system is shown in Fig. 11.11. For maximum mode of operation, the pin $MN/\overline{MX}$ of the 8086 processor is tied to the ground.

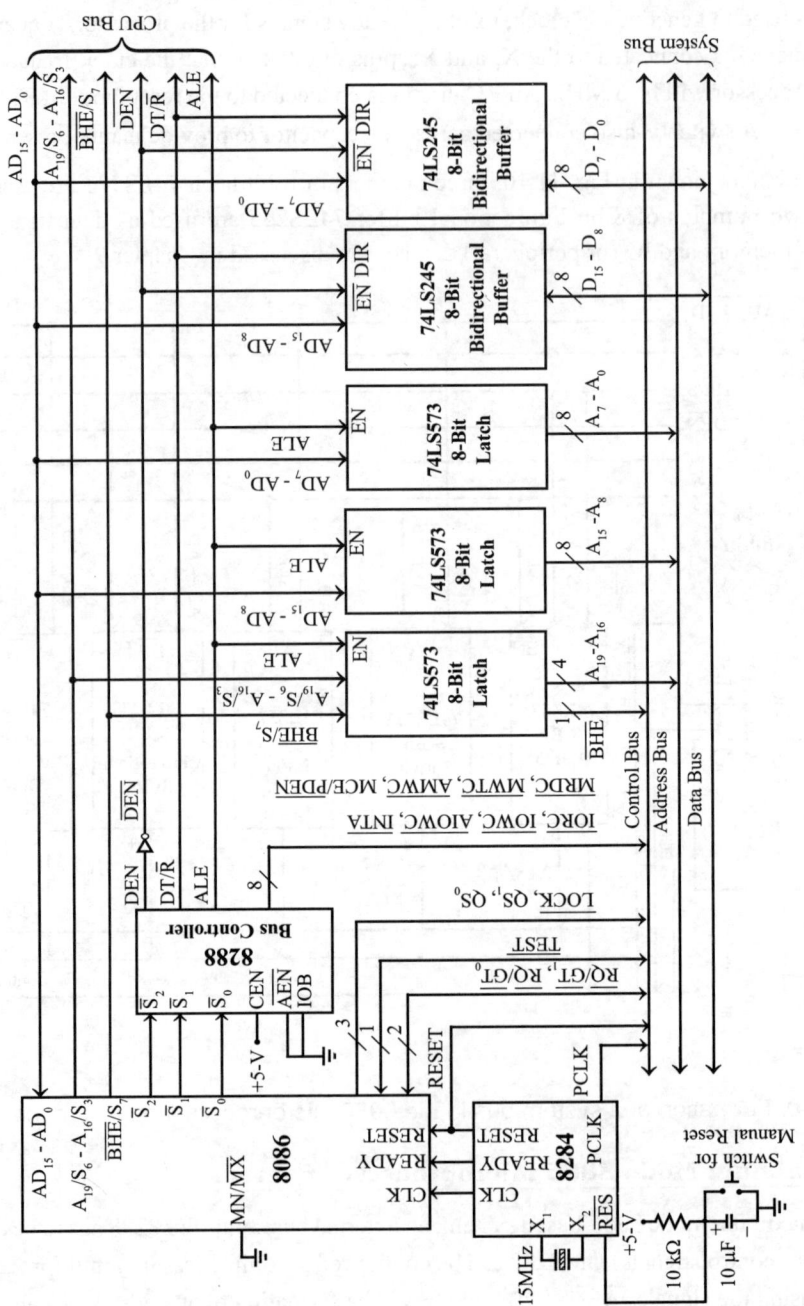

Fig. 11.11: Formation of a system bus in the 8086 microprocessor maximum mode system.

The system shown in Fig. 11.11 employs a bus controller 8288 to generate bus-control signals. Here, the bus controller is configured for the uniprocessor mode of operation by grounding $\overline{AEN}$ and IOB, and by applying +5 V to CEN. (For multiprocessor mode of operation, IOB should be tied to +5 V and the signals $\overline{AEN}$, and CEN are supplied by a bus arbiter such as INTEL 8289.)

In the 8086 processor, the address is multiplexed with the data or status signals. In a system, multiplexing is not allowed and so the multiplexed address lines of the CPU bus has to be demultiplexed by using external latches. In the system shown in Fig. 11.11, three 8-bit latches 74LS573 are employed to demultiplex the address lines. The signal ALE generated by the bus controller is used as enable for the latches.

In the 8086-based system, the data bus should be provided with data transceivers to drive the data on the bus. In the system shown in Fig. 11.11, two bidirectional buffers 74LS245 are employed as data transceivers. The signals DEN and $DT/\overline{R}$ generated by the bus controller are used as enable and direction control of buffers, respectively.

The system employs a clock generator INTEL 8284 to generate the clock, reset and ready signals for the 8086 processor. A quartz crystal of 15 MHz frequency is connected to the X_1 and X_2 pins of 8284 so that the clock frequency supplied to the 8086 processor will be 5 MHz. An RC circuit is connected to the reset input of the 8284 to provide power-ON reset. A switch is also connected across the capacitor to provide manual reset. The bus controller generates separate read and write controls for memory and IO devices. It also generates extended write control signal for memory and IO devices requiring higher write time.

11.6 MULTIPROCESSOR CONFIGURATIONS OF 8086 MICROPROCESSOR-BASED SYSTEM

A multiprocessor system will have two or more processors that can execute instructions (or perform operations) simultaneously. In multiprocessor systems, the extra or added processors can be special-purpose processors which are specifically designed to perform certain tasks efficiently or can be other general-purpose processors. For example, a multiprocessor system can be formed by using an 8086 microprocessor and an 8087 coprocessor in order to impart efficient floating-point arithmetic capability to the 8086-based system.

The multiprocessor systems offers the following advantages over single-processor design:

1. Several low-cost processors may be combined to fit the needs of an application while avoiding the expense of the unneeded capabilities of a centralized system.
2. The multiprocessor system provides room for expansion because it is easy to add more processors as the need arises.
3. In a multiprocessor system, implementation of modular processing of tasks can be achieved.
4. When a failure occurs, it is easier to replace only the faulty part/processor.

The two major issues in multiprocessor system design are bus contention and interprocessor communication. In a multiprocessor system, more than one processor will share the system memory and IO devices through a common system bus and so extra logic must be included to ensure that only one processor has access to the system bus at any one time.

Also, there should be an unambiguous way of interprocessor communication so that one processor can dispatch a task or return a result to another processor unambiguously. The maximum mode 8086 microprocessor has features for designing a multiprocessor system. Two types of multiprocessor configurations can be formed using 8086 processor: closely coupled (or tightly coupled) configuration and loosely coupled configuration.

11.6.1 Closely Coupled Configuration

In an 8086-based system, only two or three processors can be connected to work in closely coupled configuration and in this, 8086 is the host/master processor and the other/supporting processor is the slave processor. The slave processor can be a coprocessor or an IO processor or general purpose processor. Two numbers of 8086 processors cannot work in closely coupled configurations. In closely coupled configuration, both the master and slave processors share the same bus control logic, clock generator, system memory and IO. In closely coupled configurations, each processor will not have its own local memory or IO. The general block diagram of a closely coupled configuration is shown in Fig. 11.12.

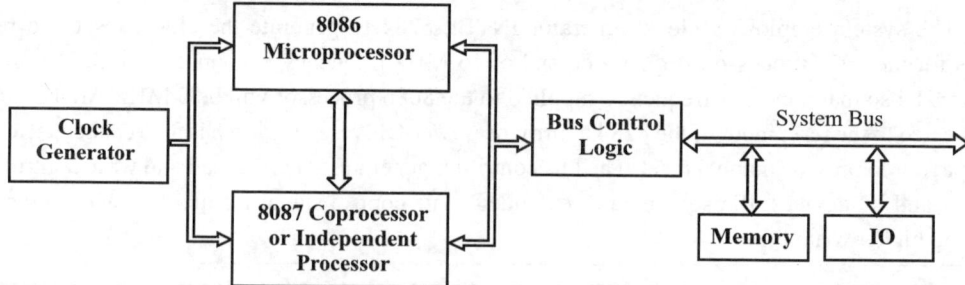

Fig. 11.12: Closely coupled configuration.

In a closely coupled configuration, the bus access control is provided by the master/host processor and so the bus request of the supporting/slave processor is connected to the master (In case of 8086/8088 the signal RQ, GT is used for bus request and grant.) In a closely coupled configuration, when the slave is a coprocessor, it interacts directly with the master and to a certain extent its functioning depends on the master. But when the slave is another processor then it can work independently.

11.6.2 Loosely Coupled Configuration

In a loosely coupled configuration, a number of modules of microprocessor-based system (or masters) can be interfaced through a common system bus to work as a multiprocessor system. Each module in the loosely coupled configuration is an independent microprocessor-based system with its own clock source, and its own memory and IO devices interfaced through a local bus. Each module can also be a closely coupled configuration of a processor and coprocessor. The block diagram of a loosely coupled configuration using three modules of 8086 is shown in Fig. 11.13.

Each module in a loosely coupled configuration functions independently and there is no direct connection between them. The modules can have access to system resources (system memory and IO) through the system bus. Each module has a local and system bus control logic which takes care of interprocessor communication and system bus allocation to the masters/modules competing for system resources.

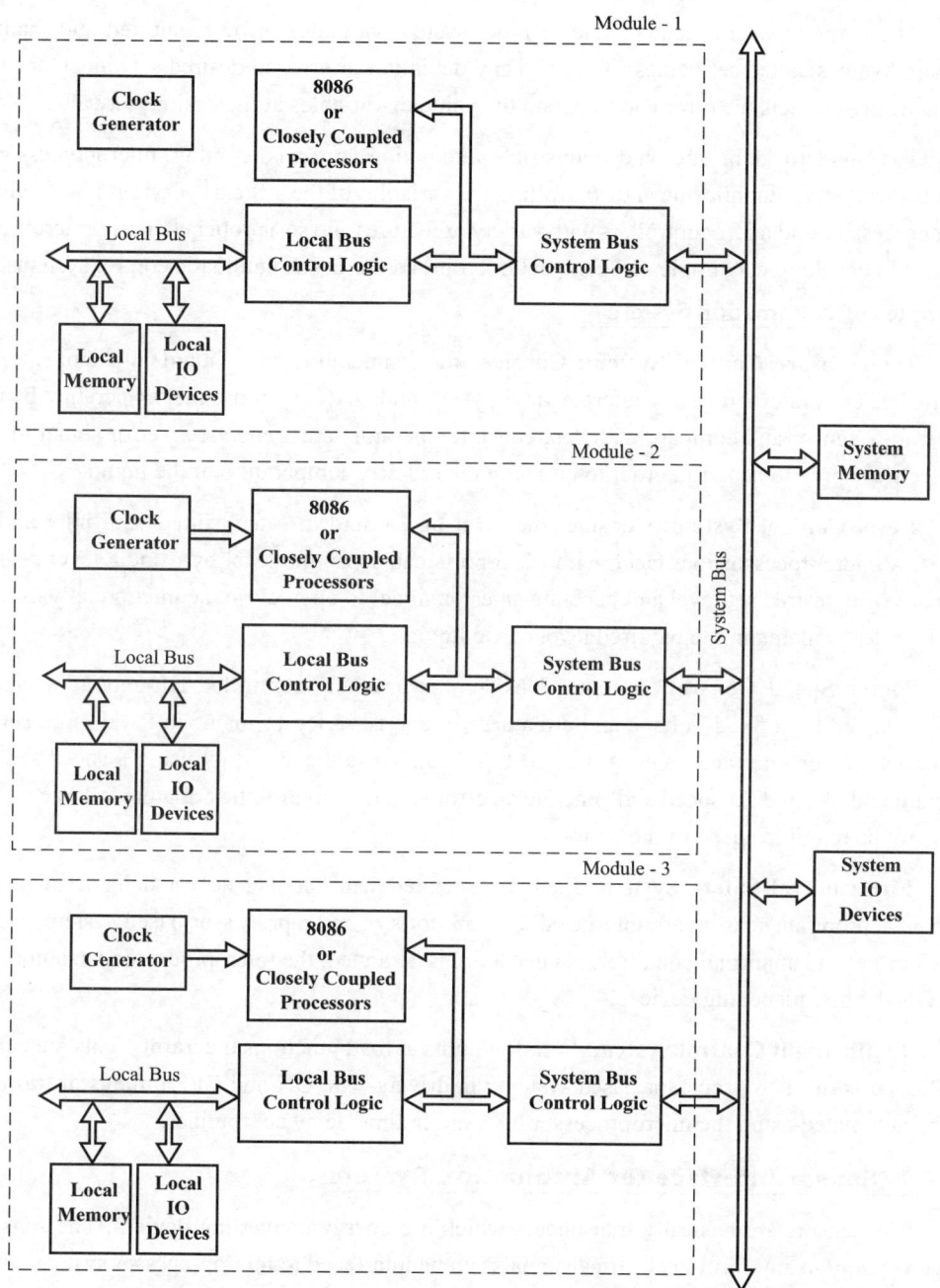

Fig. 11.13: Loosely coupled configuration.

11.7 APPLICATION TO AUTOMATION SYSTEMS

Automation systems require one or more system variables to be monitored and maintained at desired values called set values. If there is any deviation or error in desired set values, the system automatically corrects the error and maintain the value of variables at the desired set values.

The modern domestic and industrial automation systems employ microprocessors and microcontrollers for monitoring and controlling the variables of the systems. There is a wide choice of microprocessors and microcontrollers with variety of features and so a proper choice of microprocessor can lead to development simple and compact microprocessor based automatic control systems.

Examples of Automation Systems

Temperature Control System: Consider the temperature of a liquid in a tank fitted with electric heater. A microprocessor interfaced with ADC and DAC can sense the temperature by using a temperature sensor and compare with desired set temperature and generate an error signal to control the electric supply to heater in order to maintain the required temperature of the liquid.

Level Control System: Consider the level of a liquid in a tank fitted with inlet and outlet valves. A microprocessor interfaced with 8255 ports can sense the level by using a level sensor and compare with desired set level and generate an error signal to open/close the inlet/outlet valves of the tank in order to maintain the required level of the liquid.

Motor Speed Control System: Consider the armature-controlled DC motor in which the field is excited by a fixed voltage and armature voltage is varied to maintain the desired set speed. A microprocessor interfaced with ADC and DAC can sense the speed by using a speed sensor and compare with desired set speed and generate an error signal to control the armature voltage in order to maintain the required speed of the motor.

Fire Control Safety System: Fire is associated with smoke, heat and light. In fire safety systems appropriate sensors are interfaced to 8255 ports of microprocessor based system in order to sense normal and abnormal conditions. When a fire is deducted, the microprocessor can automatically turn ON the fire quenching devices.

Traffic Light Control System: In traffic lights at road junctions the traffic lights are connected to 8255 ports of microprocessor based system via drivers. The ON and OFF timings of traffic lights can be automated using the microprocessor by running time delay subroutines.

11.7.1 Sensor Interface for Automation System

The sensors are basically transducers which are energy-converting devices. The sensors that convert energy in any form to electrical signals can be interfaced to microprocessor system with ADC and ports. These sensors generate electrical signals proportional to the quantity being sensed and so can be used for measurement of the physical quantity being sensed. Examples of such applications are measurement of temperature, humidity, pressure, flow, etc.

When a sensor is used for measurement, the magnitude of the electrical signal from the sensor should be calibrated in order to relate the quantity being measured with electrical output. Also, the electrical signal has to be amplified and scaled by using operational amplifiers to a range of voltage acceptable by ADC (Analog-to-Digital-Converter). This process is also known as signal conditioning.

The output signal from the signal conditioning unit will be an analog voltage. Using ADC, this analog voltage can be converted to digital data (8-bit/10-bit/12-bit data) and fed to a microprocessor.

In certain applications, the sensors can be used to sense normal and abnormal conditions of a physical quantity or presence and absence of a physical quantity. Examples of such applications are gas-leakage detection, fire detection, etc.

In such cases, the analog signal from the sensor can be converted to digital signal using comparator and fed to one of the interrupts of a microprocessor or fed to one of the IO port pins of the 8255 ports interfaced with microprocessor. The comparator will compare two analog inputs and generate a high or low signal that can be used as a digital signal. One input to the comparator is a preset analog voltage and another input is the signal conditioned voltage from the sensor.

Temperature Sensors

The various types of temperature sensors are thermistors, thermocouples, bimetallic strips, integrated circuits (IC), temperature sensors, etc. Since the output voltage of an IC version of temperature sensors is internally linearized, they may not require calibration and so they are superior versions of temperature sensors for interfacing with microprocessors.

Some examples of the IC version of temperature sensors are AD590, LM34, LM35, etc. In an IC version of the temperature sensor, the change in Base-Emitter voltage of a transistor Base-Emitter junction is used to sense the temperature.

The IC includes circuits for amplification and linearization of this voltage in order to generate an output voltage proportional to temperature. Therefore, no additional hardware is required for interfacing the IC version of the temperature sensor to a microprocessor via an ADC.

Ultrasonic Sensors

The ultrasonic sensors are widely used for distance and speed measurement. The other applications include humidifiers, sonar, medical ultrasonography, burglar alarms, non-destructive testing and wireless charging.

The ultrasonic sensor basically employs a transducer-like piezoelectric crystal which generates high-frequency sound waves. Piezoelectric crystals change size and shape when an AC voltage is applied which creates sound waves whose frequency is same as that of the applied AC voltage. Since piezoelectric materials generate a voltage when force is applied to them, they can also work as ultrasonic detectors.

The ultrasonic sensors for distance measurement are commercially available as a module consisting of an ultrasonic transmitter and receiver. The transmitter is basically a sound-wave generator, that generates high-frequency sound waves, above 20 kHz (typically values are in the range

28 kHz to 40 kHz). When the ultrasonic signals are transmitted in a medium, they get reflected at the interface where there is a change in the medium. The reflected wave can be detected by an ultrasonic receiver. The time difference between the transmitted wave and reflected wave can be used to estimate the distance between the transmitter and reflected object.

Humidity Sensor

Humidity is the presence of water in air. The amount of water vapour in air can affect human comfort as well as many manufacturing processes in industries. Electronic-type humidity sensors are of two types: capacitive type and resistive type.

The capacitive-type humidity sensor is based on the change in dielectric due to change in humidity. Humidity sensors relying on this principle consist of a hygroscopic dielectric material sandwiched between a pair of electrodes forming a small capacitor. The change in capacitance can be converted to a change in analog voltage by an appropriate circuit.

Resistive-type humidity sensors depend on change in the resistance of the sensor element due to change in humidity. The humidity-sensitive resistor is composed of an organic polymer, such as polyvinyl chloride or polyethylene or a metal oxide. The change in resistance can be converted to a change in analog voltage by an appropriate circuit.

Humidity sensors are available as modules that give an electrical voltage output which can be amplified and scaled using signal-conditioning circuit and applied to the input of an ADC interfaced to the microprocessor.

Gas Sensor

Gas sensors can be used to detect combustible, flammable and toxic gases, and non-availability of oxygen. The gas sensors are mostly part of a safety system in industrial plants, refineries, waste water treatment facilities, vehicles and homes to detect potentially hazardous gas leakage.

The various types of gas sensors are infrared gas sensors, ultrasonic gas sensors, electrochemical gas sensors and semiconductor gas sensors.

Semiconductor sensors employ chemicals such as tin dioxide to detect gases. The resistance of tin dioxide decreases with increase in the concentration of gas surrounding it. This change in resistance is used to estimate the gas concentration. Semiconductor sensors are commonly used to detect hydrogen, oxygen, alcohol vapour, and harmful gases such as carbon monoxide.

Infra-Red(IR) Sensors

Infrared (IR) rays are invisible electromagnetic waves which have a wide frequency range between microwaves and visible light in the electromagnetic spectrum. The frequency range of IR waves will be from 300 GHz to 430 THz (with wavelength in the range 1000 micrometre to 0.75 micrometre). The wavelength region of 0.75 μm to 3 μm is called near infrared, the region from 3 μm to 6 μm is called mid infrared, and the region higher than 6 μm is called far infrared.

The infrared sensors basically employ IR diodes as sources to generate IR waves and photo transistors that work in the IR frequency range to deduct IR waves. The IR diodes can emit infrared waves, when forward biased. The conduction of an IR photo transistor will depend on the strength of an IR wave incident on its base-emitter junction.

One of the major applications of an IR sensor is object sensing. When IR waves are transmitted in a medium, they will get reflected when they encounter an object in their path. The reflected wave can be sensed by the IR receiver to recognize the presence of an object.

11.8 TEMPERATURE CONTROL SYSTEM

11.8.1 Temperature Control System using 8085/8086 Processor

The microprocessor-based temperature control system can be used for automatic control of the temperature of a plant. A simplified block diagram of an 8085/8086 microprocessor-based temperature control system is shown in Fig. 11.14.

The system consist of 8085/8086 microprocessor as CPU, EPROM memory for program storage, RAM memory for stack and data storage, INTEL 8279 for keyboard and display interface, ADC, DAC, INTEL 8255 for IO ports, amplifiers, signal conditioning circuit, temperature sensor and supply control circuit. In this system, the temperature is controlled by controlling the power input to the heating element.

The EPROM memory is provided for storing the system program, and RAM memory for temporary data storage and stack operation. Using INTEL 8279, a keyboard and six numbers of 7-segment LEDs are interfaced to the system. The system has been designed to accept the desired temperature and various control commands through the keyboard. The 7-segment display has been provided to display the temperature of the plant at any time instant.

The temperature of the plant is measured using a temperature sensor. The different types of temperature sensors that can be used for temperature measurement are thermo-couples, thermistors, PN-junctions, IC sensors like AD590, etc. These sensors will convert the input temperature to proportional analog voltage or current. The output signal of the sensor will be a weak signal and so it has to be amplified using high input impedance operational amplifier. Then the analog signal is scaled to a suitable level by the signal conditioning circuit.

The microprocessor can process only digital signals and so the analog signal from signal conditioning circuit cannot be read by the processor directly. The system has an **A**nalog-to-**D**igital **C**onverter (ADC) to convert the analog signal to proportional digital data. In this system, the ADC is interfaced to 8085/8086 processor through port-A and port-C of 8255. The 8085/8086 processor sends signal to ADC through port-C to start conversion and at the end of conversion it reads the digital data from the port-A of 8255.

The 8085/8086 processor calculates the actual temperature using the input data and displays it on the 7-segment LED. The processor also compares the desired temperature with actual temperature (the operator can enter the desired temperature through the keyboard) and calculate the error (the difference between actual temperature and desired temperature.)

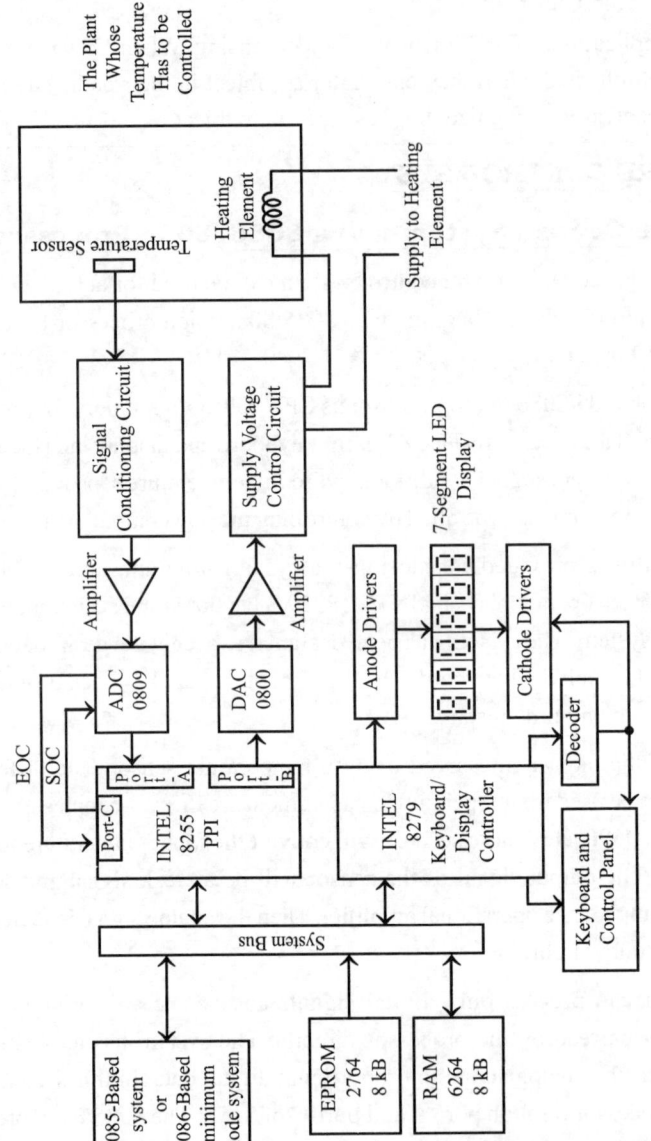

Fig. 11.14: An 8085/8086 microprocessor-based temperature control system.

The error is used to compute a digital control signal, which is converted to analog control signal by DAC. The DAC is interfaced to the system through port-B of 8255. The analog control signal produced by DAC is used to control the power supply of the heating element of the plant.

The digital control signal can be computed by the 8085/8086 processor using different digital control algorithms (P/PI/PID/FUZZY logic control algorithms).

The control circuit for power supply can be either a thyristor-based circuit or relay. In case of thyristor control circuits, the firing angle can be varied by the control signal to control the power input to the heater. In case of relay, the control signal can switch ON/OFF the relay to control the power input to the heater.

The sequence of operations performed by the microprocessor based system are shown in the flowchart of Fig. 11.15.

11.8.2 Interfacing Temperature Sensor with the 8051 Microcontroller

The interfacing of the IC version of a temperature sensor, LM35 with the 8051 controller is shown in Fig. 11.16. The LM35 devices are precision integrated-circuit temperature sensors, with an output voltage linearly proportional to the Centigrade temperature and can be used to measure temperature in the range $-55°C$ to $150°C$, with typical accuracies of $\pm \frac{1}{4}$ °C at room temperature and $\pm \frac{3}{4}$ °C over a full temperature range.

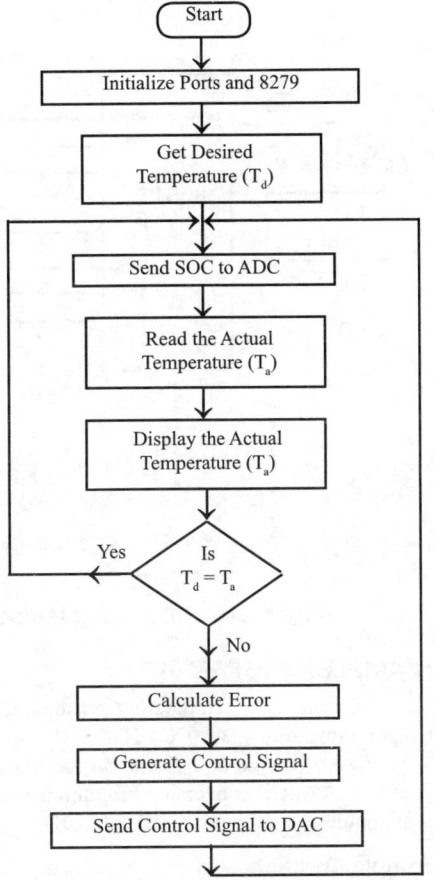

Fig. 11.15: Flowchart for temperature control system.

The device has only three pins. Two pins for connecting a DC power supply in the range +4 V to +20 V and one pin for output voltage proportional to temperature with respect to power supply ground. The output voltage will be 10 mV/°C. In the interface circuit shown in Fig. 11.16, a capacitor is connected across the output for noise immunity. The sensor output is connected to a high input impedance operational amplifier, to avoid loading of the sensor. Also, the operational amplifier is designed to provide a gain in order to scale the output value suitable for ADC input.

The ADC program developed in Chapter-9, Section 9.15.5 can be used to read the temperature sensor output through ADC as digital data. The digital data read from ADC should be converted to temperature value by multiplying by an appropriate constant and then converted to ASCII value and send to LCD for display. The LCD program developed in Chapter-9, Section 9.16.4 can be used to display the temperature value on the LCD. Refer Example Program 51.10 in Chapter 8 for ASCII conversion.

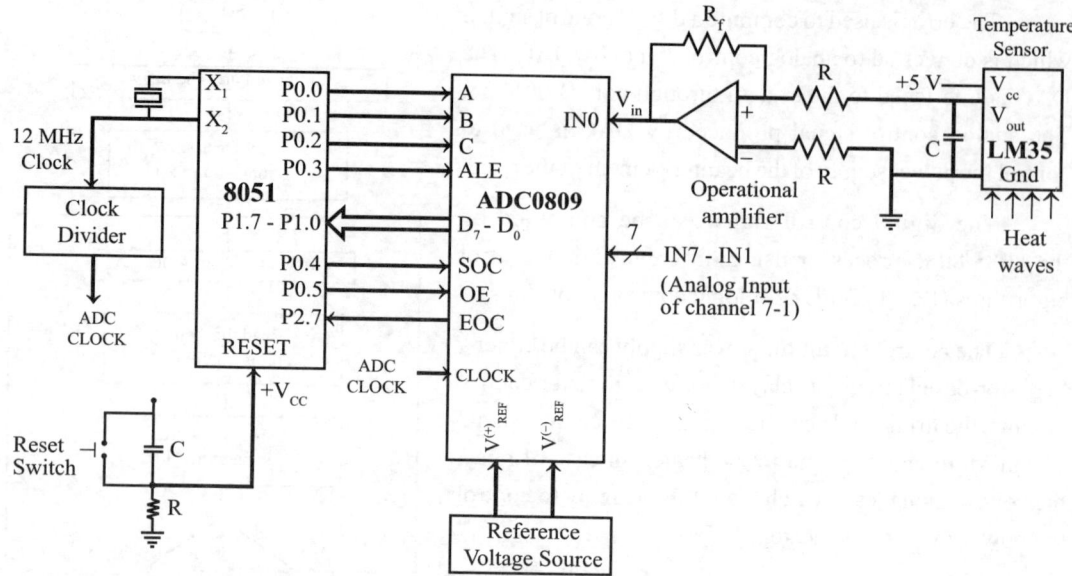

Fig. 11.16: Interfacing temperature sensor LM35 to 8051 microcontroller via ADC.

EXAMPLE PROGRAM 11.1

Consider the temperature sensor interface shown in Fig. 11.16. Let the interface be designed to read temperature in the range 0°C - 100°C. Let the signal conditioning circuit using operational amplifier be designed to give zero voltage at 0°C and an increment of 19.61 mV for every one degree rise in temperature. Let us assume that ASCII conversion program is available as a subroutine called ASCICON and LCD display program is available as a subroutine called LCDDISP.

Sample Calculation of Temperature

The 8-bit ADC data for the input analog voltage, V_{in}, when $+V_{ref}$ is $+5$ V and $-V_{ref}$ is 0 V is given by,

$$\text{ADC data} = \frac{V_{in}}{+V_{ref}} \times 255$$

The ADC data is calculated for various values of input analog voltage, V_{in} and tabulated in the following table:

Temperature	V_{in}	ADC data
0°C	0 V	0 $= 00_H$
1°C	19.61 mV	1 $= 01_H$
2°C	39.22 mV	2 $= 02_H$
.	.	.
25°C	490.25 mV	25 $= 19_H$
.	.	.
50°C	980.5 mV	50 $= 32_H$
.	.	.
75°C	1470.75 mV	75 $= 4B_H$
.	.	.
100°C	1961 mV	100 $= 64_H$

Flowchart

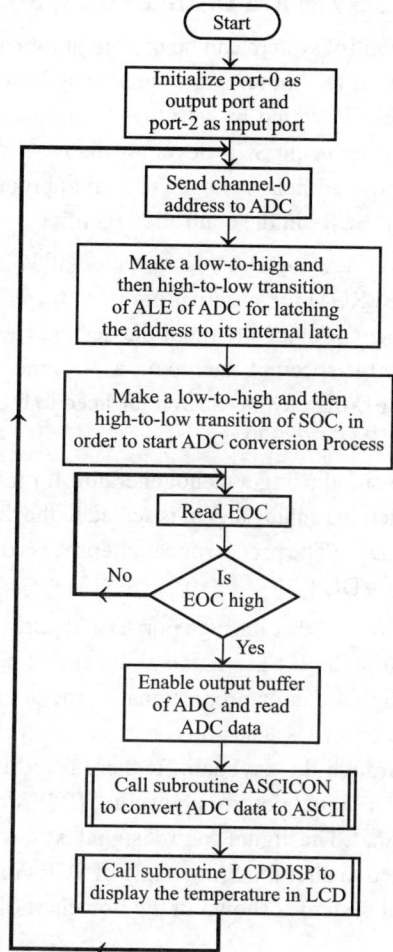

Assembly Language Program

```
;Program to measure temperature using LM35 interfaced to 8051 via ADC

START:   MOV    P0,#00H        ;Initialize port-0 as output port
         MOV    P2,#FFH        ;Initialize port-2 as input port
AGAIN:   MOV    P0,#00H        ;Send channel-0 address to port-0
         SETB   P0.3           ;Set ALE high
         NOP                   ;Delay and make ALE low
         NOP
         CLR    P0.3

         SETB   P0.4           ;Set SOC high
         NOP                   ;Delay and make SOC low
         NOP
         CLR    P0.4
WAIT:    JNB    P2.7,WAIT      ;Wait until EOC is high

         SETB   P0.5           ;Set OE high, to enable ADC output buffer
         MOV    A,P1           ;Get ADC data in A register

         ACALL  ASCICON        ;Call subroutine ASCICON to convert ADC data to ASCII
         ACALL  LCDDISP        ;Call subroutine LCDDISP to display the temperature value in LCD

         SJMP   AGAIN
         END                   ;Assembly end
```

11.9 MOTOR SPEED CONTROL SYSTEM USING 8085/8086/8051

The microprocessor-based speed control system can be used to automatically control the speed of a motor. A typical 8085/8086/8051-based DC motor speed control system is shown in Fig. 11.18. In this system, the speed of the DC motor is varied by varying the armature voltage and the field voltage is kept constant. A controlled rectifier using SCR develops the required armature voltage and the uncontrolled rectifier generates the required field voltage. The microprocessor controls the speed of the motor by varying the firing angle of SCRs in the controlled rectifier.

The speed control system has been developed using the 8085/8086 microprocessor or 8051 microcontroller as CPU. The system has EPROM for system program storage, and RAM for temporary data storage and stack. A keyboard has been provided to input the desired speed and other commands to operate the system. In order to display the speed of the motor, a 7-segment LED display has been provided. The keyboard and 7-segment LED display has been interfaced to the 8085/8086/8051-based system using keyboard/display controller INTEL 8279.

The speed of the DC motor is measured using a tachogenerator. It produces an analog voltage proportional to the speed of the motor. Then the analog signal is scaled to the desired level by the signal conditioning circuit and digitized using ADC. (The processor cannot process the analog signal directly, hence the analog signal is digitized using ADC.)

The ADC is interfaced to the 8085/8086/8051 through port-B and port-C of 8255. The processor can send a start of conversion to ADC through port-C pin and at the end of conversion it can read the digital data from port-B of 8255. This digital data is proportional to the actual speed. The processor calculates the actual speed and displays it on LEDs. The processor also compares the actual speed with desired speed entered by the operator through the keyboard. If there is a difference then the error is estimated. The error can be modified by a digital control algorithm, (P/PI/PID/FUZZY logic control algorithm) to produce a digital control signal. The digital control signal is converted to analog signal by the DAC. The analog control signal is used to alter the firing angle of SCRs in the controlled rectifiers. The operational flow of the speed control system is shown in the flowchart of Fig. 11.17.

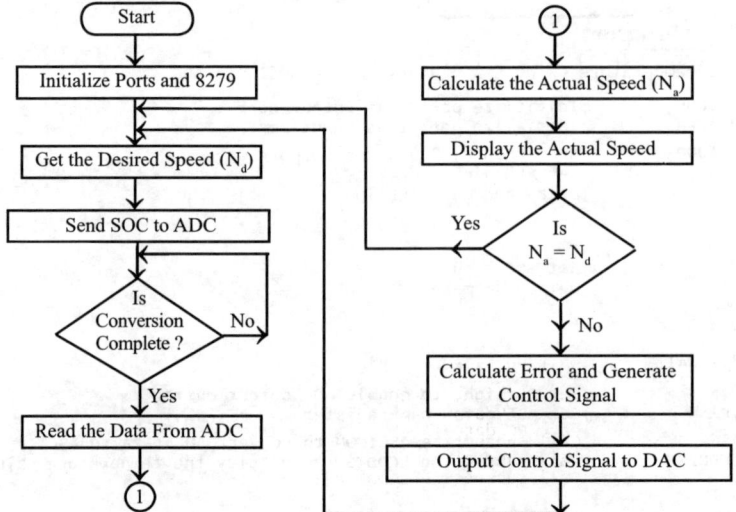

Fig. 11.17: Flowchart for a DC motor speed control system.

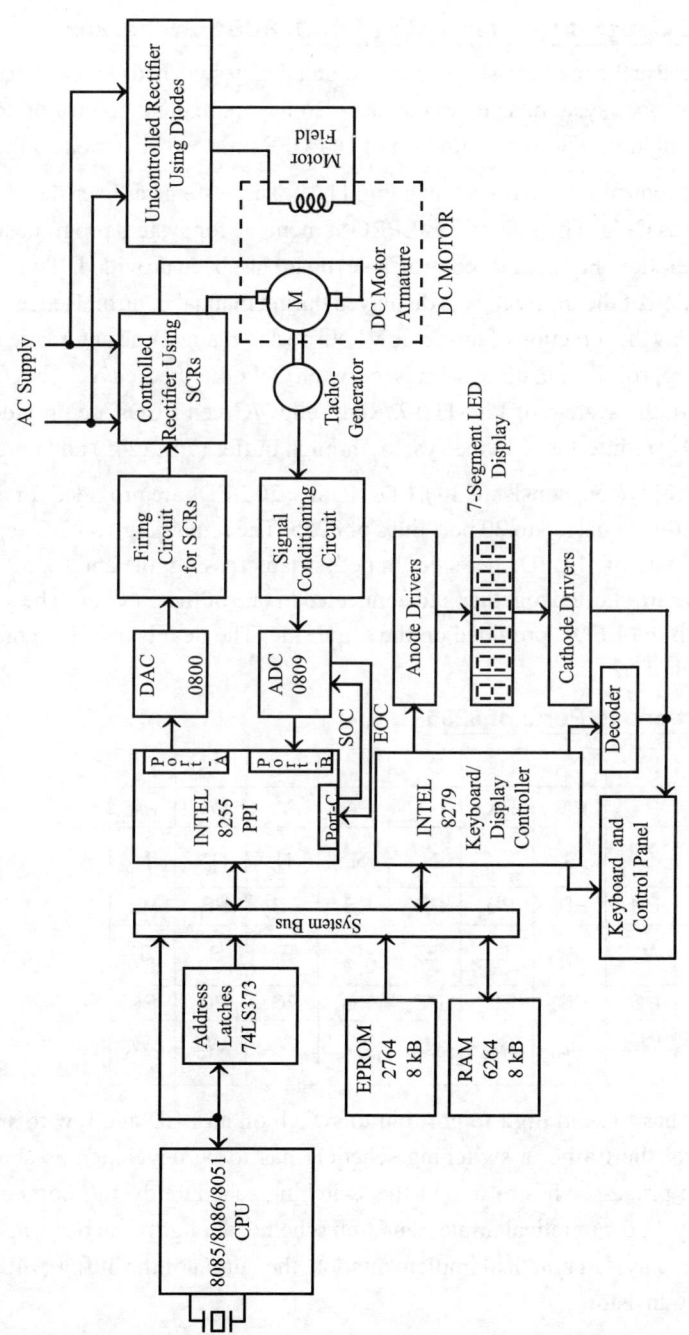

Fig. 11.18: An 8085/8086/8051-based DC motor speed control system.

11.10 TRAFFIC LIGHT CONTROL SYSTEM

11.10.1 Traffic Light Control System using 8085/8086 Processor

The traffic lights placed at the road crossings can be automatically switched ON/OFF in the desired sequence using a microprocessor system. The system can also have a manual control option, so that during heavy traffic (or during traffic jam) the duration of ON/OFF time can be varied by the operator.

A typical traffic light control system is shown in Fig. 11.19. The system has been developed using 8085/8086 microprocessor as CPU. The system has EPROM memory for system program storage and RAM memory for stack operation. For manual control, a keyboard has been provided. It will be helpful for the operator if the direction of the traffic flow is displayed during manual control. Hence, 7-segment LEDs are interfaced to display the direction of traffic flow both during manual and automatic mode. The primary function of the microprocessor in the system is to switch ON/OFF the Red/Yellow/Green lights in the specified sequence. In the system of Fig. 11.19, Red/Yellow/Green LEDs are provided instead of lights (lamps). The LEDs are interfaced to the system through buffer (74LS245) and ports of 8255.

The traffic light control system consist of 36 LEDs. In this, 20 LEDs are provided for controlling vehicle traffic and are directly connected to 20 port lines of 8255. The remaining 16 LEDs are used for pedestrian crossings, at the rate of 4 LEDs per side. In pedestrian crossing, the anodes of two green LEDs provided on each side are shorted together and connected to one of the port line. The same signal is inverted and given to two red LEDs provided on the same side. The details of LED connection to port lines are shown in Table 11.2.

Table 11.2: LED Connections to Ports of 8255

Data Bus	D_7	D_6	D_5	D_4	D_3	D_2	D_1	D_0
Port-A Pins	PA_7	PA_6	PA_5	PA_4	PA_3	PA_2	PA_1	PA_0
LEDs Connected to port-A	S_G	S_Y	S_R	N_{FL}	N_{FR}	N_G	N_Y	N_R
Port-B Pins	PB_7	PB_6	PB_5	PB_4	PB_3	PB_2	PB_1	PB_0
LEDs Connected to port-B	W_R	E_{FL}	E_{FR}	E_G	E_Y	E_R	S_{FL}	S_{FR}
Port-C Pins	PC_7	PC_6	PC_5	PC_4	PC_3	PC_2	PC_1	PC_0
LEDs Connected to port-C	W_{PG}	E_{PG}	S_{PG}	N_{PG}	W_{FL}	W_{FR}	W_G	W_Y

The microprocessor has to send **high** to port pin to switch on an LED and **low** to switch OFF an LED. In order to control the traffic, a switching schedule has to be developed as shown in the following example and the processor has to output the switching schedule data to ports one by one after the required time delay. In the practical implementation scheme, the lights can be turned ON/OFF using driver transistors and relays. In practical implementation, the output of the buffer (74LS245) can be connected to the driver transistor.

A relay placed at the collector of the transistor can be used to switch ON/OFF the light as shown in Fig. 11.20. A reverse-biased diode is connected across the relay coil to prevent relay chattering (for free-wheeling action).

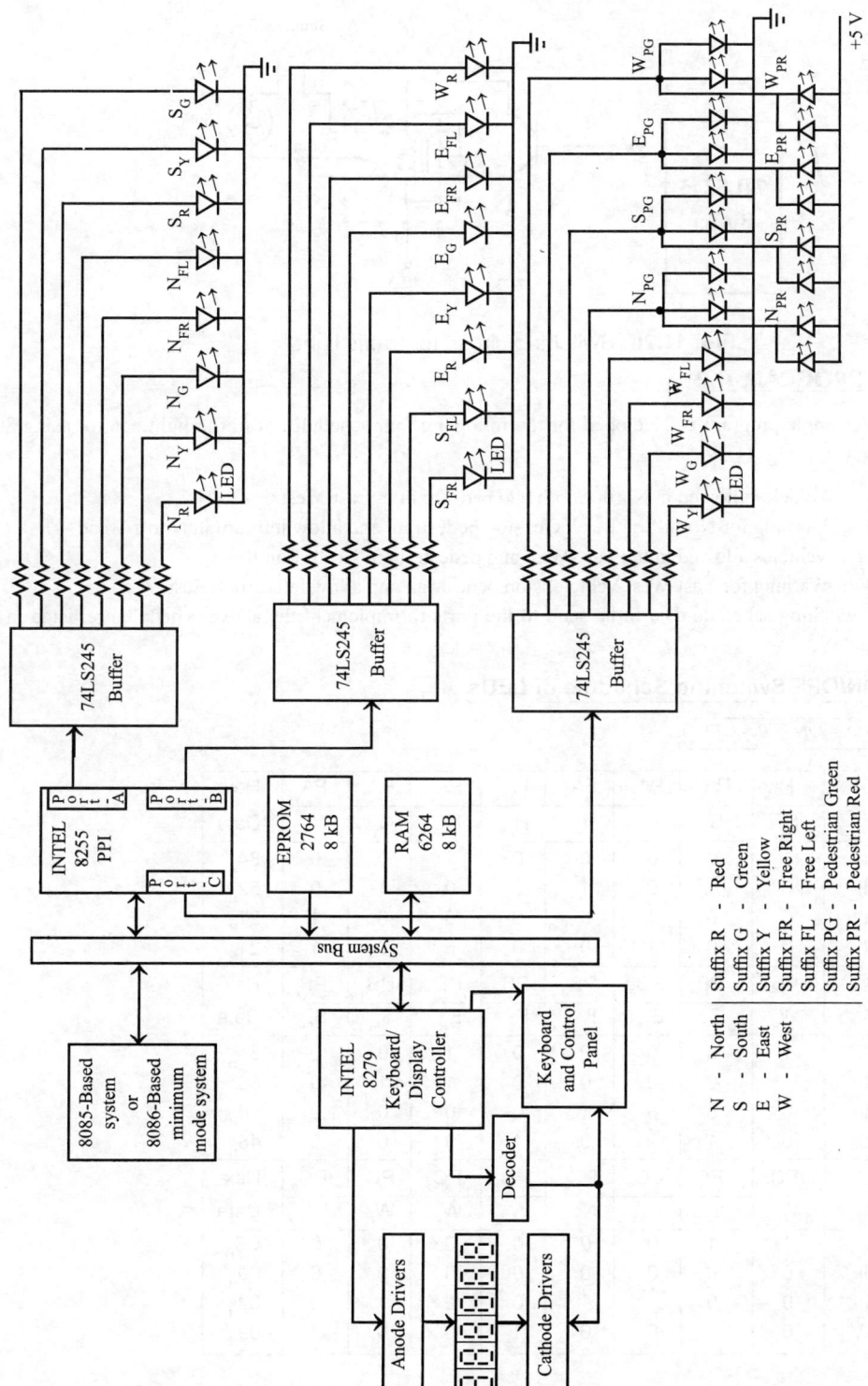

Fig. 11.19: An 8085/8086 microprocessor-based traffic light control system.

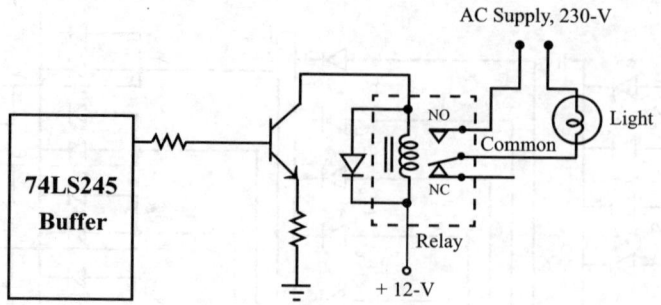

Fig. 11.20: Switching circuit for traffic light.

EXAMPLE PROGRAM 11.2

This example program is developed for the following four schedules of traffic-light control using 8085 processor shown in Fig. 11.19.

Schedule I : Vehicles allowed in North-South and pedestrain in East-West.
Schedule II : Warning for North-South vehicle, stop pedestrain and allow left turn in North-West.
Schedule III : Vehicles allowed in the East-West and pedestrain in North-South.
Schedule IV : Warning for East-West Vehicle, stop pedestrian and allow left turn in East-West.

The switching schedule data to be send to the ports to implement the above schedule are listed in the following Table 1.

Table 1: ON/OFF Switching Schedule of LEDs

> *Note:* 1 = ON ; 0 = OFF

Port-A Schedule	PA_7 S_G	PA_6 S_Y	PA_5 S_R	PA_4 N_{FL}	PA_3 N_{FR}	PA_2 N_G	PA_1 N_Y	PA_0 N_R	Hex Data
Schedule I	1	0	0	0	0	1	0	0	84_H
Schedule II	0	1	0	1	0	0	1	0	52_H
Schedule III	0	0	1	0	0	0	0	1	21_H
Schedule IV	0	0	1	0	0	0	0	1	21_H
Port-B Schedule	PB_7 W_R	PB_6 E_{FL}	PB_5 E_{FR}	PB_4 E_G	PB_3 E_Y	PB_2 E_R	PB_1 S_{FL}	PB_0 S_{FR}	Hex Data
Schedule I	1	0	0	0	0	1	0	0	84_H
Schedule II	1	0	0	0	0	1	1	0	86_H
Schedule III	0	0	0	1	0	0	0	0	10_H
Schedule IV	0	1	0	0	1	0	0	0	48_H
Port-C Schedule	PC_7 W_{PG}	PC_6 E_{PG}	PC_5 S_{PG}	PC_4 N_{PG}	PC_3 W_{FL}	PC_2 W_{FR}	PC_1 W_G	PC_0 W_Y	Hex Data
Schedule I	1	1	0	0	0	0	0	0	$C0_H$
Schedule II	0	0	0	0	0	0	0	0	00_H
Schedule III	0	0	1	1	0	0	1	0	32_H
Schedule IV	0	0	0	0	1	0	0	1	09_H

Flowchart

Flowchart for Main Program

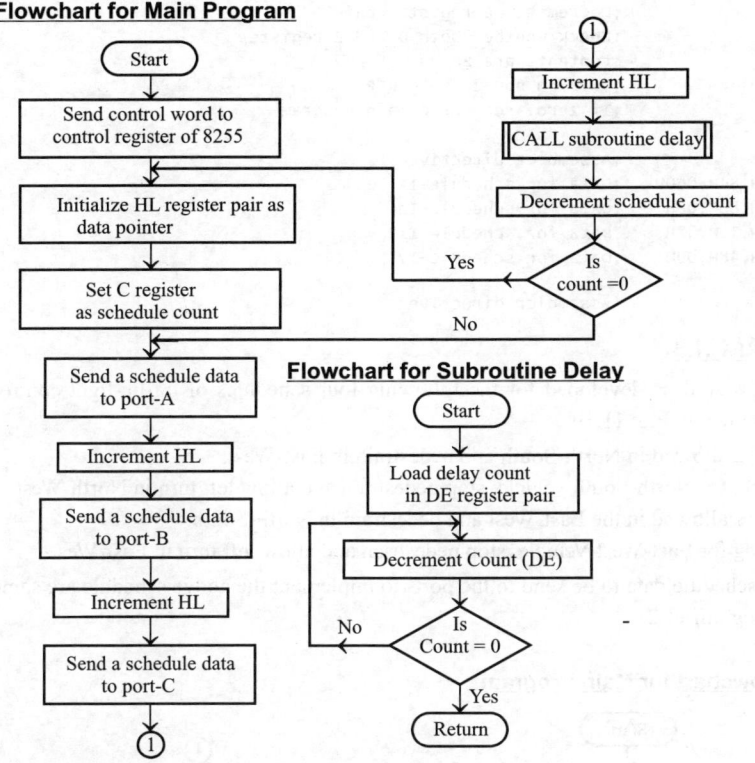

Flowchart for Subroutine Delay

Assembly Language Program

```
;Program for traffic light control
        PORTA    EQU  10H        ;Port-A address, 8255 is IO mapped with 8-bit IO addresses
        PORTB    EQU  11H        ;Port-B address
        PORTC    EQU  12H        ;Port-C address
        CNTPORT  EQU  13H        ;Control Port address
        CNTWORD  EQU  80H        ;Control Word to initialize ports in mode-0 output
        ORG  1000H               ;Assembler directive
        MVI   A,CNTWORD          ;Load control word in A-register to set 8255 ports in mode-0 output
        OUT   CNTPORT            ;Send control word to control register

START:  LXI   H,1100H            ;Initialize HL register pair as address pointerfor schedule data
        MVI   C,04H              ;Set C-register as schedule counter

AGAIN:MVI   A,M                 ;Load A-register with port-A schedule data
        OUT   PORTA              ;Send data to port-A
        INXH                     ;Increment address pointer
        MVI   A,M                ;Load A-register with port-B schedule data
        OUT   PORTB              ;Send data to port-B
        INXH                     ;Increment address pointer
        MVI   A,M                ;Load A-register with port-C schedule data
        OUT   PORTC              ;Send data to port-C
        INXH                     ;Increment address pointer
        CALL  DELAY              ;Call delay subroutine
        DCR   C                  ;Decrement schedule count
        JNZ   AGAIN              ;Repeat until schedule count is zero
        JMP   START              ;Jump to START, to repeat 4 schedules again and again
```

```
;DELAY ROUTINE
DELAY: LXI D,0F429H      ;Load DE register pair with delay count
REPT:  DCX D            ;Decrement DE register pair
       MOV A,E          ;Check whether both D and E register
       CMP D            ;contents are zeros
       JNZ REPT         ;Jump on non-zero to REPT
       RET              ;If zero, return to main program

ORG 1100H               ;Assembler directive
       DB  84H,84H,0C0H ;Data for schedule-I
       DB  52H,86H,00H  ;Data for schedule-II
       DB  21H,10H,32H  ;Data for schedule-III
       DB  21H,48H,09H  ;Data for schedule-IV

END                     ;Assembler directive
```

EXAMPLE PROGRAM 11.3

This example program is developed for the following four schedules of traffic-light control sequence using 8086 processer shown in Fig. 11.19.

Schedule I   :   Vehicles allowed in North-South and pedestrain in East-West.

Schedule II   :   Warning for North-South vehicle, stop pedestrain and allow left turn in North-West.

Schedule III :   Vehicles allowed in the East-West and pedestrain in North-South.

Schedule IV :   Warning for East-West Vehicle, stop pedestrian and allow left turn in East-West.

The switching schedule data to be send to the ports to implement the above schedule are same as that of Table-1 in Example Program 11.2.

Flowchart

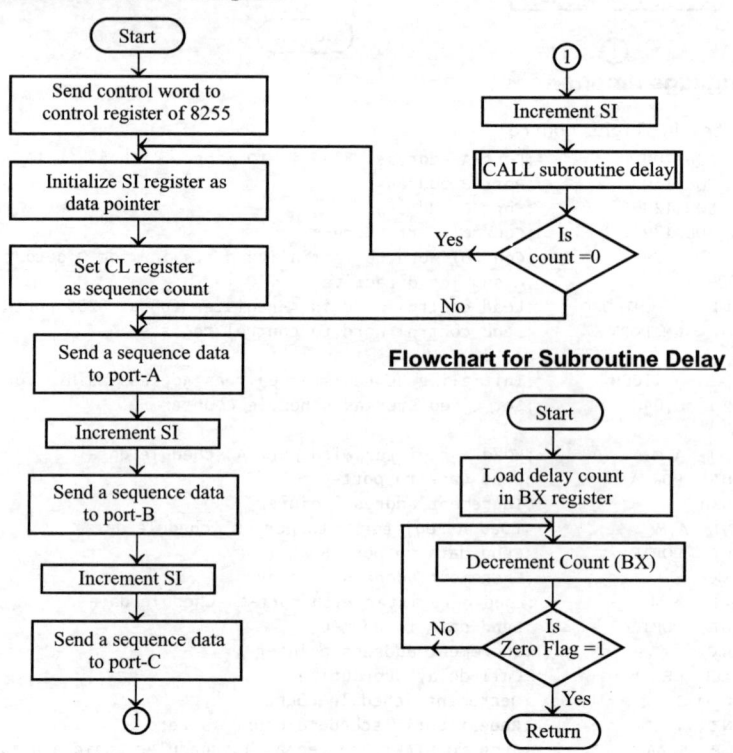

Flowchart for Main Program

Flowchart for Subroutine Delay

Assembly Language Program

```
;Program for traffic light control
PORTA    EQU 0000H              ;Port-A address
PORTB    EQU 0002H              ;Port-B address
PORTC    EQU 0004H              ;Port-C address
CNTPORT EQU 0006H               ;Control Port address
CNTWORD EQU 80H                 ;Control Word to initialize ports in mode-0 output

CODE SEGMENT                    ;Assembler directive
         ASSUME CS:CODE         ;Assembler directive
         ORG 1000H              ;Assembler directive
         MOV DX,CNTPORT         ;Get control register address in DX register
         MOV AL,CNTWORD         ;Load control word in AL to set 8255 ports in mode-0 output
         OUT [DX],AL            ;Send control word to control register
START:   MOV SI,1100H           ;Initialize SI register with offset address of schedule data
         MOV CL,04H             ;Set CL register as sequence counter
AGAIN:   MOV DX,PORTA           ;Get port-A address in DX register
         MOV AL,[SI]            ;Load AL register with sequence data
         OUT [DX],AL            ;Send data to port-A
         INC SI                 ;Increment SI register
         MOV DX,PORTB           ;Get port-B address in DX register
         MOV AL,[SI]            ;Load AL register with sequence data
         OUT [DX],AL            ;Send data to port-B
         INC SI                 ;Increment SI register
         MOV DX,PORTC           ;Get port-c address in DX register
         MOV AL,[SI]            ;Load AL register with sequence data
         OUT [DX],AL            ;Send data to port-C
         INC SI                 ;Increment SI register
         CALL DELAY             ;Call delay subroutine
         LOOP AGAIN             ;Repeat until the count in CL register is zero
         JMP START              ;Jump to START
;DELAY ROUTINE
DELAY PROC NEAR                 ;Assembler directive
         MOV BX,0F429H          ;Load BX register with delay count
REPT:    DEC BX                 ;Decrement BX register
         JNZ REPT               ;Jump on non-zero to REPT
         RET                    ;Return to main program
DELAY ENDP                      ;Assembler directive

         ORG 1100H              ;Assembler directive
         DB 84H,84H,0C0H        ;Data for schedule-I
         DB 52H,86H,00H         ;Data for schedule-II
         DB 21H,10H,32H         ;Data for schedule-III
         DB 21H,48H,09H         ;Data for schedule-IV

CODE ENDS                       ;Assembler directive
END                             ;Assembler directive
```

> **Note:** *In the assembly language program, it is assumed that 8255 ports are IO mapped in the system with 16-bit IO port address.*

11.10.2 Traffic Light Control System using 8051 Controller

A typical traffic light control demonstration system using an 8051 microcontroller with LEDs interfaced to ports of 8051 is shown in Fig. 11.21. Here, the LEDs are interfaced to ports of 8051 microcontroller through buffers. The LED connections and control logic are similar to that 8085/8086 based traffic light control system. In 8051 based system the ports-0, 1, 2 are used instead of ports-A, B, C of 8255 in 8085/8086 based system. The details of LED connections to ports of 8051 controller is shown in Table 11.3.

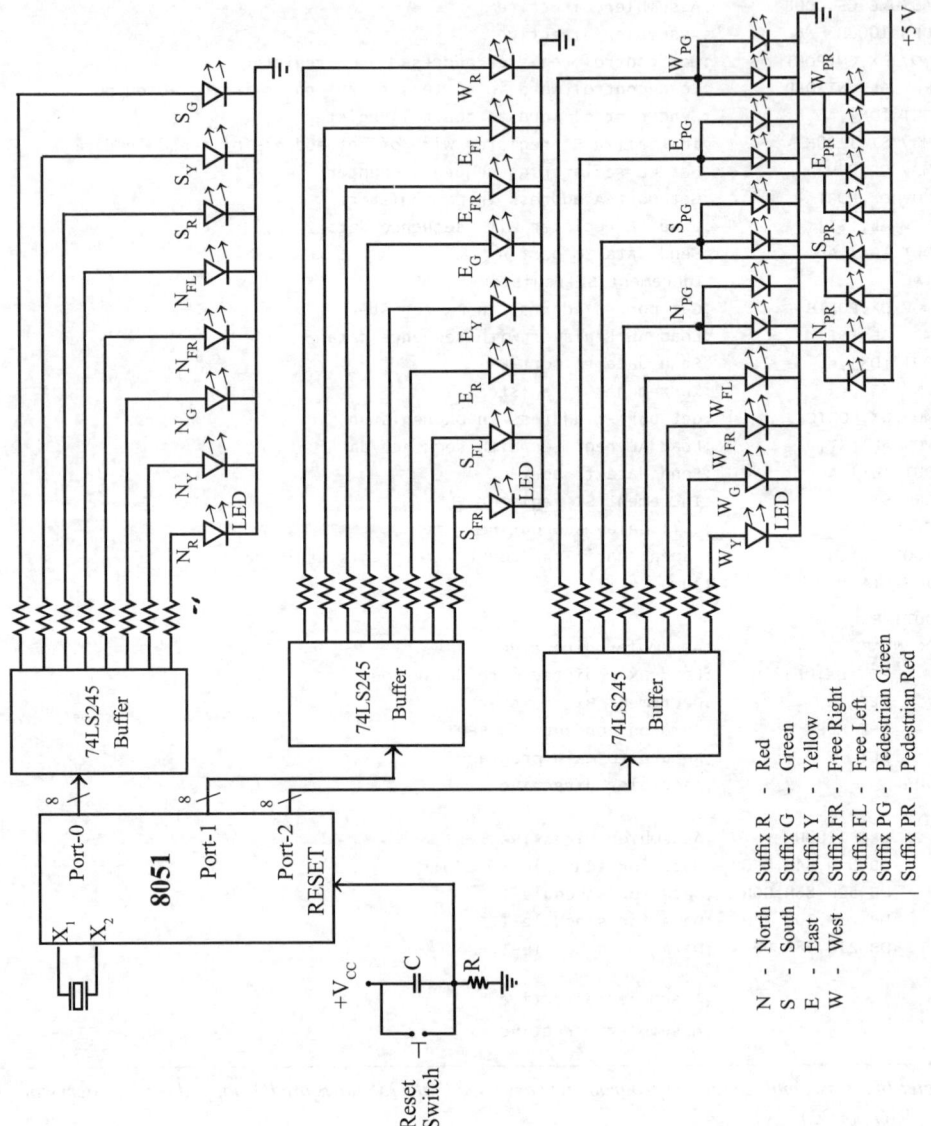

N - North
S - South
E - East
W - West

Suffix R - Red
Suffix G - Green
Suffix Y - Yellow
Suffix FR - Free Right
Suffix FL - Free Left
Suffix PG - Pedestrian Green
Suffix PR - Pedestrian Red

Fig. 11.21: An 8051 microcontroller-based traffic light control system.

Table 11.3: LED Connections to Ports of 8051

Port-0 Pins	P0.7	P0.6	P0.5	P0.4	P0.3	P0.2	P0.1	P0.0
LEDs Connected to port-0	S_G	S_Y	S_R	N_{FL}	N_{FR}	N_G	N_Y	N_R
Port-1 Pins	P1.7	P1.6	P1.5	P1.4	P1.3	P1.2	P1.1	P1.0
LEDs Connected to port-1	W_R	E_{FL}	E_{FR}	E_G	E_Y	E_R	S_{FL}	S_{FR}
Port-2 Pins	P2.7	P2.6	P2.5	P2.4	P2.3	P2.2	P2.1	P2.0
LEDs Connected to port-2	W_{PG}	E_{PG}	S_{PG}	N_{PG}	W_{FL}	W_{FR}	W_G	W_Y

EXAMPLE PROGRAM 11.4

This example program is developed for the following four schedules of traffic-light control using 8051 controller shown in Fig. 11.21.

Schedule I   : Vehicles allowed in North-South and pedestrain in East-West.
Schedule II : Warning for North-South vehicle, stop pedestrain and allow left turn in North-West.
Schedule III : Vehicles allowed in the East-West and pedestrain in North-South.
Schedule IV : Warning for East-West Vehicle, stop pedestrian and allow left turn in East-West.

The switching schedule data to be send to the ports to implement the above schedule are listed in the following Table 1.

Table 1: ON/OFF Switching Schedule of LEDs

Note: 1 = ON ; 0 = OFF

Port-0 Schedule	P0.7 S_G	P0.6 S_Y	P0.5 S_R	P0.4 N_{FL}	P0.3 N_{FR}	P0.2 N_G	P0.1 N_Y	P0.0 N_R	Hex Data
Schedule I	1	0	0	0	0	1	0	0	84_H
Schedule II	0	1	0	1	0	0	1	0	52_H
Schedule III	0	0	1	0	0	0	0	1	21_H
Schedule IV	0	0	1	0	0	0	0	1	21_H

Port-1 Schedule	P1.7 W_R	P1.6 E_{FL}	P1.5 E_{FR}	P1.4 E_G	P1.3 E_Y	P1.2 E_R	P1.1 S_{FL}	P1.0 S_{FR}	Hex Data
Schedule I	1	0	0	0	0	1	0	0	84_H
Schedule II	1	0	0	0	0	1	1	0	86_H
Schedule III	0	0	0	1	0	0	0	0	10_H
Schedule IV	0	1	0	0	1	0	0	0	48_H

Port-2 Schedule	P2.7 W_{PG}	P2.6 E_{PG}	P2.5 S_{PG}	P2.4 N_{PG}	P2.3 W_{FL}	P2.2 W_{FR}	P2.1 W_G	P2.0 W_Y	Hex Data
Schedule I	1	1	0	0	0	0	0	0	$C0_H$
Schedule II	0	0	0	0	0	0	0	0	00_H
Schedule III	0	0	1	1	0	0	1	0	32_H
Schedule IV	0	0	0	0	1	0	0	1	09_H

Flowchart

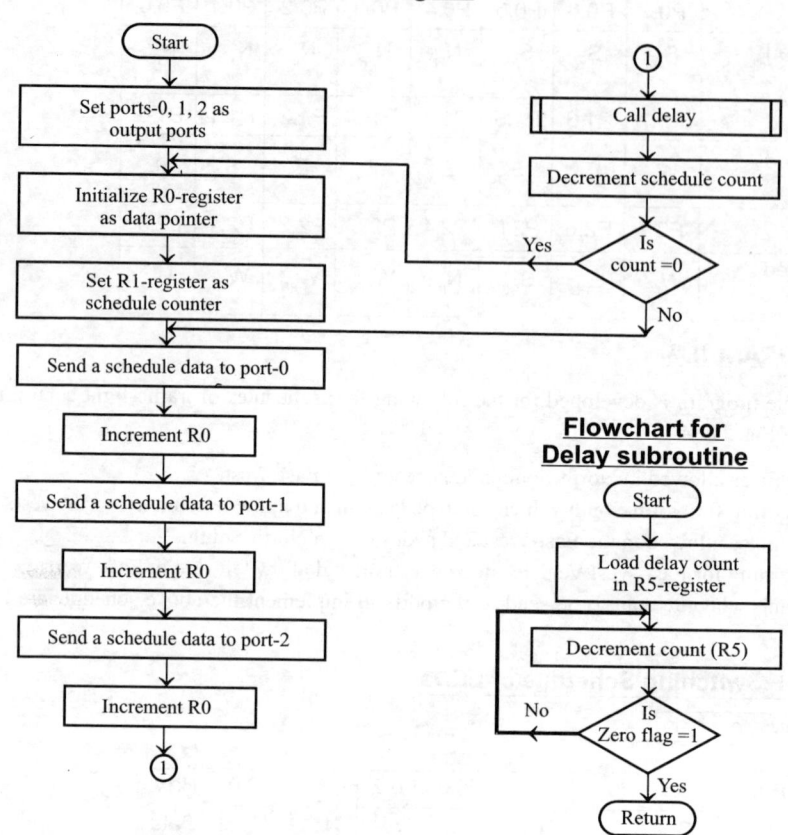

Flowchart for Main Program

Flowchart for Delay subroutine

Assembly Language Program

```
;Program for traffic light control

ORG 2000H                       ;Assembler directive
        MOV P0, #00H            ;Set port-0 as output port
        MOV P1, #00H            ;Set port-1 as output port
        MOV P2, #00H            ;Set port-2 as output port

START:  MOV R0, #2100H          ;Initialize R0-register with address of schedule data
        MOV R1, #04H            ;Set R1-register as schedule counter
NEXT:   MOVX A, @R0             ;Load A-register with schedule data
        MOV P0, A               ;Send data to port-0
        INC R0                  ;Increment address pointer
        MOVX A, @R0             ;Load A-register with schedule data
        MOV P1, A               ;Send data to port-1
        INC R0                  ;Increment address pointer
        MOVX A, @R0             ;Load A-register with schedule data
        MOV P2, A               ;Send data to port-2
        INC R0                  ;Increment address pointer
        ACALL DELAY             ;Wait for some time to change schedule
        DJNZ R1, NEXT           ;Decrement R1-register and Jump on non-zero to next schedule
        SJMP START              ;Repeat the schedules
```

```
;DELAY ROUTINE
DELAY:     MOV R5, #FFH        ;Load R5-register with delay count
AGAIN:     DJNZ R5, AGAIN      ;Decrement R5-register and Jump on non-zero to AGAIN
           RET                 ;Return to main program

ORG 2100H                      ;Assembler directive
DB 84H,84H,0C0H                ;Data for schedule-I
DB 52H,86H,00H                 ;Data for schedule-II
DB 21H,10H,32H                 ;Data for schedule-III
DB 21H,48H,09H                 ;Data for schedule-IV

END                            ;Assembly end
```

11.11 STEPPER MOTOR CONTROL SYSTEM

Stepper motors are popularly used in computer peripherals, plotters, robots and machine tools for precise incremental rotation. In a stepper motor, the stator windings are excited by electrical pulses and for each pulse, the motor shaft advances by one angular step. (Since the stepper motor can be driven by digital pulses, it is also called digital motor.) The step size in the motor is determined by the numer of poles in the rotor and the number of pairs of stator windings (one pair of stator winding is called one phase). The stator windings are also called control windings.

The motor is controlled by switching ON/OFF the control winding. The popular stepper motor used for demonstration in laboratories has a step size of 1.8° (i.e., 200 steps per revolution). This motor consists of four stator windings and requires four switching sequences as shown in Table 11.4.

The basic step size of the motor is called full-step. By altering the switching sequence, the motor can be made to run with incremental motion of half the full-step value. The switching sequence for half step rotation is shown in Table 11.5.

11.11.1 Stepper Motor Control System using 8085/8086 Processor

A typical stepper motor control system is shown in Fig. 11.22, for a two-phase of four winding stepper motor. The system consists of an 8085/8086 microprocessor as CPU, EPROM and RAM memory for program and data storage and for stack.

Using INTEL 8279, a keyboard and six number of 7-segment LED display have been interfaced in the system. Through the keyboard, the operator can issue commands to control the system. The LED display have been provided to display messages to the operator.

The windings of the stepper motor are connected to the collector of the darlington pair transistors. The transistors are switched ON/OFF by the microprocessor through the ports of 8255 and buffer (74LS245). A free-wheeling diode is connected across each winding for fast switching.

The processor has to output a switching sequence and wait for 1 to 5 milliseconds before sending the next switching sequence. (The delay is necessary to allow the motor transients to die out.)

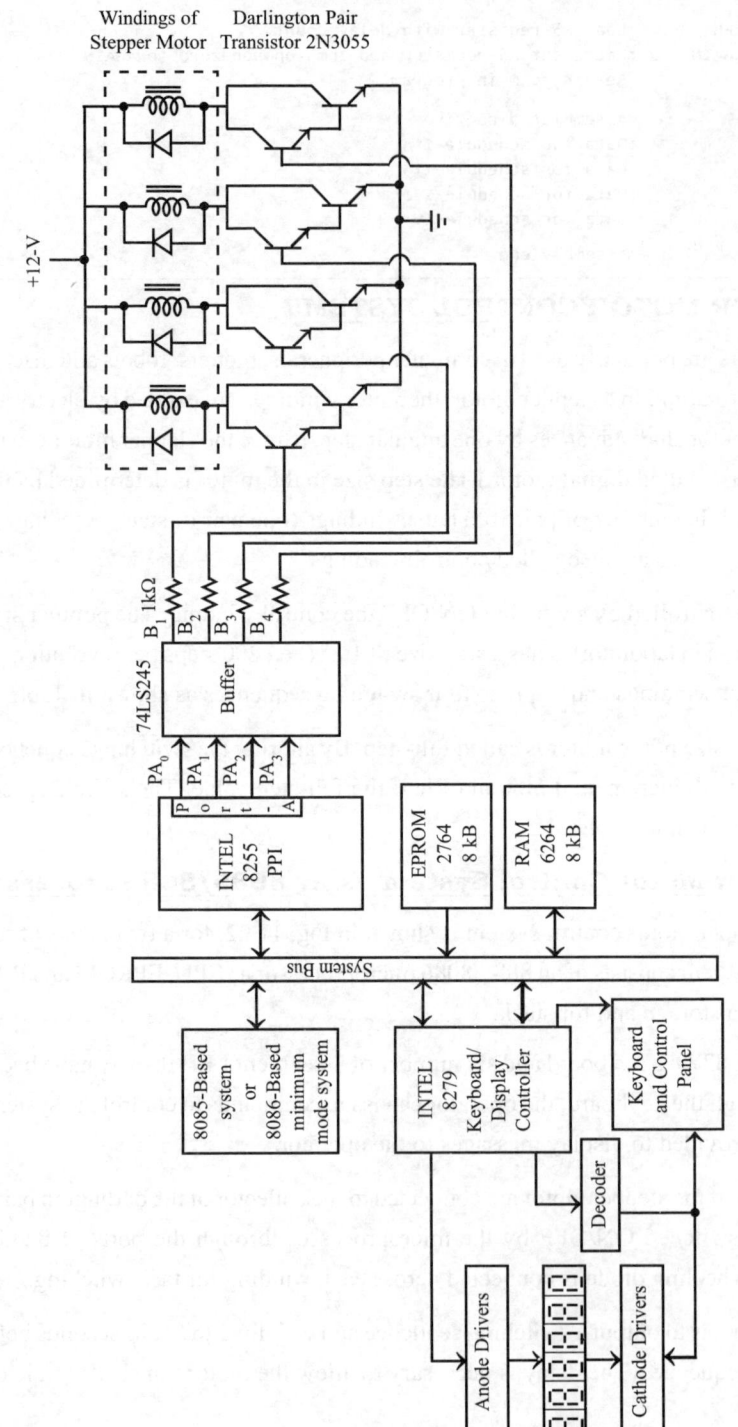

Fig. 11.22: An 8085/8086 microprocessor-based stepper motor control system.

Table 11.4: Switching Sequence for Full-Step Rotation

Switching sequence	Clockwise rotation				Anticlockwise rotation			
	PA_3	PA_2	PA_1	PA_0	PA_3	PA_2	PA_1	PA_0
Sequence-1	1	1	0	0	0	0	1	1
Sequence-2	0	1	1	0	0	1	1	0
Sequence-3	0	0	1	1	1	1	0	0
Sequence-4	1	0	0	1	1	0	0	1

Table 11.5: Switching Sequence for Half-Step Rotation

Switching sequence	Clockwise rotation				Anticlockwise rotation			
	PA_3	PA_2	PA_1	PA_0	PA_3	PA_2	PA_1	PA_0
Sequence-1	1	1	0	0	0	0	1	1
Sequence-2	0	1	0	0	0	0	1	0
Sequence-3	0	1	1	0	0	1	1	0
Sequence-4	0	0	1	0	0	1	0	0
Sequence-5	0	0	1	1	1	1	0	0
Sequence-6	0	0	0	1	1	0	0	0
Sequence-7	1	0	0	1	1	0	0	1
Sequence-8	1	0	0	0	0	0	0	1

EXAMPLE PROGRAM 11.5

Write a program to run the stepper motor interfaced with 8085 microprocessor in clockwise direction. Assume that 8255 is IO mapped and address of Port-A as 10_H and control register address as 13_H. The switching sequence for clockwise rotation is shown in following table.

Table 1: Switching Sequence for Clockwise Rotation

Sequence No	PA_7	PA_6	PA_5	PA_4	PA_3	PA_2	PA_1	PA_0	Hex Value
Sequence - 1	*	*	*	*	1	1	0	0	$0C_H$
Sequence - 2	*	*	*	*	0	1	1	0	06_H
Sequence - 3	*	*	*	*	0	0	1	1	03_H
Sequence - 4	*	*	*	*	1	0	0	1	09_H

Flowchart

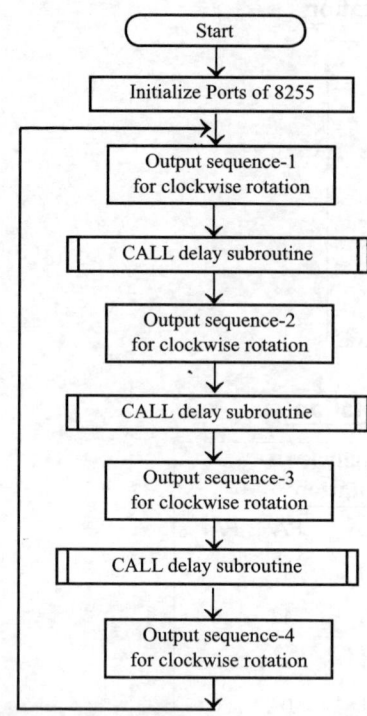

Assembly Language Program

```
;Program for stepper motor to rotate in clockwise direction
;The control word to initialize 8255 ports as output port in mode-0 is 80H
        ORG 1000H       ;Assembler directive

        MVI   A,80H     ;Load control word in A-register
        OUT   13H       ;Send control word to control register

START:  MVI   A,0CH     ;Load A-register with sequence-1 data
        OUT   10H       ;Send the data to port-A
        CALL  DELAY     ;Call DELAY subroutine
        MVI   A,06H     ;Load A-register with sequence-2 data
        OUT   10H       ;Send the data to port-A
        CALL  DELAY     ;Call DELAY subroutine
        MVI   A,03H     ;Load A-register with sequence-3 data
        OUT   10H       ;Send the data to port-A
        CALL  DELAY     ;Call DELAY subroutine

        MVI   A,09H     ;Load A-register with sequence-4 data
        OUT   10H       ;Send the data to port-A
        CALL  DELAY     ;Call DELAY subroutine
        JMP   START     ;Jump to START

;DELAY ROUTINE
DELAY:  LXI   D,0F429H  ;Load DE register pair with delay count
REPT:   DCX   D         ;Decrement DE register pair
        MOV   A,E       ;Check whether both D and E register
        CMP   D         ;contents are zeros
        JNZ   REPT      ;Jump on non-zero to REPT
        RET             ;If zero, return to main program
        END             ;Assembler directive
```

EXAMPLE PROGRAM 11.6

Write a program to run the stepper motor interfaced with 8086 microprocessor in clockwise direction. Assume that 8255 is IO mapped and address of Port-A as 0020_H and control register address as 0026_H. The switching sequence for clockwise rotation is same as that of Table-1 in Example Program 11.5.

Assembly Language Program

```
;Program for stepper motor to rotate in clockwise direction
;The control word to initialize 8255 ports as output port in mode-0 is 80H
CODE SEGMENT              ;Assembler directive
       ASSUME CS:CODE ;Assembler directive
       ORG 1000H         ;Assembler directive
       MOV DX,0026H      ;Load control register address in DX
       MOV AL,80H        ;Load control word in AL
       OUT [DX],AL       ;Send control word to control register
       MOV DX,0020H      ;Load the port-A address in DX register
START: MOV AL,0CH        ;Load AL register with sequence-1 data
       OUT [DX],AL       ;Send the data to port-A
       CALL DELAY        ;Call DELAY subroutine
       MOV AL,06H        ;Load AL register with sequence-2 data
       OUT [DX],AL       ;Send the data to port-A
       CALL DELAY        ;Call DELAY subroutine
       MOV AL,03H        ;Load AL reg.with sequence-3 data
       OUT [DX],AL       ;Send the data to port-A
       CALL DELAY        ;Call DELAY subroutine
       MOV AL,09H        ;Load AL reg.with sequence-4 data
       OUT [DX],AL       ;Send the data to port-A
       CALL DELAY        ;Call DELAY subroutine
       JMP START         ;Jump to START
;DELAY ROUTINE
DELAY PROC NEAR          ;Assembler directive
       MOV AX,0F80H      ;Load AX register with delay value
LP:    DEC AX            ;Decrement AX register
       JNZ LP            ;Jump on non-zero to LP
       RET               ;Return to main program
DELAY ENDP               ;Assembler directive
CODE ENDS                ;Assembler directive
END                      ;Assembler directive
```

11.11.2 Stepper Motor Control using the 8051 Microcontroller

The stepper motor interfaced to the 8051 microcontroller is shown in Fig. 11.23. The stepper motor interface and control logic are similar to that of 8085/8086-based system. In 8051 based system the port-0 of 8051 is used to interface motor instead of port-A of 8255 in 8085/8086-based system. The switching sequence for full step and half step rotation are same as that Table 11.4 and 11.5 in Section 11.11.1.

The stepper motor interface has four keys interfaced through port-2 pins in order to control the motor operation. The keys can be used to start and stop the motor and also used to select forward or reverse rotation of the motor. Normally, the port pins are pulled high and pressing a key will make the port pin low, and this low signal is sensed by the microcontroller to initiate the operation selected by the key press.

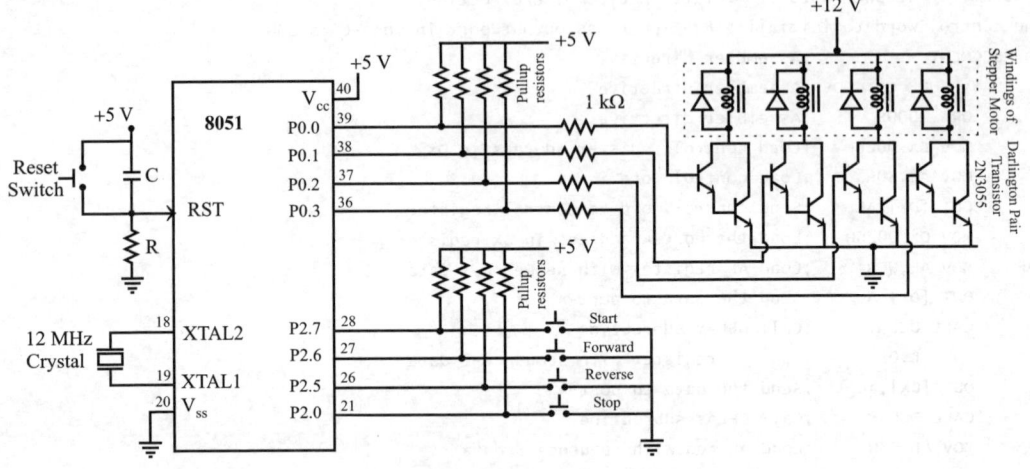

Fig. 11.23: Stepper motor interface to 8051

EXAMPLE PROGRAM 11.7

This example program is developed for the stepper motor interface shown in Fig. 11.23. The switching sequence data for forward and reverse are shown in Table 1.

Table 1: Forward/Reverse Sequence Data

Direction	Sequence No	P0.7	P0.6	P0.5	P0.4	P0.3	P0.2	P0.1	P0.0	Hex Value
Forward	Sequence - 1	*	*	*	*	1	1	0	0	$0C_H$
	Sequence - 2	*	*	*	*	0	1	1	0	06_H
	Sequence - 3	*	*	*	*	0	0	1	1	03_H
	Sequence - 4	*	*	*	*	1	0	0	1	09_H
Reverse	Sequence - 1	*	*	*	*	0	0	1	1	03_H
	Sequence - 2	*	*	*	*	0	1	1	0	06_H
	Sequence - 3	*	*	*	*	1	1	0	0	$C0_H$
	Sequence - 4	*	*	*	*	1	0	0	1	09_H

Flowchart

Flowchart for Main Program

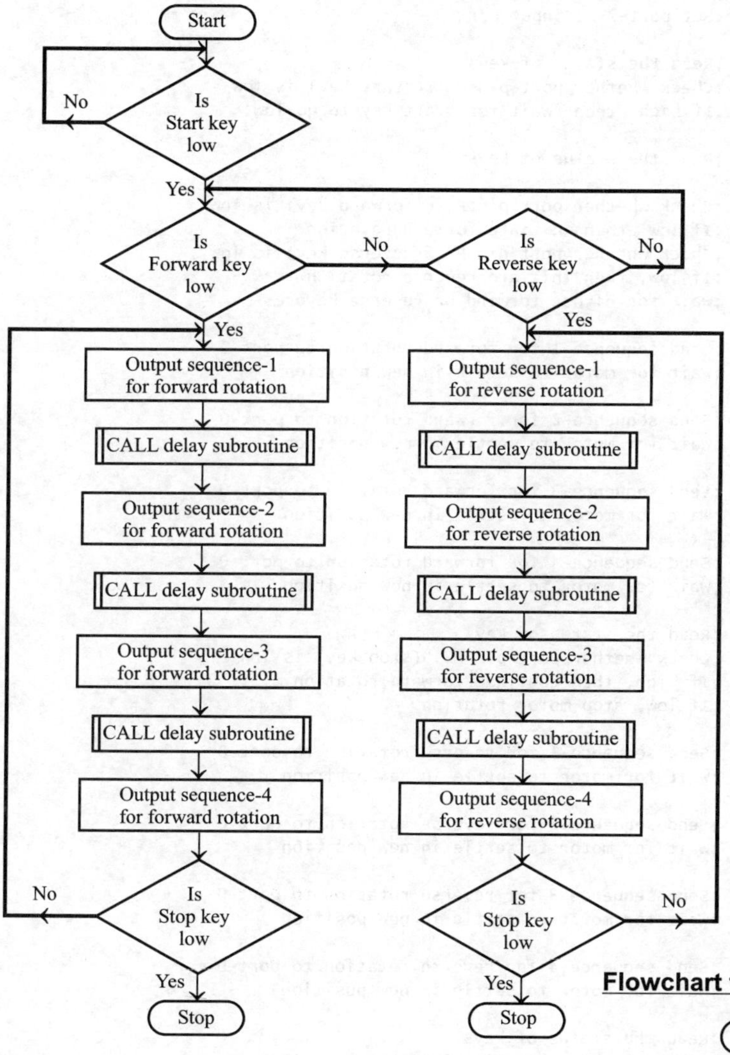

Start

```
Is Start key low
No →
Yes ↓
```

Is Forward key low — No → Is Reverse key low — No

Yes ↓ (Forward)   Yes ↓ (Reverse)

Forward	Reverse
Output sequence-1 for forward rotation	Output sequence-1 for reverse rotation
CALL delay subroutine	CALL delay subroutine
Output sequence-2 for forward rotation	Output sequence-2 for reverse rotation
CALL delay subroutine	CALL delay subroutine
Output sequence-3 for forward rotation	Output sequence-3 for reverse rotation
CALL delay subroutine	CALL delay subroutine
Output sequence-4 for forward rotation	Output sequence-4 for reverse rotation

Is Stop key low — No (Forward)   Is Stop key low — No (Reverse)

Yes ↓ Stop   Yes ↓ Stop

Flowchart for Subroutine Delay

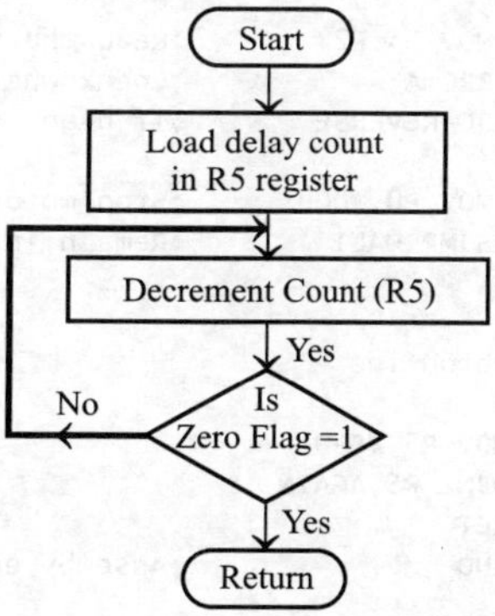

Start

Load delay count in R5 register

Decrement Count (R5)

Is Zero Flag =1 — No

Yes ↓

Return

Assembly Language Program

```
;Program for Stepper Motor
        MOV  P0,#00H    ;Set port-0 as output port
        MOV  P2,#FFH    ;Set port-2 as input port

START:  MOV  A,P2       ;Read the status of keys
        RLC A           ;Check whether port pin P2.7(start key) is low
        JC START        ;If high, then  wait for start key to go low

WAIT:   MOV  A,P2       ;Read the status of keys
        RLC A
        RLC A           ;Check whether port pin P2.6(forward key) is low
        JNC FORWARD     ;If low, then initiate forward rotation
        RLC A           ;Check whether port pin P2.5(reverse key) is low
        JNC REVERSE     ;If low, then initiate reverse rotation
        SJMP WAIT       ;Wait for either forward or reverse keypress

FORWARD: MOV  P0,#0CH   ;Send sequence-1 for forward rotation to port-0
         ACALL DELAY    ;Wait for motor to settle in new position

         MOV  P0,#06H   ;Send sequence-2 for forward rotation to port-0
         ACALL DELAY    ;Wait for motor to settle in new position

         MOV  P0,#03H   ;Send sequence-3 for forward rotation to port-0
         ACALL DELAY    ;Wait for motor to settle in new position

         MOV  P0,#09H   ;Send sequence-4 for forward rotation to port-0
         ACALL DELAY    ;Wait for motor to settle in new position

         MOV  A,P2      ;Read the status of keys
         RRC A          ;Check whether port pin P2.0(stop key) is low
         JC FORWARD     ;If high, then continue forward rotation
         SJMP STOP      ;If low, stop motor rotation

REVERSE: MOV  P0,#03H   ;Send sequence-1 for reverse rotation to port-0
         ACALL DELAY    ;Wait for motor to settle in new position

         MOV  P0,#06H   ;Send sequence-2 for reverse rotation to port-0
         ACALL DELAY    ;Wait for motor to settle in new position

         MOV  P0,#C0H   ;Send sequence-3 for reverse rotation to port-0
         ACALL DELAY    ;Wait for motor to settle in new position

         MOV  P0,#09H   ;Send sequence-4 for reverse rotation to port-0
         ACALL DELAY    ;Wait for motor to settle in new position

         MOV  A,P2      ;Read the status of keys
         RRC A          ;Check whether port pin P2.0(stop key) is low
         JC REVERSE     ;If high, then continue reverse rotation

STOP:   MOV  P0,#00H    ;Stop motor rotation
HALT:   SJMP HALT       ;Remain in infinite loop until reset, program end


;Delay Subroutine

DELAY:  MOV R5,#FFH
AGAIN:  DJNZ R5,AGAIN
        RET
        END             ;Assemby end
```

11.12 ALARM CONTROLLER

11.12.1 Alarm Controller using 8085/8086 Processor

The alarm controllers are microprocessor-based systems designed to sense some abnormal conditions using sensors and turn on a prerecorded voice or generate a sound to indicate the abnormality.

Some of the abnormal conditions that can be sensed by alarm controllers are gas leakage, fire, leakage in flow pipes, etc. The sensors are basically transducers which are energy-converting devices. The sensors generate electrical signals proportional to quantity being sensed, and the magnitude of this signal can be used to differentiate the normal and abnormal conditions.

These electrical signals are amplified and scaled by using operational amplifiers. This process is also known as **signal conditioning**. The output signal from signal conditioning unit will be an analog signal. Using ADC (Analog-to-Digital-Converter), this analog can be converted to digital data (8-bit/10-bit/12-bit data) and fed to a microprocessor based system.

Alternatively, the analog signal can be converted to digital signal using comparator and fed to one of the interrupt of microprocessor or fed to one of the input pin of an IO port interfaced to the microprocessor. The comparator will compare two analog inputs and generate a high or low signal that can be used as digital signal. One of the inputs to comparator is a preset analog voltage and another input is the signal conditioned voltage from the sensor.

When ADC is employed, the microprocessor will read the digital value from ADC and compare it with a known value to determine the normal and abnormal conditions. When digital signal is fed to a microprocessor, the change of signal level from high to low (or low to high), is used to sense the abnormal condition. Whenever the abnormality is sensed, the microprocessor will generate an output signal to turn ON a sound system that is interfaced via IO port.

A typical alarm controller using the 8085/8086 microprocessor is shown in Fig. 11.24. In this system, a gas sensor and a fire sensor are interfaced to 8085/8086 microprocessor via ports of 8255. The pin assignment of 8255 ports are shown in Table 11.6.

The signal from the gas sensor is amplified and scaled by a signal conditioning circuit to 0 - 5 volt range analog signal and fed to ADC. Depending on the density of gas the analog signal will vary from 0 to 5 volts, and this value is used to sense conditions like light gas leakage (when near to 0 volt) and severe gas leakage (when near to 5 volt).

The ADC is interfaced to 8085/8086 processor via an input port. The processor will read the ADC periodically and keeps on monitoring the amount of gas leakage, and if the amount exceeds a predefined value, then the processor will send an output signal to turn ON an audio signal generator to create an electromagnetic signal in the audio frequency range and it is amplified and fed to a speaker in order to convert it to a sound signal which is the required alarm for abnormal gas leakage.

The signal from the fire sensor is amplified by a signal conditioning circuit and fed to one of the input of a comparator. Another input of the comparator is an analog voltage called preset voltage. The preset voltage is decided such that during normal condition (or when no fire) the signal from the fire sensor is less than the preset voltage and when fire is sensed, the signal from the fire sensor is greater than the preset voltage. Therefore, when there is no fire, the output of comparator is low and whenever fire is sensed the output of comparator will be high.

The comparator output is connected to the input pin of an IO port interfaced to microprocessor. The processor will read the IO port periodically and keeps on monitoring the level of input signal, and if the signal goes high, then the processor will send an output signal to turn ON an alarm system similar to that of gas leakage detection. In the system shown in Fig. 11.24, the same alarm system is used for gas leakage and fire.

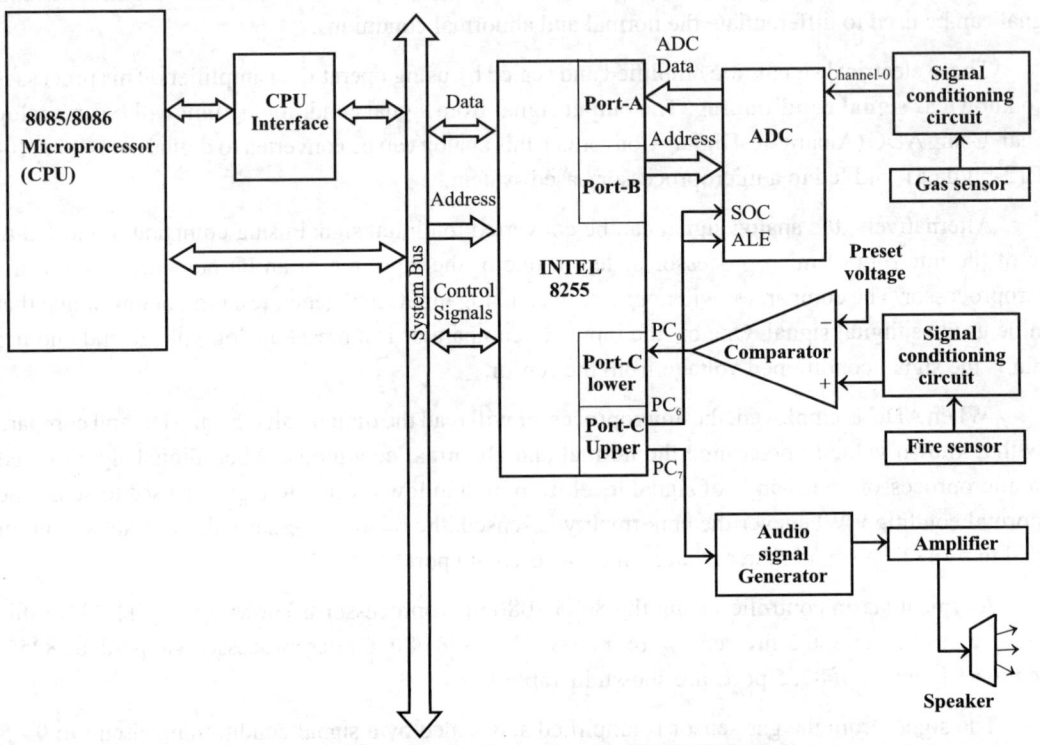

Fig. 11.24: A typical alarm controller using the 8085/8086 microprocessor.

Table 11.6: Pin Assignment of 8255 Ports

Port-A	PA_7	PA_6	PA_5	PA_4	PA_3	PA_2	PA_1	PA_0
	8-bit ADC Data							
Port-B	PB_7	PB_6	PB_5	PB_4	PB_3	PB_2	PB_1	PB_0
	*	*	*	*	*	3-bit ADC Channel Address		
Port-C	PC_7	PC_6	PC_5	PC_4	PC_3	PC_2	PC_1	PC_0
	Alarm ON/OFF	SOC ALE	*	*	*	*	*	Fire Sense

EXAMPLE PROGRAM 11.8

This example program is developed for the alarm controller using the 8085 microporcessor shown in Fig. 11.24. It is assumed the 8255 IO ports are IO mapped in the system. The formation of control words are listed in the following Table 1.

Table 1: Control Words of 8255

Control word/ Port-B	Feature selected	Binary code								Hex code	Variable name in program
Mode set Control set	Mode-0, Port-B & C output, Port-A input	B_7	B_6	B_5	B_4	B_3	B_2	B_1	B_0	90_H	MCSW
		1	0	0	1	0	0	0	0		
Port-B	ADC channel-0 Address	PB_7	PB_6	PB_5	PB_4	PB_3	PB_2	PB_1	PB_0	00_H	CH0AD
		*	*	*	*	*	0	0	0		
Bit Set/Reset control word	Set SOC and ALE	B_7	B_6	B_5	B_4	B_3	B_2	B_1	B_0	$0D_H$	BSWC6
		0	*	*	*	1	1	0	1		
	Reset SOC and ALE	0	*	*	*	1	1	0	0	$0C_H$	BRWC6
	Alarm ON	0	*	*	*	1	1	1	1	$0F_H$	BSWC7
	Alarm OFF	0	*	*	*	1	1	1	0	$0E_H$	BRWC7

Flowchart

Flowchart for Main Program

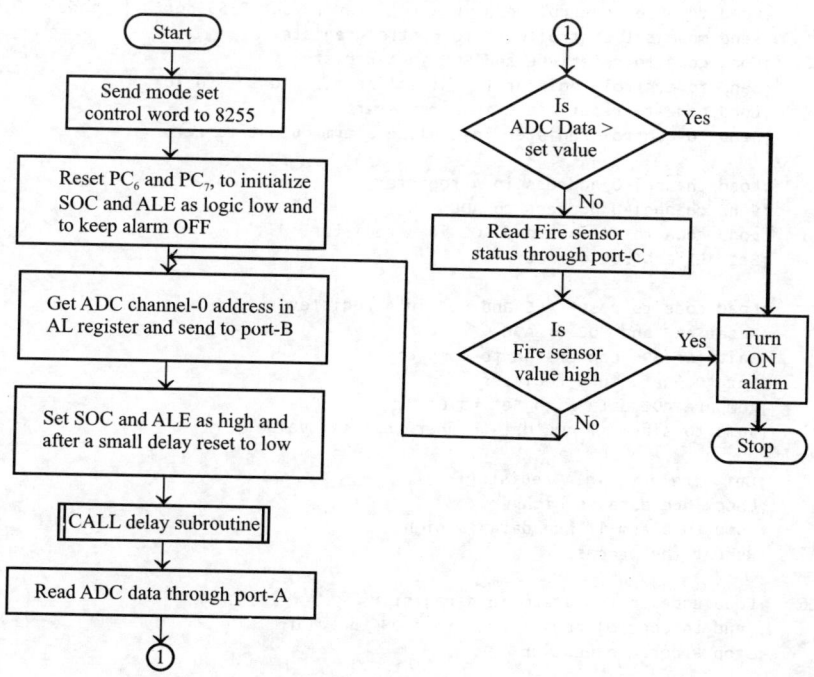

Flowchart for
Delay subroutine

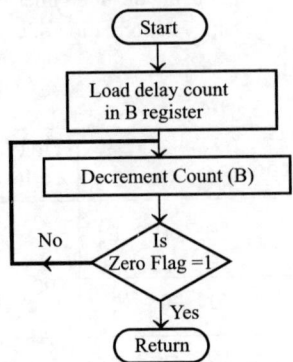

Assembly Language Program

```
;PROGRAM FOR ALARM CONTROLLER
PORTA EQU 10H          ;Address of port-A of 8255
PORTB EQU 11H          ;Address of port-B of 8255
PORTC EQU 12H          ;Address of port-C of 8255
CNREG EQU 13H          ;Address of 8255 control register
MSCW  EQU 80H          ;Mode set control set control word of 8255
BSWC6 EQU 0DH          ;Bit set Control Word of port-C pin-6
BRWC6 EQU 0CH          ;Bit reset Control Word of port-C pin-6
BSWC7 EQU 0FH          ;Bit set Control Word of port-C pin-7
BRWC7 EQU 0EH          ;Bit reset Control Word of port-C pin-7
CH0AD EQU 00H          ;ADC Channel-0 address to be send via Port-B
GSET  EQU 7FH          ;Gas sensor set value
      ORG  1000H       ;Assembler directive
      MVI  A,MSCW       ;Load mode set control word in A-register to set 8255 ports in mode-0
      OUT  CNREG        ;Send mode set control word to control register
      MVI  A,BRWC6      ;Load code to reset ALE and SOC in A-register
      OUT  CNREG        ;Send to control register to initialize ALE and SOC to low
      MVI  A,BRWC7      ;Load code to reset Alarm in A-register
      OUT  CNREG        ;Send to control register initialize alarm output to zero

AGAIN:MVI  A,CH0AD      ;Load Channel-0 address in A-register
      OUT  PORTB        ;Send channel-0 address to ADC
      MVI  A,BSWC6      ;Load code to set ALE and SOC in A-register
      OUT  CNREG        ;Set ALE and SOC of ADC
      NOP
      MVI  A,BRWC6      ;Load code to reset ALE and SOC in A-register
      OUT  CNREG        ;Reset ALE and SOC of ADC
      CALL DELAY        ;Wait for ADC conversion to complete
      IN   PORTA        ;Get ADC data in A-register
      CPI  GSET         ;Compare ADC data with Set value
      JNC  ALARM        ;Jump to alarm if ADC data higher than set value

      IN   PORTC        ;Get Fire data in A-register
      RRC               ;Check fire data is high
      JC   ALARM        ;Jump to alarm if fire data is high
      JMP  AGAIN        ;Repeat the process

ALARM:MVI  A,BSWC6      ;Load code to set alarm in A-register
      OUT  CNREG        ;Send to control register to turn ON the alarm
      HLT               ;Stop Program execution
```

```
;DELAYROUTINE
DELAY:MVI  B,7FH      ;Load B-register with delay count
REPT: DCR  B          ;Decrement B-register
      JNZ  REPT       ;Jump on non-zero to REPT
      RET             ;Return to main program

      END             ;Assembler directive
```

EXAMPLE PROGRAM 11.9

This example program is developed for the alarm controller using the 8086 microprocessor shown in Fig. 11.24. It is assumed the 8255 IO ports are IO mapped in the system. The formation of control words are same as that of Table-1 in Example Program 11.8.

Flowchart

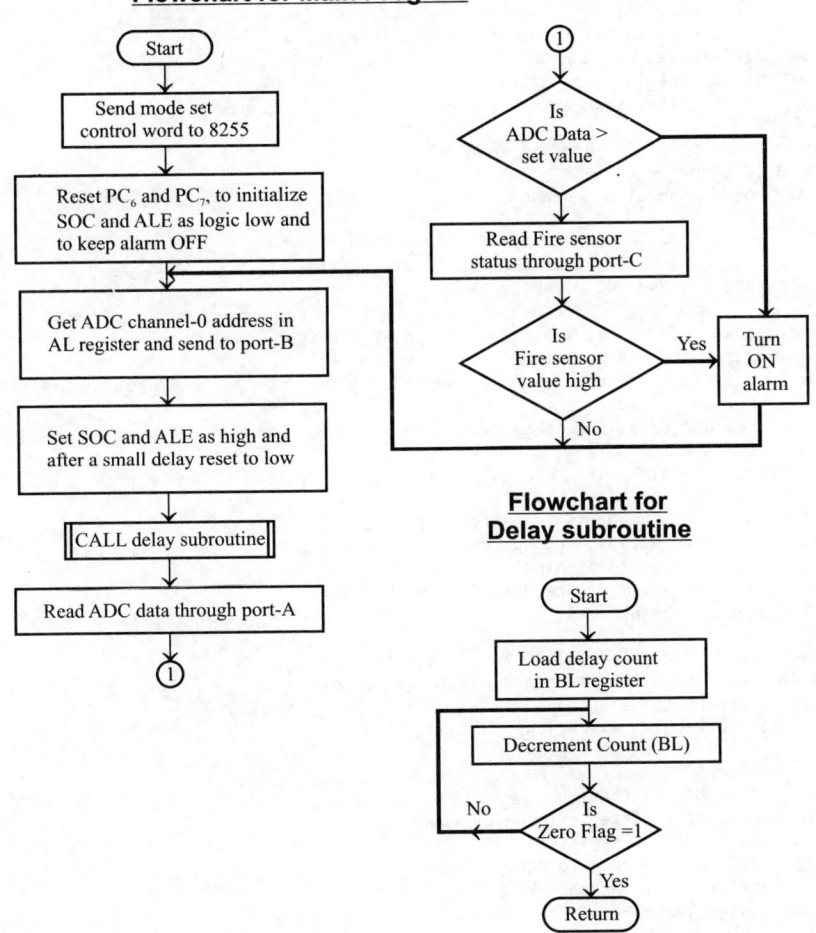

Flowchart for Main Program

Start

Send mode set control word to 8255

Reset PC_6 and PC_7, to initialize SOC and ALE as logic low and to keep alarm OFF

Get ADC channel-0 address in AL register and send to port-B

Set SOC and ALE as high and after a small delay reset to low

CALL delay subroutine

Read ADC data through port-A

Is ADC Data > set value

Read Fire sensor status through port-C

Is Fire sensor value high — Yes → Turn ON alarm

No

Flowchart for Delay subroutine

Start

Load delay count in BL register

Decrement Count (BL)

Is Zero Flag =1 — No

Yes

Return

Assembly Language Program

```
;PROGRAM FOR ALARM CONTROLLER

PORTA EQU 0000H     ;Address of port-A of 8255
PORTB EQU 0002H     ;Address of port-B of 8255
PORTC EQU 0004H     ;Address of port-C of 8255
CNREG EQU 0006H     ;Address of 8255 control register
MSCW  EQU 80H       ;Mode set control set control word of 8255
BSWC6 EQU 0DH       ;Bit set Control Word of port-C pin-6
BRWC6 EQU 0CH       ;Bit reset Control Word of port-C pin-6
BSWC7 EQU 0FH       ;Bit set Control Word of port-C pin-7
BRWC7 EQU 0EH       ;Bit reset Control Word of port-C pin-7
CH0AD EQU 00H       ;ADC Channel-0 address to be send via Port-B
GSET  EQU 7FH       ;Gas sensor set value
CODE SEGMENT        ;Assembler directive
ASSUME CS:CODE      ;Assembler directive
     ORG 1000H      ;Assembler directive
     MOV DX,CNREG   ;Get control register address in DX register
     MOV AL,MSCW    ;Set 8255 ports in mode-0
     OUT [DX],AL    ;Send mode set control word to control register
     MOV AL,BRWC6   ;Load code to reset ALE and SOC in AL register
     OUT [DX],AL    ;Initialize ALE and SOC to low
     MOV AL,BRWC7   ;Load code to reset Alarm
     OUT [DX],AL    ;Initialize alarm output to zero

AGAIN:MOV AL,CH0AD  ;Load Channel-0 address in AL register
     MOV DX,PORTB   ;Get port-B address in DX register
     OUT [DX],AL    ;Send channel-0 address to ADC

     MOV AL,BSWC6   ;Load code to set ALE and SOC in AL register
     MOV DX,CNREG   ;Get control register address in DX register
     OUT [DX],AL    ;Set ALE and SOC of ADC
     NOP
     MOV AL,BRWC6   ;Load code to reset ALE and SOC in AL register
     OUT [DX],AL    ;Reset ALE and SOC of ADC
     CALL DELAY     ;Wait for ADC conversion to complete
     MOV DX,PORTA   ;Get port-A address in DX register
     IN AL,[DX]     ;Get ADC data in AL register
     CMP AL,GSET    ;Compare ADC data with Set value
     JNC ALARM      ;Jump to alarm if ADC data higher than set value

     MOV DX,PORTC   ;Get port-C address in DX register
     IN AL,[DX]     ;Get Fire data in AL register
     ROR AL,1       ;Check fire data is high
     JC ALARM       ;Jump to alarm if fire data is high
     JMP AGAIN      ;Repeat the process

ALARM:MOV AL,BSWC6  ;Load code to set alarm in AL register
     MOV DX,CNREG   ;Get control register address in DX register
     OUT [DX],AL    ;Turn ON the alarm
     HLT            ;Stop Program execution

;DELAY ROUTINE
DELAY PROC NEAR     ;Assembler directive
     MOV BL,7FH     ;Load BL register with delay count
REPT: DEC BL        ;Decrement BL register
     JNZ REPT       ;Jump on non-zero to REPT
     RET            ;Return to main program
DELAY ENDP          ;Assembler directive

CODE ENDS           ;Assembler directive
END                 ;Assembler directive
```

11.12.2 Alarm Controller using 8051 Controller

A typical alarm controller using an 8051 microcontroller is shown in Fig. 11.25. The 8051 based system is similar to that of 8085/8086 based alarm controller except 8255. The 8051 has internal ports and so 8255 ports are not required in 8051 based system. In 8051-based system, the ADC, fire sensor and speaker are directly interfaced to ports of 8051. The working of 8051 based alarm controller and control logic are similar to that 8085/8086 based system. The pin assignments of ports of 8051 microcontroller are shown in Table 11.7.

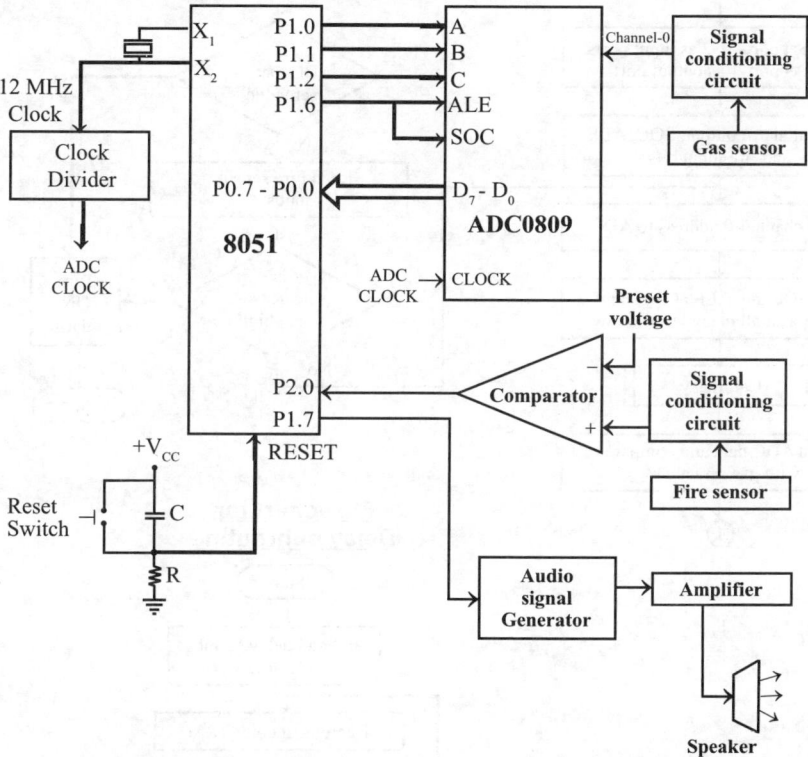

Fig. 11.25: A typical alarm controller using the 8051 microcontroller.

Table 11.7: Pin Assignment of 8051 Ports

Port-0	P0.7	P0.6	P0.5	P0.4	P0.3	P0.2	P0.1	P0.0
	8-bit ADC Data							
Port-1	P1.7	P1.6	P1.5	P1.4	P1.3	P1.2	P1.1	P1.0
	Alarm ON/OFF	SOC ALE	*	*	*	3-bit ADC Channel Address		
Port-2	P2.7	P2.6	P2.5	P2.4	P2.3	P2.2	P2.1	P2.0
	*	*	*	*	*	*	*	Fire Sense

EXAMPLE PROGRAM 11.10

This example program is developed for the alarm controller using the 8051 microcontroller shown in Fig. 11.25.

Flowchart

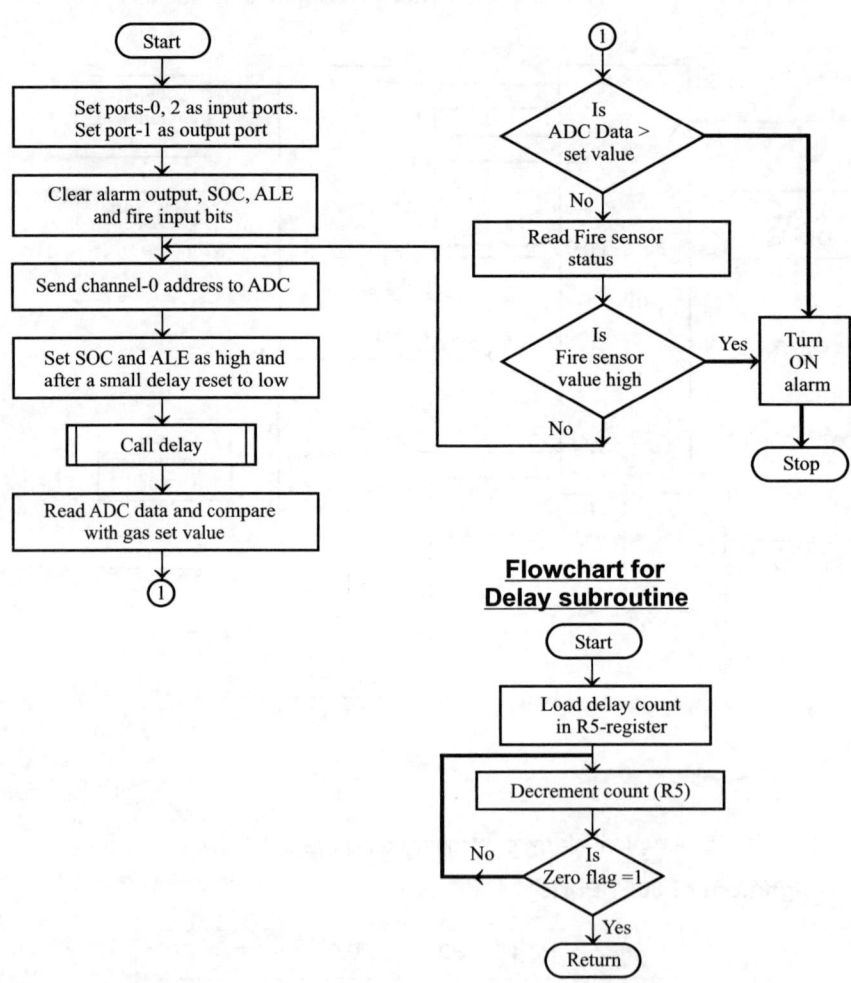

Flowchart for Main Program

Start

Set ports-0, 2 as input ports. Set port-1 as output port

Clear alarm output, SOC, ALE and fire input bits

Send channel-0 address to ADC

Set SOC and ALE as high and after a small delay reset to low

Call delay

Read ADC data and compare with gas set value

①

Is ADC Data > set value
No
Read Fire sensor status
Is Fire sensor value high
Yes → Turn ON alarm
No
Stop

Flowchart for Delay subroutine

Start

Load delay count in R5-register

Decrement count (R5)

No
Is Zero flag =1
Yes
Return

Assembly Language Program

```
;Program for alarm controller

ORG 2000H                    ;Assembler directive
INIT:   MOV P0, #FFH         ;Set port-0 as input port
        MOV P1, #00H         ;Set port-1 as output port
        MOV P2, #FFH         ;Set port-2 as input port
        CLR P1.7             ;Reset output of alarm
```

```
          CLR P1.6            ;Reset SOC and ALE of ADC
          CLR P2.0            ;Clear fire input bit

REPT:     MOV P1, #CH0AD      ;Send channel-0 address to ADC
          SETB P1.6           ;Set SOC and ALE of ADC
          NOP
          NOP
          CLR P1.6            ;After a small delay reset SOC and ALE of ADC
          ACALL DELAY         ;Wait for ADC conversion to complete
          MOV A. P0           ;Read ADC data in A-register
          CLR C               ;Clear carry
          SUBB A, #GSET       ;Compare gas set value with ADC value
          JNC ALARM           ;If ADC value is higher turn ON alarm
          JB P2.0, ALARM      ;Check for fire input bit, if set turn ON alarm
          SJMP REPT           ;If no fault then repeat the process

ALARM:    SETB P1.7           ;Turn ON alarm
STOP:     SJMP STOP           ;Halt program execution

;DELAY ROUTINE
DELAY:    MOV R5, #FFH        ;Load R5-register with delay count
AGAIN:    DJNZ R5, AGAIN      ;Decrement R5-register and Jump on non-zero to AGAIN
          RET                 ;Return to main program

ORG 2100H                     ;Assembler directive
CH0AD EQU 00H                 ;ADC Channel-0 address
GSET  EQU 7FH                 ;Gas sensor set value
END                           ;Assembly end
```

11.13 SERVO MOTOR

The motors that are used in automatic position control systems are called Servomotors. When the objective of the system is to control the position of an object then the system is called Servomechanism. The servomotors are used to convert an electrical signal (control voltage) applied to them into an angular displacement of the shaft.

They can either operate in a continuous duty or step duty depending on construction. Depending on the supply required to run the motor, they are broadly classified as DC servomotors and AC servomotors. The DC motors are expensive than AC motors. But the DC servomotors have linear characteristics and so it is easier to control.

The DC servomotors are generally used for large power applications such as in machine tools and robotics. Advantages of AC motors are lower cost, higher efficiency and less maintenance since there is no commutator and brushes.

The disadvantage of AC motor is that the characteristics are quite non-linear and these motors are more difficult to control especially for positioning applications (servomechanisms). The AC motor are best suited for low power applications such as instrument servo (Example: Control of pen in X-Y recorders) and computer related equipment (Example: Disk drives, Tape drives, Printers, etc.)

11.13.1 Speed Control of Servo Motor

The DC servo motors are specially designed DC motors to satisfy the characteristics of servo applications. A normal DC motor can be converted to servo motor by adding speed reduction gear system to alter the speed torque ratio. Power output a motor is directly proportional to product of speed and torque. Therefore, at lesser speeds it is possible to achieve very high torque. Most of servo motors are used to provide small incremental motions proportional to input voltage and so they can offer very high driving torque.

The speed and hence angular rotation of servo motor is directly proportional to armature voltage when field excitation is constant. Hence in servo applications the field of DC motor is excited by a constant DC voltage. Alternatively, field is constructed with high retentivity permanent magnet. For continuous rotation the speed is controlled by varying armature voltage by PWM (Pulse Width Modulation) technique. In PWM technique a pulsating DC voltage is applied to armature so that the average DC voltage is varied by varying the ON time.

If the ON time is 10% then average DC voltage is 10%.

If the ON time is 20% then average DC voltage is 20% and so on.

For incremental angular motion, the armature is supplied with time-based pulses. The angular motion is limited to fixed number of revolutions. The resolution of angular motion and the duration of pulses required to achieve angular motions depends on the type and construction of the servo motor. For example, in a DC servo motor, if 0.05 millisecond pulse can produce an angular motion of 2°, then for 180° motion (180°/2 = 90°), 90 pulses of 0.05 millisecond duration is required.

In DC servo motor the reversal of direction of rotation can be achieved by reversing the polarity or sign of DC voltage applied to armature. For accurate speed or motion control, closed loop control is required. In continuous type DC servo motor, speed sensor is employed for sensing the speed and compare with desired set speed and error is generated to correct the speed if there is a difference between actual and set speed. In incremental motion DC servo motor the angular position is sensed by using potentiometer and feedback signal from potentiometer is used to correct error in desired angular position.

11.13.2 Servo Motors with Inbuilt Control Circuit

Servo motors with inbuilt control circuit are popularly used in wireless control toys, robots, etc, are they are also called RC Servo motors. A RC servo motor is permanent magnet type DC servo motor fitted with a train of reduction gears, a feedback potentiometer to sense angular position and driver circuit for forward and reverse angular motion control. The angular motion is limited to 0 to 180° (or – 90° to +90°) The RC servo motors are popularly used in microcontroller-based applications due to simplicity in controlling the angular motion using time-based pulses. A periodic pulse train of fixed ON time duration can turn the motor shaft in a fixed angular motion.

In RC servo motors, the field is constructed using permanent magnet and so field flux is constant and speed or angular motion is directly proportional to armature voltage. The armature of a typical servo motor is excited by a periodic pulse train with time period of 20 milliseconds. The ON time of the periodic pulse train decides the angular motion of the shaft.

Microprocessor or microcontroller-based closed loop position control system of a typical DC servo motor is shown in Fig. 11.26. The system consists of a DC motor fitted with gear wheels specially designed for servo applications, a potentiometer to generate a voltage called feedback voltage proportional to shaft position and driver transistors to supply the required volage to armature of DC motor.

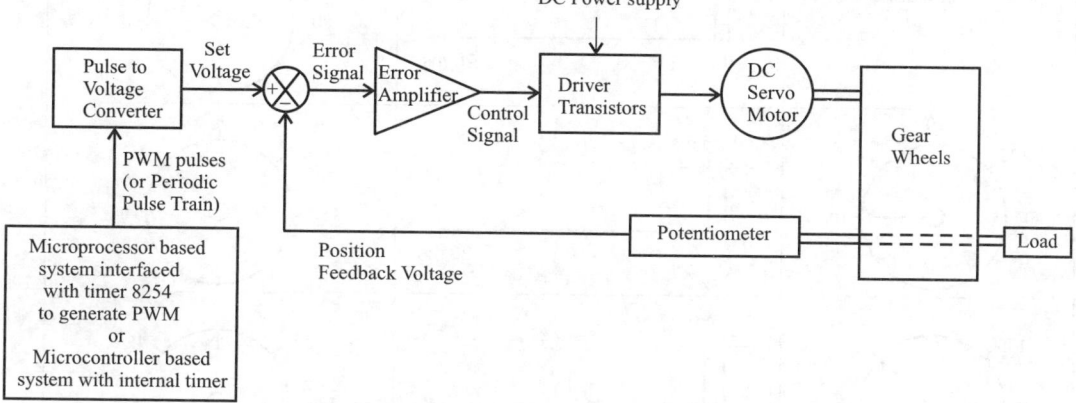

Fig. 11.26: Microprocessor or microcontroller-based position control system using DC servo motor.

The microprocessor/microcontroller-based system will generate the required PWM pulses which is converted to a voltage called set voltage for the desired position of motor shaft. This set voltage is compared with feedback voltage from potentiometer and an error signal is generated. The amplified error signal is used as control signal for driver transistors so that motor will be ON for the required time duration and move the shaft to desired position.

Most of the commercial motors require a response time of 15 to 20 milliseconds for the motor shaft to move from one position to another when there is a change in pulse width and so the time period of PWM signal or the periodic pulse train is normally selected as 20 milliseconds.

11.13.3 Angular Motion Control of Servo Motor

0° to 180° Motion Control

For 0 to 180° angular motion, a periodic pulse train with 1 millisecond ON time will move the motor shaft to one extreme end referred to as 0° position and a periodic pulse train with 2 millisecond ON time will move the motor shaft to the other extreme end referred to as 180° position.

Hence, in a time period of 20 milliseconds, the ON pulse time duration is 1 millisecond to 2 milliseconds, with a total time difference of 1 millisecond (1000 microseconds) for 180° rotation. Most of the servo motors are available with a resolution of 1° motion.

ON time for 1° motion = 1000 + 1000/180 = 1000 + 5.5556 microseconds.

ON time for 2° motion = 1000 + (2 × 1000/180) = 1000 + (2 × 5.5556) microseconds.

ON time for N° motion = 1000 + (N × 1000/180) = 1000 + (N × 5.5556) microseconds.

Table 11.8: 0° to 180° Rotation in Forward Direction (45° Steps)

Initial position	PWM/Periodic pulse train	Final position
Any position	1 ms, 0, 20 ms	90°, 135°, 45°, 180°, 0°
90°, 135°, 45°, 180°, 0°	1.25 ms, 0, 20 ms	90°, 135°, 45°, 180°, 0°
90°, 135°, 45°, 180°, 0°	1.5 ms, 0, 20 ms	90°, 135°, 45°, 180°, 0°
90°, 135°, 45°, 180°, 0°	1.75 ms, 0, 20 ms	90°, 135°, 45°, 180°, 0°
90°, 135°, 45°, 180°, 0°	2 ms, 0, 20 ms	90°, 135°, 45°, 180°, 0°

An example of PWM pulses and motor shaft position of servo motor in 0 to 180° position control for forward rotation with 45° incremental motion is shown in Table 11.8.

The servo motor can be operated in forward and reverse direction. For forward rotation the ON time of periodic pulses are can be estimated as shown above. But for reverse rotation, the ON time should be decreased from current value (Current Ton).

ON time for N° reverse motion = Current Ton − (N × 1000/180) microseconds.

An example of PWM pulses and motor shaft position of servo motor in 0 to 180° position control for reverse rotation with 45° incremental motion is shown in Table 11.9.

Table 11.9: 0° to 180° Rotation in Reverse Direction (45° Steps)

Initial position	PWM/Periodic pulse train	Final position
Any position	2 ms, 0, 20 ms	90°, 135°, 45°, 180°, 0°
90°, 135°, 45°, 180°, 0°	1.75 ms, 0, 20 ms	90°, 135°, 45°, 180°, 0°
90°, 135°, 45°, 180°, 0°	1.5 ms, 0, 20 ms	90°, 135°, 45°, 180°, 0°
90°, 135°, 45°, 180°, 0°	1.25 ms, 0, 20 ms	90°, 135°, 45°, 180°, 0°
90°, 135°, 45°, 180°, 0°	1 ms, 0, 20 ms	90°, 135°, 45°, 180°, 0°

− 90° to + 90° Motion Control

For − 90° to + 90° angular motion, 1.5 millisecond pulse will move the motor shaft to midpoint referred to as neutral or 0° position, 1 millisecond pulse will move the motor shaft to one extreme end referred to as − 90° position and 2 millisecond pulse will move the motor shaft to the other extreme end referred to as + 90° position.

0° to + 90° Motion Control

In a time period of 20 milliseconds, for 0° to + 90° angular motion, the ON pulse time duration is 1.5 millisecond to 2 milliseconds, with a total time difference of 0.5 millisecond (500 microseconds) for 90° rotation. Therefore, for a servo motor with 1° resolution,

Timing pulse for 1° motion = 1500 + 500/90 = 1500 + 5.5556 microseconds.

Timing pulse for 2° motion = 1500 + (2 × 500/90) = 1500 + (2 × 5.5556) microseconds.

Timing pulse for N° motion = 1500 + (N × 500/90) = 1500 + (N × 5.5556) microseconds.

For forward rotation the ON time of periodic pulses are can be estimated as shown above. But for reverse rotation, the ON time should be decreased from current value (Current Ton).

ON time for N° reverse motion = Current Ton – (N × 500/90) microseconds.

An example of PWM pulses and motor shaft position of servo motor in 0° to + 90° position control for forward and reverse rotation with 45° incremental motion is shown in Table 11.10.

Table 11.10: 0° to +90° Rotation in Forward and Reverse Direction (45° Steps)

Initial position	PWM/Periodic pulse train	Final position
Any position	1.5 ms 0 20 ms	90°, 45°, 0°, –45°, –90°
90°, 45°, 0°, –45°, –90°	1.75 ms 0 20 ms	90°, 45°, 0°, –45°, –90°
90°, 45°, 0°, –45°, –90°	2 ms 0 20 ms	90°, 45°, 0°, –45°, –90°
90°, 45°, 0°, –45°, –90°	1.75 ms 0 20 ms	90°, 45°, 0°, –45°, –90°

Table 11.10 continued...

Initial position	PWM/Periodic pulse train	Final position

0° to − 90° Motion Control

In a time period of 20 milliseconds, for 0° to -90° angular motion, the ON pulse time duration is 1.5 millisecond to 1 milliseconds, with a total time difference of 0.5 millisecond (500 microseconds) for 90° rotation. Therefore, for a servo motor with 1° resolution,

Timing pulse for 1° motion = 1500 − 500/90 = 1500 − 5.5556 microseconds.

Timing pulse for 2° motion = 1500 − (2 × 500/90) = 1500 − (2 × 5.5556) microseconds.

Timing pulse for N° motion = 1500 − (N × 500/90) = 1500 − (N × 5.5556) microseconds.

For reverse rotation the ON time of periodic pulses are can be estimated as shown above. But for forward rotation, the ON time should be increased from current value (Current Ton).

ON time for N° forward motion = Current Ton + (N × 500/90) microseconds.

An example of PWM pulses and motor shaft position of servo motor in 0° to − 90° position control for forward and reverse rotation with 45° incremental motion is shown in Table 11.11.

Table 11.11: 0° to − 90° Rotation in Forward and Reverse Direction (45° Steps)

Initial position	PWM/Periodic pulse train	Final position
Any position		

Table 11.11 continued...

Initial position	PWM/Periodic pulse train	Final position
	 1 ms ⊣⊢ 0 ... 20 ms	
	 1.25 ms ⊣⊢ 0 ... 20 ms	
	 1.5 ms ⊣⊢ 0 ... 20 ms	

11.14 SHORT-ANSWER QUESTIONS

Q11.1 Write a short note on Clock generator-INTEL 8284.

The Clock generator-INTEL 8284 is specially designed for the 8086 microprocessor. It generates the clock signal, READY and RESET signals required for 8086 and a TTL level peripheral clock signal, which can be used for other peripheral devices in the system.

Q11.2 How is the clock signal generated in 8284?

The 8284 has an internal oscillator circuit. An external quartz crystal has to be connected to the 8284 to generate a clock signal.

Q11.3 Write a short note on Bus Controller-INTEL 8288.

The Bus controller-INTEL 8288 is specially designed for the 8086 microprocessor for use in maximum mode to generate bus control signals. It reads the status signal from the processor and generates bus control signal for memory and IO devices.

Q11.4 List the control signals of 8086 that are common to minimum and maximum modes.

The control signals that are common to minimum mode and maximum mode are $\overline{\text{BHE}}$, RD, MN/$\overline{\text{MX}}$, $\overline{\text{TEST}}$, READY and RESET.

Q11.5 How do bus request signals work in 8086-based maximum mode system?

1. When a local bus master requires system bus control, it sends a low pulse to the 8086.
2. At the end of the current bus cycle, the processor (8086) drives its pins to high impedance state and sends an acknowledgement signal as a low pulse on the same pin to the device which requested the bus control.
3. On receiving the acknowledge signal, the local master will take control of the system bus. After completing its work, at the end, the local bus master sends a low signal on the same pin to the 8086 to inform the end of control. Now 8086 regains the control of the bus.

Q11.6 "Driving a system at the maximum clock is advantageous". Justify the statement.

The above statement is true, because the execution time will be minimum if the clock is maximum

Q11.7 What is minimum mode system?

The minimum mode is the uniprocessor mode in which the 8086 processor itself generates all bus control signals and so there is no need for an external bus controller. The signal $MN/\overline{MX}$ of the 8086 microprocessor has to be permanently tied high, for minimum mode system.

Q11.8 What is maximum mode system?

The maximum mode is a multiprocessor mode in which the 8086 processor is interfaced with the 8087 coprocessor to enhance the mathematical capability of the system. The signal $MN/\overline{MX}$ of the 8086 microprocessor is permanently tied low, for maximum mode system. In the maximum mode system, an external bus controller 8288 has to be employed to generate the bus control signals.

Q11.9 What is multiprocessor configuration?

The multiprocessor configuration is a system bus structure in which two or more processors can execute instructions (or perform operations) simultaneously. In multiprocessor systems, the added processors can be special-purpose processors or general-purpose processors.

Q11.10 Differentiate between the minimum and maximum mode of the 8086 system.

Minimum mode	Maximum mode
1. External bus controller, not required	1. External bus controller 8288 has to be employed.
2. $MN/\overline{MX}$ pin of 8086 is connected to Vcc(+5)	2. $MN/\overline{MX}$ pin of 8086 is connected to ground.
3. Suitable for uniprocessor system.	3. Suitable for multiprocessor system.

Q11.11 What are the advantages of multiprocessor configuration?

1. Multiple low-cost processors can be interfaced to form larger systems to handle heavy tasks.
2. Possible to expand the system resources whenever the need arises.
3. Modular programming can be easily implemented.
4. In case of failure, it will be easier to replace the faulty part alone.

Q11.12 What is a coprocessor and what is its need?

A coprocessor is an additional device/IC that can be integrated with the main processor to enhance its working capability. INTEL 8087 is a coprocessor that has been developed to work with 8086 processor in maximum mode to take care of mathematical calculations involving integer and floating-point data.

Q11.13 What are the features of the 8087 coprocessor?

1. It has 16-bit external data bus and 20-bit external address bus.
2. It has a numeric execution unit to take care of mathematical operations.
3. It has 8 numbers of 80-bit registers to store large-sized data.
4. It can work with binary, BCD, integer and floating-point data.

Q11.14 What is loosely coupled configuration?

The loosely coupled configuration is a multiprocessor system in which a number of independent microprocessor-based systems (or masters) are interfaced through a common system bus to share system resources. Each independent microprocessor system has its own clock source, memory and IO devices interfaced through a local bus.

Q11.15 What is closely coupled configuration?

The closely coupled configuration is a multiprocessor system, in which two processors will be connected through the system bus such that they share the same bus control logic, clock generator, system memory and IO.

Q11.16 Compare closely coupled and loosely coupled configurations.

Closely coupled configuration	Loosely coupled configuration
1. Two processors called master and slave processors connected to a system bus to work together.	1. Multiple independent microprocessor based systems connected to a common bus to share resources.
2. Master and slave processors share a bus, memory, clock generator and IO.	2. Each module have its own clock, memory and IO.

Q11.17 List the specification of microprocessor system?

1. Input device   5. Peripheral devices required
2. Output device   6. Type of CPU (Microprocessor)
3. Memory requirement   7. Applications or Nature of work.
4. System clock frequency

Q11.18 List of peripheral devices that can be interfaced to 8085-based system ?

1. Programmable Interval Timer - INTEL 8253/8254   5. Programmable Interrupt Controller - INTEL 8259
2. USART - INTEL 8251   6. DMA Controller - INTEL 8237/8257
3. Programmable peripheral Interface - INTEL 8255   7. ADC
4. Keyboard / Display controller - INTEL 8279   8. DAC.

Q11.19 Give some examples of Microprocessor-based automation systems

1. Temperature control system   4. Fire control safety system
2. Level Control system   5. Traffic light control system.
3. Motor speed control system

Q11.20 List the various type of temperature sensors?

1. Thermistors   4. Integrated circuits
2. Thermo couples   5. Temperature sensors
3. Bimetallic strips

Q11.21 What are the types of electronic-type humidity sensors?

1. **Capacitive type**
2. **Resistive type**

Q11.22 List the types of gas sensors?

1. **Infrared gas sensors.**
2. **Ultrasonic gas sensors.**
3. **Electrochemical gas sensors.**
4. **Semiconductor gas sensors**

Q11.23 Write short notes on Infra-Red(IR) Sensors ?

Infrared (IR) rays are invisible electromagnetic waves which have a wide frequency range between microwaves and visible light in the electromagnetic spectrum. The frequency range of IR waves will be from 300 GHz to 430 THz (with wavelength in the range 1000 micrometre to 0.75 micrometre). The wavelength region of 0.75 μm to 3 μm is called near infrared, the region from 3 μm to 6 μm is called mid infrared, and the region higher than 6 μm is called far infrared.

Q11.24 Define signal conditioning ?

Some of the abnormal conditions that can be sensed by alarm controllers are gas leakage, fire, leakage in flow pipes, etc. The sensors are basically transducers which are energy-converting devices. The sensors generate electrical signals proportional to quantity being sensed, and the magnitude of this signal can be used to differentiate the normal and abnormal conditions. These electrical signals are amplified and scaled by using operational amplifiers. This process is also known as **signal conditioning**.

Q11.25 Write the applications of ultrasonic sensors?

1. **Humidifiers**
2. **Sonar**
3. **Medical ultrasonography**
4. **Burgar alarms**
5. **Non-destructive testing**
6. **Testing and wireless charging**

11.15 EXERCISES

I. Fill in the blanks with appropriate words

1. The most widely used input device in a microprocessor trainer kit is _____ .
2. The _____ chip provides RAM memory for 8085-based minimum system.
3. The method in which unused address lines are directly connected CS pin peripheral ICs is called _____.
4. The control signal which is required to control the supply of the heating element should be _____.
5. The _____ converts the temperature into proportional analog voltage or current.
6. The _____ voltage of the DC motor is variable while the _____ voltage is kept constant for varying the DC motor speed.
7. The stepper motor which has step size of 1.8° can rotate _____ number of steps per revolution.
8. The _____ pin of 8284A provides same frequency as that of crystal.
9. The _____ and _____ signals of 8284A are used for providing extended IO write and Memory write time for IO and memory devices respectively.
10. The most significant five bits of 8087 instructions have the _____ binary bit pattern.
11. A system which has two or more processors which execute instructions simultaneously is called _____ system.

12. In _____ closely coupled configurations, both master and slave processors share the same bus control logic.

13. The _____ device is used to generate bus control signals in maximum mode in 8086 based systems.

14. The coprocessor which is used with 8086 inorder to perform floating point arithmetic operations effieciently is _____ .

Answers

1. hex-keyboard	5. temperature sensor	9. AIOWC, AMWC	13. bus controller-8288
2. 8155	6. armature, field	10. 11011	14. 8087-math processor
3. linear address selection	7. 200	11. multiprocessor	
4. analog	8. OSC	12. closely	

II. State whether the following statements are True/False.

1. The INTEL 8279 keyboard and display controller can be used to generate maximum 256 key-codes.
2. The interrupt vector table of 8085 is stored in RAM
3. In minimum systems, both monitor program and system program are stored in permanent memory.
4. The clock generator INTEL 8284 has an on-chip internal oscillator circuit and quartz crystal.
5. The temperature of a plant is kept under control by varying the supply voltage of heating element.
6. An SCR circuit is used to generate the required armature voltage in dc motor speed control systems.
7. The firing angle of SCR should be constant for controlling the speed of dc motor.
8. A forward biased diode is connected across relay coil to prevent free-wheeling action in traffic light control system.
9. The clock generator 8284A can also generate peripheral clock signal.
10. The bus controller-INTEL 8288 is used to generate bus control signals in 8086 minimum mode.
11. The coprocessor is also called as math processor or Numeric Data Processor(NDP).
12. The INTEL 80286 has on-chip co-processor.
13. The co-processor 8087 works with 8086 only in maximum mode.
14. The co-processors can work on BCD numbers directly.
15. The letter "F" is used to differentiate 8086 and 8087 instructions.
16. The 8087 treats all the 8086 instructions encountered as NOP.
17. The 8086 processor has an on-chip latches to demultiplex the address/data lines.
18. In 8086 maximum mode, the signals ALE, DEN and DT/$\overline{R}$ are generated by bus controller.
19. Two numbers of 8086 processors cannot work in closely coupled configurations.
20. Each module in loosely coupled configuration can work as an independent microprocessor based system.
21. Maximum mode of 8086 can support only uniprocessor mode of operation.

Answers

1. True	4. False	7. False	10. False	13. True	16. True	19. True
2. False	5. True	8. False	11. True	14. True	17. False	20. True
3. True	6. True	9. True	12. False	15. True	18. True	21. False

III. Choose the right answer for the following questions.

1. *The INTEL 8279 can be used to interface _____.*

   a) hex-keyboard   b) LED display   c) both a& b   d) neither a nor b

2. *Which of the following chip is not used by 8085-based minimum system?*

   a) 8155   b) 8255   c) 8355   d) 8755

3. *The following Programmable Peripheral Interface (PPI) ICs demultiplex the address/data lines internally.*

   a) 8155   b) 8355   c) 8755   d) all the three

4. *Which of the following INTEL IC is used to generate the clock required for 8086?*

   a) 8255   b) 8234   c) 8279   d) 8284

5. *Which of the following circuit is used to strengthen the weak signal?*

   a) signal conditioner b) amplifier   c) ADC   d) DAC

6. *Which of the following is not a temperature sensor?*

   a) thermo couples   b) thermistor   c) AD590   d) AD0808

7. *Which of the following circuit is used to generated the required field voltage to control the speed of the dc motor?*

   a) SCR   b) controlled rectifier   c) uncontrolled rectifier d) none of the above

8. *The _____ is used for precise incremental rotation in most of the computer peripherals.*

   a) stepper motor   b) DC motor   c) a and b   d) neither a nor b

9. *In stepper motor control system the free-wheeling diode is connected across each winding for _____.*

   a) fast switchig   b) fast halt   c) both a and b   d) neither a nor b

10. *The clock generator 8284A is used to generate the following signal/s*

    a) clock   b) READY   c) RESET   d) all the three

11. *Which of the following 8284A pin provides clock signal for peripherals?*

    a) CLK   b) PCLK   c) RESET   d) CSYNC

12. *Which of the following processor haave on-chip coprocessors?*

    a) 80186   b) 80286   c) 80386   d) 80486

13. *Which of the following data types are supported by coprocessor?*

    a) 80-bit BCD   b) 32-bit real   c) 64-bit real   d) all the three

14. *In following configuration, the bus request of the slave processor is connected to the master.*

    a) loosely coupled   b) dual coupled   c) closely coupled   d) all the three

Answers						
1. c	3. d	5. b	7. c	9. a	11. b	13. d
2. b	4. d	6. d	8. a	10. d	12. d	14. c

IV. Answer the following questions.

E11.1 What are the required devices/units for designing a microprocessor based system?

E11.2 Comment on input devices required for designing a microprocessor based system

E11.3 Discuss about output devices required for designig a microprocessor based system

E11.4 Discuss the memory requirement of a microprocessor based system.

E11.5 Write short notes on system clock frequecy selection for a microprocessor based system.

E11.6 What is meant by minimum system?

E11.7 What are the requirements for designing a 8085 based minimum system?

E11.8 List the functional units of temperature control system.

E11.9 Briefly explain the steps involved in controlling the temperature of a plant.

E11.10 List the various types temperature sensors.

E11.11 What is the need of signal conditioning circuit in temperature control system?

E11.12 What are the uses of ADC and DAC in temperature control system?

E11.13 Write down the steps involved in controlling the speed of a dc motor.

E11.14 Write down the stepper motor switching sequence for full-step rotation

E11.15 Write down the stepper motor switching sequence fo half-step rotation

E11.16 Write short notes on clock generator-INTEL 8284A.

E11.17 What is the use of bus controller-INTEL 8288?

E11.18 Define coprocessor

E11.19 List few coprocessors used for INTEL 80×86 family processors.

E11.20 Write down the type of operands on which the coprocessor can work.

E11.21 How is the main-processor and co-processor instructions are differentiated?

E11.22 Write short notes on minimum mode 8086 microprocessor system.

E11.23 Write short notes on maximum mode 8086 microprocessor system.

E11.24 What is meant by multiprocessor system?Give examples.

E11.25 What are the advantages of multiprocessor system?

E11.26 What are the major issues of multiprocessor system?

E11.27 Mention the two types of multiprocessor configurations.

E11.28 Discuss about closely coupled configuration.

E11.29 Write short notes on loosely coupled configurations

E11.30 What are the advantages and disadvantages of closely coupled configurations?

E11.31 What are the advantages and disadvantages of loosely coupled configurations?

APPENDIX I : LIST OF MICROPROCESSORS RELEASED BY INTEL

MICROPROCESSOR	DATE OF INTRODUCTION	NUMBER OF TRANSISTORS	CLOCK SPEED
4004	15th Nov, 1971	2,300	400 kHz
8008	Apr, 1972	3,500	500-800 kHz
8080	Apr, 1974	4,500	2 MHz
8085	Mar, 1976	6,500	5 MHz
8086	8th Jun, 1978	29,000	5/8/10 MHz
8088	Jun, 1979	29,000	5/8 MHz
80186	1982	....	10/12 MHz
80286	Feb, 1982	134,000	6/10/12 MHz
INTEL386 DX	17th Oct, 1985	275,000	16/20/25/33 MHz
INTEL386 SX	16th Jun, 1988	275,000	16/20/25/33 MHz
INTEL386 SL	15th Oct, 1990	855,000	20/25 MHz
INTEL486 DX	10th Apr, 1989	1.2 million	25/33/50 MHz
INTEL486 SX	16th Sep, 1991	900,000	16/20/25 MHz
INTEL486 SX	21st Sep, 1992	1.185 million	33 MHz
INTEL486 SL	4th Nov, 1992	1.4 million	20/25/33 MHz
INTELDX 2	3rd Mar, 1992	1.2 million	50/66 MHz
INTELDX 4	7th Mar, 1994	3.2 million	75/100 MHz
Pentium	22nd Mar, 1993	3.1 million	60/66 MHz
Pentium	7th Mar, 1994	3.2 million	75/90/100/120 MHz
Pentium	Jun, 1995	3.3 million	133/150/166/200 MHz
Pentium Pro	1st Nov, 1995	5.5 million	150/166/180/200 MHz
Pentium (MMX)	8th Jan, 1997	4.5 million	166/200/233 MHz
Mobile Pentium (MMX)	9th Sep, 1997	4.5 million	200/233/266/300 MHz
Pentium II	7th May, 1997	7.5 million	233/266/300/333/350/400/ 450 MHz
Mobile Pentium II	2nd Apr, 1998	7.5 million	233/266/300 MHz
Mobile Pentium II	25th Jan, 1999	27.4 million	333/366/400 MHz
Pentium II Xeon	29th Jun, 1998	7.5 million	400/450 MHz
Celeron	15th Apr, 1998	7.5 million	266/300 MHz
Celeron	24th Aug, 1998	19 million	333 MHZ to 2.7 GHz
Mobile Celeron	25th Jan, 1999	18.9 million	266 MHz to 2.4 GHz
Pentium III	26th Feb, 1999	9.5 million	450/500/550/600 MHZ

Appendix I continued...

MICROPROCESSOR	DATE OF INTRODUCTION	NUMBER OF TRANSISTORS	CLOCK SPEED
Pentium III	25th Oct, 1999	28 million	500 MHz to 1 GHz
Pentium III Xeon	17th Mar, 1999	9.5 million	500/550 MHz
Pentium III Xeon	25th Oct, 1999	28 million	600 to 900 MHz
Mobile Pentium III	25th Oct, 1999	28 million	400 MHz to 1 GHz
Mobile Pentium III	30th Jul, 2001	44 million	1/1.06/1.13/1.2/1.33 GHz
Pentium 4	20th Nov, 2000	42 million	1.4/1.5/1.6/1.7/1.8/1.9/2 GHz
Pentium 4	27th Aug, 2001	55 million	2 to 2.8 GHz
Pentium 4 (HT Technology)	14th Nov, 2002	55 million	2.4 to 3.3 GHz
Mobile Pentium 4	4th Mar, 2002	55 million	1.5 to 3.2 GHz
INTEL Xeon	21st May, 2001	42 million	1.4/1.5/1.7/2 GHz
INTEL Xeon	9th Jan, 2002	52 million	1.8/2/2.2/2.4/2.6/2.8 GHz
INTEL Xeon	18th Nov, 2002	108 million	1.4 to 3.2 GHz
INTEL Itanium	May, 2001	25 million	733/800 MHz
INTEL Itanium 2	8th Jul, 2002	220 million	900 MHz/1 GHz
INTEL Itanium 2	30th Jun, 2003	410 million	1/1.4/1.5 GHz
INTEL Pentium-M	12th Mar, 2003	77 million	900 MHz to 1.7 GHz

Note : The date mentioned here is the date of introduction of the lowest clock version of the processor. For the date of introduction of higher clock version of a processor please refer to INTEL website www.intel.com.

APPENDIX II : 8051 INSTRUCTIONS IN HEXADECIMAL ORDER

OPCODE IN HEX	MNEMONIC		OPCODE IN HEX	MNEMONIC		OPCODE IN HEX	MNEMONIC	
00	NOP		2B	ADD	A,R3	56	ANL	A,@R0
01	AJMP	addr11	2C	ADD	A,R4	57	ANL	A,@R1
02	LJMP	addr16	2D	ADD	A,R5	58	ANL	A,R0
03	RR	A	2E	ADD	A,R6	59	ANL	A,R1
04	INC	A	2F	ADD	A,R7	5A	ANL	A,R2
05	INC	DPTR	30	JNB	bit,offset	5B	ANL	A,R3
06	INC	@R0	31	ACALL	addr11	5C	ANL	A,R4
07	INC	@R1	32	RETI		5D	ANL	A,R5
08	INC	R0	33	RLC	A	5E	ANL	A,R6
09	INC	R1	34	ADDC	A,#data	5F	ANL	A,R7
0A	INC	R2	35	ADDC	A,direct	60	JZ	offset
0B	INC	R3	36	ADDC	A,@R0	61	AJMP	addr11
0C	INC	R4	37	ADDC	A,@R1	62	XRL	direct,A
0D	INC	R5	38	ADDC	A,R0	63	XRL	direct,#data
0E	INC	R6	39	ADDC	A,R1	64	XRL	A,#data
0F	INC	R7	3A	ADDC	A,R2	65	XRL	A,#direct
10	JBC	bit,offset	3B	ADDC	A,R3	66	XRL	A,@R0
11	ACALL	addr11	3C	ADDC	A,R4	67	XRL	A,@R1
12	LCALL	addr16	3D	ADDC	A,R5	68	XRL	A,R0
13	RRC	A	3E	ADDC	A,R6	69	XRL	A,R1
14	DEC	A	3F	ADDC	A,R7	6A	XRL	A,R2
15	DEC	direct	40	JC	offset	6B	XRL	A,R3
16	DEC	@R0	41	AJMP	addr11	6C	XRL	A,R4
17	DEC	@R1	42	ORL	direct,A	6D	XRL	A,R5
18	DEC	R0	43	ORL	direct,#data	6E	XRL	A,R6
19	DEC	R1	44	ORL	A,#data	6F	XRL	A,R7
1A	DEC	R2	45	ORL	A,direct	70	JNZ	offset
1B	DEC	R3	46	ORL	A,@R0	71	ACALL	addr11
1C	DEC	R4	47	ORL	A,@R1	72	ORL	C,bitaddr
1D	DEC	R5	48	ORL	A,R0	73	JMP	@A+DPTR
1E	DEC	R6	49	ORL	A,R1	74	MOV	A,#data
1F	DEC	R7	4A	ORL	A,R2	75	MOV	direct,#data
20	JB	bit,offset	4B	ORL	A,R3	76	MOV	@R0,#data
21	AJMP	addr11	4C	ORL	A,R4	77	MOV	@R1,#data
22	RET		4D	ORL	A,R5	78	MOV	R0,#data
23	RL	A	4E	ORL	A,R6	79	MOV	R1,#data
24	ADD	A,#data	4F	ORL	A,R7	7A	MOV	R2,#data
25	ADD	A,direct	50	JNC	offset	7B	MOV	R3,#data
26	ADD	A,@R0	51	ACALL	addr11	7C	MOV	R4,#data
27	ADD	A,@R1	52	ANL	direct,A	7D	MOV	R5,#data
28	ADD	A,R0	53	ANL	direct,#data	7E	MOV	R6,#data
29	ADD	A,R1	54	ANL	A,#data	7F	MOV	R7,#data
2A	ADD	A,R2	55	ANL	A,direct	80	SJMP	offset

Appendix II continued...

OPCODE IN HEX	MNEMONIC		OPCODE IN HEX	MNEMONIC		OPCODE IN HEX	MNEMONIC	
81	AJMP	addr11	AC	MOV	R4,direct	D7	XCHD	A,@R1
82	ANL	C,bit	AD	MOV	R5,direct	D8	DJNZ	R0,offset
83	MOVC	A,@A+PC	AE	MOV	R6,direct	D9	DJNZ	R1,offset
84	DIV	AB	AF	MOV	R7,direct	DA	DJNZ	R2,offset
85	MOV	direct,direct	B0	ANL	C,/bit	DB	DJNZ	R3,offset
86	MOV	direct,@R0	B1	ACALL	addr11	DC	DJNZ	R4,offset
87	MOV	direct,@R1	B2	CPL	bit	DD	DJNZ	R5,offset
88	MOV	direct,R0	B3	CPL	C	DE	DJNZ	R6,offset
89	MOV	direct,R1	B4	CJNE	A,#data,offset	DF	DJNZ	R7,offset
8A	MOV	direct,R2	B5	CJNE	A,direct,offset	E0	MOVX	A,@DPTR
8B	MOV	direct,R3	B6	CJNE	@R0,#data,offset	E1	AJMP	addr11
8C	MOV	direct,R4	B7	CJNE	@R1,#data,offset	E2	MOVX	A,@R0
8D	MOV	direct,R5	B8	CJNE	R0,#data,offset	E3	MOVX	A,@R1
8E	MOV	direct,R6	B9	CJNE	R1,#data,offset	E4	CLR	A
8F	MOV	direct,R7	BA	CJNE	R2,#data,offset	E5	MOV	A,direct
90	MOV	DPTR,#data16	BB	CJNE	R3,#data,offset	E6	MOV	A,@R0
91	ACALL	addr11	BC	CJNE	R4,#data,offset	E7	MOV	A,@R1
92	MOV	bit,C	BD	CJNE	R5,#data,offset	E8	MOV	A,R0
93	MOVC	A,@A+DPTR	BE	CJNE	R6,#data,offset	E9	MOV	A,R1
94	SUBB	A,#data	BF	CJNE	R7,#data,offset	EA	MOV	A,R2
95	SUBB	A,#direct	C0	PUSH	direct	EB	MOV	A,R3
96	SUBB	A,@R0	C1	AJMP	addr11	EC	MOV	A,R4
97	SUBB	A,@R1	C2	CLR	bit	ED	MOV	A,R5
98	SBBB	A,R0	C3	CLR	C	EE	MOV	A,R6
99	SBBB	A,R1	C4	SWAP	A	EF	MOV	A,R7
9A	SBBB	A,R2	C5	XCH	A,direct	F0	MOVX	@DPTR,A
9B	SBBB	A,R3	C6	XCH	A,@R0	F1	ACALL	addr11
9C	SBBB	A,R4	C7	XCH	A,@R1	F2	MOVX	@R0,A
9D	SBBB	A,R5	C8	XCH	A,R0	F3	MOVX	@R1,A
9E	SBBB	A,R6	C9	XCH	A,R1	F4	CPL	A
9F	SBBB	A,R7	CA	XCH	A,R2	F5	MOV	direct,A
A0	ORL	C,/bit	CB	XCH	A,R3	F6	MOV	@R0,A
A1	AJMP	addr11	CC	XCH	A,R4	F7	MOV	@R1,A
A2	MOV	C,bit	CD	XCH	A,R5	F8	MOV	R0,A
A3	INC	DPTR	CE	XCH	A,R6	F9	MOV	R1,A
A4	MUL	AB	CF	XCH	A,R7	FA	MOV	R2,A
A5	UNUSED		D0	POP	direct	FB	MOV	R3,A
A6	MOV	@R0,direct	D1	ACALL	addr11	FC	MOV	R4,A
A7	MOV	@R1,direct	D2	SETB	bit	FD	MOV	R5,A
A8	MOV	R0,direct	D3	SETB	C	FE	MOV	R6,A
A9	MOV	R1,direct	D4	DA	A	FF	MOV	R7,A
AA	MOV	R2,direct	D5	DJNZ	direct,offset			
AB	MOV	R3,direct	D6	XCHD	A,@R0			

APPENDIX III : 8051 INSTRUCTIONS IN ALPHABETICAL ORDER

HEX CODE	MNEMONIC		HEX CODE	MNEMONIC		HEX CODE	MNEMONIC	
11	ACALL	addr11	5F	ANL	A,R7	DE	DJNZ	R6,offset
31	ACALL	addr11	55	ANL	A,direct	DF	DJNZ	R7,offset
51	ACALL	addr11	56	ANL	A,@R0	D5	DJNZ	direct,offset
71	ACALL	addr11	57	ANL	A,@R1	04	INC	A
91	ACALL	addr11	54	ANL	A,#data	08	INC	R0
B1	ACALL	addr11	52	ANl	direct,A	09	INC	R1
D1	ACALL	addr11	53	ANl	direct,#data	0A	INC	R2
F1	ACALL	addr11	82	ANL	C,bit	0B	INC	R3
28	ADD	A,R0	B0	ANL	C,/bit	0C	INC	R4
29	ADD	A,R1	B5	CJNE	A,direct,offset	0D	INC	R5
2A	ADD	A,R2	B4	CJNE	A,#data,offset	0E	INC	R6
2B	ADD	A,R3	B8	CJNE	R0,#data,offset	0F	INC	R7
2C	ADD	A,R4	B9	CJNE	R1,#data,offset	05	INC	direct
2D	ADD	A,R5	BA	CJNE	R2,#data,offset	06	INC	@R0
2E	ADD	A,R6	BB	CJNE	R3,#data,offset	07	INC	@R1
2F	ADD	A,R7	BC	CJNE	R4,#data,offset	A3	INC	DPTR
25	ADD	A,direct	BD	CJNE	R5,#data,offset	20	JB	bit,offset
26	ADD	A,@R0	BE	CJNE	R6,#data,offset	10	JBC	bit,offset
27	ADD	A,@R1	BF	CJNE	R7,#data,offset	40	JC	offset
24	ADD	A,#data	B6	CJNE	@R0,#data,offset	73	JMP	@A+DPTR
38	ADDC	A,R0	B7	CJNE	@R1,#data,offset	30	JNB	bit,offset
39	ADDC	A,R1	E4	CLR	A	50	JNC	offset
3A	ADDC	A,R2	C3	CLR	C	70	JNZ	offset
3B	ADDC	A,R3	C2	CLR	bit	60	JZ	offset
3C	ADDC	A,R4	F4	CPL	A	12	LCALL	addr16
3D	ADDC	A,R5	B3	CPL	C	02	LJMP	addr16
3E	ADDC	A,R6	B2	CPL	bit	E8	MOV	A,R0
3F	ADDC	A,R7	D4	DA	A	E9	MOV	A,R1
35	ADDC	A,direct	14	DEC	A	EA	MOV	A,R2
36	ADDC	A,@R0	18	DEC	R0	EB	MOV	A,R3
37	ADDC	A,@R1	19	DEC	R1	EC	MOV	A,R4
34	ADDC	A,#data	1A	DEC	R2	ED	MOV	A,R5
01	AJMP	addr11	1B	DEC	R3	EE	MOV	A,R6
21	AJMP	addr11	1C	DEC	R4	EF	MOV	A,R7
41	AJMP	addr11	1D	DEC	R5	E5	MOV	A,direct
61	AJMP	addr11	1E	DEC	R6	E6	MOV	A,@R0
81	AJMP	addr11	1F	DEC	R7	E7	MOV	A,@R1
A1	AJMP	addr11	15	DEC	direct	74	MOV	A,#data
C1	AJMP	addr11	16	DEC	@R0	F8	MOV	R0,A
E1	AJMP	addr11	17	DEC	@R1	F9	MOV	R1,A
58	ANL	A,R0	84	DIV	AB	FA	MOV	R2,A
59	ANL	A,R1	D8	DJNZ	R0,offset	FB	MOV	R3,A
5A	ANL	A,R2	D9	DJNZ	R1,offset	FC	MOV	R4,A
5B	ANL	A,R3	DA	DJNZ	R2,offset	FD	MOV	R5,A
5C	ANL	A,R4	DB	DJNZ	R3,offset	FE	MOV	R6,A
5D	ANL	A,R5	DC	DJNZ	R4,offset	FF	MOV	R7,A
5E	ANL	A,R6	DD	DJNZ	R5,offset			

Appendix III continued...

HEX CODE	MNEMONIC		HEX CODE	MNEMONIC		HEX CODE	MNEMONIC	
A8	MOV	R0,direct	A4	MUL	AB	CC	XCH	A,R4
A9	MOV	R1,direct	00	NOP		CD	XCH	A,R5
AA	MOV	R2,direct	48	ORL	A,R0	CE	XCH	A,R6
AB	MOV	R3,direct	49	ORL	A,R1	CF	XCH	A,R7
AC	MOV	R4,direct	4A	ORL	A,R2	C5	XCH	A,direct
AD	MOV	R5,direct	4B	ORL	A,R3	C6	XCH	A,@R0
AE	MOV	R6,direct	4C	ORL	A,R4	C7	XCH	A,@R1
AF	MOV	R7,direct	4D	ORL	A,R5	D6	XCHD	A,@R0
78	MOV	R0,#data	4E	ORL	A,R6	D7	XCHD	A,@R1
79	MOV	R1,#data	4F	ORL	A,R7	68	XRL	A,R0
7A	MOV	R2,#data	45	ORL	A,direct	69	XRL	A,R1
7B	MOV	R3,#data	46	ORL	A,@R0	6A	XRL	A,R2
7C	MOV	R4,#data	47	ORL	A,@R1	6B	XRL	A,R3x
7D	MOV	R5,#data	44	ORL	A,#data	6C	XRL	A,R4
7E	MOV	R6,#data	42	ORL	direct,A	6D	XRL	A,R5
7F	MOV	R7,#data	43	ORL	direct,#data	6E	XRL	A,R6
F5	MOV	direct,A	72	ORL	C,bit	6F	XRL	A,R7
88	MOV	direct,R0	A0	ORL	C,/bit	65	XRL	A,direct
89	MOV	direct,R1	D0	POP	direct	66	XRL	A,@R0
8A	MOV	direct,R2	C0	PUSH	direct	67	XRL	A,@R1
8B	MOV	direct,R3	22	RET		64	XRL	A,#data
8C	MOV	direct,R4	32	RETI		62	XRL	direct,A
8D	MOV	direct,R5	23	RL	A	63	XRL	direct,#data
8E	MOV	direct,R6	33	RLC	A			
8F	MOV	direct,R7	03	RR	A			
85	MOV	direct,direct	13	RRC	A			
86	MOV	direct,@R0	D3	SETB	C			
87	MOV	direct,@R1	D2	SETB	bit			
75	MOV	direct,#data	80	SJMP	offset			
F6	MOV	@R0,A	98	SUBB	A,R0			
F7	MOV	@R1,A	99	SUBB	A,R1			
A6	MOV	@R0,direct	9A	SUBB	A,R2			
A7	MOV	@R1,direct	9B	SUBB	A,R3			
76	MOV	@R0,#data	9C	SUBB	A,R4			
77	MOV	@R1,#data	9D	SUBB	A,R5			
A2	MOV	C,bit	9E	SUBB	A,R6			
92	MOV	bit,C	9F	SUBB	A,R7			
90	MOV	DPTR,#data16	95	SUBB	A,direct			
93	MOVC	A, @A+DPTR	96	SUBB	A,@R0			
83	MOVC	A,@A+PC	97	SUBB	A,@R1			
E2	MOVX	A,@R0	94	SUBB	A,#data			
E3	MOVX	A,@R1	C4	SWAP	A			
E0	MOVX	A,@DPTR	C8	XCH	A,R0			
F2	MOVX	@R0,A	C9	XCH	A,R1			
F3	MOVX	@R1,A	CA	XCH	A,R2			
F0	MOVX	@DPTR,A	CB	XCH	A,R3			

APPENDIX IV : 8085 INSTRUCTIONS IN HEXADECIMAL ORDER

OPCODE IN HEX	MNEMONIC		OPCODE IN HEX	MNEMONIC		OPCODE IN HEX	MNEMONIC	
00	NOP		2B	DCX	H	56	MOV	D, M
01	LXI	B, d16	2C	INR	L	57	MOV	D, A
02	STAX	B	2D	DCR	L	58	MOV	E, B
03	INX	B	2E	MVI	L, d8	59	MOV	E, C
04	INR	B	2F	CMA		5A	MOV	E, D
05	DCR	B	30	SIM		5B	MOV	E, E
06	MVI	B, d8	31	LXI	SP, d16	5C	MOV	E, H
07	RLC		32	STA	addr16	5D	MOV	E, L
08	---		33	INX	SP	5E	MOV	E, M
09	DAD	B	34	INR	M	5F	MOV	E, A
0A	LDAX	B	35	DCR	M	60	MOV	H, B
0B	DCX	B	36	MVI	M, d8	61	MOV	H, C
0C	INR	C	37	STC		62	MOV	H, D
0D	DCR	C	38	---		63	MOV	H, E
0E	MVI	C, d8	39	DAD	SP	64	MOV	H, H
0F	RRC		3A	LDA	addr16	65	MOV	H, L
10	---		3B	DCX	SP	66	MOV	H, M
11	LXI	D, d16	3C	INR	A	67	MOV	H, A
12	STAX	D	3D	DCR	A	68	MOV	L, B
13	INX	D	3E	MVI	A, d8	69	MOV	L, C
14	INR	D	3F	CMC		6A	MOV	L, D
15	DCR	D	40	MOV	B, B	6B	MOV	L, E
16	MVI	D, d8	41	MOV	B, C	6C	MOV	L, H
17	RAL		42	MOV	B, D	6D	MOV	L, L
18	---		43	MOV	B, E	6E	MOV	L, M
19	DAD	D	44	MOV	B, H	6F	MOV	L, A
1A	LDAX	D	45	MOV	B, L	70	MOV	M, B
1B	DCX	D	46	MOV	B, M	71	MOV	M, C
1C	INR	E	47	MOV	B, A	72	MOV	M, D
1D	DCR	E	48	MOV	C, B	73	MOV	M, E
1E	MVI	E, d8	49	MOV	C, C	74	MOV	M, H
1F	RAR		4A	MOV	C, D	75	MOV	M, L
20	RIM		4B	MOV	C, E	76	HLT	
21	LXI	H, d16	4C	MOV	C, H	77	MOV	M, A
22	SHLD	addr16	4D	MOV	C, L	78	MOV	A, B
23	INX	H	4E	MOV	C, M	79	MOV	A, C
24	INR	H	4F	MOV	C, A	7A	MOV	A, D
25	DCR	H	50	MOV	D, B	7B	MOV	A, E
26	MVI	H, d8	51	MOV	D, C	7C	MOV	A, H
27	DAA		52	MOV	D, D	7D	MOV	A, L
28	---		53	MOV	D, E	7E	MOV	A, M
29	DAD	H	54	MOV	D, H	7F	MOV	A, A
2A	LHLD	addr16	55	MOV	D, L	80	ADD	B

Appendix IV continued...

OPCODE IN HEX	MNEMONIC		OPCODE IN HEX	MNEMONIC		OPCODE IN HEX	MNEMONIC	
81	ADD	C	AC	XRA	H	D7	RST	2
82	ADD	D	AD	XRA	L	D8	RC	
83	ADD	E	AE	XRA	M	D9	---	
84	ADD	H	AF	XRA	A	DA	JC	addr16
85	ADD	L	B0	ORA	B	DB	IN	addr8
86	ADD	M	B1	ORA	C	DC	CC	addr16
87	ADD	A	B2	ORA	D	DD	---	
88	ADC	B	B3	ORA	E	DE	SBI	d8
89	ADC	C	B4	ORA	H	DF	RST	3
8A	ADC	D	B5	ORA	L	E0	RPO	
8B	ADC	E	B6	ORA	M	E1	POP	H
8C	ADC	H	B7	ORA	A	E2	JPO	addr16
8D	ADC	L	B8	CMP	B	E3	XTHL	
8E	ADC	M	B9	CMP	C	E4	CPO	addr16
8F	ADC	A	BA	CMP	D	E5	PUSH	H
90	SUB	B	BB	CMP	E	E6	ANI	d8
91	SUB	C	BC	CMP	H	E7	RST	4
92	SUB	D	BD	CMP	L	E8	RPE	
93	SUB	E	BE	CMP	M	E9	PCHL	
94	SUB	H	BF	CMP	A	EA	JPE	addr16
95	SUB	L	C0	RNZ		EB	XCHG	
96	SUB	M	C1	POP	B	EC	CPE	addr16
97	SUB	A	C2	JNZ	addr16	ED	---	
98	SBB	B	C3	JMP	addr16	EE	XRI	d8
99	SBB	C	C4	CNZ	addr16	EF	RST	5
9A	SBB	D	C5	PUSH	B	F0	RP	
9B	SBB	E	C6	ADI	d8	F1	POP	PSW
9C	SBB	H	C7	RST	0	F2	JP	addr16
9D	SBB	L	C8	RZ		F3	DI	
9E	SBB	M	C9	RET		F4	CP	addr16
9F	SBB	A	CA	JZ	addr16	F5	PUSH	PSW
A0	ANA	B	CB	---		F6	ORI	d8
A1	ANA	C	CC	CZ	addr16	F7	RST	6
A2	ANA	D	CD	CALL	addr16	F8	RM	
A3	ANA	E	CE	ACI	d8	F9	SPHL	
A4	ANA	H	CF	RST	1	FA	JM	addr16
A5	ANA	L	D0	RNC		FB	EI	
A6	ANA	M	D1	POP	D	FC	CM	addr16
A7	ANA	A	D2	JNC	addr16	FD	---	
A8	XRA	B	D3	OUT	addr8	FE	CPI	d8
A9	XRA	C	D4	CNC	addr16	FF	RST	7
AA	XRA	D	D5	PUSH	D	--	---	
AB	XRA	E	D6	SUI	d8	--	---	

d8	→	8-bit data	addr16	→	16-bit address
d16	→	16-bit data	M	→	Memory
addr8	→	8-bit address	PSW	→	Program Status Word

APPENDIX V : 8085 INSTRUCTIONS IN ALPHABETICAL ORDER

OPCODE IN HEX	MNEMONIC		OPCODE IN HEX	MNEMONIC		OPCODE IN HEX	MNEMONIC	
CE	ACI	d8	E4	CPO	addr16	0A	LDAX	B
8F	ADC	A	CC	CZ	addr16	1A	LDAX	D
88	ADC	B	27	DAA		2A	LHLD	addr16
89	ADC	C	09	DAD	B	01	LXI	B,d16
8A	ADC	D	19	DAD	D	11	LXI	D,d16
8B	ADC	E	29	DAD	H	21	LXI	H,d16
8C	ADC	H	39	DAD	SP	31	LXI	SP,d16
8D	ADC	L	3D	DCR	A	7F	MOV	A,A
8E	ADC	M	05	DCR	B	78	MOV	A,B
87	ADD	A	0D	DCR	C	79	MOV	A,C
80	ADD	B	15	DCR	D	7A	MOV	A,D
81	ADD	C	1D	DCR	E	7B	MOV	A,E
82	ADD	D	25	DCR	H	7C	MOV	A,H
83	ADD	E	2D	DCR	L	7D	MOV	A,L
84	ADD	H	35	DCR	M	7E	MOV	A,M
85	ADD	L	0B	DCX	B	47	MOV	B,A
86	ADD	M	1B	DCX	D	40	MOV	B,B
C6	ADI	d8	2B	DCX	H	41	MOV	B,C
A7	ANA	A	3B	DCX	SP	42	MOV	B,D
A0	ANA	B	F3	DI		43	MOV	B,E
A1	ANA	C	FB	EI		44	MOV	B,H
A2	ANA	D	76	HLT		45	MOV	B,L
A3	ANA	E	DB	IN	addr8	46	MOV	B,M
A4	ANA	H	3C	INR	A	4F	MOV	C,A
A5	ANA	L	04	INR	B	48	MOV	C,B
A6	ANA	M	0C	INR	C	49	MOV	C,C
E6	ANI	d8	14	INR	D	4A	MOV	C,D
CD	CALL	addr16	1C	INR	E	4B	MOV	C,E
DC	CC	addr16	24	INR	H	4C	MOV	C,H
FC	CM	addr16	2C	INR	L	4D	MOV	C,L
2F	CMA		34	INR	M	4E	MOV	C,M
3F	CMC		03	INX	B	57	MOV	D,A
BF	CMP	A	13	INX	D	50	MOV	D,B
B8	CMP	B	23	INX	H	51	MOV	D,C
B9	CMP	C	33	INX	SP	52	MOV	D,D
BA	CMP	D	DA	JC	addr16	53	MOV	D,E
BB	CMP	E	FA	JM	addr16	54	MOV	D,H
BC	CMP	H	C3	JMP	addr16	55	MOV	D,L
BD	CMP	L	D2	JNC	addr16	56	MOV	D,M
BE	CMP	M	C2	JNZ	addr16	5F	MOV	E,A
D4	CNC	addr16	F2	JP	addr16	58	MOV	E,B
C4	CNZ	addr16	EA	JPE	addr16	59	MOV	E,C
F4	CP	addr16	E2	JPO	addr16	5A	MOV	E,D
EC	CPE	addr16	CA	JZ	addr16	5B	MOV	E,E
FE	CPI	d8	3A	LDA	addr16	5C	MOV	E,H

Appendix V continued...

OPCODE IN HEX	MNEMONIC		OPCODE IN HEX	MNEMONIC		OPCODE IN HEX	MNEMONIC	
5D	MOV	E,L	C1	POP	B	97	SUB	A
5E	MOV	E,M	D1	POP	D	90	SUB	B
67	MOV	H,A	E1	POP	H	91	SUB	C
60	MOV	H,B	F1	POP	PSW	92	SUB	D
61	MOV	H,C	C5	PUSH	B	93	SUB	E
62	MOV	H,D	D5	PUSH	D	94	SUB	H
63	MOV	H,E	E5	PUSH	H	95	SUB	L
64	MOV	H,H	F5	PUSH	PSW	96	SUB	M
65	MOV	H,L	17	RAL		D6	SUI	d8
66	MOV	H,M	1F	RAR		EB	XCHG	
6F	MOV	L,A	D8	RC		AF	XRA	A
68	MOV	L,B	C9	RET		A8	XRA	B
69	MOV	L,C	20	RIM		A9	XRA	C
6A	MOV	L,D	07	RLC		AA	XRA	D
6B	MOV	L,E	F8	RM		AB	XRA	E
6C	MOV	L,H	D0	RNC		AC	XRA	H
6D	MOV	L,L	C0	RNZ		AD	XRA	L
6E	MOV	L,M	F0	RP		AE	XRA	M
77	MOV	M,A	E8	RPE		EE	XRI	d8
70	MOV	M,B	E0	RPO		E3	XTHL	
71	MOV	M,C	0F	RRC				
72	MOV	M,D	C7	RST	0			
73	MOV	M,E	CF	RST	1			
74	MOV	M,H	D7	RST	2			
75	MOV	M,L	DF	RST	3			
3E	MVI	A, d8	E7	RST	4			
06	MVI	B, d8	EF	RST	5			
0E	MVI	C, d8	F7	RST	6			
16	MVI	D, d8	FF	RST	7			
1E	MVI	E, d8	C8	RZ				
26	MVI	H, d8	98	SBB	B			
2E	MVI	L, d8	99	SBB	C			
36	MVI	M, d8	9A	SBB	D			
00	NOP		9B	SBB	E			
B7	ORA	A	9C	SBB	H			
B0	ORA	B	9D	SBB	L			
B1	ORA	C	9E	SBB	M			
B2	ORA	D	DE	SBI	d8			
B3	ORA	E	22	SHLD	addr16			
B4	ORA	H	30	SIM				
B5	ORA	L	F9	SPHL				
B6	ORA	M	32	STA	addr16			
F6	ORI	d8	02	STAX	B			
D3	OUT	addr8	12	STAX	D			
E9	PCHL		37	STC				

d8	→	8-bit data	addr16	→	16-bit address
d16	→	16-bit data	M	→	Memory
addr8	→	8-bit address	PSW	→	Program Status Word

APPENDIX VI : TEMPLATES FOR 8086 INSTRUCTIONS

S.No.	Mnemonic	Templates
Group - I Data Transfer Instructions		
1.	Mov reg2/mem, reg1/mem	
	a) MOV reg2, reg1	`1000 10dw` `mod reg r/m`
	b) MOV mem, reg1	
	c) MOV reg2, mem	`1000 10dw` `mod reg r/m` `l.b.disp` `h.b.disp`
2.	MOV reg/mem, data	
	a) MOV reg, data	`1100 011w` `mod 000 r/m` `l.b.data` `h.b.data`
	b) MOV mem, data	`1100 011w` `mod 000 r/m` `l.b.disp` `h.b.disp` `l.b.data` `h.b.disp`
3.	MOV reg, data	`1011 wreg` `l.b.data` `h.b.data`
4.	MOV A, mem	`1010 000 w` `l.b.disp` `h.b.disp`
	a) MOV AL, mem	
	b) MOV AX, mem	
5.	MOV mem, A	`1010 001w` `l.b.disp` `h.b.disp`
	a) MOV mem, AL	
	b) MOV mem, AX	
6.	MOV seg reg, reg16/mem	
	a) MOV seg reg, reg16	`1000 1110` `mod 0 sr r/m`
	b) MOV seg reg, mem	`1000 1110` `mod 0 sr r/m` `l.b.disp` `h.b.disp`
7.	MOV reg16/mem, seg reg	
	a) MOV reg16, segreg	`1000 1100` `mod 0sr r/m`
	b) MOV mem, segreg	`1000 1100` `mod 0 sr r/m` `l.b disp` `h.b.disp`
8.	PUSH reg16/mem	
	a) PUSH reg16	`1111 1111` `mod 110r/m`
	b) PUSH mem	`1111 1111` `mod 110 r/m` `l.b disp` `h.b.disp`
9.	PUSH reg16	`0101 0 reg`
10.	PUSH segreg	`000 sr 110`
11.	PUSHF	`1001 1100`
12.	POP reg16/mem	
	a) POP reg16	`1000 1111` `mod 000r/m`
	b) POP mem	`1000 1111` `mod 000 r/m` `l.b disp` `h.b.disp`
13.	POP reg16	`0101 1 reg`
14.	POP segreg	`000 sr 111`
15.	POPF	`1001 1101`

Appendix VI continued ...

S.No.	Mnemonic	Templates					
16.	XCHG reg2/mem, reg1						
	a) XCHG reg2, reg1	1000 011w	mod reg r/m				
	b) XCHG mem, reg1	1000 011w	mod reg r/m	l.b disp	h.b.disp		
17.	XCHG AX, reg16	1001 0 reg					
18.	XLAT	1101 0111					
19.	IN A, [DX]	1110 110w					
	a) IN AL, [DX]						
	b) IN AX, [DX]						
20.	IN A, addr8	1110 010w	addr8				
	a) IN AL, addr8						
	b) IN AX, addr8						
21.	OUT [DX], A	1110 111w					
	a) OUT [DX], AL						
	b) OUT [DX], AX						
22.	OUT addr8, A	1110 011w	addr8				
	a) OUT addr8, AL						
	b) OUT addr8, AX						
23.	LEA reg16, mem	1000 1101	mod reg r/m	l.b.disp	h.b. disp		
24.	LDS reg16, mem	1100 0101	mod reg r/m	l.b.disp	h.b. disp		
25.	LES reg16, mem	1100 0100	mod reg r/m	l.b.disp	h.b. disp		
26.	LAHF	1001 1111					
27.	SAHF	1001 1110					
Group - II Arithmetic Instructions							
28.	ADD reg2/mem,reg1/mem						
	a) ADD reg2, reg1	0000 00dw	mod reg r/m				
	b) ADD reg2, mem } c) ADD mem, reg1	0000 00dw	mod reg r/m	l.b. disp	h.b. disp		
29.	ADD reg/mem, data						
	a) ADD reg, data	1000 00sw	mod 000 r/m	l.b. data	h.b. data		
	b) ADD mem, data	1000 00sw	mod 000 r/m	l.b. disp	h.b. disp	l.b. data	h.b. data
30.	ADD A, data	0000 010w	l.b. data	h.b. data			
	a) ADD AL, data8						
	b) ADD AX, data16						
31.	ADC reg2/mem,reg1/mem						
	a) ADC reg2, reg1	0001 00dw	mod reg r/m				
	b) ADC reg2, mem } c) ADC mem, reg1	001 00dw	mod reg r/m	l.b. disp	h.b.disp		

Appendix VI continued ...

S.No.	Mnemonic	Templates					
32.	ADC reg/mem, data						
	a) ADC reg, data	`1000 00sw`	`mod 010 r/m`	`l.b.data`	`h.b.data`		
	b) ADC mem, data	`1000 00sw`	`mod 010 r/m`	`l.b.disp`	`h.b.disp`	`l.b.data`	`h.b.data`
33.	ADC A, data	`0001 010w`	`l.b.data`	`h.b.data`			
	a) ADC AL, data8						
	b) ADC AX, data16						
34.	AAA	`0011 0111`					
35.	DAA	`0010 0111`					
36.	SUB reg2/mem,reg1/mem						
	a) SUB reg2, reg1	`0010 10dw`	`mod reg r/m`				
	b) SUB reg2, mem ⎫						
	c) SUB mem, reg1 ⎬	`0010 10dw`	`mod reg r/m`	`l.b.disp`	`h.b.disp`		
37.	SUB reg/mem, data						
	a) SUB reg, data	`1000 00sw`	`mod 101 r/m`	`l.b. data`	`h.b.data`		
	b) SUB mem, data	`000 00sw`	`mod 101 'r/m`	`l.b.disp`	`h.b.disp`	`l.b.data`	`h.b.data`
38.	SUB A, data	`0010 110w`	`l.b.data`	`h.b.data`			
	a) SUB AL, data8						
	b) SUB AX, data16						
39.	SBB reg2/mem,reg1/mem						
	a) SBB reg2, reg1	`0001 10dw`	`mod reg r/m`				
	b) SBB reg2, mem ⎫						
	0001 10dw	`mod reg r/m`	`l.b.disp`	`h.b.disp`			
	c) SBB mem, reg1 ⎬						
40.	SBB reg/mem, data						
	a) SBB reg, data	`1000 00sw`	`mod 011 r/m`	`l.b.data`	`h.b.data`		
	b) SBB mem, data	`1000 00sw`	`mod 011 r/m`	`l.b.disp`	`h.b.disp`	`l.b.data`	`h.b.data`
41.	SBB A, data	`0001 110w`	`l.b.data`	`h.b.data`			
	a) SBB AL, data8						
	b) SBB AX, data16						
42.	AAS	`0011 1111`					
43.	DAS	`0010 1111`					
44.	MUL reg/mem						
	a) MUL reg	`1111 011w`	`mod 100 r/m`				
	b) MUL mem	`1111 011w`	`mod 100 r/m`	`l.b.disp`	`h.b.disp`		
45.	IMUL reg/mem						
	a) IMUL reg	`1111 011w`	`mod 101 r/m`				
	b) IMUL mem	`1111 011w`	`mod 101 r/m`	`l.b.disp`	`h.b.disp`		

Appendix VI continued ...

S.No.	Mnemonic	Templates				
46.	AAM	1101 0100	0000 1010			
47.	DIV reg/mem					
	a) DIV reg	1111 011w	mod 110 r/m			
	b) DIV mem	1111 011w	mod 110 r/m	l.b.disp	h.b.disp	
48.	IDIV reg/mem					
	a) IDIV reg	1111 011w	mod 111 r/m			
	b) IDIV mem	1111 011w	mod 111 r/m	l.b.disp	h.b.disp	
49.	AAD	1101 0101	0000 1010			
50.	NEG mem/reg					
	a) NEG reg	1111 011w	mod 011 r/m			
	b) NEG mem	1111 011w	mod 011 r/m	l.b.disp	h.b.disp	
51.	INC reg8/mem					
	a) INC reg8	1111 111w	mod 000 r/m			
	b) INC mem	1111 111w	mod 000 r/m	l.b.disp	h.b.disp	
52.	INC reg16	0100 0 reg				
53.	DEC reg8/mem					
	a) DEC reg8	1111 111w	mod 001 r/m			
	b) DEC mem	1111 111w	mod 001 r/m	l.b.disp	h.b.disp	
54.	DEC reg16	0100 1 reg				
55.	CBW	1001 1000				
56.	CWD	1001 1001				
57.	CMP reg2/mem, reg1/mem					
	a) CMP reg2, reg1	0011 10dw	mod reg r/m			
	b) CMP reg2, mem	0011 10dw	mod reg r/m	l.b.disp	h.b.disp	
	c) CMP mem, reg1					
58.	CMP reg/mem, data					
	a) CMP reg, data	1000 00sw	mod 111 r/m	l.b.data	h.b.data	
	b) CMP mem, data	1000 00sw	mod 111 r/m	l.b.disp	h.b.disp	l.b.data	h.b.data
59.	CMP A, data	0001 110w	l.b.data	h.b.data		
	a) CMP AL, data8					
	b) CMP AX, data16					
Group - III Logical Instructions						
60.	AND reg2/mem, reg1/mem					
	a) AND reg2, reg1	0010 00dw	mod reg r/m			
	b) AND reg2, mem	0010 00dw	mod reg r/m	l.b.disp	h.b.disp	
	c) AND mem, reg1					

Appendix VI continued ...

S.No.	Mnemonic	Templates					
61.	AND reg/mem, data						
	a) AND reg, data	`1000 000w`	`mod 100 r/m`	`l.b.data`	`h.b.data`		
	b) AND mem, data	`1000 000w`	`mod 100 r/m`	`l.b.disp`	`h.b.disp`	`l.b.data`	`h.b.data`
62.	AND A, data a) AND AL, data8 b) AND AX, data16	`0010 010w`	`l.b.data`	`h.b.data`			
63.	OR reg2/mem, reg1/mem						
	a) OR reg2, reg1	`0000 10dw`	`mod reg r/m`				
	b) OR reg2, mem c) OR mem, reg1	`0000 10dw`	`mod reg r/m`	`l.b.disp`	`h.b.disp`		
64.	OR reg/mem, data						
	a) OR reg, data	`1000 000w`	`mod 001 r/m`	`l.b.data`	`h.b.data`		
	b) OR mem, data	`1000 000w`	`mod 001 r/m`	`l.b.disp`	`h.b.disp`	`l.b.data`	`h.b.data`
65.	OR A, data a) OR AL, data8 b) OR AX, data16	`0000 110w`	`l.b.data`	`h.b.data`			
66.	XOR reg2/mem, reg1/mem						
	a) XOR reg2, reg1	`0011 00dw`	`mod reg r/m`				
	b) XOR reg2, mem c) XOR mem, reg1	`0011 00dw`	`mod reg r/m`	`l.b.disp`	`h.b.disp`		
67.	XOR reg/mem, data						
	a) XOR reg, data	`1000 000w`	`mod 110 r/m`	`l.b.data`	`h.b.data`		
	b) XOR mem, data	`1000 000w`	`mod 110 r/m`	`l.b.disp`	`h.b.disp`	`l.b.data`	`h.b.data`
68.	XOR A, data a) XOR AL, data8 b) XOR AX, data16	`0011 010w`	`l.b.data`	`h.b.data`			
69.	TEST reg2/mem, reg1/mem						
	a) TEST reg2, reg1	`1000 010w`	`mod reg r/m`				
	b) TEST reg2, mem c) TEST mem, reg1	`1000 010w`	`mod reg r/m`	`l.b.disp`	`h.b.disp`		
70.	TEST reg/mem, data						
	a) TEST reg, data	`1111 011w`	`mod 000 r/m`	`l.b.data`	`h.b.data`		
	b) TEST mem, data	`1111 011w`	`mod 000 r/m`	`l.b.disp`	`h.b.disp`	`l.b.data`	`h.b.data`
71.	TEST A, data a) TEST AL, data8 b) TEST AX, data16	`1010 100w`	`l.b.data`	`h.b.data`			

Appendix VI continued ...

S.No.	Mnemonic	Templates
72.	NOT reg/mem a) NOT reg	`1111 011w` `mod 010 r/m`
	b) NOT mem	`1111 011w` `mod 010 r/m` `l.b.disp` `h.b.disp`
73.	SHL reg/mem or SAL reg/mem a) SHL reg or SAL reg i) SHL reg, 1 or SAL reg, 1 ii) SHL reg, CL or SAL reg, CL	`1101 00vw` `mod 100 r/m`
	b) SHL mem or SAL mem i) SHL mem, 1 or SAL mem, 1 ii) SHL mem, CL or SAL mem, CL	`1101 00vw` `mod 100 r/m` `l.b.disp` `h.b.disp`
74.	SHR reg/mem a) SHR reg i) SHR reg, 1 ii) SHR reg, CL	`1101 00vw` `mod 101 r/m`
	b) SHR mem i) SHR mem,1 ii) SHR mem,CL	`1101 00vw` `mod 101 r/m` `l.b.disp` `h.b.disp`
75.	SAR reg/mem a) SAR reg i) SAR reg, 1 ii) SAR reg, CL	`1101 00vw` `mod 111 r/m`
	b) SAR mem i) SAR mem, 1 ii) SAR mem, CL	`1101 00vw` `mod 111 r/m` `l.b.disp` `h.b.disp`
76.	ROL reg/mem a) ROL reg i) ROL reg, 1 ii) ROL reg, CL	`1101 00vw` `mod 000 r/m`
	b) ROL mem i) ROL mem, 1 ii) ROL mem, CL	`1101 00vw` `mod 000 r/m` `l.b.disp` `h.b.disp`
77.	RCL reg/mem a) RCL reg i) RCL reg, 1 ii) RCL reg, CL	`1101 00vw` `mod 010 r/m`
	b) RCL mem i) RCL mem, 1 ii) RCL mem, CL	`1101 00vw` `mod 010 r/m` `l.b.disp` `h.b.disp`

Appendix VI continued ...

S.No.	Mnemonic	Templates
78.	ROR reg/mem a) ROR reg i) ROR reg, 1 ii) ROR reg, CL	`1101 00vw` `mod 001 r/m`
	b) ROR mem i) ROR mem, 1 ii) ROR mem, CL	`1101 00vw` `mod 001 r/m` `l.b.disp` `h.b.disp`
79.	RCR reg/mem a) RCR reg i) RCR reg, 1 ii) RCR reg, CL	`1101 00vw` `mod 011 r/m`
	b) RCR mem i) RCR mem, 1 ii) RCR mem, CL	`1101 00vw` `mod 011 r/m` `l.b.disp` `h.b.disp`
Group - IV String Manipulation Instructions		
80.	REP a) REPZ/REPE b) REPNZ/REPNE	`1111  001z`
81.	MOVS a) MOVSB b) MOVSW	`1010 010w`
82.	CMPS a) CMPSB b) CMPSW	`1010 011w`
83.	SCAS a) SCASB b) SCASW	`1010 111w`
84.	LODS a) LODSB b) LODSW	`1010 110w`
85.	STOS a) STOSB b) STOSW	`1010 101w`
Group - V Control Transfer Instructions		
86.	CALL disp16 (Call near - direct within segment)	`1110 1000` `l.b.disp` `h.b.disp`
87.	CALL reg/mem (Call near - indirect within segment)	
	a) CALL reg	`1111 1111` `mod 010 r/m`
	b) CALL mem	`1111 1111` `mod 010 r/m` `l.b.disp` `h.b.disp`

Appendix VI continued ...

S.No.	Mnemonic	Templates				
88.	CALL addr$_{offset}$, addr$_{base}$ (Call far-direct intersegment)	1001 1010	l.b.offset	h.b.offset	l.b.base	h.b.base
89.	CALL mem (Call far-indirect intersegment)	1111 1111	mod 011 r/m	l.b.disp	h.b.disp	
90.	RET (Return from call within segment)	1100 0011				
91.	RET data16 (Return from call within segment adding immediate data to SP)	1100 0010	l.b. data	h.b.data		
92.	RET (Return from intersegment call)	1100 1011				
93.	RET data16 (Return from intersegment call adding immediate data to SP)	1100 1010	l.b. data	h.b.data		
94.	JMP disp16 (Unconditional jump near-direct within segment)	1110 1001	l.b. disp	h.b.disp		
95.	JMP disp8 (Unconditional jump short-direct within segment)	1110 1011	disp8			
96.	JMP reg/mem (Unconditional jump near-indirect within segment) a) JMP reg	1111 1111	mod 100 r/m			
	b) JMP mem	1111 1111	mod 100 r/m	l.b.disp	h.b.disp	
97.	JMP addr$_{offset}$, addr$_{base}$ (Unconditional jump far-direct intersegment)	1110 1010	l.b.offset	h.b.offset	l.b.base	h.b.base
98.	JMP mem (Unconditional jump far-indirect intersegment)	1111 1111	mod 101 r/m	l.b.disp	h.b.disp	
99.	JE/JZ disp8	0111 0100	disp8			
100.	JL/JNGE disp8	0111 1100	disp8			
101.	JLE/JNG disp8	0111 1110	disp8			
102.	JB/JNAE/JC disp8	0111 0010	disp8			
103.	JBE/JNA disp8	0111 0110	disp8			
104.	JP/JPE disp8	0111 1010	disp8			
105.	JNB/JAE/JNC disp8	0111 0011	disp8			
106.	JNBE/JA disp8	0111 0111	disp8			
107.	JNP/JPO disp8	0111 1011	disp8			

Appendix VI continued ...

S.No.	Mnemonic	Templates
108.	JNO disp8	`0111 0001` `disp8`
109.	JNS disp8	`0111 1001` `disp8`
110.	JO disp8	`0111 0000` `disp8`
111.	JS disp8	`0111 1000` `disp8`
112.	JNE/JNZ disp8	`0111 0101` `disp8`
113.	JNL/JGE disp8	`0111 1101` `disp8`
114.	JNLE/JG disp8	`0111 1111` `disp8`
115.	JCXZ disp8	`1110 0011` `disp8`
116.	LOOP disp8	`1110 0010` `disp8`
117.	LOOPZ/LOOPE disp8	`1110 0001` `disp8`
118.	LOOPNZ/LOOPNE disp8	`1110 0000` `disp8`
119.	INT type	`1100 1101` `type`
120.	INT 3	`1100 1100`
121.	INTO	`1100 1110`
122.	IRET	`1100 1111`
Group - VI Processor Control Instructions		
123.	CLC	`1111 1000`
124.	CMC	`1111 0101`
125.	STC	`1111 1001`
126.	CLD	`1111 1100`
127.	STD	`1111 1101`
128.	CLI	`1111 1010`
129.	STI	`1111 1011`
130.	HLT	`1111 0100`
131.	WAIT	`1001 1011`
132.	ESC opcode, mem/reg	
	a) ESC opcode, mem	`1101 1opc` `mod opc r/m` `l.b.disp` `h.b.disp`
	b) ESC opcode, reg	`1101 1opc` `mod opc r/m`
133.	LOCK	`1111 0000`
134.	NOP	`1001 0000`
135.	Segment override prefix	`001 sr 110`

Table A-1: ONE Bit Special Indicator

Special bit value	Meaning
w = 0	8-bit operation.
w = 1	16-bit operation.
d = 0	The register specified by reg field is source operand.
d = 1	The register specified by reg field is destination operand.
sw = 00	8-bit operation with an 8-bit immediate data.
sw = 01	16-bit operation with a 16-bit immediate data.
sw = 11	16-bit operation with a sign extended 8-bit immediate operand.
v = 0	Shift/rotate operation is performed one time.
v = 1	The content of CL is count value for number of shift/rotate operations to be performed.
z = 0	Repeat execution of string instruction until ZF = 0.
z = 1	Repeat execution of string instruction until ZF = 1.

Table A-2: Codes for "mod" Field

Code for mod field	Name of the mode
00	Memory mode with no displacement
01	Memory mode with 8-bit displacement
10	Memory mode with 16-bit displacement
11	Register mode

Table A-3: Codes for "reg" Field

Code for reg field	Name of the register represented by the code when w = 0 or 1	
	When w = 0	When w = 1
000	AL	AX
001	CL	CX
010	DL	DX
011	BL	BX
100	AH	SP
101	CH	BP
110	DH	SI
111	BH	DI

Table A-4: Codes for "sr" Field

Code for sr field	Segment register
00	E S
01	C S
10	SS
11	DS

Table A-5: Codes for "r/m" Field

Code for r/m field	Effective address calculation when mod = 00/01/10		
	mod = 00	mod = 01	mod = 10
000	[BX + SI]	[BX + SI + disp8]	[BX + SI + disp16]
001	[BX + DI]	[BX + DI + disp8]	[BX + DI + disp16]
010	[BP + SI]	[BP + SI + disp8]	[BP + SI + disp16]
011	[BP + DI]	[BP + DI + disp8]	[BP + DI + disp16]
100	[SI]	[SI + disp8]	[SI + disp16]
101	[DI]	[DI + disp8]	[DI + disp16]
110	[disp16]	[BP + disp8]	[BP + disp16]
111	[BX]	[BX + disp8]	[BX + disp16]

Table A-6: Meanings of Various Terms used in the Operand Field of Instructions and Templates

Term	Meaning
reg/reg1/reg2	8-bit or 16-bit register
reg8	8-bit register
reg16	16-bit register
segreg, sr	segment register
mem	8-bit or 16-bit memory
mem8	8-bit memory
mem16	16-bit memory
data	8-bit or 16-bit data

Term	Meaning
data8	8-bit data
data16	16-bit data
addr8	8-bit address
addr$_{offset}$	16-bit offset/effective address
addr$_{base}$	16-bit base address
disp8	8-bit signed displacement
disp16	16-bit displacement
opc	opcode

APPENDIX VII : DOS AND BIOS INTERRUPTS

Table 1: DOS Interrupts

Interrupt number	Function code	Dedicated operation
INT 20H	-	Program Terminate
Character Input/Output		
INT 21H	01H	Read character from standard input device.
INT 21H	02H	Write character to standard output device.
INT 21H	03H	Read character from auxiliary input device.
INT 21H	04H	Write character to auxiliary output device.
INT 21H	05H	Write character to printer.
INT 21H	06H	Console input or output.
INT 21H	07H	Unfiltered character input without echo.
INT 21H	08H	Read character without echo.
INT 21H	09H	Display string.
INT 21H	0AH	Buffered string input.
INT 21H	0BH	Get input status.
INT 21H	0CH	Clear input buffer and read.
File Operations		
INT 21H	0FH	Open file using FCB (File Control Block).
INT 21H	10H	Close file using FCB.
INT 21H	11H	Find first matching file using FCB.
INT 21H	12H	Find next matching file using FCB.
INT 21H	13H	Delete file using FCB.
INT 21H	16H	Create/Truncate file using FCB.
INT 21H	17H	Rename file using FCB.
INT 21H	23H	Get file size in records using FCB.
INT 21H	29H	Parse file name.
INT 21H	3CH	Create/Truncate file.
INT 21H	3DH	Open file.
INT 21H	3EH	Close file.
INT 21H	41H	Delete file.
INT 21H	43H	Set/Get file attributes - CHMOD.
INT 21H	45H	Duplicate a file handle.
INT 21H	46H	Force duplication of handle.
INT 21H	4EH	DOS first find (matching file).
INT 21H	4FH	DOS next find (matching file).
INT 21H	56H	Rename file.
INT 21H	57H	Get/Set file time and date.
INT 21H	5AH	Create unique temporary file.
INT 21H	5BH	Create new file.
Record Operations		
INT 21H	14H	Read file sequentially using FCB.
INT 21H	15H	Write file sequentially using FCB.
INT 21H	1AH	Set disk transfer area address.

Appendix VII continued...

Interrupt number	Function code	Dedicated operation
INT 21H	21H	Random record read using FCB.
INT 21H	22H	Random file read/write using FCB.
INT 21H	24H	Set random record number.
INT 21H	27H	Read random file block.
INT 21H	28H	Write random file block.
INT 21H	2FH	Get current disk transfer area address.
INT 21H	3FH	Read file or device.
INT 21H	40H	Write to file or device.
INT 21H	42H	Move file pointer.
INT 21H	5CH	Lock/unlock file access.
Directory Operations		
INT 21H	39H	Create new subdirectory.
INT 21H	3AH	Delete subdirectory.
INT 21H	3BH	Set current directory.
INT 21H	47H	Get present working directory.
Disk Management		
INT 21H	0DH	Reset disk.
INT 21H	0EH	Set DOS default disk.
INT 21H	19H	Get DOS default disk drive.
INT 21H	1BH	Get FAT information for default drive.
INT 21H	1CH	Get FAT information for specified drive.
INT 21H	2EH	Set or reset verify flag.
INT 21H	36H	Get disk free space.
INT 21H	54H	Get DOS verify switch.
Process Management		
INT 21H	00H	Program terminate.
INT 21H	26H	Create program segment prefix.
INT 21H	31H	Terminate and stay resident.
INT 21H	4BH	Execute or load program.
INT 21H	4CH	Terminate with return code.
INT 21H	4DH	Get return code.
INT 21H	62H	Get PSP (Program Segment Prefix) pointer.
Memory Management		
INT 21H	48H	Allocate memory.
INT 21H	49H	Release memory.
INT 21H	4AH	Modify memory allocation.
INT 21H	58H	Get or set allocation strategy.
Network Functions		
INT 21H	5EH	Get machine name and Get/Set printer setup.
INT 21H	5FH	Get redirection entry.
Time and Date Functions		
INT 21H	2AH	Get DOS system date.
INT 21H	2BH	Set DOS system date.
INT 21H	2CH	Get DOS system time.
INT 21H	2DH	Set DOS system time.

Appendix VII continued...

Interrupt number	Function code	Dedicated operation
	Miscellaneous System Functions	
INT 21H	25H	Set interrupt vector.
INT 21H	30H	Get DOS version number.
INT 21H	32H	Get DOS disk information.
INT 21H	33H	Get/set Ctrl-Break flag, Get boot drive.
INT 21H	34H	DOS re-entrance status address.
INT 21H	35H	Get interrupt vector.
INT 21H	37H	Set/Get switch character.
INT 21H	38H	Get/Set country dependent information.
INT 21H	44H	Device I/O control.
INT 21H	59H	Get extended error information.
INT 21H	63H	Get load byte table.
INT 22H	-	Terminate handler pointer.
INT 23H	-	Ctrl-C handler pointer.
INT 24H	-	Critical error handler pointer.
INT 25H	-	Absolute disk read.
INT 26H	-	Absolute disk write.
INT 27H	-	Terminate and stay resident.
INT 28H to INT 2EH	-	Reserved.
INT 2FH	01H	Print spooler.
INT 2FH	10H	Share.

Table 2: BIOS Interrupts

Interrupt number	Function code	Dedicated operation
	BIOS Video Driver Services	
INT 10H	00H	Set video mode.
INT 10H	01H	Set cursor shape.
INT 10H	02H	Set cursor position.
INT 10H	03H	Read cursor position.
INT 10H	04H	Read light pen position.
INT 10H	05H	Set active video page.
INT 10H	06H	Scroll/Initialize rectangle window up.
INT 10H	07H	Scroll/Initialize rectangle window down.
INT 10H	08H	Read character and attribute at cursor.
INT 10H	09H	Write character and attribute at cursor.
INT 10H	0AH	Write character only at cursor.
INT 10H	0BH	Set colour palette.
INT 10H	0CH	Set pixel.
INT 10H	0DH	Get pixel.
INT 10H	0EH	Write text in teletype mode.
INT 10H	0FH	Get video mode.
INT 10H	10H	Set colour palette registers.

Appendix VII continued...

Interrupt number	Function code	Dedicated operation
INT 10H	13H	Display string.
INT 10H	0FEH	Get video buffer pointer.
INT 10H	0FFH	Update video buffer.
INT 11H	-	Get machine configuration.
INT 12H	-	Get conventional memory size.
BIOS Floppy Disk Services		
INT 13H	00H	Reset disk system.
INT 13H	01H	Get disk system status.
INT 13H	02H	Read disk sector.
INT 13H	03H	Write disk sector.
INT 13H	04H	Verify disk sectors.
INT 13H	05H	Format disk track.
BIOS Serial Communication Port Services		
INT 14H	00H	Initialize communication port.
INT 14H	01H	Write communication port.
INT 14H	02H	Read communication port.
INT 14H	03H	Read communication port status.
INT 15H	- AT services.	
BIOS Keyboard Driver Services		
INT 16H	00H	Read keyboard character.
INT 16H	01H	Read keyboard status.
INT 16H	02H	Read keyboard flags.
BIOS Printer Driver Services		
INT 17H	00H	Write to printer.
INT 17H	01H	Initialize printer port.
INT 17H	02H	Read printer status.

INDEX

CHIP INDEX

8755

ADC0809/0808

DAC0800

MAX232

PENTIUM

Z80